Functional Analysis, Calculus of Variations and Numerical Methods for Models in Physics and Engineering

Fabio Silva Botelho, *Adjunct Professor*

Department of Mathematics
Universidade Federal de Santa Catarina
Florianópolis, SC, Brazil

CRC Press
Taylor & Francis Group
Boca Raton   London   New York

CRC Press is an imprint of the
Taylor & Francis Group, an **informa** business

A SCIENCE PUBLISHERS BOOK

CRC Press
Taylor & Francis Group
6000 Broken Sound Parkway NW, Suite 300
Boca Raton, FL 33487-2742

First issued in paperback 2022

Version Date: 20200427

ISBN: 978-0-367-51003-9 (pbk)
ISBN: 978-0-367-35674-3 (hbk)

DOI: 10.1201/9780429343315

Visit the Taylor & Francis Web site at
http://www.taylorandfrancis.com

and the CRC Press Web site at
http://www.crcpress.com

Preface

The first part of this book develops in details the basic tools of functional analysis, measure and integration and Sobolev spaces theory, which will be used in the development of the subsequent chapters. The results are established in a rigorous and precise fashion, however we believe the proofs presented are relatively easy to follow, since all steps are developed in a very transparent and clear way.

Of course, in general the results in this first part are not new and may be found in similar format in many other excellent books on functional analysis, such as those by Bachman and Narici [6], Brezis [24], Rudin [67] and others.

In many cases we give our own version of some proofs we believe the original ones are not clear enough, always aiming to improve their clarity and preciseness.

In the second book part, we develop the basic concepts on the calculus of variations, convex analysis, duality theory and constrained optimization in a Banach spaces context.

We start with a more basic approach on the calculus of variations which gradually evolves to the development of more advanced results on convex analysis and optimization. The text includes a formal proof of the Lagrange multiplier theorem, in a first step, for the equality-restriction case and in the subsequent sections we address the general case of constrained optimization also in a Banach spaces context.

Finally, in the third part we address the development of several important examples which apply the previous theoretical approaches exposed.

Applications include duality principles for non-linear plate and shell models, duality principles for problems in non-linear elasticity, duality for shape optimization models, existence and duality for the Ginzburg-Landau system in superconductivity, duality for a semi-linear model in micro-magnetism and others.

About such applications we highlight the results on the Generalized Method of Lines, which is a numerical method for which the domain of the partial differential equation in question is discretized in lines, and the concerning solution is written on these lines as function of the boundary conditions and boundary shape. Some important models are addressed and among them we highlight the applications of such a method to the Navier-Stokes system.

Also as an application of the Generalized Method of Lines, we finish the book with a chapter which presents a numerical procedure for the solution of an inverse optimization problem.

At this point, it is worth emphasizing that the Chapter 21 is co-authored by myself and my colleague Alexandre Molter and Chapter 23 is co-authored by myself and my graduate student Eduardo Pandini Barros, to whom I would like to express my gratitude for their contributions.

As a final note we highlight that the content of the present book overlaps something with the ones of my previous books "Topics on Functional Analysis, Calculus of Variations and Duality" published by Academic Publications, Sofia, 2011 and, "Functional Analysis and Applied Optimization in Banach Spaces" published by Springer in 2014, but only relating a part on basic standard mathematics, so that, among some other new chapters, the applications presented from Chapters 17 and 30 are almost all new developments.

Acknowledgements

I would like to thank some people of Virginia Tech—USA, where I got my Ph.D. degree in Mathematics in 2009. I am especially grateful to Professor Robert C Rogers for his excellent work as advisor. I would like to thank the Department of Mathematics for its constant support and this opportunity of studying mathematics in advanced level. Among many other Professors, I particularly thank Martin Day (Calculus of Variations), William Floyd and Peter Haskell (Elementary Real Analysis), James Thomson (Real Analysis), Peter Linnell (Abstract Algebra) and George Hagedorn (Functional Analysis) for the excellent lectured courses. Finally, special thanks to all my Professors at I.T.A. (Instituto Tecnológico de Aeronáutica, SP-Brasil) my undergraduate and masters school. Specifically about I.T.A., among many others I would like to express my gratitude to Professors Leo H Amaral, Tânia Rabelo and my master thesis advisor Antônio Marmo de Oliveira, also for their valuable work.

January 2020

Fabio Silva Botelho
Florianópolis, SC, Brazil

Contents

SECTION I: FUNCTIONAL ANALYSIS

SECTION II: CALCULUS OF VARIATIONS, CONVEX ANALYSIS AND RESTRICTED OPTIMIZATION

SECTION III: APPLICATIONS TO MODELS IN PHYSICS AND ENGINEERING

SECTION I
FUNCTIONAL ANALYSIS

Chapter 1

Metric Spaces

1.1 Introduction

In this chapter we present the main basic definitions and results relating to the concept of metric spaces.

We may recall that any Banach space is a metric one, so that the framework here introduced is suitable for a very large class of spaces.

The main references for this chapter are [14, 82].

1.2 The main definitions

We begin this section by presenting the metric definitions and some concerning examples.

Definition 1.2.1 (Metric space) *Let V be a non-empty set. We say that V is a metric space as it is possible to define a function $d : V \times V \to \mathbb{R}^+ = [0, +\infty)$ such that*

1. $d(u,v) > 0$ if $u \neq v$ and $d(u,u) = 0$, $\forall u,v \in V$.

2. $d(u,v) = d(v,u)$, $\forall u,v \in V$.

3. $d(u,w) \leq d(u,v) + d(v,w)$, $\forall u,v,w \in V$.

Such a function d is said to be a metric for V, so that the metric space in question is denoted by (V,d).

Example 1.2.2 *$V = \mathbb{R}$ is a metric space with the metric $d : \mathbb{R} \times \mathbb{R} \to \mathbb{R}^+$, where*

$$d(u,v) = |u-v|, \forall u,v \in \mathbb{R}.$$

Example 1.2.3 $V = \mathbb{R}^2$ *is a metric space with the metric* $d : V \times V \to \mathbb{R}^+$, *where*

$$d(\mathbf{u}, \mathbf{v}) = \sqrt{(u_1 - v_1)^2 + (u_2 - v_2)^2}, \ \forall \mathbf{u} = (u_1, u_2), \ \mathbf{v} = (v_1, v_2) \in \mathbb{R}^2.$$

Example 1.2.4 $V = \mathbb{R}^n$ *is a metric space with the metric* $d : V \times V \to \mathbb{R}^+$ *where*

$$d(\mathbf{u}, \mathbf{v}) = \sqrt{(u_1 - v_1)^2 + \cdots + (u_n - v_n)^2}, \ \forall \mathbf{u} = (u_1, \ldots, u_n), \ \mathbf{v} = (v_1, \ldots, v_n) \in \mathbb{R}^n.$$

Example 1.2.5 $V = C([a,b])$, *where* $C([a,b])$ *is the metric space of continuous functions* $u : [a,b] \to \mathbb{R}$ *with the metric* $d : V \times V \to \mathbb{R}^+$, *where*

$$d(u, v) = \max_{x \in [a,b]} \{|u(x) - v(x)|\} \equiv \|u - v\|_\infty, \ \forall u, v \in V.$$

Example 1.2.6 $V = C([a,b])$, *is a metric space with the metric* $d : V \times V \to \mathbb{R}^+$ *where*

$$d(u, v) = \int_a^b |u(x) - v(x)| \, dx, \ \forall u, v \in V.$$

1.2.1 The space l^∞

In this subsection we shall define some important classes of metric spaces.

The first definition presented is about the l^∞ space of sequences.

Definition 1.2.7 *We define the space* l^∞ *as*

$$l^\infty = \{\mathbf{u} = \{u_n\}_{n \in \mathbb{N}} \ : \ u_n \in \mathbb{C} \text{ and there exists } M > 0 \text{ such that } |u_n| < M, \ \forall n \in \mathbb{N}\}.$$

A metric for l^∞ *may be defined by*

$$d(\mathbf{u}, \mathbf{v}) = \sup_{j \in \mathbb{N}} \{|u_j - v_j|\},$$

where $\mathbf{u} = \{u_n\}$ *e* $\mathbf{v} = \{v_n\} \in l^\infty$.

1.2.2 Discrete metric

At this point we introduce the definition of discrete metric.

Definition 1.2.8 *Let V be a non-empty set. We define the discrete metric for V by*

$$d(u, v) = \begin{cases} 0, & \text{if } u = v, \\ 1, & \text{if } u \neq v. \end{cases} \tag{1.1}$$

In such a case we say that (V, d) is a discrete metric space.

Exercise 1.2.9 *Let $V = \mathbb{R}$ and let $d : V \times V \to \mathbb{R}$ be defined by*

$$d(u, v) = \sqrt{|u - v|}.$$

Show that d is a metric for V.

1.2.3   The metric space *s*

In the next lines we define one more metric space of sequences, namely, the space *s*.

Definition 1.2.10 (The metric space *s*) *We define the metric space s as* $s = (V,d)$, *where*

$$V = \{\mathbf{u} = \{u_n\}, : u_n \in \mathbb{C}, \forall n \in \mathbb{N}\},$$

with the metric

$$d(\mathbf{u},\mathbf{v}) = \sum_{n=1}^{\infty} \frac{1}{2^n} \frac{|u_n - v_n|}{(1 + |u_n - v_n|)},$$

$\forall \mathbf{u} = \{u_n\}$ *and* $\mathbf{v} = \{v_n\} \in V$.

Exercise 1.2.11 *Show that this last function d is indeed a metric.*

1.2.4   The space $B(A)$

Another important metric space is the space of bounded functions defined on a set A, and denoted by $B(A)$.

Definition 1.2.12 *Let A be a non-empty set and define*

$$B(A) = \{u : A \to \mathbb{R}, \text{ such that there exists } M > 0 \text{ such that } |u(x)| < M, \forall x \in A\}.$$

$B(A)$ *is said to be the space of bounded functions defined on A.*

Exercise 1.2.13 *Show that $B(A)$ is a metric space with the metric*

$$d(u,v) = \sup_{x \in A} \{|u(x) - v(x)|\}.$$

1.2.5   The space l^p

Finally, one of the most important metric space of sequences is the l^p one, whose definition is presented in the next lines.

Definition 1.2.14 *Let $p \geq 1$, $p \in \mathbb{R}$.*
   We define the space l^p by

$$l^p = \left\{ \mathbf{u} = \{u_n\} : u_n \in \mathbb{C} \text{ and } \sum_{n=1}^{\infty} |u_n|^p < \infty \right\}$$

   with the metric

$$d(\mathbf{u},\mathbf{v}) = \left(\sum_{n=1}^{\infty} |u_n - v_n|^p \right)^{1/p},$$

where $\mathbf{u} = \{u_n\}$ and $\mathbf{v} = \{v_n\} \in l^p$.

At this point we shall show that d is indeed a metric.

Let $p > 1$, $p \in \mathbb{R}$. Let $q > 1$ be such that

$$\frac{1}{p} + \frac{1}{q} = 1,$$

that is

$$q = \frac{p}{p-1}.$$

Let $x, y \geq 0$, $x, y \in \mathbb{R}$.

We are going to show that

$$xy \leq \frac{1}{p}x^p + \frac{1}{q}y^q.$$

Observe that if $x = 0$ or $y = 0$ the inequality is immediate. Thus, suppose $x > 0$ and $y > 0$. Fix $y > 0$ and define

$$h(x) = \frac{1}{p}x^p + \frac{1}{q}y^q - xy, \ \forall x > 0.$$

Observe that

$$h'(x) = x^{p-1} - y$$

and

$$h''(x) = (p-1)x^{p-2} > 0, \forall x > 0.$$

Therefore h is convex and its minimum on $(0, +\infty)$ is attained through the equation

$$h'(x) = x^{p-1} - y = 0,$$

that is, at $x_0 = y^{1/(p-1)}$.

Hence,

$$
\begin{aligned}
\min_{x \in (0,+\infty)} h(x) &= h(x_0) \\
&= \frac{1}{p}(x_0)^p + \frac{1}{q}y^q - x_0 y \\
&= \frac{1}{p}y^{p/(p-1)} - y^{1/(p-1)}y + \frac{1}{q}y^q \\
&= (1/p - 1)y^q + \frac{1}{q}y^q \\
&= -\frac{1}{q}y^q + \frac{1}{q}y^q \\
&= 0.
\end{aligned}
\tag{1.2}
$$

Thus,

$$h(x) = \frac{1}{p}x^p + \frac{1}{q}y^q - xy \geq h(x_0) = 0, \ \forall x > 0.$$

Therefore, since $y > 0$ is arbitrary, we obtain

$$xy \leq \frac{1}{p}x^p + \frac{1}{q}y^q, \ \forall x, y > 0.$$

so that

$$xy \leq \frac{1}{p}x^p + \frac{1}{q}y^q, \ \forall x, y \geq 0. \tag{1.3}$$

Let $\mathbf{u} = \{u_n\} \in l^p$ and $\mathbf{v} = \{v_n\} \in l^q$.
Denote

$$\|\mathbf{u}\|_p = \left(\sum_{n=1}^{\infty} |u_n|^p \right)^{1/p}$$

and

$$\|\mathbf{v}\|_q = \left(\sum_{n=1}^{\infty} |v_n|^q \right)^{1/q}.$$

Define also

$$\hat{\mathbf{u}} = \frac{\mathbf{u}}{\|\mathbf{u}\|_p} = \left\{ \frac{u_n}{\left(\sum_{n=1}^{\infty} |u_n|^p \right)^{1/p}} \right\},$$

and

$$\hat{\mathbf{v}} = \frac{\mathbf{v}}{\|\mathbf{v}\|_q} = \left\{ \frac{v_n}{\left(\sum_{n=1}^{\infty} |v_n|^q \right)^{1/q}} \right\}.$$

From this and (1.3) we obtain,

$$\begin{aligned}
\sum_{n=1}^{\infty} |\hat{u}_n \hat{v}_n| &\leq \frac{1}{p} \sum_{n=1}^{\infty} |\hat{u}_n|^p + \frac{1}{q} \sum_{n=1}^{\infty} |\hat{v}_n|^q \\
&= \frac{1}{p} + \frac{1}{q} \\
&= 1.
\end{aligned} \tag{1.4}$$

Thus,

$$\sum_{n=1}^{\infty} |u_n v_n| \leq \|\mathbf{u}\|_p \|\mathbf{v}\|_q, \ \forall \mathbf{u} \in l^p, \ \mathbf{v} \in l^q.$$

This last inequality is well known as the Hölder one.

Exercise 1.2.15 *Prove the Minkowski inequality, namely*

$$\|\mathbf{u} + \mathbf{v}\|_p \leq \|\mathbf{u}\|_p + \|\mathbf{v}\|_p, \ \forall \mathbf{u}, \mathbf{v} \in l^p.$$

Hint

$$\begin{aligned}
\|\mathbf{u} + \mathbf{v}\|_p^p &= \sum_{n=1}^{\infty} |u_n + v_n|^p \\
&\leq \sum_{n=1}^{\infty} |u_n + v_n|^{p-1} (|u_n| + |v_n|).
\end{aligned} \tag{1.5}$$

Apply the Hölder inequality to each part of the right hand side of the last inequality.
Use such an inequality to prove the triangle inequality concerning the metrics definition.
Prove also the remaining properties relating to the metric definition and conclude that $d : l^p \times l^p \to \mathbb{R}^+$, where

$$d(\mathbf{u}, \mathbf{v}) = \|\mathbf{u} - \mathbf{v}\|_p, \ \forall \mathbf{u}, \mathbf{v} \in l^p$$

is indeed a metric for the space l^p.

1.2.6 Some fundamental definitions

Definition 1.2.16 (neighborhood) *Let (U,d) be a metric space. Let $u \in U$ and $r > 0$. We define the neighborhood of center u and radius r, denoted by $V_r(u)$, by*

$$V_r(u) = \{v \in U \mid d(u,v) < r\}.$$

Definition 1.2.17 (limit point) *Let (U,d) be a metric space and $E \subset U$. A point $u \in U$ is said to be a limit point of E if for each $r > 0$ there exists $v \in V_r(u) \cap E$ such that $v \neq u$.*

We shall denote by E' the set of all limit points of E.

Example 1.2.18 $U = \mathbb{R}^2$, $E = B_r(0)$. *Thus $E' = \overline{B}_r(0)$.*

Remark 1.2.19 *In the next definitions U shall denote a metric space with a metric d.*

Definition 1.2.20 (Isolated point) *Let $u \in E \subset U$. We say that u is an isolated point of E if it is not a limit point of E.*

Example 1.2.21

$U = \mathbb{R}^2$, $E = B_1((0,0)) \cup \{(3,3)\}$. *Thus $(3,3)$ is an isolated point of E.*

Definition 1.2.22 (Closed set) *Let $E \subset U$ and let E' be the set of limit points of E. We say that E is closed if $E \supset E'$.*

Example 1.2.23

Let $U = \mathbb{R}^2$ and $r > 0$, thus $E = \overline{B}_r((0,0))$ is closed.

Definition 1.2.24 *A point $u \in E \subset U$ is said to be an interior point of E if there exists $r > 0$ such that $V_r(u) \subset E$, where*

$$V_r(u) = \{v \in U \mid d(u,v) < r\}.$$

Example 1.2.25

For $U = \mathbb{R}^2$, let $E = B_1((0,0)) \cup \{(3,3)\}$, for example $u = (0.25, 0.25)$ is an interior point of E, in fact, for $r = 0.5$, if $v \in V_r(u)$ then $d(u,v) < 0.5$ so that $d(v,(0,0)) \leq d((0,0),u) + d(u,v) \leq \sqrt{1/8} + 0.5 < 1$ that is, $v \in B_1((0,0))$ and thus $V_r(u) \subset B_1((0,0))$. We may conclude that u is an interior point of $B_1((0,0))$. In fact all points of $B_1((0,0))$ are interior.

Definition 1.2.26 (Open set) *$E \subset U$ is said to be open if all its points are interior.*

Example 1.2.27

For $U = \mathbb{R}^2$, the ball $B_1(0,0)$ is open.

Definition 1.2.28 *Let $E \subset U$, we define its complement, denoted by E^c, by:*

$$E^c = \{v \in U \mid v \notin E\}.$$

Definition 1.2.29 *A set $E \subset U$ is said to bounded if there exists $M > 0$ such that*

$$\sup\{d(u,v) \mid u,v \in E\} \leq M.$$

Definition 1.2.30 *A set $E \subset U$ is said to be dense in U if each point of U is either a point of E or it is a limit point of E, that is, $U = E \cup E'$.*

Example 1.2.31

The set $\mathbb{Q}$ is dense in $\mathbb{R}$. Let $u \in \mathbb{R}$ and let $r > 0$. Thus, from a well known result in elementary analysis there exists $v \in \mathbb{Q}$ such that $u < v < u + r$, that is, $v \in \mathbb{Q} \cap V_r(u)$ and $v \neq u$, where $V_r(u) = (u - r, u + r)$. Therefore u is a limit point of $\mathbb{Q}$. Since $u \in \mathbb{R}$ is arbitrary, we may conclude that $\mathbb{R} \subset \mathbb{Q}'$, that is, $\mathbb{Q}$ is dense in $\mathbb{R}$.

Theorem 1.2.32 *Let (U, d) be a metric space. Let $u \in U$ and $r > 0$. Then $V_r(u)$ is open.*

Proof 1.1   First we recall that
$$V_r(u) = \{v \in U \mid d(u, v) < r\}.$$
Let $v \in V_r(u)$. We have to show that v is an interior point of $V_r(u)$. Define $r_1 = r - d(u, v) > 0$. We shall show that $V_{r_1}(v) \subset V_r(u)$.

Let $w \in V_{r_1}(v)$, thus $d(v, w) < r_1$. Hence
$$d(u, w) \leq d(u, v) + d(v, w) < d(u, v) + r_1 = r.$$

Therefore $w \in V_r(u), \forall w \in V_{r_1}(v)$, that is $V_{r_1}(v) \subset V_r(u)$, so that we may conclude that v is an interior point of $V_r(u), \forall v \in V_r(u)$, thus, $V_r(u)$ is open. The proof is complete.

Theorem 1.2.33 *Let u be a limit point of $E \subset U$, where (U, d) is a metric space. Then each neighborhood of u has an infinite number of points of E, distinct from u.*

Proof 1.2   Suppose to obtain contradiction, that there exists $r > 0$ such that $V_r(u)$ has a finite number of points of E distinct from u. Let $\{v_1, ..., v_n\}$ be such points of $V_r(u) \cap E$ distinct from u. Choose $0 < r_1 < \min\{d(u, v_1), d(u, v_2), ..., d(u, v_n)\}$. Hence $V_{r_1}(u) \subset V_r(u)$ and $v_i \notin V_{r_1}(u), \forall i \in \{1, 2, ..., n\}$. Therefore either $V_{r_1}(u) \cap E = \{u\}$ or $V_{r_1} \cap E = \emptyset$, which contradicts the fact that u is a limit point of E.

The proof is complete.

Corollary 1.2.34 *Let $E \subset U$ be a finite set. Then E has no limit points.*

1.2.7   Properties of open and closed sets in a metric space

In this section we present some basic properties of open and closed sets.

Proposition 1.2.35 *Let $\{E_\alpha, \alpha \in L\}$ be a collection of sets. Then*
$$(\cup_{\alpha \in L} E_\alpha)^c = \cap_{\alpha \in L} E_\alpha^c.$$

Proof 1.3   Observe that
$$
\begin{aligned}
u \in (\cup_{\alpha \in L} E_\alpha)^c \quad &\Leftrightarrow \quad u \notin \cup_{\alpha \in L} E_\alpha \\
&\Leftrightarrow \quad u \notin E_\alpha, \forall \alpha \in L \\
&\Leftrightarrow \quad u \in E_\alpha^c, \forall \alpha \in L \\
&\Leftrightarrow \quad u \in \cap_{\alpha \in L} E_\alpha^c. \tag{1.6}
\end{aligned}
$$

Exercise 1.2.36

Prove that
$$(\cap_{\alpha \in L} E_\alpha)^c = \cup_{\alpha \in L} E_\alpha^c.$$

Theorem 1.2.37 *Let (U, d) be a metric space and $E \subset U$. Thus, E is open if and only if E^c is closed.*

Proof 1.4   Suppose E^c is closed. Choose $u \in E$, thus $u \notin E^c$ and therefore u is not a limit point E^c. Hence there exists $r > 0$ such that $V_r(u) \cap E^c = \emptyset$. Hence, $V_r(u) \subset E$, that is, u is an interior point of E, $\forall u \in E$, so that E is open.

Reciprocally, suppose E is open. Let $u \in (E^c)'$. Thus for each $r > 0$ there exists $v \in V_r(u) \cap E^c$ such that $v \neq u$, so that

$$V_r(u) \nsubseteq E, \ \forall r > 0.$$

Therefore u is not an interior point of E. Since E is open we have that $u \notin E$, that is, $u \in E^c$. Hence $(E^c)' \subset E^c$, that is, E^c is closed.

The proof is complete.

Corollary 1.2.38 *Let (U, d) be a metric space, $F \subset U$ is closed if and only if F^c is open.*

Theorem 1.2.39 *Let (U, d) be a metric space.*

1. *If $G_\alpha \subset U$ and G_α is open $\forall \alpha \in L$, then*

$$\cup_{\alpha \in L} G_\alpha$$

is open.

2. *If $F_\alpha \subset U$ and F_α is closed $\forall \alpha \in L$, then*

$$\cap_{\alpha \in L} F_\alpha$$

is closed.

3. *If $G_1, ..., G_n \subset U$ and G_i is open $\forall i \in \{1, ..., n\}$, then*

$$\cap_{i=1}^n G_i$$

is open.

4. *If $F_1, ..., F_n \subset U$ and F_i is closed $\forall i \in \{1, ..., n\}$, then*

$$\cup_{i=1}^n F_i$$

is closed.

Proof 1.5

1. Let $G_\alpha \subset U$, where G_α is open $\forall \alpha \in L$. Let $u \in \cup_{\alpha \in L} G_\alpha$. Thus $u \in G_{\alpha_0}$ for some $\alpha_0 \in L$. Since G_{α_0} is open, there exists $r > 0$ such that $V_r(u) \subset G_{\alpha_0} \subset \cup_{\alpha \in L} G_\alpha$. Hence, u is an interior point, $\forall u \in \cup_{\alpha \in L} G_\alpha$. Thus $\cup_{\alpha \in L} G_\alpha$ is open.

2. Let $F_\alpha \subset U$, where F_α is closed $\forall \alpha \in L$. Thus, F_α^c is open $\forall \alpha \in L$. From the last item, we have $\cup_{\alpha \in L} F_\alpha^c$ is open so that

$$\cap_{\alpha \in L} F_\alpha = (\cup_{\alpha \in L} F_\alpha^c)^c$$

is closed.

3. Let $G_1, ..., G_n \subset U$ be open sets. Let

$$u \in \cap_{i=1}^n G_i.$$

Thus,

$$u \in G_i, \forall i \in \{1, ..., n\}.$$

Since G_i is open, there exists $r_i > 0$ such that $V_{r_i}(u) \subset G_i$.

Define $r = \min\{r_1, ..., r_n\}$. Hence, $V_r(u) \subset V_{r_i}(u) \subset G_i, \forall i \in \{1, ..., n\}$ and therefore

$$V_r(u) \subset \cap_{i=1}^{n} G_i.$$

This means that u is an interior point of $\cap_{i=1}^{n} G_i$, and being $u \in \cap_{i=1}^{n} G_i$ arbitrary we obtain that $\cap_{i=1}^{n} G_i$ is open.

4. Let $F_1, ..., F_n \subset U$ be closed sets. Thus, $F_1^c, ..., F_n^c$ are open. Thus, from the last item, we obtain:

$$\cap_{i=1}^{n} F_i^c$$

is open, so that

$$\cup_{i=1}^{n} F_i = (\cap_{i=1}^{n} F_i^c)^c$$

is closed.

The proof is complete.

Exercise 1.2.40

Let (U, d) be a metric space and let $u_0 \in U$. Show that $A = \{u_0\}$ is closed. Let $B = \{u_1, ..., u_n\} \subset U$. Show that B is closed.

Definition 1.2.41 (Closure) *Let (U, d) be a metric space and let $E \subset U$. Denote the set of limit points of E by E'. We define the closure of E, denoted by $\overline{E}$, by:*

$$\overline{E} = E \cup E'.$$

Examples 1.2.42

1. *Let $U = \mathbb{R}^2$, $E = B_1(0,0)$, we have that $E' = \overline{B}_1(0,0)$, so that in this example $\overline{E} = E \cup E' = E'$.*

2. *Let $U = \mathbb{R}$, $A = \{1/n \, : \, n \in \mathbb{N}\}$, we have that $A' = \{0\}$, and thus $\overline{A} = A \cup A' = A \cup \{0\}$.*

Theorem 1.2.43 *Let (U, d) be a metric space and $E \subset U$. Thus,*

1. *$\overline{E}$ is closed.*

2. *$E = \overline{E} \Leftrightarrow E$ is closed.*

3. *If $F \supset E$ and F is closed, then $F \supset \overline{E}$.*

Proof 1.6

1. Observe that $\overline{E} = E \cup E'$. Let $u \in \overline{E}^c$. Thus $u \notin E$ and $u \notin E'$ (u is not a limit point of E). Therefore, there exists $r > 0$ such that $V_r(u) \cap E = \emptyset$, that is, $V_r(u) \subset E^c$, thus, u is an interior point of E^c.

   We shall prove that $V_r(u) \cap \overline{E} = \emptyset$. Let $v \in V_r(u)$ and define $r_1 = r - d(u,v) > 0$. We shall show that

   $$V_{r_1}(v) \subset V_r(u).$$

   Let $w \in V_{r_1}(v)$, thus $d(v,w) < r_1$ and therefore

   $$d(u,w) \leq d(u,v) + d(v,w) < d(u,v) + r_1 = r,$$

   that is, $w \in V_r(u)$. Hence,

   $$V_{r_1}(v) \subset V_r(u),$$

   and thus v is not a limit point of E, that is, $v \in \overline{E}^c, \forall v \in V_r(u)$. Thus, $V_r(u) \subset \overline{E}^c$ which means that u is an interior point of $\overline{E}^c$, so that $\overline{E}^c$ is open, and hence $\overline{E}$ is closed.

2. Observe that $E \subset \overline{E} = E \cup E'$. Suppose that E is closed. Thus $E \supset E'$, that is $E \supset E \cup E' = \overline{E}$. Hence $E = \overline{E}$. Suppose $E = \overline{E}$. From the last item $\overline{E}$ is closed, and thus E is closed.

3. Let F be a closed set such that $F \supset E$. Thus, $F' \supset E'$.

   Hence

$$F = \overline{F} = F \cup F' \supset E \cup E' = \overline{E}.$$

The proof is complete.

1.2.8 Compact sets

Definition 1.2.44 (Open covering) *Let (U,d) be a metric space. We say that a collection of sets $\{G_\alpha, \alpha \in L\} \subset U$ is an open covering of $A \subset U$ if*

$$A \subset \cup_{\alpha \in L} G_\alpha$$

and G_α is open, $\forall \alpha \in L$.

Definition 1.2.45 (Compact set) *Let (U,d) be a metric space and $K \subset U$. We say that K is compact if each open covering $\{G_\alpha, \ \alpha \in L\}$ of K admits a finite sub-covering. That is, if $K \subset \cup_{\alpha \in L} G_\alpha$, and G_α is open $\forall \alpha \in L$, then there exist $\alpha_1, \alpha_2, ..., \alpha_n \in L$ such that $K \subset \cup_{i=1}^n G_{\alpha_i}$.*

Theorem 1.2.46 *Let (U,d) be a metric space. Let $K \subset U$ where K is compact. Then K is closed.*

Proof 1.7   Let us show that K^c is open. Let $u \in K^c$. For convenience, let us generically denote in this proof $V_r(u) = V(u,r)$.

For each $v \in K$ we have $d(u,v) > 0$. Define $r_v = d(u,v)/2$. Thus,

$$V(u,r_v) \cap V(v,r_v) = \emptyset, \forall v \in K. \tag{1.7}$$

Observe that

$$\cup_{v \in K} V(v,r_v) \supset K.$$

since K is compact, there exist $v_1, ..., v_n \in K$ such that

$$K \subset \cup_{i=1}^n V(v_i, r_{v_i}). \tag{1.8}$$

Define $r_0 = \min\{r_{v_1}, ..., r_{v_n}\}$, thus

$$V(u,r_0) \subset V(u,r_{v_i}), \forall i \in \{1,...,n\},$$

so that from this and (1.7) we get

$$V(u,r_0) \cap V(v_i, r_{v_i}) = \emptyset, \forall i \in \{1,2,...,n\}.$$

Hence,

$$V(u,r_0) \cap (\cup_{i=1}^n V(v_i, r_{v_i})) = \emptyset.$$

From this and (1.8) we obtain, $V(u,r_0) \cap K = \emptyset$, that is $V(u,r_0) \subset K^c$. Therefore u is an interior point of K^c and being $u \in K^c$ arbitrary, K^c is open so that K is closed.

The proof is complete.

Theorem 1.2.47 *Let (U,d) be a metric space. If $F \subset K \subset U$, K is compact and F is closed, then F is compact.*

Proof 1.8 Let $\{G_\alpha, \alpha \in L\}$ be an open covering of F, that is

$$F \subset \cup_{\alpha \in L} G_\alpha.$$

Observe that $U = F \cup F^c \supset K$, and thus,

$$F^c \cup (\cup_{\alpha \in L} G_\alpha) \supset K.$$

Therefore, since F^c is open $\{F^c, G_\alpha, \alpha \in L\}$ is an open covering of K, and since K is compact, there exist $\alpha_1, ..., \alpha_n \in L$ such that

$$F^c \cup G_{\alpha_1} \cup ... \cup G_{\alpha_n} \supset K \supset F.$$

Therefore

$$G_{\alpha_1} \cup ... \cup G_{\alpha_n} \supset F,$$

so that F is compact.

Theorem 1.2.48 *If $\{K_\alpha, \alpha \in L\}$ is a collection of compact sets in a metric space (U, d) such that the intersection of each finite sub-collection is non-empty, then*

$$\cap_{\alpha \in L} K_\alpha \neq \emptyset.$$

Proof 1.9 Suppose, to obtain contradiction, that

$$\cap_{\alpha \in L} K_\alpha = \emptyset. \tag{1.9}$$

Fix $\alpha_0 \in L$ and denote $L_1 = L \setminus \{\alpha_0\}$. From (1.9) we obtain

$$K_{\alpha_0} \cap (\cap_{\alpha \in L_1} K_\alpha) = \emptyset.$$

Hence

$$K_{\alpha_0} \subset (\cap_{\alpha \in L_1} K_\alpha)^c,$$

that is,

$$K_{\alpha_0} \subset \cup_{\alpha \in L_1} K_\alpha^c.$$

Since, K_{α_0} is compact and K_α^c is open, $\forall \alpha \in L$, there exist $\alpha_1, \alpha_2, ..., \alpha_n \in L_1$ such that

$$K_{\alpha_0} \subset \cup_{j=1}^n K_{\alpha_j}^c = \left(\cap_{j=1}^n K_{\alpha_j}\right)^c,$$

therefore,

$$K_{\alpha_0} \cap \left(\cap_{j=1}^n K_{\alpha_j}\right) = K_{\alpha_0} \cap K_{\alpha_1} \cap ... \cap K_{\alpha_n} = \emptyset,$$

which contradicts the hypotheses. The proof is complete.

Corollary 1.2.49 *Let (U, d) be a metric space. If $\{K_n, n \in \mathbb{N}\} \subset U$ is a sequence of compact non-empty sets such that $K_n \supset K_{n+1}, \forall n \in \mathbb{N}$ then $\cap_{n=1}^\infty K_n \neq \emptyset$.*

Theorem 1.2.50 *Let (U, d) be a metric space. If $E \subset K \subset U$, K is compact and E is infinite, then E has at least one limit point in K.*

Proof 1.10 Suppose, to obtain contradiction, that no point of K is a limit point of E. Then, for each $u \in K$ there exists $r_u > 0$ such that $V(u, r_u)$ has at most one point of E, namely, u if $u \in E$. Observe that $\{V(u, r_u), u \in K\}$ is an open covering of K and therefore of E. Since each $V(u, r_u)$ has at most one point of E which is infinite, no finite sub-covering (relating the open cover in question), covers E, and hence no finite sub-covering covers $K \supset E$, which contradicts the fact that K is compact. This completes the proof.

Theorem 1.2.51 *Let $\{I_n\}$ be a sequence of bounded closed non-empty real intervals, such that $I_n \supset I_{n+1}, \forall n \in \mathbb{N}$. Thus, $\cap_{n=1}^{\infty} I_n \neq \emptyset$.*

Proof 1.11   Let $I_n = [a_n, b_n]$ and let $E = \{a_n, n \in \mathbb{N}\}$. Thus, $E \neq \emptyset$ and E is upper bounded by b_1. Let $x = \sup E$.

Observe that, given $m, n \in \mathbb{N}$ we have that

$$a_n \leq a_{n+m} \leq b_{n+m} \leq b_m,$$

so that

$$\sup_{n \in \mathbb{N}} a_n \leq b_m, \forall m \in \mathbb{N},$$

that is, $x \leq b_m, \forall m \in \mathbb{N}$. Hence,

$$a_m \leq x \leq b_m, \forall m \in \mathbb{N},$$

that is,

$$x \in [a_m, b_m], \forall m \in \mathbb{N},$$

so that $x \in \cap_{m=1}^{\infty} I_m$.

The proof is complete.

Theorem 1.1

Let $I = [a, b] \subset \mathbb{R}$ be a bounded closed non-empty real interval. Under such hypotheses, I is compact.

Proof 1.12   Observe that if $x, y \in [a, b]$ then $|x - y| \leq (b - a)$.. Suppose there exists an open covering of I, denoted by $\{G_\alpha, \alpha \in L\}$ for which there is no finite sub-covering.

Let $c = (a + b)/2$. Thus, either $[a, c]$ or $[c, b]$ has no finite sub-covering related to $\{G_\alpha, \alpha \in L\}$. Denote such an interval by I_1. Dividing I_1 into two connected closed sub-intervals of same size, we get an interval I_2 for which there is no finite sub-covering related to $\{G_\alpha, \alpha \in L\}$.

Proceeding in this fashion, we may obtain a sequence of closed intervals $\{I_n\}$ such that

1. $I_n \supset I_{n+1}, \forall n \in \mathbb{N}$.

2. No finite sub-collection of $\{G_\alpha, \alpha \in L\}$ covers $I_n, \forall n \in \mathbb{N}$.

3. If $x, y \in I_n$ then $|x - y| \leq 2^{-n}(b - a)$.

From the last theorem, there exists $x^* \in \mathbb{R}$ such that $x^* \in \cap_{n=1}^{\infty} I_n \subset I \subset \cup_{\alpha \in L} G_\alpha$. Hence, there exists $\alpha_0 \in L$ such that $x^* \in G_{\alpha_0}$. Since G_{α_0} is open, there exists $r > 0$ such that

$$V_r(x^*) = (x^* - r, x^* + r) \subset G_{\alpha_0}.$$

Choose $n_0 \in \mathbb{N}$ such that

$$2^{-n_0}(b - a) < r/2.$$

Hence, since $x^* \in I_{n_0}$, if $y \in I_{n_0}$ then from item 3 above, $|y - x^*| \leq 2^{-n_0}(b - a) < r/2$, that is $y \in V_r(x^*) \subset G_{\alpha_0}$.

Therefore

$$y \in I_{n_0} \Rightarrow y \in G_{\alpha_0},$$

so that $I_{n_0} \subset G_{\alpha_0}$, which contradicts the item 2 above indicated.

The proof is complete.

Theorem 1.2 Heine-Borel

Let $E \subset \mathbb{R}$, thus the following three properties are equivalent:

1. *E is closed and bounded.*

2. *E is compact.*

3. *Each infinite subset of E has a limit point of E.*

Proof 1.13

■ 1 implies 2: Let $E \subset \mathbb{R}$ be a closed and bounded. Thus, since E is bounded there exists $[a,b]$, a bounded closed interval such that $E \subset [a,b]$. From the last theorem $[a,b]$ is compact and since E is closed, from Theorem 1.2.47 we may infer that E is compact.

■ 2 implies 3: This follows from Theorem 1.2.50.

■ 3 implies 1: We prove the contrapositive, that is, the negation of 1 implies the negation of 3.

The negation of 1 is: E is not bounded or E is not closed. If $E \subset \mathbb{R}$ is not bounded, choosing $x_1 \in E$, for each $n \in \mathbb{N}$ there exists $x_{n+1} \in E$ such that $|x_{n+1}| > n + |x_n| \geq n$. Hence $\{x_n\}$ has no limit points so that we have got the negation of 3.

On the other hand, suppose E is not closed. Thus there exists $x_0 \in \mathbb{R}$ such that $x_0 \in E'$ and $x_0 \notin E$.

Since $x_0 \in E'$, for each $n \in \mathbb{N}$ there exists $x_n \in E$ such that $|x_n - x_0| < 1/n$ $(x_n \in V_{1/n}(x_0))$.

Let $y \in E$, we are going to show that y is not limit point $\{x_n\} \subset E$. Observe that

$$
\begin{aligned}
|x_n - y| &\geq |x_0 - y| - |x_n - x_0| \\
&> |x_0 - y| - 1/n \\
&> |x_0 - y|/2 > 0
\end{aligned}
\tag{1.10}
$$

for all n sufficiently big.

Hence, y is not a limit point of $\{x_n\}$, $\forall y \in E$. Therefore $\{x_n\} \subset E$ is a infinite set with no limit point in E.

In any case, we have got the negation of 3. This completes the proof.

Theorem 1.2.52 (Weierstrass) *Any real set which is bounded and infinite has a limit point in $\mathbb{R}$.*

Proof 1.14 Let $E \subset \mathbb{R}$ be a bounded infinite set. Thus, there exists $r > 0$ such that $E \subset [-r,r] = I_r$. Since E is infinite and I_r is compact, from Theorem 1.2.50, E has a limit point in $I_r \subset \mathbb{R}$. The proof is complete.

1.2.9 Separable metric spaces

Definition 1.2.53 (Separable metric space) *Let (V,d) be a metric space. We say that a set $M \subset V$ is dense in V if*

$$
\overline{M} = M \cup M' = V.
$$

If V has a dense subset which is countable, we say that V is separable.

Example 1.2.54 *$V = \mathbb{R}$ is separable. Indeed, $\mathbb{Q}$, the set of rational number, is dense in $\mathbb{R}$ and countable.*

Example 1.2.55 *The space l^∞ is not separable.*
In fact, let $A \subset l^\infty$ be the set of all real sequences whose entries are only 0 and 1.
From elementary analysis it is well known that A is non-countable.
Let $0 < \varepsilon < 1/4$.

Suppose, to obtain contradiction, that

$$B = \{u_n\}_{n \in \mathbb{N}} \subset l^\infty$$

is dense in l^∞.

Thus, for each $v \in A$, we may select a $n_v \in \mathbb{N}$ such that

$$d(v, u_{n_v}) < \varepsilon.$$

Let $v, w \in A$ be such that $v \neq w$. Therefore,

$$d(v, w) = 1,$$

so that

$$d(v, w) \leq d(v, u_{n_v}) + d(u_{n_v}, w),$$

and thus

$$
\begin{aligned}
d(u_{n_v}, w) &\geq 1 - d(v, u_{n_v}) \\
&\geq 1 - \varepsilon \\
&> 1 - 1/4 \\
&= 3/4 \\
&> \varepsilon.
\end{aligned}
\tag{1.11}
$$

So to summarize, if $v \neq w$, then

$$u_{n_v} \neq u_{n_w}.$$

Let $T : A \to B$, where

$$T(v) = u_{n_v}.$$

Thus T is a bijection on $Im(T) \subset B$.

Therefore, generically denoting $C \sim D$ if, and only if, there exists a bijection between the sets C and D, we have

$$A \sim Im(T) \sim B \sim \mathbb{N}.$$

This contradicts A to be non-countable.
So, we may infer that l^∞ is non-separable.

Exercise 1.2.56 *Let $1 \leq p < +\infty$. Prove that l^p is separable.*

1.2.10 Complete metric spaces

Definition 1.1 Let $\{u_n\} \subset V$, where (V, d) is a metric space.

We say that $u_0 \in V$ is the limit of $\{u_n\}$ as n goes to infinity (∞), if for each $\varepsilon > 0$ there exists $n_0 \in \mathbb{N}$ such that if $n > n_0$, then

$$d(u_n, u_0) < \varepsilon.$$

In such a a case we denote

$$\lim_{n \to \infty} u_n = u_0,$$

or

$$u_n \to u_0, \text{ as } n \to \infty$$

and say that the sequence $\{u_n\}$ is convergent.

Thus

$$\lim_{n\to\infty} d(u_n, v_n) = \lim_{n\to\infty} d(u'_n, v'_n), \ \forall \{u'_n\} \in \hat{u}, \ \{v'_n\} \in \hat{v}.$$

Therefore, the candidate to metric in question is well defined.

Furthermore,

$$\begin{aligned}
\hat{d}(\hat{u}, \hat{v}) = 0 \quad &\Leftrightarrow \quad \lim_{n\to\infty} d(u_n, v_n) = 0 \\
&\Leftrightarrow \quad \{v_n\} \in \hat{u} \\
&\Leftrightarrow \quad \hat{u} = \hat{v}.
\end{aligned} \tag{1.12}$$

Finally, let $\hat{u}, \hat{v}$ and $\hat{w} \in \hat{V}$.

Thus,

$$\begin{aligned}
\hat{d}(\hat{u}, \hat{w}) &= \lim_{n\to\infty} d(u_n, w_n) \\
&\leq \lim_{n\to\infty} [d(u_n, v_n) + d(v_n, w_n)] \\
&= \lim_{n\to\infty} d(u_n, v_n) + \lim_{n\to\infty} d(v_n, w_n) \\
&= d(\hat{u}, \hat{v}) + d(\hat{v}, \hat{w}).
\end{aligned} \tag{1.13}$$

From this we may conclude that $\hat{d}$ is in fact a metric for $\hat{V}$.

2. We shall show now that V is isometric of a dense subspace of $\hat{V}$.

Let $b \in V$. Define $\hat{b}$ by its representative

$$\{u_n\} = \{b, b, b, \ldots\}.$$

Define

$$W = \{\hat{b} = \widehat{\{b, b, b, \ldots\}} \ : \ b \in V\}.$$

Define also $T : V \to W$ by

$$T(b) = \hat{b} = \widehat{\{b, b, b, \ldots\}}.$$

Thus,

$$\hat{d}(\hat{b}, \hat{c}) = \lim_{n\to\infty} d(b, c) = d(b, c).$$

Therefore, T is an isometry.

We are going to show that W is dense in $\hat{V}$.

Let $\hat{u} \in \hat{V}$ and $\{u_n\} \in \hat{u}$. Let $\varepsilon > 0$.

Since $\{u_n\}$ is a Cauchy sequence, there exists $n_0 \in \mathbb{N}$ such that if $m, n > n_0$, then

$$d(u_n, u_m) < \varepsilon/2.$$

Choose $N > n_0$.

Thus, $d(u_n, u_N) < \varepsilon/2, \forall n > n_0$.

Observe that

$$\hat{u}_N = \widehat{\{u_N, u_N, u_N, \ldots\}} \in W,$$

and

$$\hat{d}(\hat{u}, \hat{u}_N) = \lim_{n\to\infty} d(u_n, u_N) \leq \varepsilon/2 < \varepsilon.$$

Since $\varepsilon > 0$ is arbitrary, we may conclude that

$$\hat{u} \in W' \cup W, \forall \hat{u} \in \hat{V},$$

that is, W is dense in $\hat{V}$.

3. Now we shall show that $\hat{V}$ complete.

Let $\{\hat{u}_n\}$ be a Cauchy sequence in $\hat{V}$.

Since W is dense in $\hat{V}$, for each $n \in \mathbb{N}$ there exists $\hat{z}_n \in W$ such that

$$\hat{d}(\hat{u}_n, \hat{z}_n) < \frac{1}{n}.$$

Observe that

$$
\begin{aligned}
\hat{d}(\hat{z}_n, \hat{z}_m) &\leq \hat{d}(\hat{z}_m, \hat{u}_m) + d(\hat{u}_m, \hat{u}_n) + \hat{d}(\hat{u}_n, \hat{z}_n) \\
&\leq \frac{1}{n} + d(\hat{u}_m, \hat{u}_n) + \frac{1}{m}.
\end{aligned}
\tag{1.14}
$$

Let $\varepsilon > 0$ (a new one). Hence, there exists $n_0 \in \mathbb{N}$ such that if $m, n > n_0$, then

$$\hat{d}(\hat{u}_n, \hat{u}_m) < \frac{\varepsilon}{3}.$$

Thus, if

$$m, n > \max\left\{\frac{3}{\varepsilon}, n_0\right\},$$

then

$$\hat{d}(\hat{z}_n, \hat{z}_m) < \varepsilon.$$

Therefore, $\{\hat{z}_m\}$ is a Cauchy sequence and since $T : V \to W$ is an isometry it follows that

$$\{z_m\} = \{T^{-1}(\hat{z}_m)\},$$

is also a Cauchy one.

Let $\hat{u}$ be the class of $\{z_m\}$. We will show that

$$\lim_{n \to \infty} \hat{d}(\hat{u}_n, \hat{u}) = 0.$$

Indeed,

$$
\begin{aligned}
\hat{d}(\hat{u}_n, \hat{u}) &\leq \hat{d}(\hat{u}_n, \hat{z}_n) + \hat{d}(\hat{z}_n, \hat{u}) \\
&\leq \frac{1}{n} + \lim_{m \to \infty} d(z_n, z_m).
\end{aligned}
\tag{1.15}
$$

Therefore,

$$\lim_{n \to \infty} \hat{d}(\hat{u}_n, \hat{u}) = 0.$$

Thus, $\hat{V}$ is complete.

The proof is complete.

1.4   Advanced topics on compactness in metric spaces

Definition 1.4.1 (Diameter of a set) *Let (U,d) be a metric space and $A \subset U$. We define the diameter of A, denoted by $diam(A)$ by*

$$diam(A) = \sup\{d(u,v) \mid u,v \in A\}.$$

Definition 1.4.2 *Let (U,d) be a metric space. We say that $\{F_k\} \subset U$ is a nested sequence of sets if*

$$F_1 \supset F_2 \supset F_3 \supset \ldots.$$

Theorem 1.4.3 *If (U,d) is a complete metric space then every nested sequence of non-empty closed sets $\{F_k\}$ is such that*

$$\lim_{k \to +\infty} diam(F_k) = 0$$

has non-empty intersection, that is

$$\cap_{k=1}^{\infty} F_k \neq \emptyset.$$

Proof 1.16   Suppose $\{F_k\}$ is a nested sequence and $\lim_{k \to \infty} diam(F_k) = 0$. For each $n \in \mathbb{N}$ select $u_n \in F_n$. Suppose given $\varepsilon > 0$. Since

$$\lim_{n \to \infty} diam(F_n) = 0,$$

there exists $N \in \mathbb{N}$ such that if $n \geq N$ then

$$diam(F_n) < \varepsilon.$$

Thus, if $m,n > N$ we have $u_m, u_n \in F_N$ so that

$$d(u_n, u_m) < \varepsilon.$$

Hence, $\{u_n\}$ is a Cauchy sequence. Being U complete, there exists $u \in U$ such that

$$u_n \to u \text{ as } n \to \infty.$$

Choose $m \in \mathbb{N}$. We have that $u_n \in F_m, \forall n > m$, so that

$$u \in \bar{F}_m = F_m.$$

Since $m \in \mathbb{N}$ is arbitrary we obtain

$$u \in \cap_{m=1}^{\infty} F_m.$$

The proof is complete.

Theorem 1.4.4 *Let (U,d) be a metric space. If $A \subset U$ is compact then it is closed and bounded.*

Proof 1.17   We have already proved that A is closed. Suppose, to obtain contradiction that A is not bounded. Thus for each $K \in \mathbb{N}$ there exists $u, v \in A$ such that

$$d(u,v) > K.$$

Observe that

$$A \subset \cup_{u \in A} B_1(u).$$

Since A is compact there exists $u_1, u_2, ..., u_n \in A$ such that

$$A = \subset \cup_{k=1}^{n} B_1(u_k).$$

Define
$$R = \max\{d(u_i, u_j) \mid i, j \in \{1, ..., n\}\}.$$

Choose $u, v \in A$ such that

$$d(u, v) > R + 2. \tag{1.16}$$

Observe that there exist $i, j \in \{1, ..., n\}$ such that

$$u \in B_1(u_i), \ v \in B_1(u_j).$$

Thus

$$
\begin{aligned}
d(u, v) &\leq d(u, u_i) + d(u_i, u_j) + d(u_j, v) \\
&\leq 2 + R,
\end{aligned} \tag{1.17}
$$

which contradicts (1.16). This completes the proof.

Definition 1.4.5 (Relative compactness) *In a metric space (U, d) a set $A \subset U$ is said to be relatively compact if $\overline{A}$ is compact.*

Definition 1.4.6 (ε - nets) *Let (U, d) be a metric space. A set $N \subset U$ is said to be a ε-net with respect to a set $A \subset U$ if for each $u \in A$ there exists $v \in N$ such that*

$$d(u, v) < \varepsilon.$$

Definition 1.4.7 *Let (U, d) be a metric space. A set $A \subset U$ is said to be totally bounded if for each $\varepsilon > 0$ there exists a finite ε-net with respect to A.*

Proposition 1.4.8 *Let (U, d) be a metric space. If $A \subset U$ is totally bounded then it is bounded.*

Proof 1.18   Choose $u, v \in A$. Let $\{u_1, ..., u_n\}$ be the $1 - net$ with respect to A. Define

$$R = \max\{d(u_i, u_j) \mid i, j \in \{1, ..., n\}\}.$$

Observe that there exist $i, j \in \{1, ..., n\}$ such that

$$d(u, u_i) < 1, \ d(v, u_j) < 1.$$

Thus

$$
\begin{aligned}
d(u, v) &\leq d(u, u_i) + d(u_i, u_j) + d(u_j, v) \\
&\leq R + 2.
\end{aligned} \tag{1.18}
$$

Since $u, v \in A$ are arbitrary, A is bounded.

Theorem 1.4.9 *Let (U, d) be a metric space. If from each sequence $\{u_n\} \subset A$ we can select a convergent subsequence $\{u_{n_k}\}$, then A is totally bounded.*

Proof 1.19   Suppose, to obtain contradiction, that A is not totally bounded. Thus there exists $\varepsilon_0 > 0$ such that there exists no ε_0-net with respect to A. Choose $u_1 \in A$, hence $\{u_1\}$ is not a ε_0-net, that is, there exists $u_2 \in A$ such that

$$d(u_1, u_2) > \varepsilon_0.$$

Again $\{u_1, u_2\}$ is not a ε_0-net for A, so that there exists $u_3 \in A$ such that

$$d(u_1, u_3) > \varepsilon_0 \text{ and } d(u_2, u_3) > \varepsilon_0.$$

Proceeding in this fashion we can obtain a sequence $\{u_n\}$ such that

$$d(u_n, u_m) > \varepsilon_0, \text{ if } m \neq n. \tag{1.19}$$

Clearly, we cannot extract a convergent subsequence of $\{u_n\}$, otherwise such a subsequence would be Cauchy contradicting (1.19). The proof is complete.

Definition 1.4.10 (Sequentially compact sets) *Let (U, d) be a metric space. A set $A \subset U$ is said to be sequentially compact if for each sequence $\{u_n\} \subset A$ there exists a subsequence $\{u_{n_k}\}$ and $u \in A$ such that*

$$u_{n_k} \to u, \text{ as } k \to \infty.$$

Theorem 1.4.11 *A subset A of a metric space (U, d) is compact if and only if it is sequentially compact.*

Proof 1.20   Suppose A is compact. By Proposition 2.4.8, A is countably compact. Let $\{u_n\} \subset A$ be a sequence. We have two situations to consider.

1. $\{u_n\}$ has infinitely many equal terms, that is in this case we have

$$u_{n_1} = u_{n_2} = \dots = u_{n_k} = \dots = u \in A.$$

   Thus the result follows trivially.

2. $\{u_n\}$ has infinitely many distinct terms. In such a case, being A countably compact, $\{u_n\}$ has a limit point in A, so that there exists a subsequence $\{u_{n_k}\}$ and $u \in A$ such that

$$u_{n_k} \to u, \text{ as } k \to \infty.$$

In both cases we may find a subsequence converging to some $u \in A$.

Thus, A is sequentially compact.

Conversely suppose A is sequentially compact, and suppose $\{G_\alpha, \ \alpha \in L\}$ is an open cover of A. For each $u \in A$ define

$$\delta(u) = \sup\{r \mid B_r(u) \subset G_\alpha, \text{ for some } \alpha \in L\}.$$

First we prove that $\delta(u) > 0, \forall u \in A$. Choose $u \in A$. Since $A \subset \cup_{\alpha \in L} G_\alpha$, there exists $\alpha_0 \in L$ such that $u \in G_{\alpha_0}$. Being G_{α_0} open, there exists $r_0 > 0$ such that $B_{r_0}(u) \subset G_{\alpha_0}$.

Thus,

$$\delta(u) \geq r_0 > 0.$$

Now define δ_0 by

$$\delta_0 = \inf\{\delta(u) \mid u \in A\}.$$

Therefore, there exists a sequence $\{u_n\} \subset A$ such that

$$\delta(u_n) \to \delta_0 \text{ as } n \to \infty.$$

Since A is sequentially compact, we may obtain a subsequence $\{u_{n_k}\}$ and $u_0 \in A$ such that

$$\delta(u_{n_k}) \to \delta_0 \text{ and } u_{n_k} \to u_0,$$

as $k \to \infty$. Therefore, we may find $K_0 \in \mathbb{N}$ such that if $k > K_0$ then

$$d(u_{n_k}, u_0) < \frac{\delta(u_0)}{4}. \tag{1.20}$$

We claim that

$$\delta(u_{n_k}) \geq \frac{\delta(u_0)}{4}, \text{ if } k > K_0.$$

To prove the claim, suppose

$$z \in B_{\frac{\delta(u_0)}{4}}(u_{n_k}), \forall k > K_0,$$

(observe that in particular from (1.20)

$$u_0 \in B_{\frac{\delta(u_0)}{4}}(u_{n_k}), \forall k > K_0).$$

Since

$$\frac{\delta(u_0)}{2} < \delta(u_0),$$

there exists some $\alpha_1 \in L$ such that

$$B_{\frac{\delta(u_0)}{2}}(u_0) \subset G_{\alpha_1}.$$

However, since

$$d(u_{n_k}, u_0) < \frac{\delta(u_0)}{4}, \text{ if } k > K_0,$$

we obtain

$$B_{\frac{\delta(u_0)}{2}}(u_0) \supset B_{\frac{\delta(u_0)}{4}}(u_{n_k}), \text{ if } k > K_0,$$

so that

$$\delta(u_{n_k}) \geq \frac{\delta(u_0)}{4}, \forall k > K_0.$$

Therefore,

$$\lim_{k \to \infty} \delta(u_{n_k}) = \delta_0 \geq \frac{\delta(u_0)}{4}.$$

Choose $\varepsilon > 0$ such that

$$\delta_0 > \varepsilon > 0.$$

From the last theorem, since A it is sequentially compact, it is totally bounded. For the $\varepsilon > 0$ chosen above, consider an ε-net contained in A (the fact that the ε-net may be chosen contained in A is also a consequence of last theorem) and denote it by N, that is,

$$N = \{v_1, ..., v_n\} \in A.$$

Since $\delta_0 > \varepsilon$, there exists

$$\alpha_1, ..., \alpha_n \in L$$

such that

$$B_\varepsilon(v_i) \subset G_{\alpha_i}, \forall i \in \{1, ..., n\},$$

considering that

$$\delta(v_i) \geq \delta_0 > \varepsilon > 0, \forall i \in \{1, ..., n\}.$$

For $u \in A$, since N is an ε-net we have

$$u \in \cup_{i=1}^n B_\varepsilon(v_i) \subset \cup_{i=1}^n G_{\alpha_i}.$$

Since $u \in U$ is arbitrary, we obtain

$$A \subset \cup_{i=1}^{n} G_{\alpha_i}.$$

Thus

$$\{G_{\alpha_1}, ..., G_{\alpha_n}\}$$

is a finite subcover for A of

$$\{G_\alpha, \ \alpha \in L\}.$$

Hence A is compact.

The proof is complete;

Theorem 1.4.12 *Let (U, d) be a metric space. Thus, $A \subset U$ is relatively compact if and only if for each sequence in A, we may select a convergent subsequence.*

Proof 1.21 Suppose A is relatively compact. Thus $\overline{A}$ is compact so that from the last Theorem, $\overline{A}$ is sequentially compact.

Thus from each sequence in $\overline{A}$ we may select a subsequence which converges to some element of $\overline{A}$. In particular, for each sequence in $A \subset \overline{A}$, we may select a subsequence that converges to some element of $\overline{A}$.

Conversely, suppose that for each sequence in A we may select a convergent subsequence. It suffices to prove that $\overline{A}$ is sequentially compact. Let $\{v_n\}$ be a sequence in $\overline{A}$. Since A is dense in $\overline{A}$, there exists a sequence $\{u_n\} \subset A$ such that

$$d(u_n, v_n) < \frac{1}{n}.$$

From the hypothesis we may obtain a subsequence $\{u_{n_k}\}$ and $u_0 \in \overline{A}$ such that

$$u_{n_k} \to u_0, \ \text{as } k \to \infty.$$

Thus,

$$v_{n_k} \to u_0 \in \overline{A}, \ \text{as } k \to \infty.$$

Therefore, $\overline{A}$ is sequentially compact so that it is compact.

Theorem 1.4.13 *Let (U, d) be a metric space.*

1. *If $A \subset U$ is relatively compact then it is totally bounded.*

2. *If (U, d) is a complete metric space and $A \subset U$ is totaly bounded then A is relatively compact.*

Proof 1.22

1. Suppose $A \subset U$ is relatively compact. From the last theorem, from each sequence in A, we can extract a convergent subsequence. From Theorem 1.4.9, A is totally bounded.

2. Let (U, d) be a metric space and let A be a totally bounded subset of U.

   Let $\{u_n\}$ be a sequence in A. Since A is totally bounded for each $k \in \mathbb{N}$, we find a ε_k-net where $\varepsilon_k = 1/k$, denoted by N_k, where

   $$N_k = \{v_1^{(k)}, v_2^{(k)}, ..., v_{n_k}^{(k)}\}.$$

   In particular for $k = 1$ $\{u_n\}$ is contained in the 1-net N_1. Thus at least one ball of radius 1 of N_1 contains infinitely many points of $\{u_n\}$. Let us select a subsequence $\{u_{n_k}^{(1)}\}_{k \in \mathbb{N}}$ of this infinite set (which is contained in a ball of radius 1). Similarly, we may select a subsequence here just partially relabeled $\{u_{n_l}^{(2)}\}_{l \in \mathbb{N}}$ of $\{u_{n_k}^{(1)}\}$ which is contained in one of the balls of the $\frac{1}{2}$-net. Proceeding in this fashion for each $k \in \mathbb{N}$, we may find a subsequence denoted by $\{u_{n_m}^{(k)}\}_{m \in \mathbb{N}}$ of the original sequence contained in a ball of radius $1/k$.

Now consider the diagonal sequence denoted by $\{u_{n_k}^{(k)}\}_{k \in \mathbb{N}} = \{z_k\}$. Thus

$$d(z_n, z_m) < \frac{2}{k}, \text{ if } m, n > k,$$

that is, $\{z_k\}$ is a Cauchy sequence, and since (U, d) is complete, there exists $u \in U$ such that

$$z_k \to u \text{ as } k \to \infty.$$

From Theorem 1.4.12, A is relatively compact.

The proof is complete.

1.5 The Arzela-Ascoli theorem

In this section we present a classical result in analysis, namely the Arzela-Ascoli theorem.

Definition 1.5.1 (Equi-continuity) *Let $\mathscr{F}$ be a collection of complex functions defined on a metric space (U, d). We say that $\mathscr{F}$ is equicontinuous if for each $\varepsilon > 0$, there exists $\delta > 0$ such that if $u, v \in U$ and $d(u, v) < \delta$ then*

$$|f(u) - f(v)| < \varepsilon, \forall f \in \mathscr{F}.$$

Furthermore, we say that $\mathscr{F}$ is point-wise bounded if for each $u \in U$ there exists $M(u) \in \mathbb{R}$ such that

$$|f(u)| < M(u), \forall f \in \mathscr{F}.$$

Theorem 1.5.2 (Arzela-Ascoli) *Suppose $\mathscr{F}$ is a point-wise bounded equicontinuous collection of complex functions defined on a metric space (U, d). Also suppose that U has a countable dense subset E. Thus, each sequence $\{f_n\} \subset \mathscr{F}$ has a subsequence that converges uniformly on every compact subset of U.*

Proof 1.23   Let $\{u_n\}$ be a countable dense set in (U, d). By hypothesis, $\{f_n(u_1)\}$ is a bounded sequence, therefore it has a convergent subsequence, which is denoted by $\{f_{n_k}(u_1)\}$. Let us denote

$$f_{n_k}(u_1) = \tilde{f}_{1,k}(u_1), \forall k \in \mathbb{N}.$$

Thus there exists $g_1 \in \mathbb{C}$ such that

$$\tilde{f}_{1,k}(u_1) \to g_1, \text{ as } k \to \infty.$$

Observe that $\{f_{n_k}(u_2)\}$ is also bounded and also it has a convergent subsequence, which similarly as above we will denote by $\{\tilde{f}_{2,k}(u_2)\}$. Again there exists $g_2 \in \mathbb{C}$ such that

$$\tilde{f}_{2,k}(u_1) \to g_1, \text{ as } k \to \infty.$$

$$\tilde{f}_{2,k}(u_2) \to g_2, \text{ as } k \to \infty.$$

Proceeding in this fashion for each $m \in \mathbb{N}$ we may obtain $\{\tilde{f}_{m,k}\}$ such that

$$\tilde{f}_{m,k}(u_j) \to g_j, \text{ as } k \to \infty, \forall j \in \{1, ..., m\},$$

where the set $\{g_1, g_2, ..., g_m\}$ is obtained as above. Consider the diagonal sequence

$$\{\tilde{f}_{k,k}\},$$

and observe that the sequence

$$\{\tilde{f}_{k,k}(u_m)\}_{k>m}$$

is such that

$$\tilde{f}_{k,k}(u_m) \to g_m \in \mathbb{C}, \text{ as } k \to \infty, \forall m \in \mathbb{N}.$$

Therefore we may conclude that from $\{f_n\}$ we may extract a subsequence also denoted by

$$\{f_{n_k}\} = \{\tilde{f}_{k,k}\}$$

which is convergent in

$$E = \{u_n\}_{n \in \mathbb{N}}.$$

Now suppose $K \subset U$, being K compact. Suppose given $\varepsilon > 0$. From the equi-continuity hypothesis there exists $\delta > 0$ such that if $u, v \in U$ and $d(u,v) < \delta$ we have

$$|f_{n_k}(u) - f_{n_k}(v)| < \frac{\varepsilon}{3}, \forall k \in \mathbb{N}.$$

Observe that

$$K \subset \cup_{u \in K} B_{\frac{\delta}{2}}(u),$$

and being K compact we may find $\{\tilde{u}_1, ..., \tilde{u}_M\}$ such that

$$K \subset \cup_{j=1}^{M} B_{\frac{\delta}{2}}(\tilde{u}_j).$$

Since E is dense in U, there exists

$$v_j \in B_{\frac{\delta}{2}}(\tilde{u}_j) \cap E, \forall j \in \{1, ..., M\}.$$

Fixing $j \in \{1, ..., M\}$, from $v_j \in E$ we obtain that

$$\lim_{k \to \infty} f_{n_k}(v_j)$$

exists as $k \to \infty$. Hence there exists $K_{0_j} \in \mathbb{N}$ such that if $k, l > K_{0_j}$ then

$$|f_{n_k}(v_j) - f_{n_l}(v_j)| < \frac{\varepsilon}{3}.$$

Pick $u \in K$, thus

$$u \in B_{\frac{\delta}{2}}(\tilde{u}_{\hat{j}})$$

for some $\hat{j} \in \{1, ..., M\}$, so that

$$d(u, v_{\hat{j}}) < \delta.$$

Therefore, if

$$k, l > \max\{K_{0_1}, ..., K_{0_M}\},$$

then

$$\begin{aligned}
|f_{n_k}(u) - f_{n_l}(u)| &\leq |f_{n_k}(u) - f_{n_k}(v_{\hat{j}})| + |f_{n_k}(v_{\hat{j}}) - f_{n_l}(v_{\hat{j}})| \\
&\quad + |f_{n_l}(v_{\hat{j}}) - f_{n_l}(u)| \\
&\leq \frac{\varepsilon}{3} + \frac{\varepsilon}{3} + \frac{\varepsilon}{3} = \varepsilon.
\end{aligned} \tag{1.21}$$

Since $u \in K$ is arbitrary, we conclude that $\{f_{n_k}\}$ is uniformly Cauchy on K.

The proof is complete.

Chapter 2

Topological Vector Spaces

2.1 Introduction

The main objective of this chapter is to present an outline of the basic tools of analysis necessary to develop the subsequent chapters. We assume the reader has a background in linear algebra and elementary real analysis at an undergraduate level. The main references for this chapter are the excellent books on functional analysis, Rudin [67], Bachman and Narici [6] and Reed and Simon [62]. All proofs are developed in detail.

2.2 Vector spaces

We denote by $\mathbb{F}$ a scalar field. In practice this is either $\mathbb{R}$ or $\mathbb{C}$, the set of real or complex numbers.

Definition 2.2.1 (Vector spaces) *A vector space over $\mathbb{F}$ is a set which we will denote by U whose elements are called vectors, for which are defined two operations, namely, addition denoted by $(+) : U \times U \to U$, and scalar multiplication denoted by $(\cdot) : \mathbb{F} \times U \to U$, so that the following relations are valid*

1. *$u + v = v + u, \forall u, v \in U,$*

2. *$u + (v + w) = (u + v) + w, \forall u, v, w \in U,$*

3. *there exists a vector denoted by θ such that $u + \theta = u, \forall u \in U,$*

4. *for each $u \in U$, there exists a unique vector denoted by $-u$ such that $u + (-u) = \theta,$*

5. *$\alpha \cdot (\beta \cdot u) = (\alpha \cdot \beta) \cdot u, \forall \alpha, \beta \in \mathbb{F}, u \in U,$*

6. *$\alpha \cdot (u + v) = \alpha \cdot u + \alpha \cdot v, \forall \alpha \in \mathbb{F}, u, v \in U,$*

7. *$(\alpha + \beta) \cdot u = \alpha \cdot u + \beta \cdot u, \forall \alpha, \beta \in \mathbb{F}, u \in U,$*

8. *$1 \cdot u = u, \forall u \in U.$*

Remark 2.2.2 *From now on we may drop the dot $(\cdot)$ in scalar multiplications and denote $\alpha \cdot u$ simply as αu.*

Definition 2.2.3 (Vector subspace) *Let U be a vector space. A set $V \subset U$ is said to be a vector subspace of U if V is also a vector space with the same operations as those of U. If $V \neq U$ we say that V is a proper subspace of U.*

Definition 2.2.4 (Finite dimensional space) *A vector space is said to be of finite dimension if there exists fixed $u_1, u_2, ..., u_n \in U$ such that for each $u \in U$ there are corresponding $\alpha_1, ..., \alpha_n \in \mathbb{F}$ for which*

$$u = \sum_{i=1}^{n} \alpha_i u_i. \tag{2.1}$$

Definition 2.2.5 (Topological spaces) *A set U is said to be a topological space if it is possible to define a collection σ of subsets of U called a topology in U, for which are valid the following properties:*

1. *$U \in \sigma$,*

2. *$\emptyset \in \sigma$,*

3. *if $A \in \sigma$ and $B \in \sigma$ then $A \cap B \in \sigma$, and*

4. *arbitrary unions of elements in σ also belong to σ.*

Any $A \in \sigma$ is said to be an open set.

Remark 2.2.6 *When necessary, to clarify the notation, we shall denote the vector space U endowed with the topology σ by (U, σ).*

Definition 2.2.7 (Closed sets) *Let U be a topological space. A set $A \subset U$ is said to be closed if $U \setminus A$ is open. We also denote $U \setminus A = A^c = \{u \in U \mid u \notin A\}$.*

Remark 2.2.8 *For any sets $A, B \subset U$ we denote*

$$A \setminus B = \{u \in A \mid u \notin B\}.$$

Sometimes, as the meaning is clear, we may also denote $A \setminus B = A - B$.

Proposition 2.2.9 *For closed sets we have the following properties:*

1. *U and $\emptyset$ are closed,*

2. *If A and B are closed sets then $A \cup B$ is closed,*

3. *Arbitrary intersections of closed sets are closed.*

Proof 2.1

1. Since $\emptyset$ is open and $U = \emptyset^c$, by Definition 2.2.7, U is closed. Similarly, since U is open and $\emptyset = U \setminus U = U^c$, $\emptyset$ is closed.

2. A, B closed implies that A^c and B^c are open, and by Definition 2.2.5, $A^c \cap B^c$ is open, so that $A \cup B = (A^c \cap B^c)^c$ is closed.

3. Consider $A = \cap_{\lambda \in L} A_\lambda$, where L is a collection of indices and A_λ is closed, $\forall \lambda \in L$. We may write $A = (\cup_{\lambda \in L} A_\lambda^c)^c$ and since A_λ^c is open $\forall \lambda \in L$ we have, by Definition 2.2.5, that A is closed.

Definition 2.2.10 (Closure) *Given $A \subset U$, we define the closure of A, denoted by $\bar{A}$, as the intersection of all closed sets that contain A.*

Remark 2.2.11 *From Proposition 2.2.9 Item 3 we have that $\bar{A}$ is the smallest closed set that contains A, in the sense that, if C is closed and $A \subset C$ then $\bar{A} \subset C$.*

Definition 2.2.12 (Interior) *Given $A \subset U$ we define its interior, denoted by A°, as the union of all open sets is contained in A.*

Remark 2.2.13 *It is not difficult to prove that if A is open then $A = A^\circ$.*

Definition 2.2.14 (Neighborhood) *Given $u_0 \in U$ we say that $\mathscr{V}$ is a neighborhood of u_0 if such a set is open and contains u_0. We denote such neighborhoods by $\mathscr{V}_{u_0}$.*

Proposition 2.2.15 *If $A \subset U$ is a set such that for each $u \in A$ there exists a neighborhood $\mathscr{V}_u \ni u$ such that $\mathscr{V}_u \subset A$, then A is open.*

Proof 2.2 This follows from the fact that $A = \cup_{u \in A} \mathscr{V}_u$ and any arbitrary union of open sets is open.

Definition 2.2.16 (Function) *Let U and V be two topological spaces. We say that $f : U \to V$ is a function if f is a collection of pairs $(u,v) \in U \times V$ such that for each $u \in U$ there exists only one $v \in V$ such that $(u,v) \in f$.*
  In such a case we denote
$$v = f(u).$$

Definition 2.2.17 (Continuity at a point) *A function $f : U \to V$ is continuous at $u \in U$ if for each neighborhood $\mathscr{V}_{f(u)} \subset V$ of $f(u)$ there exists a neighborhood $\mathscr{V}_u \subset U$ of u such that $f(\mathscr{V}_u) \subset \mathscr{V}_{f(u)}$.*

Definition 2.2.18 (Continuous function) *A function $f : U \to V$ is continuous if it is continuous at each $u \in U$.*

Proposition 2.2.19 *A function $f : U \to V$ is continuous if and only if $f^{-1}(\mathscr{V})$ is open for each open $\mathscr{V} \subset V$, where*
$$f^{-1}(\mathscr{V}) = \{u \in U \mid f(u) \in \mathscr{V}\}. \tag{2.2}$$

Proof 2.3 Suppose $f^{-1}(\mathscr{V})$ is open whenever $\mathscr{V} \subset V$ is open. Pick $u \in U$ and any open $\mathscr{V}$ such that $f(u) \in \mathscr{V}$. Since $u \in f^{-1}(\mathscr{V})$ and $f(f^{-1}(\mathscr{V})) \subset \mathscr{V}$, we have that f is continuous at $u \in U$. Since $u \in U$ is arbitrary we have that f is continuous. Conversely, suppose f is continuous and pick $\mathscr{V} \subset V$ open. If $f^{-1}(\mathscr{V}) = \emptyset$ we are done, since $\emptyset$ is open. Thus, suppose $u \in f^{-1}(\mathscr{V})$, since f is continuous, there exists $\mathscr{V}_u$ a neighborhood of u such that $f(\mathscr{V}_u) \subset \mathscr{V}$. This means $\mathscr{V}_u \subset f^{-1}(\mathscr{V})$ and therefore, from Proposition 2.2.15, $f^{-1}(\mathscr{V})$ is open.

Definition 2.2.20 *We say that (U, σ) is a Hausdorff topological space if, given $u_1, u_2 \in U$, $u_1 \neq u_2$, there exists $\mathscr{V}_1, \mathscr{V}_2 \in \sigma$ such that*
$$u_1 \in \mathscr{V}_1 \,,\; u_2 \in \mathscr{V}_2 \text{ and } \mathscr{V}_1 \cap \mathscr{V}_2 = \emptyset. \tag{2.3}$$

Definition 2.2.21 (Base) *A collection $\sigma' \subset \sigma$ is said to be a base for σ if every element of σ may be represented as a union of elements of σ'.*

Definition 2.2.22 (Local base) *A collection $\hat{\sigma}$ of neighborhoods of a point $u \in U$ is said to be a local base at u if each neighborhood of u contains a member of $\hat{\sigma}$.*

Definition 2.2.23 (Topological vector spaces) *A vector space endowed with a topology, denoted by (U, σ), is said to be a topological vector space if and only if*

  1. *Every single point of U is a closed set,*

  2. *The vector space operations (addition and scalar multiplication) are continuous with respect to σ.*

More specifically, addition is continuous if, given $u, v \in U$ and $\mathcal{V} \in \sigma$ such that $u + v \in \mathcal{V}$ then there exists $\mathcal{V}_u \ni u$ and $\mathcal{V}_v \ni v$ such that $\mathcal{V}_u + \mathcal{V}_v \subset \mathcal{V}$. On the other hand, scalar multiplication is continuous if given $\alpha \in \mathbb{F}$, $u \in U$ and $\mathcal{V} \ni \alpha \cdot u$, there exists $\delta > 0$ and $\mathcal{V}_u \ni u$ such that, $\forall \beta \in \mathbb{F}$ satisfying $|\beta - \alpha| < \delta$ we have $\beta \mathcal{V}_u \subset \mathcal{V}$.

Given (U, σ), let us associate with each $u_0 \in U$ and $\alpha_0 \in \mathbb{F}$ ($\alpha_0 \neq 0$) the functions $T_{u_0} : U \to U$ and $M_{\alpha_0} : U \to U$ defined by

$$T_{u_0}(u) = u_0 + u \tag{2.4}$$

and

$$M_{\alpha_0}(u) = \alpha_0 \cdot u. \tag{2.5}$$

The continuity of such functions is a straightforward consequence of the continuity of vector space operations (addition and scalar multiplication). It is clear that the respective inverse maps, namely T_{-u_0} and M_{1/α_0} are also continuous. So if $\mathcal{V}$ is open, then $u_0 + \mathcal{V}$, that is, $(T_{-u_0})^{-1}(\mathcal{V}) = T_{u_0}(\mathcal{V}) = u_0 + \mathcal{V}$ is open. By analogy $\alpha_0 \mathcal{V}$ is open. Thus σ is completely determined by a 'local base', so that the term local base will be understood henceforth as a local base at θ. So to summarize, a local base of a topological vector space is a collection Ω of neighborhoods of θ, such that each neighborhood of θ contains a member of Ω.

Now we present some simple results, namely:

Proposition 2.2.24 *If $A \subset U$, then $\forall u \in A$, there exists a neighborhood $\mathcal{V}$ of θ such that $u + \mathcal{V} \subset A$*

Proof 2.4   Just take $\mathcal{V} = A - u$.

Proposition 2.2.25 *Given a topological vector space (U, σ), any element of σ may be expressed as a union of translates of members of Ω, so that the local base Ω generates the topology σ.*

Proof 2.5   Let $A \subset U$ open and $u \in A$. $\mathcal{V} = A - u$ is a neighborhood of θ and by definition of local base, there exists a set $\mathcal{V}_{\Omega_u} \subset \mathcal{V}$ such that $\mathcal{V}_{\Omega_u} \in \Omega$. Thus, we may write

$$A = \cup_{u \in A}(u + \mathcal{V}_{\Omega_u}). \tag{2.6}$$

2.3   Some properties of topological vector spaces

In this section we study some fundamental properties of topological vector spaces. We start with the following proposition:

Proposition 2.3.1 *Any topological vector space U is a Hausdorff space.*

Proof 2.6   Pick $u_0, u_1 \in U$ such that $u_0 \neq u_1$. Thus $\mathcal{V} = U \setminus \{u_1 - u_0\}$ is an open neighborhood of zero. As $\theta + \theta = \theta$, by the continuity of addition, there exist $\mathcal{V}_1$ and $\mathcal{V}_2$ neighborhoods of θ such that

$$\mathcal{V}_1 + \mathcal{V}_2 \subset \mathcal{V} \tag{2.7}$$

define $\mathcal{U} = \mathcal{V}_1 \cap \mathcal{V}_2 \cap (-\mathcal{V}_1) \cap (-\mathcal{V}_2)$, thus $\mathcal{U} = -\mathcal{U}$ (symmetric) and $\mathcal{U} + \mathcal{U} \subset \mathcal{V}$ and hence

$$u_0 + \mathcal{U} + \mathcal{U} \subset u_0 + \mathcal{V} \subset U \setminus \{u_1\} \tag{2.8}$$

so that

$$u_0 + v_1 + v_2 \neq u_1, \quad \forall v_1, v_2 \in \mathcal{U}, \tag{2.9}$$

or

$$u_0 + v_1 \neq u_1 - v_2, \quad \forall v_1, v_2 \in \mathscr{U}, \tag{2.10}$$

and since $\mathscr{U} = -\mathscr{U}$

$$(u_0 + \mathscr{U}) \cap (u_1 + \mathscr{U}) = \emptyset. \tag{2.11}$$

Definition 2.3.2 (Bounded sets) *A set $A \subset U$ is said to be bounded if to each neighborhood of zero $\mathscr{V}$ there corresponds a number $s > 0$ such that $A \subset t\mathscr{V}$ for each $t > s$.*

Definition 2.3.3 (Convex sets) *A set $A \subset U$ such that*

$$if \ u, v \in A \ then \ \lambda u + (1 - \lambda)v \in A, \quad \forall \lambda \in [0, 1], \tag{2.12}$$

is said to be convex.

Definition 2.3.4 (Locally convex spaces) *A topological vector space U is said to be locally convex if there is a local base Ω whose elements are convex.*

Definition 2.3.5 (Balanced sets) *A set $A \subset U$ is said to be balanced if $\alpha A \subset A$, $\forall \alpha \in \mathbb{F}$ is such that $|\alpha| \leq 1$.*

Theorem 2.3.6 *In a topological vector space U we have:*

1. *Every neighborhood of zero contains a balanced neighborhood of zero,*

2. *Every convex neighborhood of zero contains a balanced convex neighborhood of zero.*

Proof 2.7

1. Suppose $\mathscr{U}$ is a neighborhood of zero. From the continuity of scalar multiplication, there exist $\mathscr{V}$ (neighborhood of zero) and $\delta > 0$, such that $\alpha\mathscr{V} \subset \mathscr{U}$ whenever $|\alpha| < \delta$. Define $\mathscr{W} = \cup_{|\alpha|<\delta}\alpha\mathscr{V}$, thus $\mathscr{W} \subset \mathscr{U}$ is a balanced neighborhood of zero.

2. Suppose $\mathscr{U}$ is a convex neighborhood of zero in U. Define

$$A = \{\cap \alpha\mathscr{U} \mid \alpha \in \mathbb{C}, \ |\alpha| = 1\}. \tag{2.13}$$

As $0 \cdot \theta = \theta$ (where $\theta \in U$ denotes the zero vector) from the continuity of scalar multiplication there exists $\delta > 0$ and there is a neighborhood of zero $\mathscr{V}$ such that if $|\beta| < \delta$ then $\beta\mathscr{V} \subset \mathscr{U}$. Define $\mathscr{W}$ as the union of all such $\beta\mathscr{V}$. Thus, $\mathscr{W}$ is balanced and $\alpha^{-1}\mathscr{W} = \mathscr{W}$ as $|\alpha| = 1$, so that $\mathscr{W} = \alpha\mathscr{W} \subset \alpha\mathscr{U}$, and hence $\mathscr{W} \subset A$, which implies that the interior A° is a neighborhood of zero. Also, $A^{\circ} \subset \mathscr{U}$. Since A is an intersection of convex sets, it is convex and so is A°. Now will show that A° is balanced and complete the proof. For this, it suffices to prove that A is balanced. Choose r and β such that $0 \leq r \leq 1$ and $|\beta| = 1$. Then

$$r\beta A = \cap_{|\alpha|=1} r\beta\alpha\mathscr{U} = \cap_{|\alpha|=1} r\alpha\mathscr{U}. \tag{2.14}$$

Since $\alpha\mathscr{U}$ is a convex set that contains zero, we obtain $r\alpha\mathscr{U} \subset \alpha\mathscr{U}$, so that $r\beta A \subset A$, which completes the proof.

Proposition 2.3.7 *Let U be a topological vector space and $\mathscr{V}$ a neighborhood of zero in U. Given $u \in U$, there exists $r \in \mathbb{R}^+$ such that $\beta u \in \mathscr{V}$, $\forall \beta$ such that $|\beta| < r$.*

Proof 2.8   Observe that $u + \mathscr{V}$ is a neighborhood of $1 \cdot u$, then by the continuity of scalar multiplication, there exists $\mathscr{W}$ neighborhood of u and $r > 0$ such that

$$\beta \mathscr{W} \subset u + \mathscr{V}, \forall \beta \text{ such that } |\beta - 1| < r, \tag{2.15}$$

so that

$$\beta u \in u + \mathscr{V}, \tag{2.16}$$

or

$$(\beta - 1)u \in \mathscr{V}, \text{ where } |\beta - 1| < r, \tag{2.17}$$

and thus

$$\hat{\beta} u \in \mathscr{V}, \forall \hat{\beta} \text{ such that } |\hat{\beta}| < r, \tag{2.18}$$

which completes the proof.

Corollary 2.3.8 *Let $\mathscr{V}$ be a neighborhood of zero in U, if $\{r_n\}$ is a sequence such that $r_n > 0$, $\forall n \in \mathbb{N}$ and $\lim\limits_{n \to \infty} r_n = \infty$, then $U \subset \cup_{n=1}^{\infty} r_n \mathscr{V}$.*

Proof 2.9   Let $u \in U$, then $\alpha u \in \mathscr{V}$ for any α sufficiently small, from the last proposition $u \in \frac{1}{\alpha} \mathscr{V}$. As $r_n \to \infty$ we have that $r_n > \frac{1}{\alpha}$ for n sufficiently big, so that $u \in r_n \mathscr{V}$, which completes the proof.

Proposition 2.3.9 *Suppose $\{\delta_n\}$ is sequence such that $\delta_n \to 0$, $\delta_n < \delta_{n-1}$, $\forall n \in \mathbb{N}$ and $\mathscr{V}$ a bounded neighborhood of zero in U, then $\{\delta_n \mathscr{V}\}$ is a local base for U.*

Proof 2.10   Let $\mathscr{U}$ be a neighborhood of zero; as $\mathscr{V}$ is bounded, there exists $t_0 \in \mathbb{R}^+$ such that $\mathscr{V} \subset t \mathscr{U}$ for any $t > t_0$. As $\lim\limits_{n \to \infty} \delta_n = 0$, there exists $n_0 \in \mathbb{N}$ such that if $n \geq n_0$ then $\delta_n < \frac{1}{t_0}$, so that $\delta_n \mathscr{V} \subset \mathscr{U}$, $\forall n$ such that $n \geq n_0$.

Definition 2.3.10 (Convergence in topological vector spaces) *Let U be a topological vector space. We say $\{u_n\}$ converges to $u_0 \in U$, if for each neighborhood $\mathscr{V}$ of u_0 then, there exists $N \in \mathbb{N}$ such that*

$$u_n \in \mathscr{V}, \forall n \geq N.$$

Definition 2.3.11 (Dense set) *Let (V, σ) be a topological vector space (T.V.E.). Let $A, B \subset V$. We say that A is dense in B as*

$$B \subset \overline{A}.$$

Definition 2.3.12 *We say that topological vector space V is separable as it has a dense and countable set.*

2.3.1   Nets and convergence

Definition 2.3.13 *A directed system is a set of indices I, with an order relation $\prec$, which satisfies the following properties:*

1. If α, $\beta \in I$, then there exists $\gamma \in I$ such that

$$\alpha \prec \gamma \text{ and } \beta \prec \gamma.$$

2. $\prec$ is a partial order relation.

Definition 2.3.14 (Net) *Let (V, σ) be a topological space. A net in (V, σ) is a function defined on a directed system I with range in V, where we denote such a net by*

$$\{u_\alpha\}_{\alpha \in I},$$

and where

$$u_\alpha \in V, \, \forall \alpha \in I.$$

Definition 2.3.15 (Convergent net) *Let (V, σ) be a topological space and let $\{u_\alpha\}_{\alpha \in I}$ be a net in V.*
    We say that such a net converges to $u \in V$ as for each neighborhood $W \in \sigma$ of u there exists $\beta \in I$ such that if $\alpha \succ \beta$, then

$$u_\alpha \in W.$$

Definition 2.3.16 *Let (V, σ) be a topological space and let $\{u_\alpha\}_{\alpha \in I}$ be a net in V. We say that $u \in V$ is a cluster point of the net in question as for each neighborhood $W \in \sigma$ of u and each $\beta \in I$, there exists $\alpha \succ \beta$ such that*

$$u_\alpha \in W.$$

Definition 2.3.17 (Limit point) *Let (V, σ) be a topological space and let $A \subset V$. We say that $u \in V$ is a limit point of A as for each neighborhood $W \in \sigma$ of u, there exists $v \in W \cap A$ such that $v \neq u$.*

Theorem 2.3.18 *Let (V, d) be a topological space and let $A \subset V$.*
    Under such hypotheses,

$$\overline{A} = A \cup A'$$

where A' denotes the set of limit points of A.

Proof 2.11   Let $u \in A \cup A'$.
    If $u \in A$, then $u \in \overline{A}$.
    Thus, suppose $u \in A' \setminus A$.
    Hence, for each neighborhood $W \in \sigma$ of u, there exists $u_w \in A \setminus \{u\}$ such that $u_w \in W$.
    Denote by I the set of all neighborhoods of u, partially ordered by the relation

$$W_1 \prec W_2 \Leftrightarrow W_2 \subset W_1.$$

From the exposed above we may obtain a net $\{u_w\}_{w \in I}$ such that

$$u_w \to u.$$

Assume, to obtain contradiction, that $u \notin \overline{A}$.
Hence $u \in \overline{A}^c$ which is an open set. Since $u_w \to u$, there exists $W_1 \in I$ such that if $W_2 \succ W_1$, then

$$u_{w_2} \in \overline{A}^c.$$

In particular, $u_{w_2} \in A^c$, if $W_2 \succ W_1$, which contradicts

$$u_{w_2} \in A \setminus \{u\}.$$

Summarizing,

$$u \in \overline{A}, \, \forall u \in A \cup A'.$$

Therefore,

$$A \cup A' \subset \overline{A}. \tag{2.19}$$

Reciprocally, suppose $u \in \overline{A}$.

If $u \in A$, then $u \in A \cup A'$.

Suppose, to obtain contradiction, that $u \notin A$ and $u \notin A'$.

Thus there exists a neighborhood $W \in \sigma$ of u such that

$$W \cap A = \emptyset.$$

Thus, $A \subset W^c$ and W^c is closed, so that

$$\overline{A} \subset W^c.$$

From this and $u \in W$ we get

$$u \notin \overline{A},$$

a contradiction. Therefore $u \in A$ or $u \in A'$, $\forall u \in \overline{A}$.

Thus

$$\overline{A} \subset A \cup A'. \tag{2.20}$$

From (2.19) and (2.20), we obtain

$$\overline{A} = A \cup A'.$$

This complete the proof.

Theorem 2.3.19 *Let (V_1, σ_1) and (V_2, σ_2) be topological spaces.*

Let $f : V_1 \to V_2$ be a function.

Let $u \in V_1$. Thus f is continuous at u if, and only if, for each net $\{u_\alpha\}_{\alpha \in I} \subset V_1$ such that $u_\alpha \to u$, we have that

$$f(u_\alpha) \to f(u).$$

Proof 2.12   Suppose f is continuous at u. Let $\{u_\alpha\}_{\alpha \in I}$ be a net such that

$$u_\alpha \to u.$$

Let $W_{f(u)} \in \sigma_2$ be such that $f(u) \in W_{f(u)}$.

From the hypotheses, there exists $V_u \in \sigma_1$ such that $u \in V_u$ and

$$f(V_u) \subset W_{f(u)}.$$

From $u_\alpha \to u$, there exists $\beta \in I$ such that if $\alpha \succ \beta$, then

$$u_\alpha \in V_u.$$

Therefore,

$$f(u_\alpha) \in W_{f(u)}, \ \ \text{if } \alpha \succ \beta.$$

Since $W_{f(u)}$ is arbitrary, it follows that

$$f(u_\alpha) \to f(u).$$

Reciprocally, suppose

$$f(u_\alpha) \to f(u)$$

whenever

$$u_\alpha \to u.$$

Suppose, to obtain contradiction, that f is not continuous at u.

Thus there exists $W_{f(u)} \in \sigma_2$ such that $f(u) \in W_{f(u)}$ and so that for each neighborhood $W \in \sigma_1$ of u there exists $u_W \in W$ such that

$$f(u_W) \notin W_{f(u)}.$$

Denote by I the set of all neighborhoods of u, partially ordered by the relation

$$W_1 \prec W_2 \Leftrightarrow W_2 \subset W_1.$$

Thus, the net $\{u_W\}_{W \in I}$ is such that

$$u_W \to u.$$

However,

$$f(u_W) \notin W_{f(u)}, \ \forall W \in I.$$

Hence, $\{f(u_W)\}$ does not converges to $f(u)$, which contradicts the reciprocal hypothesis. The proof is complete.

2.4 Compactness in topological vector spaces

We start this section with the definition of open covering.

Definition 2.4.1 (Open covering) *Given $B \subset U$ we say that $\{\mathcal{O}_\alpha, \ \alpha \in A\}$ is a covering of B if $B \subset \cup_{\alpha \in A} \mathcal{O}_\alpha$ If $\mathcal{O}_\alpha$ is open $\forall \alpha \in A$ then $\{\mathcal{O}_\alpha\}$ is said to be an open covering of B.*

Definition 2.4.2 (Compact sets) *A set $B \subset U$ is said to be compact if each open covering of B has a finite sub-covering. More explicitly, if $B \subset \cup_{\alpha \in A} \mathcal{O}_\alpha$, where $\mathcal{O}_\alpha$ is open $\forall \alpha \in A$, then, there exist $\alpha_1, ..., \alpha_n \in A$ such that $B \subset \mathcal{O}_{\alpha_1} \cup ... \cup \mathcal{O}_{\alpha_n}$, for some n, a finite positive integer.*

Theorem 2.4.3 *Let (V, σ) be a topological space. Let $K \subset V$.*
Under such hypotheses, K is compact if, and only if, each net $\{u_\alpha\}_{\alpha \in I} \subset K$ has a limit point in K.

Proof 2.13 Suppose K is compact. Let $\{u_\alpha\}_{\alpha \in I} \subset K$ be a net with infinite distinct terms (otherwise the result is immediate).

Denote $E = \{u_\alpha\}_{\alpha \in I}$. Suppose, to obtain contradiction, that no point of K is a limit point of E.

Hence, for each $u \in K$, there exists a neighborhood W_u of u such that

$$W_u \cap E = \emptyset,$$

or

$$W_u \cap E = \{u\} \text{ if } u \in E.$$

In any case each W_u has no more than one point of E.

Observe that $\cup_{u \in K} W_u \supset K$. Since K is compact, there exist $u_1, \ldots, u_n \in K$ such that

$$E \subset K \subset \cup_{j=1}^n W_{u_j}.$$

From this we may conclude that E has no more than n distinct elements, which contradicts E to have infinity distinct terms.

Reciprocally, suppose that each net $\{u_\alpha\}_{\alpha \in I} \subset K$ has at least one limit point in K.

Suppose, to obtain contradiction K is not compact.

Thus there exists an open covering $\{G_\alpha, \ \alpha \in L\}$ of K which admits no finite sub-covering.

Denote by F the finite sub-collections of $\{G_\alpha, \ \alpha \in L\}$.

Hence, for a $W \in F$ we may select a $u_W \notin W$ where $u_W \in K$.

Let us partially order F through the relation

$$W_1 \prec W_2 \Leftrightarrow W_1 \subset W_2.$$

From the hypotheses, the net $\{u_W\}_{W \in F}$ has a limit point $u \in K$.

Observe that

$$u \in K \subset \cup_{\alpha \in L} G_\alpha.$$

Thus, there exists $\alpha_0 \in L$ such that

$$u \in G_{\alpha_0}.$$

Since u is a limit point of $\{u_W\}_{W \in F} \subset K$, there exists $W_1 \succ G_{\alpha_0}$ such that

$$u_{W_1} \in G_{\alpha_0} \subset W_1.$$

This contradicts $u_{W_1} \notin W_1$. Therefore, K is compact.
The proof is complete.

Proposition 2.4.4 *A compact subset of a Hausdorff space is closed.*

Proof 2.14   Let U be a Hausdorff space and consider $A \subset U$, A compact. Given $x \in A$ and $y \in A^c$, there exist open sets $\mathcal{O}_x$ and $\mathcal{O}_y^x$ such that $x \in \mathcal{O}_x$, $y \in \mathcal{O}_y^x$ and $\mathcal{O}_x \cap \mathcal{O}_y^x = \emptyset$. It is clear that $A \subset \cup_{x \in A} \mathcal{O}_x$ and since A is compact, we may find $\{x_1, x_2, ..., x_n\}$ such that $A \subset \cup_{i=1}^n \mathcal{O}_{x_i}$. For the selected $y \in A^c$ we have $y \in \cap_{i=1}^n \mathcal{O}_y^{x_i}$ and $(\cap_{i=1}^n \mathcal{O}_y^{x_i}) \cap (\cup_{i=1}^n \mathcal{O}_{x_i}) = \emptyset$. Since $\cap_{i=1}^n \mathcal{O}_y^{x_i}$ is open, and y is an arbitrary point of A^c we have that A^c is open, so that A is closed, which completes the proof. The next result is very useful.

Theorem 2.4.5 *Let $\{K_\alpha, \ \alpha \in L\}$ be a collection of compact subsets of a Hausdorff topological vector space U, such that the intersection of every finite sub-collection (of $\{K_\alpha, \ \alpha \in L\}$) is non-empty.*
Under such hypotheses

$$\cap_{\alpha \in L} K_\alpha \neq \emptyset.$$

Proof 2.15   Fix $\alpha_0 \in L$. Suppose, to obtain contradiction that

$$\cap_{\alpha \in L} K_\alpha = \emptyset.$$

That is,

$$K_{\alpha_0} \cap [\cap_{\substack{\alpha \in L \\ \alpha \neq \alpha_0}} K_\alpha] = \emptyset.$$

Thus,

$$\cap_{\substack{\alpha \in L \\ \alpha \neq \alpha_0}} K_\alpha \subset K_{\alpha_0}^c,$$

so that

$$K_{\alpha_0} \subset [\cap_{\substack{\alpha \in L \\ \alpha \neq \alpha_0}} K_\alpha]^c,$$

$$K_{\alpha_0} \subset [\cup_{\substack{\alpha \in L \\ \alpha \neq \alpha_0}} K_\alpha^c].$$

However K_{α_0} is compact and K_α^c is open, $\forall \alpha \in L$.
Hence, there exist $\alpha_1, ..., \alpha_n \in L$ such that

$$K_{\alpha_0} \subset \cup_{i=1}^n K_{\alpha_i}^c.$$

From this we may infer that

$$K_{\alpha_0} \cap [\cap_{i=1}^n K_{\alpha_i}] = \emptyset,$$

which contradicts the hypotheses.
The proof is complete.

Proposition 2.4.6 *Let U be a topological Hausdorff space and let $A \subset B$ where A is closed and B is compact. Under such hypotheses, A is compact.*

Proof 2.16 Consider $\{\mathcal{O}_\alpha, \alpha \in L\}$ an open cover of A. Thus $\{A^c, \mathcal{O}_\alpha, \alpha \in L\}$ is a cover of U, so that it is a cover of B. As B is compact, there exist $\alpha_1, \alpha_2, ..., \alpha_n$ such that $A^c \cup (\cup_{i=1}^n \mathcal{O}_{\alpha_i}) \supset B \supset A$, so that $\{\mathcal{O}_{\alpha_i}, i \in \{1, ..., n\}\}$ covers A. From this we may infer that A is compact. The proof is complete.

Definition 2.4.7 (Countably compact sets) *A set A is said to be countably compact if every infinite subset of A has a limit point in A.*

Proposition 2.4.8 *Every compact subset of a Hausdorff topological space U is countably compact.*

Proof 2.17 Let B an infinite subset of A compact and suppose B has no limit point in A, so that there is no any limit point. Choose a countable infinite set $\{x_1, x_2, x_3, ...\} \subset B$ and define $F = \{x_1, x_2, x_3, ...\}$. It is clear that F has no limit point. Thus for each $n \in \mathbb{N}$, there exist $\mathcal{O}_n$ open such that $\mathcal{O}_n \cap F = \{x_n\}$. Also, for each $x \in A \setminus F$, there exist $\mathcal{O}_x$ such that $x \in \mathcal{O}_x$ and $\mathcal{O}_x \cap F = \emptyset$. Thus $\{\mathcal{O}_x, x \in A \setminus F; \mathcal{O}_1, \mathcal{O}_2, ...\}$ is an open cover of A without a finite subcover, which contradicts the fact that A is compact.

2.4.1 A note on convexity in topological vector spaces

Definition 2.4.9 *Let (V, σ) be a topological vector space. Let $A \subset V$ be such that $A \neq \emptyset$. We define the convex hull of A, denoted by $Conv(A)$, as*

$$Conv(A) = \left\{ \sum_{k=1}^n \lambda_k u_k : n \in \mathbb{N}, \lambda_k \geq 0, u_k \in A, \forall k \in \{1, ..., n\} \text{ and } \sum_{k=1}^n \lambda_k = 1 \right\}.$$

Theorem 2.4.10 *let (V, σ) be a topological vector space. Let $A \subset V$ be such that $A \neq \emptyset$. Under such hypotheses, $Conv(A)$ is convex.*

Proof 2.18 Let $u, v \in Conv(A)$ and let $\lambda \in [0, 1]$. Thus, there exist $n_1, n_2 \in \mathbb{N}$ such that

$$u = \sum_{k=1}^{n_1} \lambda_k u_k \text{ and } v = \sum_{k=1}^{n_2} \tilde{\lambda}_k v_k,$$

where $u_k \in A$ and $\lambda_k \geq 0, \forall k \in \{1, ... n_1\}$ and also $\sum_{k=1}^{n_1} \lambda_k = 1$.

Moreover, $v_k \in A$, $\tilde{\lambda}_k \geq 0, \forall k \in \{1, ... n_2\}$ and $\sum_{k=1}^{n_2} \tilde{\lambda}_k = 1$.

Thus, we have that

$$\lambda u + (1 - \lambda) v = \sum_{k=1}^{n_1} \lambda \lambda_k u_k + \sum_{k=1}^{n_2} (1 - \lambda) \tilde{\lambda}_k v_k,$$

where

$$\lambda \lambda_k \geq 0, u_k \in A, \forall k \in \{1, ..., n_1\}$$

and

$$(1 - \lambda) \tilde{\lambda}_k \geq 0, v_k \in A, \forall k \in \{1, ..., n_2\}$$

so that

$$\sum_{k=1}^{n_1} \lambda \lambda_k + \sum_{k=1}^{n_2} (1 - \lambda) \tilde{\lambda}_k = \lambda + (1 - \lambda) = 1.$$

Therefore,

$$\lambda u + (1 - \lambda v) \in Conv(A), \forall u, v \in Conv(A), \lambda \in [0, 1].$$

Hence, $Conv(A)$ is convex.
The proof is complete.

Theorem 2.4.11 *Let (V, σ) be a topological vector space. Let $A \subset V$ be such that $A \neq \emptyset$.*
Under such hypotheses, A is convex if, and only if, $Conv(A) = A$.

Proof 2.19 Suppose that A is convex. We shall prove that

$$A = B_n \equiv \left\{ \sum_{k=1}^{n} \lambda_k u_k \; : \; \lambda_k \geq 0, \; u_k \in A, \; \forall k \in \{1, \ldots, n\} \text{ and } \sum_{k=1}^{n} \lambda_k = 1 \right\}, \forall n \in \mathbb{N}.$$

We shall do it by induction on n.
Observe that for $n = 1$ and $n = 2$, from the convexity of A we obtain $A = B_1$ and $A = B_2$.
Let $n \in \mathbb{N}$. Suppose $A = B_n$. We are going to prove that $A = B_{n+1}$ which will complete the induction.
Clearly $B_n \subset B_{n+1}$, so that $A \subset B_{n+1}$.
Reciprocally, let $u \in B_{n+1}$. Thus, there exist $u_1, \ldots, u_{n+1} \in A$ and $\lambda_1, \ldots \lambda_{n+1}$ such that $\lambda_k \geq 0, \forall k \in \{1, \ldots, n+1\}$, $\sum_{k=1}^{n+1} \lambda_k = 1$, and

$$u = \sum_{k=1}^{n+1} \lambda_k u_k.$$

With no loss in generality, assume $0 < \lambda_{n+1} < 1$ (otherwise the conclusion is immediate).
Thus,

$$\lambda_1 + \cdots + \lambda_n = (1 - \lambda_{n+1}) > 0.$$

Hence,

$$\frac{\lambda_1}{1 - \lambda_{n+1}} + \cdots + \frac{\lambda_n}{1 - \lambda_{n+1}} = 1.$$

Therefore, defining

$$\tilde{\lambda}_k = \frac{\lambda_k}{1 - \lambda_{n+1}} \geq 0, \; \forall k \in \{1, \ldots, n\}$$

we have that

$$\sum_{k=1}^{n} \tilde{\lambda}_k = 1,$$

so that

$$w = \sum_{k=1}^{n} \tilde{\lambda}_k u_k \in B_n = A.$$

Since A convex, we obtain

$$w_1 = (1 - \lambda_{n+1})w + \lambda_{n+1} u_{n+1} \in A,$$

that is,

$$w_1 = \sum_{k=1}^{n+1} \lambda_k u_k = u \in A, \forall u \in B_{n+1}.$$

Thus,

$$B_{n+1} \subset A,$$

and hence

$$B_{n+1} = A.$$

This completes the induction, that is,

$$A = B_n, \; \forall n \in \mathbb{N}.$$

Hence,

$$A = \cup_{n=1}^{\infty} B_n = Conv(A).$$

Reciprocally, assume $A = Conv(A)$. Since $Conv(A)$ is convex, A is convex.
The proof is complete.

Remark 2.4.12 *Let $A \subset B \subset V$. Clearly $Conv(A) \subset Conv(B)$. In particular, if B is convex, then*

$$Conv(A) \subset B = Conv(B).$$

Proposition 2.4.13 *Let (V, σ) be a topological vector space. Suppose that a non-empty $A \subset V$ is open. Under such hypotheses, $Conv(A)$ is open.*

Proof 2.20 Let $u \in Conv(A)$. Thus, there exist $n \in \mathbb{N}$, $u_k \in A$, $\lambda_k \geq 0$, $\forall k \in \{1, \ldots, n\}$ such that $\sum_{k=1}^n \lambda_k = 1$, and $u = \sum_{k=1}^n \lambda_k u_k$.

With no loss in generality, assume $\lambda_1 \neq 0$ (redefine the indices, if necessary).

Since $u_1 \in A$ and A is open, there exists a neighborhood V_{u_1} of u_1 such that $V_{u_1} \subset A$.

Thus, $W = \lambda_1 V_{u_1} + \lambda_2 u_2 + \cdots + \lambda_n u_n \subset Conv(A)$.

Observe that W is open and $u \in W \subset Conv(A)$.

Therefore u is an interior point of $Conv(A), \forall u \in Conv(A)$. Thus, $Conv(A)$ is open.

This completes the proof.

Proposition 2.4.14 *Let (V, σ) be a topological vector space. Suppose $A \subset V$ is convex and $A^\circ \neq \emptyset$. Under such hypotheses, A° is convex.*

Proof 2.21 Let $u, v \in A^\circ$ and $\lambda \in [0, 1]$. Thus, there exist neighborhoods V_u of u and V_v of v such that $V_u \subset A$ and $V_v \subset A$. Hence,

$$B \equiv V_u \cup V_v \subset A.$$

Therefore, since A is convex, we obtain

$$Conv(B) \subset Conv(A) = A.$$

From the last proposition $Conv(B)$ is open and moreover $Conv(B) \subset A^\circ$. Thus,

$$\lambda u + (1 - \lambda)v \in Conv(B) \subset A^\circ, \ \forall u, v \in A^\circ, \ \lambda \in [0, 1].$$

From this we may infer that A° is convex.

The proof is complete.

Remark 2.4.15 *Let (V, σ) be a topological vector space a let $A \subset V$ be a non-empty open set..*
Thus, tA is open, $\forall t \in \mathbb{F}$ such that $t \neq 0$.
Let $B \subset V$ be a balanced set such that $\mathbf{0} \in B^\circ$.
Let $\alpha \in \mathbb{F}$ be such that $0 < |\alpha| \leq 1$. Thus,

$$\alpha B^\circ \subset \alpha B \subset B.$$

Since αB° is open, we have that $\alpha B^\circ \subset B^\circ, \forall \alpha \in \mathbb{F}$ such that $|\alpha| \leq 1$.
From this we may infer that B° is balanced.

2.5 Normed and metric spaces

The idea here is to prepare a route for the study of Banach spaces defined below. We start with the definition of norm.

Definition 2.5.1 (Norm) *A vector space U is said to be a normed space, if it is possible to define a function $\| \cdot \|_U : U \to \mathbb{R}^+ = [0, +\infty)$, called a norm, which satisfies the following properties:*

1. $\|u\|_U > 0$, if $u \neq \theta$ and $\|u\|_U = 0 \Leftrightarrow u = \theta$

2. $\|u+v\|_U \leq \|u\|_U + \|v\|_U, \forall\ u,v \in U,$

3. $\|\alpha u\|_U = |\alpha| \|u\|_U, \forall u \in U, \alpha \in \mathbb{F}.$

Now we recall the definition of metric.

Definition 2.5.2 (Metric space) *A vector space U is said to be a metric space if it is possible to define a function $d : U \times U \to \mathbb{R}^+$, called a metric on U, such that*

1. $0 \leq d(u,v),\ \forall u,v \in U,$

2. $d(u,v) = 0 \Leftrightarrow u = v,$

3. $d(u,v) = d(v,u),\ \forall u,v \in U,$

4. $d(u,w) \leq d(u,v) + d(v,w), \forall u,v,w \in U.$

A metric can be defined through a norm, that is

$$d(u,v) = \|u-v\|_U. \tag{2.21}$$

In this case we say that the metric is induced by the norm.

The set $B_r(u) = \{v \in U \mid d(u,v) < r\}$ is called the open ball with center at u and radius r. A metric $d : U \times U \to \mathbb{R}^+$ is said to be invariant if

$$d(u+w, v+w) = d(u,v), \forall u,v,w \in U. \tag{2.22}$$

The following are some basic definitions concerning metric and normed spaces:

Definition 2.5.3 (Convergent sequences) *Given a metric space U, we say that $\{u_n\} \subset U$ converges to $u_0 \in U$ as $n \to \infty$, if for each $\varepsilon > 0$, there exists $n_0 \in \mathbb{N}$, such that if $n \geq n_0$ then $d(u_n, u_0) < \varepsilon$. In this case we write $u_n \to u_0$ as $n \to +\infty$.*

Definition 2.5.4 (Cauchy sequence) *$\{u_n\} \subset U$ is said to be a Cauchy sequence if for each $\varepsilon > 0$ there exists $n_0 \in \mathbb{N}$ such that $d(u_n, u_m) < \varepsilon, \forall m,n \geq n_0$*

Definition 2.5.5 (Completeness) *A metric space U is said to be complete if each Cauchy sequence related to $d : U \times U \to \mathbb{R}^+$ converges to an element of U.*

Definition 2.5.6 (Limit point) *Let (U,d) be a metric space and let $E \subset U$. We say that $v \in U$ is a limit point of E if for each $r > 0$ there exists $w \in B_r(v) \cap E$ such that $w \neq v$.*

Definition 2.5.7 (Interior point, topology for (U,d)) *Let (U,d) be a metric space and let $E \subset U$. We say that $u \in E$ is interior point if there exists $r > 0$ such that $B_r(u) \subset E$. If a point of E is not a limit point is said to be an isolated one. We may define a topology for a metric space (U,d), by declaring as open all set $E \subset U$ such that all its points are interior. Such a topology is said to be induced by the metric d.*

Proposition 2.5.8 *Let (U,d) be a metric space. The set σ of all open sets, defined through the last definition, is indeed a topology for (U,d).*

Proof 2.22

1. Obviously $\emptyset$ and U are open sets.

2. Assume A and B are open sets and define $C = A \cap B$. Let $u \in C = A \cap B$, thus from $u \in A$, there exists $r_1 > 0$ such that $B_{r_1}(u) \subset A$. Similarly from $u \in B$ there exists $r_2 > 0$ such that $B_{r_2}(u) \subset B$.

   Define $r = \min\{r_1, r_2\}$. Thus $B_r(u) \subset A \cap B = C$, so that u is an interior point of C. Since $u \in C$ is arbitrary, we may conclude that C is open.

3. Suppose $\{A_\alpha, \ \alpha \in L\}$ is a collection of open sets. Define $E = \cup_{\alpha \in L} A_\alpha$ and we shall show that E is open.

Choose $u \in E = \cup_{\alpha \in L} A_\alpha$. Thus there exists $\alpha_0 \in L$ such that $u \in A_{\alpha_0}$. Since A_{α_0} is open there exists $r > 0$ such that $B_r(u) \subset A_{\alpha_0} \subset \cup_{\alpha \in L} A_\alpha = E$. Hence u is an interior point of E, since $u \in E$ is arbitrary, $E = \cup_{\alpha \in L} A_\alpha$ is open.

The proof is complete.

Definition 2.5.9 *Let (U,d) be a metric space and let $E \subset U$. We define E' as the set of all the limit points of E.*

Theorem 2.5.10 *Let (U,d) be a metric space and let $E \subset U$. Then E is closed if and only if $E' \subset E$.*

Proof 2.23   Suppose $E' \subset E$. Let $u \in E^c$, thus $u \notin E$ and $u \notin E'$. Therefore there exists $r > 0$ such that $B_r(u) \cap E = \emptyset$, so that $B_r(u) \subset E^c$. Therefore u is an interior point of E^c. Since $u \in E^c$ is arbitrary we may infer that E^c is open, so that $E = (E^c)^c$ is closed.

Conversely, suppose that E is closed, that is E^c is open.

If $E' = \emptyset$ we are done.

Thus assume $E' \neq \emptyset$ and choose $u \in E'$. Thus for each $r > 0$ there exists $v \in B_r(u) \cap E$ such that $v \neq u$. Thus $B_r(u) \not\subset E^c, \forall r > 0$ so that u is not a interior point of E^c. Since E^c is open, we have that $u \notin E^c$ so that $u \in E$. We have thus obtained, $u \in E, \forall u \in E'$, so that $E' \subset E$.

The proof is complete.

Remark 2.5.11 *From this last result, we may conclude that in a metric space $E \subset U$ is closed if and only if $E' \subset E$.*

At this point we recall the definition of Banach space.

Definition 2.5.12 (Banach spaces) *A normed vector space U is said to be a Banach space if each Cauchy sequence related to the metric induced by the norm converges to an element of U.*

Remark 2.5.13 *Let (U,σ) be a topological space. We say that the topology σ is compatible with a metric $d : U \times U \to \mathbb{R}^+$ if σ coincides with the topology generated by such a metric. In this case we say that $d : U \times U \to \mathbb{R}^+$ induces the topology σ.*

Definition 2.5.14 (Metrizable spaces) *A topological vector space (U,σ) is said to be metrizable if σ is compatible with some metric d.*

Definition 2.5.15 (Normable spaces) *A topological vector space (U,σ) is said to be normable if the induced metric (by this norm) is compatible with σ.*

2.6   Linear mappings

Given U, V topological vector spaces, a function (mapping) $f : U \to V, A \subset U$ and $B \subset V$, we define:

$$f(A) = \{f(u) \mid u \in A\}, \tag{2.23}$$

and the inverse image of B, denoted $f^{-1}(B)$ as

$$f^{-1}(B) = \{u \in U \mid f(u) \in B\}. \tag{2.24}$$

Definition 2.6.1 (Linear functions) *A function $f : U \to V$ is said to be linear if*

$$f(\alpha u + \beta v) = \alpha f(u) + \beta f(v), \forall u, v \in U, \ \alpha, \beta \in \mathbb{F}. \tag{2.25}$$

Definition 2.6.2 (Null space and range) *Given $f : U \to V$, we define the null space and the range of f, denoted by $N(f)$ and $R(f)$ respectively, as*

$$N(f) = \{ u \in U \mid f(u) = \theta \}. \tag{2.26}$$

and

$$R(f) = \{ f(u) \; : \; u \in U \}. \tag{2.27}$$

Note that if f is linear then $N(f)$ and $R(f)$ are subspaces of U and V respectively.

Proposition 2.6.3 *Let U, V be topological vector spaces. If $f : U \to V$ is linear and continuous at θ, then it is continuous everywhere.*

Proof 2.24 Since f is linear we have $f(\theta) = \theta$. Since f is continuous at θ, given $\mathscr{V} \subset V$ a neighborhood of zero, there exists $\mathscr{U} \subset U$ neighborhood of zero, such that

$$f(\mathscr{U}) \subset \mathscr{V}. \tag{2.28}$$

Thus

$$v - u \in \mathscr{U} \Rightarrow f(v - u) = f(v) - f(u) \in \mathscr{V}, \tag{2.29}$$

or

$$v \in u + \mathscr{U} \Rightarrow f(v) \in f(u) + \mathscr{V}, \tag{2.30}$$

which means that f is continuous at u. Since u is arbitrary, f is continuous everywhere.

2.7 Linearity and continuity

Definition 2.7.1 (Bounded functions) *A function $f : U \to V$ is said to be bounded if it maps bounded sets into bounded sets.*

Proposition 2.7.2 *A set E is bounded if and only if the following condition is satisfied: whenever $\{u_n\} \subset E$ and $\{\alpha_n\} \subset \mathbb{F}$ are such that $\alpha_n \to 0$ as $n \to \infty$ we have $\alpha_n u_n \to \theta$ as $n \to \infty$.*

Proof 2.25 Suppose E is bounded. Let $\mathscr{U}$ be a balanced neighborhood of θ in U, then $E \subset t\mathscr{U}$ for some t. For $\{u_n\} \subset E$, as $\alpha_n \to 0$, there exists N such that if $n > N$ then $t < \frac{1}{|\alpha_n|}$. Since $t^{-1}E \subset \mathscr{U}$ and $\mathscr{U}$ is balanced, we have that $\alpha_n u_n \in \mathscr{U}$, $\forall n > N$, and thus $\alpha_n u_n \to \theta$. Conversely, if E is not bounded, there is a neighborhood $\mathscr{V}$ of θ and $\{r_n\}$ such that $r_n \to \infty$ and E is not contained in $r_n \mathscr{V}$, that is, we can choose u_n such that $r_n^{-1} u_n$ is not in $\mathscr{V}$, $\forall n \in \mathbb{N}$, so that $\{r_n^{-1} u_n\}$ does not converge to θ.

Proposition 2.7.3 *Let $f : U \to V$ be a linear function. Consider the following statements*

1. *f is continuous,*

2. *f is bounded,*

3. *If $u_n \to \theta$ then $\{f(u_n)\}$ is bounded,*

4. *If $u_n \to \theta$ then $f(u_n) \to \theta$.*

Then,

- ■ 1 *implies* 2,

- ■ 2 *implies* 3,

- ■ *if U is metrizable with invariant metric,then* 3 *implies* 4, *which implies* 1.

Proof 2.26

1. 1 implies 2: Suppose f is continuous, for $\mathscr{W} \subset V$ neighborhood of zero, there exists a neighborhood of zero in U, denoted by $\mathscr{V}$, such that

$$f(\mathscr{V}) \subset \mathscr{W}. \tag{2.31}$$

   If E is bounded, there exists $t_0 \in \mathbb{R}^+$ such that $E \subset t\mathscr{V}, \forall t \geq t_0$, so that

$$f(E) \subset f(t\mathscr{V}) = tf(\mathscr{V}) \subset t\mathscr{W}, \ \forall t \geq t_0, \tag{2.32}$$

   and thus f is bounded.

2. 2 implies 3: Suppose $u_n \to \theta$ and let $\mathscr{W}$ be a neighborhood of zero. Then there exists $N \in \mathbb{N}$ such that if $n \geq N$ then $u_n \in \mathscr{V} \subset \mathscr{W}$ where $\mathscr{V}$ is a balanced neighborhood of zero. On the other hand, for $n < N$, there exists K_n such that $u_n \in K_n\mathscr{V}$. Define $K = \max\{1, K_1, ..., K_n\}$. Then $u_n \in K\mathscr{V}, \forall n \in \mathbb{N}$ and hence $\{u_n\}$ is bounded. Finally from 2, we have that $\{f(u_n)\}$ is bounded.

3. 3 implies 4: Suppose U is metrizable with invariant metric and let $u_n \to \theta$. Given $K \in \mathbb{N}$, there exists $n_K \in \mathbb{N}$ such that if $n > n_K$ then $d(u_n, \theta) < \frac{1}{K^2}$. Define $\gamma_n = 1$ if $n < n_1$ and $\gamma_n = K$, if $n_K \leq n < n_{K+1}$ so that

$$d(\gamma_n u_n, \theta) = d(Ku_n, \theta) \leq Kd(u_n, \theta) < K^{-1}. \tag{2.33}$$

   Thus since 2 implies 3 we have that $\{f(\gamma_n u_n)\}$ is bounded so that, by Proposition 2.7.2 $f(u_n) = \gamma_n^{-1}f(\gamma_n u_n) \to \theta$ as $n \to \infty$.

4. 4 implies 1: suppose 1 fails. Thus there exists a neighborhood of zero $\mathscr{W} \subset V$ such that $f^{-1}(\mathscr{W})$ contains no neighborhood of zero in U. Particularly, we can select $\{u_n\}$ such that $u_n \in B_{1/n}(\theta)$ and $f(u_n)$ not in $\mathscr{W}$ so that $\{f(u_n)\}$ does not converge to zero. Thus 4 fails.

2.8 Continuity of operators in Banach spaces

Let U, V be Banach spaces. We call a function $A : U \to V$ an operator.

Proposition 2.8.1 *Let U, V be Banach spaces. A linear operator $A : U \to V$ is continuous if and only if there exists $K \in \mathbb{R}^+$ such that*

$$\|A(u)\|_V < K\|u\|_U, \forall u \in U.$$

Proof 2.27   Suppose A is linear and continuous. From Proposition 2.7.3,

$$\text{if } \{u_n\} \subset U \text{ is such that } u_n \to \theta \text{ then } A(u_n) \to \theta. \tag{2.34}$$

We claim that for each $\varepsilon > 0$ there exists $\delta > 0$ such that if $\|u\|_U < \delta$ then $\|A(u)\|_V < \varepsilon$.
Suppose, to obtain contradiction that the claim is false.

Thus there exists $\varepsilon_0 > 0$ such that for each $n \in \mathbb{N}$ there exists $u_n \in U$ such that $\|u_n\|_U \leq \frac{1}{n}$ and $\|A(u_n)\|_V \geq \varepsilon_0$.

Therefore $u_n \to \theta$ and $A(u_n)$ does not converge to θ, which contradicts (2.34).

Thus the claim holds.

In particular, for $\varepsilon = 1$ there exists $\delta > 0$ such that if $\|u\|_U < \delta$ then $\|A(u)\|_V < 1$. Thus given an arbitrary not relabeled $u \in U$, $u \neq \theta$, for

$$w = \frac{\delta u}{2\|u\|_U}$$

we have

$$\|A(w)\|_V = \frac{\delta \|A(u)\|_V}{2\|u\|_U} < 1,$$

that is,

$$\|A(u)\|_V < \frac{2\|u\|_U}{\delta}, \forall u \in U.$$

Defining

$$K = \frac{2}{\delta}$$

the first part of the proof is complete. Reciprocally, suppose there exists $K > 0$ such that

$$\|A(u)\|_V < K\|u\|_U, \forall u \in U.$$

Hence $u_n \to \theta$ implies $\|A(u_n)\|_V \to \theta$, so that from Proposition 2.7.3, A is continuous.

The proof is complete.

2.9 Some classical results on Banach spaces

In this section we present some important results in Banach spaces. We start with the following theorem.

2.9.1 The Baire Category theorem

Theorem 2.9.1 *Let U and V be Banach spaces and let $A : U \to V$ be a linear operator. Then A is bounded if and only if the set $C \subset U$ has at least one interior point, where*

$$C = A^{-1}[\{v \in V \mid \|v\|_V \leq 1\}].$$

Proof 2.28   Suppose there exists $u_0 \in U$ in the interior of C. Thus, there exists $r > 0$ such that

$$B_r(u_0) = \{u \in U \mid \|u - u_0\|_U < r\} \subset C.$$

Fix $u \in U$ such that $\|u\|_U < r$. Thus, we have

$$\|A(u)\|_V \leq \|A(u + u_0)\|_V + \|A(u_0)\|_V.$$

Observe also that

$$\|(u + u_0) - u_0\|_U < r,$$

so that $u + u_0 \in B_r(u_0) \subset C$ and thus

$$\|A(u + u_0)\|_V \leq 1$$

and hence

$$\|A(u)\|_V \le 1 + \|A(u_0)\|_V, \tag{2.35}$$

$\forall u \in U$ such that $\|u\|_U < r$. Fix an arbitrary not relabeled $u \in U$ such that $u \ne \theta$. From (2.35)

$$w = \frac{u}{\|u\|_U} \frac{r}{2}$$

is such that

$$\|A(w)\|_V = \frac{\|A(u)\|_V}{\|u\|_U} \frac{r}{2} \le 1 + \|A(u_0)\|_V,$$

so that

$$\|A(u)\|_V \le (1 + \|A(u_0)\|_V)\|u\|_U \frac{2}{r}.$$

Since $u \in U$ is arbitrary, A is bounded.

Reciprocally, suppose A is bounded. Thus

$$\|A(u)\|_V \le K\|u\|_U, \forall u \in U,$$

for some $K > 0$. In particular

$$D = \left\{ u \in U \mid \|u\|_U \le \frac{1}{K} \right\} \subset C.$$

The proof is complete.

Definition 2.9.2 *A set S in a metric space U is said to be nowhere dense if $\overline{S}$ has an empty interior.*

Theorem 2.9.3 (Baire Category theorem) *A complete metric space is never the union of a countable number of nowhere dense sets.*

Proof 2.29 Suppose, to obtain contradiction, that U is a complete metric space and

$$U = \cup_{n=1}^{\infty} A_n$$

where each A_n is nowhere dense. Since A_1 is nowhere dense, there exist $u_1 \in U$ which is not in $\overline{A}_1$, otherwise we would have $U = \overline{A}_1$, which is not possible since U is open. Furthermore, $\overline{A}_1^c$ is open, so that we may obtain $u_1 \in A_1^c$ and $0 < r_1 < 1$ such that

$$B_1 = B_{r_1}(u_1)$$

satisfies

$$B_1 \cap A_1 = \emptyset.$$

Since A_2 is nowhere dense we have B_1 is not contained in $\overline{A}_2$. Therefore we may select $u_2 \in B_1 \setminus \overline{A}_2$ and since $B_1 \setminus \overline{A}_2$ is open, there exists $0 < r_2 < 1/2$ such that

$$\overline{B}_2 = \overline{B}_{r_2}(u_2) \subset B_1 \setminus \overline{A}_2,$$

that is

$$B_2 \cap A_2 = \emptyset.$$

Proceeding inductively in this fashion, for each $n \in \mathbb{N}$ we may obtain $u_n \in B_{n-1} \setminus \overline{A}_n$ such that we may choose an open ball $B_n = B_{r_n}(u_n)$ such that

$$\overline{B}_n \subset B_{n-1},$$

$$B_n \cap A_n = \emptyset$$

and

$$0 < r_n < 2^{1-n}.$$

Observe that $\{u_n\}$ is a Cauchy sequence, considering that if $m, n > N$ then $u_n, u_m \in B_N$, so that

$$d(u_n, u_m) < 2(2^{1-N}).$$

Define

$$u = \lim_{n \to \infty} u_n.$$

Since

$$u_n \in B_N, \forall n > N,$$

we get

$$u \in \bar{B}_N \subset B_{N-1}.$$

Therefore, u is not in $A_{N-1}, \forall N > 1$, which means u is not in $\cup_{n=1}^{\infty} A_n = U$, a contradiction.

The proof is complete.

2.9.2   The Principle of Uniform Boundedness

Theorem 2.9.4 (The Principle of Uniform Boundedness) *Let U be a Banach space. Let $\mathscr{F}$ be a family of linear bounded operators from U into a normed linear space V. Suppose for each $u \in U$ there exists a $K_u \in \mathbb{R}$ such that*

$$\|T(u)\|_V < K_u, \forall T \in \mathscr{F}.$$

Then, there exists $K \in \mathbb{R}$ such that

$$\|T\| < K, \forall T \in \mathscr{F}.$$

Proof 2.30   Define

$$B_n = \{u \in U \mid \|T(u)\|_V \leq n, \forall T \in \mathscr{F}\}.$$

By the hypotheses, given $u \in U$, $u \in B_n$ for all n sufficiently big. Thus,

$$U = \cup_{n=1}^{\infty} B_n.$$

Moreover, each B_n is closed. By the Baire category theorem there exists $n_0 \in \mathbb{N}$ such that B_{n_0} has non-empty interior. That is, there exists $u_0 \in U$ and $r > 0$ such that

$$B_r(u_0) \subset B_{n_0}.$$

Thus, fixing an arbitrary $T \in \mathscr{F}$, we have

$$\|T(u)\|_V \leq n_0, \forall u \in B_r(u_0).$$

Thus if $\|u\|_U < r$ then $\|(u + u_0) - u_0\|_U < r$, so that

$$\|T(u + u_0)\|_V \leq n_0,$$

that is,

$$\|T(u)\|_V - \|T(u_0)\|_V \leq n_0.$$

Thus

$$\|T(u)\|_V \leq 2n_0, \text{ if } \|u\|_U < r. \tag{2.36}$$

For $u \in U$ arbitrary, $u \neq \theta$, define

$$w = \frac{ru}{2\|u\|_U},$$

from (2.36) we obtain

$$\|T(w)\|_V = \frac{r\|T(u)\|_V}{2\|u\|_U} \leq 2n_0,$$

so that

$$\|T(u)\|_V \leq \frac{4n_0\|u\|_U}{r}, \forall u \in U.$$

Hence

$$\|T\| \leq \frac{4n_0}{r}, \forall T \in \mathscr{F}.$$

The proof is complete.

2.9.3 The Open Mapping theorem

Theorem 2.9.5 (The Open Mapping theorem) *Let U and V be Banach spaces and let $A : U \to V$ be a bounded onto linear operator. Thus if $\mathscr{O} \subset U$ is open then $A(\mathscr{O})$ is open in V.*

Proof 2.31   First we will prove that given $r > 0$, there exists $r' > 0$ such that

$$A(B_r(\theta)) \supset B_{r'}^V(\theta). \tag{2.37}$$

Here $B_{r'}^V(\theta)$ denotes a ball in V of radius r' with center in θ. Since A is onto

$$V = \cup_{n=1}^{\infty} A(nB_1(\theta)).$$

By the Baire Category theorem, there exists $n_0 \in \mathbb{N}$ such that the closure of $A(n_0 B_1(\theta))$ has non-empty interior, so that $\overline{A(B_1(\theta))}$ has a non-empty interior. We will show that there exists $r' > 0$ such that

$$B_{r'}^V(\theta) \subset \overline{A(B_1(\theta))}.$$

Observe that there exists $y_0 \in V$ and $r_1 > 0$ such that

$$B_{r_1}^V(y_0) \subset \overline{A(B_1(\theta))}. \tag{2.38}$$

Define $u_0 \in U$ which satisfies $A(u_0) = y_0$. We claim that

$$\overline{A(B_{r_2}(\theta))} \supset B_{r_1}^V(\theta),$$

where $r_2 = 1 + \|u_0\|_U$. To prove the claim, pick

$$y \in A(B_1(\theta))$$

thus there exists $u \in U$ such that $\|u\|_U < 1$ and $A(u) = y$. Therefore

$$A(u) = A(u - u_0 + u_0) = A(u - u_0) + A(u_0).$$

But observe that

$$\begin{aligned} \|u - u_0\|_U &\leq &\|u\|_U + \|u_0\|_U \\ &< &1 + \|u_0\|_U \\ &= &r_2, \end{aligned} \tag{2.39}$$

so that

$$A(u - u_0) \in A(B_{r_2}(\theta)).$$

This means

$$y = A(u) \in A(u_0) + A(B_{r_2}(\theta)),$$

and hence

$$A(B_1(\theta)) \subset A(u_0) + A(B_{r_2}(\theta)).$$

That is, from this and (2.38), we obtain

$$A(u_0) + \overline{A(B_{r_2}(\theta))} \supset \overline{A(B_1(\theta))} \supset B_{r_1}^V(y_0) = A(u_0) + B_{r_1}^V(\theta),$$

and therefore

$$\overline{A(B_{r_2}(\theta))} \supset B_{r_1}^V(\theta).$$

Since

$$A(B_{r_2}(\theta)) = r_2 A(B_1(\theta)),$$

we have, for some not relabeled $r_1 > 0$ that

$$\overline{A(B_1(\theta))} \supset B_{r_1}^V(\theta).$$

Thus it suffices to show that

$$\overline{A(B_1(\theta))} \subset A(B_2(\theta)),$$

to prove (2.37). Let $y \in \overline{A(B_1(\theta))}$, since A is continuous we may select $u_1 \in B_1(\theta)$ such that

$$y - A(u_1) \in B_{r_1/2}^V(\theta) \subset \overline{A(B_{1/2}(\theta))}.$$

Now select $u_2 \in B_{1/2}(\theta)$ so that

$$y - A(u_1) - A(u_2) \in B_{r_1/4}^V(\theta).$$

By induction, we may obtain

$$u_n \in B_{2^{1-n}}(\theta),$$

such that

$$y - \sum_{j=1}^{n} A(u_j) \in B_{r_1/2^n}^V(\theta).$$

Define

$$u = \sum_{n=1}^{\infty} u_n,$$

we have that $u \in B_2(\theta)$, so that

$$y = \sum_{n=1}^{\infty} A(u_n) = A(u) \in A(B_2(\theta)).$$

Therefore,

$$\overline{A(B_1(\theta))} \subset A(B_2(\theta)).$$

The proof of (2.37) is complete.

To finish the proof of this theorem, assume $\mathcal{O} \subset U$ is open. Let $v_0 \in A(\mathcal{O})$. Let $u_0 \in \mathcal{O}$ be such that $A(u_0) = v_0$. Thus there exists $r > 0$ such that

$$B_r(u_0) \subset \mathcal{O}.$$

From (2.37),

$$A(B_r(\theta)) \supset B_{r'}^V(\theta),$$

for some $r' > 0$. Thus

$$A(\mathcal{O}) \supset A(u_0) + A(B_r(\theta)) \supset v_0 + B_{r'}^V(\theta).$$

This means that v_0 is an interior point of $A(\mathcal{O})$. Since $v_0 \in A(\mathcal{O})$ is arbitrary, we may conclude that $A(\mathcal{O})$ is open.

The proof is complete.

Theorem 2.9.6 (The Inverse Mapping theorem) *A continuous linear bijection of one Banach space onto another has a continuous inverse.*

Proof 2.32   Let $A : U \to V$ satisfying the theorem hypotheses. Since A is open, A^{-1} is continuous.

2.9.4   The Closed Graph theorem

Definition 2.9.7 (Graph of a mapping) *Let $A : U \to V$ be an operator, where U and V are normed linear spaces. The **graph** of A denoted by $\Gamma(A)$ is defined by*

$$\Gamma(A) = \{(u,v) \in U \times V \mid v = A(u)\}.$$

Theorem 2.9.8 (The Closed Graph theorem) *Let U and V be Banach spaces and let $A : U \to V$ be a linear operator. Then A is bounded if and only if its graph is closed.*

Proof 2.33   Suppose $\Gamma(A)$ is closed. Since A is linear $\Gamma(A)$ is a subspace of $U \oplus V$. Also, being $\Gamma(A)$ closed, it is a Banach space with the norm

$$\|(u,A(u))\| = \|u\|_U + \|A(u)\|_V.$$

Consider the continuous mappings

$$\Pi_1(u,A(u)) = u$$

and

$$\Pi_2(u,A(u)) = A(u).$$

Observe that Π_1 is a bijection, so that by the inverse mapping theorem Π_1^{-1} is continuous. As

$$A = \Pi_2 \circ \Pi_1^{-1},$$

it follows that A is continuous. The converse is immediate.

2.10   A note on finite dimensional normed spaces

We start this section with the theorem.

Theorem 2.10.1 *Let V be a complex normed vector space (not necessarily of finite-dimension). Suppose $\{u_1,\ldots,u_n\} \subset V$ be a linearly independent set. Under such hypotheses, there exists $c > 0$ such that*

$$\|\alpha_1 u_1 + \cdots + \alpha_n u_n\|_V \geq c(|\alpha_1| + \cdots + |\alpha_n|), \ \forall \alpha_1, \ldots, \alpha_n \in \mathbb{C}. \tag{2.40}$$

Proof 2.34   For $\alpha_1, \ldots, \alpha_n \in \mathbb{C}$, let us denote

$$s = |\alpha_1| + \cdots + |\alpha_n|.$$

Thus, if $s = 0$, then $\alpha_1 = \ldots = \alpha_n = 0$ and (2.40) holds.

Suppose then $s > 0$

Denoting $\beta_j = \frac{\alpha_j}{s}$, $\forall j \in \{1,\ldots,n\}$ we have that (2.40) is equivalent to

$$\|\beta_1 u_1 + \cdots + \beta_n u_n\|_V \geq c, \ \forall \beta_1,\ldots,\beta_n \in \mathbb{C}, \ \text{such that} \ \sum_{j=1}^{n} |\beta_j| = 1. \tag{2.41}$$

Suppose to obtain, contradiction, there is no $c > 0$ such that (2.41) holds.

Thus there exist sequences $\{v_m\} \subset V$ and $\{\beta_j^m\} \subset \mathbb{C}$, $\forall j \in \{1,\ldots,n\}$ such that

$$v_m = \beta_1^m u_1 + \ldots + \beta_n^m u_n$$

such that

$$\sum_{j=1}^{n} |\beta_j^m| = 1, \ \forall m \in \mathbb{N}$$

and

$$\|v_m\|_V \to 0, \ \text{as} \ m \to \infty.$$

In particular

$$|\beta_j^m| \leq 1, \ \forall m \in \mathbb{N}, \ j \in \{1,\ldots,n\}.$$

It is a well-known result in elementary analysis that a bounded sequence $\mathbb{C}^n$ has a convergent subsequence.

Hence, there exists a subsequence $\{m_k\}$ of $\mathbb{N}$ and

$$\beta_j^0 \in \mathbb{C}, \ \forall j \in \{1,\ldots,n\}$$

such that

$$\beta_j^{m_k} \to \beta_j^0, \ \forall j \in \{1,\ldots,n\},$$

and

$$\sum_{j=1}^{n} |\beta_j^0| = 1.$$

Thus

$$v_{m_k} \to \sum_{j=1}^{n} \beta_j^0 u_j.$$

Since

$$\sum_{j=1}^{n} |\beta_j^0| = 1,$$

this contradicts

$$v_{m_k} \to \mathbf{0}.$$

Therefore, there exists $c > 0$ such that (2.40) holds.

The proof is complete.

Now we present the following result about the completeness of finite dimensional subspaces in a normed vector space.

Theorem 2.10.2 *Let V be a complex normed vector space and let M be finite dimensional subspace of V.*

Under such hypotheses, M is complete (closed). In particular, each normed finite dimensional vector space is complete.

Proof 2.35   Let $\{v_m\} \subset M$ be a Cauchy sequence. Let $n \in \mathbb{N}$ be the dimension of M. Let $\{u_1, \ldots, u_n\}$ be a basis for M.

Hence, there exists a sequence $\{\alpha_j^m\} \subset \mathbb{C}$ such that

$$v_m = \alpha_1^m u_1 + \cdots + \alpha_n^m u_n.$$

Let $\varepsilon > 0$. Since $\{v_m\}$ is a Cauchy sequence, there exists $n_0 \in \mathbb{N}$ such that, if $m, l > n_0$, then

$$\|v_m - v_l\|_V < \varepsilon.$$

Hence, from this and the last theorem, there exists $c > 0$ such that

$$
\begin{aligned}
\varepsilon \;>\; & \left\| \sum_{j=1}^{n} (\alpha_j^m - \alpha_j^l) u_j \right\|_V \\
\geq\; & c \sum_{i=1}^{n} |\alpha_j^m - \alpha_j^l| \\
\geq\; & c|\alpha_j^m - \alpha_j^l|, \; \forall m, l > n_0, \; \forall j \in \{1, \cdots, n\}.
\end{aligned}
\tag{2.42}
$$

Thus $\{\alpha_j^m\} \subset \mathbb{C}$ is a Cauchy sequence.

Therefore, there exists a $\alpha_j^0 \in \mathbb{C}$ such that

$$\alpha_j^m \to \alpha_j^0, \; \forall j \in \{1, \ldots, n\}.$$

From this, denoting $v_0 = \sum_{j=1}^{n} \alpha_j^0 u_j \in M$, we get

$$\|v_m - v_0\|_V \leq \sum_{j=1}^{n} |\alpha_j^m - \alpha_j^0| \|u_j\|_V \to 0, \text{ as } m \to \infty.$$

From this last result, we may infer that M is complete.

Definition 2.10.3 (Equivalence between two norms) *Let V be a vector space. Two norms*

$$\|\cdot\|_0, \|\cdot\|_1 : V \to \mathbb{R}^+$$

are said to be equivalent, if there exists $\alpha, \beta > 0$ such that

$$\alpha \|u\|_0 \leq \|u\|_1 \leq \beta \|u\|_0, \; \forall u \in V.$$

Theorem 2.10.4 *Let V be a finite dimensional vector space. Under such hypotheses, any two norms defined on V are equivalent.*

Proof 2.36   Assume the dimension of V is n.

Let $\{u_1, \ldots, u_n\} \subset V$ be a basis for V. Let $\|\cdot\|_0, \|\cdot\|_1 : V \to \mathbb{R}^+$ be two norms in V.

Let $u \in V$. Hence there exists $\alpha_1, \ldots, \alpha_n \in \mathbb{C}$ such that

$$u = \sum_{j=1}^{n} \alpha_j u_j,$$

so that there exists $c > 0$ which does not depend on u, such that

$$\|u\|_1 \geq c(|\alpha_1| + \cdots + |\alpha_n|).$$

On the other hand,

$$\|u\|_0 \leq \sum_{j=1}^{n} \|\alpha_j\| \|u_j\|_0 \leq K \sum_{j=1}^{n} |\alpha_j| \leq \frac{K}{c} \|u\|_1, \ \forall u \in V,$$

where $K = \max_{j \in \{1,\dots n\}} \{\|u_j\|_0\}$.

Interchanging the roles of $\|\cdot\|_0$ and $\|\cdot\|_1$ we may obtain $K_1, c_1 > 0$ such that

$$\|u_1\|_1 \leq \frac{K_1}{c_1} \|u\|_0, \ \forall u \in V.$$

Denoting $\alpha = \frac{c}{K}$ and $\beta = \frac{K_1}{c_1}$, we have obtained,

$$\alpha \|u\|_0 \leq \|u\|_1 \leq \beta \|u\|_0, \ \forall u \in V.$$

The proof is complete.

Chapter 3

Hilbert Spaces

3.1 Introduction

At this point we introduce an important class of spaces namely, the Hilbert spaces, which are a special class of metric spaces.

The main references for this chapter are [62, 24].

3.2 The main definitions and results

Definition 3.2.1 *Let H be a vector space. We say that H is a real pre-Hilbert space if there exists a function $(\cdot,\cdot)_H : H \times H \to \mathbb{R}$ such that*

1. $(u,v)_H = (v,u)_H,\ \forall u,v \in H,$

2. $(u+v,w)_H = (u,w)_H + (v,w)_H,\ \forall u,v,w \in H,$

3. $(\alpha u,v)_H = \alpha(u,v)_H,\ \forall u,v \in H,\ \alpha \in \mathbb{R},$

4. $(u,u)_H \geq 0,\ \forall u \in H,$ *and* $(u,u)_H = 0,$ *if and only if* $u = \theta.$

Remark 3.2.2 *The function $(\cdot,\cdot)_H : H \times H \to \mathbb{R}$ is called an inner-product.*

Proposition 3.2.3 (Cauchy-Schwartz inequality) *Let H be a pre-Hilbert space. Defining*

$$\|u\|_H = \sqrt{(u,u)_H}, \forall u \in H,$$

we have

$$|(u,v)_H| \leq \|u\|_H \|v\|_H, \forall u,v \in H.$$

Equality holds if and only if $u = \alpha v$ for some $\alpha \in \mathbb{R}$ or $v = \theta.$

Proof 3.1   If $v = \theta$ the inequality is immediate. Assume $v \neq \theta$. Given $\alpha \in \mathbb{R}$ we have

$$
\begin{aligned}
0 &\leq (u - \alpha v, u - \alpha v)_H \\
&= (u,u)_H + \alpha^2 (v,v)_H - 2\alpha(u,v)_H \\
&= \|u\|_H^2 + \alpha^2 \|v\|_H^2 - 2\alpha(u,v)_H.
\end{aligned}
\tag{3.1}
$$

In particular for $\alpha = (u,v)_H/\|v\|_H^2$, we obtain

$$0 \le \|u\|_H^2 - \frac{(u,v)_H^2}{\|v\|_H^2},$$

that is

$$|(u,v)_H| \le \|u\|_H \|v\|_H.$$

The proof of the remaining conclusions is left as an exercise.

Proposition 3.2.4 *On a pre-Hilbert space H, the function*

$$\|\cdot\|_H : H \to \mathbb{R}$$

is a norm, where as above

$$\|u\|_H = \sqrt{(u,u)}.$$

Proof 3.2   The only non-trivial property to be verified, concerning the definition of norm, is the triangle inequality.

Observe that, given $u, v \in H$, from the Cauchy-Schwartz inequality we have,

$$
\begin{aligned}
\|u+v\|_H^2 &= (u+v,u+v)_H \\
&= (u,u)_H + (v,v)_H + 2(u,v)_H \\
&\le (u,u)_H + (v,v)_H + 2|(u,v)_H| \\
&\le \|u\|_H^2 + \|v\|_H^2 + 2\|u\|_H\|v\|_H \\
&= (\|u\|_H + \|v\|_H)^2.
\end{aligned}
\tag{3.2}
$$

Therefore

$$\|u+v\|_H \le \|u\|_H + \|v\|_H, \forall u,v \in H.$$

The proof is complete.

Definition 3.2.5 *A pre-Hilbert space H is said to be a Hilbert space if it is complete, that is, if any cauchy sequence in H converges to an element of H.*

Definition 3.2.6 (Orthogonal Complement) *Let H be a Hilbert space. Considering $M \subset H$ we define its orthogonal complement, denoted by $M^\perp$, by*

$$M^\perp = \{u \in H \mid (u,m)_H = 0, \ \forall m \in M\}.$$

Theorem 3.2.7 *Let H be a Hilbert space, M a closed subspace of H and suppose $u \in H$. Under such hypotheses there exists a unique $m_0 \in M$ such that*

$$\|u - m_0\|_H = \min_{m \in M}\{\|u - m\|_H\}.$$

Moreover, $n_0 = u - m_0 \in M^\perp$ so that

$$u = m_0 + n_0,$$

where $m_0 \in M$ and $n_0 \in M^\perp$. Finally, such a representation through $M \oplus M^\perp$ is unique.

Proof 3.3 Define d by
$$d = \inf_{m \in M} \{\|u - m\|_H\}.$$

Let $\{m_i\} \subset M$ be a sequence such that

$$\|u - m_i\|_H \to d, \text{ as } i \to \infty.$$

Thus, from the parallelogram law we have

$$
\begin{aligned}
\|m_i - m_j\|_H^2 &= \|m_i - u - (m_j - u)\|_H^2 \\
&= 2\|m_i - u\|_H^2 + 2\|m_j - u\|_H^2 \\
&\quad - \| - 2u + m_i + m_j\|_H^2 \\
&= 2\|m_i - u\|_H^2 + 2\|m_j - u\|_H^2 \\
&\quad - 4\| - u + (m_i + m_j)/2\|_H^2 \\
&\leq 2\|m_i - u\|_H^2 + 2\|m_j - u\|_H^2 - 4d^2 \\
&\to 2d^2 + 2d^2 - 4d^2 = 0, \text{ as } i, j \to +\infty.
\end{aligned}
$$ (3.3)

Thus $\{m_i\} \subset M$ is a Cauchy sequence. Since M is closed, there exists $m_0 \in M$ such that

$$m_i \to m_0, \text{ as } i \to +\infty,$$

so that

$$\|u - m_i\|_H \to \|u - m_0\|_H = d.$$

Define

$$n_0 = u - m_0.$$

We will prove that $n_0 \in M^\perp$.

Pick $m \in M$ and $t \in \mathbb{R}$, thus we have

$$
\begin{aligned}
d^2 &\leq \|u - (m_0 - tm)\|_H^2 \\
&= \|n_0 + tm\|_H^2 \\
&= \|n_0\|_H^2 + 2(n_0, m)_H t + \|m\|_H^2 t^2.
\end{aligned}
$$ (3.4)

Since

$$\|n_0\|_H^2 = \|u - m_0\|_H^2 = d^2,$$

we obtain

$$2(n_0, m)_H t + \|m\|_H^2 t^2 \geq 0, \forall t \in \mathbb{R}$$

so that

$$(n_0, m)_H = 0.$$

Being $m \in M$ arbitrary, we obtain

$$n_0 \in M^\perp.$$

It remains to prove the uniqueness. Let $m \in M$, thus

$$
\begin{aligned}
\|u - m\|_H^2 &= \|u - m_0 + m_0 - m\|_H^2 \\
&= \|u - m_0\|_H^2 + \|m - m_0\|_H^2,
\end{aligned}
$$ (3.5)

since

$$(u - m_0, m - m_0)_H = (n_0, m - m_0)_H = 0.$$

From (3.5) we obtain

$$\|u-m\|_H^2 > \|u-m_0\|_H^2 = d^2,$$

if $m \neq m_0$.

Therefore m_0 is unique.

Now suppose

$$u = m_1 + n_1,$$

where $m_1 \in M$ and $n_1 \in M^\perp$. As above, for $m \in M$

$$
\begin{aligned}
\|u-m\|_H^2 &= \|u-m_1+m_1-m\|_H^2 \\
&= \|u-m_1\|_H^2 + \|m-m_1\|_H^2, \\
&\geq \|u-m_1\|_H
\end{aligned}
\tag{3.6}
$$

and thus since m_0 such that

$$d = \|u-m_0\|_H$$

is unique, we get

$$m_1 = m_0$$

and therefore

$$n_1 = u - m_0 = n_0.$$

The proof is complete.

Theorem 3.2.8 (The Riesz Lemma) *Let H be a Hilbert space and let $f : H \to \mathbb{R}$ be a continuous linear functional. Then there exists a unique $u_0 \in H$ such that*

$$f(u) = (u, u_0)_H, \forall u \in H.$$

Moreover

$$\|f\|_{H^*} = \|u_0\|_H.$$

Proof 3.4   Define N by

$$N = \{u \in H \mid f(u) = 0\}.$$

Thus, as f is a continuous and linear N is a closed subspace of H. If $N = H$, then $f(u) = 0 = (u, \theta)_H, \forall u \in H$ and the proof would be complete. Thus assume $N \neq H$. By the last theorem there exists $v \neq \theta$ such that $v \in N^\perp$.

Define

$$u_0 = \frac{f(v)}{\|v\|_H^2} v.$$

Thus if $u \in N$ we have

$$f(u) = 0 = (u, u_0)_H = 0.$$

On the other hand if $u = \alpha v$ for some $\alpha \in \mathbb{R}$, we have

$$
\begin{aligned}
f(u) &= \alpha f(v) \\
&= \frac{f(v)(\alpha v, v)_H}{\|v\|_H^2} \\
&= \left(\alpha v, \frac{f(v)v}{\|v\|_H^2} \right)_H \\
&= (\alpha v, u_0)_H.
\end{aligned}
\tag{3.7}
$$

Therefore, $f(u)$ equals $(u, u_0)_H$ in the space spanned by N and v. Now we show that this last space (then span of N and v) is in fact H. Just observe that given $u \in H$ we may write

$$u = \left(u - \frac{f(u)v}{f(v)} \right) + \frac{f(u)v}{f(v)}. \tag{3.8}$$

Since

$$u - \frac{f(u)v}{f(v)} \in N$$

we have finished the first part of the proof, that is, we have proven that

$$f(u) = (u, u_0)_H, \forall u \in H.$$

To finish the proof, assume $u_1 \in H$ is such that

$$f(u) = (u, u_1)_H, \forall u \in H.$$

Thus,

$$\begin{aligned} \|u_0 - u_1\|_H^2 &= (u_0 - u_1, u_0 - u_1)_H \\ &= (u_0 - u_1, u_0)_H - (u_0 - u_1, u_1)_H \\ &= f(u_0 - u_1) - f(u_0 - u_1) = 0. \end{aligned} \tag{3.9}$$

Hence, $u_1 = u_0$.

Let us now prove that

$$\|f\|_{H^*} = \|u_0\|_H.$$

First observe that

$$\begin{aligned} \|f\|_{H^*} &= \sup\{f(u) \mid u \in H, \|u\|_H \le 1\} \\ &= \sup\{|(u, u_0)_H| \mid u \in H, \|u\|_H \le 1\} \\ &\le \sup\{\|u\|_H \|u_0\|_H \mid u \in H, \|u\|_H \le 1\} \\ &\le \|u_0\|_H. \end{aligned} \tag{3.10}$$

On the other hand,

$$\begin{aligned} \|f\|_{H^*} &= \sup\{f(u) \mid u \in H, \|u\|_H \le 1\} \\ &\ge f\left(\frac{u_0}{\|u_0\|_H} \right) \\ &= \frac{(u_0, u_0)_H}{\|u_0\|_H} \\ &= \|u_0\|_H. \end{aligned} \tag{3.11}$$

From (3.10) and (3.11)

$$\|f\|_{H^*} = \|u_0\|_H.$$

The proof is complete.

Remark 3.2.9 *Similarly as above, we may define a Hilbert space H over $\mathbb{C}$, that is, a complex one. In this case the complex inner product $(\cdot, \cdot)_H : H \times H \to \mathbb{C}$ is defined through the following properties:*

1. $(u, v)_H = \overline{(v, u)_H}, \forall u, v \in H$,

2. $(u+v,w)_H = (u,w)_H + (v,w)_H, \ \forall u,v,w \in H,$

3. $(\alpha u,v)_H = \alpha(u,v)_H, \ \forall u,v \in H, \ \alpha \in \mathbb{C},$

4. $(u,u)_H \geq 0, \ \forall u \in H,$ and $(u,u) = 0,$ if and only if $u = \mathbf{0}.$

Observe that in this case we have

$$(u, \alpha v)_H = \overline{\alpha}(u,v)_H, \ \forall u,v \in H, \ \alpha \in \mathbb{C},$$

where for $\alpha = a + bi \in \mathbb{C},$ we have $\overline{\alpha} = a - bi.$ Finally, similar results as those proven above are valid for complex Hilbert spaces.

3.3 Orthonormal basis

In this section we study separable Hilbert spaces and the related orthonormal bases.

Definition 3.3.1 *Let H be a Hilbert space. A set $S \subset H$ is said to orthonormal if*

$$\|u\|_H = 1,$$

and

$$(u,v)_H = 0, \forall u,v \in S, \text{ such that } u \neq v.$$

If S is not properly contained in any other orthonormal set, it is said to be an orthonormal basis for H.

Theorem 3.3.2 *Let H be a Hilbert space and let $\{u_n\}_{n=1}^N$ be an orthonormal set. Then for all $u \in H,$ we have*

$$\|u\|_H^2 = \sum_{n=1}^N |(u,u_n)_H|^2 + \left\| u - \sum_{n=1}^N (u,u_n)_H u_n \right\|_H^2.$$

Proof 3.5 Observe that

$$u = \sum_{n=1}^N (u,u_n)_H u_n + \left(u - \sum_{n=1}^N (u,u_n)_H u_n \right).$$

Furthermore, we may easily obtain that

$$\sum_{n=1}^N (u,u_n)_H u_n \text{ and } u - \sum_{n=1}^N (u,u_n)_H u_n$$

are orthogonal vectors so that

$$
\begin{aligned}
\|u\|_H^2 &= (u,u)_H \\
&= \left\| \sum_{n=1}^N |(u,u_n)_H u_n \right\|_H^2 + \left\| u - \sum_{n=1}^N (u,u_n)_H u_n \right\|_H^2 \\
&= \sum_{n=1}^N |(u,u_n)_H|^2 + \left\| u - \sum_{n=1}^N (u,u_n)_H u_n \right\|_H^2.
\end{aligned}
\tag{3.12}
$$

Corollary 3.3.3 (Bessel inequality) *Let H be a Hilbert space and let $\{u_n\}_{n=1}^N$ be an orthonormal set. Then for all $u \in H,$ we have*

$$\|u\|_H^2 \geq \sum_{n=1}^N |(u,u_n)_H|^2.$$

Theorem 3.3.4 *Each Hilbert space has an orthonormal basis.*

Proof 3.6   Define by C the collection of all orthonormal sets in H. Define an order in C by stating $S_1 \prec S_2$ if $S_1 \subset S_2$. Then C is partially ordered and obviously non-empty, since

$$v/\|v\|_H \in C, \forall v \in H, v \neq \theta.$$

Now let $\{S_\alpha\}_{\alpha \in L}$ be a linearly ordered subset of C. Clearly $\cup_{\alpha \in L} S_\alpha$ is an orthonormal set which is an upper bound for $\{S_\alpha\}_{\alpha \in L}$.

Therefore, every linearly ordered subset has an upper bound, so that by Zorn's lemma C has a maximal element, that is, an orthonormal set not properly contained in any other orthonormal set.

This completes the proof.

Theorem 3.3.5 *Let H be a Hilbert space and let $S = \{u_\alpha\}_{\alpha \in L}$ be an orthonormal basis. Then for each $v \in H$ we have*

$$v = \sum_{\alpha \in L} (u_\alpha, v)_H u_\alpha,$$

and

$$\|v\|_H^2 = \sum_{\alpha \in L} |(u_\alpha, v)_H|^2.$$

Proof 3.7   Let $v \in H$. Let $L' \subset L$ a finite subset of L. From Bessel's inequality we have,

$$\sum_{\alpha \in L'} |(u_\alpha, v)_H|^2 \leq \|v\|_H^2.$$

From this, we may infer that the set $A_n = \{\alpha \in L \mid |(u_\alpha, v)_H| > 1/n\}$ is finite, so that

$$A = \{\alpha \in L \mid |(u_\alpha, v)_H| > 0\} = \cup_{n=1}^\infty A_n$$

is at most countable.

Thus $(u_\alpha, v)_H \neq 0$ for at most countably many $\alpha's \in L$, which we order by $\{\alpha_n\}_{n \in \mathbb{N}}$. Since the sequence

$$s_N = \sum_{i=1}^N |(u_{\alpha_i}, v)_H|^2,$$

is monotone and bounded, it is converging to some real limit as $N \to \infty$. Define

$$v_n = \sum_{i=1}^n (u_{\alpha_i}, v)_H u_{\alpha_i},$$

so that for $n > m$ we have

$$
\begin{aligned}
\|v_n - v_m\|_H^2 &= \left\| \sum_{i=m+1}^n (u_{\alpha_i}, v)_H u_{\alpha_i} \right\|_H^2 \\
&= \sum_{i=m+1}^n |(u_{\alpha_i}, v)_H|^2 \\
&= |s_n - s_m|.
\end{aligned}
\tag{3.13}
$$

Hence, $\{v_n\}$ is a Cauchy sequence which converges to some $v' \in H$.

Observe that

$$
\begin{aligned}
(v - v', u_{\alpha_l})_H &= \lim_{N \to \infty} (v - \sum_{i=1}^{N} (u_{\alpha_i}, v)_H u_{\alpha_i}, u_{\alpha_l})_H \\
&= (v, u_{\alpha_l})_H - (v, u_{\alpha_l})_H \\
&= 0.
\end{aligned} \tag{3.14}
$$

Also, if $\alpha \neq \alpha_l, \forall l \in \mathbb{N}$ then

$$
(v - v', u_\alpha)_H = \lim_{N \to \infty} (v - \sum_{i=1}^{\infty} (u_{\alpha_i}, v)_H u_{\alpha_i}, u_\alpha)_H = 0.
$$

Hence,

$$
v - v' \perp u_\alpha, \ \forall \alpha \in L.
$$

If

$$
v - v' \neq \theta,
$$

then we could obtain an orthonormal set

$$
\left\{ u_\alpha, \ \alpha \in L, \frac{v - v'}{\|v - v'\|_H} \right\}
$$

which would properly contain the complete orthonormal set

$$
\{u_\alpha, \ \alpha \in L\},
$$

a contradiction.

Therefore, $v - v' = \theta$, that is

$$
v = \lim_{N \to \infty} \sum_{i=1}^{N} (u_{\alpha_i}, v)_H u_{\alpha_i}.
$$

3.3.1 The Gram-Schmidt orthonormalization

Let H be a Hilbert space and and $\{u_n\} \subset H$ be a sequence of linearly independent vectors. Consider the procedure:

$$
w_1 = u_1, \ v_1 = \frac{w_1}{\|w_1\|_H},
$$

$$
w_2 = u_2 - (v_1, u_2)_H v_1, \ v_2 = \frac{w_2}{\|w_2\|_H},
$$

and inductively,

$$
w_n = u_n - \sum_{k=1}^{n-1} (v_k, u_n)_H v_k, \ v_n = \frac{w_n}{\|w_n\|_H}, \forall n \in \mathbb{N}, n > 2.
$$

Observe that clearly $\{v_n\}$ is an orthonormal set and for each $m \in \mathbb{N}$, $\{v_k\}_{k=1}^{m}$ and $\{u_k\}_{k=1}^{m}$ span the same vector subspace of H.

Such a process of obtaining the orthonormal set $\{v_n\}$ is known as the Gram-Schmidt orthonormalization. We finish this section with the following theorem.

Theorem 3.3.6 *A Hilbert space H is separable if and only if has a countable orthonormal basis. If $dim(H) = N < \infty$, the H is isomorphic to $\mathbb{C}^N$. If $dim(H) = +\infty$ then H is isomorphic to l^2, where*

$$
l^2 = \left\{ \{y_n\} \mid y_n \in \mathbb{C}, \forall n \in \mathbb{N} \text{ and } \sum_{n=1}^{\infty} |y_n|^2 < +\infty \right\}.
$$

Proof 3.8 Suppose H is separable and let $\{u_n\}$ be a countable dense set in H. To obtain an orthonormal basis it suffices to apply the Gram-Schmidt orthonormalization procedure to the greatest linearly independent subset of $\{u_n\}$.

Conversely, if $B = \{v_n\}$ is an orthonormal basis for H, the set of all finite linear combinations of elements of B with rational coefficients are dense in H, so that H is separable.

Moreover, if $dim(H) = +\infty$ consider the isomorphism $F : H \to l^2$ given by

$$F(u) = \{(u_n, u)_H\}_{n \in \mathbb{N}}.$$

Finally, if $dim(H) = N < +\infty$, consider the isomorphism $F : H \to \mathbb{C}^N$ given by

$$F(u) = \{(u_n, u)_H\}_{n=1}^{N}.$$

The proof is complete.

3.4 Projection on a convex set

Theorem 3.4.1 *Let H be a Hilbert space and let $K \subset H$ be a non-empty, closed and convex set. Under such hypotheses, for each $f \in H$ there exists a unique $u \in K$ such that*

$$\|f - u\|_H = \min_{v \in K} \|f - v\|_H.$$

Moreover, $u \in K$ is such that

$$(f - u, v - u)_H \leq 0, \ \forall v \in K.$$

Proof 3.9 Define

$$d = \inf_{v \in K} \|f - v\|_H.$$

Hence, for each $n \in \mathbb{N}$ there exists $v_n \in K$ such that

$$d \leq \|f - v_n\|_H < d + 1/n.$$

Let $m, n \in \mathbb{N}$. Define $a = f - v_n$ and $b = f - v_m$. From the parallelogram law, we have

$$\|a + b\|_H^2 + \|a - b\|_H^2 = 2(\|a\|_H^2 + \|b\|_H^2),$$

that is,

$$\|2f - (v_n + v_m)\|_H^2 + \|v_n - v_m\|_H^2 = 2(\|f - v_n\|_H^2 + \|f - v_m\|_H^2),$$

so that

$$
\begin{aligned}
\|v_n - v_m\|_H^2 &= -4\left(\left\|f - \frac{v_n + v_m}{2}\right\|_H^2\right) + 2(\|f - v_n\|_H^2 + \|f - v_m\|_H^2) \\
&\leq -4d^2 + 2(d + 1/n)^2 + (2(d + 1/m)^2 \\
&\to -4d^2 + 2d^2 + 2d^2 \\
&= 0, \text{ as } m, n \to \infty.
\end{aligned}
\tag{3.15}
$$

Hence $\{v_n\}$ is a Cauchy sequence so that there exists $u \in K$ such that

$$\|v_n - u\|_H \to 0, \text{ as } n \to \infty.$$

Therefore,

$$\|f - v_n\|_H \to \|f - u\|_H = d, \text{ as } n \to \infty.$$

Now let $v \in K$ and $t \in [0, 1]$. Define

$$w = (1 - t)u + tv.$$

Observe that, since K is convex, $w \in K$, $\forall t \in [0, 1]$.
 Hence,

$$
\begin{aligned}
\|f - u\|_H^2 &\leq \|f - w\|_H^2 \\
&= \|f - (1 - t)u - tv\|_H^2 \\
&= \|(f - u) + t(u - v)\|_H^2 \\
&= \|f - u\|_H^2 + 2(f - u, u - v)_H t + t^2 \|u - v\|_H^2.
\end{aligned}
\tag{3.16}
$$

From this, we obtain

$$(f - u, v - u)_H \leq \frac{t}{2} \|u - v\|_H, \, \forall t \in [0, 1].$$

Letting $t \to 0^+$ we get

$$(f - u, v - u)_H \leq 0, \, \forall v \in K.$$

The proof is complete.

Corollary 3.4.2 *In the context of the last theorem, assume*

$$(f - u, v - u) \leq 0, \forall v \in K.$$

Under such hypotheses,

$$\|f - u\|_H = \min_{v \in K} \|f - v\|_H.$$

Finally, such a $u \in K$ is unique.

Proof 3.10 Let $v \in K$. Thus,

$$
\begin{aligned}
\|f - u\|_H^2 - \|f - v\|_H^2 &= \|f\|_H^2 - 2(f, u)_H + \|u\|_H^2 \\
&\quad - \|f\|_H^2 + 2(f, v)_H - \|v\|_H^2 \\
&\quad - \|u - v\|_H^2 + \|u\|_H^2 - 2(u, v)_H + \|v\|_H^2 \\
&= 2(u, u)_H - 2(u, v)_H + 2(f, v)_H - 2(f, u)_H - \|u - v\|_H^2 \\
&= 2(f - u, v - u)_H - \|u - v\|_H^2 \\
&\leq 0.
\end{aligned}
\tag{3.17}
$$

Summarizing,

$$\|f - v\|_H \geq \|f - u\|_H, \, \forall v \in K.$$

Suppose now that $u_1, u_2 \in K$ be such that

$$(f - u_1, v - u_1)_H \leq 0, \, \forall v \in K,$$

$$(f - u_2, v - u_2)_H \leq 0, \, \forall v \in K.$$

With $v = u_2$ in the last first inequality and $v = u_1$ in the last second one, we get

$$(f - u_1, u_2 - u_1)_H \leq 0$$

and

$$(f - u_2, u_1 - u_2)_H \leq 0.$$

Adding these two last inequalities, we obtain

$$(f, u_2 - u_1)_H + (f, u_1 - u_2)_H - (u_1, u_2 - u_1)_H + (u_2, u_2 - u_1)_H \leq 0,$$

that is,

$$\|u_2 - u_1\|_H^2 \leq 0,$$

so that

$$\|u_1 - u_2\|_H = 0,$$

and therefore,

$$u_1 = u_2.$$

Hence, the $u \in K$ in question is unique.
The proof is complete.

Proposition 3.4.3 *Let H be a Hilbert space and $K \subset H$ a non-empty, closed and convex set. Let $f \in H$. Define $P_K(f) = u$ where $u \in K$ is such that*

$$\|f - u\|_H = \min_{v \in K} \|f - v\|_H.$$

Under such hypotheses,

$$\|P_K f_1 - P_K f_2\|_H \leq \|f_1 - f_2\|_H, \ \forall f_1, f_2 \in H.$$

Proof 3.11   Let $f_1, f_2 \in H$ and $u_1 = P_K f_1$ and $u_2 = P_K f_2$.
From the last proposition,

$$(f_1 - u_1, v - u_1)_H \leq 0, \ \forall v \in K$$

and

$$(f_2 - u_2, v - u_2)_H \leq 0, \ \forall v \in K.$$

With $v = u_2$ in the first last inequality and $v = u_1$ in the last second one, we obtain

$$(f_1 - u_1, u_2 - u_1)_H \leq 0,$$

$$(f_2 - u_2, u_1 - u_2)_H \leq 0.$$

Adding these two last inequalities, we get

$$(f_1, u_2 - u_1)_H - (f_2, u_2 - u_1)_H + (u_2 - u_1, u_2 - u_1)_H \leq 0,$$

that is,

$$\begin{aligned}
\|u_2 - u_1\|_H^2 &\leq (f_2 - f_1, u_2 - u_1)_H \\
&\leq \|f_2 - f_1\|_H \|u_2 - u_1\|_H,
\end{aligned} \tag{3.18}$$

so that

$$\|u_2 - u_1\|_H \leq \|f_2 - f_1\|_H.$$

This completes the proof.

Corollary 3.4.4 *Let H be a Hilbert space and let $M \subset H$ be a closed vector subspace of H. Let $f \in H$. Thus, $u = P_M(f)$ is such that $u \in M$ and*

$$(f - u, v)_H = 0, \forall v \in M.$$

Proof 3.12 In the previous results, we have got,

$$(f - u, v - u)_H \leq 0, \ \forall v \in M.$$

Let $v \in M$ be such that $v \neq \mathbf{0}$.
Thus,

$$(f - u, tv - u)_H \leq 0, \ \forall t \in \mathbb{R}.$$

Hence,

$$t(f - u, v)_H \leq (f - u, u)_H, \ \forall t \in \mathbb{R}.$$

From this we obtain

$$(f - u, v)_H = 0, \ \forall v \in M.$$

Remark 3.4.5 *Reciprocally, if*

$$(f - u, v)_H = 0, \ \forall v \in M,$$

then,

$$(f - u, v - u)_H = 0, \ \forall v \in M,$$

so that

$$u = P_M(f).$$

3.5   The theorems of Stampacchia and Lax-Milgram

In this section we present the statement and proof of two well-known results, namely, the Stampacchia and Lax-Milgram theorems.

Definition 3.5.1 *Let $a : H \times H \to \mathbb{R}$ be a bilinear form.*

1. *We say that a is bounded if there exists $c > 0$ such that*

$$|a(u, v)| \leq c \|u\|_H \|v\|_H, \ \forall u, v \in H.$$

2. *We say that a is coercive if there exists $\alpha > 0$ such that*

$$|a(v, v)| \geq \alpha \|v\|_H^2, \ \forall v \in H.$$

Theorem 3.5.2 (Stampacchia) *Let H be a Hilbert space and let $a : H \times H \to \mathbb{R}$ be a bounded and coercive bilinear form.*

Let $K \subset H$ be a non-empty, closed and convex set. Under such hypotheses, for each $f \in H$ there exists a unique $u \in K$ such that

$$a(u, v - u) \geq (f, v - u)_H, \ \forall v \in K. \tag{3.19}$$

Moreover, if a is symmetric, that is, $a(u, v) = a(v, u)$, $\forall u, v \in H$, such $u \in K$ in question is also such that

$$\frac{1}{2} a(u, u) - (f, u)_H = \min_{v \in K} \left\{ \frac{a(v, v)}{2} - (f, v)_H \right\}. \tag{3.20}$$

Proof 3.13 Fix $u \in H$. The function

$$v \mapsto a(u,v), \ \forall v \in H,$$

is continuous and linear.

Hence from the Riesz representation theorem, there exists a unique vector denoted by $A(u) \in H$ such that

$$(A(u),v)_H = a(u,v), \ \forall v \in H.$$

Clearly such an operator A is linear, and

$$|(A(u),v)_H| = |a(u,v)| \le c\|u\|_H\|v\|_H,$$

for some $c > 0$, so that

$$\|A(u)\|_H = \sup_{v \in H}\{|(A(u),v)_H| \ : \ \|v\|_H \le 1\} \le c\|u\|_H$$

Moreover,

$$(Av,v)_H = a(v,v) \ge \alpha\|v\|_H^2, \ \forall v \in H,$$

for some $\alpha > 0$.

Let $\rho > 0$ to be specified. Let $f \in H$.
Define $T(v) = P_K(\rho f - \rho A(v) + v)$.
Observe that

$$
\begin{aligned}
\|T(v_1) - T(v_2)\|_H^2 &= \|P_K(\rho f - \rho Av_1 + v_1) - P_K(\rho f - \rho v_2 + v_2)\|_H^2 \\
&\le \| - \rho A(v_1) + \rho A(v_2) + (v_1 - v_2)\|_H^2 \\
&= \|v_1 - v_2\|_H^2 - 2\rho(A(v_1 - v_2), v_1 - v_2)_H + \rho^2\|A(v_1 - v_2)\|_H^2 \\
&\le (1 - 2\alpha\rho + c\rho^2)\|v_1 - v_2\|_H^2.
\end{aligned}
\tag{3.21}
$$

Let $F(\rho) = 1 - 2\alpha\rho + c\rho^2$.
Thus, if $F'(\rho_0) = 0$ then

$$-2\alpha + 2\rho_0 c = 0,$$

that is,

$$\rho_0 = \frac{\alpha}{c}.$$

Therefore,

$$F(\rho_0) = 1 - 2\frac{\alpha^2}{c} + c\frac{\alpha^2}{c^2} = 1 - \frac{\alpha^2}{c}.$$

Observe that we may redefine a larger $c > 0$ such that

$$0 < 1 - \frac{\alpha^2}{c} < 1.$$

Hence,

$$\|T(v_1) - T(v_2)\|_H \le \lambda\|v_1 - v_2\|_H,$$

where

$$\lambda = \sqrt{1 - \frac{\alpha^2}{c}} < 1.$$

From this and Banach fixed point theorem, there exists $u \in K$ such that

$$T(u) = u,$$

that is,

$$P_K(\rho f - \rho Au + u) = u.$$

From this and Theorem 3.4.1, we obtain

$$(\rho f - \rho A(u) + u - u, v - u)_H \leq 0, \ \forall v \in K.$$

Thus,

$$a(u, v - u) = (A(u), v - u)_H \geq (f, v - u)_H, \ \forall v \in K.$$

Assume now that $a(u, v)$ is also symmetric. Thus, $a(u, v)$ define a inner product in H, inducing a norm

$$\sqrt{a(u, u)}$$

which is equivalent to $\|u\|_H$, since

$$\sqrt{c}\|u\|_H \geq \sqrt{a(u, u)} \geq \sqrt{\alpha}\|u\|_H, \ \forall u \in H.$$

From the Riesz representation theorem, there exists a unique $g \in H$ such that

$$(f, v)_H = a(g, v), \ \forall v \in H.$$

Similarly to the indicated above we may obtain $u \in K$ such that

$$u = P_K(g)$$

so that

$$a(g - u, v - u) \leq 0, \ \forall v \in K.$$

Hence,

$$a(g, v - u) - a(u, v - u) \leq 0,$$

that is,

$$a(u, v - u) \geq a(g, v - u) = (f, v - u), \ \forall v \in K.$$

Moreover from $u = P_k(g)$ we obtain

$$a(g - u, g - u) = \min_{v \in K} a(g - v, g - v).$$

Therefore,

$$a(g, g) - 2a(g, u) + a(u, u) \leq a(g, g) - 2a(g, v) + a(v, v),$$

that is,

$$\frac{a(u, u)}{2} - a(g, u) \leq \frac{a(v, v)}{2} - a(g, v),$$

so that

$$\frac{a(u, u)}{2} - (f, u)_H \leq \frac{a(v, v)}{2} - (f, v)_H, \ \forall v \in K.$$

The proof is complete.

Corollary 3.5.3 (Lax-Milgram theorem) *Assume $a(u, v)$ is a bounded, coercive and symmetric bilinear form on H. Under such hypotheses, there exists a unique $u \in H$ such that*

$$a(u, v) = (f, v), \ \forall v \in H.$$

Moreover, such a $u \in H$ is such that

$$\frac{1}{2}a(u, u) - (f, u)_H = \min_{v \in H} \frac{1}{2}a(v, v) - (f, v)_H.$$

Proof 3.14   The proof follows from the Stampacchia theorem with $K = H$ and Remark 3.4.5.

Chapter 4

The Hahn-Banach Theorems and the Weak Topologies

4.1 Introduction

In this chapter we present the Hahn-Banach theorems and some important applications. Also, a study on weak topologies is developed in details.

Finally, we highlight the main reference for this chapter is Brezis [24].

4.2 The Hahn-Banach theorems

In this chapter, U always denotes a Banach space.

Theorem 4.2.1 (The Hahn-Banach theorem) *Consider a functional* $p : U \to \mathbb{R}$ *such that*

$$p(\lambda u) = \lambda p(u), \forall u \in U, \lambda > 0, \tag{4.1}$$

and

$$p(u+v) \leq p(u) + p(v), \forall u, v \in U. \tag{4.2}$$

Let $V \subset U$ *be a proper subspace of* U *and let* $g : V \to \mathbb{R}$ *be a linear functional such that*

$$g(u) \leq p(u), \forall u \in V. \tag{4.3}$$

Under such hypotheses, there exists a linear functional $f : U \to \mathbb{R}$ *such that*

$$g(u) = f(u), \forall u \in V, \tag{4.4}$$

and

$$f(u) \leq p(u), \forall u \in U. \tag{4.5}$$

Proof 4.1   Choose $z \in U \setminus V$. Denote by $\tilde{V}$ the space spanned by V and z, that is,

$$\tilde{V} = \{v + \alpha z \mid v \in V \text{ and } \alpha \in \mathbb{R}\}. \tag{4.6}$$

We may define an extension of g from V to $\tilde{V}$, denoted by $\tilde{g}$, by

$$\tilde{g}(\alpha z + v) = \alpha \tilde{g}(z) + g(v), \tag{4.7}$$

where $\tilde{g}(z)$ will be properly specified in the next lines.

Let $v_1, v_2 \in V$, $\alpha > 0$, $\beta > 0$. Thus,

$$
\begin{aligned}
\beta g(v_1) + \alpha g(v_2) &= g(\beta v_1 + \alpha v_2) \\
&= (\alpha + \beta) g\left(\frac{\beta}{\alpha + \beta} v_1 + \frac{\alpha}{\alpha + \beta} v_2 \right) \\
&\leq (\alpha + \beta) p\left(\frac{\beta}{\alpha + \beta}(v_1 - \alpha z) + \frac{\alpha}{\alpha + \beta}(v_2 + \beta z) \right) \\
&\leq \beta p(v_1 - \alpha z) + \alpha p(v_2 + \beta z)
\end{aligned}
\tag{4.8}
$$

and therefore,

$$\frac{1}{\alpha}[-p(v_1 - \alpha z) + g(v_1)] \leq \frac{1}{\beta}[p(v_2 + \beta z) - g(v_2)],$$

$\forall v_1, v_2 \in V$, $\alpha, \beta > 0$. Thus, there exists $a \in \mathbb{R}$ such that

$$\sup_{v \in V, \alpha > 0} \left[\frac{1}{\alpha}(-p(v - \alpha z) + g(v)) \right] \leq a \leq \inf_{v \in V, \alpha > 0} \left[\frac{1}{\alpha}(p(v + \alpha z) - g(v)) \right]. \tag{4.9}$$

We shall define $\tilde{g}(z) = a$. Therefore, if $\alpha > 0$, then

$$
\begin{aligned}
\tilde{g}(\alpha z + v) &= a\alpha + g(v) \\
&\leq \left[\frac{1}{\alpha}(p(v + \alpha z) - g(v)) \right] \alpha + g(v) \\
&= p(v + \alpha z).
\end{aligned}
\tag{4.10}
$$

On the other hand, if $\alpha < 0$, then $-\alpha > 0$. Thus,

$$a \geq \frac{1}{-\alpha}(-p(v - (-\alpha)z) + g(v)),$$

so that

$$
\begin{aligned}
\tilde{g}(\alpha z + v) &= a\alpha + g(v) \\
&\leq \left[\frac{1}{-\alpha}(-p(v + \alpha z) + g(v)) \right] \alpha + g(v) \\
&= p(v + \alpha z)
\end{aligned}
\tag{4.11}
$$

and hence,

$$\tilde{g}(u) \leq p(u), \forall u \in \tilde{V}.$$

Define now by $\mathscr{E}$ the set of all extensions e of g, which satisfy $e(u) \leq p(u)$ on the domain of e, where such a domain is always a subspace of U. We shall also define a partial order for $\mathscr{E}$ denoting $e_1 \prec e_2$, as the domain of e_2 contains the domain of e_1 and $e_1 = e_2$ on the domain of e_1. Let $\{e_\alpha\}_{\alpha \in A}$ be an ordered subset of $\mathscr{E}$. Let V_α be the domain of e_α, $\forall \alpha \in A$. Define e on $\cup_{\alpha \in A} V_\alpha$ by setting $e = e_\alpha$ on V_α. Clearly, $e_\alpha \prec e$, $\forall \alpha \in A$ so that each ordered subset of $\mathscr{E}$ has an upper bound. From this and Zorn Lemma, $\mathscr{E}$ has a maximal element f defined on some subspace $\tilde{U} \subset U$ such that $f(u) \leq p(u), \forall u \in \tilde{U}$. Suppose, to obtain contradiction, that $\tilde{U} \neq U$ and let $z_1 \in U \setminus \tilde{U}$. As indicated above we may obtain an extension f_1 from $\tilde{U}$ to the subspace spanned by z_1 and $\tilde{U}$, which contradicts the maximality of f.

The proof is complete.

Definition 4.2.2 (Topological dual spaces) *Let U be a Banach space. We will define its dual topological space, as the set of all linear continuous functionals defined on U. We suppose such a dual space of U may be represented by another vector space U^*, through a bilinear form $\langle \cdot, \cdot \rangle_U : U \times U^* \to \mathbb{R}$ (here we are referring to standard representations of dual spaces of Sobolev and Lebesgue spaces, to be addressed in the subsequent chapters). Thus, given $f : U \to \mathbb{R}$ linear and continuous, we assume the existence of a unique $u^* \in U^*$ such that*

$$f(u) = \langle u, u^* \rangle_U, \forall u \in U. \tag{4.12}$$

The norm of f, denoted by $\|f\|_{U^*}$, is defined as

$$\|f\|_{U^*} = \sup_{u \in U}\{|\langle u, u^* \rangle_U| \ : \ \|u\|_U \leq 1\} \equiv \|u^*\|_{U^*}. \tag{4.13}$$

Corollary 4.2.3 *Let $V \subset U$ be a proper subspace of U and let $g : V \to \mathbb{R}$ be a linear and continuous functional with norm*

$$\|g\|_{V^*} = \sup_{u \in V}\{|g(u)| \ | \ \|u\|_U \leq 1\}. \tag{4.14}$$

Under such hypotheses, there exists u^* in U^* such that

$$\langle u, u^* \rangle_U = g(u), \forall u \in V, \tag{4.15}$$

and

$$\|u^*\|_{U^*} = \|g\|_{V^*}. \tag{4.16}$$

Proof 4.2   It suffices to apply Theorem 4.2.1 with $p(u) = \|g\|_{V^*}\|u\|_V$. Indeed, from such a theorem, there exists a linear functional $f : U \to \mathbb{R}$ such that

$$f(u) = g(u), \ \forall u \in V$$

and

$$f(u) \leq p(u) = \|g\|_{V^*}\|u\|_U,$$

that is,

$$|f(u)| \leq p(u) = \|g\|_{V^*}\|u\|_U, \forall u \in U.$$

Therefore,

$$\|f\|_{U^*} = \sup_{u \in U}\{|f(u)| \ : \ \|u\|_U \leq 1\} \leq \|g\|_{V^*}.$$

On the other hand,

$$\|f\|_{U^*} \geq \sup_{u \in V}\{|f(u)| \ : \ \|u\|_U \leq 1\} = \|g\|_{V^*}.$$

Thus,

$$\|f\|_{U^*} = \|g\|_{V^*}.$$

Finally, since f is linear and continuous, there exists $u^* \in U^*$ such that

$$f(u) = \langle u, u^* \rangle_U, \ \forall u \in U,$$

and hence

$$\langle u, u^* \rangle_U = f(u) = g(u), \ \forall u \in V.$$

Moreover,

$$\|u^*\|_{U^*} = \|f\|_{U^*} = \|g\|_{V^*}.$$

The proof is complete.

Corollary 4.2.4 *Let $u_0 \in U$. Under such hypotheses, there exists $u_0^* \in U^*$ such that*

$$\|u_0^*\|_{U^*} = \|u_0\|_U \text{ and } \langle u_0, u_0^* \rangle_U = \|u_0\|_U^2. \tag{4.17}$$

Proof 4.3 It suffices to apply the Corollary 4.2.3 with $V = \{\alpha u_0 \mid \alpha \in \mathbb{R}\}$ and $g(tu_0) = t\|u_0\|_U^2$ so that $\|g\|_{V^*} = \|u_0\|_U$.

Indeed, from the last corollary, there exists $u_0^* \in U^*$ such that

$$\langle tu_0, u_0^* \rangle_U = g(tu_0), \, \forall t \in \mathbb{R},$$

and

$$\|u_0^*\|_{U^*} = \|g\|_{V^*},$$

where,

$$\|g\|_{V^*} = \sup_{t \in \mathbb{R}}\{t\|u_0\|_U^2 \; : \; \|tu_0\|_U \le 1\} = \|u_0\|_U.$$

Moreover, also from the last corollary,

$$\|u_0^*\|_{U^*} = \|g\|_{V^*} = \|u_0\|_U.$$

Finally,

$$\langle tu_0, u_0^* \rangle_U = g(tu_0) = t\|u_0\|_U^2, \, \forall t \in \mathbb{R},$$

so that

$$\langle u_0, u_0^* \rangle_U = \|u_0\|_U^2.$$

This completes the proof.

Corollary 4.2.5 *Let $u \in U$. Under such hypotheses*

$$\|u\|_U = \sup_{u^* \in U^*} \{|\langle u, u^* \rangle_U| \mid \|u^*\|_{U^*} \le 1\}. \tag{4.18}$$

Proof 4.4 Suppose $u \ne \mathbf{0}$, otherwise the result is immediate. Since

$$|\langle u, u^* \rangle_U| \le \|u\|_U \|u^*\|_{U^*}, \forall u \in U, u^* \in U^*$$

we have

$$\sup_{u^* \in U^*} \{|\langle u, u^* \rangle_U| \mid \|u^*\|_{U^*} \le 1\} \le \|u\|_U. \tag{4.19}$$

However, from the last corollary, there exists $u_0^* \in U^*$ such that $\|u_0^*\|_{U^*} = \|u\|_U$ and $\langle u, u_0^* \rangle_U = \|u\|_U^2$. Define $u_1^* = \|u\|_U^{-1} u_0^*$. Thus, $\|u_1^*\|_U = 1$ and $\langle u, u_1^* \rangle_U = \|u\|_U$.

The proof is complete.

Definition 4.2.6 (Affine hyperplane) *Let U be a Banach space. An affine hyperplane H is a set defined by*

$$H = \{u \in U \mid \langle u, u^* \rangle_U = \alpha\} \tag{4.20}$$

for some $u^* \in U^*$ and $\alpha \in \mathbb{R}$.

Proposition 4.2.7 *An affine hyperplane H defined as above indicated is closed.*

Proof 4.5 The result follows directly from the continuity of $\langle u, u^* \rangle_U$ as functional on U.

Definition 4.2.8 (Separation) *Let $A, B \subset U$. We say that a hyperplane H, as above indicated separates A and B, as there exist $\alpha \in \mathbb{R}$ and $u^* \in U^*$ such that*

$$\langle u, u^* \rangle_U \leq \alpha, \forall u \in A, \text{ and } \langle u, u^* \rangle_U \geq \alpha, \forall u \in B. \tag{4.21}$$

We say that H separates A and B strictly if there exists $\varepsilon > 0$ such that

$$\langle u, u^* \rangle_U \leq \alpha - \varepsilon, \forall u \in A, \text{ and } \langle u, u^* \rangle_U \geq \alpha + \varepsilon, \forall u \in B, \tag{4.22}$$

Theorem 4.2.9 (The Hahn-Banach theorem, the geometric form) *Let $A, B \subset U$ be two non-empty, convex sets such that $A \cap B = \emptyset$ and A is open. Under such hypotheses, there exists a closed hyperplane which separates A and B, that is, there exist $\alpha \in \mathbb{R}$ and $u^* \in U^*$ such that*

$$\langle u, u^* \rangle_U \leq \alpha \leq \langle v, u^* \rangle_U, \; \forall u \in A, \; v \in B.$$

To prove such a theorem, we need two lemmas.

Lemma 4.2.10 *Let $C \subset U$ be a convex set such that $\mathbf{0} \in C$. For each $u \in U$ define*

$$p(u) = \inf\{\alpha > 0, \; \alpha^{-1} u \in C\}. \tag{4.23}$$

Under such hypotheses, p is such that there exists $M \in \mathbb{R}^+$ such that

$$0 \leq p(u) \leq M \|u\|_U, \forall u \in U, \tag{4.24}$$

and

$$C = \{u \in U \mid p(u) < 1\}. \tag{4.25}$$

Moreover,

$$p(u + v) \leq p(u) + p(v), \forall u, v \in U.$$

Proof 4.6 Let $r > 0$ be such that $B(\mathbf{0}, r) \subset C$. Let $u \in U$ such that $u \neq \mathbf{0}$. Thus,

$$\frac{u}{\|u\|_U} r \in \overline{B(\mathbf{0}, r)} \subset \overline{C},$$

and therefore

$$p(u) \leq \frac{\|u\|_U}{r}, \forall u \in U \tag{4.26}$$

which proves (4.24). Suppose now $u \in C$. Since C is open there exists $\varepsilon > 0$ sufficiently small such that $(1 + \varepsilon)u \in C$. Thus, $p(u) \leq \frac{1}{1+\varepsilon} < 1$. Reciprocally, if $p(u) < 1$, there exists $0 < \alpha < 1$ such that $\alpha^{-1} u \in C$ and hence, since C is convex, we get $u = \alpha(\alpha^{-1}u) + (1 - \alpha)\mathbf{0} \in C$.

Finally, let $u, v \in C$ and $\varepsilon > 0$. Thus, $\frac{u}{p(u)+\varepsilon} \in C$ and $\frac{v}{p(v)+\varepsilon} \in C$ so that $\frac{tu}{p(u)+\varepsilon} + \frac{(1-t)v}{p(v)+\varepsilon} \in C, \forall t \in [0, 1]$. Particularly, for $t = \frac{p(u)+\varepsilon}{p(u)+p(v)+2\varepsilon}$ we obtain $\frac{u+v}{p(u)+p(v)+2\varepsilon} \in C$, and thus,

$$p(u + v) \leq p(u) + p(v) + 2\varepsilon, \forall \varepsilon > 0.$$

The proof of this lemma is complete.

Lemma 4.2.11 *Let $C \subset U$ be an non-empty, open and convex set and let $u_0 \in U$ such that $u_0 \notin C$. Under such hypotheses, there exists $u^* \in U^*$ such that $\langle u, u^* \rangle_U < \langle u_0, u^* \rangle_U, \forall u \in C$*

Proof 4.7   By translation, if necessary, there is no loos in generality in assuming $0 \in C$. Consider the functional p defined in the last lema. Define $V = \{\alpha u_0 \mid \alpha \in \mathbb{R}\}$. Define also g on V, by

$$g(tu_0) = t, \ \forall t \in \mathbb{R}. \tag{4.27}$$

Let $t \in \mathbb{R}$ be such that $t \neq 0$. Since

$$\frac{tu_0}{t} = u_0 \notin C,$$

we have

$$g(tu_0) = t \leq p(tu_0)$$

and therefore

$$g(u) \leq p(u), \forall u \in V.$$

From the Hahn-Banach theorem, there exists a linear functional f defined on U which extends g such that

$$f(u) \leq p(u) \leq M\|u\|_U. \tag{4.28}$$

Here, we have applied the Lemma 4.2.10. In particular, $f(u_0) = g(u_0) = g(1u_0) = 1$, also from the last lemma, $f(u) < 1, \forall u \in C$. The existence of u^* satisfying this lemma conclusion follows from the continuity of f, indicated in (4.28).

Proof of Theorem 4.2.9. Define $C = A + (-B)$ so that C is convex and $0 \notin C$. From Lemma 4.2.11, there exists $u^* \in U^*$ such that $\langle w, u^* \rangle_U < 0, \forall w \in C$, and thus,

$$\langle u, u^* \rangle_U < \langle v, u^* \rangle_U, \forall u \in A, \ v \in B. \tag{4.29}$$

Therefore, there exists $\alpha \in \mathbb{R}$ such that

$$\sup_{u \in A} \langle u, u^* \rangle_U \leq \alpha \leq \inf_{v \in B} \langle v, u^* \rangle_U, \tag{4.30}$$

which completes the proof.

Proposition 4.2.12 *Let U be a Banach space and let $A, B \subset U$ be such that A is compact, B is closed and $A \cap B = \emptyset$.*

Under such hypotheses, there exists $\varepsilon_1 > 0$ such that

$$[A + B_{\varepsilon_1}(\mathbf{0})] \cap [B + B_{\varepsilon_1}(\mathbf{0})] = \emptyset.$$

Proof 4.8   Suppose, to obtain contradiction, the proposition conclusion is false.

Thus, for each $n \in \mathbb{N}$ there exists $u_n \in U$ such that $d(u_n, A) < \frac{1}{n}$ and $d(u_n, B) < \frac{1}{n}$.

Therefore, there exist $v_n \in A$ and $w_n \in B$ such that

$$\|u_n - v_n\|_U < \frac{1}{n} \tag{4.31}$$

and

$$\|u_n - w_n\|_U < \frac{1}{n}, \ \forall n \in \mathbb{N}. \tag{4.32}$$

Since $\{v_n\} \subset A$ and A is compact, there exist a subsequence $\{v_{n_j}\}$ of $\{v_n\}$ and $v_0 \in A$, such that

$$\|v_{n_j} - v_0\|_U \to 0, \text{ as } j \to \infty.$$

Thus, from this, (4.31) and (4.32) we obtain,

$$\|u_{n_j} - v_0\|_U \to 0, \text{ as } j \to \infty,$$

and

$$\|w_{n_j} - v_0\|_U \to 0, \text{ as } j \to \infty.$$

Since A and B are closed we may infer that

$$v_0 \in A \cap B,$$

which contradicts $A \cap B = \emptyset$.

The proof is complete.

Theorem 4.2.13 (The Hahn-Banach theorem, the second geometric form) *Let $A, B \subset U$ be two non-empty, convex sets such that $A \cap B = \emptyset$. Suppose A is compact and B is closed. Under such hypotheses, there exists an hyperplane which separates A and B strictly.*

Proof 4.9   Observe that, from the last proposition, there exists $\varepsilon > 0$ sufficiently small such that $A_\varepsilon = A + B(0, \varepsilon)$ and $B_\varepsilon = B + B(0, \varepsilon)$ are disjoint and convex sets. From Theorem 4.2.9, there exists $u^* \in U^*$ such that $u^* \neq \mathbf{0}$ and

$$\langle u + \varepsilon w_1, u^* \rangle_U \leq \langle u + \varepsilon w_2, u^* \rangle_U, \forall u \in A, \, v \in B, \, w_1, w_2 \in B(0, 1). \tag{4.33}$$

Thus, there exists $\alpha \in \mathbb{R}$ such that

$$\langle u, u^* \rangle_U + \varepsilon \|u^*\|_{U^*} \leq \alpha \leq \langle v, u^* \rangle_U - \varepsilon \|u^*\|_{U^*}, \forall u \in A, \, v \in B. \tag{4.34}$$

The proof is complete.

Corollary 4.2.14 *Suppose $V \subset U$ is a vector subspace such that $\overline{V} \neq U$. Under such hypotheses, there exists $u^* \in U^*$ such that $u^* \neq \mathbf{0}$ and*

$$\langle u, u^* \rangle_U = 0, \forall u \in V. \tag{4.35}$$

Proof 4.10   Let $u_0 \in U$ be such that $u_0 \notin \overline{V}$. Applying Theorem 4.2.9 to $A = \overline{V}$ and $B = \{u_0\}$ we obtain $u^* \in U^*$ and $\alpha \in \mathbb{R}$ such that $u^* \neq \mathbf{0}$ e

$$\langle u, u^* \rangle_U < \alpha < \langle u_0, u^* \rangle_U, \forall u \in V. \tag{4.36}$$

Since V is a subspace, we must have $\langle u, u^* \rangle_U = 0, \forall u \in V$.

4.3 The weak topologies

Definition 4.3.1 (Weak neighborhoods) *Let U be a Banach space and let $u_0 \in U$. We define a weak neighborhood of u_0, denoted by $\mathcal{V}_w(u_0)$, as*

$$\mathcal{V}_w(u_0) = \{u \in U \mid |\langle u - u_0, u_i^* \rangle_U| < \varepsilon_i, \forall i \in \{1, ..., m\}\}, \tag{4.37}$$

for some $m \in \mathbb{N}$, $\varepsilon_i > 0$, and $u_i^* \in U^*$, $\forall i \in \{1, ..., m\}$.

Let $A \subset U$. We say that $u_0 \in A$ is weakly interior to A, as there exists a weak neighborhood $V_w(u_0)$ of u_0 contained in A.

If all points of A are weakly interior, we say that A is weakly open.

Finally, we define the weak topology $\sigma(U, U^*)$ for U, as the set of all subsets weakly open of U.

Proposition 4.3.2 *A Banach space U is Hausddorff as endowed with the weak topology $\sigma(U, U^*)$.*

Proof 4.11   Choose $u_1, u_2 \in U$ such that $u_1 \neq u_2$. From the Hahn-Banach theorem, second geometric form, there exists an hyperplane separating $\{u_1\}$ and $\{u_2\}$ strictly, thats is, there exist $u^* \in U^*$ and $\alpha \in \mathbb{R}$ such that

$$\langle u_1, u^* \rangle_U < \alpha < \langle u_2, u^* \rangle_U. \tag{4.38}$$

Define

$$\mathcal{V}_{w1}(u_1) = \{u \in U \mid |\langle u - u_1, u^* \rangle| < \alpha - \langle u_1, u^* \rangle_U\}, \tag{4.39}$$

and

$$\mathcal{V}_{w2}(u_2) = \{u \in U \mid |\langle u - u_2, u^* \rangle_U| < \langle u_2, u^* \rangle_U - \alpha\}. \tag{4.40}$$

We claim that
$$V_{w_1}(u_1) \cap V_{w_2}(u_2) = \emptyset.$$

Suppose, to obtain contradiction, there exists $u \in V_{w_1}(u_1) \cap V_{w_2}(u_2)$.
Thus,

$$\langle u - u_1, u^* \rangle_U < \alpha - \langle u_1, u^* \rangle_U,$$

and therefore

$$\langle u, u^* \rangle_U < \alpha.$$

Also

$$-\langle u - u_2, u^* \rangle_U < \langle u_2, u^* \rangle_U - \alpha,$$

and hence

$$\langle u, u^* \rangle_U > \alpha.$$

We have got

$$\langle u, u^* \rangle_U < \alpha < \langle u, u^* \rangle_U,$$

a contradiction.

Summarizing, we have obtained $u_1 \in \mathcal{V}_{w_1}(u_1)$, $u_2 \in \mathcal{V}_{w_2}(u_2)$ and $\mathcal{V}_{w_1}(u_1) \cap \mathcal{V}_{w_2}(u_2) = \emptyset$.
The proof is complete.

Remark 4.3.3 *If $\{u_n\} \in U$ is such that u_n converges to u in $\sigma(U, U^*)$, then we write $u_n \rightharpoonup u$, weakly.*

Proposition 4.3.4 *Let U be a Banach space. For a sequence $\{u_n\} \subset U$, we have*

1. *$u_n \rightharpoonup u$, for $\sigma(U, U^*) \Leftrightarrow \langle u_n, u^* \rangle_U \to \langle u, u^* \rangle_U, \forall u^* \in U^*$,*

2. *If $u_n \to u$ strongly (in norm), then $u_n \rightharpoonup u$ weakly,*

3. *If $u_n \rightharpoonup u$ weakly, then $\{\|u_n\|_U\}$ is bounded and $\|u\|_U \leq \liminf_{n \to \infty} \|u_n\|_U$,*

4. *If $u_n \rightharpoonup u$ weakly and $u_n^* \to u^*$ strongly (in norm) in U^*, then $\langle u_n, u_n^* \rangle_U \to \langle u, u^* \rangle_U$.*

Proof 4.12

1. The result follows from the definition of $\sigma(U, U^*)$.

   Indeed, suppose that $\{u_n\} \subset U$ and $u_n \rightharpoonup u$, weakly.

   Let $u^* \in U^*$ and let $\varepsilon > 0$.

   Define

   $$V_w(u) = \{v \in U \ : \ |\langle v - u, u^* \rangle_U| < \varepsilon\}.$$

From the hypotheses, there exists $n_0 \in \mathbb{N}$ such that if $n > n_0$, then

$$u_n \in V_w(u).$$

That is,

$$|\langle u_n - u, u^* \rangle_U| < \varepsilon,$$

if $n > n_0$.

Therefore,

$$\langle u_n, u^* \rangle_U \to \langle u, u^* \rangle_U, \text{ as } n \to \infty$$

$\forall u^* \in U^*$.

Reciprocally, suppose that

$$\langle u_n, u^* \rangle_U \to \langle u, u^* \rangle_U, \text{ as } n \to \infty$$

$\forall u^* \in U^*$.

Let $V(u) \in \sigma(U, U^*)$ be a set which contains $\{u\}$.

Thus, there exists a weak neighborhood a $V_w(u)$ such that $u \in V_w(u) \subset V(u)$, where there exist $m \in \mathbb{N}$, $\varepsilon_i > 0$ and $u_i^* \in U^*$ such that

$$V_w(u) = \{v \in U \ : \ |\langle v - u, u_i^* \rangle_U| < \varepsilon_i, \ \forall i \in \{1, \cdots, m\}\}.$$

From the hypotheses, for each $i \in \{1, \cdots m\}$, there exists $n_i \in \mathbb{N}$ such that if $n > n_i$, then

$$|\langle u_n - u, u_i^* \rangle_U| < \varepsilon_i.$$

Define $n_0 = \max\{n_1, \cdots, n_m\}$.

Thus,

$$u_n \in V_w(u) \subset V(u), \text{ if } n > n_0.$$

From this we may infer that $u_n \rightharpoonup u$ for $\sigma(U, U^*)$.

2. This follows from the inequality

$$|\langle u_n, u^* \rangle_U - \langle u, u^* \rangle_U| \leq \|u^*\|_{U^*} \|u_n - u\|_U. \tag{4.41}$$

3. For each $u^* \in U^*$ the sequence $\{\langle u_n, u^* \rangle_U\}$ is convergent for some bounded sequence. From this and the Uniform Boundedness Principle, there exists $M > 0$ such that $\|u_n\|_U \leq M, \forall n \in \mathbb{N}$. Moreover, for $u^* \in U^*$, we have

$$|\langle u_n, u^* \rangle_U| \leq \|u^*\|_{U^*} \|u_n\|_U, \tag{4.42}$$

and letting $n \to \infty$, we obtain

$$|\langle u, u^* \rangle_U| \leq \liminf_{n \to \infty} \|u^*\|_{U^*} \|u_n\|_U. \tag{4.43}$$

Thus,

$$\|u\|_U = \sup_{u^* \in U^*} \{|\langle u, u^* \rangle_U| \ : \ \|u\|_{U^*} \leq 1\} \leq \liminf_{n \to \infty} \|u_n\|_U. \tag{4.44}$$

4. Just observe that

$$
\begin{aligned}
|\langle u_n, u_n^* \rangle_U - \langle u, u^* \rangle_U| &\leq |\langle u_n, u_n^* - u^* \rangle_U| \\
&\quad + |\langle u - u_n, u^* \rangle_U| \\
&\leq \|u_n^* - u^*\|_{U^*} \|u_n\|_U \\
&\quad + |\langle u_n - u, u^* \rangle_U| \\
&\leq M \|u_n^* - u^*\|_{U^*} \\
&\quad + |\langle u_n - u, u^* \rangle_U| \\
&\rightarrow 0, \text{ as } n \rightarrow \infty.
\end{aligned}
\tag{4.45}
$$

Theorem 4.3.5 *Let U be a Banach space and let $A \subset U$ be a non-empty convex set. Under such hypotheses, A is closed for the topology $\sigma(U, U^*)$ if, and only if A is closed for the topology induced by $\|\cdot\|_U$.*

Proof 4.13 If $A = U$, the result is immediate. Thus, assume $A \neq U$. Suppose that A is strongly closed. Let $u_0 \notin A$. From the Hahn-Banach theorem there exists a closed hyperplane which separates u_0 and A strictly, that is, there exist $\alpha \in \mathbb{R}$ and $u^* \in U^*$ such that

$$
\langle u_0, u^* \rangle_U < \alpha < \langle v, u^* \rangle_U, \forall v \in A.
\tag{4.46}
$$

Define

$$
\mathscr{V} = \{u \in U \mid \langle u, u^* \rangle_U < \alpha\},
\tag{4.47}
$$

so that $u_0 \in \mathscr{V}$, $\mathscr{V} \subset U \setminus A$.
Let
$$
V_w(u_0) = \{v \in U : |\langle v - u_0, u^* \rangle_U| < \alpha - \langle u_0, u^* \rangle_U.
$$

Let $v \in V_w(u_0)$.
Thus,

$$
\begin{aligned}
\langle v, u^* \rangle_U &= \langle v - u_0 + u_0, u^* \rangle_U \\
&= \langle v - u_0, u^* \rangle_U + \langle u_0, u^* \rangle_U \\
&\leq |\langle v - u_0, u^* \rangle_U| + \langle u_0, u^* \rangle_U \\
&< \alpha - \langle u_0, u^* \rangle_U + \langle u_0, u^* \rangle_U \\
&= \alpha.
\end{aligned}
\tag{4.48}
$$

From this we may infer that $V_w(u_0) \subset \mathscr{V} \subset U \setminus A$, that is, u_0 is an interior point for $\sigma(U, U^*)$ of $U \setminus A$, $\forall u_0 \in U \setminus A$
Therefore, $\mathscr{V}$ is weakly open.
Summarizing, $U \setminus A$ is open in $\sigma(U, U^*)$ and thus A is closed for $\sigma(U, U^*)$ (weakly closed).
Finally, the reciprocal is immediate.

Theorem 4.3.6 *Let (Z, σ) be a topological space and let U be a Banach space. Let $\phi : Z \rightarrow U$ be a function, considering U with the weak topology $\sigma(U, U^*)$.*
  Under such hypotheses, ϕ is continuous if, and only if, $f_{u^*} : Z \rightarrow \mathbb{R}$, where

$$
f_{u^*}(z) = \langle \phi(z), u^* \rangle_U
$$

is continuous, $\forall u^* \in U^*$.

Proof 4.14   Assume ϕ is continuous. Let $z_0 \in Z$ and let $\{z_\alpha\}_{\alpha \in I}$ be a net such that

$$z_\alpha \to z_0.$$

From the hypotheses,

$$\phi(z_\alpha) \rightharpoonup \phi(z_0), \text{ in } \sigma(U, U^*).$$

Therefore,

$$\langle \phi(z_\alpha), u^* \rangle_U \to \langle \phi(z_0), u^* \rangle_U, \forall u^* \in U^*.$$

Thus, f_{u^*} is continuous at z_0, $\forall u^* \in U^*$, $\forall z_0 \in Z$.
Reciprocally, assume $f_{u^*} : Z \to \mathbb{R}$, where

$$f_{u^*}(z) = \langle \phi(z), u^* \rangle_U$$

is continuous, $\forall u^* \in U^*$.

Suppose, to obtain contradiction, that ϕ is not continuous.
Thus, there exists $z_0 \in Z$ such that ϕ is not continuous at z_0.
In particular, there exists a net $\{z_\alpha\}_{\alpha \in I}$ such that $z_\alpha \to z_0$ and we do not have

$$\phi(z_\alpha) \rightharpoonup \phi(z_0), \text{ in } \sigma(U, U^*).$$

Hence, there exists $u^* \in U^*$ such that we do not have

$$\langle \phi(z_\alpha), u^* \rangle_U \to \langle \phi(z_0), u^* \rangle_U,$$

and thus f_{u^*} is not continuous at z_0, a contradiction.
Therefore, ϕ is continuous.
The proof is complete.

4.4   The weak-star topology

Definition 4.4.1 (Reflexive spaces) *Let U be a Banach space. We say that U is reflexive, if the canonical injection*

$$J : U \to U^{**}$$

is onto, where

$$\langle u, u^* \rangle_U = \langle u^*, J(u) \rangle_{U^*}, \forall u \in U, u^* \in U^*.$$

Thus, if U is reflexive, we may identify the bi-dual space of U, U^{**}, with U.
The weak topology for U^* may be defined similarly to $\sigma(U, U^*)$ and it is denoted by $\sigma(U^*, U^{**})$.
We define as well, the weak-star topology for U^*, denoted by $\sigma(U^*, U)$, as it follows.
Firstly, we define weak-star neighborhoods.
Let $u_0^* \in U^*$. We define a weak-star neighborhood for u_0^*, denoted by $V_w(u_0^*)$, as

$$V_w(u_0^*) = \{u^* \in U^* : |\langle u_i, u^* - u_0^* \rangle_U| < \varepsilon_i, \forall i \in \{1, \cdots, m\}\},$$

where $m \in \mathbb{N}$, $\varepsilon_i > 0$ and $u_i \in U$, $\forall i \in \{1, \cdots, m\}$.
Let $A \subset U^*$. We say that $u_0^* \in A$ is weakly-star interior to A, as there exists a weak-star neighborhood $V_w(u_0^*)$ contained in A.
If all point of A are weakly-star interior, we say that A weakly-star open.
Finally, we define the weak-star topology $\sigma(U^*, U)$ for U^*, as the set of all subsets weakly-star open of U^*.
Observe that $\sigma(U^*, U^{**})$ and $\sigma(U^*, U)$ coincide if U is reflexive.

4.5   Weak-star compactness

Theorem 4.5.1 (Banach and Alaoglu) *Let U be a Banach space. Denote*

$$B_{U^*} = \{u^* \in U^* : \|u^*\|_{U^*} \leq 1\}.$$

Under such hypotheses, B_{U^*} is compact for U^* with the weak-star topology $\sigma(U^*, U)$.

Proof 4.15   For each $u \in U$, we shall associate a real number ω_u and denote

$$\omega = \prod_{u \in U} \omega_u \in \mathbb{R}^U,$$

and consider the projections

$$P_u : \mathbb{R}^U \to \mathbb{R}$$

where

$$P_u(\omega) = \omega_u, \ \forall \omega \in \mathbb{R}^U, \ u \in U.$$

We shall define a topology for $\mathbb{R}^U$, which is induced by the weak neighborhoods specified in the next lines.

Let $\tilde{\omega} \in \mathbb{R}^U$. We define a weak neighborhood $\tilde{V}(\tilde{\omega})$ of $\tilde{\omega}$ as

$$\tilde{V}(\tilde{w}) = \{\omega \in \mathbb{R}^U : |P_{u_i}(\omega) - P_{u_i}(\tilde{\omega})| < \varepsilon_i, \ \forall \in \{1, \cdots, m\}\},$$

where $m \in \mathbb{N}$, $\varepsilon_i > 0$ and $u_i \in U$, $\forall i \in \{1, \cdots, m\}$.

Let $A \subset \mathbb{R}^U$. We say that $\tilde{\omega} \in A$ is interior to A, as there exists a neighborhood $\tilde{V}_w(\tilde{\omega})$ contained in A. If all points of A are interior, we say that A is weakly open.

Finally, we define the weak topology σ para $\mathbb{R}^U$, as the set of all subsets weakly open of $\mathbb{R}^U$.

Now consider U^* with the topology $\sigma(U^*, U)$ and let $\phi : U^* \to \mathbb{R}^U$ where

$$\phi(u^*) = \prod_{u \in U} \langle u, u^* \rangle_U.$$

We shall show that ϕ is continuous. Suppose, to obtain contradiction that ϕ is not continuous. Thus, there exists $u^* \in U^*$ such that ϕ is not continuous at u^*.

Hence there exist a net $\{u_\alpha^*\}_{\alpha \in I}$ such that

$$u_\alpha^* \to u^* \text{ in } \sigma(U^*, U),$$

but we do not have

$$\phi(u_\alpha^*) \to \phi(u^*) \text{ in } \sigma.$$

Therefore, there exists a weak neighborhood $\tilde{V}(\phi(u^*))$ such that for each $\beta \in I$ there exists $\alpha_\beta \in I$ such that $\alpha_\beta \succ \beta$ and

$$\phi(u_{\alpha_\beta}^*) \notin \tilde{V}(\phi(u^*)),$$

with with no loss of generality, we may assume

$$\tilde{V}(\phi(u^*)) = \{\omega \in \mathbb{R}^U : |P_{u_i}(w) - P_{u_i}(\phi(u^*))| < \varepsilon_i, \ \forall i \in \{1, \cdots, m\}\},$$

where $m \in \mathbb{N}$, $\varepsilon_i > 0$ and $u_i \in U$, $\forall i \in \{1, \cdots, m\}$.

From this, we get $j \in \{1, \cdots, m\}$ and a sub-net $\{u_{\alpha_\beta}^*\}$ also denoted by $\{u_{\alpha_\beta}^*\}$ such that

$$|P_{u_j}(\phi(u_{\alpha_\beta}^*)) - P_{u_j}(\phi(u^*))| \geq \varepsilon_j, \forall \alpha_\beta \in I.$$

Thus,

$$\begin{aligned} |P_{u_j}(\phi(u_{\alpha_\beta}^*)) - P_{u_j}(\phi(u^*))| &= |\langle u_j, u_{\alpha_\beta}^* - u^* \rangle_U| \\ &\geq \varepsilon_j, \ \forall \alpha_\beta \in I. \end{aligned} \tag{4.49}$$

Therefore, we do not have,

$$\langle u_j, u_{\alpha_\beta}^* \rangle_U \to \langle u_j, u^* \rangle_U,$$

that is, we do not have,

$$u_\alpha^* \to u^*, \ \text{em} \ \sigma(U^*, U),$$

a contradiction.

Hence, ϕ is continuous with $\mathbb{R}^U$ with the topology σ above specified.

We shall prove now that

$$\phi^{-1} : \phi(U^*) \to U^*$$

is also continuous.

This follows from a little adaptation of the last proposition, considering that

$$f_u(\omega) = \langle u, \phi^{-1}(w) \rangle_U = \omega_u = P_u(\omega),$$

on $\phi(U^*)$ so that f_u, is continuous on $\phi(U^*)$, for all $u \in U$.

Thus, from the last proposition, ϕ^{-1} is continuous.

On the other hand, observe that

$$\phi(B_{U^*}) = K$$

where

$$\begin{aligned} K &= \{\omega \in \mathbb{R}^U \ : \ |\omega_u| \leq \|u\|_U, \omega_{u+v} = \omega_u + \omega_v, \\ &\quad \omega_{\lambda u} = \lambda \omega_u, \forall u, v \in U, \ \lambda \in \mathbb{R}\}. \end{aligned} \tag{4.50}$$

To finish this proof, it suffices, from the continuity of ϕ^{-1}, to show that $K \subset \mathbb{R}^U$ is compact with $\mathbb{R}^U$ with the topology σ.

Observe that $K = K_1 \cap K_2$ where

$$K_1 = \{\omega \in \mathbb{R}^U \ : \ |\omega_u| \leq \|u\|_U, \forall u \in U\}, \tag{4.51}$$

and

$$K_2 = \{\omega \in \mathbb{R}^U \ : \ \omega_{u+v} = \omega_u + \omega_v, \ \omega_{\lambda u} = \lambda \omega_u, \forall u, v \in U, \ \lambda \in \mathbb{R}\}. \tag{4.52}$$

The set $K_3 = \prod_{u \in U} [-\|u\|_U, \|u\|_U]$ is compact as a Cartesian product of compact real intervals.

Since $K_1 \subset K_3$ and K_1 is closed, we have that K_1 is compact concerning the topology in question.

On the other hand, K_2 is closed, since defining the closed sets $A_{u,v}$ e $B_{\lambda,u}$ (these sets are closed from the continuity of projections P_u com $\mathbb{R}^U$ for the topology σ, as inverse images of closed sets in $\mathbb{R}$) by

$$A_{u,v} = \{\omega \in \mathbb{R}^U \ : \ \omega_{u+v} - \omega_u - \omega_v = 0\}, \tag{4.53}$$

and

$$B_{\lambda,u} = \{\omega \in \mathbb{R}^U \ : \ \omega_{\lambda u} - \lambda \omega_u = 0\} \tag{4.54}$$

we have

$$K_2 = (\cap_{u,v \in U} A_{u,v}) \cap (\cap_{(\lambda,u) \in \mathbb{R} \times U} B_{\lambda,u}). \tag{4.55}$$

Recall that K_2 is closed as an intersection of closed sets.

Finally, we have that $K_1 \cap K_2 \subset K_1$ is compact.

This completes the proof.

Theorem 4.5.2 (Kakutani) *Let U be a Banach space. Then U is reflexive if and only if*

$$B_U = \{u \in U \mid \|u\|_U \le 1\} \tag{4.56}$$

is compact for the weak topology $\sigma(U, U^*)$.

Proof 4.16 Suppose U is reflexive, then $J(B_U) = B_{U^{**}}$. From the last theorem $B_{U^{**}}$ is compact for the topology $\sigma(U^{**}, U^*)$. Therefore it suffices to verify that $J^{-1} : U^{**} \to U$ is continuous from U^{**} with the topology $\sigma(U^{**}, U^*)$ to U, with the topology $\sigma(U, U^*)$.

From Theorem 4.3.6 it is sufficient to show that the function $u \mapsto \langle J^{-1}u, f \rangle_U$ is continuous for the topology $\sigma(U^{**}, U^*)$, for each $f \in U^*$. Since

$$\langle J^{-1}u, f \rangle_U = \langle f, u \rangle_{U^*},$$

we have completed the first part of the proof.

For the second we need two lemmas.

Lemma 4.5.3 (Helly) *Let U be a Banach space, $f_1, ..., f_n \in U^*$ and $\alpha_1, ..., \alpha_n \in \mathbb{R}$, then 1 and 2 are equivalent, where:*

1.

$$\text{Given } \varepsilon > 0, \text{ there exists } u_\varepsilon \in U \text{ such that } \|u_\varepsilon\|_U \le 1 \text{ and}$$

$$|\langle u_\varepsilon, f_i \rangle_U - \alpha_i| < \varepsilon, \forall i \in \{1, ..., n\}.$$

2.

$$\left| \sum_{i=1}^n \beta_i \alpha_i \right| \le \left\| \sum_{i=1}^n \beta_i f_i \right\|_{U^*}, \forall \beta_1, ..., \beta_n \in \mathbb{R}. \tag{4.57}$$

Proof. $1 \Rightarrow 2$: *Fix $\beta_1, ..., \beta_n \in \mathbb{R}$, $\varepsilon > 0$ and define $S = \sum_{i=1}^n |\beta_i|$. From 1, we have*

$$\left| \sum_{i=1}^n \beta_i \langle u_\varepsilon, f_i \rangle_U - \sum_{i=1}^n \beta_i \alpha_i \right| < \varepsilon S \tag{4.58}$$

and therefore

$$\left| \sum_{i=1}^n \beta_i \alpha_i \right| - \left| \sum_{i=1}^n \beta_i \langle u_\varepsilon, f_i \rangle_U \right| < \varepsilon S \tag{4.59}$$

or

$$\left| \sum_{i=1}^n \beta_i \alpha_i \right| < \left\| \sum_{i=1}^n \beta_i f_i \right\|_{U^*} \|u_\varepsilon\|_U + \varepsilon S \le \left\| \sum_{i=1}^n \beta_i f_i \right\|_{U^*} + \varepsilon S \tag{4.60}$$

so that

$$\left| \sum_{i=1}^n \beta_i \alpha_i \right| \le \left\| \sum_{i=1}^n \beta_i f_i \right\|_{U^*} \tag{4.61}$$

since ε is arbitrary.

Now let us show that $2 \Rightarrow 1$. Define $\vec{\alpha} = (\alpha_1, ..., \alpha_n) \in \mathbb{R}^n$ and consider the function $\varphi(u) = (\langle f_1, u \rangle_U, ..., \langle f_n, u \rangle_U)$. Item 1 is equivalent to $\vec{\alpha}$ belongs to the closure of $\varphi(B_U)$. Let us suppose that $\vec{\alpha}$ does not belong to the closure of $\varphi(B_U)$ and obtain a contradiction. Thus we can separate $\vec{\alpha}$ and the closure of $\varphi(B_U)$ strictly, that is there exists $\vec{\beta} = (\beta_1, ..., \beta_n) \in \mathbb{R}^n$ and $\gamma \in \mathbb{R}$ such that

$$\varphi(u) \cdot \vec{\beta} < \gamma < \vec{\alpha} \cdot \vec{\beta}, \forall u \in B_U \tag{4.62}$$

Taking the supremum in u we contradict 2.

Also we need the lemma.

Lemma 4.5.4 *Let U be a Banach space. Then $J(B_U)$ is dense in $B_{U^{**}}$ for the topology $\sigma(U^{**}, U^*)$.*

Proof 4.17 Let $u^{**} \in B_{U^{**}}$ and consider $\mathcal{V}_{u^{**}}$ a neighborhood of u^{**} for the topology $\sigma(U^{**}, U^*)$. It suffices to show that $J(B_U) \cap \mathcal{V}_{u^{**}} \neq \emptyset$. As $\mathcal{V}_{u^{**}}$ is a weak neighborhood, there exists $f_1, ..., f_n \in U^*$ and $\varepsilon > 0$ such that

$$\mathcal{V}_{u^{**}} = \{\eta \in U^{**} \mid \langle f_i, \eta - u^{**} \rangle_{U^*} \mid < \varepsilon, \forall i \in \{1, ..., n\}\}. \tag{4.63}$$

Define $\alpha_i = \langle f_i, u^{**} \rangle_{U^*}$ and thus for any given $\beta_1, ..., \beta_n \in \mathbb{R}$ we have

$$\left| \sum_{i=1}^{n} \beta_i \alpha_i \right| = \left| \langle \sum_{i=1}^{n} \beta_i f_i, u^{**} \rangle_{U^*} \right| \leq \left\| \sum_{i=1}^{n} \beta_i f_i \right\|_{U^*}, \tag{4.64}$$

so that from Helly lemma, there exists $u_\varepsilon \in U$ such that $\|u_\varepsilon\|_U \leq 1$ and

$$|\langle u_\varepsilon, f_i \rangle_U - \alpha_i| < \varepsilon, \forall i \in \{1, ..., n\} \tag{4.65}$$

or,

$$|\langle f_i, J(u_\varepsilon) - u^{**} \rangle_{U^*}| < \varepsilon, \forall i \in \{1, ..., n\} \tag{4.66}$$

and hence

$$J(u_\varepsilon) \in \mathcal{V}_{u^{**}}. \tag{4.67}$$

Now we will complete the proof of Kakutani theorem. Suppose B_U is weakly compact (that is, compact for the topology $\sigma(U, U^*)$). Observe that $J : U \to U^{**}$ is weakly continuous, that is, it is continuous with U endowed with the topology $\sigma(U, U^*)$ and U^{**} endowed with the topology $\sigma(U^{**}, U^*)$. Thus, as B_U is weakly compact, we have that $J(B_U)$ is compact for the topology $\sigma(U^{**}, U^*)$. From the last lemma, $J(B_U)$ is dense $B_{U^{**}}$ for the topology $\sigma(U^{**}, U^*)$. Hence $J(B_U) = B_{U^{**}}$, or $J(U) = U^{**}$, which completes the proof. $\square$

Proposition 4.5.5 *Let U be a reflexive Banach space. Let $K \subset U$ be a convex closed bounded set. Then K is weakly compact.*

Proof 4.18 From Theorem 4.3.5, K is weakly closed (closed for the topology $\sigma(U, U^*)$). Since K is bounded, there exists $\alpha \in \mathbb{R}^+$ such that $K \subset \alpha B_U$. Since K is weakly closed and $K = K \cap \alpha B_U$, we have that it is weakly compact.

Proposition 4.5.6 *Let U be a reflexive Banach space and $M \subset U$ a closed subspace. Then M with the norm induced by U is reflexive.*

Proof 4.19   We can identify two weak topologies in M, namely:

$$\sigma(M, M^*) \text{ and the trace of } \sigma(U, U^*). \tag{4.68}$$

It can be easily verified that these two topologies coincide (through restrictions and extensions of linear forms). From the Kakutani Theorem it suffices to show that B_M is compact for the topology $\sigma(M, M^*)$. But B_U is compact for $\sigma(U, U^*)$ and $M \subset U$ is closed (strongly) and convex so that it is weakly closed, thus from last proposition, B_M is compact for the topology $\sigma(U, U^*)$, and therefore it is compact for $\sigma(M, M^*)$.

4.6   Separable sets

Definition 4.6.1 (Separable spaces)   *A metric space U is said to be separable if there exist a set $K \subset U$ such that K is countable and dense in U.*

The next proposition is proved in [24].

Proposition 4.6.2   *Let U be a separable metric space. If $V \subset U$ then V is separable.*

Theorem 4.6.3   *Let U be a Banach space such that U^* is separable. Then U is separable.*

Proof 4.20   Consider $\{u_n^*\}$ a countable dense set in U^*. Observe that

$$\|u_n^*\|_{U^*} = \sup\{|\langle u_n^*, u\rangle_U| \mid u \in U \text{ and } \|u\|_U = 1\} \tag{4.69}$$

so that for each $n \in \mathbb{N}$, there exists $u_n \in U$ such that $\|u_n\|_U = 1$ and $\langle u_n^*, u_n\rangle_U \geq \frac{1}{2}\|u_n^*\|_{U^*}$.

Define U_0 as the vector space on $\mathbb{Q}$ spanned by $\{u_n\}$, and U_1 as the vector space on $\mathbb{R}$ spanned by $\{u_n\}$. It is clear that U_0 is dense in U_1 and we will show that U_1 is dense in U, so that U_0 is a dense set in U. For, suppose u^* is such that $\langle u, u^*\rangle_U = 0, \forall u \in U_1$. Since $\{u_n^*\}$ is dense in U^*, given $\varepsilon > 0$, there exists $n \in \mathbb{N}$ such that $\|u_n^* - u^*\|_{U^*} < \varepsilon$, so that

$$
\begin{aligned}
\frac{1}{2}\|u_n^*\|_{U^*} \leq \langle u_n, u_n^*\rangle_U &= \langle u_n, u_n^* - u^*\rangle_U + \langle u_n, u^*\rangle_U \\
&\leq \|u_n^* - u^*\|_{U^*}\|u_n\|_U + 0 \\
&< \varepsilon
\end{aligned}
\tag{4.70}
$$

or

$$\|u^*\|_{U^*} \leq \|u_n^* - u^*\|_{U^*} + \|u_n^*\|_{U^*} < \varepsilon + 2\varepsilon = 3\varepsilon. \tag{4.71}$$

Therefore, since ε is arbitrary, $\|u^*\|_{U^*} = 0$, that is $u^* = \theta$. By Corollary 4.2.14 this completes the proof.

Proposition 4.6.4   *U is reflexive if and only if U^* is reflexive.*

Proof 4.21   Suppose U is reflexive, as B_{U^*} is compact for $\sigma(U^*, U)$ and $\sigma(U^*, U) = \sigma(U^*, U^{**})$ we have that B_{U^*} is compact for $\sigma(U^*, U^{**})$, which means that U^* is reflexive.

Suppose U^* is reflexive, from above U^{**} is reflexive. Since $J(U)$ is a closed subspace of U^{**}, from Proposition 4.5.6, J(U) is reflexive. From the Kakutani Theorem $J(B_U)$ is weakly compact. At this point we shall prove that $J^{-1} : J(U) \to V$ is continuous from $J(U)$ with the topology $\sigma(U^{**}, U^*)$ to U with the topology $\sigma(U, U^*)$.

Let $\{u_\alpha^{**}\}_{\alpha \in I} \subset J(B_U)$ be a net such that

$$u_\alpha^{**} \rightharpoonup u_0^{**}$$

weakly in $\sigma(U^{**}, U^*)$.

Let $u^* \in U^*$. Thus,

$$\langle u^*, u_\alpha^{**} \rangle_{U^*} \to \langle u^*, u_0^{**} \rangle_{U^*}.$$

From this

$$\langle u^*, J(J^{-1}(u_\alpha^{**})) \rangle_{U^*} \to \langle u^*, J(J^{-1}(u_0^{**})) \rangle_{U^*},$$

so that

$$\langle (J^{-1}(u_\alpha^{**})), u^* \rangle_U \to \langle (J^{-1}(u_0^{**})), u^* \rangle_U.$$

Since the net is question and $u^* \in U^*$ have been arbitrary, we may infer that J^{-1} is weakly continuous for the concerning topology.

Hence $J^{-1}(J(B_U))$ is also weakly compact so that from this, from the fact that B_U is weakly closed and

$$B_U \subset J^{-1}J(B_U),$$

it follows that B_U is compact for the topology $\sigma(U, U^*)$.

From such a result and from the Kakutani Theorem we may infer that U is reflexive.

The proof is complete.

Proposition 4.6.5 *Let U be a Banach space. Then U is reflexive and separable if and only if U^* is reflexive and separable.*

Our final result in this section refers to the metrizability of B_{U^*}.

Theorem 4.6.6 *Let U be separable Banach space. Under such hypotheses B_{U^*} is metrizable with respect to the weak-star topology $\sigma(U^*, U)$. Conversely, if B_{U^*} is mertizable in $\sigma(U^*, U)$ then U is separable.*

Proof 4.22 Let $\{u_n\}$ be a dense countable set in B_U. For each $u^* \in U^*$ define

$$\|u^*\|_w = \sum_{n=1}^{\infty} \frac{1}{2^n} |\langle u_n, u^* \rangle_U|.$$

It may be easily verified that $\| \cdot \|_w$ is a norm in U^* and

$$\|u^*\|_w \le \|u^*\|_U.$$

So, we may define a metric in U^* by

$$d(u^*, v^*) = \|u^* - v^*\|_w.$$

Now we shall prove that the topology induced by d coincides with $\sigma(U^*, U)$ in U^*.

For, let $u_0^* \in B_{U^*}$ and let V be neighborhood of u_0^* in $\sigma(U^*, U)$.

We need to prove that there exists $r > 0$ such that

$$V_w = \{u^* \in B_{U^*} \mid d(u_0^*, u^*) < r\} \subset V.$$

Observe that for V we may assume the general format

$$V = \{u^* \in U^* \mid |\langle v_i, u^* - u_0^* \rangle_U| < \varepsilon,$$

for some $\varepsilon > 0$ and $v_1, ..., v_k \in U$.

There is no loss in generality in assuming

$$\|v_i\|_U \le 1, \forall i \in \{1, ..., k\}.$$

Since $\{u_n\}$ is dense in U, for each $i \in \{1,...,k\}$ there exists $n_i \in \mathbb{N}$ such that

$$\|u_{n_i} - v_i\|_U < \frac{\varepsilon}{4}.$$

Choose $r > 0$ small enough such that

$$2^{n_i} r < \frac{\varepsilon}{2}, \forall i \in \{1,...,k\}.$$

We are going to show that $V_w \subset V$, where

$$V_w = \{u^* \in B_{U^*} \mid d(u_0^*, u^*) < r\} \subset V.$$

Observe that, if $u^* \in V_w$ then

$$d(u_0^*, u^*) < r,$$

so that

$$\frac{1}{2^{n_i}} |\langle u_{n_i}, u^* - u_0^* \rangle_U| < r, \forall i \in \{1,...,k\},$$

so that

$$
\begin{aligned}
|\langle v_i, u^* - u_0^* \rangle_U| &\leq |\langle v_i - u_{n_i}, u^* - u_0^* \rangle_U| + |\langle u_{n_i}, u^* - u_0^* \rangle_U| \\
&\leq (\|u^*\|_{U^*} + \|u_0^*\|_{U^*}) \|v_i - u_{n_i}\|_U + |\langle u_{n_i}, u^* - u_0^* \rangle_U| \\
&< 2\frac{\varepsilon}{4} + \frac{\varepsilon}{2} = \varepsilon.
\end{aligned}
\tag{4.72}
$$

Therefore $u^* \in V$, so that $V_w \subset V$.

Now let $u_0 \in B_{U^*}$ and fix $r > 0$. We have to obtain a neighborhood $V \in \sigma(U^*U)$ such that

$$V \subset V_w = \{u^* \in B_{U^*} \mid d(u_0^*, u^*) < r\}.$$

We shall define $k \in \mathbb{N}$ and $\varepsilon > 0$ in the next lines so that $V \subset V_w$, where

$$V = \{u^* \in B_{U^*} \mid |\langle u_i, u^* - u_0^* \rangle_U| < \varepsilon, \forall i \in \{1,...,k\}\}.$$

For $u^* \in V_w$ we have

$$
\begin{aligned}
d(u^*, u_0^*) &= \sum_{n=1}^{k} \frac{1}{2^n} |\langle u_n, u^* - u_0^* \rangle_U| \\
&\quad + \sum_{n=k+1}^{\infty} \frac{1}{2^n} |\langle u_n, u^* - u_0^* \rangle_U| \\
&< \varepsilon + 2 \sum_{n=k+1}^{\infty} \frac{1}{2^n} \\
&= \varepsilon + \frac{1}{2^{k-1}}.
\end{aligned}
\tag{4.73}
$$

Hence, it suffices to take $\varepsilon = r/2$, and k sufficiently big such that

$$\frac{1}{2^{k-1}} < r/2.$$

The first part of the proof is finished.

Conversely, assume B_U is metrizable in $\sigma(U^*, U)$. We are going to show that U is separable.

Define,

$$\tilde{V}_n = \left\{ u^* \in B_{U^*} \mid d(u^*, \theta) < \frac{1}{n} \right\}.$$

From the first part, we may find V_n a neighborhood of zero in $\sigma(U^*, U)$ such that

$$V_n \subset \tilde{V}_n.$$

Moreover, we may assume that V_n has the form

$$V_n = \{ u^* \in B_{U^*} \mid |\langle u, u^* - \theta \rangle_U| < \varepsilon_n, \forall u \in C_n \},$$

where C_n is a finite set.

Define

$$D = \cup_{i=1}^{\infty} C_n.$$

Thus D is countable and we are going to prove that such a set is dense in U.

For, suppose $u^* \in U^*$ is such that

$$\langle u, u^* \rangle_U = 0, \forall u \in D.$$

Hence,

$$u^* \in V_n \subset \tilde{V}_n, \forall n \in \mathbb{N},$$

so that $u^* = \theta$.

The proof is complete.

4.7 Uniformly convex spaces

Definition 4.7.1 (Uniformly convex spaces) *A Banach space U is said to be uniformly convex if for each $\varepsilon > 0$, there exists $\delta > 0$ such that:*

If $u, v \in U$, $\|u\|_U \le 1$, $\|v\|_U \le 1$, and $\|u - v\|_U > \varepsilon$ then $\frac{\|u+v\|_U}{2} < 1 - \delta$.

Theorem 4.7.2 (Milman Pettis) *Every uniformly convex Banach space is reflexive.*

Proof 4.23 Let $\eta \in U^{**}$ be such that $\|\eta\|_{U^{**}} = 1$. It suffices to show that $\eta \in J(B_U)$. Since $J(B_U)$ is closed in U^{**}, we have only to show that for each $\varepsilon > 0$ there exists $u \in U$ such that $\|\eta - J(u)\|_{U^{**}} < \varepsilon$.

Thus, suppose given $\varepsilon > 0$. Let $\delta > 0$ be the corresponding constant relating the uniformly convex property.

Choose $f \in U^*$ such that $\|f\|_{U^*} = 1$ and

$$\langle f, \eta \rangle_{U^*} > 1 - \frac{\delta}{2}. \tag{4.74}$$

Define

$$V = \left\{ \zeta \in U^{**} \mid |\langle f, \zeta - \eta \rangle_{U^*}| < \frac{\delta}{2} \right\}.$$

Observe that V is a neighborhood of η in $\sigma(U^{**}, U^*)$. Since $J(B_U)$ is dense in $B_{U^{**}}$ concerning the topology $\sigma(U^{**}, U^*)$, we have that $V \cap J(B_U) \neq \emptyset$ and thus there exists $u \in B_U$ such that $J(u) \in V$. Suppose, to obtain contradiction, that

$$\|\eta - J(u)\|_{U^{**}} > \varepsilon.$$

Therefore, defining

$$W = (J(u) + \varepsilon B_{U^{**}})^c,$$

we have that $\eta \in W$, where W is also a weak neighborhood of η in $\sigma(U^{**}, U^*)$, since $B_{U^{**}}$ is closed in $\sigma(U^{**}, U^*)$.

Hence $V \cap W \cap J(B_U) \neq \emptyset$, so that there exists some $v \in B_U$ such that $J(v) \in V \cap W$. Thus, $J(u) \in V$ and $J(v) \in V$, so that

$$|\langle u, f \rangle_U - \langle f, \eta \rangle_{U^*}| < \frac{\delta}{2},$$

and

$$|\langle v, f \rangle_U - \langle f, \eta \rangle_{U^*}| < \frac{\delta}{2}.$$

Hence,

$$\begin{aligned} 2\langle f, \eta \rangle_{U^*} &< \langle u + v, f \rangle_U + \delta \\ &\leq \|u + v\|_U + \delta. \end{aligned} \tag{4.75}$$

From this and (4.74) we obtain

$$\frac{\|u + v\|_U}{2} > 1 - \delta,$$

and thus from the definition of uniform convexity, we obtain

$$\|u - v\|_U \leq \varepsilon. \tag{4.76}$$

On the other hand, since $J(v) \in W$, we have

$$\|J(u) - J(v)\|_{U^{**}} = \|u - v\|_U > \varepsilon,$$

which contradicts (4.76). The proof is complete.

Chapter 5

Topics on Linear Operators

The main references for this chapter are Reed and Simon [62] and Bachman and Narici [6].

5.1 Topologies for bounded operators

Let U, Y be Banach spaces. First we recall that the set of all bounded linear operators $A : U \to Y$, denoted by $\mathscr{L}(U,Y)$, is a Banach space with the norm

$$\|A\| = \sup\{\|Au\|_Y \mid \|u\|_U \leq 1\}.$$

The topology related to the metric induced by this norm is called the uniform operator topology.

Let us introduce now the strong operator topology, which is defined as the weakest topology for which the functions

$$E_u : \mathscr{L}(U,Y) \to Y$$

are continuous where

$$E_u(A) = Au, \forall A \in \mathscr{L}(U,Y).$$

For such a topology a base at origin is given by sets of the form

$$\{A \mid A \in \mathscr{L}(U,Y), \, \|Au_i\|_Y < \varepsilon, \forall i \in \{1,,...,n\}\},$$

where $u_1,...,u_n \in U$ and $\varepsilon > 0$.

Observe that a sequence $\{A_n\} \subset \mathscr{L}(U,Y)$ converges to A concerning this last topology if

$$\|A_n u - Au\|_Y \to 0, \text{ as } n \to \infty, \forall u \in U.$$

In the next lines we describe the weak operator topology in $\mathscr{L}(U,Y)$. Such a topology is weakest one such that the functions

$$E_{u,v} : \mathscr{L}(U,Y) \to \mathbb{C}$$

are continuous, where

$$E_{u,v}(A) = \langle Au, v \rangle_Y, \forall A \in \mathscr{L}(U,Y), u \in U, \, v \in Y^*.$$

For such a topology, a base at origin is given by sets of the form

$$\{A \in \mathscr{L}(U,Y) \mid |\langle Au_i, v_j \rangle_Y| < \varepsilon, \forall i \in \{1,...,n\}, \, j \in \{1,...,m\}\}.$$

where $\varepsilon > 0$, $u_1, ..., u_n \in U$, $v_1, ..., v_m \in Y^*$.

A sequence $\{A_n\} \subset \mathscr{L}(U,Y)$ converges to $A \in \mathscr{L}(U,Y)$ if

$$|\langle A_n u, v \rangle_Y - \langle Au, v \rangle_Y| \to 0,$$

as $n \to \infty$, $\forall u \in U$, $v \in Y^*$.

5.2 Adjoint operators

We start this section recalling the definition of adjoint operator.

Definition 5.2.1 *Let U, Y be Banach spaces. Given a bounded linear operator $A : U \to Y$ and $v^* \in Y^*$, we have that $T(u) = \langle Au, v^* \rangle_Y$ is such that*

$$|T(u)| \leq \|Au\|_Y \cdot \|v^*\| \leq \|A\| \|v^*\|_{Y^*} \|u\|_U.$$

Hence, $T(u)$ is a continuous linear functional on U and considering our fundamental representation hypothesis, there exists $u^* \in U^*$ such that

$$T(u) = \langle u, u^* \rangle_U, \forall u \in U.$$

We define A^* by setting $u^* = A^* v^*$, so that

$$T(u) = \langle u, u^* \rangle_U = \langle u, A^* v^* \rangle_U$$

that is,

$$\langle u, A^* v^* \rangle_U = \langle Au, v^* \rangle_Y, \forall u \in U, \ v^* \in Y^*.$$

We call $A^* : Y^* \to U^*$ the adjoint operator relating $A : U \to Y$.

Theorem 5.2.2 *Let U, Y be Banach spaces and let $A : U \to Y$ be a bounded linear operator. Then*

$$\|A\| = \|A^*\|.$$

Proof 5.1 Observe that

$$
\begin{aligned}
\|A\| &= \sup_{u \in U} \{\|Au\| \mid \|u\|_U = 1\} \\
&= \sup_{u \in U} \{ \sup_{v^* \in Y^*} \{\langle Au, v^* \rangle_Y \mid \|v^*\|_{Y^*} = 1\}, \|u\|_U = 1\} \\
&= \sup_{(u,v^*) \in U \times Y^*} \{\langle Au, v^* \rangle_Y \mid \|v^*\|_{Y^*} = 1, \|u\|_U = 1\} \\
&= \sup_{(u,v^*) \in U \times Y^*} \{\langle u, A^* v^* \rangle_U \mid \|v^*\|_{Y^*} = 1, \|u\|_U = 1\} \\
&= \sup_{v^* \in Y^*} \{\sup_{u \in U} \{\langle u, A^* v^* \rangle_U \mid \|u\|_U = 1\}, \|v^*\|_{Y^*} = 1\} \\
&= \sup_{v^* \in Y^*} \{\|A^* v^*\|, \|v^*\|_{Y^*} = 1\} \\
&= \|A^*\|.
\end{aligned}
\tag{5.1}
$$

In particular, if $U = Y = H$ where H is Hilbert space, we have

Theorem 5.2.3 *Given the bounded linear operators $A, B : H \to H$ we have*

1. $(AB)^* = B^* A^*$,

2. $(A^*)^* = A$,

3. *If A has a bounded inverse A^{-1} then A^* has a bounded inverse and*

$$(A^*)^{-1} = (A^{-1})^*.$$

4. $\|AA^*\| = \|A\|^2.$

Proof 5.2

1. Observe that

$$(ABu, v)_H = (Bu, A^*v)_H = (u, B^*A^*v)_H, \forall u, v \in H.$$

2. Observe that

$$(Au, v)_H = (u, A^*v)_H = \overline{(A^*v, u)_H} = \overline{(v, A^{**}u)_H} = (A^{**}u, v)_H, \forall u, v \in H.$$

3. We have that

$$I = AA^{-1} = A^{-1}A,$$

so that

$$I = I^* = (AA^{-1})^* = (A^{-1})^*A^* = (A^{-1}A)^* = A^*(A^{-1})^*.$$

4. Observe that

$$\|A^*A\| \leq \|A\|\|A^*\| = \|A\|^2,$$

and

$$
\begin{aligned}
\|A^*A\| &\geq \sup_{u \in U}\{(u, A^*Au)_H \mid \|u\|_U = 1\} \\
&= \sup_{u \in U}\{(Au, Au)_H \mid \|u\|_U = 1\} \\
&= \sup_{u \in U}\{\|Au\|_H^2 \mid \|u\|_U = 1\} = \|A\|^2,
\end{aligned}
\tag{5.2}
$$

and hence

$$\|A^*A\| = \|A\|^2.$$

Definition 5.2.4 *Given $A \in \mathscr{L}(H)$ we say that A is self-adjoint if*

$$A = A^*.$$

Theorem 5.2.5 *Let U and Y be Banach spaces and let $A : U \to Y$ be a bounded linear operator. Then*

$$[R(A)]^\perp = N(A^*),$$

where

$$[R(A)]^\perp = \{v^* \in Y^* \mid \langle Au, v^* \rangle_Y = 0, \ \forall u \in U\}.$$

Proof 5.3   Let $v^* \in N(A^*)$. Choose $v \in R(A)$. Thus there exists u in U such that $Au = v$ so that

$$\langle v, v^* \rangle_Y = \langle Au, v^* \rangle_Y = \langle u, A^*v^* \rangle_U = 0.$$

Since $v \in R(A)$ is arbitrary, we have obtained

$$N(A^*) \subset [R(A)]^\perp.$$

Suppose $v^* \in [R(A)]^\perp$. Choose $u \in U$. Thus,

$$\langle Au, v^* \rangle_Y = 0,$$

so that

$$\langle u, A^* v^* \rangle_U, \forall u \in U.$$

Therefore, $A^* v^* = \theta$, that is, $v^* \in N(A^*)$. Since $v^* \in [R(A)]^\perp$ is arbitrary, we get

$$[R(A)]^\perp \subset N(A^*).$$

This completes the proof.

The next result is relevant for subsequent developments.

Lemma 5.1

Let U, Y be Banach spaces and let $A : U \to Y$ be a bounded linear operator. Suppose also that $R(A) = \{A(u) : u \in U\}$ is closed. Under such hypotheses, there exists $K > 0$ such that for each $v \in R(A)$ there exists $u_0 \in U$ such that

$$A(u_0) = v$$

and

$$\|u_0\|_U \leq K \|v\|_Y.$$

Proof 5.4 Define $L = N(A) = \{u \in U : A(u) = \theta\}$ (the null space of A). Consider the space U/L, where

$$U/L = \{\bar{u} : u \in U\},$$

where

$$\bar{u} = \{u + w : w \in L\}.$$

Define $\bar{A} : U/L \to R(A)$, by

$$\bar{A}(\bar{u}) = A(u).$$

Observe that $\bar{A}$ is one-to-one, linear, onto and bounded. Moreover $R(A)$ is closed so that it is a Banach space. Hence by the inverse mapping theorem we have that $\bar{A}$ has a continuous inverse. Thus, for any $v \in R(A)$ there exists $\bar{u} \in U/L$ such that

$$\bar{A}(\bar{u}) = v$$

so that

$$\bar{u} = \bar{A}^{-1}(v),$$

and therefore

$$\|\bar{u}\| \leq \|\bar{A}^{-1}\| \|v\|_Y.$$

Recalling that

$$\|\bar{u}\| = \inf_{w \in L} \{\|u + w\|_U\},$$

we may find $u_0 \in \bar{u}$ such that

$$\|u_0\|_U \leq 2\|\bar{u}\| \leq 2\|\bar{A}^{-1}\| \|v\|_Y,$$

and so that

$$A(u_0) = \bar{A}(\overline{u_0}) = \bar{A}(\bar{u}) = v.$$

Taking $K = 2\|\bar{A}^{-1}\|$, we have completed the proof.

Theorem 5.1

Let U, Y be Banach spaces and let $A : U \to Y$ be a bound linear operator. Assume $R(A)$ is closed. Under such hypotheses

$$R(A^*) = [N(A)]^{\perp}.$$

Proof 5.5   Let $u^* \in R(A^*)$. Thus there exists $v^* \in Y^*$ such that

$$u^* = A^*(v^*).$$

Let $u \in N(A)$. Hence,

$$\langle u, u^* \rangle_U = \langle u, A^*(v^*) \rangle_U = \langle A(u), v^* \rangle_Y = 0.$$

Since $u \in N(A)$ is arbitrary, we get $u^* \in [N(A)]^{\perp}$, so that

$$R(A^*) \subset [N(A)]^{\perp}.$$

Now suppose $u^* \in [N(A)]^{\perp}$. Thus,

$$\langle u, u^* \rangle_U = 0, \ \forall u \in N(A).$$

Fix $v \in R(A)$. From the Lemma 5.1, there exists $K > 0$ (which does not depend on v) and $u_v \in U$ such that

$$A(u_v) = v$$

and

$$\|u_v\|_U \leq K \|v\|_Y.$$

Define $f : R(A) \to \mathbb{R}$ by

$$f(v) = \langle u_v, u^* \rangle_U.$$

Observe that

$$|f(v)| \leq \|u_v\|_U \|u^*\|_{U^*} \leq K \|v\|_Y \|u^*\|_{U^*},$$

so that f is a bounded linear functional. Hence by a Hahn-Banach theorem corollary, there exists $v^* \in Y^*$ such that

$$f(v) = \langle v, v^* \rangle_Y \equiv F(v), \ \forall v \in R(A),$$

that is, F is an extension of f from $R(A)$ to Y.

In particular

$$f(v) = \langle u_v, u^* \rangle_U = \langle v, v^* \rangle_Y = \langle A(u_v), v^* \rangle_Y \ \forall v \in R(A),$$

where $A(u_v) = v$, so that

$$\langle u_v, u^* \rangle_U = \langle A(u_v), v^* \rangle_Y \ \forall v \in R(A).$$

Now let $u \in U$ and define $A(u) = v_0$. Observe that

$$u = (u - u_{v_0}) + u_{v_0},$$

and

$$A(u - u_{v_0}) = A(u) - A(u_{v_0}) = v_0 - v_0 = \theta.$$

Since $u^* \in [N(A)]^{\perp}$, we get

$$\langle u - u_{v_0}, u^* \rangle_U = 0$$

so that

$$
\begin{aligned}
\langle u, u^* \rangle_U &= \langle (u - u_{v_0}) + u_{v_0}, u^* \rangle_U \\
&= \langle u_{v_0}, u^* \rangle_U \\
&= \langle A(u_{v_0}), v^* \rangle_Y \\
&= \langle A(u - u_{v_0}) + A(u_{v_0}), v^* \rangle_Y \\
&= \langle A(u), v^* \rangle_Y.
\end{aligned}
\tag{5.3}
$$

Hence,

$$
\langle u, u^* \rangle_U = \langle A(u), v^* \rangle_Y, \ \forall u \in U.
$$

We may conclude that $u^* = A^*(v^*) \in R(A^*)$. Since $u^* \in [N(A)]^\perp$ is arbitrary we obtain

$$
[N(A)]^\perp \subset R(A^*).
$$

The proof is complete.

We finish this section with the following result.

Definition 5.2.6 *Let U be a Banach space and $S \subset U$. We define the positive conjugate cone of S, denoted by $S^\oplus$ by*

$$
S^\oplus = \{ u^* \in U^* \ : \ \langle u, u^* \rangle_U \geq 0, \ \forall u \in S \}.
$$

Similarly, we define the the negative cone of S, denoted by denoted by $S^\ominus$ by

$$
S^\ominus = \{ u^* \in U^* \ : \ \langle u, u^* \rangle_U \leq 0, \ \forall u \in S \}.
$$

Theorem 5.2.7 *Let U, Y be Banach spaces and $A : U \to Y$ be a bounded linear operator. Let $S \subset U$. Then*

$$
[A(S)]^\oplus = (A^*)^{-1}(S^\oplus),
$$

where

$$
(A^*)^{-1}(S^\oplus) = \{ v^* \in Y^* \ : \ A^* v^* \in S^\oplus \}.
$$

Proof 5.6 Let $v^* \in [A(S)]^\oplus$ and $u \in S$. Thus,

$$
\langle A(u), v^* \rangle_Y \geq 0,
$$

so that

$$
\langle u, A^*(v^*) \rangle_U \geq 0.
$$

Since $u \in S$ is arbitrary, we get

$$
v^* \in (A^*)^{-1}(S^\oplus).
$$

From this,

$$
[A(S)]^\oplus = (A^*)^{-1}(S^\oplus).
$$

Reciprocally, let $v^* \in (A^*)^{-1}(S^\oplus)$. Hence $A^*(v^*) \in S^\oplus$ so that, for $u \in S$ we obtain

$$
\langle u, A^*(v^*) \rangle_U \geq 0,
$$

and therefore

$$
\langle A(u), v^* \rangle_Y \geq 0.
$$

Since $u \in S$ is arbitrary, we get $v^* \in [A(S)]^\oplus$, that is,

$$
(A^*)^{-1}(S^\oplus) \subset [A(S)]^\oplus.
$$

The proof is complete.

5.3 Compact operators

We start this section defining compact operators.

Definition 5.3.1 *Let U and Y be Banach spaces. An operator $A \in \mathscr{L}(U,Y)$ (linear and bounded) is said to compact if A takes bounded sets into pre-compact sets. Summarizing, A is compact if for each bounded sequence $\{u_n\} \subset U$, $\{Au_n\}$ has a convergent subsequence in Y.*

Theorem 5.3.2 *A compact operator maps weakly convergent sequences into norm convergent sequences.*

Proof 5.7 Let $A : U \to Y$ be a compact operator. Suppose

$$u_n \rightharpoonup u \text{ weakly in U.}$$

By the uniform boundedness theorem, $\{\|u_n\|\}$ is bounded. Thus, given $v^* \in Y^*$ we have

$$
\begin{aligned}
\langle Au_n, v^* \rangle_Y &= \langle u_n, A^* v^* \rangle_U \\
&\to \langle u, A^* v^* \rangle_U \\
&= \langle Au, v^* \rangle_Y.
\end{aligned}
\tag{5.4}
$$

Being $v^* \in Y^*$ arbitrary, we get that

$$Au_n \rightharpoonup Au \text{ weakly in } Y. \tag{5.5}$$

Suppose Au_n does not converge in norm to Au. Thus there exists $\varepsilon > 0$ and a subsequence $\{Au_{n_k}\}$ such that

$$\|Au_{n_k} - Au\|_Y \geq \varepsilon, \forall k \in \mathbb{N}.$$

As $\{u_{n_k}\}$ is bounded and A is compact, $\{Au_{n_k}\}$ has a subsequence converging para $\tilde{v} \neq Au$. But then such a sequence converges weakly to $\tilde{v} \neq Au$, which contradicts (5.5). The proof is complete.

Theorem 5.3.3 *Let H be a separable Hilbert space. Thus each compact operator in $\mathscr{L}(H)$ is the limit in norm of a sequence of finite rank operators.*

Proof 5.8 Let A be a compact operator in H. Let $\{\phi_j\}$ be an orthonormal basis in H. For each $n \in \mathbb{N}$ define

$$\lambda_n = \sup\{\|A\psi\|_H \mid \psi \in [\phi_1,...,\phi_n]^\perp \text{ and } \|\psi\|_H = 1\}.$$

It is clear that $\{\lambda_n\}$ is a non-increasing sequence that converges to a limit $\lambda \geq 0$. We will show that $\lambda = 0$. Choose a sequence $\{\psi_n\}$ such that

$$\psi_n \in [\phi_1,...,\phi_n]^\perp,$$

$\|\psi_n\|_H = 1$ and $\|A\psi_n\|_H \geq \lambda/2$. Now we will show that

$$\psi_n \rightharpoonup \theta, \text{ weakly in } H.$$

Let $\psi^* \in H^* = H$, thus there exists a sequence $\{a_j\} \subset \mathbb{C}$ such that

$$\psi^* = \sum_{j=1}^{\infty} a_j \phi_j.$$

Suppose given $\varepsilon > 0$. We may find $n_0 \in \mathbb{N}$ such that

$$\sum_{j=n_0}^{\infty} |a_j|^2 < \varepsilon.$$

Choose $n > n_0$. Hence there exists $\{b_j\}_{j>n}$ such that

$$\psi_n = \sum_{j=n+1}^{\infty} b_j \phi_j,$$

and

$$\sum_{j=n+1}^{\infty} |b_j|^2 = 1.$$

Therefore,

$$
\begin{aligned}
|(\psi_n, \psi^*)_H| &= \left| \sum_{j=n+1}^{\infty} (\phi_j, \phi_j)_H a_j \cdot b_j \right| \\
&= \left| \sum_{j=n+1}^{\infty} a_j \cdot b_j \right| \\
&\leq \sqrt{\sum_{j=n+1}^{\infty} |a_j|^2} \sqrt{\sum_{j=n+1}^{\infty} |b_j|^2} \\
&\leq \sqrt{\varepsilon},
\end{aligned}
\tag{5.6}
$$

if $n > n_0$. Since $\varepsilon > 0$ is arbitrary,

$$(\psi_n, \psi^*)_H \to 0, \text{ as } n \to \infty.$$

Since $\psi^* \in H$ is arbitrary, we get

$$\psi_n \rightharpoonup \theta, \text{ weakly in } H.$$

Hence, as A is compact, we have

$$A\psi_n \to \theta \text{ in norm },$$

so that $\lambda = 0$. Finally, we may define $\{A_n\}$ by

$$A_n(u) = A \left(\sum_{j=1}^{n} (u, \phi_j)_H \phi_j \right) = \sum_{j=1}^{n} (u, \phi_j)_H A\phi_j,$$

for each $u \in H$. Thus

$$\|A - A_n\| = \lambda_n \to 0, \text{ as } n \to \infty.$$

The proof is complete.

5.4 The square root of a positive operator

Definition 5.4.1 *Let H be a Hilbert space. A mapping $E : H \to H$ is said to be a projection on $M \subset H$ if for each $z \in H$ we have*

$$Ez = x$$

where $z = x + y$, $x \in M$ and $y \in M^{\perp}$.

Observe that

1. E is linear,

2. E is idempotent, that is $E^2 = E$,

3. $R(E) = M$,

4. $N(E) = M^\perp$.

Also observe that from

$$Ez = x$$

we have

$$\|Ez\|_H^2 = \|x\|_H^2 \le \|x\|_H^2 + \|y\|_H^2 = \|z\|_H^2,$$

so that

$$\|E\| \le 1.$$

Definition 5.4.2 *Let* $A, B \in \mathscr{L}(H)$. *We write*

$$A \ge \theta$$

if

$$(Au, u)_H \ge 0, \forall u \in H,$$

and in this case we say that A *is positive. Finally, we denote*

$$A \ge B$$

if

$$A - B \ge \theta.$$

Theorem 5.4.3 *Let* A *and* B *be bounded self-adjoint operators such that* $A \ge \theta$ *and* $B \ge \theta$. *If* $AB = BA$ *then*

$$AB \ge \theta.$$

Proof 5.9 If $A = \theta$, the result is obvious. Assume $A \ne \theta$ and define the sequence

$$A_1 = \frac{A}{\|A\|}, \quad A_{n+1} = A_n - A_n^2, \forall n \in \mathbb{N}.$$

We claim that

$$\theta \le A_n \le I, \forall n \in \mathbb{N}.$$

We prove the claim by induction.
For $n = 1$, it is clear that $A_1 \ge \theta$. And since $\|A_1\| = 1$, we get

$$(A_1 u, u)_H \le \|A_1\| \|u\|_H \|u\|_H = (Iu, u)_H, \forall u \in H,$$

so that

$$A_1 \le I.$$

Thus,

$$\theta \le A_1 \le I.$$

Now suppose $\theta \le A_n \le I$. Since A_n is self adjoint we have,

$$\begin{aligned}
(A_n^2(I - A_n)u, u)_H &= ((I - A_n)A_n u, A_n u)_H \\
&= ((I - A_n)v, v)_H \ge 0, \forall u \in H
\end{aligned} \tag{5.7}$$

where $v = A_n u$. Therefore

$$A_n^2(I - A_n) \ge \theta.$$

Similarly, we may obtain

$$A_n(I - A_n)^2 \geq \theta,$$

so that

$$\theta \leq A_n^2(I - A_n) + A_n(I - A_n)^2 = A_n - A_n^2 = A_{n+1}.$$

So we also have,

$$\theta \leq I - A_n + A_n^2 = I - A_{n+1},$$

that is,

$$\theta \leq A_{n+1} \leq I,$$

so that

$$\theta \leq A_n \leq I, \, \forall n \in \mathbb{N}.$$

Observe that,

$$
\begin{aligned}
A_1 &= A_1^2 + A_2 \\
&= A_1^2 + A_2^2 + A_3 \\
&\cdots \quad \cdots\cdots\cdots\cdots \\
&= A_1^2 + \ldots + A_n^2 + A_{n+1}.
\end{aligned}
\tag{5.8}
$$

Since $A_{n+1} \geq \theta$, we obtain

$$A_1^2 + A_2^2 + \ldots + A_n^2 = A_1 - A_{n+1} \leq A_1. \tag{5.9}$$

From this, for a fixed $u \in H$, we have

$$
\begin{aligned}
\sum_{j=1}^{n} \|A_j u\|^2 &= \sum_{j=1}^{n} (A_j u, A_j u)_H \\
&= \sum_{j=1}^{n} (A_j^2 u, u)_H \\
&\leq (A_1 u, u)_H.
\end{aligned}
\tag{5.10}
$$

Since $n \in \mathbb{N}$ is arbitrary, we get

$$\sum_{j=1}^{\infty} \|A_j u\|^2$$

is a converging series, so that

$$\|A_n u\| \to 0,$$

that is,

$$A_n u \to \theta, \text{ as } n \to \infty.$$

From this and (5.9), we get

$$\sum_{j=1}^{n} A_j^2 u = (A_1 - A_{n+1})u \to A_1 u, \text{ as } n \to \infty.$$

Finally, we may write,

$$
\begin{aligned}
(ABu, u)_H &= \|A\|(A_1 Bu, u)_H \\
&= \|A\|(BA_1 u, u)_H \\
&= \|A\|(B \lim_{n \to \infty} \sum_{j=1}^{n} A_j^2 u, u)_H \\
&= \|A\| \lim_{n \to \infty} \sum_{j=1}^{n} (BA_j^2 u, u)_H \\
&= \|A\| \lim_{n \to \infty} \sum_{j=1}^{n} (BA_j u, A_j u)_H \\
&\geq 0.
\end{aligned}
\tag{5.11}
$$

Hence,

$$(ABu, u)_H \geq 0, \forall u \in H.$$

The proof is complete.

Theorem 5.4.4 *Let $\{A_n\}$ be a sequence of self-adjoint commuting operators in $\mathscr{L}(H)$. Let $B \in \mathscr{L}(H)$ be a self adjoint operator such that*

$$A_i B = BA_i, \forall i \in \mathbb{N}.$$

Suppose also that

$$A_1 \leq A_2 \leq A_3 \leq \dots \leq A_n \leq \dots \leq B.$$

Under such hypotheses there exists a self-adjoint, bounded, linear operator A such that

$$A_n \to A \text{ in norm },$$

and

$$A \leq B.$$

Proof 5.10 Consider the sequence $\{C_n\}$ where

$$C_n = B - A_n \geq 0, \forall n \in \mathbb{N}.$$

Fix $u \in H$. First, we show that $\{C_n u\}$ converges. Observe that

$$C_i C_j = C_j C_i, \forall i, j \in \mathbb{N}.$$

Also, if $n > m$ then

$$A_n - A_m \geq \theta$$

so that

$$C_m = B - A_m \geq B - A_n = C_n.$$

Therefore, from $C_m \geq \theta$ and $C_m - C_n \geq \theta$ we obtain

$$(C_m - C_n)C_m \geq \theta, \text{ if } n > m$$

and also

$$C_n(C_m - C_n) \geq \theta.$$

Thus,

$$(C_m^2 u, u)_H \geq (C_n C_m u, u)_H \geq (C_n^2 u, u)_H,$$

and we may conclude that

$$(C_n^2 u, u)_H$$

is a monotone non-increasing sequence of real numbers, bounded below by 0, so that there exists $\alpha \in \mathbb{R}$ such that

$$\lim_{n \to \infty} (C_n^2 u, u)_H = \alpha.$$

Since each C_n is self-adjoint we obtain

$$
\begin{aligned}
\|(C_n - C_m)u\|_H^2 &= ((C_n - C_m)u, (C_n - C_m)u)_H \\
&= ((C_n - C_m)(C_n - C_m)u, u)_H \\
&= (C_n^2 u, u)_H - 2(C_n C_m u, u) + (C_m^2 u, u)_H \\
&\to \alpha - 2\alpha + \alpha = 0, \tag{5.12}
\end{aligned}
$$

as

$$m, n \to \infty.$$

Therefore, $\{C_n u\}$ is a Cauchy sequence in norm, so that there exists the limit

$$\lim_{n \to \infty} C_n u = \lim_{n \to \infty} (B - A_n)u,$$

and hence there exists

$$\lim_{n \to \infty} A_n u, \forall u \in H.$$

Now define A by

$$Au = \lim_{n \to \infty} A_n u.$$

Since the limit

$$\lim_{n \to \infty} A_n u, \forall u \in H$$

exists, we have that

$$\sup_{n \in \mathbb{N}} \{\|A_n u\|_H\}$$

is finite for all $u \in H$. By the principle of uniform boundedness

$$\sup_{n \in \mathbb{N}} \{\|A_n\|\} < \infty$$

so that there exists $K > 0$ such that

$$\|A_n\| \leq K, \forall n \in \mathbb{N}.$$

Therefore,

$$\|A_n u\|_H \leq K \|u\|_H,$$

so that

$$\|Au\| = \lim_{n \to \infty} \{\|A_n u\|_H\} \leq K \|u\|_H, \forall u \in H$$

which means that A is bounded. Fixing $u, v \in H$, we have

$$(Au, v)_H = \lim_{n \to \infty} (A_n u, v)_H = \lim_{n \to \infty} (u, A_n v)_H = (u, Av)_H,$$

and thus A is self-adjoint. Finally

$$(A_n u, u)_H \leq (Bu, u)_H, \forall n \in \mathbb{N},$$

so that

$$(Au, u) = \lim_{n \to \infty} (A_n u, u)_H \leq (Bu, u)_H, \forall u \in H.$$

Hence $A \leq B$.

The proof is complete.

Definition 5.4.5 *Let $A \in \mathscr{L}(A)$ be a positive operator. The self-adjoint operator $B \in \mathscr{L}(H)$ such that*

$$B^2 = A$$

is called the square root of A. If $B \geq \theta$ we denote

$$B = \sqrt{A}.$$

Theorem 5.4.6 *Suppose $A \in \mathscr{L}(H)$ is positive. Then there exists $B \geq \theta$ such that*

$$B^2 = A.$$

Furthermore B commutes with any $C \in \mathscr{L}(H)$ such that commutes with A.

Proof 5.11 There is no loss of generality in considering

$$\|A\| \leq 1,$$

which means $\theta \leq A \leq I$, because we may replace A by

$$\frac{A}{\|A\|}$$

so that if

$$C^2 = \frac{A}{\|A\|}$$

then

$$B = \|A\|^{1/2} C.$$

Let

$$B_0 = \theta,$$

and consider the sequence of operators given by

$$B_{n+1} = B_n + \frac{1}{2}(A - B_n^2), \forall n \in \mathbb{N} \cup \{0\}.$$

Since each B_n is polynomial in A, we have that B_n is self-adjoint and commute with any operator with commutes with A. In particular

$$B_i B_j = B_j B_i, \forall i, j \in \mathbb{N}.$$

First we show that

$$B_n \leq I, \forall n \in \mathbb{N} \cup \{0\}.$$

Since $B_0 = \theta$, and $B_1 = \frac{1}{2}A$, the statement holds for $n = 1$. Suppose $B_n \leq I$. Thus,

$$\begin{aligned} I - B_{n+1} &= I - B_n - \frac{1}{2}A + \frac{1}{2}B_n^2 \\ &= \frac{1}{2}(I - B_n)^2 + \frac{1}{2}(I - A) \geq \theta \end{aligned} \tag{5.13}$$

so that

$$B_{n+1} \leq I.$$

The induction is complete, that is,

$$B_n \leq I, \forall n \in \mathbb{N}.$$

Now we prove the monotonicity also by induction. Observe that

$$B_0 \leq B_1,$$

and supposing

$$B_{n-1} \leq B_n,$$

we have

$$
\begin{aligned}
B_{n+1} - B_n &= B_n + \frac{1}{2}(A - B_n^2) - B_{n-1} - \frac{1}{2}(A - B_{n-1}^2) \\
&= B_n - B_{n-1} - \frac{1}{2}(B_n^2 - B_{n-1}^2) \\
&= B_n - B_{n-1} - \frac{1}{2}(B_n + B_{n-1})(B_n - B_{n-1}) \\
&= (I - \frac{1}{2}(B_n + B_{n-1}))(B_n - B_{n-1}) \\
&= \frac{1}{2}((I - B_{n-1}) + (I - B_n))(B_n - B_{n-1}) \geq \theta.
\end{aligned}
$$

The induction is complete, that is

$$\theta = B_0 \leq B_1 \leq B_2 \leq \ldots \leq B_n \leq \ldots \leq I.$$

By the last theorem there exists a self-adjoint operator B such that

$$B_n \to B \text{ in norm}.$$

Fixing $u \in H$ we have

$$B_{n+1}u = B_n u + \frac{1}{2}(A - B_n^2)u,$$

so that taking the limit in norm as $n \to \infty$, we get

$$\theta = (A - B^2)u.$$

Being $u \in H$ arbitrary we obtain

$$A = B^2.$$

It is also clear that

$$B \geq \theta$$

The proof is complete.

Chapter 6

Spectral Analysis, a General Approach in Normed Spaces

6.1 Introduction

We start by presenting some results about the spectrum and resolvent sets for a bounded operator defined on a normed space. The main reference for this chapter is Bachman and Narici [6].

Definition 6.1.1 *Let V be a complex normed vector space and let $A : D \subset V \to V$ be a linear operator, where D is dense on V. We say that $A^{-1} : R(A) \to D$ is the inverse operator related to A, as A is a bijection from D to $R(A)$ and*

$$A^{-1}y = u \text{ if, and only if, } Au = y, \ \forall u \in D, y \in R(A),$$

where $R(A) = \{Au : u \in D\}$, is the range of A.

In such case we have,

$$A^{-1}Au = u, \ \forall u \in D$$

and

$$AA^{-1}y = y, \ \forall y \in R(A).$$

Let $\lambda \in \mathbb{C}$.

1. *If $R(\lambda I - A)$ is dense in V and $\lambda I - A$ has a bounded inverse, we write $\lambda \in \rho(A)$, where $\rho(A)$ denotes the resolvent set of A.*

2. *If $R(\lambda I - A)$ is dense in V and $(\lambda I - A)^{-1}$ exists but it is not bounded, we write $\lambda \in C\sigma(A)$, where $C\sigma(A)$ denotes the continuous spectrum of A.*

3. *If $R(\lambda I - A)$ is not dense in V and $\lambda I - A$ has an inverse either bounded or unbounded, we write $\lambda \in R\sigma(A)$, where $R\sigma(A)$ denotes the residual spectrum of A.*

4. *If $(\lambda I - A)^{-1}$ does not exist, we write $\lambda \in P\sigma(A)$ where $P\sigma(A)$ denotes the point spectrum of A.*

   In such a case there exists $u \in V$ such that

$$Au - \lambda u = \mathbf{0}$$

where $u \neq \mathbf{0}$, so that u is said to be an eigenvector of A and λ the corresponding eigenvalue.

Table 6.1 summarizes such results.

Table 6.1: About the spectrum and resolvent sets of A.

$(\lambda I - A)^{-1}$	Boundedness of $(\lambda I - A)^{-1}$	$R(\lambda I - A)$	Set
exists	bounded	dense in V	$\rho(A)$
exists	unbounded	dense in V	$C\sigma(A)$
exists	bounded or not	not dense in V	$R\sigma(A)$
not exists	—	dense or not in V	$P\sigma(A)$

Remark 6.1.2 *Observe that*

$$\mathbb{C} = \rho(A) \cup C\sigma(A) \cup R\sigma(A) \cup P\sigma(A)$$

and such union is disjoint.

The spectrum of A, denoted by $\sigma(A)$, is defined by

$$\sigma(A) = C\sigma(A) \cup R\sigma(A) \cup P\sigma(A).$$

Theorem 6.1.3 (Riesz) *Let V be a normed vector space and let $0 < \alpha < 1$. Let M be a proper closed vector subspace of V.*

Under such hypotheses, there exists $u_\alpha \in V$ such that

$$\|u_\alpha\|_V = 1$$

and

$$\|u - u_\alpha\|_V \geq \alpha, \ \forall u \in M.$$

Proof 6.1   Since $M \subset V$ is a proper closed subspace of V, we may select a $v \in V \setminus M$.
Define

$$d = \inf_{u \in M} \|u - v\|_V.$$

Observe that, since M is closed, we have $d > 0$, otherwise if we had $d = 0$ we had $v \in \overline{M} = M$, which contradicts $v \notin M$.

Also,

$$d/\alpha > d.$$

Hence, there exists $u_0 \in M$ such that

$$0 < d \leq \|u_0 - v\|_V < d/\alpha.$$

Define

$$u_\alpha = \frac{v - u_0}{\|v - u_0\|_V}.$$

Thus, $\|u_\alpha\|_V = 1$ and also for $u \in M$ we have

$$
\begin{aligned}
\|u - u_\alpha\|_V &= \left\| u - \frac{v}{\|v - u_0\|_V} + \frac{u_0}{\|v - u_0\|_V} \right\|_V \\
&= \| u(\|v - u_0\|_V) + u_0 - v \|_V \frac{1}{\|v - u_0\|_V}.
\end{aligned}
\tag{6.1}
$$

From this, since $u\|v - u_0\|_V + u_0 \in M$, we have

$$\|u - u_\alpha\|_V \geq \frac{d}{\|v - u_0\|_V} > \alpha, \ \forall u \in M.$$

The proof is complete.

Theorem 6.1.4 *Let V be a complex normed vector space and let $A : D \subset V \to V$ be a linear compact operator. Under such hypotheses, $P\sigma(A)$ is countable and 0 is its unique possible limit point.*

Proof 6.2 Let $\varepsilon > 0$. We shall prove that there exists at most a finite number of points in P_ε where

$$P_\varepsilon = \{\lambda \in P\sigma(A) \; : \; |\lambda| \geq \varepsilon\}.$$

Observe that in such a case

$$P\sigma(A) \setminus \{0\} = \cup_{n=1}^{\infty} P_{1/n}$$

and such a set is countable and has 0 as the unique possible limit point.

Suppose, to obtain contradiction, there exists $\varepsilon > 0$ such that P_ε has infinite points. Hence, there exists a sequence $\{\lambda_n\}_{n \in \mathbb{N}} \subset P_\varepsilon$ and a sequence of linearly independent eigenvectors $\{u_n\}$ such that

$$Au_n = \lambda_n u_n, \; \forall n \in \mathbb{N}.$$

Define

$$M_n = \text{Span}\{u_1, \ldots, u_n\},$$

so that

$$M_{n-1} \subset M_n$$

and M_n is finite dimensional, $\forall n \in \mathbb{N}$. Observe that $M_{n-1} \subset M_n$ properly.

From the Riesz theorem, there exists $y_n \in M_n$ such that $\|y_n\|_V = 1$ and

$$\|y_n - u\| \geq 1/2, \; \forall u \in M_{n-1}, \; \forall n > 1.$$

Let

$$u = \sum_{i=1}^{n} \alpha_i u_i \in M_n.$$

Thus,

$$Au = \sum_{i=1}^{n} \alpha_i Au_i = \sum_{i=1}^{n} \alpha_i \lambda_i u_i.$$

Therefore,

$$
\begin{aligned}
(\lambda_n - A)u &= \lambda_n u - Au \\
&= \sum_{i=1}^{n-1} \alpha_i(\lambda_n - \lambda_i)u_i \in M_{n-1}.
\end{aligned}
\tag{6.2}
$$

Therefore,

$$(\lambda_n - A)(M_n) \subset M_{n-1}$$

and from this

$$A(M_n) \subset M_n, \forall n \in \mathbb{N}.$$

Let $1 < m < n$. Thus,

$$w = (\lambda_n - A)y_n + Ay_m \in M_{n-1},$$

so that

$$Ay_n - Ay_m = \lambda_n y_n - w = \lambda_n(y_n - \lambda_n^{-1}w).$$

Since, $\lambda_n^{-1}w \in M_{n-1}$ we get

$$\|Ay_n - Ay_m\| = |\lambda_n| \|y_n - \lambda_n^{-1}w\| \geq \frac{|\lambda_n|}{2} \geq \frac{\varepsilon}{2},$$

$\forall 1 \leq m < n \in \mathbb{N}$.

Therefore, $\{y_n\}$ is a bounded sequence and such that $\{Ay_n\}$ has no Cauchy subsequence, that is, $\{Ay_n\}$ has no convergent subsequence, which contradicts A be compact.

The proof is complete.

Definition 6.1.5 *Let V be a normed vector space and let $A : D \subset V \to V$ be a linear operator, where D is dense in V.*

We say that $\lambda \in \mathbb{C}$ is a proper approximate value of A if for each $\varepsilon > 0$ there exists $u \in D$ such that

$$\|u\|_V = 1 \text{ and } \|(\lambda I - A)u\|_V < \varepsilon.$$

In such a case we denote $\lambda \in \pi(A)$, where $\pi(A)$ is the approximate spectrum of A.

Theorem 6.1.6 *Considering the statements of the last definition, we have that $\lambda \in \pi(A)$ if, and only if, $\lambda I - A$ has no a bounded inverse.*

Proof 6.3 Suppose $\lambda \in \pi(A)$. Thus, for each $n \in \mathbb{N}$, there exists $u_n \in D$ such that $\|u_n\|_V = 1$ and

$$\|(\lambda I - A)u_n\|_V < 1/n. \tag{6.3}$$

Suppose, to obtain contradiction, there exists $K > 0$ such that

$$\|(\lambda I - A)u\|_V \geq K\|u\|_V, \ \forall u \in D.$$

In particular we have

$$\|(\lambda I - A)u_n\|_V \geq 1, \ \forall n \in \mathbb{N},$$

which contradicts (6.3).

Reciprocally, suppose $\lambda I - A$ has no bounded inverse.

Thus, there is no $K > 0$ such that

$$\|(\lambda I - A)u\|_V \geq K\|u\|_V, \ \forall u \in D.$$

Hence, for each $\varepsilon > 0$ we may find $u \in D$ such that

$$\|(\lambda I - A)u\|_V < \varepsilon\|u\|_V.$$

From this, for each $\varepsilon > 0$ we may find $u \in D$ such that

$$\|u\|_V = 1 \text{ and } \|(\lambda I - A)u\|_V < \varepsilon.$$

Therefore $\lambda \in \pi(A)$.

6.2 Sesquilinear functionals

Definition 6.2.1 *Let V be a complex vector space. A functional $f : V \times V \to \mathbb{C}$ is said to be a sesquilinear functional, as*

1. $f(u_1 + u_2, v) = f(u_1, v) + f(u_2, v), \ \forall u_1, u_2, v \in V$.

2. $f(\alpha u, v) = \alpha f(u, v), \ \forall u, v \in V, \ \forall \alpha \in \mathbf{C}$.

3. $f(u, v_1 + v_2) = f(u, v_1) + f(u, v_2), \ \forall u, v_1, v_2 \in V$.

4. $f(u, \alpha v) = \overline{\alpha} f(u, v), \ \forall u, v \in V, \ \alpha \in \mathbb{C}$.

Remark 6.2.2 *Let H be a complex Hilbert space and let $A : H \to H$ be a linear operator. Hence $f : H \times H \to \mathbb{C}$ defined by*

$$f(u, v) = (Au, v)_H, \forall u, v \in H$$

is a sequilinear functional.

Remark 6.2.3 *Given a sesquilinear functional*

$$f : H \times H \to \mathbb{C}$$

we shall define $\hat{f} : H \to \mathbb{C}$ *by*

$$\hat{f}(u) = f(u,u), \ \forall u \in H.$$

Exercise 6.2.4 *In the context of the last definitions, prove that*

$$
\begin{aligned}
f(u,v) \quad = \quad & \hat{f}\left(\frac{1}{2}(u+v)\right) - \hat{f}\left(\frac{1}{2}(u-v)\right) \\
& + i\hat{f}\left(\frac{1}{2}(u+iv)\right) - i\hat{f}\left(\frac{1}{2}(u-iv)\right), \ \forall u,v \in H.
\end{aligned}
\tag{6.4}
$$

Conclude that if $\hat{f}_1 = \hat{f}_2$, *then* $f_1 = f_2$.

Theorem 6.2.5 *Let V be a complex vector space and let $f : V \times V \to \mathbb{C}$ be a symmetric sesquilinear functional, that is, assume f is such that $f(u,v) = \overline{f(v,u)}$, $\forall u,v \in V$, where $\overline{f(v,u)}$ denotes the complex conjugate of $f(v,u)$.*

Under such hypotheses,

$$\hat{f}(u) \in \mathbb{R}, \ \forall u \in V.$$

Moreover, the reciprocal also holds.

Proof 6.4 Suppose

$$f(u,v) = \overline{f(v,u)}, \ \forall u,v \in V.$$

Thus,

$$\hat{f}(u) = f(u,u) = \overline{f(u,u)} = \overline{\hat{f}(u)},$$

so that

$$\hat{f}(u) \in \mathbb{R}, \forall u \in V.$$

Reciprocally, suppose $\hat{f}$ is real.
Define

$$g(u,v) = \overline{f(v,u)}, \ \forall u,v \in V.$$

Hence,

$$\hat{g}(u) = g(u,u) = \overline{f(u,u)} = f(u,u) = \hat{f}(u), \ \forall u \in V.$$

From this and the last exercise, we obtain $g = f$, so that

$$f(u,v) = \overline{f(v,u)}, \ \forall u,v \in V.$$

Remark 6.2.6 *Let $A : D \subset V \to V$ be a symmetric operator, that is, such that*

$$(Au,v)_V = (u,Av)_V, \ \forall u,v \in V,$$

where V is a space with inner product.

Thus,

$$
\begin{aligned}
f(u,v) \quad = \quad & (Au,v)_V \\
= \quad & (u,Av)_V \\
= \quad & \overline{(Av,u)_V} \\
= \quad & \overline{f(v,u)}, \ \forall u,v \in V,
\end{aligned}
\tag{6.5}
$$

so that f is symmetric.

Definition 6.2.7 *Let V be a normed vector space. A sesquilinear functional $f : V \times V \to \mathbb{C}$ is said to be bounded if there exists $K > 0$ such that*

$$|f(u,v)| \leq K\|u\|_V\|v\|_V, \ \forall u,v \in V. \tag{6.6}$$

Defining

$$B = \{K > 0 \text{ such that (6.6) is satisfied }\}$$

we also define the norm of f, denoted by $\|f\|$ as

$$\|f\| = \inf\{K \ : \ K \in B\}.$$

Moreover, defining

$$C = \{K > 0 \text{ such that } |\hat{f}(u)| \leq K\|u\|_V^2, \ \forall u \in V\},$$

we define also the norm of $\hat{f}$, denoted by $\|\hat{f}\|$, as

$$\|\hat{f}\| = \inf\{K > 0 \ : \ K \in C\}.$$

Proposition 6.2.8 *Considering the context of the last definition,*

$$\|f\| = \sup_{(u,v)\in V\times V} \{|f(u,v)| \ : \ \|u\|_V = \|v\|_V = 1\}$$

and

$$\|\hat{f}\| = \sup_{u\in V}\{|\hat{f}(u)| \ : \ \|u\|_V = 1\}.$$

Proof 6.5   We firstly denote

$$\alpha = \sup_{(u,v)\in V\times V} \{|f(u,v)| \ : \ \|u\|_V = \|v\|_V = 1\}.$$

Observe that

$$|f(u,v)| \leq \|f\|\|u\|_V\|v\|_V, \ \forall u,v \in V,$$

so that

$$\alpha \leq \|f\|. \tag{6.7}$$

On the other hand,

$$
\begin{aligned}
|f(u,v)| &= \left| f\left(u\frac{\|u\|_V}{\|u\|_V}, v\frac{\|v\|_V}{\|v\|_V} \right) \right| \\
&= \left| f\left(\frac{u}{\|u\|_V}, \frac{v}{\|v\|_V} \right) \right| \|u\|_V\|v\|_V \\
&\leq \alpha\|u\|_V\|v\|_V, \ \forall u \neq \mathbf{0}, v \neq \mathbf{0}. \tag{6.8}
\end{aligned}
$$

Hence $\alpha \in B$ so that

$$\alpha \geq \inf B = \|f\|. \tag{6.9}$$

From (6.7) and (6.9) we may infer that

$$\alpha = \|f\|.$$

Similarly, the second result may be proven.
The proof is complete.

Theorem 6.2.9 *Let V be a complex normed vector space and let $F : V \times V \to \mathbb{C}$ be a sesquilinear, bounded and symmetric functional. Under such hypotheses,*

$$\|f\| = \|\hat{f}\|.$$

Proof 6.6   Observe that

$$
\begin{aligned}
f(u,v) \quad = \quad & \hat{f}\left(\frac{1}{2}(u+v)\right) - \hat{f}\left(\frac{1}{2}(u-v)\right) \\
& + i\hat{f}\left(\frac{1}{2}(u+iv)\right) - i\hat{f}\left(\frac{1}{2}(u-iv)\right), \ \forall u,v \in H.
\end{aligned}
\tag{6.10}
$$

Since $\hat{f}(u) \in \mathbb{R}$, $\forall u \in V$, we have that

$$
\begin{aligned}
|Re[f(u,v)]| \quad &\leq \quad \left| \hat{f}\left(\frac{1}{2}(u+v)\right) \right| + \left| \hat{f}\left(\frac{1}{2}(u-v)\right) \right| \\
&\leq \quad \frac{1}{4}\|\hat{f}\| \|u+v\|_V^2 + \frac{1}{4}\|\hat{f}\| \|u-v\|_V^2 \\
&= \quad \frac{1}{4}\|\hat{f}\| \left(2\|u\|_V^2 + 2\|v\|_V^2\right), \ \forall u,v \in V.
\end{aligned}
\tag{6.11}
$$

Thus, if $\|u\|_V = \|v\|_V = 1$, we get

$$|Re[f(u,v)]| \leq \|\hat{f}\|.$$

Observe that in its polar form, we have

$$f(u,v) = re^{i\theta}.$$

Denoting $\alpha = e^{-i\theta}$, we obtain

$$f(\alpha u, v) = \alpha f(u,v) = r = |f(u,v)| = |Re[f(\alpha u, v)]| \leq \|\hat{f}\|.$$

Thus,

$$\|f\| = \sup\{|f(u,v)| \ : \ \|u\|_V = \|v\|_V = 1\} \leq \|\hat{f}\|.$$

However, from the definitions, $\|f\| \geq \|\hat{f}\|$.
From these last two lines, we may infer that

$$\|f\| = \|\hat{f}\|.$$

The proof is complete.

Definition 6.2.10 (Normal operator)   *Let H be a complex Hilbert space. We say that a bounded linear operator $A : H \to H$ is normal as*

$$A^*A = AA^*.$$

Theorem 6.2.11 *Let H be a complex Hilbert space. and let $A : H \to H$ be a bounded linear operator. Under such hypotheses, A is normal if, and only if,*

$$\|A^*u\|_H = \|Au\|_H, \ \forall u \in H.$$

Proof 6.7   Suppose A is normal. Thus,

$$(A^*Au, u)_H = (AA^*u, u)_H$$

so that

$$\|Au\|_H^2 = (Au, Au)_H = (A^*u, A^*u)_H = \|A^*u\|_H^2,$$

that is,

$$\|Au\|_H = \|A^*u\|_H, \ \forall u \in H.$$

Reciprocally, suppose that

$$\|Au\|_H = \|A^*u\|_H, \ \forall u \in H.$$

Hence

$$(Au, Au)_H = (A^*u, A^*u)_H$$

so that

$$(A^*Au, u)_H = (A^{**}A^*u, u)_H = (AA^*u, u)_H, \ \forall u \in H.$$

From this, denoting

$$f_1(u, v) = (A^*Au, v)_H$$

and

$$f_2(u, v) = (AA^*u, v)_H$$

we obtain

$$\hat{f}_1(u) = \hat{f}_2(u), \ \forall u \in H.$$

Thus, $f_1 = f_2$, so that

$$(A^*Au, v)_H = (AA^*u, v)_H, \ \forall u, v \in H.$$

From this, we may infer that

$$A^*A = AA^*.$$

Theorem 6.2.12 *Let H be a complex Hilbert space and let $A \in L(H)$. Under such hypotheses the following proprieties are equivalent.*

1. *There exists $\lambda \in \pi(A)$ such that $|\lambda| = \|A\|$.*

2.
$$\|A\| = \sup_{u \in H}\{|(Au, u)_H| : \|u\|_H = 1\}.$$

Proof 6.8

■ 1 implies 2: Suppose $\lambda \in \pi(A)$ is such that

$$|\lambda| = \|A\|.$$

We shall prove that

$$\lambda \in \overline{\{(Au, u)_H : u \in H, \|u\|_H = 1\}}.$$

From this we may obtain

$$
\begin{aligned}
\|A\| &= |\lambda| \\
&\leq \sup_{u \in H}\{|(Au, u)_H| : \|u\|_H = 1\} \\
&\leq \sup_{(u,v) \in H \times H}\{|(Au, v)_H| : \|u\|_H = 1, \|v\|_H = 1\} \\
&= \|A\|.
\end{aligned}
\tag{6.12}
$$

which would complete the first part of the proof.

From $\lambda \in \pi(A)$, there exists $\{u_n\}_{n \in \mathbb{N}} \subset H$ such that

$$\|u_n\|_H = 1$$

and

$$\|Au_n - \lambda u_n\|_H \to 0.$$

Thus,

$$
\begin{aligned}
|(Au_n, u_n)_H - \lambda| &= |(Au_n, u_n)_H - \lambda(u_n, u_n)_H| \\
&= |(Au_n - \lambda u_n, u_n)_H| \\
&\leq \|Au_n - \lambda u_n\|_H \|u_n\|_H \\
&\to 0, \text{ as } n \to \infty.
\end{aligned}
\tag{6.13}
$$

Thus,

$$\lambda \in \overline{\{(Au, u)_H \ : \ u \in H, \ \|u\|_H = 1\}}.$$

The first part of the proof is complete.

■ 2 implies 1:

Reciprocally, suppose

$$\|A\| = \sup_{u \in H}\{|(Au, u)_H| \ : \ \|u\|_H = 1\}.$$

Hence, there exists a sequence $\{u_n\} \subset H$ such that $\|u_n\|_H = 1$, $\forall n \in \mathbb{N}$, and

$$|(Au_n, u_n)_H| \to \|A\|_H.$$

Thus $\{(Au_n, u_n)_H\} \subset \mathbb{C}$ is a bounded sequence. From this there exists a subsequence $\{(Au_{n_k}, u_{n_k})_H\}$ of $\{(Au_n, u_n)_H\}$ and $\lambda \in \mathbb{C}$ such that

$$(Au_{n_k}, u_{n_k})_H \to \lambda, \text{ as } k \to \infty.$$

Therefore,

$$
\begin{aligned}
\|Au_{n_k} - \lambda u_{n_k}\|_H^2 &= \|Au_{n_k}\|_H^2 - \overline{\lambda}(Au_{n_k}, u_{n_k})_H - \lambda(u_{n_k}, Au_{n_k})_H + |\lambda|^2 \\
&\leq \|A\|^2 \|u_{n_k}\|_H^2 - \overline{\lambda}(Au_{n_k}, u_{n_k})_H - \lambda\overline{(Au_{n_k}, u_{n_k})_H} + |\lambda|^2 \\
&\to |\lambda|^2 - \overline{\lambda}\lambda - \lambda\overline{\lambda} + |\lambda|^2 = 0.
\end{aligned}
\tag{6.14}
$$

Summarizing,

$$\|Au_{n_k} - \lambda u_{n_k}\|_H \to 0, \text{ as } k \to \infty,$$

so that $\lambda \in \pi(A)$ and $|\lambda| = \|A\|$.

The proof is complete.

Theorem 6.2.13 *Let H be a Hilbert space and let $A \in L(H)$ be a self-adjoint operator.*
Define

$$M = \sup_{u \in H}\{(Au, u)_H \ : \ \|u\|_H = 1\}$$

and

$$m = \inf_{u \in H}\{(Au, u)_H \ : \ \|u\|_H = 1\}.$$

Under such hypotheses, $m \in \sigma(A)$ and $M \in \sigma(A)$.

Proof 6.9   Choose $\alpha \in \mathbb{R}$ such that
$$M - \alpha \geq m - \alpha > 0.$$

Define $\hat{M} = M - \alpha$ and $\hat{A} = A - \alpha I$. Since $\hat{A}$ is self-adjoint (see Theorem 6.4.2 for details), we have that
$$\|\hat{A}\| = \hat{M}.$$

Thus there exists a subsequence $\{u_n\} \subset H$ such that $\|u_n\|_H = 1$ and
$$(\hat{A}u_n, u_n)_H \to \hat{M}, \text{ as } n \to \infty.$$

Thus,
$$(\hat{A}u_n - \hat{M}u_n, u_n)_H \to 0, \text{ as } n \to \infty.$$

Hence,
$$
\begin{aligned}
\|\hat{A}u_n - \hat{M}u_n\|_H^2 &= (\hat{A}u_n - \hat{M}u_n, \hat{A}u_n - \hat{M}u_n)_H \\
&= \|\hat{A}u_n\|_H^2 - 2\hat{M}(\hat{A}u_n, u_n)_H + \hat{M}^2 \\
&\leq \|\hat{A}\|_H^2 \|u_n\|_H^2 - 2\hat{M}(\hat{A}u_n, u_n)_H + \hat{M}^2 \\
&\to \hat{M}^2 - 2\hat{M}^2 + \hat{M}^2 \\
&= 0.
\end{aligned}
\tag{6.15}
$$

Summarizing,
$$\|\hat{A}u_n - \hat{M}u_n\|_H \to 0, \text{ as } n \to \infty,$$

so that
$$\|Au_n - Mu_n\|_H \to 0, \text{ as } n \to \infty.$$

From this we may infer that
$$M \in \pi(A) \subset \sigma(A).$$

Similarly, select $\beta \in \mathbb{R}$ such that
$$-m + \beta \geq -M + \beta > 0.$$

Define $\hat{A} = -A + \beta I$. The remaining parts of the proof are similar to those of the previous case. This completes the proof.

Theorem 6.2.14 *Let H be a complex Hilbert space. Let $U : H \to H$ be a linear bounded operator. Under such hypotheses, U is a isometry if, and only if,*
$$U^*U = I,$$

where I denotes the identity operator.

Proof 6.10   Observe that
$$(Uu, Uv)_H = (u, v)_H, \ \forall u, v \in H,$$

if, and only if,
$$(U^*Uu, v)_H = (u, v), \ \forall u, v \in H,$$

if, and only if,
$$U^*U = I.$$

The proof is complete.

Theorem 6.2.15 *Let H be a complex Hilbert space. Under such hypothesis, $U : H \to H$ is a bijective isometry if, and only if,*

$$U^*U = UU^* = I, \text{ in } H.$$

Proof 6.11   Assume $U : H \to H$ is a bijective isometry.

From the last theorem $U^*U = I$, and U is a bijection, we obtain $U^{-1} = U^*$ so that $UU^* = I$ in H.

On the other hand, if $U^*U = UU^* = I$ in H, we have that $U^* = U^{-1}$ and the domain of $U^* = U^{-1}$ is H, so that the range of U is also H.

From this U is a bijection.

Moreover, $(u,v)_H = (U^*Uu,v)_H = (Uu,Uv)_H, \forall u, v \in H$, and thus U is a bijective isometry.

Theorem 6.2.16 *Let H be a complex Hilbert space. Suppose $U : H \to H$ is a linear operator such that*

$$\|Uu\|_H = \|u\|_H, \forall u \in H$$

(in such a case we say that U is unitary).

Under such hypotheses, U is a isometry.

Proof 6.12   From the hypotheses,

$$(Uu,Uu)_H = (u,u)_H, \forall u \in H.$$

Thus,

$$(U^*Uu,u)_H = (u,u)_H, \forall u \in H,$$

so that

$$((U^*U - I)u,u)_H = 0, \forall u \in H.$$

Since $U^*U - I$ is self adjoint, it follows that

$$\|U^*U - I\| = \sup_{u \in H}\{|((U^*U - I)u,u)_H| : \|u\|_H = 1\} = 0,$$

so that $U^*U = I$.

From this and Theorem 6.2.14, we have that U is a isometry.

Corollary 6.2.17 *Let $U : H \to H$ be a unitary operator.*

Under such hypotheses, if $\lambda \in \mathbb{C}$ is an eigenvalue of U then

$$|\lambda| = 1.$$

Proof 6.13   Suppose $Uu = \lambda u$ and $\|u\|_H = 1$. Thus,

$$1 = (u,u)_H = (Uu,Uu)_H = (\lambda u, \lambda u)_H = \lambda \overline{\lambda}(u,u)_H = |\lambda|^2.$$

The proof is complete.

Theorem 6.2.18 *Let H be a complex Hilbert space. Let $A : D_A \subset H \to H$ be a linear self-adjoint operator but not necessarily bounded, where D_A is dense in H, that is, $\overline{D_A} = H$. Let U be the Cayley transform of A, that is $U = (A - i)(A + i)^{-1}$.*

Under such hypotheses, U is unitary.

Proof 6.14   First, we shall prove that $A \pm i$ is injective, so that its inverse is well defined on $R(A \pm i)$.
Since A is self-adjoint, we have that

$$
\begin{aligned}
\|(A \pm i)u\|_H^2 &= ((A \pm i)u, (A \pm i))_H \\
&= (Au, Au)_H \pm (Au, iu)_H \pm (iu, Au)_H + (iu, iu)_H \\
&= \|Au\|_H^2 + (\pm i \mp i)(u, Au)_H + \|u\|_H^2 \\
&= \|Au\|_H^2 + \|u\|_H^2 \geq \|u\|_H^2. \tag{6.16}
\end{aligned}
$$

Thus, if $(A \pm i)u = 0$ then $u = 0$, so that $(A \pm i)$ is injective.

Now, we are going to show that

$$
R(A \pm i) = H.
$$

Let $z \perp R(A + i)$. Hence,

$$
((A + i)u, z)_H = (Au, z)_H + i(u, z)_H = 0, \forall u \in D_A.
$$

From this

$$
(Au, z)_H = ((u, iz), \forall u \in H,
$$

so that

$$
A^* z = Az = iz,
$$

that is,

$$
(A - i)z = 0.
$$

Thus, $z = 0$.

Summarizing these last results, if $z \perp R(A + i)$ then $z = 0$, so that

$$
\overline{R(A + i)} = H.
$$

Now we are going to show that $\overline{R(A + i)} = R(A + i) = H$. Let $v \in H$. Thus there exists a sequence $\{v_n\} \subset R(A + i)$ such that

$$
v_n \to v, \text{ in norm, as } n \to \infty.
$$

Therefore there exists a sequence $\{u_n\} \subset D_A$ such that

$$
Au_n + iu_n = v_n \to v, \text{ as } n \to \infty.
$$

Similarly as above, we may obtain,

$$
\|v_n - v_m\|_H^2 = \|Au_n + iu_n - Au_m - iu_m\|_H^2 = \|A(u_n - u_m)\|_H^2 + \|u_n - u_m\|_H^2, \forall m, n \in \mathbb{N}.
$$

From this, since $\{v_n\}$ is a Cauchy sequence, we may infer that $\{u_n\}$ and $\{Au_n\}$ are Cauchy sequences, so that there exists $u \in H$, and $w \in H$ such that

$$
Au_n \to w
$$

and

$$
u_n \to u, \text{ as } n \to \infty.
$$

Since $A = A^*$ is closed (see Theorem 6.6.7 for details), we may infer that $w = Au$ and

$$
Au_n + iu_n \to Au + iu.
$$

since A is closed we get $(u, Au) \in Gr(A)$ where $Gr(A)$ denotes the graph of A, so that $Au + iu = v \in R(A + i)$.

Summarizing, if $v \in H$ then $v \in R(A + i)$, so that $R(A + i) = H = \overline{R(A + i)}$.

A similar result we may obtain for $A - i$, that is

$$R(A - i) = H.$$

Observe that

$$U = (A - i)(A + i)^{-1}$$

and

$$R((A + i)^{-1}) = D_A = D_{(A+i)},$$

and $R(A - i) = H$.

Thus $R(U) = H$, that is $U : H \to H$ is linear and onto (recalling that $D_{(A+i)^{-1}} = H$).
At this point, we shall prove that U is unitary.
Let $v \in H$. Since $R(A + i) = H$, there exists $u \in D_A$ such that

$$(A + i)u = v.$$

Hence

$$Uv = (A - i)(A + i)^{-1}v = (A - i)u,$$

so that

$$
\begin{aligned}
\|Uv\|_H^2 &= \|(A - i)u\|_H^2 \\
&= \|Au\|_H^2 + \|u\|_H^2 \\
&= \|(A + i)u\|_H^2 \\
&= \|v\|_H^2, \ \forall v \in H.
\end{aligned}
\tag{6.17}
$$

Summarizing,

$$\|Uv\|_H = \|v\|_H, \ \forall v \in H,$$

so that U is unitary.
The proof is complete.

Remark 6.2.19 *Let $v \in H$. Since $R(A + i) = H$, we may obtain $u \in H$ such that $v = (A + i)u$.
From this we have*

$$Uv = (A - i)(A + i)^{-1}v = (A - i)u,$$

so that

$$(I + U)v = 2Au$$

and

$$(I - U)v = 2iu.$$

Thus, if $(I - U)v = 0$, then $u = 0$ so that $v = (A + i)u = 0$.
Therefore, $I - U$ is injective and its inverse exists on $R(I - U)$.
Moreover, for $u \in R(I - U)$ as above, we have,

$$(I + U)[(I - U)^{-1}](2iu) = (I + U)[(I - U)^{-1}](I - U)v = (I + U)v = 2Au,$$

so that

$$Au = i(I + U)(I - U)^{-1}u, \ \forall u \in R(I - U).$$

In the next lines we shall show that $R(I - U) = D_A$.
Indeed, let $v \in H$. Thus, from the last lines above, there exists $u \in D_A$ such that

$$(I - U)v = 2iu$$

so that $R(I - U) \subset D_A$.

Reciprocally, let $u \in D_A$ *and define* $v = (A + i)u$.

Therefore,

$$2iu = (I - U)v$$

so that $u \in R(I - U)$, $\forall u \in D_A$. *Thus,*

$$D_A \subset R(I - U)$$

so that

$$R(I - U) = D_A.$$

From such last results, we may infer that

$$A = i(I + U)(I - U)^{-1},$$

in D_A.

6.3   About the spectrum of a linear operator defined on a Banach space

Definition 6.3.1 *Let* U *be a Banach space and let* $A \in \mathscr{L}(U)$. *We recall that a complex number* λ *is said to be in the resolvent set* $\rho(A)$ *of* A, *if*

$$\lambda I - A$$

is a bijection with a bounded inverse. As previously indicated, we call

$$R_\lambda(A) = (\lambda I - A)^{-1}$$

the resolvent of A *in* λ.

If $\lambda \notin \rho(A)$, *we write*

$$\lambda \in \sigma(A) = \mathbb{C} - \rho(A),$$

where $\sigma(A)$ *is said to be the spectrum of* A.

Definition 6.3.2 *Let* $A \in \mathscr{L}(U)$.

1. *If* $u \neq \theta$ *and* $Au = \lambda u$ *for some* $\lambda \in \mathbb{C}$ *then* u *is said to be an eigenvector of* A *and* λ *the corresponding eigenvalue. If* λ *is an eigenvalue, then* $(\lambda I - A)$ *is not injective and therefore* $\lambda \in \sigma(A)$.

   The set of eigenvalues is said to be the point spectrum of A.

2. *If* λ *is not an eigenvalue but*

   $$R(\lambda I - A)$$

   is not dense in U *and therefore* $\lambda I - A$ *is not a bijection, we have that* $\lambda \in \sigma(A)$. *In this case we say that* λ *is in the residual spectrum of* A, *or briefly* $\lambda \in Res[\sigma(A)]$.

Theorem 6.3.3 *Let* U *be a Banach space and suppose that* $A \in \mathscr{L}(U)$. *Then* $\rho(A)$ *is an open subset of* $\mathbb{C}$ *and*

$$F(\lambda) = R_\lambda(A)$$

is an analytic function with values in $\mathscr{L}(U)$ *on each connected component of* $\rho(A)$. *For* λ, $\mu \in \sigma(A)$, $R_\lambda(A)$ *and* $R_\mu(A)$ *commute and*

$$R_\lambda(A) - R_\mu(A) = (\mu - \lambda)R_\mu(A)R_\lambda(A).$$

Proof 6.15 Let $\lambda_0 \in \rho(A)$. We will show that λ_0 is an interior point of $\rho(A)$.

Observe that symbolically we may write

$$
\begin{aligned}
\frac{1}{\lambda - A} &= \frac{1}{\lambda - \lambda_0 + (\lambda_0 - A)} \\
&= \frac{1}{\lambda_0 - A} \left[\frac{1}{1 - \left(\frac{\lambda_0 - \lambda}{\lambda_0 - A} \right)} \right] \\
&= \frac{1}{\lambda_0 - A} \left(1 + \sum_{n=1}^{\infty} \left(\frac{\lambda_0 - \lambda}{\lambda_0 - A} \right)^n \right).
\end{aligned}
\tag{6.18}
$$

Define,

$$
\hat{R}_\lambda(A) = R_{\lambda_0}(A) \{ I + \sum_{n=1}^{\infty} (\lambda - \lambda_0)^n (R_{\lambda_0})^n \}.
\tag{6.19}
$$

Observe that

$$
\| (R_{\lambda_0})^n \| \leq \| R_{\lambda_0} \|^n.
$$

Thus, the series indicated in (6.19) will converge in norm if

$$
|\lambda - \lambda_0| < \| R_{\lambda_0} \|^{-1}.
\tag{6.20}
$$

Hence, for λ satisfying (6.20), $\hat{R}(A)$ is well defined and we can easily check that

$$
(\lambda I - A)\hat{R}_\lambda(A) = I = \hat{R}_\lambda(A)(\lambda I - A).
$$

Therefore

$$
\hat{R}_\lambda(A) = R_\lambda(A), \text{ if } |\lambda - \lambda_0| < \| R_{\lambda_0} \|^{-1},
$$

so that λ_0 is an interior point. Since $\lambda_0 \in \rho(A)$ is arbitrary, we have that $\rho(A)$ is open. Finally, observe that

$$
\begin{aligned}
R_\lambda(A) - R_\mu(A) &= R_\lambda(A)(\mu I - A)R_\mu(A) - R_\lambda(A)(\lambda I - A)R_\mu(A) \\
&= R_\lambda(A)(\mu I)R_\mu(A) - R_\lambda(A)(\lambda I)R_\mu(A) \\
&= (\mu - \lambda)R_\lambda(A)R_\mu(A)
\end{aligned}
\tag{6.21}
$$

Interchanging the roles of λ and μ we may conclude that R_λ and R_μ commute.

Corollary 6.3.4 *Let U be a Banach space and $A \in \mathscr{L}(U)$. Then the spectrum of A is non-empty.*

Proof 6.16 Observe that if

$$
\frac{\|A\|}{|\lambda|} < 1
$$

we have

$$
\begin{aligned}
(\lambda I - A)^{-1} &= [\lambda(I - A/\lambda)]^{-1} \\
&= \lambda^{-1}(I - A/\lambda)^{-1} \\
&= \lambda^{-1} \left(I + \sum_{n=1}^{\infty} \left(\frac{A}{\lambda} \right)^n \right).
\end{aligned}
\tag{6.22}
$$

Therefore we may obtain

$$R_\lambda(A) = \lambda^{-1}\left(I + \sum_{n=1}^{\infty}\left(\frac{A}{\lambda}\right)^n\right).$$

In particular

$$\|R_\lambda(A)\| \to 0, \text{ as } |\lambda| \to \infty. \tag{6.23}$$

Suppose, to obtain contradiction, that

$$\sigma(A) = \emptyset.$$

In such a case $R_\lambda(A)$ would be a entire bounded analytic function. From Liouville's theorem, $R_\lambda(A)$ would be constant, so that from (6.23) we would have

$$R_\lambda(A) = \theta, \forall \lambda \in \mathbb{C},$$

which is a contradiction.

Proposition 6.3.5 *Let H be a Hilbert space and $A \in \mathcal{L}(H)$.*

1. *If $\lambda \in Res[\sigma(A)]$ then $\overline{\lambda} \in P\sigma(A^*)$.*

2. *If $\lambda \in P\sigma(A)$ then $\overline{\lambda} \in P\sigma(A^*) \cup Res[\sigma(A^*)]$.*

Proof 6.17

1. If $\lambda \in Res[\sigma(A)]$ then

$$R(A - \lambda I) \neq H.$$

   Therefore there exists $v \in (R(A - \lambda I))^\perp, v \neq \theta$ such that

$$(v, (A - \lambda I)u)_H = 0, \forall u \in H$$

   that is

$$((A^* - \overline{\lambda}I)v, u)_H = 0, \forall u \in H$$

   so that

$$(A^* - \overline{\lambda}I)v = \theta,$$

   which means that $\overline{\lambda} \in P\sigma(A^*)$.

2. Suppose there exists $v \neq \theta$ such that

$$(A - \lambda I)v = \theta,$$

   and

$$\overline{\lambda} \notin P\sigma(A^*).$$

   Thus,

$$(u, (A - \lambda I)v))_H = 0, \forall u \in H,$$

   so that

$$((A^* - \overline{\lambda}I)u, v)_H = 0, \forall u \in H.$$

   Since

$$(A^* - \overline{\lambda}I)u \neq \theta, \forall u \in H, u \neq \theta,$$

   we get $v \in (R(A^* - \overline{\lambda}I))^\perp$, so that $R(A^* - \overline{\lambda}I) \neq H$.
   Hence, $\overline{\lambda} \in Res[\sigma(A^*)]$.

Theorem 6.3.6 *Let $A \in \mathscr{L}(H)$ be a self-adjoint operator. then*

1. $\sigma(A) \subset \mathbb{R}$.

2. Eigenvectors corresponding to distinct eigenvalues of A are orthogonal.

Proof 6.18 Let $\mu, \lambda \in \mathbb{R}$. Thus, given $u \in H$ we have

$$\|(A - (\lambda + \mu i))u\|^2 = \|(A - \lambda)u\|^2 + \mu^2 \|u\|^2,$$

so that

$$\|(A - (\lambda + \mu i))u\|^2 \geq \mu^2 \|u\|^2.$$

Therefore, if $\mu \neq 0$, $A - (\lambda + \mu i)$ has a bounded inverse on its range, which is closed. If $R(A - (\lambda + \mu i)) \neq H$ then by the last result $(\lambda - \mu i)$ would be in the point spectrum of A, which contradicts the last inequality. Hence, if $\mu \neq 0$ then $\lambda + \mu i \in \rho(A)$. To complete the proof, suppose

$$Au_1 = \lambda_1 u_1,$$

and

$$Au_2 = \lambda_2 u_2,$$

where

$$\lambda_1, \lambda_2 \in \mathbb{R}, \ \lambda_1 \neq \lambda_2 \text{ and } u_1, u_2 \neq \theta.$$

Thus,

$$
\begin{aligned}
(\lambda_1 - \lambda_2)(u_1, u_2)_H &= \lambda_1 (u_1, u_2)_H - \lambda_2 (u_1, u_2)_H \\
&= (\lambda_1 u_1, u_2)_H - (u_1, \lambda_2 u_2)_H \\
&= (Au_1, u_2)_H - (u_1, Au_2)_H \\
&= (u_1, Au_2)_H - (u_1, Au_2)_H \\
&= 0. \tag{6.24}
\end{aligned}
$$

Since $\lambda_1 - \lambda_2 \neq 0$ we get

$$(u_1, u_2)_H = 0.$$

We finish this section with an exercise and its solution.

Exercise 6.3.7 *Let H be a complex Hilbert space e let $A \in L(H)$ be a self-adjoint operator. Prove that $\lambda \in \sigma(A)$ if, and only if, there exists a sequence $\{u_n\} \subset H$ such that $\|u_n\|_H = 1$, $\forall n \in \mathbb{N}$ and*

$$\|Au_n - \lambda u_n\|_H \to 0, \text{ as } n \to \infty.$$

Solution: Suppose $\lambda \in \mathbb{C}$ is such that there exists $\{u_n\} \subset H$ such that $\|u_n\| = 1$, $\forall n \in \mathbb{N}$ and

$$\|Au_n - \lambda u_n\|_H \to 0, \text{ as } n \to \infty. \tag{6.25}$$

Suppose, to obtain contradiction, that $\lambda \in \rho(A)$. Thus, $(A - \lambda I)^{-1}$ exists and it is bounded, so that there exists $K > 0$ such that

$$\|Au - \lambda Iu\|_H \geq K\|u\|_H, \forall u \in H.$$

From this we obtain

$$\|Au_n - \lambda u_n\|_H \geq K, \ \forall n \in \mathbb{N}$$

which contradicts (6.25).

Thus $\lambda \notin \rho(A)$ so that $\lambda \in \sigma(A)$.

Reciprocally, suppose $\lambda \in \sigma(A)$. Suppose, to obtain contradiction, that there exists $K > 0$ such that

$$\|Au - \lambda u\|_H \geq K\|u\|_H, \ \forall u \in H. \tag{6.26}$$

Thus, $(A - \lambda I)^{-1}$ exists and it is bounded. Since $\lambda \in \sigma(A)$, we must have that $R(A - \lambda I)$ is not dense, so that $\lambda \in Res[\sigma(A)]$.

From Proposition 6.3.5, we have $\overline{\lambda} \in P\sigma(A^*)$.

Since $A = A^*$, from this we obtain $\lambda = \overline{\lambda} \in P(\sigma(A))$ which contradicts $(A - I\lambda)^{-1}$ to exist.

Thus, we may infer that it does not exist $K > 0$ such that (6.26) holds.

From this, for each $n \in \mathbb{N}$ there exists $u_n \in H$ such that

$$\|u_n\|_H = 1$$

and

$$\|Au_n - \lambda u_n\|_H < 1/n$$

so that

$$\|Au_n - \lambda u_n\|_H \to 0.$$

The solution is complete.

6.4   The spectral theorem for bounded self-adjoint operators

Let H be a complex Hilbert space. Consider $A : H \to H$ a linear bounded operator, that is $A \in \mathscr{L}(H)$, and suppose also that such an operator is self-adjoint. Define

$$m = \inf_{u \in H} \{(Au, u)_H \mid \|u\|_H = 1\},$$

and

$$M = \sup_{u \in H} \{(Au, u)_H \mid \|u\|_H = 1\}.$$

Remark 6.4.1 *It is possible to prove that for a linear self-adjoint operator $A : H \to H$ we have*

$$\|A\| = \sup\{|(Au, u)_H| \mid u \in H, \|u\|_H = 1\}.$$

This propriety, which prove in the next lines, is crucial for the subsequent results, since for example for A, B linear and self-adjoint and $\varepsilon > 0$ we have

$$-\varepsilon I \leq A - B \leq \varepsilon I,$$

we also would have

$$\|A - B\| < \varepsilon.$$

So, we present the following basic result.

Theorem 6.4.2 *Let H be a real Hilbert space and let $A : H \to H$ be a bounded linear self-adjoint operator. Define*

$$\alpha = \max\{|m|, |M|\},$$

where

$$m = \inf_{u \in H} \{(Au, u)_H \mid \|u\|_H = 1\},$$

and

$$M = \sup_{u \in H}\{(Au,u)_H \mid \|u\|_H = 1\}.$$

Then,

$$\|A\| = \alpha.$$

Proof 6.19　Observe that

$$(A(u+v),u+v)_H = (Au,u)_H + (Av,v)_H + 2(Au,v)_H,$$

and

$$(A(u-v),u-v)_H = (Au,u)_H + (Av,v)_H - 2(Au,v)_H.$$

Thus,

$$4(Au,v) = (A(u+v),u+v)_H - (A(u-v),u-v)_H \le M\|u+v\|_U^2 - m\|u-v\|_U^2,$$

so that

$$4(Au,v)_H \le \alpha(\|u+v\|_U^2 + \|u-v\|_U^2).$$

Hence, replacing v by $-v$ we obtain

$$-4(Au,v)_H \le \alpha(\|u+v\|_U^2 + \|u-v\|_U^2),$$

and therefore

$$4|(Au,v)_H| \le \alpha(\|u+v\|_U^2 + \|u-v\|_U^2).$$

Replacing v by βv, we get

$$4|(A(u),v)_H| \le 2\alpha(\|u\|_U^2/\beta + \beta\|v\|_U^2).$$

Minimizing the last expression in $\beta > 0$, for the optimal

$$\beta = \|u\|_U/\|v\|_U,$$

we obtain

$$|(Au,v)_H| \le \alpha\|u\|_U\|v\|_U, \forall u,v \in U.$$

Thus

$$\|A\| \le \alpha.$$

On the other hand,

$$|(Au,u)_H| \le \|A\|\|u\|_U^2,$$

so that

$$|M| \le \|A\|$$

and

$$|m| \le \|A\|,$$

so that

$$\alpha \le \|A\|.$$

The proof is complete.

Remark 6.4.3 *A similar result is valid as H is a complex Hilbert space.*

At this point we start to develop the spectral theory. Define by P the set of all real polynomials defined in $\mathbb{R}$. Define

$$\Phi_1 : P \to \mathscr{L}(H),$$

by

$$\Phi_1(p(\lambda)) = p(A), \forall p \in P.$$

Thus we have

1. $\Phi_1(p_1 + p_2) = p_1(A) + p_2(A),$

2. $\Phi_1(p_1 \cdot p_2) = p_1(A)p_2(A),$

3. $\Phi_1(\alpha p) = \alpha p(A), \forall \alpha \in \mathbb{R},\ p \in P$

4. if $p(\lambda) \geq 0$, on $[m, M]$, then $p(A) \geq \theta,$

We will prove (4):

Consider $p \in P$. Denote the real roots of $p(\lambda)$ less or equal to m by $\alpha_1, \alpha_2, ..., \alpha_n$ and denote those that are greater or equal to M by $\beta_1, \beta_2, ..., \beta_l$. Finally denote all the remaining roots, real or complex by

$$\nu_1 + i\mu_1, ..., \nu_k + i\mu_k.$$

Observe that if $\mu_i = 0$ then $\nu_i \in (m, M)$. The assumption that $p(\lambda) \geq 0$ on $[m, M]$ implies that any real root in (m, M) must be of even multiplicity.

Since complex roots must occur in conjugate pairs, we have the following representation for $p(\lambda)$:

$$p(\lambda) = a \prod_{i=1}^{n} (\lambda - \alpha_i) \prod_{i=1}^{l} (\beta_i - \lambda) \prod_{i=1}^{k} ((\lambda - \nu_i)^2 + \mu_i^2),$$

where $a \geq 0$. Observe that

$$A - \alpha_i I \geq \theta,$$

since,

$$(Au, u)_H \geq m(u, u)_H \geq \alpha_i(u, u)_H, \forall u \in H,$$

and by analogy

$$\beta_i I - A \geq \theta.$$

On the other hand, since $A - \nu_k I$ is self-adjoint, its square is positive and hence since the sum of positive operators is positive, we obtain

$$(A - \nu_k I)^2 + \mu_k^2 I \geq \theta.$$

Therefore,

$$p(A) \geq \theta.$$

The idea is now to extend de domain of Φ_1 to the set of upper semi-continuous functions, and such set we will denote by C^{up}.

Observe that if $f \in C^{up}$, there exists a sequence of continuous functions $\{g_n\}$ such that

$$g_n \downarrow f, \text{ pointwise },$$

that is,

$$g_n(\lambda) \downarrow f(\lambda), \forall \lambda \in \mathbb{R}.$$

Considering the Weierstrass theorem, since $g_n \in C([m, M])$ we may obtain a sequence of polynomials $\{p_n\}$ such that

$$\left\| \left(g_n + \frac{1}{2^n} \right) - p_n \right\|_\infty < \frac{1}{2^n},$$

where the norm $\|\cdot\|_\infty$ refers to $[m,M]$. Thus,

$$p_n(\lambda) \downarrow f(\lambda), \text{ on } [m,M].$$

Therefore,

$$p_1(A) \geq p_2(A) \geq p_3(A) \geq \ldots \geq p_n(A) \geq \ldots$$

Since $p_n(A)$ is self-adjoint for all $n \in \mathbb{N}$, we have

$$p_j(A)p_k(A) = p_k(A)p_j(A), \forall j,k \in \mathbb{N}.$$

Then the $\lim_{n \to \infty} p_n(A)$ (in norm) exists, and we denote

$$\lim_{n \to \infty} p_n(A) = f(A).$$

Now recall the Dini's theorem:

Theorem 6.4.4 (Dini) *Let $\{g_n\}$ be a sequence of continuous functions defined on a compact set $K \subset \mathbb{R}$. Suppose $g_n \to g$ point-wise and monotonically on K. Under such assumptions the convergence in question is also uniform.*

Now suppose that $\{p_n\}$ and $\{q_n\}$ are sequences of polynomial such that

$$p_n \downarrow f, \text{ and } q_n \downarrow f,$$

we will show that

$$\lim_{n \to \infty} p_n(A) = \lim_{n \to \infty} q_n(A).$$

First, observe that being $\{p_n\}$ and $\{q_n\}$ sequences of continuous functions we have that

$$\hat{h}_{nk}(\lambda) = \max\{p_n(\lambda), q_k(\lambda)\}, \forall \lambda \in [m,M]$$

is also continuous, $\forall n,k \in \mathbb{N}$. Now fix $n \in \mathbb{N}$ and define

$$h_k(\lambda) = \max\{p_k(\lambda), q_n(\lambda)\}.$$

observe that

$$h_k(\lambda) \downarrow q_n(\lambda), \forall \lambda \in \mathbb{R},$$

so that by Dini's theorem

$$h_k \to q_n, \text{ uniformly on } [m,M].$$

It follows that for each $n \in \mathbb{N}$ there exists $k_n \in \mathbb{N}$ such that if $k > k_n$ then

$$h_k(\lambda) - q_n(\lambda) \leq \frac{1}{n}, \forall \lambda \in [m,M].$$

Since

$$p_k(\lambda) \leq h_k(\lambda), \forall \lambda \in [m,M],$$

we obtain

$$p_k(\lambda) - q_n(\lambda) \leq \frac{1}{n}, \forall \lambda \in [m,M].$$

By analogy, we may show that for each $n \in \mathbb{N}$ there exists $\hat{k}_n \in \mathbb{N}$ such that if $k > \hat{k}_n$ then

$$q_k(\lambda) - p_n(\lambda) \leq \frac{1}{n}.$$

From above we obtain

$$\lim_{k \to \infty} p_k(A) \le q_n(A) + \frac{1}{n}.$$

Since the self adjoint $q_n(A) + 1/n$ commutes with the

$$\lim_{k \to \infty} p_k(A)$$

we obtain

$$
\begin{aligned}
\lim_{k \to \infty} p_k(A) &\le \lim_{n \to \infty} \left(q_n(A) + \frac{1}{n} \right) \\
&\le \lim_{n \to \infty} q_n(A).
\end{aligned}
\tag{6.27}
$$

Similarly, we may obtain

$$\lim_{k \to \infty} q_k(A) \le \lim_{n \to \infty} p_n(A),$$

so that

$$\lim_{n \to \infty} q_n(A) = \lim_{n \to \infty} p_n(A) = f(A).$$

Hence, we may extend $\Phi_1 : P \to \mathscr{L}(H)$ to $\Phi_2 : C^{up} \to \mathscr{L}(H)$ where C^{up} as earlier indicated, denotes the set of upper semi-continuous functions, where

$$\Phi_2(f) = f(A).$$

Observe that Φ_2 has the following properties:

1. $\Phi_2(f_1 + f_2) = \Phi_2(f_1) + \Phi_2(f_2)$,

2. $\Phi_2(f_1 \cdot f_2) = f_1(A) f_2(A)$,

3. $\Phi_2(\alpha f) = \alpha \Phi_2(f), \forall \alpha \in \mathbb{R}, \ \alpha \ge 0$.

4. if $f_1(\lambda) \ge f_2(\lambda), \forall \lambda \in [m, M]$, then

$$f_1(A) \ge f_2(A).$$

The next step is to extend Φ_2 to $\Phi_3 : C_-^{up} \to \mathscr{L}(H)$, where

$$C_-^{up} = \{ f - g \mid f, g \in C^{up} \}.$$

For $h = f - g \in C_-^{up}$ we define

$$\Phi_3(h) = f(A) - g(A).$$

Now we will show that Φ_3 is well defined. Suppose that $h \in C_-^{up}$ and

$$h = f_1 - g_1 \text{ and } h = f_2 - g_2.$$

Thus,

$$f_1 - g_1 = f_2 - g_2,$$

that is,

$$f_1 + g_2 = f_2 + g_1,$$

so that from the definition of Φ_2 we obtain

$$f_1(A) + g_2(A) = f_2(A) + g_1(A),$$

that is,

$$f_1(A) - g_1(A) = f_2(A) - g_2(A).$$

Therefore, Φ_3 is well defined. Finally, observe that for $\alpha < 0$

$$\alpha(f - g) = -\alpha g - (-\alpha)f,$$

where $-\alpha g \in C^{up}$ and $-\alpha f \in C^{up}$. Thus,

$$\Phi_3(\alpha f) = \alpha f(A) = \alpha \Phi_3(f), \forall \alpha \in \mathbb{R}.$$

6.4.1 The spectral theorem

Consider the upper semi-continuous function

$$h_\mu(\lambda) = \begin{cases} 1, & \text{if } \lambda \leq \mu, \\ 0, & \text{if } \lambda > \mu. \end{cases} \tag{6.28}$$

Denote

$$E(\mu) = \Phi_3(h_\mu) = h_\mu(A).$$

Observe that

$$h_\mu(\lambda)h_\mu(\lambda) = h_\mu(\lambda), \forall \lambda \in \mathbb{R},$$

so that

$$[E(\mu)]^2 = E(\mu), \forall \mu \in \mathbb{R}.$$

Therefore,

$$\{E(\mu) \mid \mu \in \mathbb{R}\}$$

is a family of orthogonal projections. Also observe that if $\nu \geq \mu$ we have

$$h_\nu(\lambda)h_\mu(\lambda) = h_\mu(\lambda)h_\nu(\lambda) = h_\mu(\lambda),$$

so that

$$E(\nu)E(\mu) = E(\mu)E(\nu) = E(\mu), \forall \nu \geq \mu.$$

If $\mu < m$, then $h_\mu(\lambda) = 0$, on $[m, M]$, so that

$$E(\mu) = 0, \text{ if } \mu < m.$$

Similarly, if $\mu \geq M$ them $h_\mu(\lambda) = 1$, on $[m, M]$, so that

$$E(\mu) = I, \text{ if } \mu \geq M.$$

Next we show that the family $\{E(\mu)\}$ is strongly continuous from the right. First we will establish a sequence of polynomials $\{p_n\}$ such that

$$p_n \downarrow h_\mu,$$

and

$$p_n(\lambda) \geq h_{\mu + \frac{1}{n}}(\lambda), \text{ on } [m, M].$$

Observe that for any fixed n there exists a sequence of polynomials $\{p_j^n\}$ such that

$$p_j^n \downarrow h_{\mu + 1/n}, \text{ point-wise .}$$

Consider the monotone sequence

$$g_n(\lambda) = \min\{p_s^r(\lambda) \mid r, s \in \{1, ..., n\}\}.$$

Thus,

$$g_n(\lambda) \geq h_{\mu+\frac{1}{n}}(\lambda), \forall \lambda \in \mathbb{R},$$

and we obtain

$$\lim_{n\to\infty} g_n(\lambda) \geq \lim_{n\to\infty} h_{\mu+\frac{1}{n}}(\lambda) = h_\mu(\lambda).$$

On the other hand,

$$g_n(\lambda) \leq p_n^r(\lambda), \forall \lambda \in \mathbb{R}, \forall r \in \{1,...,n\},$$

so that

$$\lim_{n\to\infty} g_n(\lambda) \leq \lim_{n\to\infty} p_n^r(\lambda).$$

Therefore,

$$
\begin{aligned}
\lim_{n\to\infty} g_n(\lambda) &\leq \lim_{r\to\infty}\lim_{n\to\infty} p_n^r(\lambda) \\
&= h_\mu(\lambda).
\end{aligned}
\tag{6.29}
$$

Thus,

$$\lim_{n\to\infty} g_n(\lambda) = h_\mu(\lambda).$$

Observe that g_n are not necessarily polynomials. To set a sequence of polynomials, observe that we may obtain a sequence $\{p_n\}$ of polynomials such that

$$|g_n(\lambda) + 1/n - p_n(\lambda)| < \frac{1}{2^n}, \forall \lambda \in [m,M],\ n \in \mathbb{N}.$$

so that

$$p_n(\lambda) \geq g_n(\lambda) + 1/n - 1/2^n \geq g_n(\lambda) \geq h_{\mu+1/n}(\lambda).$$

Thus,

$$p_n(A) \to E(\mu),$$

and

$$p_n(A) \geq h_{\mu+\frac{1}{n}}(A) = E(\mu+1/n) \geq E(\mu).$$

Therefore we may write

$$E(\mu) = \lim_{n\to\infty} p_n(A) \geq \lim_{n\to\infty} E(\mu+1/n) \geq E(\mu).$$

Thus,

$$\lim_{n\to\infty} E(\mu+1/n) = E(\mu).$$

From this we may easily obtain the strong continuity from the right.

For $\mu \leq \nu$ we have

$$
\begin{aligned}
\mu(h_\nu(\lambda) - h_\mu(\lambda)) &\leq \lambda(h_\nu(\lambda) - h_\mu(\lambda)) \\
&\leq \nu(h_\nu(\lambda) - h_\mu(\lambda)).
\end{aligned}
\tag{6.30}
$$

To verify this observe that if $\lambda < \mu$ or $\lambda > \nu$ then all terms involved in the above inequalities are zero. On the other hand if

$$\mu \leq \lambda \leq \nu$$

then

$$h_\nu(\lambda) - h_\mu(\lambda) = 1,$$

so that in any case (6.30) holds. From the monotonicity property we have

$$
\begin{aligned}
\mu(E(\nu) - E(\mu)) &\leq A(E(\nu) - E(\mu)) \\
&\leq \nu(E(\nu) - E(\mu)).
\end{aligned}
\tag{6.31}
$$

Now choose $a, b \in \mathbb{R}$ such that

$$
a < m \text{ and } b \geq M.
$$

Suppose given $\varepsilon > 0$. Choose a partition P_0 of $[a,b]$, that is

$$
P_0 = \{a = \lambda_0, \lambda_1, ..., \lambda_n = b\},
$$

such that

$$
\max_{k \in \{1,...,n\}} \{|\lambda_k - \lambda_{k-1}|\} < \varepsilon.
$$

Hence,

$$
\begin{aligned}
\lambda_{k-1}(E(\lambda_k) - E(\lambda_{k-1})) &\leq A(E(\lambda_k) - E(\lambda_{k-1})) \\
&\leq \lambda_k(E(\lambda_k) - E(\lambda_{k-1})).
\end{aligned}
\tag{6.32}
$$

Summing up on k and recalling that

$$
\sum_{k=1}^{n} E(\lambda_k) - E(\lambda_{k-1}) = I,
$$

we obtain

$$
\begin{aligned}
\sum_{k=1}^{n} \lambda_{k-1}(E(\lambda_k) - E(\lambda_{k-1})) &\leq A \\
&\leq \sum_{k=1}^{n} \lambda_k(E(\lambda_k) - E(\lambda_{k-1})).
\end{aligned}
\tag{6.33}
$$

Let $\lambda_k^0 \in [\lambda_{k-1}, \lambda_k]$. Since $(\lambda_k - \lambda_k^0) \leq (\lambda_k - \lambda_{k-1})$ from (6.32) we obtain

$$
\begin{aligned}
A - \sum_{k=1}^{n} \lambda_k^0(E(\lambda_k) - E(\lambda_{k-1})) &\leq \varepsilon \sum_{k=1}^{n} (E(\lambda_k) - E(\lambda_{k-1})) \\
&= \varepsilon I.
\end{aligned}
\tag{6.34}
$$

By analogy

$$
-\varepsilon I \leq A - \sum_{k=1}^{n} \lambda_k^0(E(\lambda_k) - E(\lambda_{k-1})).
\tag{6.35}
$$

Since

$$
A - \sum_{k=1}^{n} \lambda_k^0(E(\lambda_k) - E(\lambda_{k-1}))
$$

is self-adjoint we obtain

$$
\left\| A - \sum_{k=1}^{n} \lambda_k^0(E(\lambda_k) - E(\lambda_{k-1})) \right\| < \varepsilon.
$$

Being $\varepsilon > 0$ arbitrary, we may write

$$
A = \int_a^b \lambda \, dE(\lambda),
$$

that is,

$$
A = \int_{m^-}^{M} \lambda \, dE(\lambda).
$$

Remark 6.4.5 *Consider again the function $h_\mu : \mathbb{R} \to \mathbb{R}$ where*

$$h_\mu(\lambda) = \begin{cases} 1, & \text{if } \lambda \le \mu \\ 0, & \text{if } \lambda > \mu. \end{cases} \tag{6.36}$$

Let H be a complex Hilbert space and let $A \in L(H)$, where A is a self-adjoint operator.
  Suppose $f \in C([m, M])$ where

$$m = \inf_{u \in H} \{(Au, u)_H \ : \ \|u\|_H = 1\},$$

and

$$M = \sup_{u \in H} \{(Au, u)_H \ : \ \|u\|_H = 1\}.$$

  Let $\varepsilon > 0$. Since f is uniformly continuous on the compact set $[m, M]$, there exists $\delta > 0$ such that if $x, y \in [m, M]$ and $|x - y| < \delta$, then

$$|f(x) - f(y)| < \varepsilon. \tag{6.37}$$

  Let $P = \{\lambda_0 = m, \lambda_1, \ldots, \lambda_n = M\}$ be a partition of $[m, M]$, such that $\|P\| = \max\{\lambda_k - \lambda_{k-1} \ : \ k \in \{1, \ldots, n\}\} < \delta$.
  Choose

$$\lambda_k^0 \in (\lambda_{k-1}, \lambda_k), \forall k \in \{1, \ldots, n\}$$

and observe that

$$h_{\lambda_k}(\lambda) - h_{\lambda_{k-1}}(\lambda) = \begin{cases} 1, & \text{if } \lambda_{k-1} < \lambda \le \lambda_k \\ 0, & \text{otherwise.} \end{cases} \tag{6.38}$$

From this and (6.37), we may obtain

$$\left| f(\lambda) - \sum_{k=1}^n f(\lambda_k^0)[h_{\lambda_k}(\lambda) - h_{\lambda_{k-1}}(\lambda)] \right| < \varepsilon, \ \forall \lambda \in [m, M].$$

  Therefore, for the corresponding operators, we have got

$$\left\| f(A) - \sum_{k=1}^n f(\lambda_k^0)[E(\lambda_k) - E(\lambda_{k-1})] \right\| < \varepsilon.$$

Since $\varepsilon > 0$, the partition P and $\{\lambda_k^0\}$ have been arbitrary, we may denote

$$f(A) = \int_{m^-}^M f(\lambda) \, dE(\lambda).$$

6.5   The spectral decomposition of unitary transformations

Definition 6.5.1 *Let H be a Hilbert space. A transformation $U : H \to H$ is said to be unitary if*

$$(Uu, Uu)_H = (u, u)_H, \forall u, u \in H.$$

Observe that in this case

$$U^* U = U U^* = I,$$

so that

$$U^{-1} = U^*.$$

Theorem 6.5.2 *Every Unitary transformation U has a spectral decomposition*

$$U = \int_{0^-}^{2\pi} e^{i\phi} dE(\phi),$$

where $\{E(\phi)\}$ is a spectral family on $[0, 2\pi]$. Furthermore $E(\phi)$ is continuous at 0 and it is the limit of polynomials in U and U^{-1}.

We present just a sketch of the proof. For the trigonometric polynomials

$$p(e^{i\phi}) = \sum_{k=-n}^{n} c_k e^{ik\phi},$$

consider the transformation

$$p(U) = \sum_{k=-n}^{n} c_k U^k,$$

where $c_k \in \mathbb{C}, \forall k \in \{-n, ..., 0, ..., n\}$.

Observe that

$$\overline{p(e^{i\phi})} = \sum_{k=-n}^{n} \overline{c}_k e^{-ik\phi},$$

so that the corresponding operator is

$$p(U)^* = \sum_{k=-n}^{n} \overline{c}_k U^{-k} = \sum_{k=-n}^{n} \overline{c}_k (U^*)^k.$$

Also if

$$p(e^{i\phi}) \geq 0$$

there exists a polynomial q such that

$$p(e^{i\phi}) = |q(e^{i\phi})|^2 = \overline{q(e^{i\phi})} q(e^{i\phi}),$$

so that

$$p(U) = [q(U)]^* q(U).$$

Therefore

$$(p(U)v, v)_H = (q(U)^* q(U)v, v)_H = (q(U)v, q(U)v)_H \geq 0, \forall v \in H,$$

which means

$$p(U) \geq 0.$$

Define the function $h_\mu(\phi)$ by

$$h_\mu(\phi) = \begin{cases} 1, & \text{if } 2k\pi < \phi \leq 2k\pi + \mu, \\ 0, & \text{if } 2k\pi + \mu < \phi \leq 2(k+1)\pi, \end{cases} \tag{6.39}$$

for each $k \in \{0, \pm 1, \pm 2, \pm 3, ...\}$. Define $E(\mu) = h_\mu(U)$. Observe that the family $\{E(\mu)\}$ are projections and in particular

$$E(0) = 0,$$

$$E(2\pi) = I$$

and if $\mu \leq \nu$, since

$$h_\mu(\phi) \leq h_\nu(\phi),$$

we have

$$E(\mu) \leq E(\nu).$$

Suppose given $\varepsilon > 0$. Let P_0 be a partition of $[0, 2\pi]$ that is,

$$P_0 = \{0 = \phi_0, \phi_1, ..., \phi_n = 2\pi\}$$

such that

$$\max_{j \in \{1,...,n\}} \{|\phi_j - \phi_{j-1}|\} < \varepsilon.$$

For fixed $\phi \in [0, 2\pi]$, let $j \in \{1, ..., n\}$ be such that

$$\phi \in [\phi_{j-1}, \phi_j].$$

$$
\begin{aligned}
\left| e^{i\phi} - \sum_{k=1}^{n} e^{i\phi_k} (h_{\phi_k}(\phi) - h_{\phi_{k-1}}(\phi)) \right| &= \left| e^{i\phi} - e^{i\phi_j} \right| \\
&\leq |\phi - \phi_j| < \varepsilon.
\end{aligned}
\tag{6.40}
$$

Thus,

$$0 \leq \left| e^{i\phi} - \sum_{k=1}^{n} e^{i\phi_k} (h_{\phi_k}(\phi) - h_{\phi_{k-1}}(\phi)) \right|^2 \leq \varepsilon^2$$

so that, for the corresponding operators

$$
\begin{aligned}
0 &\leq \left[U - \sum_{k=1}^{n} e^{i\phi_k} (E(\phi_k) - E(\phi_{k-1})) \right]^* \left[U - \sum_{k=1}^{n} e^{i\phi_k} (E(\phi_k) - E(\phi_{k-1})) \right] \\
&\leq \varepsilon^2 I
\end{aligned}
\tag{6.41}
$$

and hence

$$\left\| U - \sum_{k=1}^{n} e^{i\phi_k} (E(\phi_k) - E(\phi_{k-1})) \right\| < \varepsilon.$$

Being $\varepsilon > 0$ arbitrary, we may infer that

$$U = \int_0^{2\pi} e^{i\phi} dE(\phi).$$

6.6   Unbounded operators

6.6.1   Introduction

Let H be a Hilbert space. Let $A : D(A) \to H$ be an operator, where unless otherwise indicated $D(A)$ is a dense subset of H. We consider in this section the special case where A is unbounded.

Definition 6.6.1 *Given $A : D \to H$ we define the graph of A, denoted by $\Gamma(A)$ by,*

$$\Gamma(A) = \{(u, Au) \mid u \in D\}.$$

Definition 6.6.2 *An operator $A : D \to H$ is said to be closed if $\Gamma(A)$ is closed.*

Definition 6.6.3 *Let $A_1 : D_1 \to H$ and $A_2 : D_2 \to H$ operators. We write $A_2 \supset A_1$ if $D_2 \supset D_1$ and*

$$A_2 u = A_1 u, \forall u \in D_1.$$

In this case we say that A_2 is an extension of A_1.

Definition 6.6.4 *A linear operator $A : D \to H$ is said to be closable if it has a linear closed extension. The smallest closed extension of A is denote by $\overline{A}$ and is called the closure of A.*

Proposition 6.6.5 *Let $A : D \to H$ be a linear operator. If A is closable then*

$$\Gamma(\overline{A}) = \overline{\Gamma(A)}.$$

Proof 6.20 Suppose B is a closed extension of A. Then

$$\overline{\Gamma(A)} \subset \overline{\Gamma(B)} = \Gamma(B),$$

so that if $(\theta, \phi) \in \overline{\Gamma(A)}$ then $(\theta, \phi) \in \Gamma(B)$, and hence $\phi = \theta$. Define the operator C by

$$D(C) = \{\psi \mid (\psi, \phi) \in \overline{\Gamma(A)} \text{ for some } \phi\},$$

and $C(\psi) = \phi$, where ϕ is the unique point such that $(\psi, \phi) \in \overline{\Gamma(A)}$. Hence

$$\Gamma(C) = \overline{\Gamma(A)} \subset \Gamma(B),$$

so that

$$A \subset C.$$

However, $C \subset B$ and since B is an arbitrary closed extension of A we have

$$C = \overline{A}$$

so that

$$\Gamma(C) = \Gamma(\overline{A}) = \overline{\Gamma(A)}.$$

Definition 6.6.6 *Let $A : D \to H$ be a linear operator where D is dense in H. Define $D(A^*)$ by*

$$D(A^*) = \{\phi \in H \mid (A\psi, \phi)_H = (\psi, \eta)_H, \forall \psi \in D \text{ for some } \eta \in H\}.$$

In this case we denote

$$A^* \phi = \eta.$$

A^* defined in this way is called the adjoint operator related to A.

Observe that by the Riesz lemma, $\phi \in D(A^*)$ if and only if there exists $K > 0$ such that

$$|(A\psi, \phi)_H| \leq K\|\psi\|_H, \forall \psi \in D.$$

Also note that if

$$A \subset B \text{ then } B^* \subset A^*.$$

Finally, as D is dense in H then

$$\eta = A^*(\phi)$$

is uniquely defined. However the domain of A^* may not be dense, and in some situations we may have $D(A^*) = \{\theta\}$.

If $D(A^*)$ is dense we define

$$A^{**} = (A^*)^*.$$

Theorem 6.6.7 *Let $A : D \to H$ a linear operator, being D dense in H. Then*

1. A^* is closed,

2. *A is closable if and only if $D(A^*)$ is dense and in this case*

$$\overline{A} = A^{**}.$$

3. *If A is closable then $(\overline{A})^* = A^*$.*

Proof 6.21

1. We define the operator $V : H \times H \to H \times H$ by

$$V(\phi, \psi) = (-\psi, \phi).$$

Let $E \subset H \times H$ be a subspace. Thus if $(\phi_1, \psi_1) \in V(E^\perp)$ then there exists $(\phi, \psi) \in E^\perp$ such that

$$V(\phi, \psi) = (-\psi, \phi) = (\phi_1, \psi_1).$$

Hence

$$\psi = -\phi_1 \text{ and } \phi = \psi_1,$$

so that for $(\psi_1, -\phi_1) \in E^\perp$ and $(w_1, w_2) \in E$ we have

$$((\psi_1, -\phi_1), (w_1, w_2))_{H \times H} = 0 = (\psi_1, w_1)_H + (-\phi_1, w_2)_H.$$

Thus,

$$(\phi_1, -w_2)_H + (\psi_1, w_1)_H = 0,$$

and therefore

$$((\phi_1, \psi_1), (-w_2, w_1))_{H \times H} = 0,$$

that is

$$((\phi_1, \psi_1), V(w_1, w_2))_{H \times H} = 0, \forall (w_1, w_2) \in E.$$

This means that

$$(\phi_1, \psi_1) \in (V(E))^\perp,$$

so that

$$V(E^\perp) \subset (V(E))^\perp.$$

It is easily verified that the implications from which the last inclusion results are in fact equivalences, so that

$$V(E^\perp) = (V(E))^\perp.$$

Suppose $(\phi, \eta) \in H \times H$. Thus $(\phi, \eta) \in V(\Gamma(A))^\perp$ if and only if

$$((\phi, \eta), (-A\psi, \psi))_{H \times H} = 0, \forall \psi \in D,$$

which holds if and only if

$$(\phi, A\psi)_H = (\eta, \psi)_H, \forall \psi \in D,$$

that is, if and only if

$$(\phi, \eta) \in \Gamma(A^*).$$

Thus,

$$\Gamma(A^*) = V(\Gamma(A))^\perp.$$

Since $(V(\Gamma(A))^\perp$ is closed, A^* is closed.

2. Observe that $\Gamma(A)$ is a linear subset of $H \times H$ so that

$$
\begin{aligned}
\overline{\Gamma(A)} &= [\Gamma(A)^\perp]^\perp \\
&= V^2[\Gamma(A)^\perp]^\perp \\
&= [V[V(\Gamma(A))^\perp]]^\perp \\
&= [V(\Gamma(A^*))]^\perp
\end{aligned}
\tag{6.42}
$$

so that from the proof of item 1, if A^* is densely defined we get

$$
\overline{\Gamma(A)} = \Gamma[(A^*)^*].
$$

Conversely, suppose $D(A^*)$ is not dense. Thus there exists $\psi \in [D(A^*)]^\perp$ such that $\psi \neq \theta$. Le $(\phi, A^*\phi) \in \Gamma(A^*)$. Hence

$$
((\psi, \theta), (\phi, A^*\phi))_{H \times H} = (\psi, \phi)_H = 0,
$$

so that

$$
(\psi, \theta) \in [\Gamma(A^*)]^\perp.
$$

Therefore, $V[\Gamma(A^*)]^\perp$ is not the graph of a linear operator. Since $\overline{\Gamma(A)} = V[\Gamma(A^*)]^\perp$ A is not closable

3. Observe that if A is closable then

$$
A^* = \overline{(A^*)} = A^{***} = (\overline{A})^*.
$$

6.7 Symmetric and self-adjoint operators

Definition 6.7.1 *Let $A : D \to H$ be a linear operator, where D is dense in H. A is said to be symmetric i $A \subset A^*$, that is if $D \subset D(A^*)$ and*

$$
A^*\phi = A\phi, \forall \phi \in D.
$$

Equivalently, A is symmetric if and only if

$$
(A\phi, \psi)_H = (\phi, A\psi)_H, \forall \phi, \psi \in D.
$$

Definition 6.7.2 *Let $A : D \to H$ be a linear operator. We say that A is self-adjoint if $A = A^*$, that is if A i symmetric and $D = D(A^*)$.*

Definition 6.7.3 *Let $A : D \to H$ be a symmetric operator. We say that A is essentially self-adjoint if its closur $\overline{A}$ is self-adjoint. If A is closed, a subset $E \subset D$ is said to be a core for A if $\overline{A|_E} = A$.*

Theorem 6.7.4 *Let $A : D \to H$ be a symmetric operator. Then the following statements are equivalent*

1. *A is self-adjoint.*

2. *A is closed and $N(A^* \pm iI) = \{\theta\}$.*

3. *$R(A \pm iI) = H$.*

Proof 6.22

■ 1 implies 2:

Suppose A is self-adjoint let $\phi \in D = D(A^*)$ be such that

$$
A\phi = i\phi
$$

so that
$$A^*\phi = i\phi.$$

Observe that

$$
\begin{aligned}
i(\phi,\phi)_H &= (i\phi,\phi)_H \\
&= (A\phi,\phi)_H \\
&= (\phi,A\phi)_H \\
&= (\phi,i\phi)_H \\
&= -i(\phi,\phi)_H,
\end{aligned}
\tag{6.43}
$$

so that $(\phi,\phi)_H = 0$, that is $\phi = \theta$. Thus,

$$N(A - iI) = \{\theta\}.$$

Similarly, we prove that $N(A + iI) = \{\theta\}$. Finally, since $\overline{A^*} = A^* = A$, we get that $A = A^*$ is closed.

■ **2 implies 3:**

Suppose 2 holds. Thus the equation
$$A^*\phi = -i\phi$$

has no non trivial solution. We will prove that $R(A - iI)$ is dense in H. If $\psi \in R(A - iI)^{\perp}$ then

$$((A - iI)\phi, \psi)_H = 0, \forall \phi \in D,$$

so that $\psi \in D(A^*)$ and

$$(A - iI)^*\psi = (A^* + iI)\psi = \theta,$$

and hence by above $\psi = \theta$. Now we will prove that $R(A - iI)$ is closed and conclude that

$$R(A - iI) = H.$$

Given $\phi \in D$ we have
$$\|(A - iI)\phi\|_H^2 = \|A\phi\|_H^2 + \|\phi\|_H^2. \tag{6.44}$$

Let $\psi_0 \in H$ be a limit point of $R(A - iI)$. Thus we may find $\{\phi_n\} \subset D$ such that

$$(A - iI)\phi_n \to \psi_0.$$

From (6.44)
$$\|\phi_n - \phi_m\|_H \leq \|(A - iI)(\phi_n - \phi_m)\|_H, \forall m,n \in \mathbb{N}$$

so that $\{\phi_n\}$ is a Cauchy sequence, therefore converging to some $\phi_0 \in H$. Also from (6.44)

$$\|A\phi_n - A\phi_m\|_H \leq \|(A - iI)(\phi_n - \phi_m)\|_H, \forall m,n \in \mathbb{N}$$

so that $\{A\phi_n\}$ is a Cauchy sequence, hence also a converging one. Since A is closed, we get $\phi_0 \in D$ and

$$(A - iI)\phi_0 = \psi_0.$$

Therefore, $R(A - iI)$ is closed, so that

$$R(A - iI) = H.$$

Similarly,

$$R(A + iI) = H.$$

■ 3 implies 1: Let $\phi \in D(A^*)$. Since, $R(A - iI) = H$, there is an $\eta \in D$ such that

$$(A - iI)\eta = (A^* - iI)\phi,$$

and since $D \subset D(A^*)$ we obtain $\phi - \eta \in D(A^*)$, and

$$(A^* - iI)(\phi - \eta) = \theta.$$

Since $R(A + iI) = H$ we have $N(A^* - iI) = \{\theta\}$. Therefore, $\phi = \eta$, so that $D(A^*) = D$. The proof is complete.

6.7.1 The spectral theorem using Cayley transform

In this section H is a complex Hilbert space. We suppose A is defined on a dense subspace of H, being A self-adjoint but possibly unbounded. We have shown that $(A + i)$ and $(A - i)$ are onto H and it is possible to prove that

$$U = (A - i)(A + i)^{-1},$$

exists on all H and it is unitary. Furthermore on the domain of A,

$$A = i(I + U)(I - U)^{-1}.$$

The operator U is called the Cayley transform of A. We have already proven that

$$U = \int_0^{2\pi} e^{i\phi} dF(\phi),$$

where $\{F(\phi)\}$ is a monotone family of orthogonal projections, strongly continuous from the right and we may consider it such that

$$F(\phi) = \begin{cases} 0, & \text{if } \phi \leq 0, \\ I, & \text{if } \phi \geq 2\pi. \end{cases} \tag{6.45}$$

Since $F(\phi) = 0$, for all $\phi \leq 0$ and

$$F(0) = F(0^+)$$

we obtain

$$F(0^+) = 0 = F(0^-),$$

that is, $F(\phi)$ is continuous at $\phi = 0$. We claim that F is continuous at $\phi = 2\pi$. Observe that $F(2\pi) = F(2\pi^+)$ so that we need only to show that

$$F(2\pi^-) = F(2\pi).$$

Suppose

$$F(2\pi) - F(2\pi^-) \neq \theta.$$

Thus there exists some $u, v \in H$ such that

$$(F(2\pi) - F(2(\pi^-)))u = v \neq \theta.$$

Therefore,

$$F(\phi)v = F(\phi)[(F(2\pi) - F(2\pi^-))u],$$

so that

$$F(\phi)v = \begin{cases} 0, & \text{if } \phi < 2\pi, \\ v, & \text{if } \phi \geq 2\pi. \end{cases} \tag{6.46}$$

Observe that

$$U - I = \int_0^{2\pi} (e^{i\phi} - 1) dF(\phi),$$

and

$$U^* - I = \int_0^{2\pi} (e^{-i\phi} - 1) dF(\phi).$$

Let $\{\phi_n\}$ be a partition of $[0, 2\pi]$. From the monotonicity of $[0, 2\pi]$ and pairwise orthogonality of

$$\{F(\phi_n) - F(\phi_{n-1})\}$$

we can show that (this is not proved in details here)

$$(U^* - I)(U - I) = \int_0^{2\pi} (e^{-i\phi} - 1)(e^{i\phi} - 1) dF(\phi),$$

so that, given $z \in H$ we have

$$((U^* - I)(U - I)z, z)_H = \int_0^{2\pi} |e^{i\phi} - 1|^2 (dF(\phi)z, z)_H,$$

thus, for v defined above

$$
\begin{aligned}
\|(U - I)v\|^2 &= ((U - I)v, (U - I)v)_H \\
&= ((U - I)^*(U - I)v, v)_H \\
&= \int_0^{2\pi} |e^{i\phi} - 1|^2 (dF(\phi)v, v)_H \\
&= \int_0^{2\pi^-} |e^{i\phi} - 1|^2 (dF(\phi)v, v)_H \\
&= 0 \quad\quad\quad (6.47)
\end{aligned}
$$

The last two equalities results from $e^{2\pi i} - 1 = 0$ and $dF(\phi)v = \theta$ on $[0, 2\pi)$. Since $v \neq \theta$, the last equation implies that $1 \in P\sigma(U)$, which contradicts the existence of

$$(I - U)^{-1}.$$

Thus, F is continuous at $\phi = 2\pi$.

Now choose a sequence of real numbers $\{\phi_n\}$ such that $\phi_n \in (0, 2\pi)$, $n = 0, \pm 1, \pm 2, \pm 3, \ldots$ such that

$$-\cot\left(\frac{\phi_n}{2}\right) = n.$$

Now define $T_n = F(\phi_n) - F(\phi_{n-1})$. Since U commutes with $F(\phi)$, U commutes with T_n. since

$$A = i(I + U)(I - U)^{-1},$$

this implies that the range of T_n is invariant under U and A. Observe that

$$
\begin{aligned}
\sum_n T_n &= \sum_n (F(\phi_n) - F(\phi_{n-1})) \\
&= \lim_{\phi \to 2\pi} F(\phi) - \lim_{\phi \to 0} F(\phi) \\
&= I - \theta = I. \quad\quad\quad (6.48)
\end{aligned}
$$

Hence,

$$\sum_n R(T_n) = H.$$

Also, for $u \in H$ we have that

$$F(\phi)T_n u = \begin{cases} 0, \text{ if } \phi < \phi_{n-1}, \\ (F(\phi) - F(\phi_{n-1}))u, \text{ if } \phi_{n-1} \leq \phi \leq \phi_n, \\ F(\phi_n) - F(\phi_{n-1}))u, \text{ if } \phi > \phi_n, \end{cases} \tag{6.49}$$

so that

$$
\begin{aligned}
(I - U)T_n u &= \int_0^{2\pi} (1 - e^{i\phi})dF(\phi)T_n u \\
&= \int_{\phi_{n-1}}^{\phi_n} (1 - e^{i\phi})dF(\phi)u.
\end{aligned} \tag{6.50}
$$

Therefore,

$$
\begin{aligned}
&\int_{\phi_{n-1}}^{\phi_n} (1 - e^{i\phi})^{-1}dF(\phi)(I - U)T_n u \\
&= \int_{\phi_{n-1}}^{\phi_n} (1 - e^{i\phi})^{-1}dF(\phi) \int_{\phi_{n-1}}^{\phi_n} (1 - e^{i\phi})dF(\phi)u \\
&= \int_{\phi_{n-1}}^{\phi_n} (1 - e^{i\phi})^{-1}(1 - e^{i\phi})dF(\phi)u \\
&= \int_{\phi_{n-1}}^{\phi_n} dF(\phi)u \\
&= \int_0^{2\pi} dF(\phi)T_n u = T_n u.
\end{aligned} \tag{6.51}
$$

Hence,

$$\left[(I - U)|_{R(T_n)}\right]^{-1} = \int_{\phi_{n-1}}^{\phi_n} (1 - e^{i\phi})^{-1}dF(\phi).$$

From this, from above and as

$$A = i(I + U)(I - U)^{-1}$$

we obtain

$$AT_n u = \int_{\phi_{n-1}}^{\phi_n} i(1 + e^{i\phi})(1 - e^{i\phi})^{-1}dF(\phi)u.$$

Therefore, defining

$$\lambda = -cot\left(\frac{\phi}{2}\right),$$

and

$$E(\lambda) = F(-2cot^{-1}\lambda),$$

we get

$$i(1 + e^{i\phi})(1 - e^{i\phi})^{-1} = -cot\left(\frac{\phi}{2}\right) = \lambda.$$

Hence,

$$AT_n u = \int_{n-1}^{n} \lambda dE(\lambda)u.$$

Finally, from

$$u = \sum_{n=-\infty}^{\infty} T_n u,$$

we can obtain

$$
\begin{aligned}
Au &= A\left(\sum_{n=-\infty}^{\infty} T_n u\right) \\
&= \sum_{n=-\infty}^{\infty} AT_n u \\
&= \sum_{n=-\infty}^{\infty} \int_{n-1}^{n} \lambda \, dE(\lambda) u.
\end{aligned}
\tag{6.52}
$$

Being the convergence in question in norm, we may write

$$
Au = \int_{-\infty}^{\infty} \lambda \, dE(\lambda) u.
$$

Since $u \in H$ is arbitrary, we may denote

$$
A = \int_{-\infty}^{\infty} \lambda \, dE(\lambda).
\tag{6.53}
$$

Chapter 7

Basic Results on Measure and Integration

The main references for this chapter are Rudin [68], Royden [69] and Stein and Shakarchi [72], where more details may be found. All these three books are excellent and we strongly recommend their reading.

7.1 Basic concepts

In this chapter U denotes a topological space.

Definition 7.1.1 (σ-Algebra) *A collection $\mathcal{M}$ of subsets of U is said to be a σ-Algebra if $\mathcal{M}$ has the following properties:*

1. $U \in \mathcal{M}$,

2. if $A \in \mathcal{M}$ then $U \setminus A \in \mathcal{M}$,

3. if $A_n \in \mathcal{M}, \forall n \in \mathbb{N}$, then $\cup_{n=0}^{\infty} A_n \in \mathcal{M}$.

Definition 7.1.2 (Measurable spaces) *If $\mathcal{M}$ is a σ-algebra in U we say that U is a measurable space. The elements of $\mathcal{M}$ are called the measurable sets of U.*

Definition 7.1.3 (Measurable function) *If U is a measurable space and V is a topological space, we say that $f : U \to V$ is a measurable function if $f^{-1}(\mathcal{V})$ is measurable whenever $\mathcal{V} \subset V$ is an open set.*

Remark 7.1.4   *1. Observe that $\emptyset = U \setminus U$ so that from 1 and 2 in Definition 7.1.1, we have that $\emptyset \in \mathcal{M}$.*

2. From 1 and 3 from Definition 7.1.1, it is clear that $\cup_{i=1}^{n} A_i \in \mathcal{M}$ whenever $A_i \in \mathcal{M}, \forall i \in \{1, ..., n\}$.

3. Since $\cap_{i=1}^{\infty} A_i = (\cup_{i=1}^{\infty} A_i^c)^c$ also from Definition 7.1.1, it is clear that $\mathcal{M}$ is closed under countable intersections.

4. Since $A \setminus B = B^c \cap A$ we obtain: if $A, B \in \mathcal{M}$ then $A \setminus B \in \mathcal{M}$.

Theorem 7.1.5 *Let $\mathcal{F}$ be any collection of subsets of U. Then there exists a smallest σ-algebra $\mathcal{M}_0$ in U such that $\mathcal{F} \subset M_0$.*

Proof 7.1   Let Ω be the family of all σ-Algebras that contain $\mathscr{F}$. Since the set of all subsets in U is a σ-algebra, Ω is non-empty.

Let $\mathscr{M}_0 = \cap_{\mathscr{M}_\lambda \in \Omega} \mathscr{M}_\lambda$, it is clear that $\mathscr{M}_0 \supset \mathscr{F}$, it remains to prove that in fact $\mathscr{M}_0$ is a σ-algebra. Observe that:

1. $U \in \mathscr{M}_\lambda, \forall \mathscr{M}_\lambda \in \Omega$, so that, $U \in \mathscr{M}_0$,

2. $A \in \mathscr{M}_0$ implies $A \in \mathscr{M}_\lambda, \forall \mathscr{M}_\lambda \in \Omega$, so that $A^c \in \mathscr{M}_\lambda, \forall \mathscr{M}_\lambda \in \Omega$, which means $A^c \in \mathscr{M}_0$,

3. $\{A_n\} \subset \mathscr{M}_0$ implies $\{A_n\} \subset \mathscr{M}_\lambda, \forall \mathscr{M}_\lambda \in \Omega$, so that $\cup_{n=1}^\infty A_n \in \mathscr{M}_\lambda, \forall \mathscr{M}_\lambda \in \Omega$, which means $\cup_{n=1}^\infty A_n \in \mathscr{M}_0$.

From Definition 7.1.1 the proof is complete.

Definition 7.1.6 (Borel sets) *Let U be a topological space, considering the last theorem there exists a smallest σ-algebra in U, denoted by $\mathscr{B}$, which contains the open sets of U. The elements of $\mathscr{B}$ are called the Borel sets.*

Theorem 7.1.7 *Suppose $\mathscr{M}$ is a σ-algebra in U and V is a topological space. For $f : U \to V$, we have:*

1. If $\Omega = \{E \subset V \mid f^{-1}(E) \in \mathscr{M}\}$, then Ω is a σ-algebra.

2. If $V = [-\infty, \infty]$, and $f^{-1}((\alpha, \infty]) \in \mathscr{M}$, for each $\alpha \in \mathbb{R}$, then f is measurable.

Proof 7.2

1. (a) $V \in \Omega$ since $f^{-1}(V) = U$ and $U \in \mathscr{M}$.
   (b) $E \in \Omega \Rightarrow f^{-1}(E) \in \mathscr{M} \Rightarrow U \setminus f^{-1}(E) \in \mathscr{M} \Rightarrow f^{-1}(V \setminus E) \in \mathscr{M} \Rightarrow V \setminus E \in \Omega$.
   (c) $\{E_i\} \subset \Omega \Rightarrow f^{-1}(E_i) \in \mathscr{M}, \forall i \in \mathbb{N} \Rightarrow \cup_{i=1}^\infty f^{-1}(E_i) \in \mathscr{M} \Rightarrow f^{-1}(\cup_{i=1}^\infty E_i) \in \mathscr{M} \Rightarrow \cup_{i=1}^\infty E_i \in \Omega$.
   Thus Ω is a σ-algebra.

2. Define $\Omega = \{E \subset [-\infty, \infty] \mid f^{-1}(E) \in \mathscr{M}\}$ from above Ω is a σ- algebra. Given $\alpha \in \mathbb{R}$, let $\{\alpha_n\}$ be a real sequence such that $\alpha_n \to \alpha$ as $n \to \infty$, $\alpha_n < \alpha, \forall n \in \mathbb{N}$. Since $(\alpha_n, \infty] \in \Omega$ for each n and

$$[-\infty, \alpha) = \cup_{n=1}^\infty [-\infty, \alpha_n] = \cup_{n=1}^\infty (\alpha_n, \infty]^C, \tag{7.1}$$

we obtain, $[-\infty, \alpha) \in \Omega$. Furthermore, we have $(\alpha, \beta) = [-\infty, \beta) \cap (\alpha, \infty] \in \Omega$. Since every open set in $[-\infty, \infty]$ may be expressed as a countable union of intervals (α, β) we have that Ω contains all the open sets. Thus, $f^{-1}(E) \in \mathscr{M}$ whenever E is open, so that f is measurable.

Proposition 7.1.8 *If $\{f_n : U \to [-\infty, \infty]\}$ is a sequence of measurable functions and $g = \sup_{n \geq 1} f_n$ and $h = \limsup_{n \to \infty} f_n$ then g and h are measurable.*

Proof 7.3   Observe that $g^{-1}((\alpha, \infty]) = \cup_{n=1}^\infty f_n^{-1}((\alpha, \infty])$. From last theorem g is measurable. By analogy $h = \inf_{k \geq 1} \{\sup_{i \geq k} f_i\}$ is measurable.

7.2   Simple functions

Definition 7.2.1 (Simple functions) *A function $f : U \to \mathbb{C}$ is said to be a simple function if its range $(R(f))$ has only finitely many points. If $\{\alpha_1, ..., \alpha_n\} = R(f)$ and we set $A_i = \{u \in U \mid f(u) = \alpha_i\}$, clearly we have: $f = \sum_{i=1}^n \alpha_i \chi_{A_i}$, where*

$$\chi_{A_i}(u) = \begin{cases} 1, & \text{if } u \in A_i, \\ 0, & \text{otherwise.} \end{cases} \tag{7.2}$$

Theorem 7.2.2 *Let $f : U \to [0, \infty]$ be a measurable function. Thus there exists a sequence of simple functions $\{s_n : U \to [0, \infty]\}$ such that*

1. *$0 \leq s_1 \leq s_2 \leq \ldots \leq f$,*

2. *$s_n(u) \to f(u)$ as $n \to \infty, \forall u \in U$.*

Proof 7.4 Define $\delta_n = 2^{-n}$. To each $n \in \mathbb{N}$ and each $t \in \mathbb{R}^+$, there corresponds a unique integer $K = K_n(t)$ such that

$$K\delta_n \leq t \leq (K+1)\delta_n. \tag{7.3}$$

Defining

$$\varphi_n(t) = \begin{cases} K_n(t)\delta_n, & if \ 0 \leq t < n, \\ n, & if \ t \geq n, \end{cases} \tag{7.4}$$

we have that each φ_n is a Borel function on $[0, \infty]$, such that

1. *$t - \delta_n < \varphi_n(t) \leq t$ if $0 \leq t \leq n$,*

2. *$0 \leq \varphi_1 \leq \ldots \leq t$,*

3. *$\varphi_n(t) \to t$ as $n \to \infty, \forall t \in [0, \infty]$.*

It follows that the sequence $\{s_n = \varphi_n \circ f\}$ corresponds to the results indicated above.

7.3   Measures

Definition 7.3.1 (Measure) *Let $\mathscr{M}$ be a σ-algebra on a topological space U. A function $\mu : \mathscr{M} \to [0, \infty]$ is said to be a measure if $\mu(\emptyset) = 0$ and μ is countably additive, that is, given $\{A_i\} \subset U$, a sequence of pairwise disjoint sets then*

$$\mu(\cup_{i=1}^{\infty} A_i) = \sum_{i=1}^{\infty} \mu(A_i). \tag{7.5}$$

In this case $(U, \mathscr{M}, \mu)$ is called a measure space.

Proposition 7.3.2 *Let $\mu : \mathscr{M} \to [0, \infty]$, where $\mathscr{M}$ is a σ-algebra of U. Then we have the following.*

1. *$\mu(A_1 \cup \ldots \cup A_n) = \mu(A_1) + \ldots + \mu(A_n)$ for any given $\{A_i\}$ of pairwise disjoint measurable sets of $\mathscr{M}$.*

2. *If $A, B \in \mathscr{M}$ and $A \subset B$ then $\mu(A) \leq \mu(B)$.*

3. *If $\{A_n\} \subset \mathscr{M}, A = \cup_{n=1}^{\infty} A_n$ and*

$$A_1 \subset A_2 \subset A_3 \subset \ldots \tag{7.6}$$

then, $\lim_{n \to \infty} \mu(A_n) = \mu(A)$.

4. *If $\{A_n\} \subset \mathscr{M}, A = \cap_{n=1}^{\infty} A_n, A_1 \supset A_2 \supset A_3 \supset \ldots$ and $\mu(A_1)$ is finite then,*

$$\lim_{n \to \infty} \mu(A_n) = \mu(A). \tag{7.7}$$

Proof 7.5

1. Take $A_{n+1} = A_{n+2} = ... = \emptyset$ in Definition 7.1.1 item 1,

2. Observe that $B = A \cup (B-A)$ and $A \cap (B-A) = \emptyset$ so that by above $\mu(A \cup (B-A)) = \mu(A) + \mu(B-A) \geq \mu(A)$,

3. Let $B_1 = A_1$ and let $B_n = A_n - A_{n-1}$ then $B_n \in \mathcal{M}$, $B_i \cap B_j = \emptyset$ if $i \neq j$, $A_n = B_1 \cup ... \cup B_n$ and $A = \cup_{i=1}^{\infty} B_i$. Thus

$$\mu(A) = \mu(\cup_{i=1}^{\infty} B_i) = \sum_{n=1}^{\infty} \mu(B_i) = \lim_{n \to \infty} \sum_{i=1}^{n} \mu(B_i) = \lim_{n \to \infty} \mu(A_n). \tag{7.8}$$

4. Let $C_n = A_1 \setminus A_n$. Then $C_1 \subset C_2 \subset ...$, $\mu(C_n) = \mu(A_1) - \mu(A_n)$, $A_1 \setminus A = \cup_{n=1}^{\infty} C_n$.

   Thus, by 3 we have

$$\mu(A_1) - \mu(A) = \mu(A_1 \setminus A) = \lim_{n \to \infty} \mu(C_n) = \mu(A_1) - \lim_{n \to \infty} \mu(A_n). \tag{7.9}$$

7.4 Integration of simple functions

Definition 7.4.1 (Integral for simple functions) *For $s : U \to [0, \infty]$, a measurable simple function, that is,*

$$s = \sum_{i=1}^{n} \alpha_i \chi_{A_i}, \tag{7.10}$$

where

$$\chi_{A_i}(u) = \begin{cases} 1, & \text{if } u \in A_i, \\ 0, & \text{otherwise,} \end{cases} \tag{7.11}$$

we define the integral of s over $E \subset \mathcal{M}$, denoted by $\int_E s \, d\mu$ as

$$\int_E s \, d\mu = \sum_{i=1}^{n} \alpha_i \mu(A_i \cap E). \tag{7.12}$$

The convention $0.\infty = 0$ is used here.

Definition 7.4.2 (Integral for non-negative measurable functions) *If $f : U \to [0, \infty]$ is measurable, for $E \in \mathcal{M}$, we define the integral of f on E, denoted by $\int_E f d\mu$, as*

$$\int_E f d\mu = \sup_{s \in A} \left\{ \int_E s d\mu \right\}, \tag{7.13}$$

where

$$A = \{s \text{ simple and measurable} \mid 0 \leq s \leq f\}. \tag{7.14}$$

Definition 7.4.3 (Integrals for measurable functions) *For a measurable $f : U \to [-\infty, \infty]$ and $E \in \mathcal{M}$, we define $f^+ = \max\{f, 0\}$, $f^- = \max\{-f, 0\}$ and the integral of f on E, denoted by $\int_E f \, d\mu$, as*

$$\int_E f \, d\mu = \int_E f^+ \, d\mu - \int_E f^- \, d\mu.$$

Theorem 7.4.4 (Lebesgue's monotone convergence theorem) *Let $\{f_n\}$ be a sequence of real measurable functions on U and suppose that*

1. $0 \leq f_1(u) \leq f_2(u) \leq ... \leq \infty, \forall u \in U,$

2. $f_n(u) \to f(u)$ *as* $n \to \infty, \forall u \in U.$

Then,

(a) *f is measurable,*

(b) *$\int_U f_n d\mu \to \int_U f d\mu$ as $n \to \infty$.*

Proof 7.6   Since $\int_U f_n d\mu \leq \int_U f_{n+1} d\mu, \forall n \in \mathbb{N}$, there exists $\alpha \in [0, \infty]$ such that

$$\int_U f_n d\mu \to \alpha, \text{ as } n \to \infty, \tag{7.15}$$

By Proposition 7.1.8, f is measurable, and since $f_n \leq f$ we have

$$\int_U f_n d\mu \leq \int_U f \, d\mu. \tag{7.16}$$

From (7.15) and (7.16), we obtain

$$\alpha \leq \int_U f d\mu. \tag{7.17}$$

Let s be any simple function such that $0 \leq s \leq f$, and let $c \in \mathbb{R}$ such that $0 < c < 1$. For each $n \in \mathbb{N}$ we define

$$E_n = \{u \in U \mid f_n(u) \geq cs(u)\}. \tag{7.18}$$

Clearly E_n is measurable and $E_1 \subset E_2 \subset ...$ and $U = \cup_{n \in \mathbb{N}} E_n$. Observe that

$$\int_U f_n \, d\mu \geq \int_{E_n} f_n \, d\mu \geq c \int_{E_n} s d\mu. \tag{7.19}$$

Letting $n \to \infty$ and applying Proposition 7.3.2, we obtain

$$\alpha = \lim_{n \to \infty} \int_U f_n \, d\mu \geq c \int_U s d\mu, \tag{7.20}$$

so that

$$\alpha \geq \int_U s d\mu, \forall s \text{ simple and measurable such that } 0 \leq s \leq f. \tag{7.21}$$

This implies

$$\alpha \geq \int_U f d\mu. \tag{7.22}$$

From (7.17) and (7.22) the proof is complete.

The next result we do not prove it (it is a direct consequence of last theorem). For a proof see [68].

Corollary 7.4.5 *Let $\{f_n\}$ be a sequence of non-negative measurable functions defined on U ($f_n : U \to [0, \infty], \forall n \in \mathbb{N}$). Defining $f(u) = \sum_{n=1}^{\infty} f_n(u), \forall u \in U$, we have*

$$\int_U f \, d\mu = \sum_{n=1}^{\infty} \int_U f_n \, d\mu.$$

Theorem 7.4.6 (Fatou's lemma) *If* $\{f_n : U \to [0,\infty]\}$ *is a sequence of measurable functions, then*

$$\int_U \liminf_{n\to\infty} f_n \, d\mu \leq \liminf_{n\to\infty} \int_U f_n d\mu. \tag{7.23}$$

Proof 7.7 For each $k \in \mathbb{N}$ define $g_k : U \to [0,\infty]$ by

$$g_k(u) = \inf_{i \geq k} \{f_i(u)\}. \tag{7.24}$$

Then

$$g_k \leq f_k \tag{7.25}$$

so that

$$\int_U g_k \, d\mu \leq \int_U f_k \, d\mu, \forall k \in \mathbb{N}. \tag{7.26}$$

Also $0 \leq g_1 \leq g_2 \leq ...$, each g_k is measurable, and

$$\lim_{k\to\infty} g_k(u) = \liminf_{n\to\infty} f_n(u), \forall u \in U. \tag{7.27}$$

From the Lebesgue monotone convergence theorem

$$\liminf_{k\to\infty} \int_U g_k \, d\mu = \lim_{k\to\infty} \int_U g_k \, d\mu = \int_U \liminf_{n\to\infty} f_n \, d\mu. \tag{7.28}$$

From (7.26) we have

$$\liminf_{k\to\infty} \int_U g_k \, d\mu \leq \liminf_{k\to\infty} \{\int_U f_k \, d\mu\}. \tag{7.29}$$

Thus, from (7.28) and (7.29) we obtain

$$\int_U \liminf_{n\to\infty} f_n \, d\mu \leq \liminf_{n\to\infty} \int_U f_n \, d\mu. \tag{7.30}$$

Theorem 7.4.7 (Lebesgue's dominated convergence theorem) *Suppose* $\{f_n\}$ *is sequence of complex measurable functions on U such that*

$$\lim_{n\to\infty} f_n(u) = f(u), \forall u \in U. \tag{7.31}$$

If there exists a measurable function $g : U \to \mathbb{R}^+$ *such that* $\int_U g \, d\mu < \infty$ *and* $|f_n(u)| \leq g(u), \forall u \in U, n \in \mathbb{N}$, *then*

1. $\int_U |f| \, d\mu < \infty$,

2. $\lim_{n\to\infty} \int_U |f_n - f| \, d\mu = 0$.

Proof 7.8

1. This inequality holds since f is measurable and $|f| \leq g$.

2. Since $2g - |f_n - f| \geq 0$ we may apply the Fatou's lemma and obtain:

$$\int_U 2g d\mu \leq \liminf_{n \to \infty} \int_U (2g - |f_n - f|) d\mu, \tag{7.32}$$

so that

$$\limsup_{n \to \infty} \int_U |f_n - f| \, d\mu \leq 0. \tag{7.33}$$

Hence,

$$\lim_{n \to \infty} \int_U |f_n - f| \, d\mu = 0. \tag{7.34}$$

This completes the proof.

We finish this section with an important remark:

Remark 7.4.8 *In a measurable space U we say that a property holds almost everywhere (a.e.) if it holds on U except for a set of measure zero. Finally, since integrals are not changed by the redefinition of the functions in question on sets of zero measure, the proprieties of items (1) and (2) of the Lebesgue's monotone convergence may be considered a.e. in U, instead of in all U. Similar remarks are valid for the Fatou's lemma and the Lebesgue's dominated Convergence theorem.*

7.5 Signed measures

In this section we study signed measures. We start with the following definition:

Definition 7.5.1 *Let $(U, \mathcal{M})$ be a measurable space. We say that a measure μ is finite if $\mu(U) < \infty$. On the other hand, we say that μ is σ-finite if there exists a sequence $\{U_n\} \subset U$ such that $U = \cup_{n=1}^{\infty} U_n$ and $\mu(U_n) < \infty, \forall n \in \mathbb{N}$.*

Definition 7.5.2 (Signed measure) *Let $(U, \mathcal{M})$ be a measurable space. We say that $\nu : \mathcal{M} \to [-\infty, +\infty]$ is a signed measure if*

- *ν may assume at most one the values $-\infty, +\infty$.*

- *$\nu(\emptyset) = 0$.*

- *$\nu\left(\sum_{n=1}^{\infty} E_n\right) = \sum_{n=1}^{\infty} \nu(E_n)$ for all sequence of measurable disjoint sets $\{E_n\}$*

We say that $A \in \mathcal{M}$ is a positive set with respect to ν if A is measurable and $\nu(E) \geq 0$ for all E measurable such that $E \subset A$.

Similarly, We say that $B \in \mathcal{M}$ is a negative set with respect to ν if B is measurable and $\nu(E) \leq 0$ for all E measurable such that $E \subset B$.

Finally, if $A \in \mathcal{M}$ is both positive and negative with respect to ν, it is said to be a null set.

Lemma 7.5.3 *Considering the last definitions, we have that a countable union of positive measurable sets is positive.*

Proof 7.9 Let $A = \cup_{n=1}^{\infty} A_n$ where A_n is positive, $\forall n \in \mathbb{N}$. Choose a measurable set $E \subset A$. Set

$$E_n = (E \cap A_n) \setminus (\cup_{i=1}^{n-1} A_i).$$

Thus, E_n is a measurable subset of A_n so that $v(E_n) \geq 0$. Observe that

$$E = \cup_{n=1}^{\infty} E_n$$

where $\{E_n\}$ is a sequence of measurable disjoint sets.

Therefore $v(E) = \sum_{n=1}^{\infty} v(E_n) \geq 0$.

Since $E \subset A$ is arbitrary, A is positive.

The proof is complete.

Lemma 7.5.4 *Considering the last definitions, let E be a measurable set such that*

$$0 < v(E) < \infty.$$

Then there exists a positive set $A \subset E$ such that $v(A) > 0$.

Proof 7.10　Observe that if E is not positive then it contains a set of negative measure. In such a case, let n_1 be the smallest positive integer such that there exists a measurable set $E_1 \subset E$ such that

$$v(E_1) < -1/n_1.$$

Reasoning inductively, if $E \setminus \left(\cup_{j=1}^{k-1} E_j \right)$ is not positive, let n_k be the smallest positive integer such that there exists a measurable set

$$E_k \subset E \setminus \left(\cup_{j=1}^{k-1} E_j \right)$$

such that

$$v(E_k) < -1/n_k.$$

Define

$$A = E \setminus (\cup_{k=1}^{\infty} E_k).$$

Then

$$E = A \cup (\cup_{k=1}^{\infty} E_k).$$

Since such a union is disjoint, we have,

$$v(E) = v(A) + \sum_{k=1}^{\infty} v(E_k),$$

so that, since $v(E) < \infty$, this last series is convergent.

Also, since

$$1/n_k < -v(E_k),$$

we have that

$$\sum_{k=1}^{\infty} 1/n_k$$

is convergent so that $n_k \to \infty$ as $k \to \infty$.

From $v(E) > 0$ we must have $v(A) > 0$.

Now, we will show that A is positive. Let $\varepsilon > 0$. Choose k sufficiently big such that $1/(n_k - 1) < \varepsilon$.

Since

$$A \subset E \setminus \left(\cup_{j=1}^{k} E_j \right),$$

A contains no measurable set with measure less than

$$-1/(n_k - 1) > -\varepsilon,$$

that is, A contains no measurable set with measure less than $-\varepsilon$.

Since $\varepsilon > 0$ is arbitrary, A contains no measurable negative set. Thus, A is positive.

This completes the proof.

Proposition 7.5.5 (The Hahn decomposition) *Let v be a signed measure on a measurable space $(U, \mathcal{M})$. Then there exist a positive set A and a negative set B such that $U = A \cup B$ and $A \cap B = \emptyset$.*

Proof 7.11   Without loosing generality, suppose v does not assume the value $+\infty$ (the other case may be dealt similarly). Define

$$\lambda = \sup\{v(A) \mid A \text{ is positive}\}.$$

Since the empty set $\emptyset$ is positive, we obtain $\lambda \geq 0$.
  Let $\{A_n\}$ be a sequence of positive sets such that

$$\lim_{n \to \infty} v(A_n) = \lambda.$$

Define

$$A = \cup_{i=1}^{\infty} A_i.$$

  From Lemma 7.5.3, A is a positive set, so that

$$\lambda \geq v(A).$$

  On the other hand,

$$A \setminus A_n \subset A$$

so that

$$v(A - A_n) \geq 0, \forall n \in \mathbb{N}.$$

  Therefore

$$v(A) = v(A_n) + v(A \setminus A_n) \geq v(A_n), \forall n \in \mathbb{N}.$$

  Hence

$$v(A) \geq \lambda,$$

so that $\lambda = v(A)$.
  Let $B = U \setminus A$. Suppose $E \subset B$, so that E is positive. Hence,

$$
\begin{aligned}
\lambda &\geq v(E \cup A) \\
&= v(E) + v(A) \\
&= v(E) + \lambda,
\end{aligned}
\tag{7.35}
$$

so that $v(E) = 0$.
  Thus, B contains no positive set of positive measure, so that, by Lemma 7.5.4, B contains no subsets of positive measure, that is, B is negative.
  The proof is complete.

Remark 7.5.6 *Denoting the Hahn decomposition of U relating v by $\{A, B\}$, we may define the measures v^+ and v^- by*

$$v^+(E) = v(E \cap A),$$

and-

$$v^-(E) = -v(E \cap B),$$

so that

$$v = v^+ - v^-.$$

  We recall that two measures v_1 and v_2 are mutually singular if there are disjoint measurable sets such that

$$U = A \cup B$$

and

$$v_1(A) = v_2(B) = 0.$$

Observe that the measures v^+ and v^- above defined are mutually singular. The decomposition

$$v = v^+ - v^-$$

is called the Jordan one of v. The measures v^+ and v^- are called the positive and negative parts of v respectively.

Observe that either v^+ or v^- is finite since only one of the values $+\infty, -\infty$ may be assumed by v. We may also define

$$|v|(E) = v^+(E) + v^-(E),$$

which is called the absolute value or total variation of v.

7.6   The Radon-Nikodym theorem

We start this section with the definition of absolutely continuous measures.

Definition 7.6.1 (Absolutely continuous measures) *We say that a measure v is absolutely continuous with respect to a measure μ and write $v \ll \mu$, if $v(A) = 0$ for all set such that $\mu(A) = 0$. In case of a signed measure we write $v \ll \mu$ if $|v| \ll |\mu|$.*

Theorem 7.6.2 (The Radon-Nikodym theorem) *Let $(U, \mathcal{M}, \mu)$ be a σ-finite measure space. Let v be a measure defined on $\mathcal{M}$ which is absolutely continuous with respect to μ, that is, $v \ll \mu$.*

Then there exists a non-negative measurable function f such that

$$v(E) = \int_E f \, d\mu, \forall E \in \mathcal{M}.$$

The function f is unique up to usual representatives.

Proof 7.12   First assume v and μ are finite.

Define $\lambda = v + \mu$. Also define the functional F by

$$F(f) = \int_U f \, d\mu.$$

We recall that $f \in L^2(\mu)$ if f is measurable and

$$\int_U |f|^2 \, d\mu < \infty.$$

The space $L^2(\mu)$ is a Hilbert one with inner product

$$(f,g)_{L^2(\mu)} = \int_U fg \, d\mu.$$

Observe that, from the Cauchy-Schwartz inequality, we may write,

$$
\begin{aligned}
|F(f)| &= |(f,1)_{L^2(\mu)}| \\
&\leq \|f\|_{L^2(\mu)} [\mu(U)]^{1/2} \\
&\leq \|f\|_{L^2(\lambda)} [\mu(U)]^{1/2},
\end{aligned}
\tag{7.36}
$$

since

$$\|f\|_{L^2(\mu)}^2 = \int_U |f|^2 \, d\mu \le \int_U |f|^2 \, d\lambda = \|f\|_{L^2(\lambda)}^2.$$

Thus F is a bounded linear functional on $L^2(\lambda)$, where $f \in L^2(\lambda)$, if f is measurable and

$$\int_U |f|^2 \, d\lambda < \infty.$$

Since $L^2(\lambda)$ is also a Hilbert space with the inner product

$$(f,g)_{L^2(\lambda)} = \int_U fg \, d\lambda,$$

from the Riesz representation theorem, there exists $g \in L^2(\lambda)$, such that

$$F(f) = \int_U fg \, d\lambda.$$

Thus,

$$\int_U f \, d\mu = \int_U fg \, d\lambda,$$

and in particular,

$$\int_U f \, d\mu = \int_U fg \, (d\mu + d\nu).$$

Hence,

$$\int_U f(1-g) \, d\mu = \int_U fg \, d\nu. \tag{7.37}$$

Assume, to obtain contradiction, that $g < 0$ in a set A such that $\mu(A) > 0$.
Thus,

$$\int_A (1-g) \, d\mu > 0,$$

so that from this and (7.37) with $f = \chi_A$ we get

$$\int_A g \, d\nu > 0.$$

Since $g < 0$ on A we have a contradiction. Thus $g \ge 0$, a.e. $[\mu]$ on U.
Now, assume, also to obtain contradiction that $g > 1$ on set B such that $\mu(B) > 0$.
Thus

$$\int_B (1-g) \, d\mu \le 0,$$

so that from this and (7.37) with $f = \chi_B$ we obtain

$$\nu(B) \le \int_B g \, d\nu \le 0$$

and hence

$$\nu(B) = 0.$$

Thus, $\int_B g \, d\nu = 0$ so that

$$\int_B (1-g) \, d\mu = 0,$$

which implies that $\mu(B) = 0$, a contradiction.

From above we conclude that
$$0 \leq g \leq 1, \text{ a.e. } [\mu] \text{ in } U.$$

On the other hand, for a fixed E μ-measurable again from (7.37) with $f = \chi_E$ we get
$$\int_E (1-g)\, d\mu = \int_E g\, dv,$$

so that
$$\int_E (1-g)\, d\mu = \int_E g\, dv - \int_E dv + v(E),$$

and therefore
$$v(E) = \int_E (1-g)\, (d\mu + dv),$$

that is,
$$v(E) = \int_E (1-g)\, d\lambda, \forall E \in \mathcal{M}$$

Define
$$B = \{u \in U \; : \; g(u) = 0\}.$$

Hence, $\mu(B) = \int_B g\, d\lambda = 0$.
From this, since $\lambda \ll \mu$, we obtain
$$g^{-1}g = 1, \text{ a.e. } [\lambda].$$

Therefore, for a not relabeled $E \in \mathcal{M}$ we have
$$\lambda(E) = \int_E g^{-1}g\, d\lambda = \int_E g^{-1}\, d\mu.$$

Finally, observe that
$$
\begin{aligned}
\mu(E) + v(E) &= \lambda(E) \\
&= \int_E d\lambda \\
&= \int_E g^{-1}\, d\mu.
\end{aligned}
\tag{7.38}
$$

Thus,
$$
\begin{aligned}
v(E) &= \int_E g^{-1}\, d\mu - \mu(E) \\
&= \int_E (g^{-1} - 1)\, d\mu \\
&= \int_E (1-g)g^{-1}\, d\mu, \forall E \in \mathcal{M}.
\end{aligned}
\tag{7.39}
$$

The proof for the finite case in complete. The proof for σ-finite is developed in the next lines.

Since U is σ-finite, there exists a sequence $\{U_n\}$ such that $U = \cup_{n=1}^{\infty} U_n$, and $\mu(U_n) < \infty$ and $v(U_n) < \infty, \forall n \in \mathbb{N}$.

Define
$$F_n = U_n \setminus \left(\cup_{j=1}^{n-1} U_j \right),$$

thus $U = \cup_{n=1}^{\infty} F_n$ and $\{F_n\}$ is a sequence of disjoint sets, such that $\mu(F_n) < \infty$ and $v(F_n) < \infty, \forall n \in \mathbb{N}$.

Let $E \in \mathcal{M}$. For each $n \in \mathbb{N}$ from above we may obtain f_n such that
$$v(E \cap F_n) = \int_{E \cap F_n} f_n\, d\mu, \forall E \in \mathcal{M}.$$

From this and the monotone convergence theorem corollary we may write

$$
\begin{aligned}
v(E) &= \sum_{n=1}^{\infty} v(E \cap F_n) \\
&= \sum_{n=1}^{\infty} \int_{E \cap F_n} f_n \, d\mu \\
&= \sum_{n=1}^{\infty} \int_E f_n \chi_{F_n} \, d\mu \\
&= \int_E \sum_{n=1}^{\infty} f_n \chi_{F_n} \, d\mu \\
&= \int_E f \, d\mu,
\end{aligned}
\tag{7.40}
$$

where

$$
f = \sum_{n=1}^{\infty} f_n \chi_{F_n}.
$$

The proof is complete.

Theorem 7.6.3 (The Lebesgue decomposition) *Let $(U, \mathcal{M}, \mu)$ be a σ-finite measure space and let v be a σ-finite measure defined on $\mathcal{M}$.*

Then we may find a measure v_0, singular with respect to μ, and a measure v_1, absolutely continuous with respect to μ, such that

$$
v = v_0 + v_1.
$$

Furthermore, the measures v_0 and v_1 are unique.

Proof 7.13 Since μ and v are σ-finite measures, so is

$$
\lambda = v + \mu.
$$

Observe that v and μ are absolutely continuous with respect to λ. Hence, by the Radon-Nikodym theorem, there exist non-negative measurable functions f and g such that

$$
\mu(E) = \int_E f \, d\lambda, \ \forall E \in \mathcal{M}
$$

and

$$
v(E) = \int_E g \, d\lambda \ \forall E \in \mathcal{M}.
$$

Define

$$
A = \{u \in U \mid f(u) > 0\},
$$

and

$$
B = \{u \in U \mid f(u) = 0\}.
$$

Thus,

$$
U = A \cup B,
$$

and

$$
A \cap B = \emptyset.
$$

Also define

$$
v_0(E) = v(E \cap B), \forall E \in \mathcal{M}.
$$

We have that $\nu_0(A) = 0$ so that

$$\nu_0 \perp \mu.$$

Define

$$\nu_1(E) = \nu(E \cap A)$$
$$= \int_{E \cap A} g \, d\lambda. \tag{7.41}$$

Therefore,

$$\nu = \nu_0 + \nu_1.$$

To finish the proof, we have only to show that

$$\nu_1 \ll \mu.$$

Let $E \in \mathcal{M}$ such that $\mu(E) = 0$. Thus

$$0 = \mu(E) = \int_E f \, d\lambda,$$

and in particular

$$\int_{E \cap A} f \, d\lambda = 0.$$

Since $f > 0$ on $A \cap E$ we conclude that

$$\lambda(A \cap E) = 0.$$

Therefore, since $\nu \ll \lambda$, we obtain

$$\nu(E \cap A) = 0,$$

so that

$$\nu_1(E) = \nu(E \cap A) = 0.$$

From this we may infer that

$$\nu_1 \ll \mu.$$

The proof of uniqueness is left to the reader.

7.7 Outer measure and measurability

Let U be a set. Denote by $\mathscr{P}$ the set of all subsets of U. An outer measure $\mu^* : \mathscr{P} \to [0, +\infty]$ is a set function such that

1. $\mu^*(\emptyset) = 0$,

2. if $A \subset B$ then $\mu^*(A) \le \mu^*(B), \forall A, B \subset U$,

3. if $E \subset \cup_{n=1}^\infty E_n$ then

$$\mu^*(E) \le \sum_{n=1}^\infty \mu^*(E_n).$$

The outer measure is called finite if $\mu^*(U) < \infty$.

Definition 7.7.1 (measurable set) *A set $E \subset U$ is said to be measurable with respect to μ^* if*

$$\mu^*(A) = \mu^*(A \cap E) + \mu^*(A \cap E^c), \forall A \subset U.$$

Theorem 7.7.2 *The set $\mathcal{B}$ of μ^*-measurable sets is a σ-algebra. If $\overline{\mu}$ is defined to be μ^* restricted to $\mathcal{B}$, then $\overline{\mu}$ is a complete measure on $\mathcal{B}$.*

Proof 7.14   Let $E = \emptyset$ and let $A \subset U$.
Thus,

$$\mu^*(A) = \mu^*(A \cap \emptyset) + \mu^*(A \cap \emptyset^c) = \mu^*(A \cap U) = \mu^*(A).$$

Therefore $\emptyset$ is μ^*-measurable.
Let $E_1, E_2 \in U$ be μ^*-measurable sets. Let $A \subset U$. Thus,

$$\mu^*(A) = \mu^*(A \cap E_2) + \mu^*(A \cap E_2^c),$$

so that from the measurability of E_1 we get

$$\mu^*(A) = \mu^*(A \cap E_2) + \mu^*(A \cap E_2^c \cap E_1) + \mu^*(A \cap E_2^c \cap E_1^c). \tag{7.42}$$

Since,

$$A \cap (E_1 \cup E_2) = (A \cap E_2) \cup (A \cap E_1 \cap E_2^c),$$

we obtain

$$\mu^*(A \cap (E_1 \cup E_2)) \leq \mu^*(A \cap E_2) + \mu^*(A \cap E_2^c \cap E_1). \tag{7.43}$$

From this and (7.42) we obtain

$$\begin{aligned}
\mu^*(A) &\geq \mu^*(A \cap (E_1 \cup E_2)) + \mu^*(A \cap E_1^c \cap E_2^c) \\
&= \mu^*(A \cap (E_1 \cup E_2)) + \mu^*(A \cap (E_1 \cup E_2)^c). \tag{7.44}
\end{aligned}$$

Hence $E_1 \cup E_2$ is μ^*-measurable.
By induction, the union of a finite number of μ^*- measurable sets is μ^*-measurable.
Assume $E = \cup_{i=1}^{\infty} E_i$ where $\{E_i\}$ is a sequence of disjoint μ^*-measurable sets.
Define $G_n = \cup_{i=1}^{n} E_i$. Then G_n is μ^*-measurable and for a given $A \subset U$ we have

$$\begin{aligned}
\mu^*(A) &= \mu^*(A \cap G_n) + \mu^*(A \cap G_n^c) \\
&\geq \mu^*(A \cap G_n) + \mu^*(A \cap E^c), \tag{7.45}
\end{aligned}$$

since $E^c \subset G_n^c, \ \forall n \in \mathbb{N}$.
Observe that

$$G_n \cap E_n = E_n$$

and

$$G_n \cap E_n^c = G_{n-1}.$$

Thus, from the measurability of E_n we may get

$$\begin{aligned}
\mu^*(A \cap G_n) &= \mu^*(A \cap G_n \cap E_n) + \mu^*(A \cap G_n \cap E_n^c) \\
&= \mu^*(A \cap E_n) + \mu^*(A \cap G_{n-1}). \tag{7.46}
\end{aligned}$$

By induction we obtain

$$\mu^*(A \cap G_n) = \sum_{i=1}^{n} \mu^*(A \cap E_i),$$

so that

$$\mu^*(A) \geq \mu^*(A \cap E^c) + \sum_{i=1}^{n} \mu^*(A \cap E_i), \forall n \in \mathbb{N},$$

that is, considering that

$$A \cap E \subset \cup_{i=1}^{\infty}(A \cap E_i),$$

we get

$$
\begin{aligned}
\mu^*(A) &\geq \mu^*(A \cap E^c) + \sum_{i=1}^{\infty} \mu^*(A \cap E_i) \\
&\geq \mu^*(A \cap E^c) + \mu^*(A \cap E) \tag{7.47}
\end{aligned}
$$

Since $A \subset U$ is arbitrary we may conclude that $E = \cup_{i=1}^{\infty} E_i$ is μ^*-measurable. Therefore $\mathscr{B}$ is a σ-algebra. Finally, we prove that $\overline{\mu}$ is a measure.

Let $E_1, E_2 \subset U$ be two disjoint μ^*-measurable sets.

Thus

$$
\begin{aligned}
\overline{\mu}(E_1 \cup E_2) &= \mu^*(E_1 \cup E_2) \\
&= \mu^*((E_1 \cup E_2) \cap E_2) + \mu^*((E_1 \cup E_2) \cap E_2^c) \\
&= \mu^*(E_2) + \mu^*(E_1). \tag{7.48}
\end{aligned}
$$

By induction we obtain the finite additivity.

Also, if

$$E = \cup_{i=1}^{\infty} E_i$$

where $\{E_i\}$ is a sequence of disjoint measurable sets.

Thus,

$$\overline{\mu}(E) \geq \overline{\mu}(\cup_{i=1}^{n} E_i) = \sum_{i=1}^{n} \overline{\mu}(E_i), \ \forall n \in \mathbb{N}.$$

Therefore,

$$\overline{\mu}(E) \geq \sum_{i=1}^{\infty} \overline{\mu}(E_i).$$

Now observe that

$$\overline{\mu}(E) = \mu^*(\cup_{i=1}^{\infty} E_i) \leq \sum_{i=1}^{\infty} \mu^*(E_i) = \sum_{i=1}^{\infty} \overline{\mu}(E_i),$$

and thus

$$\overline{\mu}(E) = \sum_{i=1}^{\infty} \overline{\mu}(E_i).$$

The proof is complete.

Definition 7.7.3 *A measure on an Algebra $\mathscr{A} \subset U$ is a set function $\mu : \mathscr{A} \to [0, +\infty)$ such that*

1. $\mu(\emptyset) = 0$,

2. *if $\{E_i\}$ is a sequence of disjoint sets in $\mathscr{A}$ so that $E = \cup_{i=1}^{\infty} E_i \in \mathscr{A}$, then*

$$\mu(E) = \sum_{i=1}^{\infty} \mu(E_i).$$

We may define an outer measure in U, by

$$\mu^*(E) = \inf \left\{ \sum_{i=1}^{\infty} \mu(A_i) \mid E \subset \cup_{i=1}^{\infty} A_i \right\},$$

where $A_i \in \mathscr{A}$, $\forall i \in \mathbb{N}$.

Proposition 7.7.4 *Suppose $A \in \mathscr{A}$ and $\{A_i\} \subset \mathscr{A}$ is such that*

$$A \subset \cup_{i=1}^{\infty} A_i.$$

Under such hypotheses,

$$\mu(A) \leq \sum_{i=1}^{\infty} \mu(A_i).$$

Proof 7.15 Define

$$B_n = (A \cap A_n) \setminus (\cup_{i=1}^{n-1} A_i).$$

Thus,

$$B_n \subset A_n, \; \forall n \in \mathbb{N},$$

$B_n \in \mathscr{B}, \; \forall n \in \mathbb{N}$, and

$$A = \cup_{i=1}^{\infty} B_i.$$

Moreover, $\{B_n\}$ is a sequence of disjoint sets, so that

$$\mu(A) = \sum_{i=1}^{\infty} \mu(B_i) \leq \sum_{i=1}^{\infty} \mu(A_i).$$

Corollary 7.7.5 *If $A \in \mathscr{A}$ then $\mu^*(A) = \mu(A)$.*

Theorem 7.7.6 *The set function μ^* is an outer measure.*

Proof 7.16 The only not immediate property to be proven is the countably subadditivity.
Suppose $E \subset \cup_{i=1}^{\infty} E_i$. If $\mu^*(E_i) = +\infty$ for some $i \in \mathbb{N}$ the result holds.
Thus, assume $\mu^*(E_i) < +\infty, \; \forall i \in \mathbb{N}$.
Let $\varepsilon > 0$. Thus for each $i \in \mathbb{N}$ there exists $\{A_{ij}\} \subset \mathscr{A}$ such that $E_i \subset \cup_{j=1}^{\infty} A_{ij}$, and

$$\sum_{j=1}^{\infty} \mu(A_{ij}) \leq \mu^*(E_i) + \frac{\varepsilon}{2^i}.$$

Therefore,

$$\mu^*(E) \leq \sum_{i=1}^{\infty} \sum_{j=1}^{\infty} \mu(A_{ij}) \leq \sum_{i=1}^{\infty} \mu^*(E_i) + \varepsilon.$$

Since $\varepsilon > 0$ is arbitrary, we get

$$\mu^*(E) \leq \sum_{i=1}^{\infty} \mu^*(E_i).$$

Proposition 7.7.7 *Suppose $A \in \mathscr{A}$. Then A is μ^*-measurable.*

Proof 7.17 Let $E \in U$ such that $\mu^*(E) < +\infty$. Let $\varepsilon > 0$.
Thus, there exists $\{A_i\} \subset \mathscr{A}$ such that $E \subset \cup_{i=1}^{\infty} A_i$ and

$$\sum_{i=1}^{\infty} \mu(A_i) < \mu^*(E) + \varepsilon.$$

Observe that

$$\mu(A_i) = \mu(A_i \cap A) + \mu(A_i \cap A^c),$$

so that, from the fact that

$$E \cap A \subset \cup_{i=1}^{\infty} (A_i \cap A),$$

and

$$(E \cap A^c) \subset \cup_{i=1}^{\infty} (A_i \cap A^c),$$

we obtain

$$\mu^*(E) + \varepsilon > \sum_{i=1}^{\infty} (A_i \cap A) + \sum_{i=1}^{\infty} \mu(A_i \cap A^c)$$
$$\geq \mu^*(E \cap A) + \mu^*(E \cap A^c). \tag{7.49}$$

Since $\varepsilon > 0$ is arbitrary, we get

$$\mu^*(E) \geq \mu^*(E \cap A) + \mu^*(E \cap A^c).$$

The proof is complete.

Proposition 7.7.8 *Suppose μ is a measure on an Algebra $\mathscr{A} \subset U$, μ^* is the outer measure induced by μ and $E \subset U$ is a set. Then, for each $\varepsilon > 0$, there is a set $A \in \mathscr{A}_\sigma$ with $E \subset A$ and*

$$\mu^*(A) \leq \mu^*(E) + \varepsilon.$$

Also, there is a set $B \in \mathscr{A}_{\sigma\delta}$ such that $E \subset B$ and

$$\mu^*(E) = \mu^*(B).$$

Proof 7.18   Let $\varepsilon > 0$. Thus, there is a sequence $\{A_i\} \subset \mathscr{A}$ such that

$$E \subset \cup_{i=1}^{\infty} A_i$$

and

$$\sum_{i=1}^{\infty} \mu(A_i) \leq \mu^*(E) + \varepsilon.$$

Define $A = \cup_{i=1}^{\infty} A_i$, then

$$\mu^*(A) \leq \sum_{i=1}^{\infty} \mu^*(A_i)$$
$$= \sum_{i=1}^{\infty} \mu(A_i)$$
$$\leq \mu^*(E) + \varepsilon. \tag{7.50}$$

Now, observe that we write $A \in \mathscr{A}_\sigma$ if $A = \cup_{i=1}^{\infty} A_i$ where $A_i \in \mathscr{A}$, $\forall i \in \mathbb{N}$. Also, we write $B \in \mathscr{A}_{\sigma\delta}$ if $B = \cap_{n=1}^{\infty} A_n$, where $A_n \in \mathscr{A}_\sigma$, $\forall n \in \mathbb{N}$. From above, for each $n \in \mathbb{N}$ there exists $A_n \in \mathscr{A}_\sigma$ such that

$$E \subset A_n$$

and

$$\mu^*(A_n) \leq \mu(E) + 1/n.$$

Define $B = \cap_{n=1}^{\infty} A_n$. Thus $B \in \mathscr{A}_{\sigma\delta}$, $E \subset B$ and

$$\mu^*(B) \leq \mu^*(A_n) \leq \mu^*(E) + 1/n, \forall n \in \mathbb{N}.$$

Hence,

$$\mu^*(B) = \mu^*(E).$$

The proof is complete.

Proposition 7.7.9 *Suppose* μ *is a* σ-*finite measure on a* σ-*algebra* $\mathscr{A}$, *and let* μ^* *be the outer measure induced by* μ.

Under such hypotheses, a set E *is* μ^* *measurable if and only if* $E = A \setminus B$ *where* $A \in \mathscr{A}_{\sigma\delta}$, $B \subset A$, $\mu^*(B) = 0$.

Finally, for each set B *such that* $\mu^*(B) = 0$, *there exists* $C \in \mathscr{A}_{\sigma\delta}$ *such that* $B \subset C$ *and* $\mu^*(C) = 0$.

Proof 7.19 The if part is obvious.

Now suppose E is μ^*-measurable. Let $\{U_i\}$ be a countable collection of disjoint sets of finite measure such that

$$U = \cup_{i=1}^{\infty} U_i.$$

Observe that

$$E = \cup_{i=1}^{\infty} E_i$$

where

$$E_i = E \cap U_i,$$

is μ^*-measurable for each $i \in \mathbb{N}$.

Let $\varepsilon > 0$. From the last proposition for each $i, n \in \mathbb{N}$ there exists $A_{ni} \in \mathscr{A}_\sigma$ such that

$$\mu(A_{ni}) < \mu^*(E_i) + \frac{1}{n2^i}.$$

Define

$$A_n = \cup_{i=1}^{\infty} A_{ni}.$$

Thus,

$$E \subset A_n$$

and

$$A_n \setminus E \subset \cup_{i=1}^{\infty}(A_{ni} \setminus E_i),$$

and therefore,

$$\mu(A_n \setminus E) \leq \sum_{i=1}^{\infty} \mu(A_{ni} \setminus E_i) \leq \sum_{i=1}^{\infty} \frac{1}{n2^i} = \frac{1}{n}.$$

Since $A_n \in \mathscr{A}_\sigma$, defining

$$A = \cap_{n=1}^{\infty} A_n,$$

we have that $A \in \mathscr{A}_{\sigma\delta}$, and,

$$A \setminus E \subset A_n \setminus E$$

so that

$$\mu^*(A \setminus E) \leq \mu^*(A_n \setminus E) \leq \frac{1}{n}, \forall n \in \mathbb{N}.$$

Hence $\mu^*(A \setminus E) = 0$.

The proof is complete.

Theorem 7.7.10 (Carathéodory) *Let* μ *be a measure on a algebra* $\mathscr{A}$ *and* μ^* *the respective induced outer measure.*

Then the restriction $\overline{\mu}$ *of* μ^* *to the* μ^*-*measurable sets is an extension of* μ *to an* σ-*algebra containing* $\mathscr{A}$. *If* μ *is finite or* σ-*finite, so is* $\overline{\mu}$. *In particular, if* μ *is* σ-*finite, then* $\overline{\mu}$ *is the only measure on the smallest* σ-*algebra containing* $\mathscr{A}$ *which is an extension of* μ.

Proof 7.20   From the Theorem 7.7.2, $\overline{\mu}$ is an extension of μ to a σ-algebra containing $\mathscr{A}$, that is, $\overline{\mu}$ is a measure on such a set.

Observe that, from the last results, if μ is σ-finite so is $\overline{\mu}$.

Now assume μ is σ-finite. We will prove the uniqueness of $\overline{\mu}$.

Let $\mathscr{B}$ be the smallest σ-algebra containing $\mathscr{A}$ and let $\tilde{\mu}$ be another measure on $\mathscr{B}$ which extends μ on $\mathscr{A}$.

Since each set $\mathscr{A}_\sigma$ may be expressed as a disjoint countable union of sets in $\mathscr{A}$, the measure $\tilde{\mu}$ equals $\overline{\mu}$ on $\mathscr{A}_\sigma$. Let B be a μ^*-measurable set such that $\mu^*(B) < \infty$.

Let $\varepsilon > 0$. By the Proposition 7.7.9 there exists an $A \in \mathscr{A}_\sigma$ such that $B \subset A$ and

$$\mu^*(A) < \mu^*(B) + \varepsilon.$$

Since $B \subset A$, we obtain

$$\tilde{\mu}(B) \le \tilde{\mu}(A) = \mu^*(A) \le \mu^*(B) + \varepsilon.$$

Considering that $\varepsilon > 0$ is arbitrary, we get

$$\tilde{\mu}(B) \le \mu^*(B).$$

Observe that the class of μ^*-measurable sets is an σ-algebra containing $\mathscr{A}$.

Therefore as above indicated, we have obtained $A \in \mathscr{A}_\sigma$ such that $B \subset A$ and

$$\mu^*(A) \le \mu^*(B) + \varepsilon$$

so that

$$\mu^*(A) = \mu^*(B) + \mu^*(A \setminus B),$$

from this and above

$$\tilde{\mu}(A \setminus B) \le \mu^*(A \setminus B) \le \varepsilon,$$

if $\mu^*(B) < \infty$.

Therefore,

$$
\begin{aligned}
\mu^*(B) \;\le\;& \mu^*(A) = \tilde{\mu}(A) \\
=\;& \tilde{\mu}(B) + \tilde{\mu}(A \setminus B) \le \tilde{\mu}(B) + \varepsilon.
\end{aligned}
\tag{7.51}
$$

Since $\varepsilon > 0$ is arbitrary we have

$$\mu^*(B) \le \tilde{\mu}(B),$$

so that

$\mu^*(B) = \tilde{\mu}(B)$.

Finally, since μ is σ-finite, there exists a sequence of countable disjoint sets $\{U_i\}$ such that $\mu(U_i) < \infty$, $\forall i \in \mathbb{N}$ and $U = \cup_{i=1}^\infty U_i$.

If $B \in \mathscr{B}$, then

$$B = \cup_{i=1}^\infty (U_i \cap B).$$

Thus, from above

$$
\begin{aligned}
\tilde{\mu}(B) \;=\;& \sum_{i=1}^\infty \tilde{\mu}(U_i \cap B) \\
=\;& \sum_{i=1}^\infty \overline{\mu}(U_i \cap B) \\
=\;& \overline{\mu}(B).
\end{aligned}
\tag{7.52}
$$

The proof is complete.

Remark 7.7.11 *We may start the process of construction of a measure by the action of a set function on a semi-algebra. Here, a semi-algebra $\mathscr{C}$ is a collection of subsets of U such that the intersection of any two sets in $\mathscr{C}$ is in $\mathscr{C}$ and, the complement of any set in $\mathscr{C}$ is a finite disjoint union of sets in $\mathscr{C}$.*

If $\mathscr{C}$ is any semi-algebra of sets, then the collection consisting of the empty set and all finite disjoint unions of sets in $\mathscr{C}$ is an algebra, which is said to be generated by $\mathscr{C}$. We denote such algebra by $\mathscr{A}$.

If we have a set function acting on $\mathscr{C}$, we may extend it to $\mathscr{A}$ by defining

$$\mu(A) = \sum_{i=1}^{n} \mu(E_i),$$

where, $A = \cup_{i=1}^{n} E_i$ and $E_i \in \mathscr{C}$, $\forall i \in \{1,...,n\}$, so that this last union is disjoint. We recall that any $A \in \mathscr{A}$ admits such a representation.

7.8 Fubini's theorem

We start this section with the definition of complete measure space.

Definition 7.8.1 *We say that a measure space $(U, \mathscr{M}, \mu)$ is complete if $\mathscr{M}$ contains all subsets of sets of zero measure. That is, if $A \in \mathscr{M}$, $\mu(A) = 0$ and $B \subset A$ then $B \in \mathscr{M}$.*

In the next lines we recall the formal definition of a semi-algebra.

Definition 7.8.2 *We say (in fact recall) that $\mathscr{C} \subset U$ is a semi-algebra in U if the two conditions below are valid.*

1. if $A, B \in \mathscr{C}$ then $A \cap B \in \mathscr{C}$,

2. For each $A \in \mathscr{C}$, A^c is a finite disjoint union of elements in $\mathscr{C}$.

7.8.1 Product measures

Let $(U, \mathscr{M}_1, \mu_1)$ and $(V, \mathscr{M}_2, \mu_2)$ be two complete measure spaces. We recall that the cartesian product between U and V, denoted by $U \times V$ is defined by

$$U \times V = \{(u,v) \mid u \in U \text{ and } v \in V\}.$$

If $A \subset U$ and $B \subset V$ we call $A \times B$ a rectangle. If $A \in \mathscr{M}_1$ and $B \in \mathscr{M}_2$ we say that $A \times B$ is a measurable rectangle. The collection $\mathscr{R}$ of measurable rectangles is a semi-algebra since

$$(A \times B) \cap (C \times D) = (A \cap C) \times (B \cap D),$$

and

$$(A \times B)^c = (A^c \times B) \cup (A \times B^c) \cup (A^c \times B^c).$$

We define $\lambda : \mathscr{M}_1 \times \mathscr{M}_2 \to \mathbb{R}^+$ by

$$\lambda(A \times B) = \mu_1(A)\mu_2(B).$$

Lemma 7.8.3 *Let $\{A_i \times B_i\}_{i \in \mathbb{N}}$ be a countable disjoint collection of measurable rectangles whose the union is the rectangle $A \times B$. Then*

$$\lambda(A \times B) = \sum_{i=1}^{\infty} \mu_1(A_i)\mu_2(B_i).$$

Proof 7.21   Let $u \in A$. Thus each $v \in B$ is such that (u, v) is exactly in one $A_i \times B_i$. Therefore

$$\chi_{A \times B}(u, v) = \sum_{i=1}^{\infty} \chi_{A_i}(u) \chi_{B_i}(v).$$

Hence for the fixed u in question, from the corollary of Lebesgue monotone convergence theorem we may write

$$\int_V \chi_{A \times B}(u, v) d\mu_2(v) = \int_V \sum_{i=1}^{\infty} \chi_{A_i}(u) \chi_{B_i}(v) d\mu_2(v)$$

$$= \sum_{i=1}^{\infty} \chi_{A_i}(u) \mu_2(B_i) \tag{7.53}$$

so that also from the mentioned corollary

$$\int_U d\mu_1(u) \int_V \chi_{A \times B}(u, v) d\mu_2(v) = \sum_{i=1}^{\infty} \mu_1(A_i) \mu_2(B_i).$$

Observe that

$$\int_U d\mu_1(u) \int_V \chi_{A \times B}(u, v) d\mu_2(v) = \int_U d\mu_1(u) \int_V \chi_A(u) \chi_B(v) d\mu_2(v)$$

$$= \mu_1(A) \mu_2(B).$$

From the last two equations we may write

$$\lambda(A \times B) = \mu_1(A) \mu_2(B) = \sum_{i=1}^{\infty} \mu_1(A_i) \mu_2(B_i).$$

Definition 7.8.4 *Let $E \subset U \times V$. We define E_u and E_v by*

$$E_u = \{v \mid (u, v) \in E\},$$

and

$$E_v = \{u \mid (u, v) \in E\}.$$

Observe that

$$\chi_{E_u}(v) = \chi_E(u, v),$$
$$(E^c)_u = (E_u)^c,$$

and

$$(\cup E_\alpha)_u = \cup (E_\alpha)_u,$$

for any collection $\{E_\alpha\}$.

   We denote by $\mathscr{R}_\sigma$ as the collection of sets which are countable unions of measurable rectangles. Also, $\mathscr{R}_{\sigma\delta}$ will denote the collection of sets which are countable intersections of elements of $\mathscr{R}_\sigma$.

Lemma 7.8.5 *Let $u \in U$ and $E \in \mathscr{R}_{\sigma\delta}$. Then E_u is a measurable subset of V.*

Proof 7.22   If $E \in \mathscr{R}$ the result is trivial. Let $E \in \mathscr{R}_\sigma$. Then E may be expressed as a disjoint union

$$E = \cup_{i=1}^{\infty} E_i,$$

where $E_i \in \mathcal{R}, \forall i \in \mathbb{N}$. Thus,

$$
\begin{aligned}
\chi_{E_u}(v) &= \chi_E(u,v) \\
&= \sup_{i \in \mathbb{N}} \chi_{E_i}(u,v) \\
&= \sup_{i \in \mathbb{N}} \chi_{(E_i)_u}(v).
\end{aligned}
\tag{7.54}
$$

Since each $(E_i)_u$ is measurable we have that

$$\chi_{(E_i)_u}(v)$$

is a measurable function of v, so that

$$\chi_{E_u}(v)$$

is measurable, which implies that E_u is measurable. Suppose now

$$E = \cap_{i=1}^{\infty} E_i,$$

where $E_{i+1} \subset E_i, \forall i \in \mathbb{N}$. Then

$$
\begin{aligned}
\chi_{E_u}(v) &= \chi_E(u,v) \\
&= \inf_{i \in \mathbb{N}} \chi_{E_i}(u,v) \\
&= \inf_{i \in \mathbb{N}} \chi_{(E_i)_u}(v).
\end{aligned}
\tag{7.55}
$$

Thus as from above $\chi_{(E_i)_u}(v)$ is measurable for each $i \in \mathbb{N}$, we have that χ_{E_u} is also measurable so that E_u is measurable.

Lemma 7.8.6 *Let E be a set in $\mathcal{R}_{\sigma\delta}$ with $(\mu_1 \times \mu_2)(E) < \infty$. Then the function g defined by*

$$g(u) = \mu_2(E_u)$$

is a measurable function and

$$\int_U g \, d\mu_1(u) = (\mu_1 \times \mu_2)(E).$$

Proof 7.23   The lemma is true if E is a measurable rectangle. Let $\{E_i\}$ be a disjoint sequence of measurable rectangles and $E = \cup_{i=1}^{\infty} E_i$. Set

$$g_i(u) = \mu_2((E_i)_u).$$

Then each g_i is a non-negative measurable function and

$$g = \sum_{i=1}^{\infty} g_i.$$

Thus g is measurable, and by the corollary of the Lebesgue monotone convergence theorem, we have

$$
\begin{aligned}
\int_U g(u) d\mu_1(u) &= \sum_{i=1}^{\infty} \int_U g_i(u) d\mu_1(u) \\
&= \sum_{i=1}^{\infty} (\mu_1 \times \mu_2)(E_i) \\
&= (\mu_1 \times \mu_2)(E).
\end{aligned}
\tag{7.56}
$$

Let E be a set of finite measure in $\mathscr{R}_{\sigma\delta}$. Then there is a sequence in $\mathscr{R}_{\sigma}$ such that

$$E_{i+1} \subset E_i$$

and

$$E = \cap_{i=1}^{\infty} E_i.$$

Let $g_i(u) = \mu_2((E_i)_u)$, since

$$\int_U g_1(u) = (\mu_1 \times \mu_2)(E_1) < \infty,$$

we have that

$$g_1(u) < \infty \text{ a.e. in } E_1.$$

For an $u \in E_1$ such that $g_1(u) < \infty$ we have that $\{(E_i)_u\}$ is a sequence of measurable sets of finite measure whose intersection is E_u. Thus

$$g(u) = \mu_2(E_u) = \lim_{i \to \infty} \mu_2((E_i)_u) = \lim_{i \to \infty} g_i(u), \tag{7.57}$$

that is,

$$g_i \to g, \text{ a.e. in } E.$$

We may conclude that g is also measurable. Since

$$0 \le g_i \le g, \forall i \in \mathbb{N}$$

the Lebesgue dominated convergence theorem implies that

$$\begin{aligned}
\int_U g(u)\, d\mu_1(u) &= \lim_{i \to \infty} \int_U g_i\, d\mu_1(u) \\
&= \lim_{i \to \infty} (\mu_1 \times \mu_2)(E_i) \\
&= (\mu_1 \times \mu_2)(E).
\end{aligned}$$

Lemma 7.8.7 *Let E be a set such that $(\mu_1 \times \mu_2)(E) = 0$. then for almost all $u \in U$ we have*

$$\mu_2(E_u) = 0.$$

Proof 7.24   Observe that there is a set in $\mathscr{R}_{\sigma\delta}$ such that $E \subset F$ and

$$(\mu_1 \times \mu_2)(F) = 0.$$

From the last lemma

$$\mu_2(F_u) = 0$$

fora almost all u. From $E_u \subset F_u$ we obtain

$$\mu_2(E_u) = 0$$

for almost all u, since μ_2 is complete.

Proposition 7.8.8 *Let E be a measurable subset of $U \times V$ such that $(\mu_1 \times \mu_2)(E)$ is finite. The for almost all u the set E_u is a measurable subset of V. The function g defined by*

$$g(u) = \mu_2(E_u)$$

is measurable and

$$\int g\, d\mu_1(u) = (\mu_1 \times \mu_2)(E).$$

Proof 7.25 First observe that there is a set $F \in \mathscr{R}_{\sigma\delta}$ such that $E \subset F$ and

$$(\mu_1 \times \mu_2)(F) = (\mu_1 \times \mu_2)(E).$$

Let $G = F \setminus E$. Since F and E are measurable, G is measurable, and

$$(\mu_1 \times \mu_2)(G) = 0.$$

By the last lemma we obtain

$$\mu_2(G_u) = 0,$$

for almost all u so that

$$g(u) = \mu_2(E_u) = \mu_2(F_u), \text{ a.e. in } U.$$

By Lemma 7.8.6 we may conclude that g is measurable and

$$\int_U g \, d\mu_1(u) = (\mu_1 \times \mu_2)(E).$$

Theorem 7.8.9 (Fubini) *Let $(U, \mathscr{M}_1, \mu_1)$ and $(V, \mathscr{M}_2, \mu_2)$ be two complete measure spaces and f an integrable function on $U \times V$. Then*

1. *$f_u(v) = f(u,v)$ is measurable and integrable for almost all u.*

2. *$f_v(u) = f(u,v)$ is measurable and integrable for almost all v.*

3. *$h_1(u) = \int_V f(u,v) \, d\mu_2(v)$ is integrable on U.*

4. *$h_2(v) = \int_U f(u,v) \, d\mu_1(u)$ is integrable on V.*

5.

$$
\int_U \left[\int_V f \, d\mu_2(v) \right] d\mu_1(u) = \int_V \left[\int_U f \, d\mu_1(u) \right] d\mu_2(v)
$$
$$
= \int_{U \times V} f d(\mu_1 \times \mu_2). \tag{7.58}
$$

Proof 7.26 It suffices to consider the case where f is non-negative (we can then apply the result to $f^+ = \max(f, 0)$ and $f^- = \max(-f, 0)$). The last proposition asserts that the theorem is true if f is a simple function which vanishes outside a set of finite measure. Similarly as in in Theorem 7.2.2, we may obtain a sequence of non-negative simple functions $\{\phi_n\}$ such that

$$\phi_n \uparrow f.$$

Observe that, given $u \in U$, f_u is such that

$$(\phi_n)_u \uparrow f_u, \text{ a.e. } .$$

By the Lebesgue monotone convergence theorem we get

$$\int_V f(u,v) \, d\mu_2(v) = \lim_{n \to \infty} \int_V \phi_n(u,v) \, d\mu_2(v),$$

so that this last resulting function is integrable in U. Again by the Lebesgue monotone convergence theorem, we obtain

$$
\int_U \left[\int_V f \, d\mu_2(v) \right] d\mu_1(u) = \lim_{n \to \infty} \int_U \left[\int_V \phi_n \, d\mu_2(v) \right] d\mu_1(u)
$$
$$
= \lim_{n \to \infty} \int_{U \times V} \phi_n \, d(\mu_1 \times \mu_2)
$$
$$
= \int_{U \times V} f \, d(\mu_1 \times \mu_2). \tag{7.59}
$$

Chapter 8

The Lebesgue Measure in $\mathbb{R}^n$

8.1 Introduction

In this chapter we will define the Lebesgue measure and the concept of Lebesgue measurable set. We show that the set of Lebesgue measurable sets is a $\sigma-$ algebra so that the earlier results, proven for more general measure spaces, remain valid in the present context (such as the Lebesgue monotone and dominated convergence theorems).

The main references for this chapter are [72, 69].

We start with the following theorems without proofs.

Theorem 8.1.1 *Every open set $A \subset \mathbb{R}$ may be expressed as a countable union of disjoint open intervals.*

Remark 8.1.2 *In this text Q_j denotes a closed cube in $\mathbb{R}^n$ and $|Q_j|$ its volume, that is, $|Q_j| = \prod_{i=1}^{n}(b_i - a_i)$, where $Q_j = \prod_{i=1}^{n}[a_i, b_i]$. Also we assume that if two Q_1 and Q_2 closed or not, have the same interior, then $|Q_1| = |Q_2| = |\bar{Q}_1|$. We recall that two cubes $Q_1, Q_2 \subset \mathbb{R}^n$ are said to be quasi-disjoint if their interiors are disjoint.*

Theorem 8.1.3 *Every open set $A \subset \mathbb{R}^n$, where $n \geq 1$ may be expressed as a countable union of quasi-disjoint closed cubes.*

Definition 8.1.4 (Outer measure) *Let $E \subset \mathbb{R}^n$. The outer measure of E, denoted by $m^*(E)$, is defined by*

$$m^*(E) = \inf \left\{ \sum_{j=1}^{\infty} |Q_j| \ : \ E \subset \cup_{j=1}^{\infty} Q_j \right\},$$

where Q_j is a closed cube, $\forall j \in \mathbb{N}$.

8.2 Properties of the outer measure

First observe that given $\varepsilon > 0$, there exists a sequence $\{Q_j\}$ such that

$$E \subset \cup_{j=1}^{\infty} Q_j$$

and

$$\sum_{j=1}^{\infty} |Q_j| \leq m^*(E) + \varepsilon.$$

1. Monotonicity: If $E_1 \subset E_2$ then $m^*(E_1) \leq m^*(E_2)$. This follows from the fact that if $E_2 \subset \cup_{j=1}^\infty Q_j$ then $E_1 \subset \cup_{j=1}^\infty Q_j$.

2. Countable sub-additivity : If $E \subset \cup_{j=1}^\infty E_j$, then $m^*(E) \leq \sum_{j=1}^\infty m^*(E_j)$.

Proof 8.1   First assume that $m^*(E_j) < \infty, \forall j \in \mathbb{N}$, otherwise the result is obvious. Thus, given $\varepsilon > 0$ for each $j \in \mathbb{N}$ there exists a sequence $\{Q_{k,j}\}_{k \in \mathbb{N}}$ such that

$$E_j \subset \cup_{k=1}^\infty Q_{k,j}$$

and

$$\sum_{k=1}^\infty |Q_{k,j}| < m^*(E_j) + \frac{\varepsilon}{2^j}.$$

Hence

$$E \subset \cup_{j,k=1}^\infty Q_{k,j}$$

and therefore

$$
\begin{aligned}
m^*(E) &\leq \sum_{j,k=1}^\infty |Q_{k,j}| = \sum_{j=1}^\infty \left(\sum_{k=1}^\infty |Q_{k,j}| \right) \\
&\leq \sum_{j=1}^\infty \left(m^*(E_j) + \frac{\varepsilon}{2^j} \right) \\
&= \sum_{j=1}^\infty m^*(E_j) + \varepsilon.
\end{aligned}
\tag{8.1}
$$

Being $\varepsilon > 0$ arbitrary, we obtain

$$m^*(E) \leq \sum_{j=1}^\infty m^*(E_j).$$

3. If

$$E \subset \mathbb{R}^n,$$

and

$$\alpha = \inf\{m^*(A) \,|\, A \text{ is open and } E \subset A\},$$

then

$$m^*(E) = \alpha.$$

Proof 8.2   From the monotonicity, we have

$$m^*(E) \leq m^*(A), \forall A \supset E, A \text{ open .}$$

Thus,

$$m^*(E) \leq \alpha.$$

Suppose given $\varepsilon > 0$. Choose a sequence $\{Q_j\}$ of closed cubes such that

$$E \subset \cup_{j=1}^\infty Q_j$$

and

$$\sum_{j=1}^{\infty} |Q_j| \leq m^*(E) + \varepsilon.$$

Let $\{\tilde{Q}_j\}$ be a sequence of open cubes such that $\tilde{Q}_j \supset Q_j$

$$|\tilde{Q}_j| \leq |Q_j| + \frac{\varepsilon}{2^j}, \forall j \in \mathbb{N}.$$

Define

$$A = \cup_{j=1}^{\infty} \tilde{Q}_j,$$

hence A is open, $A \supset E$ and

$$
\begin{aligned}
m^*(A) \quad &\leq \quad \sum_{j=1}^{\infty} |\tilde{Q}_j| \\
&\leq \quad \sum_{j=1}^{\infty} \left(|Q_j| + \frac{\varepsilon}{2^j} \right) \\
&= \quad \sum_{j=1}^{\infty} |Q_j| + \varepsilon \\
&\leq \quad m^*(E) + 2\varepsilon.
\end{aligned}
\tag{8.2}
$$

therefore

$$\alpha \leq m^*(E) + 2\varepsilon.$$

Being $\varepsilon > 0$ arbitrary, we have

$$\alpha \leq m^*(E).$$

The proof is complete.

4. If $E = E_1 \cup E_2$ and $d(E_1, E_2) > 0$, then

$$m^*(E) = m^*(E_1) + m^*(E_2).$$

Proof 8.3 First observe that being $E = E_1 \cup E_2$ we have

$$m^*(E) \leq m^*(E_1) + m^*(E_2).$$

Let $\varepsilon > 0$. Choose $\{Q_j\}$ a sequence of closed cubes such that

$$E \subset \cup_{j=1}^{\infty} Q_j,$$

and

$$\sum_{j=1}^{\infty} |Q_j| \leq m^*(E) + \varepsilon.$$

Let $\delta > 0$ such that

$$d(E_1, E_2) > \delta > 0.$$

Dividing the cubes Q_j if necessary, we may assume that the diameter of each cube Q_j is smaller than δ. Thus each Q_j intersects just one of the sets E_1 and E_2. Denote by J_1 and J_2 the sets of indices j such that Q_j intersects E_1 and E_2 respectively. thus,

$$E_1 \subset \cup_{j \in J_1} Q_j \text{ and } E_2 \subset \cup_{j \in J_2} Q_j.$$

hence,

$$
\begin{aligned}
m^*(E_1) + m^*(E_2) &\leq \sum_{j \in J_1} |Q_j| + \sum_{j \in J_2} |Q_j| \\
&\leq \sum_{j=1}^{\infty} |Q_j| \leq m^*(E) + \varepsilon.
\end{aligned}
\tag{8.3}
$$

Being $\varepsilon > 0$ arbitrary,

$$
m^*(E_1) + m^*(E_2) \leq m^*(E).
$$

This completes the proof.

5. If a set E is a countable union of cubes quasi disjoints, that is

$$
E = \cup_{j=1}^{\infty} Q_j
$$

then

$$
m^*(E) = \sum_{j=1}^{\infty} |Q_j|.
$$

Proof 8.4   Let $\varepsilon > 0$.

Let $\{\tilde{Q}_j\}$ be open cubes such that $\tilde{Q}_j \subset\subset Q_j^{\circ}$ (that is, the closure of $\tilde{Q}_j$ is contained in the interior of Q_j) and

$$
|Q_j| \leq |\tilde{Q}_j| + \frac{\varepsilon}{2^j}.
$$

Thus, for each $N \in \mathbb{N}$, the cubes $\tilde{Q}_1, ..., \tilde{Q}_N$ are disjoint and each pair have a finite distance. Hence,

$$
m^*(\cup_{j=1}^{N} \tilde{Q}_j) = \sum_{j=1}^{N} |\tilde{Q}_j| \geq \sum_{j=1}^{N} \left(|Q_j| - \frac{\varepsilon}{2^j} \right).
$$

Being

$$
\cup_{j=1}^{N} \tilde{Q}_j \subset E
$$

we obtain

$$
m^*(E) \geq \sum_{j=1}^{N} |\tilde{Q}_j| \geq \sum_{j=1}^{N} |Q_j| - \varepsilon.
$$

Therefore,

$$
\sum_{j=1}^{\infty} |Q_j| \leq m^*(E) + \varepsilon.
$$

Being $\varepsilon > 0$ arbitrary, we may conclude that

$$
\sum_{j=1}^{\infty} |Q_j| \leq m^*(E).
$$

The proof is complete.

8.3   The Lebesgue measure

Definition 8.3.1 *Let* $E \subset \mathbb{R}^n$. *We say that* E *is measurable if for each* $A \subset \mathbb{R}^n$ *we have*

$$m^*(A) = m^*(A \cap E) + m^*(A \cap E^c).$$

In such a case we define the Lebesgue measure of E, *denoted by* $m(E)$, *by*

$$m(E) = m^*(E).$$

Proposition 8.3.2 *Let* $E \subset \mathbb{R}^n$ *be such that* $m^*(E) = 0$. *Under such hypotheses,* E *is measurable.*

Proof 8.5   Let $A \subset \mathbb{R}^n$. Observe that $A \cap E \subset E$ so that

$$m^*(A \cap E) = 0.$$

From this, we obtain

$$
\begin{aligned}
m^*(A) &\leq m^*(A \cap E) + m^*(A \cap E^c) \\
&= m^*(A \cap E^c) \\
&\leq m^*(A).
\end{aligned}
\tag{8.4}
$$

Thus,

$$m^*(A) = m^*(A \cap E) + m^*(A \cap E^c).$$

Since A is arbitrary, we may infer that E is measurable.

Theorem 8.3.3 *Let* $E = \cup_{k=1}^n E_k$ *where* $E_k \subset \mathbb{R}^n$ *is measurable,* $\forall k \in \{1, \ldots, n\}$. *Under such hypotheses,* E *is measurable.*

Proof 8.6   First we prove that $E_1 \cup E_2$ is measurable.
Let $A \subset \mathbb{R}^n$. Since E_1 and E_2 are measurable, we have

$$
\begin{aligned}
m^*(A) &= m^*(A \cap E_1) + m^*(A \cap E_1^c) \\
&= m^*(A \cap E_1) + m^*((A \cap E_1^c) \cap E_2) + m^*((A \cap E_1^c) \cap E_2^c).
\end{aligned}
\tag{8.5}
$$

Now observe that

$$(A \cap E_1^c) \cap E_2^c = A \cap (E_1^c \cap E_2^c) = A \cap [E_1 \cup E_2]^c.$$

Moreover,

$$
\begin{aligned}
(A \cap E_1) \cup [A \cap E_1^c \cap E_2] &= (A \cap E_1) \cup (A \cap E_2) \cap E_1^c \\
&= A \cap (E_1 \cup E_2) \cap [A \cap (E_1 \cup E_1^c)] \\
&= A \cap [E_1 \cup E_2] \cap A \\
&= A \cap (E_1 \cup E_2).
\end{aligned}
\tag{8.6}
$$

From these last results, we obtain,

$$
\begin{aligned}
m^*(A) &= m^*(A \cap E_1) + m^*((A \cap E_1^c) \cap E_2) + m^*((A \cap E_1^c) \cap E_2^c) \\
&= m^*(A \cap E_1) + m^*((A \cap E_1^c) \cap E_2) + m^*(A \cap (E_1 \cup E_2)^c) \\
&\geq m^*(A \cap (E_1 \cup E_2)) + m^*(A \cap (E_1 \cup E_2)^c).
\end{aligned}
\tag{8.7}
$$

Since A is arbitrary, from this we may infer that $E_1 \cup E_2$ is measurable.

To complete the proof, we reason by induction. Let $k \in \{2, \ldots n-1\}$.

Assume $\cup_{j=1}^{k} E_j$ is measurable,

Observe that $\cup_{j=1}^{k+1} E_j = (\cup_{j=1}^{k} E_j)) \cup E_k$, which from the last result is measurable. In particular,

$$\cup_{j=1}^{n} E_j$$

is measurable.

The proof is complete.

Theorem 8.3.4 *Let $A \subset \mathbb{R}^n$ and let $\{E_k\}_{k=1}^{n}$ be finite collection of measurable disjoint sets. Under such hypotheses,*

$$m^*(A \cap [\cup_{k=1}^{n} E_k]) = \sum_{k=1}^{n} m^*(A \cap E_k).$$

In particular,

$$m^*(\cup_{k=1}^{n} E_k) = \sum_{k=1}^{n} m^*(E_k).$$

Proof 8.7   We prove the result by induction on n. The result is immediate for $n = 1$.

Assume the result holds for $n - 1$.

Since the collection $\{E_k\}_{k=1}^{n}$ is disjoint, we have

$$A \cap [\cup_{k=1}^{n} E_k] \cap E_n = A \cap E_n,$$

and

$$A \cap [\cup_{k=1}^{n} E_k] \cap E_n^c = A \cap [\cup_{k=1}^{n-1} E_k].$$

Thus, from the measurability of E_n and the induction assumption, we get

$$
\begin{aligned}
m^*(A \cap [\cup_{k=1}^{n} E_k]) &= m^*(A \cap [\cup_{k=1}^{n} E_k] \cap E_n) + m^*(A \cap [\cup_{k=1}^{n} E_k] \cap E_n^c) \\
&= m^*(A \cap E_n) + m^*(A \cap [\cup_{k=1}^{n-1} E_k]) \\
&= m^*(A \cap E_n) + \sum_{k=1}^{n-1} m^*(A \cap E_k) \\
&= \sum_{k=1}^{n} m^*(A \cap E_k).
\end{aligned}
$$

(8.8

The proof is complete.

8.4   Outer and inner approximations of Lebesgue measurable sets

We start this section with the following remark:

Remark 8.4.1 *Let $A \subset B \subset \mathbb{R}^n$ be sets such that A is a measurable set which has a finite outer measure. In such a case*

$$
\begin{aligned}
m^*(B) &= m^*(B \cap A) + m^*(B \cap A^c) \\
&= m^*(A) + m^*(B \setminus A),
\end{aligned}
$$

(8.9

so that

$$m^*(B \setminus A) = m^*(B) - m^*(A).$$

Theorem 8.4.2 *Let* $E \subset \mathbb{R}^n$. *Under such hypothesis, the following properties are equivalent to the measurability of* E:

1. *For each* $\varepsilon > 0$ *there exists an open set* $\mathscr{O} \subset \mathbb{R}^n$ *such that*

$$E \subset \mathscr{O}$$

   and

$$m^*(\mathscr{O} \setminus E) < \varepsilon.$$

2. *There is a* G_δ *set G such that*

$$E \subset G$$

   and $m^*(G \setminus E) = 0$, *where a* G_δ *set is one which may be expressed as a countable intersection of open sets.*

3. *For each* $\varepsilon > 0$ *there exists a closed set* $F \subset \mathbb{R}^n$ *such that*

$$F \subset E$$

   and

$$m^*(E \setminus F) < \varepsilon.$$

4. *There is a* F_σ *set G such that*

$$F \subset E$$

   and $m^*(E \setminus F) = 0$, *where a* F_σ *set is one which may be expressed as a countable union of closed sets.*

Proof 8.8 Assume E is measurable. Let $\varepsilon > 0$. Assume first $m^*(E) < \infty$. Hence, there exists a sequence of open blocks $\{Q_j\} \subset \mathbb{R}^n$ such that

$$E \subset \cup_{j=1}^{\infty} Q_j$$

and

$$\sum_{j=1}^{\infty} m^*(Q_j) < m^*(E) + \varepsilon.$$

Define $\mathscr{O} = \cup_{j=1}^{\infty} Q_j$. Thus,

$$E \subset \mathscr{O}$$

and

$$m^*(\mathscr{O}) \leq \sum_{j=1}^{\infty} m^*(Q_j) < m^*(E) + \varepsilon.$$

Thus

$$m^*(\mathscr{O}) - m^*(E) < \varepsilon.$$

From the measurability of E we get,

$$\begin{aligned} m^*(\mathscr{O}) &= m^*(\mathscr{O} \cap E) + m^*(\mathscr{O} \cap E^c) \\ &= m^*(\mathscr{O} \cap E) + m^*(\mathscr{O} \setminus E), \end{aligned} \tag{8.10}$$

so that

$$m(\mathscr{O} \setminus E) = m^*(\mathscr{O}) - m^*(E) < \varepsilon.$$

Assume now $m^*(E) = \infty$. Let $F_k = E \cap B_k(\mathbf{0})$, $\forall k \in \mathbb{R}^n$, where $B_k(\mathbf{0})$ denotes the open ball of center $\mathbf{0}$ and radius k.

Define also, $E_1 = F_1$ and

$$E_k = F_k \setminus \cup_{j=1}^{k-1} F_k, \ \forall k \geq 2.$$

Hence, $E = \cup_{k=1}^{\infty} E_k$ and such union is disjoint.

Observe that, from the last previous lines, for each $k \in \mathbb{N}$ there exists an open set $\mathscr{O}_k$ such that $E_k \subset \mathscr{O}_k$ and

$$m^*(\mathscr{O}_k \setminus E_k) < \frac{\varepsilon}{2^k}.$$

Define now $\mathscr{O} = \cup_{k=1}^{\infty} \mathscr{O}_k$. Thus such a set is open and

$$\mathscr{O} \setminus E = (\cup_{k=1}^{\infty} \mathscr{O}_k) \setminus E = \cup_{k=1}^{\infty} (\mathscr{O}_k \setminus E) \subset \cup_{k=1}^{\infty} (\mathscr{O}_k \setminus E_k).$$

Thus,

$$\begin{aligned} m^*(\mathscr{O} \setminus E) &\leq \sum_{k=1}^{\infty} m^*(\mathscr{O}_k \setminus E_k) \\ &\leq \sum_{k=1}^{\infty} \frac{\varepsilon}{2^k} \\ &= \varepsilon. \end{aligned} \tag{8.11}$$

From this 1 holds.

Hence from 1, for each $k \in \mathbb{N}$ there exists $\mathscr{O}_k$ open such that $E \subset \mathscr{O}_k$ and

$$m^*(\mathscr{O}_k \setminus E) < \frac{1}{k}.$$

Define $G = \cap_{k=1}^{\infty} \mathscr{O}_k$. Thus G is a G_δ set such that $E \subset G$ and

$$m^*(G \setminus E) \leq m^*(\mathscr{O}_k \setminus E) < \frac{1}{k}, \ \forall k \in \mathbb{N}.$$

Therefore, $m^*(G \setminus E) = 0$.

We have proven the measurability of E implies 1, which implies 2.

Now, suppose 2 holds. Thus, there exists a G_δ set G such that $E \subset G$ and

$$m^*(G \setminus E) = 0.$$

Hence, $G \setminus E$ is measurable.

We now are going to show that

$$E = G \cap (G \setminus E)^c.$$

Indeed, since $E \subset G$, we have,

$$\begin{aligned} x \in G \cap (G \setminus E)^c &\Leftrightarrow x \in G \cap (G \cap E^c)^c \\ &\Leftrightarrow x \in G \cap (G^c \cup E) \\ &\Leftrightarrow x \in (G \cap G^c) \cup (G \cap E) \\ &\Leftrightarrow x \in E. \end{aligned} \tag{8.12}$$

Summarizing

$$E = G \cap (G \setminus E)^c,$$

so that E is measurable.

From this we may infer that the measurability of E is equivalent of 1 and equivalent to 2.

Reasoning with the complements, we may obtain that 1 is equivalent to 3 and 2 is equivalent to 4.

This completes the proof.

8.5   Some other properties of measurable sets

1. Each open set is measurable. This results immediately from the last theorem.

2. If $E \subset \mathbb{R}^n$ is measurable, then E^c is measurable. This is obvious from the definition of measurable sets.

3. Closed sets are measurable. This results from the last two items

4. A countable intersection of measurable sets is measurable.

Proof 8.9   This follows just from observing that

$$\cap_{j=1}^\infty E_j = (\cup_{j=1}^\infty E_j^c)^c.$$

Theorem 8.5.1 *If* $\{E_i\}$ *is sequence of measurable pairwise disjoint sets and* $E = \cup_{j=1}^\infty E_i$ *then*

$$m(E) = \sum_{j=1}^\infty m(E_j).$$

Proof 8.10   First assume that E_j is bounded. Being E_j^c measurable, given $\varepsilon > 0$ there exists an open $H_j \supset E_j^c$ such that

$$m^*(H_j - E_j^c) < \frac{\varepsilon}{2^j}, \forall j \in \mathbb{N}.$$

Denoting $F_j = H_j^c$ we have that $F_j \subset E_j$ is closed and

$$m^*(E_j - F_j) < \frac{\varepsilon}{2^j}, \forall j \in \mathbb{N}.$$

For each $N \in \mathbb{N}$ the sets $F_1, ..., F_N$ are compact and disjoint, so that

$$m(\cup_{j=1}^N F_j) = \sum_{j=1}^N m(F_j).$$

As

$$\cup_{j=1}^N F_j \subset E$$

we have

$$m(E) \geq \sum_{j=1}^N m(F_j) \geq \sum_{j=1}^N m(E_j) - \varepsilon.$$

Hence,

$$m(E) \geq \sum_{j=1}^\infty m(E_j) - \varepsilon.$$

being $\varepsilon > 0$ arbitrary, we obtain

$$m(E) \geq \sum_{j=1}^\infty m(E_j).$$

As the reverse inequality is always valid, we have

$$m(E) = \sum_{j=1}^\infty m(E_j).$$

For the general case, select a sequence of cubes $\{Q_k\}$ such that

$$\mathbb{R}^n = \cup_{k=1}^{\infty} Q_k$$

and $Q_k \subset Q_{k+1} \forall k \in \mathbb{N}$. Define $S_1 = Q_1$ e $S_k = Q_k - Q_{k-1}, \forall k \geq 2$. Also define

$$E_{j,k} = E_j \cap S_k, \forall j, k \in \mathbb{N}.$$

Thus,

$$E = \cup_{j=1}^{\infty} \left(\cup_{k=1}^{\infty} E_{j,k} \right) = \cup_{j,k=1}^{\infty} E_{j,k},$$

where such an union is disjoint and each $E_{j,k}$ is bounded. Through the last result, we get

$$
\begin{aligned}
m(E) &= \sum_{j,k=1}^{\infty} m(E_{j,k}) \\
&= \sum_{j=1}^{\infty} \sum_{k=1}^{\infty} m(E_{j,k}) \\
&= \sum_{j=1}^{\infty} m(E_j).
\end{aligned}
\tag{8.13}
$$

The proof is complete.

Theorem 8.5.2 *Suppose $E \subset \mathbb{R}^n$ is a measurable set. Then for each $\varepsilon > 0$:*

1. *If $m(E)$ is finite, there exists a compact set $K \subset E$ such that*

$$m(E \setminus K) < \varepsilon.$$

2. *If $m(E)$ is finite there exist a finite union of closed cubes*

$$F = \cup_{j=1}^{N} Q_j$$

such that

$$m(E \triangle F) \leq \varepsilon,$$

where

$$E \triangle F = (E \setminus F) \cup (F \setminus E).$$

Proof 8.11

1. Choose a closed set such that $F \subset E$ and

$$m(E \setminus F) < \frac{\varepsilon}{2}.$$

Let B_n be a closed ball with center at origin and radius n. Define $K_n = F \cap B_n$ and observe that K_n is compact, $\forall n \in \mathbb{N}$. Thus

$$E \setminus K_n \searrow E \setminus F.$$

Being $m(E) < \infty$ we have

$$m(E \setminus K_n) < \varepsilon,$$

for all n sufficiently big.

2. Choose a sequence of closed cubes $\{Q_j\}$ such that

$$E \subset \cup_{j=1}^{\infty} Q_j,$$

and

$$\sum_{j=1}^{\infty} |Q_j| \leq m(E) + \frac{\varepsilon}{2}.$$

Being $m(E) < \infty$ the series converges and there exists $N_0 \in \mathbb{N}$ such that

$$\sum_{N_0+1}^{\infty} |Q_j| < \frac{\varepsilon}{2}.$$

Defining $F = \cup_{j=1}^{N_0} Q_j$, we have

$$
\begin{aligned}
m(E \triangle F) &= m(E - F) + m(F - E) \\
&\leq m\left(\cup_{j=N_0+1}^{\infty} Q_j\right) + m\left(\cup_{j=1}^{\infty} Q_j - E\right) \\
&\leq \sum_{j=N_0+1}^{\infty} |Q_j| + \sum_{j=1}^{\infty} |Q_j| - m(E) \\
&\leq \frac{\varepsilon}{2} + \frac{\varepsilon}{2} = \varepsilon.
\end{aligned}
\tag{8.14}
$$

8.6 Lebesgue measurable functions

Definition 8.6.1 *Let* $E \subset \mathbb{R}^n$ *be a measurable set. A function* $f : E \to [-\infty, +\infty]$ *is said to be Lebesgue measurable if for each* $a \in \mathbb{R}$*, the set*

$$f^{-1}([-\infty, a)) = \{x \in E \mid f(x) < a\}$$

is measurable.

Observe that:

1. If

$$f^{-1}([-\infty, a))$$

is measurable for each $a \in \mathbb{R}$ then

$$f^{-1}([-\infty, a]) = \cap_{k=1}^{\infty} f^{-1}([-\infty, a + 1/k))$$

is measurable for each $a \in \mathbb{R}$.

2. If

$$f^{-1}([-\infty, a])$$

is measurable for each $a \in \mathbb{R}$ then

$$f^{-1}([-\infty, a)) = \cup_{k=1}^{\infty} f^{-1}([-\infty, a - 1/k])$$

is also measurable for each $a \in \mathbb{R}$.

3. Given $a \in \mathbb{R}$, observe that

$$f^{-1}([-\infty, a)) \text{ is measurable} \Leftrightarrow E - f^{-1}([-\infty, a)) \text{ is measurable}$$
$$\Leftrightarrow f^{-1}(\mathbb{R}) - f^{-1}([-\infty, a)) \Leftrightarrow f^{-1}(\mathbb{R} - [-\infty, a)) \text{ is measurable}$$
$$\Leftrightarrow f^{-1}([a, +\infty]) \text{ is measurable} . \tag{8.15}$$

4. From above, we can prove that

$$f^{-1}([-\infty, a))$$

is measurable $\forall a \in \mathbb{R}$ if, and only if

$$f^{-1}((a, b))$$

is measurable for each $a, b \in \mathbb{R}$ such that $a < b$. Therefore f is measurable if and only if $f^{-1}(\mathcal{O})$ is measurable whenever $\mathcal{O} \subset \mathbb{R}$ is open.

5. Thus, f is measurable if $f^{-1}(\mathcal{F})$ is measurable whenever $\mathcal{F} \subset \mathbb{R}$ is closed.

Proposition 8.6.2 *If f is continuous in $\mathbb{R}^n$, then f is measurable. If f is measurable and real and ϕ is continuous, then $\phi \circ f$ is measurable.*

Proof 8.12   The first implication is obvious. For the second, being ϕ continuous

$$\phi^{-1}([-\infty, a))$$

is open, and therefore

$$(\phi \circ f)^{-1}(([-\infty, a)) = f^{-1}(\phi^{-1}([-\infty, a)))$$

is measurable, $\forall a \in \mathbb{R}$.

Proposition 8.6.3 *Suppose $\{f_k\}$ is a sequence of measurable functions. Then*

$$\sup_{k \in \mathbb{N}} f_k(x), \quad \inf_{k \in \mathbb{N}} f_k(x),$$

and

$$\limsup_{k \to \infty} f_k(x), \quad \liminf_{k \to \infty} f_k(x)$$

are measurable.

Proof 8.13   We will prove only that $\sup_{n \in \mathbb{N}} f_n(x)$ is measurable. The remaining proofs are analogous. Let

$$f(x) = \sup_{n \in \mathbb{N}} f_n(x).$$

Thus

$$f^{-1}((a, +\infty]) = \cup_{n=1}^{\infty} f_n^{-1}((a, +\infty]).$$

Being each f_n measurable, such a set is measurable, $\forall a \in \mathbb{R}$. By analogy

$$\inf_{k \in \mathbb{N}} f_k(x)$$

is measurable,

$$\limsup_{k \to \infty} f_k(x) = \inf_{k \geq 1} \sup_{j \geq k} f_j(x),$$

and

$$\liminf_{k \to \infty} f_k(x) = \sup_{k \geq 1} \inf_{j \geq k} f_j(x)$$

are measurable.

Proposition 8.6.4 *Let* $\{f_k\}$ *be a sequence of measurable functions such that*

$$\lim_{k \to \infty} f_k(x) = f(x).$$

Then f is measurable.

Proof 8.14   Just observe that

$$f(x) = \lim_{k \to \infty} f_k(x) = \limsup_{k \to \infty} f_k(x).$$

The next result we do not prove it. For a proof see [72].

Proposition 8.6.5 *If f and g are measurable functions, then*

1. f^2 *is measurable.*

2. $f + g$ *and* $f \cdot g$ *are measurable if both assume finite values.*

Proposition 8.6.6 *Let* $E \subset \mathbb{R}^n$ *a measurable set. Suppose* $f : E \to \mathbb{R}$ *is measurable. Thus, if* $g : E \to \mathbb{R}$ *is such that*

$$g(x) = f(x), \; a.e. \; in \; E$$

then g is measurable.

Proof 8.15   Define

$$A = \{x \in E \mid f(x) \neq g(x)\},$$

and

$$B = \{x \in E \mid f(x) = g(x)\}.$$

A is measurable since $m^*(A) = m(A) = 0$ and therefore $B = E - A$ is also measurable. Let $a \in \mathbb{R}$. Hence,

$$g^{-1}((a, +\infty]) = \left(g^{-1}((a, +\infty]) \cap A\right) \cup \left(g^{-1}((a, +\infty]) \cap B\right).$$

Observe that

$$
\begin{aligned}
x \in g^{-1}((a, +\infty]) \cap B &\Leftrightarrow x \in B \text{ and } g(x) \in (a, +\infty] \\
&\Leftrightarrow x \in B \text{ and } f(x) \in (a, +\infty] \\
&\Leftrightarrow x \in B \cap f^{-1}((a, +\infty]). \tag{8.16}
\end{aligned}
$$

Thus $g^{-1}((a, +\infty]) \cap B$ is measurable. As $g^{-1}((a, +\infty]) \cap A \subset A$ we have $m^*(g^{-1}((a, +\infty]) \cap A) = 0$, that is, such a set is measurable. Hence being $g^{-1}((a, +\infty])$ the union of two measurable sets is also measurable. Being $a \in \mathbb{R}$ arbitrary, g is measurable.

Theorem 8.6.7 *Suppose f is a non-negative measurable function on* $\mathbb{R}^n$. *Then there exists a increasing sequence of non-negative simple functions* $\{\varphi_k\}$ *such that*

$$\lim_{k \to \infty} \varphi_k(x) = f(x), \forall x \in \mathbb{R}^n.$$

Proof 8.16   Let $N \in \mathbb{N}$. Let Q_N be the cube with center at origin and side of measure N. Define

$$F_N(x) = \begin{cases} f(x), & \text{if } x \in Q_N \text{ and } f(x) \leq N, \\ N, & \text{if } x \in Q_N \text{ and } f(x) > N, \\ 0, & \text{otherwise.} \end{cases}$$

Thus, $F_N(x) \to f(x)$ as $N \to \infty, \forall x \in \mathbb{R}^n$. Fixing $M, N \in \mathbb{N}$ define

$$E_{l,M} = \left\{ x \in Q_N : \frac{l}{M} \le F_N(x) \le \frac{l+1}{M} \right\},$$

for $0 \le l \le N \cdot M$. Defining

$$F_{N,M} = \sum_{l=0}^{NM} \frac{l}{M} \chi_{E_{l,M}},$$

we have that $F_{N,M}$ is a simple function and

$$0 \le F_N(x) - F_{N,M}(x) \le \frac{1}{M}.$$

If $\varphi_K(x) = F_{K,K}(x)$ we obtain

$$0 \le |F_K(x) - \varphi_K(x)| \le \frac{1}{K}.$$

Hence,

$$|f(x) - \varphi_K(x)| \le |f(x) - F_K(x)| + |F_K(x) - \varphi_K(x)|.$$

Therefore,

$$\lim_{K \to \infty} |f(x) - \varphi_K(x)| = 0, \forall x \in \mathbb{R}^n.$$

The proof is complete.

Theorem 8.6.8 *Suppose that f is a measurable function defined on $\mathbb{R}^n$. Then there exists a sequence of simple functions $\{\varphi_k\}$ such that*

$$|\varphi_k(x)| \le |\varphi_{k+1}(x)|, \forall x \in \mathbb{R}^n, k \in \mathbb{N},$$

and

$$\lim_{k \to \infty} \varphi_k(x) = f(x), \forall x \in \mathbb{R}^n.$$

Proof 8.17   Write

$$f(x) = f^+(x) - f^-(x),$$

where

$$f^+(x) = \max\{f(x), 0\}$$

and

$$f^-(x) = \max\{-f(x), 0\}.$$

Thus f^+ and f^- are non-negative measurable functions so that from the last theorem there exist increasing sequences of non-negative simple functions such that

$$\varphi_k^{(1)}(x) \to f^+(x), \forall x \in \mathbb{R}^n,$$

and

$$\varphi_k^{(2)}(x) \to f^-(x), \forall x \in \mathbb{R}^n,$$

as $k \to \infty$. Defining

$$\varphi_k(x) = \varphi_k^{(1)}(x) - \varphi_k^{(2)}(x),$$

we obtain

$$\varphi_k(x) \to f(x), \forall x \in \mathbb{R}^n$$

as $k \to \infty$ and

$$|\varphi_k(x)| = \varphi_k^{(1)}(x) + \varphi_k^{(2)}(x) \nearrow |f(x)|, \forall x \in \mathbb{R}^n,$$

as $k \to \infty$.

Theorem 8.6.9 *Suppose* f *is a measurable function in* $\mathbb{R}^n$. *Then there exists a sequence of step functions* $\{\varphi_k\}$ *which converges to* f *a.e. in* $\mathbb{R}^n$.

Proof 8.18 From the last theorem, it suffices to prove that if E is measurable and $m(E) < \infty$ then χ_E may be approximated almost everywhere in E by step functions. Suppose given $\varepsilon > 0$. Observe that from Proposition 8.5.2, there exist cubes $Q_1, ..., Q_N$ such that

$$m(E \triangle \cup_{j=1}^{N} Q_j) < \varepsilon.$$

We may obtain almost disjoints rectangles $\tilde{R}_j$ such that $\cup_{j=1}^{M} \tilde{R}_j = \cup_{j=1}^{N} Q_j$ and disjoints rectangles $R_j \subset \tilde{R}_j$ such that

$$m(E \triangle \cup_{j=1}^{M} R_j) < 2\varepsilon.$$

Thus,

$$f(x) = \sum_{j=1}^{M} \chi_{R_j},$$

possibly except in a set of measure $< 2\varepsilon$. Hence, for each $k > 0$, there exists a step function φ_k such that $m(E_k) < 2^{-k}$ where

$$E_k = \{x \in \mathbb{R}^n \mid f(x) \neq \varphi_k(x)\}.$$

Defining

$$F_k = \cup_{j=k+1}^{\infty} E_j$$

we have

$$
\begin{aligned}
m(F_k) &\leq \sum_{j=k+1}^{\infty} m(E_j) \\
&\leq \sum_{j=k+1}^{\infty} 2^{-j} \\
&= \frac{2^{-(k+1)}}{1 - 1/2} \\
&= 2^{-k}.
\end{aligned}
\tag{8.17}
$$

Therefore also defining

$$F = \cap_{k=1}^{\infty} F_k$$

we have $m(F) = 0$ considering that

$$m(F) \leq 2^{-k}, \forall k \in \mathbb{N}.$$

Finally, observe that

$$\varphi_k(x) \to f(x), \forall x \in F^c.$$

The proof is complete.

Theorem 8.6.10 (Egorov) *Suppose that* $\{f_k\}$ *is a sequence of measurable functions defined in a measurable set* E *such that* $m(E) < \infty$. *Assume that* $f_k \to f$, *a.e in* E. *Thus given* $\varepsilon > 0$ *we may find a closed set* $A_\varepsilon \subset E$ *such that* $f_k \to f$ *uniformly in* A_ε *and* $m(E - A_\varepsilon) < \varepsilon$.

Proof 8.19 Without losing generality we may assume that

$$f_k \to f, \forall x \in E.$$

For each $N, k \in \mathbb{N}$ define

$$E_k^N = \{x \in E \mid |f_j(x) - f(x)| < 1/N, \forall j \geq k\}.$$

Fixing $N \in \mathbb{N}$, we may observe that

$$E_k^N \subset E_{k+1}^N,$$

and that $\cup_{k=1}^{\infty} E_k^N = E$. Thus we may obtain k_N such that

$$m(E - E_{k_N}^N) < \frac{1}{2^N}.$$

Observe that

$$|f_j(x) - f(x)| < \frac{1}{N}, \forall j \geq k_N, \ x \in E_{k_N}^N.$$

Choose $M \in \mathbb{N}$ such that

$$\sum_{k=M}^{\infty} 2^{-k} \leq \frac{\varepsilon}{2}.$$

Define

$$\tilde{A}_\varepsilon = \cap_{N \geq M}^{\infty} E_{k_N}^N.$$

Thus,

$$m(E - \tilde{A}_\varepsilon) \leq \sum_{N=M}^{\infty} m(E - E_{k_N}^N) < \frac{\varepsilon}{2}.$$

Suppose given $\delta > 0$. Let $N \in \mathbb{N}$ be such that $N > M$ and $1/N < \delta$. Thus, if $x \in \tilde{A}_\varepsilon$ then $x \in E_{k_N}^N$ so that

$$|f_j(x) - f(x)| < \delta, \forall j > k_N.$$

Hence, $f_k \to f$ uniformly in $\tilde{A}_\varepsilon$. Observe that $\tilde{A}_\varepsilon$ is measurable and thus there exists a closed set $A_\varepsilon \subset \tilde{A}_\varepsilon$ such that

$$m(\tilde{A}_\varepsilon - A_\varepsilon) < \frac{\varepsilon}{2}.$$

That is:

$$m(E - A_\varepsilon) \leq m(E - \tilde{A}_\varepsilon) + m(\tilde{A}_\varepsilon - A_\varepsilon) < \frac{\varepsilon}{2} + \frac{\varepsilon}{2} = \varepsilon,$$

and

$$f_k \to f$$

uniformly in A_ε. The proof is complete.

Definition 8.6.11 *We say that $f : \mathbb{R}^n \to [-\infty, +\infty] \in L^1(\mathbb{R}^n$ if f is measurable and*

$$\int_{\mathbb{R}^n} |f| \, dx < \infty.$$

Definition 8.6.12 *We say that a set $A \subset L^1(\mathbb{R}^n)$ is dense in $L^1(\mathbb{R}^n)$, if for each $f \in L^1(\mathbb{R}^n)$ and each $\varepsilon > 0$ there exists $g \in A$ such that*

$$\|f - g\|_{L^1(\mathbb{R}^n)} = \int_{\mathbb{R}^n} |f - g| \, dx < \varepsilon.$$

Theorem 8.6.13 *About dense sets in $L^1(\mathbb{R}^n)$ we have:*

1. *The set of simple functions is dense in $L^1(\mathbb{R}^n)$.*

2. *The set of step functions is dense in $L^1(\mathbb{R}^n)$.*

3. *the set of continuous functions with compact support is dense in $L^1(\mathbb{R}^n)$.*

Proof 8.20

1. From the last theorems given $f \in L^1(\mathbb{R}^n)$ there exists a sequence of simple functions such that

$$\varphi_k(x) \to f(x) \text{ a.e. in } \mathbb{R}^n.$$

Since $\{\varphi_k\}$ may be also such that

$$|\varphi_k| \le |f|, \forall k \in \mathbb{N}$$

from the Lebesgue dominated converge theorem, we have

$$\|\varphi_k - f\|_{L^1(\mathbb{R}^n)} \to 0,$$

as $k \to \infty$.

2. From the last item , it suffices to show that simple functions may be approximated by step functions. As a simple function is a linear combination of characteristic functions of sets of finite measure, it suffices to prove that given $\varepsilon > 0$ and a set of finite measure, there exists φ a step function such that

$$\|\chi_E - \varphi\|_{L^1(\mathbb{R}^n)} < \varepsilon.$$

This may be made similar as in Theorem 8.6.9.

3. From the last item, it suffices to establish the result as f is a characteristic function of a rectangle in $\mathbb{R}^n$. First consider the case of a interval $[a,b]$. we may approximate $f = \chi_{[a,b]}$ by $g(x)$, where g is continuous, and linear on $(a-\varepsilon, a)$ e $(b, b+\varepsilon)$ and

$$g(x) = \begin{cases} 1, & \text{if } a \le x \le b, \\ 0, & \text{if } x \le a - \varepsilon \text{ or } x \ge b + \varepsilon. \end{cases}$$

Thus,

$$\|f - g\|_{L^1(\mathbb{R}^n)} < 2\varepsilon.$$

for the general case of a rectangle in $\mathbb{R}^n$, we just recall that in this case f is the product of the characteristic functions of n intervals. Therefore we may approximate f by the product of n functions similar to g defined above.

Chapter 9

Other Topics in Measure and Integration

In this chapter we present some important results which may be found in similar form at Chapters 2, 6 and ? in the excellent book Real and Complex Analysis [68] by W. Rudin, where more details may be found.

9.1 Some preliminary results

In the next results μ is a measure on U. We start with the following theorem.

Theorem 9.1.1 *Let $f : U \to [0, \infty]$ be a measurable function. If $E \in \mathcal{M}$ and*

$$\int_E f \, d\mu = 0$$

then

$$f = 0, \text{ a.e. in } E.$$

Proof 9.1 Define

$$A_n = \{u \in E \mid f(u) > 1/n\}, \forall n \in \mathbb{N}.$$

Thus,

$$\mu(A_n)/n \le \int_{A_n} f \, d\mu \le \int_E f \, d\mu = 0.$$

Therefore, $\mu(A_n) = 0, \forall n \in \mathbb{N}$.

Define

$$A = \{u \in E \mid f(u) > 0\}.$$

Hence,

$$A = \cup_{n=1}^\infty A_n,$$

so that $\mu(A) = 0$.

Thus,

$$f = 0, \text{ a.e. in } E.$$

Theorem 9.1.2 *Assume $f \in L^1(\mu)$ and $\int_E f \, d\mu = 0, \forall E \in \mathcal{M}$. Under such hypotheses, $f = 0$, a.e. in U.*

Proof 9.2 Consider first the case $f : U \to [-\infty, +\infty]$. Define

$$A_n = \{u \in U \mid f(u) > 1/n\}, \forall n \in \mathbb{N}.$$

Thus,

$$\mu(A_n)/n \le \int_{A_n} f \, d\mu = 0.$$

Hence, $\mu(A_n) = 0, \forall n \in \mathbb{N}$.

Define

$$A = \{u \in E \mid f(u) > 0\}.$$

Therefore,

$$A = \cup_{n=1}^{\infty} A_n,$$

so that $\mu(A) = 0$.

Thus,

$$f \le 0, \text{ a.e. in } U.$$

By analogy we get

$$f \ge 0, \text{ a.e. in } U,$$

so that

$$f = 0, \text{ a.e. in } U.$$

To complete the proof, just apply this last result to the real and imaginary parts of a complex f.

Theorem 9.1.3 *Suppose $\mu(U) < \infty$ and $f \in L^1(\mu)$. Moreover, assume*

$$\frac{\int_E |f| \, d\mu}{\mu(E)} \le \alpha \in [0, \infty), \forall E \in \mathcal{M}.$$

Under such hypothesis we have

$$|f| \le \alpha, \text{ a.e. in } U.$$

Proof 9.3 Define

$$A_n = \{u \in U \mid |f(u)| > \alpha + 1/n\}, \forall n \in \mathbb{N}.$$

Thus, if $\mu(A_n) > 0$ we get

$$1/n \le \frac{\int_{A_n} (|f| - \alpha) \, d\mu}{\mu(A_n)} = \frac{\int_{A_n} |f| \, d\mu}{\mu(A_n)} - \alpha \le 0,$$

a contradiction. Hence, $\mu(A_n) = 0, \forall n \in \mathbb{N}$.

Define

$$A = \{u \in U \mid |f(u)| > \alpha\}.$$

Therefore,

$$A = \cup_{n=1}^{\infty} A_n,$$

so that $\mu(A) = 0$.

Thus,

$$|f(u)| \le \alpha, \text{ a.e. in } U.$$

The proof is complete.

At this point we present some preliminary results to the development of the well known Urysohn's lemma.

Theorem 9.1.4 *Let U be a Hausdorff space and $K \subset U$ compact. Let $v \in K^c$. Then there exists open sets V and $W \subset U$ such that $v \in V$, $K \subset W$ and $V \cap W = \emptyset$.*

Proof 9.4   For each $u \in K$ there exists open sets $W_u, V_v^u \subset U$ such that $u \in W_u$, $v \in W_v^u$ and $W_u \cap V_v^u = \emptyset$. Observe that $K \subset \cup_{u \in K} W_u$ so that, since K is compact there exist $u_1, u_2, ..., u_n \in K$ such that

$$K \subset \cup_{i=1}^n W_{u_i}.$$

Finally, defining the open sets

$$V = \cap_{i=1}^n V_v^{u_i},$$

and

$$W = \cup_{i=1}^n W_{u_i},$$

we get

$$V \cap W = \emptyset,$$

$v \in V$ and $K \subset W$.

The proof is complete.

Theorem 9.1.5 *Let $\{K_\alpha, \ \alpha \in L\}$ be a collection of compact subsets of a Hausdorff space U.*

Assume $\cap_{\alpha \in L} K_\alpha = \emptyset$. Under such hypotheses some finite sub-collection of $\{K_\alpha, \ \alpha \in L\}$ has empty inter sections.

Proof 9.5   Define $V_\alpha = K_\alpha^c$, $\forall \alpha \in L$. Fix $\alpha_0 \in L$. From the hypotheses

$$K_{\alpha_0} \cap [\cap_{\alpha \in L \setminus \{\alpha_0\}} K_\alpha] = \emptyset.$$

Hence,

$$K_{\alpha_0} \subset [\cap_{\alpha \in L \setminus \{\alpha_0\}} K_\alpha]^c,$$

that is,

$$K_{\alpha_0} \subset \cup_{\alpha \in L \setminus \{\alpha_0\}} K_\alpha^c = \cup_{\alpha \in L \setminus \{\alpha_0\}} V_\alpha.$$

Since K_{α_0} is compact, there exists $\alpha_1, ..., \alpha_n \in L$ such that

$$K_{\alpha_0} \subset V_{\alpha_1} \cup ... \cup V_{\alpha_n} = (K_{\alpha_1} \cap ... \cap K_{\alpha_n})^c,$$

so that,

$$K_{\alpha_0} \cap K_{\alpha_1} \cap ... \cap K_{\alpha_n} = \emptyset.$$

The proof is complete.

Definition 9.1.6 *We say that a space U is locally compact if each $u \in U$ has a neighborhood whose closure is compact.*

Theorem 9.1.7 *Let U be a locally compact Hausdorff space. Suppose $W \subset U$ is open and $K \subset W$, where K is compact. Then there exists an open set $V \subset U$ whose closure is compact and such that*

$$K \subset V \subset \overline{V} \subset W.$$

Proof 9.6   Let $u \in K$. Since U is locally compact there exists an open $V_u \subset U$ such that $u \in V_u$ and $\overline{V}_u$ is compact.

Observe that

$$K \subset \cup_{u \in K} V_u$$

and since K is compact there exists $u_1, u_2, .., u_n \in K$ such that

$$K \subset \cup_{j=1}^n V_{u_j}.$$

Hence, defining, $G = \cup_{j=1}^n V_{u_j}$ we get

$$K \subset G$$

where $\overline{G}$ is compact.

If $W = U$ define $V = G$ and the proof would be complete.

Otherwise, if $W \neq U$ define $C = U \setminus W$. From Theorem 9.1.4, for each $v \in C$ there exists an open W_v such that $K \subset W_v$ and $v \notin \overline{W}_v$.

Hence, $\{C \cap \overline{G} \cap \overline{W}_v : v \in C\}$ is a collection of compact sets with empty intersection.

From Theorem 9.1.5 there are points $v_1, .., v_n \in C$ such that

$$C \cap \overline{G} \cap \overline{W}_{v_1} \cap ... \cap \overline{W}_{v_n} = \emptyset.$$

Defining

$$V = G \cap W_{v_1} \cap ... \cap W_{v_n}$$

we obtain

$$\overline{V} \subset \overline{G} \cap \overline{W}_{v_1} \cap ... \cap \overline{W}_{v_n}.$$

Also,

$$K \subset V \subset \overline{V} \subset W.$$

This completes the proof.

Definition 9.1.8 *Let $f : U \to [-\infty, +\infty]$ be a function on a topological space U.*

We say that f is lower semi-continuous if $A_\alpha = \{u \in U : f(u) > \alpha\}$ is open for all $\alpha \in \mathbb{R}$. Similarly, we say that f is upper semi-continuous if $B_\alpha = \{u \in U : f(u) < \alpha\}$ is open for all $\alpha \in \mathbb{R}$.

Observe that from this last definition f is continuous if and only if, it is both lower and upper semi-continuous.

Here we state and prove a very important result namely, the Uryshon's lemma.

Lemma 9.1.9 (Urysohn's lemma) *Assume U is a locally compact Hausdorff space, $V \subset U$ is an open set which contains a compact set K. Under such assumptions, there exists a function $f \in C_c(V)$ such that*

- $0 \leq f(u) \leq 1, \forall u \in V$,

- $f(u) = 1, \forall u \in K$.

Proof 9.7 Set $r_1 = 0$ and $r_2 = 1$, and let $r_3, r_4, r_5, ...$ be an enumeration of the rational numbers in $(0, 1)$. Observe that we may find open sets V_0 and V_1 such that $\overline{V}_0$ is compact and

$$K \subset V_1 \subset \overline{V}_1 \subset V_0 \subset \overline{V}_0 \subset V.$$

Reasoning by induction, suppose $n \geq 2$ and that $V_{r_1}, ..., V_{r_n}$ have been chosen so that if $r_i < r_j$ then $\overline{V}_{r_j} \subset V_{r_i}$. Denote

$$r_i = \max\{r_k \mid k \in \{1, ..., n\} \text{ and } r_k < r_{n+1}\},$$

and

$$r_j = \min\{r_k, \mid k \in \{1, ..., n\} \text{ and } r_k > r_{n+1}\}.$$

We may find again an open set $V_{r_{n+1}}$ such that

$$\overline{V}_{r_j} \subset V_{r_{n+1}} \subset \overline{V}_{r_{n+1}} \subset V_{r_i}.$$

Thus, we have obtained a sequence V_r of open sets such that for every r rational in $(0,1)$ $\overline{V}_r$ is compact and if $s > r$ then $\overline{V}_s \subset V_r$. Define,

$$f_r(u) = \begin{cases} r, & \text{if } u \in V_r, \\ 0, & \text{otherwise,} \end{cases}$$

and

$$g_s(u) = \begin{cases} 1, & \text{if } u \in \overline{V}_s, \\ s, & \text{otherwise.} \end{cases}$$

Also define

$$f(u) = \sup_{r \in \mathbb{Q} \cap (0,1)} f_r(u), \forall u \in V,$$

and

$$g(u) = \inf_{s \in \mathbb{Q} \cap (0,1)} g_s(u), \forall u \in V.$$

Observe that f is lower semi-continuous and g is upper semi-continuous. Moreover,

$$0 \leq f \leq 1,$$

and

$$f = 1, \text{ if } u \in K.$$

Observe also that the support of f is contained in $\overline{V}_0$
To complete the proof, it suffices to show that

$$f = g.$$

The inequality

$$f_r(u) > g_s(u)$$

is possible only if $r > s$, $u \in V_r$ and $u \notin \overline{V}_s$.

But if $r > s$ then $V_r \subset V_s$, and hence $f_r \leq g_s, \forall r, s \in \mathbb{Q} \cap (0,1)$, so that $f \leq g$. Suppose there exists $u \in V$ such that

$$f(u) < g(u).$$

Thus there exists rational numbers r, s such that

$$f(u) < r < s < g(u).$$

Since $f(u) < r$, $u \notin V_r$. Since $g(u) > s$, $u \in \overline{V}_s$.

As $\overline{V}_s \subset V_r$, we have a contradiction. Hence, $f = g$, and such a function is continuous.

The proof is complete.

Theorem 9.1.10 *[Partition of unity] Let U be a locally compact Hausdorff space. Assume $K \subset U$ is compact so that*

$$K \subset \cup_{i=1}^n V_i,$$

where V_i is open $\forall i \in \{1,...n\}$. Under such hypotheses, there exists functions $h_1,...,h_n$ such that

$$\sum_{i=1}^n h_i = 1, \text{ on } K,$$

$$h_i \in C_c(V_i) \text{ and } 0 \leq h_i \leq 1, \forall i \in \{1,...,n\}.$$

Proof 9.8   Let $u \in K \subset \cup_{i=1}^{n} V_i$. Thus there exists $j \in \{1,...,n\}$ such that $u \in V_j$. We may select an open set W_u such that $u \in W_u$, $\overline{W}_u$ is compact and $\overline{W}_u \subset V_j$.

Observe that
$$K \subset \cup_{u \in K} W_u.$$

From this, since K is compact, there exist $u_1,...,u_N$ such that
$$K \subset \cup_{j=1}^{N} W_{u_j}.$$

For each $i \in \{1,...,n\}$ define by $\tilde{W}_i$ the union of those $\overline{W_{u_j}}$, contained in V_i. By the Uryshon's lemma we may find continuous functions g_i such that

$$g_i = 1, \text{ on } \tilde{W}_i,$$

$$g_i \in C_c(V_i),$$

$$0 \le g_i \le 1, \forall i \in \{1,...,n\}.$$

Define,

$$
\begin{aligned}
h_1 &= g_1 \\
h_2 &= (1-g_1)g_2 \\
h_3 &= (1-g_1)(1-g_2)g_3 \\
&\cdots \quad \cdots\cdots\cdots\cdots \\
h_n &= (1-g_1)(1-g_2)...(1-g_{n-1})g_n.
\end{aligned}
\tag{9.1}
$$

Thus,
$$0 \le h_i \le 1 \text{ and } h_i \in C_c(V_i), \ \forall i \in \{1,..,n\}.$$

Furthermore, by induction, we may obtain

$$h_1 + h_2 + ... + h_n = 1 - (1-g_1)(1-g_2)...(1-g_n).$$

Finally, if $u \in K$ then $u \in \tilde{W}_i$ for some $i \in \{1,..,n\}$, so that $g_i(u) = 1$, and hence

$$(h_1 + ... + h_n)(u) = 1, \forall u \in K.$$

The set $\{h_1,...,h_n\}$ is said to be a partition of unity on K subordinate to the open cover $\{V_1,...,V_n\}$. The proof is complete.

9.2   The Riesz representation theorem

In the next lines we introduce the main result in this section, namely, the Riesz representation theorem.

Theorem 9.2.1 (Riesz representation theorem) *Let U be a locally compact Hausdorff space and let F be a positive linear functional on $C_c(U)$. Then there exists a σ-algebra $\mathcal{M}$ in U which contains all the Borel sets and there exists a unique positive measure μ on $\mathcal{M}$ such that*

1. $F(f) = \int_U f \, d\mu, \forall f \in C_c(U)$.

2. $\mu(K) < \infty$, for every compact $K \subset U$.

3. $\mu(E) = \inf\{\mu(V) \mid E \subset V, V \text{ open }\}, \forall E \in \mathcal{M}$

4. $\mu(E) = \sup\{\mu(K) \mid K \subset E, K \text{ compact}\}$ *holds for all open E and all* $E \in \mathcal{M}$ *such that* $\mu(E) < \infty$.

5. *If* $E \in \mathcal{M}$, $A \subset E$ *and* $\mu(E) = 0$ *then* $A \in \mathcal{M}$.

Proof 9.9   We start by proving the uniqueness of μ. If μ satisfies (3) and (4), then μ is determined by its values on compact sets. Then, if μ_1 and μ_2 are two measures for which the theorem holds, to prove uniqueness, it suffices to show that

$$\mu_1(K) = \mu_2(K)$$

for every compact $K \subset U$. Let $\varepsilon > 0$. Fix a compact $K \subset U$. By (2) and (3), there exists an open $V \supset K$ such that

$$\mu_2(V) < \mu_2(K) + \varepsilon.$$

By the Urysohn's lema, there exists a $f \in C_c(V)$ such that

$$0 \leq f(u) \leq 1, \forall u \in V,$$

and

$$f(u) = 1, \ \forall u \in K.$$

Thus,

$$
\begin{aligned}
\mu_1(K) &= \int_U \chi_K \, d\mu_1 \\
&\leq \int_U f \, d\mu_1 \\
&= F(f) \\
&= \int_U f \, d\mu_2 \\
&\leq \int_U \chi_V \, d\mu_2 \\
&= \mu_2(V) \\
&< \mu_2(K) + \varepsilon.
\end{aligned}
\tag{9.2}
$$

Since $\varepsilon > 0$ is arbitrary, we get,

$$\mu_1(K) \leq \mu_2(K).$$

Interchanging the roles of μ_1 and μ_2 we similarly obtain

$$\mu_2(K) \leq \mu_1(K),$$

so that

$$\mu_1(K) = \mu_2(K).$$

The proof of uniqueness is complete.

Now for every open $V \subset U$, define

$$\mu(V) = \sup\{F(f) \mid f \in C_c(V) \text{ and } 0 \leq f \leq 1\}.$$

If V_1, V_2 are open and $V_1 \subset V_2$, then

$$\mu(V_1) \leq \mu(V_2).$$

Hence,

$$\mu(E) = \inf\{\mu(V) \mid E \subset V, V \text{ open}\},$$

if E is an open set. Define,

$$\mu(E) = \inf\{\mu(V) \mid E \subset V, V \text{ open}\},$$

$\forall E \subset U$. Define by $\mathcal{M}_F$ the collection of all $E \subset U$ such that $\mu(E) < \infty$ and

$$\mu(E) = \sup\{\mu(K) \mid K \subset E, K \text{ compact}\}.$$

Finally, define by $\mathcal{M}$ the collection of all sets such that $E \subset U$ and $E \cap K \in \mathcal{M}_F$ for all compact $K \subset U$. Since

$$\mu(A) \leq \mu(B),$$

if $A \subset B$ we have that $\mu(E) = 0$ implies $E \cap K \in \mathcal{M}_F$ for all K compact, so that $E \in \mathcal{M}$. Thus, (5) holds and so does (3) by definition.

Observe that if $f \geq 0$ then $F(f) \geq 0$, that is if $f \leq g$ then $F(f) \leq F(g)$.

Now we prove that if $\{E_n\} \subset U$ is a sequence then

$$\mu\left(\cup_{n=1}^{\infty} E_n\right) \leq \sum_{n=1}^{\infty} \mu(E_n). \tag{9.3}$$

First we show that

$$\mu(V_1 \cup V_2) \leq \mu(V_1) + \mu(V_2),$$

if V_1, V_2 are open sets.

Choose $g \in C_c(V_1 \cup V_2)$ such that

$$0 \leq g \leq 1.$$

By Theorem 9.1.10 there exist functions h_1 and h_2 such that $h_i \in C_c(V_i)$ and

$$0 \leq h_i \leq 1$$

and so that $h_1 + h_2 = 1$ on the support of g. Hence $h_i \in C_c(V_i)$, $0 \leq h_i g \leq 1$, and $g = (h_1 + h_2)g$ and thus

$$F(g) = F(h_1 g) + F(h_2 g) \leq \mu(V_1) + \mu(V_2).$$

Since g is arbitrary, from the definition of μ we obtain

$$\mu(V_1 \cup V_2) \leq \mu(V_1) + \mu(V_2).$$

Furthermore, if $\mu(E_n) = \infty$, for some $n \in \mathbb{N}$, then (9.3) is obviously valid. Assume then $\mu(E_n) < \infty, \forall n \in \mathbb{N}$. Let a not relabeled $\varepsilon > 0$. Therefore for each $n \in \mathbb{N}$ there exists an open $V_n \supset E_n$ such that

$$\mu(V_n) < \mu(E_n) + \frac{\varepsilon}{2^n}.$$

Define

$$V = \cup_{n=1}^{\infty} V_n,$$

and choose $f \in C_c(V)$ such that $0 \leq f \leq 1$. Since the support of f is compact, there exists $N \in \mathbb{N}$ such that

$$spt(f) \subset \cup_{n=1}^{N} V_n.$$

Therefore,

$$
\begin{aligned}
F(f) &\leq \mu\left(\cup_{n=1}^{N} V_n\right) \\
&\leq \sum_{n=1}^{N} \mu(V_n) \\
&\leq \sum_{n=1}^{\infty} \mu(E_n) + \varepsilon.
\end{aligned}
\tag{9.4}
$$

Since this holds for any $f \in C_c(V)$ with $0 \leq f \leq 1$ and $\cup_{n=1}^{\infty} E_n \subset V$, we get

$$\mu\left(\cup_{n=1}^{\infty} E_n\right) \leq \mu(V) \leq \sum_{i=1}^{\infty} \mu(E_n) + \varepsilon.$$

Since $\varepsilon > 0$ is arbitrary, we have proven (9.3).

In the next lines we prove that if K is compact then $K \in \mathcal{M}_F$ and

$$\mu(K) = \inf\{F(f) \mid f \in C_c(U),\ f = 1 \text{ on } K\}. \tag{9.5}$$

For, if $f \in C_c(U)$, $f = 1$ on K, and $0 < \alpha < 1$ define

$$V_\alpha = \{u \in U \mid f(u) > \alpha\}.$$

Thus, $K \subset V_\alpha$ and if $g \in C_c(V_\alpha)$ and $0 \leq g \leq 1$ we get

$$\alpha g \leq f.$$

Hence,

$$\begin{aligned}
\mu(K) &\leq \mu(V_\alpha) \\
&= \sup\{F(g) \mid g \in C_c(V_\alpha),\ 0 \leq g \leq 1\} \\
&\leq \alpha^{-1} F(f). \tag{9.6}
\end{aligned}$$

Letting $\alpha \to 1$ we obtain

$$\mu(K) \leq F(f).$$

Thus, $\mu(K) < \infty$, and obviously $K \in \mathcal{M}_F$.

Also there exists an open $V \supset K$ such that

$$\mu(V) < \mu(K) + \varepsilon.$$

By the Urysohn's lemma, we may find $f \in C_c(V)$ such that $f = 1$ on K and $0 \leq f \leq 1$. Thus,

$$F(f) \leq \mu(V) < \mu(K) + \varepsilon.$$

Since $\varepsilon > 0$ is arbitrary, (9.5) holds.

At this point we prove that for every open V we have

$$\mu(V) = \sup\{\mu(K) \mid K \subset V,\ K \text{ compact}\} \tag{9.7}$$

and hence $\mathcal{M}_F$ contains every open set such that $\mu(V) < \infty$.

Let $V \subset U$ be an open set such that $\mu(V) < \infty$.

Let $\alpha \in \mathbb{R}$ be such that $\alpha < \mu(V)$. Therefore there exists $f \in C_c(V)$ such that $0 \leq f \leq 1$ and such that $\alpha < F(f)$.

If $W \subset U$ is an open set such that $K = spt(f) \subset W$, we have that $f \in C_c(W)$ and $0 \leq f \leq 1$ so that

$$F(f) \leq \mu(W).$$

Thus, since $W \supset K$ is arbitrary we obtain

$$F(f) \leq \mu(K),$$

so that

$$\alpha < \mu(K),$$

where $K \subset U$ is a compact set.

Hence (9.7) holds.

Suppose that

$$E = \cup_{n=1}^{\infty} E_n,$$

where $\{E_n\}$ is a sequence of disjoint sets in $\mathcal{M}_F$.

We are going to show that

$$\mu(E) = \sum_{n=1}^{\infty} \mu(E_n). \tag{9.8}$$

In addition, if $\mu(E) < \infty$ then also $E \subset \mathcal{M}_F$.

First we show that if $K_1, K_2 \subset U$ are compact disjoint sets then,

$$\mu(K_1 \cup K_2) = \mu(K_1) + \mu(K_2). \tag{9.9}$$

From the Uryshon's lemma there exists $f \in C_c(U)$ such that $f = 1$ on K_1, $f = 0$ on K_2 and,

$$0 \le f \le 1.$$

From (9.5) there exists $g \in C_c(U)$ such that $g = 1$ on $K_1 \cup K_2$ and

$$F(g) < \mu(K_1 \cup K_2) + \varepsilon.$$

Observe that $fg = 1$ on K_1 and $(1-f)g = 1$ on K_2 and also fg, $(1-f)g \in C_c(U)$ and $0 \le fg \le 1$ and $0 \le (1-f)g \le 1$ so that

$$
\begin{aligned}
\mu(K_1) + \mu(K_2) &\le F(fg) + F((1-f)g) \\
&= F(g) \\
&\le \mu(K_1 \cup K_2) + \varepsilon. \tag{9.10}
\end{aligned}
$$

Since $\varepsilon > 0$ is arbitrary we obtain

$$\mu(K_1) + \mu(K_2) \le \mu(K_1 \cup K_2).$$

From this (9.9) holds.

Also, if $\mu(E) = \infty$, (9.8) follows from (9.3).

Thus assume $\mu(E) < \infty$.

Since $E_n \in \mathcal{M}_F, \forall n \in \mathbb{N}$ we may obtain compact sets $H_n \subset E_n$ such that

$$\mu(H_n) > \mu(E_n) - \frac{\varepsilon}{2^n}, \forall n \in \mathbb{N}.$$

Defining $K_N = \cup_{n=1}^{N} H_n$, by (3) we get

$$
\begin{aligned}
\mu(E) &\ge \mu(K_N) \\
&= \sum_{n=1}^{N} \mu(H_n) \\
&\ge \sum_{n=1}^{N} \mu(E_n) - \varepsilon, \forall N \in \mathbb{N}. \tag{9.11}
\end{aligned}
$$

Since $N \in \mathbb{N}$ and $\varepsilon > 0$ are arbitrary we get,

$$\mu(E) \ge \sum_{n=1}^{\infty} \mu(E_n).$$

From this and (9.3) we obtain

$$\mu(E) = \sum_{n=1}^{\infty} \mu(E_n).\tag{9.12}$$

Let $\varepsilon_0 > 2\varepsilon$. If $\mu(E) < \infty$, there exists $N_0 \in \mathbb{N}$ such that $\mu(K_{N_0}) > \sum_{n=1}^{\infty} \mu(E_n) - \varepsilon_0$.
From this and (9.12) we obtain

$$\mu(E) \leq \mu(K_{N_0}) + \varepsilon_0.$$

Therefore, since $\varepsilon > 0$ and $\varepsilon_0 > 2\varepsilon$ are arbitrary we may conclude that E satisfies (4) so that $E \in \mathcal{M}_F$.
Now we prove the following.
If $E \in \mathcal{M}_F$ there is a compact $K \subset U$ and an open $V \subset U$ such that $K \subset E \subset V$ and

$$\mu(V \setminus K) < \varepsilon.$$

For, from above, there exists a compact K and an open V such that

$$K \subset E \subset V$$

and

$$\mu(V) - \frac{\varepsilon}{2} < \mu(E) < \mu(K) + \frac{\varepsilon}{2}.$$

Since $V \setminus K$ is open and of finite measure, it is in $\mathcal{M}_F$. From the last chain of inequalities we obtain

$$\mu(K) + \mu(V \setminus K) = \mu(V) < \mu(K) + \varepsilon,$$

so that

$$\mu(V \setminus K) < \varepsilon.$$

In the next lines we prove that if $A, B \in \mathcal{M}_F$ then

$$A \setminus B, A \cup B \text{ and } A \cap B \in \mathcal{M}_F.$$

By above there exist compact sets K_1, K_2 and open sets V_1, V_2 such that

$$K_1 \subset A \subset V_1, \ K_2 \subset B \subset V_2$$

and

$$\mu(V_i \setminus K_i) < \varepsilon, \forall i \in \{1, 2\}.$$

Since

$$(A \setminus B) \subset (V_1 \setminus K_2) \subset (V_1 \setminus K_1) \cup (K_1 \setminus V_2) \cup (V_2 \setminus K_2),$$

we get

$$\mu(A \setminus B) < \varepsilon + \mu(K_1 \setminus V_2) + \varepsilon,$$

Since $K_1 \setminus V_2 \subset A \setminus B$ is compact and $\varepsilon > 0$ is arbitrary, we get,

$$A \setminus B \in \mathcal{M}_F.$$

Since

$$A \cup B = (A \setminus B) \cup B,$$

we obtain

$$A \cup B \in \mathcal{M}_F.$$

Since

$$A \cap B = A \setminus (A \setminus B)$$

we get

$$A \cap B \in \mathcal{M}_F.$$

At this point we prove that $\mathcal{M}$ is a σ-algebra in U which contains all the Borel sets.
Let $K \subset U$ be a compact subset. If $A \in \mathcal{M}$ then

$$A^c \cap K = K \setminus (A \cap K),$$

so that $A^c \cap K \in \mathcal{M}_F$ considering that $K \in \mathcal{M}_F$ and $A \cap K \in \mathcal{M}_F$.
Thus if $A \in \mathcal{M}$ then $A^c \in \mathcal{M}$.
Next suppose

$$A = \cup_{n=1}^{\infty} A_n$$

where $A_n \in \mathcal{M}, \forall n \in \mathbb{N}$.
Define $B_1 = A_1 \cap K$ and

$$B_n = (A_n \cap K) \setminus (B_1 \cup B_2 \cup ... \cup B_{n-1}),$$

$\forall n \geq 2, n \in \mathbb{N}$.
Then $\{B_n\}$ is disjoint sequence of sets in $\mathcal{M}_F$.
Thus,

$$A \cap K = \cup_{n=1}^{\infty} B_n \in \mathcal{M}_F.$$

Hence $A \in \mathcal{M}$. Finally, if $C \subset U$ is a closed subset, then $C \cap K$ is compact, so that $C \cap K \in \mathcal{M}_F$. Hence $C \in \mathcal{M}$.
Therefore, $\mathcal{M}$ is a σ-algebra which contains the closed sets, so that it contains the Borel sets.
Finally, we will prove that

$$\mathcal{M}_F = \{E \in \mathcal{M} \mid \mu(E) < \infty\}.$$

For, if $E \in \mathcal{M}_F$ then $E \cap K \in \mathcal{M}_F$ for all compact $K \subset U$, hence $E \in \mathcal{M}$.
Conversely, assume $E \in \mathcal{M}$ and $\mu(E) < \infty$. There is an open $V \supset E$ such that $\mu(V) < \infty$. Pick a compact $K \subset V$ such that

$$\mu(V \setminus K) < \varepsilon.$$

Since $E \cap K \in \mathcal{M}_F$ there is a compact $K_1 \subset (E \cap K)$ such that

$$\mu(E \cap K) < \mu(K_1) + \varepsilon.$$

Since

$$E \subset (E \cap K) \cup (V \setminus K),$$

it follows that

$$\mu(E) \leq \mu(E \cap K) + \mu(V \setminus K) < \mu(K_1) + 2\varepsilon.$$

This implies $E \in \mathcal{M}_F$.
To finish the proof, we show that

$$F(f) = \int_U f \, d\mu, \forall f \in C_c(U)$$

From linearity it suffices to prove the result for the case where f is real.
Let $f \in C_c(U)$. Let K be the support of f and let $[a,b] \subset \mathbb{R}$ be such that

$$R(f) \subset (a,b),$$

where $R(f)$ denotes the range of f.

Suppose given a not relabeled $\varepsilon > 0$. Choose a partition of $[a,b]$ denoted by

$$\{y_i\} = \{a = y_0 < y_1 < y_2 < .. < y_n = b\},$$

such that $y_i - y_{i-1} < \varepsilon, \forall i \in \{1,...,n\}$.

Denote

$$E_i = \{u \in K \mid y_{i-1} < f(u) \leq y_i\},$$

$\forall i \in \{1,...,n\}$.

Since f is continuous, it is Borel measurable, and the sets E_i are disjoint Borel ones such that

$$\cup_{i=1}^n E_i = K.$$

Select open sets $V_i \supset E_i$ such that

$$\mu(V_i) < \mu(E_i) + \frac{\varepsilon}{n}, \forall i \in \{1,...,n\},$$

and such that

$$f(u) < y_i + \varepsilon, \forall u \in V_i.$$

From Theorem 9.1.10 there exists a partition of unity subordinate to $\{V_i\}_{i=1}^n$ such that $h_i \in C_c(V_i)$, $0 \leq h_i \leq 1$ and

$$\sum_{i=1}^n h_i = 1, \text{ on } K.$$

Hence,

$$f = \sum_{i=1}^n h_i f,$$

and

$$\mu(K) \leq F\left(\sum_{i=1}^n h_i f\right) = \sum_{i=1}^n F(h_i f).$$

Observe that

$$\mu(E_i) + \frac{\varepsilon}{n} \quad > \quad \mu(V_i)$$
$$= \quad \sup\{F(f) \mid f \in C_c(V_i),\ 0 \leq f \leq 1\}$$
$$> \quad F(h_i), \forall i \in \{1,..,n\}. \tag{9.13}$$

Thus,

$$F(f) \quad = \quad \sum_{i=1}^n F(h_i f)$$
$$\leq \quad \sum_{i=1}^n F(h_i(y_{i-1} + 2\varepsilon))$$
$$= \quad \sum_{i=1}^n (y_{i-1} + 2\varepsilon) F(h_i)$$
$$< \quad \sum_{i=1}^n (y_{i-1} + 2\varepsilon)\left(\mu(E_i) + \frac{\varepsilon}{n}\right)$$
$$< \quad \sum_{i=1}^n y_{i-1}\mu(E_i) + \sum_{i=1}^n (y_{i-1})\frac{\varepsilon}{n} + 2\varepsilon \sum_{i=1}^n \mu(E_i) + 2\varepsilon^2$$
$$< \quad \int_U f\, d\mu + b\varepsilon + 2\varepsilon\mu(K) + 2\varepsilon^2. \tag{9.14}$$

Since $\varepsilon > 0$ is arbitrary, we obtain

$$F(f) \le \int_U f \, d\mu, \forall f \in C_c(U).$$

From this

$$F(-f) \le \int_U (-f) \, d\mu, \forall f \in C_c(U),$$

that is,

$$F(f) \ge \int_U f \, d\mu, \forall f \in C_c(U).$$

Hence,

$$F(f) = \int_U f \, d\mu, \forall f \in C_c(U).$$

The proof is complete.

9.3 The Lebesgue points

In this section we introduce a very important concept in analysis namely, the definition of Lebesgue points.

We recall that in $\mathbb{R}^n$ the open ball with center u and radius r is defined by

$$B_r(u) = \{v \in \mathbb{R}^n \mid |v - u|_2 < r\}.$$

Consider a Borel measure μ on $\mathbb{R}^n$. We may associate to μ, the function $F_{r\mu}(u)$, denoted by

$$F_{r\mu}(u) = \frac{\mu(B_r(u))}{m(B_r(u))},$$

where m denotes the Lebesgue measure.

We define the symmetric derivative of μ at u, by $(D\mu)(u)$, by

$$(D\mu)(u) = \lim_{r \to 0} F_{r\mu}(u),$$

whenever such a limit exists.

We also define the function G_μ for a positive measure μ by

$$G_\mu(u) = \sup_{0 < r < \infty} F_{r\mu}(u).$$

The function $G_\mu : \mathbb{R}^n \to [0, +\infty]$, is lower semi-continuous and hence measurable.

Lemma 9.3.1 *Let $W = \cup_{i=1}^N B_{r_i}(u_i)$ be a finite union of open balls. Then there is a set $S \subset \{1, 2, ..., N\}$ such that*

1. *The balls $B_{r_i}(u_i), i \in S$ are disjoint.*

2. *$W \subset \cup_{i \in S} B_{3r_i}(u_i)$.*

Proof 9.10 Let us first order the balls $B_{r_i}(u_i)$ so that

$$r_1 \ge r_2 \ge ... \ge r_N.$$

Set $i_1 = 1$, and discard all balls such that

$$B_{i_1} \cap B_j \ne \emptyset.$$

Let B_{i_2} be the first of the remaining balls, if any. Discard all B_j such that $j > i_2$ and $B_{i_2} \cap B_j \neq \emptyset$.

Let B_{i_3} the first of the remaining balls as long as possible. Such a process stops after a finite number of steps. Define $S = \{i_1, i_2, ...\}$. It is clear that (1) holds. Now we prove that each discarded B_j is contained in

$$\{B_{3r_i}, \ i \in S\}.$$

For, just observe that if $r' < r$ and $B_{r'}(u')$ intersects $B_r(u)$ then $B_{r'}(u') \subset B_{3r}(u)$.

The proof is complete.

Theorem 9.3.2 *Suppose μ is a finite Borel measure on $\mathbb{R}^n$ and $\lambda > 0$. Then*

$$m(A_\lambda) \leq 3^n \lambda^{-1} \|\mu\|,$$

where

$$A_\lambda = \{u \in U \mid G_\mu(u) > \lambda\},$$

and

$$\|\mu\| = |\mu|(\mathbb{R}^n).$$

Proof 9.11   Let K be a compact subset of the open set A_λ.

As $G_\mu(u) = \sup_{0 < r < \infty} \{F_{r\mu}(u)\}$, each $u \in K$ is the center of an open ball B_u such that

$$\mu(B_u) > \lambda m(B_u).$$

Since K is compact, there exists a finite number of such balls which covers K. By Lemma 9.3.1, there exists a disjoint sub-collection here denoted by $\{B_{r_1}, ..., B_{r_N}\}$ such that $K \subset \cup_{k=1}^N B_{3r_k}$, so that

$$
\begin{aligned}
m(K) \quad &\leq \quad 3^n \sum_{k=1}^N m(B_{r_k}) \\
&\leq \quad 3^n \lambda^{-1} \sum_{k=1}^N |\mu|(B_{r_k}) \\
&\leq \quad 3^n \lambda^{-1} \|\mu\|.
\end{aligned}
\tag{9.15}
$$

The result follows taking the supremum relating all compact $K \subset A_\lambda$.

Remark 9.3.3 *Observe that, if $f \in L^1(\mathbb{R}^n)$ and $\lambda > 0$, for $A_\lambda = \{u \in \mathbb{R}^n \mid |f| > \lambda\}$, we have*

$$m(A_\lambda) \leq \lambda^{-1} \|f\|_1.$$

This follows from the fact that

$$\lambda m(A_\lambda) \leq \int_{A_\lambda} |f| \, dm \leq \int_{\mathbb{R}^n} |f| \, dm = \|f\|_1.$$

Observe also that defining $d\eta = |f| \, dm$, for every $\lambda > 0$, defining

$$G_f(u) = \sup_{0 < r < \infty} \frac{\eta(B_r(u))}{m(B_r(u))},$$

and

$$A_\lambda = \{u \in U \mid G_f(u) > \lambda\},$$

we have

$$m(A_\lambda) \leq 3^n \lambda^{-1} \|f\|_1.$$

9.3.1 The main result on Lebesgue points

Finally in this section we present the main definition of Lebesgue points and some relating results.

Definition 9.3.4 *Let $f \in L^1(\mathbb{R}^n)$. A point $u \in L^1(\mathbb{R}^n)$ such that*

$$\lim_{r \to 0} \frac{1}{m(B_r(u))} \int_{B_r(u)} |f(v) - f(u)| \, dm(v) = 0,$$

is called a Lebesgue point of f.

Theorem 9.3.5 *If $f \in L^1(\mathbb{R}^n)$, then almost all $u \in \mathbb{R}^n$ is a Lebesgue point of f.*

Proof 9.12 Define

$$H_{r_f}(u) = \frac{1}{m(B_r(u))} \int_{B_r(u)} |f - f(u)| \, dm, \forall u \in \mathbb{R}^n, r > 0,$$

also define

$$H_f(u) = \limsup_{r \to 0} H_{r_f}(u).$$

We have to show that $H_f = 0$, a.e. $[m]$.

Select $y > 0$ and fix $k \in \mathbb{N}$. Observe that there exists $g \in C(\mathbb{R}^n)$ such that

$$\|f - g\|_1 < 1/k.$$

Define $h = f - g$. Since g is continuous, $H_g = 0$ in $\mathbb{R}^n$. Observe that

$$
\begin{aligned}
H_{r_h}(u) &= \frac{1}{m(B_r(u))} \int_{B_r(u)} |h - h(u)| \, dm \\
&\leq \frac{1}{m(B_r(u))} \int_{B_r(u)} |h| \, dm + |h(u)|,
\end{aligned}
\tag{9.16}
$$

so that

$$H_h < G_h + |h|.$$

Since

$$H_{r_f} \leq H_{r_g} + H_{r_h},$$

we obtain

$$H_f \leq G_h + |h|.$$

Define

$$A_y = \{u \in \mathbb{R}^n \mid H_f(u) > 2y\},$$
$$B_{y,k} = \{u \in \mathbb{R}^n \mid G_h(u) > y\}$$

and

$$C_{y,k} = \{u \in \mathbb{R}^n \mid |h| > y\}.$$

Observe that $\|h\|_1 < 1/k$, so that from remark 9.3.3 we obtain

$$m(B_{y,k}) \leq \frac{3^n}{yk},$$

and

$$m(C_{y,k}) \leq \frac{1}{yk},$$

and hence

$$m(B_{y,k} \cup C_{y,k}) \leq \frac{3^n + 1}{yk},$$

Therefore

$$m(A_y) \leq m(B_{y,k} \cup C_{y,k}) \leq \frac{3^n + 1}{yk}.$$

Since k is arbitrary, we get $m(A_y) = 0, \forall y > 0$ so that $m\{u \in \mathbb{R}^n \mid H_f(u) > 0\} = 0$.
The proof is complete.

We finish this section with the following result.

Theorem 9.3.6 *Suppose μ is a complex Borel measure on $\mathbb{R}^n$ such that $\mu \ll m$. Suppose f is the Radon-Nikodym derivative of μ with respect to m. Under such assumptions,*

$$D\mu = f, \ a.e. \ [m],$$

and

$$\mu(E) = \int_E D\mu \, dm,$$

for all Borel set $E \subset \mathbb{R}^n$.

Proof 9.13   From the Radon-Nikodym theorem we have

$$\mu(E) = \int_E f \, dm,$$

for all measurable set $E \subset \mathbb{R}^n$.
Observe that at any Lebesgue point u of f we have

$$
\begin{aligned}
f(u) &= \lim_{r \to 0} \frac{1}{m(B_r(u))} \int_{B_r(u)} f \, dm \\
&= \lim_{r \to 0} \frac{\mu(B_r(u))}{m(B_r(u))} \\
&= D\mu(u).
\end{aligned}
\tag{9.17}
$$

The proof is complete.

Chapter 10

Distributions

The main reference for this chapter is Rudin [67].

10.1 Basic definitions and results

Definition 10.1.1 (Test Functions, the Space $\mathscr{D}(\Omega)$) *Let $\Omega \subset \mathbb{R}^n$ be a nonempty open set. For each $K \subset \Omega$ compact, consider the space $\mathscr{D}_K$, the set of all $C^\infty(\Omega)$ functions with support in K. We define the space of test functions, denoted by $\mathscr{D}(\Omega)$ as*

$$\mathscr{D}(\Omega) = \cup_{K \subset \Omega} \mathscr{D}_K, \; K \; compact. \tag{10.1}$$

Thus $\phi \in \mathscr{D}(\Omega)$ if and only if $\phi \in C^\infty(\Omega)$ and the support of ϕ is a compact subset of Ω.

Definition 10.1.2 (Topology for $\mathscr{D}(\Omega)$) *Let $\Omega \subset \mathbb{R}^n$ be an open set.*

1. *For every $K \subset \Omega$ compact, σ_K denotes the topology which a local base is defined by $\{\mathscr{V}_{N,k}\}$, where $N, k \in \mathbb{N}$,*

$$\mathscr{V}_{N,k} = \{\phi \in \mathscr{D}_K \mid \|\phi\|_N < 1/k\} \tag{10.2}$$

and

$$\|\phi\|_N = \max\{|D^\alpha \phi(x)| \mid x \in \Omega, |\alpha| \leq N\}. \tag{10.3}$$

2. *$\hat{\sigma}$ denotes the collection of all convex balanced sets $\mathscr{W} \in \mathscr{D}(\Omega)$ such that $\mathscr{W} \cap \mathscr{D}_K \in \sigma_K$ for every compact $K \subset \Omega$.*

3. *We define σ in $\mathscr{D}(\Omega)$ as the collection of all unions of sets of the form $\phi + \mathscr{W}$, for $\phi \in \mathscr{D}(\Omega)$ and $\mathscr{W} \in \hat{\sigma}$.*

Theorem 10.1.3 *Concerning the last definition we have the following:*

1. *σ is a topology in $\mathscr{D}(\Omega)$.*

2. *Through σ, $\mathscr{D}(\Omega)$ is made into a locally convex topological vector space.*

Proof 10.1

1. From item 3 of Definition 10.1.2, it is clear that arbitrary unions of elements of σ are elements of σ. Let us now show that finite intersections of elements of σ also belongs to σ. Suppose $\mathscr{V}_1 \in \sigma$ and $\mathscr{V}_2 \in \sigma$, if $\mathscr{V}_1 \cap \mathscr{V}_2 = \emptyset$ we are done. Thus, suppose $\phi \in \mathscr{V}_1 \cap \mathscr{V}_2$. By the definition of σ there exist two sets of indices L_1 and L_2, such that

$$\mathscr{V}_i = \cup_{\lambda \in L_i}(\phi_{i\lambda} + \mathscr{W}_{i\lambda}), \text{ for } i = 1,2, \tag{10.4}$$

where, $W_{i\lambda} \in \hat{\sigma}, \forall \lambda \in L_i$. Moreover, since $\phi \in \mathscr{V}_1 \cap \mathscr{V}_2$ there exist $\phi_i \in \mathscr{D}(\Omega)$ and $\mathscr{W}_i \in \hat{\sigma}$ such that

$$\phi \in \phi_i + \mathscr{W}_i \subset \mathscr{V}_i, \text{ for } i = 1,2. \tag{10.5}$$

Thus there exists $K \subset \Omega$ such that $\phi_i \in \mathscr{D}_K$ for $i \in \{1,2\}$. Since $\mathscr{D}_K \cap \mathscr{W}_i \in \sigma_K$, $\mathscr{D}_K \cap \mathscr{W}_i$ is open in $\mathscr{D}_K$ so that from (10.5) there exists $0 < \delta_i < 1$ such that

$$\phi - \phi_i \in (1 - \delta_i)\mathscr{W}_i, \text{ for } i \in \{1,2\}. \tag{10.6}$$

From (10.6) and from the convexity of $\mathscr{W}_i$ we have

$$\phi - \phi_i + \delta_i \mathscr{W}_i \subset (1 - \delta_i)\mathscr{W}_i + \delta_i \mathscr{W}_i = \mathscr{W}_i \tag{10.7}$$

so that

$$\phi + \delta_i \mathscr{W}_i \subset \phi_i + \mathscr{W}_i \subset \mathscr{V}_i, \text{ for } i \in \{1,2\}. \tag{10.8}$$

Define $\mathscr{W}_\phi = (\delta_1 \mathscr{W}_1) \cap (\delta_2 \mathscr{W}_2)$ so that

$$\phi + \mathscr{W}_\phi \subset \mathscr{V}_i, \tag{10.9}$$

and therefore we may write

$$\mathscr{V}_1 \cap \mathscr{V}_2 = \cup_{\phi \in \mathscr{V}_1 \cap \mathscr{V}_2}(\phi + \mathscr{W}_\phi) \in \sigma. \tag{10.10}$$

This completes the proof.

2. It suffices to show that single points are closed sets in $\mathscr{D}(\Omega)$ and the vector space operations are continuous.

   (a) Pick $\phi_1, \phi_2 \in \mathscr{D}(\Omega)$ such that $\phi_1 \neq \phi_2$ and define

   $$\mathscr{V} = \{\phi \in \mathscr{D}(\Omega) \mid \|\phi\|_0 < \|\phi_1 - \phi_2\|_0\}. \tag{10.11}$$

   Thus, $\mathscr{V} \in \hat{\sigma}$ and $\phi_1 \notin \phi_2 + \mathscr{V}$. As $\phi_2 + \mathscr{V}$ is open is contained $\mathscr{D}(\Omega) \setminus \{\phi_1\}$ and $\phi_2 \neq \phi_1$ is arbitrary, it follows that $\mathscr{D}(\Omega) \setminus \{\phi_1\}$ is open, so that $\{\phi_1\}$ is closed.

   (b) The proof that addition is σ-continuous follows from the convexity of any element of $\hat{\sigma}$. Thus given $\phi_1, \phi_2 \in \mathscr{D}(\Omega)$ and $\mathscr{V} \in \hat{\sigma}$ we have

   $$\phi_1 + \frac{1}{2}\mathscr{V} + \phi_2 + \frac{1}{2}\mathscr{V} = \phi_1 + \phi_2 + \mathscr{V}. \tag{10.12}$$

   (c) To prove the continuity of scalar multiplication, first consider $\phi_0 \in \mathscr{D}(\Omega)$ and $\alpha_0 \in \mathbb{R}$. Then:

   $$\alpha\phi - \alpha_0\phi_0 = \alpha(\phi - \phi_0) + (\alpha - \alpha_0)\phi_0. \tag{10.13}$$

For $\mathscr{V} \in \hat{\sigma}$ there exists $\delta > 0$ such that $\delta \phi_0 \in \frac{1}{2} \mathscr{V}$. Let us define $c = \frac{1}{2(|\alpha_0| + \delta)}$. Thus if $|\alpha - \alpha_0| < \delta$ then $(\alpha - \alpha_0)\phi_0 \in \frac{1}{2} \mathscr{V}$. Let $\phi \in \mathscr{D}(\Omega)$ such that

$$\phi - \phi_0 \in c\mathscr{V} = \frac{1}{2(|\alpha_0| + \delta)} \mathscr{V}, \tag{10.14}$$

so that

$$(|\alpha_0| + \delta)(\phi - \phi_0) \in \frac{1}{2} \mathscr{V}. \tag{10.15}$$

This means

$$\alpha(\phi - \phi_0) + (\alpha - \alpha_0)\phi_0 \in \frac{1}{2} \mathscr{V} + \frac{1}{2} \mathscr{V} = \mathscr{V}. \tag{10.16}$$

Therefore, $\alpha\phi - \alpha_0\phi_0 \in \mathscr{V}$ whenever $|\alpha - \alpha_0| < \delta$ and $\phi - \phi_0 \in c\mathscr{V}$.

For the next result the proof may be found in Rudin [67].

Proposition 10.1.4 *A convex balanced set $\mathscr{V} \subset \mathscr{D}(\Omega)$ is open if and only if $\mathscr{V} \in \sigma$.*

Proposition 10.1.5 *The topology σ_K of $\mathscr{D}_K \subset \mathscr{D}(\Omega)$ coincides with the topology that $\mathscr{D}_K$ inherits from $\mathscr{D}(\Omega)$.*

Proof 10.2   From Proposition 10.1.4 we have

$$\mathscr{V} \in \sigma \text{ implies } \mathscr{D}_K \cap \mathscr{V} \in \sigma_K. \tag{10.17}$$

Now suppose $\mathscr{V} \in \sigma_K$, we must show that there exists $A \in \sigma$ such that $\mathscr{V} = A \cap \mathscr{D}_K$. The definition of σ_K implies that for every $\phi \in \mathscr{V}$, there exist $N \in \mathbb{N}$ and $\delta_\phi > 0$ such that

$$\{\varphi \in \mathscr{D}_K \mid \|\varphi - \phi\|_N < \delta_\phi\} \subset \mathscr{V}. \tag{10.18}$$

Define

$$\mathscr{U}_\phi = \{\varphi \in \mathscr{D}(\Omega) \mid \|\varphi\|_N < \delta_\phi\}. \tag{10.19}$$

Then $\mathscr{U}_\phi \in \hat{\sigma}$ and

$$\mathscr{D}_K \cap (\phi + \mathscr{U}_\phi) = \phi + (\mathscr{D}_K \cap \mathscr{U}_\phi) \subset \mathscr{V}. \tag{10.20}$$

Defining $A = \cup_{\phi \in \mathscr{V}} (\phi + \mathscr{U}_\phi)$, we have completed the proof.

The proof for the next two results may also be found in Rudin [67].

Proposition 10.1.6 *If A is a bounded set of $\mathscr{D}(\Omega)$ then $A \subset \mathscr{D}_K$ for some $K \subset \Omega$, and there are $M_N < \infty$ such that $\|\phi\|_N \leq M_N, \forall \phi \in A$, $N \in \mathbb{N}$.*

Proposition 10.1.7 *If $\{\phi_n\}$ is a Cauchy sequence in $\mathscr{D}(\Omega)$, then $\{\phi_n\} \subset \mathscr{D}_K$ for some $K \subset \Omega$ compact, and*

$$\lim_{i,j \to \infty} \|\phi_i - \phi_j\|_N = 0, \forall N \in \mathbb{N}. \tag{10.21}$$

Proposition 10.1.8 *If $\phi_n \to 0$ in $\mathscr{D}(\Omega)$, then there exists a compact $K \subset \Omega$ which contains the support of $\phi_n, \forall n \in \mathbb{N}$ and $D^\alpha \phi_n \to 0$ uniformly, for each multi-index α.*

The proof follows directly from last proposition.

Theorem 10.1.9 *Suppose* $T : \mathscr{D}(\Omega) \to V$ *is linear, where* V *is a locally convex space. Then the following statements are equivalent.*

1. *T is continuous.*

2. *T is bounded.*

3. *If $\phi_n \to \theta$ in $\mathscr{D}(\Omega)$ then $T(\phi_n) \to \theta$ as $n \to \infty$.*

4. *The restrictions of T to each $\mathscr{D}_K$ are continuous.*

Proof 10.3

- ■ $1 \Rightarrow 2$. This follows from Proposition 2.7.3 .

- ■ $2 \Rightarrow 3$. Suppose T is bounded and $\phi_n \to 0$ in $\mathscr{D}(\Omega)$, by last proposition $\phi_n \to 0$ in some $\mathscr{D}_K$ so that $\{\phi_n\}$ is bounded and $\{T(\phi_n)\}$ is also bounded. Hence by Proposition 2.7.3, $T(\phi_n) \to 0$ in V.

- ■ $3 \Rightarrow 4$. Assume 3 holds and consider $\{\phi_n\} \subset \mathscr{D}_K$. If $\phi_n \to \theta$ then, by Proposition 10.1.5, $\phi_n \to \theta$ in $\mathscr{D}(\Omega)$, so that, by above $T(\phi_n) \to \theta$ in V. Since $\mathscr{D}_K$ is metrizable, also by proposition 2.7.3 we have that 4 follows.

- ■ $4 \Rightarrow 1$. Assume 4 holds and let $\mathscr{V}$ be a convex balanced neighborhood of zero in V. Define $\mathscr{U} = T^{-1}(\mathscr{V})$. Thus $\mathscr{U}$ is balanced and convex. By Proposition 10.1.5, $\mathscr{U}$ is open in $\mathscr{D}(\Omega)$ if and only if $\mathscr{D}_K \cap \mathscr{U}$ is open in $\mathscr{D}_K$ for each compact $K \subset \Omega$, thus if the restrictions of T to each $\mathscr{D}_K$ are continuous at θ, then T is continuous at θ, hence 4 implies 1.

Definition 10.1.10 (Distribution) *A linear functional in $\mathscr{D}(\Omega)$ which is continuous with respect to σ is said to be a Distribution.*

Proposition 10.1.11 *Every differential operator is a continuous mapping from $\mathscr{D}(\Omega)$ into $\mathscr{D}(\Omega)$.*

Proof 10.4 Since $\|D^\alpha \phi\|_N \le \|\phi\|_{|\alpha|+N}, \forall N \in \mathbb{N}$, D^α is continuous on each $\mathscr{D}_K$, so that by Theorem 10.1.9 D^α is continuous on $\mathscr{D}(\Omega)$.

Theorem 10.1.12 *Denoting by $\mathscr{D}'(\Omega)$ the dual space of $\mathscr{D}(\Omega)$ we have that $T : \mathscr{D}(\Omega) \to \mathbb{R} \in \mathscr{D}'(\Omega)$ if and only if for each compact set $K \subset \Omega$ there exists an $N \in \mathbb{N}$ and $c \in \mathbb{R}^+$ such that*

$$|T(\phi)| \le c\|\phi\|_N, \forall \phi \in \mathscr{D}_K. \tag{10.22}$$

Proof 10.5 The proof follows from the equivalence of 1 and 4 in Theorem 10.1.9.

10.2 Differentiation of distributions

Definition 10.2.1 (Derivatives for Distributions) *Given $T \in \mathscr{D}'(\Omega)$ and a multi-index α, we define the D^α derivative of T as*

$$D^\alpha T(\phi) = (-1)^{|\alpha|} T(D^\alpha \phi), \forall \phi \in \mathscr{D}(\Omega). \tag{10.23}$$

Remark 10.2.2 *Observe that if $|T(\phi)| \leq c\|\phi\|_N, \forall \phi \in \mathscr{D}(\Omega)$ for some $c \in \mathbb{R}^+$, then*

$$|D^\alpha T(\phi)| \leq c\|D^\alpha \phi\|_N \leq c\|\phi\|_{N+|\alpha|}, \forall \phi \in \mathscr{D}(\Omega), \tag{10.24}$$

thus $D^\alpha T \in \mathscr{D}'(\Omega)$. Therefore, derivatives of distributions are also distributions.

Theorem 10.2.3 *Suppose $\{T_n\} \subset \mathscr{D}'(\Omega)$. Let $T : \mathscr{D}(\Omega) \to \mathbb{R}$ be defined by*

$$T(\phi) = \lim_{n \to \infty} T_n(\phi), \forall \phi \in \mathscr{D}(\Omega). \tag{10.25}$$

Then $T \in \mathscr{D}'(\Omega)$, and

$$D^\alpha T_n \to D^\alpha T \text{ in } \mathscr{D}'(\Omega). \tag{10.26}$$

Proof 10.6 Let K be an arbitrary compact subset of Ω. Since (10.25) holds for every $\phi \in \mathscr{D}_K$, the principle of uniform boundedness implies that the restriction of T to $\mathscr{D}_K$ is continuous. It follows from Theorem 10.1.9 that T is continuous in $\mathscr{D}(\Omega)$, that is, $T \in \mathscr{D}'(\Omega)$. On the other hand

$$(D^\alpha T)(\phi) = (-1)^{|\alpha|} T(D^\alpha \phi) = (-1)^{|\alpha|} \lim_{n \to \infty} T_n(D^\alpha \phi) = \lim_{n \to \infty} (D^\alpha T_n(\phi)), \forall \phi \in \mathscr{D}(\Omega). \tag{10.27}$$

10.3 Examples of distributions

10.3.1 First example

Let $\Omega \subset \mathbb{R}^n$ be an open bounded set. As a first example of distribution consider the functional

$$T : \mathscr{D}(\Omega) \to \mathbb{R}$$

given by

$$T(\phi) = \int_\Omega f\phi \, dx,$$

where $f \in L^1(\Omega)$. Observe that

$$
\begin{aligned}
|T(\phi)| &\leq \int_\Omega |f\phi| \, dx \\
&\leq \int_\Omega |f| \, dx \|\phi\|_\infty,
\end{aligned} \tag{10.28}
$$

so that T is a bounded linear functional on $\mathscr{D}(\Omega)$, that is, T is a distribution.

10.3.2 Second example

For the second example, define $\Omega = (0,1)$, and $T : \mathscr{D}(\Omega) \to \mathbb{R}$ by

$$T(\phi) = \phi(1/2) + \phi'(1/3).$$

Thus,

$$|T(\phi)| = |\phi(1/2) + \phi'(1/3)| \leq \|\phi\|_\infty + \|\phi'\|_\infty \leq 2\|\phi\|_1,$$

so that T is also a distribution (bounded and linear).

10.3.3 Third example

For the third example, consider an open bounded $\Omega \subset \mathbb{R}^n$, and $T : \mathscr{D}(\Omega) \to \mathbb{R}$ by

$$T(\phi) = \int_\Omega f\phi \, dx,$$

where, $f \in L^1(\Omega)$.

Observe that the derivative of T for the multi-index $\alpha = (\alpha_1, ..., \alpha_n)$, is defined by

$$D^\alpha T(\phi) = (-1)^{|\alpha|} T(D^\alpha \phi) = (-1)^{|\alpha|} \int_\Omega f D^\alpha \phi \, dx.$$

If there exists $g \in L^1(\Omega)$, such that

$$(-1)^{|\alpha|} \int_\Omega f D^\alpha \phi \, dx = \int_\Omega g\phi \, dx, \forall \phi \in \mathscr{D}(\Omega),$$

we say that g is the derivative D^α of f in the distributional sense.

For example, for $\Omega = (0,1)$ and $f : \overline{\Omega} \to \mathbb{R}$ given by

$$f(x) = \begin{cases} 0, & \text{if } x \in [0, 1/2], \\ 1, & \text{if } x \in (1/2, 1], \end{cases}$$

and

$$T(\phi) = \int_\Omega f\phi \, dx$$

where $\phi \in C_c^\infty(\Omega)$,we have

$$
\begin{aligned}
D_x T(\phi) &= -\int_\Omega f \frac{d\phi}{dx} \, dx \\
&= -\int_{1/2}^1 (1) \frac{d\phi}{dx} \, dx \\
&= -\phi(1) + \phi(1/2) = \phi(1/2),
\end{aligned}
\tag{10.29}
$$

that is,

$$D_x T(\phi) = \phi(1/2), \forall \phi \in C_c^\infty(\Omega).$$

Finally, defining $f : \overline{\Omega} \to \mathbb{R}$ by

$$f(x) = \begin{cases} x, & \text{if } x \in [0, 1/2], \\ -x+1, & \text{if } x \in (1/2, 1], \end{cases}$$

and,

$$T(\phi) = \int_\Omega f\phi \, dx$$

where $\phi \in C_c^\infty(\Omega)$ we have

$$
\begin{aligned}
D_x T(\phi) &= -\int_\Omega f \frac{d\phi}{dx} \, dx \\
&= -\int_0^1 f \frac{d\phi}{dx} \, dx \\
&= \int_0^1 g\phi \, dx,
\end{aligned}
\tag{10.30}
$$

where

$$g(x) = \begin{cases} 1, & \text{if } x \in [0, 1/2], \\ -1, & \text{if } x \in (1/2, 1]. \end{cases}$$

In such a case we denote $g = D_x f$ and say that g is the derivative of f in the distributional sense.

We emphasize that in this last example the classical derivative of f is not defined, since f is not differentiable at $x = 1/2$.

Chapter 11

The Lebesgue and Sobolev Spaces

Here we emphasize that the two main references for this chapter are Adams [2] and Evans [34]. We start with the definition of Lebesgue spaces, denoted by $L^p(\Omega)$, where $1 \leq p \leq \infty$ and $\Omega \subset \mathbb{R}^n$ is an open set. In this chapter, integrals always refer to the Lebesgue measure.

11.1 Definition and properties of L^p spaces

Definition 11.1.1 (L^p spaces) *For $1 \leq p < \infty$, we say that $u \in L^p(\Omega)$ if $u : \Omega \to \mathbb{R}$ is measurable and*

$$\int_\Omega |u|^p dx < \infty. \tag{11.1}$$

We also denote $\|u\|_p = [\int_\Omega |u|^p dx]^{1/p}$ and will show that $\|\cdot\|_p$ is a norm.

Definition 11.1.2 (L^∞ spaces) *We say that $u \in L^\infty(\Omega)$ if u is measurable and there exists $M \in \mathbb{R}^+$, such that $|u(x)| \leq M$, a.e. in Ω. We define*

$$\|u\|_\infty = \inf\{M > 0 \mid |u(x)| \leq M, \text{ a.e. in } \Omega\}. \tag{11.2}$$

We will show that $\|\cdot\|_\infty$ is a norm. For $1 \leq p \leq \infty$, we define q by the relations

$$q = \begin{cases} +\infty, & \text{if } p = 1, \\ \frac{p}{p-1}, & \text{if } 1 < p < +\infty, \\ 1, & \text{if } p = +\infty, \end{cases}$$

so that symbolically we have

$$\frac{1}{p} + \frac{1}{q} = 1.$$

The next result is fundamental in the proof of the Sobolev Imbedding theorem.

Theorem 11.1.3 (Hölder inequality) *Consider $u \in L^p(\Omega)$ and $v \in L^q(\Omega)$, with $1 \leq p \leq \infty$. Then $uv \in L^1(\Omega)$ and*

$$\int_\Omega |uv| dx \leq \|u\|_p \|v\|_q. \tag{11.3}$$

Proof 11.1 The result is clear if $p = 1$ or $p = \infty$. You may assume $\|u\|_p, \|v\|_q > 0$, otherwise the result is also obvious. Thus suppose $1 < p < \infty$. From the concavity of log function on $(0, \infty)$ we obtain

$$\log\left(\frac{1}{p}a^p + \frac{1}{q}b^q\right) \geq \frac{1}{p}\log a^p + \frac{1}{q}\log b^q = \log(ab). \tag{11.4}$$

Thus,

$$ab \leq \frac{1}{p}(a^p) + \frac{1}{q}(b^q), \ \forall a \geq 0, b \geq 0. \tag{11.5}$$

Therefore,

$$|u(x)||v(x)| \leq \frac{1}{p}|u(x)|^p + \frac{1}{q}|v(x)|^q, \ \text{a.e. in } \Omega. \tag{11.6}$$

Hence, $|uv| \in L^1(\Omega)$ and

$$\int_\Omega |uv|dx \leq \frac{1}{p}\|u\|_p^p + \frac{1}{q}\|v\|_q^q. \tag{11.7}$$

Replacing u by λu in (11.7) $\lambda > 0$, we obtain

$$\int_\Omega |uv|dx \leq \frac{\lambda^{p-1}}{p}\|u\|_p^p + \frac{1}{\lambda q}\|v\|_q^q. \tag{11.8}$$

For $\lambda = \|u\|_p^{-1}\|v\|_q^{q/p}$ we obtain the Hölder inequality.

The next step is to prove that $\|\cdot\|_p$ is a norm.

Theorem 11.1.4 $L^p(\Omega)$ *is a vector space and* $\|\cdot\|_p$ *is norm* $\forall p$ *such that* $1 \leq p \leq \infty$.

Proof 11.2 The only non-trivial property to be proved concerning the norm definition, is the triangle inequality. If $p = 1$ or $p = \infty$ the result is clear. Thus, suppose $1 < p < \infty$. For $u, v \in L^p(\Omega)$ we have

$$|u(x) + v(x)|^p \leq (|u(x)| + |v(x)|)^p \leq 2^p(|u(x)|^p + |v(x)|^p), \tag{11.9}$$

so that $u + v \in L^p(\Omega)$. On the other hand,

$$\begin{aligned}\|u+v\|_p^p &= \int_\Omega |u+v|^{p-1}|u+v|dx \\ &\leq \int_\Omega |u+v|^{p-1}|u|dx + \int_\Omega |u+v|^{p-1}|v|dx,\end{aligned} \tag{11.10}$$

and hence, from Hölder's inequality

$$\|u+v\|_p^p \leq \|u+v\|_p^{p-1}\|u\|_p + \|u+v\|_p^{p-1}\|v\|_p, \tag{11.11}$$

that is,

$$\|u+v\|_p \leq \|u\|_p + \|v\|_p, \forall u, v \in L^p(\Omega). \tag{11.12}$$

Theorem 11.1.5 $L^p(\Omega)$ *is a Banach space for any p such that* $1 \leq p \leq \infty$.

Proof 11.3 Suppose $p = \infty$. Suppose $\{u_n\}$ is Cauchy sequence in $L^\infty(\Omega)$. Thus, given $k \in \mathbb{N}$ there exists $N_k \in \mathbb{N}$ such that, if $m, n \geq N_k$ then

$$\|u_m - u_n\|_\infty < \frac{1}{k}. \tag{11.13}$$

Therefore, for each k, there exist a set E_k such that $m(E_k) = 0$, and

$$|u_m(x) - u_n(x)| < \frac{1}{k}, \; \forall x \in \Omega \setminus E_k, \; \forall m, n \geq N_k. \tag{11.14}$$

Observe that $E = \cup_{k=1}^\infty E_k$ is such that $m(E) = 0$. Thus $\{u_n(x)\}$ is a real Cauchy sequence at each $x \in \Omega \setminus E$. Define $u(x) = \lim_{n \to \infty} u_n(x)$ on $\Omega \setminus E$. Letting $m \to \infty$ in (11.14) we obtain

$$|u(x) - u_n(x)| < \frac{1}{k}, \; \forall x \in \Omega \setminus E, \; \forall n \geq N_k. \tag{11.15}$$

Thus, $u \in L^\infty(\Omega)$ and $\|u_n - u\|_\infty \to 0$ as $n \to \infty$.

Now suppose $1 \leq p < \infty$. Let $\{u_n\}$ a Cauchy sequence in $L^p(\Omega)$. We can extract a subsequence $\{u_{n_k}\}$ such that

$$\|u_{n_{k+1}} - u_{n_k}\|_p \leq \frac{1}{2^k}, \forall k \in \mathbb{N}. \tag{11.16}$$

To simplify the notation we write u_k in place of u_{n_k}, so that

$$\|u_{k+1} - u_k\|_p \leq \frac{1}{2^k}, \forall k \in \mathbb{N}. \tag{11.17}$$

Defining

$$g_n(x) = \sum_{k=1}^n |u_{k+1}(x) - u_k(x)|, \tag{11.18}$$

we obtain

$$\|g_n\|_p \leq 1, \; \forall n \in \mathbb{N}. \tag{11.19}$$

From the monotone convergence theorem and (11.19), $g_n(x)$ converges to a limit $g(x)$ with $g \in L^p(\Omega)$. On the other hand, for $m \geq n \geq 2$ we have

$$|u_m(x) - u_n(x)| \leq |u_m(x) - u_{m-1}(x)| + ... + |u_{n+1}(x) - u_n(x)| \leq g(x) - g_{n-1}(x), \text{ a.e. in } \Omega. \tag{11.20}$$

Hence, $\{u_n(x)\}$ is Cauchy a.e. in Ω and converges to a limit $u(x)$ so that

$$|u(x) - u_n(x)| \leq g(x), \text{ a.e. in } \Omega, \text{ for } n \geq 2, \tag{11.21}$$

which means $u \in L^p(\Omega)$. Finally, from $|u_n(x) - u(x)| \to 0$, a.e. in Ω, $|u_n(x) - u(x)|^p \leq |g(x)|^p$ and the Lebesgue dominated convergence theorem we get

$$\|u_n - u\|_p \to 0 \text{ as } n \to \infty. \tag{11.22}$$

Theorem 11.1.6 *Let* $\{u_n\} \subset L^p(\Omega)$ *and* $u \in L^p(\Omega)$ *such that* $\|u_n - u\|_p \to 0$. *Then there exists a subsequence* $\{u_{n_k}\}$ *such that*

1. $u_{n_k}(x) \to u(x)$, *a.e. in* Ω,

2. $|u_{n_k}(x)| \leq h(x)$, *a.e. in* $\Omega, \forall k \in \mathbb{N}$, *for some* $h \in L^p(\Omega)$.

Proof 11.4   The result is clear for $p = \infty$. Suppose $1 \leq p < \infty$. From the last theorem we can easily obtain that $|u_{n_k}(x) - u(x)| \to 0$ as $k \to \infty$, a.e. in Ω. To complete the proof, just take $h = |u| + g$, where g is defined in the proof of the last theorem.

Theorem 11.1.7 $L^p(\Omega)$ *is reflexive for all p such that* $1 < p < \infty$.

Proof 11.5   We divide the proof into 3 parts.

1. For $2 \leq p < \infty$ we have that

$$\left\| \frac{u+v}{2} \right\|_{L^p(\Omega)}^p + \left\| \frac{u-v}{2} \right\|_{L^p(\Omega)}^p$$

$$\leq \frac{1}{2} \left(\|u\|_{L^p(\Omega)}^p + \|v\|_{L^p(\Omega)}^p \right), \forall u, v \in L^p(\Omega). \tag{11.23}$$

Proof. Observe that

$$\alpha^p + \beta^p \leq \left(\alpha^2 + \beta^2 \right)^{p/2}, \forall \alpha, \beta \geq 0. \tag{11.24}$$

Now taking $\alpha = \left| \frac{a+b}{2} \right|$ and $\beta = \left| \frac{a-b}{2} \right|$ in (11.24), we obtain (using the convexity of $t^{p/2}$),

$$\left| \frac{a+b}{2} \right|^p + \left| \frac{a-b}{2} \right|^p \leq \left(\left| \frac{a+b}{2} \right|^2 + \left| \frac{a-b}{2} \right|^2 \right)^{p/2}$$

$$= \left(\frac{a^2}{2} + \frac{b^2}{2} \right)^{p/2}$$

$$\leq \frac{1}{2} |a|^p + \frac{1}{2} |b|^p. \tag{11.25}$$

The inequality (11.23) follows immediately.

2. $L^p(\Omega)$ is uniformly convex, and therefore reflexive for $2 \leq p < \infty$.
   Proof. Suppose given $\varepsilon > 0$ and suppose that

$$\|u\|_p \leq 1, \ \|v\|_p \leq 1 \text{ and } \|u-v\|_p > \varepsilon. \tag{11.26}$$

From part 1, we obtain

$$\left\| \frac{u+v}{2} \right\|_p^p < 1 - \left(\frac{\varepsilon}{2} \right)^p, \tag{11.27}$$

and therefore

$$\left\| \frac{u+v}{2} \right\|_p < 1 - \delta, \tag{11.28}$$

for $\delta = 1 - (1 - (\varepsilon/2)^p)^{1/p} > 0$. Thus, $L^p(\Omega)$ is uniformly convex and from Theorem 4.7.2 it is reflexive.

3. $L^p(\Omega)$ is reflexive for $1 < p \leq 2$. Let $1 < p \leq 2$, from 2 we can conclude that L^q is reflexive. We will define $T : L^p(\Omega) \to (L^q)^*$ by

$$\langle Tu, f \rangle_{L^q(\Omega)} = \int_\Omega uf dx, \forall u \in L^p(\Omega), \ f \in L^q(\Omega). \tag{11.29}$$

From the Hölder inequality, we obtain

$$|\langle Tu, f \rangle_{L^q(\Omega)}| \leq \|u\|_p \|f\|_q, \tag{11.30}$$

so that

$$\|Tu\|_{(L^q(\Omega))^*} \leq \|u\|_p. \tag{11.31}$$

Pick $u \in L^p(\Omega)$ and define $f_0(x) = |u(x)|^{p-2} u(x)$ ($f_0(x) = 0$ if $u(x) = 0$). Thus, we have that $f_0 \in L^q(\Omega)$, $\|f_0\|_q = \|u\|_p^{p-1}$ and $\langle Tu, f_0 \rangle_{L^q(\Omega)} = \|u\|_p^p$. Therefore,

$$\|Tu\|_{(L^q(\Omega))^*} \geq \frac{\langle Tu, f_0 \rangle_{L^q(\Omega)}}{\|f_0\|_q} = \|u\|_p \tag{11.32}$$

Hence from (11.31) and (11.32) we have

$$\|Tu\|_{(L^q(\Omega))^*} = \|u\|_p, \forall u \in L^p(\Omega). \tag{11.33}$$

Thus, T is an isometry from $L^p(\Omega)$ to a closed subspace of $(L^q(\Omega))^*$. Since from the first part $L^q(\Omega)$ is reflexive, we have that $(L^q(\Omega))^*$ is reflexive. Hence $T(L^p(\Omega))$ and $L^p(\Omega)$ are reflexive.

Theorem 11.1.8 (Riesz representation theorem) *Let $1 < p < \infty$ and let f be a continuous linear functional on $L^p(\Omega)$. Then there exists a unique $u_0 \in L^q$ such that*

$$f(v) = \int_\Omega vu_0 \, dx, \ \forall v \in L^p(\Omega). \tag{11.34}$$

Furthermore

$$\|f\|_{(L^p)^*} = \|u_0\|_q. \tag{11.35}$$

Proof 11.6 First we define the operator $T : L^q(\Omega) \to (L^p(\Omega))^*$ by

$$\langle Tu, v \rangle_{L^p(\Omega)} = \int_\Omega uv \, dx, \forall v \in L^p(\Omega). \tag{11.36}$$

Similarly to the last theorem, we obtain

$$\|Tu\|_{(L^p(\Omega))^*} = \|u\|_q. \tag{11.37}$$

We have to show that T is onto. Define $E = T(L^q(\Omega))$. As E is a closed subspace, it suffices to show that E is dense in $(L^p(\Omega))^*$. Suppose $h \in (L^p)^{**} = L^p$ is such that

$$\langle Tu, h \rangle_{L^p(\Omega)} = 0, \forall u \in L^q(\Omega). \tag{11.38}$$

Choosing $u = |h|^{p-2} h$ we may conclude that $h = 0$ which, by Corollary 4.2.14 completes the first part of the proof. The proof of uniqueness is left to the reader.

Definition 11.1.9 *Let $1 \leq p \leq \infty$. We say that $u \in L^p_{loc}(\Omega)$ if $u\chi_K \in L^p(\Omega)$ for all compact $K \subset \Omega$.*

11.1.1   Spaces of continuous functions

We introduce some definitions and properties concerning spaces of continuous functions. First, we recall that by a domain we mean an open set in $\mathbb{R}^n$. Thus for a domain $\Omega \subset \mathbb{R}^n$ and for any nonnegative integer m we define by $C^m(\Omega)$ the set of all functions u which the partial derivatives $D^\alpha u$ are continuous on Ω for any α such that $|\alpha| \leq m$, where if $D^\alpha = D_1^{\alpha_1} D_2^{\alpha_2} ... D_n^{\alpha_n}$ we have $|\alpha| = \alpha_1 + ... + \alpha_n$. We define $C^\infty(\Omega) = \cap_{m=0}^\infty C^m(\Omega)$ and denote $C^0(\Omega) = C(\Omega)$. Given a function $\phi : \Omega \to \mathbb{R}$, its support, denoted by $spt(\phi)$ is given by

$$spt(\phi) = \overline{\{x \in \Omega \mid \phi(x) \neq 0\}}.$$

$C_c^\infty(\Omega)$ denotes the set of functions in $C^\infty(\Omega)$ with compact support contained in Ω.

The sets $C_0(\Omega)$ and $C_0^\infty(\Omega)$ consist of the closure of $C_c(\Omega)$ (which is the set of functions in $C(\Omega)$ with compact support in Ω) and $C_c^\infty(\Omega)$ respectively, relating the uniform convergence norm. On the other hand, $C_B^m(\Omega)$ denotes the set of functions $u \in C^m(\Omega)$ for which $D^\alpha u$ is bounded on Ω for $0 \leq |\alpha| \leq m$. Observe that $C_B^m(\Omega)$ is a Banach space with the norm denoted by $\| \cdot \|_{B,m}$ given by

$$\|u\|_{B,m} = \max_{0 \leq |\alpha| \leq m} \sup_{x \in \Omega} \{|D^\alpha u(x)|\}.$$

Also, we define $C^m(\overline{\Omega})$ as the set of functions $u \in C^m(\Omega)$ for which $D^\alpha u$ is bounded and uniformly continuous on Ω for $0 \leq |\alpha| \leq m$. Observe that $C^m(\overline{\Omega})$ is a closed subspace of $C_B^m(\Omega)$ and is also a Banach space with the norm inherited from $C_B^m(\Omega)$. An important space is the one of Hölder continuous functions.

Definition 11.1.10 (Spaces of Hölder continuous functions) *If $0 < \lambda < 1$, for a nonnegative integer m we define the space of Hölder continuous functions denoted by $C^{m,\lambda}(\overline{\Omega})$, as the subspace of $C^m(\overline{\Omega})$ consisting of those functions u for which, for all $0 \leq |\alpha| \leq m$, there exists a constant K such that*

$$|D^\alpha u(x) - D^\alpha u(y)| \leq K|x-y|^\lambda, \forall x,y \in \Omega.$$

$C^{m,\lambda}(\overline{\Omega})$ is a Banach space with the norm denoted by $\| \cdot \|_{m,\lambda}$ given by

$$\|u\|_{m,\lambda} = \|u\|_{B,m} + \max_{0 \leq |\alpha| \leq m} \sup_{x,y \in \Omega} \left\{ \frac{|D^\alpha u(x) - D^\alpha u(y)|}{|x-y|^\lambda}, \quad x \neq y \right\}.$$

From now on we say that $f : \Omega \to \mathbb{R}$ is locally integrable, if it is Lebesgue integrable on any compact $K \subset \Omega$. Furthermore, we say that $f \in L_{loc}^p(\Omega)$ if $f \in L^p(K)$ for any compact $K \subset \Omega$. Finally, given an open $\Omega \subset \mathbb{R}^n$, we denote $W \subset\subset \Omega$ whenever $\overline{W}$ is compact and $\overline{W} \subset \Omega$.

Theorem 11.1.11 *The space $C_0(\Omega)$ is dense in $L^p(\Omega)$, for $1 \leq p < \infty$.*

Proof 11.7   For the proof we need the following lemma:

Lemma 11.1.12 *Let $f \in L_{loc}^1(\Omega)$ such that*

$$\int_\Omega fu \, dx = 0, \forall u \in C_0(\Omega). \tag{11.39}$$

Then $f = 0$ a.e. in Ω.

First suppose $f \in L^1(\Omega)$ and Ω bounded, so that $m(\Omega) < \infty$. Given $\varepsilon > 0$, since $C_0(\Omega)$ is dense in $L^1(\Omega)$, there exists $f_1 \in C_0(\Omega)$ such that $\|f - f_1\|_1 < \varepsilon$ and thus, from (11.39) we obtain

$$\left| \int_\Omega f_1 u \, dx \right| \leq \varepsilon \|u\|_\infty, \forall u \in C_0(\Omega). \tag{11.40}$$

Defining

$$K_1 = \{x \in \Omega \mid f_1(x) \geq \varepsilon\}, \tag{11.41}$$

and

$$K_2 = \{x \in \Omega \mid f_1(x) \leq -\varepsilon\}. \tag{11.42}$$

As K_1 and K_2 are disjoint compact sets, by the Urysohn Theorem there exists $u_0 \in C_0(\Omega)$ such that

$$u_0(x) = \begin{cases} +1, & \text{if } x \in K_1, \\ -1, & \text{if } x \in K_2 \end{cases} \tag{11.43}$$

and

$$|u_0(x)| \leq 1, \forall x \in \Omega. \tag{11.44}$$

Also defining $K = K_1 \cup K_2$, we may write

$$\int_\Omega f_1 u_0 \, dx = \int_{\Omega - K} f_1 u_0 \, dx + \int_K f_1 u_0 \, dx. \tag{11.45}$$

Observe that, from (11.40)

$$\int_K |f_1| \, dx \leq \int_\Omega |f_1 u_0| \, dx \leq \varepsilon \tag{11.46}$$

so that

$$\int_\Omega |f_1| \, dx = \int_K |f_1| \, dx + \int_{\Omega - K} |f_1| \, dx \leq \varepsilon + \varepsilon m(\Omega). \tag{11.47}$$

Hence,

$$\|f\|_1 \leq \|f - f_1\|_1 + \|f_1\|_1 \leq 2\varepsilon + \varepsilon m(\Omega). \tag{11.48}$$

Since $\varepsilon > 0$ is arbitrary, we have that $f = 0$ a.e. in Ω. Finally, if $m(\Omega) = \infty$, define

$$\Omega_n = \{x \in \Omega \mid \text{dist}(x, \Omega^c) > 1/n \text{ and } |x| < n\}. \tag{11.49}$$

It is clear that $\Omega = \cup_{n=1}^\infty \Omega_n$ and from above $f = 0$ a.e. on $\Omega_n, \forall n \in \mathbb{N}$, so that $f = 0$ a.e. in Ω.

Finally, to finish the proof of Theorem 11.1.11, suppose $h \in L^q(\Omega)$ is such that

$$\int_\Omega hu \, dx = 0, \forall u \in C_0(\Omega). \tag{11.50}$$

Observe that $h \in L^1_{loc}(\Omega)$ since $\int_K |h| \, dx \leq \|h\|_q m(K)^{1/p} < \infty$. From last lemma $h = 0$ a.e. in Ω, which by Corollary 4.2.14 completes the proof.

Theorem 11.1.13 *$L^p(\Omega)$ is separable for any $1 \leq p < \infty$.*

Proof 11.8 The result follows from last theorem and from the fact that $C_0(K)$ is separable for each $K \subset \Omega$ compact (from the Weierstrass theorem, polynomials with rational coefficients are dense $C_0(K)$). Observe that $\Omega = \cup_{n=1}^\infty \Omega_n$, Ω_n defined as in (11.49), where $\bar{\Omega}_n$ is compact, $\forall n \in \mathbb{N}$.

11.2 The Sobolev spaces

Now we define the Sobolev spaces, denoted by $W^{m,p}(\Omega)$.

Definition 11.2.1 (Sobolev Spaces) *We say that $u \in W^{m,p}(\Omega)$ if $u \in L^p(\Omega)$ and $D^\alpha u \in L^p(\Omega)$, for all α such that $0 \le |\alpha| \le m$, where the derivatives are understood in the distributional sense.*

Definition 11.2.2 *We define the norm $\| \cdot \|_{m,p}$ for $W^{m,p}(\Omega)$, where $m \in \mathbb{N}$ and $1 \le p \le \infty$, as*

$$\|u\|_{m,p} = \left\{ \sum_{0 \le |\alpha| \le m} \|D^\alpha u\|_p^p \right\}^{1/p}, \ if \ 1 \le p < \infty, \tag{11.51}$$

and

$$\|u\|_{m,\infty} = \max_{0 \le |\alpha| \le m} \|D^\alpha u\|_\infty. \tag{11.52}$$

Theorem 11.2.3 *$W^{m,p}(\Omega)$ is a Banach space.*

Proof 11.9 Consider $\{u_n\}$ a Cauchy sequence in $W^{m,p}(\Omega)$. Then $\{D^\alpha u_n\}$ is a Cauchy sequence for each $0 \le |\alpha| \le m$. Since $L^p(\Omega)$ is complete there exist functions u and u_α, for $0 \le |\alpha| \le m$, in $L^p(\Omega)$ such that $u_n \to u$ and $D^\alpha u_n \to u_\alpha$ in $L^p(\Omega)$ as $n \to \infty$. From above $L^p(\Omega) \subset L^1_{loc}(\Omega)$ and so u_n determines a distribution $T_{u_n} \in \mathscr{D}'(\Omega)$. For any $\phi \in \mathscr{D}(\Omega)$ we have, by Hölder's inequality

$$|T_{u_n}(\phi) - T_u(\phi)| \le \int_\Omega |u_n(x) - u(x)||\phi(x)|dx \le \|\phi\|_q \|u_n - u\|_p. \tag{11.53}$$

Hence $T_{u_n}(\phi) \to T_u(\phi)$ for every $\phi \in \mathscr{D}(\Omega)$ as $n \to \infty$. Similarly $T_{D^\alpha u_n}(\phi) \to T_{u_\alpha}(\phi)$ for every $\phi \in \mathscr{D}(\Omega)$. We have that

$$\begin{aligned} T_{u_\alpha}(\phi) &= \lim_{n \to \infty} T_{D^\alpha u_n}(\phi) \\ &= \lim_{n \to \infty} (-1)^{|\alpha|} T_{u_n}(D^\alpha \phi) \\ &= (-1)^{|\alpha|} T_u(D^\alpha \phi) = T_{D^\alpha u}(\phi), \end{aligned} \tag{11.54}$$

for every $\phi \in \mathscr{D}(\Omega)$. Thus $u_\alpha = D^\alpha u$ in the sense of distributions, for $0 \le |\alpha| \le m$, and $u \in W^{m,p}(\Omega)$. As $\lim_{n \to \infty} \|u - u_n\|_{m,p} = 0$, $W^{m,p}(\Omega)$ is complete.

Remark 11.2.4 *Observe that distributional and classical derivatives coincide when the latter exist and are continuous. We define $S \subset W^{m,p}(\Omega)$ by*

$$S = \{\phi \in C^m(\Omega) \mid \|\phi\|_{m,p} < \infty\} \tag{11.55}$$

Thus, the completion of S concerning the norm $\| \cdot \|_{m,p}$ is denoted by $H^{m,p}(\Omega)$.

Corollary 11.2.5 *$H^{m,p}(\Omega) \subset W^{m,p}(\Omega)$*

Proof 11.10 Since $W^{m,p}(\Omega)$ is complete we have that $H^{m,p}(\Omega) \subset W^{m,p}(\Omega)$.

Theorem 11.2.6 *$W^{m,p}(\Omega)$ is separable if $1 \le p < \infty$, and is reflexive and uniformly convex if $1 < p < \infty$. Particularly, $W^{m,2}(\Omega)$ is a separable Hilbert space with the inner product*

$$(u,v)_m = \sum_{0 \le |\alpha| \le m} \langle D^\alpha u, D^\alpha v \rangle_{L^2(\Omega)}. \tag{11.56}$$

Proof 11.11 We can see $W^{m,p}(\Omega)$ as a subspace of $L^p(\Omega, \mathbb{R}^N)$, where $N = \sum_{0 \le |\alpha| \le m} 1$. From the relevant properties for $L^p(\Omega)$, we have that $L^p(\Omega; \mathbb{R}^N)$ is a reflexive and uniformly convex for $1 < p < \infty$ and separable for $1 \le p < \infty$. Given $u \in W^{m,p}(\Omega)$, we may associate the vector $Pu \in L^p(\Omega; \mathbb{R}^N)$ defined by

$$Pu = \{D^\alpha u\}_{0 \le |\alpha| \le m}. \tag{11.57}$$

Since $\|Pu\|_{p^N} = \|u\|_{m,p}$, we have that $W^{m,p}$ is closed subspace of $L^p(\Omega; \mathbb{R}^N)$. Thus from Theorem 1.21 in Adams [1], we have that $W^{m,p}(\Omega)$ is separable if $1 \le p < \infty$ and, reflexive and uniformly convex, if $1 < p < \infty$.

Lemma 11.2.7 *Let $1 \le p < \infty$ and define $U = L^p(\Omega; \mathbb{R}^N)$. For every continuous linear functional f on U, there exists a unique $v \in L^q(\Omega; \mathbb{R}^N) = U^*$ such that*

$$f(u) = \sum_{i=1}^N \langle u_i, v_i \rangle, \forall u \in U. \tag{11.58}$$

Moreover,

$$\|f\|_{U^*} = \|v\|_{q^N}, \tag{11.59}$$

where $\|\cdot\|_{q^N} = \|\cdot\|_{L^q(\Omega, \mathbb{R}^N)}$.

Proof 11.12 For $u = (u_1, ..., u_n) \in L^p(\Omega; \mathbb{R}^N)$ we may write

$$
\begin{aligned}
f(u) &= f((u_1, 0, ..., 0)) + ... + f((0, ..., 0, u_j, 0, ..., 0)) \\
&\quad + ... + f((0, ..., 0, u_n)),
\end{aligned} \tag{11.60}
$$

and since $f((0, ..., 0, u_j, 0, ..., 0))$ is continuous linear functional on $u_j \in L^p(\Omega)$, there exists a unique $v_j \in L^q(\Omega)$ such that $f(0, ..., 0, u_j, 0, ..., 0) = \langle u_j, v_j \rangle_{L^2(\Omega)}, \forall u_j \in L^p(\Omega), \forall\, 1 \le j \le N$, so that

$$f(u) = \sum_{i=1}^N \langle u_i, v_i \rangle, \forall u \in U. \tag{11.61}$$

From Hölder's inequality we obtain

$$|f(u)| \le \sum_{j=1}^N \|u_j\|_p \|v_j\|_q \le \|u\|_{p^N} \|v\|_{q^N}, \tag{11.62}$$

and hence $\|f\|_{U^*} \le \|v\|_{q^N}$. The equality in (11.62) is achieved for $u \in L^p(\Omega, \mathbb{R}^N)$, $1 < p < \infty$ such that

$$u_j(x) = \begin{cases} |v_j|^{q-2} \bar{v}_j, & \text{if } v_j \ne 0 \\ 0, & \text{if } v_j = 0. \end{cases} \tag{11.63}$$

If $p = 1$ choose k such that $\|v_k\|_\infty = \max_{1 \le j \le N} \|v_j\|_\infty$. Given $\varepsilon > 0$, there is a measurable set A such that $m(A) > 0$ and $|v_k(x)| \ge \|v_k\|_\infty - \varepsilon, \forall x \in A$. Defining $u(x)$ as

$$u_i(x) = \begin{cases} \bar{v}_k / v_k, & \text{if } i = k, \, x \in A \text{ and } v_k(x) \ne 0 \\ 0, & \text{otherwise,} \end{cases} \tag{11.64}$$

we have

$$
\begin{aligned}
f(u_k) &= \langle u, v_k \rangle_{L^2(\Omega)} = \int_A |v_k| dx \\
&\ge (\|(v_k\|_\infty - \varepsilon) \|u_k\|_1 \\
&= (\|v\|_{\infty^N} - \varepsilon) \|u\|_{1^N}.
\end{aligned} \tag{11.65}
$$

Since ε is arbitrary, the proof is complete.

Theorem 11.2.8 *Let $1 \leq p < \infty$. Given a continuous linear functional f on $W^{m,p}(\Omega)$, there exists $v \in L^q(\Omega, \mathbb{R}^N)$ such that*

$$f(u) = \sum_{0 \leq |\alpha| \leq m} \langle D^\alpha u, v_\alpha \rangle_{L^2(\Omega)}. \tag{11.66}$$

Proof 11.13 Consider f a continuous linear operator on $U = W^{m,p}(\Omega)$. By the Hahn Banach Theorem we can extend f to $\tilde{f}$, on $L^p(\Omega; \mathbb{R}^N)$, so that $\|\tilde{f}\|_{q^N} = \|f\|_{U^*}$ and by the last theorem, there exists $\{v_\alpha\} \in L^q(\Omega; \mathbb{R}^N)$ such that

$$\tilde{f}(\hat{u}) = \sum_{0 \leq |\alpha| \leq m} \langle \hat{u}_\alpha, v_\alpha \rangle_{L^2(\Omega)}, \forall v \in L^p(\Omega; \mathbb{R}^N). \tag{11.67}$$

In particular, for $u \in W^{m,p}(\Omega)$, defining $\hat{u} = \{D^\alpha u\} \in L^p(\Omega; \mathbb{R}^N)$ we obtain

$$f(u) = \tilde{f}(\hat{u}) = \sum_{1 \leq |\alpha| \leq m} \langle D^\alpha u, v_\alpha \rangle_{L^2(\Omega)}. \tag{11.68}$$

Finally, observe that, also from the Hahn-Banach theorem $\|f\|_{U^*} = \|\tilde{f}\|_{q^N} = \|v\|_{q^N}$.

Definition 11.2.9 *Let $\Omega \subset \mathbb{R}^n$ be a domain. For m a positive integer and $1 \leq p < \infty$ we define $W_0^{m,p}(\Omega)$ a. the closure in $\|\cdot\|_{m,p}$ of $C_c^\infty(\Omega)$, where we recall that $C_c^\infty(\Omega)$ denotes the the set of $C^\infty(\Omega)$ functions with compact support contained in Ω. Finally, we also recall that the support of $\phi : \Omega \to \mathbb{R}$, denoted by $spt(\phi)$, i. given by*

$$spt(\phi) = \overline{\{x \in \Omega \mid \phi(x) \neq 0\}}.$$

11.3 The Sobolev imbedding theorem

11.3.1 The statement of Sobolev imbedding theorem

Now we present the Sobolev imbedding theorem. We recall that for normed spaces X, Y the notation

$$X \hookrightarrow Y$$

means that $X \subset Y$ and there exists a constant $K > 0$ such that

$$\|u\|_Y \leq K\|u\|_X, \forall u \in X.$$

If in addition the imbedding is compact then for any bounded sequence $\{u_n\} \subset X$ there exists a conver gent subsequence $\{u_{n_k}\}$, which converges to some u in the norm $\|\cdot\|_Y$. At this point, we first introduce th following definition:

Definition 11.3.1 *Let $\Omega \subset \mathbb{R}^n$ be an open bounded set. We say that $\partial\Omega$ is $\hat{C}^1$ if for each $x_0 \in \partial\Omega$, denotin $\hat{x} = (x_1, ..., x_{n-1})$ for a local coordinate system, there exist $r > 0$ and a function $f(x_1, ..., x_{n-1}) = f(\hat{x})$ suc that*

$$W = \overline{\Omega} \cap B_r(x_0) = \{x \in B_r(x_0) \mid x_n \leq f(x_1, ..., x_{n-1})\}.$$

Moreover, $f(\hat{x})$ is a Lipschitz continuous function, so that

$$|f(\hat{x}) - f(\hat{y})| \leq C_1 |\hat{x} - \hat{y}|_2, \text{ on its domain,}$$

for some $C_1 > 0$. Finally, we assume

$$\left\{ \frac{\partial f(\hat{x})}{\partial x_k} \right\}_{k=1}^{n-1}$$

is classically defined, almost everywhere also on its concerning domain, so that $f \in W^{1,2}$.

Theorem 11.3.2 (The Sobolev imbedding theorem) *Let Ω be an open bounded set in $\mathbb{R}^n$ such that $\partial\Omega$ is $\hat{C}^1$. Let $j \geq 0$ and $m \geq 1$ be integers and let $1 \leq p < \infty$.*

1. **Part I**

    (a) **Case A** *If either $mp > n$ or $m = n$ and $p = 1$ then*

    $$W^{j+m,p}(\Omega) \hookrightarrow C_B^j(\Omega). \tag{11.69}$$

    Moreover,

    $$W^{j+m,p}(\Omega) \hookrightarrow W^{j,q}(\Omega), \text{ for } p \leq q \leq \infty, \tag{11.70}$$

    and, in particular

    $$W^{m,p}(\Omega) \hookrightarrow L^q(\Omega), \text{ for } p \leq q \leq \infty. \tag{11.71}$$

    (b) **Case B** *If $mp = n$, then*

    $$W^{j+m,p}(\Omega) \hookrightarrow W^{j,q}(\Omega), \text{ for } p \leq q < \infty, \tag{11.72}$$

    and, in particular

    $$W^{m,p}(\Omega) \hookrightarrow L^q(\Omega), \text{ for } p \leq q < \infty. \tag{11.73}$$

    (c) **Case C** *If $mp < n$ or $p = 1$, then*

    $$W^{j+m,p}(\Omega) \hookrightarrow W^{j,q}(\Omega), \text{ for } p \leq q \leq p^* = \frac{np}{n-mp}, \tag{11.74}$$

    and, in particular

    $$W^{m,p}(\Omega) \hookrightarrow L^q(\Omega), \text{ for } p \leq q \leq p^* = \frac{np}{n-mp}. \tag{11.75}$$

2. **Part II**
   If $mp > n > (m-1)p$, then

    $$W^{j+m,p} \hookrightarrow C^{j,\lambda}(\overline{\Omega}), \text{ for } 0 < \lambda \leq m - (n/p), \tag{11.76}$$

    and if $n = (m-1)p$, then

    $$W^{j+m,p} \hookrightarrow C^{j,\lambda}(\overline{\Omega}), \text{ for } 0 < \lambda < 1. \tag{11.77}$$

    Also, if $n = m - 1$ and $p = 1$, then (11.77) holds for $\lambda = 1$ as well.

3. **Part III** *All imbeddings in Parts A and B are valid for arbitrary domains Ω if the W $-$ space undergoing the imbedding is replaced with the corresponding $W_0 -$ space.*

11.4   The proof of the Sobolev imbedding theorem

Now we present a collection of results which imply the proof of the Sobolev imbedding theorem. We start with the approximation by smooth functions.

Definition 11.4.1 *Let $\Omega \subset \mathbb{R}^n$ be an open bounded set. For each $\varepsilon > 0$ define*

$$\Omega_\varepsilon = \{x \in \Omega \mid dist(x, \partial\Omega) > \varepsilon\}.$$

Definition 11.4.2 *Define $\eta \in C_c^\infty(\mathbb{R}^n)$ by*

$$\eta(x) = \begin{cases} C\exp\left(\frac{1}{|x|_2^2 - 1}\right), & \text{if } |x|_2 < 1, \\ 0, & \text{if } |x|_2 \geq 1, \end{cases}$$

where $|\cdot|_2$ refers to the Euclidean norm in $\mathbb{R}^n$, that is for $x = (x_1, ..., x_n) \in \mathbb{R}^n$, we have

$$|x|_2 = \sqrt{x_1^2 + ... + x_n^2}.$$

Moreover, $C > 0$ is chosen so that

$$\int_{\mathbb{R}^n} \eta \, dx = 1.$$

For each $\varepsilon > 0$, set

$$\eta_\varepsilon(x) = \frac{1}{\varepsilon^n} \eta\left(\frac{x}{\varepsilon}\right).$$

The function η is said to be the fundamental mollifier. The functions $\eta_\varepsilon \in C_c^\infty(\mathbb{R}^n)$ and satisfy

$$\int_{\mathbb{R}^n} \eta_\varepsilon \, dx = 1,$$

and $spt(\eta_\varepsilon) \subset B(0, \varepsilon)$.

Definition 11.4.3 *If $f : \Omega \to \mathbb{R}$ is locally integrable, we define its mollification, denoted by $f_\varepsilon : \Omega_\varepsilon \to \mathbb{R}$ as:*

$$f_\varepsilon = \eta_\varepsilon * f,$$

that is,

$$\begin{aligned} f_\varepsilon(x) &= \int_\Omega \eta_\varepsilon(x - y) f(y) \, dy \\ &= (-1)^n \int_{B(0,\varepsilon)} \eta_\varepsilon(y) f(x - y) \, dy. \end{aligned} \tag{11.78}$$

Theorem 11.4.4 (Properties of mollifiers) *The mollifiers have the following properties:*

1. *$f_\varepsilon \in C^\infty(\Omega_\varepsilon)$,*

2. *$f_\varepsilon \to f$ a.e. as $\varepsilon \to 0$,*

3. *If $f \in C(\Omega)$ then $f_\varepsilon \to f$ uniformly on compact subsets of Ω.*

Proof 11.14

1. fix $x \in \Omega_\varepsilon$, $i \in \{1, ..., n\}$ and a $h_0 > 0$ small enough such that

$$x + he_i \in \Omega_\varepsilon,$$

$\forall 0 < |h| < h_0$. Thus

$$\frac{f_\varepsilon(x+he_i) - f_\varepsilon(x)}{h} = \frac{1}{\varepsilon^n} \int_\Omega \frac{1}{h} \left[\eta\left(\frac{x+he_i-y}{\varepsilon}\right) - \eta\left(\frac{x-y}{\varepsilon}\right) \right]$$
$$\times f(y)\, dy$$
$$= \frac{1}{\varepsilon^n} \int_V \frac{1}{h} \left[\eta\left(\frac{x+he_i-y}{\varepsilon}\right) - \eta\left(\frac{x-y}{\varepsilon}\right) \right]$$
$$\times f(y)\, dy, \tag{11.79}$$

for an appropriate set $V \subset\subset \Omega$.

Observe that

$$\frac{1}{h} \left[\eta\left(\frac{x+he_i-y}{\varepsilon}\right) - \eta\left(\frac{x-y}{\varepsilon}\right) \right] f(y) \to \frac{\partial \eta(x-y)}{\partial x_i} f(y), \text{ in } V.$$

and

$$\left| \frac{1}{h} \left[\eta\left(\frac{x+he_i-y}{\varepsilon}\right) - \eta\left(\frac{x-y}{\varepsilon}\right) \right] f(y) \right| < C|f(y)|, \text{ in } V,$$

where

$$C = \sup_{y \in \mathbb{R}^N} \left| \frac{\partial \eta(x-y)}{\partial x_i} \right|.$$

From this and the Lebesgue dominated converge theorem, we obtain

$$\frac{\partial f_\varepsilon(x)}{\partial x_i} = \lim_{h \to 0} \frac{f_\varepsilon(x+he_i) - f_\varepsilon(x)}{h}$$
$$= \lim_{h \to 0} \frac{1}{\varepsilon^n} \int_\Omega \frac{1}{h} \left[\eta\left(\frac{x+he_i-y}{\varepsilon}\right) - \eta\left(\frac{x-y}{\varepsilon}\right) \right] f(y)\, dy$$
$$= \int_\Omega \frac{\partial \eta_\varepsilon(x-y)}{\partial x_i} f(y)\, dy. \tag{11.80}$$

By analogy, we may show that

$$D^\alpha f_\varepsilon(x) = \int_\Omega D^\alpha \eta_\varepsilon(x-y) f(y)\, dy, \forall x \in \Omega_\varepsilon.$$

2. From the Lebesgue differentiation theorem we have

$$\lim_{r \to 0} \frac{1}{|B(x,r)|} \int_{B(x,r)} |f(y) - f(x)|\, dy = 0, \tag{11.81}$$

for almost all $x \in \Omega$. Fix $x \in \Omega$ such that (11.81) holds. Hence,

$$|f_\varepsilon(x) - f(x)| = \int_{B(x,\varepsilon)} \eta_\varepsilon(x-y)[f(x) - f(y)]\, dy$$
$$\leq \frac{1}{\varepsilon^n} \int_{B(x,\varepsilon)} \eta\left(\frac{x-y}{\varepsilon}\right) [f(x) - f(y)]\, dy$$
$$\leq \frac{C}{|B(x,\varepsilon)|} \int_{B(x,\varepsilon)} |f(y) - f(x)|\, dy \tag{11.82}$$

for an appropriate constant $C > 0$. From (11.81), we obtain $f_\varepsilon \to f$ as $\varepsilon \to 0$.

3. Assume $f \in C(\Omega)$. Given $V \subset\subset \Omega$ choose W such that

$$V \subset\subset W \subset\subset \Omega,$$

and note that f is uniformly continuous on $\overline{W}$ Thus the limit indicated in (11.81) holds uniformly on V, and therefore $f_\varepsilon \to f$ uniformly on V.

For the next results we denote

$$\tilde{u}(x) = \begin{cases} u(x), & \text{if } x \in \Omega, \\ 0, & \text{if } x \in \mathbb{R}^n \setminus \Omega. \end{cases}$$

Theorem 11.4.5 *Let $\Omega \subset \mathbb{R}^N$ be an open bounded set. Let $u : \Omega \to \mathbb{R}$ be such that $u \in L^p(\Omega)$, where $1 \leq p < \infty$. Then*

$$\eta_\varepsilon * \tilde{u} \in L^p(\Omega),$$

$$\|\eta_\varepsilon * \tilde{u}\|_{p,\Omega} \leq \|u\|_{p,\Omega}, \forall \varepsilon > 0$$

and

$$\lim_{\varepsilon \to 0^+} \|\eta_\varepsilon * \tilde{u} - u\|_{p,\Omega} = 0.$$

Proof 11.15   Defining $q = p/(p-1)$, from Hölder's inequality we have

$$
\begin{aligned}
|\eta_\varepsilon * \tilde{u}(x)| &= \left| \int_{\mathbb{R}^n} \eta_\varepsilon(x-y)\tilde{u}(y)\, dy \right| \\
&= \left| \int_{\mathbb{R}^n} [\eta_\varepsilon(x-y)]^{(1-1/p)}[\eta_\varepsilon(x-y)]^{1/p}\tilde{u}(y)\, dy \right| \\
&\leq \left[\int_{\mathbb{R}^n} \eta_\varepsilon(x-y)\, dy \right]^{1/q} \left[\int_{\mathbb{R}^n} \eta_\varepsilon(x-y)|\tilde{u}(y)|^p\, dy \right]^{1/p} \\
&= \left[\int_{\mathbb{R}^n} \eta_\varepsilon(x-y)|u(y)|^p\, dy \right]^{1/p}.
\end{aligned}
\tag{11.83}
$$

From this and Fubini theorem, we obtain

$$
\begin{aligned}
\int_\Omega |\eta_\varepsilon * \tilde{u}(x)|^p\, dx &\leq \int_{\mathbb{R}^n} \int_{\mathbb{R}^n} \eta_\varepsilon(x-y)|\tilde{u}(y)|^p\, dy\, dx \\
&= \int_{\mathbb{R}^n} |\tilde{u}(y)|^p \left(\int_{\mathbb{R}^n} \eta_\varepsilon(x-y)\, dx \right) dy \\
&= \int_{\mathbb{R}^N} |\tilde{u}(y)|^p\, dy \\
&= \int_\Omega |u(y)|^p\, dy \\
&= \|u\|_{p,\Omega}^p.
\end{aligned}
\tag{11.84}
$$

Suppose given $\rho > 0$. As $C_c(\Omega)$ is dense in $L^p(\Omega)$, there exists $\phi \in C_c(\Omega)$ such that

$$\|u - \phi\|_p < \rho/3.$$

From the fact that

$$\eta_\varepsilon * \phi \to \phi$$

as $\varepsilon \to 0$, uniformly in Ω we have that there exists $\delta > 0$ such that

$$\|\eta_\varepsilon * \phi - \phi\|_p < \rho/3$$

if $0 < \varepsilon < \delta$. Thus for any $0 < \varepsilon < \delta(\rho)$, we get

$$
\begin{aligned}
\|\eta_\varepsilon * \tilde{u} - u\|_p &= \|\eta_\varepsilon * \tilde{u} - \eta_\varepsilon * \phi + \eta_\varepsilon * \phi - \phi + \phi - u\|_p \\
&\leq \|\eta_\varepsilon * \tilde{u} - \eta_\varepsilon * \phi\|_p + \|\eta_\varepsilon * \phi - \phi\|_p + \|\phi - u\|_p \\
&\leq \rho/3 + \rho/3 + \rho/3 = \rho.
\end{aligned}
\tag{11.85}
$$

Since $\rho > 0$ is arbitrary, the proof is complete.

11.4.1 Relatively compact sets in $L^p(\Omega)$

Theorem 11.4.6 *Consider $1 \leq p < \infty$. A bounded set $K \subset L^p(\Omega)$ is relatively compact if and only if for each $\varepsilon > 0$, there exists $\delta > 0$ and $G \subset\subset \Omega$ (we recall that $G \subset\subset \Omega$ means that $\overline{G}$ is compact and $\overline{G} \subset \Omega$) such that for each $u \in K$ and $h \in \mathbb{R}^n$ such that $|h| < \delta$ we have*

1.

$$
\int_\Omega |\tilde{u}(x+h) - \tilde{u}(x)|^p \, dx < \varepsilon^p,
\tag{11.86}
$$

2.

$$
\int_{\Omega - \overline{G}} |u(x)|^p \, dx < \varepsilon^p.
\tag{11.87}
$$

Proof 11.16 Suppose K is relatively compact in $L^p(\Omega)$. Suppose given $\varepsilon > 0$. As $\overline{K}$ is compact we may find a finite $\varepsilon/6$-net for K. Denote such a $\varepsilon/6$-net by N where

$$
N = \{v_1, ..., v_m\} \subset L^p(\Omega).
$$

Since $C_c(\Omega)$ is dense in $L^p(\Omega)$, for each $k \in \{1, ..., m\}$ there exists $\phi_k \in C_c(\Omega)$ such that

$$
\|\phi_k - v_k\|_p < \frac{\varepsilon}{6}.
$$

Thus defining

$$
S = \{\phi_1, ..., \phi_m\},
$$

given $u \in K$, we may select $v_k \in N$ such that

$$
\|u - v_k\|_p < \frac{\varepsilon}{6},
$$

so that

$$
\begin{aligned}
\|\phi_k - u\|_p &\leq \|\phi_k - v_k\|_p + \|v_k - u\|_p \\
&\leq \frac{\varepsilon}{6} + \frac{\varepsilon}{6} = \frac{\varepsilon}{3}.
\end{aligned}
\tag{11.88}
$$

Define

$$
G = \cup_{k=1}^m spt(\phi_k),
$$

where

$$
spt(\phi_k) = \overline{\{x \in \mathbb{R}^n \mid \phi_k(x) \neq 0\}}.
$$

We have that

$$
G \subset\subset \Omega,
$$

where as above mentioned this means $\overline{G} \subset \Omega$. Observe that

$$\varepsilon^p > \|u - \phi_k\|_p^p \geq \int_{\Omega - \overline{G}} |u(x)|^p \, dx.$$

Since $u \in K$ is arbitrary, (11.87) is proven. Since ϕ_k is continuous and $spt(\phi_k)$ is compact we have that ϕ_k is uniformly continuous, that is, for the ε given above, there exists $\tilde{\delta} > 0$ such that if $|h| < \min\{\tilde{\delta}, 1\}$ then

$$|\phi_k(x+h) - \phi_k(x)| < \frac{\varepsilon}{3(|\overline{G}| + 1)}, \forall x \in \overline{G},$$

Thus,

$$\int_{\Omega} |\phi_k(x+h) - \phi_k(x)|^p \, dx < \left(\frac{\varepsilon}{3}\right)^p.$$

Also observe that since

$$\|u - \phi_k\|_p < \frac{\varepsilon}{3},$$

we have that

$$\|T_h u - T_h \phi_k\|_p < \frac{\varepsilon}{3},$$

where $T_h u = u(x+h)$. Thus if $|h| < \delta = \min\{\tilde{\delta}, 1\}$, we obtain

$$\begin{aligned}
\|T_h \tilde{u} - \tilde{u}\|_p &\leq \|T_h \tilde{u} - T_h \phi_k\|_p + \|T_h \phi_k - \phi_k\|_p \\
&\quad + \|\phi_k - u\|_p \\
&< \frac{\varepsilon}{3} + \frac{\varepsilon}{3} + \frac{\varepsilon}{3} = \varepsilon.
\end{aligned} \tag{11.89}$$

For the converse, it suffices to consider the special case $\Omega = \mathbb{R}^n$, because for the general Ω we can define $\tilde{K} = \{\tilde{u} \mid u \in K\}$. Suppose given $\varepsilon > 0$ and choose $G \subset\subset \mathbb{R}^n$ such that for all $u \in K$ we have

$$\int_{\mathbb{R}^n - \overline{G}} |u(x)|^p \, dx < \frac{\varepsilon}{3}.$$

For each $\rho > 0$ the function $\eta_\rho * u \in C^\infty(\mathbb{R}^n)$, and in particular $\eta_\rho * u \in C(\overline{G})$. Suppose $\phi \in C_0(\mathbb{R}^n)$. Fix $\rho > 0$. By Hölder's inequality we have

$$\begin{aligned}
|\eta_\rho * \phi(x) - \phi(x)|^p &= \left| \int_{\mathbb{R}^n} \eta_\rho(y)(\phi(x-y) - \phi(x)) \, dy \right|^p \\
&= \left| \int_{\mathbb{R}^n} (\eta_\rho(y))^{1-1/p} (\eta_\rho(y))^{1/p} (T_{-y}\phi(x) - \phi(x)) \, dy \right|^p \\
&\leq \int_{B_\rho(\theta)} (\eta_\rho(y)) |T_{-y}\phi(x) - \phi(x)|^p \, dy.
\end{aligned} \tag{11.90}$$

Hence, from the Fubini theorem we may write

$$\int_{\mathbb{R}^n} |\eta_\rho * \phi(x) - \phi(x)|^p \, dx \leq \int_{B_\rho(\theta)} (\eta_\rho(y)) \int_{\mathbb{R}^n} |T_{-y}\phi(x) - \phi(x)|^p \, dx \, dy, \tag{11.91}$$

so that we may write

$$\|\eta_\rho * \phi - \phi\|_p \leq \sup_{h \in B_\rho(\theta)} \{\|T_h \phi - \phi\|_p\}. \tag{11.92}$$

Fix $u \in L^p(\mathbb{R}^n)$. We may obtain a sequence $\{\phi_k\} \subset C_c(\mathbb{R}^n)$ such that

$$\phi_k \to u, \text{ in } L^p(\mathbb{R}^n).$$

Observe that

$$\eta_\rho * \phi_k \to \eta_\rho * u, \text{ in } L^p(\mathbb{R}^n),$$

as $k \to \infty$. Also

$$T_h \phi_k \to T_h u, \text{ in } L^p(\mathbb{R}^n),$$

as $k \to \infty$. Thus,

$$\|T_h \phi_k - \phi_k\|_p \to \|T_h u - u\|_p,$$

in particular

$$\limsup_{k \to \infty} \left\{ \sup_{h \in B_\rho(\theta)} \{\|T_h \phi_k - \phi_k\|\} \right\} \leq \sup_{h \in B_\rho(\theta)} \{\|T_h u - u\|_p\}.$$

Therefore, as

$$\|\eta_\rho * \phi_k - \phi_k\|_p \to \|\eta_\rho * u - u\|_p,$$

as $k \to \infty$, from (11.92) we get

$$\|\eta_\rho * u - u\|_p \leq \sup_{h \in B_\rho(\theta)} \{\|T_h u - u\|_p\}.$$

From this and (11.86) we obtain

$$\|\eta_\rho * u - u\|_p \to 0, \text{ uniformly in } K \text{ as } \rho \to 0.$$

Fix $\rho_0 > 0$ such that

$$\int_{\overline{G}} |\eta_{\rho_0} * u - u|^p \, dx < \frac{\varepsilon}{3 \cdot 2^{p-1}}, \forall u \in K.$$

Observe that

$$
\begin{aligned}
|\eta_{\rho_0} * u(x)| &= \left| \int_{\mathbb{R}^n} \eta_{\rho_0}(x-y) u(y) \, dy \right| \\
&= \left| \int_{\mathbb{R}^n} [\eta_{\rho_0}(x-y)]^{(1-1/p)} [\eta_{\rho_0}(x-y)]^{1/p} u(y) \, dy \right| \\
&\leq \left[\int_{\mathbb{R}^n} \eta_{\rho_0}(x-y) \, dy \right]^{1/q} \left[\int_{\mathbb{R}^n} \eta_{\rho_0}(x-y) |u(y)|^p \, dy \right]^{1/p} \\
&= \left[\int_{\mathbb{R}^n} \eta_{\rho_0}(x-y) |u(y)|^p \, dy \right]^{1/p}.
\end{aligned}
\tag{11.93}
$$

From this, we may write,

$$|\eta_{\rho_0} * u(x)| \leq \left(\sup_{y \in \mathbb{R}^n} \eta_{\rho_0}(y) \right)^{1/p} \|u\|_p \leq K_1, \forall x \in \mathbb{R}^n, u \in K$$

where $K_1 = K_2 K_3$,

$$K_2 = \left(\sup_{y \in \mathbb{R}^n} \eta_{\rho_0}(y) \right)^{1/p},$$

and K_3 is any constant such that

$$\|u\|_p < K_3, \forall u \in K.$$

Similarly

$$|\eta_{\rho_0} * u(x+h) - \eta_{\rho_0} u(x)| \leq \left(\sup_{y \in \mathbb{R}^n} \eta_{\rho_0}(y) \right)^{1/p} \|T_h u - u\|_p,$$

and thus from (11.86) we obtain

$$\eta_{\rho_0} * u(x+h) \to \eta_{\rho_0} * u(x), \text{ as } h \to 0$$

uniformly in $\mathbb{R}^n$ and for $u \in K$.

By the Arzela-Ascoli Theorem

$$\{\eta_{\rho_0} * u \mid u \in K\}$$

is relatively compact in $C(\overline{G})$, and it is totally bounded so that there exists a ε_0-net $N = \{v_1, ..., v_m\}$ where

$$\varepsilon_0 = \left(\frac{\varepsilon}{3 \cdot 2^{p-1}|\overline{G}|}\right)^{1/p}.$$

Thus for some $k \in \{1, ..., m\}$ we have

$$\|v_k - \eta_{\rho_0} * u\|_\infty < \varepsilon_0.$$

Hence,

$$
\begin{aligned}
\int_{\mathbb{R}^n} |u(x) - \tilde{v}_k(x)|^p \, dx &= \int_{\mathbb{R}^n - \overline{G}} |u(x)|^p \, dx + \int_G |u(x) - v_k(x)|^p \, dx \\
&\leq \frac{\varepsilon}{3} + 2^{p-1} \int_{\overline{G}} (|u(x) - (\eta_{\rho_0} * u)(x)|^p \\
&\quad + |\eta_{\rho_0} * u(x) - v_k(x)|^p) \, dx \\
&\leq \frac{\varepsilon}{3} + 2^{p-1} \left(\frac{\varepsilon}{3 \cdot 2^{p-1}} + \frac{\varepsilon|\overline{G}|}{3 \cdot 2^{p-1}|\overline{G}|}\right) \\
&= \varepsilon.
\end{aligned}
\tag{11.94}
$$

Thus K is totally bounded and therefore it is relatively compact.

The proof is complete.

11.4.2 Some approximation results

Theorem 11.4.7 *Let $\Omega \subset \mathbb{R}^n$ be an open set. Assume $u \in W^{m,p}(\Omega)$ for some $1 \leq p < \infty$, and set*

$$u_\varepsilon = \eta_\varepsilon * u \text{ in } \Omega_\varepsilon.$$

Then,

1. $u_\varepsilon \in C^\infty(\Omega_\varepsilon), \forall \varepsilon > 0$,

2. $u_\varepsilon \to u$ in $W^{m,p}_{loc}(\Omega)$, as $\varepsilon \to 0$,

Proof 11.17 Assertion 1 has been already proved. Let us prove 2. We will show that if $|\alpha| \leq m$, then

$$D^\alpha u_\varepsilon = \eta_\varepsilon * D^\alpha u, \text{ in } \Omega_\varepsilon.$$

For, let $x \in \Omega_\varepsilon$. Thus,

$$
\begin{aligned}
D^\alpha u_\varepsilon(x) &= D^\alpha \left(\int_\Omega \eta_\varepsilon(x-y)u(y) \, dy\right) \\
&= \int_\Omega D_x^\alpha \eta_\varepsilon(x-y)u(y) \, dy \\
&= (-1)^{|\alpha|} \int_\Omega D_y^\alpha(\eta_\varepsilon(x-y))u(y) \, dy.
\end{aligned}
\tag{11.95}
$$

Observe that for fixed $x \in \Omega_\varepsilon$ the function

$$\phi(y) = \eta_\varepsilon(x-y) \in C_c^\infty(\Omega).$$

Therefore,

$$\int_\Omega D_y^\alpha \left(\eta_\varepsilon(x-y) \right) u(y) \, dy = (-1)^{|\alpha|} \int_\Omega \eta_\varepsilon(x-y) D_y^\alpha u(y) \, dy,$$

and hence,

$$
\begin{aligned}
D^\alpha u_\varepsilon(x) &= (-1)^{|\alpha|+|\alpha|} \int_\Omega \eta_\varepsilon(x-y) D^\alpha u(y) \, dy \\
&= (\eta_\varepsilon * D^\alpha u)(x).
\end{aligned}
\tag{11.96}
$$

Now choose any open bounded set such that $V \subset\subset \Omega$. We have that

$$D^\alpha u_\varepsilon \to D^\alpha u, \text{ in } L^p(V) \text{ as } \varepsilon \to 0,$$

for each $|\alpha| \le m$.

Thus,

$$\|u_\varepsilon - u\|_{m,p,V}^p = \sum_{|\alpha| \le m} \|D^\alpha u_\varepsilon - D^\alpha u\|_{p,V} \to 0,$$

as $\varepsilon \to 0$.

Theorem 11.4.8 *Let $\Omega \subset \mathbb{R}^n$ be a bounded open set and suppose $u \in W^{m,p}(\Omega)$ for some $1 \le p < \infty$. Then there exists a sequence $\{u_k\} \subset C^\infty(\Omega)$ such that*

$$u_k \to u \text{ in } W^{m,p}(\Omega).$$

Proof 11.18   Observe that

$$\Omega = \cup_{i=1}^\infty \Omega_i,$$

where

$$\Omega_i = \{x \in \Omega \mid dist(x, \partial\Omega) > 1/i\}.$$

Define

$$V_i = \Omega_{i+3} - \bar{\Omega}_{i+1},$$

and choose any open set V_0 such that $V_0 \subset\subset \Omega$, so that

$$\Omega = \cup_{i=0}^\infty V_i.$$

Let $\{\zeta_i\}_{i=0}^\infty$ be a smooth partition of unit subordinate to the open sets $\{V_i\}_{i=0}^\infty$. That is,

$$\begin{cases} 0 \le \zeta_i \le 1, & \zeta_i \in C_c^\infty(V_i) \\ \sum_{i=0}^\infty \zeta_i = 1, & \text{on } \Omega, \end{cases}$$

Now suppose $u \in W^{m,p}(\Omega)$. Thus $\zeta_i u \in W^{m,p}(\Omega)$ and $\text{spt}(\zeta_i u) \subset V_i \subset \Omega$. Choose $\delta > 0$. For each $i \in \mathbb{N}$ choose $\varepsilon_i > 0$ small enough so that

$$u_i = \eta_{\varepsilon_i} * (\zeta_i u)$$

satisfies

$$\|u_i - \zeta_i u\|_{m,p,\Omega} \le \frac{\delta}{2^{i+1}},$$

and $\text{spt}(u_i) \subset W_i$ where $W_i = \Omega_{i+4} - \bar{\Omega}_i \supset V_i$. Define

$$v = \sum_{i=0}^{\infty} u_i.$$

Thus such a function belongs to $C^{\infty}(\Omega)$, since for each open $V \subset\subset \Omega$ there are at most finitely many non-zero terms in the sum. Since

$$u = \sum_{i=0}^{\infty} \zeta_i u,$$

we have that for a fixed $V \subset\subset \Omega$,

$$
\begin{aligned}
\|v - u\|_{m,p,V} &\leq \sum_{i=0}^{\infty} \|u_i - \zeta_i u\|_{m,p,V} \\
&\leq \delta \sum_{i=0}^{\infty} \frac{1}{2^{i+1}} = \delta.
\end{aligned}
\tag{11.97}
$$

Taking the supremum over sets $V \subset\subset \Omega$ we obtain

$$\|v - u\|_{m,p,\Omega} < \delta.$$

Since $\delta > 0$ is arbitrary, the proof is complete.

The next result is also relevant. For a proof see Evans, [34] page 232.

Theorem 11.4.9 *Let $\Omega \subset \mathbb{R}^n$ be a bounded set such that $\partial\Omega$ is C^1. Suppose $u \in W^{m,p}(\Omega)$ where $1 \leq p < \infty$. Thus there exists a sequence $\{u_n\} \subset C^{\infty}(\overline{\Omega})$ such that*

$$u_n \to u \text{ in } W^{m,p}(\Omega), \text{ as } n \to \infty.$$

Anyway, now we prove a more general result.

Theorem 11.4.10 *Let $\Omega \subset \mathbb{R}^n$ be an open bounded set such that $\partial\Omega$ is $\hat{C}^1$. Let $u \in W^{m,p}(\Omega)$ where m is a non-negative integer and $1 \leq p < \infty$.*

Under such assumptions, there exists $\{u_k\} \subset C^{\infty}(\overline{\Omega})$ such that

$$\|u_k - u\|_{m,p,\Omega} \to 0, \text{ as } k \to \infty.$$

Proof 11.19 Fix $x_0 \in \partial\Omega$. Since $\partial\Omega$ is $\hat{C}^1$, denoting $\hat{x} = (x_1,...,x_{n-1})$ for a local coordinate system, there exists $r > 0$ and a function $f(x_1,...,x_{n-1}) = f(\hat{x})$ such that

$$W = \overline{\Omega} \cap B_r(x_0) = \{x \in B_r(x_0) \mid x_n \leq f(x_1,...,x_{n-1})\}.$$

We emphasize $f(\hat{x})$ is a Lipschitz continuous function, so that

$$|f(\hat{x}) - f(\hat{y})| \leq C_1 |\hat{x} - \hat{y}|_2, \text{ on its domain,}$$

for some $C_1 > 0$. Furthermore,

$$\left\{ \frac{\partial f(\hat{x})}{\partial x_k} \right\}_{k=1}^{n-1}$$

is classically defined, almost everywhere also on its concerning domain.

Let $\varepsilon > 0$. For each $\delta > 0$ define $x_\delta = x + C\delta e_n$, where $C > 1$ is a fixed constant. Define $u_\delta = u(x_\delta)$. Now choose $\delta > 0$ sufficiently small such that

$$\|u_\delta - u\|_{m,p,W} < \varepsilon/2.$$

For each $n \in \mathbb{N}$, $x \in W$ define
$$v_n(x) = (\eta_{1/n} * u_\delta)(x).$$

Observe that
$$\|v_n - u\|_{m,p,W} \leq \|v_n - u_\delta\|_{m,p,W} + \|u_\delta - u\|_{m,p,W},$$

For the fixed $\delta > 0$, there exists $N_\varepsilon \in \mathbb{N}$ such that if $n > N_\varepsilon$ we have
$$\|v_n - u_\delta\|_{m,p,W} < \varepsilon/2,$$

and
$$v_n \in C^\infty(\overline{W}).$$

Hence,
$$\|v_n - u\|_{m,p,W} \leq \|v_n - u_\delta\|_{m,p,W} + \|u_\delta - u\|_{m,p,W} < \varepsilon/2 + \varepsilon/2 = \varepsilon.$$

Clarifying the dependence of r on $x_0 \in \partial\Omega$ we denote $r = r_{x_0}$. Observe that
$$\partial\Omega \subset \cup_{x_0 \in \partial\Omega} B_{r_{x_0}}(x_0)$$

so that since $\partial\Omega$ is compact, there exists $x_1, ..., x_M \in \partial\Omega$ such that
$$\partial\Omega \subset \cup_{i=1}^M B_{r_i}(x_i).$$

We denote $B_{r_i}(x_i) = B_i$ and $W_i = \Omega \cap B_i$, $\forall i \in \{1, ..., M\}$. We also choose an appropriate open set $B_0 \subset\subset \Omega$ such that
$$\Omega \subset \cup_{i=0}^M B_i.$$

Let $\{\zeta_i\}_{i=0}^M$ be a concerned partition of unity relating $\{B_i\}_{i=0}^M$.
Thus $\zeta_i \in C_c^\infty(B_i)$ and $0 \leq \zeta_i \leq 1$, $\forall i \in \{0, ..., M\}$ and also
$$\sum_{i=0}^M \zeta_i = 1 \text{ on } \Omega.$$

From above, we may find $v_i \in C^\infty(\overline{W}_i)$ such that $\|v_i - u\|_{m,p,W_i} < \varepsilon$, $\forall i \in \{1, ..., M\}$. Define $u_0 = v_0 = u$ on $B_0 \equiv W_0$,
$$u_i = \zeta_i u, \forall i \in \{0, ..., M\}$$

and
$$v = \sum_{i=0}^M \zeta_i v_i.$$

We emphasize
$$v \in C^\infty(\overline{\Omega}).$$

Therefore
$$
\begin{aligned}
\|v - u\|_{m,p,\Omega} &= \left\| \sum_{i=0}^M (\zeta_i u - \zeta_i v_i) \right\|_{m,p,\Omega} \\
&\leq C_2 \sum_{i=0}^M \|u - v_i\|_{m,p,(\Omega \cap B_i)} \\
&= C_2 \sum_{i=0}^M \|u - v_i\|_{m,p,W_i} \\
&< C_2 M \varepsilon. \quad\quad (11.98)
\end{aligned}
$$

Since neither C_2 nor M depends on $\varepsilon > 0$, the proof is complete.

11.4.3 Extensions

In this section we study extensions of Sobolev spaces from a domain $\Omega \subset \mathbb{R}^n$ to $\mathbb{R}^n$. First we enunciate a result found in Evans, [34].

Theorem 11.4.11 *Assume $\Omega \subset \mathbb{R}^n$ is an open bounded set, and that $\partial\Omega$ is C^1. Let $1 \leq p < \infty$ and let V be a bounded open set such that $\Omega \subset\subset V$. Then there exists a bounded linear operator*

$$E : W^{1,p}(\Omega) \to W^{1,p}(\mathbb{R}^n),$$

such that for each $u \in W^{1,p}(\Omega)$ we have:

1. *$Eu = u$, a.e. in Ω,*

2. *Eu has support in V, and*

3. *$\|Eu\|_{1,p,\mathbb{R}^n} \leq C\|u\|_{1,p,\Omega}$, where the constant depend only on p, Ω, and V.*

The next result, which we prove, is a more general one.

Theorem 11.4.12 *Assume $\Omega \subset \mathbb{R}^n$ is an open bounded set, and that $\partial\Omega$ is $\hat{C}^1$. Let $1 \leq p < \infty$, and let V be a bounded open set such that $\Omega \subset\subset V$. Then there exists a bounded linear operator*

$$E : W^{1,p}(\Omega) \to W^{1,p}(\mathbb{R}^n),$$

such that for each $u \in W^{1,p}(\Omega)$ we have:

1. *$Eu = u$, a.e. in Ω,*

2. *Eu has support in V, and*

3. *$\|Eu\|_{1,p,\mathbb{R}^n} \leq C\|u\|_{1,p,\Omega}$, where the constant depend only on p, Ω, and V.*

Proof 11.20   Let $u \in W^{1,p}(\Omega)$. Fix $N \in \mathbb{N}$ and select $\phi_N \in C^\infty(\overline{\Omega})$ such that

$$\|\phi_N - u\|_{1,p,\Omega} < 1/N. \tag{11.99}$$

Choose $x_0 \in \partial\Omega$. From the hypothesis we may write

$$\overline{\Omega} \cap B_r(x_0) = \{x \in B_r(x_0) \mid x_n \leq f(x_1, ..., x_{n-1})\},$$

for some $r > 0$ and so that denoting $\hat{x} = (x_1, ..., x_{n-1})$, $f(x_1, ..., x_{n-1}) = f(\hat{x})$ is a Lipschitz continuous function such that

$$\left\{ \frac{\partial f(\hat{x})}{\partial x_k} \right\}_{k=1}^{n-1}$$

is classically defined almost everywhere on its domain and

$$|f(\hat{x}) - f(\hat{y})| \leq C_1 |\hat{x} - \hat{y}|_2, \forall \hat{x}, \hat{y} \text{ on its domain,}$$

for some $C_1 > 0$.

Define the variable $y \in \mathbb{R}^n$ by $y_i = x_i, \forall i \in \{1, ..., n-1\}$, and $y_n = f(x_1, ..., x_{n-1}) - x_n$. Thus

$$\phi_N(x_1, ..., x_n) = \phi_N(y_1, ..., y_{n-1}, f(y_1, .., y_{n-1}) - y_n) = \overline{\phi}_N(y_1, ..., y_n).$$

Observe that defining $\psi(x) = y$ we have $\psi(x) = (x_1, \cdots, x_{n-1}, f(x_1, \cdots, x_{n-1}) - x_n)$ so that

$$\psi^{-1}(y) = (y_1, \cdots y_{n-1}, f(y_1, \cdots y_{n-1}) - y_n).$$

From the continuity of ψ^{-1}, there exists $r_1 > 0$ such that

$$\psi^{-1}(B_{r_1}^+(y_0)) \subset \Omega \cap B_r(x_0),$$

where $y_0 = (x_{01}, \ldots, x_{0n-1}, 0)$. We define $W^+ = \psi^{-1}(\overline{B}_{r_1}^+(y_0))$ and $W^- = \psi^{-1}(\overline{B}_{r_1}^-(y_0))$ where we denote

$$B^+ = B_{r_1}^+(y_0) = \{y \in B_{r_1}(y_0) \mid y_n \geq 0\},$$

and

$$B^- = B_{r_1}^-(y_0) = \{y \in B_{r_1}(y_0) \mid y_n < 0\}.$$

We emphasize that locally about x_0 we have that $\partial\Omega$ and $\psi(\partial\Omega)$ correspond to the the equations $x_n - f(x_1, \ldots, x_{n-1}) = 0$ and $y_n = 0$ respectively.

At this point we are going to show that $\{\overline{\phi}_N\}$ is a Cauchy sequence in $W^{1,p}(B^+)$.

Observe that, denoting $W = W^+ \cup W^-$, we have

$$\psi(x) = (x_1, \cdots, x_{n-1}, f(x_1, \cdots, x_n) - x_n), \text{ in } W.$$

Also, since $\psi'(x)$ is classically defined and $|\psi'(x)| < C_2$, a.e. in Ω, for an appropriate $C_2 > 0$ we have that $\psi' \in L^\infty(W, \mathbb{R}^{n\times n})$. Similarly,

$$\det(\psi') \in L^\infty(W),$$

$$\det[(\psi^{-1})'] \in L^\infty(B),$$

$$\frac{\partial x_k(y)}{\partial y_j} \in L^\infty(B), \forall j, k \in \{1, \cdots, n\},$$

and

$$\frac{\partial y_j(x)}{\partial x_k} \in L^\infty(W), \forall j, k \in \{1, \cdots, n\}.$$

We denote by $\hat{W} \subset W$ the set on which $\det(\psi')$ and

$$\frac{\partial y_j(x)}{\partial x_k}, \forall j, k \in \{1, \cdots, n\},$$

are classically defined.

Similarly, we denote by $\hat{B} \subset B$ the set on which $\det((\psi^{-1})')$ and

$$\frac{\partial x_k(y))}{\partial y_j}, \forall j, k \in \{1, \cdots, n\},$$

are classically defined.

Let $M, N \in \mathbb{N}$, thus

$$\int_{B^+} |\overline{\phi}_N(y) - \overline{\phi}_M(y)|^p \, dy$$

$$\leq \int_{\hat{W}^+} |\overline{\phi}_N(y(x)) - \overline{\phi}_M(y(x))|^p |\det(\psi'(x))| \, dx$$

$$\leq K_1 \int_{W^+} |\overline{\phi}_N(y(x)) - \overline{\phi}_M(y(x))|^p \, dx$$

$$= K_1 \int_{W^+} |\phi_N(x) - \phi_M(x)|^p \, dx$$

$$\leq K_1 \|\phi_N - \phi_M\|_{1,p,\Omega}^p \to 0, \text{ as } N, M \to \infty. \tag{11.100}$$

Similarly, for $j \in \{1, \cdots, n\}$, we have,

$$
\int_{B^+} \left| \frac{\partial \overline{\phi}_N(y)}{\partial y_j} - \frac{\partial \overline{\phi}_M(y)}{\partial y_j} \right|^p dy
$$

$$
\leq \int_{W^+} \left| \frac{\partial \overline{\phi}_N(y(x))}{\partial y_j} - \frac{\partial \overline{\phi}_M(y(x))}{\partial y_j} \right|^p |\det(\psi'(x))| \, dx
$$

$$
\leq \int_{W^+} \left| \left(\frac{\partial \phi_N(x)}{\partial x_k} - \frac{\partial \phi_M(x)}{\partial x_k} \right) \frac{\partial x_k(y(x))}{\partial y_k} \right|^p |\det(\psi'(x))| \, dx
$$

$$
\leq K_2 \|\phi_N - \phi_M\|_{1,p,\Omega}^p \to 0, \text{ as } N, M \to \infty. \tag{11.101}
$$

Hence, $\{\overline{\phi}_N\}$ is a Cauchy sequence.
Therefore, there exists $\tilde{u} \in W^{1,p}(B^+)$ such that

$$
\overline{\phi}_N \to \tilde{u} \text{ in norm, in } W^{1,p}(B^+).
$$

Hence, up to a not relabeled subsequence,

$$
\overline{\phi}_N \to \tilde{u} \text{ a.e. in } B^+.
$$

However, up to a not relabeled subsequence of this last subsequence, from (11.99) we obtain,

$$
\phi_N \to u, \text{ a.e. in } \Omega, \tag{11.102}
$$

so that

$$
\tilde{u}(y) = u(x(y)), \text{ a.e. in } B^+.
$$

Moreover, $\overline{\phi}_N$ is Lipschitz continuous on B^+ so that $\overline{\phi} \in W^{1,p}(B^+)$, and therefore there exists $\tilde{\phi}_N \in C^{\infty}(\overline{B}^+)$ such that

$$
\|\tilde{\phi}_N - \overline{\phi}_N\|_{1,p,B^+} < 1/N.
$$

Define $\hat{\phi}_N : B \to \mathbb{R}$ by

$$
\hat{\phi}_N(y) = \begin{cases} \tilde{\phi}_N(y) & \text{if } y \in \overline{B}^+ \\ -3\tilde{\phi}_N(y_1, ..., y_{n-1}, -y_n) + 4\tilde{\phi}_N(y_1, ..., y_{n-1}, -y_n/2) & \text{if } y \in B^-. \end{cases}
$$

It may be easily verified that $\hat{\phi}_N \in C^1(B)$. Also, there exists $C_2 > 0$ such that

$$
\begin{aligned}
\|\hat{\phi}_N\|_{1,p,B} &\leq C_2 \|\hat{\phi}_N\|_{1,p,B^+} \\
&= C_2 \|\tilde{\phi}_N\|_{1,p,B^+} \\
&\leq C_2 \|\overline{\phi}_N\|_{1,p,B^+} + C_2/N, \tag{11.103}
\end{aligned}
$$

where C_2 depends only on Ω and p.
We claim that $\{\hat{\phi}_N\}$ is a Cauchy sequence in $W^{1,p}(B)$.
For $N_1, N_2 \in \mathbb{N}$ we have,

$$
\begin{aligned}
\|\hat{\phi}_{N_1} - \hat{\phi}_{N_2}\|_{1,p,B} &\leq C_1 \|\hat{\phi}_{N_1} - \hat{\phi}_{N_2}\|_{1,p,B^+} \\
&\leq C_1 \|\tilde{\phi}_{N_1} - \overline{\phi}_{N_1} + \overline{\phi}_{N_1} - \overline{\phi}_{N_2} + \overline{\phi}_{N_2} - \hat{\phi}_{N_2}\|_{1,p,B^+} \\
&\leq C_1 \|\tilde{\phi}_{N_1} - \overline{\phi}_{N_1}\|_{1,p,B^+} + C_1 \|\overline{\phi}_{N_1} - \overline{\phi}_{N_2}\|_{1,p,B^+} \\
&\quad + C_1 \|\overline{\phi}_{N_2} - \hat{\phi}_{N_2}\|_{1,p,B^+} \\
&\leq C_1/N_1 + C_1 \|\overline{\phi}_{N_1} - \overline{\phi}_{N_2}\|_{1,p,B^+} + C_1/N_2 \\
&\to 0, \text{ as } N_1, N_2 \to \infty. \tag{11.104}
\end{aligned}
$$

Hence $\{\hat{\phi}_N\}$ is a Cauchy sequence and thus there exists $\hat{u} \in W^{1,p}(B)$ such that

$$\hat{\phi}_N \to \hat{u} \text{ in norm in } W^{1,p}(B).$$

In particular, up to a not relabeled subsequence of indices $\{N\}$ of the sequence indicated in (11.102), we have,

$$\hat{\phi}_N \to \hat{u}, \text{ a.e. in } B,$$

Observe that,

$$\|\hat{\phi}_N|_{B^+} - \overline{\phi}_N\|_{1,p,B^+} = \|\tilde{\phi}_N - \overline{\phi}_N\|_{1,p,B^+} \to 0, \tag{11.105}$$

as $N \to \infty$. From this, since $\overline{\phi}_N \to u(x(y))$, in $W^{1,p}(B^+)$, we obtain for an appropriate not relabeled subsequence

$$\hat{\phi}_N|_{B^+} \to u(x(y)), \text{ in } W^{1,p}(B^+),$$

so that

$$\hat{u}(y) = u(x(y)), \text{ a.e. in } B^+,$$

and thus,

$$\hat{u}(y(x)) = u(x), \text{ a.e. in } W^+.$$

Denoting, $\overline{u}(x) = u(y(x))$ in $W = W^+ \cup W^-$, we obtain

$$\overline{u} = u, \text{ a.e. in } W^+.$$

At this point we are going to show that

$$\|\overline{u} - \hat{\phi}_N(y(x))\|_{1,p,W} \to 0, \text{ as } N \to \infty.$$

Observe that,

$$\int_W |\overline{u}_N(x)) - \hat{\phi}_N(y(x))|^p \, dx$$

$$\leq \int_{\hat{B}} |\overline{u}(x(y)) - \hat{\phi}_N(y)|^p |\det((\psi^{-1})'(y))| \, dy$$

$$\leq K_1 \int_B |\hat{u}(y) - \hat{\phi}_N(y)|^p \, dy$$

$$\leq K_1 \|\hat{u} - \hat{\phi}_N\|_{1,p,B}^p \to 0, \text{ as } N \to \infty. \tag{11.106}$$

Similarly, for $k \in \{1, \cdots, n\}$, we have,

$$\int_W \left| \frac{\partial \overline{u}(x)}{\partial x_k} - \frac{\partial \overline{\phi}_N(y(x))}{\partial x_k} \right|^p \, dy$$

$$\leq \int_{\hat{B}} \left| \frac{\partial \overline{u}_N(x(y))}{\partial x_k} - \frac{\partial \overline{\phi}_N(y)}{\partial x_k} \right|^p |\det((\psi^{-1})'(y))| \, dy$$

$$\leq \int_{\hat{B}} \left| \left(\frac{\partial \hat{u}(y)}{\partial y_j} - \frac{\partial \phi_N(y)}{\partial y_j} \right) \frac{\partial y_j(x(y))}{\partial x_k} \right|^p |\det((\psi^{-1})'(y))| \, dy$$

$$\leq K_2 \|\hat{u} - \phi_N\|_{1,p,B}^p \to 0, \text{ as } N \to \infty. \tag{11.107}$$

Therefore,

$$\|\overline{u}(x) - \hat{\phi}(y(x))\|_{1,p,W} \to 0, \text{ as } N \to \infty.$$

Now choose $\varepsilon > 0$. Thus there exists $N_0 \in \mathbb{N}$ such that if $N > N_0$ we have

$$
\begin{aligned}
\|\bar{u}\|_{1,p,W} &\leq \|\hat{\phi}_N(y(x))\|_{1,p,W} + \varepsilon \\
&\leq C_3 \|\hat{\phi}_N(y)\|_{1,p,B} + \varepsilon \\
&\leq C_4 \|\hat{\phi}_N\|_{1,p,B^+} + \varepsilon \\
&\leq C_5 \|\overline{\phi}_N\|_{1,p,W^+} + \varepsilon
\end{aligned}
\tag{11.108}
$$

so that letting $N \to \infty$, since $\varepsilon > 0$ is arbitrary we get

$$
\|\bar{u}\|_{1,p,W} \leq C_5 \|u\|_{1,p,W^+}.
$$

Now denoting $W = W_{x_0}$ we have that $\partial\Omega \subset \cup_{x_0 \in \partial\Omega} W_{x_0}$ and since $\partial\Omega$ is compact, there exist $x_1, ..., x_M \in \partial\Omega$, such that

$$
\partial\Omega \subset \cup_{i=1}^M W_{x_i}.
$$

Hence for an appropriate open $W_0 \subset\subset \Omega$, we get

$$
\Omega \subset \cup_{i=0}^M W_i.
$$

where we have denoted $W_i = W_{x_i}, \forall i \in \{0, ..., M\}$.

Let $\{\zeta_i\}_{i=0}^M$ be a concerned partition of unity relating $\{W_i\}_{i=0}^M$, so that

$$
\sum_{i=0}^M \zeta_i = 1, \text{ in } \Omega,
$$

and $\zeta_i \in C_c^\infty(W_i)$, $0 \leq \zeta_i \leq 1$, $\forall i \in \{0, ..., M\}$.

Define

$$
u_i = \zeta_i u, \forall i \in \{0, ..., M\}.
$$

For each i we denote the extension of u from W_i^+ to W_i by $\bar{u}_i$. Also define $\bar{u}_0 = u$ in W_0, and $\bar{u} = \sum_{i=0}^M \zeta_i \bar{u}_i$.

Recalling that $u = \bar{u}_i$, a.e. on W_i^+, and that $\overline{\Omega} = \cup_{i=1}^M W_i^+ \cup W_0$, we obtain $\bar{u} = \sum_{i=0}^M \zeta_i \bar{u}_i = \sum_{i=0}^M \zeta_i u = u$, a.e. in Ω.

Furthermore

$$
\begin{aligned}
\|\bar{u}\|_{1,p,\mathbb{R}^n} &\leq \sum_{i=0}^M \|\zeta_i \bar{u}_i\|_{1,p,\mathbb{R}^n} \\
&\leq C_5 \sum_{i=0}^M \|\bar{u}_i\|_{1,p,W_i} \\
&\leq C_5 \|u\|_{1,p,W_0} + C_5 \sum_{i=1}^M \|u_i\|_{1,p,W_i^+} \\
&\leq (M+1)C_5 \|u\|_{1,p,\Omega} \\
&= C\|u\|_{1,p,\Omega},
\end{aligned}
\tag{11.109}
$$

where $C = (M+1)C_5$.

We recall that the partition of unity may be chosen so that its support is on V.

Finally, we denote $Eu = \bar{u}$.

The proof is complete.

11.4.4 The main results

Definition 11.4.13 *For* $1 \leq p < n$ *we define* $r = \frac{np}{n-p}$.

Theorem 11.4.14 (Gagliardo-Nirenberg-Sobolev inequality) *Let* $1 \leq p < n$. *Thus there exists a constant* $K > 0$ *depending only* p *and* n *such that*

$$\|u\|_{r,\mathbb{R}^n} \leq K \|Du\|_{p,\mathbb{R}^n}, \forall u \in C_c^1(\mathbb{R}^n).$$

Proof 11.21 Suppose $p = 1$. Let $u \in C_c^1(\mathbb{R}^n)$. From the fundamental theorem of calculus we have,

$$u(x) = \int_{-\infty}^{x_i} \frac{\partial u(x_1, \ldots, x_{i-1}, y_i, x_{i+1}, \ldots, x_n)}{\partial x_i}\, dy_i,$$

so that

$$|u(x)| \leq \int_{-\infty}^{\infty} |Du(x_1, \ldots, x_{i-1}, y_i, x_{i+1}, \ldots, x_n)|\, dy_i.$$

Therefore,

$$|u(x)|^{n/(n-1)} \leq \prod_{i=1}^{n} \left(\int_{-\infty}^{\infty} |Du(x_1, \ldots, x_{i-1}, y_i, x_{i+1}, \ldots, x_n)|\, dy_i \right)^{1/(n-1)}.$$

From this, we get,

$$\int_{-\infty}^{\infty} |u(x)|^{n/(n-1)}\, dx_1$$

$$\leq \int_{-\infty}^{\infty} \prod_{i=1}^{n} \left(\int_{-\infty}^{\infty} |Du|\, dy_i \right)^{1/(n-1)} dx_1$$

$$\leq \left(\int_{-\infty}^{\infty} |Du|\, dy_1 \right)^{1/(n-1)}$$

$$\times \int_{-\infty}^{\infty} \left(\prod_{i=2}^{n} \left(\int_{-\infty}^{\infty} |Du|\, dy_i \right)^{1/(n-1)} \right) dx_1. \tag{11.110}$$

From this and the generalized Hölder inequality, we obtain,

$$\int_{-\infty}^{\infty} |u(x)|^{n/(n-1)}\, dx_1$$

$$\leq \left(\int_{-\infty}^{\infty} |Du|\, dy_1 \right)^{1/(n-1)}$$

$$\times \prod_{i=2}^{n} \left(\int_{-\infty}^{\infty} \int_{-\infty}^{\infty} |Du|\, dx_1 dy_i \right)^{1/(n-1)}. \tag{11.111}$$

Integrating in x_2 we obtain,

$$\int_{-\infty}^{\infty} \int_{-\infty}^{\infty} |u(x)|^{n/(n-1)}\, dx_1 dx_2$$

$$\leq \int_{-\infty}^{\infty} \left(\left(\int_{-\infty}^{\infty} |Du|\, dy_1 \right)^{1/(n-1)} \right.$$

$$\times \prod_{i=2}^{n} \left(\int_{-\infty}^{\infty} \int_{-\infty}^{\infty} |Du|\, dx_1 dy_i \right)^{1/(n-1)} \right) dx_2,$$

so that

$$\int_{-\infty}^{\infty} \int_{-\infty}^{\infty} |u(x)|^{n/(n-1)} \, dx_1 dx_2$$
$$\leq \left(\int_{-\infty}^{\infty} \int_{-\infty}^{\infty} |Du| \, dy_2 dx_1 \right)^{1/(n-1)}$$
$$\times \int_{-\infty}^{\infty} \left(\left(\int_{-\infty}^{\infty} |Du| \, dy_1 \right)^{1/(n-1)} \right.$$
$$\left. \times \prod_{i=3}^{n} \left(\int_{-\infty}^{\infty} \int_{-\infty}^{\infty} |Du| \, dx_1 dy_i \right)^{1/(n-1)} \right) dx_2. \tag{11.112}$$

By applying the generalized Hölder inequality we get

$$\int_{-\infty}^{\infty} \int_{-\infty}^{\infty} |u(x)|^{n/(n-1)} \, dx_1 dx_2$$
$$\leq \left(\int_{-\infty}^{\infty} \int_{-\infty}^{\infty} |Du| \, dy_2 dx_1 \right)^{1/(n-1)}$$
$$\times \left(\int_{-\infty}^{\infty} \int_{-\infty}^{\infty} |Du| \, dy_1 dx_2 \right)^{1/(n-1)}$$
$$\times \prod_{i=3}^{n} \left(\int_{-\infty}^{\infty} \int_{-\infty}^{\infty} \int_{-\infty}^{\infty} |Du| \, dx_1 dx_2 dy_i \right)^{1/(n-1)}. \tag{11.113}$$

Therefore, reasoning inductively, after n steps we get

$$\int_{\mathbb{R}^n} |u(x)|^{n/(n-1)} \, dx$$
$$\leq \prod_{i=1}^{n} \left(\int_{-\infty}^{\infty} \ldots \int_{-\infty}^{\infty} |Du| \, dx \right)^{1/(n-1)}$$
$$= \left(\int_{\mathbb{R}^n} |Du| \, dx \right)^{n/(n-1)}. \tag{11.114}$$

This is the result for $p = 1$. Now suppose $1 < p < n$.
For $\gamma > 1$ apply the above result for
$$v = |u|^{\gamma},$$

to obtain

$$\left(\int_{\mathbb{R}^n} |u(x)|^{\gamma n/(n-1)} \, dx \right)^{(n-1)/n} \leq \int_{\mathbb{R}^n} |D|u|^{\gamma}|; dx$$
$$\leq \gamma \int_{\mathbb{R}^n} |u|^{\gamma-1} |Du|; dx$$
$$\leq \gamma \left(\int_{\mathbb{R}^n} |u|^{(\gamma-1)p/(p-1)} \, dx \right)^{(p-1)/p}$$
$$\times \left(\int_{\mathbb{R}^n} |Du|^p \, dx \right)^{1/p}. \tag{11.115}$$

In particular for γ such that
$$\frac{\gamma n}{n-1} = \frac{(\gamma-1)p}{p-1},$$

that is $\gamma = \frac{p(n-1)}{n-p}$, so that

$$\frac{\gamma n}{n-1} = \frac{(\gamma-1)p}{p-1} = \frac{np}{n-p},$$

we get

$$\left(\int_{\mathbb{R}^n} |u|^r \, dx\right)^{((n-1)/n-(p-1)/p)} \leq C \left(\int_{\mathbb{R}^n} |Du|^p \, dx\right)^{1/p}.$$

From this and considering that

$$\frac{n-1}{n} - \frac{p-1}{p} = \frac{n-p}{np} = \frac{1}{r},$$

we finally obtain

$$\left(\int_{\mathbb{R}^n} |u|^r \, dx\right)^{1/r} \leq C \left(\int_{\mathbb{R}^n} |Du|^p \, dx\right)^{1/p}.$$

The proof is complete.

Theorem 11.4.15 *Let $\Omega \subset \mathbb{R}^n$ be a bounded open set. Suppose $\partial\Omega$ is $\hat{C}^1$, $1 \leq p < n$ and $u \in W^{1,p}(\Omega)$. Then $u \in L^r(\Omega)$ and*

$$\|u\|_{r,\Omega} \leq K \|u\|_{1,p,\Omega},$$

where the constant depends only on p, n and Ω.

Proof 11.22 Since $\partial\Omega$ is $\hat{C}^1$, from Theorem 11.4.12, there exists an extension $Eu = \bar{u} \in W^{1,p}(\mathbb{R}^n)$ such that $\bar{u} = u$ in Ω the support of $\bar{u}$ is compact and

$$\|\bar{u}\|_{1,p,\mathbb{R}^n} \leq C \|u\|_{1,p,\Omega},$$

where C does not depend on u. As $\bar{u}$ has compact support, from Theorem 11.4.10, there exists a sequence $\{u_k\} \in C_c^\infty(\mathbb{R}^n)$ such that

$$u_k \to \bar{u} \text{ in } W^{1,p}(\mathbb{R}^n).$$

from the last theorem

$$\|u_k - u_l\|_{r,\mathbb{R}^n} \leq K \|Du_k - Du_l\|_{p,\mathbb{R}^n}.$$

Hence,

$$u_k \to \bar{u} \text{ in } L^r(\mathbb{R}^n).$$

also from the last theorem

$$\|u_k\|_{r,\mathbb{R}^n} \leq K \|Du_k\|_{p,\mathbb{R}^n}, \forall k \in \mathbb{N},$$

so that

$$\|\bar{u}\|_{r,\mathbb{R}^n} \leq K \|D\bar{u}\|_{p,\mathbb{R}^n}.$$

Therefore, we may get

$$
\begin{aligned}
\|u\|_{r,\Omega} &\leq \|\bar{u}\|_{r,\mathbb{R}^n} \\
&\leq K \|D\bar{u}\|_{p,\mathbb{R}^n} \\
&\leq K_1 \|\bar{u}\|_{1,p,\mathbb{R}^n} \\
&\leq K_2 \|u\|_{1,p,\Omega}.
\end{aligned}
\tag{11.116}
$$

The proof is complete.

Theorem 11.4.16 *Let $\Omega \subset \mathbb{R}^n$ be a bounded open set such that $\partial\Omega \in \hat{C}^1$. If $mp < n$, then $W^{m,p}(\Omega) \hookrightarrow L^q(\Omega)$ for $p \leq q \leq (np)/(n-mp)$.*

Proof 11.23 Define $q_0 = np/(n - mp)$. We first prove by induction on m that

$$W^{m,p} \hookrightarrow L^{q_0}(\Omega).$$

The last result is exactly the case for $m = 1$. Assume

$$W^{m-1,p} \hookrightarrow L^{r_1}(\Omega), \tag{11.117}$$

where

$$r_1 = np/(n - (m-1)p) = np/(n - mp + p),$$

whenever $n > (m-1)p$. If $u \in W^{m,p}(\Omega)$ where $n > mp$, then u and $D_j u$ are in $W^{m-1,p}(\Omega)$, so that from (11.117) we have $u \in W^{1,r_1}(\Omega)$ and

$$\|u\|_{1,r_1,\Omega} \leq K\|u\|_{m,p,\Omega}. \tag{11.118}$$

Since $n > mp$ we have that $r_1 = np/((n - mp) + p) < n$, from $q_0 = nr_1/(n - r_1) = np/(n - mp)$ by the last theorem we have

$$\|u\|_{q_0,\Omega} \leq K_2\|u\|_{1,r_1,\Omega},$$

where the constant K_2 does not depend on u, and therefore from this and (11.118) we obtain

$$\|u\|_{q_0,\Omega} \leq K_2\|u\|_{1,r_1,\Omega} \leq K_3\|u\|_{m,p,\Omega}. \tag{11.119}$$

The induction is complete. Now suppose $p \leq q \leq q_0$. Define

$$s = (q_0 - q)p/(q_0 - p) \text{ and } t = p/s = (q_0 - p)/(q_0 - q).$$

Through Hölder's inequality, we get

$$
\begin{aligned}
\|u\|_{q,\Omega}^q &= \int_\Omega |u(x)|^s |u(x)|^{q-s} \, dx \\
&\leq \left(\int_\Omega |u(x)|^{st} \, dx \right)^{1/t} \left(\int_\Omega |u(x)|^{(q-s)t'} \, dx \right)^{1/t'} \\
&= \|u\|_{p,\Omega}^{p/t} \|u\|_{q_0,\Omega}^{q_0/t'} \\
&\leq \|u\|_{p,\Omega}^{p/t} (K_3)^{q_0/t'} \|u\|_{m,p,\Omega}^{q_0/t'} \\
&\leq (K_3)^{q_0/t'} \|u\|_{m,p,\Omega}^{p/t} \|u\|_{m,p,\Omega}^{q_0/t'} \\
&= (K_3)^{q_0/t'} \|u\|_{m,p,\Omega}^q, \tag{11.120}
\end{aligned}
$$

since

$$p/t + q_0/t' = q.$$

This completes the proof.

Corollary 11.4.17 *If $mp = n$, then $W^{m,p}(\Omega) \hookrightarrow L^q$ for $p \leq q < \infty$.*

Proof 11.24 If $q \geq p' = p/(p-1)$ then $q = ns/(n - ms)$ where $s = pq/(p+q)$ is such that $1 \leq s \leq p$. Observe that

$$W^{m,p}(\Omega) \hookrightarrow W^{m,s}(\Omega)$$

with the imbedding constant depending only on $|\Omega|$. Since $ms < n$, by the last theorem we obtain

$$W^{m,p}(\Omega) \hookrightarrow W^{m,s}(\Omega) \hookrightarrow L^q(\Omega).$$

Now if $p \leq q \leq p'$, from above we have $W^{m,p}(\Omega) \hookrightarrow L^{p'}(\Omega)$ and the obvious imbedding $W^{m,p}(\Omega) \hookrightarrow L^p(\Omega)$. Define $s = (p'-q)p/(p'-p)$ and the result follows from a reasoning analogous to the final chain of inequalities of last theorem, indicated in (11.120).

About the next theorem, note that its hypotheses are satisfied if $\partial\Omega$ is $\hat{C}^1$ (here we do not give the details).

Theorem 11.4.18 *Let $\Omega \subset \mathbb{R}^n$ be an open bounded set, such that for each $x \in \overline{\Omega}$ there exists a convex set $C_x \subset \overline{\Omega}$ whose shape depends on x, but such that $|C_x| > \alpha$, for some $\alpha > 0$ that does not depend on x. Thus if $mp > n$, then*

$$W^{m,p}(\Omega) \hookrightarrow C_B^0(\Omega).$$

Proof 11.25 Suppose first $m = 1$ so that $p > n$. Fix $x \in \overline{\Omega}$ and pick $y \in C_x$. For $\phi \in C^\infty(\overline{\Omega})$, from the fundamental theorem of calculus, we have

$$\phi(y) - \phi(x) = \int_0^1 \frac{d(\phi(x+t(y-x)))}{dt} \, dt.$$

Thus,

$$|\phi(x)| \leq |\phi(y)| + \int_0^1 \left| \frac{d(\phi(x+t(y-x)))}{dt} \right| \, dt,$$

and hence

$$\int_{C_x} |\phi(x)| \, dy \leq \int_{C_x} |\phi(y)| \, dy + \int_{C_x} \int_0^1 \left| \frac{d(\phi(x+t(y-x)))}{dt} \right| \, dt \, dy,$$

so that, from Hölder's inequality and Fubini theorem we get,

$$
\begin{aligned}
|\phi(x)|\alpha &\leq |\phi(x)| \cdot |C_x| \\
&\leq \|\phi\|_{p,\Omega} |C_x|^{1/p'} \\
&\quad + \int_0^1 \int_{C_x} \left| \frac{d(\phi(x+t(y-x)))}{dt} \right| \, dy \, dt.
\end{aligned}
\tag{11.121}
$$

Therefore

$$|\phi(x)|\alpha \leq \|\phi\|_{p,\Omega} |\Omega|^{1/p'} + \int_0^1 \int_V |\nabla\phi(z)| \delta t^{-n} \, dz \, dt,$$

where $|V| = t^n |C_x|$ and δ denotes the diameter of Ω. From Hölder's inequality again, we obtain

$$|\phi(x)|\alpha \leq \|\phi\|_{p,\Omega} |\Omega|^{1/p'} + \delta \int_0^1 \left(\int_V |\nabla\phi(z)|^p \, dz \right)^{1/p} t^{-n} (t^n |C_x|)^{1/p'} \, dt,$$

and thus

$$|\phi(x)|\alpha \leq \|\phi\|_{p,\Omega} |\Omega|^{1/p'} + \delta |C_x|^{1/p'} \|\nabla\phi\|_{p,\Omega} \int_0^1 t^{-n(1-1/p')} \, dt.$$

Since $p > n$ we obtain

$$\int_0^1 t^{-n(1-1/p')} \, dt = \int_0^1 t^{-n/p} \, dt = \frac{1}{1-n/p}.$$

From this, the last inequality and from the fact that $|C_x| \leq |\Omega|$, we have that there exists $K > 0$ such that

$$|\phi(x)| \leq K \|\phi\|_{1,p,\Omega}, \forall x \in \overline{\Omega}, \ \phi \in C^\infty(\overline{\Omega}).$$

$$\tag{11.122}$$

Here the constant K depends only on p, n and Ω. Consider now $u \in W^{1,p}(\Omega)$.

Thus there exists a sequence $\{\phi_k\} \subset C^\infty(\overline{\Omega})$ such that

$$\phi_k \to u, \text{ in } W^{1,p}(\Omega).$$

Up to a not relabeled subsequence, we have

$$\phi_k \to u, \text{ a.e. in } \overline{\Omega}. \tag{11.123}$$

Fix $x \in \overline{\Omega}$ such that the limit indicate in (11.123) holds. Suppose given $\varepsilon > 0$. Therefore, there exists $k_0 \in \mathbb{N}$ such that

$$|\phi_{k_0}(x) - u(x)| \le \varepsilon/2$$

and

$$\|\phi_{k_0} - u\|_{1,p,\Omega} < \varepsilon/(2K).$$

Thus,

$$
\begin{aligned}
|u(x)| &\le |\phi_{k_0}(x)| + \varepsilon/2 \\
&\le K\|\phi_{k_0}\|_{1,p,\Omega} + \varepsilon/2 \\
&\le K\|u\|_{1,p,\Omega} + \varepsilon.
\end{aligned}
\tag{11.124}
$$

Since $\varepsilon > 0$ is arbitrary, the proof for $m = 1$ is complete, because for $\{\phi_k\} \in C^\infty(\overline{\Omega})$ such that $\phi_k \to u$ in $W^{1,p}(\Omega)$, from (11.122) we have that $\{\phi_k\}$ is a uniformly Cauchy sequence, so that it converges to a continuous u^*, where $u^* = u$, a.e. in $\overline{\Omega}$.

For $m > 1$ but $p > n$ we still have

$$|u(x)| \le K\|u\|_{1,p,\Omega} \le K_1\|u\|_{m,p,\Omega}, \text{ a.e. in } \Omega, \forall u \in W^{m,p}(\Omega).$$

If $p \le n \le mp$, there exists j satisfying $1 \le j \le m-1$ such that $jp \le n \le (j+1)p$. If $jp < n$ set

$$\hat{r} = np/(n - jp),$$

Let $1 \le p_1 \le n$ such that

$$\hat{r} = np_1/(n - p_1).$$

Thus we have that

$$np/(n - jp) = np_1/(n - p_1),$$

so that

$$p_1 = np/(n - (j-1)p),$$

so that by above and the last theorem:

$$\|u\|_\infty \le K_1\|u\|_{1,\hat{r},\Omega} \le K_1\|u\|_{m-j,\hat{r},\Omega} \le K_2\|u\|_{m-(j-1),p_1,\Omega}.$$

Now define

$$\hat{r}_1 = p_1 = np/(n - (j-1)p)$$

and $1 \le p_2 \le n$ such that

$$\hat{r}_1 = np_2/(n - p_2),$$

so that

$$np/(n - (j-1)p) = np_2/(n - p_2).$$

Hence $p_2 = np/(n - (j-2)p)$ so that by the last theorem

$$\|u\|_{m-(j-1),p_1,\Omega} = \|u\|_{m-(j-1),\hat{r}_1,\Omega} \le K_3\|u\|_{m-(j-2),p_2,\Omega}.$$

Proceeding inductively in this fashion, after j steps, observing that $p_j = p$, we get

$$\|u\|_\infty \le K_1 \|u\|_{1,\hat{r},\Omega} \le K_1 \|u\|_{m-j,\hat{r},\Omega} \le K_j \|u\|_{m,p,\Omega},$$

for some appropriate K_j. Finally, if $jp = n$ choosing $\hat{r} = \max\{n,p\}$ also by the last theorem we obtain the same last chain of inequalities. For, assume $\hat{r} = \max\{n,p\} = n > p$. Let p_1 be such that

$$r_1 = \frac{np_1}{n - p_1} = n,$$

That is

$$p_1 = \frac{n}{2}.$$

Since $n > p$ we have that $n \ge 2$ so that $1 \le p_1 < n$. From the last theorem we obtain

$$\|u\|_\infty \le C \|u\|_{m-j,r_1,\Omega} \le C_1 \|u\|_{m-(j-1),p_1,\Omega}.$$

Let $r_2 = p_1 = n/2$, define p_2 such that

$$r_2 = n/2 = \frac{np_2}{n - p_2},$$

that is $p_2 = n/3$.

Hence, again by the last theorem we get

$$\|u\|_\infty \le C_1 \|u\|_{m-(j-1),r_2,\Omega} \le C_2 \|u\|_{m-(j-2),p_2,\Omega}.$$

Reasoning inductively, after $j - 1$ steps, , we get $p_{j-1} = n/j = p$, so that

$$\|u\|_\infty \le C \|u\|_{m-(j-1),r_1,\Omega} \le C_3 \|u\|_{m-(j-(j-1)),p_{j-1},\Omega} \le C_4 \|u\|_{m,p,\Omega}.$$

Finally, if $r_1 = \max\{n,p\} = p \ge n$, Define p_1 such that

$$r_1 = p = \frac{np_1}{n - p_1},$$

that is,

$$p_1 = \frac{np}{n + p} \le p,$$

so that by last theorem

$$\|u\|_\infty \le \|u\|_{m-j,r_1,\Omega} \le C_5 \|u\|_{m-(j-1),p_1,\Omega} \le C_6 \|u\|_{m,p,\Omega}.$$

This completes the proof.

Theorem 11.4.19 *Let $\Omega \subset \mathbb{R}^n$ be an open and bounded set with a boundary $\hat{C}^1$. If $mp > n$, then $W^{m,p}(\Omega) \hookrightarrow L^q(\Omega)$ for $p \le q \le \infty$.*

Proof 11.26 From the proof of the last theorem, we may obtain

$$\|u\|_{\infty,\Omega} \le K \|u\|_{m,p,\Omega}, \forall u \in W^{m,p}(\Omega).$$

If $p \le q < \infty$, we have

$$
\begin{aligned}
\|u\|_{q,\Omega}^q &= \int_\Omega |u(x)|^p |u(x)|^{q-p}\, dx \\
&\le \int_\Omega |u(x)|^p \left(K \|u\|_{m,p,\Omega}\right)^{q-p} dx \\
&\le K^{q-p} \|u\|_{p,\Omega}^p \|u\|_{m,p,\Omega}^{q-p} \\
&\le K^{q-p} \|u\|_{m,p,\Omega}^p \|u\|_{m,p,\Omega}^{q-p} \\
&= K^{q-p} \|u\|_{m,p,\Omega}^q.
\end{aligned}
\tag{11.125}
$$

The proof is complete.

Theorem 11.4.20 *Let $S \subset \mathbb{R}^n$ be a n-dimensional ball of radius bigger than 3. If $n < p$, then there exists a constant C, depending only on p and n, such that*

$$\|u\|_{C^{0,\lambda}(S)} \leq C\|u\|_{1,p,S}, \forall u \in C^1(S),$$

where $0 < \lambda \leq 1 - n/p$.

Proof 11.27   First consider $\lambda = 1 - n/p$ and $u \in C^1(S)$. Let $x, y \in S$ such that $|x - y| < 1$ and define $\sigma = |x - y|$. Consider a fixed cube denoted by $R_\sigma \subset S$ such that $|R_\sigma| = \sigma^n$ and $x, y \in \bar{R}_\sigma$. For $z \in R_\sigma$, we may write:

$$u(x) - u(z) = -\int_0^1 \frac{du(x + t(z - x))}{dt}\, dt,$$

that is,

$$u(x)\sigma^n = \int_{R_\sigma} u(z)\, dz - \int_{R_\sigma} \int_0^1 \nabla u(x + t(z - x)) \cdot (z - x)\, dt\, dz.$$

Thus, denoting in the next lines V by an appropriate set such that $|V| = t^n|R_\sigma|$, we obtain

$$\left| u(x) - \int_{R_\sigma} u(z)\, dz/\sigma^n \right| \leq \sqrt{n}\sigma^{1-n} \int_{R_\sigma} \int_0^1 |\nabla u(x + t(z - x))|\, dt\, dz$$

$$\leq \sqrt{n}\sigma^{1-n} \int_0^1 t^{-n} \int_V |\nabla u(z)|\, dz\, dt$$

$$\leq \sqrt{n}\sigma^{1-n} \int_0^1 t^{-n} \|\nabla u\|_{p,S} |V|^{1/p'}\, dt$$

$$\leq \sqrt{n}\sigma^{1-n} \sigma^{n/p'} \|\nabla u\|_{p,S} \int_0^1 t^{-n} t^{n/p'}\, dt$$

$$\leq \sqrt{n}\sigma^{1-n/p} \|\nabla u\|_{p,S} \int_0^1 t^{-n/p}\, dt$$

$$\leq \sigma^{1-n/p} \|u\|_{1,p,S} K, \tag{11.126}$$

where

$$K = \sqrt{n} \int_0^1 t^{-n/p}\, dt = \sqrt{n}/(1 - n/p).$$

A similar inequality holds with y in place of x, so that

$$|u(x) - u(y)| \leq 2K|x - y|^{1-n/p} \|u\|_{1,p,S}, \forall x, y \in R_\sigma.$$

Now consider $0 < \lambda < 1 - n/p$. Observe that, as $|x - y|^\lambda \geq |x - y|^{1-n/p}$ if $|x - y| < 1$, we have,

$$\sup_{x,y \in S} \left\{ \frac{|u(x) - u(y)|}{|x - y|^\lambda} \mid x \neq y, |x - y| < 1 \right\}$$

$$\leq \sup_{x,y \in S} \left\{ \frac{|u(x) - u(y)|}{|x - y|^{1-n/p}} \mid x \neq y, |x - y| < 1 \right\} \leq K\|u\|_{1,p,S}. \tag{11.127}$$

Also,

$$\sup_{x,y \in S} \left\{ \frac{|u(x) - u(y)|}{|x - y|^\lambda} \mid |x - y| \geq 1 \right\} \leq 2\|u\|_{\infty,S} \leq 2K_1\|u\|_{1,p,S}$$

so that

$$\sup_{x,y \in S} \left\{ \frac{|u(x) - u(y)|}{|x - y|^\lambda} \mid x \neq y \right\} \leq (K + 2K_1)\|u\|_{1,p,S}, \forall u \in C^1(S).$$

The proof is complete.

Theorem 11.4.21 *Let $\Omega \subset \mathbb{R}^n$ be an open bounded set such that $\partial\Omega$ is $\hat{C}^1$. Assume $n < p \leq \infty$.*
Then
$$W^{1,p}(\Omega) \hookrightarrow C^{0,\lambda}(\overline{\Omega}),$$
for all $0 < \lambda \leq 1 - n/p$.

Proof 11.28 Fix $0 < \lambda \leq 1 - n/p$ and let $u \in W^{1,p}(\Omega)$. Since $\partial\Omega$ is $\hat{C}^1$, from Theorem 11.4.12, there exists an extension $Eu = \bar{u}$ such that $\bar{u} = u$, a.e. in Ω, and
$$\|\bar{u}\|_{1,p,\mathbb{R}^n} \leq K\|u\|_{1,p,\Omega},$$
where the constant K does not depend on u. From the proof of this same theorem, we may assume that $\mathrm{spt}(\bar{u})$ is on a n-dimensional sphere $S \supset \Omega$ with sufficiently big radius and such sphere does not depend on u. Thus, in fact, we have
$$\|\bar{u}\|_{1,p,S} \leq K\|u\|_{1,p,\Omega}.$$
Since $C^\infty(S)$ is dense in $W^{1,p}(S)$, there exists a sequence $\{\phi_k\} \subset C^\infty(S)$ such that
$$u_k \to \bar{u}, \text{ in } W^{1,p}(S). \tag{11.128}$$
Up to a not relabeled subsequence, we have
$$u_k \to \bar{u}, \text{ a.e. in } \Omega.$$
From last theorem we have
$$\|u_k - u_l\|_{C^{0,\lambda}(S)} \leq C\|u_k - u_l\|_{1,p,S},$$
so that $\{u_k\}$ is a cauchy sequence in $C^{0,\lambda}(\overline{S})$, and thus $u_k \to u^*$ for some $u^* \in C^{0,\lambda}(S)$. Hence, from this and (11.128), we have
$$u^* = \bar{u}, \text{ a.e. in } S.$$
Finally, from above and last theorem we may write:
$$\|u^*\|_{C^{0,\lambda}(\overline{\Omega})} \leq \|u^*\|_{C^{0,\lambda}(S)} \leq K_1\|\bar{u}\|_{1,p,S} \leq K_2\|u\|_{1,p,\Omega}.$$
The proof is complete.

11.5   The trace theorem

In this section we state and prove the trace theorem.

Theorem 11.5.1 *Let $1 < p < \infty$ and let $\Omega \subset \mathbb{R}^n$ be an open bounded set such that $\partial\Omega$ is $\hat{C}^1$. Then there exists a bounded linear operator*
$$T : W^{1,p}(\Omega) \to L^p(\partial\Omega),$$
such that

■ $Tu = u|_{\partial\Omega}$ *if $u \in W^{1,p}(\Omega) \cap C(\overline{\Omega})$, and*

■

$$\|Tu\|_{p,\partial\Omega} \leq C\|u\|_{1,p,\Omega}, \forall u \in W^{1,p}(\Omega),$$
where the constant C depends only on p and Ω.

Proof 11.29 Let $u \in W^{1,p}(\Omega) \cap C(\overline{\Omega})$. Choose $x_0 \in \partial\Omega$.

Since $\partial\Omega$ is $\hat{C}^1$, there exists $r > 0$ such that for a local coordinate system we may write,

$$\overline{\Omega} \cap B_r(x_0) = \{x \in B_r(x_0) \mid x_n \leq f(x_1, ..., x_{n-1})\},$$

where denoting $\hat{x} = (x_1, ..., x_{n-1})$, $f(\hat{x})$ is continuous and such that its partial derivatives are classically defined a.e. and bounded on its domain. Furthermore

$$|f(\hat{x}) - f(\hat{y})| \leq K|\hat{x} - \hat{y}|_2, \forall \hat{x}, \hat{y}$$

for some $K > 0$ also on its domain.

Define the coordinates y by

$$y_i = x_i, \forall i \in \{1, ..., n-1\},$$

and

$$y_n = f(x_1, ..., x_{n-1}) - x_n.$$

Define $\hat{u}(y)$ by,

$$u(x_1, ..., x_n) = u(y_1, ..., y_{n-1}, f(y_1, ..., y_{n-1}) - y_n) = \hat{u}(y).$$

Also define $y_0 = (x_{01}, ..., x_{0n-1}, x_{0n} - f(x_{01}, ..., x_{0n-1})) = (y_{01}, ..., y_{0n-1}, 0)$ and choose $r_1 > 0$ such that

$$\Psi^{-1}(B_{r_1}^+(y_0)) \subset \Omega \cap B_r(x_0).$$

Observe that this is possible since Ψ and Ψ^{-1} are continuous, where $y = \Psi(x)$. Here

$$B_{r_1}^+(y_0) = \{y \in B_{r_1}(y_0) \mid y_n > 0\}.$$

For each $N \in \mathbb{N}$, choose, by mollification for example, $\phi_N \in C^\infty(\overline{B}_{r_1}^+(y_0))$ such that

$$\|\phi_N - \hat{u}\|_{\infty, \overline{B}_{r_1}^+(y_0)} < \frac{1}{N}.$$

Denote $B = B_{r_1/2}(y_0)$, and $B^+ = B_{r_1/2}^+(y_0)$. Now choose $\eta \in C_c^\infty(B_{r_1}(y_0))$, such that $\eta > 0$ and $\eta \equiv 1$ on B. Also denote

$$\tilde{\Gamma} = \{y \in B \mid y_n = 0\},$$

and

$$\tilde{\Gamma}_1 = \{y \in B_{r_1}(y_0) \mid y_n = 0\}.$$

Observe that

$$\begin{aligned}
\int_{\tilde{\Gamma}} |\phi_N|^p \, d\Gamma &\leq \int_{\tilde{\Gamma}_1} \eta |\phi_N|^p \, d\Gamma \\
&= -\int_{B_{r_1}^+} (\eta |\phi_N|^p)_{y_n} \, dy \\
&\leq -\int_{B_{r_1}^+} (\eta_{y_n} |\phi_N|^p) \, dy \\
&\quad + \int_{B_{r_1}^+} (p|\phi_N|^{p-1} |(\phi_N)_{y_n}| \eta) \, dy.
\end{aligned} \tag{11.129}$$

Here we recall the Young inequality

$$ab \leq \frac{a^p}{p} + \frac{b^q}{q}, \forall a, b \geq 0, \text{ where } \frac{1}{p} + \frac{1}{q} = 1.$$

Thus,

$$(|\phi_N|^{p-1}|)(|(\phi_N)_{y_n}|\eta) \le \frac{|(\phi_N)_{y_n}|^p \eta^p}{p} + \frac{|\phi_N|^{(p-1)q}}{q},$$

so that replacing such an inequality in (11.129), since $(p-1)q = p$ we get

$$\int_{\tilde{\Gamma}} |\phi_N|^p \, d\Gamma \quad \le \quad C_1 \left(\int_{B_{r_1}^+} |\phi_N|^p \, dy + \int_{B_{r_1}^+} |D\phi_N|^p \, dy \right). \tag{11.130}$$

Letting $N \to +\infty$ we obtain

$$\int_{\Gamma} |u(x)|^p \, d\Gamma \quad \le \quad C_2 \int_{\tilde{\Gamma}} |\hat{u}(y)|^p \, d\Gamma$$

$$\le \quad C_3 \left(\int_{B_{r_1}^+} |\hat{u}|^p \, dy + \int_{B_{r_1}^+} |D\hat{u}|^p \, dy \right)$$

$$\le \quad C_4 \left(\int_{W^+} |u|^p \, dx + \int_{W^+} |Du|^p \, dx \right). \tag{11.131}$$

where $\Gamma = \psi^{-1}(\tilde{\Gamma})$ and $W^+ = \Psi^{-1}(B_{r_1}^+)$.

Observe that denoting $W = W_{x_0}$ we have that $\partial\Omega \subset \cup_{x\in\partial\Omega} W_x$, and thus, since $\partial\Omega$ is compact, we may select $x_1, ..., x_M$ such that $\partial\Omega \subset \cup_{i=1}^M W_i$. We emphasize to have denoted $W_{x_i} = W_i, \forall i \in \{1, ..., M\}$. Denoting $W_i^+ = W_i \cap \Omega$ we may obtain,

$$\int_{\partial\Omega} |u(x)|^p \, d\Gamma \quad \le \quad \sum_{i=1}^M \int_{\Gamma_i} |u(x)|^p \, d\Gamma$$

$$\le \quad \sum_{i=1}^M C_{4^i} \left(\int_{W_i^+} |u|^p \, dx + \int_{W_i^+} |Du|^p \, dx \right)$$

$$\le \quad C_5 M \left(\int_{\Omega} |u|^p \, dx + \int_{\Omega} |Du|^p \, dx \right)$$

$$= \quad C \left(\int_{\Omega} |u|^p \, dx + \int_{\Omega} |Du|^p \, dx \right). \tag{11.132}$$

At this point we denote $Tu = u|_{\partial\Omega}$.

Finally, for the case $u \in W^{1,p}(\Omega)$, select $\{u_k\} \subset C^\infty(\overline{\Omega})$ such that

$$\|u_k - u\|_{1,p,\Omega} \to 0, \text{ as } k \to \infty.$$

From above

$$\|Tu_k - Tu_l\|_{p,\partial\Omega} \le C\|u_k - u_l\|_{1,p,\Omega},$$

so that

$$\{Tu_k\}$$

is a Cauchy sequence. Hence we may define

$$Tu = \lim_{k\to\infty} Tu_k, \text{ in } L^p(\partial\Omega).$$

The proof is complete.

Remark 11.5.2 *Similar results are valid for* $W_0^{m,p}$, *however in this case the traces relative to derivatives of order up to* $m-1$ *are involved.*

11.6 Compact imbeddings

Theorem 11.6.1 *Let m be a non-negative integer and let $0 < v < \lambda \leq 1$. Then the following imbeddings exist:*

$$C^{m+1}(\overline{\Omega}) \hookrightarrow C^m(\overline{\Omega}), \tag{11.133}$$

$$C^{m,\lambda}(\overline{\Omega}) \hookrightarrow C^m(\overline{\Omega}), \tag{11.134}$$

$$C^{m,\lambda}(\overline{\Omega}) \hookrightarrow C^{m,v}(\overline{\Omega}). \tag{11.135}$$

If Ω is bounded then imbeddings (11.134) and (11.135) are compact.

Proof 11.30 The Imbeddings (11.133) and (11.134) follows from the inequalites

$$\|\phi\|_{C^m(\overline{\Omega})} \leq \|\phi\|_{C^{m+1}(\overline{\Omega})},$$

$$\|\phi\|_{C^m(\overline{\Omega})} \leq \|\phi\|_{C^{m,\lambda}(\overline{\Omega})}.$$

To establish (11.135) note that for $|\alpha| \leq m$

$$\sup_{x,y \in \Omega} \left\{ \frac{|D^\alpha \phi(x) - D^\alpha \phi(y)|}{|x-y|^v} \mid x \neq y, |x-y| < 1 \right\}$$
$$\leq \sup_{x,y \in \Omega} \left\{ \frac{|D^\alpha \phi(x) - D^\alpha \phi(y)|}{|x-y|^\lambda} \mid x \neq y, |x-y| < 1 \right\}, \tag{11.136}$$

and also,

$$\sup_{x,y \in \Omega} \left\{ \left| \frac{D^\alpha \phi(x) - D^\alpha \phi(y)}{|x-y|^v} \right| \mid |x-y| \geq 1 \right\} \leq 2 \sup_{x \in \Omega} \{|D^\alpha \phi|\}. \tag{11.137}$$

Therefore, we may conclude that

$$\|\phi\|_{C^{m,v}(\bar{\Omega})} \leq 3 \|\phi\|_{C^{m,\lambda}(\overline{\Omega})}, \forall \phi \in C^{m,v}(\overline{\Omega}).$$

Now suppose Ω is bounded. If A is a bounded set in $C^{0,\lambda}(\overline{\Omega})$ then there exists $M > 0$ such that

$$\|\phi\|_{C^{0,\lambda}(\overline{\Omega})} \leq M, \forall \phi \in A.$$

But then

$$|\phi(x) - \phi(y)| \leq M|x-y|^\lambda, \forall x,y \in \overline{\Omega}, \phi \in A,$$

so that by the Ascoli-Arzela theorem, A is pre-compact in $C(\overline{\Omega})$. This proves the compactness of (11.134) for $m = 0$.

   If $m \geq 1$ and A is bounded in $C^{m,\lambda}(\overline{\Omega})$, then A is bounded in $C^{0,\lambda}(\overline{\Omega})$. Thus, by above there is a sequence $\{\phi_k\} \subset A$ and $\phi \in C^{0,\lambda}(\overline{\Omega})$ such that

$$\phi_k \to \phi \text{ in } C(\overline{\Omega}).$$

However, $\{D_i \phi_k\}$ is also bounded in $C^{0,\lambda}(\overline{\Omega})$, so that there exist a not relabeled subsequence, also denoted by $\{\phi_k\}$ and ψ_i such that

$$D_i \phi_k \to \psi_i, \text{ in } C(\overline{\Omega}).$$

The convergence in $C(\bar{\Omega})$ being the uniform one, we have $\psi_i = D_i\phi$. We can proceed extracting (not relabeled) subsequences until obtaining

$$D^{\alpha}\phi_k \to D^{\alpha}\phi, \text{ in } C(\bar{\Omega}), \forall\, 0 \leq |\alpha| \leq m.$$

This completes the proof of compactness of (11.134). For (11.135), let S be a bounded set in $C^{m,\lambda}(\bar{\Omega})$. Observe that

$$
\begin{aligned}
\frac{|D^{\alpha}\phi(x) - D^{\alpha}\phi(y)|}{|x-y|^{\nu}} &= \left(\frac{|D^{\alpha}\phi(x) - D^{\alpha}\phi(y)|}{|x-y|^{\lambda}}\right)^{\nu/\lambda} \\
&\quad \cdot |D^{\alpha}\phi(x) - D^{\alpha}\phi(y)|^{1-\nu/\lambda} \\
&\leq K|D^{\alpha}\phi(x) - D^{\alpha}\phi(y)|^{1-\nu/\lambda},
\end{aligned}
\tag{11.138}
$$

for all $\phi \in S$. From (11.134), S has a converging subsequence in $C^m(\bar{\Omega})$. From (11.138) such a subsequence is also converging in $C^{m,\nu}(\bar{\Omega})$. The proof is complete.

Theorem 11.6.2 (Rellich-Kondrachov) *Let $\Omega \subset \mathbb{R}^n$ be an open bounded set such that $\partial\Omega$ is $\hat{C}^1$. Let j, m be integers, $j \geq 0, m \geq 1$, and let $1 \leq p < \infty$.*

1. ***Part I*** *- If $mp < n$, then the following imbeddings are compact:*

$$W^{j+m,p}(\Omega) \hookrightarrow W^{j,q}(\Omega),$$

$$\text{if } 0 < n - mp < n \text{ and } 1 \leq q < np/(n-mp), \tag{11.139}$$

$$W^{j+m,p}(\Omega) \hookrightarrow W^{j,q}(\Omega), \text{ if } n = mp, \ 1 \leq q < \infty. \tag{11.140}$$

2. ***Part II*** *- If $mp > n$, then the following imbeddings are compact:*

$$W^{j+m,p} \hookrightarrow C_B^j(\Omega), \tag{11.141}$$

$$W^{j+m,p}(\Omega) \hookrightarrow W^{j,q}(\Omega), \text{ if } 1 \leq q \leq \infty. \tag{11.142}$$

3. ***Part III*** *-The following imbeddings are compact:*

$$W^{j+m,p}(\Omega) \hookrightarrow C^j(\bar{\Omega}), \text{ if } mp > n, \tag{11.143}$$

$$W^{j+m,p}(\Omega) \hookrightarrow C^{j,\lambda}(\bar{\Omega}),$$

$$\text{if } mp > n \geq (m-1)p \text{ and } 0 < \lambda < m - n/p. \tag{11.144}$$

4. ***Part IV*** *- All the above imbeddings are compact if we replace $W^{j+m,p}(\Omega)$ by $W_0^{j+m,p}(\Omega)$.*

Remark 11.6.3 *Given X, Y, Z spaces, for which we have the imbeddings $X \hookrightarrow Y$ and $Y \hookrightarrow Z$ and if one of these imbeddings is compact then the composite imbedding $X \hookrightarrow Z$ is compact. Since the extension operator $u \to \tilde{u}$ where $\tilde{u}(x) = u(x)$ if $x \in \Omega$ and $\tilde{u}(x) = 0$ if $x \in \mathbb{R}^n - \Omega$, defines an imbedding $W_0^{j+m,p}(\Omega) \hookrightarrow W^{j+m,p}(\mathbb{R}^n)$ we have that Part-IV of above theorem follows from the application of Parts I-III to $\mathbb{R}^n$ (despite the fact we are assuming Ω bounded, the general results may be found in Adams [2]).*

Remark 11.6.4 *To prove the compactness of any of above imbeddings it is sufficient to consider the case* $j = 0$. *Suppose, for example, that the first imbedding has been proved for* $j = 0$. *For* $j \geq 1$ *and* $\{u_i\}$ *bounded sequence in* $W^{j+m,p}(\Omega)$ *we have that* $\{D^\alpha u_i\}$ *is bounded in* $W^{m,p}(\Omega)$ *for each* α *such that* $|\alpha| \leq j$. *From the case* $j = 0$ *it is possible to extract a subsequence (similarly to a diagonal process)* $\{u_{i_k}\}$ *for which* $\{D^\alpha u_{i_k}\}$ *converges in* $L^q(\Omega)$ *for each* α *such that* $|\alpha| \leq j$, *so that* $\{u_{i_k}\}$ *converges in* $W^{j,q}(\Omega)$.

Remark 11.6.5 *Since* Ω *is bounded,* $C_B^0(\Omega) \hookrightarrow L^q(\Omega)$ *for* $1 \leq q \leq \infty$. *In fact*

$$\|u\|_{0,q,\Omega} \leq \|u\|_{C_B^0}[vol(\Omega)]^{1/q}. \tag{11.145}$$

Thus the compactness of (11.142) (for $j = 0$*) follows from that of (11.141).*

Proof of Parts II and III. If $mp > n > (m-1)p$ and $0 < \lambda < (m-n)/p$, then there exists μ such that $\lambda < \mu < m - (n/p)$. Since Ω is bounded, the imbedding $C^{0,\mu}(\overline{\Omega}) \hookrightarrow C^{0,\lambda}(\overline{\Omega})$ is compact by Theorem 11.6.1. Since by the Sobolev Imbedding Theorem we have $W^{m,p}(\Omega) \hookrightarrow C^{0,\mu}(\overline{\Omega})$, we have that imbedding (11.144) is compact.

If $mp > n$, let j^* be the non-negative integer satisfying $(m-j^*)p > n \geq (m-j^*-1)p$. Thus we have the chain of imbeddings

$$W^{m,p}(\Omega) \hookrightarrow W^{m-j^*,p}(\Omega) \hookrightarrow C^{0,\mu}(\overline{\Omega}) \hookrightarrow C(\overline{\Omega}), \tag{11.146}$$

where $0 < \mu < m - j^* - (n/p)$. The last imbedding in (11.146) is compact by Theorem 11.6.1, so that (11.143) is compact for $j = 0$. By analogy (11.141) is compact for $j = 0$. Therefore from the above remarks, (11.142) is also compact. For the proof of Part I, we need the following lemma:

Lemma 11.6.6 *Let* Ω *be an bounded domain in* $\mathbb{R}^n$. *Let* $1 \leq q_1 \leq q_0$ *and suppose*

$$W^{m,p}(\Omega) \hookrightarrow L^{q_0}(\Omega), \tag{11.147}$$

$$W^{m,p}(\Omega) \hookrightarrow L^{q_1}. \tag{11.148}$$

Suppose also that (11.148) is compact. If $q_1 \leq q < q_0$, *then the imbedding*

$$W^{m,p} \hookrightarrow L^q(\Omega) \tag{11.149}$$

is compact.

Proof 11.31 Define $\lambda = q_1(q_0 - q)/(q(q_0 - q_1))$ and $\mu = q_0(q - q_1)/(q(q_0 - q_1))$. We have that $\lambda > 0$ and $\mu \geq 0$. From Hölder's inequality and (11.147) there exists $K \in \mathbb{R}^+$ such that,

$$\|u\|_{0,q,\Omega} \leq \|u\|_{0,q_1,\Omega}^\lambda \|u\|_{0,q_0,\Omega}^\mu \leq K\|u\|_{0,q_1,\Omega}^\lambda \|u\|_{m,p,\Omega}^\mu,$$

$$\forall u \in W^{m,p}(\Omega). \tag{11.150}$$

Thus considering a sequence $\{u_i\}$ bounded in $W^{m,p}(\Omega)$, since (11.148) is compact there exists a subsequence $\{u_{nk}\}$ that converges, and is therefore a Cauchy sequence in $L^{q_1}(\Omega)$. From (11.150), $\{u_{nk}\}$ is also a Cauchy sequence in $L^q(\Omega)$, so that (11.149) is compact.

Proof of Part I. Consider $j = 0$. Define $q_0 = np/(n - mp)$. To prove the imbedding

$$W^{m,p}(\Omega) \hookrightarrow L^q(\Omega), \ 1 \leq q < q_0, \tag{11.151}$$

is compact, by last lemma it suffices to do so only for $q = 1$. For $k \in \mathbb{N}$, define

$$\Omega_k = \{x \in \Omega \mid \text{dist}(x, \partial\Omega) > 2/k\}. \tag{11.152}$$

Suppose A is a bounded set of functions in $W^{m,p}(\Omega)$, that is suppose there exists $K_1 > 0$ such that

$$\|u\|_{W^{m,p}(\Omega)} < K_1, \forall u \in A.$$

Also, suppose given $\varepsilon > 0$, and define, for $u \in W^{m,p}(\Omega)$, $\tilde{u}(x) = u(x)$ if $x \in \Omega$, $\tilde{u}(x) = 0$, if $x \in \mathbb{R}^n \setminus \Omega$. Fix $u \in A$. From Hölder's inequality and considering that $W^{m,p}(\Omega) \to L^{q_0}(\Omega)$, we have

$$
\begin{aligned}
\int_{\Omega-\Omega_k} |u(x)| dx &\leq \left\{ \int_{\Omega-\Omega_k} |u(x)|^{q_0} dx \right\}^{1/q_0} \left\{ \int_{\Omega-\Omega_k} 1 dx \right\}^{1-1/q_0} \\
&\leq K_1 \|u\|_{m,p,\Omega} [vol(\Omega - \Omega_k)]^{1-1/q_0},
\end{aligned}
\tag{11.153}
$$

Thus, since A is bounded in $W^{m,p}(\Omega)$, there exists $K_0 \in \mathbb{N}$ such that if $k \geq K_0$ then

$$\int_{\Omega-\Omega_k} |u(x)| dx < \varepsilon, \forall u \in A \tag{11.154}$$

and, now fixing a not relabeled $k > K_0$, we get,

$$\int_{\Omega-\Omega_k} |\tilde{u}(x+h) - \tilde{u}(x)| dx < 2\varepsilon, \forall u \in A, \forall h \in \mathbb{R}^n. \tag{11.155}$$

Observe that if $|h| < 1/k$, then $x + th \in \Omega_{2k}$ provided $x \in \Omega_k$ and $0 \leq t \leq 1$. If $u \in C^\infty(\Omega)$ we have that

$$
\begin{aligned}
\int_{\Omega_k} |u(x+h) - u(x)| &\leq \int_{\Omega_k} dx \int_0^1 \left| \frac{du(x+th)}{dt} \right| dt \\
&\leq |h| \int_0^1 dt \int_{\Omega_{2k}} |\nabla u(y)| dy \leq |h| \|u\|_{1,1,\Omega} \\
&\leq K_2 |h| \|u\|_{m,p,\Omega}.
\end{aligned}
\tag{11.156}
$$

Since $C^\infty(\Omega)$ is dense in $W^{m,p}(\Omega)$, from above for $|h|$ sufficiently small

$$\int_\Omega |\tilde{u}(x+h) - \tilde{u}(x)| dx < 3\varepsilon, \forall u \in A. \tag{11.157}$$

From Theorem 11.4.6, A is relatively compact in $L^1(\Omega)$ and therefore the imbedding indicated (11.151) is compact for $q = 1$. This completes the proof.

Before introducing the exercises, we work out some examples.

We start start with the following J.L. Lions Lemma.

Lemma 11.1

Let U, V, W be Banach spaces. Assume $U \hookrightarrow V$, where such an imbbeding is compact, and $V \hookrightarrow W$.

Under such assumptions, for each $\varepsilon > 0$ there exists $C(\varepsilon) > 0$ such that

$$\|u\|_V \leq \varepsilon \|u\|_U + C(\varepsilon) \|u\|_W, \ \forall u \in U.$$

Proof 11.32 Suppose, to obtain contradiction, there exists $\varepsilon_0 > 0$ such that for each $n \in \mathbb{N}$ we may find $u_n \in U$ such that

$$\|u_n\|_V > \varepsilon_0 \|u_n\|_U + n \|u_n\|_W. \tag{11.158}$$

Observe that from $U \hookrightarrow V$, there exists $K_1 > 0$ such that

$$\|u\|_V \leq K_1 \|u\|_U, \forall u \in U,$$

and since from (11.158), $\|u_n\|_V > 0$ so that $u_n \neq \mathbf{0}, \forall n \in \mathbb{N}$, we also obtain,

$$\left\| \frac{u_n}{\|u_n\|_U} \right\|_V \leq K_1, \forall n \in \mathbb{N}. \tag{11.159}$$

From this and (11.158), we have,

$$K_1 \geq \left\| \frac{u_n}{\|u_n\|_U} \right\|_V > \varepsilon_0 + n \left\| \frac{u_n}{\|u_n\|_U} \right\|_W, \forall n \in \mathbb{N}. \tag{11.160}$$

From (11.159), the sequence

$$\left\{ \frac{u_n}{\|u_n\|_U} \right\}$$

is bounded in V, and since $U \hookrightarrow V$ is compact, there exists a subsequence $\{n_k\}$ of $\mathbb{N}$ and $u_0 \in V$ such that

$$\left\| \frac{u_{n_k}}{\|u_{n_k}\|_U} - u_0 \right\|_V \to 0, \text{ as } k \to \infty.$$

From this and $V \hookrightarrow W$ we obtain,

$$\left\| \frac{u_{n_k}}{\|u_{n_k}\|_U} - u_0 \right\|_W \to 0, \text{ as } k \to \infty.$$

Observe that from (11.160),

$$\left\| \frac{u_n}{\|u_n\|_U} \right\|_V > \varepsilon_0, \forall n \mathbb{N}, \tag{11.161}$$

so that

$$\|u_0\|_V \geq \varepsilon_0 > 0,$$

which means,

$$u_0 \neq \mathbf{0}.$$

Let $\varepsilon = \|u_0\|_W / 2$. From (11.161), there exists $k_0 \in \mathbb{N}$ such that if $k > k_0$, then

$$\left\| \frac{u_{n_k}}{\|u_{n_k}\|_U} - u_0 \right\|_W < \varepsilon = \|u_0\|_W / 2,$$

so that

$$\left\| \frac{u_{n_k}}{\|u_{n_k}\|_U} \right\|_W \geq \|u_0\|_W / 2, \text{ if } k > k_0.$$

From this and (11.160), we get,

$$
\begin{aligned}
K_1 &> \varepsilon_0 + n_k \left\| \frac{u_{n_k}}{\|u_{n_k}\|_U} \right\|_W \\
&\geq \varepsilon_0 + n_k \frac{\|u_0\|_W}{2}, \text{ if } k > k_0.
\end{aligned}
\tag{11.162}
$$

This contradicts $u_0 \neq \mathbf{0}$.
The proof is complete.

Example 11.6.7 *With such a result in mind, we work out the following example.*
Let $\Omega \subset \mathbb{R}^n$ be an open bounded set with a $\hat{C}^1$ class boundary.
Let $1 < p < n$ where $n \geq 2$.
Let

$$1 \leq q < \frac{np}{n-p}.$$

In such a case the imbbeding

$$W^{1,p}(\Omega) \hookrightarrow L^q(\Omega)$$

is compact, and we have also the obvious imbbeding,

$$L^q(\Omega) \hookrightarrow L^1(\Omega).$$

Let $\varepsilon > 0$. From the last lemma with $U = W^{1,p}(\Omega)$, $V = L^q(\Omega)$ and $W = L^1(\Omega)$, there exists $C(\varepsilon) > 0$ such that

$$\|u\|_q \ \leq \ \varepsilon \|u\|_{1,p} + C(\varepsilon)\|u\|_1 \ \forall u \in W^{1,p}(\Omega). \tag{11.163}$$

Exercise 11.6.8 *Let $\Omega \subset \mathbb{R}^n$ be an open bounded set with a $\hat{C}^1$ class boundary.*
Let $p > n$.
Prove that for each $\varepsilon > 0$ there exists $C(\varepsilon) > 0$ such that

$$\|Du\|_\infty \leq \varepsilon \|u\|_{2,p} + C(\varepsilon)\|u\|_\infty, \ \forall u \in W^{2,p}(\Omega).$$

Hint: From the Rellich-Kondrachov theorem, the imbbeding $W^{2,p}(\Omega) \hookrightarrow C^1(\overline{\Omega})$ is compact.

Theorem 11.6.9 *Let $\Omega \subset \mathbb{R}^n$ be an open set.*
Let $1 < p < \infty$ and let $u \in L^p(\Omega)$.
For each $j \in \{1, \cdots, n\}$ assume the exists a real $K_j > 0$ such that

$$\left| \int_\Omega u \frac{\partial \varphi}{\partial x_j} \, dx \right| \leq K_j \|\varphi\|_{p'}, \forall \varphi \in C_c^1,$$

where p' is such that

$$\frac{1}{p} + \frac{1}{p'} = 1.$$

Under such assumptions we have,

$$u \in W^{1,p}(\Omega).$$

Proof 11.33 Fix $j \in \{1, \cdots, n\}$.
Observe that, from the hypotheses, the functional $F_u : C_c^1(\Omega) \to \mathbb{R}$ where

$$F_u(\varphi) = \int_\Omega u \frac{\partial \varphi}{\partial x_j} \, dx,$$

is bounded on $C_c^1(\Omega)$, where $C_c^1(\Omega)$ is a not closed subspace of $L^{p'}$.
From the Hahn-Banach theorem, F_u may be extended to a $\tilde{F}_u : L^{p'}(\Omega) \to \mathbb{R}$ with same norm as F_u, and such that

$$\tilde{F}_u(\varphi) = F_u(\varphi), \ \forall \varphi \in C_c^1(\Omega).$$

From the Riesz representation theorem, since $L^p(\Omega)$ is reflexive, there exists $f_j \in L^p(\Omega)$ such that

$$\tilde{F}_u(v) = \int_\Omega f_j v \, dx, \ \forall v \in L^{p'},$$

and in particular,

$$F_u(\varphi) = \tilde{F}_u(\varphi) = \int_\Omega f_j \varphi \, dx, \ \forall \varphi \in C_c^1(\Omega),$$

so that

$$\int_\Omega u \frac{\partial \varphi}{\partial x_j} \, dx = F_u(\varphi) = \int_\Omega f\varphi \, dx, \forall \varphi \in C_c^1(\Omega).$$

From this, in the distributional sense,

$$\frac{\partial u}{\partial x_j} = f_j \in L^p(\Omega), \forall j \in \{1, \cdots, n\},$$

so that

$$u \in W^{1,p}(\Omega).$$

Example 11.6.10 *In this example, let $\Omega \subset \mathbb{R}^n$ be an open bounded set. Let $1 < p < \infty$.*
Assume $u \in W^{1,p}(\Omega)$.
Let $f : \mathbb{R} \to \mathbb{R}$ be such that

$$|f(x) - f(y)| \leq K|x - y|, \ \forall x, y \in \mathbb{R}.$$

We are going to show that

$$(f \circ u) \in W^{1,p}(\Omega).$$

First observe that, from the hypotheses,

$$|f(u(x)) - f(0)| \leq K|u(x)|, \ a.\ e.\ in\ \Omega,$$

so that

$$|f(u(x))| \leq K|u(x)| + |f(0)|, \ a.e.\ in\ \Omega.$$

Since $(f \circ u)$ is measurable, we may infer that

$$(f \circ u) \in L^p(\Omega).$$

Observe that

$$f(u(x)) \frac{\varphi(x + h\mathbf{e}_j) - \varphi(x)}{h} \to f(u(x)) \frac{\partial \varphi(x)}{\partial x_j}, \ as\ h \to 0, \ a.e.\ in\ \Omega,$$

$$\left| f(u(x)) \frac{\varphi(x + h\mathbf{e}_j) - \varphi(x)}{h} \right| \leq |f(u(x))|C,$$

where

$$C = \max_{x \in \Omega} \left| \frac{\partial \varphi(x)}{\partial x_j} \right|.$$

Recalling that $(f \circ u) \in L^p(\Omega) \subset L^1(\Omega)$, from the Lebesgue dominated convergence theorem, we obtain

$$\lim_{h \to 0} \int_\Omega (f(u(x)) \frac{\varphi(x + h\mathbf{e}_j) - \varphi(x)}{h} \, dx$$

$$= \int_\Omega (f(u(x)) \frac{\partial \varphi}{\partial x_j} \, dx. \tag{11.164}$$

Observe that, for any $|h| > 0$ sufficiently small,

$$\left| \int_\Omega f(u(x)) \frac{\varphi(x+h\mathbf{e}_j) - \varphi(x)}{h} \, dx \right|$$

$$= \left| \int_\Omega \frac{f(u(x-h\mathbf{e}_j)) - f(u(x))}{h} \varphi(x) \, dx \right|$$

$$\leq \int_\Omega \frac{|f(u(x-h\mathbf{e}_j)) - f(u(x))|}{|h|} |\varphi(x)| \, dx$$

$$\leq \int_\Omega \frac{K|u(x-h\mathbf{e}_j) - u(x)|}{|h|} |\varphi(x)| \, dx$$

$$\leq K \left\| \frac{u(x-h\mathbf{e}_j) - u(x)}{h} \right\|_p \|\varphi\|_{p'}$$

$$\leq KC_1 \|\varphi\|_{p'}. \tag{11.165}$$

From this and (11.164), we obtain,

$$\left| \int_\Omega f(u(x)) \frac{\partial \varphi}{\partial x_j} \, dx \right| \leq KC_1 \|\varphi\|_{p'},$$

$\forall \varphi \in C_c^1(\Omega), \forall j \in \{1, \cdots, n\}$, *so that from the last theorem,*

$$(f \circ u) \in W^{1,p}(\Omega).$$

11.7 On the regularity of Laplace equation solutions

Let $\Omega \subset \mathbb{R}^n$ be an open bounded set. Consider the functions $a_{ij} : \overline{\Omega} \to \mathbb{R}$ where we assume $a_{ij} \in C^1(\overline{\Omega})$ $\forall i, j \in \{1, \cdots, n\}$, where we also assume

$$a_{ij}\xi_i\xi_j \geq c|\xi|^2, \forall \xi \in \mathbb{R}^n.$$

Let $f \in L^2(\Omega)$ and consider the problem of minimizing $J : U = W_0^{1,2}(\Omega) \to \mathbb{R}$, where,

$$J(u) = \frac{1}{2} \int_\Omega a_{ij} \frac{\partial u}{\partial x_i} \frac{\partial u}{\partial x_j} \, dx - \int_\Omega f u \, dx,$$

$\forall u \in U$.

Observe that J is strictly convex, coercive and continuous, so that it is lower semi-continuous. Hence, from the direct method of calculus of variations, there exists a unique $u_0 \in W_0^{1,2}(\Omega)$ such that

$$J(u_0) = \min_{u \in U} J(u).$$

Moreover, from the necessary optimal condition,

$$\delta J(u_0, \varphi) = 0, \forall \varphi \in U = W_0^{1,2}(\Omega),$$

we obtain,

$$\int_\Omega a_{ij} \frac{\partial u_0}{\partial x_i} \frac{\partial \varphi}{\partial x_j} \, dx - \int_\Omega f \varphi \, dx = 0, \tag{11.166}$$

$\forall \varphi \in U$.

Let $V, W \subset \mathbb{R}^n$ be open sets such that

$$V \subset\subset W \subset\subset \Omega,$$

Chose $\eta : \mathbb{R}^n \to \mathbb{R}$ such that, $\eta \in C_c^1(\mathbb{R}^n)$ and such that

$$\eta \equiv 1, \text{ on } \overline{V},$$

$$spt(\eta) \subset W,$$

and

$$0 \le \eta(x) \le 1, \ \forall x \in \mathbb{R}^n.$$

Select $k \in \{1, \cdots, n\}$ and define

$$\varphi(x) = -D_k^{-h}[\eta^2 D_k^h u_0(x)],$$

where generically,

$$D_k^h u(x) = \frac{u(x + h\mathbf{e}_k) - u(x)}{|h|}, \forall h \ne 0.$$

Observe that, for $|h| > 0$ sufficiently small, $\varphi \in U = W_0^{1,2}(\Omega)$, so that from (11.166), we obtain,

$$-\int_\Omega a_{ij} \frac{\partial u_0}{\partial x_i} \frac{\partial D_k^{-h}[\eta^2 D_k^h u_0(x)]}{\partial x_j} \, dx$$
$$+ \int_\Omega f D_k^{-h}[\eta^2 D_k^h u_0(x)] \, dx = 0, \tag{11.167}$$

Define,

$$A = -\int_\Omega a_{ij} \frac{\partial u_0}{\partial x_i} \frac{\partial D_k^{-h}[\eta^2 D_k^h u_0(x)]}{\partial x_j} \, dx,$$

and

$$B = \int_\Omega f D_k^{-h}[\eta^2 D_k^h u_0(x)] \, dx.$$

Observe that

$$
\begin{aligned}
A &= -\int_\Omega a_{ij} \frac{\partial u_0}{\partial x_i} D_k^{-h} \left[\frac{\partial \eta^2}{\partial x_j} D_k^h u_0(x) \right] \, dx \\
&\quad - \int_\Omega a_{ij} \frac{\partial u_0}{\partial x_i} D_k^{-h} \left[\eta^2 D_k^h \frac{\partial u_0(x)}{\partial x_j} \right] \, dx \\
&= \int_\Omega D_k^h[a_{ij}] \frac{\partial u_0}{\partial x_i} \left[\frac{\partial \eta^2}{\partial x_j} D_k^h u_0(x) \right] \, dx \\
&\quad + \int_\Omega a_{ij} D_k^h \left[\frac{\partial u_0}{\partial x_i} \right] \left[\frac{\partial \eta^2}{\partial x_j} D_k^h u_0(x) \right] \, dx \\
&\quad + \int_\Omega D_k^h[a_{ij}] \frac{\partial u_0}{\partial x_i} \left[\eta^2 D_k^h \frac{\partial u_0(x)}{\partial x_j} \right] \, dx \\
&\quad + \int_\Omega a_{ij} D_k^h \left[\frac{\partial u_0}{\partial x_i} \right] \left[\eta^2 D_k^h \frac{\partial u_0(x)}{\partial x_j} \right] \, dx
\end{aligned}
\tag{11.168}
$$

On the other hand

$$
\begin{aligned}
B &= \int_{\Omega} f D_k^{-h} [\eta^2 D_k^h u_0(x)] \, dx \\
&= \leq \|f\|_2 \|D_k^{-h} [\eta^2 D_k^h u_0]\|_2 \\
&\leq \|f\|_2 \left\| \frac{\partial [\eta^2 D_k^k u_0]}{\partial x_k} \right\|_2 \\
&\leq \|f\|_2 \left\| \frac{\partial \eta^2}{\partial x_k} D_k^h u_0 \right\|_2 \\
&\quad + \|f\|_2 \left\| \eta^2 \frac{\partial D_k^h u_0}{\partial x_k} \right\|_2 \\
&\leq \|f\|_2 K_1 \|u_0\|_{1,2} + \|f\|_2 K_2 \left(\sum_{j=1}^{n} \left\| \eta \frac{\partial D_k^h u_0}{\partial x_j} \right\|_2 \right)
\end{aligned}
\tag{11.169}
$$

where

$$
K_1 = \left\| \frac{\partial \eta^2}{\partial x_k} \right\|_{\infty},
$$

and

$$
K_2 = \|\eta\|_{\infty}.
$$

From (11.167) and (11.168) we get

$$
\begin{aligned}
&\int_{\Omega} a_{ij} D_k^h \left[\frac{\partial u_0}{\partial x_i} \right] \left[\eta^2 D_k^h \frac{\partial u_0(x)}{\partial x_j} \right] \, dx \\
&= -\int_{\Omega} D_k^h [a_{ij}] \frac{\partial u_0}{\partial x_i} \left[\frac{\partial \eta^2}{\partial x_j} D_k^h u_0(x) \right] \, dx \\
&\quad - \int_{\Omega} a_{ij} D_k^h \left[\frac{\partial u_0}{\partial x_i} \right] \left[\frac{\partial \eta^2}{\partial x_j} D_k^h u_0(x) \right] \, dx \\
&\quad - \int_{\Omega} D_k^h [a_{ij}] \frac{\partial u_0}{\partial x_i} \left[\eta^2 D_k^h \frac{\partial u_0(x)}{\partial x_j} \right] \, dx \\
&\quad - B,
\end{aligned}
\tag{11.170}
$$

so that from this and (11.169), we obtain,

$$
\begin{aligned}
&c \sum_{j=1}^{n} \left\| \eta D_k^h \left(\frac{\partial u_0}{\partial x_j} \right) \right\|_2^2 \\
&\leq \int_{\Omega} a_{ij} \eta^2 D_k^h \left(\frac{\partial u_0}{\partial x_i} \right) D_k^h \left(\frac{\partial u_0}{\partial x_i} \right) \, dx \\
&\leq K_3 \|u_0\|_{1,2}^2 + K_4 \sum_{j=1}^{N} \left\| \eta D_k^h \left(\frac{\partial u_0}{\partial x_j} \right) \right\|_2 \\
&\quad + (K_5 + K_7) \sum_{j=1}^{N} \left\| \eta D_k^h \left(\frac{\partial u_0}{\partial x_j} \right) \right\|_2 + K_6
\end{aligned}
\tag{11.171}
$$

where

$$K_3 = \max_{i,j \in \{1,\cdots,n\}} \|D_k^h a_{ij}\|_\infty \max_{j \in \{1,\cdots,n\}} \left\| \eta \frac{\partial \eta}{\partial x_j} \right\|_\infty$$

$$\leq \max_{i,j \in \{1,\cdots,n\}} \left\| \frac{\partial a_{ij}}{\partial x_k} \right\|_\infty \max_{j \in \{1,\cdots,n\}} \left\| \eta \frac{\partial \eta}{\partial x_j} \right\|_\infty, \quad (11.172)$$

$$K_4 = \max_{i,j \in \{1,\cdots,n\}} \|a_{ij}\|_\infty \max_{j \in n} \left\| \eta \frac{\partial \eta}{\partial x_j} \right\|_\infty,$$

$$K_5 = \|u_0\|_{1,2} \max_{i,j \in \{1,\cdots,n\}} \|D_k^h a_{ij}\|_\infty \|\eta\|_\infty$$

$$\leq \|u_0\|_{1,2} \max_{i,j \in \{1,\cdots,n\}} \left\| \frac{\partial a_{ij}}{\partial x_k} \right\|_\infty \|\eta\|_\infty, \quad (11.173)$$

$$K_6 = \|f\|_2 K_1 \|u_0\|_{1,2},$$

and

$$K_7 = K_2 \|f\|_2.$$

We claim

$$\limsup_{h \to 0} \sum_{j=1}^n \left\| \eta D_k^h \left(\frac{\partial u_0(x)}{\partial x_j} \right) \right\|_2 \leq K_8,$$

for some $K_8 > 0$.

Suppose, to obtain contradiction, there is a sequence $\{h_m\}$ such that $h_m \to 0^+$ and

$$\sum_{j=1}^n \left\| \eta D_k^{h_m} \left(\frac{\partial u_0(x)}{\partial x_j} \right) \right\|_2 \to +\infty, \text{ as } m \to \infty.$$

From this and (11.171) we have,

$$c \limsup_{m \to \infty} \sum_{j=1}^n \left\| \eta D_k^{h_m} \left(\frac{\partial u_0(x)}{\partial x_j} \right) \right\|_2$$

$$\leq \lim_{m \to \infty} \frac{K_3 \|u_0\|_{1,2}}{\sum_{j=1}^n \left\| \eta D_k^{h_m} \left(\frac{\partial u_0(x)}{\partial x_j} \right) \right\|_2} + K_4 + K_5$$

$$+ \frac{K_6}{\sum_{j=1}^n \left\| \eta D_k^{h_m} \left(\frac{\partial u_0(x)}{\partial x_j} \right) \right\|_2} + K_7$$

$$= K_4 + K_5 + K_7, \quad (11.174)$$

which contradicts $c > 0$.

Hence,

$$\left\| D_K^h \frac{\partial u_0}{\partial x_j} \right\|_V \leq K_8, \forall j, k \in \{1, \cdots, n\},$$

for some appropriate $K_8 > 0$, so that

$$\frac{\partial^2 u_0}{\partial x_j \partial x_k} \in L^2(V), \ \forall j, k \in \{1, ..., n\},$$

that is,

$$u_0 \in W^{2,2}(V), \ \forall V \subset\subset \Omega,$$

so that

$$u_0 \in W_{loc}^{2,2}(\Omega).$$

11.7.1 Regularity, a more general result

Let us now consider the case in which,

$$\Omega = B_r(x_0) \cap \mathbb{R}^n_+,$$

for some $r > 0$ where

$$\mathbb{R}^n_+ = \{x = (x_1, \cdots, x_n) \in \mathbb{R}^n : x_n > 0\}.$$

Assume the same previous hypotheses on $\{a_{ij}\}$ and $J : U \to \mathbb{R}$ so that

$$J(u_0) = \min_{u \in U} J(u).$$

and

$$\int_\Omega a_{ij} \frac{\partial u_0}{\partial x_i} \frac{\partial \varphi}{\partial x_j} \, dx = \int_\Omega f \varphi \, dx, \tag{11.175}$$

$\forall \varphi \in U = W_0^{1,2}(\Omega)$.

Define $V = B_{r/2}(x_0) \cap \mathbb{R}^n_+$.

Choose $\eta \in C^1(\mathbb{R}^n)$ such that

$$\eta \equiv 1, \text{ on } \overline{V}$$

and

$$spt(\eta) \subset\subset B_r(x_0).$$

Select $k \in \{1, \cdots, n-1\}$, and define for $0 < |h| < r/4$

$$\varphi(x) = -D_k^{-h}[\eta^2 D_k^h u_0(x)].$$

Observe that $\varphi \in W_0^{1,2}(\Omega)$ so that with the same development presented in the last section we may obtain

$$\left\| \eta D_k^h \frac{\partial u_0(x)}{\partial x_j} \right\|_{2,\Omega} \le K_8,$$

$\forall j \in \{1, \cdots, n\}, k \in \{1, \cdots n-1\}, 0 < |h| < r/4$. so that

$$\left\| \frac{\partial^2 u_0}{\partial x_k \partial x_j} \right\|_{2,V} \le K_8, \forall j \in \{1, \cdots, n\}, k \in \{1, \cdots, n-1\}.$$

Finally, observe that in distributional sense,

$$\frac{\partial}{\partial x_j} \left(a_{ij} \frac{\partial u_0}{\partial x_i} \right) + f = 0,$$

so that

$$\begin{aligned}
c_1 \left| \frac{\partial^2 u_0}{\partial x_n^2} \right| &\le a_{nn} \left| \frac{\partial^2 u_0}{\partial x_n^2} \right| \\
&\le K \sum_{j=1}^n \sum_{k=1}^{n-1} \left| \frac{\partial^2 u_0}{\partial x_j \partial x_k} \right| + |f|, \text{ a.e in } \Omega.
\end{aligned} \tag{11.176}$$

Hence

$$\left\| \frac{\partial^2 u_0}{\partial x_n^2} \right\|_{2,V} \le K_9,$$

so that

$$u_0 \in W^{2,2}(V).$$

Our more general result is summarized by the next theorem.

Theorem 11.7.1 *Let $\Omega \subset \mathbb{R}^n$ be an open bounded set with a boundary $\partial\Omega$.*

Assume for each $x_0 \in \partial\Omega$ there exists $r > 0$ and a C^2 function f such that for a local system of coordinates

$$\Omega \cap B_r(x_0) = \{x \in B_r(x_0) \ : \ x_n < f(x_1, \cdots, x_{n-1})\}.$$

Assume also that defining

$$y = (y_1, \cdots, y_n) = \psi(x) = (x_1, \cdots, x_{n-1}, f(x_1, \cdots, x_{n-1}) - x_n),$$

so that

$$x = \psi^{-1}(y) = (y_1, \cdots, y_{n-1}, f(y_1, \cdots, y_{n-1}) - y_n),$$

we have,

$$\tilde{a}_{ij} \eta_i \eta_j \geq c_{x_0} |\eta|^2, \ \forall \eta \in \mathbb{R}^n,$$

where

$$\tilde{a}_{ij}(y) = a_{kl}(x(y)) \frac{\partial y_k(x(y))}{\partial x_i} \frac{\partial y_l(x(y))}{\partial x_j} |\det(\psi^{-1})'(y)|.$$

Let $u_0 \in U = W_0^{1,2}(\Omega)$ be such that

$$J(u_0) = \min_{u \in U} J(u),$$

where, once more, $J : U \to \mathbb{R}$ is given by

$$J(u) = \frac{1}{2} \int_\Omega a_{ij} \frac{\partial u_0}{\partial x_i} \frac{\partial u_0}{\partial x_j} \, dx - \int_\Omega f \, u \, dx,$$

and where $f \in L^2(\Omega)$.

Under such hypotheses, $u_0 \in W^{2,2}(\Omega)$.

Proof 11.34 Select $x_0 \in \partial\Omega$.

Denote $V = B_r(x_0) \cap \Omega$,

From the continuity of ψ^{-1}, there exists $r_1 > 0$ such that

$$W = \psi^{-1}(\overline{B_{r_1}^+}) \subset \overline{V} \subset \overline{\Omega},$$

where $y_0 = ((x_0)_1, \cdots, (x_0)_{n-1}, 0)$ and

$$B_{r_1}^+ = B_{r_1} \cap \mathbb{R}_+^n.$$

Let $\tilde{\varphi}(y) \in W_0^{1,2}(B_{r_1}^+)$. Hence

$$\varphi(x) = \tilde{\varphi}(y(x)) \in W_0^{1,2}(W).$$

Observe that, from the optimality condition, we have,

$$\int_W a_{ij} \frac{\partial u_0}{\partial x_i} \frac{\partial \varphi(x)}{\partial x_j} \, dx = \int_W f \, \varphi \, dx.$$

However,

$$\int_W a_{ij} \frac{\partial u_0(x)}{\partial x_i} \frac{\partial \varphi(x)}{\partial x_j} \, dx$$

$$= \int_{B_{r_1}^+} a_{ij}(x(y)) \frac{\partial u_0(x(y))}{\partial y_k} \frac{\partial y_k(x(y))}{\partial x_i} \frac{\partial \varphi(x(y))}{\partial y_l} \frac{\partial y_l(x(y))}{\partial x_j} |\det[(\psi^{-1})'(y)]| \, dy$$

$$= \int_{B_{r_1}^+} \tilde{a}_{ij}(y) \frac{\partial \tilde{u}_0(y)}{\partial y_i} \frac{\partial \tilde{\varphi}(y)}{\partial y_j} \, dy. \tag{11.177}$$

Hence,

$$\int_{B_{r_1}^+} \tilde{a}_{ij}(y) \frac{\partial \tilde{u}_0(y)}{\partial y_i} \frac{\partial \tilde{\varphi}(y)}{\partial y_j} \, dy = \int_{B_{r_1}^+} \tilde{f}(y) \tilde{\varphi}(y) \, dy,$$

where,

$$\tilde{f}(y) = f(x(y)) |\det[(\psi^{-1})'(y)]| \in L^2(B_{r_1}^+).$$

Since $\tilde{\varphi} \in W_0^{1,2}(B_{r_1}^+)$ is arbitrary, from the results of last section we may infer that,

$$\tilde{u}_0(y) = u_0(x(y)) \in W^{2,2}(B_{r_1/2}^+).$$

Now observe that, denoting

$$\tilde{W} = \psi^{-1}(B_{r_1/2}^+) \subset \Omega,$$

we have

$$\left\| \frac{\partial^2 u_0}{\partial x_i \partial x_j} \right\|_{2,\tilde{W}}^2$$

$$= \int_{B_{r_1/2}^+} \left| \frac{\partial^2 \tilde{u}_0(y)}{\partial y_k \partial y_l} \frac{\partial y_l(x(y))}{\partial x_i} \frac{\partial y_k(x(y))}{\partial x_j} + \frac{\partial \tilde{u}_0(y)}{\partial y_k} \frac{\partial^2 y_k(x(y))}{\partial x_i \partial x_j} \right|^2 |\det[(\psi^{-1})'(y)]| \, dy$$

$$\leq K_1 \sum_{j,k=1}^n \left\| \frac{\partial^2 \tilde{u}_0}{\partial y_k \partial y_l} \right\|_{2,B_{r_1/2}^+}^2$$

$$+ K_2 \sum_{i,j,k} \left\| \frac{\partial^2 \tilde{u}_0}{\partial y_i \partial y_j} \right\|_{2,B_{r_1/2}^+} \left\| \frac{\partial \tilde{u}_0}{\partial y_k} \right\|_{2,B_{r_1/2}^+}$$

$$+ K_3 \sum_{j=1}^n \left\| \frac{\partial \tilde{u}_0}{\partial y_j} \right\|_{2,B_{r_1/2}^+}^2, \tag{11.178}$$

so that

$$\frac{\partial^2 u_0}{\partial x_i \partial x_j} \in L^2(\tilde{W}).$$

At this point we remark that $x_0 \in W$, and clarifying the dependence of W relating x_0, we also denote

$$W = W_{x_0}.$$

Observe that from the compactness of $\partial \Omega$, we may find $x_1, \cdots, x_M \in \partial \Omega$ such that

$$\cup_{j=1}^M W_{x_j} \supset \partial \Omega,$$

so that we may also find $W_0 \subset\subset \Omega$ such that

$$[\cup_{j=1}^M W_{x_j}] \cup W_0 \supset \Omega,$$

and in such a case

$$W = [\cup_{j=1}^M \tilde{W}_{x_j}] \cup W_0.$$

From the previous results

$$\|u_0\|_{2,2,\tilde{W}_{x_j}} \leq K_j, \forall j \in \{1, \cdots, M\},$$

and

$$\|u_0\|_{2,2,W_0} \leq K_0,$$

for appropriate real constants $K_0 > 0$, $K_j > 0$, $\forall j \in \{1, \cdots M\}$.

Joining the pieces, we obtain

$$u_0 \in W^{2,2}(\Omega).$$

The proof is complete.

SECTION II

CALCULUS OF VARIATIONS, CONVEX ANALYSIS AND RESTRICTED OPTIMIZATION

Chapter 12

Basic Topics on the Calculus of Variations

12.1 Banach spaces

The main reference for this chapter is Troutman [79].

We start by recalling the norm definition.

Definition 12.1.1 *Let V be a vectorial space. A norm in V is a function denoted by $\|\cdot\|_V : V \to \mathbb{R}^+ = [0, +\infty)$, for which the following properties hold:*

1.
$$\|u\|_V > 0, \forall u \in V \text{ such that } u \neq \mathbf{0}$$

   and
$$\|u\|_V = 0, \text{ if, and only if } u = \mathbf{0}.$$

2. Triangular inequality, that is
$$\|u + v\|_V \leq \|u\|_V + \|v\|_V, \ \forall u, v \in V$$

3.
$$\|\alpha u\|_V = |\alpha| \|u\|, \ \forall u \in V, \ \alpha \in \mathbb{R}.$$

In such a case we say that the space V is a normed one.

Definition 12.1.2 (Convergent sequence) *Let V be a normed space and let $\{u_n\} \subset V$ be a sequence. We say that $\{u_n\}$ converges to $u_0 \in V$, if for each $\varepsilon > 0$, there exists $n_0 \in \mathbb{N}$ such that if $n > n_0$, then*

$$\|u_n - u_0\|_V < \varepsilon.$$

In such a case, we write

$$\lim_{n \to \infty} u_n = u_0, \text{ in norm.}$$

Definition 12.1.3 (Cauchy sequence in norm) *Let V be a normed space and let $\{u_n\} \subset V$ be a sequence. We say that $\{u_n\}$ is a Cauchy one, as for each $\varepsilon > 0$, there exists $n_0 \in \mathbb{N}$ such that if $m, n > n_0$, then*

$$\|u_n - u_m\|_V < \varepsilon.$$

At this point we recall the definition of Banach space.

Definition 12.1.4 (Banach space) *A normed space V is said to be a Banach space as it is complete, that is, for each Cauchy sequence $\{u_n\} \subset V$ there exists a $u_0 \in V$ such that*

$$\|u_n - u_0\|_V \to 0, \text{ as } n \to \infty.$$

Example 12.1.5 *Examples of Banach spaces:*
Consider $V = C([a,b])$, the space of continuous functions on $[a,b]$. We shall prove that such a space is a Banach one with the norm,

$$\|f\|_V = \max\{|f(x)| : x \in [a,b]\}.$$

Exercise 12.1.6 *Prove that*

$$\|f\|_V = \max\{|f(x)| : x \in [a,b]\}$$

is a norm for $V = C([a,b])$.

Solution:

1. Clearly

$$\|f\|_V \geq 0, \forall f \in V$$

   and

$$\|f\|_V = 0 \text{ if, and only if } f(x) = 0, \ \forall x \in [a,b],$$

$$\text{that is if, and only if } f = \mathbf{0}.$$

2. Let $f, g \in V$.
   Thus,

$$
\begin{aligned}
\|f + g\|_V &= \max\{|f(x) + g(x)|, x \in [a,b]\} \\
&\leq \max\{|f(x)| + |g(x)|, x \in [a,b]\} \\
&\leq \max\{|f(x)|, x \in [a,b]\} + \max\{|g(x)| \ x \in [a,b]\} \\
&= \|f\|_V + \|g\|_V.
\end{aligned}
\tag{12.1}
$$

3. Finally, let $\alpha \in \mathbb{R}$ and $f \in V$.
   Hence,

$$
\begin{aligned}
\|\alpha f\|_V &= \max\{|\alpha f(x)|, x \in [a,b]\} \\
&= \max\{|\alpha||f(x)|, x \in [a,b]\} \\
&= |\alpha| \max\{|f(x)|, x \in [a,b]\} \\
&= |\alpha| \|f\|.
\end{aligned}
\tag{12.2}
$$

From this we may infer that $\| \cdot \|_V$ is a norm for V.

The solution is complete.

Theorem 12.1.7 $V = C([a,b])$ *is a Banach space with the norm*

$$\|f\|_V = \max\{|f(x)| \; : \; x \in [a,b]\}, \; \forall f \in V.$$

Proof 12.1   The proof that $C([a,b])$ is a vector space is left as an exercise.

From the last exercise, $\| \cdot \|_V$ is a norm for V.

Let $\{f_n\} \subset V$ be a Cauchy sequence.

We shall prove that there exists $f \in V$ such that

$$\|f_n - f\|_V \to 0, \; \text{as } n \to \infty.$$

Let $\varepsilon > 0$.

Thus, there exists $n_0 \in \mathbb{N}$ such that if $m, n > n_0$, then

$$\|f_n - f_m\|_V < \varepsilon.$$

Hence

$$\max\{|f_n(x) - f_m(x)| \; : \; x \in [a,b]\} < \varepsilon,$$

that is,

$$|f_n(x) - f_m(x)| < \varepsilon, \; \forall x \in [a,b], \; m, n > n_0. \tag{12.3}$$

Let $x \in [a,b]$.

From (12.3), $\{f_n(x)\}$ is a real Cauchy sequence, therefore it is convergent.

So, define

$$f(x) = \lim_{n \to \infty} f_n(x), \; \forall x \in [a,b].$$

Also from (12.3), we have that

$$\lim_{m \to \infty} |f_n(x) - f_m(x)| = |f_n(x) - f(x)| \leq \varepsilon, \; \forall n > n_0.$$

From this we may infer that

$$\|f_n - f\|_V \to 0, \; \text{as } n \to \infty.$$

We shall prove now that f is continuous on $[a,b]$.

From the exposed above,

$$f_n \to f$$

uniformly on $[a,b]$ as $n \to \infty$.

Thus, there exists $n_1 \in \mathbb{N}$ such that if $n > n_1$, then

$$|f_n(x) - f(x)| < \frac{\varepsilon}{3}, \; \forall x \in [a,b].$$

Choose $n_2 > n_1$.

Let $x \in [a,b]$. From

$$\lim_{y \to x} f_{n_2}(y) = f_{n_2}(x),$$

there exists $\delta > 0$ such that if $y \in [a,b]$ and $|y - x| < \delta$, then

$$|f_{n_2}(y) - f_{n_2}(x)| < \frac{\varepsilon}{3}.$$

Thus, if $y \in [a,b]$ and $|y-x| < \delta$, then

$$
\begin{aligned}
|f(y) - f(x)| &= |f(y) - f_{n_2}(y) + f_{n_2}(y) - f_{n_2}(x) + f_{n_2}(x) - f(x)| \\
&\leq |f(y) - f_{n_2}(y)| + |f_{n_2}(y) - f_{n_2}(x)| + |f_{n_2}(x) - f(x)| \\
&< \frac{\varepsilon}{3} + \frac{\varepsilon}{3} + \frac{\varepsilon}{3} \\
&= \varepsilon.
\end{aligned} \tag{12.4}
$$

So, we may infer that f is continuous at x, $\forall x \in [a,b]$, that is $f \in V$.
The proof is complete.

Exercise 12.1.8 *Let $V = C^1([a,b])$ be the space of functions $f : [a,b] \to \mathbb{R}$ which the the first derivative is continuous on $[a,b]$.*
Define the function (in fact a functional) $\|\cdot\|_V : V \to \mathbb{R}^+$ by

$$
\|f\|_V = \max\{|f(x)| + |f'(x)| : x \in [a,b]\}.
$$

1. Prove that $\|\cdot\|_V$ is a norm.

2. Prove that V is a Banach space with such a norm.

Solution: The proof of item 1 is left as an exercise.
Now, we shall prove that V is complete.
Let $\{f_n\} \subset V$ be a Cauchy sequence.
Let $\varepsilon > 0$. Thus, there exists $n_0 \in \mathbb{N}$ such that if $m, n > n_0$, then

$$
\|f_n - f_m\|_V < \varepsilon/2.
$$

Therefore,

$$
|f_n(x) - f_m(x)| + |f'_n(x) - f'_m(x)| < \varepsilon/2, \ \forall x \in [a,b], \ m,n > n_0. \tag{12.5}
$$

Let $x \in [a,b]$. hence, $\{f_n(x)\}$ and $\{f'_n(x)\}$ are real Cauchy sequences, and therefore, they are convergent. Denote

$$
f(x) = \lim_{n \to \infty} f_n(x)
$$

and

$$
g(x) = \lim_{n \to \infty} f'_n(x).
$$

From this and (12.5), we obtain

$$
\begin{aligned}
|f_n(x) - f(x)| + |f'_n(x) - g(x)| &= \lim_{m \to \infty} |f_n(x) - f_m(x)| + |f'_n(x) - f'_m(x)| \\
&\leq \varepsilon/2, \ \forall x \in [a,b], \ n > n_0.
\end{aligned} \tag{12.6}
$$

Similarly to the last example, we may obtain that f and g are continuous, therefore uniformly continuous on the compact set $[a,b]$.
Thus, there exists $\delta > 0$ such that if $x, y \in [a,b]$ and $|y-x| < \delta$, then

$$
|g(y) - g(x)| < \varepsilon/2. \tag{12.7}
$$

Choose $n_1 > n_0$. Let $x \in (a,b)$.

Hence, if $0 < |h| < \delta$, then from (12.6) and (12.7) we have

$$
\begin{aligned}
&\left| \frac{f_{n_1}(x+h) - f_{n_1}(x)}{h} - g(x) \right| \\
={} & |f'_{n_1}(x+th) - g(x+th) + g(x+th) - g(x)| \\
\leq{} & |f'_{n_1}(x+th) - g(x+th)| + |g(x+th) - g(x)| \\
<{} & \varepsilon/2 + \varepsilon/2 \\
={} & \varepsilon,
\end{aligned}
\tag{12.8}
$$

where from mean value theorem, $t \in (0,1)$ (it depends on h). Therefore, letting $n_1 \to \infty$, we get

$$
\begin{aligned}
&\left| \frac{f_{n_1}(x+h) - f_{n_1}(x)}{h} - g(x) \right| \\
\to{} & \left| \frac{f(x+h) - f(x)}{h} - g(x) \right| \\
\leq{} & \varepsilon, \forall 0 < |h| < \delta.
\end{aligned}
\tag{12.9}
$$

From this we may infer that

$$
f'(x) = \lim_{h \to 0} \frac{f(x+h) - f(x)}{h} = g(x), \ \forall x \in (a,b).
$$

The cases in which $x = a$ or $x = b$ may be dealt similarly with one-sided limits. From this and (12.6), we have

$$
\|f_n - f\|_V \to 0, \ as \ n \to \infty
$$

and

$$
f \in C^1([a,b]).
$$

The solution is complete.

Definition 12.1.9 (Functional) *Let V be a Banach space. A functional F defined on V is a function whose the co-domain is $\mathbb{R}$ ($F : V \to \mathbb{R}$).*

Example 12.1.10 *Let $V = C([a,b])$ and $F : V \to \mathbb{R}$ where*

$$
F(y) = \int_a^b (\,sen^3 x + y(x)^2)\, dx, \ \forall y \in V.
$$

Example 12.1.11 *Let $V = C^1([a,b])$ and let $J : V \to \mathbb{R}$ where*

$$
J(y) = \int_a^b \sqrt{1 + y'(x)^2}\, dx, \ \forall y \in C^1([a,b]).
$$

In our first frame-work we consider functionals defined as

$$
F(y) = \int_a^b f(x, y(x), y'(x))\, dx,
$$

where we shall assume

$$
f \in C([a,b] \times \mathbb{R} \times \mathbb{R})
$$

and $V = C^1([a,b])$.

Thus, for $F : D \subset V \to \mathbb{R}$ where

$$F(y) = \int_a^b f(x, y(x), y'(x)) \, dx,$$

we assume

$$V = C^1([a,b]),$$

and

$$D = \{y \in V : y(a) = A \text{ and } y(b) = B\},$$

where $A, B \in \mathbb{R}$.

Observe that if $y \in D$, then $y + v \in D$ if, and only if, $v \in V$ and

$$v(a) = v(b) = 0.$$

Indeed, in such a case

$$y + v \in V$$

and

$$y(a) + v(a) = y(a) = A,$$

and

$$y(b) + v(b) = y(b) = B.$$

Thus, we define the space of admissible directions for F, denoted by V_a, as,

$$V_a = \{v \in V : v(a) = v(b) = 0\}.$$

Definition 12.1.12 (Global minimum) *Let V be a Banach space and let $F : D \subset V \to \mathbb{R}$ be a functional. We say that $y_0 \in D$ is a point of global minimum for F, if*

$$F(y_0) \leq F(y), \ \forall y \in D.$$

Observe that denoting $y = y_0 + v$ where $v \in V_a$, we have

$$F(y_0) \leq F(y_0 + v), \ \forall v \in V_a.$$

Example 12.1.13 *Consider $J : D \subset V \to \mathbb{R}$ where $V = C^1([a,b])$,*

$$D = \{y \in V : y(a) = 0 \text{ and } y(b) = 1\}$$

and

$$J(y) = \int_a^b (y'(x))^2 \, dx.$$

Thus,

$$V_a = \{v \in V : v(a) = v(b) = 0\}.$$

Let $y_0 \in D$ be a candidate to global minimum for F and let $v \in V_a$ be admissible direction. Hence, we must have

$$J(y_0 + v) - J(y_0) \geq 0, \tag{12.10}$$

where

$$
\begin{aligned}
J(y_0 + v) - J(y_0) &= \int_a^b (y_0'(x) + v'(x))^2 \, dx - \int_a^b y_0'(x)^2 \, dx \\
&= 2 \int_a^b y_0'(x) v'(x) \, dx + \int_a^b v'(x)^2 \, dx \\
&\geq 2 \int_a^b y_0'(x) v'(x) \, dx. \tag{12.11}
\end{aligned}
$$

Observe that if $y_0'(x) = c$ in $[a,b]$, we have (12.10) satisfied, since in such a case,

$$
\begin{aligned}
J(y_0 + v) - J(y_0) &\geq 2\int_a^b y_0'(x)v'(x)\,dx \\
&= 2c\int_a^b v'(x)\,dx \\
&= 2c[v(x)]_a^b \\
&= 2c(v(b) - v(a)) \\
&= 0.
\end{aligned}
\tag{12.12}
$$

Summarizing, if $y_0'(x) = c$ on $[a,b]$, then

$$
J(y_0 + v) \geq J(y_0), \; \forall v \in V_a.
$$

Observe that in such a case,

$$
y_0(x) = cx + d,
$$

for some $d \in \mathbb{R}$.
   However, from
$$
y(a) = 0, \; \text{we get } ca + d = 0.
$$

From $y_0(b) = 1$, we have $cb + d = 1$.
Solving this last system in c and d we obtain,

$$
c = \frac{1}{b-a},
$$

and

$$
d = \frac{-a}{b-a}.
$$

From this, we have,

$$
y_0(x) = \frac{x-a}{b-a}.
$$

Observe that the graph of y_0 corresponds to the straight line connecting the points $(a,0)$ and $(b,1)$.

12.2   The Gâteaux variation

Definition 12.2.1 *Let V be a Banach space and let $J : D \subset V \to \mathbb{R}$ be a functional. Let $y \in D$ and $v \in V_a$. We define the Gâteaux variation of J at y in the direction v, denoted by $\delta J(y;v)$, by*

$$
\delta J(y;v) = \lim_{\varepsilon \to 0} \frac{J(y + \varepsilon v) - J(y)}{\varepsilon},
$$

if such a limit exists. Equivalently,

$$
\delta J(y;v) = \frac{\partial J(y + \varepsilon v)}{\partial \varepsilon}\Big|_{\varepsilon = 0}.
$$

Example 12.2.2 *Let $V = C^1([a,b])$ and $J : V \to \mathbb{R}$ where*

$$
J(y) = \int_a^b \left(sen^3 x + y(x)^2 \right) dx.
$$

Let $y, v \in V$. Let us calculate

$$\delta J(y; v).$$

Observe that,

$$
\begin{aligned}
\delta J(y; v) &= \lim_{\varepsilon \to 0} \frac{J(y + \varepsilon v) - J(y)}{\varepsilon} \\
&= \lim_{\varepsilon \to 0} \frac{\int_a^b (sen^3 x + (y(x) + \varepsilon v(x))^2)\, dx - \int_a^b (sen^3 x + y(x)^2)\, dx}{\varepsilon} \\
&= \lim_{\varepsilon \to 0} \frac{\int_a^b (2\varepsilon y(x)v(x) + \varepsilon^2 v(x))\, dx}{\varepsilon} \\
&= \lim_{\varepsilon \to 0} \left(\int_a^b 2y(x)v(x)\, dx + \varepsilon \int_a^b v(x)^2\, dx \right) \\
&= \int_a^b 2y(x)v(x)\, dx.
\end{aligned}
\tag{12.13}
$$

Example 12.2.3 *Let $V = C^1([a,b])$ and let $J : V \to \mathbb{R}$ where*

$$J(y) = \int_a^b \rho(x)\sqrt{1 + y'(x)^2}\, dx,$$

and where $\rho : [a,b] \to (0, +\infty)$ is a fixed function.
Let $y, v \in V$.
Thus,

$$\delta J(y; v) = \frac{\partial J(y + \varepsilon v)}{\partial \varepsilon}\Big|_{\varepsilon = 0}, \tag{12.14}$$

where

$$J(y + \varepsilon v) = \int_a^b \rho(x)\sqrt{1 + (y'(x) + \varepsilon v'(x))^2}\, dx.$$

Hence,

$$
\begin{aligned}
\frac{\partial J(y + \varepsilon v)}{\partial \varepsilon}\Big|_{\varepsilon = 0} &= \frac{\partial}{\partial \varepsilon}\left(\int_a^b \rho(x)\sqrt{1 + (y'(x) + \varepsilon v'(x))^2}\, dx \right) \\
&=^{(*)} \int_a^b \rho(x)\frac{\partial}{\partial \varepsilon}\left(\sqrt{1 + (y'(x) + \varepsilon v'(x))^2} \right) dx \\
&= \int_a^b \frac{\rho(x)}{2} \frac{2(y'(x) + \varepsilon v'(x))v'(x)}{\sqrt{1 + (y'(x) + \varepsilon v'(x))^2}}\, dx.
\end{aligned}
\tag{12.15}
$$

(*): We shall prove this step is valid in the subsequent pages.

From this we get,

$$
\begin{aligned}
\delta J(y; v) &= \frac{\partial J(y + \varepsilon v)}{\partial \varepsilon}\Big|_{\varepsilon = 0} \\
&= \int_a^b \frac{\rho(x)y'(x)v'(x)}{\sqrt{1 + y'(x)^2}}\, dx.
\end{aligned}
\tag{12.16}
$$

The example is complete.

Example 12.2.4 *Let $V = C^1([a,b])$ and $f \in C^1([a,b] \times \mathbb{R} \times \mathbb{R})$. Thus f is a function of three variables, namely, $f(x,y,z)$.*

Consider the functional $F : V \to \mathbb{R}$, defined by

$$F(y) = \int_a^b f(x,y(x),y'(x))\, dx.$$

Let $y, v \in V$. Thus,

$$\delta F(y;v) = \frac{\partial}{\partial \varepsilon} F(y + \varepsilon v)|_{\varepsilon=0}.$$

Observe that

$$F(y + \varepsilon v) = \int_a^b f(x, y(x) + \varepsilon v(x), y'(x) + \varepsilon v'(x))\, dx,$$

and therefore

$$
\begin{aligned}
\frac{\partial}{\partial \varepsilon} F(y + \varepsilon v) &= \frac{\partial}{\partial \varepsilon}\left(\int_a^b f(x, y(x) + \varepsilon v(x), y'(x) + \varepsilon v'(x))\, dx \right) \\
&= \int_a^b \frac{\partial}{\partial \varepsilon}\left(f(x, y(x) + \varepsilon v(x), y'(x) + \varepsilon v'(x)) \right)\, dx \\
&= \int_a^b \left(\frac{\partial f(x, y(x) + \varepsilon v(x), y'(x) + \varepsilon v'(x))}{\partial y} v(x) \right. \\
&\qquad \left. + \frac{\partial f(x, y(x) + \varepsilon v(x), y'(x) + \varepsilon v'(x))}{\partial z} v'(x) \right)\, dx.
\end{aligned}
\tag{12.17}
$$

Thus

$$
\begin{aligned}
\delta F(y;v) &= \frac{\partial F(y + \varepsilon v)}{\partial \varepsilon}|_{\varepsilon=0} \\
&= \int_a^b \left(\frac{\partial f(x, y(x), y'(x))}{\partial y} v(x) + \frac{\partial f(x, y(x), y'(x))}{\partial z} v'(x) \right)\, dx.
\end{aligned}
\tag{12.18}
$$

12.3 Minimization of convex functionals

Definition 12.3.1 (Convex function) *A function $f : \mathbb{R}^n \to \mathbb{R}$ is said to be convex if*

$$f(\lambda x + (1-\lambda)y) \le \lambda f(x) + (1-\lambda)f(y), \ \forall x, y \in \mathbb{R}^n, \ \lambda \in [0,1].$$

Proposition 12.3.2 *Let $f : \mathbb{R}^n \to \mathbb{R}$ be a convex and differentiable function.*
Under such hypotheses,

$$f(y) - f(x) \ge \langle f'(x), y - x \rangle_{\mathbb{R}^n}, \ \forall x, y \in \mathbb{R}^n,$$

where $\langle \cdot, \cdot \rangle_{\mathbb{R}^n} : \mathbb{R}^n \times \mathbb{R}^n \to \mathbb{R}$ denotes the usual inner product for $\mathbb{R}^n$, that is,

$$\langle x, y \rangle_{\mathbb{R}^n} = x_1 y_1 + \cdots + x_n y_n,$$

$\forall x = (x_1, \cdots, x_n), \ y = (y_1, \cdots, y_n) \in \mathbb{R}^n.$

Proof 12.2 Choose $x, y \in \mathbb{R}^n$.
From the hypotheses,

$$f((1-\lambda)x + \lambda y) \le (1-\lambda)f(x) + \lambda f(y), \ \forall \lambda \in (0,1).$$

Thus,

$$\frac{f(x+\lambda(y-x))-f(x)}{\lambda} \leq f(y)-f(x), \ \forall \lambda \in (0,1).$$

Therefore,

$$\begin{aligned}
\langle f'(x),y-x\rangle_{\mathbb{R}^n} &= \lim_{\lambda \to 0^+} \frac{f(x+\lambda(y-x))-f(x)}{\lambda} \\
&\leq f(y)-f(x), \ \forall x,y \in \mathbb{R}^n.
\end{aligned} \tag{12.19}$$

The proof is complete.

Proposition 12.3.3 *Let* $f : \mathbb{R}^n \to \mathbb{R}$ *be a differentiable function on* $\mathbb{R}^n$.
Assume

$$f(y)-f(x) \geq \langle f'(x),y-x\rangle_{\mathbb{R}^n}, \ \forall x,y \in \mathbb{R}^n.$$

Under such hypotheses, f *is convex.*

Proof 12.3 Define $f^* : \mathbb{R}^n \to \mathbb{R} \cup \{+\infty\}$ by

$$f^*(x^*) = \sup_{x \in \mathbb{R}^n} \{\langle x,x^*\rangle_{\mathbb{R}^n} - f(x)\}.$$

Such a function is said to the polar function for f.
Let $x \in \mathbb{R}^n$. From the hypotheses,

$$\langle f'(x),x\rangle_{\mathbb{R}^n} - f(x) \geq \langle f'(x),y\rangle_{\mathbb{R}^n} - f(y), \ \forall y \in \mathbb{R}^n,$$

that is,

$$\begin{aligned}
f^*(f'(x)) &= \sup_{y \in \mathbb{R}^n} \{\langle f'(x),y\rangle_{\mathbb{R}^n} - f(y)\} \\
&= \langle f'(x),x\rangle_{\mathbb{R}^n} - f(x).
\end{aligned} \tag{12.20}$$

On the other hand

$$f^*(x^*) \geq \langle x,x^*\rangle_{\mathbb{R}^n} - f(x), \ \forall x,x^* \in \mathbb{R}^n,$$

and thus

$$f(x) \geq \langle x,x^*\rangle_{\mathbb{R}^n} - f^*(x^*), \ \forall x^* \in \mathbb{R}^n.$$

Hence,

$$\begin{aligned}
f(x) &\geq \sup_{x^* \in \mathbb{R}^n} \{\langle x,x^*\rangle_{\mathbb{R}^n} - f^*(x^*)\} \\
&\geq \langle f'(x),x\rangle_{\mathbb{R}^n} - f^*(f'(x)).
\end{aligned} \tag{12.21}$$

From this and (12.20), we obtain,

$$\begin{aligned}
f(x) &= \sup_{x^* \in \mathbb{R}^n} \{\langle x,x^*\rangle_{\mathbb{R}^n} - f^*(x^*)\} \\
&= \langle f'(x),x\rangle_{\mathbb{R}^n} - f^*(f'(x)).
\end{aligned} \tag{12.22}$$

Summarizing,

$$f(x) = \sup_{x^* \in \mathbb{R}^n} \{\langle x,x^*\rangle_{\mathbb{R}^n} - f^*(x^*)\}, \ \forall x \in \mathbb{R}^n.$$

Choose $x,y \in \mathbb{R}^n$ and $\lambda \in [0,1]$.

From the last equation we may write,

$$
\begin{aligned}
f(\lambda x + (1-\lambda)y) &= \sup_{x^* \in \mathbb{R}^n} \{\langle \lambda x + (1-\lambda)y, x^* \rangle_{\mathbb{R}^n} - f^*(x^*)\} \\
&= \sup_{x^* \in \mathbb{R}^n} \{\langle \lambda x + (1-\lambda)y, x^* \rangle_{\mathbb{R}^n} \\
&\quad -\lambda f^*(x^*) - (1-\lambda)f^*(x^*)\} \\
&= \sup_{x^* \in \mathbb{R}^n} \{\lambda(\langle x, x^* \rangle_{\mathbb{R}^n} - f^*(x^*)) \\
&\quad +(1-\lambda)(\langle y, x^* \rangle_{\mathbb{R}^n} - f^*(x^*))\} \\
&\leq \lambda \sup_{x^* \in \mathbb{R}^n} \{\langle x, x^* \rangle_{\mathbb{R}^n} - f^*(x^*)\} \\
&\quad +(1-\lambda) \sup_{x^* \in \mathbb{R}^n} \{\langle y, x^* \rangle_{\mathbb{R}^n} - f^*(x^*)\} \\
&= \lambda f(x) + (1-\lambda)f(y).
\end{aligned} \tag{12.23}
$$

Since $x, y \in \mathbb{R}^n$ and $\lambda \in [0,1]$ are arbitrary, we may infer that f is convex.

This completes the proof.

Definition 12.3.4 (Convex functional) *Let V be a Banach space and let $J : D \subset V \to \mathbb{R}$ be a functional. We say that J is convex if*

$$
J(y+v) - J(y) \geq \delta J(y;v), \ \forall v \in V_a(y),
$$

where

$$
V_a(y) = \{v \in V : y+v \in D\}.
$$

Theorem 12.3.5 *Let V be a Banach space and let $J : D \subset U$ be a convex functional. Thus, if $y_0 \in D$ is such that*

$$
\delta J(y_0;v) = 0, \ \forall v \in V_a(y_0),
$$

then

$$
J(y_0) \leq J(y), \ \forall y \in D,
$$

that is, y_0 minimizes J on D.

Proof 12.4   Choose $y \in D$. Let $v = y - y_0$. Thus $y = y_0 + v \in D$ so that

$$
v \in V_a(y_0).
$$

From the hypothesis,

$$
\delta J(y_0;v) = 0,
$$

and since J is convex, we obtain

$$
J(y) - J(y_0) = J(y_0 + v) - J(y_0) \geq \delta J(y_0;v) = 0,
$$

that is,

$$
J(y_0) \leq J(y), \ \forall y \in D.
$$

The proof is complete.

Example 12.3.6 *Let us see this example of convex functional. Let $V = C^1([a,b])$ and let $J : D \subset V \to \mathbb{R}$ be defined by*

$$
J(y) = \int_a^b (y'(x))^2 \, dx,
$$

where

$$D = \{y \in V \ : \ y(a) = 1 \ and \ y(b) = 5\}.$$

We shall show that J is convex.
Indeed, let $y \in D$ and $v \in V_a$ where

$$V_a = \{v \in V \ : \ v(a) = v(b) = 0\}.$$

Thus,

$$
\begin{aligned}
J(y+v) - J(y) &= \int_a^b (y'(x) + v'(x))^2 \, dx - \int_a^b y'(x)^2 \, dx \\
&= \int_a^b 2y'(x)v'(x) \, dx + \int_a^b v'(x)^2 \, dx \\
&\geq \int_a^b 2y'(x)v'(x) \, dx \\
&= \delta J(y;v). \tag{12.24}
\end{aligned}
$$

Therefore, J is convex.

12.4 Sufficient conditions of optimality for the convex case

We start this section with a remark.

Remark 12.4.1 *Consider a function $f : [a,b] \times \mathbb{R} \times \mathbb{R} \to \mathbb{R}$ where $f \in C^1([a,b] \times \mathbb{R} \times \mathbb{R})$.*
Thus, for $V = C^1([a,b])$, define $F : V \to \mathbb{R}$ by

$$F(y) = \int_a^b f(x, y(x), y'(x)) \, dx.$$

Let $y, v \in V$. We have already shown that

$$\delta F(y;v) = \int_a^b (f_y(x, y(x), y'(x))v(x) + f_z(x, y(x), y'(x))v'(x)) \, dx.$$

Suppose f is convex in (y,z) for all $x \in [a,b]$, which we denote by $f(\underline{x}, y, z)$ to be convex.
From the last section, we have that

$$
\begin{aligned}
f(x, y+v, y'+v') - f(x, y, y') &\geq \langle \underline{\nabla} f(x, y, y'), (v, v') \rangle_{\mathbb{R}^2} \\
&= f_y(x, y, y')v + f_z(x, y, y')v', \ \forall x \in [a,b] \tag{12.25}
\end{aligned}
$$

where we denote

$$\underline{\nabla} f(x, y, y') = (f_y(x, y, y'), f_z(x, y, y')).$$

Therefore,

$$
\begin{aligned}
F(y+v) - F(y) &= \int_a^b [f(x, y+v, y'+v') - f(x, y, y')] \, dx \\
&\geq \int_a^b [f_y(x, y, y')v + f_z(x, y, y')v'] \, dx \\
&= \delta J(y;v). \tag{12.26}
\end{aligned}
$$

Thus, F is convex.

Theorem 12.4.2 *Let $V = C^1([a,b])$. Let $f \in C^2([a,b] \times \mathbb{R} \times \mathbb{R})$ where $f(\underline{x}, y, z)$ is convex. Define*

$$D = \{y \in V \ : \ y(a) = a_1 \text{ and } y(b) = b_1\},$$

where $a_1, b_1 \in \mathbb{R}$.

Define also $F : D \to \mathbb{R}$ by

$$F(y) = \int_a^b f(x, y(x), y'(x)) \, dx.$$

Under such hypotheses, F is convex and if $y_0 \in D$ is such that

$$\frac{d}{dx}[f_z(x, y_0(x), y_0'(x))] = f_y(x, y_0(x), y_0'(x)), \ \forall x \in [a,b],$$

then y_0 minimizes F on D, that is,

$$F(y_0) \leq F(y), \ \forall y \in D.$$

Proof 12.5 From the last remark, F is convex. Suppose now that $y_0 \in D$ is such that

$$\frac{d}{dx}[f_z(x, y_0(x), y_0'(x))] = f_y(x, y_0(x), y_0'(x)), \ \forall x \in [a,b].$$

Let $v \in V_a = \{v \in V \ : \ v(a) = v(b) = 0\}$. Thus,

$$
\begin{aligned}
\delta F(y_0; v) &= \int_a^b (f_y(x, y_0(x), y_0'(x))v(x) + f_z(x, y_0(x), y_0'(x))v'(x)) \, dx \\
&= \int_a^b \left(\frac{d}{dx}(f_z(x, y_0(x), y_0'(x))v(x)) + f_z(x, y_0(x), y_0'(x))v'(x) \right) dx \\
&= \int_a^b \left(\frac{d}{dx}[f_z(x, y_0(x), y_0'(x))v(x)] \right) dx \\
&= [f_z(x, y_0(x), y_0'(x))v(x)]_a^b \\
&= f_z(b, y_0(b), y_0'(b))v(b) - f_z(a, y_0(a), y_0'(a))v(a) \\
&= 0, \ \forall v \in V_a.
\end{aligned}
\tag{12.27}
$$

Since F is convex, from this and Theorem 12.3.5, we may infer that y_0 minimizes J on D.

Example 12.4.3 *Let $V = C^1([a,b])$ and*

$$D = \{y \in V \ : \ y(0) = 0 \text{ and } y(1) = 1\}.$$

Define $F : D \to \mathbb{R}$ by

$$F(y) = \int_0^1 [y'(x)^2 + 5y(x)] \, dx, \ \forall y \in D.$$

Observe that

$$F(y) = \int_0^1 f(x, y, y') \, dx$$

where

$$f(x, y, z) = z^2 + 5y,$$

that is, $f(\underline{x}, y, z)$ is convex.

Thus, from the last theorem F is convex and if $y_0 \in D$ is such that

$$\frac{d}{dx} f_z(x, y_0(x), y_0'(x)) = f_y(x, y_0(x), y_0'(x)), \ \forall x \in [a,b],$$

then y_0 minimizes F on D.

Considering that $f_z(x,y,z) = 2z$ and $f_y(x,y,z) = 5$, from this last equation we get,

$$\frac{d}{dx}(2y_0'(x)) = 5,$$

that is,

$$y_0''(x) = \frac{5}{2}, \ \forall x \in [0,1].$$

Thus,

$$y_0'(x) = \frac{5}{2}x + c,$$

and

$$y_0(x) = \frac{5}{4}x^2 + cx + d.$$

From this and $y_0(0) = 0$, we obtain $d = 0$.
From this and $y_0(1) = 1$, we have

$$\frac{5}{4} + c = 1,$$

so that

$$c = -1/4$$

Therefore

$$y_0(x) = \frac{5x^2}{4} - \frac{x}{4}$$

minimizes F on D.
The example is complete.

12.5 Natural conditions, problems with free extremals

We start this section with the following theorem.

Theorem 12.5.1 *Let $V = C^1([a,b])$. Let $f \in C^2([a,b] \times \mathbb{R} \times \mathbb{R})$ be such that $f(\underline{x}, y, z)$ is convex. Define*

$$D = \{y \in V \ : \ y(a) = a_1\},$$

where $a_1 \in \mathbb{R}$.

Define also $F : D \to \mathbb{R}$ by

$$F(y) = \int_a^b f(x, y(x), y'(x)) \, dx.$$

Under such hypotheses, F is convex and if $y_0 \in D$ is such that

$$\frac{d}{dx}[f_z(x, y_0(x), y_0'(x))] = f_y(x, y_0(x), y_0'(x)), \ \forall x \in [a,b]$$

and

$$f_z(b, y_0(b), y_0'(b)) = 0$$

then y_0 minimizes F on D, that is,

$$F(y_0) \le F(y), \ \forall y \in D.$$

Proof 12.6   Since $f(\underline{x}, y, z)$ is convex from the last remark F is convex. Suppose now that $y_0 \in D$ is such that

$$\frac{d}{dx}[f_z(x, y_0(x), y_0'(x))] = f_y(x, y_0(x), y_0'(x)), \ \forall x \in [a, b]$$

and

$$f_z(b, y_0(b), y_0'(b)) = 0.$$

Let $v \in V_a = \{v \in V \ : \ v(a) = 0\}$. Thus,

$$
\begin{aligned}
\delta F(y_0; v) &= \int_a^b (f_y(x, y_0(x), y_0'(x))v(x) + f_z(x, y_0(x), y_0'(x))v'(x)) \, dx \\
&= \int_a^b \left(\frac{d}{dx}(f_z(x, y_0(x), y_0'(x))v(x)) + f_z(x, y_0(x), y_0'(x))v'(x) \right) dx \\
&= \int_a^b \left(\frac{d}{dx}[f_z(x, y_0(x), y_0'(x))v(x)] \right) dx \\
&= [f_z(x, y_0(x), y_0'(x))v(x)]_a^b \\
&= f_z(b, y_0(b), y_0'(b))v(b) - f_z(a, y_0(a), y_0'(a))v(a) \\
&= 0 v(b) - f_z(a, y_0(a), y_0'(b))0 \\
&= 0, \ \forall v \in V_a.
\end{aligned}
\tag{12.28}
$$

Since F is convex, from this and Theorem 12.3.5, we may infer that y_0 minimizes J on D.

Remark 12.5.2 *About this last theorem $y(a) = a_1$ is said to be an essential boundary condition, whereas $f_z(b, y_0(b), y_0'(b)) = 0$ is said to be a natural boundary condition.*

Theorem 12.5.3 *Let $V = C^1([a, b])$. Let $f \in C^2([a, b] \times \mathbb{R} \times \mathbb{R})$ where $f(\underline{x}, y, z)$ is convex. Define*

$$D = V$$

and $F : D \to \mathbb{R}$ by

$$F(y) = \int_a^b f(x, y(x), y'(x)) \, dx.$$

Under such hypotheses, F is convex and if $y_0 \in D$ is such that

$$\frac{d}{dx}[f_z(x, y_0(x), y_0'(x))] = f_y(x, y_0(x), y_0'(x)), \ \forall x \in [a, b],$$

$$f_z(a, y_0(a), y_0'(a)) = 0$$

and

$$f_z(b, y_0(b), y_0'(b)) = 0,$$

then y_0 minimizes F on D, that is,

$$F(y_0) \le F(y), \ \forall y \in D.$$

Proof 12.7   From the last remark F is convex. Suppose that $y_0 \in D$ is such that

$$\frac{d}{dx}[f_z(x, y_0(x), y_0'(x))] = f_y(x, y_0(x), y_0'(x)), \ \forall x \in [a, b]$$

and

$$f_z(a, y_0(a), y_0'(a)) = f_z(b, y_0(b), y_0'(b)) = 0.$$

Let $v \in D = V$. Thus,

$$
\begin{aligned}
\delta F(y_0; v) &= \int_a^b (f_y(x, y_0(x), y_0'(x))v(x) + f_z(x, y_0(x), y_0'(x))v'(x)) \, dx \\
&= \int_a^b \left(\frac{d}{dx}(f_z(x, y_0(x), y_0'(x))v(x)) + f_z(x, y_0(x), y_0'(x))v'(x) \right) dx \\
&= \int_a^b \left(\frac{d}{dx}[f_z(x, y_0(x), y_0'(x))v(x)] \right) dx \\
&= [f_z(x, y_0(x), y_0'(x))v(x)]_a^b \\
&= f_z(b, y_0(b), y_0'(b))v(b) - f_z(a, y_0(a), y_0'(a))v(a) \\
&= 0v(b) - 0v(a) \\
&= 0, \ \forall v \in D.
\end{aligned}
\tag{12.29}
$$

Since F is convex, from this and from Theorem 12.3.5, we may conclude that y_0 minimizes J on $D = V$. The proof is complete.

Remark 12.5.4 *About this last theorem, the conditions $f_z(a, y_0(a), y_0'(a)) = f_z(b, y_0(b), y_0'(b)) = 0$ are said to be natural boundary conditions and the problem in question a free extremal one.*

Exercise 12.5.5 *Show that F is convex and obtain its point of global minimum on D, D_1 and D_2, where*

$$
F(y) = \int_1^2 \frac{y'(x)^2}{x} \, dx,
$$

and where

1.

$$
D = \{y \in C^1([1,2]) : y(1) = 0, \ y(1) = 3\},
$$

2.

$$
D_1 = \{y \in C^1([1,2]) : y(2) = 3\}.
$$

3.

$$
D_2 = C^1([1,2]).
$$

Solution: Observe that

$$
F(y) = \int_1^2 f(x, y(x), y'(x)) \, dx,
$$

where $f(x, y, z) = z^2/x$, so that $f(x, y, z)$ is convex.
Therefore, F is convex.
Let $y, v \in V$, thus,

$$
\delta F(y; v) = \int_1^2 [f_y(x, y, y')v + f_z(x, y, y')v'] \, dx,
$$

where

$$
f_y(x, y, z) = 0
$$

and

$$
f_z(x, y, z) = 2z/x.
$$

Therefore,

$$
\delta F(y; v) = \int_1^2 2x^{-1}y'(x)v'(x) \, dx.
$$

For D, from Theorem 12.4.1, sufficient conditions of optimality are given by,

$$\begin{cases} \frac{d}{dx}[f_z(x,y_0(x),y_0'(x))] = f_y(x,y_0(x),y_0'(x)) \text{ in } [1,2], \\ y_0(1) = 0, \\ y_0(2) = 3. \end{cases} \tag{12.30}$$

Thus, we must have

$$\frac{d}{dx}[2x^{-1}y_0'(x)] = 0,$$

that is,

$$2x^{-1}y_0'(x) = c,$$

so that

$$y_0'(x) = \frac{cx}{2}.$$

Therefore,

$$y_0(x) = \frac{cx^2}{4} + d.$$

On the other hand, we must have also

$$y_0(1) = \frac{c}{4} + d = 0,$$

and

$$y_0(2) = c + d = 3.$$

Thus, $c = 4$ and $d = -1$ so that $y_0(x) = x^2 - 1$ minimizes F on D.

For D_1, from Theorem 12.5.1, sufficient conditions of global optimality are given by

$$\begin{cases} \frac{d}{dx}[f_z(x,y_0(x),y_0'(x))] = f_y(x,y_0(x),y_0'(x)) \text{ in } [1,2], \\ y_0(2) = 3, \\ f_z(1,y_0(1),y_0'(1)) = 0. \end{cases} \tag{12.31}$$

Thus, we must have

$$y_0(x) = \frac{cx^2}{4} + d.$$

On the other hand, we must also have

$$y_0(2) = c + d = 3,$$

and

$$f_z(1,y_0(1),y_0'(1)) = 2(1)^{-1}y_0'(1) = 0,$$

that is,

$$y_0'(1) = c/2 = 0,$$

Therefore, $c = 0$ and $d = 3$ so that $y_0(x) = 3$ minimizes F on D_1.

For D_2, from Theorem 12.5.3, sufficient conditions of global optimality are given by

$$\begin{cases} \frac{d}{dx}[f_z(x,y_0(x),y_0'(x))] = f_y(x,y_0(x),y_0'(x)) \text{ in } [1,2], \\ f_z(1,y_0(1),y_0'(1)) = 0 \\ f_z(2,y_0(2),y_0'(2)) = 0. \end{cases} \tag{12.32}$$

Thus, we have

$$y_0(x) = \frac{cx^2}{4} + d.$$

On the other hand we have also

$$f_z(1,y_0(1),y_0'(1)) = 2(1)^{-1}y_0'(1) = 0,$$

$$f_z(2,y_0(2),y_0'(2)) = 2(2)^{-1}y_0'(2) = 0,$$

that is,

$$y_0'(1) = y_0'(2) = 0,$$

where $y_0'(x) = cx/2$.
 Thus $c = 0$, *so that* $y_0(x) = d$, $\forall d \in \mathbb{R}$ *minimizes F on* D_2.

Exercise 12.5.6 *Let* $V = C^2([0,1])$ *and* $J : D \subset V \to \mathbb{R}$ *where*

$$J(y) = \frac{EI}{2} \int_0^1 y''(x)^2 \, dx - \int_0^1 P(x)y(x) \, dx,$$

represents the energy of a straight beam with rectangular cross section with inertial moment I. Here $y(x)$ *denotes the vertical displacement of the point* $x \in [0,1]$ *resulting from the action of distributed vertical load* $P(x) = \alpha x$, $\forall x \in [0,1]$, *where* $E > 0$ *is the Young modulus and* $\alpha > 0$ *is a real constant.*
 And also

$$D = \{y \in V \,:\, y(0) = y(1) = 0\}.$$

Under such hypotheses,

1. *prove that F is convex.*

2. *Prove that if* $y_0 \in D$ *is such that*

$$\begin{cases} EI\frac{d^4}{dx^4}[y_0(x)] = P(x), \; \forall x \in [0,1], \\[2mm] y_0''(0) = 0, \\[2mm] y_0''(1) = 0, \end{cases} \tag{12.33}$$

then y_0 *minimizes F on D.*

3. *Find the optimal solution* $y_0 \in D$.

Solution:
 Let $y \in D$ *and* $v \in V_a = \{v \in V \,:\, v(0) = v(1) = 0\}$.
 We recall that

$$\begin{aligned} \delta J(y;v) &= \lim_{\varepsilon \to 0} \frac{F(y+\varepsilon v) - F(y)}{\varepsilon} \\[2mm] &= \lim_{\varepsilon \to 0} \frac{(EI/2)\int_0^1 [(y'' + \varepsilon v'')^2 - (y'')^2]\, dx - \int_0^1 (P(y+\varepsilon v) - Py)\, dx}{\varepsilon} \\[2mm] &= \lim_{\varepsilon \to 0} \left(\int_0^1 (EIy''v'' - Pv)\, dx + \frac{\varepsilon EI}{2} \int_0^1 (v'')^2 \, dx \right) \\[2mm] &= \int_0^1 (EIy''v'' - Pv)\, dx. \end{aligned} \tag{12.34}$$

On the other hand,

$$
\begin{aligned}
J(y+v) - J(v) &= (EI/2) \int_0^1 [(y''+v'')^2 - (y'')^2]\, dx - \int_0^1 (P(y+v) - Py)\, dx \\
&= \int_0^1 (EIy''v'' - Pv)\, dx + \frac{EI}{2} \int_0^1 (v'')^2\, dx \\
&\geq \int_0^1 (EIy''v'' - Pv)\, dx \\
&= \delta J(y;v).
\end{aligned}
\tag{12.35}
$$

Since $y \in D$ and $v \in V_a$ are arbitrary, we may infer that J is convex.
Assume that $y_0 \in D$ is such that

$$
\begin{cases}
EI\frac{d^4}{dx^4}[y_0(x)] = P(x), \ \forall x \in [0,1], \\[2mm]
y_0''(0) = 0, \\[2mm]
y_0''(1) = 0,
\end{cases}
\tag{12.36}
$$

Thus,

$$
\begin{aligned}
\delta J(y;v) &= \int_0^1 (EIy''v'' - Pv)\, dx \\
&= \int_0^1 (EIy''v'' - EIy^{(4)}v)\, dx \\
&= \int_0^1 (EIy''v'' + EIy'''v')\, dx - [EIy'''(x)v(x)]_a^b \\
&= \int_0^1 (EIy''v'' + EIy'''v')\, dx \\
&= \int_0^1 (EIy''v'' - EIy''v'')\, dx + [EIy''(x)v'(x)]_a^b \\
&= 0
\end{aligned}
\tag{12.37}
$$

Summarizing

$$
\delta J(y_0;v) = 0, \ \forall v \in V_a.
$$

Therefore, since J is convex, we may conclude that y_0 minimizes J on D.
To obtain the solution of the ODE is question, we shall denote

$$
y_0(x) = y_p(x) + y_h(x),
$$

where a particular solution y_p is given by $y_p(x) = \frac{\alpha x^5}{120EI}$, where clearly

$$
EI\frac{d^4}{dx^4}[y_p(x)] = P(x), \forall x \in [0,1].
$$

The homogeneous associated equation

$$
EI\frac{d^4}{dx^4}[y_h(x)] = 0,
$$

has the following general solution

$$
y_h(x) = ax^3 + bx^2 + cx + d,
$$

and thus,

$$y_0(x) = y_p(x) + y_h(x) = \frac{\alpha x^5}{120EI} + ax^3 + bx^2 + cx + d.$$

From $y_0(0) = 0$, we obtain $d = 0$.
Observe that $y_0'(x) = \frac{5\alpha}{120EI}x^4 + 3ax^2 + 2bx + c$ e $y_0''(x) = \frac{\alpha}{6EI}x^3 + 6ax + 2b$.
From this and $y_0''(0) = 0$, we get $b = 0$.
From $y_0''(1) = 0$, we obtain,

$$\frac{\alpha}{6EI} 1^3 + 6a\, 1 = 0,$$

and thus

$$a = -\frac{\alpha}{36EI}.$$

From such results and from $y_0(1) = 0$, we obtain

$$\frac{\alpha}{120EI} + a\, 1^3 + c\, 1 = \frac{\alpha}{120EI} - \frac{\alpha}{36EI} + c = 0,$$

that is,

$$c = \frac{\alpha}{EI}\left(\frac{1}{36} - \frac{1}{120}\right) = \frac{7\alpha}{360EI}.$$

Finally, we have that

$$y_0(x) = \frac{\alpha x^5}{120EI} - \frac{\alpha x^3}{36EI} + \frac{7\alpha x}{360EI}$$

minimizes J on D.
The solution is complete.

12.6 The du Bois-Reymond lemma

Lemma 12.6.1 (du Bois-Reymond) *Suppose $h \in C([a,b])$ and*

$$\int_a^b h(x)v'(x)\, dx = 0, \ \forall v \in V_a,$$

where

$$V_a = \{v \in C^1([a,b]) : v(a) = v(b) = 0\}.$$

Under such hypotheses, there exists $c \in \mathbb{R}$ such that

$$h(x) = c, \ \forall x \in [a,b].$$

Proof 12.8   Let

$$c = \frac{\int_a^b h(t)\, dt}{b-a}.$$

Define

$$v(x) = \int_a^x (h(t) - c)\, dt.$$

Thus,

$$v'(x) = h(x) - c, \ \forall x \in [a,b],$$

so that $v \in C^1([a,b])$.

Moreover,

$$v(a) = \int_a^a (h(t) - c) \, dt = 0,$$

and

$$v(b) = \int_a^b (h(t) - c) \, dt = \int_a^b h(t) \, dt - c(b-a) = c(b-a) - c(b-a) = 0,$$

so that $v \in V_a$.

Observe that, from this and the hypotheses,

$$
\begin{aligned}
0 \;\leq\; & \int_a^b (h(t) - c)^2 \, dt \\
= \;& \int_a^b (h(t) - c)(h(t) - c) \, dt \\
= \;& \int_a^b (h(t) - c)v'(t) \, dt \\
= \;& \int_a^b h(t)v'(t) \, dt - c \int_a^b v'(t) \, dt \\
= \;& 0 - c(v(b) - v(a)) \\
= \;& 0.
\end{aligned}
\tag{12.38}
$$

Thus,

$$\int_a^b (h(t) - c)^2 \, dt = 0.$$

Since h is continuous, we may infer that

$$h(x) - c = 0, \; \forall x \in [a,b],$$

that is,

$$h(x) = c, \; \forall x \in [a,b].$$

The proof is complete.

Theorem 12.6.2 *Let $g,h \in C([a,b])$ and suppose*

$$\int_a^b (g(x)v(x) + h(x)v'(x)) \, dx = 0, \; \forall v \in V_a,$$

where

$$V_a = \{v \in C^1([a,b]) \, : \, v(a) = v(b) = 0\}.$$

Under such hypotheses, $h \in C^1([a,b])$ and

$$h'(x) = g(x), \; \forall x \in [a,b].$$

Proof 12.9　Define

$$G(x) = \int_a^x g(t) \, dt.$$

Thus,

$$G'(x) = g(x), \; \forall x \in [a,b].$$

Let $v \in V_a$.

From the hypotheses,

$$
\begin{aligned}
0 &= \int_a^b [g(x)v(x) + h(x)v'(x)]\, dx \\
&= \int_a^b [-G(x)v'(x) + h(x)v'(x)]\, dx + [G(x)v(x)]_a^b \\
&= \int_a^b [-G(x) + h(x)]v'(x)\, dx,\ \forall v \in V_a.
\end{aligned}
\tag{12.39}
$$

From this and from the du Bois - Reymond lemma, we may conclude that

$$
-G(x) + h(x) = c, \forall x \in [a,\ b],
$$

for some $c \in \mathbb{R}$.

Thus

$$
g(x) = G'(x) = h'(x),\ \forall x \in [a,b],
$$

so that

$$
g \in C^1([a,b]).
$$

The proof is complete.

Lemma 12.6.3 (Fundamental lemma of calculus of variation for one dimension) *Let $g \in C([a,b]) = V$.*
Assume

$$
\int_a^b g(x)v(x)\, dx = 0,\ \forall v \in V_a,
$$

where again,

$$
V_a = \{v \in C^1([a,b])\ :\ v(a) = v(b) = 0\}.
$$

Under such hypotheses,

$$
g(x) = 0,\ \forall x \in [a,b].
$$

Proof 12.10   It suffices to apply the last theorem for $h \equiv 0$.

Exercise 12.6.4 *Let $h \in C([a,b])$.*
Suppose

$$
\int_a^b h(x)w(x)\, dx = 0,\ \forall w \in D_0,
$$

where

$$
D_0 = \left\{ w \in C([a,b])\ :\ \int_a^b w(x)\, dx = 0 \right\}.
$$

Show that there exists $c \in \mathbb{R}$ such that

$$
h(x) = c,\ \forall x \in [a,b].
$$

Solution Define, as above indicated,

$$
V_a = \{v \in C^1([a,b])\ :\ v(a) = v(b) = 0\}.
$$

Let $v \in V_a$.
Let $w \in C([a,b])$ be such that

$$
w(x) = v'(x),\ \forall x \in [a,b].
$$

Observe that

$$\int_a^b w(x)\,dx = \int_a^b v'(x)\,dx = [v(x)]_a^b = v(b) - v(a) = 0.$$

From this $\int_a^b h(x)w(x)\,dx = 0$, and thus

$$\int_a^b h(x)v'(x)\,dx = 0.$$

Since $v \in V_a$ is arbitrary, from this and the du Bois-Reymond lemma, there exists $c \in \mathbb{R}$ such that

$$h(x) = c, \ \forall x \in [a,b].$$

The solution is complete.

12.7   Calculus of variations, the case of scalar functions on $\mathbb{R}^n$

Let $\Omega \subset \mathbb{R}^n$ be an open, bounded, connected with a regular boundary $\partial\Omega = S$ (Lipschitzian) (which we define as Ω to be of class $\hat{C}^1$). Let $V = C^1(\overline{\Omega})$ and let $F : D \subset V \to \mathbb{R}$, be such that

$$F(y) = \int_\Omega f(x, y(x), \nabla y(x))\,dx, \ \forall y \in V,$$

where we denote

$$dx = dx_1 \cdots dx_n.$$

Assume $f : \overline{\Omega} \times \mathbb{R} \times \mathbb{R}^n \to \mathbb{R}$ is of C^2 class. Suppose also $f(x, y, \mathbf{z})$ is convex in $(y, \mathbf{z})$, $\forall x \in \overline{\Omega}$, which we denote by $f(\underline{x}, y, \mathbf{z})$ to be convex.

Observe that for $y \in D$ and $v \in V_a$, where

$$D = \{y \in V \ : \ y = y_1 \text{ on } \partial\Omega\},$$

and

$$V_a = \{v \in V \ : \ v = 0 \text{ on } \partial\Omega\},$$

where

$$y_1 \in C^1(\overline{\Omega}),$$

we have that

$$\delta F(y; v) = \frac{\partial}{\partial \varepsilon} F(y + \varepsilon v)|_{\varepsilon=0},$$

where

$$F(y + \varepsilon v) = \int_\Omega f(x, y + \varepsilon v, \nabla y + \varepsilon \nabla v)\,dx.$$

Therefore,

$$\begin{aligned}
\frac{\partial}{\partial \varepsilon} F(y + \varepsilon v) &= \int_\Omega \left(\frac{\partial}{\partial \varepsilon}(f(x, y + \varepsilon v, \nabla y + \varepsilon \nabla v)) \right) dx \\
&= \int_\Omega [f_y(x, y + \varepsilon v, \nabla y + \varepsilon \nabla v)v + \sum_{i=1}^n f_{z_i}(x, y + \varepsilon v, \nabla y + \varepsilon \nabla v)v_{x_i}]\,dx. \quad (12.40)
\end{aligned}$$

Thus,

$$\begin{aligned}
\delta F(y; v) &= \frac{\partial}{\partial \varepsilon} F(y + \varepsilon v)|_{\varepsilon=0} \\
&= \int_\Omega [f_y(x, y, \nabla y)v + \sum_{i=1}^n f_{z_i}(x, y, \nabla y)v_{x_i}]\,dx. \quad (12.41)
\end{aligned}$$

On the other hand, since $f(\underline{x}, y, \mathbf{z})$ is convex, we have that

$$
\begin{aligned}
F(y+v) - F(y) &= \int_{\Omega} [f(x, y+v, \nabla y + \nabla v) - f(x, y, \nabla y)] \, dx \\
&\geq \langle \underline{\nabla} f(x, y, \nabla y), (v, \nabla v) \rangle_{R^{n+1}} \\
&= \int_{\Omega} [f_y(x, y, \nabla y)v + \sum_{i=1}^{n} f_{z_i}(x, y, \nabla y)v_{x_i}] \, dx \\
&= \delta F(y; v).
\end{aligned} \tag{12.42}
$$

Since $y \in D$ and $v \in V_a$ are arbitrary, w e may infer that F is convex.
Here we denote,

$$
\underline{\nabla} f(x, y, \nabla y) = (f_y(x, y, \nabla y), f_{z_1}(x, y, \nabla y), \cdots, f_{z_n}(x, y, \nabla y)).
$$

Theorem 12.7.1 *Let $\Omega \subset \mathbb{R}^n$ be a set of $\hat{C}^1$ class and let $V = C^1(\overline{\Omega})$. Let $f \in C^2(\overline{\Omega} \times \mathbb{R} \times \mathbb{R})$ where $f(\underline{x}, y, \mathbf{z})$ is convex. Define*

$$
D = \{y \in V : y = y_1 \text{ em } \partial \Omega\},
$$

where $y_1 \in C^1(\overline{\Omega})$
Define also $F : D \to \mathbb{R}$ by

$$
F(y) = \int_{\Omega} f(x, y(x), \nabla y(x)) \, dx.
$$

From such hypotheses, F is convex and if $y_0 \in D$ is such that

$$
\sum_{i=1}^{n} \frac{d}{dx_i} [f_{z_i}(x, y_0(x), \nabla y_0(x))] = f_y(x, y_0(x), \nabla y_0(x)), \ \forall x \in \overline{\Omega},
$$

then y_0 minimizes F on D, that is,

$$
F(y_0) \leq F(y), \ \forall y \in D.
$$

Proof 12.11 From the last remark, F is convex. Suppose now that $y_0 \in D$ is such that

$$
\sum_{i=1}^{n} \frac{d}{dx_i} [f_{z_i}(x, y_0(x), \nabla y_0(x))] = f_y(x, y_0(x), \nabla y_0(x)), \ \forall x \in \overline{\Omega},
$$

Let $v \in V_a = \{v \in V : v = 0 \text{ on } \partial \Omega\}$. Thus,

$$
\begin{aligned}
\delta F(y_0; v) &= \int_{\Omega} (f_y(x, y_0(x), \nabla y_0(x))v(x) + \sum_{i=1}^{n} f_{z_i}(x, y_0(x), \nabla y_0(x))v_{x_i}(x)) \, dx \\
&= \int_{\Omega} \left(\sum_{i=1}^{n} \frac{d}{dx_i} (f_{z_i}(x, y_0(x), \nabla y_0(x)))v(x) + \sum_{i=1}^{n} f_{z_i}(x, y_0(x), \nabla y_0(x))v_{x_i}(x) \right) dx \\
&= \int_{\Omega} \left(-\sum_{i=1}^{n} f_{z_i}(x, y_0(x), \nabla y_0(x))v_{x_i}(x) + \sum_{i=1}^{n} f_{z_i}(x, y_0(x), \nabla y_0(x))v_{x_i}(x) \right) dx \\
&\quad + \int_{\partial \Omega} \sum_{i=1}^{n} f_{z_i}(x, y_0(x), \nabla y_0(x)) \, n_i \, v(x) \, dS \\
&= 0, \ \forall v \in V_a,
\end{aligned} \tag{12.43}
$$

where $\mathbf{n} = (n_1, \cdots, n_n)$ denotes the outward normal field to $\partial \Omega = S$. Since F is convex, from this and Theorem 12.3.5, we have that y_0 minimizes F on D.

12.8   The second Gâteaux variation

Definition 12.8.1 *Let V be a Banach space. Let $F : D \subset V \to \mathbb{R}$ be a functional such that $\delta F(y;v)$ exists on $B_r(y_0)$ for $y_0 \in D$, $r > 0$ and for all $v \in V_a$.*

Let $y \in B_r(y_0)$ and $v, w \in V_a$. We define the second Gâteaux variation of F at the point y in the directions v and w, denoted by $\delta^2 F(y;v,w)$, as

$$\delta^2 F(y;v,w) = \lim_{\varepsilon \to 0} \frac{\delta F(y + \varepsilon w;v) - \delta F(y;v)}{\varepsilon},$$

if such a limit exists.

Remark 12.8.2 *Observe that from this last definition, if the limits in question exist, we have*

$$\delta F(y;v) = \frac{\partial}{\partial \varepsilon} F(y + \varepsilon v)|_{\varepsilon=0},$$

and

$$\delta^2 F(y;v,v) = \frac{\partial^2}{\partial \varepsilon^2} F(y + \varepsilon v)|_{\varepsilon=0}, \forall v \in V_a.$$

Thus, for example, for $V = C^1(\overline{\Omega})$ where $\Omega \subset \mathbb{R}^n$ is of $\hat{C}^1$ class and $F : V \to \mathbb{R}$ is given by

$$F(y) = \int_{\Omega} f(x, y, \nabla y) \, dx$$

and where

$$f \in C^2(\overline{\Omega} \times \mathbb{R} \times \mathbb{R}^n),$$

for $y, v \in V$, we have

$$\delta^2 F(y;v,v) = \frac{\partial^2}{\partial \varepsilon^2} F(y + \varepsilon v)|_{\varepsilon=0},$$

where

$$
\begin{aligned}
\frac{\partial^2}{\partial \varepsilon^2} F(y + \varepsilon v) &= \frac{\partial^2}{\partial \varepsilon^2} \left(\int_{\Omega} f(x, y + \varepsilon v, \nabla y + \varepsilon \nabla v) \, dx \right) \\
&= \int_{\Omega} \frac{\partial^2}{\partial \varepsilon^2} [f(x, y + \varepsilon v, \nabla y + \varepsilon \nabla v)] \, dx \\
&= \int_{\Omega} \Bigg[f_{yy}(x, y + \varepsilon v, \nabla y + \varepsilon \nabla v) v^2 + \sum_{i=1}^{n} 2 f_{yz_i}(x, y + \varepsilon v, \nabla y + \varepsilon \nabla v) v v_{x_i} \\
&\quad + \sum_{i=1}^{n} \sum_{j=1}^{n} f_{z_i z_j}(x, y + \varepsilon v, \nabla y + \varepsilon \nabla v) v_{x_i} v_{x_j} \Bigg] \, dx
\end{aligned}
$$

(12.44)

so that

$$
\begin{aligned}
\delta^2 F(y;v,v) &= \frac{\partial^2}{\partial \varepsilon^2} F(y + \varepsilon v)|_{\varepsilon=0} \\
&= \int_{\Omega} \Bigg[f_{yy}(x, y, \nabla y) v^2 + \sum_{i=1}^{n} 2 f_{yz_i}(x, y, \nabla y) v v_{x_i} \\
&\quad + \sum_{i=1}^{n} \sum_{j=1}^{n} f_{z_i z_j}(x, y, \nabla y) v_{x_i} v_{x_j} \Bigg] \, dx.
\end{aligned}
$$

(12.45)

12.9   First order necessary conditions for a local minimum

Definition 12.9.1 *Let V be a Banach space. Let $F : D \subset V \to \mathbb{R}$ be a functional. We say that $y_0 \in D$ is a point of local minimum for F on D, if there exists $\delta > 0$ such that*

$$F(y) \geq F(y_0), \ \forall y \in B_\delta(y_0) \cap D.$$

Theorem 12.9.2 *[First order necessary condition] Let V be a Banach space. Let $F : D \subset V \to \mathbb{R}$ be a functional. Suppose that $y_0 \in D$ is a point of local minimum for F on D. Let $v \in V_a$ and assume $\delta F(y_0; v)$ to exist.*

Under such hypotheses,

$$\delta F(y_0; v) = 0.$$

Proof 12.12   Define $\phi(\varepsilon) = F(y_0 + \varepsilon v)$, which from the existence of $\delta F(y_0; v)$ is well defined for all ε sufficiently small.

Also from the hypotheses, $\varepsilon = 0$ is a point of local minimum for the differentiable at 0 function ϕ.

Thus, from the standard condition for one variable calculus, we have

$$\phi'(0) = 0,$$

thst is,

$$\phi'(0) = \delta F(y_0; v) = 0.$$

The proof is complete.

Theorem 12.9.3 (Second order sufficient condition) *Let V be a Banach space. Let $F : D \subset V \to \mathbb{R}$ be a functional. Suppose $y_0 \in D$ is such that $\delta F(y_0; v) = 0$ for all $v \in V_a$ and there exists $\delta > 0$ such that*

$$\delta^2 F(y; v, v) \geq 0, \ \forall y \in B_\delta(y_0) \text{ and } v \in V_a.$$

Under such hypotheses $y_0 \in D$ is a point of local minimum for F, that is

$$F(y) \geq F(y_0), \ \forall y \in B_r(y_0) \cap D.$$

Proof 12.13   Let $y \in B_\delta(y_0) \cap D$. Define $v = y - y_0 \in V_a$.

Define also $\phi : [0, 1] \to \mathbb{R}$ by

$$\phi(\varepsilon) = F(y_0 + \varepsilon v).$$

From the Taylor Theorem for one variable, there exists $t_0 \in (0, 1)$ such that

$$\phi(1) = \phi(0) + \frac{\phi'(0)}{1!}(1 - 0) + \frac{1}{2!}\phi''(t_0)(1 - 0)^2,$$

That is,

$$
\begin{aligned}
F(y) &= F(y_0 + v) \\
&= F(y_0) + \delta F(y_0; v) + \frac{1}{2}\delta^2 F(y_0 + t_0 v; v, v) \\
&= F(y_0) + \frac{1}{2}\delta^2 F(y_0 + t_0 v; v, v) \\
&\geq F(y_0), \ \forall y \in B_\delta(y_0) \cap D.
\end{aligned}
\tag{12.46}
$$

The proof is complete.

12.10 Continuous functionals

Definition 12.10.1 *Let V be a Banach space. Let $F : D \subset V \to \mathbb{R}$ be a functional and let $y_0 \in D$.*

We say that F is continuous on $y_0 \in D$, if for each $\varepsilon > 0$ there exists $\delta > 0$ such that if $y \in D$ and $\|y - y_0\|_V < \delta$, then

$$|F(y) - F(y_0)| < \varepsilon.$$

Example 12.10.2 *Let $V = C^1([a,b])$ and $f \in C([a,b] \times \mathbb{R} \times \mathbb{R})$.*

Consider $F : V \to \mathbb{R}$ where

$$F(y) = \int_a^b f(x, y(x), y'(x)) \, dx,$$

and

$$\|y\|_V = \max\{|y(x)| + |y'(x)| \; : \; x \in [a,b]\}.$$

Let $y_0 \in V$. We shall prove that F is continuous at y_0.

Let $y \in V$ be such that

$$\|y - y_0\|_V < 1.$$

Thus,

$$\|y\|_V - \|y_0\|_V \le \|y - y_0\|_V < 1,$$

that is,

$$\|y\|_V < 1 + \|y_0\|_V \equiv \alpha.$$

Observe that f is uniformly continuous on the compact set

$$[a,b] \times [-\alpha, \alpha] \times [-\alpha, \alpha] \equiv A.$$

Let $\varepsilon > 0$. Therefore, there exists $\delta_0 > 0$ such that if (x, y_1, z_1) and $(x, y_2, z_2) \in A$ and

$$|y_1 - y_2| + |z_1 - z_2| < \delta_0,$$

then

$$|f(x, y_1, z_1) - f(x, y_2, z_2)| < \frac{\varepsilon}{b - a}. \tag{12.47}$$

Let $\delta = \min\{\delta_0, 1\}$.

Hence, if

$$\|y - y_0\|_V < \delta,$$

we have

$$\max\{|y(x) - y_0(x)| + |y'(x) - y_0'(x)| \; : \; x \in [a,b]\} < \delta \le 1,$$

so that from this and (12.47), we obtain

$$|f(x, y(x), y'(x)) - f(x, y_0(x), y_0'(x))| < \frac{\varepsilon}{b - a}, \; \forall x \in [a,b].$$

Thus,

$$
\begin{aligned}
|F(y) - F(y_0)| &= \left| \int_a^b [f(x, y(x), y'(x)) - f(x, y_0(x), y_0'(x))] \, dx \right| \\
&\le \int_a^b |f(x, y(x), y'(x)) - f(x, y_0(x), y_0'(x))| \, dx \\
&< \frac{\varepsilon(b - a)}{(b - a)} \\
&= \varepsilon.
\end{aligned}
\tag{12.48}
$$

From this we have that F is continuous at y_0, $\forall y_0 \in V$.

The example is complete.

12.11 The Gâteaux variation, the formal proof of its formula

In the previous sections we had obtained the Gâteaux variations formulas with some informality for a relatively large class of functionals.

In this section we intend to provide a formal proof for such formulas.

Our main result is summarized by the following theorem.

Theorem 12.11.1 *Let $\Omega \subset \mathbb{R}^n$ be sets of $\hat{C}^1$ class and let $V = C^1(\overline{\Omega})$.*
Let $f : \overline{\Omega} \times \mathbb{R} \times \mathbb{R}^n \to \mathbb{R}$ be a function of C^1 class.
Define $F : V \to \mathbb{R}$ by

$$F(y) = \int_{\Omega} f(x, y(x), \nabla y(x)) \, dx.$$

Let $y, v \in V$. Under such hypotheses

$$\delta F(y; v) = \int_{\Omega} \left(f_y(x, y(x), \nabla y(x))v(x) + \sum_{i=1}^{n} f_{z_i}(x, y(x), \nabla y(x))v_{x_i}(x) \right) dx.$$

Proof 12.14 Let $\{\varepsilon_n\} \subset \mathbb{R} \setminus \{0\}$ be a sequence such that

$$\varepsilon_n \to 0, \text{ as } n \to \infty.$$

Define

$$G_n(x) = \frac{f(x, y(x) + \varepsilon_n v(x), \nabla y(x) + \varepsilon_n \nabla v(x)) - f(x, y(x), \nabla y(x))}{\varepsilon_n},$$

$\forall n \in \mathbb{N}, \ x \in \overline{\Omega}$.
Define also

$$G(x) = f_y(x, y(x), \nabla y(x))v(x) + \sum_{i=1}^{n} f_{z_i}(x, y(x), \nabla y(x))v_{x_i}(x), \ \forall x \in \overline{\Omega}.$$

Observe that

$$G_n(x) \to G(x), \ \forall x \in \overline{\Omega}.$$

Now, we are going to prove that, for a not relabeled subsequence,

$$\int_{\Omega} G_n(x) \, dx \to \int_{\Omega} G(x) \, dx, \text{ as } n \to \infty$$

Define

$$c_n = \max_{x \in \overline{\Omega}} \{|G_n(x) - G(x)|\}.$$

From the continuity of the functions in question, for each $n \in \mathbb{N}$, there exists $x_n \in \overline{\Omega}$ such that

$$c_n = |G_n(x_n) - G(x_n)|.$$

Observe that $\{x_n\} \subset \overline{\Omega}$ and such a set is compact. Thus, there exist a subsequence $\{x_{n_j}\}$ of $\{x_n\}$ and $x_0 \in \overline{\Omega}$ such that

$$\lim_{j \to \infty} x_{n_j} = x_0.$$

On the other hand, from the mean value theorem, for each $j \in \mathbb{N}$ there exists $t_j \in (0, 1)$ such that

$$
\begin{aligned}
G_n(x_{n_j}) &= \frac{f(x_{n_j}, y(x_{n_j}) + \varepsilon_{n_j} v(x_{n_j}), \nabla y(x_{n_j}) + \varepsilon_{n_j} \nabla v(x_{n_j})) - f(x_{n_j}, y(x_{n_j}), \nabla y(x_{n_j}))}{\varepsilon_{n_j}} \\
&= f_y(x_{n_j}, y(x_{n_j}) + t_j \varepsilon_{n_j} v(x_{n_j}), \nabla y(x_{n_j}) + t_j \varepsilon_{n_j} \nabla v(x_{n_j}))v(x_{n_j}) \\
&\quad + \sum_{i=1}^{n} f_{z_i}(x_{n_j}, y(x_{n_j}) + t_j \varepsilon_{n_j} v(x_{n_j}), \nabla y(x_{n_j}) + t_j \varepsilon_{n_j} \nabla v(x_{n_j}))v_{x_i}(x_{n_j}) \\
&\to G(x_0), \text{ as } j \to \infty.
\end{aligned}
$$

$$\tag{12.49}$$

Hence,

$$
\begin{aligned}
c_{n_j} &= |G_{n_j}(x_{n_j}) - G(x_{n_j})| \\
&\to |G(x_0) - G(x_0)| \\
&= 0.
\end{aligned}
\tag{12.50}
$$

Let $\varepsilon > 0$. Thus, there exists $j_0 \in \mathbb{N}$ such that if $j > j_0$, then

$$
0 \le c_{n_j} < \frac{\varepsilon}{m(\Omega)},
$$

where

$$
m(\Omega) = \int_{\Omega} dx.
$$

Therefore, if $j > j_0$, then

$$
\begin{aligned}
\left| \int_{\Omega} [G_{n_j}(x) - G(x)] \, dx \right| &\le \int_{\Omega} |G_{n_j}(x) - G(x)| \, dx \\
&\le \int_{\Omega} c_{n_j} \, dx \\
&= c_{n_j} m(\Omega) \\
&< \varepsilon.
\end{aligned}
\tag{12.51}
$$

Thus,

$$
\lim_{j \to \infty} \int_{\Omega} G_{n_j}(x) \, dx = \int_{\Omega} G(x) \, dx.
$$

Suppose now, to obtain contradiction, that we do not have

$$
\lim_{\varepsilon \to 0} \int_{\Omega} G_{\varepsilon}(x) \, dx = \int_{\Omega} G(x) \, dx,
$$

where

$$
G_{\varepsilon}(x) = \frac{f(x, y(x) + \varepsilon v(x), \nabla y(x) + \varepsilon \nabla v(x)) - f(x, y(x), \nabla y(x))}{\varepsilon},
$$

$\forall \varepsilon \in \mathbb{R}$ such that $\varepsilon \ne 0$.

Hence, there exists $\varepsilon_0 > 0$ such that for each $n \in \mathbb{N}$ there exists $\tilde{\varepsilon}_n \in \mathbb{R}$ such that

$$
0 < |\tilde{\varepsilon}_n| < \frac{1}{n},
$$

and

$$
\left| \int_{\Omega} \tilde{G}_n(x) \, dx - \int_{\Omega} G(x) \, dx \right| \ge \varepsilon_0,
\tag{12.52}
$$

where

$$
\tilde{G}_n(x) = \frac{f(x, y(x) + \tilde{\varepsilon}_n v(x), \nabla y(x) + \tilde{\varepsilon}_n \nabla v(x)) - f(x, y(x), \nabla y(x))}{\tilde{\varepsilon}_n},
$$

$\forall n \in \mathbb{N}, \, x \in \overline{\Omega}$.

However, as above indicated, we may obtain a subsequence $\{\tilde{\varepsilon}_{n_j}\}$ of $\{\tilde{\varepsilon}_n\}$ such that

$$
\lim_{j \to \infty} \int_{\Omega} \tilde{G}_{n_j}(x) \, dx = \int_{\Omega} G(x) \, dx,
$$

which contradicts (12.52).

Therefore, necessarily we have that

$$\lim_{\varepsilon \to 0} \int_{\Omega} G_{\varepsilon}(x)\, dx = \int_{\Omega} G(x)\, dx,$$

that is,

$$
\begin{aligned}
\delta F(y;v) &= \lim_{\varepsilon \to 0} \frac{F(y+\varepsilon v) - F(y)}{\varepsilon} \\
&= \lim_{\varepsilon \to 0} \int_{\Omega} G_{\varepsilon}(x)\, dx \\
&= \int_{\Omega} G(x)\, dx \\
&= \int_{\Omega} \left(f_y(x, y(x), \nabla y(x)) v(x) + \sum_{i=1}^{n} f_{z_i}(x, y(x), \nabla y(x)) v_{x_i}(x) \right) dx.
\end{aligned}
\qquad (12.53)
$$

The proof is complete.

Chapter 13

More Topics on the Calculus of Variations

13.1 Introductory remarks

The main references for this chapter are [79, 44]. We start by recalling that a functional is a function whose the co-domain is the real set. We denote such functionals by $F : U \to \mathbb{R}$, where U is a Banach space. In our work format, we consider the special cases

1. $F(u) = \int_\Omega f(x, u, \nabla u)\, dx$, where $\Omega \subset \mathbb{R}^n$ is an open, bounded, connected set.

2. $F(u) = \int_\Omega f(x, u, \nabla u, D^2 u)\, dx$, here

$$Du = \nabla u = \left\{ \frac{\partial u_i}{\partial x_j} \right\}$$

and

$$D^2 u = \{D^2 u_i\} = \left\{ \frac{\partial^2 u_i}{\partial x_k \partial x_l} \right\},$$

for $i \in \{1, ..., N\}$ and $j, k, l \in \{1, ..., n\}$.

Also, $f : \overline{\Omega} \times \mathbb{R}^N \times \mathbb{R}^{N \times n} \to \mathbb{R}$ is denoted by $f(x, s, \xi)$ and we assume

1.

$$\frac{\partial f(x, s, \xi)}{\partial s}$$

and

2.

$$\frac{\partial f(x, s, \xi)}{\partial \xi}$$

are continuous $\forall (x, s, \xi) \in \overline{\Omega} \times \mathbb{R}^N \times \mathbb{R}^{N \times n}$.

Remark 13.1.1 *We also recall that the notation $\nabla u = Du$ may be used.*

Now we define our general problem, namely problem $\mathscr{P}$ where

$$\text{Problem } \mathscr{P} : \text{ minimize } F(u) \text{ on } U,$$

that is, to find $u_0 \in U$ such that

$$F(u_0) = \min_{u \in U}\{F(u)\}.$$

At this point, we introduce some essential definitions.

Theorem 13.1.2 *Consider the hypotheses stated at section 13.1 on $F : U \to \mathbb{R}$. Suppose F attains a local minimum at $u \in C^2(\bar{\Omega}; \mathbb{R}^N)$ and additionally assume that $f \in C^2(\overline{\Omega}, \mathbb{R}^N, \mathbb{R}^{N \times n})$. Then the necessary conditions for a local minimum for F are given by the Euler-Lagrange equations:*

$$\frac{\partial f(x, u, \nabla u)}{\partial s} - div\left(\frac{\partial f(x, u, \nabla u)}{\partial \xi}\right) = \theta, \text{ in } \Omega.$$

Proof 13.1   Observe that the standard first order necessary condition stands for $\delta F(u, \varphi) = 0, \forall \varphi \in \mathscr{V}$. From the related results for the Gâteaux variation expression obtained in the last chapter, after integration by parts, we get

$$\int_\Omega \left(\frac{\partial f(x, u, \nabla u)}{\partial s_i} - \sum_{\alpha=1}^n \frac{d}{dx_\alpha}\left(\frac{\partial f(x, u, \nabla u)}{\partial \xi_\alpha^i}\right)\right) \varphi^i \, dx = 0,$$

$\forall i \in \{1, \cdots, N\}$, $\forall \varphi \in C_c^\infty(\Omega, \mathbb{R}^N)$.

The result thus follows from the fundamental lemma of the calculus of variations (please see Lemma 11.1.12 for the concerning result).

13.2   The Gâteaux variation, a more general case

Theorem 13.2.1 *Consider the functional $F : U \to \mathbb{R}$, where*

$$U = \{u \in W^{1,2}(\Omega, \mathbb{R}^N) \mid u = u_0 \text{ in } \partial\Omega\}.$$

Suppose

$$F(u) = \int_\Omega f(x, u, \nabla u) \, dx,$$

where $f : \Omega \times \mathbb{R}^N \times \mathbb{R}^{N \times n}$ is such that, for each $K > 0$ there exists $K_1 > 0$ such that

$$|f(x, s_1, \xi_1) - f(x, s_2, \xi_2)| < K_1(|s_1 - s_2| + |\xi_1 - \xi_2|)$$
$$\forall s_1, s_2 \in \mathbb{R}^N, \xi_1, \xi_2 \in \mathbb{R}^{N \times n}, \text{ such that } |s_1| < K, |s_2| < K, |\xi_1| < K, |\xi_2| < K.$$

Also assume the hypotheses of section 13.1 except for the continuity of derivatives of f. Under such assumptions, for each $u \in C^1(\overline{\Omega}; \mathbb{R}^N)$ and $\varphi \in C_c^\infty(\Omega; \mathbb{R}^N)$, we have

$$\delta F(u, \varphi) = \int_\Omega \left\{\frac{\partial f(x, u, \nabla u)}{\partial s} \cdot \varphi + \frac{\partial f(x, u, \nabla u)}{\partial \xi} \cdot \nabla\varphi\right\} dx.$$

Proof 13.2   First we recall that

$$\delta F(u, \varphi) = \lim_{\varepsilon \to 0} \frac{F(u + \varepsilon\varphi) - F(u)}{\varepsilon}.$$

Observe that

$$\lim_{\varepsilon \to 0} \frac{f(x, u + \varepsilon\varphi, \nabla u + \varepsilon\nabla\varphi) - f(x, u, \nabla u)}{\varepsilon} = \frac{\partial f(x, u, \nabla u)}{\partial s} \cdot \varphi + \frac{\partial f(x, u, \nabla u)}{\partial \xi} \cdot \nabla\varphi, \text{ a.e in } \Omega.$$

Define

$$G(x,u,\varphi,\varepsilon) = \frac{f(x,u+\varepsilon\varphi,\nabla u+\varepsilon\nabla\varphi) - f(x,u,\nabla u)}{\varepsilon},$$

and

$$\tilde{G}(x,u,\varphi) = \frac{\partial f(x,u,\nabla u)}{\partial s} \cdot \varphi + \frac{\partial f(x,u,\nabla u)}{\partial \xi} \cdot \nabla\varphi.$$

Thus we have

$$\lim_{\varepsilon\to 0} G(x,u,\varphi,\varepsilon) = \tilde{G}(x,u,\varphi), \text{ a.e in } \Omega.$$

Now will show that

$$\lim_{\varepsilon\to 0} \int_\Omega G(x,u,\varphi,\varepsilon)\,dx = \int_\Omega \tilde{G}(x,u,\varphi)\,dx.$$

It suffices to show that (we do not provide details here), for an arbitrary sequence $\{\varepsilon_n\} \subset \mathbb{R}$ such that

$$\varepsilon_n \to 0, \text{ as } n \to \infty$$

we have

$$\lim_{n\to\infty} \int_\Omega G(x,u,\varphi,\varepsilon_n)\,dx = \int_\Omega \tilde{G}(x,u,\varphi)\,dx.$$

Observe that, for an appropriate $K > 0$, we have

$$|G(x,u,\varphi,\varepsilon_n)| \leq K(|\varphi| + |\nabla\varphi|), \text{ a.e. in } \Omega. \tag{13.1}$$

By the Lebesgue dominated convergence theorem, we obtain

$$\lim_{n\to\infty} \int_\Omega G(x,u,\varphi,\varepsilon_n)\,dx = \int_\Omega \tilde{G}(x,u,\varphi)\,dx,$$

that is,

$$\delta F(u,\varphi) = \int_\Omega \left\{ \frac{\partial f(x,u,\nabla u)}{\partial s} \cdot \varphi + \frac{\partial f(x,u,\nabla u)}{\partial \xi} \cdot \nabla\varphi \right\} dx.$$

13.3 Fréchet differentiability

In this section we introduce a very important definition namely, Fréchet differentiability.

Definition 13.3.1 *Let U,Y be Banach spaces and consider a transformation $T : U \to Y$. We say that T is Fréchet differentiable at $u \in U$ if there exists a bounded linear transformation $T'(u) : U \to Y$ such that*

$$\lim_{v\to\theta} \frac{\|T(u+v) - T(u) - T'(u)(v)\|_Y}{\|v\|_U} = 0, \; v \neq \theta.$$

In such a case $T'(u)$ is called the Fréchet derivative of T at $u \in U$.

13.4 The Legendre-Hadamard condition

Theorem 13.4.1 *If $u \in C^1(\bar{\Omega};\mathbb{R}^N)$ is such that*

$$\delta^2 F(u,\varphi) \geq 0, \forall\varphi \in C_c^\infty(\Omega,\mathbb{R}^N),$$

then

$$f_{\xi_\alpha^i \xi_\beta^k}(x,u(x),\nabla u(x))\rho^i\rho^k\eta_\alpha\eta_\beta \geq 0, \forall x \in \Omega, \rho \in \mathbb{R}^N, \eta \in \mathbb{R}^n.$$

Such a condition is known as the Legendre-Hadamard condition.

Proof 13.3   Suppose

$$\delta^2 F(u,\varphi) \geq 0, \forall \varphi \in C_c^\infty(\Omega;\mathbb{R}^N).$$

We denote $\delta^2 F(u,\varphi)$ by

$$\delta^2 F(u,\varphi) = \int_\Omega a(x)D\varphi(x) \cdot D\varphi(x)\, dx$$
$$+ \int_\Omega b(x)\varphi(x) \cdot D\varphi(x)\, dx + \int_\Omega c(x)\varphi(x) \cdot \varphi(x)\, dx, \tag{13.2}$$

where

$$a(x) = f_{\xi\xi}(x,u(x),Du(x)),$$
$$b(x) = 2f_{s\xi}(x,u(x),Du(x)),$$

and

$$c(x) = f_{ss}(x,u(x),Du(x)).$$

Now consider $v \in C_c^\infty(B_1(0),\mathbb{R}^N)$. Thus given $x_0 \in \Omega$ for λ sufficiently small we have that $\varphi(x) = \lambda v\left(\frac{x-x_0}{\lambda}\right)$ is an admissible direction. Now we introduce the new coordinates $y = (y^1,...,y^n)$ by setting $y = \lambda^{-1}(x-x_0)$ and multiply (13.2) by λ^{-n} to obtain

$$\int_{B_1(0)} \{a(x_0+\lambda y)Dv(y) \cdot Dv(y) + 2\lambda b(x_0+\lambda y)v(y) \cdot Dv(y) + \lambda^2 c(x_0+\lambda y)v(y) \cdot v(y)\}\, dy > 0,$$

where $a = \{a_{ij}^{\alpha\beta}\}, b = \{b_{jk}^\beta\}$ and $c = \{c_{jk}\}$. Since a,b and c are continuous, we have

$$a(x_0+\lambda y)Dv(y) \cdot Dv(y) \to a(x_0)Dv(y) \cdot Dv(y),$$
$$\lambda b(x_0+\lambda y)v(y) \cdot Dv(y) \to 0,$$

and

$$\lambda^2 c(x_0+\lambda y)v(y) \cdot v(y) \to 0,$$

uniformly on $\bar{\Omega}$ as $\lambda \to 0$. Thus this limit give us

$$\int_{B_1(0)} \tilde{f}_{jk}^{\alpha\beta} D_\alpha v^j D_\beta v^k\, dx \geq 0, \forall v \in C_c^\infty(B_1(0);\mathbb{R}^N), \tag{13.3}$$

where

$$\tilde{f}_{jk}^{\alpha\beta} = a_{jk}^{\alpha\beta}(x_0) = f_{\xi_\alpha^i \xi_\beta^k}(x_0,u(x_0),\nabla u(x_0)).$$

Now define $v = (v^1,...,v^N)$ where

$$v^j = \rho^j \cos((\eta \cdot y)t)\zeta(y)$$
$$\rho = (\rho^1,...,\rho^N) \in \mathbb{R}^N$$

and

$$\eta = (\eta_1,...,\eta_n) \in \mathbb{R}^n$$

and $\zeta \in C_c^\infty(B_1(0))$. From (13.3) we obtain

$$0 \leq \tilde{f}_{jk}^{\alpha\beta} \rho^j \rho^k \left\{ \int_{B_1(0)} (\eta_\alpha t(-\sin((\eta \cdot y)t)\zeta + \cos((\eta \cdot y)t)D_\alpha\zeta) \right.$$
$$\left. \cdot (\eta_\beta t(-\sin((\eta \cdot y)t)\zeta + \cos((\eta \cdot y)t)D_\beta\zeta)\, dy \right\} \tag{13.4}$$

By analogy for

$$v^j = \rho^j \sin((\eta \cdot y)t)\zeta(y)$$

we obtain

$$0 \leq \tilde{f}_{jk}^{\alpha\beta} \rho^j \rho^k \left\{ \int_{B_1(0)} (\eta_\alpha t (cos((\eta \cdot y)t)\zeta + sin((\eta \cdot y)t) D_\alpha \zeta) \right.$$
$$\left. \cdot (\eta_\beta t (cos((\eta \cdot y)t)\zeta + sin((\eta \cdot y)t) D_\beta \zeta) \, dy \right\} \qquad (13.5)$$

Summing up these last two equations, dividing the result by t^2 and letting $t \to +\infty$ we obtain

$$0 \leq \tilde{f}_{jk}^{\alpha\beta} \rho^j \rho^k \eta_\alpha \eta_\beta \int_{B_1(0)} \zeta^2 \, dy,$$

for all $\zeta \in C_c^\infty(B_1(0))$, which implies

$$0 \leq \tilde{f}_{jk}^{\alpha\beta} \rho^j \rho^k \eta_\alpha \eta_\beta.$$

The proof is complete.

13.5 The Weierstrass condition for $n = 1$

Here we present the Weierstrass condition for the special case $N \geq 1$ and $n = 1$. We start with a definition.

Definition 13.5.1 *We say that $u \in \hat{C}([a,b]; \mathbb{R}^N)$ if $u : [a,b] \to \mathbb{R}^N$ is Lipschitz continuous in $[a,b]$, and its derivative u' is continuous except on a finite set of points in $[a,b]$. More specifically, there exists $K > 0$ such that*

$$|u'(x^+)| \leq K, \, \forall x \in [a,b]$$

and

$$|u'(x-)| \leq K, \, \forall x \in [a,b]$$

and there exists at most a finite set of points $t_1, t_2, \cdots, t_n \in (a,b)$ such that

$$u'(t_j+) \neq u'(t_j-), \, \forall j \in \{1, \cdots, n\}.$$

Here, we have denoted

$$u'(x+) = \lim_{h \to 0^+} \frac{u(x+h) - u(x)}{h},$$

and

$$u'(x-) = \lim_{h \to 0^-} \frac{u(x+h) - u(x)}{h}, \, \forall x \in (a,b).$$

Similarly, we define $u'(a+)$ and $u'(b-)$.

Theorem 13.5.2 (Weierstrass) *Let $\Omega = (a,b)$ and $f : \bar{\Omega} \times \mathbb{R}^N \times \mathbb{R}^N \to \mathbb{R}$ be such that $f_s(x,s,\xi)$ and $f_\xi(x,s,\xi)$ are continuous on $\bar{\Omega} \times \mathbb{R}^N \times \mathbb{R}^N$.*
Define $F : U \to \mathbb{R}$ by

$$F(u) = \int_a^b f(x, u(x), u'(x)) \, dx,$$

where

$$U = \{u \in \hat{C}^1([a,b]; \mathbb{R}^N) \mid u(a) = \alpha, \, u(b) = \beta\}.$$

Suppose $u \in U$ minimizes locally F on U, that is, suppose that there exists $\varepsilon_0 > 0$ such that

$$F(u) \leq F(v), \forall v \in U, \text{ such that } \|u - v\|_\infty < \varepsilon_0.$$

Under such hypotheses, we have

$$E(x, u(x), u'(x+), w) \geq 0, \forall x \in [a, b], \ w \in \mathbb{R}^N,$$

and

$$E(x, u(x), u'(x-), w) \geq 0, \forall x \in [a, b], \ w \in \mathbb{R}^N,$$

where

$$u'(x+) = \lim_{h \to 0^+} \frac{u(x+h) - u(x)}{h},$$

and

$$u'(x-) = \lim_{h \to 0^-} \frac{u(x+h) - u(x)}{h}, \ \forall x \in (a, b)$$

and,

$$E(x, s, \xi, w) = f(x, s, w) - f(x, s, \xi) - f_\xi(x, s, \xi)(w - \xi).$$

Remark 13.5.3 *The function E is known as the Weierstrass Excess Function.*

Proof 13.4   Fix $x_0 \in (a, b)$ and $w \in \mathbb{R}^N$. Choose $0 < \varepsilon < 1$ and $h > 0$ such that $u + v \in U$ and

$$\|v\|_\infty < \varepsilon_0$$

where $v(x)$ is given by

$$v(x) = \begin{cases} (x - x_0)w, & \text{if } 0 \leq x - x_0 \leq \varepsilon h, \\ \tilde{\varepsilon}(h - x + x_0)w, & \text{if } \varepsilon h \leq x - x_0 \leq h, \\ 0, & \text{otherwise,} \end{cases}$$

where

$$\tilde{\varepsilon} = \frac{\varepsilon}{1 - \varepsilon}.$$

From

$$F(u + v) - F(u) \geq 0$$

we obtain

$$\int_{x_0}^{x_0+h} f(x, u(x) + v(x), u'(x) + v'(x)) \, dx - \int_{x_0}^{x_0+h} f(x, u(x), u'(x)) \, dx \geq 0. \tag{13.6}$$

Define

$$\tilde{x} = \frac{x - x_0}{h},$$

so that

$$d\tilde{x} = \frac{dx}{h}.$$

From (13.6) we obtain

$$h \int_0^1 f(x_0 + \tilde{x}h, u(x_0 + \tilde{x}h) + v(x_0 + \tilde{x}h), u'(x_0 + \tilde{x}h) + v'(x_0 + \tilde{x}h)) \, d\tilde{x}$$

$$- h \int_0^1 f(x_0 + \tilde{x}h, u(x_0 + \tilde{x}h), u'(x_0 + \tilde{x}h)) \, d\tilde{x} \geq 0. \tag{13.7}$$

where the derivatives are related to x.

Therefore

$$\int_0^\varepsilon f(x_0 + \tilde{x}h, u(x_0 + \tilde{x}h) + v(x_0 + \tilde{x}h), u'(x_0 + \tilde{x}h) + w) \, d\tilde{x}$$

$$- \int_0^\varepsilon f(x_0 + \tilde{x}h, u(x_0 + \tilde{x}h), u'(x_0 + \tilde{x}h)) \, d\tilde{x}$$

$$+ \int_\varepsilon^1 f(x_0 + \tilde{x}h, u(x_0 + \tilde{x}h) + v(x_0 + \tilde{x}h), u'(x_0 + \tilde{x}h) - \tilde{\varepsilon}w) \, d\tilde{x}$$

$$- \int_\varepsilon^1 f(x_0 + \tilde{x}h, u(x_0 + \tilde{x}h), u'(x_0 + \tilde{x}h)) \, d\tilde{x}$$

$$\geq 0. \tag{13.8}$$

Letting $h \to 0$ we obtain

$$\varepsilon(f(x_0, u(x_0), u'(x_0+) + w) - f(x_0, u(x_0), u'(x_0+)))$$
$$+ (1 - \varepsilon)(f(x_0, u(x_0), u'(x_0+) - \tilde{\varepsilon}w) - f(x_0, u(x_0), u'(x_0+))) \geq 0.$$

Hence, by the mean value theorem we get

$$\varepsilon(f(x_0, u(x_0), u'(x_0+) + w) - f(x_0, u(x_0), u'(x_0+)))$$
$$- (1 - \varepsilon)\tilde{\varepsilon}(f_\xi(x_0, u(x_0), u'(x_0+) + \rho(\tilde{\varepsilon})w)) \cdot w \geq 0. \tag{13.9}$$

Dividing by ε and letting $\varepsilon \to 0$, so that $\tilde{\varepsilon} \to 0$ and $\rho(\tilde{\varepsilon}) \to 0$ we finally obtain

$$f(x_0, u(x_0), u'(x_0+) + w) - f(x_0, u(x_0), u'(x_0+)) - f_\xi(x_0, u(x_0), u'(x_0+)) \cdot w \geq 0.$$

Similarly we may get

$$f(x_0, u(x_0), u'(x_0-) + w) - f(x_0, u(x_0), u'(x_0-)) - f_\xi(x_0, u(x_0), u'(x_0-)) \cdot w \geq 0.$$

Since $x_0 \in [a, b]$ and $w \in \mathbb{R}^N$ are arbitrary, the proof is complete.

13.6 The Weierstrass condition, the general case

In this section we present a proof for the Weierstrass necessary condition for $N \geq 1, n \geq 1$. Such a result may be found in similar form in [44].

Theorem 13.1
Assume $u \in C^1(\overline{\Omega}; \mathbb{R}^N)$ is a point of strong minimum for a Fréchet differentiable functional $F : U \to \mathbb{R}$ that is, in particular, there exists $\varepsilon > 0$ such that

$$F(u + \varphi) \geq F(u),$$

for all $\varphi \in C_c^\infty(\Omega; \mathbb{R}^n)$ such that

$$\|\varphi\|_\infty < \varepsilon.$$

Here

$$F(u) = \int_\Omega f(x, u, Du) \, dx,$$

where we recall to have denoted

$$Du = \nabla u = \left\{ \frac{\partial u_i}{\partial x_j} \right\}.$$

Under such hypotheses, for all $x \in \Omega$ and each rank-one matrix $\eta = \{\rho_i \beta^\alpha\} = \{\rho \otimes \beta\}$, we have that

$$E(x, u(x), Du(x), Du(x) + \rho \otimes \beta) \geq 0,$$

where

$$
\begin{aligned}
& E(x, u(x), Du(x), Du(x) + \rho \otimes \beta) \\
= \quad & f(x, u(x), Du(x) + \rho \otimes \beta) - f(x, u(x), Du(x)) \\
& - \rho^i \beta_\alpha f_{\xi^i_\alpha}(x, u(x), Du(x)).
\end{aligned}
\tag{13.10}
$$

Proof 13.5 Since u is a point of local minimum for F, we have that

$$\delta F(u; \varphi) = 0, \forall \varphi \in C_c^\infty(\Omega; \mathbb{R}^N),$$

that is

$$\int_\Omega (\varphi \cdot f_s(x, u(x), Du(x)) + D\varphi \cdot f_\xi(x, u(x), Du(x)) \, dx = 0,$$

and hence,

$$
\begin{aligned}
& \int_\Omega (f(x, u(x), Du(x) + D\varphi(x)) - f(x, u(x), Du(x)) \, dx \\
& - \int_\Omega (\varphi(x) \cdot f_s(x, u(x), Du(x)) - D\varphi(x) \cdot f_\xi(x, u(x), Du(x)) \, dx \\
\geq \quad & 0,
\end{aligned}
\tag{13.11}
$$

$\forall \varphi \in \mathcal{V}$, where

$$\mathcal{V} = \{\varphi \in C_c^\infty(\Omega; \mathbb{R}^N) : \|\varphi\|_\infty < \varepsilon\}.$$

Choose a unity vector $e \in \mathbb{R}^n$ and write

$$x = (x \cdot e)e + \bar{x},$$

where

$$\bar{x} \cdot e = 0.$$

Denote $D_e v = Dv \cdot e$, and let $\rho = (\rho_1, ..., \rho_N) \in \mathbb{R}^N$.
Also, let x_0 be any point of Ω. Without loss of generality assume $x_0 = 0$.
Choose $\lambda_0 \in (0, 1)$ such that $C_{\lambda_0} \subset \Omega$, where,

$$C_{\lambda_0} = \{x \in \mathbb{R}^n : |x \cdot e| \leq \lambda_0 \text{ and } \|\bar{x}\| \leq \lambda_0\}.$$

Let $\lambda \in (0, \lambda_0)$ and

$$\phi \in C_c((-1, 1); \mathbb{R})$$

and choose a sequence

$$\phi_k \in C_c^\infty((-\lambda^2, \lambda); \mathbb{R})$$

which converges uniformly to the Lipschitz function ϕ_λ given by

$$
\phi_\lambda = \begin{cases}
t + \lambda^2, & \text{if } -\lambda^2 \leq t \leq 0, \\
\lambda(\lambda - t), & \text{if } 0 < t < \lambda \\
0, & \text{otherwise}
\end{cases}
\tag{13.12}
$$

and such that ϕ_k' converges uniformly to ϕ_λ' on each compact subset of

$$A_\lambda = \{t : -\lambda^2 < t < \lambda, \, t \neq 0\}.$$

We emphasize the choice of $\{\phi_k\}$ may be such that for some $K > 0$ we have $\|\phi\|_\infty < K$, $\|\phi_k\|_\infty < K$ and $\|\phi_k'\|_\infty < K, \forall k \in \mathbb{N}$.

Observe that for any sufficiently small $\lambda > 0$ we have that φ_k defined by

$$\varphi_k(x) = \rho\phi_k(x \cdot e)\phi(|\overline{x}|^2/\lambda^2) \in \mathscr{V}, \forall k \in \mathbb{N}$$

so that letting $k \to \infty$ we obtain that

$$\varphi(x) = \rho\phi_\lambda(x \cdot e)\phi(|\overline{x}|^2/\lambda^2),$$

is such that (13.11) is satisfied.

Moreover,

$$D_e\varphi(x) = \rho\phi_\lambda'(x \cdot e)\phi(|\overline{x}|^2/\lambda^2),$$

and

$$\overline{D}\varphi(x) = \rho\phi_\lambda(x \cdot e)\phi'(|\overline{x}|^2/\lambda^2)2\lambda^{-2}\overline{x},$$

where $\overline{D}$ denotes the gradient relating the variable $\overline{x}$.

Note that, for such a $\varphi(x)$ the integrand of (13.11) vanishes if $x \notin C_\lambda$, where

$$C_\lambda = \{x \in \mathbb{R}^n : |x \cdot e| \leq \lambda \text{ and } \|\overline{x}\| \leq \lambda\}.$$

Define C_λ^+ and C_λ^- by

$$C_\lambda^- = \{x \in C_\lambda : x \cdot e \leq 0\},$$

and

$$C_\lambda^+ = \{x \in C_\lambda : x \cdot e > 0\}.$$

Hence, denoting

$$
\begin{aligned}
g_k(x) &= (f(x, u(x), Du(x) + D\varphi_k(x)) - f(x, u(x), Du(x))) \\
&\quad - (\varphi_k(x) \cdot f_s(x, u(x), Du(x) + D\varphi_k(x)) \cdot f_\xi(x, u(x), Du(x))
\end{aligned}
\tag{13.13}
$$

and

$$
\begin{aligned}
g(x) &= (f(x, u(x), Du(x) + D\varphi(x)) - f(x, u(x), Du(x))) \\
&\quad - (\varphi(x) \cdot f_s(x, u(x), Du(x) + D\varphi(x)) \cdot f_\xi(x, u(x), Du(x))
\end{aligned}
\tag{13.14}
$$

letting $k \to \infty$, using the Lebesgue dominated converge theorem we obtain

$$
\begin{aligned}
&\int_{C_\lambda^-} g_k(x)\,dx + \int_{C_\lambda^+} g_k(x)\,dx \\
&\quad \to \int_{C_\lambda^-} g(x)\,dx + \int_{C_\lambda^+} g(x)\,dx \geq 0,
\end{aligned}
\tag{13.15}
$$

Now define

$$y = y^e e + \overline{y},$$

where

$$y^e = \frac{x \cdot e}{\lambda^2},$$

and

$$\overline{y} = \frac{\overline{x}}{\lambda}.$$

The sets C_λ^- and C_λ^+ correspond, concerning the new variables, to the sets B_λ^- and B_λ^+, where

$$B_\lambda^- = \{y : \|\bar{y}\| \leq 1, \text{ and } -\lambda^{-1} \leq y^e \leq 0\},$$

$$B_\lambda^+ = \{y : \|\bar{y}\| \leq 1, \text{ and } 0 < y^e \leq \lambda^{-1}\}.$$

Therefore, since $dx = \lambda^{n+1} dy$, multiplying (13.15) by λ^{-n-1}, we obtain

$$\int_{B_1^-} g(x(y)) \, dy + \int_{B_\lambda^- \setminus B_1^-} g(x(y)) \, dy + \int_{B_\lambda^+} g(x(y)) \, dy \geq 0, \tag{13.16}$$

where

$$x = (x \cdot e)e + \bar{x} = \lambda^2 y^e + \lambda \bar{y} \equiv x(y).$$

Observe that

$$D_e \varphi(x) = \begin{cases} \rho \phi(\|\bar{y}\|^2) & \text{if } -1 \leq y^e \leq 0, \\ \rho \phi(\|\bar{y}\|^2)(-\lambda) & \text{if } 0 \leq y^e \leq \lambda^{-1}, \\ 0, & \text{otherwise.} \end{cases} \tag{13.17}$$

Observe also that

$$|g(x(y))| \leq o(\sqrt{|\varphi(x)|^2 + |D\varphi(x)|^2}),$$

so that from the from the expression of $\varphi(x)$ and $D\varphi(x)$ we obtain, for

$$y \in B_\lambda^+, \text{ or } y \in B_\lambda^- \setminus B_1^-,$$

that

$$|g(x(y))| \leq o(\lambda), \text{ as } \lambda \to 0.$$

Since the Lebesgue measures of B_λ^- and B_λ^+ are bounded by

$$2^{n-1}/\lambda$$

the second and third terms in (13.16) are of $o(1)$ where

$$\lim_{\lambda \to 0^+} o(1)/\lambda = 0,$$

so that letting $\lambda \to 0^+$, considering that

$$x(y) \to 0,$$

and on B_1^- (up to the limit set B)

$$\begin{aligned} g(x(y)) \quad \to \quad & f(0, u(0), Du(0) + \rho \phi(\|\bar{y}\|^2)e) \\ & -f(0, u(0), Du(0)) - \\ & \rho \phi(\|\bar{y}\|^2)ef_\xi(0, u(0), Du(0)) \end{aligned} \tag{13.18}$$

we get,

$$\begin{aligned} \int_B [f(0, u(0), Du(0) + \rho \phi(\|\bar{y}\|^2)e) - f(0, u(0), Du(0)) \\ -\rho \phi(\|\bar{y}\|^2)ef_\xi(0, u(0), Du(0))] \, d\bar{y}_2 ... d\bar{y}_n \\ \geq \quad 0, \end{aligned} \tag{13.19}$$

where B is an appropriate limit set (we do not provide more details here) such that

$$B = \{y \in \mathbb{R}^n \; : \; y^e = 0 \text{ and } \|\bar{y}\| \leq 1\}.$$

Here we have used the fact that, on the set in question,

$$D\varphi(x) \to \rho\phi(\|\bar{y}\|^2)e, \text{ as } \lambda \to 0^+.$$

Finally, inequality (13.19) is valid for a sequence $\{\phi_n\}$ (in place of ϕ) such that

$$0 \leq \phi_n \leq 1 \text{ and } \phi_n(t) = 1, \text{ if } |t| < 1 - 1/n,$$

$\forall n \in \mathbb{N}$.

Letting $n \to \infty$, from (13.19) we obtain

$$\begin{aligned} f(0, u(0), Du(0) + \rho \otimes e) - f(0, u(0), Du(0)) \\ -\rho \cdot e f_\xi(0, u(0), Du(0)) \geq 0. \end{aligned} \tag{13.20}$$

13.7 The Weierstrass-Erdmann conditions

We start with a definition.

Definition 13.7.1 *Define* $I = [a, b]$. *A function* $u \in \hat{C}([a, b]; \mathbb{R}^N)$ *is said to be a weak Lipschitz extremal of*

$$F(u) = \int_a^b f(x, u(x), u'(x)) \, dx,$$

if

$$\int_a^b (f_s(x, u(x), u'(x)) \cdot \varphi + f_\xi(x, u(x), u'(x)) \cdot \varphi'(x)) \, dx = 0,$$

$\forall \varphi \in C_c^\infty([a, b]; \mathbb{R}^N)$.

Proposition 13.7.2 *For any Lipschitz extremal of*

$$F(u) = \int_a^b f(x, u(x), u'(x)) \, dx$$

there exists a constant $c \in \mathbb{R}^N$ *such that*

$$f_\xi(x, u(x), u'(x)) = c + \int_a^x f_s(t, u(t), u'(t)) \, dt, \forall x \in [a, b]. \tag{13.21}$$

Proof 13.6 Fix $\varphi \in C_c^\infty([a, b]; \mathbb{R}^N)$. Integration by parts of the extremal condition

$$\delta F(u, \varphi) = 0,$$

implies that

$$\int_a^b f_\xi(x, u(x), u'(x)) \cdot \varphi'(x) \, dx - \int_a^b \int_a^x f_s(t, u(t), u'(t)) \, dt \cdot \varphi'(x) \, dx = 0.$$

Since φ is arbitrary, from the du Bois-Reymond lemma, there exists $c \in \mathbb{R}^N$ such that

$$f_\xi(x, u(x), u'(x)) - \int_a^x f_s(t, u(t), u'(t))\, dt = c, \forall x \in [a, b].$$

The proof is complete.

Theorem 13.7.3 (Weierstrass-Erdmann corner conditions) *Let* $I = [a, b]$. *Suppose* $u \in \hat{C}^1([a, b]; \mathbb{R}^N)$ *is such that*

$$F(u) \le F(v), \forall v \in \mathscr{C}_r,$$

for some $r > 0$. *where*

$$\mathscr{C}_r = \{v \in \hat{C}^1([a, b]; \mathbb{R}^N) \mid v(a) = u(a),\ v(b) = u(b),\ and\ \|u - v\|_\infty < r\}.$$

Let $x_0 \in (a, b)$ *be a corner point of* u. *Denoting* $u_0 = u(x_0)$, $\xi_0^+ = u'(x_0^+)$ *and* $\xi_0^- = u'(x_0^-)$, *then the following relations are valid:*

1. $f_\xi(x_0, u_0, \xi_0^-) = f_\xi(x_0, u_0, \xi_0^+)$,

2.

$$f(x_0, u_0, \xi_0^-) - \xi_0^- f_\xi(x_0, u_0, \xi_0^-)$$
$$= f(x_0, u_0, \xi_0^+) - \xi_0^+ f_\xi(x_0, u_0, \xi_0^+).$$

Remark 13.7.4 *The conditions above are known as the Weierstrass-Erdmann corner conditions.*

Proof 13.7 Condition (1) is just a consequence of equation (13.21). For (2), define

$$\tau_\varepsilon(x) = x + \varepsilon\lambda(x),$$

where $\lambda \in C_c^\infty(I)$. Observe that $\tau_\varepsilon(a) = a$ and $\tau_\varepsilon(b) = b$, $\forall \varepsilon > 0$. Also $\tau_0(x) = x$. Choose $\varepsilon_0 > 0$ sufficiently small such that for each ε satisfying $|\varepsilon| < \varepsilon_0$, we have $\tau'_\varepsilon(x) > 0$ and

$$\tilde{u}_\varepsilon(x) = (u \circ \tau_\varepsilon^{-1})(x) \in \mathscr{C}_r.$$

Define

$$\phi(\varepsilon) = F(x, \tilde{u}_\varepsilon, \tilde{u}'_\varepsilon(x)).$$

Thus ϕ has a local minimum at 0, so that $\phi'(0) = 0$, that is

$$\frac{d(F(x, \tilde{u}_\varepsilon, \tilde{u}'_\varepsilon(x)))}{d\varepsilon}\Big|_{\varepsilon=0} = 0.$$

Observe that

$$\frac{d\tilde{u}_\varepsilon}{dx} = u'(\tau_\varepsilon^{-1}(x))\frac{d\tau_\varepsilon^{-1}(x)}{dx},$$

and

$$\frac{d\tau_\varepsilon^{-1}(x)}{dx} = \frac{1}{1 + \varepsilon\lambda'(\tau_\varepsilon^{-1}(x))}.$$

Thus,

$$F(\tilde{u}_\varepsilon) = \int_a^b f\left(x, u(\tau_\varepsilon^{-1}(x)), u'(\tau_\varepsilon^{-1}(x))\left(\frac{1}{1 + \varepsilon\lambda'(\tau_\varepsilon^{-1}(x))}\right)\right) dx.$$

Defining

$$\bar{x} = \tau_\varepsilon^{-1}(x),$$

we obtain

$$d\bar{x} = \frac{1}{1 + \varepsilon \lambda'(\bar{x})} \, dx,$$

that is

$$dx = (1 + \varepsilon \lambda'(\bar{x})) \, d\bar{x}.$$

Dropping the bar for the new variable, we may write

$$F(\tilde{u}_\varepsilon) = \int_a^b f\left(x + \varepsilon \lambda(x), u(x), \frac{u'(x)}{1 + \varepsilon \lambda'(x)}\right)(1 + \varepsilon \lambda'(x)) \, dx.$$

From

$$\frac{dF(\tilde{u}_\varepsilon)}{d\varepsilon}\bigg|_{\varepsilon=0},$$

we obtain

$$\int_a^b \left(\lambda f_x(x, u(x), u'(x)) + \lambda'(x)(f(x, u(x), u'(x)) - u'(x) f_\xi(x, u(x), u'(x)))\right) dx = 0. \tag{13.22}$$

Since λ is arbitrary, from proposition 13.7.2, we obtain

$$f(x, u(x), u'(x)) - u'(x) f_\xi(x, u(x), u'(x)) - \int_a^x f_x(t, u(t), u'(t)) \, dt = c_1$$

for some $c_1 \in \mathbb{R}$.

Being $\int_a^x f_x(t, u(t), u'(t)) \, dt + c_1$ a continuous function (in fact absolutely continuous), the proof is complete.

13.8   Natural boundary conditions

Consider the functional $F : U \to \mathbb{R}$, where

$$F(u) = \int_\Omega f(x, u(x), \nabla u(x)) \, dx,$$

$$f(x, s, \xi) \in C^1(\bar{\Omega}, \mathbb{R}^N, \mathbb{R}^{N \times n}),$$

and $\Omega \subset \mathbb{R}^n$ is an open bounded connected set.

Proposition 13.8.1 *Assume*

$$U = \{u \in W^{1,2}(\Omega; \mathbb{R}^N); u = u_0 \text{ on } \Gamma_0\},$$

where $\Gamma_0 \subset \partial\Omega$ is closed and $\partial\Omega = \Gamma = \Gamma_0 \cup \Gamma_1$ being Γ_1 open in Γ and $\Gamma_0 \cap \Gamma_1 = \emptyset$. Thus if $\partial\Omega \in C^1$ $f \in C^2(\bar{\Omega}, \mathbb{R}^N, \mathbb{R}^{N \times n})$ and $u \in C^2(\bar{\Omega}; \mathbb{R}^N)$, and also

$$\delta F(u, \varphi) = 0, \forall \varphi \in C^1(\bar{\Omega}; \mathbb{R}^N), \text{ such that } \varphi = 0 \text{ on } \Gamma_0,$$

then u is a extremal of F which satisfies the following natural boundary conditions,

$$n_\alpha f_{\xi_\alpha^i}(x, u(x) \nabla u(x)) = 0, \text{ a.e. on } \Gamma_1, \forall i \in \{1, ..., N\}.$$

Proof 13.8 Observe that $\delta F(u,\varphi) = 0, \forall \varphi \in C_c^\infty(\Omega;\mathbb{R}^N)$, thus u is a extremal of F and through integration by parts and the fundamental lemma of calculus of variations, we obtain

$$L_f(u) = 0, \text{ in } \Omega,$$

where

$$L_f(u) = f_s(x,u(x),\nabla u(x)) - div(f_\xi(x,u(x),\nabla u(x))).$$

Defining

$$\mathscr{V} = \{\varphi \in C^1(\Omega;\mathbb{R}^N) \mid \varphi = 0 \text{ on } \Gamma_0\},$$

for an arbitrary $\varphi \in \mathscr{V}$, we obtain

$$
\begin{aligned}
\delta F(u,\varphi) &= \int_\Omega L_f(u) \cdot \varphi \, dx \\
&\quad + \int_{\Gamma_1} n_\alpha f_{\xi_\alpha^i}(x,u(x),\nabla u(x))\varphi^i(x) \, d\Gamma \\
&= \int_{\Gamma_1} n_\alpha f_{\xi_\alpha^i}(x,u(x),\nabla u(x))\varphi^i(x) \, d\Gamma \\
&= 0, \forall \varphi \in \mathscr{V}.
\end{aligned}
\tag{13.23}
$$

Suppose, to obtain contradiction, that

$$n_\alpha f_{\xi_\alpha^i}(x_0,u(x_0),\nabla u(x_0)) = \beta > 0,$$

for some $x_0 \in \Gamma_1$ and some $i \in \{1,...,N\}$. Defining

$$G(x) = n_\alpha f_{x_{i\alpha}^i}(x,u(x),\nabla u(x)),$$

by the continuity of G, there exists $r > 0$ such that

$$G(x) > \beta/2, \text{ in } B_r(x_0),$$

and in particular

$$G(x) > \beta/2, \text{ in } B_r(x_0) \cap \Gamma_1.$$

Choose $0 < r_1 < r$ such that $B_{r_1}(x_0) \cap \Gamma_0 = \emptyset$. This is possible since Γ_0 is closed and $x_0 \in \Gamma_1$.
    Choose $\varphi^i \in C_c^\infty(B_{r_1}(x_0))$ such that $\varphi^i \geq 0$ in $B_{r_1}(x_0)$ and $\varphi^i > 0$ in $B_{r_1/2}(x_0)$. Therefore

$$\int_{\Gamma_1} G(x)\varphi^i(x) \, dx > \frac{\beta}{2}\int_{\Gamma_1} \varphi^i \, dx > 0,$$

and this, for $\varphi = (0,\cdots,\varphi^i,\cdots,0) \in C_c^\infty(\Omega;\mathbb{R}^N)$, contradicts (13.23). Thus

$$G(x) \leq 0, \forall x \in \Gamma_1,$$

and by analogy

$$G(x) \geq 0, \forall x \in \Gamma_1,$$

so that

$$G(x) = 0, \forall x \in \Gamma_1.$$

The proof is complete.

Chapter 14

Convex Analysis and Duality Theory

14.1 Convex sets and functions

For this section the most relevant reference is Ekeland and Temam [33].

Definition 14.1.1 (Convex functional) *Let U be a vector space and let $S \subset U$ be a convex set. A functional $F : S \to \bar{\mathbb{R}} = \mathbb{R} \cup \{+\infty, -\infty\}$ is said to be convex, if*

$$F(\lambda u + (1-\lambda)v) \leq \lambda F(u) + (1-\lambda)F(v), \forall u, v \in S, \lambda \in [0,1]. \tag{14.1}$$

14.2 Weak lower semi-continuity

We start with the definition of Epigraph.

Definition 14.2.1 (Epigraph) *Let U be a Banach space and let $F : U \to \bar{\mathbb{R}}$ be a functional. We define the Epigraph of F, denoted by $Epi(F)$, by*

$$Epi(F) = \{(u,a) \in U \times \mathbb{R} \mid a \geq F(u)\}.$$

Definition 14.2.2 *Let U be a Banach space. Consider the weak topology $\sigma(U, U^*)$ for U and let $F : U \to \mathbb{R} \cup \{+\infty\}$ be a functional. Let $u \in U$. We say that F is weakly lower semi-continuous at $u \in U$ if for each $\lambda < F(u)$, there exists a weak neighborhood $V_\lambda(u) \in \sigma(U, U^*)$ such that*

$$F(v) > \lambda, \ \forall v \in V_\lambda(u).$$

If F is weakly lower semi-continuous (w.l.s.c.) on U, we write simply F is w.l.s.c..

Theorem 14.2.3 *Let U be a Banach space and let $F : U \to \mathbb{R} \cup \{+\infty\}$ be a functional. Under such hypotheses, the following properties are equivalent.*

1. F is w.l.s.c..

2. $Epi(F)$ is closed for $U \times \mathbb{R}$ with the product topology between $\sigma(U, U^*)$ and the usual topology for $\mathbb{R}$.

3. $H_\gamma^F = \{u \in U \mid F(u) \leq \gamma\}$ *is closed for* $\sigma(U, U^*)$, $\forall \gamma \in \mathbb{R}$.

4. *The set* $G_\gamma^F = \{u \in U \mid F(u) > \gamma\}$ *is open for* $\sigma(U, U^*)$, $\forall \gamma \in \mathbb{R}$.

5.
$$\liminf_{v \to u} F(v) \geq F(u), \forall u \in U,$$

where
$$\liminf_{v \to u} F(v) = \sup_{V(u) \in \sigma(U, U^*)} \inf_{v \in V(u)} F(v).$$

Proof 14.1   Assume F is w.l.s.c.. We are going to show that $Epi(F)^c$ is open for $\sigma(U, U^*) \times \mathbb{R}$. Choose $(u, r) \in Epi(F)^c$. Thus $(u, r) \notin Epi(F)$, so that $r < F(u)$. Select λ such that $r < \lambda < F(u)$. Since F is w.l.s.c. at u, there exists a a weak neighborhood $V_\lambda(u)$ such that

$$F(v) > \lambda, \forall v \in V_\lambda(u).$$

Thus,

$$V_\lambda(u) \times (-\infty, \lambda) \subset Epi(F)^c$$

so that (u, r) is an interior point of $Epi(F)^c$ and hence, since such a point is arbitrary in $Epi(F)^c$, we may infer that $Epi(F)^c$ is open so that $Epi(F)$ is closed for the topology in question. Assume now (2). Observe that

$$H_\gamma^F \times \{\gamma\} = Epi(F) \cap (U \times \{\gamma\}).$$

From the hypotheses $Epi(F)$ is closed, that is, $H_\gamma^F \times \{\gamma\}$ is closed and thus H_γ^F is closed.

Assume (3). To obtain (4), it suffices to consider the complement of H_γ^F. Suppose (4) is valid. Let $u \in U$ and let $\gamma \in \mathbb{R}$ be such that

$$\gamma < F(u).$$

Since G_γ^F is open for $\sigma(U, U^*)$ there exists a weak neighborhood $V(u)$ such that

$$V(u) \subset G_\gamma^F,$$

so that

$$F(v) > \gamma, \forall v \in V(u),$$

and hence

$$\inf_{v \in V(u)} F(v) \geq \gamma.$$

In particular, we have

$$\liminf_{v \to u} F(v) \geq \gamma.$$

Letting $\gamma \to F(u)$, we obtain

$$\liminf_{v \to u} F(v) \geq F(u).$$

Finally assume

$$\liminf_{v \to u} F(v) \geq F(u).$$

Let $\lambda < F(u)$ and let $0 < \varepsilon < F(u) - \lambda$.

Observe that

$$\liminf_{v \to u} F(v) = \sup_{V(u) \in \sigma(U, U^*)} \inf_{v \in V(u)} F(v).$$

Thus, there exists a weak neighborhood $V(u)$ such that $F(v) \geq F(u) - \varepsilon > \lambda, \forall v \in V(u)$.

The proof is complete.

Remark 14.2.4 *A similar result is valid for the strong topology (in norm) of a Banach space U so that a $F : U \to \mathbb{R} \cup \{+\infty\}$ is strongly lower semi-continuous (l.s.c.) at $u \in U$, if*

$$\liminf_{v \to u} F(v) \geq F(u). \tag{14.2}$$

Corollary 14.2.5 *All convex l.s.c. functional $F : U \to \overline{\mathbb{R}}$ is also w.l.s.c..*

Proof 14.2   The result follows from the fact for F l.s.c., its epigraph being convex and strongly closed, it is also weakly closed.

Definition 14.2.6 (Affine-continuous functionals) *Let U be a Banach space. A functional $F : U \to \mathbb{R}$ is said to be affine-continuous, if there exist $u^* \in U^*$ and $\alpha \in \mathbb{R}$ such that*

$$F(u) = \langle u, u^* \rangle_U + \alpha, \forall u \in U. \tag{14.3}$$

Definition 14.2.7 ($\Gamma(U)$) *Let U be a Banach space. We say that $F : U \to \bar{\mathbb{R}}$ is a functional in $\Gamma(U)$ and write $F \in \Gamma(U)$ if F may be represented point-wise as the supremum of a family of affine-continuous functionals. If $F \in \Gamma(U)$, and $F(u) \in \mathbb{R}$ for some $u \in U$ we write $F \in \Gamma_0(U)$.*

Definition 14.2.8 (Convex envelop) *Let U be a Banach space. Let $F : U \to \bar{\mathbb{R}}$, be a functional. we define its convex envelop, denoted by $CF : U \to \bar{\mathbb{R}}$, as*

$$CF(u) = \sup_{(u^*, \alpha) \in A^*} \{\langle u, u^* \rangle + \alpha\}, \tag{14.4}$$

where

$$A^* = \{(u^*, \alpha) \in U^* \times \mathbb{R} \mid \langle v, u^* \rangle_U + \alpha \leq F(v), \forall v \in U\} \tag{14.5}$$

14.3   Polar functionals and related topics on convex analysis

Definition 14.3.1 (Polar functional) *Let U be a Banach space and let $F : U \to \bar{\mathbb{R}}$, be a functional. We define the polar functional related to F, denoted by $F^* : U^* \to \bar{\mathbb{R}}$, by*

$$F^*(u^*) = \sup_{u \in U} \{\langle u, u^* \rangle_U - F(u)\}, \forall u^* \in U^*. \tag{14.6}$$

Definition 14.3.2 (Bipolar functional) *Let U be a Banach space and let $F : U \to \bar{\mathbb{R}}$ be a functional. We define the bi-polar functional related to F, denoted by $F^{**} : U \to \bar{\mathbb{R}}$, as*

$$F^{**}(u) = \sup_{u^* \in U^*} \{\langle u, u^* \rangle_U - F^*(u^*)\}, \forall u \in U. \tag{14.7}$$

Proposition 14.3.3 *Let U be a Banach space and let $F : U \to \bar{\mathbb{R}}$ be a functional. Under such hypotheses $F^{**}(u) = CF(u), \forall u \in U$ and in particular, if $F \in \Gamma(U)$, then $F^{**}(u) = F(u), \forall u \in U$.*

Proof 14.3   By the definition, the convex envelop of F is the supremum of affine-continuous functionals bounded by F at the point in question. In fact we need only to consider those which are maximal, that is, only those in the form

$$u \mapsto \langle u, u^* \rangle_U - F^*(u^*). \tag{14.8}$$

Thus,

$$CF(u) = \sup_{u^* \in U^*} \{\langle u, u^* \rangle_U - F^*(u^*)\} = F^{**}(u). \tag{14.9}$$

Corollary 14.3.4 *Let U be a Banach space and let $F : U \to \bar{\mathbb{R}}$ be a functional. Under such hypotheses, $F^* = F^{***}$.*

Proof 14.4 Since $F^{**} \leq F$, we obtain

$$F^* \leq F^{***}. \tag{14.10}$$

On the other hand,

$$F^{**}(u) \geq \langle u, u^* \rangle_U - F^*(u^*), \tag{14.11}$$

so that

$$F^{***}(u^*) = \sup_{u \in U} \{\langle u, u^* \rangle_U - F^{**}(u)\} \leq F^*(u^*). \tag{14.12}$$

From (14.10) and (14.12), we have $F^*(u^*) = F^{***}(u^*)$, $\forall u^* \in U^*$.

At this point, we recall the definition of Gâteaux differentiability.

Definition 14.3.5 (Gâteaux differentiability) *Let U be a Banach space. A functional $F : U \to \bar{\mathbb{R}}$ is said to be Gâteaux differentiable at $u \in U$, if there exists $u^* \in U^*$ such that*

$$\lim_{\lambda \to 0} \frac{F(u + \lambda h) - F(u)}{\lambda} = \langle h, u^* \rangle_U, \quad \forall h \in U. \tag{14.13}$$

The vector u^* is said to be the Gâteaux derivative of $F : U \to \mathbb{R}$ at u and may denoted by

$$u^* = \frac{\partial F(u)}{\partial u} \text{ or } u^* = \delta F(u) \tag{14.14}$$

Definition 14.3.6 (Sub-gradients) *Let U be a Banach space and let $F : U \to \bar{\mathbb{R}}$ be a functional. We define the set of sub-gradients of F at u, denoted by $\partial F(u)$, by*

$$\partial F(u) = \{u^* \in U^*, \text{ such that}$$
$$\langle v - u, u^* \rangle_U + F(u) \leq F(v), \forall v \in U\}. \tag{14.15}$$

Lemma 14.3.7 (Continuity of convex functions) *Let U be a Banach space and let $F : U \to \mathbb{R}$ be a convex functional. Let $u \in U$ and suppose there exists $a > 0$ and a neighborhood V of u such that*

$$F(v) < a < +\infty, \forall v \in V.$$

From the hypotheses, F is continuous at u.

Proof 14.5 Redefining the problem with $G(v) = F(v + u) - F(u)$ we need only consider the case in which $u = \mathbf{0}$ and $F(u) = 0$. Let $\mathcal{V}$ be a neighborhood of $\mathbf{0}$ such that $F(v) \leq a < +\infty, \forall v \in \mathcal{V}$. Define $\mathcal{W} = \mathcal{V} \cap (-\mathcal{V})$. Choose $\varepsilon \in (0, 1)$. Let $v \in \varepsilon \mathcal{W}$, thus

$$\frac{v}{\varepsilon} \in \mathcal{V} \tag{14.16}$$

and since F is convex, we have that

$$F(v) = F\left((1 - \varepsilon)\mathbf{0} + \varepsilon \frac{v}{\varepsilon}\right) \leq (1 - \varepsilon)F(\mathbf{0}) + \varepsilon F(v/\varepsilon) \leq \varepsilon a. \tag{14.17}$$

Also

$$\frac{-v}{\varepsilon} \in \mathscr{V}. \tag{14.18}$$

Hence,

$$F(\mathbf{0}) = F\left(\frac{v}{1+\varepsilon} + \varepsilon\frac{(-v/\varepsilon)}{1+\varepsilon}\right) \leq \frac{F(v)}{1+\varepsilon} + \frac{\varepsilon}{1+\varepsilon}F(-v/\varepsilon),$$

so that

$$F(v) \geq (1+\varepsilon)F(\mathbf{0}) - \varepsilon F(-v/\varepsilon) \geq -\varepsilon a. \tag{14.19}$$

Therefore

$$|F(v)| \leq \varepsilon a, \forall v \in \varepsilon\mathscr{W}, \tag{14.20}$$

that is, F is continuous at $u = \mathbf{0}$.

Proposition 14.3.8 *Let U be a Banach space and let $F : U \to \bar{\mathbb{R}}$ be a convex functional, which is finite and continuous at $u \in U$. Under such hypotheses, $\partial F(u) \neq \emptyset$.*

Proof 14.6 Since F is convex, $Epi(F)$ is convex. Since F is continuous at u, we have that $Epi(F)^0$ is non-empty. Observe that $(u, F(u))$ is on the boundary of $Epi(F)$. Therefore, denoting $A = Epi(F)$, from the Hahn-Banach theorem there exists a closed hyperplane H which separates $(u, F(u))$ and A^0, where H

$$H = \{(v, a) \in U \times \mathbb{R} \mid \langle v, u^* \rangle_U + \alpha a = \beta\}, \tag{14.21}$$

for some fixed $\alpha, \beta \in \mathbb{R}$ and $u^* \in U^*$, so that

$$\langle v, u^* \rangle_U + \alpha a \geq \beta, \forall (v, a) \in Epi(F), \tag{14.22}$$

and

$$\langle u, u^* \rangle_U + \alpha F(u) = \beta, \tag{14.23}$$

where $(\alpha, \beta, u^*) \neq (0, 0, \mathbf{0})$. Suppose, to obtain contradiction, that $\alpha = 0$.
Thus,

$$\langle v - u, u^* \rangle_U \geq 0, \forall v \in U, \tag{14.24}$$

and therefore we obtain, $u^* = \mathbf{0}$ and $\beta = 0$, a contradiction. Hence, we may assume $\alpha > 0$ (considering (14.22)) and thus $\forall v \in U$ we have

$$\frac{\beta}{\alpha} - \langle v, u^*/\alpha \rangle_U \leq F(v), \tag{14.25}$$

and

$$\frac{\beta}{\alpha} - \langle u, u^*/\alpha \rangle_U = F(u), \tag{14.26}$$

that is,

$$\langle v - u, -u^*/\alpha \rangle_U + F(u) \leq F(v), \forall v \in U, \tag{14.27}$$

so that

$$-u^*/\alpha \in \partial F(u). \tag{14.28}$$

The proof is complete.

Definition 14.3.9 (Carathéodory function) *Let $S \subset \mathbb{R}^n$ be an open set. We say that $g : S \times \mathbb{R}^l \to \mathbb{R}$ is a Carathéodory function if*

$$\forall \xi \in \mathbb{R}^l, \ x \mapsto g(x, \xi) \text{ is a measurable function,}$$

and

$$a.e. \text{ in } S, \ \xi \mapsto g(x, \xi) \text{ is a continuous function.}$$

The proof of next results may be found in Ekeland and Temam [33].

Proposition 14.3.10 *Let E and F be two Banach spaces, let S be a Borel subset of $\mathbb{R}^n$, and $g : S \times E \to F$ be a Carathéodory function. For each measurable function $u : S \to E$, let $G_1(u)$ be the measurable function $x \mapsto g(x, u(x)) \in F$.*
  Under such hypotheses, if G_1 maps $L^p(S, E)$ on $L^r(S, F)$ for $1 \le p, r < \infty$, then G_1 is strongly continuous.

For the functional $G : U \to \mathbb{R}$, defined by $G(u) = \int_S g(x, u(x)) dS$, where $U = U^* = [L^2(S)]^l$, we have also the following result.

Proposition 14.3.11 *Considering the statement in the last proposition we may express $G^* : U^* \to \bar{\mathbb{R}}$ by*

$$G^*(u^*) = \int_S g^*(x, u^*(x)) dx, \tag{14.29}$$

onde $g^*(x, y) = \sup_{\eta \in \mathbb{R}^l} (y \cdot \eta - g(x, \eta))$, for almost all $x \in S$.

14.4 The Legendre transform and the Legendre functional

For non-convex functionals, in some cases the global extremal through which the polar functional is obtained corresponds to a local extremal point which the analytical expression is not possible.
  This fact motivates the definition of the Legendre Transform, which is obtained through a local extremal point.

Definition 14.4.1 (Legendre transform and associated functional) *Consider the function of C^2 class, $g : \mathbb{R}^n \to \mathbb{R}$. Its Legendre transform, denoted by $g_L^* : R_L^n \to \mathbb{R}$, is expressed by*

$$g_L^*(y^*) = \sum_{i=1}^n x_{0i} \cdot y_i^* - g(x_0), \tag{14.30}$$

where x_0 is a solution of the system

$$y_i^* = \frac{\partial g(x_0)}{\partial x_i}, \tag{14.31}$$

and $R_L^n = \{y^* \in \mathbb{R}^n \text{ such an equation (14.31) has a unique solution}\}$.
  Moreover, considering the functional $G : Y \to \mathbb{R}$ defined by $G(v) = \int_S g(v) dS$, we also define the associated Legendre functional, denoted by $G_L^* : Y_L^* \to \mathbb{R}$ as

$$G_L^*(v^*) = \int_S g_L^*(v^*) \, dx, \tag{14.32}$$

where $Y_L^* = \{v^* \in Y^* \mid v^*(x) \in R_L^n, \text{ a.e. in } S\}$.

About the Legendre transform we still have the following results:

Proposition 14.4.2 *Considering the last definitions, suppose for $\hat{x}_0 \in \mathbb{R}^n$ we have*

$$\det \left\{ \frac{\partial^2 g(\hat{x}_0)}{\partial x_i \partial x_j} \right\} \neq 0.$$

Let y_0^* be such that

$$(y_0^*)_i = \frac{\partial g(\hat{x}_0)}{\partial x_i}, \ \forall i \in \{1,\dots,n\}.$$

Under such hypotheses, $y_0^* \in R_L^n$ and

$$x_0(y^*) = \left[\frac{\partial g}{\partial x} \right]^{-1} (y^*)$$

is of C^1 class in a neighborhood of y_0^*.

  Moreover, for all y^* in such a neighborhood, we have that

$$y_i^* = \frac{\partial g(x_0)}{\partial x_i}, \ \forall i \in \{1,\dots,n\} \Leftrightarrow x_{0i} = \frac{\partial g_L^*(y^*)}{\partial y_i^*}, \ \forall i \in \{1,\dots,n\}.$$

Proof 14.7 The proof that $x_0(y^*)$ is of C^1 class in a neighborhood of y_0^* results from the inverse function theorem.

Suppose now that:

$$y_i^* = \frac{\partial g(x_0)}{\partial x_i}, \ \forall \, i \in \{1,\dots,n\}, \tag{14.33}$$

thus:

$$g_L^*(y^*) = y_i^* x_{0i} - g(x_0) \tag{14.34}$$

and taking derivatives for this expression we have:

$$\frac{\partial g_L^*(y^*)}{\partial y_i^*} = y_j^* \frac{\partial x_{0j}}{\partial y_i^*} + x_{0i} - \frac{\partial g(x_0)}{\partial x_j} \frac{\partial x_{0j}}{\partial y_i^*}, \tag{14.35}$$

or

$$\frac{\partial g_L^*(y^*)}{\partial y_i^*} = \left(y_j^* - \frac{\partial g(x_0)}{\partial x_j} \right) \frac{\partial x_{0j}}{\partial y_i^*} + x_{0i} \tag{14.36}$$

which from (14.33) implies that:

$$\frac{\partial g_L^*(y^*)}{\partial y_i^*} = x_{0i}, \ \forall \, i \in \{1,\dots,n\}. \tag{14.37}$$

This completes the first half of the proof. Conversely, suppose now that:

$$x_{0i} = \frac{\partial g_L^*(y^*)}{\partial y_i^*}, \ \forall i \in \{1,\dots,n\}. \tag{14.38}$$

As $y^* \in R_L^n$ there exists $\bar{x}_0 \in \mathbb{R}^n$ such that:

$$y_i^* = \frac{\partial g(\bar{x}_0)}{\partial x_i} \ \forall i \in \{1,\dots,n\}, \tag{14.39}$$

and,

$$g_L^*(y^*) = y_i^* \bar{x}_{0i} - g(\bar{x}_0) \tag{14.40}$$

and therefore taking derivatives for this expression we can obtain:

$$\frac{\partial g_L^*(y^*)}{\partial y_i^*} = y_j^* \frac{\partial \bar{x}_{0j}}{\partial y_i^*} + \bar{x}_{0i} - \frac{\partial g(\bar{x}_0)}{\partial x_j} \frac{\partial \bar{x}_{0j}}{\partial y_i^*}, \tag{14.41}$$

$\forall i \in \{1, ..., n\}$, so that:

$$\frac{\partial g_L^*(y^*)}{\partial y_i^*} = \left(y_j^* - \frac{\partial g(\bar{x}_0)}{\partial x_j} \right) \frac{\partial \bar{x}_{0j}}{\partial y_i^*} + \bar{x}_{0i} \tag{14.42}$$

$\forall i \in \{1, ..., n\}$, which from (14.38) and (14.39), implies that:

$$\bar{x}_{0i} = \frac{\partial g_L^*(y^*)}{\partial y_i^*} = x_{0i}, \ \forall \ i \in \{1, ..., n\}, \tag{14.43}$$

from this and (14.39) we have:

$$y_i^* = \frac{\partial g(\bar{x}_0)}{\partial x_i} = \frac{\partial g(x_0)}{\partial x_i} \ \forall \ i \in \{1, ..., n\}. \tag{14.44}$$

Theorem 14.4.3 *Consider the functional $J : U \to \bar{\mathbb{R}}$ defined as $J(u) = (G \circ \Lambda)(u) - \langle u, f \rangle_U$ where $\Lambda (= \{\Lambda_i\}) : U \to Y$ ($i \in \{1, ..., n\}$) is a continuous linear operator and, $G : Y \to \mathbb{R}$ is a functional that can be expressed as $G(v) = \int_S g(v) dS$, $\forall v \in Y$ (here $g : \mathbb{R}^n \to \mathbb{R}$ is a differentiable function that admits Legendre Transform denoted by $g_L^* : R_L^n \to \mathbb{R}$. That is, the hypothesis mentioned at Proposition 14.4.2 are satisfied).*

Under these assumptions we have:

$$\delta J(u_0) = \theta \Leftrightarrow \delta(-G_L^*(v_0^*) + \langle u_0, \Lambda^* v_0^* - f \rangle_U) = \theta, \tag{14.45}$$

where $v_0^* = \frac{\partial G(\Lambda(u_0))}{\partial v}$ is supposed to be such that $v_0^*(x) \in R_L^n$, a.e. in S and in this case:

$$J(u_0) = -G_L^*(v_0^*). \tag{14.46}$$

Proof 14.8 Suppose first that $\delta J(u_0) = \theta$, that is:

$$\Lambda^* \frac{\partial G(\Lambda u_0)}{\partial v} - f = \theta \tag{14.47}$$

which, as $v_0^* = \frac{\partial G(\Lambda u_0)}{\partial v}$ implies that:

$$\Lambda^* v_0^* - f = \theta, \tag{14.48}$$

and

$$v_{0i}^* = \frac{\partial g(\Lambda u_0)}{\partial x_i}. \tag{14.49}$$

Thus from the last proposition we can write:

$$\Lambda_i(u_0) = \frac{\partial g_L^*(v_0^*)}{\partial y_i^*}, \text{ for } i \in \{1, .., n\} \tag{14.50}$$

which means:

$$\Lambda u_0 = \frac{\partial G_L^*(v_0^*)}{\partial v^*}. \tag{14.51}$$

Therefore from (14.48) and (14.51) we have:

$$\delta(-G_L^*(v_0^*) + \langle u_0, \Lambda^* v_0^* - f \rangle_U) = \theta. \tag{14.52}$$

This completes the first part of the proof.

Conversely, suppose now that:

$$\delta(-G_L^*(v_0^*) + \langle u_0, \Lambda^* v_0^* - f \rangle_U) = \theta, \tag{14.53}$$

that is:

$$\Lambda^* v_0^* - f = \theta \tag{14.54}$$

and

$$\Lambda u_0 = \frac{\partial G_L^*(v_0^*)}{\partial v^*}. \tag{14.55}$$

Clearly, from (14.55), the last proposition and (14.54) we can write:

$$v_0^* = \frac{\partial G(\Lambda(u_0))}{\partial v} \tag{14.56}$$

and

$$\Lambda^* \frac{\partial G(\Lambda u_0)}{\partial v} - f = \theta, \tag{14.57}$$

which implies:

$$\delta J(u_0) = \theta. \tag{14.58}$$

Finally, we have:

$$J(u_0) = G(\Lambda u_0) - \langle u_0, f \rangle_U \tag{14.59}$$

From this, (14.54) and (14.56) we have

$$J(u_0) = G(\Lambda u_0) - \langle u_0, \Lambda^* v_0^* \rangle_U = G(\Lambda u_0) - \langle \Lambda u_0, v_0^* \rangle_Y = -G_L^*(v_0^*). \tag{14.60}$$

14.5 Duality in convex optimization

Let U be a Banach space. Given $F : U \to \bar{\mathbb{R}}$ ($F \in \Gamma_0(U)$) we define the problem $\mathscr{P}$ as

$$\mathscr{P}: \text{ minimize } F(u) \text{ on } U. \tag{14.61}$$

We say that $u_0 \in U$ is a solution of problem $\mathscr{P}$ if $F(u_0) = \inf_{u \in U} F(u)$. Consider a function $\phi(u, p)$ (ϕ $U \times Y \to \bar{\mathbb{R}}$) such that

$$\phi(u, 0) = F(u), \tag{14.62}$$

we define the problem $\mathscr{P}^*$, as

$$\mathscr{P}^*: \text{ maximize } -\phi^*(0, p^*) \text{ on } Y^*. \tag{14.63}$$

Observe that

$$\phi^*(0, p^*) = \sup_{(u,p) \in U \times Y} \{\langle 0, u \rangle_U + \langle p, p^* \rangle_Y - \phi(u, p)\} \geq -\phi(u, 0), \tag{14.64}$$

or

$$\inf_{u \in U} \{\phi(u, 0)\} \geq \sup_{p^* \in Y^*} \{-\phi^*(0, p^*)\}. \tag{14.65}$$

Proposition 14.5.1 *Consider* $\phi \in \Gamma_0(U \times Y)$. *If we define*

$$h(p) = \inf_{u \in U}\{\phi(u,p)\}, \tag{14.66}$$

then h is convex.

Proof 14.9 We have to show that given $p, q \in Y$ and $\lambda \in (0,1)$, we have

$$h(\lambda p + (1-\lambda)q) \leq \lambda h(p) + (1-\lambda)h(q). \tag{14.67}$$

If $h(p) = +\infty$ or $h(q) = +\infty$ we are done. Thus let us assume $h(p) < +\infty$ and $h(q) < +\infty$. For each $a > h(p)$ there exists $u \in U$ such that

$$h(p) \leq \phi(u,p) \leq a, \tag{14.68}$$

and, if $b > h(q)$, there exists $v \in U$ such that

$$h(q) \leq \phi(v,q) \leq b. \tag{14.69}$$

Thus

$$h(\lambda p + (1-\lambda)q) \leq \inf_{w \in U}\{\phi(w, \lambda p + (1-\lambda)q)\}$$
$$\leq \phi(\lambda u + (1-\lambda)v, \lambda p + (1-\lambda)q) \leq \lambda \phi(u,p) + (1-\lambda)\phi(v,q)$$
$$\leq \lambda a + (1-\lambda)b. \tag{14.70}$$

Letting $a \to h(p)$ and $b \to h(q)$ we obtain

$$h(\lambda p + (1-\lambda)q) \leq \lambda h(p) + (1-\lambda)h(q). \quad \square \tag{14.71}$$

Proposition 14.5.2 *For h as above, we have* $h^*(p^*) = \phi^*(0, p^*), \forall p^* \in Y^*$, *so that*

$$h^{**}(0) = \sup_{p^* \in Y^*}\{-\phi^*(0, p^*)\}. \tag{14.72}$$

Proof. Observe that

$$h^*(p^*) = \sup_{p \in Y}\{\langle p, p^*\rangle_Y - h(p)\} = \sup_{p \in Y}\{\langle p, p^*\rangle_Y - \inf_{u \in U}\{\phi(u,p)\}\}, \tag{14.73}$$

so that

$$h^*(p^*) = \sup_{(u,p) \in U \times Y}\{\langle p, p^*\rangle_Y - \phi(u,p)\} = \phi^*(0, p^*). \tag{14.74}$$

Proposition 14.5.3 *The set of solutions of the problem* $\mathscr{P}^*$ *(the dual problem) is identical to* $\partial h^{**}(0)$.

Proof 14.10 Consider $p_0^* \in Y^*$ a solution of Problem $\mathscr{P}^*$, that is,

$$-\phi^*(0, p_0^*) \geq -\phi^*(0, p^*), \forall p^* \in Y^*, \tag{14.75}$$

which is equivalent to

$$-h^*(p_0^*) \geq -h^*(p^*), \forall p^* \in Y^*, \tag{14.76}$$

which is equivalent to

$$-h(p_0^*) = \sup_{p^* \in Y^*}\{\langle 0, p^*\rangle_Y - h^*(p^*)\} \Leftrightarrow -h^*(p_0^*) = h^{**}(0) \Leftrightarrow p_0^* \in \partial h^{**}(0). \tag{14.77}$$

Theorem 14.5.4 *Consider* $\phi : U \times Y \to \bar{\mathbb{R}}$ *convex. Assume* $\inf_{u \in U}\{\phi(u,0)\} \in \mathbb{R}$ *and there exists* $u_0 \in U$ *such that* $p \mapsto \phi(u_0, p)$ *is finite and continuous at* $0 \in Y$, *then*

$$\inf_{u \in U}\{\phi(u,0)\} = \sup_{p^* \in Y^*}\{-\phi^*(0,p^*)\}, \tag{14.78}$$

and the dual problem has at least one solution.

Proof 14.11 By hypothesis $h(0) \in \mathbb{R}$ and as was shown above, h is convex. As the function $p \mapsto \phi(u_0, p)$ is convex and continuous at $0 \in Y$, there exists a neighborhood $\mathscr{V}$ of zero in Y such that

$$\phi(u_0, p) \leq M < +\infty, \forall p \in \mathscr{V}, \tag{14.79}$$

for some $M \in \mathbb{R}$. Thus, we may write

$$h(p) = \inf_{u \in U}\{\phi(u,p)\} \leq \phi(u_0, p) \leq M, \forall p \in \mathscr{V}. \tag{14.80}$$

Hence, from Lemma 14.3.7, h is continuous at 0. Thus by Proposition 14.3.8, h is sub-differentiable at 0 which means $h(0) = h^{**}(0)$. Therefore by Proposition 14.5.3, the dual problem has solutions and

$$h(0) = \inf_{u \in U}\{\phi(u,0)\} = \sup_{p^* \in Y^*}\{-\phi^*(0,p^*)\} = h^{**}(0). \tag{14.81}$$

Now we apply the last results to $\phi(u,p) = G(\Lambda u + p) + F(u)$, where $\Lambda : U \to Y$ is a continuous linear operator whose adjoint operator is denoted by $\Lambda^* : Y^* \to U^*$. We may enunciate the following theorem.

Theorem 14.5.5 *Suppose* U *is a reflexive Banach space and define* $J : U \to \mathbb{R}$ *by*

$$J(u) = G(\Lambda u) + F(u) = \phi(u,0), \tag{14.82}$$

where $\lim J(u) = +\infty$ *as* $\|u\|_U \to \infty$ *and* $F \in \Gamma_0(U)$, $G \in \Gamma_0(Y)$. *Also suppose there exists* $\hat{u} \in U$ *such that* $J(\hat{u}) < +\infty$ *with the function* $p \mapsto G(p)$ *continuous at* $\Lambda\hat{u}$. *Under such hypothesis, there exist* $u_0 \in U$ *and* $p_0^* \in Y^*$ *such that*

$$J(u_0) = \min_{u \in U}\{J(u)\} = \max_{p^* \in Y^*}\{-G^*(p^*) - F^*(-\Lambda^*p^*)\} = -G^*(p_0^*) - F^*(-\Lambda^*p_0^*). \tag{14.83}$$

Proof 14.12 The existence of solutions for the primal problem follows from the direct method of calculus of variations. That is, considering a minimizing sequence, from above (coercivity hypothesis), such a sequence is bounded and has a weakly convergent subsequence to some $u_0 \in U$. Finally, from the lower semi-continuity of primal formulation, we may conclude that u_0 is a minimizer. The other conclusions follow from Theorem 14.5.4 just observing that

$$\phi^*(0, p^*) = \sup_{u \in U, p \in Y}\{\langle p, p^*\rangle_Y - G(\Lambda u + p) - F(u)\} = \sup_{u \in U, q \in Y}\{\langle q, p^*\rangle - G(q) - \langle \Lambda u, p^*\rangle - F(u)\}, \tag{14.84}$$

so that

$$\phi^*(0, p^*) = G^*(p^*) + \sup_{u \in U}\{-\langle u, \Lambda^*p^*\rangle_U - F(u)\} = G^*(p^*) + F^*(-\Lambda^*p^*). \tag{14.85}$$

Thus,

$$\inf_{u \in U}\{\phi(u,0)\} = \sup_{p^* \in Y^*}\{-\phi^*(0,p^*)\} \tag{14.86}$$

and solutions u_0 and p_0^* for the primal and dual problems, respectively, imply that

$$J(u_0) = \min_{u \in U}\{J(u)\} = \max_{p^* \in Y^*}\{-G^*(p^*) - F^*(-\Lambda^*p^*)\} = -G^*(p_0^*) - F^*(-\Lambda^*p_0^*). \tag{14.87}$$

14.6   The min-max theorem

Our main objective in this section is to state and prove the min-max theorem.

Definition 14.1   Let U, Y be Banach spaces, $A \subset U$ and $B \subset Y$ and let $L : A \times B \to \mathbb{R}$ be a functional. We say that $(u_0, v_0) \in A \times B$ is a saddle point for L if

$$L(u_0, v) \leq L(u_0, v_0) \leq L(u, v_0), \ \forall u \in A, \ v \in B.$$

Proposition 14.1
Let U, Y be Banach spaces, $A \subset U$ and $B \subset Y$. A functional $L : U \times Y \to \mathbb{R}$ has a saddle point if and only if

$$\max_{v \in B} \inf_{u \in A} L(u, v) = \min_{u \in A} \sup_{v \in B} L(u, v).$$

Proof 14.13   Suppose $(u_0, v_0) \in A \times B$ is a saddle point of L.
   Thus,

$$L(u_0, v) \leq L(u_0, v_0) \leq L(u, v_0), \forall u \in A, \ v \in B. \tag{14.88}$$

Define

$$F(u) = \sup_{v \in B} L(u, v).$$

Observe that

$$\inf_{u \in A} F(u) \leq F(u_0),$$

so that

$$\inf_{u \in A} \sup_{v \in B} L(u, v) \leq \sup_{v \in B} L(u_0, v). \tag{14.89}$$

Define

$$G(v) = \inf_{u \in A} L(u, v).$$

Thus

$$\sup_{v \in B} G(v) \geq G(v_0),$$

so that

$$\sup_{v \in B} \inf_{u \in A} L(u, v) \geq \inf_{u \in A} L(u, v_0). \tag{14.90}$$

From (14.88), (14.89) and (14.90) we obtain

$$
\begin{aligned}
\inf_{u \in A} \sup_{v \in B} L(u, v) \quad &\leq \quad \sup_{v \in B} L(u_0, v) \\
&\leq \quad L(u_0, v_0) \\
&\leq \quad \inf_{u \in A} L(u, v_0) \\
&\leq \quad \sup_{v \in B} \inf_{u \in A} L(u, v). \tag{14.91}
\end{aligned}
$$

Hence

$$
\begin{aligned}
\inf_{u \in A} \sup_{v \in B} L(u, v) \quad &\leq \quad L(u_0, v_0) \\
&\leq \quad \sup_{v \in B} \inf_{u \in A} L(u, v). \tag{14.92}
\end{aligned}
$$

On the other hand

$$\inf_{u \in A} L(u, v) \leq L(u, v), \forall u \in A, \ v \in B,$$

so that

$$\sup_{v \in B} \inf_{u \in A} L(u, v) \leq \sup_{v \in B} L(u, v), \forall u \in A,$$

and hence

$$\sup_{v \in B} \inf_{u \in A} L(u, v) \leq \inf_{u \in A} \sup_{v \in B} L(u, v). \tag{14.93}$$

From (14.88), (14.92), (14.93) we obtain

$$\begin{aligned}
\inf_{u \in A} \sup_{v \in B} L(u, v) &= \sup_{v \in B} L(u_0, v) \\
&= L(u_0, v_0) \\
&= \inf_{u \in A} L(u, v_0) \\
&= \sup_{v \in B} \inf_{u \in A} L(u, v). \tag{14.94}
\end{aligned}$$

Conversely suppose

$$\max_{v \in B} \inf_{u \in A} L(u, v) = \min_{u \in A} \sup_{v \in B} L(u, v).$$

As above defined,

$$F(u) = \sup_{v \in B} L(u, v),$$

and

$$G(v) = \inf_{u \in A} L(u, v).$$

From the hypotheses, there exists $(u_0, v_0) \in A \times B$ such that

$$\sup_{v \in B} G(v) = G(v_0) = F(u_0) = \inf_{u \in A} F(u).$$

so that

$$F(u_0) = \sup_{v \in B} L(u_0, v) = \inf_{u \in U} L(u, v_0) = G(v_0).$$

In particular

$$L(u_0, v_0) \leq \sup_{v \in B} L(u_0, v) = \inf_{u \in U} L(u, v_0) \leq L(u_0, v_0).$$

Therefore

$$\sup_{v \in B} L(u_0, v) = L(u_0, v_0) = \inf_{u \in U} L(u, v_0).$$

The proof is complete.

Proposition 14.2

Let U, Y be Banach spaces, $A \subset U$, $B \subset Y$ and let $L : A \times B \to \mathbb{R}$ be a functional. Assume there exist $u_0 \in A$ $v_0 \in B$ and $\alpha \in \mathbb{R}$ such that

$$L(u_0, v) \leq \alpha, \ \forall v \in B,$$

and

$$L(u, v_0) \geq \alpha, \ \forall u \in A.$$

Under such hypotheses (u_0, v_0) is a saddle point of L, that is,

$$L(u_0, v) \leq L(u_0, v_0) \leq L(u, v_0), \ \forall u \in A, \ v \in B.$$

Proof 14.14   Observe that, from the hypotheses we have

$$L(u_0, v_0) \leq \alpha,$$

and

$$L(u_0, v_0) \geq \alpha,$$

so that

$$L(u_0, v) \leq \alpha = L(u_0, v_0) \leq L(u, v_0), \ \forall u \in A, \ v \in B.$$

This completes the proof.

In the next lines we state and prove the min − max theorem.

Theorem 14.1

Let U, Y be reflexive Banach spaces, $A \subset U$, $B \subset Y$ and let $L : A \times B \to \mathbb{R}$ be a functional. Suppose that

1. *$A \subset U$ is convex, closed and non-empty.*

2. *$B \subset Y$ is convex, closed and non-empty.*

3. *For each $u \in A$, $F_u(v) = L(u, v)$ is concave and upper semi-continuous.*

4. *For each $v \in B$, $G_v(u) = L(u, v)$ is convex and lower semi-continuous.*

5. *The set A and B are bounded.*

Under such hypotheses L has at least one saddle point $(u_0, v_0) \in A \times B$ such that

$$
\begin{aligned}
L(u_0, v_0) &= \min_{u \in A} \max_{v \in B} L(u, v) \\
&= \max_{v \in B} \min_{u \in A} L(u, v).
\end{aligned}
\tag{14.95}
$$

Proof 14.15   Fix $v \in B$. Observe that $G_v(u) = L(u, v)$ is convex and lower semi-continuous, therefore it is weakly lower semi-continuous on the weakly compact set A. At first we assume the additional hypothesis that $G_v(u)$ is strictly convex, $\forall v \in B$. Hence $G_v(u)$ attains a unique minimum on A. We denote the optimal $u \in A$ by $u(v)$

Define

$$G(v) = \min_{u \in A} G_v(u) = \min_{u \in U} L(u, v).$$

Thus,

$$G(v) = L(u(v), v).$$

The function $G(v)$ is expressed as the minimum of a family of concave weakly upper semi-continuous functions, and hence it is also concave and upper semi-continuous.

Moreover, $G(v)$ is bounded above on the weakly compact set B, so that there exists $v_0 \in B$ such that

$$G(v_0) = \max_{v \in B} G(v) = \max_{v \in B} \min_{u \in A} L(u,v).$$

Observe that

$$G(v_0) = \min_{u \in A} L(u,v_0) \le L(u,v_0), \ \forall u \in U.$$

Observe also that, from the concerned concavity, for $u \in A$, $v \in B$ and $\lambda \in (0,1)$ we have

$$L(u,(1-\lambda)v_0 + \lambda v) \ge (1-\lambda)L(u,v_0) + \lambda L(u,v).$$

In particular denote $u((1-\lambda)v_0 + \lambda v) = u_\lambda$, where u_λ is such that

$$
\begin{aligned}
G((1-\lambda)v_0 + \lambda v) &= \min_{u \in A} L(u,(1-\lambda)v_0 + \lambda v) \\
&= L(u_\lambda, (1-\lambda)v_0 + \lambda v).
\end{aligned}
\tag{14.96}
$$

Therefore,

$$
\begin{aligned}
G(v_0) &= \max_{v \in B} G(v) \\
&\ge G((1-\lambda)v_0 + \lambda v) \\
&= L(u_\lambda, (1-\lambda)v_0 + \lambda v) \\
&\ge (1-\lambda)L(u_\lambda, v_0) + \lambda L(u_\lambda, v) \\
&\ge (1-\lambda)\min_{u \in A} L(u,v_0) + \lambda L(u_\lambda, v) \\
&= (1-\lambda)G(v_0) + \lambda L(u_\lambda, v).
\end{aligned}
\tag{14.97}
$$

From this, we obtain

$$G(v_0) \ge L(u_\lambda, v). \tag{14.98}$$

Let $\{\lambda_n\} \subset (0,1)$ be such that $\lambda_n \to 0$.
Let $\{u_n\} \subset A$ be such that

$$
\begin{aligned}
G((1-\lambda_n)v_0 + \lambda_n v) &= \min_{u \in A} L(u,(1-\lambda_n)v_0 + \lambda_n v) \\
&= L(u_n, (1-\lambda_n)v_0 + \lambda_n v).
\end{aligned}
\tag{14.99}
$$

Since A is weakly compact, there exists a subsequence $\{u_{n_k}\} \subset \{u_n\} \subset A$ and $u_0 \in A$ such that

$$u_{n_k} \rightharpoonup u_0, \text{ weakly in } U, \text{ as } k \to \infty.$$

Observe that

$$
\begin{aligned}
(1-\lambda_{n_k})L(u_{n_k},v_0) + \lambda_{n_k}L(u_{n_k},v) &\le L(u_{n_k},(1-\lambda_{n_k})v_0 + \lambda_{n_k}v) \\
&= \min_{u \in A} L(u,(1-\lambda_{n_k})v_0 + \lambda_{n_k}v) \\
&\le L(u,(1-\lambda_{n_k})v_0 + \lambda_{n_k}v),
\end{aligned}
\tag{14.100}
$$

$\forall u \in A$, $k \in \mathbb{N}$.
Recalling that $\lambda_{n_k} \to 0$, from this and (14.100) we obtain

$$
\begin{aligned}
L(u_0,v_0) &\le \liminf_{k \to \infty} L(u_{n_k}, v_0) \\
&= \liminf_{k \to \infty}((1-\lambda_{n_k})L(u_{n_k},v_0) + \lambda_{n_k}L(u,v)) \\
&\le \limsup_{k \to \infty} L(u,(1-\lambda_{n_k})v_0 + \lambda_{n_k}v) \\
&\le L(u,v_0), \ \forall u \in U.
\end{aligned}
\tag{14.101}
$$

Hence, $L(u_0, v_0) = \min_{u \in A} L(u, v_0)$.

Observe that from (14.98) we have

$$G(v_0) \geq L(u_{n_k}, v),$$

so that

$$G(v_0) \geq \liminf_{k \to \infty} L(u_{n_k}, v) \geq L(u_0, v), \forall v \in B.$$

Denoting $\alpha = G(v_0)$ we have

$$\alpha = G(v_0) \geq L(u_0, v), \forall v \in B,$$

and

$$\alpha = G(v_0) = \min_{u \in U} L(u, v_0) \leq L(u, v_0), \ \forall u \in A.$$

From these last two results and Proposition 14.2 we have that (u_0, v_0) is a saddle point for L. Now assume that

$$G_v(u) = L(u, v)$$

is convex but not strictly convex $\forall v \in B$.

For each $n \in \mathbb{N}$ Define L_n by

$$L_n(u, v) = L(u, v) + \|u\|_U / n.$$

In such a case

$$(F_v)_n(u) = L_n(u, v)$$

is strictly convex for all $n \in \mathbb{N}$.

From above we main obtain $(u_n, v_n) \in A \times B$ such that

$$
\begin{aligned}
L(u_n, v) + \|u_n\|_U / n &\leq L(u_n, v_n) + \|u_n\|_U / n \\
&\leq L(u, v_n) + \|u\| / n. \quad (14.102)
\end{aligned}
$$

Since $A \times B$ is weakly compact and $\{(u_n, v_n)\} \subset A \times B$, up to subsequence not relabeled, there exists $(u_0, v_0) \in A \times B$ such that

$$u_n \rightharpoonup u_0, \text{ weakly in } U,$$

$$v_n \rightharpoonup v_0, \text{ weakly in } Y,$$

so that,

$$
\begin{aligned}
L(u_0, v) &\leq \liminf_{n \to \infty} (L(u_n, v) + \|u_n\|_U / n) \\
&\leq \limsup_{n \to \infty} L(u, v_n) + \|u\|_U / n \\
&\leq L(u, v_0). \quad (14.103)
\end{aligned}
$$

Hence,

$$L(u_0, v) \leq L(u, v_0), \forall u \in A, \ v \in B,$$

so that

$$L(u_0, v) \leq L(u_0, v_0) \leq L(u, v_0), \forall u \in A, \ v \in B.$$

This completes the proof.

In the next result we deal with more general situations.

Theorem 14.2
*Let U, Y be reflexive Banach spaces, $A \subset U$, $B \subset Y$ and let $L : A \times B \to \mathbb{R}$ be a functional.
Suppose that*

1. *$A \subset U$ is convex, closed and non-empty.*

2. *$B \subset Y$ is convex, closed and non-empty.*

3. *For each $u \in A$, $F_u(v) = L(u, v)$ is concave and upper semi-continuous.*

4. *For each $v \in B$, $G_v(u) = L(u, v)$ is convex and lower semi-continuous.*

5. *Either the set A is bounded or there exists $\tilde{v} \in B$ such that*

$$L(u, \tilde{v}) \to +\infty, \text{ as } \|u\| \to +\infty, \ u \in A.$$

6. *Either the set B is bounded or there exists $\tilde{u} \in A$ such that*

$$L(\tilde{u}, v) \to -\infty, \text{ as } \|v\| \to +\infty, \ v \in B.$$

Under such hypotheses L has at least one saddle point $(u_0, v_0) \in A \times B$.

Proof 14.16   We prove the result just for the special case such that there exists $\tilde{v} \in B$ such that

$$L(u, \tilde{v}) \to +\infty, \text{ as } \|u\| \to +\infty, \ u \in A,$$

and B is bounded. The proofs of remaining cases are similar.
For each $n \in \mathbb{N}$ denote

$$A_n = \{u \in A \ : \ \|u\|_U \le n\}.$$

Fix $n \in \mathbb{N}$. The sets A_n and B are closed, convex and bounded, so that from the last Theorem 14.1 there exists a saddle point $(u_n, v_n) \in A_n \times B$ for

$$L : A_n \times B \to \mathbb{R}.$$

Hence,

$$L(u_n, v) \le L(u_n, v_n) \le L(u, v_n), \forall u \in A_n, \ v \in B.$$

For a fixed $\tilde{u} \in A_1$ we have

$$
\begin{aligned}
L(u_n, \tilde{v}) \ &\le \ L(u_n, v_n) \\
&\le \ L(\tilde{u}, v_n) \\
&\le \ \sup_{v \in B} L(\tilde{u}, v) \equiv b \in \mathbb{R}.
\end{aligned}
\tag{14.104}
$$

On the other hand, from the hypotheses,

$$G_{\tilde{v}}(u) = L(u, \tilde{v})$$

is convex, lower semi-continuous and coercive, so that it is bounded below. Thus there exists $a \in \mathbb{R}$ such that

$$-\infty < a \le G_{\tilde{v}}(u) = L(u, \tilde{v}), \ \forall u \in A.$$

Hence

$$a \le L(u_n, \tilde{v}) \le L(u_n, v_n) \le b, \forall n \in \mathbb{N}.$$

Therefore $\{L(u_n, v_n)\}$ is bounded.

Moreover, from the coercivity hypotheses and

$$a \le L(u_n, \tilde{v}) \le b, \forall n \in \mathbb{N},$$

we may infer that $\{u_n\}$ is bounded.

Summarizing, $\{u_n\}, \{v_n\}$, and $\{L(u_n, v_n)\}$ are bounded sequences, and thus there exists a subsequence $\{n_k\}, u_0 \in A, v_0 \in B$ and $\alpha \in \mathbb{R}$ such that

$$u_{n_k} \rightharpoonup u_0, \text{ weakly in } U,$$

$$v_{n_k} \rightharpoonup v_0, \text{ weakly in } Y,$$

$$L(u_{n_k}, v_{n_k}) \to \alpha \in \mathbb{R},$$

as $k \to \infty$. Fix $(u, v) \in A \times B$. Observe that if $n_k > n_0 = \|u\|_U$, then

$$L(u_{n_k}, v) \le L(u_{n_k}, v_{n_k}) \le L(u, v_{n_k}),$$

so that letting $k \to \infty$, we obtain

$$
\begin{aligned}
L(u_0, v) &\le \liminf_{k \to \infty} L(u_{n_k}, v) \\
&\le \lim_{k \to \infty} L(u_{n_k}, v_{n_k}) = \alpha \\
&\le \limsup_{k \to \infty} L(u, v_{n_k}) \\
&\le L(u, v_0),
\end{aligned}
\tag{14.105}
$$

that is,

$$L(u_0, v) \le \alpha \le L(u, v_0), \ \forall u \in A, \ v \in B.$$

From this and Proposition 14.2 we may conclude that (u_0, v_0) is a saddle point for $L : A \times B \to \mathbb{R}$. The proof is complete.

14.7 Relaxation for the scalar case

In this section, $\Omega \subset \mathbb{R}^N$ denotes a bounded open set with a locally Lipschitz boundary. That is, for each point $x \in \partial\Omega$ there exists a neighborhood $\mathcal{U}_x$ whose the intersection with $\partial\Omega$ is the graph of a Lipschitz continuous function.

We start with the following definition.

Definition 14.7.1 *A function $u : \Omega \to \mathbb{R}$ is said to be affine if ∇u is constant on Ω. Furthermore, we say that $u : \Omega \to \mathbb{R}$ is piecewise affine if it is continuous and there exists a partition of Ω into a set of zero measure and finite number of open sets on which u is affine.*

The proof of next result is found in [33].

Theorem 14.7.2 *Let $r \in \mathbb{N}$ and let u_k, $1 \le k \le r$ be piecewise affine functions from Ω into $\mathbb{R}$ and $\{\alpha_k\}$ such that $\alpha_k > 0, \forall k \in \{1, ..., r\}$ and $\sum_{k=1}^{r} \alpha_k = 1$. Given $\varepsilon > 0$, there exists a locally Lipschitz function $u : \Omega \to \mathbb{R}$ and r disjoint open sets Ω_k, $1 \le k \le r$, such that*

$$|m(\Omega_k) - \alpha_k m(\Omega)| < \alpha_k \varepsilon, \ \forall k \in \{1, ..., r\},
\tag{14.106}$$

$$\nabla u(x) = \nabla u_k(x), \ a.e. \ on \ \Omega_k, \tag{14.107}$$

$$|\nabla u(x)| \le \max_{1 \le k \le r} \{|\nabla u_k(x)|\}, \ a.e. \ on \ \Omega, \tag{14.108}$$

$$\left| u(x) - \sum_{k=1}^{r} \alpha_k u_k \right| < \varepsilon, \ \forall x \in \Omega, \tag{14.109}$$

$$u(x) = \sum_{k=1}^{r} \alpha_k u_k(x), \forall x \in \partial\Omega. \tag{14.110}$$

The next result is also found in [33].

Proposition 14.7.3 *Let $r \in \mathbb{N}$ and let u_k, $1 \le k \le r$ be piecewise affine functions from Ω into $\mathbb{R}$. Consider a Carathéodory function $f : \Omega \times \mathbb{R}^N \to \mathbb{R}$ and a positive function $c \in L^1(\Omega)$ which satisfy*

$$c(x) \ge \sup\{|f(x,\xi)| \ | \ |\xi| \le \max_{1 \le k \le r} \{\|\nabla u_k\|_\infty\}\}. \tag{14.111}$$

Given $\varepsilon > 0$, there exists a locally Lipschitz function $u : \Omega \to \mathbb{R}$ such that

$$\left| \int_\Omega f(x, \nabla u) dx - \sum_{k=1}^{r} \alpha_k \int_\Omega f(x, \nabla u_k) dx \right| < \varepsilon, \tag{14.112}$$

$$|\nabla u(x)| \le \max_{1 \le k \le r} \{|\nabla u_k(x)|\}, \ a.e. \ in \ \Omega, \tag{14.113}$$

$$\left| u(x) - \sum_{k=1}^{r} \alpha_k u_k(x) \right| < \varepsilon, \forall x \in \Omega \tag{14.114}$$

$$u(x) = \sum_{k=1}^{r} \alpha_k u_k(x), \forall x \in \partial\Omega. \tag{14.115}$$

Proof 14.17 It is sufficient to establish the result for functions u_k affine over Ω, since Ω can be divided into pieces on which u_k are affine, and such pieces can be put together through (14.115). Let $\varepsilon > 0$ be given. We know that simple functions are dense in $L^1(\Omega)$, concerning the L^1 norm. Thus there exists a partition of Ω into a finite number of open sets $\mathcal{O}_i$, $1 \le i \le N_1$ and a negligible set, and there exists $\bar{f}_k$ constant functions over each $\mathcal{O}_i$ such that

$$\int_\Omega |f(x, \nabla u_k(x)) - \bar{f}_k(x)| dx < \varepsilon, \ 1 \le k \le r. \tag{14.116}$$

Now choose $\delta > 0$ such that

$$\delta \le \frac{\varepsilon}{N_1(1 + \max_{1 \le k \le r}\{\|\bar{f}_k\|_\infty\})} \tag{14.117}$$

and if B is a measurable set

$$m(B) < \delta \Rightarrow \int_B c(x) dx \le \varepsilon/N_1. \tag{14.118}$$

Now we apply Theorem 14.7.2, to each of the open sets $\mathcal{O}_i$, therefore there exists a locally Lipschitz function $u : \mathcal{O}_i \to \mathbb{R}$ and there exist r open disjoints spaces Ω_k^i, $1 \leq k \leq r$, such that

$$|m(\Omega_k^i) - \alpha_k m(\mathcal{O}_i)| \leq \alpha_k \delta, \text{ for } 1 \leq k \leq r, \tag{14.119}$$

$$\nabla u = \nabla u_k, \text{ a.e. in } \Omega_k^i, \tag{14.120}$$

$$|\nabla u(x)| \leq \max_{1 \leq k \leq r} \{|\nabla u_k(x)|\}, \text{ a.e. } \mathcal{O}_i, \tag{14.121}$$

$$\left| u(x) - \sum_{k=1}^{r} \alpha_k u_k(x) \right| \leq \delta, \forall x \in \mathcal{O}_i \tag{14.122}$$

$$u(x) = \sum_{k=1}^{r} \alpha_k u_k(x), \forall x \in \partial \mathcal{O}_i. \tag{14.123}$$

We can define $u = \sum_{k=1}^{r} \alpha_k u_k$ on $\Omega - \cup_{i=1}^{N_1} \mathcal{O}_i$. Therefore u is continuous and locally Lipschitz. Now observe that

$$\int_{\mathcal{O}_i} f(x, \nabla u(x)) dx - \sum_{k=1}^{r} \int_{\Omega_k^i} f(x, \nabla u_k(x)) dx = \int_{\mathcal{O}_i - \cup_{k=1}^{r} \Omega_k^i} f(x, \nabla u(x)) dx. \tag{14.124}$$

From $|f(x, \nabla u(x))| \leq c(x)$, $m(\mathcal{O}_i - \cup_{k=1}^{r} \Omega_k^i) \leq \delta$ and (14.118) we obtain

$$\left| \int_{\mathcal{O}_i} f(x, \nabla u(x)) dx - \sum_{k=1}^{r} \int_{\Omega_k^i} f(x, \nabla u_k(x)) dx \right| = \left| \int_{\mathcal{O}_i - \cup_{k=1}^{r} \Omega_k^i} f(x, \nabla u(x)) dx \right| \leq \varepsilon/N_1. \tag{14.125}$$

Considering that $\bar{f}_k$ is constant in $\mathcal{O}_i$, from (14.117), (14.118) and (14.119) we obtain

$$\sum_{k=1}^{r} \left| \int_{\Omega_k^i} \bar{f}_k(x) dx - \alpha_k \int_{\mathcal{O}_i} \bar{f}_k(x) dx \right| < \varepsilon/N_1. \tag{14.126}$$

We recall that $\Omega_k = \cup_{i=1}^{N_1} \Omega_k^i$ so that

$$\left| \int_{\Omega} f(x, \nabla u(x)) dx - \sum_{k=1}^{r} \alpha_k \int_{\Omega} f(x, \nabla u_k(x)) dx \right|$$

$$\leq \left| \int_{\Omega} f(x, \nabla u(x)) dx - \sum_{k=1}^{r} \int_{\Omega_k} f(x, \nabla u_k(x)) dx \right|$$

$$+ \sum_{k=1}^{r} \int_{\Omega_k} |f(x, \nabla u_k(x) - \bar{f}_k(x)| dx$$

$$+ \sum_{k=1}^{r} \left| \int_{\Omega_k} \bar{f}_k(x) dx - \alpha_k \int_{\Omega} \bar{f}_k(x) dx \right|$$

$$+ \sum_{k=1}^{r} \alpha_k \int_{\Omega} |\bar{f}_k(x) - f(x, \nabla u_k(x))| dx. \tag{14.127}$$

From (14.125), (14.116), (14.126) and (14.116) again, we obtain

$$\left| \int_{\Omega} f(x, \nabla u(x)) dx - \sum_{k=1}^{r} \alpha_k \int_{\Omega} f(x, \nabla u_k) dx \right| < 4\varepsilon. \tag{14.128}$$

The next result we do not prove it. It is a well known result from the finite element theory.

Proposition 14.7.4 *If $u \in W_0^{1,p}(\Omega)$ there exists a sequence $\{u_n\}$ of piecewise affine functions over Ω, null on $\partial\Omega$, such that*

$$u_n \to u, \text{ in } L^p(\Omega) \tag{14.129}$$

and

$$\nabla u_n \to \nabla u, \text{ in } L^p(\Omega; \mathbb{R}^N). \tag{14.130}$$

Proposition 14.7.5 *For p such that $1 < p < \infty$, suppose that $f : \Omega \times \mathbb{R}^N \to \mathbb{R}$ is a Carathéodory function , for which there exist $a_1, a_2 \in L^1(\Omega)$ and constants $c_1 \geq c_2 > 0$ such that*

$$a_2(x) + c_2 |\xi|^p \leq f(x, \xi) \leq a_1(x) + c_1 |\xi|^p, \forall x \in \Omega, \, \xi \in \mathbb{R}^N. \tag{14.131}$$

Then, given $u \in W^{1,p}(\Omega)$ piecewise affine, $\varepsilon > 0$ and a neighborhood $\mathcal{V}$ of zero in the topology $\sigma(L^p(\Omega, \mathbb{R}^N), L^q(\Omega, \mathbb{R}^N))$ there exists a function $v \in W^{1,p}(\Omega)$ such that

$$\nabla v - \nabla u \in \mathcal{V}, \tag{14.132}$$

$$u = v \text{ on } \partial\Omega,$$

$$\|v - u\|_{\infty} < \varepsilon, \tag{14.133}$$

and

$$\left| \int_{\Omega} f(x, \nabla v(x)) dx - \int_{\Omega} f^{**}(x, \nabla u(x)) dx \right| < \varepsilon. \tag{14.134}$$

Proof 14.18 Suppose given $\varepsilon > 0$, $u \in W^{1,p}(\Omega)$ piecewise affine continuous, and a neighborhood $\mathcal{V}$ of zero, which may be expressed as

$$\mathcal{V} = \left\{ w \in L^p(\Omega, \mathbb{R}^N) \mid \left| \int_{\Omega} h_m \cdot w dx \right| < \eta, \forall m \in \{1, ..., M\} \right\}, \tag{14.135}$$

where $M \in \mathbb{N}$, $h_m \in L^q(\Omega, \mathbb{R}^N)$, $\eta \in \mathbb{R}^+$. By hypothesis, there exists a partition of Ω into a negligible set Ω_0 and open subspaces Δ_i, $1 \leq i \leq r$, over which $\nabla u(x)$ is constant. From standard results of convex analysis in $\mathbb{R}^N$, for each $i \in \{1, ..., r\}$ we can obtain $\{\alpha_k \geq 0\}_{1 \leq k \leq N+1}$, and ξ_k such that $\sum_{k=1}^{N+1} \alpha_k = 1$ and

$$\sum_{k=1}^{N+1} \alpha_k \xi_k = \nabla u, \forall x \in \Delta_i, \tag{14.136}$$

and

$$\sum_{k=1}^{N+1} \alpha_k f(x, \xi_k) = f^{**}(x, \nabla u(x)). \tag{14.137}$$

Define $\beta_i = \max_{k \in \{1,\ldots,N+1\}} \{|\xi_k| \ on \ \Delta_i\}$, and $\rho_1 = \max_{i \in \{1,\ldots,r\}} \{\beta_i\}$, and $\rho = \max\{\rho_1, \|\nabla u\|_\infty\}$. Now, observe that we can obtain functions $\hat{h}_m \in C_0^\infty(\Omega; \mathbb{R}^N)$ such that

$$\max_{m \in \{1,\ldots,M\}} \|\hat{h}_m - h_m\|_{L^q(\Omega, \mathbb{R}^N)} < \frac{\eta}{4\rho m(\Omega)}. \tag{14.138}$$

Define $C = \max_{m \in \{1,\ldots,M\}} \|div(\hat{h}_m)\|_{L^q(\Omega)}$ and we can also define

$$\varepsilon_1 = \min\{\varepsilon/4, 1/(m(\Omega)^{1/p}), \eta/(2Cm(\Omega)^{1/p}), 1/m(\Omega)\} \tag{14.139}$$

We recall that ρ does not depend on ε. Furthermore, for each $i \in \{1,\ldots,r\}$ there exists a compact subset $K_i \subset \Delta_i$ such that

$$\int_{\Delta_i - K_i} [a_1(x) + c_1(x) \max_{|\xi| \le \rho} \{|\xi|^p\}] dx < \frac{\varepsilon_1}{r}. \tag{14.140}$$

Also, observe that the sets K_i may be obtained such that the restrictions of f and f^{**} to $K_i \times \rho B$ are continuous, so that from this and from the compactness of ρB, for all $x \in K_i$, we can find an open ball ω_x with center in x and contained in Ω, such that

$$|f^{**}(y, \nabla u(x)) - f^{**}(x, \nabla u(x))| < \frac{\varepsilon_1}{m(\Omega)}, \forall y \in \omega_x \cap K_i, \tag{14.141}$$

and

$$|f(y, \xi) - f(x, \xi)| < \frac{\varepsilon_1}{m(\Omega)}, \forall y \in \omega_x \cap K_i, \forall \xi \in \rho B. \tag{14.142}$$

Therefore, from this and (14.137) we may write

$$\left| f^{**}(y, \nabla u(x)) - \sum_{k=1}^{N+1} \alpha_k f(y, \xi_k) \right| < \frac{2\varepsilon_1}{m(\Omega)}, \forall y \in \omega_x \cap K_i. \tag{14.143}$$

We can cover the compact set K_i with a finite number of those open balls ω_x, denoted by ω_j, $1 \le j \le l$. Consider the open sets $\omega_j' = \omega_j - \cup_{i=1}^{j-1} \bar{\omega}_i$, we have that $\cup_{j=1}^l \bar{\omega}_j' = \cup_{j=1}^l \bar{\omega}_j$. Defining functions u_k, for $1 \le k \le N+1$ such that $\nabla u_k = \xi_k$ and $u = \sum_{k=1}^{N+1} \alpha_k u_k$ we may apply Proposition 14.7.3 to each of the open sets ω_j', so that we obtain functions $v_i \in W^{1,p}(\Omega)$ such that

$$\left| \int_{\omega_j'} f(x, \nabla v_i(x)) dx - \sum_{k=1}^{N+1} \alpha_k \int_{\omega_j'} f(x, \xi_k) dx \right| < \frac{\varepsilon_1}{rl}, \tag{14.144}$$

$$|\nabla v_i| < \rho, \forall x \in \omega_j', \tag{14.145}$$

$$|v_i(x) - u(x)| < \varepsilon_1, \forall x \in \omega_j', \tag{14.146}$$

and

$$v_i(x) = u(x), \forall x \in \partial \omega_j'. \tag{14.147}$$

Finally we set

$$v_i = u \ \text{on} \ \Delta_i - \cup_{j=1}^l \omega_j. \tag{14.148}$$

We may define a continuous mapping $v : \Omega \to \mathbb{R}$ by

$$v(x) = v_i(x), \text{ if } x \in \Delta_i, \tag{14.149}$$

$$v(x) = u(x), \text{ if } x \in \Omega_0. \tag{14.150}$$

We have that $v(x) = u(x), \forall x \in \partial\Omega$ and $\|\nabla v\|_\infty < \rho$. Also, from (14.140)

$$\int_{\Delta_i - K_i} |f^{**}(x, \nabla u(x)| dx < \frac{\varepsilon_1}{r} \tag{14.151}$$

and

$$\int_{\Delta_i - K_i} |f(x, \nabla v(x)| dx < \frac{\varepsilon_1}{r}. \tag{14.152}$$

On the other hand, from (14.143) and (14.144)

$$\left| \int_{K_i \cap \omega_j'} f(x, \nabla v(x)) dx - \int_{K_i \cap \omega_j'} f^{**}(x, \nabla u(x)) dx \right| \leq \frac{\varepsilon_1}{rl} + \frac{\varepsilon_1 m(\omega_j' \cap K_i)}{m(\Omega)} \tag{14.153}$$

so that

$$\left| \int_{K_i} f(x, \nabla v(x)) dx - \int_{K_i} f^{**}(x, \nabla u(x)) dx \right| \leq \frac{\varepsilon_1}{r} + \frac{\varepsilon_1 m(K_i)}{m(\Omega)}. \tag{14.154}$$

Now summing up in i and considering (14.151) and (14.152) we obtain (14.134), that is

$$\left| \int_\Omega f(x, \nabla v(x)) dx - \int_\Omega f^{**}(x, \nabla u(x)) dx \right| < 4\varepsilon_1 \leq \varepsilon. \tag{14.155}$$

Also, observe that from above, we have

$$\|v - u\|_\infty < \varepsilon_1, \tag{14.156}$$

and thus

$$
\begin{aligned}
\left| \int_\Omega \hat{h}_m \cdot (\nabla v(x) - \nabla u(x)) dx \right| &= \left| -\int_\Omega div(\hat{h}_m)(v(x) - u(x)) dx \right| \\
&\leq \|div(\hat{h}_m)\|_{L^q(\Omega)} \|v - u\|_{L^p(S)} \\
&\leq C\varepsilon_1 m(\Omega)^{1/p} \\
&< \frac{\eta}{2}.
\end{aligned}
\tag{14.157}
$$

Also we have that

$$\left| \int_\Omega (\hat{h}_m - h_m) \cdot (\nabla v - \nabla u) dx \right| \leq \|\hat{h}_m - h_m\|_{L^q(\Omega, \mathbb{R}^N)} \|\nabla v - \nabla u\|_{L^p(\Omega, \mathbb{R}^N)} \leq \frac{\eta}{2}. \tag{14.158}$$

Thus

$$\left| \int_\Omega h_m \cdot (\nabla v - \nabla u) dx \right| < \eta, \forall m \in \{1, ..., M\}. \tag{14.159}$$

Theorem 14.7.6 *Assuming the hypothesis of last theorem, given a function* $u \in W_0^{1,p}(\Omega)$, *given* $\varepsilon > 0$ *and a neighborhood of zero* $\mathcal{V}$ *in* $\sigma(L^p(\Omega, \mathbb{R}^N), L^q(\Omega, \mathbb{R}^N))$, *we have that there exists a function* $v \in W_0^{1,p}(\Omega)$ *such that*

$$\nabla v - \nabla u \in \mathcal{V}, \tag{14.160}$$

and

$$\left| \int_\Omega f(x, \nabla v(x)) dx - \int_\Omega f^{**}(x, \nabla u(x)) dx \right| < \varepsilon. \tag{14.161}$$

Proof 14.19 We can approximate u by a function w which is piecewise affine and null on the boundary. Thus, there exists $\delta > 0$ such that we can obtain $w \in W_0^{1,p}(\Omega)$ piecewise affine such that

$$\|u - w\|_{1,p} < \delta \tag{14.162}$$

so that

$$\nabla w - \nabla u \in \frac{1}{2} \mathcal{V}, \tag{14.163}$$

and

$$\left| \int_\Omega f^{**}(x, \nabla w(x)) dx - \int_\Omega f^{**}(x, \nabla u(x)) dx \right| < \frac{\varepsilon}{2}. \tag{14.164}$$

From Proposition 14.7.5 we may obtain $v \in W_0^{1,p}(\Omega)$ such that

$$\nabla v - \nabla w \in \frac{1}{2} \mathcal{V}, \tag{14.165}$$

and

$$\left| \int_\Omega f^{**}(x, \nabla w(x)) dx - \int_\Omega f(x, \nabla v(x)) dx \right| < \frac{\varepsilon}{2}. \tag{14.166}$$

From (14.164) and (14.166)

$$\left| \int_\Omega f^{**}(x, \nabla u(x)) dx - \int_\Omega f(x, \nabla v(x)) dx \right| < \varepsilon. \tag{14.167}$$

Finally, from (14.163), (14.165) and from the fact the weak neighborhoods are convex, we have

$$\nabla v - \nabla u \in \mathcal{V}. \tag{14.168}$$

To finish this chapter, we present two theorems which summarize the last results.

Theorem 14.7.7 *Let* f *be a Carathéodory function from* $\Omega \times \mathbb{R}^N$ *into* $\mathbb{R}$ *which satisfies*

$$a_2(x) + c_2|\xi|^p \leq f(x, \xi) \leq a_1(x) + c_1|\xi|^p \tag{14.169}$$

where $a_1, a_2 \in L^1(\Omega)$, $1 < p < +\infty$, $b \geq 0$ *and* $c_1 \geq c_2 > 0$. *Under such assumptions, defining* $\hat{U} = W_0^{1,p}(\Omega)$, *we have*

$$\inf_{u \in \hat{U}} \left\{ \int_\Omega f(x, \nabla u) dx \right\} = \min_{u \in \hat{U}} \left\{ \int_\Omega f^{**}(x, \nabla u) dx \right\} \tag{14.170}$$

The solutions of relaxed problem are weak cluster points in $W_0^{1,p}(\Omega)$ *of the minimizing sequences of primal problem.*

Proof 14.20   The existence of solutions for the convex relaxed formulation is a consequence of the reflexivity of U and coercivity hypothesis, which allows an application of the direct method of calculus of variations. That is, considering a minimizing sequence, from above (coercivity hypothesis), such a sequence is bounded and has a weakly convergent subsequence to some $\hat{u} \in W^{1,p}(\Omega)$. Finally, from the lower semi-continuity of relaxed formulation, we may conclude that $\hat{u}$ is a minimizer. The relation (14.170) follows from last theorem.

Theorem 14.7.8 *Let f be a Carathéodory function from $\Omega \times \mathbb{R}^N$ into $\mathbb{R}$ which satisfies*

$$a_2(x) + c_2|\xi|^p \leq f(x,\xi) \leq a_1(x) + c_1|\xi|^p \tag{14.171}$$

where $a_1, a_2 \in L^1(\Omega)$, $1 < p < +\infty$, $b \geq 0$ and $c_1 \geq c_2 > 0$. Let $u_0 \in W^{1,p}(\Omega)$. Under such assumptions, defining $\hat{U} = \{u \mid u - u_0 \in W_0^{1,p}(\Omega)\}$, we have

$$\inf_{u \in \hat{U}} \left\{ \int_\Omega f(x, \nabla u)dx \right\} = \min_{u \in \hat{U}} \left\{ \int_\Omega f^{**}(x, \nabla u)dx \right\} \tag{14.172}$$

   The solutions of relaxed problem are weak cluster points in $W^{1,p}(\Omega)$ of the minimizing sequences of primal problem.

Proof 14.21   Just apply the last theorem to the integrand $g(x,\xi) = f(x, \xi + \nabla u_0)$. For details see [33].

14.8   Duality suitable for the vectorial case

14.8.1   The Ekeland variational principle

In this section we present and prove the Ekeland variational principle. This proof may be found in Giusti [47], pages 160–161.

Theorem 14.8.1 (Ekeland variational principle) *Let (U,d) be a complete metric space and let $F : U \to \overline{\mathbb{R}}$ be a lower semi continuous bounded below functional taking a finite value at some point.*
   Let $\varepsilon > 0$. Assume for some $u \in U$ we have

$$F(u) \leq \inf_{u \in U}\{F(u)\} + \varepsilon.$$

Under such hypotheses, there exists $v \in U$ such that

   1. $d(u,v) \leq 1$,

   2. $F(v) \leq F(u)$,

   3. $F(v) \leq F(w) + \varepsilon d(v,w), \forall w \in U$.

Proof 14.22   Define the sequence $\{u_n\} \subset U$ by:

$$u_1 = u,$$

and having $u_1, ..., u_n$, select u_{n+1} as specified in the next lines. First, define

$$S_n = \{w \in U \mid F(w) \leq F(u_n) - \varepsilon d(u_n, w)\}.$$

   Observe that $u_n \in S_n$ so that S_n in non-empty.

On the other hand, from the definition of infimum, we may select $u_{n+1} \in S_n$ such that

$$F(u_{n+1}) \leq \frac{1}{2} \left\{ F(u_n) + \inf_{w \in S_n} \{F(w)\} \right\}. \tag{14.173}$$

Since $u_{n+1} \in S_n$ we have

$$\varepsilon d(u_{n+1}, u_n) \leq F(u_n) - F(u_{n+1}). \tag{14.174}$$

and hence

$$\varepsilon d(u_{n+m}, u_n) \leq \varepsilon \sum_{i=1}^{m} d(u_{n+i}, u_{n+m-i}) \leq F(u_n) - F(u_{n+m}). \tag{14.175}$$

From (14.174) $\{F(u_n)\}$ is decreasing sequence bounded below by $\inf_{u \in U} F(u)$ so that there exists $\alpha \in \mathbb{R}$ such that

$$F(u_n) \to \alpha \text{ as } n \to \infty.$$

From this and (14.175), $\{u_n\}$ is a Cauchy sequence, converging to some $v \in U$.

Since F is lower semi-continuous we get,

$$\alpha = \liminf_{m \to \infty} F(u_{n+m}) \geq F(v),$$

so that letting $m \to \infty$ in (14.175) we obtain

$$\varepsilon d(u_n, v) \leq F(u_n) - F(v), \tag{14.176}$$

and, in particular for $n = 1$ we get

$$0 \leq \varepsilon d(u, v) \leq F(u) - F(v) \leq F(u) - \inf_{u \in U} F(u) \leq \varepsilon.$$

Thus, we have proven 1 and 2.

Suppose, to obtain contradiction, that 3 does not hold.

Hence, there exists $w \in U$ such that

$$F(w) < F(v) - \varepsilon d(w, v).$$

In particular we have

$$w \neq v. \tag{14.177}$$

Thus, from this and (14.176) we have

$$F(w) < F(u_n) - \varepsilon(u_n, v) - \varepsilon d(w, v) \leq F(u_n) - \varepsilon d(u_n, w), \forall n \in \mathbb{N}.$$

Now observe that $w \in S_n, \forall n \in \mathbb{N}$ so that

$$\inf_{w \in S_n} \{F(w)\} \leq F(w), \forall n \in \mathbb{N}.$$

From this and (14.173) we obtain,

$$2F(u_{n+1}) - F(u_n) \leq F(w) < F(v) - \varepsilon d(v, w),$$

so that

$$2 \liminf_{n \to \infty} \{F(u_{n+1})\} \leq F(v) - \varepsilon d(v, w) + \liminf_{n \to \infty} \{F(u_n)\}.$$

Hence,

$$F(v) \leq \liminf_{n \to \infty} \{F(u_{n+1})\} \leq F(v) - \varepsilon d(v, w),$$

so that

$$0 \leq -\varepsilon d(v, w),$$

which contradicts (14.177).

Thus 3 holds.

Remark 14.8.2 *We may introduce in U a new metric given by $d_1 = \varepsilon^{1/2}d$. We highlight that the topology remains the same and also F remains lower semi-continuous. Under the hypotheses of the last theorem, if there exists $u \in U$ such that $F(u) < \inf_{u \in U} F(u) + \varepsilon$, then there exists $v \in U$ such that*

1. *$d(u,v) \leq \varepsilon^{1/2}$,*

2. *$F(v) \leq F(u)$,*

3. *$F(v) \leq F(w) + \varepsilon^{1/2}d(u,w), \forall w \in U$.*

Remark 14.8.3 *Observe that, if U is a Banach space,*

$$F(v) - F(v+tw) \leq \varepsilon^{1/2}t\|w\|_U, \forall t \in [0,1], \; w \in U, \tag{14.178}$$

so that, if F is Gâteaux differentiable, we obtain

$$-\langle \delta F(v), w \rangle_U \leq \varepsilon^{1/2}\|w\|_U. \tag{14.179}$$

Similarly

$$F(v) - F(v+t(-w)) \leq \varepsilon^{1/2}t\|w\|_U \leq, \forall t \in [0,1], \; w \in U, \tag{14.180}$$

so that, if F is Gâteaux differentiable, we obtain

$$\langle \delta F(v), w \rangle_U \leq \varepsilon^{1/2}\|w\|_U. \tag{14.181}$$

Thus

$$\|\delta F(v)\|_{U^*} \leq \varepsilon^{1/2}. \tag{14.182}$$

We have thus obtained, from the last theorem and remarks, the following result.

Theorem 14.8.4 *Let U be a Banach space. Let $F : U \to \mathbb{R}$ be a lower semi-continuous Gâteaux differentiable functional. Given $\varepsilon > 0$ suppose that $u \in U$ is such that*

$$F(u) \leq \inf_{u \in U}\{F(u)\} + \varepsilon. \tag{14.183}$$

Then there exists $v \in U$ such that

$$F(v) \leq F(u), \tag{14.184}$$

$$\|u - v\|_U \leq \sqrt{\varepsilon}, \tag{14.185}$$

and

$$\|\delta F(v)\|_{U^*} \leq \sqrt{\varepsilon}. \tag{14.186}$$

The next theorem easily follows from above results.

Theorem 14.8.5 *Let $J : U \to \mathbb{R}$, be defined by*

$$J(u) = G(\nabla u) - \langle f, u \rangle_{L^2(S;\mathbb{R}^N)}, \tag{14.187}$$

where

$$U = W_0^{1,2}(S;\mathbb{R}^N), \tag{14.188}$$

We suppose G is a l.s.c and Gâteaux-differentiable so that J is bounded below. Then, given $\varepsilon > 0$, there exists $u_\varepsilon \in U$ such that

$$J(u_\varepsilon) < \inf_{u \in U}\{J(u)\} + \varepsilon, \tag{14.189}$$

and

$$\|\delta J(u_\varepsilon)\|_{U^*} < \sqrt{\varepsilon}. \; \square \tag{14.190}$$

14.9 Some examples of duality theory in convex and non-convex analysis

We start with a well known result of Toland, published in 1979.

Theorem 14.9.1 (Toland, 1979) *Let U be a Banach space and let $F, G : U \to \mathbb{R}$ be functionals such that*

$$\inf_{u \in U} \{G(u) - F(u)\} = \alpha \in \mathbb{R}.$$

Under such hypotheses

$$F^*(u^*) - G^*(u^*) \geq \alpha, \ \forall u^* \in U^*.$$

Moreover, suppose that $u_0 \in U$ is such that

$$G(u_0) - F(u_0) = \min_{u \in U} \{G(u) - F(u)\} = \alpha.$$

Assume also $u_0^* \in \partial F(u_0)$.
Under such hypotheses,

$$F^*(u_0^*) - G^*(u_0^*) = \alpha,$$

so that

$$
\begin{aligned}
G(u_0) - F(u_0) &= \min_{u \in U} \{G(u) - F(u)\} \\
&= \min_{u^* \in U^*} \{F^*(u^*) - G^*(u^*)\} \\
&= F^*(u_0^*) - G^*(u_0^*).
\end{aligned}
\tag{14.191}
$$

Proof 14.23 Under such hypotheses,

$$\inf_{u \in U} \{G(u) - F(u)\} = \alpha \in \mathbb{R}.$$

Thus

$$G(u) - F(u) \geq \alpha, \ \forall u \in U.$$

Therefore, for $u^* \in U^*$, we have

$$-\langle u, u^* \rangle_U + G(u) + \langle u, u^* \rangle_U - F(u) \geq \alpha, \ \forall u \in U.$$

Thus,

$$-\langle u, u^* \rangle_U + G(u) + \sup_{u \in U} \{\langle u, u^* \rangle_U - F(u)\} \geq \alpha, \ \forall u \in U,$$

that is,

$$-\langle u, u^* \rangle_U + G(u) + F^*(u^*) \geq \alpha, \ \forall u \in U,$$

so that

$$\inf_{u \in U} \{-\langle u, u^* \rangle_U + G(u)\} + F^*(u^*) \geq \alpha,$$

that is,

$$-G^*(u^*) + F^*(u^*) \geq \alpha, \ \forall u^* \in U^*.
\tag{14.192}$$

Also from the hypotheses,

$$G(u_0) - F(u_0) \leq G(u) - F(u), \ \forall u \in U.
\tag{14.193}$$

On the other hand, from $u_0^* \in \partial F(u_0)$, we obtain

$$\langle u_0, u_0^* \rangle_U - F(u_0) \geq \langle u, u_0^* \rangle_U - F(u), \forall u \in U$$

so that

$$-F(u) \leq \langle u_0 - u, u_0^* \rangle_U - F(u_0), \forall u \in U.$$

From this and (14.193), we get,

$$G(u_0) - F(u_0) \leq G(u) + \langle u_0 - u, u_0^* \rangle_U - F(u_0), \forall u \in U. \tag{14.194}$$

so that,

$$\langle u_0, u_0^* \rangle_U - G(u_0) \geq \langle u, u_0^* \rangle_U - G(u), \forall u \in U,$$

that is

$$\begin{aligned} G^*(u_0^*) &= \sup_{u \in U} \{\langle u, u_0^* \rangle_U - G(u)\} \\ &= \langle u_0, u_0^* \rangle_U - G(u_0). \end{aligned} \tag{14.195}$$

Summarizing, we have got

$$G^*(u_0^*) = \langle u_0, u_0^* \rangle_U - G(u_0),$$

and

$$F^*(u_0^*) = \langle u_0, u_0^* \rangle_U - F(u_0).$$

Hence,

$$F^*(u_0) - G^*(u_0^*) = G(u_0) - F(u_0) = \alpha.$$

From this and (14.192), we have

$$\begin{aligned} G(u_0) - F(u_0) &= \min_{u \in U} \{G(u) - F(u)\} \\ &= \min_{u^* \in U^*} \{F^*(u^*) - G^*(u^*)\} \\ &= F^*(u_0^*) - G^*(u_0^*). \end{aligned} \tag{14.196}$$

The proof is complete.

Exercise 14.9.2 *Let $\Omega \subset \mathbb{R}^2$ be a set of $\hat{C}^1$ class. Let $V = C^1(\overline{\Omega})$ and let $J : D \subset V \to \mathbb{R}$ where*

$$J(u) = \frac{\gamma}{2} \int_\Omega \nabla u \cdot \nabla u \, dx - \int_\Omega fu \, dx, \forall u \in U$$

and where

$$D = \{u \in V : u = 0 \text{ on } \partial\Omega\}.$$

1. *Prove that J is convex.*

2. *Prove that $u_0 \in D$ such that*

$$\gamma \nabla^2 u_0 + f = 0, \text{ in } \Omega,$$

   minimizes J on D.

3. *Prove that*

$$\inf_{u \in U} J(u) \geq \sup_{v^* \in Y^*} \{-G^*(v^*) - F^*(-\Lambda^* v^*)\},$$

where

$$G(\nabla u) = \frac{1}{2} \int_\Omega \nabla u \cdot \nabla u \, dx,$$

and

$$G^*(v^*) = \sup_{v \in Y} \{\langle v, v^* \rangle_Y - G(v)\},$$

where $Y = Y^* = L^2(\Omega)$.

4. *Defining* $\Lambda : U \to Y$ *by*

$$\Lambda u = \nabla u,$$

and

$$F : D \to \mathbb{R}$$

as

$$F(u) = \int_\Omega f u \, dx,$$

so that

$$
\begin{aligned}
F^*(-\Lambda^* v^*) &= \sup_{u \in D} \{-\langle \nabla u, v^* \rangle_Y - F(u)\} \\
&= \sup_{u \in D} \left\{ \langle u, div\, v^* \rangle_Y + \int_\Omega f u \, dx \right\} \\
&= \sup_{u \in D} \left\{ \int_\Omega (div\, v^* + f) \, u \, dx \right\} \\
&= \begin{cases} 0, & \text{if } div(v^*) + f = 0, \text{ in } \Omega \\ +\infty, & \text{otherwise,} \end{cases}
\end{aligned}
\tag{14.197}
$$

Prove that $v_0^* = \gamma \nabla u_0$ *is such that*

$$
\begin{aligned}
J(u_0) &= \min_{u \in D} J(u) \\
&= \min_{u \in D} \{G(\Lambda u) + F(u)\} \\
&= \max_{v^* \in Y^*} \{-G^*(v^*) - F^*(-\Lambda^* v^*)\} \\
&= -G^*(v_0^*) - F^*(-\Lambda^* v_0^*).
\end{aligned}
\tag{14.198}
$$

Solution: Let $u \in D$ *and*

$$v \in V_a = \{v \in V \; : \; v = 0 \text{ on } \partial\Omega\}.$$

Thus,

$$
\begin{aligned}
&\delta J(u; v) \\
&= \lim_{\varepsilon \to 0} \frac{J(u + \varepsilon v) - J(u)}{\varepsilon} \\
&= \lim_{\varepsilon \to 0} \frac{(\gamma/2) \int_\Omega (\nabla u + \varepsilon \nabla v) \cdot (\nabla u + \varepsilon \nabla v) \, dx - (\gamma/2) \int_\Omega \nabla u \cdot \nabla u \, dx - \int_\Omega (u + \varepsilon v - u) f \, dx}{\varepsilon} \\
&= \lim_{\varepsilon \to 0} \left(\gamma \int_\Omega \nabla u \cdot \nabla v \, dx - \int_\Omega f v \, dx + \varepsilon (\gamma/2) \int_\Omega \nabla v \cdot \nabla v \, dx \right) \\
&= \gamma \int_\Omega \nabla u \cdot \nabla v \, dx - \int_\Omega f v \, dx.
\end{aligned}
\tag{14.199}
$$

Hence,

$$
\begin{aligned}
J(u+v) - J(u) &= (\gamma/2) \int_\Omega (\nabla u + \nabla v) \cdot (\nabla u + \nabla v) \, dx - (\gamma/2) \int_\Omega \nabla u \cdot \nabla u \, dx \\
&\quad - \int_\Omega (u+v-u) f \, dx \\
&= \gamma \int_\Omega \nabla u \cdot \nabla v \, dx - \int_\Omega f v \, dx + (\gamma/2) \int_\Omega \nabla v \cdot \nabla v \, dx \\
&\geq \gamma \int_\Omega \nabla u \cdot \nabla v \, dx - \int_\Omega f v \, dx \\
&= \delta J(u;v) \qquad\qquad\qquad\qquad\qquad\qquad\qquad (14.200)
\end{aligned}
$$

$\forall u \in D, \ v \in V_a.$

From this we may infer that J is convex.

From the hypotheses $u_0 \in D$ is such that

$$
\gamma \nabla^2 u_0 + f = 0, \ in \ \Omega.
$$

Let $v \in V_a$.

Therefore, we have

$$
\begin{aligned}
\delta J(u_0; v) &= \gamma \int_\Omega \nabla u_0 \cdot \nabla v \, dx - \int_\Omega f v \, dx \\
&= \gamma \int_\Omega \nabla u_0 \cdot \nabla v \, dx + \gamma \int_\Omega \nabla^2 u_0 v \, dx \\
&= \gamma \int_\Omega \nabla u_0 \cdot \nabla v \, dx - \gamma \int_\Omega \nabla u_0 \cdot \nabla v \, dx + \int_{\partial\Omega} \nabla u_0 \cdot \mathbf{n} \, v \, ds \\
&= 0 \qquad\qquad\qquad\qquad\qquad\qquad\qquad\qquad\qquad (14.201)
\end{aligned}
$$

where $\mathbf{n}$ denotes the unit outward field to $\partial\Omega$.

Summarizing, we got $\delta J(u_0; v) = 0, \ \forall v \in V_a$.

Since J is convex,from this we may conclude that u_0 minimizes J on D.

Observe now that,

$$
\begin{aligned}
J(u) &= G(\nabla u) + F(u) \\
&= -\langle \nabla u, v^* \rangle_Y + G(\nabla u) + \langle \nabla u \cdot v^* \rangle_Y + F(u) \\
&\geq \inf_{v \in Y} \{ -\langle v, v^* \rangle_Y + G(v) \} \\
&\quad + \inf_{u \in U} \{ \langle \nabla u, v^* \rangle_Y + F(u) \} \\
&= -G^*(v^*) - F^*(-\Lambda^* v^*), \ \forall u \in U, \ v^* \in Y^*. \qquad (14.202)
\end{aligned}
$$

Summarizing,

$$
\inf_{u \in D} J(u) \geq \sup_{v^* \in Y^*} \{ -G^*(v^*) - F^*(-\Lambda^* v^*) \}. \qquad\qquad (14.203)
$$

Also from the hypotheses we have $v_0^* = \gamma \nabla u_0$.

Thus,

$$
v_0^* = \frac{\partial G(\nabla u_0)}{\partial v},
$$

so that

$$
\begin{aligned}
G^*(v_0^*) &= \sup_{v \in Y} \{ \langle v, v_0^* \rangle_Y - G(v) \} \\
&= \langle \nabla u_0, v_0^* \rangle_Y - G(\nabla u_0) \\
&= -\langle u_0, div \, v_0^* \rangle_{L^2} - G(\nabla u_0). \qquad\qquad (14.204)
\end{aligned}
$$

On the other hand, from de $v_0^* = \gamma \nabla u_0$, we have

$$div\ v_0^* = \gamma div(\nabla u_0) = \gamma \nabla^2 u_0 = -f$$

From this and (14.204), we obtain,

$$G^*(v_0^*) = -\langle u_0, f \rangle_{L^2} - G(\nabla u_0),$$

and from

$$div\ v_0^* + f = 0$$

we get

$$F^*(-\Lambda^* v_0^*) = 0.$$

Hence

$$G(\nabla u_0) - \langle u_0, f \rangle_{L^2} = -G^*(v_0^*) - F^*(-\Lambda^* v_0^*),$$

so that from this and (14.203) we have

$$
\begin{aligned}
J(u_0) &= \min_{u \in D} J(u) \\
&= \min_{u \in D} \{ G(\Lambda u) + F(u) \} \\
&= \max_{v^* \in Y^*} \{ -G^*(v^*) - F^*(-\Lambda^* v^*) \} \\
&= -G^*(v_0^*) - F^*(-\Lambda^* v_0^*).
\end{aligned}
\tag{14.205}
$$

The solution is complete.

Chapter 15

Constrained Variational Optimization

15.1 Basic concepts

For this chapter the most relevant reference is the excellent book of Luenberger [57], where more details may be found. We start with the definition of cone:

Definition 15.1.1 (Cone) *Given U a Banach space, we say that $C \subset U$ is a cone with vertex at origin, if given $u \in C$, we have that $\lambda u \in C$, $\forall \lambda \geq 0$. By analogy we define a cone with vertex at $p \in U$ as $P = p + C$, where C is any cone with vertex at origin.*

Definition 15.1.2 *Let P be a convex cone in U. For $u, v \in U$ we write $u \geq v$ (with respect to P) if $u - v \in P$. In particular $u \geq \theta$ if and only if $u \in P$. Also*

$$P^+ = \{u^* \in U^* \mid \langle u, u^* \rangle_U \geq 0, \forall u \in P\}. \tag{15.1}$$

If $u^* \in P^+$ we write $u^* \geq \theta^*$.

Proposition 15.1.3 *Let U be a Banach space and P be a convex closed cone in U. If $u \in U$ satisfies $\langle u, u^* \rangle_U \geq 0$, $\forall u^* \geq \theta^*$, then $u \geq \theta$.*

Proof 15.1 We prove the contrapositive. Assume $u \notin P$. Then by the separating hyperplane theorem there is an $u^* \in U^*$ such that $\langle u, u^* \rangle_U < \langle p, u^* \rangle_U, \forall p \in P$. Since P is a cone, given any $p \in P$ we must have $\langle p, u^* \rangle_U \geq 0$, otherwise we would have $\langle u, u^* \rangle > \langle \alpha p, u^* \rangle_U$ for some $\alpha > 0$. Thus $u^* \in P^+$. Finally, since $\inf_{p \in P}\{\langle p, u^* \rangle_U\} = 0$, we obtain $\langle u, u^* \rangle_U < 0$ which completes the proof.

Definition 15.1.4 (Convex mapping) *Let U, Z be vector spaces. Let $P \subset Z$ be a convex cone. A mapping $G : U \to Z$ is said to be convex if*

$$G(\alpha u_1 + (1 - \alpha)u_2) \leq \alpha G(u_1) + (1 - \alpha)G(u_2), \forall u_1, u_2 \in U, \alpha \in [0, 1]. \tag{15.2}$$

Consider the problem $\mathscr{P}$, defined as

$$\text{Problem } \mathscr{P} : \text{ Minimize } F : U \to \mathbb{R} \text{ subject to } u \in \Omega, \text{ and } G(u) \leq \theta$$

Define

$$\omega(z) = \inf\{F(u) \mid u \in \Omega \text{ and } G(u) \le z\}. \tag{15.3}$$

For such a functional we have the following result.

Proposition 15.1.5 *If F is a real convex functional and G is convex, then ω is convex.*

Proof 15.2 Observe that

$$\omega(\alpha z_1 + (1-\alpha)z_2) = \inf\{F(u) \mid u \in \Omega$$
$$\text{and } G(u) \le \alpha z_1 + (1-\alpha)z_2\} \tag{15.4}$$

$$\le \inf\{F(u) \mid u = \alpha u_1 + (1-\alpha)u_2 \ u_1, u_2 \in \Omega$$
$$\text{and } G(u_1) \le z_1, \ G(u_2) \le z_2\} \tag{15.5}$$

$$\le \alpha \inf\{F(u_1) \mid u_1 \in \Omega, \ G(u_1) \le z_1\}$$
$$+ (1-\alpha)\inf\{F(u_2) \mid u_2 \in \Omega, \ G(u_2) \le z_2\} \tag{15.6}$$

$$\le \alpha \omega(z_1) + (1-\alpha)\omega(z_2). \tag{15.7}$$

Now we establish the Lagrange multiplier theorem for convex global optimization.

Theorem 15.1.6 *Let U be a vector space, Z a Banach space, Ω a convex subset of U, P a positive convex closed cone of Z. Assume that P contains an interior point. Let F be a real convex functional on Ω and G a convex mapping from Ω into Z. Assume the existence of $u_1 \in \Omega$ such that $G(u_1) < \theta$. Defining*

$$\mu_0 = \inf_{u \in \Omega}\{F(u) \mid G(u) \le \theta\}, \tag{15.8}$$

then there exists $z_0^* \ge \theta$, $z_0^* \in Z^*$ such that

$$\mu_0 = \inf_{u \in \Omega}\{F(u) + \langle G(u), z_0^*\rangle z\}. \tag{15.9}$$

Furthermore, if the infimum in (15.8) is attained by $u_0 \in U$ such that $G(u_0) \le \theta$, it is also attained in (15.9) by the same u_0 and also $\langle G(u_0), z_0^*\rangle z = 0$. We refer to z_0^* as the Lagrangian Multiplier.

Proof 15.3 Consider the space $W = \mathbb{R} \times Z$ and the sets A, B where

$$A = \{(r, z) \in \mathbb{R} \times Z \mid r \ge F(u), \ z \ge G(u) \ for \ some \ u \in \Omega\}, \tag{15.10}$$

and

$$B = \{(r, z) \in \mathbb{R} \times Z \mid r \le \mu_0, \ z \le \theta\}, \tag{15.11}$$

where $\mu_0 = \inf_{u \in \Omega}\{F(u) \mid G(u) \le \theta\}$. Since F and G are convex, A and B are convex sets. It is clear that A contains no interior point of B, and since $N = -P$ contains an interior point, the set B contains an interior point. Thus, from the separating hyperplane theorem, there is a non-zero element $w_0^* = (r_0, z_0^*) \in W^*$ such that

$$r_0 r_1 + \langle z_1, z_0^*\rangle z \ge r_0 r_2 + \langle z_2, z_0^*\rangle z, \forall(r_1, z_1) \in A, \ (r_2, z_2) \in B. \tag{15.12}$$

From the nature of B it is clear that $w_0^* \geq \theta$. That is, $r_0 \geq 0$ and $z_0^* \geq \theta$. We will show that $r_0 > 0$. The point $(\mu_0, \theta) \in B$, hence

$$r_0 r + \langle z, z_0^* \rangle_Z \geq r_0 \mu_0, \forall (r, z) \in A. \tag{15.13}$$

If $r_0 = 0$ then $\langle G(u_1), z_0^* \rangle_Z \geq 0$ and $z_0^* \neq \theta$. Since $G(u_1) < \theta$ and $z_0^* \geq \theta$ we have a contradiction. Therefore $r_0 > 0$ and, without loss of generality we may assume $r_0 = 1$. Since the point (μ_0, θ) is arbitrarily close to A and B, we have

$$\mu_0 = \inf_{(r,z) \in A} \{r + \langle z, z_0^* \rangle_Z\} \leq \inf_{u \in \Omega} \{F(u) + \langle G(u), z_0^* \rangle_Z\} \leq \inf\{F(u) \mid u \in \Omega, \ G(u) \leq \theta\} = \mu_0. \tag{15.14}$$

Also, if there exists u_0 such that $G(u_0) \leq \theta$, $\mu_0 = F(u_0)$, then

$$\mu_0 \leq F(u_0) + \langle G(u_0), z_0^* \rangle_Z \leq F(u_0) = \mu_0. \tag{15.15}$$

Hence

$$\langle G(u_0), z_0^* \rangle_Z = 0. \tag{15.16}$$

Corollary 15.1.7 *Let the hypothesis of the last theorem hold. Suppose*

$$F(u_0) = \inf_{u \in \Omega} \{F(u) \mid G(u) \leq \theta\}. \tag{15.17}$$

Then there exists $z_0^* \geq \theta$ such that the Lagrangian $L : U \times Z^* \to \mathbb{R}$ defined by

$$L(u, z^*) = F(u) + \langle G(u), z^* \rangle_Z \tag{15.18}$$

has a saddle point at (u_0, z_0^*). That is

$$L(u_0, z^*) \leq L(u_0, z_0^*) \leq L(u, z_0^*), \forall u \in \Omega, z^* \geq \theta. \tag{15.19}$$

Proof 15.4 For z_0^* obtained in the last theorem, we have

$$L(u_0, z_0^*) \leq L(u, z_0^*), \forall u \in \Omega. \tag{15.20}$$

As $\langle G(u_0), z_0^* \rangle_Z = 0$, we have

$$L(u_0, z^*) - L(u_0, z_0^*) = \langle G(u_0), z^* \rangle_Z - \langle G(u_0), z_0^* \rangle_Z = \langle G(u_0), z^* \rangle_Z \leq 0. \tag{15.21}$$

We now prove two theorems relevant to develop the subsequent section.

Theorem 15.1.8 *Let $F : \Omega \subset U \to \mathbb{R}$ and $G : \Omega \to Z$. Let $P \subset Z$ be a convex closed cone. Suppose there exists $(u_0, z_0^*) \in U \times Z^*$ where $z_0^* \geq \theta$ and $u_0 \in \Omega$ are such that*

$$F(u_0) + \langle G(u_0), z_0^* \rangle_Z \leq F(u) + \langle G(u), z_0^* \rangle_Z, \forall u \in \Omega. \tag{15.22}$$

Then

$$F(u_0) = \inf\{F(u) \mid u \in \Omega \text{ and } G(u) \leq G(u_0)\}. \tag{15.23}$$

Proof 15.5 Suppose there is a $u_1 \in \Omega$ such that $F(u_1) < F(u_0)$ and $G(u_1) \leq G(u_0)$. Thus

$$\langle G(u_1), z_0^* \rangle_Z \leq \langle G(u_0), z_0^* \rangle_Z \tag{15.24}$$

so that

$$F(u_1) + \langle G(u_1), z_0^* \rangle_Z < F(u_0) + \langle G(u_0), z_0^* \rangle_Z, \tag{15.25}$$

which contradicts the hypothesis of the theorem.

Theorem 15.1.9 *Let F be a convex real functional and $G : \Omega \to Z$ convex and let u_0 and u_1 be solutions to the problems $\mathscr{P}_0$ and $\mathscr{P}_1$ respectively, where*

$$\mathscr{P}_0 : \ minimize \ F(u) \ subject \ to \ u \in \Omega \ and \ G(u) \leq z_0, \tag{15.26}$$

and

$$\mathscr{P}_1 : \ minimize \ F(u) \ subject \ to \ u \in \Omega \ and \ G(u) \leq z_1. \tag{15.27}$$

Suppose z_0^* and z_1^* are the Lagrange multipliers related to these problems. Then

$$\langle z_1 - z_0, z_1^* \rangle_Z \leq F(u_0) - F(u_1) \leq \langle z_1 - z_0, z_0^* \rangle_Z. \tag{15.28}$$

Proof 15.6   For u_0, z_0^* we have

$$F(u_0) + \langle G(u_0) - z_0, z_0^* \rangle_Z \leq F(u) + \langle G(u) - z_0, z_0^* \rangle_Z, \forall u \in \Omega, \tag{15.29}$$

and, particularly for $u = u_1$ and considering that $\langle G(u_0) - z_0, z_0^* \rangle_Z = 0$, we obtain

$$F(u_0) - F(u_1) \leq \langle G(u_1) - z_0, z_0^* \rangle_Z \leq \langle z_1 - z_0, z_0^* \rangle_Z. \tag{15.30}$$

A similar argument applied to u_1, z_1^* provides us the other inequality.

15.2   Duality

Consider the basic convex programming problem:

$$Minimize \ F(u) \ subject \ to \ G(u) \leq \theta, \ u \in \Omega, \tag{15.31}$$

where $F : U \to \mathbb{R}$ is a convex functional, $G : U \to Z$ is convex mapping, and Ω is a convex set. We define $\varphi : Z^* \to \mathbb{R}$ by

$$\varphi(z^*) = \inf_{u \in \Omega} \{F(u) + \langle G(u), z^* \rangle_Z\}. \tag{15.32}$$

Proposition 15.2.1 *φ is concave and*

$$\varphi(z^*) = \inf_{z \in \Gamma} \{\omega(z) + \langle z, z^* \rangle_Z\}, \tag{15.33}$$

where

$$\omega(z) = \inf_{u \in \Omega} \{F(u) \mid G(u) \leq z\}, \tag{15.34}$$

and

$$\Gamma = \{z \in Z \mid G(u) \leq z \ for \ some \ u \in \Omega\}.$$

Proof 15.7   Observe that

$$\begin{aligned}
\varphi(z^*) &= \inf_{u \in \Omega} \{F(u) + \langle G(u), z^* \rangle_Z\} \\
&\leq \inf_{u \in \Omega} \{F(u) + \langle z, z^* \rangle_Z \mid G(u) \leq z\} \\
&= \omega(z) + \langle z, z^* \rangle_Z, \forall z^* \geq \theta, z \in \Gamma.
\end{aligned} \tag{15.35}$$

On the other hand, for any $u_1 \in \Omega$, defining $z_1 = G(u_1)$, we obtain

$$F(u_1) + \langle G(u_1), z^* \rangle_Z \geq \inf_{u \in \Omega} \{ F(u) + \langle z_1, z^* \rangle_Z \mid G(u) \leq z_1 \} = \omega(z_1) + \langle z_1, z^* \rangle_Z, \qquad (15.36)$$

so that

$$\varphi(z^*) \geq \inf_{z \in \Gamma} \{ \omega(z) + \langle z, z^* \rangle_Z \}. \qquad (15.37)$$

Theorem 15.2.2 (Lagrange duality) *Consider $F : \Omega \subset U \to \mathbb{R}$ a convex functional, Ω a convex set, and $G : U \to Z$ a convex mapping. Suppose there exists a u_1 such that $G(u_1) < \theta$ and that $\inf_{u \in \Omega} \{ F(u) \mid G(u) \leq \theta \} < \infty$, with such order related to a convex closed cone in Z. Under such assumptions, we have*

$$\inf_{u \in \Omega} \{ F(u) \mid G(u) \leq \theta \} = \max_{z^* \geq \theta} \{ \varphi(z^*) \}. \qquad (15.38)$$

If the infimum on the left side in (15.38) is achieved at some $u_0 \in U$ and the max on the right side at $z_0^* \in Z^*$, then

$$\langle G(u_0), z_0^* \rangle_Z = 0 \qquad (15.39)$$

and u_0 minimizes $F(u) + \langle G(u), z_0^* \rangle_Z$ on Ω.

Proof 15.8 For $z^* \geq \theta$ we have

$$\inf_{u \in \Omega} \{ F(u) + \langle G(u), z^* \rangle_Z \} \leq \inf_{u \in \Omega, G(u) \leq \theta} \{ F(u) + \langle G(u), z^* \rangle_Z \} \leq \inf_{u \in \Omega, G(u) \leq \theta} F(u) \leq \mu_0. \qquad (15.40)$$

or

$$\varphi(z^*) \leq \mu_0. \qquad (15.41)$$

The result follows from Theorem 15.1.6.

15.3   The Lagrange multiplier theorem

Remark 15.3.1 *This section was published in similar form by the journal "Computational and Applied Mathematics, SBMAC-Springer", reference [42].*

In this section we develop a new and simpler proof of the Lagrange multiplier theorem in a Banach space context. In particular, we address the problem of minimizing a functional $F : U \to \mathbb{R}$ subject to $G(u) = \theta$, where θ denotes the zero vector and $G : U \to Z$ is a Fréchet differentiable transformation. Here U, Z are Banach spaces. General results in Banach spaces may be found in [2, 34], for example. For the theorem in question, among others we would cite [57, 54, 19]. Specially the proof given in [57] is made through the generalized inverse function theorem. We emphasize such a proof is extensive and requires the continuous Fréchet differentiability of F and G. Our approach here is different and the results are obtained through other hypotheses.

The main result is summarized by the following theorem.

Theorem 15.3.2 *Let U and Z be Banach spaces. Assume u_0 is a local minimum of $F(u)$ subject to $G(u) = \theta$ where $F : U \to \mathbb{R}$ is a Gâteaux differentiable functional and $G : U \to Z$ is a Fréchet differentiable transformation such that $G'(u_0)$ maps U onto Z. Finally, assume there exist $\alpha > 0$ and $K > 0$ such that if $\|\varphi\|_U < \alpha$ then,*

$$\|G'(u_0 + \varphi) - G'(u_0)\| \leq K\|\varphi\|_U.$$

Under such assumptions, there exists $z_0^* \in Z^*$ such that

$$F'(u_0) + [G'(u_0)]^*(z_0^*) = \theta,$$

that is,

$$\langle \varphi, F'(u_0) \rangle_U + \langle G'(u_0)\varphi, z_0^* \rangle_Z = 0, \forall \varphi \in U.$$

Proof 15.9 First observe that there is no loss of generality in assuming $0 < \alpha < 1$. Also from the generalized mean value inequality and our hypothesis, if $\|\varphi\|_U < \alpha$, then

$$
\begin{aligned}
&\|G(u_0 + \varphi) - G(u_0) - G'(u_0) \cdot \varphi\| \\
\leq\ & \sup_{h \in [0,1]} \{\|G'(u_0 + h\varphi) - G'(u_0)\|\} \|\varphi\|_U \\
\leq\ & K \sup_{h \in [0,1]} \{\|h\varphi\|_U\} \|\varphi\|_U \leq K\|\varphi\|_U^2.
\end{aligned}
\tag{15.42}
$$

For each $\varphi \in U$, define $H(\varphi)$ by

$$G(u_0 + \varphi) = G(u_0) + G'(u_0) \cdot \varphi + H(\varphi),$$

that is,

$$H(\varphi) = G(u_0 + \varphi) - G(u_0) - G'(u_0) \cdot \varphi.$$

Let $L_0 = N(G'(u_0))$ where $N(G'(u_0))$ denotes the null space of $G'(u_0)$. Observe that U/L_0 is a Banach space for which we define $A : U/L_0 \to Z$ by

$$A(\bar{u}) = G'(u_0) \cdot u,$$

where $\bar{u} = \{u + v \mid v \in L_0\}$.

Since $G'(u_0)$ is onto, so is A, so that by the inverse mapping theorem A has a continuous inverse A^{-1}.

Let $\varphi \in U$ be such that $G'(u_0) \cdot \varphi = \theta$. For a given t such that $0 < |t| < \frac{\alpha}{1+\|\varphi\|_U}$, let $\psi_0 \in U$ be such that

$$G'(u_0) \cdot \psi_0 + \frac{H(t\varphi)}{t^2} = \theta,$$

Observe that, from (15.42),

$$\|H(t\varphi)\| \leq Kt^2 \|\varphi\|_U^2,$$

and thus from the boundedness of A^{-1}, $\|\psi_0\|$ as a function of t may be chosen uniformly bounded relating t (that is, despite the fact that ψ_0 may vary with t, there exists $K_1 > 0$ such that $\|\psi_0\|_U < K_1, \forall t$ such that $0 < |t| < \frac{\alpha}{1+\|\varphi\|_U}$).

Now choose $0 < r < 1/4$ and define $g_0 = \theta$.

Also define

$$\varepsilon = \frac{r}{4(\|A^{-1}\| + 1)(K + 1)(K_1 + 1)(\|\varphi\|_U + 1)}.$$

Since from the hypotheses $G'(u)$ is continuous at u_0, we may choose $0 < \delta < \alpha$ such that if $\|v\|_U < \delta$ then

$$\|G'(u_0 + v) - G'(u_0)\| < \varepsilon.$$

Fix $t \in \mathbb{R}$ such that

$$0 < |t| < \frac{\delta}{2(1 + \|\varphi\|_U + K_1)}.$$

Observe that $\psi \in U$ is such that $G(u_0 + t\varphi + t^2\psi) = \theta$ if and only if

$$G'(u_0) \cdot \psi + \frac{H(t\varphi + t^2\psi)}{t^2} = \theta.$$

Define

$$L_1 = A^{-1}\left[G'(u_0) \cdot (\psi_0 - g_0) + \frac{H(t\varphi + t^2(\psi_0 - g_0))}{t^2}\right],$$

so that

$$
\begin{aligned}
L_1 &= A^{-1}[A(\overline{\psi_0 - g_0})] + A^{-1}\left(\frac{H(t\varphi + t^2(\psi_0 - g_0))}{t^2}\right) \\
&= \overline{\psi_0 - g_0} + \overline{w_1} \\
&= \overline{\psi_0 + w_1} \\
&= \{\psi_0 + w_1 + v \mid v \in L_0\}.
\end{aligned}
$$

Here $w_1 \in U$ is such that

$$\overline{w_1} = A^{-1}\left(\frac{H(t\varphi + t^2(\psi_0 - g_0))}{t^2}\right),$$

that is,

$$A(\overline{w_1}) = \frac{H(t\varphi + t^2(\psi_0 - g_0))}{t^2},$$

so that

$$G'(u_0) \cdot w_1 = \frac{H(t\varphi + t^2(\psi_0 - g_0))}{t^2}.$$

Select $g_1 \in L_1$ such that

$$\|g_1 - g_0\|_U \le 2\|L_1 - L_0\|.$$

This is possible since

$$\|L_1 - L_0\| = \inf_{g \in L_1}\{\|g - g_0\|_U\}.$$

So we have that

$$L_1 = A^{-1}\left[-\frac{H(t\varphi)}{t^2} + \frac{H(t\varphi + t^2(\psi_0 - g_0))}{t^2}\right]. \tag{15.43}$$

However

$$
\begin{aligned}
& H(t\varphi + t^2(\psi_0 - g_0)) - H(t\varphi) \\
={} & G(u_0 + t\varphi + t^2(\psi_0)) - G(u_0) \\
& -G'(u_0) \cdot (t\varphi + t^2(\psi_0)) \\
& -G(u_0 + t\varphi) + G(u_0) \\
& +G'(u_0) \cdot (t\varphi) \\
={} & G(u_0 + t\varphi + t^2(\psi_0)) - G(u_0 + t\varphi) \\
& -G'(u_0) \cdot (t^2(\psi_0)), \tag{15.44}
\end{aligned}
$$

so that by the generalized mean value inequality we may write

$$
\begin{aligned}
& \|H(t\varphi + t^2(\psi_0 - g_0)) - H(t\varphi)\| \\
\le{} & \sup_{h \in [0,1]} \|G'(u_0 + t\varphi + ht^2(\psi_0)) - G'(u_0)\| \|t^2\psi_0\|_U \\
<{} & \varepsilon t^2 \|\psi_0\|_U. \tag{15.45}
\end{aligned}
$$

From this and (15.43) we get

$$
\begin{aligned}
\|L_1\| \ &\leq \ \|A^{-1}\|\|H(t\varphi + t^2(\psi_0 - g_0)) - H(t\varphi)\|/t^2 \\
&< \ \|A^{-1}\|\varepsilon\|\psi_0\|_U \\
&< \ \|A^{-1}\|K_1 \frac{r}{4(\|A^{-1}\| + 1)(K + 1)(K_1 + 1)(\|\varphi\|_U + 1)} \\
&< \ \frac{r}{4}.
\end{aligned}
\tag{15.46}
$$

Hence

$$\|g_1\|_U < 2\|L_1\| < r/2.$$

Now reasoning by induction, for $n \geq 2$ assume that $\|g_{n-1}\|_U < r$ and $\|g_{n-2}\|_U < r$ and define L_n by

$$L_n - L_{n-1} = A^{-1}\left[G'(u_0) \cdot (\psi_0 - g_{n-1}) + \frac{H(t\varphi + t^2(\psi_0 - g_{n-1}))}{t^2}\right].$$

Observe that

$$
\begin{aligned}
L_n \ &= \ A^{-1}\left[G'(u_0) \cdot (\psi_0 - g_{n-1}) + \frac{H(t\varphi + t^2(\psi_0 - g_{n-1}))}{t^2}\right] + L_{n-1} \\
&= \ A^{-1}A(\overline{\psi_0 - g_{n-1}}) + A^{-1}\left[\frac{H(t\varphi + t^2(\psi_0 - g_{n-1}))}{t^2}\right] + \overline{g}_{n-1} \\
&= \ \overline{\psi_0 - g_{n-1}} + A^{-1}\left[\frac{H(t\varphi + t^2(\psi_0 - g_{n-1}))}{t^2}\right] + \overline{g}_{n-1} \\
&= \ \overline{\psi_0} + A^{-1}\left[\frac{H(t\varphi + t^2(\psi_0 - g_{n-1}))}{t^2}\right] \\
&= \ \{\psi_0 + w_n + v \mid v \in L_0\}.
\end{aligned}
$$

Here $w_n \in U$ is such that

$$\overline{w_n} = A^{-1}\left[\frac{H(t\varphi + t^2(\psi_0 - g_{n-1}))}{t^2}\right],$$

that is,

$$A(\overline{w_n}) = \left[\frac{H(t\varphi + t^2(\psi_0 - g_{n-1}))}{t^2}\right],$$

so that

$$G'(u_0) \cdot w_n = \left[\frac{H(t\varphi + t^2(\psi_0 - g_{n-1}))}{t^2}\right].$$

Choose $g_n \in L_n$ such that

$$\|g_n - g_{n-1}\|_U \leq 2\|L_n - L_{n-1}\|.$$

This is possible since

$$\|L_n - L_{n-1}\| = \inf_{g \in L_n} \{\|g - g_{n-1}\|_U\}.$$

Observe that we may write

$$L_{n-1} = A^{-1}[A(\overline{g}_{n-1})] = A^{-1}[G'(u_0) \cdot g_{n-1}].$$

Thus

$$L_n = A^{-1}\left[G'(u_0) \cdot (\psi_0 - g_{n-1}) + \frac{H(t\varphi + t^2(\psi_0 - g_{n-1}))}{t^2} + G'(u_0) \cdot g_{n-1}\right].$$

By analogy

$$L_{n-1} = A^{-1} \left[G'(u_0) \cdot (\psi_0 - g_{n-2}) + \frac{H(t\varphi + t^2(\psi_0 - g_{n-2}))}{t^2} + G'(u_0) \cdot g_{n-2} \right].$$

Observe that

$$
\begin{aligned}
& H(t\varphi + t^2(\psi_0 - g_{n-1})) - H(t\varphi + t^2(\psi_0 - g_{n-2})) \\
= \ & G(u_0 + t\varphi + t^2(\psi_0 - g_{n-1})) - G(u_0) \\
& - G'(u_0) \cdot (t\varphi + t^2(\psi_0 - g_{n-1})) \\
& - G(u_0 + t\varphi + t^2(\psi_0 - g_{n-2})) + G(u_0) \\
& + G'(u_0) \cdot (t\varphi + t^2(\psi_0 - g_{n-2})) \\
= \ & G(u_0 + t\varphi + t^2(\psi_0 - g_{n-1})) - G(u_0 + t\varphi + t^2(\psi_0 - g_{n-2})) \\
& - G'(u_0) \cdot (t^2(-g_{n-1} + g_{n-2})),
\end{aligned}
\tag{15.47}
$$

so that by the generalized mean value inequality we may write

$$
\begin{aligned}
& \| H(t\varphi + t^2(\psi_0 - g_{n-1})) - H(t\varphi + t^2(\psi_0 - g_{n-2})) \| \\
\leq \ & \sup_{h \in [0,1]} \| G'(u_0 + t\varphi + t^2\psi_0 - t^2(h(g_{n-1}) + (1-h)g_{n-2})) - G'(u_0) \| \\
& \times \| t^2(-g_{n-1} + g_{n-2}) \|_U \\
< \ & \varepsilon t^2 \| g_{n-1} - g_{n-2} \|_U.
\end{aligned}
$$

Therefore, similarly as above,

$$
\begin{aligned}
\| L_n - L_{n-1} \| \ & \leq \ \frac{\|A^{-1}\|}{t^2} \| H(t\varphi + t^2(\psi_0 - g_{n-1})) - H(t\varphi + t^2(\psi_0 - g_{n-2})) \| \\
& < \ \varepsilon \|A^{-1}\| \| g_{n-1} - g_{n-2} \|_U \\
& < \ (r/4) \| g_{n-1} - g_{n-2} \|_U \\
& < \ \frac{1}{4} \| g_{n-1} - g_{n-2} \|_U.
\end{aligned}
\tag{15.48}
$$

Thus,

$$\| g_n - g_{n-1} \|_U \leq 2 \| L_n - L_{n-1} \| < \frac{1}{2} \| g_{n-1} - g_{n-2} \|_U.$$

Finally

$$
\begin{aligned}
\| g_n \|_U \ & = \ \| g_n - g_{n-1} + g_{n-1} - g_{n-2} + g_{n-2} - \dots + g_1 - g_0 \|_U \\
& \leq \ \| g_1 \|_U \left(1 + \frac{1}{2} + \dots + \frac{1}{2^n} \right) < 2 \| g_1 \|_U < r.
\end{aligned}
\tag{15.49}
$$

Thus $\| g_n \|_U < r$ and

$$\| g_n - g_{n-1} \|_U < \frac{1}{2} \| g_{n-1} - g_{n-2} \|_U, \forall n \in \mathbb{N},$$

so that $\{g_n\}$ is a Cauchy sequence, and since U is a Banach space there exists $g \in U$ such that

$$g_n \to g, \text{ in norm, as } n \to \infty.$$

Hence

$$L_n \to L = \bar{g}, \text{ in norm, as } n \to \infty,$$

so that,

$$L_n - L_{n-1} \to L - L = \theta = A^{-1} \left[G'(u_0) \cdot (\psi_0 - g) + \frac{H(t\varphi + t^2(\psi_0 - g))}{t^2} \right].$$

Since A^{-1} is a bijection, denoting $\tilde{\psi}_0 = (\psi_0 - g)$, we get

$$G'(u_0) \cdot \tilde{\psi}_0 + \frac{H(t\varphi + t^2(\tilde{\psi}_0))}{t^2} = \theta$$

Clarifying the dependence on t we denote $\tilde{\psi}_0 = \tilde{\psi}_0(t)$ where as above mentioned, $t \in \mathbb{R}$ is such that

$$0 < |t| < \frac{\delta}{2(1 + \|\varphi\|_U + K_1)}.$$

Therefore

$$G(u_0 + t\varphi + t^2\tilde{\psi}_0(t)) = \theta.$$

Observe also that $\|t^2\tilde{\psi}_0(t)\|_U = \|t^2(\psi_0(t) - g)\|_U \le t^2(K_1 + r) \le t^2(K_1 + 1)$ so that $t^2\tilde{\psi}_0(t) \to \theta$ as $t \to 0$. Thus, by defining $t^2\tilde{\psi}_0(t)|_{t=0} = \theta$ (observe that in principle such a function would not be defined at $t = 0$), we obtain

$$\frac{d(t^2\tilde{\psi}_0(t))}{dt}\Big|_{t=0} = \lim_{t \to 0} \left(\frac{t^2\tilde{\psi}_0(t) - \theta}{t} \right) = \theta,$$

considering that

$$\|t\tilde{\psi}_0(t)\|_U \le |t|(K_1 + 1) \to 0, \text{ as } t \to 0.$$

Finally, defining

$$\phi(t) = F(u_0 + t\varphi + t^2\tilde{\psi}_0(t)),$$

from the hypotheses we have that there exists a suitable $\tilde{t}_2 > 0$ such that

$$\phi(0) = F(u_0) \le F(u_0 + t\varphi + t^2\tilde{\psi}_0(t)) = \phi(t), \forall |t| < \tilde{t}_2,$$

also from the hypothesis we get

$$\phi'(0) = \delta F(u_0, \varphi) = 0,$$

that is,

$$\langle \varphi, F'(u_0) \rangle_U = 0, \forall \varphi \text{ such that } G'(u_0) \cdot \varphi = \theta.$$

In the next lines as usual $N[G'(u_0)]$ and $R[G'(u_0)]$ denote the null space and the range of $G'(u_0)$, respectively. Thus $F'(u_0)$ is orthogonal to the null space of $G'(u_0)$, which we denote by

$$F'(u_0) \perp N[G'(u_0)].$$

Since $R[G'(u_0)]$ is closed, we get $F'(u_0) \in R([G'(u_0)]^*)$, that is, there exists $z_0^* \in Z^*$ such that

$$F'(u_0) = [G'(u_0)]^*(-z_0^*).$$

The proof is complete.

15.4 Some examples concerning inequality constraints

In this section we assume the hypotheses of last theorem for F and G below specified. As an application of this same result, consider the problem of locally minimizing $F(u)$ subject to $G_1(u) = \theta$ and $G_2(u) \leq \theta$, where $F : U \to \mathbb{R}$, U being a function Banach space, $G_1 : U \to [L^p(\Omega)]^{m_1}$, $G_2 : U \to [L^p(\Omega)]^{m_2}$ where $1 < p < \infty$ and Ω is an appropriate subset of $\mathbb{R}^N$. We refer to the simpler case in which the partial order in $[L^p(\Omega)]^{m_2}$ is defined by $u = \{u_i\} \geq \theta$ if and only if $u_i \in L^p(\Omega)$ and $u_i(x) \geq 0$ a.e. in $\Omega, \forall i \in \{1, ..., m_2\}$.

Observe that defining

$$\tilde{F}(u,v) = F(u),$$

$$G(u,v) = \left(\{(G_1)_i(u)\}_{m_1 \times 1}, \{(G_2)_i(u) + v_i^2\}_{m_2 \times 1}\right)$$

it is clear that (locally) minimizing $\tilde{F}(u,v)$ subject to $G(u,v) = (\theta, \theta)$ is equivalent to the original problem. We clarify the domain of $\tilde{F}$ is denoted by $U \times Y$, where

$$Y = \{v \text{ measurable such that } v_i^2 \in L^p(\Omega), \forall i \in \{1, ..., m_2\}\}.$$

Therefore, if u_0 is a local minimum for the original constrained problem, then for an appropriate and easily defined v_0 we have that (u_0, v_0) is a point of local minimum for the extended constrained one, so that by the last theorem there exists a Lagrange multiplier $z_0^* = (z_1^*, z_2^*) \in [L^q(\Omega)]^{m_1} \times [L^q(\Omega)]^{m_2}$ where $1/p + 1/q = 1$ and

$$\tilde{F}'(u_0, v_0) + [G'(u_0, v_0)]^*(z_0^*) = (\theta, \theta),$$

that is,

$$F'(u_0) + [G_1'(u_0)]^*(z_1^*) + [G_2'(u_0)]^*(z_2^*) = \theta, \tag{15.50}$$

and

$$(z_2^*)_i v_{0i} = \theta, \forall i \in \{1, ..., m_2\}.$$

In particular for almost all $x \in \Omega$, if x is such that $v_{0i}(x)^2 > 0$ then $z_{2i}^*(x) = 0$, and if $v_{0i}(x) = 0$ then $(G_2)_i(u_0(x)) = 0$, so that $(z_2^*)_i(G_2)_i(u_0) = 0$, a.e. in $\Omega, \forall i \in \{1, ..., m_2\}$.

Furthermore, consider the problem of minimizing $F_1(v) = \tilde{F}(u_0, v) = F(u_0)$ subject $\{G_{2i}(u_0) + v_i^2\} = \theta$. From above such a local minimum is attained at v_0. Thus, from the stationarity of $F_1(v) + \langle z_2^*, \{(G_2)_i(u_0) - v_i^2\}\rangle_{[L^p(\Omega)]^{m_2}}$ at v_0 and the standard necessary conditions for the case of convex (in fact quadratic) constraints we get $(z_2^*)_i \geq 0$ a.e. in $\Omega, \forall i \in \{1, ..., m_2\}$, that is, $z_2^* \geq \theta$.

Summarizing, for the order in question the first order necessary optimality conditions are given by (15.50), $z_2^* \geq \theta$ and $(z_2^*)_i(G_2)_i(u_0) = \theta, \forall i \in \{1, ..., m_2\}$ (so that $\langle z_2^*, G_2(u_0)\rangle_{[L^p(\Omega)]^{m_2}} = 0$), $G_1(u_0) = \theta$, and $G_2(u_0) \leq \theta$.

Remark 15.4.1 *For the case $U = \mathbb{R}^n$ and $\mathbb{R}^{m_k}$ replacing $[L^p(\Omega)]^{m_k}$, for $k \in \{1, 2\}$ the conditions $(z_2^*)_i v_i = \theta$ means that for the constraints not active (for example $v_i \neq 0$) the corresponding coordinate $(z_2^*)_i$ of the Lagrange multiplier is 0. If $v_i = 0$ then $(G_2)_i(u_0) = 0$, so that in any case $(z_2^*)_i(G_2)_i(u_0) = 0$.*

Summarizing, for this last mentioned case we have obtained the standard necessary optimality conditions: $(z_2^*)_i \geq 0$, and $(z_2^*)_i(G_2)_i(u_0) = 0, \forall i \in \{1, ..., m_2\}$.

15.5 The Lagrange multiplier theorem for equality and inequality constraints

In this section we develop more rigorous results concerning the Lagrange multiplier theorem for the case involving equalities and inequalities.

Theorem 15.1
Let U, Z_1, Z_2 be Banach spaces. Consider a cone C in Z_2 as above specified and such that if $z_1 \leq \theta$ and $z_2 < \theta$ then $z_1 + z_2 < \theta$, where $z \leq \theta$ means that $z \in -C$ and $z < \theta$ means that $z \in (-C)^\circ$. The concerned order

supposed to be also that if $z < \theta$, $z^* \geq \theta^*$ and $z^* \neq \theta$ then $\langle z, z^* \rangle_{Z_2} < 0$. Furthermore, assume $u_0 \in U$ is a point of local minimum for $F : U \to \mathbb{R}$ subject to $G_1(u) = \theta$ and $G_2(u) \leq \theta$, where $G_1 : U \to Z_1$, $G_2 : U \to Z_2$ and F are Fréchet differentiable at $u_0 \in U$. Suppose also $G_1'(u_0)$ is onto and that there exist $\alpha > 0, K > 0$ such that if $\|\varphi\|_U < \alpha$ then

$$\|G_1'(u_0 + \varphi) - G_1'(u_0)\| \leq K\|\varphi\|_U.$$

Finally, suppose there exists $\varphi_0 \in U$ such that

$$G_1'(u_0) \cdot \varphi_0 = \theta$$

and

$$G_2'(u_0) \cdot \varphi_0 < \theta.$$

Under such hypotheses, there exists a Lagrange multiplier $z_0^* = (z_1^*, z_2^*) \in Z_1^* \times Z_2^*$ such that

$$F'(u_0) + [G_1'(u_0)]^*(z_1^*) + [G_2'(u_0)]^*(z_2^*) = \theta,$$

$$z_2^* \geq \theta^*,$$

and

$$\langle G_2(u_0), z_2^* \rangle_{Z_2} = 0.$$

Proof 15.10   Let $\varphi \in U$ be such that

$$G_1'(u_0) \cdot \varphi = \theta$$

and

$$G_2'(u_0) \cdot \varphi = v - \lambda G_2(u_0),$$

for some $v \leq \theta$ and $\lambda \geq 0$.

Select $\alpha \in (0, 1)$ and define

$$\varphi_\alpha = \alpha \varphi_0 + (1 - \alpha)\varphi.$$

Observe that $G_1(u_0) = \theta$ and $G_1'(u_0) \cdot \varphi_\alpha = \theta$ so that as in the proof of the Lagrange multiplier theorem 15.3.2 we may find $K_1 > 0$, $\varepsilon > 0$ and $\psi_0(t)$ such that

$$G_1(u_0 + t\varphi_\alpha + t^2 \psi_0(t)) = \theta, \ \forall |t| < \varepsilon,$$

and

$$\|\psi_0(t)\|_U < K_1, \forall |t| < \varepsilon.$$

Observe that

$$
\begin{aligned}
& G_2'(u_0) \cdot \varphi_\alpha \\
= \ & \alpha G_2'(u_0) \cdot \varphi_0 + (1 - \alpha) G_2'(u_0) \cdot \varphi \\
= \ & \alpha G_2'(u_0) \cdot \varphi_0 + (1 - \alpha)(v - \lambda G_2(u_0)) \\
= \ & \alpha G_2'(u_0) \cdot \varphi_0 + (1 - \alpha)v - (1 - \alpha)\lambda G_2(u_0) \\
= \ & v_0 - \lambda_0 G_2(u_0),
\end{aligned}
\tag{15.51}
$$

where,

$$\lambda_0 = (1 - \alpha)\lambda,$$

and

$$v_0 = \alpha G_2'(u_0) \cdot \varphi_0 + (1 - \alpha)v < \theta.$$

Hence, for $t > 0$

$$G_2(u_0 + t\varphi_\alpha + t^2 \psi_0(t)) = G_2(u_0) + G_2'(u_0) \cdot (t\varphi_\alpha + t^2 \psi_0(t)) + r(t),$$

where

$$\lim_{t \to 0^+} \frac{\|r(t)\|}{t} = 0.$$

Therefore from (15.51) we obtain

$$G_2(u_0 + t\varphi_\alpha + t^2\psi_0(t)) = G_2(u_0) + tv_0 - t\lambda_0 G_2(u_0) + r_1(t),$$

where

$$\lim_{t \to 0^+} \frac{\|r_1(t)\|}{t} = 0.$$

Observe that there exists $\varepsilon_1 > 0$ such that if $0 < t < \varepsilon_1 < \varepsilon$, then

$$v_0 + \frac{r_1(t)}{t} < \theta,$$

and

$$G_2(u_0) - t\lambda_0 G_2(u_0) = (1 - t\lambda_0)G_2(u_0) \leq \theta.$$

Hence

$$G_2(u_0 + t\varphi_\alpha + t^2\psi_0(t)) < \theta, \ \text{ if } 0 < t < \varepsilon_1.$$

From this there exists $0 < \varepsilon_2 < \varepsilon_1$ such that

$$\begin{aligned}
&F(u_0 + t\varphi_\alpha + t^2\psi_0(t)) - F(u_0) \\
&= \langle t\varphi_\alpha + t^2\psi_0(t), F'(u_0)\rangle_U + r_2(t) \geq 0,
\end{aligned} \tag{15.52}$$

where

$$\lim_{t \to 0^+} \frac{|r_2(t)|}{t} = 0.$$

Dividing the last inequality by $t > 0$ we get

$$\langle \varphi_\alpha + t\psi_0(t), F'(u_0)\rangle_U + r_2(t)/t \geq 0, \forall 0 < t < \varepsilon_2.$$

Letting $t \to 0^+$ we obtain

$$\langle \varphi_\alpha, F'(u_0)\rangle_U \geq 0.$$

Letting $\alpha \to 0^+$, we get

$$\langle \varphi, F'(u_0)\rangle_U \geq 0,$$

if

$$G_1'(u_0) \cdot \varphi = \theta,$$

and

$$G_2'(u_0) \cdot \varphi = v - \lambda G_2(u_0),$$

for some $v \leq \theta$ and $\lambda \geq 0$. Define

$$\begin{aligned}
A =\ & \{((\langle \varphi, F'(u_0)\rangle_U + r, G_1'(u_0) \cdot \varphi, G_2'(u_0)\varphi - v + \lambda G(u_0)), \\
& \varphi \in U, r \geq 0, v \leq \theta, \lambda \geq 0\}.
\end{aligned} \tag{15.53}$$

Observe that A is a convex set with a non-empty interior.

If

$$G_1'(u_0) \cdot \varphi = \theta,$$

and

$$G_2'(u_0) \cdot \varphi - v + \lambda G_2(u_0) = \theta,$$

with $v \leq \theta$ and $\lambda \geq 0$ then

$$\langle \varphi, F'(u_0) \rangle_U \geq 0,$$

so that

$$\langle \varphi, F'(u_0) \rangle_U + r \geq 0.$$

Moreover, if

$$\langle \varphi, F'(u_0) \rangle + r = 0,$$

with $r \geq 0$,

$$G'_1(u_0) \cdot \varphi = \theta,$$

and

$$G'_2(u_0) \cdot \varphi - v + \lambda G_2(u_0) = \theta,$$

with $v \leq \theta$ and $\lambda \geq 0$, then we have

$$\langle \varphi, F'(u_0) \rangle_U \geq 0,$$

so that

$$\langle \varphi, F'(u_0) \rangle_U = 0,$$

and $r = 0$. Hence $(0, \theta, \theta)$ is on the boundary of A. Therefore, by the Hahn-Banach theorem, geometric form, there exists

$$(\beta, z_1^*, z_2^*) \in \mathbb{R} \times Z_1^* \times Z_2^*$$

such that

$$(\beta, z_1^*, z_2^*) \neq (0, \theta, \theta)$$

and

$$\beta(\langle \varphi, F'(u_0) \rangle_U + r) \quad + \quad \langle G'_1(u_0) \cdot \varphi, z_1^* \rangle_{Z_1}$$
$$+ \quad \langle G'_2(u_0) \cdot \varphi - v + \lambda G_2(u_0), z_2^* \rangle_{Z_2} \geq 0, \tag{15.54}$$

$\forall \varphi \in U, r \geq 0, v \leq \theta, \lambda \geq 0$. Suppose $\beta = 0$. Fixing all variable except v we get $z_2^* \geq \theta$. Thus, for $\varphi = c\varphi_0$ with arbitrary $c \in \mathbb{R}, v = \theta, \lambda = 0$, if $z_2^* \neq \theta$, then $\langle G'_2(u_0) \cdot \varphi_0, z_2^* \rangle_{Z_2} < 0$, so that we get $z_2^* = \theta$. Since $G'_1(u_0)$ is onto, a similar reasoning lead us to $z_1^* = \theta$, which contradicts $(\beta, z_1^*, z_2^*) \neq (0, \theta, \theta)$.

Hence, $\beta \neq 0$, and fixing all variables except r we obtain $\beta > 0$. There is no loss of generality in assuming $\beta = 1$.

Again fixing all variables except v, we obtain $z_2^* \geq \theta$. Fixing all variables except λ, since $G_2(u_0) \leq \theta$ we get

$$\langle G_2(u_0), z_2^* \rangle_{Z_2} = 0.$$

Finally, for $r = 0, v = \theta, \lambda = 0$, we get

$$\langle \varphi, F'(u_0) \rangle_U + \langle G'_1(u_0)\varphi, z_1^* \rangle_{Z_1} + \langle G'_2(u_0) \cdot \varphi, z_2^* \rangle_{Z_2} \geq 0, \ \forall \varphi \in U,$$

that is, since obviously such an inequality is valid also for $-\varphi, \forall \varphi \in U$, we obtain

$$\langle \varphi, F'(u_0) \rangle_U + \langle \varphi, [G'_1(u_0)]^*(z_1^*) \rangle_U + \langle \varphi, [G'_2(u_0)]^*(z_2^*) \rangle_U = 0, \ \forall \varphi \in U,$$

so that

$$F'(u_0) + [G'_1(u_0)]^*(z_1^*) + [G'_2(u_0)]^*(z_2^*) = \theta.$$

The proof is complete.

15.6 Second order necessary conditions

In this section we establish second order necessary conditions for a class of constrained problems in Banach spaces. We highlight the next result is particularly applicable to optimization in $\mathbb{R}^n$.

Theorem 15.2
Let U, Z_1, Z_2 be Banach spaces. Consider a cone C in Z_2 as above specified and such that if $z_1 \leq \theta$ and $z_2 < \theta$ then $z_1 + z_2 < \theta$, where $z \leq \theta$ means that $z \in -C$ and $z < \theta$ means that $z \in (-C)^\circ$. The concerned order is supposed to be also that if $z < \theta$, $z^* \geq \theta^*$ and $z^* \neq \theta$ then $\langle z, z^* \rangle_{Z_2} < 0$. Furthermore, assume $u_0 \in U$ is a point of local minimum for $F : U \to \mathbb{R}$ subject to $G_1(u) = \theta$ and $G_2(u_0) \leq \theta$, where $G_1 : U \to Z_1$, $G_2 : U \to (Z_2)^k$ and F are twice Fréchet differentiable at $u_0 \in U$. Assume $G_2(u) = \{(G_2)_i(u)\}$ where $(G_2)_i : U \to Z_2$, $\forall i \in \{1, ..., k\}$ and define

$$A = \{i \in \{1, ..., k\} \; : \; (G_2)_i(u_0) = \theta\},$$

and also suppose that $(G_2)_i(u_0) < \theta$, if $i \notin A$. Moreover, suppose $\{G_1'(u_0), \{(G_2)_i'(u_0)\}_{i \in A}\}$ is onto and that there exist $\alpha > 0, K > 0$ such that if $\|\varphi\|_U < \alpha$ then

$$\|\tilde{G}'(u_0 + \varphi) - \tilde{G}'(u_0)\| \leq K \|\varphi\|_U,$$

where

$$\tilde{G}(u) = \{G_1(u), \{(G_2)_i(u)\}_{i \in A}\}.$$

Finally, suppose there exists $\varphi_0 \in U$ such that

$$G_1'(u_0) \cdot \varphi_0 = \theta$$

and

$$G_2'(u_0) \cdot \varphi_0 < \theta.$$

Under such hypotheses, there exists a Lagrange multiplier $z_0^* = (z_1^*, z_2^*) \in Z_1^* \times (Z_2^*)^k$ such that

$$F'(u_0) + [G_1'(u_0)]^*(z_1^*) + [G_2'(u_0)]^*(z_2^*) = \theta,$$

$$z_2^* \geq (\theta^*, ..., \theta^*) \equiv \theta_k^*,$$

and

$$\langle (G_2)_i(u_0), (z_2^*)_i \rangle_Z = 0, \forall i \in \{1, ..., k\},$$

$$(z_2^*)_i = \theta^*, \; if \; i \notin A,$$

Moreover, defining

$$L(u, z_1^*, z_2^*) = F(u) + \langle G_1(u), z_1^* \rangle_{Z_1} + \langle G_2(u), z_2^* \rangle_{Z_2},$$

we have that

$$\delta_{uu}^2 L(u_0, z_1^*, z_2^*; \varphi) \geq 0, \forall \varphi \in \mathcal{V}_0,$$

where

$$\mathcal{V}_0 = \{\varphi \in U \; : \; G_1'(u_0) \cdot \varphi = \theta, \; (G_2)_i'(u_0) \cdot \varphi = \theta, \; \forall i \in A\}.$$

Proof 15.11 Observe that A is defined by

$$A = \{i \in \{1, ..., k\} \; : \; (G_2)_i(u_0) = \theta\}.$$

Observe also that $(G_2)_i(u_0) < \theta$, if $i \notin A$.

Hence the point $u_0 \in U$ is a local minimum for $F(u)$ under the constraints

$$G_1(u) = \theta, \text{ and } (G_2)_i(u) \le \theta, \forall i \in A.$$

From the last Theorem 15.1 for such an optimization problem there exists a Lagrange multiplier $(z_1^*, \{(z_2^*)_{i \in A}\})$ such that $(z_2^*)_i \ge \theta^*$, $\forall i \in A$, and

$$F'(u_0) + [G_1'(u_0)]^*(z_1^*) + \sum_{i \in A} [(G_2)_i'(u_0)]^*((z_2^*)_i) = \theta. \tag{15.55}$$

The choice $(z_2^*)_i = \theta$, if $i \notin A$ leads to the existence of a Lagrange multiplier $(z_1^*, z_2^*) = (z_1^*, \{(z_2^*)_{i \in A}, (z_2^*)_{i \notin A}\})$ such that

$$z_2^* \ge \theta_k^*$$

and

$$\langle (G_2)_i(u_0), (z_2^*)_i \rangle_Z = 0, \forall i \in \{1, ..., k\}.$$

Let $\varphi \in \mathcal{V}_0$, that is, $\varphi \in U$,

$$G_1'(u_0)\varphi = \theta$$

and

$$(G_2)_i'(u_0) \cdot \varphi = \theta, \forall i \in A.$$

Recall that $\tilde{G}(u) = \{G_1(u), (G_2)_{i \in A}(u)\}$ and therefore, similarly as in the proof of the Lagrange multiplier theorem 15.3.2, we may obtain $\psi_0(t), K > 0$ and $\varepsilon > 0$ such that

$$\tilde{G}(u_0 + t\varphi + t^2\psi_0(t)) = \theta, \forall |t| < \varepsilon,$$

and

$$\|\psi_0(t)\| \le K, \forall |t| < \varepsilon.$$

Also, if $i \notin A$, we have that $(G_2)_i(u_0) < \theta$, so that

$$(G_2)_i(u_0 + t\varphi + t^2\psi_0(t)) = (G_2)_i(u_0) + G_i'(u_0) \cdot (t\varphi + t^2\psi_0(t)) + r(t),$$

where

$$\lim_{t \to 0} \frac{\|r(t)\|}{t} = 0,$$

that is,

$$(G_2)_i(u_0 + t\varphi + t^2\psi_0(t)) = (G_2)_i(u_0) + t(G_2)_i'(u_0) \cdot \varphi + r_1(t),$$

where,

$$\lim_{t \to 0} \frac{\|r_1(t)\|}{t} = 0,$$

and hence there exists, $0 < \varepsilon_1 < \varepsilon$, such that

$$(G_2)_i(u_0 + t\varphi + t^2\psi_0(t)) < \theta, \forall |t| < \varepsilon_1 < \varepsilon.$$

Therefore, since u_0 is a point of local minimum under the constraint $G(u) \le \theta$, there exists $0 < \varepsilon_2 < \varepsilon_1$, such that

$$F(u_0 + t\varphi + t^2\psi_0(t)) - F(u_0) \ge 0, \forall |t| < \varepsilon_2,$$

so that,

$$
\begin{aligned}
& F(u_0 + t\varphi + t^2\psi_0(t)) - F(u_0) \\
={} & F(u_0 + t\varphi + t^2\psi_0(t)) - F(u_0) \\
& + \langle G_1(u_0 + t\varphi + t^2\psi_0(t)), z_1^* \rangle_{Z_1} + \sum_{i \in A} \left\{ \langle (G_2)_i(u_0 + t\varphi + t^2\psi_0(t)), (z_2^*)_i \rangle_{Z_2} \right\} \\
& - \langle G_1(u_0), z_1^* \rangle_{Z_1} - \sum_{i \in A} \left\{ \langle (G_2)_i(u_0), (z_2^*)_i \rangle_{Z_2} \right\} \\
={} & F(u_0 + t\varphi + t^2\psi_0(t)) - F(u_0) \\
& + \langle G_1(u_0 + t\varphi + t^2\psi_0(t)), z_1^* \rangle_{Z_1} - \langle G_1(u_0), z_1^* \rangle_{Z_1} \\
& + \langle G_2(u_0 + t\varphi + t^2\psi_0(t)), z_2^* \rangle_{Z_2} - \langle G_2(u_0), z_2^* \rangle_{Z_2} \\
={} & L(u_0 + t\varphi + t^2\psi_0(t)), z_1^*, z_2^*) - L(u_0, z_1^*, z_2^*) \\
={} & \delta_u L(u_0, z_1^*, z_2^*; t\varphi + t^2\psi_0(t)) + \frac{1}{2}\delta_{uu}^2 L(u_0, z_1^*, z_2^*; t\varphi + t^2\psi_0(t)) + r_2(t) \\
={} & \frac{t^2}{2}\delta_{uu}^2 L(u_0, z_1^*, z_2^*; \varphi + t\psi_0(t)) + r_2(t) \geq 0, \forall |t| < \varepsilon_2.
\end{aligned}
$$

where

$$
\lim_{t \to 0} |r_2(t)|/t^2 = 0.
$$

To obtain the last inequality we have used

$$
\delta_u L(u_0, z_1^*, z_2^*; t\varphi + t^2\psi_0(t)) = 0
$$

Dividing the last inequality by $t^2 > 0$ we obtain

$$
\frac{1}{2}\delta_{uu}^2 L(u_0, z_1^*, z_2^*; \varphi + t\psi_0(t)) + r_2(t)/t^2 \geq 0, \forall 0 < |t| < \varepsilon_2,
$$

and finally, letting $t \to 0$ we get

$$
\frac{1}{2}\delta_{uu}^2 L(u_0, z_1^*, z_2^*; \varphi) \geq 0.
$$

The proof is complete.

15.7 On the Banach fixed point theorem

Now we recall a classical definition namely the Banach fixed theorem also known as the contraction mapping theorem

Definition 15.7.1 *Let C be a subset of a Banach space U and let $T : C \to C$ be an operator. Thus T is said to be a contraction mapping if there exists $0 \leq \alpha < 1$ such that*

$$
\|T(u) - T(v)\|_U \leq \alpha \|u - v\|_U, \forall u, v \in C.
$$

Remark 15.7.2 *Observe that if $\|T'(u)\|_U \leq \alpha < 1$, on a convex set C then T is a contraction mapping, since by the mean value inequality,*

$$
\|T(u) - T(v)\|_U \leq \sup_{u \in C} \{\|T'(u)\|\} \|u - v\|_U, \forall u, v \in C.
$$

The next result is the base of our generalized method of lines.

Theorem 15.7.3 (Contraction mapping theorem) *Let C be a closed subset of a Banach space U. Assume T is contraction mapping on C, then there exists a unique $u_0 \in C$ such that $u_0 = T(u_0)$. Moreover, for an arbitrary $u_1 \in C$ defining the sequence*

$$u_2 = T(u_1) \text{ and } u_{n+1} = T(u_n), \forall n \in \mathbb{N}$$

we have

$$u_n \to u_0, \text{ in norm, as } n \to +\infty.$$

Proof 15.12   Let $u_1 \in C$. Let $\{u_n\} \subset C$ be defined by

$$u_{n+1} = T(u_n), \forall n \in \mathbb{N}.$$

Hence, reasoning inductively

$$
\begin{aligned}
\|u_{n+1} - u_n\|_U &= \|T(u_n) - T(u_{n-1})\|_U \\
&\leq \alpha \|u_n - u_{n-1}\|_U \\
&\leq \alpha^2 \|u_{n-1} - u_{n-2}\|_U \\
&\leq \quad \ldots\ldots \\
&\leq \alpha^{n-1} \|u_2 - u_1\|_U, \forall n \in \mathbb{N}.
\end{aligned}
\tag{15.56}
$$

Thus, for $p \in \mathbb{N}$ we have

$$
\begin{aligned}
&\|u_{n+p} - u_n\|_U \\
={}& \|u_{n+p} - u_{n+p-1} + u_{n+p-1} - u_{n+p-2} + \ldots - u_{n+1} + u_{n+1} - u_n\|_U \\
\leq{}& \|u_{n+p} - u_{n+p-1}\|_U + \|u_{n+p-1} - u_{n+p-2}\|_U + \ldots + \|u_{n+1} - u_n\|_U \\
\leq{}& (\alpha^{n+p-2} + \alpha^{n+p-3} + \ldots + \alpha^{n-1})\|u_2 - u_1\|_U \\
\leq{}& \alpha^{n-1}(\alpha^{p-1} + \alpha^{p-2} + \ldots + \alpha^0)\|u_2 - u_1\|_U \\
\leq{}& \alpha^{n-1}\left(\sum_{k=0}^{\infty} \alpha^k\right)\|u_2 - u_1\|_U \\
\leq{}& \frac{\alpha^{n-1}}{1-\alpha}\|u_2 - u_1\|_U
\end{aligned}
\tag{15.57}
$$

Denoting $n + p = m$, we obtain

$$\|u_m - u_n\|_U \leq \frac{\alpha^{n-1}}{1-\alpha}\|u_2 - u_1\|_U, \forall m > n \in \mathbb{N}.$$

Let $\varepsilon > 0$. Since $0 \leq \alpha < 1$ there exists $n_0 \in \mathbb{N}$ such that if $n > n_0$ then

$$0 \leq \frac{\alpha^{n-1}}{1-\alpha}\|u_2 - u_1\|_U < \varepsilon,$$

so that

$$\|u_m - u_n\|_U < \varepsilon, \text{ if } m > n > n_0.$$

From this we may infer that $\{u_n\}$ is a Cauchy sequence, and since U is a Banach space, there exists $u_0 \in U$ such that

$$u_n \to u_0, \text{ in norm, as } n \to \infty.$$

Observe that

$$
\begin{aligned}
\|u_0 - T(u_0)\|_U &= \|u_0 - u_n + u_n - T(u_0)\|_U \\
&\leq \|u_0 - u_n\|_U + \|u_n - T(u_0)\|_U \\
&\leq \|u_0 - u_n\|_U + \alpha\|u_{n-1} - u_0\|_U \\
&\to 0, \text{ as } n \to \infty.
\end{aligned}
\tag{15.58}
$$

Thus $\|u_0 - T(u_0)\|_U = 0$.

Finally, we prove the uniqueness. Suppose $u_0, v_0 \in C$ are such that

$$
u_0 = T(u_0) \text{ and } v_0 = T(v_0).
$$

Hence,

$$
\begin{aligned}
\|u_0 - v_0\|_U &= \|T(u_0) - T(v_0)\|_U \\
&\leq \alpha\|u_0 - v_0\|_U.
\end{aligned}
\tag{15.59}
$$

From this we get

$$
\|u_0 - v_0\|_U \leq 0,
$$

that is

$$
\|u_0 - v_0\|_U = 0.
$$

The proof is complete.

15.8   Sensitivity analysis

15.8.1   *Introduction*

In this section we state and prove the implicit function theorem for Banach spaces. A similar result may be found in Ito and Kunisch [54], page 31.

We emphasize the result found in [54] is more general however, the proof present here is almost the same for a simpler situation. The general result found in [54] is originally from Robinson [63].

15.9   The implicit function theorem

Theorem 15.9.1 *Let V, U, W be Banach spaces. Let $F : V \times U \to W$ be a functions such that*

$$
F(x_0, u_0) = \mathbf{0},
$$

where $(x_0, u_0) \in V \times U$.

Assume there exists $r > 0$ such that F is Fréchet differentiable and $F_x(x, u)$ is continuous in (x, u) in $B_r(x_0, u_0)$.

Suppose also $[F_x(x_0, u_0)]^{-1}$ exists and is bounded so that there exists $\rho > 0$ such that

$$
0 < \|[F_x(x_0, u_0)]^{-1}\| \leq \rho.
$$

Under such hypotheses, there exist $0 < \varepsilon_1 < r/2$ and $0 < \varepsilon_2 < 1$ such that for each $u \in B_{\varepsilon_1}(u_0)$, there exists $x \in B_{\varepsilon_2}(x_0)$ such that

$$
F(x, u) = \mathbf{0},
$$

where we denote $x = x(u)$ so that

$$
F(x(u), u) = \mathbf{0}.
$$

Moreover, there exists $\delta > 0$ *such that* $0 < \delta\rho < 1$, *such that for each* $u, v \in B_{\varepsilon_1}(u_0)$ *we have*

$$\|x(u) - x(v)\| \leq \frac{\rho^2 \delta}{1 - \rho\delta} \|F(x(v), u) - F(x(v), v)\|.$$

Finally, if there exists $K > 0$ *such that*

$$\|F_u(x, u)\| \leq K, \ \forall (x, u) \in B_{\varepsilon_2}(x_0) \times B_{\varepsilon_1}(u_0)$$

so that

$$\|F(x, u) - F(x, v)\| \leq K\|u - v\|, \ \forall (x, u) \in B_{\varepsilon_2}(x_0) \times B_{\varepsilon_1}(u_0),$$

then

$$\|x(u) - x(v)\| \leq K_1 \|u - v\|,$$

where

$$K_1 = K \frac{\rho^2 \delta}{1 - \delta\rho}.$$

Proof 15.13   Let $0 < \varepsilon < r/2$. Choose $\delta > 0$ such that

$$0 < \rho\delta < 1.$$

Define
$$T(x) = F(x_0, u_0) + F_x(x_0, u_0)(x - x_0) = F_x(x_0, u_0)(x - x_0),$$

and
$$h(x, u) = F(x_0, u_0) + F_x(x_0, u_0)(x - x_0) - F(x, u).$$

Choose $0 < \varepsilon_3 < r/2$ and $0 < \varepsilon_2 < 1$ such that

$$B_{\varepsilon_2}(u_0) \times B_{\varepsilon_3}(x_0) \subset B_r(x_0, u_0)$$

and if $(x, u) \in B_{\varepsilon_2}(x_0) \times B_{\varepsilon_3}(u_0)$ then

$$\|F_x(x, u) - F_x(x_0, u_0)\| < \frac{\delta}{2}.$$

Select $0 < \varepsilon_1 < \varepsilon_3$ such that if $u \in B_{\varepsilon_1}(u_0)$, then

$$\rho\|F(x_0, u) - F(x_0, u_0)\| < (1 - \rho\delta)\varepsilon_2.$$

For each $u \in B_{\varepsilon_1}(u_0)$ define
$$\phi_u(x) = T^{-1}(h(x, u)).$$

Fix $u \in B_{\varepsilon_1}(u_0)$.
Observe that for $x_1, x_2 \in B_{\varepsilon_2}(x_0)$ we have

$$
\begin{aligned}
\|\phi_u(x_1) - \phi_u(x_2)\| &\leq \|[F_x(x_0, u_0)]^{-1}\|\|h(x_1, u) - h(x_2, u)\| \\
&\leq \rho\|h(x_1, u) - h(x_2, u)\| \\
&\leq \rho \sup_{t \in [0,1]} \|h_x(tx_1 + (1-t)x_2, u)\|\|x_1 - x_2\| \\
&\leq \rho 2 \frac{\delta}{2}\|x_1 - x_2\| \\
&= \rho\delta\|x_1 - x_2\|.
\end{aligned}
\tag{15.60}
$$

Observe that since $0 < \rho\delta < 1$ we have that ϕ_u is a candidate to be a contraction mapping. Observe also that

$$x_0 = T^{-1}(\mathbf{0}),$$

so that

$$T(x_0) = \mathbf{0} = h(x_0, u_0)$$

and

$$x_0 = T^{-1}(h(x_0, u_0)).$$

Thus,

$$
\begin{aligned}
\|\phi_u(x_0) - x_0\| &\leq \rho\|h(x_0, u) - h(x_0, u_0)\| \\
&= \rho\|h(x_0, u)\| \\
&\leq \rho\|F(x_0, u) - F(x_0, u_0)\| \\
&\leq (1 - \rho\delta)\varepsilon_2.
\end{aligned}
\tag{15.61}
$$

On the other hand, for each $x \in B_{\varepsilon_2}(x_0)$, we have that

$$
\begin{aligned}
\|\phi_u(x) - x_0\| &= \|\phi_u(x) - \phi_u(x_0) + \phi_u(x_0) - x_0\| \\
&\leq \|\phi_u(x) - \phi_u(x_0)\| + \|\phi_u(x_0) - x_0\| \\
&\leq \rho\delta\|x - x_0\| + (1 - \rho\delta)\varepsilon_2 \\
&< \rho\delta\varepsilon_2 + (1 - \rho\delta)\varepsilon_2 \\
&= \varepsilon_2.
\end{aligned}
\tag{15.62}
$$

From this we may infer that

$$\phi_u(x) \in B_{\varepsilon_2}(x_0), \ \forall x \in B_{\varepsilon_2}(x_0)$$

so that indeed ϕ_u is a contraction mapping.

Therefore, from the Banach fixed point theorem, there exists a unique fixed point $x = x(u)$ for $\phi_u(x)$, so that

$$
\begin{aligned}
x(u) &= \phi_u(x(u)) \\
&= T^{-1}(h(x(u), u)) \\
&= T^{-1}(F_x(x_0, u_0)(x(u) - x_0) - F(x(u), u))) \\
&= [F_x(x_0, u_0)]^{-1}[F_x(x_0, u_0)(x(u) - x_0) - F(x(u), u)] + x_0 \\
&= x(u) - x_0 + x_0 - [F_x(x_0, u_0)]^{-1}(F(x(u), u)) \\
&= x(u) - [F_x(x_0, u_0)]^{-1}(F(x(u), u)).
\end{aligned}
\tag{15.63}
$$

From this, we have

$$[F_x(x_0, u_0)]^{-1}(F(x(u), u)) = \mathbf{0},$$

so that

$$F(x(u), u) = F_x(x_0, u_0)\mathbf{0} = \mathbf{0},$$

that is,

$$F(x(u), u) = \mathbf{0}.$$

Let $x \in B_{\varepsilon_2}(x_0)$ and $u \in B_{\varepsilon_1}(u_0)$. Thus

$$\|\phi_u(x) - x\| \leq \|\phi_u(x)\| + \|x\| < 2\varepsilon_2.$$

Moreover,

$$\|\phi_u(x_1) - \phi_u(x_2)\| < \rho\delta\|x_2 - x_1\|, \forall x_1, x_2 \in B_{\varepsilon_2}(x_0).$$

From these last two results we obtain

$$\|\phi_u^2(x) - \phi_u(x)\| < 2\varepsilon_2(\rho\delta),$$

and reasoning inductively,

$$\|\phi_u^{n+1}(x) - \phi_u^n(x)\| < 2\varepsilon_2(\rho\delta)^n, \ \forall n \in \mathbb{N}.$$

Therefore

$$
\begin{aligned}
\|\phi_u^{n+1}(x) - x\| &= \|\phi_u^{n+1}(x) - \phi_u^n(x) + \phi_u^n(x) - \cdots + \phi_u(x) - x\| \\
&\leq \|\phi_u^{n+1}(x) - \phi_u^n(x)\| + \|\phi_u^n(x) - \phi_u^{n-1}(x)\| + \cdots + \|\phi_u(x) - x\| \\
&\leq \sum_{j=1}^{n+1} (\rho\delta)^j \|\phi_u(x) - x\| \\
&\leq \frac{\rho\delta}{1 - \rho\delta} \|\phi_u(x) - x\|, \ \forall n \in \mathbb{N}
\end{aligned}
\tag{15.64}
$$

Letting $n \to \infty$, we obtain

$$\|x(u) - x\| \leq \frac{\rho\delta}{1 - \rho\delta} \|\phi_u(x) - x\|.$$

In particular for $v \in B_{\varepsilon_1}(u_0)$ and $x = x(v)$, we get

$$
\begin{aligned}
\|x(u) - x(v)\| &\leq \frac{\rho\delta}{1 - \rho\delta} \|\phi_u(x(v)) - x(v)\| \\
&= \frac{\rho\delta}{1 - \rho\delta} \|\phi_u(x(v)) - \phi_v(x(v))\| \\
&\leq \frac{\rho^2\delta}{1 - \rho\delta} \|h(x(v), u) - h(x(v), v)\| \\
&\leq \frac{\rho^2\delta}{1 - \rho\delta} \|F(x(v), u) - F(x(v), v)\| \\
&\leq K\frac{\rho^2\delta}{1 - \rho\delta} \|u - v\| \\
&= K_1 \|u - v\|, \ \forall u, v \in B_{\varepsilon_1}(u_0).
\end{aligned}
\tag{15.65}
$$

The proof is complete.

Corollary 15.9.2 *Consider the hypotheses and statements of the last theorem. Moreover, assume $F_x : V \times U \to W$ is such that $[F_x(x, u)]^{-1}$ exists and it is bounded in $B_r(x_0, u_0)$.*

Suppose also, F is Fréchet differentiable in $B_r(x_0, u_0)$.

Let $\varphi \in U$.

Under such hypotheses,

$$x'(u, \varphi) = -[F_x(x(u), u)]^{-1}[F_u(x(u), u)](\varphi),$$

where

$$x'(u, \varphi) = \lim_{t \to 0} \frac{x(u + t\varphi) - x(u)}{t}.$$

Proof 15.14   Observe that

$$F(x(u), u) = \mathbf{0}, \text{ in } B_{\varepsilon_1}(u_0).$$

Let $u \in B_{\varepsilon_1}(u_0)$.

Let $t_0 > 0$ be such that

$$u + t\varphi \in B_{\varepsilon_1}(u_0), \ \forall |t| < t_0.$$

Observe that

$$F(x(u+t\varphi), u+t\varphi) - F(x(u), u) = \mathbf{0}, \ \forall |t| < t_0.$$

From the Fréchet differentiability of F at $(x(u), u)$, for $0 < |t| < t_0$, we obtain

$$\begin{aligned}F_x(x(u), u) \cdot (x(u+t\varphi) - x(u)) + F_u(x(u), u)(t\varphi) \\ + W(u, \varphi, t)(\|x(u+t\varphi) - x(u)\| + |t|\|\varphi\|) = \mathbf{0},\end{aligned} \tag{15.66}$$

where W is such that

$$W(u, \varphi, t) \to \mathbf{0}, \text{ as } t \to 0,$$

since

$$x(u+t\varphi) - x(u) \to \mathbf{0}$$

and

$$t\varphi \to \mathbf{0}, \text{ as } t \to 0.$$

From this we obtain

$$\begin{aligned}\frac{x(u+t\varphi) - x(u)}{t} &= -[F_x(x(u), u)]^{-1}[[F_u(x(u), u)](\varphi) + \mathbf{r}(u, \varphi, t)] \\ &\to -[F_x(x(u), u)]^{-1}[F_u(x(u), u)](\varphi), \text{ as } t \to 0,\end{aligned} \tag{15.67}$$

since

$$\begin{aligned}\|\mathbf{r}(u, \varphi, t)\| &\leq \|W(u, \varphi, t)\| \left| \frac{\|x(u+t\varphi) - x(u)\|}{t} + \|\varphi\| \right| \\ &\leq \|W(u, \varphi, t)\|(K_1\|\varphi\| + \|\varphi\|) \\ &\to 0, \text{ as } t \to 0.\end{aligned} \tag{15.68}$$

Summarizing,

$$\begin{aligned}x'(u, \varphi) &= \lim_{t \to 0} \frac{x(u+t\varphi) - x(u)}{t} \\ &= -[F_x(x(u), u)]^{-1}[F_u(x(u), u)](\varphi).\end{aligned} \tag{15.69}$$

The proof is complete.

15.9.1 The main results about Gâteaux differentiability

Again let V, U be Banach spaces and let $F : V \times U \to \mathbb{R}$ be a functional. Fix $u \in U$ and consider the problem of minimizing $F(x, u)$ subject to $G(x, u) \leq \theta$ and $H(x, u) = \theta$. Here the order and remaining details on the primal formulation are the same as those indicated in section 15.4.

Hence, for the specific case in which

$$G : V \times U \to [L^p(\Omega)]^{m_1}$$

and

$$H : V \times U \to [L^p(\Omega)]^{m_2},$$

(the cases in which the co-domains of G and H are $\mathbb{R}^{m_1}$ and $\mathbb{R}^{m_2}$ respectively are dealt similarly) we redefine the concerned optimization problem, again for a fixed $u \in U$, by minimizing $F(x,u)$ subject to

$$\{G_i(x,u) + v_i^2\} = \theta,$$

and

$$H(x,u) = \theta.$$

At this point we assume $F(x,u)$, $\tilde{G}(x,u,v) = \{G_i(x,u) + v_i^2\} \equiv G(u) + v^2$ (from now on we use this general notation) and $H(x,u)$ satisfy the hypotheses of the Lagrange multiplier theorem 15.3.2.

Hence, for the fixed $u \in U$ we assume there exists an optimal $x \in V$ which locally minimizes $F(x,u)$ under the mentioned constraints.

From Theorem 15.3.2 there exist Lagrange multipliers λ_1, λ_2 such that denoting $[L^p(\Omega)]^{m_1}$ and $[L^p(\Omega)]^{m_2}$ simply by L^p, and defining

$$\tilde{F}(x,u,\lambda_1,\lambda_2,v) = F(x,u) + \langle \lambda_1, G(u) + v^2 \rangle_{L^p} + \langle \lambda_2, H(x,u) \rangle_{L^p},$$

the following necessary conditions hold,

$$\tilde{F}_x(x,u) = F_x(x,u) + \lambda_1 \cdot G_x(x,u) + \lambda_2 \cdot H_x(x,u) = \theta, \tag{15.70}$$

$$G(x,u) + v^2 = \theta, \tag{15.71}$$

$$\lambda_1 \cdot v = \theta, \tag{15.72}$$

$$\lambda_1 \geq \theta, \tag{15.73}$$

$$H(x,u) = \theta. \tag{15.74}$$

Clarifying the dependence on u, we denote the solution $x, \lambda_1, \lambda_2, v$ by $x(u), \lambda_1(u), \lambda_2(u), v(u)$, respectively. In particular, we assume that for a $u_0 \in U$, $x(u_0), \lambda_1(u_0), \lambda_2(u_0), v(u_0)$ satisfy the hypotheses of the implicit function theorem. Thus, for any u in an appropriate neighborhood of u_0, the corresponding $x(u), \lambda_1(u), \lambda_2(u), v(u)$ are uniquely defined.

We emphasize that from now on the main focus of our analysis is to evaluate variations of the optimal $x(u), \lambda_1(u), \lambda_2(u), v(u)$ with variations of u in a neighborhood of u_0.

For such an analysis, the main tool is the implicit function theorem and its main hypothesis is satisfied through the invertibility of the matrix of Fréchet second derivatives.

Hence, denoting, $x_0 = x(u_0), (\lambda_1)_0 = \lambda_1(u_0), (\lambda_2)_0 = \lambda_2(u_0), v_0 = v(u_0)$, and

$$A_1 = F_x(x_0,u_0) + (\lambda_1)_0 \cdot G_x(x_0,u_0) + (\lambda_2)_0 \cdot H_x(x_0,u_0),$$

$$A_2 = G(x_0,u_0) + v_0^2$$

$$A_3 = H(x_0,u_0),$$

$$A_4 = (\lambda_1)_0 \cdot v_0,$$

we reiterate to assume that

$$A_1 = \theta, \ A_2 = \theta, \ A_3 = \theta, \ A_4 = \theta,$$

and M^{-1} to represent a bounded linear operator, where

$$M = \begin{bmatrix} (A_1)_x & (A_1)_{\lambda_1} & (A_1)_{\lambda_2} & (A_1)_v \\ (A_2)_x & (A_2)_{\lambda_1} & (A_2)_{\lambda_2} & (A_2)_v \\ (A_3)_x & (A_3)_{\lambda_1} & (A_3)_{\lambda_2} & (A_3)_v \\ (A_4)_x & (A_4)_{\lambda_1} & (A_4)_{\lambda_2} & (A_4)_v \end{bmatrix} \tag{15.75}$$

where the derivatives are evaluated at $(x_0, u_0, (\lambda_1)_0, (\lambda_2)_0, v_0)$, so that,

$$M = \begin{bmatrix} A & G_x(x_0, u_0) & H_x(x_0, u_0) & \theta \\ G_x(x_0, u_0) & \theta & \theta & 2v_0 \\ H_x(x_0, u_0) & \theta & \theta & \theta \\ \theta & v_0 & \theta & (\lambda_1)_0 \end{bmatrix} \tag{15.76}$$

where

$$A = F_{xx}(x_0, u_0) + (\lambda_1)_0 \cdot G_{xx}(x_0, u_0) + (\lambda_2)_0 \cdot H_{xx}(x_0, u_0).$$

Moreover, also from the implicit function theorem,

$$\|(x(u), \lambda_1(u), \lambda_2(u), v(u)) - (x(u_0), \lambda_1(u_0), \lambda_2(u_0), v(u_0))\| \le K\|u - u_0\|, \tag{15.77}$$

for some appropriate $K > 0$, $\forall u \in B_r(u_0)$, for some $r > 0$.

We highlight to have denoted $\lambda(u) = (\lambda_1(u), \lambda_2(u))$.

Let $\varphi \in [C^\infty(\Omega)]^k \cap U$, where k depends on the vectorial expression of U.

At this point we will be concerned with the following Gâteaux variation evaluation

$$\delta_u \tilde{F}(x(u_0), u_0, \lambda(u_0), v(u_0); \varphi).$$

Observe that

$$\delta_u \tilde{F}(x(u_0), u_0, \lambda(u_0), v(u_0); \varphi) =$$
$$\lim_{\varepsilon \to 0} \left\{ \frac{\tilde{F}(x(u_0 + \varepsilon\varphi), u_0 + \varepsilon\varphi, \lambda(u_0 + \varepsilon\varphi), v(u_0 + \varepsilon\varphi))}{\varepsilon} \right.$$
$$\left. - \frac{\tilde{F}(x(u_0), u_0, \lambda(u_0), v(u_0))}{\varepsilon} \right\},$$

so that

$$\delta_u \tilde{F}(x(u_0), u_0, \lambda(u_0), v(u_0); \varphi) =$$
$$\lim_{\varepsilon \to 0} \left\{ \frac{\tilde{F}(x(u_0 + \varepsilon\varphi), u_0 + \varepsilon\varphi, \lambda(u_0 + \varepsilon\varphi), v(u_0 + \varepsilon\varphi))}{\varepsilon} \right.$$
$$- \frac{\tilde{F}(x(u_0), u_0 + \varepsilon\varphi, \lambda(u_0 + \varepsilon\varphi), v(u_0 + \varepsilon\varphi))}{\varepsilon}$$
$$+ \frac{\tilde{F}(x(u_0), u_0 + \varepsilon\varphi, \lambda(u_0 + \varepsilon\varphi), v(u_0 + \varepsilon\varphi))}{\varepsilon}$$
$$\left. - \frac{\tilde{F}(x(u_0), u_0, \lambda(u_0), v(u_0))}{\varepsilon} \right\}.$$

However,

$$\left| \frac{\tilde{F}(x(u_0 + \varepsilon\varphi), u_0 + \varepsilon\varphi, \lambda(u_0 + \varepsilon\varphi), v(u_0 + \varepsilon\varphi))}{\varepsilon} \right.$$
$$\left. - \frac{\tilde{F}(x(u_0), u_0 + \varepsilon\varphi, \lambda(u_0 + \varepsilon\varphi), v(u_0 + \varepsilon\varphi))}{\varepsilon} \right|$$
$$\le \|\tilde{F}_x(x(u_0 + \varepsilon\varphi), u_0 + \varepsilon\varphi, \lambda(u_0 + \varepsilon\varphi), v(u_0 + \varepsilon\varphi))\| K\|\varphi\|$$
$$+ K_1 \|x(u_0 + \varepsilon\varphi) - x(u_0)\|$$
$$\le K_1 K \|\varphi\| \varepsilon$$
$$\to 0, \text{ as } \varepsilon \to 0.$$

In these last inequalities we have used

$$\limsup_{\varepsilon \to 0} \left\| \frac{x(u_0 + \varepsilon\varphi) - x(u_0)}{\varepsilon} \right\| \leq K\|\varphi\|,$$

and

$$\tilde{F}_x(x(u_0 + \varepsilon\varphi), u_0 + \varepsilon\varphi, \lambda(u_0 + \varepsilon\varphi), v(u_0 + \varepsilon\varphi)) = \theta.$$

On the other hand,

$$\left\{ \frac{\tilde{F}(x(u_0), u_0 + \varepsilon\varphi, \lambda(u_0 + \varepsilon\varphi), v(u_0 + \varepsilon\varphi))}{\varepsilon} \right. \\ \left. - \frac{\tilde{F}(x(u_0), u_0, \lambda(u_0), v(u_0))}{\varepsilon} \right\} \\ = \left\{ \frac{\tilde{F}(x(u_0), u_0 + \varepsilon\varphi, \lambda(u_0 + \varepsilon\varphi), v(u_0 + \varepsilon\varphi))}{\varepsilon} \right. \\ - \frac{\tilde{F}(x(u_0), u_0 + \varepsilon\varphi, \lambda(u_0), v(u_0))}{\varepsilon} \\ + \frac{\tilde{F}(x(u_0), u_0 + \varepsilon\varphi, \lambda(u_0), v(u_0))}{\varepsilon} \\ \left. - \frac{\tilde{F}(x(u_0), u_0, \lambda(u_0), v(u_0))}{\varepsilon} \right\}$$

Now observe that

$$\frac{\tilde{F}(x(u_0), u_0 + \varepsilon\varphi, \lambda(u_0 + \varepsilon\varphi), v(u_0 + \varepsilon\varphi))}{\varepsilon} \\ - \frac{\tilde{F}(x(u_0), u_0 + \varepsilon\varphi, \lambda(u_0), v(u_0))}{\varepsilon} \\ = \frac{\langle \lambda_1(u_0 + \varepsilon\varphi), G(x(u_0), u_0 + \varepsilon\varphi) + v(u_0 + \varepsilon\varphi)^2 \rangle_{L^p}}{\varepsilon} \\ - \frac{\langle \lambda_1(u_0), G(x(u_0), u_0 + \varepsilon\varphi) + v(u_0)^2 \rangle_{L^p}}{\varepsilon} \\ + \frac{\langle \lambda_2(u_0 + \varepsilon\varphi) - \lambda_2(u_0), H(x(u_0), u_0 + \varepsilon\varphi) \rangle_{L^p}}{\varepsilon}. \tag{15.78}$$

Also,

$$\left| \frac{\langle \lambda_1(u_0 + \varepsilon\varphi), G(x(u_0), u_0 + \varepsilon\varphi) + v(u_0 + \varepsilon\varphi)^2 \rangle_{L^p}}{\varepsilon} \right.$$
$$\left. - \frac{\langle \lambda_1(u_0), G(x(u_0), u_0 + \varepsilon\varphi) + v(u_0)^2 \rangle_{L^p}}{\varepsilon} \right|$$

$$\leq \left| \frac{\langle \lambda_1(u_0 + \varepsilon\varphi), G(x(u_0), u_0 + \varepsilon\varphi) + v(u_0 + \varepsilon\varphi)^2 \rangle_{L^p}}{\varepsilon} \right.$$
$$\left. - \frac{\langle \lambda_1(u_0), G(x(u_0), u_0 + \varepsilon\varphi) + v(u_0 + \varepsilon\varphi)^2 \rangle_{L^p}}{\varepsilon} \right|$$

$$+ \left| \frac{\langle \lambda_1(u_0), G(x(u_0), u_0 + \varepsilon\varphi) + v(u_0 + \varepsilon\varphi)^2 \rangle_{L^p}}{\varepsilon} \right.$$
$$\left. - \frac{\langle \lambda_1(u_0), G(x(u_0), u_0 + \varepsilon\varphi) + v(u_0)^2 \rangle_{L^p}}{\varepsilon} \right|$$

$$\leq \varepsilon \frac{K\|\varphi\|}{\varepsilon} \|G(x(u_0), u_0 + \varepsilon\varphi) + v(u_0 + \varepsilon\varphi)^2\|$$

$$+ \|\lambda_1(u_0)(v(u_0 + \varepsilon\varphi) + v(u_0))\| \frac{K\|\varphi\|\varepsilon}{\varepsilon}$$

$$\to 0 \text{ as } \varepsilon \to 0.$$

To obtain the last inequalities we have used

$$\limsup_{\varepsilon \to 0} \left\| \frac{\lambda_1(u_0 + \varepsilon\varphi) - \lambda_1(u_0)}{\varepsilon} \right\| \leq K\|\varphi\|,$$

$$\lambda_1(u_0)v(u_0) = \theta,$$
$$\lambda_1(u_0)v(u_0 + \varepsilon\varphi) \to \theta, \text{ as } \varepsilon \to 0,$$

and

$$\left\| \frac{\lambda_1(u_0)(v(u_0 + \varepsilon\varphi)^2 - v(u_0)^2)}{\varepsilon} \right\|$$

$$= \left\| \frac{\lambda_1(u_0)(v(u_0 + \varepsilon\varphi) + v(u_0))(v(u_0 + \varepsilon\varphi) - v(u_0))}{\varepsilon} \right\|$$

$$\leq \frac{\|\lambda_1(u_0)(v(u_0 + \varepsilon\varphi) + v(u_0))\|K\|\varphi\|\varepsilon}{\varepsilon}$$

$$\to 0, \text{ as } \varepsilon \to 0. \tag{15.79}$$

Finally,

$$\left| \frac{\langle \lambda_2(u_0 + \varepsilon\varphi) - \lambda_2(u_0), H(x(u_0), u_0 + \varepsilon\varphi) \rangle_{L^p}}{\varepsilon} \right|$$

$$\leq \frac{K\varepsilon\|\varphi\|}{\varepsilon} \|H(x(u_0), u_0 + \varepsilon\varphi)\|$$
$$\to 0, \text{ as } \varepsilon \to 0.$$

To obtain the last inequalities we have used

$$\limsup_{\varepsilon \to 0} \left\| \frac{\lambda_2(u_0 + \varepsilon\varphi) - \lambda_2(u_0)}{\varepsilon} \right\| \leq K\|\varphi\|,$$

and

$$H(x(u_0), u_0 + \varepsilon\varphi) \to \theta, \text{ as } \varepsilon \to 0.$$

From these last results, we get

$$\delta_u \tilde{F}(x(u_0), u_0, \lambda(u_0), v(u_0); \varphi)$$

$$=$$

$$\lim_{\varepsilon \to 0} \left\{ \frac{\tilde{F}(x(u_0), u_0 + \varepsilon\varphi, \lambda(u_0), v(u_0))}{\varepsilon} \right.$$

$$\left. - \frac{\tilde{F}(x(u_0), u_0, \lambda(u_0), v(u_0))}{\varepsilon} \right\}$$

$$=$$

$$\langle F_u(x(u_0), u_0), \varphi \rangle_U + \langle \lambda_1(u_0) \cdot G_u(x(u_0), u_0), \varphi \rangle_{L^p}$$
$$+ \langle \lambda_2(u_0) \cdot H_u(x(u_0), u_0), \varphi \rangle_{L^p}.$$

In the last lines we have proven the following corollary of the implicit function theorem,

Corollary 15.1

Suppose $(x_0, u_0, (\lambda_1)_0, (\lambda_2)_0, v_0)$ *is a solution of the system (15.70), (15.71),(15.72), (15.74), and assume the corresponding hypotheses of the implicit function theorem are satisfied. Also assume* $\tilde{F}(x, u, \lambda_1, \lambda_2, v)$ *is such that the Fréchet second derivative* $\tilde{F}_{xx}(x, u, \lambda_1, \lambda_2)$ *is continuous in a neighborhood of*

$$(x_0, u_0, (\lambda_1)_0, (\lambda_2)_0).$$

Under such hypotheses, for a given $\varphi \in [C^\infty(\Omega)]^k$, *denoting*

$$F_1(u) = \tilde{F}(x(u), u, \lambda_1(u), \lambda_2(u), v(u)),$$

we have

$$\delta(F_1(u); \varphi)|_{u=u_0}$$

$$=$$

$$\langle F_u(x(u_0), u_0), \varphi \rangle_U + \langle \lambda_1(u_0) \cdot G_u(x(u_0), u_0), \varphi \rangle_{L^p}$$
$$+ \langle \lambda_2(u_0) \cdot H_u(x(u_0), u_0), \varphi \rangle_{L^p}.$$

Chapter 16

On Central Fields in the Calculus of Variations

16.1 Introduction

In this short communication we develop sufficient conditions of local optimality for a relatively large class of problems in the calculus of variations. The concerning approach is developed through a generalization of some theoretical results about central fields presented in [79]. We address both the scalar and vectorial cases for a domain in $\mathbb{R}^n$. Finally the Weierstrass Excess function has a fundamental role in the formal proofs of the main results.

16.2 Central fields for the scalar case in the calculus of variations

Let $\Omega \subset \mathbb{R}^n$ be an open, bounded, simply connected set with a regular (Lipschitzian) boundary denoted by $\partial \Omega$.

Let $f \in C^1(\overline{\Omega} \times \mathbb{R} \times \mathbb{R}^n)$ ($f(x,y,\mathbf{z})$) and $V = C^1(\overline{\Omega})$. We suppose f is convex in $\mathbf{z}$, which we denote by $f(\underline{x}, \underline{y}, \mathbf{z})$ be convex.

Choose $\tilde{x}_1, \ldots, \tilde{x}_n, \tilde{y}_1, \ldots, \tilde{y}_n \in \mathbb{R}$ such that if $x = (x_1, \ldots, x_n) \in \Omega$, then

$$\tilde{x}_k < x_k < \tilde{y}_k \ \forall k \in \{1, \ldots, n\}$$

Suppose that $y_0 \in V_1 = C^1(B_0)$ is stationary for f in $B_0 = \prod_{k=1}^n [\tilde{x}_k, \tilde{y}_k]$, that is, suppose

$$\sum_{k=1}^n \frac{d}{dx_k} f_{z_k}(x, y_0(x), \nabla y_0(x)) = f_y(x, y_0(x), \nabla y_0(x)), \ \forall x \in B_0.$$

We also assume f to be of C^1 class in $B_0 \times \mathbb{R} \times \mathbb{R}^n$.

Consider the problem of minimizing $F : D \to \mathbb{R}$ where

$$F(y) = \int_\Omega f(x, y(x), \nabla y(x)) \, dx,$$

and where

$$D = \{y \in V \ : \ y = y_0 \text{ in } \partial \Omega\}.$$

Assume there exists a family of functions $\mathscr{F}$, such that for each $(x,y) \in D_1 \subset \mathbb{R}^{n+1}$, there exists a unique stationary function $y_3 \in V_1 = C^1(B_0)$ for f in $\mathscr{F}$, such that

$$(x,y) = (x, y_3(x)).$$

More specifically, we define $\mathscr{F}$ as a subset of $\mathscr{F}_1$, where

$$\mathscr{F}_1 = \{\phi(t, \Lambda(x,y)) \text{ stationary for } f \text{ such that } \phi(x, \Lambda(x,y)) = y \text{ for a } \Lambda \in B_r(0) \subset \mathbb{R}\}$$

where $B_r(0) = (0 - r, 0 + r)$ for some $r > 0$.

Here, ϕ is stationary and

$$\phi((t_1, \ldots, t_k = \tilde{x}_k, \ldots, t_n), \Lambda(x,y)) = y_0(t_1, \ldots, t_k = \tilde{x}_k, \ldots, t_n), \text{ on } \partial B_0$$

and

$$\nabla \phi((t_1, \ldots, t_k = \tilde{x}_k, \ldots, t_n), \Lambda(x,y)) \cdot \mathbf{n} = \Lambda \in \mathbb{R}, \text{ on } \partial B_0, \forall k \in \{1, \ldots, n\},$$

where $\mathbf{n}$ denotes the outward normal field to ∂B_0 so that, clarifying the dependence of Λ on (x,y), we have denoted $\Lambda = \Lambda(x,y)$.

Define the field $\theta : D_1 \to \mathbb{R}^n$ by

$$\theta(x,y) = \nabla y_3(x),$$

where as above indicated y_3 is such that

$$(x,y) = (x, y_3(x)),$$

so that

$$\theta(x, y_3(x)) = \nabla y_3(x), \forall x \in B_0.$$

At this point we assume the hypotheses of the implicit function theorem so that $\phi(x, \Lambda(x,y))$ is of C^1 class and $\theta(x,y)$ is continuous (in fact, since ϕ is stationary, the partial derivatives of $\theta(x,y)$ are well defined).

Define also

$$h(x,y) = f(x, y, \theta(x,y)) - \sum_{j=1}^{n} f_{z_j}(x, y, \theta(x,y)) \theta_j(x,y)$$

and

$$P_j(x,y) = f_{z_j}(x, y, \theta(x,y)), \forall j \in \{1, \ldots, n\}.$$

Observe that

$$
\begin{aligned}
h_y(x, y_3(x)) &= f_y(x, y_3(x), \theta(x, y_3(x))) + \sum_{j=1}^{n} f_{z_j}(x, y_3(x), \theta(x, y_3(x)))(\theta_j)_y(x, y_3(x)) \\
&\quad - \sum_{j=1}^{n} \{(P_j)_y(x, y_3(x))\theta_j(x, y_3(x)) + P_j(x, y_3(x))(\theta_j)_y(x, y_3(x))\} \\
&= f_y(x, y_3(x), \theta(x, y_3(x))) + \sum_{j=1}^{n} P_j(x, y_3(x))(\theta_j)_y(x, y_3(x)) \\
&\quad - \sum_{j=1}^{n} \{(P_j)_y(x, y_3(x))\theta_j(x, y_3(x)) + P_j(x, y_3(x))(\theta_j)_y(x, y_3(x))\} \\
&= f_y(x, y_3(x), \theta(x, y_3(x))) - \sum_{j=1}^{n} (P_j)_y(x, y_3(x))\theta_j(x, y_3(x)). \quad (16.1)
\end{aligned}
$$

Thus,

$$h_y(x,y) = f_y(x, y, \theta(x,y)) - \sum_{j=1}^{n} (P_j)_y(x,y)\theta_j(x,y), \forall(x,y) \in D_1.$$

On the other hand, since $y_3(x)$ is stationary, we obtain

$$
\begin{aligned}
0 &= f_y(x, y_3(x), \nabla y_3(x)) - \sum_{j=1}^{n} \frac{d}{dx_j} f_{z_j}(x, y_3(x), \nabla y_3(x)) \\
&= h_y(x, y_3(x)) + \sum_{j=1}^{n} (P_j)_y(x, y) \theta_j(x, y) \\
&\quad - \sum_{j=1}^{n} \left(\frac{\partial P_j(x, y_3(x))}{\partial x_j} + (P_j)_y(x, y_3(x)) \frac{\partial y_3(x)}{\partial x_j} \right) \\
&= h_y(x, y_3(x)) + \sum_{j=1}^{n} (P_j)_y(x, y_3(x)) \theta_j(x, y_3(x)) \\
&\quad - \sum_{j=1}^{n} \left(\frac{\partial P_j(x, y_3(x))}{\partial x_j} + (P_j)_y(x, y_3(x)) \theta_j(x, y_3(x)) \right) \\
&= h_y(x, y_3(x)) - \sum_{j=1}^{n} \left(\frac{\partial P_j(x, y_3(x))}{\partial x_j} \right).
\end{aligned}
\tag{16.2}
$$

Therefore,

$$
h_y(x, y) = \sum_{j=1}^{n} \left(\frac{\partial P_j(x, y)}{\partial x_j} \right), \forall (x, y) \in D_1
$$

Let $H_j(x, y)$ be such that

$$
\frac{\partial H_j(x, y)}{\partial y} = P_j(x, y).
$$

From these two last lines, we get

$$
h_y(x, y) = \sum_{j=1}^{n} \left(\frac{\partial (H_j)_y(x, y)}{\partial x_j} \right) = \left(\sum_{j=1}^{n} \frac{\partial H_j(x, y)}{\partial x_j} \right)_y, \forall (x, y) \in D_1
$$

so that

$$
h(x, y) = \sum_{j=1}^{n} \frac{\partial H_j(x, y)}{\partial x_k} + W(x),
$$

for some $W : B_0 \to \mathbb{R}$.

Hence, assuming D_1 contains an open which contains $\overline{\mathscr{C}_0}$, where $\mathscr{C}_0 = \{(x, y_0(x)) : x \in \Omega\}$, for $y \in L$ sufficiently close to y_0 in L^∞ norm, the generalized Hilbert integral, denoted by $I(y)$, will be defined by

$$
\begin{aligned}
I(y) &= \int_\Omega h(x, y(x)) \, dx + \sum_{j=1}^{n} P_j(x, y(x)) \frac{\partial y(x)}{\partial x_j} \, dx \\
&= \int_\Omega \sum_{j=1}^{n} \left(\frac{\partial H_j(x, y(x))}{\partial x_j} + (H_j)_y(x, y(x)) \frac{\partial y(x)}{\partial x_j} \right) dx + \int_\Omega W(x) \, dx \\
&= \int_\Omega \sum_{j=1}^{n} \frac{dH_j(x, y(x))}{dx_j} \, dx + \int_\Omega W(x) \, dx \\
&= \int_{\partial\Omega} \sum_{j=1}^{n} (-1)^{j+1} H_j(x, y(x)) dx_1 \wedge \cdots \wedge \widehat{dx_j} \wedge \cdots \wedge dx_n + \int_\Omega W(x) \, dx \\
&= W_1(y|_{\partial\Omega}) \\
&= W_1((y_0)|_{\partial\Omega}),
\end{aligned}
\tag{16.3}
$$

so that such an integral is invariant, that is, it does not depend on y.

Finally, observe that

$$F(y) - F(y_0) = \int_\Omega f(x, y(x), \nabla y(x)) \, dx - \int_\Omega f(x, y_0(x), \nabla y_0(x)) \, dx.$$

On the other hand,

$$
\begin{aligned}
I(y) &= \int_\Omega f(x, y(x), \theta(x, y)) \, dx \\
&\quad + \sum_{j=1}^n \int_\Omega f_{z_j}(x, y(x), \theta(x, y(x)))(\theta_j(x, y(x)) - y_{x_j}(x)) \, dx \\
&= I(y_0) \\
&= \int_\Omega f(x, y_0(x), \theta(x, y_0(x))) \, dx \\
&= \int_\Omega f(x, y_0(x), \nabla y_0(x)) \, dx \\
&= F(y_0).
\end{aligned}
\tag{16.4}
$$

Therefore,

$$
\begin{aligned}
F(y) - F(y_0) &= \int_\Omega [f(x, y(x), \nabla y(x)) - f(x, y(x), \theta(x, y))] \, dx \\
&\quad - \sum_{j=1}^n \int_\Omega f_{z_j}(x, y(x), \theta(x, y(x)))(y_{x_j}(x) - \theta_j(x, y(x))) \, dx \\
&= \int_\Omega \mathscr{E}(x, y(x), \theta(x, y(x)), \nabla y(x)) \, dx,
\end{aligned}
\tag{16.5}
$$

where

$$
\begin{aligned}
\mathscr{E}(x, y(x), \theta(x, y(x)), \nabla y(x)) &= f(x, y(x), \nabla y(x)) \\
&\quad - f(x, y(x), \theta(x, y)) \\
&\quad - \sum_{j=1}^n f_{z_j}(x, y(x), \theta(x, y(x)))(y_{x_j}(x) - \theta_j(x, y(x))),
\end{aligned}
\tag{16.6}
$$

is the Weierstrass Excess function.

With such results, we may prove the following theorem.

Theorem 16.2.1 *Let $\Omega \subset \mathbb{R}^n$ be an open, bounded and simply connected set, with a regular (Lipschitzian) boundary denoted by $\partial\Omega$. Let $V = C^1(\overline{\Omega})$ and let $f \in C^1(\overline{\Omega} \times \mathbb{R} \times \mathbb{R}^n)$ be such that $f(\underline{x}, \underline{y}, \mathbf{z})$ is convex. Let $F : D \to \mathbb{R}$ be defined by*

$$F(y) = \int_\Omega f(x, y(x), \nabla y(x)) \, dx,$$

where

$$D = \{y \in V \ : \ y = y_1 \ em \ \partial\Omega\}.$$

Let $y_0 \in D$ be a stationary function for f which may be extended to B_0, being kept stationary in B_0, where B_0 has been specified above in this section. Suppose we may define a field $\theta : D_1 \to \mathbb{R}^n$, also as it has been specified above in this section. Assume $D_1 \subset \mathbb{R}^{n+1}$ contains an open set which contains $\overline{\mathscr{C}_0}$, where

$$\mathscr{C}_0 = \{(x, y_0(x)) \ : \ x \in \Omega\}.$$

Under such hypotheses, there exists $\delta > 0$ such that

$$F(y) \geq F(y_0), \ \forall y \in B_\delta(y_0) \cap D,$$

where

$$B_\delta(y_0) = \{y \in V \ : \ \|y - y_0\|_\infty < \delta\}.$$

Proof 16.1 From the hypotheses and from the exposed above in this section, there exists $\delta > 0$ such that $\theta(x, y(x))$ is well defined for all $y \in D$ such that

$$\|y - y_0\|_\infty < \delta.$$

Let $y \in B_\delta(y_0) \cap D$. Thus, since $f(\underline{x}, \underline{y}, \mathbf{z})$ is convex, we have

$$
\begin{aligned}
\mathscr{E}(x, y(x), \theta(x, y(x)), \nabla y(x)) &= f(x, y(x), \nabla y(x)) \\
&\quad - f(x, y(x)\theta(x, y)) \\
&\quad - \sum_{j=1}^{n} f_{z_j}(x, y(x), \theta(x, y(x)))(y_{x_j}(x) - \theta_j(x, y(x))) \\
&\geq 0, \text{ in } \Omega
\end{aligned}
\tag{16.7}
$$

so that,

$$F(y) - F(y_0) = \int_\Omega \mathscr{E}(x, y(x), \theta(x, y(x)), \nabla y(x)) \, dx \geq 0, \ \forall y \in B_\delta(y_0) \cap D.$$

The proof is complete.

16.3 Central fields and the vectorial case in the calculus of variations

Let $\Omega \subset \mathbb{R}^n$ be an open, bounded, simply connected set with a regular (Lipschitzian) boundary denoted by $\partial\Omega$.

Let $f \in C^1(\overline{\Omega} \times \mathbb{R}^N \times \mathbb{R}^{Nn})$ ($f(x, \mathbf{y}, \mathbf{z})$) and $V = C^1(\overline{\Omega}; \mathbb{R}^N)$. We suppose f is convex in $\mathbf{z}$, which we denote by $f(\underline{x}, \underline{y}, \mathbf{z})$ be convex.

Choose $\tilde{x}_1, \ldots, \tilde{x}_n, \tilde{y}_1, \ldots, \tilde{y}_n \in \mathbb{R}$ such that if $x = (x_1, \ldots, x_n) \in \Omega$, then

$$\tilde{x}_k < x_k < \tilde{y}_k \ \forall k \in \{1, \ldots, n\}$$

Suppose that $\mathbf{y}_0 \in V_1 = C^1(B_0; \mathbb{R}^N)$ is stationary for f in $B_0 = \prod_{k=1}^{n}[\tilde{x}_k, \tilde{y}_k]$, that is, suppose that

$$\sum_{k=1}^{n} \frac{d}{dx_k} f_{z_{jk}}(x, \mathbf{y}_0(x), \nabla \mathbf{y}_0(x)) = f_{y_j}(x, \mathbf{y}_0(x), \nabla \mathbf{y}_0(x)), \ \forall x \in B_0, \ \forall j \in \{1, \ldots, N\}.$$

We also assume f to be of C^1 class in $B_0 \times \mathbb{R}^N \times \mathbb{R}^{Nn}$.

Consider the problem of minimizing $F : D \to \mathbb{R}$ where

$$F(\mathbf{y}) = \int_\Omega f(x, \mathbf{y}(x), \nabla \mathbf{y}(x)) \, dx,$$

and where

$$D = \{\mathbf{y} \in V \ : \ \mathbf{y} = \mathbf{y}_0 \text{ in } \partial\Omega\}.$$

Assume there exists a family of stationary functions $\mathscr{F}$, such that for each $(x,\mathbf{y}) \in D_1 \subset \mathbb{R}^{n+N}$, there exists a unique stationary function $\mathbf{y}_3 \in V_1$ for f in $\mathscr{F}$, such that

$$(x,\mathbf{y}) = (x,\mathbf{y}_3(x)).$$

More specifically, we define $\mathscr{F}$ as a subset of $\mathscr{F}_1$, where

$$\mathscr{F}_1 = \{\Phi(t,\Lambda(x,\mathbf{y})) \text{ stationary for } f \text{ such that } \Phi(x,\Lambda(x,\mathbf{y})) = \mathbf{y} \text{ for a } \Lambda \in B_r(\mathbf{0}) \subset \mathbb{R}^N\}$$

such that $B_r(\mathbf{0})$ is an open ball of center $\mathbf{0}$ and radius r, for some $r > 0$.

Here, Φ is stationary and

$$\Phi((t_1,\ldots,t_k=\tilde{x}_k,\ldots,t_n),\Lambda(x,\mathbf{y})) = \mathbf{y}_0(t_1,\ldots,t_k=\tilde{x}_k,\ldots,t_n), \text{ on } \partial B_0$$

and

$$\nabla(\Phi)_j((t_1,\ldots,t_k=\tilde{x}_k,\ldots,t_n),\Lambda(x,\mathbf{y})) \cdot \mathbf{n} = \Lambda_j, \text{ on } \partial B_0, \forall k \in \{1,\ldots,n\}, j \in \{1,\ldots,N\}.$$

Here, $\mathbf{n}$ denotes the outward normal field to ∂B_0.

Define the field $\theta : D_1 \subset \mathbb{R}^{n+N} \to \mathbb{R}^{Nn}$ by

$$\theta_j(x,\mathbf{y}) = \nabla(y_3)_j(x),$$

where as above indicated, $\mathbf{y}_3$ is such that

$$(x,\mathbf{y}) = (x,\mathbf{y}_3(x)),$$

and thus

$$\theta_j(x,(\mathbf{y}_3)(x)) = \nabla(y_3)_j(x), \forall x \in B_0.$$

At this point we assume the hypotheses of the implicit function theorem so that $\Phi_j(x,\Lambda(x,\mathbf{y}))$ is of C^1 class and $\theta_j(x,\mathbf{y})$ is continuous (in fact, since Φ_j is stationary, the partial derivatives of $\theta_j(x,\mathbf{y})$ are well defined, $\forall j \in \{1,\ldots,N\}$).

Define also

$$h(x,\mathbf{y}) = f(x,\mathbf{y},\theta(x,\mathbf{y})) - \sum_{j=1}^{n}\sum_{k=1}^{n} f_{z_{jk}}(x,y,\theta(x,\mathbf{y}))\theta_{jk}(x,\mathbf{y})$$

and

$$P_{jk}(x,\mathbf{y}) = f_{z_{jk}}(x,\mathbf{y},\theta(x,\mathbf{y})), \forall j \in \{1,\ldots,N\}, k \in \{1,\ldots,n\}.$$

Observe that

$$
\begin{aligned}
h_{y_j}(x,\mathbf{y}_3(x)) &= f_{y_j}(x,\mathbf{y}_3(x),\theta(x,\mathbf{y}_3(x))) + \sum_{l=1}^{N}\sum_{k=1}^{n} f_{z_{lk}}(x,\mathbf{y}_3(x),\theta(x,\mathbf{y}_3(x)))(\theta_{lk})_{y_j}(x,\mathbf{y}_3(x)) \\
&\quad - \sum_{l=1}^{N}\sum_{k=1}^{n}\{(P_{lk})_{y_j}(x,\mathbf{y}_3(x))\theta_{lk}(x,\mathbf{y}_3(x)) + P_{lk}(x,\mathbf{y}_3(x))(\theta_{lk})_{y_j}(x,\mathbf{y}_3(x))\} \\
&= f_{y_j}(x,\mathbf{y}_3(x),\theta(x,\mathbf{y}_3(x))) + \sum_{l=1}^{N}\sum_{k=1}^{n} P_{lk}(x,\mathbf{y}_3(x))(\theta_{lk})_{y_j}(x,\mathbf{y}_3(x)) \\
&\quad - \sum_{j=1}^{n}\{(P_j)_y(x,\mathbf{y}_3(x))\theta_j(x,\mathbf{y}_3(x)) + P_j(x,\mathbf{y}_3(x))(\theta_j)_y(x,\mathbf{y}_3(x))\} \\
&= f_{y_j}(x,\mathbf{y}_3(x),\theta(x,\mathbf{y}_3(x))) - \sum_{l=1}^{N}\sum_{k=1}^{n}(P_{lk})_{y_j}(x,\mathbf{y}_3(x))\theta_{lk}(x,\mathbf{y}_3(x)). \quad (16.8)
\end{aligned}
$$

Thus,

$$h_{y_j}(x,\mathbf{y}) = f_{y_j}(x,\mathbf{y},\theta(x,\mathbf{y})) - \sum_{l=1}^{N}\sum_{k=1}^{n}(P_{lk})_{y_j}(x,\mathbf{y})\theta_{lk}(x,\mathbf{y}),\ \forall(x,\mathbf{y}) \in D_1.$$

On the other hand, considering that $\mathbf{y}_3(x)$ is stationary, we obtain $H_k(x,\mathbf{y})$ such that

$$P_{jk}(x,\mathbf{y}) = \frac{\partial H_k(x,\mathbf{y})}{\partial y_j},\ \forall j \in \{1,\dots,N\},\ k \in \{1,\dots,n\}.$$

Indeed, we define

$$H_k(x,\mathbf{y}) = sta_{\phi \in \hat{D}_k} \int_{\tilde{x}_k}^{x_k} f(x_1,\dots,t_k,\dots,x_n,\phi_1(t_k),\dots,\phi_N(t_k),\nabla\tilde{\phi}(x,t_k))\,dt_k,$$

where, denoting $\hat{t}_k = (x_1,\dots,t_k,\dots,x_n)$, we have that

$$\nabla\tilde{\phi}(x,t_k) = \begin{bmatrix} ((y_3)_1)_{x_1}(\hat{t}_k) & ((y_3)_1)_{x_2}(\hat{t}_k) & \cdots & \frac{\partial\phi_1(t_k)}{\partial t_k} & \cdots & ((y_3)_1)_{x_n}(\hat{t}_k) \\ ((y_3)_2)_{x_1}(\hat{t}_k) & ((y_3)_2)_{x_2}(\hat{t}_k) & \cdots & \frac{\partial\phi_2(t_k)}{\partial t_k} & \cdots & ((y_3)_2)_{x_n}(\hat{t}_k) \\ \vdots & \vdots & \cdots & \vdots & \ddots & \vdots \\ ((y_3)_N)_{x_1}(\hat{t}_k) & ((y_3)_N)_{x_2}(\hat{t}_k) & \cdots & \frac{\partial\phi_N(t_k)}{\partial t_k} & \cdots & ((y_3)_N)_{x_n}(\hat{t}_k) \end{bmatrix}_{N\times n}, \tag{16.9}$$

and where

$$\hat{D}_k = \{\phi \in C^1([\tilde{x}_k,x_k];\mathbb{R}^N)\ :\ \phi(\tilde{x}_k) = \mathbf{y}_3(\hat{\tilde{x}}_k)\ \text{and}\ \phi(x_k) = \mathbf{y}_3(\hat{x}_k)\},$$

$\forall k \in \{1,\dots,n\}$.

Observe that, since $\phi(t_k)$ is stationary, we have

$$f_{y_j}[\phi(t_k)] - \frac{d}{dt_k}f_{z_{jk}}[\phi(t_k)] = 0,\ \text{in}\ [\tilde{x}_k,x_k],\ \forall j \in \{1,\dots,N\} \tag{16.10}$$

where generically, we denote

$$f[\phi(t_k)] = f(x,\phi(t_k),\nabla\tilde{\phi}(t_k)).$$

Observe that from these Euler-Lagrange equations we may obviously obtain

$$\phi(t_k) = \mathbf{y}_3(x_1,\dots,t_k,\dots,x_n).$$

At this point we shall also denote

$$\phi_j(t_k) \equiv \phi_j(t_k,\tilde{\Lambda}(x,\mathbf{y})),$$

where

$$\phi_j(\tilde{x}_k,\tilde{\Lambda}(x,\mathbf{y})) = (y_3)_j(x_1,\dots,\tilde{x}_k,\dots,x_n),$$

$$\frac{\partial\phi_j(\tilde{x}_k,\tilde{\Lambda}(x,\mathbf{y}))}{\partial t_k} = \tilde{\Lambda}_j,$$

and where $\tilde{\Lambda}(x,\mathbf{y})$ is such that

$$\phi_j(x_k,\tilde{\Lambda}(x,\mathbf{y})) = (y_3)_j(x_1,\dots,x_k,\dots,x_n).$$

Therefore,

$$H_k(x,\mathbf{y}) = \int_{\tilde{x}_k}^{x_k} f[\phi(t_k,\tilde{\Lambda}(x,\mathbf{y})]\,dt_k,$$

and thus,

$$
\begin{aligned}
[H_k(x,\mathbf{y})]_{y_j} &= \int_{\tilde{x}_k}^{x_k} f_{y_l}[\phi(t_k,\tilde{\Lambda}(x,\mathbf{y})](\phi_l)_{\tilde{\Lambda}}\tilde{\Lambda}_{y_j}(x,\mathbf{y})\, dt_k \\
&\quad + \int_{\tilde{x}_k}^{x_k} f_{z_{lk}}[\phi(t_k,\tilde{\Lambda}(x,\mathbf{y})](\phi_l)_{t_k\tilde{\Lambda}}\tilde{\Lambda}_{y_j}(x,\mathbf{y})\, dt_k \\
&= \int_{\tilde{x}_k}^{x_k} f_{y_l}[\phi(t_k,\tilde{\Lambda}(x,\mathbf{y})](\phi_l)_{\tilde{\Lambda}}\tilde{\Lambda}_{y_j}(x,\mathbf{y})\, dt_k \\
&\quad + \int_{\tilde{x}_k}^{x_k} f_{z_{lk}}[\phi(t_k,\tilde{\Lambda}(x,\mathbf{y})][(\phi_l)_{\tilde{\Lambda}}]_{t_k}\tilde{\Lambda}_{y_j}(x,\mathbf{y})\, dt_k.
\end{aligned}
\tag{16.11}
$$

From this and (16.10), we obtain

$$
\begin{aligned}
[H_k(x,\mathbf{y})]_{y_j} &= \int_{\tilde{x}_k}^{x_k} \frac{d}{dt_k} f_{z_{lk}}[\phi(t_k,\tilde{\Lambda}(x,\mathbf{y})](\phi_l)_{\tilde{\Lambda}}\tilde{\Lambda}_{y_j}(x,\mathbf{y})\, dt_k \\
&\quad + \int_{\tilde{x}_k}^{x_k} f_{z_{lk}}[\phi(t_k,\tilde{\Lambda}(x,\mathbf{y})][(\phi_l)_{\tilde{\Lambda}}]_{t_k}\tilde{\Lambda}_{y_j}(x,\mathbf{y})\, dt_k \\
&= \int_{\tilde{x}_k}^{x_k} \frac{d}{dt_k}\{f_{z_{lk}}[\phi(t_k,\tilde{\Lambda}(x,\mathbf{y})](\phi_l)_{\tilde{\Lambda}}]\}\, dt_k\tilde{\Lambda}_{y_j}(x,\mathbf{y}) \\
&= \{f_{z_{lk}}[\phi(t_k,\tilde{\Lambda}(x,\mathbf{y})](\phi_l)_{\tilde{\Lambda}}]\}|_{t_k=\tilde{x}_k}^{t_k=x_k}\tilde{\Lambda}_{y_j}(x,\mathbf{y})
\end{aligned}
\tag{16.12}
$$

On the other hand,

$$
\phi_l(\tilde{x}_k,\tilde{\Lambda}(x,\mathbf{y})) = (y_3)_l(x_1,\ldots,\tilde{x}_k,\ldots,x_n) = (y_0)_l(x_1,\ldots,\tilde{x}_k,\ldots,x_n),
$$

which does not depend on $\tilde{\Lambda}$, so that

$$
(\phi_l)_\Lambda(\tilde{x}_k,\tilde{\Lambda}(x,\mathbf{y})) = \mathbf{0}.
$$

Also,

$$
\phi_l(x_k,\tilde{\Lambda}(x,\mathbf{y})) = y_l
$$

and thus

$$
[\phi_l(x_k,\tilde{\Lambda}(x,\mathbf{y}))]_{y_j} = \frac{\partial y_l}{\partial y_j} = \delta_{lj}.
$$

Hence,

$$
(\phi_l)_{\tilde{\Lambda}}(x_k,\tilde{\Lambda}(x,\mathbf{y}))\tilde{\Lambda}_{y_j} = \delta_{lj}.
$$

From these last results and from (16.12), we obtain,

$$
\begin{aligned}
[H_k(x,\mathbf{y})]_{y_j} &= f_{z_{lk}}[\phi(x_k,\tilde{\Lambda}(x,\mathbf{y}))]\delta_{lj} \\
&= f_{z_{jk}}[\phi(x_k,\tilde{\Lambda}(x,\mathbf{y}))] \\
&= f_{z_{jk}}(x,\mathbf{y},\theta(x,\mathbf{y})) \\
&= P_{jk}(x,\mathbf{y}),\ \forall j \in \{1,\ldots,N\},\ k \in \{1,\ldots,n\}.
\end{aligned}
\tag{16.13}
$$

Therefore,

$$
\begin{aligned}
0 &= f_{y_j}(x,\mathbf{y}_3(x),\nabla\mathbf{y}_3(x)) - \sum_{k=1}^{n}\frac{d}{dx_k}f_{z_{jk}}(x,\mathbf{y}_3(x),\nabla\mathbf{y}_3(x)) \\
&= h_{y_j}(x,\mathbf{y}_3(x)) + \sum_{l=1}^{N}\sum_{k=1}^{n}(P_{lk})_{y_j}(x,\mathbf{y}_3(x))\theta_{lk}(x,\mathbf{y}_3(x)) \\
&\quad - \sum_{k=1}^{n}\left(\frac{\partial P_{jk}(x,\mathbf{y}_3(x))}{\partial x_k} + \sum_{l=1}^{N}(P_{jk})_{y_l}(x,\mathbf{y}_3(x))\frac{\partial (y_3)_l(x)}{\partial x_k}\right) \\
&= h_{y_j}(x,\mathbf{y}_3(x)) + \sum_{l=1}^{N}\sum_{k=1}^{n}(P_{lk})_{y_j}(x,\mathbf{y}_3(x))\theta_{lk}(x,\mathbf{y}_3(x)) \\
&\quad - \sum_{k=1}^{n}\left(\frac{\partial P_{jk}(x,\mathbf{y}_3(x))}{\partial x_k} + \sum_{l=1}^{N}(P_{jk})_{y_l}(x,\mathbf{y}_3(x))\theta_{lk}(x,\mathbf{y}_3(x))\right) \\
&= h_{y_j}(x,\mathbf{y}_3(x)) - \sum_{k=1}^{n}\left(\frac{\partial (H_k)_{y_j}(x,\mathbf{y}_3(x))}{\partial x_k}\right). \qquad (16.14)
\end{aligned}
$$

Thus,

$$
h_{y_j}(x,\mathbf{y}) = \sum_{k=1}^{n}\left(\frac{\partial H_k(x,\mathbf{y})}{\partial x_k}\right)_{y_j}, \forall(x,\mathbf{y}) \in D_1
$$

so that

$$
h(x,\mathbf{y}) = \sum_{k=1}^{n}\frac{\partial H_k(x,\mathbf{y})}{\partial x_k} + W(x),
$$

for some $W : B_0 \to \mathbb{R}$.

Hence, assuming D_1 contains an open set which contains $\overline{\mathscr{C}_0}$, where

$$
\mathscr{C}_0 = \{(x,\mathbf{y}_0(x)) \,:\, x \in \Omega\},
$$

for $\mathbf{y} \in D$ sufficiently close to $\mathbf{y}_0$ in L^∞ norm, the generalized Hilbert integral, denoted by $I(\mathbf{y})$, will be defined as

$$
\begin{aligned}
I(\mathbf{y}) &= \int_\Omega h(x,\mathbf{y}(x))\,dx + \sum_{j=1}^{N}\sum_{k=1}^{n}P_{jk}(x,\mathbf{y}(x))\frac{\partial y_j(x)}{\partial x_k}\,dx \\
&= \int_\Omega \sum_{k=1}^{n}\left(\frac{\partial H_k(x,\mathbf{y}(x))}{\partial x_k} + \sum_{j=1}^{n}(H_k)_{y_j}(x,\mathbf{y}(x))\frac{\partial y_j(x)}{\partial x_k}\right)dx + \int_\Omega W(x)\,dx \\
&= \int_\Omega \sum_{k=1}^{n}\frac{dH_k(x,\mathbf{y}(x))}{dx_k}\,dx + \int_\Omega W(x)\,dx \\
&= \int_{\partial\Omega}\sum_{k=1}^{n}(-1)^{k+1}H_k(x,\mathbf{y}(x))dx_1 \wedge \cdots \wedge \widehat{dx_k} \wedge \cdots \wedge dx_n + \int_\Omega W(x)\,dx \\
&= W_1(\mathbf{y}|_{\partial\Omega}) \\
&= W_1((\mathbf{y}_1)|_{\partial\Omega}), \qquad (16.15)
\end{aligned}
$$

so that such an integral is invariant, that is, it does not depend on $\mathbf{y}$.

Finally, observe that

$$
F(\mathbf{y}) - F(\mathbf{y}_0) = \int_\Omega f(x,\mathbf{y}(x),\nabla\mathbf{y}(x))\,dx - \int_\Omega f(x,\mathbf{y}_0(x),\nabla\mathbf{y}_0(x))\,dx.
$$

On the other hand,

$$
\begin{aligned}
I(\mathbf{y}) &= \int_{\Omega} f(x, \mathbf{y}(x), \theta(x, \mathbf{y}(x)))\, dx \\
&\quad + \sum_{j=1}^{N} \sum_{k=1}^{n} \int_{\Omega} f_{z_{jk}}(x, \mathbf{y}(x), \theta(x, \mathbf{y}(x)))(\theta_{jk}(x, \mathbf{y}(x)) - (y_j)_{x_k}(x))\, dx \\
&= I(\mathbf{y}_0) \\
&= \int_{\Omega} f(x, \mathbf{y}_0(x), \theta(x, \mathbf{y}_0(x)))\, dx \\
&= \int_{\Omega} f(x, \mathbf{y}_0(x), \nabla \mathbf{y}_0(x))\, dx \\
&= F(\mathbf{y}_0).
\end{aligned}
\tag{16.16}
$$

Thus,

$$
\begin{aligned}
F(\mathbf{y}) - F(\mathbf{y}_0) &= \int_{\Omega} [f(x, \mathbf{y}(x), \nabla \mathbf{y}(x)) - f(x, \mathbf{y}(x), \theta(x, \mathbf{y}(x)))]\, dx \\
&\quad - \sum_{j=1}^{N} \sum_{k=1}^{n} \int_{\Omega} f_{z_{jk}}(x, \mathbf{y}(x), \theta(x, \mathbf{y}(x)))((y_j)_{x_k}(x) - \theta_{jk}(x, \mathbf{y}(x)))\, dx \\
&= \int_{\Omega} \mathcal{E}(x, \mathbf{y}(x), \theta(x, \mathbf{y}(x)), \nabla \mathbf{y}(x))\, dx,
\end{aligned}
\tag{16.17}
$$

where

$$
\begin{aligned}
\mathcal{E}(x, \mathbf{y}(x), \theta(x, \mathbf{y}(x)), \nabla \mathbf{y}(x)) &= f(x, \mathbf{y}(x), \nabla \mathbf{y}(x)) \\
&\quad - f(x, \mathbf{y}(x), \theta(x, \mathbf{y}(x))) \\
&\quad - \sum_{j=1}^{N} \sum_{k=1}^{n} f_{z_{jk}}(x, \mathbf{y}(x), \theta(x, \mathbf{y}(x)))((y_j)_{x_k}(x) - \theta_{jk}(x, \mathbf{y}(x)))
\end{aligned}
$$

is the Weierstrass Excess function.

With such results, we may prove the following result.

Theorem 16.3.1 *Let $\Omega \subset \mathbb{R}^n$ be an open, bounded, simply connected set with a regular (Lipschitzian) boundary denoted by $\partial \Omega$. Let $V = C^1(\overline{\Omega}; \mathbb{R}^N)$ and let $f \in C^1(\overline{\Omega} \times \mathbb{R}^N \times \mathbb{R}^{Nn})$ be such that $f(\underline{x}, \underline{y}, \mathbf{z})$ is convex. Let $F : D \to \mathbb{R}$ be defined by*

$$
F(\mathbf{y}) = \int_{\Omega} f(x, \mathbf{y}(x), \nabla \mathbf{y}(x))\, dx,
$$

where

$$
D = \{\mathbf{y} \in V : \mathbf{y} = \mathbf{y}_1 \text{ on } \partial \Omega\}.
$$

Let $\mathbf{y}_0 \in D$ be an stationary function for f which may be extended to B_0, keeping it stationary in B_0, where B_0 has been specified above in this section. Suppose we may define a field $\theta : D_1 \to \mathbb{R}^{Nn}$, also as specified above in this section. Assume $D_1 \subset \mathbb{R}^{n+N}$ contains an open set which contains $\mathscr{C}_0$, where

$$
\mathscr{C}_0 = \{(x, \mathbf{y}_0(x)) : x \in \Omega\}.
$$

Under such hypotheses, there exists $\delta > 0$ such that

$$
F(\mathbf{y}) \geq F(\mathbf{y}_0), \ \forall \mathbf{y} \in B_\delta(\mathbf{y}_0) \cap D,
$$

where

$$
B_\delta(\mathbf{y}_0) = \{\mathbf{y} \in V : \|\mathbf{y} - \mathbf{y}_0\|_\infty < \delta\}.
$$

Proof 16.2   From the hypotheses and from the exposed above in this section, there exists $\delta > 0$ such that $\theta(x, \mathbf{y}(x))$ is well defined for each $\mathbf{y} \in D$ such that

$$\|\mathbf{y} - \mathbf{y}_0\|_\infty < \delta.$$

Let $\mathbf{y} \in B_\delta(\mathbf{y}_0) \cap D$. Thus, since $f(\underline{x}, \underline{y}, \mathbf{z})$ is convex, we have

$$
\begin{aligned}
\mathscr{E}(x, \mathbf{y}(x), \theta(x, \mathbf{y}(x)), \nabla \mathbf{y}(x)) &= f(x, \mathbf{y}(x), \nabla \mathbf{y}(x)) \\
&\quad - f(x, \mathbf{y}(x), \theta(x, \mathbf{y}(x))) \\
&\quad - \sum_{j=1}^{N} \sum_{k=1}^{n} f_{z_{jk}}(x, \mathbf{y}(x), \theta(x, \mathbf{y}(x)))((y_j)_{x_j}(x) - \theta_{jk}(x, \mathbf{y}(x))) \\
&\geq 0, \text{ in } \Omega
\end{aligned}
\tag{16.18}
$$

so that,

$$F(\mathbf{y}) - F(\mathbf{y}_0) = \int_\Omega \mathscr{E}(x, \mathbf{y}(x), \theta(x, \mathbf{y}(x)), \nabla \mathbf{y}(x)) \, dx \geq 0, \ \forall \mathbf{y} \in B_\delta(\mathbf{y}_0) \cap D.$$

The proof is complete.

SECTION III

APPLICATIONS TO MODELS IN PHYSICS AND ENGINEERING

Chapter 17

Global Existence Results and Duality for Non-Linear Models of Plates and Shells

17.1   Introduction

In this chapter, in a first step, we develop a new existence proof and a dual variational formulation for the Kirchhoff-Love thin plate model. Previous results on existence in mathematical elasticity and related models may be found in [29, 30, 31].

At this point we refer to the exceptionally important article "A contribution to contact problems for a class of solids and structures" by W.R. Bielski and J.J. Telega [11], published in 1985, as the first one to successfully apply and generalize the convex analysis approach to a model in non-convex and non-linear mechanics.

The present work is, in some sense, a kind of extension of this previous work [11] and others such as [10], which greatly influenced and inspired my work and recent book [14].

Here we highlight that such earlier results establish the complementary energy under the hypothesis of positive definiteness of the membrane force tensor at a critical point (please see [11, 10, 39] for details).

We have obtained a dual variational formulation which allows the global optimal point in question not to be positive definite (for related results see F. Botelho [14]), but also not necessarily negative definite. The approach developed also includes sufficient conditions of optimality for the primal problem. It is worth mentioning that the standard tools of convex analysis used in this text may be found in [33, 14, 64], for example.

At this point we start to describe the primal formulation.

Let $\Omega \subset \mathbb{R}^2$ be an open, bounded, connected set which represents the middle surface of a plate of thickness h. The boundary of Ω, which is assumed to be regular (Lipschitzian), is denoted by $\partial\Omega$. The vectorial basis related to the cartesian system $\{x_1, x_2, x_3\}$ is denoted by $(\mathbf{a}_\alpha, \mathbf{a}_3)$, where $\alpha = 1, 2$ (in general Greek indices stand for 1 or 2), and where $\mathbf{a}_3$ is the vector normal to Ω, whereas $\mathbf{a}_1$ and $\mathbf{a}_2$ are orthogonal vectors parallel to Ω. Also, $\mathbf{n}$ is the outward normal to the plate surface.

The displacements will be denoted by

$$\hat{\mathbf{u}} = \{\hat{u}_\alpha, \hat{u}_3\} = \hat{u}_\alpha \mathbf{a}_\alpha + \hat{u}_3 \mathbf{a}_3.$$

The Kirchhoff-Love relations are

$$\hat{u}_\alpha(x_1, x_2, x_3) = u_\alpha(x_1, x_2) - x_3 w(x_1, x_2)_{,\alpha}$$
$$\text{and } \hat{u}_3(x_1, x_2, x_3) = w(x_1, x_2). \tag{17.1}$$

Here $-h/2 \le x_3 \le h/2$ so that we have $u = (u_\alpha, w) \in U$ where

$$\begin{aligned} U &= \left\{ (u_\alpha, w) \in W^{1,2}(\Omega; \mathbb{R}^2) \times W^{2,2}(\Omega), \right. \\ &\qquad \left. u_\alpha = w = \frac{\partial w}{\partial \mathbf{n}} = 0 \text{ on } \partial\Omega \right\} \\ &= W_0^{1,2}(\Omega; \mathbb{R}^2) \times W_0^{2,2}(\Omega). \end{aligned}$$

It is worth emphasizing that the boundary conditions here specified refer to a clamped plate.

We define the operator $\Lambda : U \to Y \times Y$, where $Y = Y^* = L^2(\Omega; \mathbb{R}^{2\times2})$, by

$$\Lambda(u) = \{\gamma(u), \kappa(u)\},$$

$$\gamma_{\alpha\beta}(u) = \frac{u_{\alpha,\beta} + u_{\beta,\alpha}}{2} + \frac{w_{,\alpha} w_{,\beta}}{2},$$

$$\kappa_{\alpha\beta}(u) = -w_{,\alpha\beta}.$$

The constitutive relations are given by

$$N_{\alpha\beta}(u) = H_{\alpha\beta\lambda\mu} \gamma_{\lambda\mu}(u), \tag{17.2}$$

$$M_{\alpha\beta}(u) = h_{\alpha\beta\lambda\mu} \kappa_{\lambda\mu}(u), \tag{17.3}$$

where: $\{H_{\alpha\beta\lambda\mu}\}$ and $\{h_{\alpha\beta\lambda\mu} = \frac{h^2}{12} H_{\alpha\beta\lambda\mu}\}$, are symmetric positive definite fourth order tensors. From now on, we denote $\{\overline{H}_{\alpha\beta\lambda\mu}\} = \{H_{\alpha\beta\lambda\mu}\}^{-1}$ and $\{\overline{h}_{\alpha\beta\lambda\mu}\} = \{h_{\alpha\beta\lambda\mu}\}^{-1}$.

Furthermore $\{N_{\alpha\beta}\}$ denote the membrane force tensor and $\{M_{\alpha\beta}\}$ the moment one. The plate stored energy, represented by $(G \circ \Lambda) : U \to \mathbb{R}$ is expressed by

$$(G \circ \Lambda)(u) = \frac{1}{2} \int_\Omega N_{\alpha\beta}(u) \gamma_{\alpha\beta}(u) \, dx + \frac{1}{2} \int_\Omega M_{\alpha\beta}(u) \kappa_{\alpha\beta}(u) \, dx \tag{17.4}$$

and the external work, represented by $F : U \to \mathbb{R}$, is given by

$$F(u) = \langle w, P \rangle_{L^2(\Omega)} + \langle u_\alpha, P_\alpha \rangle_{L^2(\Omega)}, \tag{17.5}$$

where $P, P_1, P_2 \in L^2(\Omega)$ are external loads in the directions $\mathbf{a}_3$, $\mathbf{a}_1$ and $\mathbf{a}_2$ respectively. The potential energy denoted by $J : U \to \mathbb{R}$ is expressed by:

$$J(u) = (G \circ \Lambda)(u) - F(u)$$

Finally, we also emphasize from now on, as their meaning are clear, we may denote $L^2(\Omega)$ and $L^2(\Omega; \mathbb{R}^{2\times2})$ simply by L^2, and the respective norms by $\|\cdot\|_2$. Moreover derivatives are always understood in the distributional sense, **0** may denote the zero vector in appropriate Banach spaces and, the following and relating notations are used:

$$w_{,\alpha\beta} = \frac{\partial^2 w}{\partial x_\alpha \partial x_\beta},$$

$$u_{\alpha,\beta} = \frac{\partial u_\alpha}{\partial x_\beta},$$

$$N_{\alpha\beta,1} = \frac{\partial N_{\alpha\beta}}{\partial x_1},$$

and

$$N_{\alpha\beta,2} = \frac{\partial N_{\alpha\beta}}{\partial x_2}.$$

17.2   On the existence of a global minimizer

At this point we present an existence result concerning the Kirchhoff-Love plate model.

We start with the following two remarks.

Remark 17.1   Let $\{P_\alpha\} \in L^\infty(\Omega; \mathbb{R}^2)$. We may easily obtain by appropriate Lebesgue integration $\{\tilde{T}_{\alpha\beta}\}$ symmetric and such that

$$\tilde{T}_{\alpha\beta,\beta} = -P_\alpha, \text{ in } \Omega.$$

Indeed, extending $\{P_\alpha\}$ to zero outside Ω if necessary, we may set

$$\tilde{T}_{11}(x,y) = -\int_0^x P_1(\xi,y)\, d\xi,$$

$$\tilde{T}_{22}(x,y) = -\int_0^y P_2(x,\xi)\, d\xi,$$

and

$$\tilde{T}_{12}(x,y) = \tilde{T}_{21}(x,y) = 0, \text{ in } \Omega.$$

Thus, we may choose a $C > 0$ sufficiently big, such that

$$\{T_{\alpha\beta}\} = \{\tilde{T}_{\alpha\beta} + C\delta_{\alpha\beta}\}$$

is positive definite in Ω, so that

$$T_{\alpha\beta,\beta} = \tilde{T}_{\alpha\beta,\beta} = -P_\alpha,$$

where

$$\{\delta_{\alpha\beta}\}$$

is the Kronecker delta.

So, for the kind of boundary conditions of the next theorem, we do NOT have any restriction for the $\{P_\alpha\}$ norm.

Summarizing, the next result is new and it is really a step forward concerning the previous one in Ciarlet [30]. We emphasize this result and its proof through such a tensor $\{T_{\alpha\beta}\}$ are new, even though the final part of the proof is established through a standard procedure in the calculus of variations.

About the other existence result for plates, its proof through the tensor well specified $\{(T_0)_{\alpha\beta}\}$ is also new, even though the final part of such a proof is also performed through a standard procedure.

A similar remark is valid for the existence result for the model of shells, which is also established through a tensor T_0 properly specified.

Finally, the duality principles and concerning optimality conditions are established through new functionals. Similar results may be found in [14]. ■

Remark 17.2 Specifically about the existence of the tensor T_0 relating Theorem 17.3.1, we recall the following well known duality principle of the calculus of variations

$$\inf_{T=\{T_{\alpha\beta}\}\in B^*} \left\{ \frac{1}{2}\|T\|_2^2 \right\}$$

$$= \sup_{\{u_\alpha\}\in\tilde{U}} \left\{ -\frac{1}{2}\int_\Omega \nabla u_\alpha \cdot \nabla u_\alpha \, dx + \langle u_\alpha, P_\alpha \rangle_{L^2(\Omega)} + \langle u_\alpha, P_\alpha^t \rangle_{L^2(\Gamma_t)} \right\}. \tag{17.6}$$

Here

$$B^* = \{T \in L^2(\Omega;\mathbb{R}^4) : T_{\alpha\beta,\beta} + P_\alpha = 0, \text{ in } \Omega, T_{\alpha\beta}n_\beta - P_\alpha^t = 0, \text{ on } \Gamma_t\},$$

and

$$\tilde{U} = \{\{u_\alpha\} \in W^{1,2}(\Omega;\mathbb{R}^2) : u_\alpha = 0 \text{ on } \Gamma_0\}.$$

We also recall that the existence of a unique solution for both these primal and dual convex formulations is a well known result of the duality theory in the calculus of variations. Please, see related results in [33].

A similar duality principle may be established for the case of Theorem 17.6.1. ■

Theorem 17.2.1 *Let $\Omega \subset \mathbb{R}^2$ be an open, bounded, connected set with a Lipschitzian boundary denoted by $\partial\Omega = \Gamma$. Suppose $(G \circ \Lambda) : U \to \mathbb{R}$ is defined by*

$$G(\Lambda u) = G_1(\gamma(u)) + G_2(\kappa(u)), \ \forall u \in U,$$

where

$$G_1(\gamma u) = \frac{1}{2}\int_\Omega H_{\alpha\beta\lambda\mu}\gamma_{\alpha\beta}(u)\gamma_{\lambda\mu}(u) \, dx,$$

and

$$G_2(\kappa u) = \frac{1}{2}\int_\Omega h_{\alpha\beta\lambda\mu}\kappa_{\alpha\beta}(u)\kappa_{\lambda\mu}(u) \, dx,$$

where

$$\Lambda(u) = (\gamma(u), \kappa(u)) = (\{\gamma_{\alpha\beta}(u)\}, \{\kappa_{\alpha\beta}(u)\}),$$

$$\gamma_{\alpha\beta}(u) = \frac{u_{\alpha,\beta} + u_{\beta,\alpha}}{2} + \frac{w_{,\alpha}w_{,\beta}}{2},$$

$$\kappa_{\alpha\beta}(u) = -w_{,\alpha\beta},$$

and where

$$U = \{u = (u_1, u_2, w) \in W^{1,2}(\Omega;\mathbb{R}^2) \times W^{2,2}(\Omega) : u_1 = u_2 = w = 0 \text{ on } \partial\Omega\}.$$

We also define,

$$F_1(u) = \langle w, P \rangle_{L^2} + \langle u_\alpha, P_\alpha \rangle_{L^2}$$

$$\equiv \langle u, \mathbf{f} \rangle_{L^2}, \tag{17.7}$$

where

$$\mathbf{f} = (P_\alpha, P) \in L^\infty(\Omega;\mathbb{R}^3).$$

Let $J : U \to \mathbb{R}$ be defined by

$$J(u) = G(\Lambda u) - F_1(u), \ \forall u \in U.$$

Assume there exists $\{c_{\alpha\beta}\} \in \mathbb{R}^{2\times2}$ such that $c_{\alpha\beta} > 0, \ \forall \alpha, \beta \in \{1,2\}$ and

$$G_2(\kappa(u)) \geq c_{\alpha\beta} \|w_{,\alpha\beta}\|_2^2, \ \forall u \in U.$$

Under such hypotheses, there exists $u_0 \in U$ such that

$$J(u_0) = \min_{u \in U} J(u).$$

Proof 17.1 Observe that we may find $\mathbf{T}_\alpha = \{(T_\alpha)_\beta\}$ such that

$$div\mathbf{T}_\alpha = T_{\alpha\beta,\beta} = -P_\alpha$$

an also such that $\{T_{\alpha\beta}\}$ is positive definite and symmetric (please, see Remark 17.1).
Thus defining

$$v_{\alpha\beta}(u) = \frac{u_{\alpha,\beta} + u_{\beta,\alpha}}{2} + \frac{1}{2}w_{,\alpha}w_{,\beta}, \tag{17.8}$$

we obtain

$$
\begin{aligned}
J(u) &= G_1(\{v_{\alpha\beta}(u)\}) + G_2(\kappa(u)) - \langle P_\alpha, u_\alpha\rangle_{L^2} - \langle w, P\rangle_{L^2} \\
&= G_1(\{v_{\alpha\beta}(u)\}) + G_2(\kappa(u)) + \langle T_{\alpha\beta,\beta}, u_\alpha\rangle_{L^2} - \langle w, P\rangle_{L^2} \\
&= G_1(\{v_{\alpha\beta}(u)\}) + G_2(\kappa(u)) - \left\langle T_{\alpha\beta}, \frac{u_{\alpha,\beta} + u_{\beta,\alpha}}{2}\right\rangle_{L^2} - \langle w, P\rangle_{L^2} \\
&= G_1(\{v_{\alpha\beta}(u)\}) + G_2(\kappa(u)) - \left\langle T_{\alpha\beta}, v_{\alpha\beta}(u) - \frac{1}{2}w_{,\alpha}w_{,\beta}\right\rangle_{L^2} - \langle w, P\rangle_{L^2} \\
&\geq c_{\alpha\beta}\|w_{,\alpha\beta}\|_2^2 + \frac{1}{2}\langle T_{\alpha\beta}, w_{,\alpha}w_{,\beta}\rangle_{L^2} - \langle w, P\rangle_{L^2} + G_1(\{v_{\alpha\beta}(u)\}) \\
&\quad - \langle T_{\alpha\beta}, v_{\alpha\beta}(u)\rangle_{L^2}. \tag{17.9}
\end{aligned}
$$

From this, since $\{T_{\alpha\beta}\}$ is positive definite, clearly J is bounded below.
Let $\{u_n\} \in U$ be a minimizing sequence for J. Thus there exists $\alpha_1 \in \mathbb{R}$ such that

$$\lim_{n\to\infty} J(u_n) = \inf_{u\in U} J(u) = \alpha_1.$$

From (17.9), there exists $K_1 > 0$ such that

$$\|(w_n)_{,\alpha\beta}\|_2 < K_1, \forall \alpha, \beta \in \{1,2\}, \ n \in \mathbb{N}.$$

Therefore, there exists $w_0 \in W^{2,2}(\Omega)$ such that, up to a subsequence not relabeled,

$$(w_n)_{,\alpha\beta} \rightharpoonup (w_0)_{,\alpha\beta}, \ \text{weakly in } L^2,$$

$\forall \alpha, \beta \in \{1,2\}$, as $n \to \infty$.
Moreover, also up to a subsequence not relabeled,

$$(w_n)_{,\alpha} \to (w_0)_{,\alpha}, \ \text{strongly in } L^2 \text{ and } L^4, \tag{17.10}$$

$\forall \alpha, \in \{1,2\}$, as $n \to \infty$.
Also from (17.9), there exists $K_2 > 0$ such that,

$$\|(v_n)_{\alpha\beta}(u)\|_2 < K_2, \forall \alpha, \beta \in \{1,2\}, \ n \in \mathbb{N},$$

and thus, from this, (17.8) and (17.10), we may infer that there exists $K_3 > 0$ such that

$$\|(u_n)_{\alpha,\beta} + (u_n)_{\beta,\alpha}\|_2 < K_3, \forall \alpha, \beta \in \{1,2\}, \, n \in \mathbb{N}.$$

From this and Korn's inequality, there exists $K_4 > 0$ such that

$$\|u_n\|_{W^{1,2}(\Omega;\mathbb{R}^2)} \le K_4, \, \forall n \in \mathbb{N}.$$

So, up to a subsequence not relabeled, there exists $\{(u_0)_\alpha\} \in W^{1,2}(\Omega,\mathbb{R}^2)$, such that

$$(u_n)_{\alpha,\beta} + (u_n)_{\beta,\alpha} \rightharpoonup (u_0)_{\alpha,\beta} + (u_0)_{\beta,\alpha}, \text{ weakly in } L^2,$$

$\forall \alpha, \beta \in \{1,2\}$, as $n \to \infty$, and,

$$(u_n)_\alpha \to (u_0)_\alpha, \text{ strongly in } L^2,$$

$\forall \alpha \in \{1,2\}$, as $n \to \infty$.

Moreover, the boundary conditions satisfied by the subsequences are also satisfied for w_0 and u_0 in a trace sense, so that

$$u_0 = ((u_0)_\alpha, w_0) \in U.$$

From this, up to a subsequence not relabeled, we get

$$\gamma_{\alpha\beta}(u_n) \rightharpoonup \gamma_{\alpha\beta}(u_0), \text{ weakly in } L^2,$$

$\forall \alpha, \beta \in \{1,2\}$, and

$$\kappa_{\alpha\beta}(u_n) \rightharpoonup \kappa_{\alpha\beta}(u_0), \text{ weakly in } L^2,$$

$\forall \alpha, \beta \in \{1,2\}$.

Therefore, from the convexity of G_1 in γ and G_2 in κ we obtain

$$
\begin{aligned}
\inf_{u \in U} J(u) &= \alpha_1 \\
&= \liminf_{n \to \infty} J(u_n) \\
&\ge J(u_0).
\end{aligned}
\tag{17.11}
$$

Thus,

$$J(u_0) = \min_{u \in U} J(u).$$

The proof is complete.

17.3 Existence of a minimizer for the plate model for a more general case

At this point we present an existence result for a more general case.

Theorem 17.3.1 *Consider the statements and assumptions concerning the plate model described in the last section.*

More specifically, consider the functional $J : U \to \mathbb{R}$ given, as above described, by

$$
\begin{aligned}
J(u) = \; & W(\gamma(u), \kappa(u)) - \langle P_\alpha u_\alpha \rangle_{L^2} \\
& - \langle w, P \rangle_{L^2} - \langle P_\alpha^t, u_\alpha \rangle_{L^2(\Gamma_t)} \\
& - \langle P^t, w \rangle_{L^2(\Gamma_t)},
\end{aligned}
\tag{17.12}
$$

where,

$$U = \{u = (u_\alpha, w) = (u_1, u_2, w) \in W^{1,2}(\Omega; \mathbb{R}^2) \times W^{2,2}(\Omega) :$$
$$u_\alpha = w = \frac{\partial w}{\partial \mathbf{n}} = 0, \ on \ \Gamma_0\}, \tag{17.13}$$

where $\partial\Omega = \Gamma_0 \cup \Gamma_t$ and the Lebesgue measures

$$m_\Gamma(\Gamma_0 \cap \Gamma_t) = 0,$$

and

$$m_\Gamma(\Gamma_0) > 0.$$

Let T_0 be such that,

$$\|T_0\|_2^2 = \min_{T \in L^2(\Omega; \mathbb{R}^{2\times2})} \{\|T\|_2^2\},$$

subject to

$$T_{\alpha\beta,\beta} + P_\alpha = 0 \ in \ \Omega,$$

$$(T_0)_{\alpha\beta}\mathbf{n}_\beta - P_\alpha^t = 0, \ on \ \Gamma_t.$$

Assume $\|P_\alpha\|_\infty$ and $\|P_\alpha^t\|_\infty$ are small enough so that

$$J_1(u) \to +\infty, \ as \ \langle w_{,\alpha\beta}, w_{,\alpha\beta}\rangle_{L^2} \to +\infty, \tag{17.14}$$

where

$$J_1(u) = G_2(\kappa(u)) + \frac{1}{2}\langle (T_0)_{\alpha\beta}, w_{,\alpha}w_{,\beta}\rangle_{L^2}$$
$$- \langle P, w\rangle_{L^2} - \langle P^t, w\rangle_{L^2(\Gamma_t)}. \tag{17.15}$$

Under such hypotheses, there exists $u_0 \in U$ such that,

$$J(u_0) = \min_{u \in U}\{J(u)\}.$$

Proof 17.2

Observe that defining

$$v_{\alpha\beta}(u) = \frac{u_{\alpha,\beta} + u_{\beta,\alpha}}{2} + \frac{1}{2}w_{,\alpha}w_{,\beta}, \tag{17.16}$$

we have,

$$
\begin{aligned}
J(u) &= G_1(v(u)) + G_2(\kappa(u)) \\
&\quad - \langle P_\alpha, u_\alpha \rangle_{L^2} - \langle P, w \rangle_{L^2} \\
&\quad - \langle P_\alpha^t, u_\alpha \rangle_{L^2(\Gamma_t)} - \langle P^t, w \rangle_{L^2(\Gamma_t)} \\
&= G_1(v(u)) + G_2(\kappa(u)) + \langle (T_0)_{\alpha\beta,\beta}, u_\alpha \rangle_{L^2} \\
&\quad - \langle P, w \rangle_{L^2} \\
&\quad - \langle P_\alpha^t, u_\alpha \rangle_{L^2(\Gamma_t)} - \langle P^t, w \rangle_{L^2(\Gamma_t)} \\
&= G_1(v(u)) + G_2(\kappa(u)) - \left\langle (T_0)_{\alpha\beta}, \frac{u_{\alpha,\beta} + u_{\beta,\alpha}}{2} \right\rangle_{L^2} \\
&\quad + \langle (T_0)_{\alpha\beta} \mathbf{n}_\beta, u_\alpha \rangle_{L^2(\Gamma_t)} - \langle P, w \rangle_{L^2} \\
&\quad - \langle P_\alpha^t, u_\alpha \rangle_{L^2(\Gamma_t)} - \langle P^t, w \rangle_{L^2(\Gamma_t)} \\
&= G_1(v(u)) + G_2(\kappa(u)) - \left\langle (T_0)_{\alpha\beta}, \frac{u_{\alpha,\beta} + u_{\beta,\alpha}}{2} \right\rangle_{L^2} \\
&\quad - \langle P, w \rangle_{L^2} - \langle P^t, w \rangle_{L^2(\Gamma_t)} \\
&= G_1(v(u)) + G_2(\kappa(u)) - \left\langle (T_0)_{\alpha\beta}, v_{\alpha\beta}(u) - \frac{1}{2} w_{,\alpha} w_{,\beta} \right\rangle_{L^2} \\
&\quad - \langle P, w \rangle_{L^2} - \langle P^t, w \rangle_{L^2(\Gamma_t)} \\
&= J_1(u) + G_1(v(u)) - \langle (T_0)_{\alpha\beta}, v_{\alpha\beta}(u) \rangle_{L^2} \tag{17.17}
\end{aligned}
$$

From this and the hypothesis (17.14), J is bounded below. So, there exists $\alpha_1 \in \mathbb{R}$ such that

$$
\alpha_1 = \inf_{u \in U} J(u).
$$

Let $\{u_n\}$ be a minimizing sequence for J.

From (17.17) and also from the hypothesis (17.14), $\{\|w_n\|_{2,2}\}$ is bounded.

So there exists $w_0 \in U$ such that, up to a not relabeled subsequence,

$$
(w_n)_{,\alpha\beta} \rightharpoonup (w_0)_{,\alpha\beta} \text{ weakly in } L^2(\Omega), \ \forall \alpha, \beta \in \{1,2\},
$$

$$
(w_n)_{,\alpha} \to (w_0)_{,\alpha} \text{ strongly in } L^2(\Omega), \ \forall \alpha \in \{1,2\}, \tag{17.18}
$$

Also from (17.9), there exists $K_2 > 0$ such that,

$$
\|(v_n)_{\alpha\beta}(u)\|_2 < K_2, \forall \alpha, \beta \in \{1,2\}, \ n \in \mathbb{N},
$$

and thus, from this, (17.16) and (17.18), we may infer that there exists $K_3 > 0$ such that

$$
\|(u_n)_{\alpha,\beta}(u) + (u_n)_{\beta,\alpha}(u)\|_2 < K_3, \forall \alpha, \beta \in \{1,2\}, \ n \in \mathbb{N}.
$$

From this and Korn's inequality, there exists $K_4 > 0$ such that

$$
\|u_n\|_{W^{1,2}(\Omega;\mathbb{R}^2)} \leq K_4, \ \forall n \in \mathbb{N}.
$$

So, up to a subsequence not relabeled, there exists $\{(u_0)_\alpha\} \in W^{1,2}(\Omega, \mathbb{R}^2)$, such that

$$
(u_n)_{\alpha,\beta} + (u_n)_{\beta,\alpha} \rightharpoonup (u_0)_{\alpha,\beta} + (u_0)_{\beta,\alpha}, \text{ weakly in } L^2,
$$

$\forall \alpha, \beta \in \{1,2\}$, as $n \to \infty$, and,

$$
(u_n)_\alpha \to (u_0)_\alpha, \text{ strongly in } L^2,
$$

$\forall \alpha \in \{1,2\}$, as $n \to \infty$.

Moreover, the boundary conditions satisfied by the subsequences are also satisfied for w_0 and u_0 in a trace sense, so that

$$u_0 = ((u_0)_\alpha, w_0) \in U.$$

From this, up to a subsequence not relabeled, we get

$$\gamma_{\alpha\beta}(u_n) \rightharpoonup \gamma_{\alpha\beta}(u_0), \text{ weakly in } L^2,$$

$\forall \alpha, \beta \in \{1,2\}$, and

$$\kappa_{\alpha\beta}(u_n) \rightharpoonup \kappa_{\alpha\beta}(u_0), \text{ weakly in } L^2,$$

$\forall \alpha, \beta \in \{1,2\}$.

Therefore, from the convexity of G_1 in γ and G_2 in κ we obtain

$$
\begin{aligned}
\inf_{u \in U} J(u) &= \alpha_1 \\
&= \liminf_{n \to \infty} J(u_n) \\
&\geq J(u_0).
\end{aligned}
\tag{17.19}
$$

Thus,

$$J(u_0) = \min_{u \in U} J(u).$$

The proof is complete.

17.4   The main duality principle

In this section we present a duality principle for the plate model in question.

For such a result, we emphasize the dual variational formulation is concave.

Remark 17.3   In the proofs relating our duality principles we apply a very well known result found in Toland [78].

Indeed, for

$$\{N_{\alpha\beta}\} \in L^2(\Omega; \mathbb{R}^{2\times 2}),$$

assume

$$F_5(w) - G_5(\{w_\alpha\}) > 0, \ \forall w \in W_0^{2,2}(\Omega) \text{ such that } w \neq \mathbf{0},$$

where here

$$F_5(w) = \frac{1}{2} \int_\Omega h_{\alpha\beta\lambda\mu} w_{,\alpha\beta} w_{,\lambda\mu} \, dx + \frac{K}{2} \langle w_{,\alpha}, w_{,\alpha} \rangle_{L^2},$$

and

$$G_5(\{w_\alpha\}) = -\frac{1}{2} \int_\Omega N_{\alpha\beta} w_{,\alpha} w_{,\beta} \, dx + \frac{K}{2} \langle w_{,\alpha}, w_{,\alpha} \rangle_{L^2},$$

where $K > 0$ is supposed to be sufficiently big so that G_5 is convex in w.

Thus,

$$
\begin{aligned}
F_5(w) - G_5(\{w_\alpha\}) &= \frac{1}{2} \int_\Omega h_{\alpha\beta\lambda\mu} w_{,\alpha\beta} w_{,\lambda\mu} \, dx \\
&\quad + \frac{1}{2} \int_\Omega N_{\alpha\beta} w_{,\alpha} w_{,\beta} \, dx > 0,
\end{aligned}
\tag{17.20}
$$

$\forall w \in W_0^{2,2}(\Omega)$ such that $w \neq \mathbf{0}$.

Therefore,

$$-\langle w_\alpha, z_\alpha^* \rangle_{L^2} + F_5(w) + \sup_{v_2 \in L^2} \{ \langle (v_2)_\alpha, z_\alpha^* \rangle_{L^2} - G_5(\{(v_2)_\alpha\})$$

$$= -\langle w_\alpha, z_\alpha^* \rangle_{L^2} + F_5(w) + \frac{1}{2} \int_\Omega \overline{(-N_{\alpha\beta})^K} z_\alpha^* z_\beta^* \, dx > 0, \tag{17.21}$$

$$\forall w \in W_0^{2,2}(\Omega) \text{ such that } w \neq \mathbf{0},$$

so that

$$\inf_{w \in W_0^{2,2}(\Omega)} \{ -\langle w_\alpha, z_\alpha^* \rangle_{L^2} + F_5(w) \} + \frac{1}{2} \int_\Omega \overline{(-N_{\alpha\beta})^K} z_\alpha^* z_\beta^* \, dx$$

$$= -F_5^*(z^*) + \frac{1}{2} \int_\Omega \overline{(-N_{\alpha\beta})^K} z_\alpha^* z_\beta^* \, dx \geq 0, \tag{17.22}$$

$\forall z^* \in L^2$.

Indeed, from the general result in Toland [78], we have

$$\inf_{z^* \in L^2} \left\{ -F_5^*(z^*) + \frac{1}{2} \int_\Omega \overline{(-N_{\alpha\beta})^K} z_\alpha^* z_\beta^* \, dx \right\}$$

$$= \inf_{w \in W_0^{2,2}} \{ F_5(w) - G_5(\{w_{,\alpha}\}) \}$$

$$\leq F_5(w) - G_5(\{w_{,\alpha}\})$$

$$= \frac{1}{2} \int_\Omega h_{\alpha\beta\lambda\mu} w_{,\alpha\beta} w_{,\lambda\mu} \, dx$$

$$+ \frac{1}{2} \int_\Omega N_{\alpha\beta} w_{,\alpha} w_{,\beta} \, dx, \ \forall w \in W_0^{2,2}(\Omega). \tag{17.23}$$

∎

At this point we enunciate and prove our main duality principle.

Theorem 17.4.1 *Let $\Omega \subset \mathbb{R}^2$ be an open, bounded, connected set with a Lipschitzian boundary denoted by $\partial\Omega = \Gamma$. Suppose $(G \circ \Lambda) : U \to \mathbb{R}$ is defined by*

$$G(\Lambda u) = G_1(\gamma(u)) + G_2(\kappa(u)), \ \forall u \in U,$$

where

$$G_1(\gamma(u)) = \frac{1}{2} \int_\Omega H_{\alpha\beta\lambda\mu} \gamma_{\alpha\beta}(u) \gamma_{\lambda\mu}(u) \, dx,$$

and

$$G_2(\kappa(u)) = \frac{1}{2} \int_\Omega h_{\alpha\beta\lambda\mu} \kappa_{\alpha\beta}(u) \kappa_{\lambda\mu}(u) \, dx,$$

where

$$\Lambda(u) = (\gamma(u), \kappa(u)) = (\{\gamma_{\alpha\beta}(u)\}, \{\kappa_{\alpha\beta}(u)\}),$$

$$\gamma_{\alpha\beta}(u) = \frac{u_{\alpha,\beta} + u_{\beta,\alpha}}{2} + \frac{w_{,\alpha} w_{,\beta}}{2},$$

$$\kappa_{\alpha\beta}(u) = -w_{\alpha\beta}.$$

Here,

$$u = (u_1, u_2, w) = (u_\alpha, w) \in U = W_0^{1,2}(\Omega; \mathbb{R}^2) \times W_0^{2,2}(\Omega).$$

We also define,

$$
\begin{aligned}
F_1(u) &= \langle w, P \rangle_{L^2} + \langle u_\alpha, P_\alpha \rangle_{L^2} \\
&\equiv \langle u, \mathbf{f} \rangle_{L^2},
\end{aligned}
\tag{17.24}
$$

where

$$
\mathbf{f} = (P_\alpha, P) \in L^2(\Omega; \mathbb{R}^3).
$$

Let $J : U \to \mathbb{R}$ be defined by

$$
J(u) = G(\Lambda u) - F_1(u), \ \forall u \in U.
$$

Under such hypotheses,

$$
\begin{aligned}
&\inf_{u \in U} J(u) \\
&\geq \sup_{v^* \in A^*} \{ \inf_{z^* \in Y_2^*} \{ -F^*(z^*, Q) + G^*(z^*, N) \} \},
\end{aligned}
\tag{17.25}
$$

where,

$$
F(u) = G_2(\kappa(u)) + \frac{K}{2} \langle w_{,\alpha}, w_{,\alpha} \rangle_{L^2}, \forall u \in U.
$$

Moreover, $F^* : [Y_2^*]^2 \to \mathbb{R}$ is defined by,

$$
F^*(z^*, Q) = \sup_{u \in U} \{ \langle z_\alpha^* + Q_\alpha, w_{,\alpha} \rangle_{L^2} - F(u) \}, \ \forall z^* \in [Y_2^*]^2.
$$

Also,

$$
\begin{aligned}
G(v) &= -\frac{1}{2} \int_\Omega H_{\alpha\beta\lambda\mu} \left[(v_1)_{\alpha\beta} + \frac{(v_2)_\alpha (v_2)_\beta}{2} \right] \left[(v_1)_{\lambda\mu} + \frac{(v_2)_\lambda (v_2)_\mu}{2} \right] dx \\
&+ \frac{K}{2} \langle (v_2)_\alpha, (v_2)_\alpha \rangle_{L^2},
\end{aligned}
\tag{17.26}
$$

$$
\begin{aligned}
&G^*(z^*, N) \\
&= \sup_{v_2 \in Y_2} \{ \inf_{v_1 \in Y_1} \{ \langle N_{\alpha\beta}, (v_1)_{\alpha\beta} \rangle_{L^2} + \langle Q_\alpha + z_\alpha^*, (v_2)_\alpha \rangle_{L^2} - G(v) \} \} \\
&= \frac{1}{2} \int_\Omega \overline{(-N_{\alpha\beta})^K} z_\alpha^* z_\beta^* \, dx \\
&\quad - \frac{1}{2} \int_\Omega \overline{H}_{\alpha\beta\lambda\mu} N_{\alpha\beta} N_{\lambda\mu} \, dx,
\end{aligned}
\tag{17.27}
$$

if $v^* = (Q, N) \in A_3$, where

$$
A_3 = \{ v^* \in Y^* \ : \ \{(-N_{\alpha\beta})^K\} \text{ is positive definite in } \Omega \},
$$

$$
\{(-N_{\alpha\beta})^K\} = \left\{ \begin{array}{cc} -N_{11} + K & -N_{12} \\ -N_{21} & -N_{22} + K \end{array} \right\},
\tag{17.28}
$$

and

$$
\overline{(-N_{\alpha\beta})^K} = \{(-N_{\alpha\beta})^K\}^{-1}.
$$

Moreover,

$$
A^* = A_1 \cap A_2 \cap A_3 \cap A_4 \cap A_5,
$$

where

$$
A_1 = \{ v^* = (N, Q) \in Y^* \ : \ N_{\alpha\beta,\beta} + P_\alpha = 0, \text{ in } \Omega, \ \forall \alpha \in \{1, 2\} \}
$$

and

$$A_2 = \{v^* \in Y^* \ : \ Q_{\alpha,\alpha} + P = 0, \ in \ \Omega\}.$$

Also,

$$\begin{aligned} A_4 \ &= \ \{v^* = (Q,N) \in Y^* \ : \ \hat{J}^*(z^*) > 0, \\ &\quad \forall z^* \in Y_2^*, \ such \ that \ z^* \neq \mathbf{0}\}, \end{aligned} \tag{17.29}$$

and

$$A_5 = \left\{v^* = (Q,N) \in Y^* \ : \ \{N_{\alpha\beta} + K\delta_{\alpha\beta}\} \geq \frac{K}{2}\{\delta_{\alpha\beta}\}\right\},$$

where $K > 0$ is supposed to be such that, in an appropriate matrix sense,

$$\{\overline{H}_{\alpha\beta\lambda\mu}\} > \frac{2}{K}\{\delta_{\alpha\beta}\}.$$

Furthermore,

$$\begin{aligned} \hat{J}^*(z^*) \ &= \ -F^*(z^*,\mathbf{0}) + G^*(z^*,\mathbf{0}) \\ &= \ -F^*(z^*,\mathbf{0}) + \frac{1}{2}\int_\Omega \overline{(-N_{\alpha\beta})^K} \, z_\alpha^* z_\beta^* \, dx, \ \forall z^* \in Y_2^*. \end{aligned}$$

Here,

$$Y^* = Y = L^2(\Omega;\mathbb{R}^2) \times L^2(\Omega;\mathbb{R}^{2\times2}),$$
$$Y_1^* = Y_1 = L^2(\Omega;\mathbb{R}^{2\times2}),$$

and

$$Y_2^* = Y_2 = L^2(\Omega;\mathbb{R}^2),$$

Finally, denoting

$$J^*(v^*,z^*) = -F^*(z^*,Q) + G^*(z^*,N), \forall(v^*,z^*) \in A^* \times Y_2^*,$$

and

$$\tilde{J}^*(v^*) = \inf_{z^* \in Y_2^*} J^*(v^*,z^*), \forall v^* \in A^*,$$

suppose there exist $v_0^* = (N_0,Q_0) \in A^*$, $z_0^* \in Y_2^*$ and $u_0 \in U$ such that

$$\delta\{J^*(v_0^*,z_0^*) - \langle(u_0)_\alpha, (N_0)_{\alpha\beta,\beta} + P_\alpha\rangle_{L^2} - \langle w_0, (Q_0)_{\alpha,\alpha} + P\rangle_{L^2}\} = 0.$$

Under such hypotheses,

$$\begin{aligned} J(u_0) \ &= \ \min_{u \in U} J(u) \\ &= \ \max_{v^* \in A^*} \tilde{J}^*(v^*) \\ &= \ \tilde{J}^*(v_0^*) \\ &= \ J^*(v_0^*,z_0^*). \end{aligned} \tag{17.30}$$

Proof 17.3   Observe that, from the general result in Toland [78], we have

$$
\begin{aligned}
\inf_{z^*\in Y_2^*} J^*(v^*,z^*) &= \inf_{z^*\in Y_2^*}\left\{-F^*(z^*,Q)+G^*(z^*,N)\right\}\\
&= \inf_{z^*\in Y_2^*}\left\{-F^*(z^*,Q)+\frac{1}{2}\int_\Omega \overline{(-N_{\alpha\beta})^K}\,z_\alpha^*\,z_\beta^*\,dx\right.\\
&\qquad\left.-\frac{1}{2}\int_\Omega \overline{H}_{\alpha\beta\lambda\mu}N_{\alpha\beta}N_{\lambda\mu}\,dx\right\}\\
&\le -\langle Q_\alpha+z_\alpha^*\,,\,w_{,\alpha}\rangle_{L^2}+F(u)+\langle z_\alpha^*\,,\,w_{,\alpha}\rangle_{L^2}+\frac{1}{2}\langle N_{\alpha\beta}-K\delta_{\alpha\beta},w_{,\alpha}w_{,\beta}\rangle_{L^2}\\
&\qquad-\frac{1}{2}\int_\Omega \overline{H}_{\alpha\beta\lambda\mu}N_{\alpha\beta}N_{\lambda\mu}\,dx\\
&= -\langle Q_\alpha,w_{,\alpha}\rangle_{L^2}+F(u)-\frac{1}{2}\langle N_{\alpha\beta}-K\delta_{\alpha\beta},w_{,\alpha}w_{,\beta}\rangle_{L^2}\\
&\qquad-\frac{1}{2}\int_\Omega \overline{H}_{\alpha\beta\lambda\mu}N_{\alpha\beta}N_{\lambda\mu}\,dx.
\end{aligned}
\tag{17.31}
$$

From this,

$$
\begin{aligned}
\inf_{z^*\in Y_2^*} J^*(v^*,z^*) &\le -\langle P,w\rangle_{L^2}+G_2(\kappa(u))+\sup_{N\in Y_1^*}\left\{\frac{1}{2}\langle N_{\alpha\beta},w_{,\alpha}w_{,\beta}\rangle_{L^2}\right.\\
&\qquad\left.-\frac{1}{2}\int_\Omega \overline{H}_{\alpha\beta\lambda\mu}N_{\alpha\beta}N_{\lambda\mu}\,dx-\langle u_\alpha,N_{\alpha\beta,\beta}+P_\alpha\rangle_{L^2}\right\}\\
&= -\langle P,w\rangle_{L^2}+G_2(\kappa(u))+\sup_{N\in Y_1^*}\left\{\frac{1}{2}\langle N_{\alpha\beta},u_{\alpha,\beta}+u_{\beta,\alpha}+w_{,\alpha}w_{,\beta}\rangle_{L^2}\right.\\
&\qquad\left.-\frac{1}{2}\int_\Omega \overline{H}_{\alpha\beta\lambda\mu}N_{\alpha\beta}N_{\lambda\mu}\,dx-\langle u_\alpha,P_\alpha\rangle_{L^2}\right\}\\
&= G_2(\kappa(u))+G_1(\gamma(u))-\langle w,P\rangle_{L^2}-\langle u_\alpha,P_\alpha\rangle_{L^2}\\
&= J(u),\quad \forall u\in U,\ v^*=(Q,N)\in A^*.
\end{aligned}
\tag{17.32}
$$

Thus,

$$
\begin{aligned}
J(u) &\ge \inf_{z^*\in Y_2^*}\left\{-F^*(z^*,Q)+G^*(z^*,N)\right\}\\
&= \inf_{z^*\in Y_2^*} J^*(v^*,z^*)\\
&= \tilde{J}^*(v^*),
\end{aligned}
\tag{17.33}
$$

$\forall v^*\in A^*,\ u\in U.$

Summarizing,

$$
J(u)\ge \tilde{J}^*(v^*),\ \forall u\in U,\ v^*\in A^*,
$$

so that,

$$
\inf_{u\in U} J(u)\ge \sup_{v^*\in A^*}\tilde{J}^*(v^*).
\tag{17.34}
$$

Finally, assume $v_0^*=(N_0,Q_0)\in A^*$, $z_0^*\in Y_2^*$ and $u_0\in U$ are such that

$$
\delta\{J^*(v_0^*,z_0^*)-\langle(u_0)_\alpha,(N_0)_{\alpha\beta,\beta}+P_\alpha\rangle_{L^2}-\langle w_0,(Q_0)_{\alpha,\alpha}+P\rangle_{L^2}\}=0.
$$

From the variation in Q^* we get

$$\frac{\partial F^*(z_0^*, Q_0)}{\partial Q_\alpha} = (w_0)_{,\alpha},$$

so that from this and the variation in z^*, we get

$$
\begin{aligned}
\frac{\partial G^*(z_0^*, N_0)}{\partial z_\alpha^*} &= \frac{\partial F^*(z_0^*, Q_0)}{\partial z_\alpha^*} \\
&= \frac{\partial F^*(z_0^*, Q_0)}{\partial Q_\alpha} \\
&= (w_0)_{,\alpha} \\
&= \overline{N_{\alpha\beta}^K}(z_0^*)_\beta, \text{ in } \Omega.
\end{aligned}
\tag{17.35}
$$

Hence,

$$F^*(z_0^*, Q_0) = \langle (z_0^*)_\alpha + (Q_0)_\alpha, (w_0)_{,\alpha} \rangle_{L^2} - F(u_0).$$

Also, from (17.35) we have,

$$(z_0^*)_\alpha = (N_0)_{\alpha\beta}(w_0)_{,\beta} + K(w_0)_{,\alpha}.
\tag{17.36}$$

From such results and the variation in N we obtain

$$\frac{(u_0)_{\alpha,\beta} + (u_0)_{\beta,\alpha}}{2} + \frac{(w_0)_{,\alpha}(w_0)_{,\beta}}{2} - \overline{H}_{\alpha\beta\gamma\mu}(N_0)_{\lambda\mu} = 0,$$

so that

$$(N_0)_{\alpha\beta} = H_{\alpha\beta\lambda\mu}\gamma_{\lambda\mu}(u_0).$$

From these last results we have,

$$
\begin{aligned}
G^*(z_0^*, N_0) &= \langle (z_0^*)_\alpha, (w_0)_\alpha \rangle_{L^2} + \frac{1}{2}\langle (N_0)_{\alpha\beta} - K\delta_{\alpha\beta}, (w_0)_\alpha(w_0)_\beta \rangle_{L^2} \\
&\quad - \frac{1}{2}\int_\Omega \overline{H_{\alpha\beta\lambda\mu}}(N_0)_{\alpha\beta}(N_0)_{\lambda\mu}\, dx \\
&= \langle (z_0^*)_\alpha, (w_0)_\alpha \rangle_{L^2} + \frac{1}{2}\left\langle (N_0)_{\alpha\beta}, \frac{(u_0)_{\alpha,\beta} + (u_0)_{\beta,\alpha}}{2} + (w_0)_\alpha(w_0)_\beta \right\rangle_{L^2} \\
&\quad - \langle P_\alpha, (u_0)_\alpha \rangle_{L^2} - \frac{1}{2}\int_\Omega \overline{H_{\alpha\beta\lambda\mu}}(N_0)_{\alpha\beta}(N_0)_{\lambda\mu}\, dx \\
&\quad - \frac{K}{2}\langle (w_0)_{,\alpha}, (w_0)_{,\alpha} \rangle_{L^2} \\
&= \langle (z_0^*)_\alpha, (w_0)_\alpha \rangle_{L^2} + G_1(\gamma(u_0)) - \frac{K}{2}\langle (w_0)_{,\alpha}, (w_0)_{,\alpha} \rangle_{L^2} \\
&\quad - \langle P_\alpha, (u_0)_\alpha \rangle_{L^2}.
\end{aligned}
\tag{17.37}
$$

Joining the pieces, we obtain

$$
\begin{aligned}
J^*(v_0, z_0^*) &= -F^*(z_0^*, Q_0) + G^*(z_0^*, N_0) \\
&= F(u_0) + G_1(\gamma(u_0)) + \frac{K}{2}\langle (w_0)_\alpha, (w_0)_\alpha \rangle_{L^2} \\
&\quad - \langle P, w_0 \rangle_{L^2} - \langle P_\alpha, (u_0)_\alpha \rangle_{L^2} \\
&= G_1(\gamma(u_0)) + G_2(\kappa(u_0)) \\
&\quad - \langle P, w_0 \rangle_{L^2} - \langle P_\alpha, (u_0)_\alpha \rangle_{L^2} \\
&= J(u_0).
\end{aligned}
\tag{17.38}
$$

From this and from $v_0^* \in A_4$, we have

$$J(u_0) = J^*(v_0^*, z_0^*) = \inf_{z^* \in Y_2^*} J^*(v_0^*, z^*) = \tilde{J}^*(v_0^*).$$

Therefore, from such a last equality and (17.34), we may infer that

$$
\begin{aligned}
J(u_0) &= \min_{u \in U} J(u) \\
&= \max_{v^* \in A^*} \tilde{J}^*(v^*) \\
&= \tilde{J}^*(v_0^*) \\
&= J^*(v_0^*, z_0^*).
\end{aligned}
\tag{17.39}
$$

The proof is complete.

17.5   Existence and duality for a non-linear shell model

In this section we present the primal variational formulation concerning the shell model presented in [10]. Indeed, in some sense, the duality principle here presented extends the results developed in [10]. In fact, through a generalization of some ideas developed in [78], we have obtained a duality principle for which the membrane force tensor, concerning the global optimal solution of the primal formulation, may not be necessarily either positive or negative definite. We emphasize details on the Sobolev spaces involved may be found in [1]. About the fundamental concepts of convex analysis and duality here used, we would cite [64, 33, 14]. Similar problems are addressed in [14, 76, 77, 43].

At this point we start to describe the shell model in question. Let $D \subset \mathbb{R}^2$ be an open, bounded, connected set with a C^3 class boundary.

Let $S \subset \mathbb{R}^3$ be a C^3 class manifold, where

$$S = \{\mathbf{r}(\xi) \, : \, \xi = (\xi_1, \xi_2) \in \overline{D}\},$$

$\mathbf{r} : \overline{D} \subset \mathbb{R}^2 \to \mathbb{R}^3$ is a C^3 class function and where,

$$\mathbf{r}(\xi) = X_1(\xi)\mathbf{e}_1 + X_2(\xi)\mathbf{e}_2 + X_3(\xi)\mathbf{e}_3.$$

Here, $\{\mathbf{e}_1, \mathbf{e}_2, \mathbf{e}_3\}$ is the canonical basis of $\mathbb{R}^3$.

We assume S is the middle surface of a shell of constant thickness h so that we denote,

$$\mathbf{a}_\alpha = \frac{\partial \mathbf{r}}{\partial \xi_\alpha}, \; \forall \alpha \in \{1, 2\},$$

$$a_{\alpha\beta} = \mathbf{a}_\alpha \cdot \mathbf{a}_\beta.$$

Let

$$\mathbf{n} = \frac{\mathbf{a}_\alpha \times \mathbf{a}_\beta}{|\mathbf{a}_\alpha \times \mathbf{a}_\beta|},$$

be the unit normal to S, so that we define the covariant components $b_{\alpha\beta}$ of the curvature tensor

$$b = \{b_{\alpha\beta}\},$$

by

$$b_{\alpha\beta} = \mathbf{n} \cdot \mathbf{a}_{\alpha,\beta} = \mathbf{n} \cdot \mathbf{r}_{,\alpha\beta}.$$

Observe that

$$\mathbf{n} \cdot \mathbf{a}_\alpha = 0,$$

so that

$$\mathbf{n}_\beta \cdot \mathbf{a}_\alpha + \mathbf{n} \cdot \mathbf{a}_{\alpha,\beta} = 0,$$

and thus we obtain,

$$b_{\alpha\beta} = \mathbf{n} \cdot \mathbf{a}_{\alpha,\beta} = -\mathbf{n}_\beta \cdot \mathbf{a}_\alpha.$$

The Christofell symbols relating S, would be,

$$\Gamma_{\alpha\beta\gamma} = \frac{1}{2}(a_{\alpha\beta,\gamma} + a_{\alpha\gamma,\beta} - a_{\beta\gamma,\alpha}), \ \forall \alpha, \beta, \gamma \in \{1,2\}$$

and

$$\Gamma^\lambda_{\alpha\beta} = a^{\lambda\gamma}\Gamma_{\gamma\alpha\beta}, \ \forall \alpha, \beta, \lambda \in \{1,2\},$$

where

$$\{a^{\alpha\beta}\} = \{a_{\alpha\beta}\}^{-1}.$$

Let us denote with a bar the quantities relating the deformed middle surface.
So, the middle surface strain tensor $\gamma = \{\gamma_{\alpha\beta}\}$ is given by

$$\gamma_{\alpha\beta} = \frac{(\bar{a}_{\alpha\beta} - a_{\alpha\beta})}{2},$$

while the tensor relating change in curvature is given by

$$\kappa_{\alpha\beta} = -(\bar{b}_{\alpha\beta} - b_{\alpha\beta}), \ \forall \alpha, \beta \in \{1,2\}.$$

We also denote,

$$\begin{aligned} U \ &= \ \{u = (u_\alpha, w) = (u_1, u_2, w) \in W^{1,2}(S; \mathbb{R}^2) \times W^{2,2}(S) : \\ & u_\alpha = w = \frac{\partial w}{\partial \mathbf{n}} = 0, \text{ on } \partial S\} \\ &= \ W_0^{1,2}(S; \mathbb{R}^2) \times W_0^{2,2}(S), \end{aligned} \tag{17.40}$$

where, in order to simplify the analysis, the boundary conditions in question refer to a clamped shell and where ∂S denotes the boundary of S.

Also from reference [10], for moderately large rotations around tangents, the strain displacements relations are given by,

1.

$$\gamma_{\alpha\beta}(u) = \theta_{\alpha\beta}(u) + \frac{1}{2}\varphi_\alpha(u)\varphi_\beta(u),$$

2.

$$\begin{aligned} \kappa_{\alpha\beta}(u) \ = \ &-w_{|\alpha\beta} - b^\lambda_{\alpha|\beta}u_\lambda \\ &-b^\lambda_\alpha u_{\lambda|\beta} - b^\lambda_\beta u_{\lambda|\alpha} + b^\lambda_\alpha b_{\lambda\beta}w, \end{aligned} \tag{17.41}$$

where,

$$u_{\alpha|\beta} = u_{\alpha,\beta} - \Gamma^\lambda_{\alpha\beta}u_\lambda,$$

$$w_{|\alpha\beta} = w_{,\alpha\beta} - \Gamma^\lambda_{\alpha\beta}w_{,\lambda},$$

$$\theta_{\alpha\beta}(u) = \frac{1}{2}(u_{\alpha|\beta} + u_{\beta|\alpha}) - b_{\alpha\beta}w,$$

$$\varphi_\alpha(u) = w_{,\alpha} + b_\alpha^\beta u_\beta, \tag{17.42}$$

and

$$b_\alpha^\beta = b_{\alpha\lambda} a^{\lambda\beta}.$$

The primal shell inner energy is defined by

$$\begin{aligned}
W(\gamma(u), \kappa(u)) &= \frac{1}{2} \int_S H^{\alpha\beta\lambda\mu} \gamma_{\alpha\beta}(u) \gamma_{\lambda\mu}(u) \, dS \\
&\quad + \frac{1}{2} \int_S h^{\alpha\beta\lambda\mu} \kappa_{\alpha\beta}(u) \kappa_{\lambda\mu}(u) \, dS,
\end{aligned} \tag{17.43}$$

where

$$H^{\alpha\beta\lambda\mu} = \frac{Eh}{2(1+\nu)} \left(a^{\alpha\lambda} a^{\beta\mu} + a^{\alpha\mu} a^{\beta\lambda} + \frac{2\nu}{1-\nu} a^{\alpha\beta} a^{\lambda\mu} \right),$$

$$h^{\alpha\beta\lambda\mu} = \frac{h^2}{12} H^{\alpha\beta\lambda\mu},$$

and E denotes the Young's modulus and ν the Poisson ratio.
The constitutive relations are,

$$N^{\alpha\beta} = H^{\alpha\beta\lambda\mu} \gamma_{\lambda\mu}(u),$$

and

$$M^{\alpha\beta} = h^{\alpha\beta\lambda\mu} \kappa_{\lambda\mu}(u),$$

where $\{N^{\alpha\beta}\}$ is the membrane force tensor and $\{M^{\alpha\beta}\}$ is the moment one.
We assume

$$H = \{H^{\alpha\beta\lambda\mu}\}$$

to be positive definite in the sense that there exists $c_0 > 0$ such that

$$H^{\alpha\beta\lambda\mu} t_{\alpha\beta} t_{\lambda\mu} \geq c_0 t_{\alpha\beta} t_{\alpha\beta} \geq 0, \ \forall t = \{t_{\alpha\beta}\} \in \mathbb{R}^4.$$

Finally, the primal variational formulation for this model will be given by

$$J : U \to \mathbb{R}$$

where,

$$J(u) = W(\gamma(u), \kappa(u)) - \langle u, \mathbf{f} \rangle_{L^2},$$

$$W(\gamma(u), \kappa(u)) = G_1(\gamma(u)) + G_2(\kappa(u)),$$

$$G_1(\gamma(u)) = \frac{1}{2} \int_S H^{\alpha\beta\lambda\mu} \gamma_{\alpha\beta}(u) \gamma_{\lambda\mu}(u) \, dS,$$

$$G_2(\kappa(u)) = \frac{1}{2} \int_S h^{\alpha\beta\lambda\mu} \kappa_{\alpha\beta}(u) \kappa_{\lambda\mu}(u) \, dS,$$

and

$$\langle u, \mathbf{f} \rangle_{L^2} = \int_S (P^\alpha u_\alpha + Pw) \, dS.$$

Here

$$\mathbf{f} = (P^\alpha, P) \in L^2(S; \mathbb{R}^3),$$

are the external loads distributed on S, P^α relating the directions $\mathbf{a}_\alpha$ and P relating the direction $\mathbf{n}$, respectively.
Moreover, generically for $f_1, f_2 \in L^2(S)$, we denote,

$$\langle f_1, f_2 \rangle_{L^2} = \int_S f_1 f_2 \, dS,$$

where $dS = \sqrt{a} \, d\xi_1 \, d\xi_2$, and $a = det\{a_{\alpha\beta}\}$.

17.6 Existence of a minimizer

Theorem 17.6.1 *Consider the statements and assumptions concerning the shell model described in the last section.*

More specifically, consider the functional $J : U \to \mathbb{R}$ given, as above described, by

$$J(u) = W(\gamma(u), \kappa(u)) - \langle u, \mathbf{f} \rangle_{L^2},$$

where,

$$
\begin{aligned}
U &= \{u = (u_\alpha, w) = (u_1, u_2, w) \in W^{1,2}(S; \mathbb{R}^2) \times W^{2,2}(S) : \\
&\quad u_\alpha = w = \frac{\partial w}{\partial \mathbf{n}} = 0, \text{ on } \partial S\} \\
&= W_0^{1,2}(S; \mathbb{R}^2) \times W_0^{2,2}(S).
\end{aligned}
\tag{17.44}
$$

Let T_0 be such that,

$$\|T_0\|_2^2 = \min_{T \in L^2(S; \mathbb{R}^{2 \times 2})} \{\|T\|_2^2\},$$

subject to

$$\frac{(\sqrt{g} T_{\alpha\beta})_{,\beta}}{\sqrt{g}} + \Gamma^\alpha_{\lambda\beta} T_{\lambda\beta} + P_\alpha = 0 \text{ in } D.$$

Assume

$$J_1(u) \to +\infty, \text{ as } \langle w_{,\alpha\beta}, w_{,\alpha\beta} \rangle_{L^2} + \langle u_{\alpha,\beta}, u_{\alpha,\beta} \rangle_{L^2} \to +\infty, \tag{17.45}$$

where

$$J_1(u) = G_2(\kappa(u)) + \frac{1}{2} \langle (T_0)_{\alpha\beta}, \varphi_\alpha(u)\varphi_\beta(u) \rangle - \langle (T_0)_{\alpha\beta} b_{\alpha\beta} + P, w \rangle_{L^2}.$$

Under such hypotheses, there exists $u_0 \in U$ such that,

$$J(u_0) = \min_{u \in U} \{J(u)\}.$$

Proof 17.4

Observe that

$$
\begin{aligned}
\langle P_\alpha, u_\alpha \rangle_{L^2} &= -\int_D \left(\frac{(\sqrt{g}(T_0)_{\alpha\beta})_{,\beta}}{\sqrt{g}} + \Gamma^\alpha_{\lambda\beta} (T_0)_{\lambda\beta} \right) u_\alpha \sqrt{g} \, d\xi \\
&= \left\langle (T_0)_{\alpha\beta}, \frac{u_{\alpha,\beta} + u_{\beta,\alpha}}{2} - \Gamma^\lambda_{\alpha\beta} u_\lambda \right\rangle_{L^2} \\
&= \langle (T_0)_{\alpha\beta}, \theta_{\alpha\beta}(u) + b_{\alpha\beta} w \rangle_{L^2}, \tag{17.46}
\end{aligned}
$$

so that defining

$$v_{\alpha\beta}(u) = \theta_{\alpha\beta}(u) + \frac{1}{2} \varphi_\alpha(u)\varphi_\beta(u), \tag{17.47}$$

we obtain

$$\langle P_\alpha, u_\alpha \rangle_{L^2} = \left\langle (T_0)_{\alpha\beta}, v_{\alpha\beta}(u) - \frac{1}{2} \varphi_\alpha(u)\varphi_\beta(u) + b_{\alpha\beta} w \right\rangle_{L^2}.$$

Thus,

$$
\begin{aligned}
J(u) &= G_1(v(u)) - \langle (T_0)_{\alpha\beta}, v_{\alpha\beta}(u) \rangle_{L^2} \\
&\quad + G_2(\kappa(u)) + \frac{1}{2} \langle (T_0)_{\alpha\beta}, \varphi_\alpha(u)\varphi_\alpha(u) \rangle_{L^2} \\
&\quad - \langle (T_0)_{\alpha\beta} b_{\alpha\beta} + P, w \rangle_{L^2}, \tag{17.48}
\end{aligned}
$$

From this and hypothesis (17.45), J is bounded below. So, there exists $\alpha_1 \in \mathbb{R}$ such that

$$\alpha_1 = \inf_{u \in U} J(u).$$

Let $\{u_n\}$ be a minimizing sequence for J.

From (17.48) and also from the hypotheses (17.45), $\{\|w_n\|_{2,2}\}$ and $\{\|(u_n)_\alpha\|_{1,2}\}$ are bounded. So there exists $w_0 \in W_0^{2,2}(S)$ and $\{(u_0)_\alpha\} \in W_0^{1,2}(S;\mathbb{R}^2)$ such that, up to a not relabeled subsequence,

$$(w_n)_{,\alpha\beta} \rightharpoonup (w_0)_{,\alpha\beta} \text{ weakly in } L^2(S), \ \forall \alpha, \beta \in \{1,2\},$$

$$(w_n)_{,\alpha} \to (w_0)_{,\alpha} \text{ strongly in } L^2(S), \ \forall \alpha \in \{1,2\}, \tag{17.49}$$

$$(u_n)_{\alpha,\beta} \rightharpoonup (u_0)_{\alpha,\beta} \text{ weakly in } L^2(S), \ \forall \alpha, \beta \in \{1,2\}, \tag{17.50}$$

$$(u_n)_\alpha \to (u_0)_\alpha \text{ strongly in } L^2(S), \ \forall \alpha \in \{1,2\}. \tag{17.51}$$

Also from (17.48), there exists $K_2 > 0$ such that,

$$\|(v_n)_{\alpha\beta}(u)\|_2 < K_2, \forall \alpha, \beta \in \{1,2\}, \ n \in \mathbb{N},$$

and thus, from this, (17.42), (17.47), (17.49) and (17.51), we may infer that there exists $K_3 > 0$ such that

$$\|\theta_{\alpha\beta}(u_n)\|_2 < K_3, \forall \alpha, \beta \in \{1,2\}, \ n \in \mathbb{N}.$$

Thus, from this, (17.51) and (17.50), up to a subsequence not relabeled,

$$\theta_{\alpha\beta}(u_n) \rightharpoonup \theta_{\alpha\beta}(u_0), \text{ weakly in } L^2,$$

$\forall \alpha, \beta \in \{1,2\}$, as $n \to \infty$.

From this, also up to a subsequence not relabeled, we have

$$\gamma_{\alpha\beta}(u_n) \rightharpoonup \gamma_{\alpha\beta}(u_0), \text{ weakly in } L^2,$$

$\forall \alpha, \beta \in \{1,2\}$, and

$$\kappa_{\alpha\beta}(u_n) \rightharpoonup \kappa_{\alpha\beta}(u_0), \text{ weakly in } L^2,$$

$\forall \alpha, \beta \in \{1,2\}$.

Therefore, from the convexity of G_1 in γ and G_2 in κ we obtain

$$
\begin{aligned}
\inf_{u \in U} J(u) &= \alpha_1 \\
&= \liminf_{n \to \infty} J(u_n) \\
&\geq J(u_0).
\end{aligned}
\tag{17.52}
$$

Thus,

$$J(u_0) = \min_{u \in U} J(u).$$

The proof is complete.

17.7 The duality principle for the shell model

At this point we present the main duality principle, which is summarized by the next theorem.

Theorem 17.7.1 *Consider the statements and assumptions concerning the shell model described in the last two sections.*

More specifically, consider the functional $J : U \to \mathbb{R}$ given, as above described by,

$$J(u) = W(\gamma(u), \kappa(u)) - \langle u, \mathbf{f} \rangle_{L^2},$$

where,

$$
\begin{aligned}
U &= \{ u = (u_\alpha, w) = (u_1, u_2, w) \in W^{1,2}(S; \mathbb{R}^2) \times W^{2,2}(S) \; : \\
& \quad u_\alpha = w = \frac{\partial w}{\partial \mathbf{n}} = 0, \; on \; \partial S \} \\
&= W_0^{1,2}(S; \mathbb{R}^2) \times W_0^{2,2}(S).
\end{aligned}
\tag{17.53}
$$

Under such hypotheses,

$$
\begin{aligned}
& \inf_{u \in U} J(u) \\
& \geq \sup_{v^* \in A^*} \{ \inf_{z^* \in Y_2^*} \{ -F^*(z^*, Q) + G^*(z^*, N) \} \},
\end{aligned}
\tag{17.54}
$$

where,

$$F(u) = G_2(\kappa(u)) + \frac{K}{2} \langle \varphi_\alpha(u), \varphi_\alpha(u) \rangle_{L^2}, \forall u \in U.$$

Moreover, $F^* : [Y_2^*]^2 \to \mathbb{R}$ is defined by,

$$F^*(z^*, Q) = \sup_{u \in U} \{ \langle z_\alpha^* + Q_\alpha, \varphi_\alpha(u) \rangle_{L^2} - F(u) \}, \; \forall (z^*, Q) \in [Y_2^*]^2.$$

Also,

$$
\begin{aligned}
G(v) &= -\frac{1}{2} \int_S H_{\alpha\beta\lambda\mu} \left[(v_1)_{\alpha\beta} + \frac{(v_2)_\alpha (v_2)_\beta}{2} \right] \left[(v_1)_{\lambda\mu} + \frac{(v_2)_\lambda (v_2)_\mu}{2} \right] dS \\
& \quad + \frac{K}{2} \langle (v_2)_\alpha, (v_2)_\alpha \rangle_{L^2},
\end{aligned}
\tag{17.55}
$$

$$
\begin{aligned}
G^*(z^*, N) &= \sup_{v_2 \in Y_2} \{ \inf_{v_1 \in Y_1} \{ \langle N^{\alpha\beta}, (v_1)_{\alpha\beta} \rangle_{L^2} + \langle z_\alpha^*, (v_2)_\alpha \rangle_{L^2} - G(v) \} \} \\
&= \frac{1}{2} \int_S \overline{(-N^{\alpha\beta})^K} z_\alpha^* z_\beta^* \, dS \\
& \quad - \frac{1}{2} \int_S \overline{H}_{\alpha\beta\lambda\mu} N^{\alpha\beta} N^{\lambda\mu} \, dS,
\end{aligned}
\tag{17.56}
$$

if $v^* = (Q, N) \in A_3$, where

$$A_3 = \{ v^* \in Y^* \; : \; \{ (-N^{\alpha\beta})_K \} \; is \; positive \; definite \; in \; \Omega \},$$

$$
\{ (-N^{\alpha\beta})_K \} = \left\{
\begin{array}{cc}
-N^{11} + K & -N^{12} \\
-N^{21} & -N^{22} + K
\end{array}
\right\},
\tag{17.57}
$$

and

$$\overline{(-N^{\alpha\beta})_K} = \{(-N^{\alpha\beta})_K\}^{-1}.$$

Moreover, defining

$$Y^* = Y = L^2(S;\mathbb{R}^2) \times L^2(S;\mathbb{R}^{2\times2}),$$
$$Y_1^* = Y_1 = L^2(S;\mathbb{R}^{2\times2}),$$

and

$$Y_2^* = Y_2 = L^2(S;\mathbb{R}^2),$$

also,

$$A^* = A_1 \cap A_2 \cap A_3 \cap A_4 \cap A_5,$$

where,

$$A_1 = \{v^* = (Q,N) \in Y^* : -N^{\alpha\beta}|_\beta + b_\lambda^\alpha Q^\lambda - P^\alpha = 0, \text{ in } S\}, \tag{17.58}$$

$$A_2 = \{v^* = (Q,N) \in Y^* : -b_{\alpha\beta}N^{\alpha\beta} - Q_{|\alpha}^\alpha - P = 0, \text{ in } S\}. \tag{17.59}$$

Moreover,

$$\{\overline{H}_{\alpha\beta\lambda\mu}\} = \{H^{\alpha\beta\lambda\mu}\}^{-1},$$

and,

$$\{\overline{h}_{\alpha\beta\lambda\mu}\} = \{h^{\alpha\beta\lambda\mu}\}^{-1}.$$

Also,

$$A_4 = \{v^* = (Q,N) \in Y^* : \hat{J}^*(z^*) > 0,$$
$$\forall z^* \in Y_2^*, \text{ such that } z^* \neq \mathbf{0}\}, \tag{17.60}$$

and

$$A_5 = \left\{v^* = (Q,N) \in Y^* : \{N_{\alpha\beta} + K\delta_{\alpha\beta}\} \geq \frac{K}{2}\{\delta_{\alpha\beta}\}\right\},$$

where $K > 0$ is supposed to be such that, in an appropriate matrix sense,

$$\{\overline{H}_{\alpha\beta\lambda\mu}\} > \frac{2}{K}\{\delta_{\alpha\beta}\}.$$

Furthermore,

$$\hat{J}^*(z^*) = -F^*(z^*,\mathbf{0}) + G^*(z^*,\mathbf{0})$$
$$= -F^*(z^*,\mathbf{0}) + \frac{1}{2}\int_S \overline{(-N^{\alpha\beta})_K}z_\alpha^* z_\beta^* \, dS, \forall z^* \in Y_2^*.$$

Finally, denoting

$$J^*(v^*,z^*) = -F^*(z^*,Q) + G^*(z^*,N), \forall(v^*,z^*) \in A^* \times Y_2^*,$$

and

$$\tilde{J}^*(v^*) = \inf_{z^* \in Y_2^*} J^*(v^*,z^*), \forall v^* \in A^*,$$

suppose there exist $v_0^* = (N_0,Q_0) \in A^*$, $z_0^* \in Y_2^*$ and $u_0 \in U$ such that

$$\delta\{J^*(v_0^*,z_0^*) - \langle\theta_{\alpha\beta}(u_0),(N_0)^{\alpha\beta}\rangle_{L^2}$$
$$+ \langle(u_0)_\alpha,P_\alpha\rangle_{L^2} - \langle(\varphi_0)_\alpha,(Q_0)_\alpha\rangle_{L^2} + \langle w_0,P\rangle_{L^2}\} = 0. \tag{17.61}$$

Under such hypotheses,

$$
\begin{aligned}
J(u_0) &= \min_{u \in U} J(u) \\
&= \max_{v^* \in A^*} \tilde{J}^*(v^*) \\
&= \tilde{J}^*(v_0^*) \\
&= J^*(v_0^*, z_0^*).
\end{aligned}
$$ (17.62)

Proof 17.5

Observe that, from the general result in Toland [78], we have

$$
\begin{aligned}
\inf_{z^* \in Y_2^*} J^*(v^*, z^*) &= \inf_{z^* \in Y_2^*} \{-F^*(z^*, Q) + G^*(z^*, N)\} \\
&= \inf_{z^* \in Y_2^*} \left\{ -F^*(z^*, Q) + \frac{1}{2} \int_S \overline{(-N^{\alpha\beta})_K} z_\alpha^* z_\beta^* \; dx \right. \\
&\quad \left. - \frac{1}{2} \int_S \overline{H}_{\alpha\beta\lambda\mu} N^{\alpha\beta} N^{\lambda\mu} \, dS \right\} \\
&\leq -\langle Q_\alpha + z_\alpha^*;, \varphi_\alpha(u) \rangle_{L^2} + F(u) \\
&\quad + \langle z_\alpha^*, \varphi_\alpha(u) \rangle_{L^2} + \frac{1}{2} \langle N^{\alpha\beta} - K\delta_{\alpha\beta}, \varphi_\alpha(u)\varphi_\beta(u) \rangle_{L^2} \\
&\quad - \frac{1}{2} \int_S \overline{H}_{\alpha\beta\lambda\mu} N^{\alpha\beta} N^{\lambda\mu} \, dS \\
&= -\langle Q_\alpha, \varphi_\alpha(u) \rangle_{L^2} + F(u) - \frac{1}{2} \langle N^{\alpha\beta} - K\delta_{\alpha\beta}, \varphi_\alpha(u)\varphi_\beta(u) \rangle_{L^2} \\
&\quad - \frac{1}{2} \int_\Omega \overline{H}_{\alpha\beta\lambda\mu} N^{\alpha\beta} N^{\lambda\mu} \, dS
\end{aligned}
$$ (17.63)

From this, we have

$$
\begin{aligned}
\inf_{z^* \in Y_2^*} J^*(v^*, z^*) &\leq -\langle P, w \rangle_{L^2} + G_2(\kappa(u)) + \sup_{N \in Y_1^*} \left\{ \frac{1}{2} \langle N^{\alpha\beta}, \varphi_\alpha(u)\varphi_\beta(u) \rangle_{L^2} \right. \\
&\quad \left. - \frac{1}{2} \int_\Omega \overline{H}_{\alpha\beta\lambda\mu} N^{\alpha\beta} N^{\lambda\mu} \, dS - \langle \theta_{\alpha\beta}(u), N^{\alpha\beta} \rangle_{L^2} - \langle u_\alpha, P_\alpha \rangle_{L^2} \rangle_{L^2} \right\} \\
&= -\langle P, w \rangle_{L^2} + G_2(\kappa(u)) + \sup_{N \in Y_1^*} \left\{ \frac{1}{2} \langle N^{\alpha\beta}, \theta_{\alpha\beta}(u) + \varphi_\alpha(u)\varphi_\beta(u) \rangle_{L^2} \right. \\
&\quad \left. - \frac{1}{2} \int_S \overline{H}_{\alpha\beta\lambda\mu} N^{\alpha\beta} N^{\lambda\mu} \, dS - \langle u_\alpha, P_\alpha \rangle_{L^2} \right\} \\
&= G(\kappa(u)) + G_1(\gamma(u)) - \langle w, P \rangle_{L^2} - \langle u_\alpha, P_\alpha \rangle_{L^2} \\
&= J(u), \forall u \in U, \, v^* = (Q, N) \in A^*.
\end{aligned}
$$ (17.64)

Thus,

$$
\begin{aligned}
J(u) &\geq \inf_{z^* \in Y_2^*} \{-F^*(z^*, Q) + G^*(z^*, N)\} \\
&= \inf_{z^* \in Y_2^*} J^*(v^*, z^*) \\
&= \tilde{J}^*(v^*),
\end{aligned}
$$ (17.65)

$\forall v^* \in A^*, \, u \in U.$

Summarizing,

$$J(u) \geq \tilde{J}^*(v^*), \; \forall u \in U, \; v^* \in A^*,$$

so that,

$$\inf_{u \in U} J(u) \geq \sup_{v^* \in A^*} \tilde{J}^*(v^*). \tag{17.66}$$

Finally, assume $v_0^* = (N_0, Q_0) \in A^*$, $z_0^* \in Y_2^*$ and $u_0 \in U$ are such that

$$\delta\{J^*(v_0^*, z_0^*) - \langle \theta_{\alpha\beta}(u_0), (N_0)^{\alpha\beta} \rangle_{L^2}$$
$$+ \langle (u_0)_\alpha, P_\alpha \rangle_{L^2} - \langle (\varphi_0)_\alpha, (Q_0)_\alpha \rangle_{L^2} + \langle w_0, P \rangle_{L^2}\} = 0. \tag{17.67}$$

From the variation in Q^* we get

$$\frac{\partial F^*(z_0^*, Q_0)}{\partial Q_\alpha} = \varphi_\alpha(u_0), \tag{17.68}$$

so that from this and the variation in z^*, we get

$$\begin{aligned}
\frac{\partial G^*(z_0^*, Q)}{\partial z_\alpha^*} &= \frac{\partial F^*(z_0^*, Q_0)}{\partial z_\alpha^*} \\
&= \frac{\partial F^*(z_0^*, Q_0)}{\partial Q_\alpha} \\
&= \varphi_\alpha(u_0) \\
&= \overline{N_K^{\alpha\beta}}(z_0^*)_\beta, \text{ in } \Omega. \tag{17.69}
\end{aligned}$$

Hence,

$$F^*(z_0^*, Q_0) = \langle (z_0^*)_\alpha + (Q_0)_\alpha, \varphi_\alpha(u_0) \rangle_{L^2} - F(u_0). \tag{17.70}$$

Also, from (17.69) we have,

$$(z_0^*)_\alpha = (N_0)^{\alpha\beta} \varphi_\beta(u_0) + K\varphi_\alpha(u_0). \tag{17.71}$$

From such results and the variation in N we obtain

$$\frac{(u_0)_{\alpha,\beta} + (u_0)_{\beta,\alpha}}{2} + \frac{\varphi_\alpha(u_0)\varphi_\beta(u_0)}{2} - \overline{H}_{\alpha\beta\gamma\mu}(N_0)^{\lambda\mu} = 0,$$

so that

$$(N_0)^{\alpha\beta} = H_{\alpha\beta\lambda\mu}\gamma_{\lambda\mu}(u_0).$$

From these last results we obtain,

$$\begin{aligned}
G^*(z_0^*, N_0) &= \langle (z_0^*)_\alpha, \varphi_\alpha(u_0)_\alpha \rangle_{L^2} + \frac{1}{2}\left\langle (N_0)^{\alpha\beta} - K\delta_{\alpha\beta}, \varphi_\alpha(u_0)\varphi_\beta(u_0) \right\rangle_{L^2} \\
&\quad - \frac{1}{2}\int_\Omega \overline{H_{\alpha\beta\lambda\mu}}(N_0)^{\alpha\beta}(N_0)^{\lambda\mu} \, dS \\
&= \langle (z_0^*)_\alpha, \varphi_\alpha(u_0)_\alpha \rangle_{L^2} + \left\langle (N_0)^{\alpha\beta}, \theta_{\alpha\beta}(u_0) + \frac{1}{2}\varphi_\alpha(u_0)\varphi_\beta(u_0) \right\rangle_{L^2} \\
&\quad - \langle P_\alpha, (u_0)_\alpha \rangle_{L^2} - \frac{1}{2}\int_\Omega \overline{H_{\alpha\beta\lambda\mu}}(N_0)^{\alpha\beta}(N_0)^{\lambda\mu} \, dS - \frac{K}{2}\langle \varphi_\alpha(u_0), \varphi_\alpha(u_0) \rangle_{L^2} \\
&= \langle (z_0^*)_\alpha, \varphi_\alpha(u_0) \rangle_{L^2} + G_1(\gamma(u_0)) - \frac{K}{2}\langle \varphi_\alpha(u_0), \varphi_\alpha(u_0) \rangle_{L^2} \\
&\quad - \langle P_\alpha, (u_0)_\alpha \rangle_{L^2}. \tag{17.72}
\end{aligned}$$

Joining the pieces, we obtain

$$
\begin{aligned}
J^*(v_0, z_0^*) &= -F^*(z_0^*, Q_0) + G^*(z_0^*, N_0) \\
&= F(u_0) + G_1(\gamma(u_0)) + \frac{K}{2} \langle \varphi_\alpha(u_0), \varphi_\alpha(u_0) \rangle_{L^2} \\
&\quad - \langle P, w_0 \rangle_{L^2} - \langle P_\alpha, (u_0)_\alpha \rangle_{L^2} \\
&= G_1(\gamma(u_0)) + G_2(\kappa(u_0)) \\
&\quad - \langle P, w_0 \rangle_{L^2} - \langle P_\alpha, (u_0)_\alpha \rangle_{L^2} \\
&= J(u_0).
\end{aligned}
\tag{17.73}
$$

From this and from $v_0^* \in A_4$, we obtain

$$
J(u_0) = J^*(v_0^*, z_0^*) = \inf_{z^* \in Y_2^*} J^*(v_0^*, z^*) = \tilde{J}^*(v_0^*).
$$

Therefore, from such a last equality and (17.66), we may infer that

$$
\begin{aligned}
J(u_0) &= \min_{u \in U} J(u) \\
&= \max_{v^* \in A^*} \tilde{J}^*(v^*) \\
&= \tilde{J}^*(v_0^*) \\
&= J^*(v_0^*, z_0^*).
\end{aligned}
\tag{17.74}
$$

The proof is complete.

17.8 Conclusion

In this article, in a first step, we have developed new proofs of global existence of minimizers for the Kirchhoff-Love plate and a shell model presented in [10].

In a second step, we have developed duality principles for these same models. In [10], the authors developed a duality principle valid for the special case in which the membrane force tensor at a critical point is positive definite. We have generalized such a result, considering that in our approach, such a previous case is included but here we do not request the optimal membrane force to be either positive or negative definite. Thus, in some sense, we have complemented the important work developed in [10]. We would emphasize sufficient optimality conditions are presented and the results here developed are applicable to a great variety of problems, including other shell models.

Chapter 18

A Primal Dual Formulation and a Multi-Duality Principle for a Non-Linear Model of Plates

18.1   Introduction

In this chapter we develop a new primal dual variational formulation for the Kirchhoff-Love non-linear plate model. We emphasize the results here presented may be applied to a large class of non-convex variational problems.

At this point we start to describe the primal formulation.

Let $\Omega \subset \mathbb{R}^2$ be an open, bounded, connected set which represents the middle surface of a plate of thickness h. The boundary of Ω, which is assumed to be regular (Lipschitzian), is denoted by $\partial\Omega$. The vectorial basis related to the cartesian system $\{x_1, x_2, x_3\}$ is denoted by $(\mathbf{a}_\alpha, \mathbf{a}_3)$, where $\alpha = 1, 2$ (in general Greek indices stand for 1 or 2), and where $\mathbf{a}_3$ is the vector normal to Ω, whereas $\mathbf{a}_1$ and $\mathbf{a}_2$ are orthogonal vectors parallel to Ω. Also, $\mathbf{n}$ is the outward normal to the plate surface.

The displacements will be denoted by

$$\hat{\mathbf{u}} = \{\hat{u}_\alpha, \hat{u}_3\} = \hat{u}_\alpha \mathbf{a}_\alpha + \hat{u}_3 \mathbf{a}_3.$$

The Kirchhoff-Love relations are

$$\hat{u}_\alpha(x_1, x_2, x_3) = u_\alpha(x_1, x_2) - x_3 w(x_1, x_2)_{,\alpha}$$
$$\text{and } \hat{u}_3(x_1, x_2, x_3) = w(x_1, x_2). \tag{18.1}$$

Here $-h/2 \le x_3 \le h/2$ so that we have $u = (u_\alpha, w) \in U$ where

$$U = \{(u_\alpha, w) \in W^{1,2}(\Omega; \mathbb{R}^2) \times W^{2,2}(\Omega),$$
$$u_\alpha = w = \frac{\partial w}{\partial \mathbf{n}} = 0 \text{ on } \partial\Omega\}$$
$$= W_0^{1,2}(\Omega; \mathbb{R}^2) \times W_0^{2,2}(\Omega).$$

It is worth emphasizing that the boundary conditions here specified refer to a clamped plate.

We define the operator $\Lambda : U \to Y_1 \times Y_1$, where $Y_1 = Y_1^* = L^2(\Omega; \mathbb{R}^{2 \times 2})$, by

$$\Lambda(u) = \{\gamma(u), \kappa(u)\},$$

$$\gamma_{\alpha\beta}(u) = \frac{u_{,\alpha,\beta} + u_{\beta,\alpha}}{2} + \frac{w_{,\alpha}w_{,\beta}}{2},$$

$$\kappa_{\alpha\beta}(u) = -w_{,\alpha\beta}.$$

The constitutive relations are given by

$$N_{\alpha\beta}(u) = H_{\alpha\beta\lambda\mu}\gamma_{\lambda\mu}(u), \tag{18.2}$$

$$M_{\alpha\beta}(u) = h_{\alpha\beta\lambda\mu}\kappa_{\lambda\mu}(u), \tag{18.3}$$

where: $\{H_{\alpha\beta\lambda\mu}\}$ and $\{h_{\alpha\beta\lambda\mu} = \frac{h^2}{12} H_{\alpha\beta\lambda\mu}\}$, are symmetric positive definite fourth order tensors. From now on, we denote $\{\overline{H}_{\alpha\beta\lambda\mu}\} = \{H_{\alpha\beta\lambda\mu}\}^{-1}$ and $\{\bar{h}_{\alpha\beta\lambda\mu}\} = \{h_{\alpha\beta\lambda\mu}\}^{-1}$.

Furthermore $\{N_{\alpha\beta}\}$ denote the membrane stress tensor and $\{M_{\alpha\beta}\}$ the moment one. The plate stored energy, represented by $(G \circ \Lambda) : U \to \mathbb{R}$ is expressed by

$$(G \circ \Lambda)(u) = \frac{1}{2} \int_\Omega N_{\alpha\beta}(u)\gamma_{\alpha\beta}(u)\,dx + \frac{1}{2} \int_\Omega M_{\alpha\beta}(u)\kappa_{\alpha\beta}(u)\,dx \tag{18.4}$$

and the external work, represented by $F_1 : U \to \mathbb{R}$, is given by

$$F_1(u) = \langle w, P \rangle_{L^2(\Omega)} + \langle u_\alpha, P_\alpha \rangle_{L^2(\Omega)}, \tag{18.5}$$

where $P, P_1, P_2 \in L^2(\Omega)$ are external loads in the directions $\mathbf{a}_3$, $\mathbf{a}_1$ and $\mathbf{a}_2$ respectively. The potential energy, denoted by $J : U \to \mathbb{R}$ is expressed by:

$$J(u) = (G \circ \Lambda)(u) - F_1(u)$$

Finally, we also emphasize from now on, as their meaning are clear, we may denote $L^2(\Omega)$ and $L^2(\Omega; \mathbb{R}^{2 \times 2})$ simply by L^2, and the respective norms by $\|\cdot\|_2$. Moreover derivatives are always understood in the distributional sense, $\mathbf{0}$ may denote the zero vector in appropriate Banach spaces and, the following and relating notations are used:

$$w_{,\alpha\beta} = \frac{\partial^2 w}{\partial x_\alpha \partial x_\beta},$$

$$u_{\alpha,\beta} = \frac{\partial u_\alpha}{\partial x_\beta},$$

$$N_{\alpha\beta,1} = \frac{\partial N_{\alpha\beta}}{\partial x_1},$$

and

$$N_{\alpha\beta,2} = \frac{\partial N_{\alpha\beta}}{\partial x_2}.$$

Here we emphasize the general Einstein convention of sum of repeated indices holds throughout the text unless otherwise indicated.

Remark 18.1   About the references, details on the Sobolev involved may be found in [1]. Mandatory references are the original results of Telega and his co-workers in [10, 11, 76, 77]. About convex analysis the results here developed follow in some extent [14], for which the main references are [33, 78].

We emphasize, our results complement, in some sense, the original ones presented in [10, 11, 76, 77].

Finally, existence results for models in elasticity including the plate model here addressed are developed in [29, 30, 31]. Similar problems are addressed in [39, 43].   ■

18.2   The first duality principle

Theorem 18.2.1 *Let* $\Omega \subset \mathbb{R}^2$ *be an open, bounded, connected set with a regular (Lipschitzian) boundary denoted by* $\partial\Omega$. *Let* $U = U_1 \times U_2$ *where* $U_1 = W_0^{1,2}(\Omega; \mathbb{R}^2)$ *and* $U_2 = W_0^{2,2}(\Omega)$. *We recall that* $Y_1 = Y_1^* = L^2(\Omega; \mathbb{R}^{2 \times 2})$ *and define* $Y_2 = Y_2^* = L^2(\Omega; \mathbb{R}^2)$,

$$\gamma_{\alpha\beta}(u) = \frac{1}{2}(u_{\alpha,\beta} + u_{\beta,\alpha}) + \frac{1}{2}w_{,\alpha}w_{,\beta},$$

$$\kappa_{\alpha\beta}(u) = -w_{,\alpha\beta}, \quad \forall u \in U, \ \alpha,\beta \in \{1,2\}$$

and $J : U \to \mathbb{R}$ *by,*

$$J(u) = G_1(\gamma(u)) + G_2(\kappa(u)) - \langle u, f \rangle_{L^2},$$

where

$$G_1(\gamma(u)) = \frac{1}{2}\int_{\Omega} H_{\alpha\beta\lambda\mu}\gamma_{\alpha\beta}(u)\gamma_{\lambda\mu}(u)\,dx,$$

$$G_2(\kappa(u)) = \frac{1}{2}\int_{\Omega} h_{\alpha\beta\lambda\mu}\kappa_{\alpha\beta}(u)\kappa_{\lambda\mu}(u)\,dx,$$

and

$$\langle u, f \rangle_{L^2} = \langle w, P \rangle_{L^2} + \langle u_\alpha, P_\alpha \rangle_{L^2},$$

where

$$f = (P_1, P_2, P) \in L^2(\Omega; \mathbb{R}^3).$$

In the next lines we shall denote

$$w = -(h_{\alpha\beta\lambda\mu}D_{\lambda\mu}^*D_{\alpha\beta})^{-1}(\ div\ Q + \ div\ z^* - P),$$

if

$$div\ Q + \ div\ z^* - P = -h_{\alpha\beta\lambda\mu}w_{,\alpha\beta\lambda\mu}$$

and

$$w \in U_2.$$

Define also $G_2^* : Y_2^* \times Y_2^* \to \mathbb{R}$, *by*

$$
\begin{aligned}
&G_2^*(z^*, Q) \\
&= \sup_{w \in U}\{\langle w_{,\alpha}, z_\alpha^* + Q_\alpha \rangle_{L^2} - G_2(\kappa(u)) + \langle P, w \rangle_{L^2}\} \\
&= \frac{1}{2}\int_{\Omega}\{(h_{\alpha\beta\lambda\mu}D_{\lambda\mu}^*D_{\alpha\beta})^{-1}(\ div\ Q + \ div\ z^* - P)\}\{(\ div\ Q + \ div\ z^* - P)\}\,dx,
\end{aligned}
$$

and $\tilde{G}_1^* : Y_2^* \times Y_1^* \to \mathbb{R} \cup \{+\infty\}$ *where*

$$
\begin{aligned}
\tilde{G}_1^*(-Q, N) &= \sup_{(v_1, v_2) \in Y_1 \times Y_2}\Big\{-\langle (v_2)_\alpha, Q_\alpha \rangle_{L^2} + \langle (v_1)_{\alpha\beta}, N_{\alpha\beta} \rangle_{L^2} \\
&\quad - G_1\left(\left\{(v_1)_{\alpha\beta} + \frac{1}{2}(v_2)_\alpha(v_2)_\beta\right\}\right) - \frac{1}{2}\int_{\Omega} K_{\alpha\beta}(v_2)_\alpha(v_2)_\beta\,dx\Big\} \\
&= \frac{1}{2}\int_{\Omega} \overline{N_{\alpha\beta} + K_{\alpha\beta}}Q_\alpha Q_\beta\,dx \\
&\quad + \frac{1}{2}\int_{\Omega} \overline{H}_{\alpha\beta\lambda\mu}N_{\alpha\beta}N_{\lambda\mu}\,dx
\end{aligned}
$$

$$\tag{18.6}$$

if

$$\{N_{\alpha\beta} + K_{\alpha\beta}\}$$

is positive definite.

  Here we have denoted,

$$\{\overline{N_{\alpha\beta} + K_{\alpha\beta}}\} = \{N_{\alpha\beta} + K_{\alpha\beta}\}^{-1}.$$

  We denote also,

$$
\begin{aligned}
F^*(z^*) &= \sup_{v_2 \in Y_2} \left\{ \langle (v_2)_\alpha, (z_2^*)_\alpha \rangle_{L^2} - \frac{1}{2} \int_\Omega K_{\alpha\beta}(v_2)_\alpha (v_2)_\beta \, dx \right. \\
&= \frac{1}{2} \int_\Omega \overline{K}_{\alpha\beta}(z_2^*)_\alpha (z_2^*)_\beta \, dx \\
&= \left. \frac{1}{2} \int_\Omega \overline{-N_{\alpha\beta} + \varepsilon\delta_{\alpha\beta}}(z_2^*)_\alpha (z_2^*)_\beta \, dx \right\},
\end{aligned}
\tag{18.7}
$$

if $\{K_{\alpha\beta}\}$ is positive definite, where

$$\{\overline{K}_{\alpha\beta}\} = \{K_{\alpha\beta}\}^{-1},$$

$$\{K_{\alpha\beta}\} = \{-N_{\alpha\beta} + \varepsilon\delta_{\alpha\beta}\},$$

$$\{\overline{H}_{\alpha\beta\lambda\mu}\} = \{H_{\alpha\beta\lambda\mu}\}^{-1},$$

and

$$F(\{w_\alpha\}) = \frac{1}{2} \int_\Omega K_{\alpha\beta} w_{,\alpha} w_{,\beta} \, dx,$$

for some

$$\varepsilon > 0.$$

  At this point we also define,

$$B^* = \{N \in Y_1^* \ : \ \{K_{\alpha\beta}\} \text{ is positive definite and } \hat{J}^*(N,z^*) > 0, \forall z^* \in Y_2^* \text{ such that } z^* \neq \mathbf{0}\}$$

where,

$$
\begin{aligned}
\hat{J}^*(N,z^*) &= F^*(z^*) - G_2^*(z^*, \mathbf{0}) \\
&= \frac{1}{2} \int_\Omega \overline{(-N_{\alpha\beta} + \varepsilon\delta_{\alpha\beta})} z_\alpha^* z_\beta^* \, dx \\
&\quad - \frac{1}{2} \int_\Omega [(h_{\alpha\beta\lambda\mu} D_{\lambda\mu}^* D_{\alpha\beta})^{-1}(\operatorname{div} z^*)][\operatorname{div} z^*] \, dx.
\end{aligned}
\tag{18.8}
$$

Moreover, we denote,

$$C^* = \{N \in Y_1^* \ : \ N_{\alpha\beta,\beta} + P_\alpha = 0, \text{ in } \Omega\}$$

and define

$$A^* = B^* \cap C^*.$$

Assume $u_0 \in U$ is such that $\delta J(u_0) = \mathbf{0}$ and $N_0 \in B^*$, where

$$\{(N_0)_{\alpha\beta}\} = \{H_{\alpha\beta\lambda\mu}\gamma_{\lambda\mu}(u_0)\}.$$

  Under such hypotheses,

$$
\begin{aligned}
J(u_0) &= \inf_{u \in U} J(u) \\
&= \sup_{(Q,N) \in Y_2^* \times A^*} \tilde{J}^*(Q,N) \\
&= \tilde{J}^*(Q_0, N_0) \\
&= J^*(z_0^*, Q_0, N_0),
\end{aligned}
\tag{18.9}
$$

where

$$J^*(z^*,Q,N) = F^*(z^*) - G_2^*(z^*,Q) - \tilde{G}_1^*(-Q,N),$$
$$\tilde{J}^*(Q,N) = \inf_{z^* \in Y_2^*} J^*(z^*,Q,N),$$

and where

$$(z_0^*)_\alpha = (-(N_0)_{\alpha\beta} + \varepsilon\delta_{\alpha\beta})(w_0)_{,\beta},$$

and

$$Q_0 = -\varepsilon\nabla w_0.$$

Proof 18.1 From the general result in Toland [78], we have

$$\inf_{z^* \in Y_2^*} J^*(z^*,Q,N)$$

$$= \inf_{z^* \in Y_2^*} \{F^*(z^*) - G_2^*(z^*,Q) - \tilde{G}_1^*(-Q,N)\}$$

$$\leq -\langle w_{,\alpha}, z_\alpha^* \rangle_{L^2} - \langle w_{,\alpha}, Q_\alpha \rangle_{L^2} + \frac{1}{2}\int_\Omega h_{\alpha\beta\lambda\mu} w_{,\alpha\beta} w_{,\lambda\mu}\, dx$$

$$+ \langle w_{,\alpha}, z_\alpha^* \rangle_{L^2} - \frac{1}{2}\int_\Omega (-N_{\alpha\beta} + \varepsilon\delta_{\alpha\beta}) w_{,\alpha} w_{,\beta}\, dx - \langle w, P \rangle_{L^2}$$

$$- \frac{1}{2\varepsilon}\int_\Omega \delta_{\alpha\beta} Q_\alpha Q_\beta\, dx$$

$$- \frac{1}{2}\int_\Omega \overline{H}_{\alpha\beta\lambda\mu} N_{\alpha\beta} N_{\lambda\mu}\, dx - \langle u_\alpha, N_{\alpha\beta,\beta} + P_\alpha \rangle_{L^2}$$

$$\leq -\langle w_{,\alpha}, z_\alpha^* \rangle_{L^2} - \langle w_{,\alpha}, Q_\alpha \rangle_{L^2} + \frac{1}{2}\int_\Omega h_{\alpha\beta\lambda\mu} w_{,\alpha\beta} w_{,\lambda\mu}\, dx$$

$$+ \langle w_{,\alpha}, z_\alpha^* \rangle_{L^2} - \frac{1}{2}\int_\Omega (-N_{\alpha\beta} + \varepsilon\delta_{\alpha\beta}) w_{,\alpha} w_{,\beta}\, dx - \langle w, P \rangle_{L^2}$$

$$+ \langle w_{,\alpha}, Q_\alpha^* \rangle_{L^2} + \frac{1}{2}\int_\Omega \varepsilon\delta_{\alpha\beta} w_{,\alpha} w_{,\beta}\, dx$$

$$- \frac{1}{2}\int_\Omega \overline{H}_{\alpha\beta\lambda\mu} N_{\alpha\beta} N_{\lambda\mu}\, dx - \langle u_\alpha, N_{\alpha_\beta,\beta} + P_\alpha \rangle_{L^2}, \tag{18.10}$$

$\forall u \in U,\ Q \in Y_2^*,\ N \in A^*.$

Therefore,

$$
\inf_{z^* \in Y_2^*} J^*(z^*, Q, N)
$$

$$
\leq \quad \frac{1}{2} \int_\Omega h_{\alpha\beta\lambda\mu} w_{,\alpha\beta} w_{,\lambda\mu} \, dx + \frac{1}{2} \int_\Omega N_{\alpha\beta} w_{,\alpha} w_{,\beta} \, dx
$$

$$
+ \left\langle \left(\frac{u_{\alpha,\beta} + u_{\beta,\alpha}}{2} \right), N_{\alpha\beta} \right\rangle_{L^2} - \frac{1}{2} \int_\Omega \overline{H}_{\alpha\beta\lambda\mu} N_{\alpha\beta} N_{\lambda\mu} \, dx
$$

$$
- \langle w, P \rangle_{L^2} - \langle u_\alpha, P_\alpha \rangle_{L^2}
$$

$$
\leq \quad \sup_{N \in Y_1^*} \left\{ \left\langle \frac{u_{\alpha,\beta} + u_{\beta,\alpha}}{2} + \frac{1}{2} w_{,\alpha} w_{,\beta}, N_{\alpha\beta} \right\rangle_{L^2} - \frac{1}{2} \int_\Omega \overline{H}_{\alpha\beta\lambda\mu} N_{\alpha\beta} N_{\lambda\mu} \, dx \right\}
$$

$$
+ \frac{1}{2} \int_\Omega h_{\alpha\beta\lambda\mu} w_{,\alpha\beta} w_{,\lambda\mu} \, dx
$$

$$
- \langle w, P \rangle_{L^2} - \langle u_\alpha, P_\alpha \rangle_{L^2}
$$

$$
= \quad \frac{1}{2} \int_\Omega \gamma_{\alpha\beta}(u) \gamma_{\lambda\mu}(u) \, dx + \frac{1}{2} \int_\Omega h_{\alpha\beta\lambda\mu} \kappa_{\alpha\beta}(u) \kappa_{\lambda\mu}(u) \, dx
$$

$$
- \langle w, P \rangle_{L^2} - \langle u_\alpha, P_\alpha \rangle_{L^2}
$$

$$
= \quad J(u), \tag{18.11}
$$

$\forall u \in U,\ Q \in Y_2^*,\ N \in A^*$.

Summarizing,

$$
J(u) \quad \geq \quad \inf_{z^* \in Y_2^*} J^*(z^*, Q, N)
$$

$$
= \quad \tilde{J}^*(Q, N) \tag{18.12}
$$

$\forall u \in U,\ Q \in Y_2^*,\ N \in A^*$, so that

$$
\inf_{u \in U} J(u) \geq \sup_{(Q,N) \in Y_2^* \cap A^*} \tilde{J}^*(Q, N). \tag{18.13}
$$

Suppose now $u_0 \in U$ is such that

$$
\delta J(u_0) = \mathbf{0}, \tag{18.14}
$$

and $N_0 \in B^*$.

Observe that from (18.14), we get

$$
(N_0)_{\alpha\beta,\beta} + P_\alpha = 0, \text{ in } \Omega,
$$

so that

$$
N_0 \in C^*.
$$

Hence,

$$
N_0 \in A^* = B^* \cap C^*.
$$

Moreover, from $\delta J(u_0) = \mathbf{0}$, we obtain

$$
- \operatorname{div}(Q_0 + z_0^*) = h_{\alpha\beta\lambda\mu}(w_0)_{\alpha\beta\lambda\mu} - P \text{ in } \Omega, \tag{18.15}
$$

where, as above indicated

$$
(z_0^*)_\alpha = (-(N_0)_{\alpha\beta} + \varepsilon\delta_{\alpha\beta})(w_0)_{,\beta}. \tag{18.16}
$$

and

$$
Q_0 = -\varepsilon \nabla w_0.
$$

From (18.15),

$$
w_0 = -(h_{\alpha\beta\lambda\mu} D_{\lambda\mu}^* D_{\alpha\beta})^{-1}(\operatorname{div}(z_0^* + Q_0) - P),
$$

so that from this and the inversion of (18.16), we have

$$
\begin{aligned}
(w_0),_\rho &= [(h_{\alpha\beta\lambda\mu}D^*_{\lambda\mu}D_{\alpha\beta})^{-1}(-\operatorname{div}(z^*_0 + Q_0) + P)],_\rho \\
&= \overline{((-N_0)_{\rho\beta} + \varepsilon\delta_{\rho\beta})}(z^*_0)_\beta,
\end{aligned}
\tag{18.17}
$$

so that

$$
[(h_{\alpha\beta\lambda\mu}D^*_{\lambda\mu}D_{\alpha\beta})^{-1}(\operatorname{div}(z^*_0 + Q_0) - P)],_\rho + \overline{((-N_0)_{\rho\beta} + \varepsilon\delta_{\rho\beta})}(z^*_0)_\beta = 0, \text{ in } \Omega,
$$

that is,

$$
\frac{\partial \hat{J}^*_1(z^*_0, Q_0, w_0, u_0)}{\partial z^*} = \mathbf{0},
$$

where

$$
\hat{J}^*_1(z^*, Q, N, u) = J^*(z^*, Q, N) + \langle u_\alpha, N_{\alpha\beta,\beta} + P_\alpha \rangle_{L^2}.
$$

Also, from (18.15) and (18.16) we obtain

$$
\begin{aligned}
-\operatorname{div} Q_0 &= \varepsilon \operatorname{div} \nabla w_0 \\
&= \varepsilon \nabla^2 w_0 \\
&= \varepsilon \delta_{\alpha\beta} w,_{\alpha\beta} \\
&= h_{\alpha\beta\lambda\mu}(w_0)_{\alpha\beta\lambda\mu} - [((N_0)_{\alpha\beta} - \varepsilon\delta_{\alpha\beta})(w_0),_\beta],_\alpha - P \\
&= h_{\alpha\beta\lambda\mu}(w_0)_{\alpha\beta\lambda\mu} + \operatorname{div} z^*_0 - P.
\end{aligned}
\tag{18.18}
$$

Furthermore,

$$
\begin{aligned}
\frac{(Q_0)_\rho}{\varepsilon} &= -(w_0),_\rho \\
&= [(h_{\alpha\beta\lambda\mu}D^*_{\lambda\mu}D_{\alpha\beta})^{-1}(\operatorname{div}(z^*_0 + Q_0) - P)],_\rho.
\end{aligned}
\tag{18.19}
$$

This last equation corresponds to

$$
\frac{\partial \hat{J}^*_1(z^*_0, Q_0, N_0, u_0)}{\partial Q_\alpha} = 0.
$$

Moreover,

$$
\overline{H}_{\alpha\beta\lambda\mu}(N_0)_{\lambda\mu} = \left(\frac{(u_0)_{\alpha,\beta} + (u_0)_{\beta,\alpha}}{2}\right) + \frac{1}{2}(w_0),_\alpha(w_0),_\beta,
$$

so that

$$
\begin{aligned}
&\overline{H}_{\alpha\beta\lambda\mu}(N_0)_{\lambda\mu} \\
&= \left(\frac{(u_0)_{\alpha,\beta} + (u_0)_{\beta,\alpha}}{2}\right) \\
&\quad + \left(\frac{1}{2}\overline{(-(N_0)_{\alpha\rho} + \varepsilon\delta_{\alpha\rho})}(z^*_0)_\rho\right)\left(\overline{(-(N_0)_{\beta\eta} + \varepsilon\delta_{\beta\eta})}(z^*_0)_\eta\right),
\end{aligned}
\tag{18.20}
$$

which means

$$
\frac{\partial \hat{J}^*_1(z^*_0, Q_0, N_0, u_0)}{\partial N_{\alpha\beta}} = 0.
$$

Finally, from

$$
N_{\alpha\beta,\beta} + P_\alpha = 0, \text{ in } \Omega,
$$

we get

$$
\frac{\partial \hat{J}^*_1(z^*_0, Q_0, N_0, u_0)}{\partial u_\alpha} = 0.
$$

Summarizing, we have obtained

$$\delta \hat{J}_1^*(z_0^*, Q_0, N_0, u_0) = \mathbf{0}.$$

At this point we shall obtain a standard correspondence between the primal and dual formulations. First, we recall that from

$$(z_0^*)_\alpha = (-N_{\alpha\beta} + \varepsilon \delta_{\alpha\beta})(w_0)_{,\beta},$$

we have

$$F^*(z_0^*) = \langle (w_0)_\alpha, (z_0^*)_\alpha \rangle_{L^2} - \frac{1}{2}(-N_{\alpha\beta} + \varepsilon \delta_{\alpha\beta})(w_0)_{,\alpha}(w_0)_{,\beta}.$$

From

$$\text{div } Q_0 + \text{div } z_0^* = -h_{\alpha\beta\lambda\mu}(w_0)_{\alpha\beta\lambda\mu} + P$$

we obtain

$$G_2^*(z_0^*, Q_0) = \langle (w_0)_\alpha, (Q_0)_\alpha \rangle_{L^2} + \langle (w_0)_\alpha, (z_0^*)_\alpha \rangle_{L^2} - G_2(\kappa(u_0)) + \langle w_0, P \rangle_{L^2}.$$

Finally, from

$$\frac{-(Q_0)_\alpha}{\varepsilon} = (w_0)_{,\alpha}$$

and

$$(N_0)_{\alpha\beta} = H_{\alpha\beta\lambda\mu}\gamma_{\lambda\mu}(u_0),$$

we get

$$
\begin{aligned}
\tilde{G}_1^*(-Q_0, N_0) &= -\langle (w_0)_\alpha, (Q_0)_\alpha \rangle_{L^2} + \left\langle \frac{(u_0)_{\alpha,\beta} + (u_0)_{\beta,\alpha}}{2} + \frac{(w_0)_{,\alpha}(w_0)_{,\beta}}{2}, (N_0)_{\alpha\beta} \right\rangle_{L^2} \\
&\quad - G_1(\gamma(u_0)) - \frac{\varepsilon}{2}\int_\Omega \delta_{\alpha\beta}(w_0)_{,\alpha}(w_0)_{,\beta}\, dx \\
&= -\langle (w_0)_\alpha, (Q_0)_\alpha \rangle_{L^2} + \left\langle \frac{(w_0)_{,\alpha}(w_0)_{,\beta}}{2}, (N_0)_{\alpha\beta} \right\rangle_{L^2} \\
&\quad - \frac{\varepsilon}{2}\int_\Omega \delta_{\alpha\beta}(w_0)_{,\alpha}(w_0)_{,\beta}\, dx + \langle u_\alpha, P_\alpha \rangle_{L^2} \\
&\quad - \frac{1}{2}\int_\Omega H_{\alpha\beta\lambda\mu}\gamma_{\alpha\beta}(u_0)\gamma_{\lambda\mu}(u_0)\, dx.
\end{aligned}
\tag{18.21}
$$

Joining the pieces, we obtain

$$
\begin{aligned}
\hat{J}_1^*(z_0^*, Q_0, N_0, u_0) &= J^*(z_0, Q_0, N_0) \\
&= F^*(z_0^*) - G_2^*(z_0^*, Q_0) - \tilde{G}_1^*(-Q_0, N_0) \\
&= G_2(\kappa u_0) + G_1(\gamma(u_0)) - \langle w_0, P \rangle_{L^2} - \langle (u_0)_\alpha, P_\alpha \rangle_{L^2} \\
&= J(u_0).
\end{aligned}
\tag{18.22}
$$

Moreover, since $N_0 \in A^*$, we have

$$
\begin{aligned}
J^*(z_0^*, Q_0, N_0) &= \inf_{z^* \in Y_2^*} J^*(z^*, Q_0, N_0) \\
&= \tilde{J}^*(Q_0, N_0).
\end{aligned}
\tag{18.23}
$$

From this, (26.34) and (18.22), we obtain

$$
\begin{aligned}
J(u_0) &= \inf_{u \in U} J(u) \\
&= \sup_{(Q,N) \in Y_2^* \times A^*} \tilde{J}^*(Q, N) \\
&= \tilde{J}^*(Q_0, N_0) \\
&= J^*(z_0^*, Q_0, N_0),
\end{aligned}
\tag{18.24}
$$

The proof is complete.

18.3 The primal dual formulation and related duality principle

At this point we present the main result of this article, which is summarized by the next theorem.

Theorem 18.3.1 *Consider the notation and context of the last theorem. Assume those hypotheses, more specifically suppose* $\delta J(u_0) = \mathbf{0}$ *and* $N_0 \in B^*$, *where*

$$\{(N_0)_{\alpha\beta}\} = \{H_{\alpha\beta\lambda\mu}\gamma_{\lambda\mu}(u_0)\},$$

$$(z_0^*)_\alpha = (-(N_0)_{\alpha\beta} + \varepsilon\delta_{\alpha\beta})(w_0)_{,\beta},$$

and

$$Q_0 = -\varepsilon\nabla w_0.$$

Recall also that

$$B^* = \{N \in Y_1^* \ : \ \{K_{\alpha\beta}\} \text{ is positive definite and } \hat{J}^*(N, z^*) > 0, \forall z^* \in Y_2^* \text{ such that } z^* \neq \mathbf{0}\}$$

where,

$$
\begin{aligned}
\hat{J}^*(N, z^*) &= F^*(z^*) - G_2^*(z^*, \mathbf{0}) \\
&= \frac{1}{2}\int_\Omega \overline{(-N_{\alpha\beta} + \varepsilon\delta_{\alpha\beta})} z_\alpha^* z_\beta^* \, dx \\
&\quad - \frac{1}{2}\int_\Omega [(h_{\alpha\beta\lambda\mu}D_{\lambda\mu}^* D_{\alpha\beta})^{-1}(\,\mathrm{div}\, z^*)][\,\mathrm{div}\, z^*] \, dx.
\end{aligned}
\tag{18.25}
$$

Moreover,

$$C^* = \{N \in Y_1^* \ : \ N_{\alpha\beta,\beta} + P_\alpha = 0, \text{ in } \Omega\}$$

and

$$A^* = B^* \cap C^*.$$

Under such assumptions and notation, denoting

$$\hat{Y}_2^* = \{Q \in Y_2^* \ : \ Q = \nabla v, \text{ for some } v \in W_0^{2,2}(\Omega)\},$$

we have

$$
\begin{aligned}
J(u_0) &= \inf_{u \in U} J(u) \\
&= \sup_{(Q,N) \in \hat{Y}_2^* \times A^*} \tilde{J}^*(Q, N) \\
&= \tilde{J}^*(Q_0, N_0) \\
&= J^*(z_0^*, Q_0, N_0) \\
&= J_3(w_0, N_0) \\
&= \sup_{(w,N) \in U \times A^*} J_3(w, N)
\end{aligned}
\tag{18.26}
$$

where,

$$
\begin{aligned}
J_3(w, N) &= -\frac{1}{2}\int_\Omega h_{\alpha\beta\lambda\mu} w_{,\alpha\beta} w_{,\lambda\mu} \, dx \\
&\quad -\frac{1}{2}\int_\Omega (N_{\alpha\beta} - \varepsilon\delta_{\alpha\beta}) w_{,\alpha} w_{,\beta} \, dx \\
&\quad -\frac{1}{2\varepsilon}\int_\Omega \left[(-\nabla^2)^{-1}\left(h_{\alpha\beta\lambda\mu} w_{,\alpha\beta\lambda\mu} - [(N_{\alpha\beta} - \varepsilon\delta_{\alpha\beta}) w_{,\beta}]_{,\alpha} - P\right)\right] \\
&\quad \times \left(h_{\alpha\beta\lambda\mu} w_{,\alpha\beta\lambda\mu} - [(N_{\alpha\beta} - \varepsilon\delta_{\alpha\beta}) w_{,\beta}]_{,\alpha} - P\right) \, dx \\
&\quad -\frac{1}{2}\int_\Omega \overline{H}_{\alpha\beta\lambda\mu} N_{\alpha\beta} N_{\lambda\mu} \, dx,
\end{aligned}
\tag{18.27}
$$

where generically we have denoted

$$w = (\nabla^2)^{-1}\eta \text{ for } \eta \in L^2(\Omega),$$

if $\eta = \nabla^2 w$ and $w \in W_0^{1,2}(\Omega)$.

Proof 18.2 Observe that

$$\tilde{J}^*(Q,N) = \inf_{z^* \in Y_2^*}\{F^*(z^*) - G_2^*(z^*,Q) - \tilde{G}_1^*(-Q,N)\},$$

$\forall Q \in \hat{Y}_2^*,\ N \in A^*$.

Also, such an infimum is attained through the equation

$$\frac{\partial F^*(z^*)}{\partial z^*} = \frac{\partial G_2^*(z^*,Q)}{\partial z^*},$$

that is,

$$
\begin{aligned}
&-\nabla[(h_{\alpha\beta\lambda\mu}D_{\lambda\mu}^*D_{\alpha\beta})^{-1}(\text{ div }(z^*+Q)-P)]\\
=\ &\{\overline{(-N_{\alpha\beta}+\varepsilon\delta_{\alpha\beta})(z_2^*)}_\beta\},
\end{aligned}
\tag{18.28}
$$

that is,

$$
\begin{aligned}
&-\text{ div }\left(\nabla(h_{\alpha\beta\lambda\mu}D_{\lambda\mu}^*D_{\alpha\beta})^{-1}(\text{ div }(z^*+Q)-P)\right)\\
=\ &[\overline{(-N_{\alpha\beta}+\varepsilon\delta_{\alpha\beta})(z_2^*)}_\beta]_{,\alpha}\\
=\ &\nabla^2 w,
\end{aligned}
\tag{18.29}
$$

where

$$w = -(h_{\alpha\beta\lambda\mu}D_{\lambda\mu}^*D_{\alpha\beta})^{-1}(\text{ div }(z^*+Q)-P).$$

Hence,

$$F^*(z^*) = \langle w_{,\alpha},z_\alpha^*\rangle_{L^2} - \frac{1}{2}\langle(-N_{\alpha\beta}+\varepsilon\delta_{\alpha\beta})w_{,\alpha}w_{,\beta}\rangle_{L^2}. \tag{18.30}$$

From

$$-\text{ div }(z^*+Q) = h_{\alpha\beta\lambda\mu}w_{,\alpha\beta\lambda\mu} - P,$$

and

$$(z_\alpha^*)_{,\alpha} = [(-N_{\alpha\beta}+\varepsilon\delta_{\alpha\beta})(w_{,\beta})]_{,\alpha},$$

we obtain

$$
\begin{aligned}
-\text{ div } Q &= h_{\alpha\beta\lambda\mu}w_{,\alpha\beta\lambda\mu} - P + \text{ div }(z^*)\\
&= h_{\alpha\beta\lambda\mu}w_{,\alpha\beta\lambda\mu} - [(N_{\alpha\beta}-\varepsilon\delta_{\alpha\beta})(w_{,\beta})]_{,\alpha} - P.
\end{aligned}
\tag{18.31}
$$

Let $v \in W_0^{2,2}(\Omega)$ be such that $Q = \nabla v$
From the last equation

$$-\nabla^2 v = -\text{ div }(\nabla v) = -\text{ div } Q = h_{\alpha\beta\lambda\mu}w_{,\alpha\beta\lambda\mu} - [(N_{\alpha\beta}-\varepsilon\delta_{\alpha\beta})(w_{,\beta})]_{,\alpha} - P,$$

so that

$$v = -(\nabla^2)^{-1}(h_{\alpha\beta\lambda\mu}w_{,\alpha\beta\lambda\mu} - [(N_{\alpha\beta}-\varepsilon\delta_{\alpha\beta})(w_{,\beta})]_{,\alpha} - P)$$

Thus,

$$Q = \{[-(\nabla^2)^{-1}(h_{\alpha\beta\lambda\mu}w_{,\alpha\beta\lambda\mu} - [(N_{\alpha\beta} - \varepsilon\delta_{\alpha\beta})(w_{,\beta})]_{,\alpha} - P]_{,\rho}\}$$

so that

$$
\begin{aligned}
&\frac{1}{2\varepsilon}\int_{\Omega}\delta_{\alpha\beta}Q_{\alpha}Q_{\beta}\,dx \\
&= \frac{1}{2\varepsilon}\int_{\Omega}\left[(-\nabla^2)^{-1}\left(h_{\alpha\beta\lambda\mu}w_{,\alpha\beta\lambda\mu} - [(N_{\alpha\beta} - \varepsilon\delta_{\alpha\beta})w_{,\beta}]_{,\alpha} - P\right)\right] \\
&\quad \times \left(h_{\alpha\beta\lambda\mu}w_{,\alpha\beta\lambda\mu} - [(N_{\alpha\beta} - \varepsilon\delta_{\alpha\beta})w_{,\beta}]_{,\alpha} - P\right)\,dx.
\end{aligned}
\tag{18.32}
$$

Thus,

$$
\begin{aligned}
\tilde{J}^*(Q,N) &= -\frac{1}{2}\int_{\Omega}h_{\alpha\beta\lambda\mu}w_{,\alpha\beta}w_{,\lambda\mu}\,dx \\
&\quad -\frac{1}{2}\int_{\Omega}(N_{\alpha\beta} - \varepsilon\delta_{\alpha\beta})w_{,\alpha}w_{,\beta}\,dx \\
&\quad -\frac{1}{2\varepsilon}\int_{\Omega}\left[(-\nabla^2)^{-1}\left(h_{\alpha\beta\lambda\mu}w_{,\alpha\beta\lambda\mu} - [(N_{\alpha\beta} - \varepsilon\delta_{\alpha\beta})w_{,\beta}]_{,\alpha} - P\right)\right] \\
&\quad \times \left(h_{\alpha\beta\lambda\mu}w_{,\alpha\beta\lambda\mu} - [(N_{\alpha\beta} - \varepsilon\delta_{\alpha\beta})w_{,\beta}]_{,\alpha} - P\right)\,dx \\
&\quad -\frac{1}{2}\int_{\Omega}\overline{H}_{\alpha\beta\lambda\mu}N_{\alpha\beta}N_{\lambda\mu}\,dx \\
&= J_3(w,N).
\end{aligned}
\tag{18.33}
$$

Moreover, from $\delta J(u_0) = \mathbf{0}$ we have

$$h_{\alpha\beta\lambda\mu}(w_0)_{,\alpha\beta\lambda\mu} - [((N_0)_{\alpha\beta} - \varepsilon\delta_{\alpha\beta})(w_0)_{,\beta}]_{,\alpha} - \varepsilon\delta_{\alpha\beta}(w_0)_{\alpha\beta} - P = 0,\ \text{in}\ \Omega,$$

so that

$$\hat{w}_0 \equiv w_0 = (\nabla^2)^{-1}(h_{\alpha\beta\lambda\mu}(w_0)_{,\alpha\beta\lambda\mu} - [((N_0)_{\alpha\beta} - \varepsilon\delta_{\alpha\beta})(w_0)_{,\beta}]_{,\alpha} - P)/\varepsilon$$

and therefore

$$
\begin{aligned}
&-h_{\alpha\beta\lambda\mu}(w_0)_{,\alpha\beta\lambda\mu} + [((N_0)_{\alpha\beta} - \varepsilon\delta_{\alpha\beta})(w_0)_{,\beta}]_{,\alpha} \\
&+h_{\alpha\beta\lambda\mu}(\hat{w}_0)_{,\alpha\beta\lambda\mu} - [((N_0)_{\alpha\beta} - \varepsilon\delta_{\alpha\beta})(\hat{w}_0)_{,\beta}]_{,\alpha} \\
&= 0.
\end{aligned}
\tag{18.34}
$$

Also,

$$\overline{H}_{\alpha\beta\lambda\mu}(N_0)_{\lambda\mu} = \frac{1}{2}\left(\frac{(u_0)_{\alpha,\beta} + (u_0)_{\beta,\alpha}}{2}\right) + \frac{1}{2}(w_0)_{,\alpha}(w_0)_{,\beta},$$

so that

$$\delta\left\{J_3(w_0,N_0) + \left\langle\frac{1}{2}\left(\frac{(u_0)_{\alpha,\beta} + (u_0)_{\beta,\alpha}}{2}\right),(N_0)_{\alpha\beta}\right\rangle_{L^2} - \langle(u_0)_{,\alpha},P_{\alpha}\rangle_{L^2}\right\} = \mathbf{0}.$$

From these last results and from the last theorem, we may obtain

$$
\begin{aligned}
J_3(w_0,N_0) &= J(w_0) \\
&= J^*(z_0,Q_0,N_0) \\
&= \tilde{J}^*(Q_0,N_0).
\end{aligned}
\tag{18.35}
$$

From this, also from the last theorem and from (18.33), we finally get

$$
\begin{aligned}
J(u_0) &= \inf_{u \in U} J(u) \\
&= \sup_{(Q,N) \in \hat{Y}_2^* \times A^*} \tilde{J}^*(Q,N) \\
&= \tilde{J}^*(Q_0,N_0) \\
&= J^*(z_0^*,Q_0,N_0) \\
&= J_3(w_0,N_0) \\
&= \sup_{(w,N) \in U \times A^*} J_3(w,N).
\end{aligned}
\tag{18.36}
$$

The proof is complete.

18.4 A multi-duality principle for non-convex optimization

Our final result is a multi-duality principle, which is summarized by the following theorem.

Theorem 18.4.1 *Considering the notation and statements of the plate model addressed in the last sections, assuming a not relabeled finite dimensional approximate model, in a finite elements or finite difference context, let $J_1 : U \times Y_1^* \times Y_2^* \to \mathbb{R}$ be a functional where*

$$
\begin{aligned}
J_1(u,Q,N) &= \frac{1}{2} \int_\Omega h_{\alpha\beta\lambda\mu} w_{,\alpha\beta} w_{,\lambda\mu}\, dx - \langle P,w \rangle_{L^2} \\
&\quad + \frac{1}{2} \int_\Omega (\overline{-N_{\alpha\beta}^K}) Q_\alpha Q_\beta\, dx - \langle w_{,\alpha}, Q_\alpha \rangle_{L^2} \\
&\quad + \frac{K}{2} \int_\Omega w_{,\alpha} w_{,\alpha}\, dx - \frac{1}{2} \int_\Omega \overline{H}_{\alpha\beta\lambda\mu} N_{\alpha\beta} N_{\lambda\mu}\, dx \\
&\quad - \langle N_{\alpha\beta,\beta} + P_\alpha, u_\alpha \rangle_{L^2},
\end{aligned}
\tag{18.37}
$$

and where

$$
\{\overline{-N_{\alpha\beta}^K}\} = \{-N_{\alpha\beta} + K\delta_{\alpha\beta}\}^{-1}.
$$

Define also,

$$
C^* = \left\{ N \in Y_1^* \ : \ \{-N_{\alpha\beta} + K\delta_{\alpha\beta}\} > \left\{ \frac{K}{2} \delta_{\alpha\beta} \right\} \right\},
$$

$$
B^* = \{ N \in Y_1^* \ : \ N_{\alpha\beta,\beta} + P_\alpha = 0, \ in\ \Omega \},
$$

$$
D^+ = \{ N \in Y_1^* \ : \ \hat{J}_1^*(Q,N) > 0, \ \forall Q \in Y_2^* \ such\ that\ Q \neq \mathbf{0} \},
$$

$$
D^- = \{ N \in Y_1^* \ : \ \hat{J}_2^*(Q,N) < 0, \ \forall Q \in Y_2^* \ such\ that\ Q \neq \mathbf{0} \},
$$

where

$$
\hat{J}_1^*(Q,N) = -F_K^*(Q) + \frac{1}{2} \int_\Omega (\overline{-N_{\alpha\beta}^K}) Q_\alpha Q_\beta\, dx,
$$

and where

$$
F_K^*(Q) = \sup_{u \in U} \left\{ \langle w_{,\alpha}, Q_\alpha \rangle_{L^2} - \frac{1}{2} \int_\Omega h_{\alpha\beta\lambda\mu} w_{,\alpha\beta} w_{,\lambda\mu}\, dx - \frac{K}{2} \int_\Omega w_{,\alpha} w_{,\alpha}\, dx \right\}.
$$

Moreover

$$
\hat{J}_2^*(Q,N) = -F_K^*(Q) + H_K^*(Q,N),
$$

where

$$H_K^*(Q,N) = \sup_{u \in U} \left\{ \langle w_{,\alpha}, Q_{,\alpha} \rangle_{L^2} - \frac{1}{2} \int_\Omega (-N_{\alpha\beta} + K\delta_{\alpha\beta}) w_{,\alpha} w_{,\beta} \, dx \right\}.$$

Moreover, we also define,

$$A_+^* = B^* \cap C^* \cap D^+$$
$$A_-^* = B^* \cap C^* \cap D^-,$$

and

$$E^* = B^* \cap C^*.$$

Let $u_0 \in U$ be such that $\delta J(u_0) = \mathbf{0}$ and define

$$(N_0)_{\alpha\beta} = H_{\alpha\beta\lambda\mu} \gamma_{\lambda\mu}(u_0),$$

and

$$(Q_0)_\alpha = ((N_0)_{\alpha\beta} + K\delta_{\alpha\beta})(w_0)_{,\beta}.$$

Under such hypotheses,

1. *if $\delta^2 J(u_0) > \mathbf{0}$ and $N_0 \in E^*$, defining*

$$J_2(u,Q) = \sup_{N \in E^*} J_1(u,Q,N),$$

   and

$$\tilde{J}^*(Q) = \inf_{u \in B_{r_1}(u_0)} J_2(u,Q),$$

   where $r_1 > 0$ is such that

$$\delta^2 J(u) > \mathbf{0}$$

   in $B_{r_1}(u_0)$, we have

$$J(u_0) = \tilde{J}^*(Q_0),$$
$$\delta \tilde{J}^*(Q_0) = \mathbf{0}$$

   and if $K > 0$ is sufficiently big,

$$\delta^2 \tilde{J}^*(Q_0) \geq \mathbf{0}$$

   and there exist $r, r_2 > 0$ such that

$$\begin{aligned}
J(u_0) &= \inf_{u \in B_r(u_0)} J(u) \\
&= \inf_{Q \in B_{r_2}(Q_0)} \tilde{J}^*(Q) \\
&= \tilde{J}^*(Q_0).
\end{aligned} \tag{18.38}$$

2. *If $N_0 \in A_+^*$, defining*

$$J_3(u,Q) = \sup_{N \in A_+^*} J_1(u,Q,N),$$

   and

$$\tilde{J}_3^*(Q) = \inf_{u \in U} J_3(u,Q),$$

   then

$$\delta \tilde{J}_3^*(Q_0) = \mathbf{0},$$
$$\delta^2 \tilde{J}_3^*(Q_0) \geq \mathbf{0},$$

and

$$
\begin{aligned}
J(u_0) &= \inf_{u \in U} J(u) \\
&= \inf_{Q \in Y_2^*} \tilde{J}_3^*(Q) \\
&= \tilde{J}_3^*(Q_0).
\end{aligned}
\tag{18.39}
$$

3. *If* $\delta^2 J(u_0) < \mathbf{0}$ *so that* $N_0 \in A_-^*$, *defining*

$$
\hat{J}^*(Q,N) = -\hat{F}_K^*(Q) + H_K^*(Q,N) - \frac{1}{2} \int_\Omega \overline{H}_{\alpha\beta\lambda\mu} N_{\alpha\beta} N_{\lambda\mu} \, dx,
$$

where

$$
\hat{F}_K^*(Q) = \sup_{u \in U} \left\{ \langle w_{,\alpha}, Q_\alpha \rangle_{L^2} - \frac{1}{2} \int_\Omega h_{\alpha\beta\lambda\mu} w_{,\alpha\beta} w_{,\lambda\mu} \, dx - \frac{K}{2} \int_\Omega w_{,\alpha} w_{,\alpha} \, dx + \langle w, P \rangle_{L^2} \right\}.
$$

we have that

$$
\hat{J}^*(Q_0, N_0) = J(u_0),
$$

$$
\delta\{\hat{J}^*(Q_0, N_0) - \langle (N_0)_{\alpha\beta,\beta} + P_\alpha, (u_0)_\alpha \rangle_{L^2}\} = \mathbf{0},
$$

and there exist r, r_1, $r_2 > 0$ *such that*

$$
\begin{aligned}
J(u_0) &= \sup_{u \in B_r(u_0)} J(u) \\
&= \sup_{Q \in B_{r_1}(Q_0)} \left\{ \sup_{N \in B_{r_2}(N_0) \cap E^*} \hat{J}^*(Q,N) \right\} \\
&= \hat{J}^*(Q_0, N_0).
\end{aligned}
\tag{18.40}
$$

Proof 18.3 From the assumption $N_0 \in E^*$ we have that

$$
J_2(u_0, Q_0) = \sup_{N \in E^*} J_1(u_0, Q_0, N) = J_1(u_0, Q_0, N_0),
$$

where such a supremum is attained through the equation

$$
\frac{\partial J_1(u_0, Q_0, N_0)}{\partial N} = \mathbf{0}.
$$

Moreover, there exists $r > 0$, $r_2 > 0$ such that

$$
J(u_0) = \inf_{u \in B_r(u_0)} J(u),
$$

and (we justify that the first infimum Q in this equation (18.41) is well defined in the next lines)

$$
\begin{aligned}
J(u_0) &= \inf_{u \in B_r(u_0)} \inf_{Q \in B_{r_2}(Q_0)} \sup_{N \in E^*} J_1(u, Q, N) \\
&= \inf_{Q \in B_{r_2}(Q_0)} \inf_{u \in B_{r_1}(u_0)} \sup_{N \in E^*} J_1(u, Q, N) \\
&= \inf_{Q \in B_{r_2}(Q_0)} \tilde{J}^*(Q) \\
&= \tilde{J}^*(Q_0).
\end{aligned}
\tag{18.4}
$$

Observe the concerning extremal in Q is attained through the equation,

$$\frac{\partial J_1(u_0, Q_0, N_0)}{\partial Q} = \mathbf{0}.$$

Hence, from

$$\frac{\partial J(u_0)}{\partial u} = \mathbf{0},$$

from the implicit function theorem and chain rule, we get

$$
\begin{aligned}
\mathbf{0} &= \frac{\partial J(u_0)}{\partial u} \\
&= \frac{\partial J_1(u_0, Q_0, N_0)}{\partial u} \\
&\quad + \frac{\partial J_1(u_0, Q_0, N_0)}{\partial Q} \frac{\partial Q_0}{\partial u} \\
&\quad + \frac{\partial J_1(u_0, Q_0, N_0)}{\partial N} \frac{\partial N_0}{\partial u} \\
&= \frac{\partial J_1(u_0, Q_0, N_0)}{\partial u}.
\end{aligned}
\tag{18.42}
$$

Therefore,

$$
\begin{aligned}
&\frac{\partial \tilde{J}^*(Q_0)}{\partial Q} \\
&= \frac{\partial J_1(u_0, Q_0, N_0)}{\partial Q} \\
&\quad + \frac{\partial J_1(u_0, Q_0, N_0)}{\partial u} \frac{\partial u_0}{\partial Q} \\
&\quad + \frac{\partial J_1(u_0, Q_0, N_0)}{\partial N} \frac{\partial N_0}{\partial Q} \\
&= \mathbf{0}.
\end{aligned}
\tag{18.43}
$$

From this we shall denote

$$\delta \tilde{J}^*(Q_0) = \mathbf{0}.$$

Let us now show that the first infimum in Q in (18.41) is well defined.
Recall again that,

$$J_2(u_0, Q_0) = \sup_{N \in E^*} J_1(u_0, Q_0, N),$$

where such a supremum is attained through the equation

$$\frac{\partial J_1(u_0, Q_0, N_0)}{\partial N} = \mathbf{0},$$

that is,

$$
\begin{aligned}
&(Q_0)_\lambda \overline{(-N_0)^K_{\alpha\lambda} \, (-N_0)^K_{\beta\mu}} (Q_0)_\mu \\
&- \overline{H}_{\alpha\beta\lambda\mu}(N_0)_{\lambda\mu} + \frac{(u_0)_{\alpha,\beta} + (u_0)_{\beta,\alpha}}{2} \\
&= 0, \text{ in } \Omega.
\end{aligned}
\tag{18.44}
$$

Taking the variation of this last equation in Q_ρ we have

$$
\begin{aligned}
&-\overline{(-N_0)_{\alpha\rho}^K}\,\overline{(-N_0)_{\beta\mu}^K}(Q_0)_\mu \\
&+(w_0)_\lambda\overline{(-N_0)_{\alpha\beta}^K}(w_0)_\mu\frac{\partial(N_0)_{\lambda\mu}}{\partial Q_\rho} \\
&-\overline{H}_{\alpha\beta\lambda\mu}\frac{\partial(N_0)_{\lambda\mu}}{\partial Q_\rho} \\
&= \quad 0, \text{ in } \Omega,
\end{aligned}
\tag{18.45}
$$

that is,

$$
\begin{aligned}
&-\overline{(-N_0)_{\alpha\rho}^K}(w_0)_\beta \\
&+(w_0)_\lambda\overline{(-N_0)_{\alpha\beta}^K}(w_0)_\mu\frac{\partial(N_0)_{\lambda\mu}}{\partial Q_\rho} \\
&-\overline{H}_{\alpha\beta\lambda\mu}\frac{\partial(N_0)_{\lambda\mu}}{\partial Q_\rho} \\
&= \quad 0, \text{ in } \Omega,
\end{aligned}
\tag{18.46}
$$

so that

$$
\begin{aligned}
&\frac{\overline{\partial(N_0)_{\alpha\beta}}}{\partial Q_\rho} \\
&= \quad \overline{\overline{H}_{\alpha\beta\lambda\mu} - (w_0)_\lambda\overline{(-N_0)_{\alpha\beta}^K}(w_0)_\mu}\big((-N_0)_{\lambda\mu}^K(w_0)_\rho\big).
\end{aligned}
\tag{18.47}
$$

Hence, if $K > 0$ is sufficiently big, we obtain

$$
\begin{aligned}
&\left\{\frac{\partial^2 J_2(u_0,Q_0)}{\partial Q_\alpha\partial Q_\beta}\right\} \\
&= \left\{\frac{\partial^2 J_1(u_0,Q_0,N_0)}{\partial Q_\alpha\partial Q_\beta}+\right. \\
&\quad \left.+\frac{\partial^2 J_1(u_0,Q_0,N_0)}{\partial Q_\alpha\partial N_{\lambda\mu}}\frac{\partial(N_0)_{\lambda\mu}}{\partial Q_\beta}\right\} \\
&= \{\overline{(-N_0)_{\alpha\beta}^K}\}+\overline{(-N_0)_{\alpha\eta}^K}(w_0)_\rho \\
&\quad \times\left[\overline{H}_{\eta\rho\lambda\mu}-(w_0)_\lambda\overline{(-N_0)_{\eta\rho}^K}(w_0)_\mu\right]\big((-N_0)_{\lambda\mu}^K(w_0)_\beta\big) \\
&= \{\overline{(-N_0)_{\alpha\beta}^K}\}+\mathscr{O}(1/K^2) \\
&> \mathbf{0}.
\end{aligned}
\tag{18.48}
$$

Therefore, the first infimum in Q in (18.41) is well defined.
Also, from (18.41) and the second order necessary condition for a local minimum, we obtain

$$
\delta^2\tilde{J}^*(Q_0) \geq \mathbf{0}.
$$

Assume now again $\delta J(u_0) = \mathbf{0}$ and $N_0 \in A_+^*$.
Recall that

$$
J_3(u,Q) = \sup_{N\in A_+^*} J_1(u,Q,N).
$$

Observe that if $N \in A_+^*$, by direct computation we may obtain

$$\delta_{uQ}^2 J_1(u,Q,N) > \mathbf{0}.$$

Therefore, $J_3(u,Q)$ is convex since is the supremum of a family of convex functionals. Similarly as above we may obtain

$$\delta J_3(u_0,Q_0) = \mathbf{0},$$

and

$$J_3(u_0,Q_0) = J(u_0)$$

so that

$$
\begin{aligned}
J(u_0) &= J_3(u_0,Q_0) \\
&= \inf_{(u,Q) \in U \times Y_2^*} J_3(u,Q) \\
&\leq \inf_{Q \in Y_2^*} J_3(u,Q) \\
&= J(u), \ \forall u \in U.
\end{aligned}
\tag{18.49}
$$

Moreover,

$$
\begin{aligned}
\tilde{J}_3^*(Q_0) &= \inf_{u \in U} J_3(u,Q_0) \\
&= J_3(u_0,Q_0) \\
&\leq J_3(u,Q), \forall u \in U, \ Q \in Y_2^*.
\end{aligned}
\tag{18.50}
$$

Hence,

$$J(u_0) = \tilde{J}_3^*(Q_0) \leq \inf_{u \in U} J_3(u,Q) = \tilde{J}_3^*(Q), \ \forall Q \in Y_2^*.$$

From these last results we may write,

$$
\begin{aligned}
J(u_0) &= \inf_{u \in U} J(u) \\
&= \inf_{Q \in Y_2^*} \tilde{J}_3^*(Q) \\
&= \tilde{J}_3^*(Q_0).
\end{aligned}
\tag{18.51}
$$

From this, similarly as above, we may obtain

$$\delta^2 \tilde{J}_3^*(Q_0) \geq \mathbf{0}.$$

Finally, suppose now $\delta^2 J(u_0) < \mathbf{0}$, so that $N_0 \in A_-^*$.
From this we obtain

$$\frac{\partial^2 \hat{J}^*(Q_0,N_0)}{\partial (Q_{\alpha,\alpha})^2} < \mathbf{0}$$

where, as previously indicated

$$\hat{J}^*(Q,N) = -\hat{F}_K^*(Q) + H_K^*(Q,N) - \frac{1}{2} \int_\Omega \overline{H}_{\alpha\beta\lambda\mu} N_{\alpha\beta} N_{\lambda\mu} \, dx,$$

Here,

$$H_K^*(Q,N) = \sup_{u \in U} \left\{ \langle w_{,\alpha}, Q_{,\alpha} \rangle_{L^2} - \frac{1}{2} \int_\Omega (-N_{\alpha\beta} + K\delta_{\alpha\beta}) w_{,\alpha} w_{,\beta} \, dx \right\}.$$

Denoting,

$$\hat{J}(u,N) = \frac{1}{2}\int_\Omega h_{\alpha\beta\lambda\mu}w_{,\alpha\beta}w_{,\lambda\mu}\,dx$$
$$+\frac{1}{2}\int_\Omega N_{\alpha\beta}w_{,\alpha}w_{,\beta}\,dx - \frac{1}{2}\int_\Omega \overline{H}_{\alpha\beta\lambda\mu}N_{\alpha\beta}N_{\lambda\mu}\,dx$$
$$-\langle N_{\alpha\beta,\beta} - P_\alpha, u_\alpha\rangle_{L^2}, \tag{18.52}$$

also from $N_0 \in A_-^*$ and from

$$\hat{J}^*(Q_0,N_0) = J(u_0),$$

$$\delta\{\hat{J}^*(Q_0,N_0) - \langle (N_0)_{\alpha\beta,\beta} + P_\alpha, (u_0)_\alpha\rangle_{L^2}\} = \mathbf{0},$$

(the proofs of such results are very similar to those of the corresponding cases developed above), there exist $r, r_1, r_2 > 0$ such that for $N \in A_-^* \cap B_{r_2}(N_0)$, we have

$$\sup_{u\in B_r(u)} \hat{J}(u,N) = \sup_{Q\in B_{r_1}(Q_0)} \hat{J}^*(Q,N), \tag{18.53}$$

and

$$J(u_0) = \sup_{u\in B_r(u_0)} J(u)$$

$$= \sup_{u\in B_r(u_0)} \left\{ \sup_{N\in B_{r_2}(N_0)\cap E^*} \hat{J}(u,N) \right\}$$

$$= \sup_{Q\in B_{r_1}(Q_0)} \left\{ \sup_{N\in B_{r_2}(N_0)\cap E^*} \hat{J}^*(Q,N) \right\}$$

$$= \hat{J}^*(Q_0,N_0). \tag{18.54}$$

The proof is complete.

18.5   Conclusion

In this chapter we have developed a new primal dual variational formulation and a multi-duality principle applied to a non-linear model of plates.

About the primal dual formulation, we emphasize such a formulation is concave so that it is very interesting from a numerical analysis point of view.

Finally, the results here presented may be also developed in a similar fashion for a large class of problems, including non-linear models in elasticity and other non-linear models of plates and shells.

Chapter 19

On Duality Principles for One and Three-Dimensional Non-Linear Models in Elasticity

19.1 Introduction

In this chapter, we develop duality principles applicable to primal variational formulations found in the non-linear elasticity theory. As a first application, we establish the concerning results in details for one and three-dimensional models. We emphasize such duality principles are applicable to a larger class of variational optimization problems, such as non-linear models of plates and shells and other models in elasticity. Finally, we formally prove there is no duality gap between the primal and dual formulations, in a local extremal context.

About the references, this article in some sense extends and complements the original works of Telega, Bielski and their co-workers [11, 10, 76, 77]. In particular in [11], published in 1985 and in [77], for three-dimensional elasticity and related models, the authors established duality principles and concerning global optimality conditions, for the special case in which the stress tensor is positive definite at a critical point. In this specific sense, the present work complements such previous ones, considering we establish a sufficient condition for local minimality which does not require the stress tensor to be either positive or negative defined along the concerning domain. Such an optimality condition is summarized by the condition $\|u_x\|_\infty < 1/4$ at a critical point.

The tools of convex analysis and duality theory here used may be found in [14, 33, 64]. Existence of results in non-linear elasticity and related models may be found in [29, 30, 31].

Finally, details on the function spaces addressed may be found in [1].

At this point, we start to describe the primal variational formulation for the one-dimensional model.

Let $\Omega = [0, L] \subset \mathbb{R}$ be an interval which represents the axis of a straight bar of length L and constant cross section area A.

We denote by $u : [0, L] \to \mathbb{R}$ the field of axial displacements for such a bar, resulting from the application of an axial load field $P \in C([0, L])$.

We also denote

$$U = \{u \in C^1([0, L]) \ : \ u(0) = u(L) = 0\},$$

and
$$\hat{U} = \{u \in U \ : \ \|u_x\|_\infty < 1/4\}.$$

The energy for such a system, denoted by $J : U \to \mathbb{R}$, is expressed as

$$J(u) = \frac{EA}{2} \int_0^L \left(u_x + \frac{1}{2}u_x^2 \right)^2 dx - \int_0^L Pu \, dx, \ \forall u \in U,$$

where $E > 0$ is the Young modulus.

We shall also define

$$G(u_x) = \frac{EA}{2} \int_0^L \left(u_x + \frac{1}{2}u_x^2 \right)^2 dx.$$

Finally, generically we shall denote, for $u \in U$ and $r > 0$,

$$B_r(u) = \{v \in U \ : \ \|v - u\|_U < r\}$$

where

$$\|v\|_U = \max_{x \in [0,L]} \{|v(x)| + |v_x(x)|\}, \ \forall v \in U.$$

Moreover, defining $V = C([0,L])$, for $z^* \in V$ and $r_1 > 0$, we shall generically also denote

$$B_{r_1}(z^*) = \{v \in V \ : \ \|v - z^*\|_V < r_1\},$$

where

$$\|v\|_V = \|v\|_\infty = \max_{x \in [0,L]} |v(x)|, \ \forall v \in V.$$

Similar corresponding standard notations are valid for $V \times V$ and the 3-dimensional model.

19.2 The main duality principle for the one-dimensional model

Our first duality principle is summarized by the following theorem:

Theorem 19.2.1 *Let $J : U \to \mathbb{R}$ be defined by*

$$J(u) = \frac{EA}{2} \int_0^L \left(u_x + \frac{1}{2}u_x^2 \right)^2 dx - \int_0^L Pu \, dx.$$

Assume $u_0 \in \hat{U}$ is such that

$$\delta J(u_0) = \mathbf{0}.$$

Define $F : U \to \mathbb{R}$ by

$$F(u_x) = \frac{K}{2} \int_0^L u_x^2 \, dx,$$

$G_K : U \to \mathbb{R}$ by

$$G_K(u_x) = G(u_x) + \frac{K}{2} \int_0^L u_x^2 \, dx$$

and $J^* : V \times V \times V \to \mathbb{R}$ by

$$J^*(v^*, z^*) = F^*(z^*) - G_K^*(v^*, z^*),$$

$$F^*(z^*) = \sup_{v \in V}\{\langle v, z^* \rangle_{L^2} - F(v)\}$$

$$= \frac{1}{2K} \int_0^L (z^*)^2 \, dx \tag{19.1}$$

and

$$
\begin{aligned}
G_K^*(v^*, z^*) &= \sup_{(v_1, v_2) \in V \times V} \left\{ \langle v_1, z^* + v_2^* \rangle_{L^2} + \langle v_2, v_1^* \rangle_{L^2} \right. \\
&\quad \left. - \frac{EA}{2} \int_0^L \left(v_1 + \frac{1}{2} v_2^2 \right)^2 dx - \frac{K}{2} \int_0^L v_2^2 \, dx \right\} \\
&= \frac{1}{2} \int_0^L \frac{(v_1^*)^2}{z^* + v_2^* + K} \, dx + \frac{1}{2EA} \int_0^L (v_2^* + z^*)^2 \, dx,
\end{aligned}
\tag{19.2}
$$

if $v_2^* + z^* + K > 0$, in Ω.

   Define also,

$$
A^* = \{ v^* = (v_1^*, v_2^*) \in V \times V \ : \ (v_1^*)_x + (v_2^*)_x + P = 0, \ in \ \Omega \},
$$

$$
K = EA/2,
$$

$$
\hat{z}^* = K(u_0)_x,
$$

$$
\hat{v}_2^* = EA \left((u_0)_x + \frac{1}{2}(u_0)_x^2 \right) - \hat{z}^*,
$$

$$
\hat{v}_1^* = (\hat{z}^* + \hat{v}_2^* + K)(u_0)_x.
$$

Under hypotheses and definitions, we have

$$
\delta^2 J(u_0, \varphi, \varphi) \geq 0, \ \forall \varphi \in C_c^1((0, L))
\tag{19.3}
$$

and there exist $r, r_1, r_2 > 0$ such that

$$
\begin{aligned}
J(u_0) &= \inf_{u \in B_r(u_0)} J(u) \\
&= \sup_{v^* \in B_{r_2}(\hat{v}^*) \cap A^*} \left\{ \inf_{z^* \in B_{r_1}(\hat{z}^*)} \{ J^*(v^*, z^*) \} \right\} \\
&= J^*(\hat{v}^*, \hat{z}^*).
\end{aligned}
\tag{19.4}
$$

Proof 19.1   Observe that

$$
\begin{aligned}
\frac{\partial^2 J^*(\hat{v}^*, \hat{z}^*)}{\partial (z^*)^2} &= \frac{2}{EA} - \frac{(\hat{v}_1^*)^2}{(\hat{v}_2^* + \hat{z}^* + \frac{EA}{2})^3} - \frac{1}{EA} \\
&= \frac{1}{EA} - \frac{(u_0)_x^2}{\hat{v}_2^* + \hat{z}^* + \frac{EA}{2}}.
\end{aligned}
\tag{19.5}
$$

Also,

$$
\begin{aligned}
\hat{v}_2^* + \hat{z}^* + \frac{EA}{2} &= EA \left((u_0)_x + \frac{1}{2}(u_0)_x^2 \right) + \frac{EA}{2} \\
&> \frac{EA}{2} - EA \left(\frac{1}{4} + \frac{1}{2}\frac{1}{16} \right) \\
&= EA \left(\frac{1}{4} - \frac{1}{32} \right) \\
&= EA \frac{7}{32}.
\end{aligned}
\tag{19.6}
$$

From this and (19.5), we obtain

$$
\begin{aligned}
\frac{\partial^2 J^*(\hat{v}^*, \hat{z}^*)}{\partial (z^*)^2} \quad &> \quad \frac{1}{EA} - \frac{1}{16} \frac{32}{(7\,EA)} \\
&= \quad \frac{1}{EA}\left(1 - \frac{2}{7}\right) \\
&= \quad \frac{5}{7EA} \\
&> \quad 0, \text{ in } \Omega.
\end{aligned}
\tag{19.7}
$$

Thus, we may infer that there exists $r_1, r_2 > 0$ such that $J^*(v^*, z^*)$ is convex in z^* and concave in v^*, on

$$
B_{r_1}(\hat{z}^*) \times B_{r_2}(\hat{v}^*).
$$

Now, denoting

$$
\hat{J}(v^*, z^*, u) = J^*(v^*, z^*) - \langle u, (v_1^*)_x + (v_2^*)_x + P \rangle_{L^2}
$$

we obtain

$$
\begin{aligned}
\frac{\partial \hat{J}^*(\hat{v}^*, \hat{z}^*, u_0)}{\partial z^*} \quad &= \quad \frac{z^*}{K} + \frac{1}{2} \frac{(\hat{v}_1^*)^2}{(\hat{v}_2^* + \hat{z}^* + K)^2} - \frac{\hat{v}_2^* + \hat{z}^*}{EA} \\
&= \quad (u_0)_x + \frac{1}{2}(u_0)_x^2 - \frac{EA\left((u_0)_x + \frac{1}{2}(u_0)_x^2\right)}{EA} \\
&= \quad 0, \text{ in } \Omega.
\end{aligned}
\tag{19.8}
$$

Also,

$$
\begin{aligned}
\frac{\partial \hat{J}^*(\hat{v}^*, \hat{z}^*, u_0)}{\partial v_1^*} \quad &= \quad -\frac{(\hat{v}_1^*)}{(\hat{v}_2^* + \hat{z}^* + K)} + (u_0)_x \\
&= \quad 0, \text{ in } \Omega,
\end{aligned}
\tag{19.9}
$$

and

$$
\begin{aligned}
\frac{\partial \hat{J}^*(\hat{v}^*, \hat{z}^*, u_0)}{\partial v_2^*} \quad &= \quad \frac{1}{2} \frac{(v_1^*)^2}{(\hat{v}_2^* + \hat{z}^* + K)^2} - \frac{\hat{v}_2^* + \hat{z}^*}{EA} + (u_0)_x \\
&= \quad (u_0)_x + \frac{1}{2}(u_0)_x^2 - \frac{EA\left((u_0)_x + \frac{1}{2}(u_0)_x^2\right)}{EA} \\
&= \quad 0, \text{ in } \Omega.
\end{aligned}
\tag{19.10}
$$

Finally,

$$
\begin{aligned}
\frac{\partial \hat{J}^*(\hat{v}^*, \hat{z}^*, u_0)}{\partial u} \quad &= \quad -(\hat{v}_1^*)_x - (\hat{v}_2^*)_x - P \\
&= \quad \delta J(u_0) \\
&= \quad \mathbf{0}, \text{ in } \Omega.
\end{aligned}
\tag{19.11}
$$

These last four results may be summarized by the equation

$$
\delta \hat{J}^*(\hat{v}^*, \hat{z}^*, u_0) = \mathbf{0}.
$$

Since above we have obtained that $J^*(v^*, z^*)$ is convex in z^* and concave in v^*, on

$$
B_{r_1}(\hat{z}^*) \times B_{r_2}(\hat{v}^*),
$$

from this, the last result and from the min-max theorem, we have

$$
\begin{aligned}
J^*(\hat{v}^*, \hat{z}^*) &= \hat{J}^*(\hat{v}^*, \hat{z}^*, u_0) \\
&= \sup_{v^* \in B_{r_2}(\hat{v}^*)} \left\{ \inf_{z^* \in B_{r_1}(\hat{z}^*)} \left\{ \hat{J}^*(v^*, z^*, u_0) \right\} \right\} \\
&= \sup_{v^* \in B_{r_2}(\hat{v}^*) \cap A^*} \left\{ \inf_{z^* \in B_{r_1}(\hat{z}^*)} \hat{J}^*(v^*, z^*) \right\}.
\end{aligned}
\tag{19.12}
$$

At this point we observe that

$$
\begin{aligned}
J^*(\hat{v}^*, \hat{z}^*) &= \hat{J}^*(\hat{v}^*, \hat{z}^*, u_0) \\
&= F^*(\hat{z}^*) - G_K^*(\hat{v}^*, \hat{z}^*) \\
&\quad + \langle (u_0)_x, \hat{v}_1^* + \hat{v}_2^* \rangle_{L^2} - \langle u_0, P \rangle_{L^2} \\
&= \frac{K}{2} \int_0^L (u_0)_x^2 \, dx - \langle (u_0)_x, \hat{z}^* \rangle_{L^2} \\
&\quad - \langle (u_0)_x, \hat{v}_1^* + \hat{v}_2^* + \hat{z}^* \rangle_{L^2} \\
&\quad + G((u_0)_x) + \frac{K}{2} \int_0^L (u_0)_x^2 \, dx \\
&\quad + \langle (u_0)_x, \hat{v}_1^* + \hat{v}_2^* \rangle_{L^2} - \langle u_0, P \rangle_{L^2} \\
&= G((u_0)_x) - \langle u_0, P \rangle_{L^2} \\
&= J(u_0).
\end{aligned}
\tag{19.13}
$$

On the other hand, since $\hat{v}^* \in A^*$, we may write

$$
\begin{aligned}
J^*(\hat{v}^*, \hat{z}^*) &= \hat{J}^*(\hat{v}^*, \hat{z}^*) \\
&= \inf_{z^* \in B_{r_1}(\hat{z}^*)} \hat{J}^*(\hat{v}^*, z^*) \\
&\leq F^*(z^*) - G_K^*(\hat{v}^*, z^*) \\
&\quad + \langle (u)_x, \hat{v}_1^* + \hat{v}_2^* \rangle_{L^2} - \langle u, P \rangle_{L^2} \\
&\leq F^*(z^*) - \langle (u)_x, \hat{v}_1^* + \hat{v}_2^* + z^* \rangle_{L^2} \\
&\quad + G(u_x) + K \int_0^L \frac{u_x^2}{2} \, dx \\
&\quad + \langle (u)_x, \hat{v}_1^* + \hat{v}_2^* \rangle_{L^2} - \langle u, P \rangle_{L^2},
\end{aligned}
\tag{19.14}
$$

$\forall u \in U, \; z^* \in B_{r_1}(\hat{z}^*)$.

In particular, there exists $r > 0$ such that if $u \in B_r(u_0)$ then $z^* = K u_x \in B_{r_1}(\hat{z}^*)$, so that from this and (19.14), we obtain

$$
\begin{aligned}
J^*(\hat{v}^*, \hat{z}^*) &\leq -K \int_0^L \frac{u_x^2}{2} \, dx + G(u_x) + K \int_0^L \frac{u_x^2}{2} \, dx - \langle u, P \rangle_{L^2} \\
&= G(u_x) - \langle u, P \rangle_{L^2} \\
&= J(u), \; \forall u \in B_r(u_0).
\end{aligned}
\tag{19.15}
$$

Finally, from (19.12), (19.13), and (19.15), we may infer that

$$
\begin{aligned}
J(u_0) &= \inf_{u \in B_r(u_0)} J(u) \\
&= \sup_{v^* \in B_{r_2}(\hat{v}^*) \cap A^*} \left\{ \inf_{z^* \in B_{r_1}(\hat{z}^*)} \left\{ J^*(v^*, z^*) \right\} \right\} \\
&= J^*(\hat{v}^*, \hat{z}^*).
\end{aligned}
\tag{19.16}
$$

From the first equation in such a result we may also obtain the standard second order necessary condition indicated in (19.3).

The proof is complete.

19.3 The primal variational formulation for the three-dimensional model

At this point we start to describe the primal formulation for the three-dimensional model.

Consider $\Omega \subset \mathbb{R}^3$ an open, bounded, connected set, which represents the reference volume of an elastic solid under the loads $f \in C(\overline{\Omega}; \mathbb{R}^3)$ and the boundary loads $\hat{f} \in C(\Gamma; \mathbb{R}^3)$, where Γ denotes the regular (Lipschitzian) boundary of Ω. The field of displacements resulting from the actions of f and $\hat{f}$ is denoted by $u \equiv (u_1, u_2, u_3) \in U$, where u_1, u_2, and u_3 denotes the displacements relating the directions x, y, and z respectively, in the cartesian system (x, y, z).

Here, U is defined by

$$U = \{u = (u_1, u_2, u_3) \in C^1(\overline{\Omega}; \mathbb{R}^3) \mid u = (0, 0, 0) \equiv \mathbf{0} \text{ on } \Gamma_0\} \tag{19.17}$$

and $\Gamma = \Gamma_0 \cup \Gamma_1$, $\Gamma_0 \cap \Gamma_1 = \emptyset$. We assume $|\Gamma_0| > 0$ where $|\Gamma_0|$ denotes the Lebesgue measure of Γ_0.

The stress tensor is denoted by $\{\sigma_{ij}\}$, where

$$\sigma_{ij} = H_{ijkl}\left(\frac{1}{2}(u_{k,l} + u_{l,k} + u_{m,k}u_{m,l})\right), \tag{19.18}$$

$$\{H_{ijkl}\} = \{\lambda\delta_{ij}\delta_{kl} + \mu(\delta_{ik}\delta_{jl} + \delta_{il}\delta_{jk})\},$$

$\{\delta_{ij}\}$ is the Kronecker delta and $\lambda, \mu > 0$ are the Lamé constants (we assume they are such that $\{H_{ijkl}\}$ is a symmetric constant positive definite forth order tensor). Here, $i, j, k, l \in \{1, 2, 3\}$.

The boundary value form of the non-linear elasticity model is given by

$$\begin{cases} \sigma_{ij,j} + (\sigma_{mj}u_{i,m})_{,j} + f_i = 0, & \text{in } \Omega, \\ u = \mathbf{0}, & \text{on } \Gamma_0, \\ \sigma_{ij}n_j + \sigma_{mj}u_{i,m}n_j = \hat{f}_i, & \text{on } \Gamma_1, \end{cases} \tag{19.19}$$

where $\mathbf{n} = (n_1, n_2, n_3)$ denotes the outward normal to the surface Γ.

The corresponding primal variational formulation is represented by $J : U \to \mathbb{R}$, where

$$\begin{aligned} J(u) = & \frac{1}{2}\int_\Omega H_{ijkl}\left(\frac{1}{2}(u_{i,j} + u_{j,i} + u_{m,i}u_{m,j})\right)\left(\frac{1}{2}(u_{k,l} + u_{l,k} + u_{m,k}u_{m,l})\right) dx \\ & - \langle u, f \rangle_{L^2(\Omega;\mathbb{R}^3)} - \int_{\Gamma_1} \hat{f}_i u_i \, d\Gamma \end{aligned} \tag{19.20}$$

where

$$\langle u, f \rangle_{L^2(\Omega;\mathbb{R}^3)} = \int_\Omega f_i u_i \, dx.$$

Remark 19.3.1 *By a regular Lipschitzian boundary Γ of Ω we mean regularity enough so that the standard Gauss-Green formulas of integrations by parts to hold. Also, we denote by $\mathbf{0}$ the zero vector in appropriate function spaces.*

About the references, similarly as for the one-dimensional case, we refer to [76, 77, 10, 11] as the first articles to deal with the convex analysis approach applied to non-convex and non-linear mechanics models. Indeed, the present work complements such important original publications, since in these previous results the complementary energy is established as a perfect duality principle for the case of positive definiteness of the stress tensor (or the membrane force tensor, for plates and shells models) at a critical point.

We have relaxed such constraints, allowing to some extent, the stress tensor to not be necessarily either positive or negative definite in Ω. Similar problems and models are addressed in [14].

Moreover, we highlight again that existence results for models in elasticity are addressed in [29, 30, 31]. Finally, the standard tools of convex analysis here used may be found in [33, 78, 64, 14].

19.4 The main duality principle for the three-dimensional model

In this section we present the main duality principle for the 3-Dimensional model.

The main result is summarized by the following theorem.

Theorem 19.4.1 *Let $J : U \to \mathbb{R}$ be defined by*

$$
\begin{aligned}
J(u) \;=\; & \int_\Omega H_{ijkl} \left(\frac{1}{2}(u_{i,j}+u_{j,i}+u_{m,i}u_{m,j}) \right) \left(\frac{1}{2}(u_{k,l}+u_{k,l}+u_{q,k}u_{q,l}) \right) \, dx \\
& - \langle u_i f_i \rangle_{L^2(\Omega)} - \langle u_i, \hat{f}_i \rangle_{L^2(\Gamma_1)}.
\end{aligned}
\tag{19.21}
$$

Assume $u_0 \in \hat{U}$ is such that

$$
\delta J(u_0) = \mathbf{0},
$$

where

$$
\hat{U} = \{ u \in U \;:\; \|u_{i,j}\|_\infty < 1/8, \ \forall i,j \in \{1,2,3\} \}.
$$

Define $F : U \to \mathbb{R}$ by

$$
F(u_{i,j}) = \frac{K}{2} \int_\Omega \left(\frac{u_{i,j}+u_{j,i}}{2} \right) \left(\frac{u_{i,j}+u_{j,i}}{2} \right) \, dx,
$$

$G_K : U \to \mathbb{R}$ by

$$
G_K(\{u_{i,j}\}) = G(u) + \frac{K}{2} \int_\Omega \left(\frac{u_{i,j}+u_{j,i}}{2} \right) \left(\frac{u_{i,j}+u_{j,i}}{2} \right) \, dx
$$

where

$$
G(u) = \int_\Omega H_{ijkl} \left(\frac{1}{2}(u_{i,j}+u_{j,i}+u_{m,i}u_{m,j}) \right) \left(\frac{1}{2}(u_{k,l}+u_{k,l}+u_{q,k}u_{q,l}) \right) \, dx,
$$

and $J^* : V \times V \times V \to \mathbb{R}$ by

$$
J^*(v^*, z^*) = F^*(z^*) - G_K^*(v^*, z^*),
$$

where

$$
V = C(\overline{\Omega}; \mathbb{R}^{3 \times 3}),
$$

$$
\begin{aligned}
F^*(z^*) \;=\; & \sup_{v \in V} \{ \langle v_{ij}, z_{ij}^* \rangle_{L^2} - F(v) \} \\
=\; & \frac{1}{2K} \int_\Omega z_{ij}^* z_{ij}^* \, dx
\end{aligned}
\tag{19.22}
$$

and

$$
\begin{aligned}
G_K^*(v^*, z^*) &= \sup_{(v_1, v_2) \in V \times V} \Big\{ \langle (v_1)_{ij}, z_{ij}^* + (v_2^*)_{ij} \rangle_{L^2} + \langle (v_2)_{ij}, (v_1^*)_{ij} \rangle_{L^2} \\
&\quad - \frac{1}{2} \int_\Omega H_{ijkl} \left((v_1)_{ij} + \frac{1}{2} (v_2)_{mi}(v_2)_{mj} \right) \left((v_1)_{kl} + \frac{1}{2} (v_2)_{qk}(v_2)_{ql} \right) dx \\
&\quad - \frac{K}{2} \int_\Omega (v_2)_{ij}(v_2)_{ij}\, dx \Big\} \\
&= \frac{1}{2} \int_\Omega \overline{((v_2^*)_{ij} + (z^*)_{ij} + K\delta_{ij})} (v_1^*)_{mi}(v_1^*)_{mj}; dx \\
&\quad + \frac{1}{2} \int_\Omega \overline{H}_{ijkl} ((v_2^*)_{ij} + z_{ij}^*)((v_2^*)_{kl} + z_{kl}^*)\, dx,
\end{aligned}
\tag{19.23}
$$

if $\{(v_2^*)_{ij} + (z^*)_{ij} + K\delta_{ij}\}$ *is positive definite in* Ω.

  Here

$$
\{\overline{(v_2^*)_{ij} + (z^*)_{ij} + K\delta_{ij}}\} = \{(v_2^*)_{ij} + (z^*)_{ij} + K\delta_{ij}\}^{-1}.
$$

  Define also, $A^* = A_1 \cap A_2$, *where*

$$
A_1 = \{ v^* = (v_1^*, v_2^*) \in V \times V : (v_1^*)_{ij,j} + (v_2^*)_{ij,j} + f_i = 0, \text{ in } \Omega \},
$$

and

$$
A_2 = \{ v^* = (v_1^*, v_2^*) \in V \times V : (v_1^*)_{ij} n_j + (v_2^*)_{ij} n_j - \hat{f}_i = 0, \text{ on } \Gamma_t \},
$$

and let $K > 0$ *be such that*

$$
M = \left\{ \frac{D_{ijkl}}{K} - \frac{3}{32K} \delta_{ij} - \overline{H}_{ijkl} \right\}
$$

is a positive definite tensor, where

$$
D_{ijkl} = \begin{cases} 1, & \text{if } i = k \text{ and } j = l, \\ 0, & \text{otherwise} \end{cases}
\tag{19.24}
$$

and, in an appropriate sense,

$$
\{\overline{H}_{ijkl}\} = \{H_{ijkl}\}^{-1}.
$$

  Assume also,

$$
\hat{z}_{ij}^* = K \left(\frac{(u_0)_{i,j} + (u_0)_{j,i}}{2} \right),
$$

$$
(\hat{v}_2^*)_{ij} = H_{ijkl} \left(\frac{(u_0)_{k,l} + (u_0)_{l,k}}{2} + \frac{1}{2} (u_0)_{m,k}(u_0)_{m,l} \right) - (\hat{z}^*)_{ij},
$$

$$
(\hat{v}_1^*)_{ij} = ((\hat{z}^*)_{im} + (\hat{v}_2^*)_{im} + K\delta_{im})(u_0)_{m,j},
$$

$\forall i, j \in \{1, 2, 3\}$ *and* $K > 0$ *is also such that*

$$
\{(\hat{v}_2^*)_{ij} + (\hat{z}^*)_{ij} + K\delta_{ij}\} \geq K\{\delta_{ij}\}/2
$$

Under hypotheses and definitions, there exist $r, r_1, r_2 > 0$ *such that*

$$
\begin{aligned}
J(u_0) &= \inf_{u \in B_r(u_0)} J(u) \\
&= \sup_{v^* \in B_{r_2}(\hat{v}^*) \cap A^*} \left\{ \inf_{z^* \in B_{r_1}(\hat{z}^*)} \{J^*(v^*, z^*)\} \right\} \\
&= J^*(\hat{v}^*, \hat{z}^*).
\end{aligned}
\tag{19.25}
$$

Proof 19.2   Observe that, denoting

$$\overline{\{\overline{(\hat{v}_2^*)_{ij} + \hat{z}_{ij}^* + K\delta_{ij}\}}} = \{(\hat{v}_2^*)_{ij} + \hat{z}_{ij}^* + K\delta_{ij}\}^{-2}$$

and

$$\overline{\overline{\{\overline{(\hat{v}_2^*)_{ij} + \hat{z}_{ij}^* + K\delta_{ij}\}}}} = \{(\hat{v}_2^*)_{ij} + \hat{z}_{ij}^* + K\delta_{ij}\}^{-3}$$

we have

$$
\left\{ \frac{\partial^2 J^*(\hat{v}^*, \hat{z}^*)}{\partial z_{ij}^* \partial z_{kl}^*} \right\}
= \left\{ \frac{D_{ijkl}}{K} - (\overline{\overline{(\hat{v}_2^*)_{ij} + (\hat{z}^*)_{ij} + K\delta_{ij}}})(\hat{v}_1^*)_{mk}(\hat{v}_1^*)_{ml} - \overline{H}_{ijkl} \right\}
$$

$$
= \left\{ \frac{D_{ijkl}}{K} - (\overline{\overline{(\hat{v}_2^*)_{ij} + (\hat{z}^*)_{ij} + K\delta_{ij}}})(u_0)_{mk}(u_0)_{ml} - \overline{H}_{ijkl} \right\}
$$

$$
\geq \left\{ \frac{D_{ijkl}}{K} - \frac{3}{32K}\delta_{ij} - \overline{H}_{ijkl} \right\}
$$

$$
> 0. \tag{19.26}
$$

Thus, there exist $r_1, r_2 > 0$ such that $J^*(v^*, z^*)$ is convex in z^* and concave in v^* on

$$B_{r_1}(\hat{z}^*) \times B_{r_2}(\hat{v}^*).$$

Now, denoting

$$
\hat{J}(v^*, z^*, u) = J^*(v^*, z^*) - \langle u, (v_1^*)_{ij,j} + (v_2^*)_{ij,j} + f_i \rangle_{L^2(\Omega)}
$$

$$
+ \langle u, (v_1^*)_{ij} n_j + (v_2^*)_{ij} n_j - \hat{f}_i \rangle_{L^2(\Gamma_t)} \tag{19.27}
$$

we obtain

$$
\frac{\partial \hat{J}^*(\hat{v}^*, \hat{z}^*, u_0)}{\partial (z^*)_{ij}} = \frac{z_{ij}^*}{K} + \frac{1}{2}\overline{(\hat{v}_2^*)_{ij} + (\hat{z}^*)_{ij} + K\delta_{ij}}(\hat{v}_1^*)_{mi}(\hat{v}_1^*)_{mj} - \overline{H}_{ijkl}((\hat{v}_2^*)_{kl} + z_{kl}^*)
$$

$$
= \frac{(u_0)_{i,j} + (u_0)_{j,i}}{2} + \frac{1}{2}(u_0)_{mi}(u_0)_{mj} - \overline{H}_{ijkl}((\hat{v}_2^*)_{kl} + z_{kl}^*)
$$

$$
= 0, \text{ in } \Omega. \tag{19.28}
$$

Also,

$$
\frac{\partial \hat{J}^*(\hat{v}^*, \hat{z}^*, u_0)}{\partial (v_1^*)_{ij}} = -\overline{(\hat{v}_2^*)_{im} + (\hat{z}^*)_{im} + K\delta_{im}}(\hat{v}_1^*)_{mj} + (u_0)_{i,j}
$$

$$
= 0, \text{ in } \Omega, \tag{19.29}
$$

and

$$
\frac{\partial \hat{J}^*(\hat{v}^*, \hat{z}^*, u_0)}{\partial (v_2^*)_{ij}} = \frac{1}{2}\overline{(\hat{v}_2^*)_{ij} + (\hat{z}^*)_{ij} + K\delta_{ij}}(\hat{v}_1^*)_{mi}(\hat{v}_1^*)_{mj} - \overline{H}_{ijkl}((\hat{v}_2^*)_{kl} + \hat{z}_{kl}^*)
$$

$$
+ \frac{(u_0)_{i,j} + (u_0)_{j,i}}{2}
$$

$$
= \frac{(u_0)_{i,j} + (u_0)_{j,i}}{2} + \frac{1}{2}(u_0)_{mi}(u_0)_{mj} - \overline{H}_{ijkl}((\hat{v}_2^*)_{kl} + \hat{z}_{kl}^*)
$$

$$
= 0, \text{ in } \Omega. \tag{19.30}
$$

Finally,

$$
\frac{\partial \hat{J}^*(\hat{v}^*, \hat{z}^*, u_0)}{\partial u} = \begin{cases} -(\hat{v}_1^*)_{ij,j} - (\hat{v}_2^*)_{ij,j} - f_i, & \text{in } \Omega, \\ (\hat{v}_1^*)_{ij} n_j + (\hat{v}_2^*)_{ij} n_j - \hat{f}_i, & \text{on } \Gamma_1, \end{cases} \tag{19.31}
$$

Hence,

$$\frac{\partial \hat{J}^*(\hat{v}^*, \hat{z}^*, u_0)}{\partial u} = \delta J(u_0) = \mathbf{0}.$$

These last four results may be summarized by the equation

$$\delta \hat{J}^*(\hat{v}^*, \hat{z}^*, u_0) = \mathbf{0}.$$

Since above we have obtained that $J^*(v^*, z^*)$ is convex in z^* and concave in v^*, on

$$B_{r_1}(\hat{z}^*) \times B_{r_2}(\hat{v}^*),$$

from this, the last result and from the min-max theorem, we have

$$
\begin{aligned}
J^*(\hat{v}^*, \hat{z}^*) &= \hat{J}^*(\hat{v}^*, \hat{z}^*, u_0) \\
&= \sup_{v^* \in B_{r_2}(\hat{v}^*)} \left\{ \inf_{z^* \in B_{r_1}(\hat{z}^*)} \left\{ \hat{J}^*(v^*, z^*) \right\} \right\} \\
&= \sup_{v^* \in B_{r_2}(\hat{v}^*) \cap A^*} \left\{ \inf_{z^* \in B_{r_1}(\hat{z}^*)} \hat{J}^*(v^*, z^*) \right\}.
\end{aligned}
\tag{19.32}
$$

At this point we observe that

$$
\begin{aligned}
J^*(\hat{v}^*, \hat{z}^*) &= \hat{J}^*(\hat{v}^*, \hat{z}^*, u_0) \\
&= F^*(\hat{z}^*) - G_K^*(\hat{v}^*, \hat{z}^*) \\
&\quad + \langle (u_0)_{i,j}, (\hat{v}_1^*)_{ij} + (\hat{v}_2^*)_{ij} \rangle_{L^2} - \langle (u_0)_i, f_i \rangle_{L^2(\Omega)} - \langle (u_0)_i, \hat{f}_i \rangle_{L^2(\Gamma_1)} \\
&= -F(\{(u_0)_{i,j}\}) + \langle (u_0)_{i,j}, (\hat{z}^*)_{ij} \rangle_{L^2} \\
&\quad - \langle (u_0)_{i,j}, (\hat{v}_1^*)_{ij} + (\hat{v}_2^*)_{ij} + (\hat{z}^*)_{ij} \rangle_{L^2} \\
&\quad + G((u_0)) + \frac{K}{2} \int_\Omega \left(\frac{(u_0)_{i,j} + (u_0)_{j,i}}{2} \right) \left(\frac{(u_0)_{i,j} + (u_0)_{j,i}}{2} \right) dx \\
&\quad + \langle (u_0)_{i,j}, (\hat{v}_1^*)_{ij} + (\hat{v}_2^*)_{ij} \rangle_{L^2} - \langle (u_0)_i, f_i \rangle_{L^2(\Omega)} - \langle (u_0)_i, \hat{f}_i \rangle_{L^2(\Gamma_1)} \\
&= G((u_0)) - \langle (u_0)_i, f_i \rangle_{L^2(\Omega)} - \langle (u_0)_i, \hat{f}_i \rangle_{L^2(\Gamma_1)} \\
&= J(u_0).
\end{aligned}
\tag{19.33}
$$

On the other hand, since $\hat{v}^* \in A^*$, we may write

$$
\begin{aligned}
J^*(\hat{v}^*, \hat{z}^*) &= \inf_{z^* \in B_{r_1}(\hat{z}^*)} \hat{J}^*(\hat{v}^*, z^*) \\
&\leq F^*(z^*) - G_K^*(\hat{v}^*, z^*) \\
&\quad + \langle (u)_{i,j}, (\hat{v}_1^*)_{ij} + (\hat{v}_2^*)_{ij} \rangle_{L^2} - \langle u_i, f_i \rangle_{L^2(\Omega)} - \langle u_i, \hat{f}_i \rangle_{L^2(\Gamma_1)} \\
&\leq F^*(z^*) - \langle u_{ij}, (\hat{v}_1^*)_{ij} + (\hat{v}_2^*)_{ij} + (z^*)_{ij} \rangle_{L^2} \\
&\quad + G(u) + \frac{K}{2} \int_\Omega \left(\frac{u_{i,j} + u_{j,i}}{2} \right) \left(\frac{u_{i,j} + u_{j,i}}{2} \right) dx \\
&\quad + \langle u_{i,j}, (\hat{v}_1^*)_{ij} + (\hat{v}_2^*)_{ij} \rangle_{L^2} - \langle u_i, f_i \rangle_{L^2(\Omega)} - \langle u_i, \hat{f}_i \rangle_{L^2(\Gamma_1)} \\
&= F^*(z^*) - \langle u_{i,j}, z_{ij}^* \rangle_{L^2} \\
&\quad + G(u) + \frac{K}{2} \int_\Omega \left(\frac{u_{i,j} + u_{j,i}}{2} \right) \left(\frac{u_{i,j} + u_{j,i}}{2} \right) dx \\
&\quad - \langle u_i, f_i \rangle_{L^2(\Omega)} - \langle u_i, \hat{f}_i \rangle_{L^2(\Gamma_1)},
\end{aligned}
\tag{19.34}
$$

$\forall u \in U,\ z^* \in B_{r_1}(\hat{z}^*).$

In particular, there exists $r > 0$ such that if $u \in B_r(u_0)$ then $\{z_{ij}^*\} = K\left\{\frac{u_{i,j}+u_{j,i}}{2}\right\} \in B_{r_1}(\hat{z}^*)$, so that from this and (19.34), we obtain

$$
\begin{aligned}
J^*(\hat{v}^*, \hat{z}^*) \quad \leq \quad & -\frac{K}{2}\int_\Omega \left(\frac{u_{i,j}+u_{j,i}}{2}\right)\left(\frac{u_{i,j}+u_{j,i}}{2}\right)\,dx + G(u) \\
& +\frac{K}{2}\int_\Omega \left(\frac{u_{i,j}+u_{j,i}}{2}\right)\left(\frac{u_{i,j}+u_{j,i}}{2}\right)\,dx - \langle u_i, f_i\rangle_{L^2(\Omega)} - \langle u_i, \hat{f}_i\rangle_{L^2(\Gamma_1)} \\
= \quad & G(u) - \langle u_i, f_i\rangle_{L^2(\Omega)} - \langle u_i, \hat{f}_i\rangle_{L^2(\Gamma_1)} \\
= \quad & J(u),\ \forall u \in B_r(u_0).
\end{aligned}
\tag{19.35}
$$

Finally, from (19.32), (19.33), and (19.35), we may infer that

$$
\begin{aligned}
J(u_0) \quad &= \quad \inf_{u \in B_r(u_0)} J(u) \\
\\
&= \quad \sup_{v^* \in B_{r_2}(\hat{v}^*) \cap A^*} \left\{ \inf_{z^* \in B_{r_1}(\hat{z}^*)} \{J^*(v^*, z^*)\} \right\} \\
&= \quad J^*(\hat{v}^*, \hat{z}^*).
\end{aligned}
\tag{19.36}
$$

The proof is complete.

19.5 Conclusion

In this chapter we have developed some theoretical results on duality for a class of non-convex optimization problems in elasticity. In this first approach we have developed in details duality principles and sufficient optimality conditions for local minimality for one and three-dimensional models in elasticity. It is worth mentioning the results may be extended to other models in elasticity and to other models of plates and shells.

Chapter 20

A Primal Dual Variational Formulation Suitable for a Large Class of Non-Convex Problems in Optimization

20.1 Introduction

In this article we develop a new primal dual variational formulation suitable for a large class of non-convex problems in the calculus of variations.

The results are obtained through basic tools of convex analysis, duality theory, the Legendre transform concept and the respective relations between the primal and dual variables. The novelty here is that the dual formulation is established also for the primal variables, however with a large domain region of concavity about a critical point.

We formally prove there is no duality gap between the primal and dual formulations in a local extremal context.

We emphasize, our work, in some sense, generalizes, extends and complements the original Telega, Bielski and their co-workers results in the articles [10, 11, 77, 76].

The convex analysis results here used may be found in [33, 78, 64, 14], for example. Similar results for other problems may be found in [14].

Finally, details on the function spaces addressed may be found in [1].

At this point we start to describe the primal formulation.

Let $\Omega \subset \mathbb{R}^3$ be an open, bounded and connected set with a regular (Lipschitzian) boundary denoted by $\partial\Omega$.

Consider the functional $J : V \rightarrow \mathbb{R}$ where

$$
\begin{aligned}
J(u) &= G_0(\nabla u) + G_1(u) + G_2(u) \\
&\quad - F(u) - \langle u, f \rangle_{L^2} + G_3(u),
\end{aligned}
\tag{20.1}
$$

where

$$
G_0(\nabla u) = \frac{\gamma}{2} \int_\Omega \nabla u \cdot \nabla u \, dx,
$$

$$G_1(u) = \frac{\alpha}{4} \int_\Omega u^4 \, dx,$$

$$G_2(u) = \frac{(-\beta + K - \varepsilon)}{2} \int_\Omega u^2 \, dx,$$

$$F(u) = \frac{K}{2} \int_\Omega u^2 \, dx$$

and

$$G_3(u) = \frac{\varepsilon}{2} \int_\Omega u^2 \, dx$$

so that

$$
\begin{aligned}
J(u) &= \frac{\gamma}{2} \int_\Omega \nabla u \cdot \nabla u \, dx + \frac{\alpha}{4} \int_\Omega u^4 \, dx \\
&\quad - \frac{\beta}{2} \int_\Omega u^2 \, dx - \langle u, f \rangle_{L^2}, \ \forall u \in V.
\end{aligned}
\tag{20.2}
$$

Here $dx = dx_1 \, dx_2 \, dx_3$, $\alpha > 0, \beta > 0, \gamma > 0, \varepsilon > 0$, $K > \beta + \varepsilon$, $f \in C(\overline{\Omega})$ and

$$V = \{u \in C^2(\overline{\Omega}) \ : \ u = 0, \text{ on } \partial\Omega\}.$$

Moreover, we recall that V is a Banach space with the norm $\|\cdot\|_V$, where

$$
\begin{aligned}
\|u\|_V &= \max_{\mathbf{x} \in \overline{\Omega}} \{|u(\mathbf{x})| + |u_x(\mathbf{x})| + |u_y(\mathbf{x})| + |u_z(\mathbf{x})| + |u_{xy}(\mathbf{x})| + |u_{xz}(\mathbf{x})| + |u_{yz}(\mathbf{x})| \\
&\quad + |u_{xx}(\mathbf{x})| + |u_{yy}(\mathbf{x})| + |u_{zz}(\mathbf{x})|\}, \forall u \in V,
\end{aligned}
\tag{20.3}
$$

and generically we denote

$$\langle u, v \rangle_{L^2} = \int_\Omega u \, v \, dx, \ \forall u, v \in L^2(\Omega) \equiv L^2,$$

and

$$\langle \mathbf{u}, \mathbf{v} \rangle_{L^2} = \int_\Omega \mathbf{u} \cdot \mathbf{v} \, dx, \ \forall \mathbf{u}, \mathbf{v} \in L^2(\Omega; \mathbb{R}^3) \equiv L^2.$$

Now observe that

$$
\begin{aligned}
\inf_{u \in U} J(u) &\leq \inf_{u \in U} \{G_0(\nabla u) - \langle \nabla u, z_0^* \rangle_{L^2} \\
&\quad + G_1(u) - \langle u, z_1^* \rangle_{L^2} \\
&\quad + G_2(u) - \langle u, z_2^* \rangle_{L^2} \\
&\quad + G_3(u) - \langle u, f \rangle_{L^2} \\
&\quad + \sup_{u \in U} \{\langle \nabla u, z_0^* \rangle_{L^2} + \langle u, z_1^* \rangle_{L^2} + \langle u, z_2^* \rangle_{L^2} - F(u)\}\} \\
&= \inf_{u \in U} \{G_0(\nabla u) - \langle \nabla u, z_0^* \rangle_{L^2} \\
&\quad + G_1(u) - \langle u, z_1^* \rangle_{L^2} \\
&\quad + G_2(u) - \langle u, z_2^* \rangle_{L^2} \\
&\quad + G_3(u) - \langle u, f \rangle_{L^2}\} \\
&\quad + F^*(-\operatorname{div} z_0^* + z_1^* + z_1^*) \\
&= \sup_{v^* \in Y_1 \times V \times V} \{-G_0^*(v_0^* + z_0^*) - G_1^*(v_1^* + z_1^*) \\
&\quad - G_2^*(v_2^* + z_2^*) - G_3^*(\operatorname{div} v_0^* - v_1^* - v_2^* + f)\} \\
&\quad + F^*(-\operatorname{div} z_0^* + z_1^* + z_2^*), \ \forall z^* = (z_0^*, z_1^*, z_2^*) \in Y_1 \times V \times V,
\end{aligned}
\tag{20.4}
$$

where

$$F^*(-\operatorname{div} z_0^* + z_1^* + z_2^*) = \frac{1}{2K} \int_\Omega (-\operatorname{div} z_0^* + z_1^* + z_2^*)^2 \, dx,$$

$v^* = (v_0^*, v_1^*, v_2^*) \in Y_1 \times V \times V,\ Y_1 = C^1(\overline{\Omega}; \mathbb{R}^3)$ and

$$
\begin{aligned}
G_0^*(v_0^*, z_0^*) &= \sup_{v_0 \in Y_1} \{\langle v_0, v_0^* + z_0^* \rangle_{L^2} - G_0(v_0)\} \\
&= \sup_{v_0 \in Y_1} \left\{ \langle v_0, v_0^* + z_0^* \rangle_{L^2} - \frac{\gamma}{2} \int_\Omega |v_0|^2 \, dx \right\} \\
&= \frac{1}{2\gamma} \int_\Omega |v_0^* + z_0^*|^2 \, dx.
\end{aligned}
\tag{20.5}
$$

Also,

$$
\begin{aligned}
G_1^*(v_1^*, z_1^*) &= \sup_{u \in V} \{\langle u, v_1^* + z_1^* \rangle_{L^2} - G_1(u)\} \\
&= \sup_{u \in V} \left\{ \langle u, v_1^* + z_1^* \rangle_{L^2} - \frac{\alpha}{4} \int_\Omega u^4 \, dx \right\} \\
&= \frac{3}{4\alpha^{1/3}} \int_\Omega |v_1^* + z_1^*|^{4/3} \, dx,
\end{aligned}
\tag{20.6}
$$

$$
\begin{aligned}
G_2^*(v_2^*, z_2^*) &= \sup_{u \in V} \{\langle u, v_2^* + z_2^* \rangle_{L^2} - G_2(u)\} \\
&= \sup_{u \in V} \left\{ \langle u, v_2^* + z_2^* \rangle_{L^2} - \frac{(-\beta + K - \varepsilon)}{2} \int_\Omega u^2 \, dx \right\} \\
&= \frac{1}{2(K - \beta - \varepsilon)} \int_\Omega (v_2^* + z_2^*)^2 \, dx
\end{aligned}
\tag{20.7}
$$

and

$$
\begin{aligned}
G_3^*(\operatorname{div} v_0^* - v_1^* - v_2^* + f) &= \sup_{u \in V} \{-\langle \nabla u, v_0^* \rangle_{L^2} - \langle u, v_1^* \rangle_{L^2} \\
&\quad - \langle u, v_2^* \rangle_{L^2} + \langle u, f \rangle_{L^2} - G_3(u)\} \\
&= \sup_{u \in V} \{-\langle \nabla u, v_0^* \rangle_{L^2} - \langle u, v_1^* \rangle_{L^2} \\
&\quad - \langle u, v_2^* \rangle_{L^2} + \langle u, f \rangle_{L^2} - \frac{\varepsilon}{2} \int_\Omega u^2 \, dx\} \\
&= \frac{1}{2\varepsilon} \int_\Omega (\operatorname{div} v_0^* - v_1^* - v_2^* + f)^2 \, dx.
\end{aligned}
\tag{20.8}
$$

At this point, we denote,

$$
\begin{aligned}
J^*(v^*, z^*) &= -G_0^*(v_0^* + z_0^*) - G_1^*(v_1^* + z_1^*) \\
&\quad - G_2^*(v_2^* + z_2^*) - G_3^*(\operatorname{div} v_0^* - v_1^* - v_2^* + f) \\
&\quad + F^*(-\operatorname{div} z_0^* + z_1^* + z_2^*).
\end{aligned}
\tag{20.9}
$$

The extremal equation,

$$\frac{\partial J^*(v^*, z^*)}{\partial z^*} = \mathbf{0}$$

gives the following system.

Specifically from

$$\frac{\partial J^*(v^*, z^*)}{\partial z_0^*} = \mathbf{0},$$

we get

$$-\frac{\partial G_0^*(v_0^* + z_0^*)}{\partial z_0^*} + \nabla \left(\frac{\partial F^*(-\operatorname{div} z_0^* + z_1^* + z_2^*)}{\partial w_0^*} \right) = \mathbf{0},$$

where

$$w_0^* = -\operatorname{div} z_0^*$$

so that

$$\frac{v_0^* + z_0^*}{\gamma} - \nabla \left(\frac{-divz_0^* + z_1^* + z_2^*}{K} \right) = \mathbf{0}.$$

Hence, defining

$$\hat{u} = \frac{-divz_0^* + z_1^* + z_2^*}{K},$$

we have

$$v_0^* = -z_0^* + \gamma \nabla \hat{u}. \tag{20.10}$$

From

$$\frac{\partial J^*(v^*, z^*)}{\partial z_1^*} = \mathbf{0},$$

we get

$$-\frac{\partial G_1^*(v_1^* + z_1^*)}{\partial z_1^*} + \frac{\partial F^*(-\operatorname{div} z_0^* + z_1^* + z_2^*)}{\partial z_1^*} = \mathbf{0},$$

so that

$$\frac{(v_1^* + z_1^*)^{1/3}}{\alpha^{1/3}} = \hat{u},$$

that is,

$$v_1^* = -z_1^* + \alpha \hat{u}^3. \tag{20.11}$$

Finally, from

$$\frac{\partial J^*(v^*, z^*)}{\partial z_2^*} = \mathbf{0},$$

we get

$$-\frac{\partial G_2^*(v_2^* + z_2^*)}{\partial z_2^*} + \frac{\partial F^*(-\operatorname{div} z_0^* + z_1^* + z_2^*)}{\partial z_2^*} = \mathbf{0},$$

so that

$$\frac{(v_2^* + z_2^*)}{K - \beta - \varepsilon} = \hat{u},$$

that is,

$$v_2^* = -z_2^* + (K - \beta - \varepsilon)\hat{u}. \tag{20.12}$$

Replacing (20.10), (20.11) and (20.12) into (20.9), we obtain

$$
\begin{aligned}
J^*(v^*,z^*) &= -\frac{\gamma}{2}\int_\Omega |\nabla \hat{u}|^2\, dx - \frac{3}{4\alpha^{1/3}}\int_\Omega \alpha^{4/3}\hat{u}^4\, dx \\
&\quad -\frac{(K-\beta-\varepsilon)}{2}\int_\Omega \hat{u}^2\, dx + \frac{K}{2}\int_\Omega \hat{u}^2\, dx \\
&\quad -\frac{1}{2\varepsilon}\int_\Omega (\,\text{div}\, z_0^* - z_1^* - z_2^* - \gamma\nabla^2\hat{u} + \alpha\hat{u}^3 + (K-\beta-\varepsilon)\hat{u} - f)^2\, dx \\
&= -\frac{\gamma}{2}\int_\Omega |\nabla \hat{u}|^2\, dx - \frac{3}{4}\alpha\int_\Omega \hat{u}^4\, dx + \frac{\beta+\varepsilon}{2}\int_\Omega \hat{u}^2\, dx \\
&\quad -\frac{1}{2\varepsilon}\int_\Omega (-\gamma\nabla^2\hat{u} + \alpha\hat{u}^3 - (\beta+\varepsilon)\hat{u} - f)^2\, dx \\
&\equiv \hat{J}_\varepsilon(\hat{u}). \tag{20.13}
\end{aligned}
$$

20.2 The main duality principle

With such statements, definitions and results in mind we prove the following theorem.

Theorem 20.2.1 *Considering the context of the last section statements, definitions and results, let $J : V \to \mathbb{R}$ be defined by*

$$
\begin{aligned}
J(u) &= \frac{\gamma}{2}\int_\Omega \nabla u \cdot \nabla u\, dx + \frac{\alpha}{4}\int_\Omega u^4\, dx \\
&\quad -\frac{\beta}{2}\int_\Omega u^2\, dx - \langle u, f\rangle_{L^2},\ \forall u \in V. \tag{20.14}
\end{aligned}
$$

Here, $f \in C(\overline{\Omega})$. Let $\hat{J}_\varepsilon : V \to \mathbb{R}$ be defined by

$$
\begin{aligned}
\hat{J}_\varepsilon(\hat{u}) &= -\frac{\gamma}{2}\int_\Omega |\nabla \hat{u}|^2\, dx - \frac{3}{4}\alpha\int_\Omega \hat{u}^4\, dx + \frac{\beta+\varepsilon}{2}\int_\Omega \hat{u}^2\, dx \\
&\quad -\frac{1}{2\varepsilon}\int_\Omega (-\gamma\nabla^2\hat{u} + \alpha\hat{u}^3 - (\beta+\varepsilon)\hat{u} - f)^2\, dx. \tag{20.15}
\end{aligned}
$$

Assume $u_0 \in V$ is such that

$$
\delta J(u_0) = \mathbf{0},
$$

and

$$
\delta^2 J(u_0) > \mathbf{0}.
$$

Under such hypotheses,

$$
\delta \hat{J}_\varepsilon(u_0) = \mathbf{0},
$$
$$
J(u_0) = \hat{J}_\varepsilon(u_0)
$$

and for $\varepsilon > 0$ sufficiently small,

$$
\begin{aligned}
\delta^2 \hat{J}_\varepsilon(u_0) &= -(\delta^2 J(u_0) - \varepsilon) - \frac{(\delta^2 J(u_0) - \varepsilon)^2}{\varepsilon} \\
&\approx -\mathscr{O}\left(\frac{1}{\varepsilon}\right) \\
&< \mathbf{0}, \tag{20.16}
\end{aligned}
$$

so that there exist $r, r_1 > 0$ such that

$$
\begin{aligned}
J(u_0) &= \inf_{u \in B_r(u_0)} J(u) \\
&= \sup_{\hat{u} \in B_{r_1}(u_0)} \hat{J}_\varepsilon(\hat{u}) \\
&= \hat{J}_\varepsilon(u_0).
\end{aligned}
\tag{20.17}
$$

Proof 20.1   From $\delta J(u_0) = \mathbf{0}$ we have

$$
-\gamma \nabla^2 u_0 + \alpha u_0^3 - \beta u_0 - f = 0, \text{ in } \Omega.
$$

Thus defining $\hat{u} = u_0$ we have

$$
\begin{aligned}
&-\gamma \nabla^2 u_0 + \alpha u_0^3 - (\beta + \varepsilon) u_0 - f \\
&= -\varepsilon u_0 \\
&= -\varepsilon \hat{u}, \text{ in } \Omega.
\end{aligned}
\tag{20.18}
$$

Hence

$$
\hat{u} = -\frac{-\gamma \nabla^2 u_0 + \alpha u_0^3 - (\beta + \varepsilon) u_0 - f}{\varepsilon}.
$$

From such an expression for $\hat{u}$ we obtain

$$
\delta \hat{J}_\varepsilon(u_0) = \gamma \nabla^2 u_0 - 3\alpha u_0^3 + (\beta + \varepsilon) u_0 - \gamma \nabla^2 \hat{u} + 3\alpha \hat{u} u_0^2 - (\beta + \varepsilon) \hat{u}.
\tag{20.19}
$$

From this, since $\hat{u} = u_0$, we get

$$
\delta \hat{J}_\varepsilon(u_0) = \mathbf{0}.
$$

Also, we may define v^*, z^* such that

$$
v_0^* = -z_0^* + \gamma \nabla \hat{u},
\tag{20.20}
$$

$$
v_1^* = -z_1^* + \alpha \hat{u}^3,
\tag{20.21}
$$

$$
v_2^* = -z_2^* + (K - \beta - \varepsilon) \hat{u},
\tag{20.22}
$$

$$
-\operatorname{div} z_0^* + z_1^* + z_2^* = K \hat{u},
$$

so that

$$
\begin{aligned}
\operatorname{div} v_0^* - v_1^* - v_2^* &= -\operatorname{div} z_0^* + z_1^* + z_2^* \\
&\quad + \gamma \nabla^2 \hat{u} - \alpha \hat{u}^3 - (K - \beta - \varepsilon) \hat{u} \\
&= \gamma \nabla^2 \hat{u} - \alpha \hat{u}^3 + (\beta + \varepsilon) \hat{u} \\
&= \varepsilon \hat{u} - f.
\end{aligned}
\tag{20.23}
$$

From such relations we obtain,

$$
G_0^*(v_0^*, z_0^*) = \langle \nabla \hat{u}, v_0^* + z_0^* \rangle_{L^2} - G_0(\nabla \hat{u}),
$$

$$
G_1^*(v_1^*, z_1^*) = \langle \hat{u}, v_1^* + z_1^* \rangle_{L^2} - G_1(\hat{u}),
$$

$$
G_2^*(v_2^*, z_2^*) = \langle \hat{u}, v_2^* + z_2^* \rangle_{L^2} - G_3(\hat{u}),
$$

$$
G_3^*(\operatorname{div} v_0^* - v_1^* - v_2^* + f) = \langle \hat{u}, \operatorname{div} v_0^* - v_1^* - v_2^* \rangle_{L^2} + \langle \hat{u}, f \rangle_{L^2} - G_3(\hat{u})
$$

and

$$
F^*(\operatorname{div} z_0^* - z_1^* - z_2^*) = \langle \hat{u}, -\operatorname{div} z_0^* + z_1^* + z_2^* \rangle_{L^2} - F(\hat{u}).
$$

From such results, we obtain

$$
\begin{aligned}
\hat{J}_\varepsilon(\hat{u}) &= J^*(v^*, z^*) \\
&= -G_0^*(v_0^* + z_0^*) - G_1^*(v_1^* + z_1^*) \\
&\quad - G_2^*(v_2^* + z_2^*) - G_3^*(\operatorname{div} v_0^* - v_1^* - v_2^* + f) + F^*(-\operatorname{div} z_0^* + z_1^* + z_2^*) \\
&= G_0(\nabla \hat{u}) + G_1(\hat{u}) + G_2(\hat{u}) + G_3(\hat{u}) - F(\hat{u}) - \langle \hat{u}, f \rangle_{L^2} \\
&= J(\hat{u}) \\
&= J(u_0).
\end{aligned}
\tag{20.24}
$$

Finally, for $\varepsilon > 0$ sufficiently small,

$$
\begin{aligned}
\delta^2 \hat{J}_\varepsilon(u_0) &= \gamma \nabla^2 - 9\alpha u_0^2 + \beta + \varepsilon \\
&= -\frac{(-\gamma \nabla^2 + 3\alpha u_0^2 - (\beta + \varepsilon))^2}{\varepsilon} + 6\alpha u_0^2 \\
&= \gamma \nabla^2 - 3\alpha u_0^2 + \beta + \varepsilon - \frac{(-\gamma \nabla^2 + 3\alpha u_0^2 - (\beta + \varepsilon))^2}{\varepsilon} \\
&= -(\delta^2 J(u_0) - \varepsilon) - \frac{(\delta^2 J(u_0) - \varepsilon)^2}{\varepsilon} \\
&\approx -\mathscr{O}\left(\frac{1}{\varepsilon}\right) \\
&< 0.
\end{aligned}
\tag{20.25}
$$

From these last results, there exist $r, r_1 > 0$ such that

$$
\begin{aligned}
J(u_0) &= \inf_{u \in B_r(u_0)} J(u) \\
&= \sup_{\hat{u} \in B_{r_1}(u_0)} \hat{J}_\varepsilon(\hat{u}) \\
&= \hat{J}_\varepsilon(u_0).
\end{aligned}
\tag{20.26}
$$

The proof is complete.

20.3   Conclusion

In this chapter we have developed a primal dual variational formulation for a large class of problems in the calculus of variations.

We emphasize the dual functional obtained has a large domain region of concavity about a critical point, which makes such a formulation very interesting from a numerical analysis point of view.

Finally, it has been formally proven there is no duality gap between the primal and dual formulations in a local extremal context.

Chapter 21

A Duality Principle and Concerning Computational Method for a Class of Optimal Design Problems in Elasticity

Fabio Silva Botelho and Alexandre Molter

21.1 Introduction

Consider an elastic solid which the volume corresponds to an open, bounded, connected set, denoted by $\Omega \subset \mathbb{R}^3$ with a regular (Lipschitzian) boundary denoted by $\partial \Omega = \Gamma_0 \cup \Gamma_t$ where $\Gamma_0 \cap \Gamma_t = \emptyset$. Consider also the problem of minimizing the functional $\hat{J} : U \times B \to \mathbb{R}$ where

$$\hat{J}(u,t) = \frac{1}{2} \langle u_i, f_i \rangle_{L^2(\Omega)} + \frac{1}{2} \langle u_i, \hat{f}_i \rangle_{L^2(\Gamma_t)},$$

subject to

$$\begin{cases} (H_{ijkl}(t)e_{kl}(u))_{,j} + f_i = 0 \text{ in } \Omega, \\ \\ H_{ijkl}(t)e_{kl}(u)n_j - \hat{f}_i = 0, \text{ on } \Gamma_t, \ \forall i \in \{1,2,3\}. \end{cases} \quad (21.1)$$

Here, $\mathbf{n} = (n_1, n_2, n_3)$ denotes the outward normal to $\partial \Omega$ and

$$U = \{u = (u_1, u_2, u_3) \in W^{1,2}(\Omega; \mathbb{R}^3) \ : \ u = (0,0,0) = \mathbf{0} \text{ on } \Gamma_0\},$$

$$B = \left\{ t : \Omega \to [0,1] \text{ measurable } : \ \int_\Omega t(x) \, dx = t_1 |\Omega| \right\},$$

where

$$0 < t_1 < 1$$

and $|\Omega|$ denotes the Lebesgue measure of Ω.

Moreover, $u = (u_1, u_2, u_3) \in W^{1,2}(\Omega; \mathbb{R}^3)$ is the field of displacements relating the cartesian system $(0, x_1, x_2, x_3)$, resulting from the action of the external loads $f \in L^2(\Omega; \mathbb{R}^3)$ and $\hat{f} \in L^2(\Gamma_t; \mathbb{R}^3)$.

We also define the stress tensor $\{\sigma_{ij}\} \in Y^* = Y = L^2(\Omega; \mathbb{R}^{3 \times 3})$, by

$$\sigma_{ij}(u) = H_{ijkl}(t)e_{kl}(u),$$

and the strain tensor $e : U \to L^2(\Omega; \mathbb{R}^{3 \times 3})$ by

$$e_{ij}(u) = \frac{1}{2}(u_{i,j} + u_{j,i}), \ \forall i, j \in \{1, 2, 3\}.$$

Finally,

$$\{H_{ijkl}(t)\} = \{tH^0_{ijkl} + (1-t)H^1_{ijkl}\},$$

where H^0 corresponds to a strong material and H^1 to a very soft material, intending to simulate voids along the solid structure.

The variable t is the design one, which the optimal distribution values along the structure are intended to minimize its inner work with a volume restriction indicated through the set B.

The duality principle obtained is developed inspired by the works in [11, 10]. Similar theoretical results have been developed in [14], however we believe the proof here presented, which is based on the min-max theorem is easier to follow (indeed we thank an anonymous referee for his suggestion about applying the min-max theorem to complete the proof). We highlight throughout this text we have used the standard Einstein sum convention of repeated indices.

Moreover, details on the Sobolev spaces addressed may be found in [1]. In addition, the primal variational development of the topology optimization problem has been described also in [14].

The main contributions of this work are to present the detailed development, through duality theory, for such a kind of optimization problems. We emphasize that to avoid the check-board standard and obtain appropriate robust optimized structures without the use of filters, it is necessary to discretize more in the load direction, in which the displacements are much larger.

Finally, it is worth mentioning the numerical examples presented have been developed in a Finite Element (FE) context, based on the work of [70].

21.2 Mathematical formulation of the topology optimization problem

Our mathematical topology optimization problem is summarized by the following theorem.

Theorem 21.2.1 *Consider the statements and assumptions indicated in the last section, in particular those refereing to Ω and the functional $\hat{J} : U \times B \to \mathbb{R}$.*
Define $J_1 : U \times B \to \mathbb{R}$ by

$$J_1(u,t) = -G(e(u),t) + \langle u_i, f_i \rangle_{L^2(\Omega)} + \langle u_i, \hat{f}_i \rangle_{L^2(\Gamma_t)},$$

where

$$G(e(u),t) = \frac{1}{2} \int_\Omega H_{ijkl}(t)e_{ij}(u)e_{kl}(u) \, dx,$$

and where

$$dx = dx_1 dx_2 dx_3.$$

Define also $J^* : U \to \mathbb{R}$ by

$$
\begin{aligned}
J^*(u) &= \inf_{t \in B}\{J_1(u,t)\} \\
&= \inf_{t \in B}\{-G(e(u),t) + \langle u_i, f_i \rangle_{L^2(\Omega)} + \langle u_i, \hat{f}_i \rangle_{L^2(\Gamma_t)}\}.
\end{aligned}
\tag{21.2}
$$

Assume there exists $c_0, c_1 > 0$ such that

$$H^0_{ijkl} z_{ij} z_{kl} > c_0 z_{ij} z_{ij}$$

and

$$H^1_{ijkl} z_{ij} z_{kl} > c_1 z_{ij} z_{ij}, \; \forall z = \{z_{ij}\} \in \mathbb{R}^{3 \times 3}, \; \text{such that } z \neq \mathbf{0}.$$

Finally, define $J : U \times B \to \mathbb{R} \cup \{+\infty\}$ by

$$J(u,t) = \hat{J}(u,t) + Ind(u,t),$$

where

$$Ind(u,t) = \begin{cases} 0, & \text{if } (u,t) \in A^*, \\ +\infty, & \text{otherwise}, \end{cases} \tag{21.3}$$

where $A^* = A_1 \cap A_2$,

$$A_1 = \{(u,t) \in U \times B \; : \; (\sigma_{ij}(u))_{,j} + f_i = 0, \; in \; \Omega, \; \forall i \in \{1,2,3\}\}$$

and

$$A_2 = \{(u,t) \in U \times B \; : \; \sigma_{ij}(u) n_j - \hat{f}_i = 0, \; on \; \Gamma_t, \; \forall i \in \{1,2,3\}\}.$$

Under such hypotheses, there exists $(u_0, t_0) \in U \times B$ such that

$$
\begin{aligned}
J(u_0, t_0) &= \inf_{(u,t) \in U \times B} J(u,t) \\
&= \sup_{\hat{u} \in U} J^*(\hat{u}) \\
&= J^*(u_0) \\
&= \hat{J}(u_0, t_0) \\
&= \inf_{(t,\sigma) \in B \times C^*} G^*(\sigma, t) \\
&= G^*(\sigma(u_0), t_0),
\end{aligned}
\tag{21.4}
$$

where

$$
\begin{aligned}
G^*(\sigma, t) &= \sup_{v \in Y} \{\langle v_{ij}, \sigma_{ij} \rangle_{L^2(\Omega)} - G(v,t)\} \\
&= \frac{1}{2} \int_{\Omega} \overline{H}_{ijkl}(t) \sigma_{ij} \sigma_{kl} \, dx,
\end{aligned}
\tag{21.5}
$$

$$\{\overline{H}_{ijkl}(t)\} = \{H_{ijkl}(t)\}^{-1}$$

and $C^* = C_1 \cap C_2$, where

$$C_1 = \{\sigma \in Y^* \; : \; \sigma_{ij,j} + f_i = 0, \; in \; \Omega, \; \forall i \in \{1,2,3\}\}$$

and

$$C_2 = \{\sigma \in Y^* \; : \; \sigma_{ij} n_j - \hat{f}_i = 0, \; on \; \Gamma_t, \; \forall i \in \{1,2,3\}\}.$$

Proof 21.1   Observe that

$$
\begin{aligned}
\inf_{(u,t)\in U\times B} J(u,t) &= \inf_{t\in B}\left\{\inf_{u\in U} J(u,t)\right\}\\[4pt]
&= \inf_{t\in B}\left\{\sup_{\hat{u}\in U}\left\{\inf_{u\in U}\left\{\frac{1}{2}\int_\Omega H_{ijkl}(t)e_{ij}(u)e_{kl}(u)\,dx\right.\right.\right.\\
&\quad\left.\left.\left.+\langle \hat{u}_i,(H_{ijkl}(t)e_{kl}(u))_{,j}+f_i\rangle_{L^2(\Omega)}\right.\right.\right.\\
&\quad\left.\left.\left.-\langle \hat{u}_i,H_{ijkl}(t)e_{kl}(u)n_j-\hat{f}_i\rangle_{L^2(\Gamma_t)}\right\}\right\}\right\}\\[4pt]
&= \inf_{t\in B}\left\{\sup_{\hat{u}\in U}\left\{\inf_{u\in U}\left\{\frac{1}{2}\int_\Omega H_{ijkl}(t)e_{ij}(u)e_{kl}(u)\,dx\right.\right.\right.\\
&\quad\left.\left.\left.-\int_\Omega H_{ijkl}(t)e_{ij}(\hat{u})e_{kl}(u)\,dx\right.\right.\right.\\
&\quad\left.\left.\left.+\langle \hat{u}_i,f_i\rangle_{L^2(\Omega)}+\langle \hat{u}_i,\hat{f}_i\rangle_{L^2(\Gamma_t)}\right\}\right\}\right\}\\[4pt]
&= \inf_{t\in B}\left\{\sup_{\hat{u}\in U}\left\{-\int_\Omega H_{ijkl}(t)e_{ij}(\hat{u})e_{kl}(\hat{u})\,dx\right.\right.\\
&\quad\left.\left.\langle \hat{u}_i,f_i\rangle_{L^2(\Omega)}+\langle \hat{u}_i,\hat{f}_i\rangle_{L^2(\Gamma_t)}\right\}\right\}\\[4pt]
&= \inf_{t\in B}\left\{\inf_{\sigma\in C^*} G^*(\sigma,t)\right\}. \tag{21.6}
\end{aligned}
$$

Also, from this and the min-max theorem, there exist $(u_0,t_0)\in U\times B$ such that

$$
\begin{aligned}
\inf_{(u,t)\in U\times B} J(u,t) &= \inf_{t\in B}\left\{\sup_{\hat{u}\in U} J_1(u,t)\right\}\\[4pt]
&= \sup_{u\in U}\left\{\inf_{t\in B} J_1(u,t)\right\}\\[4pt]
&= J_1(u_0,t_0)\\[4pt]
&= \inf_{t\in B} J_1(u_0,t)\\[4pt]
&= J^*(u_0). \tag{21.7}
\end{aligned}
$$

Finally, from the extremal necessary condition

$$
\frac{\partial J_1(u_0,t_0)}{\partial u}=\mathbf{0}
$$

we obtain

$$
(H_{ijkl}(t_0)e_{kl}(u_0))_{,j}+f_i=0 \text{ in } \Omega,
$$

and

$$
H_{ijkl}(t_0)e_{kl}(u_0)n_j-\hat{f}_i=0 \text{ on } \Gamma_t,\ \forall i\in\{1,2,3\},
$$

so that

$$
G(e(u_0))=\frac{1}{2}\langle(u_0)_i,f_i\rangle_{L^2(\Omega)}+\frac{1}{2}\langle(u_0)_i,\hat{f}_i\rangle_{L^2(\Gamma_t)}.
$$

Hence, $(u_0,t_0)\in A^*$ so that $Ind(u_0,t_0)=0$ and $\sigma(u_0)\in C^*$.

Moreover

$$
\begin{aligned}
J^*(u_0) &= -G(e(u_0)) + \langle (u_0)_i, f_i \rangle_{L^2(\Omega)} + \langle (u_0)_i, \hat{f}_i \rangle_{L^2(\Gamma_t)} \\
&= G(e(u_0)) \\
&= G(e(u_0)) + Ind(u_0, t_0) \\
&= J(u_0, t_0) \\
&= G^*(\sigma(u_0), t_0).
\end{aligned}
\tag{21.8}
$$

This completes the proof.

21.3   About the computational method

The continuous topology optimization problem described in the previous section is discretized using the FE method, considering in plane deformations. The FE discretization is performed taking into account the bilinear isoparametric element as a master one, in similar way as in [70].

To obtain computational results, we have defined the following algorithm.

1. Set $n = 1$.

2. Set $t_1(x) = t_1$, in Ω.

3. Calculate $u_n \in U$ as the solution of equation

$$
\frac{\partial J_1(u, t_n)}{\partial u} = \mathbf{0},
$$

   that is,

$$
\begin{cases}
(H_{ijkl}(t_n)e_{kl}(u_n))_{,j} + f_i = 0 \text{ in } \Omega, \\[2mm]
H_{ijkl}(t_n)e_{kl}(u_n)n_j - \hat{f}_i = 0, \text{ on } \Gamma_t, \ \forall i \in \{1, 2, 3\}.
\end{cases}
\tag{21.9}
$$

4. Obtain t_{n+1} by

$$
t_{n+1} = \arg\min_{t \in B} J_1(u_n, t).
$$

5. Set $n := n + 1$ and go to step 3 up to the satisfaction of an appropriate convergence criterion.

In the FE formulation, equations indicated in (21.9) stands for

$$
\mathbf{H}(t)\mathbf{U} = \mathbf{f},
\tag{21.10}
$$

where $\mathbf{H}(t)$ is the global stiffness matrix, $\mathbf{U}$ is the global deflection and $\mathbf{f}$ is the global forces vector.

Thus, for such a FE models (N elements where $e \in \{1, ..., N\}$), the primal optimization problem can be written in matrix form as

$$
\begin{aligned}
\min \quad & \hat{J}(u, t) = \frac{1}{2}\mathbf{U}^T \mathbf{H}(t)\mathbf{U} \\
&= \frac{1}{2}\sum_{e=1}^{N}(t_e)^P \mathbf{u}_e^T \mathbf{H}_e \mathbf{u}_e \\
s.t. \quad & (t_e)^P \mathbf{H}_e \mathbf{u}_e = \mathbf{f}_e \\
& \sum_{e=1}^{N} t_e V_e = t_1 |\Omega| \\
& 0 \le t \le 1 \\
& e = 1, 2, 3, ..., N,
\end{aligned}
\tag{21.11}
$$

On the other hand, the dual problem may be expressed by

$$max \quad J^*(u), \text{ where}$$

$$J^*(u) = \min_{t \in B} \left(-\frac{1}{2} \sum_{e=1}^{N} \left((t_e)^p \mathbf{u}_e^T \mathbf{H}_e \mathbf{u}_e + \mathbf{f}_e \mathbf{u}_e \right) \right)$$

where $t \in B$ if and only if

$$\sum_{e=1}^{N} t_e V_e = t_1 |\Omega|$$

$$0 \leq t_e \leq 1,$$

$$e = 1, 2, 3, \dots, N,$$

(21.12)

and where V_e is the area of element e.

Finally, the last minimization indicated corresponds to item 4 in the concerning algorithm. Indeed, such a procedure refers to minimize at each iteration, through the Matlab Linprog routine, the function

$$\sum_{e=1}^{N} \frac{\partial J_1(u_n, \{t_e^n\})}{\partial t_e} t_e = \sum_{e=1}^{N} \left(-p(t_e^n)^{p-1} t_e \mathbf{u}_e^T \mathbf{H}_e \mathbf{u}_e \right)$$

subject to $t \in B$, where p is a penalization parameter (typically, $p = 3$).

21.4 Computational simulations and results

We present numerical results in an analogous two-dimensional context, more specifically for two-dimensional beams of dimensions $1 \times l$ (units refer to the international system) represented by $\Omega = [0,1] \times [0,l]$, with $l = 0.5$, $F = -10^6$ for the first case, $l = 0.5$ and $F = -10^7$ for the second one, $l = 0.6$ and $F = -10^6$ for the third case and, $l = 1$ and $F = -10^8$ for the fourth one. F is in the y-direction and corresponds to f of the theoretical formulation presented above.

We consider the strain tensor as

$$e(\mathbf{u}) = (e_x(\mathbf{u}), e_y(\mathbf{u}), e_{xy}(\mathbf{u}))^T,$$

where $\mathbf{u} = (u,v) \in W^{1,2}(\Omega; \mathbb{R}^2)$, $e_x(\mathbf{u}) = u_x$, $e_y(\mathbf{u}) = v_y$ and $e_{xy}(\mathbf{u}) = \frac{1}{2}(u_y + v_x)$.

Moreover, the stress tensor $\sigma(e(\mathbf{u}))$ is given by

$$\sigma(e(\mathbf{u})) = He(\mathbf{u}),$$

where

$$H = \frac{E(t)}{1 - v^2} \left\{ \begin{matrix} 1 & v & 0 \\ v & 1 & 0 \\ 0 & 0 & \frac{1}{2}(1-v) \end{matrix} \right\}$$

(21.13)

and

$$E(t) = tE_0 + (1-t)E_1,$$

where $E_0 = 210 * 10^9$ (the modulus of Young) and $E_1 \ll E_0$. Moreover, $v = 0.33$.

As previously mentioned, we present four numerical simulations.

Case 1. For the first case see Figure 21.1, on the left, for the concerning case, Figure 21.1, in the middle for the optimal topology for this case with no filter, Figure 21.1, on the right, for the optimal topology for this first case with filter. The objective function as function of iteration numbers also for such a case with no filter, Figure 21.2, on the left, and the objective function as function of iteration numbers also for this first case with filter, Figure 21.2, on the right.

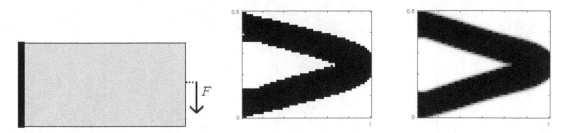

Figure 21.1: On the left, a clamped beam at $x = 0$ (cantilever beam). In the middle, the optimal topology for $t_1 = 0.5$, for the case with no filter. On the right, the optimal topology for $t_1 = 0.5$, for the case with filter. The FE mesh was 60 x 50.

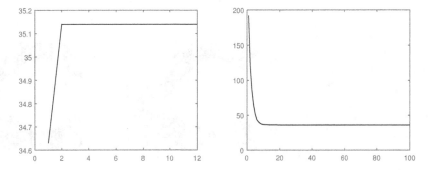

Figure 21.2: On the left, the objective function by iteration numbers for $t_1 = 0.5$, for the case with no filter. On the right, the objective function by iteration numbers for $t_1 = 0.5$, for the case with filter.

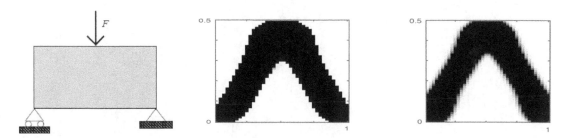

Figure 21.3: On the left, a simply supported beam at $x = 0$ and $x = 1$. In the middle the optimal topology for $t_1 = 0.5$, for the case with no filter. On the right the optimal topology for $t_1 = 0.5$, for the case with filter. The FE mesh was 40 x 50.

Case 2. For the second case see Figure 21.3, on the left, for the concerning case, Figure 21.3, in the middle, for the optimal topology for this case with no filter, Figure 21.3, on the right, for the optimal topology for this second case with filter. The objective function as function of iteration numbers also for such a case with no filter, Figure 21.4, on the left, and the objective function as function of iteration numbers also for this second case with filter, Figure 21.4, on the right.

Case 3. For the third case see Figure 21.5, on the left, for the concerning case, Figure 21.5, in the middle, for the optimal topology for this case with no filter, Figure 21.5, on the right, for the optimal topology for this third case with filter. The objective function as function of iteration numbers also for such a case with no

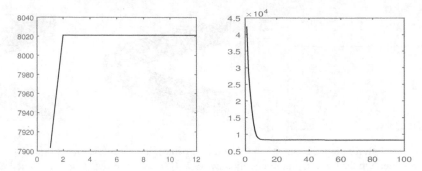

Figure 21.4: On the left the objective function by iteration numbers for $t_1 = 0.5$, for the case with no filter. On the right the objective function by iteration numbers for $t_1 = 0.5$, for the case with filter.

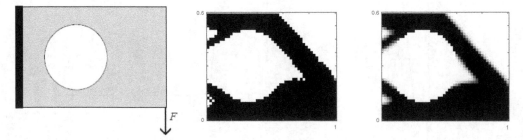

Figure 21.5: On the left a beam with a hole clamped at $x = 0$. In the middle the optimal topology for $t_1 = 0.5$, for the case with no filter. On the right the optimal topology for $t_1 = 0.5$, for the case with filter. The FE mesh was 50 x 40.

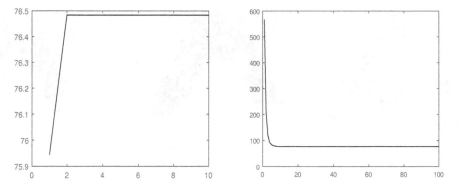

Figure 21.6: On the left the objective function by iteration numbers for $t_1 = 0.5$, for the case with no filter. On the right the objective function by iteration numbers for $t_1 = 0.5$, for the case with filter.

filter, Figure 21.6, on the left, and the objective function as function of iteration numbers also for this third case with filter, Figure 21.6, on the right.

Case 4. For the fourth case see Figure 21.7, on the left, for the concerning case, Figure 21.7, in the middle for the optimal topology for this case with no filter, Figure 21.7, on the right, for the optimal topology for the fourth case with filter. The objective function as function of iteration numbers also for such a case with no

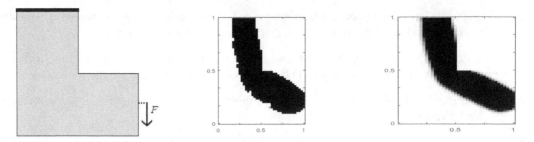

Figure 21.7: On the left a L shape beam clamped at $y = 1$. In the middle the optimal topology for $t_1 = 0.5$, for the case with no filter. On the right the optimal topology for $t_1 = 0.5$, for the case with filter. The FE mesh was 40x60.

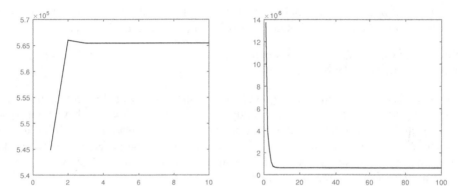

Figure 21.8: On the left the objective function by iteration numbers for $t_1 = 0.5$, for the case with no filter. On the right the objective function by iteration numbers for $t_1 = 0.5$, for the case with filter.

filter, Figure 21.8, on the left, and the objective function as function of iteration numbers also for this fourth case with filter, Figure 21.8, on the right.

We emphasize to have obtained in both optimized structures, without filter and with filter, robust topology from a structural point of view. One can note also in the figures that in all cases the objective functions, without filter and with filter, have similar final values, which indicates the results obtained are consistent.

21.5 Final remarks and conclusions

In this chapter we have developed a duality principle and relating computational method for a class of structural optimization problems in elasticity. It is worth mentioning we have not used a filter to post-process the results, having obtained a solution $t : \Omega \to \{0, 1\}$ (that is, $t(x, y) = 0$ or $t(x, y) = 1$ in Ω), by finding a critical point $(u_0, t_0) \in U \times B$ for the functional $J_1 : U \times B \to \mathbb{R}$. This corresponds, in some sense, to solving the dual problem.

We address some final remarks and conclusions on the results obtained.

■ For all examples, in a first step, we have obtained numerical results through our algorithm with a software which uses the Matlab-Linprog as optimizer at each iteration without any filter. In a second step, we obtain numerical results using the OC optimizer with filter, with a software developed based in the article [70] by Sigmund, 2001.

- We emphasize, to obtain good and consistent results, it is necessary to discretize more in the direction y, that is, the load direction, in which the displacements are much larger.

- If we do not discretize enough in the load direction, for the software with no filter, a check-board standard in the material distribution is obtained in some parts of the concerning structure.

- Summarizing, with no filter, the check-board problem is solved by increasing the discretization in the load direction.

- Moreover, with the OC optimizer with filter, the volume fraction of material is kept constant in 0.5 at each iteration during the optimization process, whereas for the case with no filter we start with a volume fraction of 0.95 which is gradually decreased to the value 0.5, using as the initial solution for a iteration with a specific volume fraction, the solution of the previous one.

- We also highlight the result obtained with no filter is indeed a critical point for the original optimization problem, whereas there is some heuristic in the procedure with filter.

- Once more we emphasize to have obtained more robust and consistent shapes by properly discretizing the approximate model in a FE context.

- Finally, it is also worth mentioning, we have obtained similar final objective function values without and with filter in all examples, even though without filter such values have been something smaller as expected. The qualitative differences between the graphs without and with filter, for the objective function as function of the number of iterations, refer to the differences between the optimization processes, where in the case with filter the volume fraction is kept 0.5 and without filter it is gradually decreased from 0.95 to 0.5, as above described.

We highlight the results obtained may be applied to other problems, such as other models of plates, shell and elasticity.

Chapter 22

Existence and Duality Principles for the Ginzburg-Landau System in Superconductivity

22.1 Introduction

In this work we present three theorems which represent duality principles suitable for a large class of non-convex variational problems.

At this point we refer to the exceptionally important article, "A contribution to contact problems for a class of solids and structures" by Bielski and Telega [11], published in 1985, as the first one to successfully apply and generalize the convex analysis approach to a model in non-convex and non-linear mechanics.

The present work is, in some sense, a kind of extension of this previous work [11] and others such as [10], which greatly influenced and inspired my work and recent book [14].

We extend and generalize the approaches in [11] and [78] and develop two multi-duality principles through which we classify qualitatively the critical points.

Thus, we emphasize the first multi-duality principle generalizes some Toland results found in [78] and in some appropriate sense, such a work complements the results presented in [11], now applied to a Ginzburg-Landau type model context.

On the other hand, the conclusions of the second multi-duality principle may be qualitatively found in similar form in the triality approach found in [83] and other references therein, even though the construction of the present result and the respective proofs be substantially different.

About the model in physics involved, we recall that about the year 1950, Ginzburg and Landau introduced a theory to model the super-conducting behavior of some types of materials below a critical temperature T_c, which depends on the material in question. They postulated the free density energy may be written close to T_c as

$$F_s(T) = F_n(T) + \frac{\hbar}{4m} \int_\Omega |\nabla \psi|_2^2 \, dx + \frac{\alpha(T)}{4} \int_\Omega |\psi|^4 \, dx - \frac{\beta(T)}{2} \int_\Omega |\psi|^2 \, dx,$$

where ψ is a complex parameter, $F_n(T)$ and $F_s(T)$ are the normal and super-conducting free energy densities, respectively (see [4, 55] for details). Here $\Omega \subset \mathbb{R}^3$ denotes the super-conducting sample with a boundary denoted by $\partial \Omega = \Gamma$. The complex function $\psi \in W^{1,2}(\Omega; \mathbb{C})$ is intended to minimize $F_s(T)$ for a fixed temperature T.

Denoting $\alpha(T)$ and $\beta(T)$ simply by α and β, the corresponding Euler-Lagrange equations are given by:

$$\begin{cases} -\frac{\hbar}{2m}\nabla^2\psi + \alpha|\psi|^2\psi - \beta\psi = 0, & \text{in } \Omega \\ \frac{\partial\psi}{\partial\mathbf{n}} = 0, & \text{on } \partial\Omega. \end{cases} \tag{22.1}$$

This last system of equations is well known as the Ginzburg-Landau (G-L) one in the absence of a magnetic field and respective potential.

Remark 22.1 About the notation, for an open bounded subset $\Omega \subset \mathbb{R}^3$, we denote the $L^2(\Omega)$ norm by $\|\cdot\|_{L^2(\Omega)}$ or simply by $\|\cdot\|_2$. Similar remark is valid for the $L^2(\Omega;\mathbb{R}^3)$ norm, which is denoted by $\|\cdot\|_{L^2(\Omega;\mathbb{R}^3)}$ or simply by $\|\cdot\|_2$, when its meaning is clear. On the other hand, by $|\cdot|_2$ we denote the standard Euclidean norm in $\mathbb{R}^3$ or $\mathbb{C}^3$.

Moreover derivatives are always understood in the distributional sense. Also, by a regular Lipschitzian boundary $\partial\Omega = \Gamma$ of Ω, we mean regularity enough so that the standard Sobolev Imbedding theorems, the trace theorem and Gauss-Green formulas of integration by parts to hold. Details about such results may be found in [1].

Finally, in general $\delta F(u,v)$ will denote the Fréchet derivative of the functional $F(u,v)$ at (u,v),

$$\delta_u F(u,v) \text{ or } \frac{\partial F(u,v)}{\partial u}$$

denotes the first Fréchet derivative of F relating the variable u and

$$\delta^2 F_{u,v}(u,v) \text{ or } \frac{\partial^2 F(u,v)}{\partial u\partial v}$$

denotes the second one relating the variables u and v, always at (u,v).

At some point of our analysis, we shall assume a finite-dimensional approximate model version. In such a context, we remark that generically in a matrix sense the notation

$$\frac{1}{K+\gamma\nabla^2}$$

will indicate the inverse

$$(KI_d + \gamma\nabla^2)^{-1},$$

where I_d denotes the identity matrix and ∇^2 is the matrix originated by a discretized version of the Laplac operator. We also emphasize that, as the meaning is clear, other similar notations may be used to indicate th inverse of matrices or operators. ■

Remark 22.2 For an appropriate set $\Omega \subset \mathbb{R}^3$ and a space U, our primal functional $J : U \to \mathbb{R}$ is specifie by

$$J(u) = \frac{\gamma}{2}\int_\Omega \nabla u \cdot \nabla u \, dx + \frac{\alpha}{2}\int_\Omega (u^2 - \beta)^2 \, dx - \langle u, f\rangle_{L^2}$$

$\forall u \in U$, where $\alpha, \beta, \gamma > 0$, $f \in L^2(\Omega)$.

We define,

$$F(u) = -\frac{\gamma}{2}\int_\Omega \nabla u \cdot \nabla u \, dx + \frac{K}{2}\int_\Omega u^2 \, dx,$$

where in a finite dimensional discretized model version, in a finite elements or finite differences contex $K > 0$ is such that

$$F(u) > 0, \ \forall u \in U \text{ such that } u \neq \mathbf{0},$$

and

$$G(u,v) = \frac{\alpha}{2} \int_\Omega (u^2 - \beta + v)^2 \, dx + \frac{K}{2} \int_\Omega u^2 \, dx - \langle u, f \rangle_{L^2},$$

so that

$$J(u) = G(u,0) - F(u), \ \forall u \in U.$$

Let $F^* : Y^* \to \mathbb{R}$ denote the polar functional related to F, that is,

$$F^*(v_1^*) = \sup_{u \in U} \{ \langle u, v_1^* \rangle_{L^2} - F(u) \}.$$

Since the optimization in question is quadratic, we have

$$F^*(v_1^*) = \langle \tilde{u}, v_1^* \rangle_{L^2} - F(\tilde{u}), \tag{22.2}$$

where $\tilde{u} \in U$ is such that

$$v_1^* = \frac{\partial F(\tilde{u})}{\partial u},$$

that is,

$$v_1^* = \gamma \nabla^2 \tilde{u} + K \tilde{u},$$

so that

$$\tilde{u} = \frac{v_1^*}{K + \gamma \nabla^2}.$$

Replacing such a $\tilde{u}$ into (22.2), we obtain

$$F^*(v_1^*) = \frac{1}{2} \int_\Omega v_1^* [(K + \gamma \nabla^2)^{-1} v_1^*] \, dx.$$

Similarly, for $G : U \times Y \to \mathbb{R}$, for $v_0^* \in Y^*$ such that

$$2 v_0^* + K > 0, \ \text{in } \overline{\Omega},$$

we also define

$$G^*(v_1^*, v_0^*) = \sup_{(u,v) \in U \times Y} \{ \langle u, v_1^* \rangle_{L^2} + \langle v, v_0^* \rangle_{L^2} - G(u,v) \},$$

where, as indicated above

$$G(u,v) = \frac{\alpha}{2} \int_\Omega (u^2 - \beta + v)^2 \, dx + \frac{K}{2} \int_\Omega u^2 \, dx - \langle u, f \rangle_{L^2}.$$

Defining $w = u^2 - \beta + v$, so that

$$v = w - u^2 + \beta,$$

we may write

$$
\begin{aligned}
G^*(v_1^*, v_0^*) = \ & \sup_{(u,w) \in U \times Y} \{ \langle u, v_1^* \rangle_{L^2} + \langle w - u^2 + \beta, v_0^* \rangle_{L^2} \\
& - \frac{\alpha}{2} \int_\Omega w^2 \, dx - \frac{K}{2} \int_\Omega u^2 \, dx + \langle u, f \rangle_{L^2} \}.
\end{aligned}
\tag{22.3}
$$

Since the optimization in question is quadratic, we have

$$
\begin{aligned}
G^*(v_1^*, v_0^*) = \ & \langle \tilde{u}, v_1^* \rangle_{L^2} + \langle \tilde{w} - \tilde{u}^2 + \beta, v_0^* \rangle_{L^2} \\
& - \frac{\alpha}{2} \int_\Omega \tilde{w}^2 \, dx - \frac{K}{2} \int_\Omega \tilde{u}^2 \, dx + \langle \tilde{u}, f \rangle_{L^2} \}.
\end{aligned}
\tag{22.4}
$$

where $\tilde{u} \in L^2$ and $\tilde{w} \in Y$ are such that

$$v_1^* = 2\tilde{u}v_0^* + K\tilde{u} - f,$$

and

$$v_0^* = \alpha\tilde{w},$$

so that

$$\tilde{u} = \frac{v_1^* + f}{2v_0^* + K},$$

and

$$\tilde{w} = \frac{v_0^*}{\alpha}.$$

Replacing such results into (22.4), we get

$$
\begin{aligned}
G^*(v_1^*, v_0^*) &= \frac{1}{2}\int_\Omega \frac{(v_1^* + f)^2}{2v_0^* + K}\, dx + \frac{1}{2\alpha}\int_\Omega (v_0^*)^2\, dx \\
&\quad + \beta\int_\Omega v_0^*\, dx.
\end{aligned}
\tag{22.5}
$$

22.2 The main result

In this section we develop a first multi-duality principle for a Ginzburg-Landau type system in a simpler real context.

About these models in physics, we refer again to [4, 55].

In the next lines we develop the main result. At this point we highlight the global optimality condition $-\gamma\nabla^2 + 2\hat{v}_0^* > \mathbf{0}$ at a critical point was presented in [83].

Theorem 22.2.1 *Let* $\Omega \subset \mathbb{R}^3$ *be an open, bounded, connected set with a regular (Lipschitzian) boundary denoted by* $\partial\Omega$. *Suppose* $J : U \to \mathbb{R}$ *is a functional defined by*

$$J(u) = \frac{\gamma}{2}\int_\Omega \nabla u \cdot \nabla u\, dx + \frac{\alpha}{2}\int_\Omega (u^2 - \beta)^2\, dx - \langle u, f \rangle_{L^2}$$

$\forall u \in U$, *where* $\alpha, \beta, \gamma > 0$, $f \in L^2(\Omega)$ *and* $U = W_0^{1,2}(\Omega)$.

At this point, we assume a discretized finite dimensional model version (in a finite elements or finite differences context, so that from now on, the not relabeled spaces, functions and operators refer to such a finite dimensional approximation) and suppose

$$\delta J(u_0) = \mathbf{0}$$

where in an appropriate matrices sense, we have

$$\delta^2 J(u_0) = -\gamma\nabla^2 + 6\alpha\{u_0(i)^2\} - 2\alpha\beta I_d.$$

Here $\{u_0(i)^2\}$ *denotes the diagonal matrix which the diagonal is given by the vector* $[u_0(i)^2]$.

Also, from now on, as the meaning is clear, we shall denote such a second Fréchet derivative simply by

$$\delta^2 J(u_0) = -\gamma\nabla^2 + 6\alpha u_0^2 - 2\alpha\beta.$$

Define

$$F(u) = -\frac{\gamma}{2}\int_\Omega \nabla u \cdot \nabla u\, dx + \frac{K}{2}\int_\Omega u^2\, dx,$$

where $K > 0$ is such that

$$F(u) > 0, \ \forall u \in U \ such \ that \ u \neq \mathbf{0},$$

and

$$G(u,v) = \frac{\alpha}{2} \int_\Omega (u^2 - \beta + v)^2 \, dx + \frac{K}{2} \int_\Omega u^2 \, dx - \langle u, f \rangle_{L^2},$$

so that

$$J(u) = G(u,0) - F(u), \ \forall u \in U.$$

Define also $Y = Y^* = L^2(\Omega)$, $F^* : Y^* \to \mathbb{R}$ by

$$
\begin{aligned}
F^*(v_1^*) &= \sup_{u \in U} \{ \langle u, v_1^* \rangle_{L^2} - F(u) \} \\
&= \frac{1}{2} \int_\Omega v_1^* [(K + \gamma \nabla^2)^{-1} v_1^*] \, dx,
\end{aligned}
\tag{22.6}
$$

and $G^* : Y^* \times Y^* \to \overline{\mathbb{R}} = \mathbb{R} \cup \{+\infty\}$ by

$$
\begin{aligned}
G^*(v_1^*, v_0^*) &= \sup_{(v_1,v) \in Y \times Y} \{ \langle v, v_0^* \rangle_Y + \langle v_1, v_1^* \rangle_Y - G(v_1, v) \} \\
&= \frac{1}{2} \int_\Omega \frac{(v_1^* + f)^2}{2v_0^* + K} \, dx + \frac{1}{2\alpha} \int_\Omega (v_0^*)^2 \, dx \\
&\quad + \beta \int_\Omega v_0^* \, dx,
\end{aligned}
\tag{22.7}
$$

if $v_0^* \in A^* = \{ v_0^* \in Y^* : 2v_0^* + K > 0, \ in \ \overline{\Omega} \}$.
    Let $J^* : Y^* \times Y^* \to \mathbb{R} \cup \{-\infty\}$ be such that

$$J^*(v_1^*, v_0^*) = -G^*(v_1^*, v_0^*) + F^*(v_1^*),$$

so that $\tilde{J}^* : Y^* \to \mathbb{R}$ is expressed by

$$\tilde{J}^*(v_1^*) = \sup_{v_0^* \in A^*} J^*(v_1^*, v_0^*).$$

    Define

$$\hat{v}_0^* = \alpha(u_0^2 - \beta),$$

and

$$\hat{v}_1^* = (2\hat{v}_0^* + K)u_0 - f.$$

    Suppose $\hat{v}_0^* \in A^*$.
    Under such hypotheses,

$$\delta \tilde{J}^*(\hat{v}_1^*) = \mathbf{0},$$

and

$$\tilde{J}^*(\hat{v}_1^*) = J(u_0).$$

Moreover,

    1. if $\delta^2 J(u_0) > \mathbf{0}$, then

$$\delta^2 \tilde{J}^*(\hat{v}_1^*) > \mathbf{0},$$

    so that there exist $r > 0$ and $r_1 > 0$ such that

$$
\begin{aligned}
J(u_0) &= \min_{u \in B_r(u_0)} J(u) \\
&= \min_{v_1^* \in B_{r_1}(\hat{v}_1^*)} \tilde{J}^*(v_1^*) \\
&= \tilde{J}^*(\hat{v}_1^*) \\
&= J^*(\hat{v}_1^*, \hat{v}_0^*).
\end{aligned}
\tag{22.8}
$$

2. *If* $-\gamma \nabla^2 + 2\hat{v}_0^* > 0$ *so that* $\delta^2 J(u_0) > 0$, *then defining*

$$B^* = \{v_0^* \in Y^* : -\gamma \nabla^2 + 2v_0^* > 0\},$$

we have

$$\delta J_2^*(\hat{v}_1^*) = 0,$$

$$\delta^2 J_2^*(\hat{v}_1^*) > 0$$

and

$$
\begin{aligned}
J(u_0) &= \min_{u \in U} J(u) \\
&= \min_{v_1^* \in Y^*} J_2^*(v_1^*) \\
&= J_2^*(\hat{v}_1^*) \\
&= J^*(\hat{v}_1^*, \hat{v}_0^*),
\end{aligned}
\tag{22.9}
$$

where

$$J_2^*(v_1^*) = \sup_{v_0^* \in A^* \cap B^*} J^*(v_1^*, v_0^*).$$

3. *If* $\delta^2 J(u_0) < 0$, *then*

$$\delta^2 \tilde{J}^*(\hat{v}_1^*) < 0,$$

so that there exist $r > 0$ *and* $r_1 > 0$ *such that*

$$
\begin{aligned}
J(u_0) &= \max_{u \in B_r(u_0)} J(u) \\
&= \max_{v_1^* \in B_{r_1}(\hat{v}_1^*)} \tilde{J}^*(v_1^*) \\
&= \tilde{J}^*(\hat{v}_1^*) \\
&= J^*(\hat{v}_1^*, \hat{v}_0^*).
\end{aligned}
\tag{22.10}
$$

Proof 22.1 From $\delta J(u_0) = 0$ we have

$$-\gamma \nabla^2 u_0 + \alpha(u_0^2 - \beta)2u_0 - f = 0, \text{ in } \Omega$$

so that

$$
\begin{aligned}
\gamma \nabla^2 u_0 + K u_0 &= \alpha(u_0^2 - \beta)2u_0 + K u_0 - f \\
&= (2\hat{v}_0^* + K)u_0 - f \\
&= \hat{v}_1^*.
\end{aligned}
\tag{22.11}
$$

From this, we obtain

$$\frac{\hat{v}_1^*}{K + \gamma \nabla^2} - \frac{\hat{v}_1^* + f}{2\hat{v}_0^* + K} = u_0 - u_0 = 0, \text{ in } \Omega.$$

On the other hand,

$$\hat{v}_0^* = \alpha(u_0^2 - \beta) = \alpha(u_0^2 - \beta + 0),$$

so that

$$-\frac{\hat{v}_0^*}{\alpha} + \frac{(\hat{v}_1^* + f)^2}{(2\hat{v}_0^* + K)^2} - \beta = 0,$$

and thus

$$\frac{\partial J^*(\hat{v}_1^*, \hat{v}_0^*)}{\partial v_0^*} = 0.$$

From this, (22.11), from the definition of $\hat{v}_0^*$ and the concavity of $J^*(\hat{v}_1^*, v_0^*)$ in v_0^*, we obtain

$$
\begin{aligned}
\tilde{J}^*(\hat{v}_1^*) &= \sup_{v_0^* \in A^*} J^*(\hat{v}_1^*, v_0^*) \\
&= J^*(\hat{v}_1^*, \hat{v}_0^*) \\
&= -G^*(\hat{v}_1^*, \hat{v}_0^*) + F^*(\hat{v}_1^*) \\
&= -\langle u_0, \hat{v}_1^* \rangle_{L^2} - \langle 0, \hat{v}_0^* \rangle_{L^2} + G(u_0, 0) \\
&\quad + \langle u_0, \hat{v}_1^* \rangle_{L^2} - F(u_0) \\
&= G(u_0, 0) - F(u_0) \\
&= J(u_0).
\end{aligned}
\tag{22.12}
$$

Also, for v_1^* in a neighborhood of $\hat{v}_1^*$ we have that

$$J^*(v_1^*) = \sup_{v_0^* \in A^*} J^*(v_1^*, v_0^*) = J^*(v_1^*, \tilde{v}_0^*),$$

where such a supremum is attained through the equation,

$$\frac{\partial J^*(v_1^*, \tilde{v}_0^*)}{\partial v_0^*} = \mathbf{0},$$

so that from the implicit function theorem we have

$$
\begin{aligned}
\frac{\partial \tilde{J}^*(v_1^*)}{\partial v_1^*} &= \frac{\partial J^*(v_1^*, \tilde{v}_0^*)}{\partial v_1^*} + \frac{\partial J^*(v_1^*, \tilde{v}_0^*)}{\partial v_0^*} \frac{\partial \tilde{v}_0^*}{\partial v_1^*} \\
&= \frac{\partial J^*(v_1^*, \tilde{v}_0^*)}{\partial v_1^*}.
\end{aligned}
\tag{22.13}
$$

Moreover, from this, joining the pieces, we get

$$
\begin{aligned}
\frac{\partial \tilde{J}^*(\hat{v}_1^*)}{\partial v_1^*} &= \frac{\partial J^*(\hat{v}_1^*, \hat{v}_0^*)}{\partial v_1^*} + \frac{\partial J^*(\hat{v}_1^*, \hat{v}_0^*)}{\partial v_0^*} \frac{\partial \hat{v}_0^*}{\partial v_1^*} \\
&= \frac{\partial J^*(\hat{v}_1^*, \hat{v}_0^*)}{\partial v_1^*} \\
&= \frac{\hat{v}_1^*}{K + \gamma \nabla^2} - \frac{\hat{v}_1^* + f}{2\hat{v}_0^* + K} = u_0 - u_0 = 0, \text{ in } \Omega.
\end{aligned}
\tag{22.14}
$$

Hence, from this and (22.13), we obtain

$$
\begin{aligned}
\frac{\partial^2 \tilde{J}^*(\hat{v}_1^*)}{\partial (v_1^*)^2} &= \frac{\partial^2 J^*(\hat{v}_1^*, \hat{v}_0^*)}{\partial v_0^* \partial v_1^*} \frac{\partial \hat{v}_0^*}{\partial v_1^*} \\
&\quad + \frac{\partial^2 J^*(\hat{v}_1^*, \hat{v}_0^*)}{\partial (v_1^*)^2} \\
&= \frac{2(\hat{v}_1^* + f)}{(2\hat{v}_0^* + K)^2} \frac{\partial \hat{v}_0^*}{\partial v_1^*} \\
&\quad + \frac{1}{K + \gamma \nabla^2} - \frac{1}{2\hat{v}_0^* + K}.
\end{aligned}
\tag{22.15}
$$

At this point we may observe that

$$\frac{(\hat{v}_1^* + f)^2}{(2\hat{v}_0^* + K)^2} - \frac{\hat{v}_0^*}{\alpha} - \beta = 0,$$

so that taking the variation in v_1^* of this equation in both sides, we obtain

$$\frac{2(\hat{v}_1^* + f)}{(2\hat{v}_0^* + K)^2} - \frac{4(\hat{v}_1^* + f)^2}{(2\hat{v}_0^* + K)^3} \frac{\partial \hat{v}_0^*}{\partial v_1^*}$$
$$-\frac{1}{\alpha} \frac{\partial \hat{v}_0^*}{\partial v_1^*} = 0, \tag{22.16}$$

and thus,

$$\frac{\partial \hat{v}_0^*}{\partial v_1^*} = \frac{\frac{2(\hat{v}_1^* + f)}{(2\hat{v}_0^* + K)^2}}{\frac{1}{\alpha} + \frac{4(\hat{v}_1^* + f)^2}{(2\hat{v}_0^* + K)^3}}, \tag{22.17}$$

Replacing (22.17) into (22.15), we obtain

$$\frac{\partial^2 \tilde{J}^*(\hat{v}_1^*)}{\partial (v_1^*)^2} = \frac{\partial^2 J^*(\hat{v}_1^*, \hat{v}_0^*)}{\partial v_0^* \partial v_1^*} \frac{\partial \hat{v}_0^*}{\partial v_1^*}$$
$$+ \frac{\partial^2 J^*(\hat{v}_1^*, \hat{v}_0^*)}{\partial (v_1^*)^2}$$
$$= \frac{2(\hat{v}_1^* + f)}{(2\hat{v}_0^* + K)^2} \frac{\partial \hat{v}_0^*}{\partial v_1^*}$$
$$+ \frac{1}{K + \gamma \nabla^2} - \frac{1}{2\hat{v}_0^* + K}$$
$$= \frac{\frac{4\alpha(\hat{v}_1^* + f)^2}{(2\hat{v}_0^* + K)^4}}{1 + \alpha \frac{4(\hat{v}_1^* + f)^2}{(2\hat{v}_0^* + K)^3}}$$
$$+ \frac{1}{K + \gamma \nabla^2} - \frac{1}{2\hat{v}_0^* + K}, \tag{22.18}$$

Therefore, denoting

$$H = 1 + \alpha \frac{4(\hat{v}_1^* + f)^2}{(2\hat{v}_0^* + K)^3} = 1 + \frac{4\alpha u_0^2}{2\hat{v}_0^* + K},$$

we have

$$\frac{\partial^2 \tilde{J}^*(\hat{v}_1^*)}{\partial (v_1^*)^2}$$
$$= \left[\left(1 + \alpha \frac{4(\hat{v}_1^* + f)^2}{(2\hat{v}_0^* + K)^3} \right) \frac{1}{K + \gamma \nabla^2} - \frac{1}{2\hat{v}_0^* + K} \right] / H$$
$$= \left[\left(1 + \frac{4\alpha u_0^2}{2\hat{v}_0^* + K} \right) \frac{1}{K + \gamma \nabla^2} - \frac{1}{2\hat{v}_0^* + K} \right] / H$$
$$= (2\hat{v}_0^* + K + 4\alpha u_0^2 - K - \gamma \nabla^2) / [(K + \gamma \nabla^2)(\hat{v}_0^* + K)H]$$
$$= [-\gamma \nabla^2 + 6\alpha u_0^2 - 2\alpha \beta] / [(K + \gamma \nabla^2)(\hat{v}_0^* + K)H]$$
$$= \frac{\delta^2 J(u_0)}{[(K + \gamma \nabla^2)(\hat{v}_0^* + K)H]}. \tag{22.19}$$

Summarizing, assuming $\delta^2 J(u_0) > \mathbf{0}$, we obtain

$$\frac{\partial^2 \tilde{J}^*(\hat{v}_1^*)}{\partial(v_1^*)^2} > \mathbf{0},$$

so that $u_0 \in U$ is a point of local minimum for J and $\hat{v}_1^*$ is a point of local minimum for $\tilde{J}^*$.

Hence, there exist $r > 0$ and $r_1 > 0$ such that

$$
\begin{aligned}
J(u_0) &= \min_{u \in B_r(u_0)} J(u) \\
&= \min_{v_1^* \in B_{r_1}(\hat{v}_1^*)} \tilde{J}^*(v_1^*) \\
&= \tilde{J}^*(\hat{v}_1^*) \\
&= J^*(\hat{v}_1^*, \hat{v}_0^*).
\end{aligned}
\tag{22.20}
$$

Assume now $-\gamma\nabla^2 + 2\hat{v}_0^* > \mathbf{0}$, so that $\delta^2 J(u_0) > \mathbf{0}$.

Observe that if $v_0^* \in A^*$ and $-\gamma\nabla^2 + 2v_0^* > \mathbf{0}$, then

$$\frac{\partial^2 J^*(v_1, v_0^*)}{\partial(v_1^*)^2} = \frac{1}{K + \gamma\nabla^2} - \frac{1}{2v_0^* + K} > \mathbf{0},$$

so that $J^*(v_1^*, v_0^*)$ is convex in v_1^*, $\forall v_0^* \in A^* \cap B^*$.

Hence

$$J_2^*(v_1^*) = \sup_{v_0^* \in A^* \cap B^*} J^*(v_1^*, v_0^*),$$

is convex, as the point-wise supremum of a family of convex functionals.

As above, from $\delta J(u_0) = \mathbf{0}$, we may obtain $\delta J_2^*(\hat{v}_1^*) = \mathbf{0}$, so that, since J_2^* is convex, we may infer that

$$J(u_0) = J_2^*(\hat{v}_1^*) = \min_{v_1^* \in Y^*} J_2^*(v_1^*).$$

Moreover,

$$
\begin{aligned}
J_2^*(\hat{v}_1^*) &= \min_{v_1^* \in Y^*} J_2^*(v_1^*) \\
&\leq J_2^*(v_1^*) \\
&= \sup_{v_0^* \in A^* \cap B^*} J^*(v_1^*, v_0^*) \\
&= \sup_{v_0^* \in A^* \cap B^*} \left\{ \frac{1}{2} \int_\Omega \frac{(v_1^*)^2}{(K + \gamma\nabla^2)} \, dx \right. \\
&\qquad \left. -\frac{1}{2} \int_\Omega \frac{(v_1^* + f)^2}{2v_0^* + K} \, dx - \frac{1}{2} \int_\Omega \frac{(v_0^*)^2}{\alpha} \, dx - \beta \int_\Omega v_0^* \, dx \right\} \\
&\leq \sup_{v_0^* \in A^* \cap B^*} \left\{ \frac{1}{2} \int_\Omega \frac{(v_1^*)^2}{(K + \gamma\nabla^2)} \, dx \right. \\
&\qquad \left. -\langle u, v_1^* + f \rangle_{L^2} + \int_\Omega (2v_0^* + K) \frac{u^2}{2} \, dx - \frac{1}{2} \int_\Omega \frac{(v_0^*)^2}{\alpha} \, dx - \beta \int_\Omega v_0^* \, dx \right\} \\
&\leq \sup_{v_0^* \in Y^*} \left\{ \frac{1}{2} \int_\Omega \frac{(v_1^*)^2}{(K + \gamma\nabla^2)} \, dx \right. \\
&\qquad \left. -\langle u, v_1^* + f \rangle_{L^2} + \int_\Omega (2v_0^* + K) \frac{u^2}{2} \, dx - \frac{1}{2} \int_\Omega \frac{(v_0^*)^2}{\alpha} \, dx - \beta \int_\Omega v_0^* \, dx \right\} \\
&= \frac{1}{2} \int_\Omega \frac{(v_1^*)^2}{(K + \gamma\nabla^2)} \, dx \\
&\qquad -\langle u, v_1^* \rangle_{L^2} + \frac{\alpha}{2} \int_\Omega (u^2 - \beta)^2 \, dx + \frac{K}{2} \int_\Omega u^2 \, dx \\
&\qquad -\langle u, f \rangle_{L^2}, \forall u \in U, \ v_1^* \in Y^*.
\end{aligned}
\tag{22.21}
$$

Hence,

$$
\begin{aligned}
J_2^*(\hat{v}_1^*) &\leq \inf_{v_1^* \in Y^*} \left\{ \frac{1}{2} \int_\Omega \frac{(v_1^*)^2}{(K + \gamma \nabla^2)} \, dx \right. \\
&\qquad \left. - \langle u, v_1^* \rangle_{L^2} + \frac{\alpha}{2} \int_\Omega (u^2 - \beta)^2 \, dx + \frac{K}{2} \int_\Omega u^2 \, dx - \langle u, f \rangle_{L^2} \right\} \\
&= \frac{\gamma}{2} \int_\Omega \nabla u \cdot \nabla u \, dx - \frac{K}{2} \int_\Omega u^2 \, dx + \frac{\alpha}{2} \int_\Omega (u^2 - \beta)^2 \, dx + \frac{K}{2} \int_\Omega u^2 \, dx - \langle u, f \rangle_{L^2} \\
&= J(u), \ \forall u \in U.
\end{aligned}
\tag{22.22}
$$

From this and

$$
J(u_0) = J_2^*(\hat{v}_1^*),
$$

we obtain

$$
\begin{aligned}
J(u_0) &= \min_{u \in U} J(u) \\
&= \min_{v_1^* \in Y^*} J_2^*(v_1^*) \\
&= J_2^*(\hat{v}_1^*) \\
&= J^*(\hat{v}_1^*, \hat{v}_0^*).
\end{aligned}
\tag{22.23}
$$

The third item may be proven similarly as the first one.
This completes the proof.

22.3   A second multi-duality principle

In this section, in a similar context, we present a second multi-duality principle. This principle is significantly different from the previous one, since we invert the order of variables as evaluating the extremals.

Indeed the first part of this proof is similar to the one of the previous theorem. Important differences appears along the proof. For the sake of completeness, we present such a proof in details.

Theorem 22.3.1 *Let $\Omega \subset \mathbb{R}^3$ be an open, bounded, connected set with a regular (Lipschitzian) boundary denoted by $\partial \Omega$. Suppose $J : U \to \mathbb{R}$ is a functional defined by*

$$
J(u) = \frac{\gamma}{2} \int_\Omega \nabla u \cdot \nabla u \, dx + \frac{\alpha}{2} \int_\Omega (u^2 - \beta)^2 \, dx - \langle u, f \rangle_{L^2}
$$

$\forall u \in U$, where $\alpha, \beta, \gamma > 0$, $f \in L^2(\Omega)$ and $U = W_0^{1,2}(\Omega)$.

At this point, we assume a discretized finite dimensional model version (in a finite elements or finite differences context, so that from now on, the not relabeled spaces, functions and operators refer to such a finite dimensional approximation) and suppose

$$
\delta J(u_0) = \mathbf{0}
$$

where in an appropriate matrices sense, we have

$$
\delta^2 J(u_0) = -\gamma \nabla^2 + 6\alpha u_0^2 - 2\alpha \beta.
$$

Define,

$$F(u) = -\frac{\gamma}{2} \int_\Omega \nabla u \cdot \nabla u \, dx + \frac{K}{2} \int_\Omega u^2 \, dx,$$

where $K > 0$ is such that

$$F(u) > 0, \ \forall u \in U \text{ such that } u \neq \mathbf{0},$$

and

$$G(u,v) = \frac{\alpha}{2} \int_\Omega (u^2 - \beta + v)^2 \, dx + \frac{K}{2} \int_\Omega u^2 \, dx - \langle u, f \rangle_{L^2},$$

so that

$$J(u) = G(u,0) - F(u), \ \forall u \in U.$$

Define also $Y = Y^* = L^2(\Omega)$, $F^* : Y^* \to \mathbb{R}$ by

$$
\begin{aligned}
F^*(v_1^*) &= \sup_{u \in U}\{\langle u, v_1^* \rangle_{L^2} - F(u)\} \\
&= \frac{1}{2} \int_\Omega v_1^*[(K + \gamma \nabla^2)^{-1} v_1^*] \, dx, \quad (22.24)
\end{aligned}
$$

and $G^* : Y^* \times Y^* \to \overline{\mathbb{R}} = \mathbb{R} \cup \{+\infty\}$ by

$$
\begin{aligned}
G^*(v_1^*, v_0^*) &= \sup_{(v_1, v) \in Y \times Y}\{\langle v, v_0^* \rangle_Y + \langle v_1, v_1^* \rangle_Y - G(v_1, v)\} \\
&= \frac{1}{2} \int_\Omega \frac{(v_1^* + f)^2}{2v_0^* + K} \, dx + \frac{1}{2\alpha} \int_\Omega (v_0^*)^2 \, dx \\
&\quad + \beta \int_\Omega v_0^* \, dx, \quad (22.25)
\end{aligned}
$$

if $v_0^* \in A^* = \{v_0^* \in Y^* : v_0^* + K > 0, \text{ in } \overline{\Omega}\}$.
Let $J^* : Y^* \times Y^* \to \mathbb{R} \cup \{-\infty\}$ be such that

$$J^*(v_1^*, v_0^*) = -G^*(v_1^*, v_0^*) + F^*(v_1^*).$$

Define

$$\hat{v}_0^* = \alpha(u_0^2 - \beta),$$

and

$$\hat{v}_1^* = (2\hat{v}_0^* + K)u_0 - f.$$

Suppose $\hat{v}_0^* \in A^*$.
Under such hypotheses,

1. *if $-\gamma \nabla^2 + 2\hat{v}_0^* > \mathbf{0}$ so that $\delta^2 J(u_0) > \mathbf{0}$, then*

$$\delta J_1^*(\hat{v}_0^*) = \mathbf{0}$$

   and

$$\delta^2 J_1^*(\hat{v}_0^*) = -\frac{\delta^2 J(u_0)}{\alpha(-\gamma \nabla^2 + 2\hat{v}_0^*)} < \mathbf{0},$$

so that there exist $r, r_1 > 0$ such that

$$
\begin{aligned}
J(u_0) &= \inf_{u \in B_r(u_0)} J(u) \\
&= \sup_{v_0^* \in B_{r_1}(\hat{v}_0^*)} \left\{ \inf_{v_1^* \in Y^*} J^*(v_1^*, v_0^*) \right\} \\
&= \sup_{v_0^* \in B_{r_1}(\hat{v}_0^*)} J_1^*(v_0^*) \\
&= J_1^*(\hat{v}_0^*) \\
&= J^*(\hat{v}_1^*, \hat{v}_0^*), \quad (22.26)
\end{aligned}
$$

where

$$J_1^*(v_0^*) = \inf_{v_1^* \in Y^*} J^*(v_1^*, v_0^*).$$

2. *If* $-\gamma \nabla^2 + 2\hat{v}_0^* < 0$ *and* $\delta^2 J(u_0) > 0$, *then*

$$\delta J_2^*(\hat{v}_0^*) = 0$$

and

$$\delta^2 J_2^*(\hat{v}_0^*) = -\frac{\delta^2 J(u_0)}{\alpha(-\gamma \nabla^2 + 2\hat{v}_0^*)} > 0,$$

so that there exist $r, r_1 > 0$ *such that*

$$
\begin{aligned}
J(u_0) &= \inf_{u \in B_r(u_0)} J(u) \\
&= \inf_{v_0^* \in B_{r_1}(\hat{v}_0^*)} \left\{ \sup_{v_1^* \in Y^*} J^*(v_1^*, v_0^*) \right\} \\
&= \inf_{v_0^* \in B_{r_1}(\hat{v}_0^*)} J_2^*(v_0^*) \\
&= J_2^*(\hat{v}_0^*) \\
&= J^*(\hat{v}_1^*, \hat{v}_0^*),
\end{aligned}
$$

(22.27)

where

$$J_2^*(v_0^*) = \sup_{v_1^* \in Y^*} J^*(v_1^*, v_0^*).$$

3. *If* $\delta^2 J(u_0) < 0$ *so that* $-\gamma \nabla^2 + 2\hat{v}_0^* < 0$, *then*

$$\delta J_2^*(\hat{v}_1^*) = 0$$

and

$$\delta^2 J_2^*(\hat{v}_0^*) = -\frac{\delta^2 J(u_0)}{\alpha(-\gamma \nabla^2 + 2\hat{v}_0^*)} < 0,$$

so that there exist $r, r_1 > 0$ *such that*

$$
\begin{aligned}
J(u_0) &= \sup_{u \in B_r(u_0)} J(u) \\
&= \sup_{v_0^* \in B_{r_1}(\hat{v}_0^*)} \left\{ \sup_{v_1^* \in Y^*} J^*(v_1^*, v_0^*) \right\} \\
&= \sup_{v_0^* \in B_{r_1}(\hat{v}_0^*)} J_2^*(v_0^*) \\
&= J_2^*(\hat{v}_0^*) \\
&= J^*(\hat{v}_1^*, \hat{v}_0^*),
\end{aligned}
$$

(22.28)

where

$$J_2^*(v_0^*) = \sup_{v_1^* \in Y^*} J^*(v_1^*, v_0^*).$$

Proof 22.2 From $\delta J(u_0) = 0$ we have

$$-\gamma \nabla^2 u_0 + \alpha(u_0^2 - \beta)2u_0 - f = 0, \text{ in } \Omega$$

so that

$$
\begin{aligned}
\gamma\nabla^2 u_0 + K u_0 &= \alpha(u_0^2 - \beta)2u_0 + K u_0 - f \\
&= (2\hat{v}_0^* + K)u_0 - f \\
&= \hat{v}_1^*.
\end{aligned}
\tag{22.29}
$$

From this, we obtain

$$
\frac{\hat{v}_1^*}{K + \gamma\nabla^2} - \frac{\hat{v}_1^* + f}{2\hat{v}_0^* + K} = u_0 - u_0 = 0, \text{ in } \Omega.
$$

Thus,

$$
\frac{\partial J^*(\hat{v}_1^*, \hat{v}_0^*)}{\partial v_1^*} = 0.
$$

Define

$$
\tilde{J}(v_0^*) = J^*(\tilde{v}_1^*, v_0^*),
$$

where $\tilde{v}_1^* \in Y^*$ is such that

$$
\frac{\partial J^*(\tilde{v}_1^*, v_0^*)}{\partial v_1^*} = 0.
$$

Observe that, from

$$
\frac{\partial^2 J^*(v_1^*, v_0^*)}{\partial (v_1^*)^2} = \frac{1}{K + \gamma\nabla^2} - \frac{1}{2v_0^* + K},
$$

we have that, if

$$
-\gamma\nabla^2 + 2v_0^* > \mathbf{0},
$$

then

$$
\tilde{J}(v_0^*) = J^*(\tilde{v}_1^*, v_0^*) = \inf_{v_1^* \in Y^*} J^*(v_1^*, v_0^*),
$$

whereas if

$$
-\gamma\nabla^2 + 2v_0^* < \mathbf{0},
$$

then

$$
\tilde{J}(v_0^*) = J^*(\tilde{v}_1^*, v_0^*) = \sup_{v_1^* \in Y^*} J^*(v_1^*, v_0^*).
$$

From (22.29) and from the definition of $\hat{v}_0^*$, we obtain

$$
\begin{aligned}
\tilde{J}^*(\hat{v}_0^*) &= J^*(\hat{v}_1^*, \hat{v}_0^*) \\
&= -G^*(\hat{v}_1^*, \hat{v}_0^*) + F^*(\hat{v}_1^*) \\
&= -\langle u_0, \hat{v}_1^* \rangle_{L^2} - \langle 0, \hat{v}_0^* \rangle_{L^2} + G(u_0, 0) \\
&\quad + \langle u_0, \hat{v}_1^* \rangle_{L^2} - F(u_0) \\
&= G(u_0, 0) - F(u_0) \\
&= J(u_0).
\end{aligned}
\tag{22.30}
$$

From the implicit function theorem we have

$$
\begin{aligned}
\frac{\partial \tilde{J}^*(v_0^*)}{\partial v_0^*} &= \frac{\partial J^*(\tilde{v}_1^*, v_0^*)}{\partial v_0^*} + \frac{\partial J^*(\tilde{v}_1^*, v_0^*)}{\partial v_1^*} \frac{\partial \tilde{v}_1^*}{\partial v_0^*} \\
&= \frac{\partial J^*(\tilde{v}_1^*, v_0^*)}{\partial v_0^*}.
\end{aligned}
\tag{22.31}
$$

Moreover, from this, joining the pieces, we get

$$
\begin{aligned}
\frac{\partial \tilde{J}^*(\hat{v}_0^*)}{\partial v_0^*} &= \frac{\partial J^*(\hat{v}_1^*, \hat{v}_0^*)}{\partial v_0^*} + \frac{\partial J^*(\hat{v}_1^*, \hat{v}_0^*)}{\partial v_1^*} \frac{\partial \hat{v}_1^*}{\partial v_0^*} \\
&= \frac{\partial J^*(\hat{v}_1^*, \hat{v}_0^*)}{\partial v_0^*} \\
&= \frac{(\hat{v}_1^* + f)^2}{(2\hat{v}_0^* + K)^2} - \frac{\hat{v}_0^*}{\alpha} - \beta \\
&= u_0^2 - \frac{\hat{v}_0^*}{\alpha} - \beta \\
&= 0, \text{ in } \Omega,
\end{aligned}
\tag{22.32}
$$

since $\hat{v}_0^* = \alpha(u_0^2 - \beta)$, in Ω.

Hence, from this and (22.31), we obtain

$$
\begin{aligned}
\frac{\partial^2 \tilde{J}^*(\hat{v}_0^*)}{\partial (v_0^*)^2} &= \frac{\partial^2 J^*(\hat{v}_1^*, \hat{v}_0^*)}{\partial v_0^* \partial v_1^*} \frac{\partial \hat{v}_1^*}{\partial v_0^*} \\
&\quad + \frac{\partial^2 J^*(\hat{v}_1^*, \hat{v}_0^*)}{\partial (v_0^*)^2} \\
&= \frac{2(\hat{v}_1^* + f)}{(2\hat{v}_0^* + K)^2} \frac{\partial \hat{v}_1^*}{\partial v_0^*} \\
&\quad -4 \frac{(\hat{v}_1^* + f)^2}{(2\hat{v}_0^* + K)^3} - \frac{1}{\alpha}.
\end{aligned}
\tag{22.33}
$$

At this point we may observe that

$$
\frac{\hat{v}_1^*}{K + \gamma \nabla^2} - \frac{(\hat{v}_1^* + f)}{(2\hat{v}_0^* + K)} = 0,
$$

so that taking the variation in v_0^* of this equation in both sides, we obtain

$$
\frac{\frac{\partial \hat{v}_1^*}{\partial v_0^*}}{(K + \gamma \nabla^2)} - \frac{\frac{\partial \hat{v}_1^*}{\partial v_0^*}}{(2\hat{v}_0^* + K)} + \frac{2(\hat{v}_1^* + f)}{(2\hat{v}_0^* + K)^2} = 0,
\tag{22.34}
$$

and thus, recalling that

$$
u_0 = \frac{(\hat{v}_1^* + f)}{(2\hat{v}_0^* + K)},
$$

we get

$$
\frac{\partial \hat{v}_1^*}{\partial v_0^*} = -\frac{\frac{(K + \gamma \nabla^2)(2u_0)}{(2\hat{v}_0^* + K)}}{1 - \frac{K + \gamma \nabla^2}{2\hat{v}_0^* + K}},
\tag{22.35}
$$

Replacing (22.35) into (22.33), we obtain

$$
\begin{aligned}
\frac{\partial^2 \tilde{J}^*(\hat{v}_0^*)}{\partial (v_0^*)^2} &= \frac{\partial^2 J^*(\hat{v}_1^*, \hat{v}_0^*)}{\partial v_0^* \partial v_1^*} \frac{\partial \hat{v}_1^*}{\partial v_0^*} \\
&\quad + \frac{\partial^2 J^*(\hat{v}_1^*, \hat{v}_0^*)}{\partial (v_0^*)^2} \\
&= \frac{2(\hat{v}_1^* + f)}{(2\hat{v}_0^* + K)^2} \frac{\partial \hat{v}_1^*}{\partial v_0^*} \\
&\quad - \frac{4(\hat{v}_1^* + f)^2}{(2\hat{v}_0^* + K)^3} - \frac{1}{\alpha} \\
&= -\frac{\frac{K+\gamma\nabla^2}{(2\hat{v}_0^* + K)^2}(4u_0^2)}{1 - \frac{K+\gamma\nabla^2}{2\hat{v}_0^* + K}} - \frac{1}{\alpha} - \frac{4u_0^2}{2\hat{v}_0^* + K}.
\end{aligned}
\tag{22.36}
$$

Therefore,

$$
\begin{aligned}
&\frac{\partial^2 \tilde{J}^*(\hat{v}_0^*)}{\partial (v_0^*)^2} \\
&= -\frac{1}{\alpha} - \frac{4u_0^2}{2\hat{v}_0^* + K} \\
&\quad - \frac{K+\gamma\nabla^2}{2v_0^* + K} \frac{4u_0^2}{(2\hat{v}_0^* - \gamma\nabla^2)} \\
&= -\frac{1}{\alpha} - \frac{4u_0^2}{(2\hat{v}_0 + K)}\left(1 + \frac{K+\gamma\nabla^2}{2\hat{v}_0^* - \gamma\nabla^2}\right) \\
&= -\frac{1}{\alpha} - \frac{4u_0^2}{(2\hat{v}_0^* - \gamma\nabla^2)} \\
&= \frac{\gamma\nabla^2 - 2\hat{v}_0^* - 4\alpha u_0^2}{\alpha(-\gamma\nabla^2 + 2\hat{v}_0^*)} \\
&= \frac{\gamma\nabla^2 - 6\alpha u_0^2 + 2\alpha\beta}{\alpha(-\gamma\nabla^2 + 2\hat{v}_0^*)} \\
&= -\frac{\delta^2 J(u_0)}{\alpha(-\gamma\nabla^2 + 2\hat{v}_0^*)}.
\end{aligned}
\tag{22.37}
$$

Assume now $\delta^2 J(u_0) > \mathbf{0}$ and $-\gamma\nabla^2 + 2\hat{v}_0^* > \mathbf{0}$.
From (22.37), we obtain

$$
\delta^2 J_1^*(\hat{v}_0^*) = \frac{\partial^2 \tilde{J}^*(\hat{v}_0^*)}{\partial (v_0^*)^2} = -\frac{\delta^2 J(u_0)}{\alpha(-\gamma\nabla^2 + 2\hat{v}_0^*)} < \mathbf{0}.
$$

Summarizing,

$$
\delta^2 J_1^*(\hat{v}_0^*) < \mathbf{0},
$$

so that $u_0 \in U$ is a point of local minimum for J and $\hat{v}_0^*$ is a point of local maximum for J_1^*.

Hence, there exist $r > 0$ and $r_1 > 0$ such that

$$
\begin{aligned}
J(u_0) &= \min_{u \in B_r(u_0)} J(u) \\
&= \max_{v_0^* \in B_{r_1}(\hat{v}_0^*)} J_1^*(v_0^*) \\
&= J_1^*(\hat{v}_0^*) \\
&= J^*(\hat{v}_1^*, \hat{v}_0^*).
\end{aligned}
\tag{22.38}
$$

The remaining items may be proven similarly from (22.37).
This completes the proof.

22.4 Another duality principle for global optimization

Our next result is another duality principle suitable for global optimization.

Theorem 22.4.1 *Let $\Omega \subset \mathbb{R}^3$ be an open, bounded, connected set with a Lipschitzian boundary denoted by $\partial\Omega$. Consider the Ginzburg-Landau energy given by $J : U \to \mathbb{R}$, where*

$$J(u) = \frac{\gamma}{2} \int_\Omega \nabla u \cdot \nabla u \, dx$$
$$+ \frac{\alpha}{2} \int_\Omega (u^2 - 1)^2 \, dx - \langle u, f \rangle_{L^2}, \tag{22.39}$$

where $\alpha, \gamma > 0$, $f \in L^2(\Omega)$ and

$$U = W_0^{1,2}(\Omega) = \{u \in W^{1,2}(\Omega) : u = 0, \text{ on } \partial\Omega\}.$$

We also denote, for a finite dimensional discretized version of this problem, in a finite elements or finite differences context,

$$J(u) = -F(u) + G_1(u, 0),$$

where

$$F(u) = -\frac{\gamma}{2} \int_\Omega \nabla u \cdot \nabla u \, dx + \frac{K}{2} \int_\Omega u^2 \, dx,$$

and

$$G_1(u, v) = \frac{\alpha}{2} \int_\Omega (u^2 - 1 + v)^2 \, dx + \frac{K}{2} \int_\Omega u^2 \, dx - \langle u, f \rangle_{L^2},$$

where $K > 0$ is such that $F(u) > 0$, $\forall u \in U$, such that $u \neq \mathbf{0}$.
And where generically,

$$\langle h, g \rangle_{L^2} = \int_\Omega hg \, dx, \ \forall h, g \in L^2(\Omega).$$

We define,

$$F^*(z^*) = \sup_{u \in U} \{\langle z^*, u \rangle_{L^2} - F(u)\}$$
$$= \sup_{u \in U} \{\langle z^*, u \rangle_{L^2} + \frac{\gamma}{2} \int_\Omega \nabla u \cdot \nabla u \, dx$$
$$- \frac{K}{2} \int_\Omega u^2 \, dx\}$$
$$= \frac{1}{2} \int_\Omega z^* ((KI_d + \gamma \nabla^2)^{-1} z^*) \, dx, \tag{22.40}$$

where I_d denotes the identity matrix.

Also,

$$\begin{aligned}
G_1^*(z^*, v_1^*) &= \sup_{(u,v)\in U\times L^2} \{\langle z^*, u\rangle_{L^2} + \langle v_1^*, v\rangle_{L^2} \\
&\quad + \langle u, f\rangle_{L^2} - \frac{\alpha}{2}\int_\Omega (u^2 - 1 + v)^2\, dx - \frac{K}{2}\int_\Omega u^2\, dx\} \\
&= \frac{1}{2}\int_\Omega \frac{(z^* + f)^2}{2v_1^* + K}\, dx + \frac{1}{2\alpha}\int_\Omega (v_1^*)^2\, dx \\
&\quad + \int_\Omega v_1^*\, dx \\
&\equiv G_{1L}^*(z^*, v_1^*),
\end{aligned}$$ (22.41)

if $v_1^* \in B_1$, *where*

$$B_1 = \{v_1^* \in Y^* : 2v_1^* + K > 0, \text{ in } \overline{\Omega}\}$$

and G_{1L}^* *stands for the Legendre transform of* G_1.
  We also denote,

$$\begin{aligned}
B_2 &= \{v_1^* \in Y^* : \\
&\quad \frac{\gamma}{2}\int_\Omega \nabla u \cdot \nabla u\, dx + \int_\Omega v_1^* u^2\, dx > 0, \\
&\quad \forall u \in U \text{ such that } u \neq \mathbf{0}\},
\end{aligned}$$ (22.42)

$$C^* = B_1 \cap B_2,$$

where

$$Y = Y^* = L^2(\Omega).$$

Under such hypotheses,

$$\begin{aligned}
\inf_{u\in U} J(u) &\geq \sup_{v_1^*\in C^*} \{\inf_{z^*\in Y^*} \{J^*(z^*, v_1^*)\}\} \\
&= \sup_{v_1^*\in C^*} \tilde{J}(v_1^*),
\end{aligned}$$ (22.43)

where,

$$J^*(z^*, v_1^*) = F^*(z^*) - G_1^*(z^*, v_1^*)$$

and

$$\tilde{J}(v_1^*) = \inf_{z^*\in Y^*} J^*(z^*, v_1^*).$$

Moreover, if there exists a critical point $(z_0^*, (v_0^*)_1) \in C^* \times Y^*$, *so that*

$$\delta J^*(z_0^*, (v_0^*)_1) = 0,$$

then, denoting

$$u_0 = \frac{z_0^* + f}{2(v_0^*)_1 + K}$$

we have that

$$\begin{aligned}
J(u_0) &= \min_{u\in U} J(u) \\
&= \max_{v_1^*\in C^*} \tilde{J}^*(v_1^*) \\
&= \tilde{J}^*((v_0^*)_1) \\
&= J^*(z_0^*, (v_0^*)_1).
\end{aligned}$$ (22.44)

Proof 22.3 Observe that

$$
\begin{aligned}
& G_1^*(z^*, v_1^*) \\
\geq \;\; & \langle z^*, u \rangle_{L^2} + \langle v_1^*, v \rangle_{L^2} - G_1(u, v),
\end{aligned}
\tag{22.45}
$$

$\forall v_1^* \in C^*,\ u \in U,\ v \in Y,\ z^* \in Y^*$.

Thus,

$$
\begin{aligned}
& -\langle z^*, u \rangle_{L^2} + G_1(u, 0) \\
\geq \;\; & -G_1^*(z^*, v_1^*),
\end{aligned}
\tag{22.46}
$$

$\forall u \in U,\ v_1^* \in C^*,\ z^* \in Y^*$, so that

$$
\begin{aligned}
& F^*(z^*) - \langle z^*, u \rangle_{L^2} + G_1(u, 0) \\
\geq \;\; & F^*(z^*) - G_1^*(z^*, v_1^*),
\end{aligned}
\tag{22.47}
$$

$\forall u \in U,\ v_1^* \in C^*,\ z^* \in Y^*$.

Hence,

$$
\begin{aligned}
J(u) &= -F(u) + G_1(u, 0) \\
&= \inf_{z^* \in Y^*} \{ F^*(z^*) - \langle z^*, u \rangle_{L^2} \} + G_1(u, 0) \\
&\geq \inf_{z^* \in Y^*} \{ F^*(z^*) - G_1^*(z^*, v_1^*) \} \\
&= \inf_{z^* \in Y^*} J^*(z^*, v_1^*) \\
&= \tilde{J}^*(v_1^*),
\end{aligned}
\tag{22.48}
$$

$\forall u \in U,\ v_1^* \in C^*$.

Thus,

$$
\inf_{u \in U} J(u) \geq \sup_{v_1^* \in C^*} \tilde{J}^*(v_1^*).
\tag{22.49}
$$

Now suppose $(z_0^*, (v_0^*)_1) \in Y^* \times C^*$ is such that

$$
\delta J^*(z_0^*, (v_0^*)_1) = \mathbf{0}.
$$

From the variation in z^* we obtain,

$$
(K I_d + \gamma \nabla^2)^{-1}(z_0^*) = \frac{z_0^* + f}{2(v_0^*)_1 + K} = u_0,
$$

so that

$$
z_0^* = (K I_d + \gamma \nabla^2) u_0,
$$

and

$$
z_0^* + f = (2(v_0^*)_1 + K) u_0.
\tag{22.50}
$$

Thus,

$$
F^*(z_0^*) = \langle z_0^*, u_0 \rangle_{L^2} - F(u_0).
\tag{22.51}
$$

On the other hand, from the variation in v_1^*, we have,

$$
\frac{[z_0^* + f]^2}{[2(v_0^*)_1 + K]^2} - \frac{(v_0^*)_1}{\alpha} - 1 = 0,
$$

so that

$$(v_0^*)_1 = \alpha(u_0^2 - 1),$$

and hence, from this and (22.50) we have,

$$z_0^* + f = \alpha(u_0^2 - 1)2u_0 + Ku_0,$$

and

$$G_1^*(z_0^*, (v_0^*)_1) = \langle z_0^*, u_0 \rangle_{L^2} - G_1(u_0, 0). \tag{22.52}$$

From (22.51) and (22.52), we obtain

$$\begin{aligned} J(u_0) &= -F(u_0) + G_1(u_0, 0) \\ &= F^*(z_0^*) - G_1^*(z_0^*, (v_0^*)_1) \\ &= J^*(z_0^*, (v_0^*)_1). \end{aligned} \tag{22.53}$$

Now, let

$$v_1^* \in C^*.$$

Observe that, in such a case,

$$\frac{\gamma}{2} \int_\Omega \nabla u \cdot \nabla u \, dx + \int_\Omega v_1^* u^2 \, dx > 0,$$

$\forall u \in U$, such that $u \neq \mathbf{0}$.

Denoting

$$\alpha_1 = \inf_{u \in U} \left\{ \frac{\gamma}{2} \int_\Omega \nabla u \cdot \nabla u \, dx + \int_\Omega v_1^* u^2 \, dx - \langle u, f \rangle_{L^2} \right\}, \tag{22.54}$$

we have

$$\begin{aligned} &\int_\Omega v_1^* u^2 \, dx + \frac{K}{2} \int_\Omega u^2 \, dx - \langle u, f \rangle_{L^2} - \langle z^*, u \rangle_{L^2} \\ &\geq \quad -\frac{\gamma}{2} \int_\Omega \nabla u \cdot \nabla u \, dx + \frac{K}{2} \int_\Omega u^2 \, dx - \langle z^*, u \rangle_{L^2} + \alpha_1, \end{aligned} \tag{22.55}$$

$\forall u \in U$, so that,

$$\begin{aligned} &\inf_{u \in U} \left\{ \int_\Omega v_1^* u^2 \, dx + \frac{K}{2} \int_\Omega u^2 \, dx - \langle u, f \rangle_{L^2} - \langle z^*, u \rangle_{L^2} \right\} \\ &\geq \quad \inf_{u \in U} \left\{ -\frac{\gamma}{2} \int_\Omega \nabla u \cdot \nabla u \, dx + \frac{K}{2} \int_\Omega u^2 \, dx - \langle z^*, u \rangle_{L^2} \right\} + \alpha_1, \end{aligned} \tag{22.56}$$

and hence,

$$-\frac{1}{2} \int_\Omega \frac{(z^* + f)^2}{2v_1^* + K} \geq -F^*(z^*) + \alpha_1,$$

so that, for $v_1^* \in C^*$ fixed, we have,

$$F^*(z^*) - \frac{1}{2} \int_\Omega \frac{(z^* + f)^2}{2v_1^* + K} \, dx \geq \alpha_1,$$

$\forall z^* \in Y^*$.

And indeed, from the general result in [78], we have

$$\inf_{z^* \in Y^*} \left\{ F^*(z^*) - \frac{1}{2} \int_\Omega \frac{(z^* + f)^2}{2v_1^* + K} \, dx \right\} = \alpha_1 \in \mathbb{R}.$$

From this, since $(v_0^*)_1 \in C^*$ and the optimization in z^* in question is quadratic, we may infer that,

$$\tilde{J}^*((v_0^*)_1) = \inf_{z^* \in Y^*} J^*(z^*, (v_0^*)_1) = J^*(z_0^*, (v_0^*)_1).$$

From this, (22.49) and (22.53), we finally obtain,

$$
\begin{aligned}
J(u_0) &= \min_{u \in U} J(u) \\
&= \max_{v_1^* \in C^*} \tilde{J}^*(v_1^*) \\
&= \tilde{J}^*((v_0^*)_1) \\
&= J^*(z_0^*, (v_0^*)_1).
\end{aligned}
\tag{22.57}
$$

This completes the proof.

22.5 The existence of a global solution for the Ginzburg-Landau system in the presence of a magnetic field

In this section we develop a proof of existence of solution for the Ginzburg-Landau system in the presence of a magnetic field and concerning potential. We emphasize again that similar models, which are closely relating to those of last sections, are addressed in [12, 55].

We highlight this existence result and the next duality principle have been presented in similar form in my book, "A Classical Description of Variational Quantum Mechanics and Related Models", [22]. For the sake of completeness, we present both the results in details.

Finally, as a previous related existence result we would cite [46].

Theorem 22.5.1 *Consider the functional* $J : U \to \mathbb{R}$ *where*

$$
\begin{aligned}
J(\phi, \mathbf{A}) &= \frac{\gamma}{2} \int_\Omega |\nabla \phi - i_m \rho \mathbf{A} \phi|_2^2 \, dx \\
&\quad + \frac{\alpha}{4} \int_\Omega |\phi|^4 \, dx - \frac{\beta}{2} \int_\Omega |\phi|^2 \, dx \\
&\quad + \frac{1}{8\pi} \int_{\Omega_1} |\operatorname{curl}(\mathbf{A}) - \mathbf{B}_0|_2^2 \, dx,
\end{aligned}
\tag{22.58}
$$

where Ω, Ω_1 *are open bounded, simply connected sets such that*

$$\overline{\Omega} \subset \Omega_1.$$

We assume the boundaries $\partial\Omega$ *and* $\partial\Omega_1$ *to be regular (Lipschitzian). Here, again* i_m *denotes the imaginary unit and* γ, α, β *and* ρ *are positive constants. Also,*

$$U = W^{1,2}(\Omega; \mathbb{C}) \times L^2(\Omega_1; \mathbb{R}^3).$$

Suppose there exists a minimizing sequence $(\phi_n, \mathbf{A}_n) \subset U$ *for* J *such that*

$$\|\phi_n\|_\infty \leq K, \; \forall n \in \mathbb{N}$$

for some $K > 0$.

Under such hypotheses, there exists $(\phi_0, \mathbf{A}_0) \in U$ such that

$$J(\phi_0, \mathbf{A}_0) = \min_{(\phi, \mathbf{A}) \in U} \{J(\phi, \mathbf{A})\}.$$

Proof 22.4

Define

$$\alpha_1 = \inf_{(\phi, \mathbf{A}) \in U} \{J(\phi, \mathbf{A})\} \in \mathbb{R}.$$

From the hypotheses,

$$\lim_{n \to \infty} J(\phi_n, \mathbf{A}_n) = \alpha_1.$$

From the expression of J, there exists $K_1 > 0$ such that

$$\|\mathrm{curl}(\mathbf{A}_n)\|_2^2 \leq K_1, \ \forall n \in \mathbb{N}.$$

Given $(\phi, \mathbf{A}) \in U$, define $(\phi', \mathbf{A}') \in U$ by

$$\phi' = \phi e^{im\rho\varphi},$$

and

$$\mathbf{A}' = \mathbf{A} + \nabla\varphi,$$

where φ will be specified in the next lines.

Observe that,

$$
\begin{aligned}
|\nabla\phi' - i_m \rho \mathbf{A}' \phi'|_2 &= |\nabla(\phi e^{im\rho\varphi}) - i_m \rho(\mathbf{A} + \nabla\varphi)\phi e^{im\rho\varphi}|_2 \\
&= |\nabla\phi e^{im\rho\varphi} + \phi i_m \rho e^{im\rho\varphi}\nabla\varphi - i_m \rho \mathbf{A}\phi e^{im\rho\varphi} - i_m \rho\phi\nabla\varphi e^{im\rho\varphi}|_2 \\
&= |(\nabla\phi - i_m \rho \mathbf{A}\phi)e^{im\rho\varphi}|_2 \\
&= |\nabla\phi - i_m \rho \mathbf{A}\phi)|_2.
\end{aligned}
\tag{22.59}
$$

Moreover,

$$\mathrm{curl}(\mathbf{A}') = \mathrm{curl}(\mathbf{A}) + \mathrm{curl}(\nabla\varphi) = \mathrm{curl}(\mathbf{A}).$$

Also,

$$|\phi'| = |\phi e^{im\rho\varphi}| = |\phi|.$$

From these last calculations, we may infer the system gauge invariance, that is,

$$J(\phi, \mathbf{A}) = J(\phi', \mathbf{A}').$$

In particular, we shall choose $\varphi \in W^{1,2}(\Omega_1)$ such that

$$div(\mathbf{A}') = div(\mathbf{A}) + \nabla^2\varphi = 0,$$

and, denoting by $\mathbf{n}$ the outward normal to $\partial\Omega_1$,

$$\mathbf{A}' \cdot \mathbf{n} = \mathbf{A} \cdot \mathbf{n} + \nabla\varphi \cdot \mathbf{n} = 0, \ \text{on } \partial\Omega_1$$

that is,

$$\nabla^2\varphi = -div(\mathbf{A}), \ \text{in } \Omega_1,$$

$$\nabla\varphi \cdot \mathbf{n} = -\mathbf{A} \cdot \mathbf{n}, \ \text{on } \partial\Omega_1.$$

Observe that at first we would have,

$$\inf_{(\phi, \mathbf{A}) \in U} J(\phi, \mathbf{A}) \leq \inf_{(\phi', \mathbf{A}') \in U} J(\phi', \mathbf{A}').$$

However,

$$J(\phi'_n, \mathbf{A}'_n) = J(\phi_n, \mathbf{A}_n) \to \alpha_1, \text{ as } n \to \infty,$$

so that

$$\inf_{(\phi, \mathbf{A}) \in U} J(\phi, \mathbf{A}) = \inf_{(\phi', \mathbf{A}') \in U} J(\phi', \mathbf{A}').$$

From Friedrichs' inequality, we have,

$$\begin{aligned} K_1^2 &\geq \|\mathrm{curl}(\mathbf{A}_n)\|_2^2 \\ &= \| \mathrm{curl}(\mathbf{A}'_n)\|_2^2 + \|div(\mathbf{A}'_n)\|_2^2 \geq K_2 \|\mathbf{A}'_n\|_2^2, \ \forall n \in \mathbb{N}, \end{aligned} \tag{22.60}$$

for some $K_2 > 0$.

Hence,

$$\|\mathbf{A}'_n\|_2 \leq K_3, \forall n \in \mathbb{N}$$

for some $K_3 > 0$.

We recall that,

$$\|\phi'_n\|_\infty = \|\phi_n\|_\infty \leq K, \forall n \in \mathbb{N}.$$

Hence,

$$\begin{aligned} J(\phi'_n, \mathbf{A}'_n) &= \frac{\gamma}{2} \int_\Omega |\nabla \phi'_n - i_m \rho \mathbf{A}'_n \phi'_n|_2^2 \, dx \\ &\quad + \frac{\alpha}{4} \int_\Omega |\phi'_n|^4 \, dx - \frac{\beta}{2} \int_\Omega |\phi'_n|^2 \, dx \\ &\quad + \frac{1}{8\pi} \int_{\Omega_1} |\mathrm{curl}(\mathbf{A}'_n) - \mathbf{B}_0|_2^2 \, dx \\ &\geq \frac{\gamma}{2} \int_\Omega |\nabla \phi'_n|^2 \, dx - \gamma |\rho| \|\phi'_n\|_\infty \|\mathbf{A}'_n\|_2 \|\nabla \phi'_n\|_2 \\ &\quad + \frac{\gamma}{2} |\rho|^2 \|\mathbf{A}'_n \phi'_n\|_2^2 \\ &\quad + \frac{\alpha}{4} \int_\Omega |\phi'_n|^4 \, dx - \frac{\beta}{2} \int_\Omega |\phi'_n|^2 \, dx \\ &\quad + \frac{1}{8\pi} \int_{\Omega_1} |\mathrm{curl}(\mathbf{A}'_n) - \mathbf{B}_0|_2^2 \, dx \\ &\geq \frac{\gamma}{2} \|\nabla \phi'_n\|_2^2 \, dx - \gamma K K_3 |\rho| \|\nabla \phi'_n\|_2 \\ &\quad + \frac{\gamma}{2} |\rho|^2 \|\mathbf{A}'_n \phi'_n\|_2^2 \\ &\quad + \frac{\alpha}{4} \int_\Omega |\phi'_n|^4 \, dx - \frac{\beta}{2} \int_\Omega |\phi'_n|^2 \, dx \\ &\quad + \frac{1}{8\pi} \int_{\Omega_1} | \mathrm{curl}(\mathbf{A}'_n) - \mathbf{B}_0|_2^2 \, dx. \end{aligned} \tag{22.61}$$

Suppose, to obtain contradiction, there exists a subsequence $\{n_k\}$ such that

$$\|\nabla \phi'_{n_k}\|_2 \to +\infty, \text{ as } k \to \infty.$$

From this and (22.61) we obtain,

$$J(\phi'_{n_k}, \mathbf{A}'_{n_k}) \to +\infty, \text{ as } k \to +\infty,$$

which contradicts

$$J(\phi'_n, \mathbf{A}'_n) \to \alpha_1, \text{ as } n \to +\infty.$$

Therefore, there exists $K_4 > 0$ such that

$$\|\nabla \phi'_n\|_2 \leq K_4 \in \mathbb{R}^+, \forall n \in \mathbb{N}.$$

Hence, from the Rellich- Krondrachov theorem, there exists $\phi_0 \in W^{1,2}(\Omega; \mathbb{C})$ such that, up to a not relabeled subsequence,

$$\nabla \phi'_n \rightharpoonup \nabla \phi_0, \text{ weakly in } L^2,$$

and

$$\phi'_n \to \phi_0, \text{ strongly in } L^2.$$

Also, since

$$\|\mathrm{curl}(\mathbf{A}'_n)\|_2 \leq K_1, \forall n \in \mathbb{N},$$

there exists $\mathbf{v}_0 \in L^2(\Omega_1; \mathbb{R}^3)$ such that

$$\mathrm{curl}(\mathbf{A}'_n) \rightharpoonup \mathbf{v}_0, \text{ weakly in } L^2(\Omega_1; \mathbb{R}^3).$$

Also, since

$$\|\mathbf{A}'_n\|_2 \leq K_4, \forall n \in \mathbb{N},$$

there exists

$$\mathbf{A}_0 \in L^2(\Omega_1; \mathbb{R}^3),$$

such that, up to a not relabeled subsequence,

$$\mathbf{A}'_n \rightharpoonup \mathbf{A}_0, \text{ weakly in } L^2(\Omega_1; \mathbb{R}^3).$$

Now fix

$$\hat{\phi} \in C_c^\infty(\Omega_1; \mathbb{R}^3).$$

Thus, we have

$$
\begin{aligned}
\langle \mathbf{A}_0, \mathrm{curl}^*(\hat{\phi}) \rangle_{L^2} &= \lim_{n \to \infty} \langle \mathbf{A}'_n, \mathrm{curl}^*(\hat{\phi}) \rangle_{L^2} \\
&= \lim_{n \to \infty} \langle \mathrm{curl}(\mathbf{A}'_n), \hat{\phi} \rangle_{L^2} \\
&= \langle \mathbf{v}_0, \hat{\phi} \rangle_{L^2}.
\end{aligned}
\tag{22.62}
$$

Since $\hat{\phi} \in C_c^\infty(\Omega_1; \mathbb{R}^3)$ is arbitrary, we may infer that

$$\mathbf{v}_0 = \mathrm{curl}(\mathbf{A}_0),$$

in distributional sense.

At this point we shall prove that, up to a not relabeled subsequence, we have,

$$\mathbf{A}'_n \phi'_n \rightharpoonup \mathbf{A}_0 \phi_0, \text{ weakly in } L^2(\Omega; \mathbb{C}^3).$$

Fix $\mathbf{v} \in L^2(\Omega; \mathbb{C}^3)$. Therefore, up to a not relabeled subsequence, we have that

$$|\phi'_n \mathbf{v} - \phi_0 \mathbf{v}|_2^2 \to 0, \text{ a.e. in } \Omega.$$

Observe that

$$\|\phi'_n\|_\infty < K, \ \forall n \in \mathbb{N},$$

so that

$$|\phi'_n \mathbf{v} - \phi_0 \mathbf{v}|_2^2 \le 2K^2 |\mathbf{v}|_2^2 \in L^1(\Omega; \mathbb{R}).$$

Thus, from the Lebesgue dominated convergence theorem, we obtain

$$\|\phi'_n \mathbf{v} - \phi_0 \mathbf{v}\|_2^2 \to 0, \text{ as } n \to \infty.$$

Hence, since

$$\mathbf{A}'_n \rightharpoonup \mathbf{A}_0, \text{ weakly in } L^2(\Omega_1; \mathbb{R}^3),$$

we have,

$$\left| \int_\Omega (\mathbf{A}'_n \cdot \phi'_n \mathbf{v} - \mathbf{A}_0 \cdot \phi_0 \mathbf{v}) \, dx \right|$$

$$= \left| \int_\Omega (\mathbf{A}'_n \cdot \phi'_n \mathbf{v} - \mathbf{A}'_n \cdot \phi_0 \mathbf{v} + \mathbf{A}'_n \cdot \phi_0 \mathbf{v} - \mathbf{A}_0 \cdot \phi_0 \mathbf{v}) \, dx \right|$$

$$\le \|\mathbf{A}'_n\|_2 \|\phi'_n \mathbf{v} - \phi_0 \mathbf{v}\|_2 + \left| \int_\Omega \mathbf{A}'_n \cdot \phi_0 \mathbf{v} - \mathbf{A}_0 \cdot \phi_0 \mathbf{v}) \, dx \right|$$

$$\to 0, \text{ as } n \to \infty. \tag{22.63}$$

Since $\mathbf{v} \in L^2(\Omega; \mathbb{C}^3)$ is arbitrary, we may infer that

$$\mathbf{A}'_n \phi'_n \rightharpoonup \mathbf{A}_0 \phi_0, \text{ weakly in } L^2(\Omega; \mathbb{C}^3).$$

From this we obtain

$$\nabla \phi'_n - i_m \rho \mathbf{A}'_n \phi'_n \rightharpoonup \nabla \phi_0 - i_m \rho \mathbf{A}_0 \phi_0, \text{ weakly in } L^2(\Omega; \mathbb{C}^3),$$

so that

$$\liminf_{n \to \infty} \left\{ \int_\Omega |\nabla \phi'_n - i_m \rho \mathbf{A}'_n \phi'_n|_2^2 \, dx \right\} \ge \int_\Omega |\nabla \phi_0 - i\rho \mathbf{A}_0 \phi_0|_2^2 \, dx, \tag{22.64}$$

Also, from

$$\phi'_n \rightharpoonup \phi_0, \text{ weakly in } W^{1,2}(\Omega; \mathbb{C})$$

and

$$\mathrm{curl}(\mathbf{A}'_n) \rightharpoonup \mathrm{curl}(\mathbf{A}_0), \text{ weakly in } L^2(\Omega_1, \mathbb{R}^3),$$

from the convexity of the functional involved, we obtain,

$$\liminf_{n \to \infty} \left\{ \frac{1}{8\pi} \int_{\Omega_1} |\mathrm{curl}(\mathbf{A}'_n) - \mathbf{B}_0|_2^2 \, dx + \frac{\alpha}{4} \int_\Omega |\phi'_n|^4 \, dx \right\}$$

$$\ge \frac{1}{8\pi} \int_{\Omega_1} |\mathrm{curl}(\mathbf{A}_0) - \mathbf{B}_0|_2^2 \, dx + \frac{\alpha}{4} \int_\Omega |\phi_0|^4 \, dx, \tag{22.65}$$

so that, from these last results and from

$$\phi'_n \to \phi_0, \text{ strongly in } L^2(\Omega; \mathbb{C}),$$

we get,

$$\inf_{(\phi, \mathbf{A}) \in U} J(\phi, \mathbf{A}) = \alpha_1$$

$$= \liminf_{n \to \infty} J(\phi'_n, \mathbf{A}'_n)$$

$$\ge J(\phi_0, \mathbf{A}_0). \tag{22.66}$$

The proof is complete.

22.6   Duality for the complex Ginzburg-Landau system

In this subsection we present a duality principle and relating sufficient optimality criterion for the full complex Ginzburg-Landau system.

The basic results on convex analysis here developed may be found in [64, 78]. Our results are summarized by the following theorem.

Theorem 22.6.1 *Let $\Omega, \Omega_1 \subset \mathbb{R}^3$ be open, bounded, connected sets with regular (Lipischtzian) boundaries denoted by $\partial\Omega$ and $\partial\Omega_1$ respectively, where $\overline{\Omega} \subset \Omega_1$ and Ω corresponds to a super conducting sample. Consider the Ginzburg-Landau energy given by $J : V_1 \times V_2 \to \mathbb{R}$ where,*

$$
\begin{aligned}
J(\phi, \mathbf{A}) \;=\; & \frac{\gamma}{2} \int_\Omega |\nabla\phi - i_m \rho \mathbf{A}\phi|_2^2 \, dx \\
& + \frac{\alpha}{2} \int_\Omega (|\phi|^2 - \beta)^2 \, dx - \langle \phi, f \rangle_{L^2} \\
& + \frac{1}{8\pi} \int_{\Omega_1} |curl(\mathbf{A}) - \mathbf{B}_0|_2^2 \, dx,
\end{aligned}
\tag{22.67}
$$

and where $\alpha, \gamma, \rho > 0$, $f \in L^2(\Omega;\mathbb{C})$.

In particular, from the Ginzburg-Landau theory for the dimensionless case we have, $\gamma = 1$, $\alpha = \frac{1}{2(1+t^2)^2}$ and $\beta = 1 - t^4$, where $t = T/T_c$, T_c is the critical temperature and T is the super-conducting sample actual one. A typical value for t is $t = 0.95$. Finally, the value $1/(8\pi)$ may also vary according to type of material or type of superconductor.

Moreover,

$$
V_1 = W^{1,2}(\Omega;\mathbb{C}),
$$

$$
V_2 = W^{1,2}(\Omega_1;\mathbb{R}^3).
$$

Here, we generically denote

$$
\langle g, h \rangle_{L^2} = \int_\Omega Re[g]Re[h]\,dx - \int_\Omega Im[g]Im[h]\,dx,
$$

$\forall h, g \in L^2(\Omega;\mathbb{C})$, where $Re[a], Im[a]$ denote the real and imaginary parts of a, $\forall a \in \mathbb{C}$, respectively.

We also denote,

$$
J(\phi, \mathbf{A}) = G_0(\phi, \nabla\phi, \mathbf{A}) + G_1(\phi, 0) + G_2(\mathbf{A}),
$$

$$
G_0(\phi, \nabla\phi, \mathbf{A}) = \frac{\gamma}{2} \int_\Omega |\nabla\phi - i_m \rho \mathbf{A}\phi|_2^2 \, dx,
$$

$$
G_1(\phi, v_3) = \frac{\alpha}{2} \int_\Omega (|\phi|^2 - \beta + v_3)^2 \, dx - \langle \phi, f \rangle_{L^2},
$$

and,

$$
G_2(\mathbf{A}) = \frac{1}{8\pi} \int_{\Omega_1} |curl(\mathbf{A}) - \mathbf{B}_0|_2^2 \, dx.
$$

Moreover, we define

$$
\begin{aligned}
G_0^*(v_1^*) \;=\; & \sup_{(\phi, v_1) \in V_1 \times Y} \{ \langle v_1^*, v_1 - i_m \rho \mathbf{A}\phi \rangle_{L^2} - G_0(\phi, v_1, \mathbf{A}) \} \\
\;=\; & \frac{1}{2\gamma} \int_\Omega |v_1^*|_2^2 \, dx,
\end{aligned}
\tag{22.68}
$$

$$G_1^*(v_1^*, v_3^*, \mathbf{A}) = \sup_{(\phi, v_3) \in V_1 \times Y_1} \{\langle v_1^*, \nabla \phi - i_m \rho \mathbf{A} \phi \rangle_{L^2} + \langle v_3^*, v_3 \rangle_{L^2} - G_1(\phi, v_3)\}$$

$$= \frac{1}{2} \int_\Omega \frac{|div(v_1^*) + i_m \rho \mathbf{A} \cdot v_1^* - f|^2}{2v_3^*} \, dx + \frac{1}{2\alpha} \int_\Omega (v_3^*)^2 \, dx$$

$$+ \int_\Omega \beta(v_3^*) \, dx, \tag{22.69}$$

if $v^* \in B_1$, where

$$B_1 = \{v^* \in Y^* \times Y_1^* : v_3^* > 0 \text{ in } \overline{\Omega}\}.$$

We also denote

$$B_2 = \{v^* \in Y^* :$$

$$\frac{1}{8\pi} \int_{\Omega_1} |curl\mathbf{A}|_2^2 \, dx - \frac{1}{2} \int_\Omega \frac{|\rho v_1^* \cdot \mathbf{A}|^2}{2 v_3^*} \, dx > 0,$$

$$\forall \mathbf{A} \in D^*, \text{ such that } \mathbf{A} \neq \mathbf{0}\}, \tag{22.70}$$

$$D^* = \{\mathbf{A} \in V_2 : div(\mathbf{A}) = 0, \text{ in } \Omega_1, \text{ and } \mathbf{A} \cdot \mathbf{n} = 0 \text{ on } \partial \Omega_1\},$$

$$C^* = B_1 \cap B_2,$$

and

$$Y = Y^* = L^2(\Omega; \mathbb{C}^3) \text{ and } Y_1 = Y_1^* = L^2(\Omega).$$

Under such assumptions, we have,

$$\inf_{(\phi, \mathbf{A}) \in V_1 \times D^*} J(\phi, \mathbf{A}) \geq \sup_{v^* \in C^*} \{\inf_{\mathbf{A} \in D^*} \{J^*(v^*, \mathbf{A}) + G_2(\mathbf{A})\}\}$$

$$= \sup_{v^* \in C^*} \tilde{J}^*(v^*), \tag{22.71}$$

where

$$J^*(v^*, \mathbf{A}) = -G_0^*(v_1^*) - G_1^*(v_1^*, v_3^*, \mathbf{A}).$$

and

$$\tilde{J}^*(v^*) = \inf_{\mathbf{A} \in D^*} \{J^*(v^*, \mathbf{A}) + G_2(\mathbf{A})\}.$$

Moreover, assume there exists a critical point $(v_0^*, \mathbf{A}_0) \in C^* \times D^*$ such that

$$\delta\{J^*(v_0^*, \mathbf{A}_0) + G_2(\mathbf{A}_0)\} = \mathbf{0}.$$

Under such hypotheses, defining

$$\phi_0 = \frac{div((v_0^*)_1) + i_m \rho \mathbf{A}_0 \cdot (v_0^*)_1 - f}{2(v_0^*)_3},$$

we have

$$J(\phi_0, \mathbf{A}_0) = \min_{(\phi, \mathbf{A}) \in V_1 \times D^*} J(\phi, \mathbf{A})$$

$$= \sup_{v^* \in C^*} \{\inf_{\mathbf{A} \in D^*} \{J^*(v^*, \mathbf{A}) + G_2(\mathbf{A})\}\}$$

$$= \max_{v^* \in C^*} \tilde{J}^*(v^*)$$

$$= \tilde{J}^*(v_0^*)$$

$$= J^*(v_0^*, \mathbf{A}_0) + G_2(\mathbf{A}_0). \tag{22.72}$$

Proof 22.5   Observe that

$$
\begin{aligned}
-J^*(v^*, \mathbf{A}) - G_2(\mathbf{A}) &= G_0^*(v_1^*) + G_1^*(v_1^*, v_3^*, \mathbf{A}) - G_2(\mathbf{A}) \\
&\geq \langle v_1^*, \nabla\phi - i_m\rho\mathbf{A}\phi\rangle_{L^2} - G_0(\phi, \nabla\phi, \mathbf{A}) \\
&\quad - \langle v_1^*, \nabla\phi - i_m\rho\mathbf{A}\phi\rangle_{L^2} + \langle v_3^*, 0\rangle_{L^2} \\
&\quad - G_1(\phi, 0) - G_2(\mathbf{A}),
\end{aligned}
\tag{22.73}
$$

$\forall\phi \in V_1$, $\mathbf{A} \in D^*$, that is,

$$
\begin{aligned}
& G_0(\phi, \nabla\phi, \mathbf{A}) + G_1(\phi, 0) + G_2(\mathbf{A}) \\
&\geq -G_0^*(v_1^*) - G_1^*(v_1^*, v_3^*, \mathbf{A}) + G_2(\mathbf{A}),
\end{aligned}
\tag{22.74}
$$

so that

$$
\begin{aligned}
J(\phi, \mathbf{A}) &\geq \inf_{\mathbf{A} \in D^*} \{-G_0^*(v_1^*) - G_1^*(v_1^*, v_3^*, \mathbf{A}) + G_2(\mathbf{A})\} \\
&= \inf_{\mathbf{A} \in D^*} \{J^*(v^*, \mathbf{A}) + G_2(\mathbf{A})\} \\
&= \tilde{J}(v^*), \ \forall(\phi, \mathbf{A}) \in V_1 \times D^*, \ v^* \in C^*.
\end{aligned}
\tag{22.75}
$$

Thus,

$$
\inf_{(\phi, \mathbf{A}) \in V_1 \times D^*} J(\phi, \mathbf{A}) \geq \sup_{v^* \in C^*} \tilde{J}^*(v^*).
\tag{22.76}
$$

Now, suppose $(v_0^*, \mathbf{A}_0) \in C^* \times D^*$ is such that

$$
\delta\{J^*(v_0^*, \mathbf{A}_0) + G_2(\mathbf{A}_0)\} = \mathbf{0}.
\tag{22.77}
$$

From the variation in v_1^* we obtain,

$$
\begin{aligned}
(v_0^*)_1 &= \gamma(\nabla - i\rho\mathbf{A}_0)\left(\frac{div((v_0^*)_1) + i_m\rho\mathbf{A}_0 \cdot (v_0^*)_1 - f}{2(v_0^*)_3}\right) \\
&= \gamma(\nabla - i_m\rho\mathbf{A}_0)\phi_0,
\end{aligned}
\tag{22.78}
$$

so that,

$$
G_0^*((v_0^*)_1) = \langle(v_0^*)_1, \nabla\phi_0 - i_m\rho\mathbf{A}_0\phi_0\rangle_{L^2} - G_0(\phi_0, \nabla\phi_0, \mathbf{A}_0).
\tag{22.79}
$$

From the variation in v_3^* we obtain

$$
\frac{(div((v_0^*)_1) + i_m\rho\mathbf{A}_0 \cdot (v_0^*)_1 - f)^2}{(2(v_0^*)_3)^2} - \frac{(v_0^*)_3}{\alpha} - \beta = 0,
$$

that is,

$$
(v_0^*)_3 = \alpha(|\phi_0|^2 - \beta),
$$

so that

$$
G_1^*((v_0^*)_1, (v_0^*)_3, \mathbf{A}_0) = -\langle(v_0^*)_1, \nabla\phi_0 - i_m\rho\mathbf{A}_0\phi_0\rangle_{L^2} - G_1(\phi_0, 0).
\tag{22.80}
$$

From (22.79) and (22.80), we obtain

$$
\begin{aligned}
& J^*(v_0^*, \mathbf{A}_0) + G_2(\mathbf{A}_0) \\
&= -G_0^*((v_0^*)_1) - G_1^*((v_0^*)_1, (v_0^*)_3, \mathbf{A}_0) + G_2(\mathbf{A}_0) \\
&= G_0(\phi_0, \nabla\phi_0, \mathbf{A}_0) + G_1(\phi_0, 0) + G_2(\mathbf{A}_0) \\
&= J(\phi_0, \mathbf{A}_0)
\end{aligned}
\tag{22.81}
$$

From $v_0^* \in C^*$ and (22.77) we have,

$$J^*(v_0^*, \mathbf{A}_0) + G_2(\mathbf{A}_0) = \tilde{J}^*(v_0^*).$$

From this, (22.76) and (22.81) we obtain,

$$
\begin{aligned}
J(\phi_0, \mathbf{A}_0) &= \min_{(\phi, \mathbf{A}) \in V_1 \times D^*} J(\phi, \mathbf{A}) \\
&= \sup_{v^* \in C^*} \left\{ \inf_{\mathbf{A} \in D^*} \{ J^*(v^*, \mathbf{A}) + G_2(\mathbf{A}) \} \right\} \\
&= \max_{v^* \in C^*} \tilde{J}^*(v^*) \\
&= \tilde{J}^*(v_0^*) \\
&= J^*(v_0^*, \mathbf{A}_0) + G_2(\mathbf{A}_0).
\end{aligned}
\tag{22.82}
$$

The proof is complete.

22.7 Conclusion

In the present chapter, we have developed duality principles applicable to a large class of variational non-convex models.

In a second step we have applied such results to a Ginzburg-Landau type equation. We emphasize the main theorems here developed are a kind of generalization of the main results found in Toland [78], published in 1979 and [11, 10].

Following the approach presented in [11], we also highlight the duality principles obtained may be applied to non-linear and non-convex models of plates, shells and elasticity.

Finally, as above mentioned, in the last sections we present a global existence result, a duality principle and respective optimality conditions for the complex Ginzburg-Landau system in superconductivity in the presence of a magnetic field and concerned magnetic potential.

Chapter 23

Existence of Solution for an Optimal Control Problem Associated to the Ginzburg-Landau System in Superconductivity

Fabio Silva Botelho and Eduardo Pandini Barros

23.1   Introduction

This work develops an existence result for an optimal control problem closely related to the Ginzburg-Landau system in superconductivity. First, we recall that about the year 1950 Ginzburg and Landau introduced a theory to model the super-conducting behavior of some types of materials below a critical temperature T_c, which depends on the material in question. They postulated the free density energy may be written close to T_c as

$$F_s(T) = F_n(T) + \frac{\hbar}{4m} \int_\Omega |\nabla \psi|_2^2 \, dx + \frac{\alpha(T)}{4} \int_\Omega |\psi|^4 \, dx - \frac{\beta(T)}{2} \int_\Omega |\psi|^2 \, dx,$$

where ψ is a complex parameter, $F_n(T)$ and $F_s(T)$ are the normal and super-conducting free energy densities, respectively (see [4] for details). Here $\Omega \subset \mathbb{R}^3$ denotes the super-conducting sample with a boundary denoted by $\partial\Omega = \Gamma$. The complex function $\psi \in W^{1,2}(\Omega; \mathbb{C})$ is intended to minimize $F_s(T)$ for a fixed temperature T.

Denoting $\alpha(T)$ and $\beta(T)$ simply by α and β, the corresponding Euler-Lagrange equations are given by:

$$\begin{cases} -\frac{\hbar}{2m}\nabla^2\psi + \alpha|\psi|^2\psi - \beta\psi = 0, & \text{in } \Omega \\ \\ \frac{\partial \psi}{\partial \mathbf{n}} = 0, & \text{on } \partial\Omega. \end{cases} \tag{23.1}$$

This last system of equations is well known as the Ginzburg-Landau (G-L) one. In the physics literature is also well known the G-L energy in which a magnetic potential here denoted by $\mathbf{A}$ is included. The functional in question is given by:

$$J(\psi, \mathbf{A}) \;=\; \frac{1}{8\pi} \int_{\mathbb{R}^3} |\operatorname{curl} \mathbf{A} - \mathbf{B}_0|_2^2 \, dx + \frac{\hbar^2}{4m} \int_{\Omega} \left| \nabla \psi - \frac{2ie}{\hbar c} \mathbf{A} \psi \right|_2^2 dx$$
$$+ \frac{\alpha}{4} \int_{\Omega} |\psi|^4 \, dx - \frac{\beta}{2} \int_{\Omega} |\psi|^2 \, dx \tag{23.2}$$

Considering its minimization on the space U, where

$$U = W^{1,2}(\Omega; \mathbb{C}) \times W^{1,2}(\mathbb{R}^3; \mathbb{R}^3),$$

through the physics notation the corresponding Euler-Lagrange equations are:

$$\begin{cases} \frac{1}{2m} \left(-i\hbar\nabla - \frac{2e}{c}\mathbf{A} \right)^2 \psi + \alpha |\psi|^2 \psi - \beta\psi = 0, & \text{in } \Omega \\[2mm] \left(i\hbar\nabla\psi + \frac{2e}{c}\mathbf{A}\psi \right) \cdot \mathbf{n} = 0, & \text{on } \partial\Omega, \end{cases} \tag{23.3}$$

and

$$\begin{cases} \operatorname{curl}(\operatorname{curl} \mathbf{A}) = \operatorname{curl} \mathbf{B}_0 + \frac{4\pi}{c}\tilde{J}, & \text{in } \Omega \\[2mm] \operatorname{curl}(\operatorname{curl} \mathbf{A}) = \operatorname{curl} \mathbf{B}_0, & \text{in } \mathbb{R}^3 \setminus \overline{\Omega}, \end{cases} \tag{23.4}$$

where

$$\tilde{J} = -\frac{ie\hbar}{2m} \left(\psi^* \nabla\psi - \psi\nabla\psi^* \right) - \frac{2e^2}{mc} |\psi|^2 \mathbf{A}.$$

and

$$\mathbf{B}_0 \in L^2(\mathbb{R}^3; \mathbb{R}^3)$$

is a known applied magnetic field.

Existence of a global solution for a similar problem has been proved in [22].

23.2 An existence result for a related optimal control problem

Let $\Omega \subset \mathbb{R}^3$, $\Omega_1 \subset \mathbb{R}^3$ be open, bounded and connected sets with Lipschitzian boundaries, where $\overline{\Omega} \subset \Omega$ and Ω_1 is convex. Let $\phi_d : \Omega \to \mathbb{C}$ be a known function in $L^4(\Omega; \mathbb{C})$ and consider the problem of minimizin

$$\| |\phi|^2 - |\phi_d|^2 \|_{0,2,\Omega}^2$$

with $(\phi, \mathbf{A}, u)$ subject to the satisfaction of the Ginzburg-Landau equations, indicated in (23.5) and (23.6) the next lines.

For such a problem, the control variable is $u \in L^2(\partial\Omega; \mathbb{C})$ and the state variables are the Ginzburg-Landa order parameter $\phi \in W^{1,2}(\Omega, \mathbb{C})$ and the magnetic potential $\mathbf{A} \in W^{1,2}(\Omega_1, \mathbb{R}^3)$.

Our main existence result is summarized by the following theorem.

Theorem 23.2.1 *Consider the functional*

$$J(\phi, \mathbf{A}, u) = \frac{\varepsilon}{2} \|\nabla\phi\|_{0,2,\Omega}^2 + K_1 \| |\phi|^2 - |\phi_d|^2 \|_{0,2,\Omega}^2 + K_2 \|u\|_{0,2,\partial\Omega}^2,$$

subject to $(\phi, \mathbf{A}, u) \in \mathscr{C}$, *where*

$$\mathscr{C} = \{ (\phi, \mathbf{A}, u) \in W^{1,2}(\Omega, \mathbb{C}) \times W^{1,2}(\Omega_1, \mathbb{R}^3) \times L^2(\partial\Omega; \mathbb{C}) \; : \; \text{such that (23.5) and (23.6) hold} \},$$

where

$$\begin{cases} \frac{1}{2m}\left(-i\hbar\nabla - \frac{2e}{c}\mathbf{A}\right)^2\phi + \alpha|\phi|^2\phi - \beta\phi = 0, & in\ \Omega, \\ \\ \left(i\hbar\nabla\phi + \frac{2e}{c}\mathbf{A}\phi\right)\cdot\mathbf{n} = u, & on\ \partial\Omega, \end{cases} \tag{23.5}$$

and

$$\begin{cases} curl\ curl\ \mathbf{A} = curl\ \mathbf{B}_0 + \frac{4\pi}{c}\tilde{J}, & in\ \Omega, \\ \\ curl\ curl\ \mathbf{A} = curl\ \mathbf{B}_0, & in\ \Omega_1\setminus\Omega, \\ \\ div\ \mathbf{A} = 0, & in\ \Omega_1, \\ \\ \mathbf{A}\cdot\mathbf{n} = 0, & on\ \partial\Omega_1 \end{cases} \tag{23.6}$$

where,

$$\tilde{J} = -\frac{ie\hbar}{2m}(\phi^*\nabla\phi - \phi\nabla\phi^*) - \frac{2e^2}{mc}|\phi|^2\mathbf{A},$$

and where $\varepsilon > 0$ is a small parameter, $K_1 > 0$ and $K_2 > 0$.

Under such hypotheses, there exists $(\phi_0, \mathbf{A}_0, u_0) \in \mathscr{C}$ such that

$$J(\phi_0, \mathbf{A}_0, u_0) = \min_{(\phi,\mathbf{A},u)\in\mathscr{C}} J(\phi, \mathbf{A}, u).$$

Proof 23.1 Let $\{(\phi_n, \mathbf{A}_n, u_n)\}$ be a minimizing sequence (such a sequence exists from the existence result for $u = 0$ in [22], and from the fact that J is lower bounded by 0).

Thus, such a sequence is such that

$$J(\phi_n, \mathbf{A}_n, u_n) \to \eta = \inf_{(\phi,\mathbf{A},u)\in\mathscr{C}} J(\phi, \mathbf{A}, u).$$

From the expression of J, there exists $K > 0$ such that

$$\|\nabla\phi_n\|_{0,2,\Omega} \le K,$$
$$\|\phi_n\|_{0,4,\Omega} \le K,$$
$$\|\phi_n\|_{0,2,\Omega} \le K,$$

and

$$\|u_n\|_{0,2,\partial\Omega} \le K,\ \forall n \in \mathbb{N}$$

so that, from the Rellich-Kondrashov Theorem, there exists a not relabeled subsequence, $\phi_0 \in W^{1,2}(\Omega,\mathbb{C})$ and $u_0 \in L^2(\Omega,\mathbb{C})$ such that

$$\phi_n \rightharpoonup \phi_0,\ \text{weakly in}\ W^{1,2}(\Omega,\mathbb{C}),$$
$$\phi_n \to \phi_0,\ \text{in norm in}\ L^2(\Omega,\mathbb{C})\ \text{and}\ L^4(\Omega,\mathbb{C}),$$
$$u_n \rightharpoonup u_0,\ \text{weakly in}\ L^2(\partial\Omega,\mathbb{C}),\ \text{as}\ n \to \infty.$$

On the other hand, we have from (23.6), from the generalized Hölder inequality and for constants $\gamma = \frac{4\pi}{c}\left|\frac{-ie\hbar}{2m}\right| > 0$ and $\gamma_1 = \frac{4\pi}{c}\frac{2e^2}{mc} > 0$ that

$$\begin{aligned} 0 = \rho_{1,n} &\equiv \langle curl\ \mathbf{A}_n,\ curl\ \mathbf{A}_n\rangle_{L^2(\Omega_1;\mathbb{R}^3)} \\ &\quad - \langle curl\ \mathbf{A}_n, \mathbf{B}_0\rangle_{L^2(\Omega_1;\mathbb{R}^3)} \\ &\quad + \frac{4\pi}{c}\left\langle \frac{ie\hbar}{2m}(\phi^*\nabla\phi - \phi\nabla\phi^*) + \frac{2e^2}{mc}|\phi|^2\mathbf{A}_n, \mathbf{A}_n\right\rangle_{L^2(\Omega,\mathbb{R}^3)} \\ &\geq \langle curl\ \mathbf{A}_n,\ curl\ \mathbf{A}_n\rangle_{L^2(\Omega_1,\mathbb{R}^3)} \\ &\quad - \|curl\ \mathbf{A}_n\|_{0,2,\Omega_1}\|\mathbf{B}_0\|_{0,2,\Omega_1} - \gamma\|\mathbf{A}_n\|_{0,4,\Omega_1}\|\phi_n\|_{0,4,\Omega}\|\nabla\phi_n\|_{0,2,\Omega} \\ &\quad + \gamma_1\langle|\phi|^2, \mathbf{A}_n\cdot\mathbf{A}_n\rangle_{L^2(\Omega,\mathbb{R}^3)}. \end{aligned} \tag{23.7}$$

From the well known Friedrichs Inequality and the Sobolev Spaces Imbedding theorem for appropriate constants indicated, we obtain

$$
\begin{aligned}
\|\mathbf{A}_n\|_{0,4,\Omega_1}^2 &\leq K_3 \|\mathbf{A}_n\|_{1,2,\Omega_1}^2 \leq K_4 \left(\| \operatorname{div} \mathbf{A}_n \|_{0,2,\Omega_1} + \| \operatorname{curl} \mathbf{A}_n \|_{0,2,\Omega} \right)^2 \\
&= K_4 \| \operatorname{curl} \mathbf{A}_n \|_{0,2,\Omega_1}^2,
\end{aligned}
\tag{23.8}
$$

since from the London Gauge assumption,

$$
\operatorname{div} \mathbf{A}_n = 0, \text{ in } \Omega_1, \ \forall n \in \mathbb{N}.
$$

Summarizing, we have obtained, for some appropriate $K_5 > 0$,

$$
\begin{aligned}
0 = \rho_{1,n} &\geq K_5 \|\mathbf{A}_n\|_{0,4,\Omega_1}^2 + \frac{1}{2} \| \operatorname{curl} \mathbf{A}_n \|_{0,2,\Omega_1}^2 \\
&\quad - \| \operatorname{curl} \mathbf{A}_n \|_{0,2,\Omega_1} \| \mathbf{B}_0 \|_{0,2,\Omega_1} - \gamma \|\mathbf{A}_n\|_{0,4,\Omega_1} K^2 \\
&\quad + \gamma_1 \left\langle |\phi|^2, \mathbf{A}_n \cdot \mathbf{A}_n \right\rangle_{L^2(\Omega;\mathbb{R}^3)} \\
&\equiv \rho_{2,n}.
\end{aligned}
\tag{23.9}
$$

Now, suppose to obtain contradiction there exists a subsequence $\{n_k\} \subset \mathbb{N}$ such that

$$
\|\mathbf{A}_{n_k}\|_{0,4,\Omega_1} \to \infty, \text{ as } k \to \infty.
$$

From (23.9) we obtain

$$
\rho_{2,n_k} \to \infty, \text{ as } k \to \infty,
$$

which contradicts

$$
\rho_{2,n} \leq 0, \forall n \in \mathbb{N}.
$$

Hence, there exists $K_6 > 0$ such that

$$
\|\mathbf{A}_n\|_{0,4,\Omega_1} < K_6,
$$

and

$$
\|\mathbf{A}_n\|_{0,2,\Omega_1} < K_6, \ \forall n \in \mathbb{N}.
$$

From this and (23.7) we have,

$$
\begin{aligned}
0 = \rho_{1,n} &\geq \| \operatorname{curl} \mathbf{A}_n \|_{0,2,\Omega_1}^2 \\
&\quad - \| \operatorname{curl} \mathbf{A}_n \|_{0,2,\Omega_1} \| \mathbf{B}_0 \|_{0,2,\Omega_1} - \gamma K_6 K^2 \\
&\quad + \gamma_1 \left\langle |\phi|^2, \mathbf{A}_n \cdot \mathbf{A}_n \right\rangle_{L^2(\Omega,\mathbb{R}^3)} \\
&\equiv \rho_{3,n}.
\end{aligned}
\tag{23.10}
$$

Suppose to obtain contradiction there exists a subsequence $\{n_k\} \subset \mathbb{N}$ such that

$$
\| \operatorname{curl} \mathbf{A}_{n_k} \|_{0,2,\Omega_1} \to \infty, \text{ as } k \to \infty.
$$

From (23.10) we obtain

$$
\rho_{3,n_k} \to \infty, \text{ as } k \to \infty,
$$

which contradicts

$$
\rho_{3,n} \leq 0, \forall n \in \mathbb{N}.
$$

Hence, there exists $K_7 > 0$ such that

$$
\| \operatorname{curl} \mathbf{A}_n \|_{0,2,\Omega_1} < K_7, \ \forall n \in \mathbb{N}
$$

so that from this, the Friedrichs inequality and the London Gauge hypothesis, we obtain $K_8 > 0$ such that

$$\|\mathbf{A}_n\|_{1,2,\Omega_1} < K_8.$$

So from such a result and the Rellich-Kondrashov Theorem there exists a not relabeled subsequence and $\mathbf{A}_0 \in W^{1,2}(\Omega_1, \mathbb{R}^3)$ such that

$$\mathbf{A}_n \rightharpoonup \mathbf{A}_0 \text{ weakly } \in W^{1,2}(\Omega_1, \mathbb{R}^3)$$

$$\mathbf{A}_n \to \mathbf{A}_0 \text{ in norm in } L^2(\Omega_1, \mathbb{R}^3) \text{ and } L^4(\Omega_1, \mathbb{R}^3).$$

Moreover, from the Sobolev Imbedding Theorem, there exist real constants $\hat{K} > 0$, $\hat{K}_1 > 0$ such that

$$\|\phi_n\|_{0,6,\Omega} \leq \hat{K}\|\phi_n\|_{1,2,\Omega} < \hat{K}_1, \ \forall n \in \mathbb{N}.$$

Thus, from this and the first equation in (23.5), there exist real constants $\hat{K}_2 > 0, \ldots, \hat{K}_6 > 0$, such that

$$
\begin{aligned}
\|\nabla^2 \phi_n\|_{0,2,\Omega} \ < \ & \hat{K}_2 \|\mathbf{A}_n\|_{0,2,\Omega} \|\nabla \phi_n\|_{0,2,\Omega} \\
& \hat{K}_3 \|\mathbf{A}_n\|_{0,4,\Omega} \|\phi_n\|_{0,2,\Omega} + \hat{K}_4 \|\phi_n\|_{0,6,\Omega}^3 + \hat{K}_5 \|\phi_n\|_{0,2,\Omega} \\
\leq \ & \hat{K}_6, \ \forall n \in \mathbb{N}.
\end{aligned}
\tag{23.11}
$$

From this, up to a subsequence, we get

$$\nabla^2 \phi_n \rightharpoonup \nabla^2 \phi_0 \text{ weakly in } L^2(\Omega; \mathbb{C}).$$

Let

$$\varphi \in C_c^\infty(\Omega, \mathbb{C}), \ \varphi_1 \in C_c^\infty(\Omega, \mathbb{R}^3) \text{ and } \varphi_2 \in C_c^\infty(\Omega_1 \setminus \Omega, \mathbb{R}^3).$$

From the last results, we may easily obtain the following limits

1.
$$\langle \nabla^2 \phi_n, \varphi \rangle_{L^2} \to \langle \nabla^2 \phi_0, \varphi \rangle_{L^2},$$

2.
$$\langle \nabla \phi_n, \nabla \varphi \rangle_{L^2} \to \langle \nabla \phi_0, \nabla \varphi \rangle_{L^2},$$

3.
$$\langle \mathbf{A}_n \cdot \nabla \phi_n, \varphi \rangle_{L^2} \to \langle \mathbf{A}_0 \cdot \nabla \phi_0, \varphi \rangle_{L^2},$$

4.
$$\langle |\mathbf{A}_n|^2 \phi_n, \varphi \rangle_{L^2} \to \langle |\mathbf{A}_0|^2 \phi_0, \varphi \rangle_{L^2},$$

5.
$$\langle |\phi_n|^2 \phi_n, \varphi \rangle_{L^2} \to \langle |\phi_0|^2 \phi_0, \varphi \rangle_{L^2},$$

6.
$$\langle \text{ curl } \mathbf{A}_n, \text{ curl } \varphi_1 \rangle_{L^2} \to \langle \text{ curl } \mathbf{A}_0, \text{ curl } \varphi_1 \rangle_{L^2},$$

7.
$$\langle \phi_n^* \nabla \phi_n, \varphi_1 \rangle_{L^2} \to \langle \phi_0^* \nabla \phi_0, \varphi_1 \rangle_{L^2},$$

8.
$$\langle \phi_n \nabla \phi_n^*, \varphi_1 \rangle_{L^2} \to \langle \phi_0 \nabla \phi_0^*, \varphi_1 \rangle_{L^2},$$

9.
$$\langle |\phi_n|^2 \mathbf{A}_n, \varphi_1 \rangle_{L^2} \to \langle |\phi_0|^2 \mathbf{A}_0, \varphi_1 \rangle_{L^2}.$$

For example, for (4), for an appropriate real $\tilde{K} > 0$ we have

$$
\begin{aligned}
& |\langle |\mathbf{A}_n|^2 \phi_n, \varphi \rangle_{L^2} - \langle |\mathbf{A}_0|^2 \phi_0, \varphi \rangle_{L^2}| \\
= \ & |\langle |\mathbf{A}_n|^2 \phi_n, \varphi \rangle_{L^2} - \langle |\mathbf{A}_0|^2 \phi_n, \varphi \rangle_{L^2} + \langle |\mathbf{A}_0|^2 \phi_n, \varphi \rangle_{L^2} - \langle |\mathbf{A}_0|^2 \phi_0, \varphi \rangle_{L^2}| \\
\leq \ & |\langle |(\mathbf{A}_n|^2 - |\mathbf{A}_0|^2) \phi_n, \varphi \rangle_{L^2} + \langle |\mathbf{A}_0|^2 (\phi_n - \phi_0), \varphi \rangle_{L^2}| \\
\leq \ & |\langle |(\mathbf{A}_n| - |\mathbf{A}_0|)(|\mathbf{A}_n| + |\mathbf{A}_0|) \phi_n, \varphi \rangle_{L^2} + \langle |\mathbf{A}_0|^2 (\phi_n - \phi_0), \varphi \rangle_{L^2}| \\
\leq \ & \||(\mathbf{A}_n| + |\mathbf{A}_0|)\|_{0,4,\Omega} \||\mathbf{A}_n| - |\mathbf{A}_0|\|_{0,4,\Omega} |\phi_n|_{0,2,\Omega} \|\varphi\|_\infty + \||\mathbf{A}_0|^2\|_{0,2,\Omega} \|\phi_n - \phi_0\|_{0,2,\Omega} \|\varphi\|_\infty \\
\leq \ & \tilde{K}(\||\mathbf{A}_n| - |\mathbf{A}_0|\|_{0,4,\Omega} + \|\phi_n - \phi_0\|_{0,2,\Omega}) \\
\to \ & 0, \text{ as } n \to \infty.
\end{aligned}
\tag{23.12}
$$

The other items may be proven similarly.
Now let $\varphi \in C^\infty(\overline{\Omega}, \mathbb{C})$. Observe that

$$
\begin{aligned}
& \langle u_n, \varphi \rangle_{L^2(\partial\Omega, \mathbb{C})} \\
= \ & \left\langle \left(i\hbar\nabla\phi_n + \frac{2e}{c}\mathbf{A}_n\phi_n \right) \cdot \mathbf{n}, \varphi \right\rangle_{L^2(\partial\Omega, \mathbb{C})} \\
= \ & \left\langle i\hbar\nabla\phi_n + \frac{2e}{c}\mathbf{A}_n\phi_n, \nabla\varphi \right\rangle_{L^2(\Omega, \mathbb{C}^3)} \\
& + \left\langle \operatorname{div}\left(i\hbar\nabla\phi_n + \frac{2e}{c}\mathbf{A}_n\phi_n \right), \varphi \right\rangle_{L^2(\Omega, \mathbb{C})} \\
\to \ & \left\langle i\hbar\nabla\phi_0 + \frac{2e}{c}\mathbf{A}_0\phi_0, \nabla\varphi \right\rangle_{L^2(\Omega, \mathbb{C}^3)} \\
& + \left\langle \operatorname{div}\left(i\hbar\nabla\phi_0 + \frac{2e}{c}\mathbf{A}_0\phi_0 \right), \varphi \right\rangle_{L^2(\Omega, \mathbb{C})} \\
= \ & \left\langle \left(i\hbar\nabla\phi_0 + \frac{2e}{c}\mathbf{A}_0\phi_0 \right) \cdot \mathbf{n}, \varphi \right\rangle_{L^2(\partial\Omega, \mathbb{C})}.
\end{aligned}
\tag{23.13}
$$

From this and from

$$
\langle u_n, \varphi \rangle_{L^2(\partial\Omega, \mathbb{C})} \to \langle u_0, \varphi \rangle_{L^2(\partial\Omega, \mathbb{C})},
$$

we have

$$
\left\langle \left(i\hbar\nabla\phi_0 + \frac{2e}{c}\mathbf{A}_0\phi_0 \right) \cdot \mathbf{n} - u_0, \varphi \right\rangle_{L^2(\partial\Omega, \mathbb{C})} = 0, \ \forall \varphi \in C^\infty(\overline{\Omega}, \mathbb{C}),
$$

so that in such a distributional sense,

$$
\left(i\hbar\nabla\phi_0 + \frac{2e}{c}\mathbf{A}_0\phi_0 \right) \cdot \mathbf{n} = u_0, \text{ on } \partial\Omega.
$$

The other boundary condition may be dealt similarly. Thus, from these last results we may infer that in the distributional sense,

$$
\begin{cases}
\frac{1}{2m}\left(-i\hbar\nabla - \frac{2e}{c}\mathbf{A}_0 \right)^2 \phi_0 + \alpha|\phi_0|^2\phi_0 - \beta\phi_0 = 0, & \text{in } \Omega, \\[2mm]
\left(i\hbar\nabla\phi_0 + \frac{2e}{c}\mathbf{A}_0\phi_0 \right) \cdot \mathbf{n} = u_0, & \text{on } \partial\Omega,
\end{cases}
\tag{23.14}
$$

and

$$\begin{cases} \text{curl curl } \mathbf{A}_0 = \text{curl } \mathbf{B}_0 + \frac{4\pi}{c}\tilde{J}_0, & \text{in } \Omega, \\[2mm] \text{curl curl } \mathbf{A}_0 = \text{curl } \mathbf{B}_0, & \text{in } \Omega_1 \setminus \Omega, \\[2mm] \text{div } \mathbf{A}_0 = 0, & \text{in } \Omega_1, \\[2mm] \mathbf{A}_0 \cdot \mathbf{n} = 0, & \text{on } \partial\Omega_1 \end{cases} \tag{23.15}$$

where

$$\tilde{J}_0 = -\frac{ie\hbar}{2m}(\phi_0^*\nabla\phi_0 - \phi_0\nabla\phi_0^*) - \frac{2e^2}{mc}|\phi_0|^2\mathbf{A}_0.$$

Hence, $(\phi_0, \mathbf{A}_0, u_0) \in \mathscr{C}$.

Finally, from $\phi_n \to \phi_0$ in L^2 and L^4, $\phi_n \rightharpoonup \phi_0$ weakly in $W^{1,2}$, $u_n \rightharpoonup u_0$ weakly in $L^2(\partial\Omega)$, by continuity in ϕ and the convexity of J in $\nabla\phi$ and u, we have,

$$\eta = \liminf_{n\to\infty} J(\phi_n, \mathbf{A}_n, u_n) \geq J(\phi_0, \mathbf{A}_0, u_0).$$

The proof is complete.

23.3 A method to obtain approximate numerical solutions for a class of partial differential equations

In this section we present a new procedure to obtain approximate solutions for a large class of partial differential equations.

We emphasize there is some error in the process, so that the solutions obtained are just qualitative (but indeed of very good quality as a first approximation).

Let $\Omega = [0,1] \times [0,1]$ and consider the Ginzburg-Landau type equation

$$\begin{cases} -\varepsilon\nabla^2 u + \alpha u^3 - \beta u = f, & \text{in } \Omega, \\ u = 0, & \text{on } \partial\Omega, \end{cases} \tag{23.16}$$

where $\varepsilon > 0$, $\alpha > 0$ and $\beta > 0$.

For the generalized method of lines, we discretize the domain in x, in N vertical lines, defining $d = 1/N$, so that the system (23.16) will approximately stand for,

$$\varepsilon\frac{u_{n+1} - 2u_n + u_{n-1}}{d^2} + \varepsilon\frac{\partial^2 u_n}{\partial y^2} - \alpha u_n^3 + \beta u_n + f_n = 0,$$

$\forall n \in \{1, ..., N-1\}$.

From this, we may write,

$$u_{n+1} - 2u_n + u_{n-1} + T(u_n)\frac{d^2}{\varepsilon} + f_n\frac{d^2}{\varepsilon} = 0, \tag{23.17}$$

where

$$T(u_n) = \frac{\partial^2 u_n}{\partial y^2} - \alpha u_n^3 + \beta u_n, \forall n \in \{1, ..., N-1\}.$$

In particular, for $n = 1$, recalling that $u_0 = 0$, we obtain

$$u_2 - 2u_1 + T(u_1)\frac{d^2}{\varepsilon} + f_1\frac{d^2}{\varepsilon} = 0,$$

so that

$$u_1 = \frac{u_2}{2} + \frac{1}{2}T(u_1)\frac{d^2}{\varepsilon} + \frac{1}{2}f_1\frac{d^2}{\varepsilon},$$

that is,

$$u_1 = a_1 u_2 + b_1 T(u_1)\frac{d^2}{\varepsilon} + c_1\frac{d^2}{\varepsilon} + E_1,$$

where

$$a_1 = \frac{1}{2}, \ b_1 = \frac{1}{2}, \ c_1 = \frac{1}{2}f_1 \text{ and } E_1 = 0.$$

Reasoning inductively, having

$$u_{n-1} = a_{n-1}u_n + b_{n-1}T(u_{n-1})\frac{d^2}{\varepsilon} + c_{n-1}\frac{d^2}{\varepsilon} + E_{n-1},$$

from this and (23.17), we have

$$u_{n+1} - 2u_n + a_{n-1}u_n + b_{n-1}T(u_{n-1})\frac{d^2}{\varepsilon} + c_{n-1}\frac{d^2}{\varepsilon} + E_{n-1}$$

$$+ T(u_n)\frac{d^2}{\varepsilon} + f_n\frac{d^2}{\varepsilon} = 0, \tag{23.18}$$

so that

$$u_n = a_n u_{n+1} + b_n T(u_n)\frac{d^2}{\varepsilon} + c_n\frac{d^2}{\varepsilon} + E_n$$

where

$$a_n = \frac{1}{2 - a_{n-1}},$$

$$b_n = a_n(b_{n-1} + 1),$$

$$c_n = a_n(c_{n-1} + f_n),$$

and

$$E_n = a_n E_{n-1} + a_n b_{n-1}(T(u_{n-1}) - T(u_n))\frac{d^2}{\varepsilon},$$

$\forall n \in \{2, \ldots, N-1\}$.

In particular, for $n = N-1$, recalling that $u_N = 0$, we may obtain u_{N-1} as the solution of the approximate ordinary differential equation,

$$u_{N-1} \approx a_{N-1}u_N + b_{N-1}T(u_{N-1})\frac{d^2}{\varepsilon} + c_{N-1}\frac{d^2}{\varepsilon} \tag{23.19}$$

that is,

$$u_{N-1} \approx b_{N-1}\left(\frac{\partial^2 u_{N-1}}{\partial y^2} - \alpha u_{N-1}^3 + \beta u_{N-1}\right)\frac{d^2}{\varepsilon} + c_{N-1}\frac{d^2}{\varepsilon},$$

with the boundary conditions $u_{N-1}(0) = u_{N-1}(1) = 0$.

Similarly, having u_{N-1}, we may obtain u_{N-2} through the approximate equation,

$$u_{N-2} \approx a_{N-2}u_{N-1} + b_{N-2}T(u_{N-2})\frac{d^2}{\varepsilon} + c_{N-2}\frac{d^2}{\varepsilon},$$

that is,

$$u_{N-2} \approx a_{N-2}u_{N-1} + b_{N-2}\left(\frac{\partial^2 u_{N-2}}{\partial y^2} - \alpha u_{N-2}^3 + \beta u_{N-2}\right)\frac{d^2}{\varepsilon} + c_{N-2}\frac{d^2}{\varepsilon},$$

and so on, up to finding u_1.

The problem is thus approximately solved. We present numerical results for $\alpha = \beta = 1$ and $N = 500$ lines (mesh 500×500).

For $\varepsilon = 1, 0.1, 0.01$ and 0.001, please see the Figures 23.1, 23.2, 23.3, 23.4 and 23.5, for the respective graphs.

We observe that for small values of ε and in particular for $\varepsilon = 0.0001$, the solution $u(x,y)$ is close to the constant value 1.32 on the almost whole domain, which is the approximate solution of equation $u^3 - u - 1 = 0$.

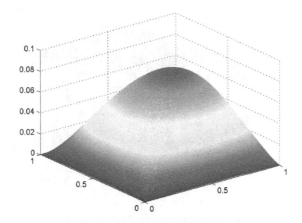

Figure 23.1: Solution $u(x,y)$ for $\varepsilon = 1$.

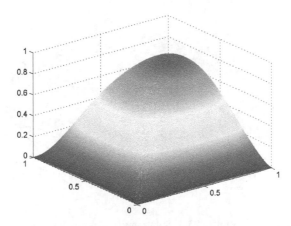

Figure 23.2: Solution $u(x,y)$ for $\varepsilon = 0.1$.

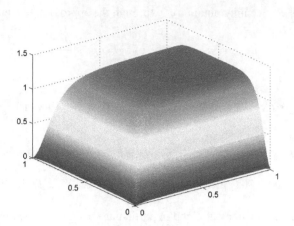

Figure 23.3: Solution $u(x,y)$ for $\varepsilon = 0.01$.

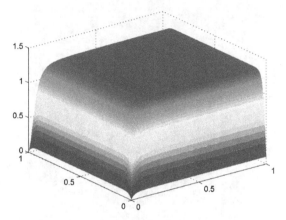

Figure 23.4: Solution $u(x,y)$ for $\varepsilon = 0.001$.

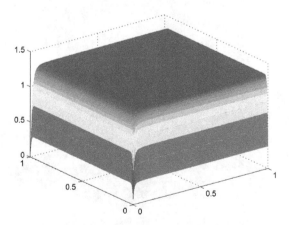

Figure 23.5: Solution $u(x,y)$ for $\varepsilon = 0.0001$.

23.4   Conclusion

In this chapter we have developed a global existence result for a control problem related to the Ginzburg-Landau system in superconductivity. We emphasize the control variable u acts on the super-conducting sample boundary, whereas the state variables, namely, the order parameter ϕ and the magnetic potential $\mathbf{A}$ are defined on Ω and Ω_1, respectively. The problem has non-linear constraints and the cost functional is non-convex. Finally, we highlight the London Gauge assumption and the Friedrichs Inequality have a fundamental role in the establishment of the main results.

Chapter 24

Duality for a Semi-Linear Model in Micro-Magnetism

24.1 Introduction

This chapter develops a dual variational formulation for a semi-linear model in micro-magnetism. For the primal formulation we refer to references [14, 15] for details. In particular we refer to the original results presented in [15], emphasizing the present work is their natural continuation and extension. We also highlight the present work develops real relevant improvements relating the previous similar results in [14].

At this point we start to describe the primal formulation.

Let $\Omega \subset \mathbb{R}^3$ be an open bounded set with a a regular (lipschitzian) boundary denoted by $\partial\Omega$. By a regular lipschitzian boundary $\partial\Omega$ we mean regularity enough so that the Sobolev imbedding theorem and relating results, the trace theorem and the standard Gauss-Green formulas of integration by parts to hold. The corresponding outward normal to $\partial\Omega$ is denoted by $\mathbf{n} = (n_1, n_2, n_3)$. Also, we denote by $\mathbf{0}$ either the zero vector in $\mathbb{R}^3$ or the zero in an appropriate function space.

Under such assumptions and notations, consider the problem of finding the magnetization $m : \Omega \to \mathbb{R}^3$ which minimizes the functional

$$
\begin{aligned}
J(m, f) &= \frac{\alpha}{2} \int_\Omega |\nabla m|_2^2 \, dx + \int_\Omega \varphi(m(x)) \, dx - \int_\Omega H(x) \cdot m \, dx \\
&\quad + \frac{1}{2} \int_{\mathbb{R}^3} |f(x)|_2^2 \, dx,
\end{aligned}
\tag{24.1}
$$

where

$$
m = (m_1, m_2, m_3) \in W^{1,2}(\Omega; \mathbb{R}^3) \equiv Y_1, \ |m(x)|_2 = 1, \text{ in } \Omega
\tag{24.2}
$$

and $f \in L^2(\mathbb{R}^3; \mathbb{R}^3) \equiv Y_2$ is the unique field determined by the simplified Maxwell's equations

$$
curl(f) = \mathbf{0}, \ div(-f + m\chi_\Omega) = 0, \text{ in } \mathbb{R}^3.
\tag{24.3}
$$

Here $H \in L^2(\Omega; \mathbb{R}^3)$ is a known external field and χ_Ω is a function defined by

$$
\chi_\Omega(x) = \begin{cases} 1, & \text{if } x \in \Omega, \\ 0, & \text{otherwise.} \end{cases}
\tag{24.4}
$$

The term

$$\frac{\alpha}{2} \int_{\Omega} |\nabla m|_2^2 \, dx$$

is called the exchange energy, where

$$|m|_2 = \sqrt{\sum_{k=1}^{3} m_k^2}$$

and

$$|\nabla m|_2^2 = \sum_{k=1}^{3} |\nabla m_k|_2^2.$$

Finally, $\varphi(m)$ represents the anisotropic contribution and is given by a multi-well functional whose minima establish the preferred directions of magnetization.

Remark 24.1.1 *Here some brief comments on the references. Relating and similar problems are addressed in [14]. The basic results on convex and variational analysis used in this text may be found in [33, 14, 64, 78]. About the duality principles, we have been greatly inspired and influenced by the work of J.J. Telega and W.R. Bielski. In particular, we would refer to [11], published in 1985, as the first article to successfully apply the convex analysis approach to non-convex and non-linear mechanics.*

Finally, an extensive study on Sobolev spaces may be found in [1].

Remark 24.1.2 *At some points of our analysis we refer to the problems in question after discretization. In such a case we are referring to their approximations in a finite element or finite differences context.*

24.2   The duality principle for the semi-linear model

We consider first the case of a uniaxial material where $\varphi(m) = \beta(1 - |m \cdot e|)$.

Observe that

$$\varphi(m) = \min\{\beta(1 + m \cdot e), \beta(1 - m \cdot e)\}$$

where $\beta > 0$ and $e \in \mathbb{R}^3$ is a unit vector.

The main duality principle is summarized by the following theorem.

Theorem 24.2.1 *Considering the previous statements and notations, define $J : Y_1 \times Y_2 \times B \to \overline{R} = \mathbb{R} \cup \{+\infty\}$ by*

$$
\begin{aligned}
J(m, f, t) \;=\; & G_0(m) - \frac{K}{2}\langle m_i, m_i \rangle_{L^2} + G_1(m, t) + G_2(f) \\
& + Ind_0(m) + Ind_1(m, f) + Ind_2(f),
\end{aligned}
\tag{24.5}
$$

where

$$G_0(m) = \frac{\alpha}{2}\langle \nabla m_i, \nabla m_i \rangle_{L^2},$$

$$G_1(m, t) = \int_{\Omega} (t g_1(m) + (1 - t) g_2(m)) \, dx - \langle H_i, m_i \rangle_{L^2} + \frac{K}{2}\langle m_i, m_i \rangle_{L^2},$$

$$G_2(f) = \frac{1}{2} \int_{\mathbb{R}^3} |f(x)|^2 \, dx,$$

$$g_1(m) = \beta(1 + m \cdot e),$$

$$g_2(m) = \beta(1 - m \cdot e),$$

$$Ind_0(m) = \begin{cases} 0, & if\ |m(x)|_2 = 1,\ in\ \Omega \\ +\infty, & otherwise, \end{cases} \tag{24.6}$$

$$Ind_1(m,f) = \begin{cases} 0, & if\ div(-f + m\chi_\Omega) = 0,\ in\ \mathbb{R}^3 \\ +\infty, & otherwise, \end{cases} \tag{24.7}$$

$$Ind_2(f) = \begin{cases} 0, & if\ curl\ f = \mathbf{0},\ in\ \mathbb{R}^3 \\ +\infty, & otherwise. \end{cases} \tag{24.8}$$

We recall the present case refers to a uniaxial material with exchange of energy, that is $\alpha > 0$.
Here, $e = (e_1, e_2, e_3) \in \mathbb{R}^3$ is a unit vector.
Under such hypotheses, we have,

$$\inf_{(m,f,t)\in Y_1\times Y_2\times B} \{J(m,f,t)\} \geq \sup_{\lambda\in A^*} \{\tilde{J}^*(\lambda)\},$$

where

$$\tilde{J}^*(\lambda) = \inf_{(z^*,t)\in Y_4^*(\lambda)\times B} J^*(\lambda, z^*, t),$$

$$J^*(\lambda, z^*) = \tilde{F}^*(z^*) - \tilde{G}^*(\lambda, z^*, t),$$

and where, for the discretized problem version,

$$\tilde{F}^*(z^*) = \sup_{m\in Y_1} \left\{ \langle z_i^*, \nabla m_i\rangle_{L^2} + G_0(m) - \frac{K}{2}\langle m_i, m_i\rangle_{L^2} \right\}. \tag{24.9}$$

Here $K > 0$ is such that

$$-G_0(m) + \frac{K}{2}\langle m_i, m_i\rangle_{L^2} > 0,\ \forall m\in Y_1,\ m \neq \mathbf{0}.$$

$$\begin{aligned} \tilde{G}^*(\lambda, z^*, t) &= G_1^*(\lambda, z^*, t) + G_2^*(\lambda) \\ &= \sup_{(m,f)\in Y_1\times Y_2} \{\langle z_i^*, \nabla m_i\rangle_{L^2} + \langle\lambda_2, div(m\chi_\Omega) - f\rangle_{L^2} \\ &\quad + \langle\lambda_1, curl\ f\rangle_{L^2} \\ &\quad - G_1(m,t) - G_0(f) - \langle\lambda_3, (\sum_{i=1}^{3} m_i^2) - 1\rangle_{L^2}\} \end{aligned} \tag{24.10}$$

where, more specifically,

$$\begin{aligned} G_1^*(\lambda, z^*, t) &= \frac{1}{2}\int_\Omega \frac{\sum_{i=1}^{3}\left(-\frac{\partial\lambda_2}{\partial x_i} + H_i + \beta(1-2t)e_i - div\ z_i^*\right)^2}{\lambda_3 + K}\ dx \\ &\quad - \frac{1}{2}\int_\Omega \lambda_3\ dx + \frac{1}{2}\int_\Omega \beta\ dx, \end{aligned} \tag{24.11}$$

$$G_2^*(\lambda) = \frac{1}{2}\|\nabla\lambda_2 - curl^*\lambda_1\|_2^2.$$

Also,

$$A_1 = \{\lambda \in Y_3 : \lambda_3 + K > 0,\ in\ \Omega\},$$

and from the standard second order sufficient optimality condition for a local minimum in m, we define

$$\begin{aligned} A_2 = \{&\lambda \in Y_3 : \\ &G_0(m) + \langle\lambda_3, \sum_{i=1}^{3} m_i^2\rangle_{L^2} > 0, \\ &\forall m\in Y_1,\ such\ that\ m \neq \mathbf{0}\} \end{aligned} \tag{24.12}$$

where

$$A^* = A_1 \cap A_2,$$

$$Y = Y^* = L^2(\Omega; \mathbb{R}^3) = L^2,$$

$$\lambda = (\lambda_1, \lambda_2, \lambda_3) \in Y_3 = W^{1,2}(\mathbb{R}^3; \mathbb{R}^3) \times W^{1,2}(\mathbb{R}^3) \times L^2(\Omega),$$

$$Y_1 = W^{1,2}(\Omega; \mathbb{R}^3),$$

$$Y_2 = W^{1,2}(\mathbb{R}^3; \mathbb{R}^3),$$

$$Y_4^*(\lambda) = \{z^* \in [Y^*]^3 \ : \ z_i^* \cdot \mathbf{n} + \lambda_2 n_i = 0, \ on \ \partial\Omega, \forall i \in \{1,2,3\}\},$$

$$B = \{t \ measurable \ : \ 0 \le t \le 1, \ in \ \Omega\}.$$

Finally, suppose there exists $(\lambda_0, z_0^*, t_0) \in A^* \times Y_4^*(\lambda_0) \times B$ *such that for an appropriate* $\lambda_4 \in L^2(\Omega)$ *we have*

$$\delta \left[J^*(\lambda_0, z_0^*, t_0) + \langle \lambda_4, t_0^2 - t_0 \rangle_{L^2} \right] = \mathbf{0},$$

$$\delta_{z^* z^*}^2 J^*(\lambda_0, z_0^*, t_0) > \mathbf{0}$$

and for the concerning Hessian

$$\det \left\{ \delta_{z^*, t}^2 \left[J^*(\lambda_0, z_0^*, t_0) + \langle \lambda_4, t_0^2 - t_0 \rangle_{L^2} \right] \right\} > 0, \ in \ \Omega.$$

Under such hypotheses,

$$
\begin{aligned}
J(m_0, f_0, t_0) &= \inf_{(m,f,t) \in Y \times Y_1 \times B} J(m, f, t) \\
&= \sup_{\lambda \in A^*} \tilde{J}^*(\lambda) \\
&= \tilde{J}^*(\lambda_0) \\
&= J^*(\lambda_0, z_0^*, t_0).
\end{aligned}
\tag{24.13}
$$

Proof 24.1 Observe that

$$
\begin{aligned}
& G_1^*(\lambda, z^*, t) + G_2^*(\lambda) \\
\ge \ & \langle z_i^*, \nabla m_i \rangle_{L^2} + \langle \lambda_2, div(m\chi_\Omega) - f) \rangle_{L^2} \\
& + \langle \lambda_1, curl f \rangle_{L^2} \\
& - \left\langle \lambda_3, \sum_{i=1}^3 m_i^2 - 1 \right\rangle_{L^2} - G_1(m, t) - G_2(f) \\
\ge \ & \langle z_i^*, \nabla m_i \rangle_{L^2} - G_1(m, t) - G_2(f) - Ind_0(m) - Ind_1(m, f) - Ind_2(f),
\end{aligned}
\tag{24.14}
$$

$\forall (m, f, t) \in Y_1 \times Y_2 \times B, z^* \in Y_4^*(\lambda), \ \lambda \in A^*$ *so that,*

$$
\begin{aligned}
& -\tilde{F}^*(z^*) + G_1^*(\lambda, z^*, t) + G_2^*(\lambda) \\
\ge \ & -\tilde{F}^*(z^*) + \langle z_i^*, \nabla m_i \rangle_{L^2} - G_1(m, t) - G_2(f) \\
& - Ind_0(m) - Ind_1(m, f) - Ind_2(f),
\end{aligned}
\tag{24.15}
$$

$\forall (m, f, t) \in Y_1 \times Y_2 \times B, z^* \in Y_4^*(\lambda), \ \lambda \in A^*$ *and hence,*

$$
\begin{aligned}
& \sup_{z^* \in Y_4^*(\lambda)} \{ -\tilde{F}^*(z^*) + G_1^*(\lambda, z^*, t) + G_2^*(\lambda) \} \\
\ge \ & \sup_{z^* \in Y_4^*(\lambda)} \{ -\tilde{F}^*(z^*) + \langle z_i^*, \nabla m_i \rangle_{L^2} \} - G_1(m, t) - G_2(f) \\
& - Ind_0(m) - Ind_1(m, f) - Ind_2(f),
\end{aligned}
\tag{24.16}
$$

$\forall (m,f,t) \in Y_1 \times Y_2 \times B$, $\lambda \in A^*$ so that that is,

$$\sup_{z^* \in Y_4^*(\lambda)} \{-\tilde{F}^*(z^*) + G_1^*(\lambda,z^*,t) + G_2^*(\lambda)\}$$

$$\geq \quad -G_0(m) + \frac{K}{2}\langle m_i, m_i \rangle_{L^2} - G_1(m,t) - G_2(f)$$

$$-Ind_0(m) - Ind_1(m,f) - Ind_2(f)$$

$$= \quad -J(m,f,t), \tag{24.17}$$

$\forall (m,f,t) \in Y_1 \times Y_2 \times B$, $\lambda \in A^*$. Thus,

$$J(m,f,t) \quad \geq \quad \inf_{z^* \in Y_4^*(\lambda)} \{\tilde{F}^*(z^*) - G_1^*(\lambda,z^*,t) + G_2^*(\lambda)\}$$

$$= \quad \inf_{z^* \in Y_4^*(\lambda)} J^*(\lambda,z^*,t), \forall (m,f,t) \in Y_1 \times Y_1 \times B, \lambda \in A^*. \tag{24.18}$$

Therefore,

$$\inf_{(m,f,t) \in Y_1 \times Y_2 \times B} J(m,f,t) \quad \geq \quad \sup_{\lambda \in A^*} \left\{ \inf_{(z^*,t) \in Y_4^*(\lambda) \times B} J^*(\lambda,z^*,t) \right\}$$

$$= \quad \sup_{\lambda \in A^*} \tilde{J}^*(\lambda). \tag{24.19}$$

By the hypotheses, $(\lambda_0, z_0^*, t_0) \in A^* \times Y_4^*(\lambda_0) \times B$ is such that

$$\delta\{J^*(\lambda_0, z_0^*, t_0) + \langle \lambda_4, t_0^2 - t_0 \rangle_{L^2}\} = \mathbf{0}.$$

From the variation in z^*,

$$\frac{\partial \tilde{F}^*(z_0^*)}{\partial z_i^*} - \nabla(m_0)_i = 0, \text{ in } \Omega, \tag{24.20}$$

where

$$(m_0)_i = \frac{-\frac{\partial(\lambda_0)_2}{\partial x_i} + H_i + \beta(1+2t_0)e_i - div[(z_0^*)_i]}{(\lambda_0)_3 + K}. \tag{24.21}$$

From the variation in λ_3, we get

$$\sum_{i=1}^{3} (m_0)_i^2 = 1, \text{ in } \Omega.$$

From the variation in λ_2, we have

$$div(m_0 \chi_\Omega - f_0) = 0, \text{ in } \mathbb{R}^3, \tag{24.22}$$

where

$$f_0 = curl(\lambda_1)_0 - \nabla(\lambda_0)_2. \tag{24.23}$$

From the variation in λ_1, we obtain,

$$curl f_0 = \mathbf{0}, \text{ in } \mathbb{R}^3. \tag{24.24}$$

From (24.20), we also have

$$\tilde{F}^*(z_0^*) = \langle (z_0^*)_i, \nabla(m_0)_i \rangle_{L^2} + G_0(m_0) - \frac{K}{2}\langle (m_0)_i, (m_0)_i \rangle_{L^2}. \tag{24.25}$$

By (24.21), (24.22), (24.23) and (24.24), we get

$$
\begin{aligned}
& G_1^*(\lambda_0, z_0^*, t_0) + G_2^*(\lambda_0) \\
=\ & \langle (z_0^*)_i, \nabla(m_0)_i \rangle_{L^2} + \langle \lambda_2, div(m_0 \chi_\Omega) - f_0 \rangle_{L^2} \\
& + \langle (\lambda_0)_1, curl\, f_0 \rangle_{L^2} \\
& - \left\langle (\lambda_0)_3, \sum_{i=1}^{3} (m_0)_i^2 - 1 \right\rangle_{L^2} - G_1(m_0, t_0) - G_2(f_0) \\
=\ & \langle (z_0^*)_i, \nabla(m_0)_i \rangle_{L^2} - G_1(m_0, t_0) - G_2(f_0) \\
& - Ind_0(m_0) - Ind_1(m_0, f_0) - Ind_2(f_0),
\end{aligned}
\tag{24.26}
$$

From (24.25) and (24.26) we obtain,

$$
\begin{aligned}
& J^*(\lambda_0, z_0^*, t_0) \\
=\ & \tilde{F}^*(z_0^*) - G_1^*(\lambda_0, z_0^*, t_0) - G_2^*(\lambda_0) \\
=\ & G_0(m_0) + G_1(m_0, t_0) + G_2(f_0) \\
& + Ind_0(m_0) + Ind_1(m_0, f_0) + Ind_2(f_0) \\
=\ & J(m_0, f_0, t_0).
\end{aligned}
\tag{24.27}
$$

Finally, from the hypotheses

$$
\delta_{z^* z^*}^2 J^*(\lambda_0, z_0^*, t_0) > \mathbf{0}
$$

and

$$
\det \left\{ \delta_{z^*, t}^2 \left[J^*(\lambda_0, z_0^*, t_0) + \langle \lambda_4, t_0^2 - t_0 \rangle_{L^2} \right] \right\} > 0, \text{ in } \Omega.
$$

Since the optimization in question in (z^*, t) is quadratic, we obtain,

$$
\tilde{J}^*(\lambda_0) = J^*(\lambda_0, z_0^*, t_0).
$$

From this, (24.19) and (24.27), we have,

$$
\begin{aligned}
J(m_0, f_0, t_0) &= \inf_{(m,f,t) \in Y \times Y_1 \times B} J(m, f, t) \\
&= \sup_{\lambda \in A^*} \tilde{J}^*(\lambda) \\
&= \tilde{J}^*(\lambda_0) \\
&= J^*(\lambda_0, z_0^*, t_0).
\end{aligned}
\tag{24.28}
$$

The proof is complete.

24.3   Conclusion

In this chapter we have developed a duality principle for a non-convex semi-linear model in micro-magnetism.

In a second step we present an optimality criterion for the dual formulation. The results are obtained through standard tools of convex analysis and duality theory.

Chapter 25

About Numerical Methods for Ordinary and Partial Differential Equations

25.1 Introduction

This chapter develops numerical methods for a large class of non-linear ordinary and partial differential equations.

More specifically, such a chapter is concerned with a kind of matrix version of the Generalized Method of Lines. Applications are developed for models in physics and engineering.

25.2 On the numerical procedures for Ginzburg-Landau type ODEs

We first apply the Newton's method to a general class of ordinary differential equations. The solution here is obtained similarly as for the generalized method of lines procedure. See the next sections for details on such a method for PDEs.

For a C^1 class function f and a continuous function g, consider the second order equation

$$\begin{cases} u'' + f(u) + g = 0, \text{ in } [0,1] \\ \\ u(0) = u_0, \ u(1) = u_f. \end{cases} \tag{25.1}$$

In finite differences we have the approximate equation:

$$u_{n+1} - 2u_n + u_{n-1} + f(u_n)d^2 + g_n d^2 = 0.$$

Assuming such an equation is non-linear and linearizing it about a first solution $\{\tilde{u}\}$, we have (in fact this an approximation),

$$u_{n+1} - 2u_n + u_{n-1} + f(\tilde{u}_n)d^2 + f'(\tilde{u}_n)(u_n - \tilde{u}_n)d^2 + g_n d^2 = 0.$$

Thus we may write

$$u_{n+1} - 2u_n + u_{n-1} + A_n u_n d^2 + B_n d^2 = 0,$$

where

$$A_n = f'(\tilde{u}_n),$$

and

$$B_n = f(\tilde{u}_n) - f'(\tilde{u}_n)\tilde{u}_n + g_n.$$

In particular, for $n = 1$ we get

$$u_2 - 2u_1 + u_0 + A_1 u_1 d^2 + B_1 d^2 = 0.$$

Solving such an equation for u_1, we get

$$u_1 = a_1 u_2 + b_1 u_0 + c_1,$$

where

$$a_1 = (2 - A_1 d^2)^{-1}, \quad b_1 = a_1, \quad c_1 = a_1 B_1 d^2.$$

Reasoning inductively, having

$$u_{n-1} = a_{n-1} u_n + b_{n-1} u_0 + c_{n-1},$$

and

$$u_{n+1} - 2u_n + u_{n-1} + A_n u_n d^2 + B_n d^2 = 0,$$

we get

$$u_{n+1} - 2u_n + a_{n-1} u_n + b_{n-1} u_0 + c_{n-1} + A_n u_n d^2 + B_n d^2 = 0,$$

so that

$$u_n = a_n u_{n+1} + b_n u_0 + c_n,$$

where

$$a_n = (2 - a_{n-1} - A_n d^2)^{-1},$$
$$b_n = a_n b_{n-1},$$

and

$$c_n = a_n(c_{n-1} + B_n d^2),$$

$\forall n \in 1, ..., N - 1$.

We have thus obtained

$$u_n = a_n u_{n+1} + b_n u_0 + c_n \equiv H_n(u_{n+1}), \forall n \in \{1, ..., N - 1\},$$

and in particular,

$$u_{N-1} = H_{N-1}(u_f),$$

so that we may calculate,

$$u_{N-2} = H_{N-2}(u_{N-1}),$$
$$u_{N-3} = H_{N-3}(u_{N-2}),$$

and so on, up to finding,

$$u_1 = H_1(u_2).$$

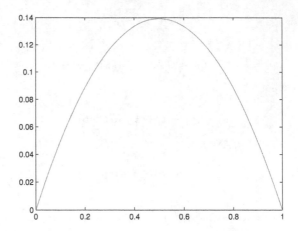

Figure 25.1: The solution $u(x)$ by Newton's method for $\varepsilon = 1$.

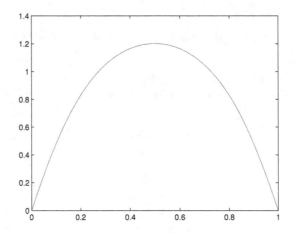

Figure 25.2: The solution $u(x)$ by Newton's method for $\varepsilon = 0.1$.

The next step is to replace $\{\tilde{u}_n\}$ by the $\{u_n\}$ calculated, and repeat the process up to the satisfaction of a appropriate convergence criterion. We present numerical results for the equation,

$$\begin{cases} u'' - \dfrac{u^3}{\varepsilon} + \dfrac{u}{\varepsilon} + g = 0, \text{ in } [0,1] \\ u(0) = 0, \ u(1) = 0, \end{cases} \tag{25.2}$$

where,

$$g(x) = \frac{1}{\varepsilon},$$

The results are obtained for $\varepsilon = 1.0$, $\varepsilon = 0.1$, $\varepsilon = 0.01$ and $\varepsilon = 0.001$. Please see Figures 25.1, 25.2, 25. and 25.4 respectively.

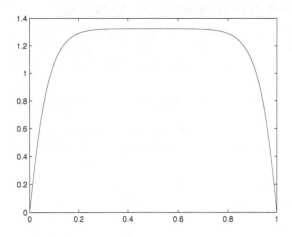

Figure 25.3: The solution $u(x)$ by Newton's method for $\varepsilon = 0.01$.

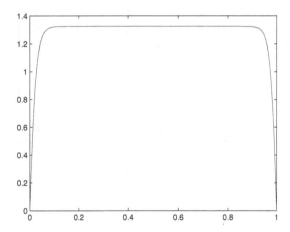

Figure 25.4: The solution $u(x)$ by Newton's method for $\varepsilon = 0.001$.

25.3 Numerical results for related P.D.E.s

25.3.1 A related P.D.E on a special class of domains

We start by describing a similar equation, but now in a two dimensional context. Let $\Omega \subset \mathbb{R}^2$ be an open, bounded, connected set with a regular boundary denoted by $\partial\Omega$. Consider a real Ginzburg-Landau type equation (see [4], [9], [55], [56] for details about such an equation), given by

$$
\begin{cases}
\varepsilon \nabla^2 u - \alpha u^3 + \beta u = f, & \text{in } \Omega \\[2mm]
u = 0, & \text{on } \partial\Omega,
\end{cases}
\tag{25.3}
$$

where α, β, $\varepsilon > 0$, $u \in U = W_0^{1,2}(\Omega)$, and $f \in L^2(\Omega)$. The corresponding primal variational formulation is represented by $J : U \to \mathbb{R}$, where

$$
J(u) = \frac{\varepsilon}{2} \int_\Omega \nabla u \cdot \nabla u \, dx + \frac{\alpha}{4} \int_\Omega u^4 \, dx - \frac{\beta}{2} \int_\Omega u^2 \, dx + \int_\Omega f u \, dx.
$$

25.3.2 About the matrix version of G.M.O.L.

The generalized method of lines was originally developed in [19]. In this work we address its matrix version. Consider the simpler case where $\Omega = [0,1] \times [0,1]$. We discretize the domain in x, that is, in $N+1$ vertical lines obtaining the following equation in finite differences (see [73] for details about finite differences schemes).

$$\frac{\varepsilon(u_{n+1} - 2u_n + u_{n-1})}{d^2} + \varepsilon M_2 u_n / d_1^2 - \alpha u_n^3 + \beta u_n = f_n, \tag{25.4}$$

$\forall n \in \{1, ..., N-1\}$, where $d = 1/N$ and u_n corresponds to the solution on the line n. The idea is to apply the Newton's method. Thus choosing a initial solution $\{(u_0)_n\}$ we linearize (25.4) about it, obtaining the linear equation:

$$u_{n+1} - 2u_n + u_{n-1} \quad + \quad \tilde{M}_2 u_n - \frac{3\alpha d^2}{\varepsilon}(u_0)_n^2 u_n$$
$$+ \quad \frac{2\alpha}{\varepsilon}(u_0)_n^3 d^2 + \frac{\beta d^2}{\varepsilon} u_n - f_n \frac{d^2}{\varepsilon} = 0, \tag{25.5}$$

where $\tilde{M}_2 = M_2 \frac{d^2}{d_1^2}$ and

$$M_2 = \begin{bmatrix} -2 & 1 & 0 & 0 & \cdots & 0 \\ 1 & -2 & 1 & 0 & \cdots & 0 \\ 0 & 1 & -2 & 1 & \cdots & 0 \\ \vdots & \vdots & \vdots & \vdots & \ddots & \vdots \\ 0 & 0 & \cdots & 1 & -2 & 1 \\ 0 & 0 & \cdots & \cdots & 1 & -2 \end{bmatrix}, \tag{25.6}$$

with N_1 lines corresponding to the discretization in the y axis. Furthermore $d_1 = 1/N_1$.

In particular, for $n = 1$ we get

$$u_2 - 2u_1 \quad + \quad \tilde{M}_2 u_1 - \frac{3\alpha d^2}{\varepsilon}(u_0)_1^2 u_1$$
$$+ \quad \frac{2\alpha}{\varepsilon}(u_0)_1^3 d^2 + \frac{\beta d^2}{\varepsilon} u_1 - f_1 \frac{d^2}{\varepsilon} = 0. \tag{25.7}$$

Denoting

$$M_{12}[1] = 2I_d - \tilde{M}_2 + 3\frac{\alpha d^2}{\varepsilon}(u_0)_1^2 I_d - \frac{\beta d^2}{\varepsilon} I_d,$$

where I_d denotes the $(N_1 - 1) \times (N_1 - 1)$ identity matrix,

$$Y_0[1] = \frac{2\alpha d^2}{\varepsilon}(u_0)_1^3 - f_1 \frac{d^2}{\varepsilon},$$

and $M_{50}[1] = M_{12}[1]^{-1}$, we obtain

$$u_1 = M_{50}[1]u_2 + z[1].$$

where

$$z[1] = M_{50}[1] \cdot Y_0[1].$$

Now for $n = 2$ we get

$$u_3 - 2u_2 + u_1 \quad + \quad \tilde{M}_2 u_2 - \frac{3\alpha d^2}{\varepsilon}(u_0)_2^2 u_2$$
$$+ \quad \frac{2\alpha}{\varepsilon}(u_0)_2^3 d^2 + \frac{\beta d^2}{\varepsilon} u_2 - f_2 \frac{d^2}{\varepsilon} = 0, \tag{25.8}$$

that is,

$$u_3 - 2u_2 + M_{50}[1]u_2 + z[1] \quad + \quad \tilde{M}_2 u_2 - \frac{3\alpha d^2}{\varepsilon}(u_0)_2^2 u_2$$
$$+ \quad \frac{2\alpha}{\varepsilon}(u_0)_2^3 d^2 + \frac{\beta d^2}{\varepsilon}u_2 - f_2\frac{d^2}{\varepsilon} = 0, \tag{25.9}$$

so that denoting

$$M_{12}[2] = 2I_d - \tilde{M}_2 - M_{50}[1] + 3\frac{\alpha d^2}{\varepsilon}(u_0)_2^2 I_d - \frac{\beta d^2}{\varepsilon}I_d,$$

$$Y_0[2] = \frac{2\alpha d^2}{\varepsilon}(u_0)_2^3 - f_2\frac{d^2}{\varepsilon},$$

and $M_{50}[2] = M_{12}[2]^{-1}$, we obtain

$$u_2 = M_{50}[2]u_3 + z[2],$$

where

$$z[2] = M_{50}[2] \cdot (Y_0[2] + z[1]).$$

Proceeding in this fashion, for the line n we obtain

$$u_{n+1} - 2u_n \quad + \quad M_{50}[n-1]u_n + z[n-1] + \tilde{M}_2 u_n - \frac{3\alpha d^2}{\varepsilon}(u_0)_n^2 u_n$$
$$+ \quad \frac{2\alpha}{\varepsilon}(u_0)_n^3 d^2 + \frac{\beta d^2}{\varepsilon}u_n - f_n\frac{d^2}{\varepsilon} = 0, \tag{25.10}$$

so that denoting

$$M_{12}[n] = 2I_d - \tilde{M}_2 - M_{50}[n-1] + 3\frac{\alpha d^2}{\varepsilon}(u_0)_n^2 I_d - \frac{\beta d^2}{\varepsilon}I_d,$$

and also denoting

$$Y_0[n] = \frac{2\alpha d^2}{\varepsilon}(u_0)_n^3 - f_n\frac{d^2}{\varepsilon},$$

and $M_{50}[n] = M_{12}[n]^{-1}$, we obtain

$$u_n = M_{50}[n]u_{n+1} + z[n],$$

where

$$z[n] = M_{50}[n] \cdot (Y_0[n] + z[n-1]).$$

Observe that we have

$$u_N = \theta,$$

where θ denotes the zero matrix $(N_1 - 1) \times 1$, so that we may calculate

$$u_{N-1} = M_{50}[N-1] \cdot u_N + z[N-1],$$

and

$$u_{N-2} = M_{50}[N-2] \cdot u_{N-1} + z[N-2],$$

and so on, up to obtaining

$$u_1 = M_{50}[1] \cdot u_2 + z[1].$$

The next step is to replace $\{(u_0)_n\}$ by $\{u_n\}$ and thus to repeat the process until convergence is achieved.

This is the Newton's Method, what seems to be relevant is the way we inverted the big matrix $((N_1 - 1) \cdot (N-1)) \times ((N_1 - 1) \cdot (N-1))$, in fact instead of inverting it directly we have inverted $N - 1$ matrices $(N_1 - 1) \times (N_1 - 1)$ through an application of the generalized method of lines.

25.3.3 Numerical results for the concerning partial differential equation

We solve the equation

$$\begin{cases} \varepsilon\nabla^2 u - \alpha u^3 + \beta u + 1 = 0, & \text{in } \Omega = [0,1] \times [0,1] \\ \\ u = 0, & \text{on } \partial\Omega, \end{cases} \tag{25.11}$$

through the algorithm specified in the last section. We consider $\alpha = \beta = 1$. For $\varepsilon = 1.0$ see Figure 25.5, for $\varepsilon = 0.0001$ see Figure 25.6.

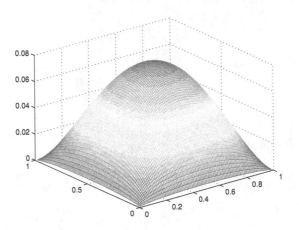

Figure 25.5: The solution $u(x,y)$ for $\varepsilon = 1.0$.

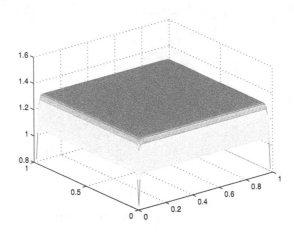

Figure 25.6: The solution $u(x,y)$ for $\varepsilon = 0.0001$.

25.4 A proximal algorithm for optimization in $\mathbb{R}^n$

In this section we develop a proximal algorithm for constrained optimization.

Let $f : \mathbb{R}^n \to \mathbb{R}$ be a C^2 class function. Consider the problem of minimizing locally f subject to $g(x) \leq 0$ where $g : \mathbb{R}^n \to \mathbb{R}$ is a given C^2 class function.

The lagrangian for this problem, denoted by $L : \mathbb{R}^{n+1} \to \mathbb{R}$, may be expressed by

$$L(x,\lambda) = f(x) + \lambda^2 g(x).$$

We define the proximal formulation for such a problem, denoted by L_p by

$$L_p(x, \lambda, x_k) = f(x) + \lambda^2 g(x) + \frac{K}{2}|x - x_k|^2.$$

25.4.1 The main result

Linearizing L_p, we propose the following procedure for looking for a critical point of such a function:
Consider

$$\begin{aligned}
\tilde{L}_p(x, \lambda, x_k) &= f(x_k) + f'(x_k) \cdot (x - x_k) + \frac{1}{2}[f''(x_k)(x - x_k)] \cdot (x - x_k) \\
&\quad + \lambda^2(g(x_k) + g'(x_k) \cdot (x - x_k)) + \frac{K}{2}|x - x_k|^2.
\end{aligned}$$

Hence from

$$\frac{\partial \tilde{L}_p(x, \lambda, x_k)}{\partial x} = 0$$

we obtain

$$f''(x_k)(x - x_k) + K(x - x_k) + f'(x_k) + \lambda^2 g'(x_k) = 0,$$

that is,

$$x - x_k = -(f''(x_k) + KI_d)^{-1}(f'(x_k) + \lambda^2 g'(x_k)),$$

and therefore

$$x(\lambda, x_k) = x_k - (f''(x_k) + KI_d)^{-1}(f'(x_k) + \lambda^2 g'(x_k)),$$

where I_d denotes the $n \times n$ identity matrix.
We define $L_1(\lambda, x_k) = \tilde{L}_p(x(\lambda, x_k), x_k, \lambda)$ so that

$$\begin{aligned}
L_1(\lambda, x_k) &= -\frac{1}{2}[(f''(x_k) + KI_d)^{-1}(f'(x_k) + \lambda^2 g'(x_k))] \cdot (f'(x_k) + \lambda^2 g'(x_k)) \\
&\quad + f(x_k) + \lambda^2 g(x_k)
\end{aligned} \tag{25.12}$$

From

$$\frac{\partial L_1(\lambda, x_k)}{\partial \lambda} = 0,$$

we get

$$[(f''(x_k) + KI_d)^{-1}(f'(x_k) + \lambda^2 g'(x_k))] \cdot g'(x_k)\lambda - \lambda g(x_k) = 0, \tag{25.13}$$

so that we have two solutions,

$$\lambda_1 = 0$$

and

$$(\lambda_2^1)^2(x_k) = -\left(\frac{[(f''(x_k) + KI_d)^{-1}f'(x_k)] \cdot g'(x_k) - g(x_k)}{[(f''(x_k) + KI_d)^{-1}g'(x_k)] \cdot g'(x_k)}\right). \tag{25.14}$$

Observe that if $(\lambda_2^1)^2(x_k) < 0$ then $\lambda_2^1(x_k)$ is complex so that, from the condition $\lambda^2 \geq 0$, we obtain

$$\lambda^2(x_k) = \max\{0, (\lambda_2^1)^2(x_k)\}.$$

Also, from the generalized inverse function theorem $\lambda^2(x)$ is locally Lipschtzian (see [54, 57, 78, 60] for details). Hence, we may infer that for a given $x_0 \in \mathbb{R}^n$ there exists $r > 0$ and $\hat{K}_3 > 0$ such that

$$|\lambda^2(x) - \lambda^2(y)| \leq \hat{K}_3|x - y|,$$

$\forall x, y \in B_r(x_0)$. With such results in mind, for such an $x_0 \in \mathbb{R}^n$, define $\{x_k\}$ by

$$x_1 = x_0 - (f''(x_0) + KI_d)^{-1}(f'(x_0) + \lambda^2(x_0)g'(x_0)),$$

$$x_{k+1} = x_k - (f''(x_k) + KI_d)^{-1}(f'(x_k) + \lambda^2(x_k)g'(x_k)), \ \forall k \in \mathbb{N}.$$

Assume

$$g(x_0) < 0 \tag{25.15}$$

and there exists $\hat{K}_1$ such that $|f''(x)| \le \hat{K}_1, \ \forall x \in B_r(x_0)$.

Define

$$K_3 = \hat{K}_3 \left(\sup_{x \in B_r(x_0)} |g'(x)| \right),$$

$\alpha_1 = 2K_3/|K - \hat{K}_1|$ and suppose

$$f''(x) + \lambda^2(y)g''(x) \ge \alpha_1 (\hat{K}_1 + K)I_d \ \forall x, y \in B_r(x_0). \tag{25.16}$$

Suppose also K is such that $K > \hat{K}_1$,

$$0 < \alpha_1 < 1,$$

$$\left(1 - \frac{\alpha_1}{4}\right) I_d \le ((f''(x) + KI_d)^{-1})(f''(y) + KI_d) \equiv H(x, y) \le \left(1 + \frac{\alpha_1}{4}\right) I_d, \tag{25.17}$$

$\forall x, y \in B_r(x_0)$ and

$$0 \le \frac{f''(x) + \lambda(y)^2 g''(x)}{K - \hat{K}_1} \le \left(1 - \frac{\alpha_1}{2}\right) I_d, \forall x, y \in B_r(x_0). \tag{25.18}$$

Observe that since $|f''(x)| \le \hat{K}_1$, we have

$$0 \le (K - \hat{K}_1)I_d \le f''(x) + KI_d,$$

so that

$$(f''(x) + KI_d)^{-1} \le \frac{1}{K - \hat{K}_1} I_d, \tag{25.19}$$

and

$$|(f''(x) + KI_d)^{-1}|K_3 \le \frac{K_3}{|K - \hat{K}_1|} = \frac{\alpha_1}{2}, \ \forall x \in B_r(x_0). \tag{25.20}$$

Assume $K > 0$ is such that

$$x_1 \in B_{r(1-\alpha_0)}(x_0)$$

and suppose the induction hypotheses

$$x_2, \ldots, x_{k+1} \in B_r(x_0).$$

where $0 < \alpha_0 < 1$ is specified in the next lines.

Note that,

$$x_{k+2} - x_{k+1} = -(f''(x_{k+1}) + KI_d)^{-1}(f'(x_{k+1}) + \lambda^2(x_{k+1})g'(x_{k+1})),$$

and

$$x_{k+1} - x_k = -(f''(x_k) + KI_d)^{-1}(f'(x_k) + \lambda^2(x_k)g'(x_k)),$$

so that,

$$(f''(x_{k+1}) + KI_d)(x_{k+2} - x_{k+1}) = -(f'(x_{k+1}) + \lambda^2(x_{k+1})g'(x_{k+1})),$$

and

$$(f''(x_k) + KI_d)(x_{k+1} - x_k) = -(f'(x_k) + \lambda^2(x_k)g'(x_k)).$$

Therefore,

$$
\begin{aligned}
&(f''(x_{k+1}) + KI_d)(x_{k+2} - x_{k+1}) \\
= \ &(f''(x_k) + KI_d)(x_{k+1} - x_k) \\
&-(f'(x_{k+1}) + \lambda^2(x_{k+1})g'(x_{k+1})) + (f'(x_k) + \lambda^2(x_k)g'(x_k)) \\
= \ &(f''(x_k) + KI_d)(x_{k+1} - x_k) - (f'(x_{k+1}) + \lambda^2(x_{k+1})g'(x_{k+1})) + (f'(x_k) + \lambda^2(x_{k+1})g'(x_k)) \\
&-(f'(x_k) + \lambda^2(x_{k+1})g'(x_k)) + (f'(x_k) + \lambda^2(x_k)g'(x_k)) \\
= \ &(f''(x_k) + KI_d)(x_{k+1} - x_k) \\
&-(f''(\tilde{x}_k) + \lambda^2(x_{k+1})g''(\tilde{x}_k))(x_{k+1} - x_k) - (\lambda^2(x_{k+1}) - \lambda^2(x_k))g'(x_k)
\end{aligned}
$$

where $\tilde{x}_k$ is on the line connecting x_k and x_{k+1}.

Thus,

$$
\begin{aligned}
x_{k+2} - x_{k+1} = \ &(f''(x_{k+1}) + KI_d)^{-1}[(f''(x_k) + KI_d)(x_{k+1} - x_k) \\
&-(f''(\tilde{x}_k) + \lambda^2(x_{k+1})g''(\tilde{x}_k))(x_{k+1} - x_k) \\
&-(\lambda^2(x_{k+1}) - \lambda^2(x_k))g'(x_k)],
\end{aligned} \tag{25.21}
$$

so that

$$
\begin{aligned}
|x_{k+2} - x_{k+1}| \leq \ &|H(x_{k+1}, x_k) - ((f''(x_{k+1}) + KI_d)^{-1})(f''(\tilde{x}_k) \\
&+ \lambda^2(x_{k+1})g''(\tilde{x}_k))||x_{k+1} - x_k| \\
&+ |(f''(x_{k+1}) + K I_d)^{-1}|K_3|x_{k+1} - x_k|.
\end{aligned} \tag{25.22}
$$

Observe that, from (25.16),

$$
f''(\tilde{x}_k) + \lambda^2(\tilde{x}_{k+1})g''(x_k) \geq \alpha_1(\hat{K}_1 + K)I_d \geq \alpha_1(f''(x_{k+1}) + K I_d),
$$

so that

$$
((f''(x_{k+1}) + K I_d)^{-1})(f''(\tilde{x}_k) + \lambda^2(x_{k+1})g''(\tilde{x}_k)) \geq \alpha_1 I_d.
$$

Hence, from this, (25.17), (25.19) and (25.18), we obtain

$$
\begin{aligned}
&I_d\left(1 + \frac{\alpha_1}{4}\right) - \alpha_1 I_d \\
\geq \ &H(x_{k+1}, x_k) - ((f''(x_{k+1}) + K I_d)^{-1})(f''(\tilde{x}_k) + \lambda^2(x_{k+1})g''(\tilde{x}_k)) \\
\geq \ &I_d\left(1 - \frac{\alpha_1}{4}\right) - (K I_d - \hat{K}_1 I_d)^{-1})(f''(\tilde{x}_k) + \lambda^2(x_{k+1})g''(\tilde{x}_k)) \\
\geq \ &I_d\left(1 - \frac{\alpha_1}{4}\right) - I_d\left(1 - \frac{\alpha_1}{2}\right) \\
= \ &\frac{\alpha_1}{4}I_d \\
\geq \ &0,
\end{aligned} \tag{25.23}
$$

and therefore,

$$
|H(x_{k+1}, x_k) - (f''(x_k) + K I_d)^{-1}(f''(\tilde{x}_k) + \lambda^2(x_{k+1})g''(\tilde{x}_k))| \leq 1 - \frac{3\alpha_1}{4}.
$$

On the other hand, from (25.20) we have,

$$
|(f''(x_k) + K I_d)^{-1}|K_3 \leq \frac{\alpha_1}{2}.
$$

From (25.22) and these last two inequalities, we obtain

$$|x_{k+2} - x_{k+1}| \leq \left(1 - \frac{3\alpha_1}{4} + \frac{\alpha_1}{2}\right)|x_{k+1} - x_k| = \left(1 - \frac{\alpha_1}{4}\right)|x_{k+1} - x_k|.$$

Thus, denoting $\alpha_0 = 1 - \alpha_1/4$, we have obtained,

$$|x_{j+2} - x_{j+1}| \leq \alpha_0 |x_{j+1} - x_j|, \forall j \in \{1, \cdots, k+1\}$$

so that

$$
\begin{aligned}
|x_{j+2} - x_{j+1}| &\leq \alpha_0 |x_{j+1} - x_j| \\
&\leq \alpha_0^2 |x_j - x_{j-1}| \\
&\leq \cdots \\
&\leq \alpha_0^{j+1} |x_1 - x_0|, \forall j \in \{1, \cdots, k\}.
\end{aligned}
\tag{25.24}
$$

Thus,

$$
\begin{aligned}
&\quad |x_{k+2} - x_1| \\
&= |x_{k+2} - x_{k+1} + x_{k+1} - x_k + x_k - x_{k-1} + \cdots + x_2 - x_1| \\
&\leq |x_{k+2} - x_{k+1}| + |x_{k+1} - x_k| + \cdots + |x_2 - x_1| \\
&\leq \sum_{j=1}^{k+1} \alpha_0^j |x_1 - x_0| \\
&\leq \sum_{j=1}^{+\infty} \alpha_0^j |x_1 - x_0| \\
&= \frac{\alpha_0}{1 - \alpha_0} |x_1 - x_0|,
\end{aligned}
\tag{25.25}
$$

so that

$$
\begin{aligned}
|x_{k+2} - x_0| &\leq |x_{k+2} - x_1| + |x_1 - x_0| \\
&\leq \frac{\alpha_0}{1 - \alpha_0} |x_1 - x_0| + |x_1 - x_0| \\
&= \frac{1}{1 - \alpha_0} |x_1 - x_0| \\
&< \frac{1}{1 - \alpha_0} r(1 - \alpha_0) \\
&= r.
\end{aligned}
\tag{25.26}
$$

Hence, $x_{k+2} \in B_r(x_0)$, and therefore the induction is complete, so that

$$x_k \in B_r(x_0), \forall k \in \mathbb{N}.$$

Moreover, $\{x_k\}$ is a Cauchy sequence, so that there exists $\tilde{x}$, such that

$$x_k \to \tilde{x}, \text{ as } k \to \infty.$$

Finally,

$$
\begin{aligned}
0 &= \lim_{k \to \infty} (x_{k+1} - x_k) \\
&= \lim_{k \to \infty} [-(f''(x_k) + KI_d)^{-1}(f'(x_k) + \lambda^2(x_k)g'(x_k))] \\
&= -(f''(\tilde{x}) + KI_d)^{-1}(f'(\tilde{x}) + \tilde{\lambda}^2 g'(\tilde{x})).
\end{aligned}
\tag{25.27}
$$

Hence, from this and

$$det(f''(\tilde{x}) + KI_d) \neq 0,$$

we obtain

$$f'(\tilde{x}) + \tilde{\lambda}^2 g'(\tilde{x}) = 0$$

In such a case, from (25.13) letting $k \to \infty$, we also obtain

$$\tilde{\lambda}^2 g(\tilde{x}) = 0.$$

Thus, if $\tilde{\lambda}^2 > 0$, then $g(\tilde{x}) = 0$.
If $\tilde{\lambda} = 0$, then $f'(\tilde{x}) = \mathbf{0}$ and

$$[(\lambda_2^1)(\tilde{x})]^2 \leq 0$$

so that from (25.14), since $(f''(\tilde{x}) + KI_d)^{-1}$ is positive definite, letting $k \to \infty$, we get

$$g(\tilde{x}) = [(\lambda_2^1)(\tilde{x})]^2 [(f''(\tilde{x}) + KI_d)^{-1} g'(\tilde{x})] \cdot g'(\tilde{x}) \leq 0.$$

That is, in any case,

$$g(\tilde{x}) \leq 0.$$

Remark 25.1   For the more general case with m_1 equality scalar constraints

$$h_j(x) = 0, \forall j \in \{1, \dots, m_1\}$$

and m_2 inequality scalar constraints

$$g_l(x) \leq 0, \ \forall l \in \{1, \dots, m_2\},$$

where $h_j, g_l : \mathbb{R}^n \to \mathbb{R}$ are C^2 class functions, $\forall j \in \{1, \dots, m_1\}$ and $\forall l \in \{1, \dots, m_2\}$, we assume $m_1 + m_2 < n$ and define the Lagrangian L_p by

$$L_p(x, \lambda, x_k) = f(x) + \sum_{j=1}^{m_1} (\lambda_h)_j h_j(x) + \sum_{l=1}^{m_2} (\lambda_g)_l^2 g_l(x) + \frac{K}{2}|x - x_k|^2.$$

Linearizing L_p, we propose the following procedure for looking for a critical point of such a function: Consider

$$
\begin{aligned}
\tilde{L}_p(x, \lambda, x_k) &= f(x_k) + f'(x_k) \cdot (x - x_k) + \frac{1}{2}[f''(x_k)(x - x_k)] \cdot (x - x_k) \\
&\quad + \sum_{j=1}^{m_1} (\lambda_h)_j (h_j(x_k) + h'_j(x_k) \cdot (x - x_k)) \\
&\quad + \sum_{l=1}^{m_2} (\lambda_g)_l^2 (g_l(x_k) + g'_l(x_k) \cdot (x - x_k)) + \frac{K}{2}|x - x_k|^2.
\end{aligned}
$$

Hence, from

$$\frac{\partial \tilde{L}_p(x, \lambda, x_k)}{\partial x} = 0,$$

we obtain,

$$f''(x_k)(x - x_k) + K(x - x_k) + f'(x_k) + \sum_{j=1}^{m_1} (\lambda_h)_j h'_j(x_k) + \sum_{l=1}^{m_2} (\lambda_g)_l^2 g'_l(x_k) = 0,$$

that is,

$$x - x_k = -(f''(x_k) + KI_d)^{-1} \left(f'(x_k) + \sum_{j=1}^{m_1} (\lambda_h)_j h'_j(x_k) + \sum_{l=1}^{m_2} (\lambda_g)_l^2 g'_l(x_k) \right),$$

and therefore,

$$x(\lambda, x_k) = x_k - (f''(x_k) + KI_d)^{-1} \left(f'(x_k) + \sum_{j=1}^{m_1} (\lambda_h)_j h'_j(x_k) + \sum_{l=1}^{m_2} (\lambda_g)_l^2 g'_l(x_k) \right), \quad (25.28)$$

where I_d denotes the $n \times n$ identity matrix.

We define $L_1(\lambda, x_k) = \tilde{L}_p(x(\lambda, x_k), x_k, \lambda)$, so that

$$\begin{aligned}
L_1(\lambda, x_k) &= -\frac{1}{2} \left[(f''(x_k) + KI_d)^{-1} \left(f'(x_k) + \sum_{j=1}^{m_1} (\lambda_h)_j h'_j(x_k) + \sum_{l=1}^{m_2} (\lambda_g)_l^2 g'_l(x_k) \right) \right. \\
&\quad \left. \cdot \left(f'(x_k) + \sum_{j=1}^{m_1} (\lambda_h)_j h'_j(x_k) + \sum_{l=1}^{m_2} (\lambda_g)_l^2 g'_l(x_k) \right) \right] \\
&\quad + f(x_k) + \sum_{j=1}^{m_1} (\lambda_h)_j h_j(x_k) + \sum_{l=1}^{m_2} (\lambda_g)_l^2 g_l(x_k).
\end{aligned} \quad (25.29)$$

From

$$\frac{\partial L_1(\lambda, x_k)}{\partial (\lambda_g)_l} = 0,$$

we get

$$\left[(f''(x_k) + KI_d)^{-1} \left(f'(x_k) + \sum_{j=1}^{m_1} (\lambda_h)_j h'_j(x_k) \right. \right. \\
\left. \left. + \sum_{l=1}^{m_2} (\lambda_g^2)_l g'_l(x_k) \right) \right] \cdot g'_l(x_k)(\lambda_g)_l - (\lambda_g)_l g_l(x_k) = 0, \quad (25.30)$$

From

$$\frac{\partial L_1(\lambda, x_k)}{\partial (\lambda_h)_j} = 0,$$

we have

$$\left[(f''(x_k) + KI_d)^{-1} \left(f'(x_k) + \sum_{j=1}^{m_1} (\lambda_h)_j h'_j(x_k) \right. \right. \\
\left. \left. + \sum_{l=1}^{m_2} (\lambda_g)_l^2 g'_l(x_k) \right) \right] \cdot h'_j(x_k) - h_j(x_k) = 0, \quad (25.31)$$

$\forall j \in \{1, \ldots, m_1\}$. Solving the linear system which comprises these last m_1 equations and the m_2 equations

$$\left[(f''(x_k) + KI_d)^{-1} \left(f'(x_k) + \sum_{j=1}^{m_1} (\lambda_h)_j h'_j(x_k) \right. \right. \\
\left. \left. + \sum_{l=1}^{m_2} (\lambda_g)_l^2 g'_l(x_k) \right) \right] \cdot g'_l(x_k) - g_l(x_k) = 0, \quad (25.32)$$

$\forall l \in \{1, \ldots, m_2\}$, we may obtain a solution

$$\left((\lambda_h)_j(x_k), (\lambda_g^1)_l^2(x_k) \right).$$

Thus, to obtain a concerning critical point, we follow the following algorithm.

1. Choose $x_0 \in \mathbb{R}^n$, $K_{max} \in \mathbb{N}$ (K_{max} is the maximum number of iterations), set $k = 0$ and $e_1 \approx 10^{-5}$.

2. Obtain a solution

$$\left((\lambda_h)_j(x_k), (\lambda_g^1)_l^2(x_k) \right)$$

   by solving the linear system (in $(\lambda_h)_j$ and $(\lambda_g)_l^2$) indicated in (25.31) and (25.32).

   Observe that if $(\lambda_g^1)_l^2 < 0$ then $(\lambda_g^1)_l$ is complex.

   To up-date λ_h and λ_g proceed as follows:

3. For each $l \in \{1, \ldots, m_2\}$ if $(\lambda_g)_l^2(x_k) \leq 0$, then set $(\lambda_g)_l(x_k) = 0$.

4. Define $J = \{l \in \{1, \ldots, m_2\}$ such that $(\lambda_g)_l^2(x_k) > 0\}$.

5. Recalculate $(\lambda_h)_j(x_k)$ and the non-zero $(\lambda_g)_l^2(x_k)$ for $l \in J$ through the solution of the linear system (in $(\lambda_h)_j$ and $(\lambda_g)_l^2$)

$$\left[(f''(x_k) + KI_d)^{-1} \left(f'(x_k) + \sum_{j=1}^{m_1} (\lambda_h)_j h'_j(x_k) \right. \right.$$

$$\left. \left. + \sum_{l \in J} (\lambda_g)_l^2 g'_l(x_k) \right) \right] \cdot h'_j(x_k) - h_j(x_k) = 0, \qquad (25.33)$$

$\forall j \in \{1, \ldots, m_1\}$ and

$$\left[(f''(x_k) + KI_d)^{-1} \left(f'(x_k) + \sum_{j=1}^{m_1} (\lambda_h)_j h'_j(x_k) \right. \right.$$

$$\left. \left. + \sum_{l \in J} (\lambda_g)_l^2 g'_l(x_k) \right) \right] \cdot g'_l(x_k) - g_l(x_k) = 0, \qquad (25.34)$$

$\forall l \in J$.

6. If $(\lambda_g)_l^2(x_k) \geq 0$, $\forall l \in \{1, \ldots, m_2\}$, then go to 7, otherwise go to item 3.

7. Up-date x_k through the equation

$$x_{k+1} = x_k - (f''(x_k) + KI_d)^{-1} \left(f'(x_k) + \sum_{j=1}^{m_1} (\lambda_h)_j(x_k) h'_j(x_k) \right.$$

$$\left. + \sum_{l=1}^{m_2} (\lambda_g)_l^2(x_k) g'_l(x_k) \right). \qquad (25.35)$$

8. If $|x_{k+1} - x_k| < e_1$ or $k > K_{max}$, then stop, otherwise $k := k+1$ and go to 2.

■

Remark 25.4.1 *In this section we have developed an algorithm for constrained optimization in $\mathbb{R}^n$. We prove the main result only for the special case of a single scalar inequality constraint. However, we highlight the proof of a more general result involving equality and inequality constraints may be developed in a similar fashion, as indicated in remark 25.1. We postpone the presentation of the formal details for such a more general case for a future work.*

Also in the context of this last result, let us consider the following example of application.

Let $\Omega \subset \mathbb{R}^2$ be an open, bounded, connected set with a regular boundary denoted by $\partial \Omega$. Consider the real Ginzburg-Landau type equation, given by

$$\begin{cases} \varepsilon \nabla^2 u - \alpha u^3 + \beta u = f, & in\ \Omega \\ u = 0, & on\ \partial \Omega, \end{cases} \tag{25.36}$$

where $\alpha, \beta, \varepsilon > 0$, $u \in U = W_0^{1,2}(\Omega)$, and $f \in L^2(\Omega)$.

Consider the sequence obtained through the algorithm:

1. *Set $n = 1$,*

2. *Choose $z_1^* \in L^2(\Omega)$.*

3. *Compute u_n by*

$$\begin{aligned} u_n &= argmin_{u \in U} \left\{ \frac{\varepsilon}{2} \int_\Omega \nabla u \cdot \nabla u\ dx + \frac{\alpha}{4} \int_\Omega u^4\ dx \right. \\ &\quad \left. - \langle u, z_n^* \rangle_{L^2} + \frac{1}{2\beta} \int_\Omega (z_n^*)^2\ dx + \int_\Omega fu\ dx \right\}, \end{aligned} \tag{25.37}$$

   which means to solve the equation

$$\begin{cases} \varepsilon \nabla^2 u - \alpha u^3 + z_n^* = f, & in\ \Omega \\ u = 0, & on\ \partial \Omega. \end{cases} \tag{25.38}$$

4. *Compute z_{n+1}^* by*

$$\begin{aligned} z_{n+1}^* &= argmin_{z^* \in L^2(\Omega)} \left\{ \frac{\varepsilon}{2} \int_\Omega \nabla u_n \cdot \nabla u_n\ dx + \frac{\alpha}{4} \int_\Omega u_n^4\ dx \right. \\ &\quad \left. - \langle u_n, z^* \rangle_{L^2} + \frac{1}{2\beta} \int_\Omega (z^*)^2\ dx + \int_\Omega fu_n\ dx \right\}, \end{aligned} \tag{25.39}$$

   that is,

$$z_{n+1}^* = \beta u_n.$$

5. *Set $n \to n+1$ and go to step 3 (up to the satisfaction of an appropriate convergence criterion).*

Assuming the hypotheses of the last result for an analogous non-restricted case, the sequence $\{u_n\}$ is such that

$$u_n \to u_0, \text{ strongly in } L^2(\Omega),$$

where

$$u_0 \in W_0^{1,2}(\Omega)$$

is a solution of equation (25.36).

Remark 25.4.2 *Observe that for each n, the procedure of evaluating u_n stands for the solution of a convex optimization problem with unique solution, given by the one of equation*

$$\varepsilon \nabla^2 u_n - \alpha u_n^3 + z_n^* + f = 0 \text{ in } \Omega,$$

which may be easily obtained, due to convexity, through the generalized method of lines (matrix version) associated with Newton's method as above described.

25.5   Conclusion

In this chapter we have develop numerical methods for large class of ODEs.

We have also introduced the matrix version of the generalized method of lines for PDEs. Further, we have developed a convergent algorithm suitable for equations that present strong variational formulation, and in particular, suitable for Ginzburg-Landau type equations. The results are rigorously proven and numerical examples are provided. We emphasize that even as the parameter ε is very small, namely $\varepsilon = 0.0001$, the results are consistent and the convergence is very fast.

Chapter 26

On the Numerical Solution of First Order Ordinary Differential Equation Systems

26.1 Introduction

In this chapter we develop an algorithm to solve a class of first order non-linear ordinary differential equations.

We start by presenting a general procedure for solving the linearized equations, and in a second step, we apply it to solve a problem in flight mechanics in a Newton's method context. In fact, a sequence of linear problems is solved intending to obtain a solution for the original non-linear problem. We emphasize the method here proposed has a performance considerably better than those so far known, particularly concerning the computation time.

At this point we present a remark on the references.

Remark 26.1   We highlight that a similar problem is addressed in [14] for a nuclear physics model. The main difference is that now our results are more general and applicable to a much larger class of problems. Specifically in the present work, we apply them to a flight mechanics model found in [80].

For the numerical results we have used finite differences. Details about finite differences schemes may be found in [73].

Finally, details on the Sobolev spaces in which the original problem is established may be found in [1].
∎

26.2 The main results

Consider the following system of difference equations in $\{(u_k)_n\}$, given by

$$(u_k)_{n+1} = \sum_{j=1}^{4} (a_{kj})_n (u_j)_n + (g_k)_n, \ \forall k \in \{1,2,3,4\}, \ n \in \{0,...,N-1\}, \qquad (26.$$

where
$$(u_k)_n, (a_{kj})_n, (g_k)_n \in \mathbb{R}, \ \forall j,k \in \{1,2,3,4\}, \ n \in \{0,...,N-1\}.$$

Assume the following boundary conditions are intended to be satisfied:

$$\begin{cases} (u_1)_0 = h_0, \\ (u_3)_0 = V_0 \\ (u_4)_0 = x_0 \\ (u_1)_N = h_f. \end{cases} \tag{26.2}$$

For $n = 0$ and $k = 1$ we obtain

$$(u_1)_1 = \sum_{j=1}^{4} (a_{1j})_0 (u_j)_0 + (g_1)_0,$$

that is,

$$(u_2)_0 = \frac{(u_1)_1 - (g_1)_0 - (a_{11})_0(u_1)_0 - (a_{13})_0(u_3)_0 - (a_{14})_0(u_4)_0}{(a_{12})_0},$$

so that we write

$$(u_2)_0 = m_2[0](u_1)_1 + \tilde{z}_2[0], \tag{26.3}$$

where

$$m_2[0] = \frac{1}{(a_{12})_0},$$

and

$$\tilde{z}_2[0] = \frac{-(g_1)_0 - (a_{11})_0(u_1)_0 - (a_{13})_0(u_3)_0 - (a_{14})_0(u_4)_0}{(a_{12})_0}.$$

Replacing (26.3) into

$$(u_2)_1 = \sum_{j=1}^{4} (a_{2j})_0 (u_j)_0 + (g_2)_0,$$

we get

$$(u_2)_1 = m_2[1](u_1)_1 + z_2[1],$$

where

$$m_2[1] = (a_{22})_0 m_2[0],$$

and,

$$z_2[1] = (a_{21})_0(u_1)_0 + (a_{22})_0\tilde{z}_2[0] + (a_{23})_0(u_3)_0 + (a_{24})_0(u_4)_0 + (g_2)_0.$$

Also, replacing (26.3) into

$$(u_3)_1 = \sum_{j=1}^{4} (a_{3j})_0 (u_j)_0 + (g_3)_0,$$

we obtain

$$(u_3)_1 = m_3[1](u_1)_1 + z_3[1],$$

where

$$m_3[1] = (a_{32})_0 m_2[0],$$

and

$$z_3[1] = (a_{31})_0(u_1)_0 + (a_{32})_0\tilde{z}_2[0] + (a_{33})_0(u_3)_0 + (a_{34})_0(u_4)_0 + (g_3)_0.$$

Finally, replacing (26.3) into

$$(u_4)_1 = \sum_{j=1}^{4} (a_{4j})_0 (u_j)_0 + (g_4)_0,$$

we may obtain

$$(u_4)_1 = m_4[1](u_1)_1 + z_4[1],$$

where

$$m_4[1] = (a_{42})_0 m_2[0],$$

and

$$z_4[1] = (a_{41})_0(u_1)_0 + (a_{42})_0 \tilde{z}_2[0] + (a_{43})_0(u_3)_0 + (a_{44})_0(u_4)_0 + (g_4)_0.$$

Reasoning inductively, having for $k \in \{2,3,4\}$,

$$(u_k)_{n-1} = m_k[n-1](u_1)_{n-1} + z_k[n-1], \tag{26.4}$$

for $n \geq 2$, replacing these last equations into (26.1) for $k = 1$, we may obtain

$$(u_1)_{n-1} = \tilde{m}_1[n-1](u_1)_n + \tilde{z}_1[n], \tag{26.5}$$

where

$$
\begin{aligned}
&\tilde{m}_1[n-1] \\
&= \{(a_{11})_{n-1} + (a_{12})_{n-1} m_2[n-1] + (a_{13})_{n-1} m_3[n-1] \\
&\quad + (a_{14})_{n-1} m_4[n-1]\}^{-1},
\end{aligned} \tag{26.6}
$$

and

$$
\begin{aligned}
&\tilde{z}_1[n-1] \\
&= -\tilde{m}_1[n-1]\{(a_{12})_{n-1} z_2[n-1] + (a_{13})_{n-1} z_3[n-1] \\
&\quad + (a_{14})_{n-1} z_4[n-1] + (g_1)_{n-1}\}.
\end{aligned} \tag{26.7}
$$

Finally, replacing (26.5) into (26.4), we may obtain

$$(u_k)_{n-1} = \tilde{m}_k[n-1](u_1)_n + \tilde{z}_k[n-1], \tag{26.8}$$

$\forall k \in \{2,3,4\}$,

where

$$\tilde{m}_k[n-1] = m_k[n-1]\tilde{m}_1[n-1],$$

$$\tilde{z}_k[n-1] = m_k[n-1]\tilde{z}_1[n-1] + z_k[n-1].$$

Replacing (26.5) and (26.8) into the system (26.1), we get

$$(u_k)_n = m_k[n](u_1)_n + z_k[n],$$

where

$$m_k[n] = (a_{k1})_n \tilde{m}_1[n-1] + (a_{k2})_n \tilde{m}_2[n-1] + (a_{k3})_n \tilde{m}_3[n-1] + (a_{k4})_n \tilde{m}_4[n-1],$$

and

$$z_k[n] = \sum_{j=1}^{4} (a_{kj})_n \tilde{z}_j[n-1] + (g_k)_n.$$

Summarizing, we have obtained,

$$(u_1)_{n-1} = \tilde{m}_1[n-1](u_1)_n + \tilde{z}_1[n-1],$$

$$(u_k)_n = m_k[n](u_1)_n + z_k[n], \quad \forall k \in \{2,3,4\}$$

$\forall n \in \{1,...,N\}$.

Therefore, having $(u_1)_N = h_f$, we may obtain

$$(u_k)_N = m_k[N](u_1)_N + z_k[N], \quad \forall k \in \{2,3,4\}$$

and

$$(u_1)_{N-1} = \tilde{m}_{N-1}(u_1)_N + \tilde{z}_1[N-1].$$

Having $(u_1)_{N-1}$, we may obtain

$$(u_k)_{N-1} = m_k[N-1](u_1)_{N-1} + z_k[N-1], \quad \forall k \in \{2,3,4\}$$

and

$$(u_1)_{N-2} = \tilde{m}_1[N-2](u_1)_{N-1} + \tilde{z}_1[N-2],$$

and so on, up to finding $(u_k)_1, \forall k \in \{1,2,3,4\}$, and finally,

$$(u_2)_0 = m_2[0](u_1)_1 + \tilde{z}_2[0].$$

At this point, the problem is solved.

26.3   Numerical results

We present numerical results for the following system of equations, which models the in plane climbing motion of an airplane (please see more details in [80]).

$$
\begin{cases}
\dot{h} = V \sin \gamma, \\
\dot{\gamma} = \frac{1}{m_f V}(T \sin(e_3) + L) - \frac{g}{V} \cos \gamma, \\
\dot{V} = \frac{1}{m_f}(T \cos(e_3) - D) - g \sin \gamma \\
\dot{x} = V \cos \gamma,
\end{cases}
\tag{26.9}
$$

with the boundary conditions,

$$
\begin{cases}
h(0) = h_0, \\
V(0) = V_0, \\
x(0) = x_0 \\
h(t_f) = h_f,
\end{cases}
\tag{26.10}
$$

where $t_f = 90s$, h is the airplane altitude, V is its speed, γ is the angle between its velocity and the horizontal axis, and finally x denotes the horizontal coordinate position.

For numerical purposes, we assume (Air bus 320)

$m_f = 120,000Kg, S_f = 260m^2, a = 0.17/10 \text{ rad}, g = 9.8m/s^2, \rho(h) = 1.225(1 - 0.0065h/288.15)^{4.225} Kg/m^3,$

$$C_L = 4.95a,$$

$$C_D = 0.0175 + 0.06C_L^2,$$

$$L = \frac{1}{2}\rho(h)V^2 C_L S_f,$$

$$D = \frac{1}{2}\rho(h)V^2 C_D S_f,$$

$$T_0 = 10000$$

and where units refer to the International System and,

$$T = D + m_f g \sin \gamma + T_0,$$

which refers to an accelerated motion.

To simplify the analysis, we redefine the variables as below indicated:

$$\begin{cases} h = u_1, \\ \gamma = u_2 \\ V = u_3 \\ x = u_4. \end{cases} \tag{26.11}$$

Thus, denoting $\mathbf{u} = (u_1, u_2, u_3, u_4) \in U = W^{1,2}([0, t_f]; \mathbb{R}^4)$, the system above indicated may be expressed by

$$\begin{cases} \dot{u}_1 = f_1(\mathbf{u}) \\ \dot{u}_2 = f_2(\mathbf{u}) \\ \dot{u}_3 = f_3(\mathbf{u}) \\ \dot{u}_4 = f_4(\mathbf{u}), \end{cases} \tag{26.12}$$

where

$$\begin{cases} f_1(\mathbf{u}) = u_3 \sin(u_2), \\ f_2(\mathbf{u}) = \frac{1}{m_f u_3}(T(\mathbf{u}) \sin(e_3) + L(\mathbf{u})) - \frac{g}{u_3} \cos(u_2), \\ f_3(\mathbf{u}) = \frac{1}{m_f}(T(\mathbf{u}) \cos(e_3) - D(\mathbf{u})) - g \sin(u_2) \\ f_4(\mathbf{u}) = u_3 \cos(u_2). \end{cases} \tag{26.13}$$

Finally,

$$L(\mathbf{u}) = \frac{1}{2} \rho_0 e^{-u_1/B_9} u_3^2 C_L S_f,$$

$$D(\mathbf{u}) = \frac{1}{2} \rho_0 e^{-u_1/B_9} u_3^2 C_D S_f,$$

$$T(\mathbf{u}) = D(\mathbf{u}) + m_f g \sin(u_2) + T_0.$$

At this point we shall write the system indicated in (26.12) in finite differences, that is,

$$\begin{cases} (u_1)_{n+1} = (u_1)_n + f_1(\mathbf{u}_n)d \\ (u_2)_{n+1} = (u_2)_n + f_2(\mathbf{u}_n)d \\ (u_3)_{n+1} = (u_3)_n + f_3(\mathbf{u}_n)d \\ (u_4)_{n+1} = (u_4)_n + f_4(\mathbf{u}_n)d, \end{cases} \tag{26.14}$$

here $d = 90/N$, we N refers to the number of nodes concerning the discretization in t (in our numerical example $N = 10000$).

Intending to apply the Newton's method we linearize the system indicated in (26.14) about a initial guess

$$\tilde{\mathbf{u}} = (\tilde{u}_1, \tilde{u}_2, \tilde{u}_3, \tilde{u}_4).$$

We obtain the following approximate system

$$\begin{cases} (u_1)_{n+1} = (u_1)_n + d \left(f_1(\tilde{\mathbf{u}}_n) + \sum_{j=1}^4 \frac{\partial f_1(\tilde{\mathbf{u}}_n)}{\partial \tilde{u}_j}((u_j)_n - (\tilde{u}_j)_n) \right) \\ (u_2)_{n+1} = (u_2)_n + d \left(f_2(\tilde{\mathbf{u}}_n) + \sum_{j=1}^4 \frac{\partial f_2(\tilde{\mathbf{u}}_n)}{\partial \tilde{u}_j}((u_j)_n - (\tilde{u}_j)_n) \right) \\ (u_3)_{n+1} = (u_3)_n + d \left(f_3(\tilde{\mathbf{u}}_n) + \sum_{j=1}^4 \frac{\partial f_3(\tilde{\mathbf{u}}_n)}{\partial \tilde{u}_j}((u_j)_n - (\tilde{u}_j)_n) \right) \\ (u_4)_{n+1} = (u_4)_n + d \left(f_4(\tilde{\mathbf{u}}_n) + \sum_{j=1}^4 \frac{\partial f_4(\tilde{\mathbf{u}}_n)}{\partial \tilde{u}_j}((u_j)_n - (\tilde{u}_j)_n) \right). \end{cases} \tag{26.15}$$

Observe that such a system is in the form,

$$(u_k)_{n+1} = (u_k)_n + \sum_{j=1}^{4} (a_{kj})_n (u_j)_n + (g_k)_n,$$

where

$$(a_{kj})_n = \frac{\partial f_k(\tilde{\mathbf{u}}_n)}{\partial u_j} \, d, \text{ for } j \neq k$$

$$(a_{jj})_n = 1 + \frac{\partial f_j(\tilde{\mathbf{u}}_n)}{\partial u_j} \, d, \text{ for } j = k,$$

and

$$(g_k)_n = f_k(\tilde{\mathbf{u}}_n) \, d - \sum_{j=1}^{4} \frac{\partial f_k(\tilde{\mathbf{u}}_n)}{\partial u_j} (\tilde{u}_j)_n \, d.$$

We solve this last system for the following boundary conditions:

$$\begin{cases} h(0) = 0 \, m, \\ V(0) = 150 m/s, \\ x(0) = 0 \, m, \\ h(t_f) = 11000 \, m. \end{cases} \tag{26.16}$$

We have obtained $\{\mathbf{u}_n\}$. In a Newton's method context, the next step is to replace $\tilde{\mathbf{u}}_n$ by $\{\mathbf{u}_n\}$ and thus to repeat the process up to the satisfaction of an appropriate convergence criterion.

We have obtained the following solutions for h, γ, V and x. Please see Figures 26.1, 26.2, 26.3 and 26.4, respectively.

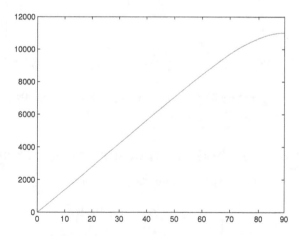

Figure 26.1: The solution h (in m) for $t_f = 90s$.

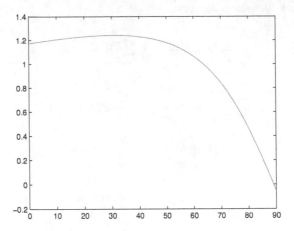

Figure 26.2: The solution γ (in rad) for $t_f = 90s$.

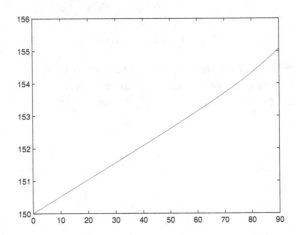

Figure 26.3: The solution V (in m/s) for $t_f = 90s$.

26.4 The Newton's method for another first order system

Consider the first order system and respective boundary conditions

$$
\begin{cases}
u' + f_1(u,v) + g_1 = 0, \text{ in } [0,1] \\[2mm]
v' + f_2(u,v) + g_2 = 0, \text{ in } [0,1] \\[2mm]
u(0) = u_0, \ \ v(1) = v_f,
\end{cases}
\tag{26.17}
$$

Linearizing the equations about the first solutions $\tilde{u}$, and $\tilde{v}$, we obtain,

$$
u' + f_1(\tilde{u}, \tilde{v}) + \frac{\partial f_1(\tilde{u}, \tilde{v})}{\partial u}(u - \tilde{u})
$$
$$
+ \frac{\partial f_1(\tilde{u}, \tilde{v})}{\partial v}(v - \tilde{v}) + g_1 = 0,
\tag{26.18}
$$

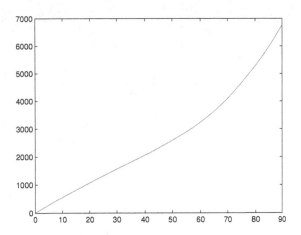

Figure 26.4: The solution x (in m) for $t_f = 90s$.

$$v' + f_2(\tilde{u}, \tilde{v}) + \frac{\partial f_2(\tilde{u}, \tilde{v})}{\partial u}(u - \tilde{u})$$

$$+ \frac{\partial f_2(\tilde{u}, \tilde{v})}{\partial v}(v - \tilde{v}) + g_2 = 0. \tag{26.19}$$

In finite differences, we could write,

$$u_n - u_{n-1} + f_1(\tilde{u}_{n-1}, \tilde{v}_{n-1})d + \frac{\partial f_1(\tilde{u}_{n-1}, \tilde{v}_{n-1})}{\partial u}(u_{n-1} - \tilde{u}_{n-1})d$$

$$+ \frac{\partial f_1(\tilde{u}_{n-1}, \tilde{v}_{n-1})}{\partial v}(v_{n-1} - \tilde{v}_{n-1})d + (g_1)_{n-1}d = 0, \tag{26.20}$$

$$v_n - v_{n-1} + f_2(\tilde{u}_{n-1}, \tilde{v}_{n-1})d + \frac{\partial f_2(\tilde{u}_{n-1}, \tilde{v}_{n-1})}{\partial u}(u_{n-1} - \tilde{u}_{n-1})d$$

$$+ \frac{\partial f_2(\tilde{u}_{n-1}, \tilde{v}_{n-1})}{\partial v}(v_{n-1} - \tilde{v}_{n-1})d + (g_2)_{n-1}d = 0. \tag{26.21}$$

Hence, we may write,

$$u_n = a_n u_{n-1} + b_n v_{n-1} + c_n,$$

$$v_n = d_n u_{n-1} + e_n v_{n-1} + f_n,$$

Where

$$a_n = -\frac{\partial f_1(\tilde{u}_{n-1}, \tilde{v}_{n-1})}{\partial u}d + 1,$$

$$b_n = -\frac{\partial f_1(\tilde{u}_{n-1}, \tilde{v}_{n-1})}{\partial v}d,$$

$$c_n = -f_1(\tilde{u}_{n-1}, \tilde{v}_{n-1})d + \frac{\partial f_1(\tilde{u}_{n-1}, \tilde{v}_{n-1})}{\partial u}\tilde{u}_{n-1}d$$

$$+ \frac{\partial f_1(\tilde{u}_{n-1}, \tilde{v}_{n-1})}{\partial v}\tilde{v}_{n-1}d - (g_1)_{n-1}d, \tag{26.22}$$

and

$$d_n = -\frac{\partial f_2(\tilde{u}_{n-1}, \tilde{v}_{n-1})}{\partial u} d,$$

$$e_n = -\frac{\partial f_2(\tilde{u}_{n-1}, \tilde{v}_{n-1})}{\partial v} d + 1,$$

$$
\begin{aligned}
f_n &= -f_2(\tilde{u}_{n-1}, \tilde{v}_{n-1}) d + \frac{\partial f_2(\tilde{u}_{n-1}, \tilde{v}_{n-1})}{\partial u} \tilde{u}_{n-1} d \\
&\quad + \frac{\partial f_2(\tilde{u}_{n-1}, \tilde{v}_{n-1})}{\partial v} \tilde{v}_{n-1} d - (g_2)_{n-1} d.
\end{aligned}
\tag{26.23}
$$

In particular, for $n = 1$, we get

$$u_1 = a_1 u_0 + b_1 v_0 + c_1, \tag{26.24}$$

and

$$v_1 = d_1 u_0 + e_1 v_0 + f_1. \tag{26.25}$$

From this last equation,

$$v_0 = (v_1 - d_1 u_0 - f_1)/e_1,$$

so that from this and equation (26.24), we get,

$$u_1 = a_1 u_0 + b_1(v_1 - d_1 u_0 - f_1)/e_1 + c_1 = F_1 v_1 + G_1,$$

where

$$F_1 = b_1/e_1, \quad G_1 = a_1 u_0 - b_1(d_1 u_0 + f_1)/e_1 + c_1.$$

Reasoning inductively, having,

$$u_{n-1} = F_{n-1} v_{n-1} + G_{n-1},$$

we also have,

$$u_n = a_n u_{n-1} + b_n v_{n-1} + c_n,$$

$$v_n = d_n u_{n-1} + e_n v_{n-1} + f_n,$$

$$v_n = d_n(F_{n-1} v_{n-1} + G_{n-1}) + e_n v_{n-1} + f_n,$$

that is,

$$v_{n-1} = H_n v_n + L_n,$$

where

$$H_n = 1/(d_n F_{n-1} + e_n),$$

$$L_n = -H_n(d_n G_{n-1} + f_n).$$

Hence,

$$u_n = a_n(F_{n-1} v_{n-1} + G_{n-1}) + b_n v_{n-1} + c_{n-1},$$

so that,

$$u_n = a_n(F_{n-1}(H_n v_n + L_n) + G_{n-1}) + b_n(H_n v_n + L_n) + c_{n-1},$$

and hence,

$$F_n = a_n F_{n-1} H_n + b_n H_n,$$

and

$$G_n = a_n(F_{n-1}L_n + G_{n-1}) + b_nL_n + c_{n-1}.$$

Thus,

$$u_n = F_nv_n + G_n,$$

so that, in particular,

$$u_N = F_Nv_f + G_N,$$

$$v_{N-1} = H_Nv_f + L_N,$$

and hence,

$$u_{N-1} = F_{N-1}v_{N-1} + G_{N-1},$$

$$v_{N-2} = H_{N-1}v_{N-1} + L_{N-1},$$

and so on, up to finding,

$$u_1 = F_1v_1 + G_1,$$

and

$$v_0 = H_0v_1 + L_0,$$

where $H_0 = 1/e_1$ and $L_0 = -(d_1u_0 + f_1)/e_1$.

The next step is to replace $\{\tilde{u}_n\}$ and $\{\tilde{v}_n\}$ by $\{u_n\}$ and $\{v_n\}$ respectively and then to repeat the process up the satisfaction of an appropriate convergence criterion.

26.4.1 An example in nuclear physics

As an application of the method above exposed we develop numerical results for the system of equations relating to the neutron kinetics of a nuclear reactor. Following [71], the system in question is given by:

$$\begin{cases} n'(t) = \frac{(\rho(T)-\beta)}{L}n(t) + \lambda C(t) \\ C'(t) = \frac{\beta}{L}n(t) - \lambda C(t) \\ T'(t) = Hn(t), \end{cases} \tag{26.26}$$

where $n(t)$ is the neutron population, $C(t)$ is the concentration of delayed neutrons, $T(t)$ is the core temperature, $\rho(T)$ is the reactivity (which depends on the temperature T), β is the delayed neutron fraction, L is the prompt reactors generation time, λ is the average decay constant of the precursors and H is the inverse of the reactor thermal capacity.

For our numerical examples we consider $T(0s) = 300K$ and $T(100s) = T_f = 350K$. Moreover we assume the relation,

$$C(0) = \frac{1}{\lambda}\frac{(\beta - \rho(0))}{L}n(0),$$

where $n(0)$ is unknown (to be numerically calculated by our method such that we have $T(100s) = T_f$).

Also we consider

$$\rho(T) = \rho(0) - \alpha(T - T(0)).$$

The remaining values are: $\beta = 0.0065$, $L = 0.0001s$, $\lambda = 0.00741s^{-1}$, $H = 0.05K/(MWs)$, $\alpha = 5 \cdot 10^{-5}K^{-1}$, and $\rho(0) = 0.2\beta$.

First we linearize the system in question about $(\tilde{n}, \tilde{T})$ obtaining (in fact its a first approximation)

$$\begin{aligned} n'(t) &= \frac{\rho(\tilde{T})-\beta}{L}n(t) + \frac{\rho(T)-\beta}{L}\tilde{n}(t) \\ &\quad -\frac{\rho(\tilde{T})-\beta}{L}\tilde{n}(t) + \lambda C(t), \end{aligned} \tag{26.27}$$

$$C'(t) = \frac{\beta}{L} n(t) - \lambda C(t),$$

$$T'(t) = Hn(t),$$

where $\rho(T) = \rho(0) - \alpha(T - T(0))$.

Discretizing such a system in finite differences, we get

$$
\begin{aligned}
(n_{i+1} - n_i)/d \;=\; & \frac{\rho(\tilde{T}_i) - \beta}{L} n_i + \frac{\rho(T_i) - \beta}{L} \tilde{n}_i \\
& - \frac{\rho(\tilde{T}_i) - \beta}{L} \tilde{n}_i + \lambda C_i,
\end{aligned}
\tag{26.28}
$$

$$(C_{i+1} - C_i)/d = \frac{\beta}{L} n_i - \lambda C_i,$$

$$(T_{i+1} - T_i)/d = Hn_i,$$

where $d = 100s/N$, where N is the number of nodes.

Hence, we may write

$$n_{i+1} = a_i n_i + b_i T_i + d_i C_i + e_i, \tag{26.29}$$

$$C_{i+1} = fn_i + gC_i, \tag{26.30}$$

$$T_{i+1} = hT_i + mn_i, \tag{26.31}$$

where

$$a_i = 1 + \frac{\rho(\tilde{T}_i) - \beta}{L} d,$$

$$b_i = \frac{-\alpha}{L} \tilde{n}_i d,$$

$$d_i = \lambda d,$$

$$e_i = \frac{(\rho(0) + \alpha T(0) - \beta)}{L} \tilde{n}_i d - \frac{(\rho(\tilde{T}_i) - \beta)}{L} \tilde{n}_i d,$$

$$f = \frac{\beta}{L} d,$$

$$g = 1 - \lambda d,$$

$$h = 1,$$

$$m = Hd.$$

Observe that

$$C_0 = \tilde{\alpha} n_0,$$

where

$$\tilde{\alpha} = \frac{\beta - \rho(0)}{L\lambda}.$$

For $i = 1$ from (26.31) we obtain

$$n_0 = \frac{T_1 - hT_0}{m} = \alpha_1 T_1 + \beta_1, \tag{26.32}$$

where $\alpha_1 = 1/m$ and $\beta_1 = -(h/m)T_0$.

Therefore,

$$C_0 = \tilde{\alpha} n_0 = \tilde{\alpha}(\alpha_1 T_1 + \beta_1).$$

Still for $i = 1$, replacing this last relation and (26.32) into (26.29), we get

$$n_1 = a_1(\alpha_1 T_1 + \beta_1) + b_1 T_0 + d_1 \tilde{\alpha}(\alpha_1 T_1 + \beta_1) + e_1,$$

so that

$$n_1 = \tilde{\alpha}_1 T_1 + \tilde{\beta}_1, \tag{26.33}$$

where

$$\tilde{\alpha}_1 = a_1 \alpha_1 + d_1 \tilde{\alpha} \alpha_1,$$

and

$$\tilde{\beta}_1 = a_1 \beta_1 + b_1 T_0 + d_1 \tilde{\alpha} \beta_1 + e_1.$$

Finally, from (26.30),

$$C_1 = f(\alpha_1 T_1 + \beta_1) + g \tilde{\alpha}(\alpha_1 T_1 + \beta_1)$$
$$= \hat{\alpha}_1 T_1 + \hat{\beta}_1, \tag{26.34}$$

where

$$\hat{\alpha}_1 = f \alpha_1 + g \tilde{\alpha} \alpha_1,$$

and

$$\hat{\beta}_1 = f \beta_1 + g \tilde{\alpha} \beta_1.$$

Reasoning inductively, having

$$n_i = \tilde{\alpha}_i T_i + \tilde{\beta}_i, \tag{26.35}$$

$$n_{i-1} = \alpha_i T_i + \beta_i, \tag{26.36}$$

$$C_i = \hat{\alpha}_i T_i + \hat{\beta}_i, \tag{26.37}$$

we are going to obtain the corresponding relations for $i + 1$, $i \geq 1$. From (26.31) and (26.35) we obtain

$$T_{i+1} = hT_i + m(\tilde{\alpha}_i T_i + \tilde{\beta}_i),$$

so that

$$T_i = \eta_i T_{i+1} + \xi_i, \tag{26.38}$$

where

$$\eta_i = (h + m\tilde{\alpha}_i)^{-1},$$

and

$$\xi_i = -(m\tilde{\beta}_i)\eta_i.$$

On the other hand, from (26.29), (26.35) and (26.37) we have

$$n_{i+1} = a_i(\tilde{\alpha}_i T_i + \tilde{\beta}_i) + b_i T_i + d_i(\hat{\alpha}_i T_i + \hat{\beta}_i) + e_i,$$

so that from this and (26.38), we obtain

$$n_{i+1} = \tilde{\alpha}_i T_{i+1} + \tilde{\beta}_{i+1},$$

where

$$\tilde{\alpha}_{i+1} = a_i \tilde{\alpha}_i \eta_i + b_i \eta_i + d_i \hat{\alpha}_i \eta_i,$$

and

$$\tilde{\beta}_{i+1} = a_i(\tilde{\alpha}_i \xi_i + \tilde{\beta}_i) + b_i \xi_i + d_i(\hat{\alpha}_i \xi_i + \hat{\beta}_i) + e_i.$$

Also from (26.35) and (26.38) we have,

$$n_i = \tilde{\alpha}_i(\eta_i T_{i+1} + \xi_i) + \tilde{\beta}_i = \alpha_{i+1} T_{i+1} + \beta_{i+1},$$

where

$$\alpha_{i+1} = \tilde{\alpha}_i \eta_i,$$

and

$$\beta_{i+1} = \tilde{\alpha}_i \xi_i + \tilde{\beta}_i.$$

Moreover,

$$
\begin{aligned}
C_{i+1} &= f n_i + g C_i \\
&= f(\alpha_{i+1} T_{i+1} + \beta_{i+1}) \\
&\quad + g(\hat{\alpha}_i T_i + \hat{\beta}_i) \\
&= f(\alpha_{i+1} T_{i+1} + \beta_{i+1}) \\
&\quad + g(\hat{\alpha}_i(\eta_i T_i + \xi_i) + \hat{\beta}_i) \\
&= \hat{\alpha}_{i+1} T_{i+1} + \hat{\beta}_{i+1},
\end{aligned}
\tag{26.39}
$$

where

$$\hat{\alpha}_{i+1} = f \alpha_{i+1} + g \hat{\alpha}_i \eta_i,$$

and

$$\hat{\beta}_i = f \beta_{i+1} + g \hat{\alpha}_i \xi_i + g \hat{\beta}_i.$$

Summarizing, we have obtained linear functions $(F_0)_i, (F_1)_i$ and $(F_2)_i$ such that

$$T_i = (F_0)_i(T_{i+1}),$$
$$n_i = (F_1)_i(T_{i+1}),$$
$$C_i = (F_2)_i(T_{i+1}),$$

$\forall i \in \{1, ..., N-1\}$.

Thus, considering the known value $T_N = T_f$ we obtain

$$T_{N-1} = (F_0)_{N-1}(T_f),$$
$$n_{N-1} = (F_1)_{N-1}(T_f),$$
$$C_{N-1} = (F_2)_{N-1}(T_f),$$

and having T_{N-1} we get,

$$T_{N-2} = (F_0)_{N-2}(T_{N-1}),$$
$$n_{N-2} = (F_1)_{N-2}(T_{N-1}),$$
$$C_{N-2} = (F_2)_{N-1}(T_{N-1}),$$

and so on, up to finding

$$T_1 = (F_0)_1(T_2),$$
$$n_1 = (F_1)_1(T_2),$$
$$C_1 = (F_2)_1(T_2),$$

and $n_0 = (F_0)_1(T_1)$.

The next step is to replace $(\tilde{n}, \tilde{T})$ by the last calculated (n, T) and then to repeat the process until an appropriate convergence criterion is satisfied.

Concerning our numerical results through such a method, for the solution $n(t)$ obtained, please see Figure 26.5. For the solution $T(t)$, see Figure 26.6.

We emphasize the numerical results here obtained are consistent with the current literature (see [71], for details).

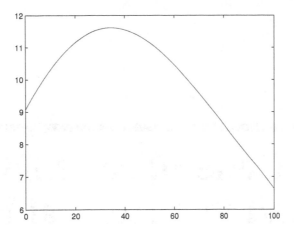

Figure 26.5: Solution $n(t)$ for $0s \leq t \leq 100s$.

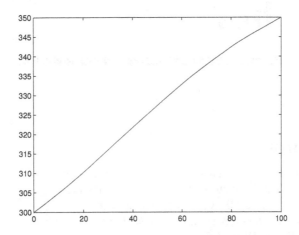

Figure 26.6: Solution $T(t)$ for $0s \leq t \leq 100s$.

26.5   Conclusion

In this chapter, we have developed a method for solving a class of first order ordinary differential equations.

The results are applied to a flight mechanics problem which models the in plane climbing of an airplane. It is worth mentioning the algorithm obtained is of easy implementation and very efficient from a computational point of view.

Finally, we would highlight the numerical results obtained are perfectly consistent with the physical problem context. In future works we intend to apply the method to solve relating optimal control problems.

Chapter 27

On the Generalized Method of Lines and its Proximal Explicit and Hyper-Finite Difference Approaches

27.1 Introduction

This chapter develops two improvements relating to the generalized method of lines. In our previous publi
cations [19, 18], we highlight the method there addressed may present a relevant error as a parameter $\varepsilon > 0$
is too small, that is, as ε is about 0.01, 0.001 or even smaller.

In the present section we develop a solution for such a problem through a proximal formulation suitable
for a large class of non-linear elliptic PDEs.

At this point we reintroduce the generalized method of lines, originally presented in F. Botelho [19]
In the present context we add new theoretical and applied results to the original presentation. Specially the
computations are all completely new. Consider first the equation

$$\varepsilon \nabla^2 u + g(u) + f = 0, \text{ in } \Omega \subset \mathbb{R}^2, \tag{27.1}$$

with the boundary conditions

$$u = 0 \text{ on } \Gamma_0 \text{ and } u = u_f, \text{ on } \Gamma_1.$$

From now on we assume that u_f, g and f are smooth functions (we mean C^∞ functions), unless otherwis
specified. Here Γ_0 denotes the internal boundary of Ω and Γ_1 the external one. Consider the simpler cas
where

$$\Gamma_1 = 2\Gamma_0,$$

and suppose there exists $r(\theta)$, a smooth function such that

$$\Gamma_0 = \{(\theta, r(\theta)) \mid 0 \le \theta \le 2\pi\},$$

being $r(0) = r(2\pi)$.

In polar coordinates the above equation may be written as

$$\frac{\partial^2 u}{\partial r^2} + \frac{1}{r}\frac{\partial u}{\partial r} + \frac{1}{r^2}\frac{\partial^2 u}{\partial \theta^2} + g(u) + f = 0, \text{ in } \Omega, \tag{27.2}$$

and

$$u = 0 \text{ on } \Gamma_0 \text{ and } u = u_f, \text{ on } \Gamma_1.$$

Define the variable t by

$$t = \frac{r}{r(\theta)}.$$

Also defining $\bar{u}$ by

$$u(r,\theta) = \bar{u}(t,\theta),$$

dropping the bar in $\bar{u}$, equation (27.1) is equivalent to

$$\begin{aligned}
\frac{\partial^2 u}{\partial t^2} \quad & + \quad \frac{1}{t}f_2(\theta)\frac{\partial u}{\partial t} \\
& + \quad \frac{1}{t}f_3(\theta)\frac{\partial^2 u}{\partial \theta \partial t} + \frac{f_4(\theta)}{t^2}\frac{\partial^2 u}{\partial \theta^2} \\
& + f_5(\theta)(g(u) + f) = 0,
\end{aligned} \tag{27.3}$$

in Ω. Here $f_2(\theta)$, $f_3(\theta)$, $f_4(\theta)$ and $f_5(\theta)$ are known functions.

More specifically, denoting

$$f_1(\theta) = \frac{-r'(\theta)}{r(\theta)},$$

we have:

$$f_2(\theta) = 1 + \frac{f_1'(\theta)}{1 + f_1(\theta)^2},$$

$$f_3(\theta) = \frac{2f_1(\theta)}{1 + f_1(\theta)^2},$$

and

$$f_4(\theta) = \frac{1}{1 + f_1(\theta)^2}.$$

Observe that $t \in [1,2]$ in Ω. Discretizing in t (N equal pieces which will generate N lines) we obtain the equation

$$\begin{aligned}
& \frac{u_{n+1} - 2u_n + u_{n-1}}{d^2} + \frac{(u_n - u_{n-1})}{d}\frac{1}{t_n}f_2(\theta) \\
& + \frac{\partial(u_n - u_{n-1})}{\partial \theta}\frac{1}{t_n d}f_3(\theta) + \frac{\partial^2 u_n}{\partial \theta^2}\frac{f_4(\theta)}{t_n^2} \\
& + f_5(\theta)\left(g(u_n)\frac{1}{\varepsilon} + f_n\frac{1}{\varepsilon}\right) = 0,
\end{aligned} \tag{27.4}$$

$\forall n \in \{1,...,N-1\}$. Here, $u_n(\theta)$ corresponds to the solution on the line n. Thus we may write

$$u_n = T_n(u_{n-1}, u_n, u_{n+1}),$$

where

$$\begin{aligned}
T_n(u_{n-1}, u_n, u_{n+1}) \quad = \quad & \left(u_{n+1} + u_n + u_{n-1} + \frac{(u_n - u_{n-1})}{d}\frac{1}{t_n}f_2(\theta)d^2\right. \\
& + \frac{\partial(u_n - u_{n-1})}{\partial \theta}\frac{1}{t_n d}f_3(\theta)d^2 + \frac{\partial^2 u_n}{\partial \theta^2}\frac{f_4(\theta)}{t_n^2}d^2 \\
& \left. + f_5(\theta)\left(g(u_n)\frac{d^2}{\varepsilon} + f_n\frac{d^2}{\varepsilon}\right)\right)/3.0.
\end{aligned} \tag{27.5}$$

27.1.1 Some preliminaries results and the main algorithm

Now we recall a classical definition.

Definition 27.1.1 Let C be a subset of a Banach space U and let $T : C \to C$ be an operator. Thus T is said to be a contraction mapping if there exists $0 \le \alpha < 1$ such that

$$\|T(x_1) - T(x_2)\|_U \le \alpha \|x_1 - x_2\|_U, \forall x_1, x_2 \in C.$$

Remark 27.1.2 Observe that if $\|T'(x)\|_U \le \alpha < 1$, on a convex set C then T is a contraction mapping, since by the mean value inequality,

$$\|T(x_1) - T(x_2)\|_U \le \sup_{x \in C} \{\|T'(x)\|\} \|x_1 - x_2\|_U, \forall x_1, x_2 \in C.$$

The next result is the base of our generalized method of lines. For a proof see Theorem 15.7.3.

Theorem 27.1.3 (Contraction mapping theorem) Let C be a closed subset of a Banach space U. Assume T is contraction mapping on C, then there exists a unique $\tilde{x} \in C$ such that $\tilde{x} = T(\tilde{x})$. Moreover, for an arbitrary $x_0 \in C$ defining the sequence

$$x_1 = T(x_0) \text{ and } x_{k+1} = T(x_k), \forall k \in \mathbb{N}$$

we have

$$x_k \to \tilde{x}, \text{ in norm, as } k \to +\infty.$$

To obtain a fixed point for each T_n indicated in (27.5) is perfectly possible if $\varepsilon \approx \mathcal{O}(1)$. However, if $\varepsilon >$ is small, the error in this process may be relevant.

To solve this problem, firstly we propose the following algorithm,

1. Choose $K \approx 30 - 80$ and set $u_0 = \mathbf{0}$.

2. Calculate $u = \{u_n\}$ by solving the equation

$$
\begin{aligned}
u_{n+1} - 2u_n + u_{n-1} &+ \frac{(u_n - u_{n-1})}{d} \frac{1}{t_n} f_2(\theta) d^2 \\
&+ \frac{\partial(u_n - u_{n-1})}{\partial \theta} \frac{1}{t_n d} f_3(\theta) d^2 + \frac{\partial^2 u_n}{\partial \theta^2} \frac{f_4(\theta)}{t_n^2} d^2 \\
&+ f_5(\theta) \left(g(u_n) \frac{d^2}{\varepsilon} + f_n \frac{d^2}{\varepsilon} \right) \\
&- K(u_n - (u_0)_n) \frac{d^2}{\varepsilon} \\
&= 0.
\end{aligned}
\tag{27.}
$$

Such an equation is solved through the Banach fixed point theorem, that is, defining

$$
\begin{aligned}
T_n(u_n, u_{n+1}, u_{n-1}) &= \left(u_{n+1} + u_n + u_{n-1} + \frac{(u_n - u_{n-1})}{d} \frac{1}{t_n} f_2(\theta) d^2 \right. \\
&+ \frac{\partial(u_n - u_{n-1})}{\partial \theta} \frac{1}{t_n d} f_3(\theta) d^2 + \frac{\partial^2 u_n}{\partial \theta^2} \frac{f_4(\theta)}{t_n^2} d^2 \\
&+ f_5(\theta) \left(g(u_n) \frac{d^2}{\varepsilon} + f_n \frac{d^2}{\varepsilon} \right) \\
&+ \left. K(u_0)_n \frac{d^2}{\varepsilon} \right) / \left(3 + K \frac{d^2}{\varepsilon} \right)
\end{aligned}
\tag{27.}
$$

equation (27.7) stands for

$$u_n = T_n(u_{n-1}, u_n, u_{n+1}),$$

so that for $n = 1$ we have

$$u_1 = T_1(0, u_1, u_2).$$

We may use the Contraction Mapping theorem to calculate u_1 as a function of u_2. The procedure would be,

(a) set $x_1 = u_2$,

(b) obtain recursively

$$x_{k+1} = T_1(0, x_k, u_2),$$

(c) and finally get

$$u_1 = \lim_{k \to \infty} x_k = g_1(u_2).$$

Thus, we have obtained

$$u_1 = g_1(u_2).$$

We can repeat the process for $n = 2$, that is, we can solve the equation

$$u_2 = T_2(u_1, u_2, u_3),$$

which from above stands for

$$u_2 = T_2(g_1(u_2), u_2, u_3).$$

The procedure would be:

(a) Set $x_1 = u_3$,

(b) calculate

$$x_{k+1} = T_2(g_1(x_k), x_k, u_3),$$

(c) obtain

$$u_2 = \lim_{k \to \infty} x_k = g_2(u_3).$$

We proceed in this fashion until obtaining

$$u_{N-1} = g_{N-1}(u_N) = g_{N-1}(u_f).$$

Being u_f known we have obtained u_{N-1}. We may then calculate

$$u_{N-2} = g_{N-2}(u_{N-1}),$$

$$u_{N-3} = g_{N-3}(u_{N-2}),$$

and so on, up to finding

$$u_1 = g_1(u_2).$$

Thus, this part of the problem is solved.

3. Set $u_0 = u$ and go to item 2 up to the satisfaction of an appropriate convergence criterion.

Remark 27.1.4 *Here we consider some points concerning the convergence of the method.*
In the next lines the norm indicated refers to the infinity one for $C([0,2\pi];\mathbb{R}^{N-1})$. In particular for $n=1$
from above we have:

$$u_1 = T_1(0,u_1,u_2),$$

that is,

$$u_2 - 2u_1 - K\frac{d^2}{\varepsilon}u_1 + \mathcal{O}\left(K\frac{d^2}{\varepsilon}\right) = 0.$$

Hence, denoting

$$a[1] = 1/\left(2 + K\frac{d^2}{\varepsilon}\right)$$

and

$$a[n] = 1/\left(2 + K\frac{d^2}{\varepsilon} - a[n-1]\right), \, \forall n \in \{2,\ldots,N-1\},$$

for N sufficiently big we may obtain

$$\|u_1 - a[1]u_2\| = \mathcal{O}\left(K\frac{d^2}{\varepsilon}\right),$$

and by induction

$$\|u_n - a[n]u_{n+1}\| = n\mathcal{O}\left(K\frac{d^2}{\varepsilon}\right),$$

so that we would have

$$\|u_n - a[n]u_{n+1}\| \leq \mathcal{O}\left(K\frac{d}{\varepsilon}\right)), \forall n \in \{1,\ldots,N-1\}$$

This last calculation is just to clarify that the procedure of obtaining the relation between consecutive line
through the contraction mapping theorem is well defined.

27.1.2 A numerical example, the proximal explicit approach

In this section we present a numerical example. Consider the equation

$$\varepsilon\nabla^2 u + g(u) + 1 = 0, \text{ in } \Omega \subset \mathbb{R}^2, \tag{27.8}$$

where, for a Ginzburg-Landau type equation (see [4, 55] for the corresponding models in physics),

$$g(u) = -u^3 + u,$$

with the boundary conditions

$$u = u_1 \text{ on } \partial\Omega = \Gamma_0 \cup \Gamma_1,$$

where $\Omega = \{(r,\theta) : 1 \leq r \leq 2, 0 \leq \theta \leq 2\pi\}$,

$$u_1 = 0, \text{ on } \Gamma_0 = \{(1,\theta) : 0 \leq \theta \leq 2\pi\},$$

$$u_1 = u_f(\theta), \text{ on } \Gamma_1 = \{(2,\theta) : 0 \leq \theta \leq 2\pi\}.$$

Through the generalized method of lines, for $N = 10$ (10 lines), $d = 1/N$ in polar coordinates and fini
differences (please see [73] for general schemes in finite differences), equation (27.8), stands for

$$(u_{n+1} - 2u_n + u_{n-1}) + \frac{1}{r_n}(u_n - u_{n-1})d + \frac{1}{r_n^2}\frac{\partial^2 u_n}{\partial\theta^2}d^2 + (-u_n^3 + u_n)\frac{d^2}{\varepsilon} + \frac{d^2}{\varepsilon} = 0,$$

$\forall n \in \{1, \ldots, N-1\}$.

At this point we present, through the generalized method of lines, the concerning algorithm which may be for the softwares maple or mathematica.

In this software, x stands for θ.

$m_8 = 10$; (number of lines)
$d = 1.0/m_8$; (thickness of the grid)
$e_1 = 0.01$; $(\varepsilon = e_1)$
$K = 70.0$;
$Clear[d_1, u, a, b, h]$;
$For[i = 1, i < m_8, i{+}{+},$
$z_1[i] = 0.0$; (vector which stores $Ku_0(i)$)
$For[k_1 = 1, k_1 < 180, k_1{+}{+},$
$Print[k_1]$;
$a = 0.0$;
$For[i = 1, i < m_8, i{+}{+},$
$Print[i]$;
$t = 1.0 + i * d$;
$b[x_-] = u[i+1][x]$;
$b_{12} = 2.0$;
$A_{18} = 5.0$;
$k = 1$;
$While[b_{12} > 10^{-4},$
$k = k+1$;
$z = (u[i+1][x] + b[x] + a + 1/t * (b[x] - a) * d * d_1^2 + 1/t^2 * D[b[x], \{x, 2\}] * d^2 * d_1^2$
$-b[x]^3 * d^2 * d_1^2/e_1 + b[x] * d^2 * d_1^2/e_1 + 1.0 * d^2/e_1 +$
$z_1[i] * d^2/e_1)/(3.0 + K * d^2/e_1)$;
$z = Series[z, \{d_1, 0, 2\}, \{u_f[x], 0, 3\}, \{u'_f[x], 0, 1\}, \{u''_f[x], 0, 1\}, \{u'''_f[x], 0, 0\}, \{u''''_f[x], 0, 0\}]$;
$z = Normal[z]$;
$z = Expand[z]$;
$b[x_-] = z$;
$u[i+1][x_-] = 0.0$;
$u_f[x_-] = 0.0$;
$d_1 = 1.0$;
$A_{19} = z$;
$b_{12} = Abs[A_{19} - A_{18}]$;
$A_{18} = A_{19}$;
$Clear[u, u_f, d_1]]$;
$a_1 = b[x]$;
$Clear[b]$;
$u[i+1][x_-] = b[x]$;
$h[i] = a_1$;
$a = a_1]$;
$b[x_-] = u_f[x]$;
$d_1 = 1.0$;

(27.9)

$$For[i = 1, i < m_8, i{+}{+},$$
$$W_1[m_8 - i] =$$
$$Series[h[m_8 - i], \{u_f[x], 0, 3\}, \{u'_f[x], 0, 1\}, \{u''_f[x], 0, 1\}, \{u'''_f[x], 0, 0\}, \{u''''_f[x], 0, 0\}];$$
$$W = Normal[W_1[m_8 - i]];$$
$$b[x_-] = Expand[W];$$
$$v[m_8 - i] = Expand[W]];$$
$$For[i = 1, i < m_8, i{+}{+},$$
$$z_1[i] = K * v[i]];$$
$$d_1 = 1.0;$$
$$Print[Expand[v[m_8/2]]];$$
$$Clear[d_1, u, b]]$$

$$(27.10)$$

At this point we present the expressions for 10 lines, firstly for $\varepsilon = 1$ and $K = 0$. In the next lines x stands for θ.

For each line $u[n]$ we have obtained,

$$\begin{aligned}
u[1] = {}& 0.0588308 + 0.167434u_f[x] - 0.00488338u_f[x]^2 - 0.00968371u_f[x]^3 + 0.0126122u''_f[x] \\
& - 0.000641358u_f[x]\,u''_f[x] - 0.00190065u_f[x]^2\,u''_f[x] + 0.0000660286u_f[x]^3\,u''_f[x]
\end{aligned}$$

$$\begin{aligned}
u[2] = {}& 0.101495 + 0.316995u_f[x] - 0.00919963u_f[x]^2 - 0.0182924u_f[x]^3 + 0.0225921u''_f[x] \\
& - 0.00113308u_f[x]u''_f[x] - 0.00336691u_f[x]^2u''_f[x] + 0.000122652u_f[x]^3\,u''_f[x]
\end{aligned}$$

$$\begin{aligned}
u[3] = {}& 0.1295 + 0.450424u_f[x] - 0.0127925u_f[x]^2 - 0.0257175u_f[x]^3 + 0.0294791u''_f[x] \\
& - 0.00142448u_f[x]\,u''_f[x] - 0.00428071u_f[x]^2\,u''_f[x] + 0.000160933u_f[x]^3u''_f[x]
\end{aligned}$$

$$\begin{aligned}
u[4] = {}& 0.143991 + 0.568538u_f[x] - 0.0153256u_f[x]^2 - 0.0315703u_f[x]^3 + 0.0331249u''_f[x] \\
& - 0.0014821u_f[x]\,u''_f[x] - 0.00456619u_f[x]^2\,u''_f[x] + 0.000168613u_f[x]^3\,u''_f[x]
\end{aligned}$$

$$\begin{aligned}
u[5] = {}& 0.146024 + 0.672307u_f[x] - 0.0164357u_f[x]^2 - 0.0352883u_f[x]^3 + 0.0336371u''_f[x] \\
& - 0.00131323u_f[x]\,u''_f[x] - 0.00421976u_f[x]^2\,u''_f[x] + 0.000141442u_f[x]^3\,u''_f[x]
\end{aligned}$$

$$\begin{aligned}
u[6] = {}& 0.136541 + 0.762571u_f[x] - 0.0158578u_f[x]^2 - 0.0361974u_f[x]^3 + 0.0312624u''_f[x] \\
& - 0.000974635u_f[x]\,u''_f[x] - 0.00333176u_f[x]^2\,u''_f[x] + 0.0000901069u_f[x]^3\,u''_f[x]
\end{aligned}$$

$$\begin{aligned}
u[7] = {}& 0.116389 + 0.840008u_f[x] - 0.0135378u_f[x]^2 - 0.0336098u_f[x]^3 + 0.0263271u''_f[x] \\
& - 0.000565842u_f[x]\,u''_f[x] - 0.00210507u_f[x]^2\,u''_f[x] + 0.0000375075u_f[x]^3\,u''_f[x]
\end{aligned}$$

$$\begin{aligned}
u[8] = {}& 0.0864095 + 0.905032u_f[x] - 0.00970893u_f[x]^2 - 0.0269167u_f[x]^3 + 0.0192033u''_f[x] \\
& - 0.000206117u_f[x]\,u''_f[x] - 0.000856701u_f[x]^2u''_f[x] + 5.78758 * 10^{-6}u_f[x]^3\,u''_f[x]
\end{aligned}$$

$$u[9] = 0.0473499 + 0.958203u_f[x] - 0.00491745u_f[x]^2 - 0.0157466u_f[x]^3 + 0.0102907u''_f[x].$$

In the next lines we present the results relating to the software indicated, with $\varepsilon = 0.01$ and $K = 70$. For each line $u[n]$ we have obtained,

$$
\begin{aligned}
u[1] \;=\; & 1.08673 + 4.6508 * 10^{-7} u_f[x] - 1.11484 * 10^{-7} u_f[x]^2 + 3.25552 * 10^{-8} u_f[x]^3 \\
& + 5.13195 * 10^{-9} u_f''[x] - 2.24741 * 10^{-9} u_f[x]\, u_f''[x] \\
& + 8.97477 * 10^{-10} u_f[x]^2\, u_f''[x] - 8.86957 * 10^{-11} u_f[x]^3\, u_f''[x]
\end{aligned}
$$

$$
\begin{aligned}
u[2] \;=\; & 1.27736 + 1.51118 * 10^{-6} u_f[x] - 4.39811 * 10^{-7} u_f[x]^2 + 1.55883 * 10^{-7} u_f[x]^3 \\
& + 1.33683 * 10^{-8} u_f''[x] - 7.50593 * 10^{-9} u_f[x]\, u_f''[x] \\
& + 3.77548 * 10^{-9} u_f[x]^2\, u_f''[x] - 4.9729 * 10^{-10} u_f[x]^3\, u_f''[x]
\end{aligned}
$$

$$
\begin{aligned}
u[3] \;=\; & 1.30559 + 6.91602 * 10^{-6} u_f[x] - 2.21813 * 10^{-6} u_f[x]^2 + 8.89891 * 10^{-7} u_f[x]^3 \\
& + 4.85851 * 10^{-8} u_f''[x] - 3.22542 * 10^{-8}\, u_f[x]\, u_f''[x] \\
& + 1.90602 * 10^{-8} u_f[x]^2\, u_f''[x] - 3.20439 * 10^{-9} u_f[x]^3\, u_f''[x]
\end{aligned}
$$

$$
\begin{aligned}
u[4] \;=\; & 1.30968 + 0.0000354152 u_f[x] - 0.0000116497 u_f[x]^2 + 5.06682 * 10^{-6} u_f[x]^3 \\
& + 1.93523 * 10^{-7} u_f''[x] - 1.42948 * 10^{-7} u_f[x]\, u_f''[x] \\
& + 9.56419 * 10^{-8} u_f[x]^2\, u_f''[x] - 2.02133 * 10^{-8} u_f[x]^3\, u_f''[x]
\end{aligned}
$$

$$
\begin{aligned}
u[5] \;=\; & 1.31014 + 0.000193272 u_f[x] - 0.0000616231 u_f[x]^2 + 0.0000278964 u_f[x]^3 \\
& + 7.96293 * 10^{-7}\, u_f''[x] - 6.22727 * 10^{-7} u_f[x]\, u_f''[x] \\
& + 4.58264 * 10^{-7} u_f[x]^2\, u_f''[x] - 1.21033 * 10^{-7} u_f[x]^3 u_f''[x]
\end{aligned}
$$

$$
\begin{aligned}
u[6] \;=\; & 1.30935 + 0.0011121 u_f[x] - 0.000331366 u_f[x]^2 + 0.000149489 u_f[x]^3 \\
& + 3.34907 * 10^{-6} u_f''[x] - 2.66647 * 10^{-6} u_f[x] u_f''[x] \\
& + 2.09295 * 10^{-6} u_f[x]^2\, u_f''[x] - 6.87318 * 10^{-7} u_f[x]^3 u_f''[x]
\end{aligned}
$$

$$
\begin{aligned}
u[7] \;=\; & 1.30383 + 0.00665186 u_f[x] - 0.00182893 u_f[x]^2 + 0.000789763 u_f[x]^3 \\
& + 0.0000140899 u_f''[x] - 0.000011257 u_f[x]\, u_f''[x] \\
& + 9.14567 * 10^{-6} u_f[x]^2 u_f''[x] - 3.70794 * 10^{-6} u_f[x]^3 u_f''[x]
\end{aligned}
$$

$$
\begin{aligned}
u[8] \;=\; & 1.26934 + 0.040481 u_f[x] - 0.0101978 u_f[x]^2 + 0.00408206 u_f[x]^3 \\
& + 0.00005612 u_f''[x] - 0.0000457826 u_f[x] \\
& + u_f''[x] + 0.000037641 u_f[x]^2\, u_f''[x] - 0.000018566 u_f[x]^3 u_f''[x]
\end{aligned}
$$

$$
\begin{aligned}
u[9] \;=\; & 1.05971 + 0.23709 u_f[x] - 0.0493175 u_f[x]^2 + 0.0173591 u_f[x]^3 \\
& + 0.000176522 u_f''[x] - 0.000150827 u_f[x] u_f''[x] \\
& + 0.000123238 u_f[x]^2\, u_f''[x] - 0.0000709253 u_f[x]^3\, u_f''[x].
\end{aligned}
$$

Remark 27.1 Observe that since $\varepsilon = 0.01$ the solution is close to the constant value 1.3247 along the domain, which is an approximate solution of equation $-u^3 + u + 1.0 = 0$. ■

27.2 The hyper-finite differences approach

In the last sections we have introduced a method to minimize the solution error for a large class of PDEs as a typical parameter $\varepsilon > 0$ is small. The main idea presented there consisted of a proximal formulation combined with the generalized method of lines.

In the present section we develop another new solution for such a problem also for a large class of non-linear elliptic PDEs, namely, the hyper-finite differences approach. We believe the result here developed is better than those of the previous sections. Indeed in the present method, in general the convergence is obtained more easily.

The idea here is to divide the domain interval, concerning variable t to be specified, into sub-domains in order to minimize the effect of the small parameter $\varepsilon > 0$. For example if we divide the domain $[1,2]$ into $10-12$ sub-domains, for $\varepsilon = 0.01$ and the example addressed in the next pages, the error applying the generalized method of lines on which the sub-domain is very small. Finally, we reconnect the sub-domains by solving the system of equations corresponding to the partial differential equation in question on each of these $N_1 - 1$ nodes. As N_1 is a small number for the amount of nodes (typically $N_1 = 10-12$), we have justified the terminology hyper-finite differences.

First observe that, in equation (27.3), $t \in [1,2]$ in Ω. Discretizing in t (N_5 equal pieces which will generate N_5 lines), we recall that such a general equation (27.3),

$$
\begin{aligned}
\frac{\partial^2 u}{\partial t^2} &+ \frac{1}{t} f_2(\theta) \frac{\partial u}{\partial t} \\
&+ \frac{1}{t} f_3(\theta) \frac{\partial^2 u}{\partial \theta \partial t} + \frac{f_4(\theta)}{t^2} \frac{\partial^2 u}{\partial \theta^2} \\
&+ f_5(\theta)(g(u) + f) = 0,
\end{aligned}
\tag{27.11}
$$

partially in finite differences, has the expression

$$
\begin{aligned}
&\frac{u_{n+1} - 2u_n + u_{n-1}}{d^2} + \frac{(u_n - u_{n-1})}{d} \frac{1}{t_n} f_2(\theta) \\
&+ \frac{\partial(u_n - u_{n-1})}{\partial \theta} \frac{1}{t_n d} f_3(\theta) + \frac{\partial^2 u_n}{\partial \theta^2} \frac{f_4(\theta)}{t_n^2} \\
&+ f_5(\theta) \left(g(u_n) \frac{1}{\varepsilon} + f_n \frac{1}{\varepsilon} \right) = 0,
\end{aligned}
\tag{27.12}
$$

$\forall n \in \{1, ..., N_5 - 1\}$. Here, $u_n(\theta)$ corresponds to the solution on the line n. Thus we may write

$$
u_n = T_n(u_{n-1}, u_n, u_{n+1}),
$$

where

$$
\begin{aligned}
T_n(u_{n-1}, u_n, u_{n+1}) &= \left(u_{n+1} + u_n + u_{n-1} + \frac{(u_n - u_{n-1})}{d} \frac{1}{t_n} f_2(\theta) d^2 \right. \\
&+ \frac{\partial(u_n - u_{n-1})}{\partial \theta} \frac{1}{t_n d} f_3(\theta) d^2 + \frac{\partial^2 u_n}{\partial \theta^2} \frac{f_4(\theta)}{t_n^2} d^2 \\
&\left. + f_5(\theta) \left(g(u_n) \frac{d^2}{\varepsilon} + f_n \frac{d^2}{\varepsilon} \right) \right) / 3.0.
\end{aligned}
\tag{27.13}
$$

27.2.1 The main algorithm

To obtain a fixed point for each T_n indicated in (27.13) is perfectly possible if $\varepsilon \approx \mathcal{O}(1)$. However, for the case in which $\varepsilon > 0$ is small, we highlight once more the error may be relevant, so that we propose the following algorithm to deal with such a situation of small $\varepsilon > 0$.

1. Choose $N = 30 - 100$ and $N_1 = 10 - 12$ (specifically for the example in the next lines). Divide the interval domain in the variable t into N_1 equal pieces (for example, for the interval $[1, 2]$ through a concerning partition $\{t_0 = 1, t_1, ..., t_{N_1} = 2\}$, where $t_k = 1 + k/N_1$ and $d = 1/(N N_1)$ is the grid thickness in t).

2. Through the generalized method of lines, solve the equation in question on the interval $[t_k, t_{k+1}]$ as function of $u(t_k)$ and $u(t_{k+1})$ and the domain shape.

   To calculate $u^k = \{u_n^k\}$ on $[t_k, t_{k+1}]$, proceed as follows. First observe that the equation in question stands for

$$u_{n+1}^k - 2u_n^k + u_{n-1}^k + \frac{(u_n^k - u_{n-1}^k)}{d} \frac{1}{t_n^k} f_2(\theta) d^2$$

$$+ \frac{\partial(u_n^k - u_{n-1}^k)}{\partial \theta} \frac{1}{t_n^k d} f_3(\theta) d^2 + \frac{\partial^2 u_n^k}{\partial \theta^2} \frac{f_4(\theta)}{(t_n^k)^2} d^2$$

$$+ f_5(\theta) \left(g(u_n^k) \frac{d^2}{\varepsilon} + f_n^k \frac{d^2}{\varepsilon} \right)$$

$$= 0, \tag{27.14}$$

   where

$$t_n^k = 1 + (k-1)/N_1 + nd, \forall k \in \{1, ..., N_1\}, \; n \in \{1, ... N - 1\}.$$

   Such an equation is solved through the Banach fixed point theorem, that is, defining

$$T_n^k(u_n^k, u_{n+1}^k, u_{n-1}^k) = \left(u_{n+1}^k + u_n^k + u_{n-1}^k + \frac{(u_n^k - u_{n-1}^k)}{d} \frac{1}{t_n^k} f_2(\theta) d^2 \right.$$

$$+ \frac{\partial(u_n^k - u_{n-1}^k)}{\partial \theta} \frac{1}{t_{nd}} f_3(\theta) d^2 + \frac{\partial^2 u_n^k}{\partial \theta^2} \frac{f_4(\theta)}{t_n^2} d^2$$

$$\left. + f_5(\theta) \left(g(u_n^k) \frac{d^2}{\varepsilon} + f_n^k \frac{d^2}{\varepsilon} \right) \right) / 3 \tag{27.15}$$

   equation (27.14) stands for

$$u_n^k = T_n^k(u_{n-1}^k, u_n^k, u_{n+1}^k),$$

   so that for $n = 1$ we have

$$u_1^k = T_1(u(t_k), u_1^k, u_2^k).$$

   We may use the Contraction Mapping Theorem to calculate u_1^k as a function of u_2^k and $u(t_k)$. The procedure would be,

   (a) set $x_1 = u_2^k$,

   (b) obtain recursively

$$x_{j+1} = T_1^k(u(t_k), x_j, u_2^k),$$

   (c) and finally get

$$u_1^k = \lim_{j \to \infty} x_j = g_1(u(t_k), u_2).$$

   Thus, we have obtained

$$u_1^k = g_1(u(t_k), u_2^k).$$

   We can repeat the process for $n = 2$, that is, we can solve the equation

$$u_2^k = T_2^k(u_1^k, u_2^k, u_3^k),$$

which from above stands for

$$u_2^k = T_2^k(g_1(u(t_k), u_2^k), u_2^k, u_3^k).$$

The procedure would be:

(a) Set $x_1 = u_3^k$,

(b) calculate

$$x_{j+1} = T_2^k(g_1(u(t_k), x_j), x_j, u_3^k),$$

(c) obtain

$$u_2^k = \lim_{j \to \infty} x_j = g_2(u(t_k), u_3^k).$$

We proceed in this fashion until obtaining

$$u_{N-1}^k = g_{N-1}(u(t_k), u_N^k) = g_{N-1}(u(t_k), u(t_{k+1})).$$

We have obtained u_{N-1}^k. We may then calculate

$$u_{N-2}^k = g_{N-2}(u(t_k), u_{N-1}^k)),$$

$$u_{N-3}^k = g_{N-3}(u(t_k), u_{N-2}^k)),$$

and so on, up to finding

$$u_1^k = g_1(u(t_k), u_2^k).$$

Thus this part of the problem is solved.

3. Calculate the solution on the lines corresponding to $u_{t_1}, \ldots, u_{t_{N_1-1}}$, by solving the system,

$$u(t_k) = \tilde{T}_k(u^{k-1}(N-1), u(t_k), u_1^{k+1}),$$

which correspond to the partial differential equation in question on the line $k\,N$, where

$$\begin{aligned} \tilde{T}_k(u_n, u_{n+1}, u_{n-1}) &= \left(u_{n+1} + u_n + u_{n-1} + \frac{(u_n - u_{n-1})}{d} \frac{1}{t_k} f_2(\theta) d^2 \right. \\ &\quad + \frac{\partial(u_n - u_{n-1})}{\partial \theta} \frac{1}{t_k d} f_3(\theta) d^2 + \frac{\partial^2 u_n}{\partial \theta^2} \frac{f_4(\theta)}{t_k^2} d^2 \\ &\quad \left. + f_5(\theta) \left(g(u_n) \frac{d^2}{\varepsilon} + f_n \frac{d^2}{\varepsilon} \right) \right) / 3. \end{aligned} \tag{27.16}$$

Here may use the Banach fixed point theorem for the final calculation as well.

The problem is then solved.

27.2.2 A numerical example

In this section we present a numerical example. Consider the equation

$$\varepsilon \nabla^2 u + g(u) + 1 = 0, \text{ in } \Omega \subset \mathbb{R}^2, \tag{27.17}$$

where, for a Ginburg-Landau type equation

$$g(u) = -u^3 + u,$$

with the boundary conditions
$$u = u_1 \text{ on } \partial\Omega = \Gamma_0 \cup \Gamma_1,$$
where $\Omega = \{(r, \theta) \ : \ 1 \leq r \leq 2, \, 0 \leq \theta \leq 2\pi\}$,
$$u_1 = 0, \text{ on } \Gamma_0 = \{(1, \theta) \ : \ 0 \leq \theta \leq 2\pi\},$$
$$u_1 = u_f(\theta), \text{ on } \Gamma_1 = \{(2, \theta) \ : \ 0 \leq \theta \leq 2\pi\}.$$

Through the generalized method of lines, for $N = 30$ (30 lines in which sub-domain), $N_1 = 10$ (10 sub-domains) and $d = 1/(N N_1)$, in polar coordinates and finite differences (please see [73] for general schemes in finite differences), equation (27.17) stands for

$$(u_{n+1} - 2u_n + u_{n-1}) + \frac{1}{r_n}(u_n - u_{n-1})d + \frac{1}{r_n^2}\frac{\partial^2 u_n}{\partial\theta^2}d^2 + (-u_n^3 + u_n)\frac{d^2}{\varepsilon} + \frac{d^2}{\varepsilon} = 0,$$

$\forall n \in \{1, \ldots, N-1\}$.

At this point we present, through the generalized method of lines, the concerning algorithm which may be for the softwares mathematica or maple.

In this software, x stands for θ.

```
ClearAll;
m8 = 30; (number of lines for each sub-domain)
N1 = 10; (number of sub-domains)
d = 1.0/m8/N1; (grid thickness)
e1 = 0.01; (ε = 0.01)
Clear[d1,u,a,b,h,U];
For[k1 = 1,k1 < N1 + 1,k1 + +,
Print[k1];
a = U[k1 − 1][x];
For[i = 1,i < m8,i++,
t = 1.0 + (k1 − 1)/N1 + i*d;
Print[i];
b[x_] = u[i + 1][x];
For[k = 1,k < 35,k++,(here we have fixed the number of iterations for this example)
z = (u[i + 1][x] + b[x] + a + 1/t*(b[x] − a)*d*d1^2 + 1/t^2*D[b[x],{x,2}]*d^2*d1^2
+(−b[x]^3*d^2*d1^2/e1 + b[x]*d^2/e1*d1^2) + 1.0*d^2/e1*d1^2)/(3.0);
z = Series[z,{d1,0,2}];
z = Normal[z];
z = Expand[z];
b[x_] = z];
a1 = b[x];
Clear[b];
u[i + 1][x_] = b[x];
h[k1,i] = Expand[a1];
Clear[d1];
a = a1];b[x_] = U[k1][x];
For[i = 1,i < m8,i++,
W1[k1,m8 − i] = Series[h[k1,m8 − i],{d1,0,2}];
W[k1,m8 − i] = Normal[W1[k1,m8 − i]];
```

$b[x_-] = Expand[W[k_1, m8 - i]];$

$v[m8 - i] = Expand[W[k_1, m8 - i]]];$

$d_1 = 1.0;$

$Print[v[m8/2]];$

$Clear[d_1, b, u]];$

$Clear[U];$

$U[0][x_-] = 0.0;$

$U[N_1][x_-] = U_f[x];$

$Clear[d_1];$

$d_1 = 1.0;$

$For[k = 1, k < N_1, k++,$

$U[k][x_-] = 0.0;$

$z_7[k] = 0.0];$

$For[k_1 = 1, k_1 < 385, k_1++,$ (here we have fixed the number of iterations)

$Print[k_1];$

$For[k = 1, k < N_1, k++,$

$t = 1 + (k)/N_1;$

$z_5[k] = (W[k, m8 - 1] + U[k][x] + W[k + 1, 1] +$

$1/t * (U[k][x] - W[k, m8 - 1]) * d * d_1^2 +$

$1/t^2 * D[U[k][x], \{x, 2\}] * d^2 * d_1^2$

$+(-U[k][x]^3 + U[k][x]) * d^2/e_1 * d1^2 + d^2/e1)/3.0];$

$For[k = 1, k < N_1, k++,$

$z_5[k] = Series[z_5[k], \{U_f[x], 0, 3\}, \{U_o[x], 0, 3\}, \{U_f'[x], 0, 1\}, \{U_o'[x], 0, 1\},$

$\{U_f''[x], 0, 1\}, \{U_o''[x], 0, 1\}, \{U_f'''[x], 0, 0\}, \{U_o'''[x], 0, 0\}, \{U_f''''[x], 0, 0\}, \{U_o''''[x], 0, 0\}];$

$z_5[k] = Normal[z_5[k]];$

$Clear[A_7];$

$A_7 = Expand[z_5[k]];$

$U[k][x_-] = A_7]]; Print[U[N_1/2][x]];$

At this point we present the expressions for $N_1 = 10$ and $\varepsilon = 0.01$. In the next lines x stands for θ ($0 \leq \theta \leq 2\pi$).

For each line $u[t_k] = U[k]$ we have obtained:

$$U[1] = 1.11698 + 3.02073 * 10^{-9} U_f[x] - 8.9816 * 10^{-10} U_f[x]^2 + 1.85216 * 10^{-10} U_f[x]^3$$
$$+ 8.284 * 10^{-11} U_f''[x] - 4.75816 * 10^{-11} U_f[x] U_f''[x]$$
$$+ 1.35882 * 10^{-11} U_f[x]^2 U_f''[x] - 1.13972 * 10^{-12} U_f[x]^3 U_f''[x]$$

$$U[2] = 1.3107 + 7.41837 * 10^{-9} U_f[x] - 3.8971 * 10^{-9} U_f[x]^2 + 1.58347 * 10^{-9} U_f[x]^3$$
$$+ 1.71087 * 10^{-10} U_f''[x] - 1.71274 * 10^{-10} U_f[x] U_f''[x]$$
$$+ 1.01077 * 10^{-10} U_f[x]^2 U_f''[x] - 8.08804 * 10^{-12} U_f[x]^3 U_f''[x]$$

$$U[3] = 1.32397 + 6.40836 * 10^{-8} U_f[x] - 4.12903 * 10^{-8} U_f[x]^2 + 1.91826 * 10^{-8} U_f[x]^3$$
$$+ 1.13579 * 10^{-9} U_f''[x] - 1.42494 * 10^{-9} U_f[x] U_f''[x]$$
$$+ 9.88305 * 10^{-10} U_f[x]^2 U_f''[x] - 9.82218 * 10^{-11} U_f[x]^3 U_f''[x]$$

$$U[4] = 1.32468 + 7.82478 * 10^{-7} U_f[x] - 5.26904 * 10^{-7} U_f[x]^2 + 2.48884 * 10^{-7} U_f[x]^3$$
$$+ 1.05836 * 10^{-8} U_f''[x] - 1.39954 * 10^{-8} U_f[x] \, U_f''[x]$$
$$+ 1.00186 * 10^{-8} \, U_f[x]^2 \, U_f''[x] - 1.26942 * 10^{-9} U_f[x]^3 \, U_f''[x]$$

$$U[5] = 1.32471 + 0.0000107587 U_f[x] - 7.21035 * 10^{-6} U_f[x]^2 + 3.33153 * 10^{-6} U_f[x]^3$$
$$+ 1.08485 * 10^{-7} U_f''[x] - 1.42477 * 10^{-7} U_f[x] \, U_f''[x]$$
$$+ 1.00683 * 10^{-7} U_f[x]^2 \, U_f''[x] - 1.65819 * 10^{-8} U_f[x]^3 \, U_f''[x]$$

$$U[6] = 1.32462 + 0.000155827 U_f[x] - 0.000102937 U_f[x]^2 + 0.00004608 U_f[x]^3$$
$$+ 1.11907 * 10^{-6} U_f''[x] - 1.43165 * 10^{-6} U_f[x] U_f''[x]$$
$$+ 9.84542 * 10^{-7} \, U_f[x]^2 \, U_f''[x] - 2.19813 * 10^{-7} U_f[x]^3 \, U_f''[x]$$

$$U[7] = 1.32331 + 0.00230353 U_f[x] - 0.00149835 U_f[x]^2 + 0.000647409 U_f[x]^3$$
$$+ 0.0000107575 \, U_f''[x] - 0.0000132399 U_f[x] \, U_f''[x]$$
$$+ 8.83908 * 10^{-6} U_f[x]^2 \, U_f''[x] - 2.87055 * 10^{-6} U_f[x]^3 \, U_f''[x]$$

$$U[8] = 1.30408 + 0.0329322 U_f[x] - 0.0200174 U_f[x]^2 + 0.00748572 U_f[x]^3$$
$$+ 0.000082938 \, U_f''[x] - 0.000092876 U_f[x] \, U_f''[x]$$
$$+ 0.0000587807 U_f[x]^2 \, U_f''[x] - 0.000027146 U_f[x]^3 \, U_f''[x]$$

$$U[9] = 1.08379 + 0.311811 U_f[x] - 0.111793 U_f[x]^2 + 0.0128827 U_f[x]^3$$
$$+ 0.000321752 U_f''[x] - 0.000274071 U_f[x] \, U_f''[x]$$
$$+ 0.000155453 U_f[x]^2 \, U_f''[x] - 0.0000707759 U_f[x]^3 \, U_f''[x]$$

Remark 27.2   Observe that since $\varepsilon = 0.01$ the solution is close to the constant value 1.3247 along the domain, which is an approximate solution of equation $-u^3 + u + 1.0 = 0$. Finally, the first output of the method is the solution on the $N_1 - 1 = 9$ nodes $U[1], \ldots, U[9]$ which, in some sense, justify the terminology hyper-finite differences, even though the solution in all the $N \cdot N_1 = 300$ lines have been obtained.   ■

27.3   Conclusion

In this chapter we have developed two improvements concerning the generalized method of lines. For a large class of models, we have solved the problem of minimizing the error as the parameter $\varepsilon > 0$ is small. In a first step we present a proximal formulation through the introduction of a parameter $K > 0$ and related equation part properly specified. In a second step, we develop the hyper-differences approach which corresponds to a domain division in smaller sub-domains so that the solution on each sub-domain is obtained through the generalized method of lines.

We highlight the methods here developed may be applied to a large class of problems, including the Ginzburg-Landau system in superconductivity in the presence of a magnetic field and respective magnetic potential.

We intend to address this kind of model and others such as the Navier-Stokes system in a future research.

Chapter 28

On the Generalized Method of Lines Applied to the Time-Independent Incompressible Navier-Stokes System

28.1 Introduction

In the first part of this article, we obtain a linear system whose the solution solves the time-independent incompressible Navier-Stokes system for the special case in which the external forces vector is a gradient. In a second step we develop approximate solutions, also for the time independent incompressible Navier-Stokes system, through the generalized method of lines. We recall that for such a method, the domain of the partial differential equation in question is discretized in lines and the concerning solution is written on these lines as functions of the boundary conditions and boundary shape. Finally, we emphasize these last main results are established through applications of the Banach fixed point theorem.

At this point we describe the system in question.

Consider $\Omega \subset \mathbb{R}^2$ an open, bounded and connected set, whose regular (Lipschitzian) internal boundary is denoted by Γ_0 and the regular external one is denoted by Γ_1. For a two-dimensional motion of a fluid on Ω, we denote by $u : \Omega \to \mathbb{R}$ the velocity field in the direction x of the Cartesian system (x,y), by $v : \Omega \to \mathbb{R}$, the velocity field in the direction y and by $p : \Omega \to \mathbb{R}$, the pressure one. We define $P = p/\rho$, where ρ is the constant fluid density. Finally, v denotes the viscosity coefficient and g denotes the gravity field. Under such notation and statements, the time-independent incompressible Navier-Stokes system of partial differential equations is expressed by,

$$
\begin{cases}
v\nabla^2 u - u\partial_x u - v\partial_y u - \partial_x P + g_x = 0, & \text{in } \Omega, \\
v\nabla^2 v - u\partial_x v - v\partial_y v - \partial_y P + g_y = 0, & \text{in } \Omega, \\
\partial_x u + \partial_y v = 0, & \text{in } \Omega,
\end{cases}
\tag{28.1}
$$

$$\begin{cases} u = v = 0, & \text{on } \Gamma_0, \\ u = u_\infty,\ v = 0,\ P = P_\infty, & \text{on } \Gamma_1 \end{cases} \tag{28.2}$$

In principle we look for solutions $(u, v, P) \in W^{2,2}(\Omega) \times W^{2,2}(\Omega) \times W^{1,2}(\Omega)$ despite the fact that less regular solutions are also possible specially concerning the weak formulation. Details about such Sobolev spaces may be found in [1]. General results on finite differences and existence theory for similar systems may be found in [73] and [75], respectively.

28.2 On the solution of the time-independent incompressible Navier-Stokes system through an associated linear one

Through the next result we obtain a linear system whose the solution also solves the time-independent incompressible Navier-Stokes system for the special case in which the external forces vector is a gradient.

Similar results for the time-independent incompressible Euler and Navier-Stokes equations have been presented in [14] and [19, 16], respectively.

Indeed in the works [19, 16], we have indicated a solution of the Navier-Stokes system given by $\mathbf{u} = (u, v)$ defined by

$$\begin{cases} u = \partial_x w_0 + \partial_x w_1, \\ v = \partial_y w_0 - \partial_y w_1, \end{cases} \tag{28.3}$$

where w_0, w_1 are solutions of the system

$$\begin{cases} \partial_{xy} w_1 = 0 & \text{in } \Omega, \\ \nabla^2 w_0 + \partial_{xx} w_1 - \partial_{yy} w_1 = 0, & \text{in } \Omega, \\ u = u_0, & \text{on } \Gamma \equiv \Gamma_0 \cup \Gamma_1, \\ v = v_0, & \text{on } \Gamma. \end{cases} \tag{28.4}$$

Thus, in such a sense, the next result complements this previous one, by introducing a new function w_2 in the solution expressions, which makes the concerning boundary conditions perfectly possible to be satisfied.

Theorem 28.2.1 *For $h = (\partial_x f, \partial_y f) \in C^1(\Omega; \mathbb{R}^2)$, consider the Navier-Stokes system similar as above indicated, that is,*

$$\begin{cases} \nu \nabla^2 u - u \partial_x u - v \partial_y u - \partial_x P + \partial_x f = 0, & \text{in } \Omega, \\ \nu \nabla^2 v - u \partial_x v - v \partial_y v - \partial_y P + \partial_y f = 0, & \text{in } \Omega, \\ \partial_x u + \partial_y v = 0, & \text{in } \Omega, \end{cases} \tag{28.5}$$

with the boundary conditions

$$\begin{cases} u = u_0, & \text{on } \Gamma \\ v = v_0, & \text{on } \Gamma, \\ P = P_0, & \text{on } \Gamma_1. \end{cases} \tag{28.6}$$

A solution for such a Navier-Stokes system is given by $\mathbf{u} = (u, v)$ defined by

$$\begin{cases} u = \partial_x w_0 + \partial_x w_1 + \partial_y w_2, \\ v = \partial_y w_0 - \partial_y w_1 - \partial_x w_2, \end{cases} \tag{28.7}$$

where w_0, w_1, w_2 are solutions of the system

$$\begin{cases} \nabla^2 w_2 + 2\partial_{xy} w_1 = 0 & in \ \Omega, \\ \nabla^2 w_0 + \partial_{xx} w_1 - \partial_{yy} w_1 = 0, & in \ \Omega, \\ u = u_0, & on \ \Gamma, \\ v = v_0, & on \ \Gamma, \end{cases} \tag{28.8}$$

and P is a solution of the system indicated in the first two lines of (28.5) with boundary conditions indicated in the third line of (28.6).

Proof 28.1   For w_0, w_1, w_2 such that $\nabla^2 w_2 + 2\partial_{xy} w_1 = 0$ in Ω, u and v as indicated above and defining

$$h_1 = u\partial_x u + v\partial_y u,$$

$$h_2 = u\partial_x v + v\partial_y v$$

and $\varphi \equiv \nabla^2 w_2 + 2\partial_{xy} w_1 = 0$, we have (you may check it using the softwares MATHEMATICA or MAPLE

$$\begin{aligned} & \frac{\partial h_1}{\partial y} - \frac{\partial h_2}{\partial x} \\ = \ & (-\partial_{yy} w_1 + \partial_{yy} w_0 - \partial_{xy} w_2)\varphi \\ & + (-\partial_y w_1 + \partial_y w_0 - \partial_x w_2)\partial_y \varphi \\ & + (\partial_{xy} w_2 + \partial_{xx} w_1 + \partial_{xx} w_0)\varphi \\ & + (\partial_y w_2 + \partial_x w_1 + \partial_x w_0)\partial_x \varphi \\ = \ & 0, \text{ in } \Omega. \end{aligned} \tag{28.9}$$

Moreover, since $\varphi = \nabla^2 w_2 + 2\partial_{xy} w_1 = 0$ in Ω, we get

$$\begin{aligned} & v(\partial_y \nabla^2 u - \partial_x \nabla^2 v) + \partial_y(\partial_x f) - \partial_x(\partial_y f) \\ = \ & v(2\partial_{xy}(\nabla^2 w_1) + \nabla^4 w_2) + \partial_y(\partial_x f) - \partial_x(\partial_y f) \\ = \ & v\nabla^2 \varphi + \partial_{yx} f - \partial_{yx} f \\ = \ & 0, \text{ in } \Omega. \end{aligned} \tag{28.10}$$

Summarizing, we have obtained

$$\begin{aligned} & \partial_y \left(v\nabla^2 u - u\partial_x u - v\partial_y u + \partial_x f \right) \\ = \ & \partial_x \left(v\nabla^2 v - u\partial_x v - v\partial_y v + \partial_y f \right). \end{aligned} \tag{28.11}$$

Also, the equation

$$\nabla^2 w_0 + \partial_{xx} w_1 - \partial_{yy} w_1 = 0, \text{ in } \Omega$$

stands for

$$\partial_x u + \partial_y v = 0, \text{ in } \Omega.$$

From these last results we may obtain P which satisfies the concerning boundary condition such that

$$\begin{cases} v\nabla^2 u - u\partial_x u - v\partial_y u - \partial_x P + \partial_x f = 0, & in \ \Omega, \\ \\ v\nabla^2 v - u\partial_x v - v\partial_y v - \partial_y P + \partial_y f = 0, & in \ \Omega, \\ \\ \partial_x u + \partial_y v = 0, & in \ \Omega, \end{cases} \tag{28.12}$$

This completes the proof.

28.3 The generalized method of lines for the Navier-Stokes system

In this section we develop the solution for the Navier-Stokes system through the generalized method of lines, which was originally introduced in [19], with further developments in [18, 14]. We consider the boundary conditions

$$u = u_0(x), \ v = v_0(x), \ P = P_0(x) \text{ on } \partial\Omega_0,$$

$$u = 0, \ v = 0, \ P = P_f(x) \text{ on } \partial\Omega_1,$$

where

$$\Omega = \{(r, \theta) : \ | \ r(\theta) \leq r \leq 2r(\theta)\}$$

and where $r(\theta)$ is a positive, smooth and periodic function with period 2π,

$$\partial\Omega = \partial\Omega_0 \cup \partial\Omega_1,$$

$$\partial\Omega_0 = \{(r(\theta), \theta) \in \mathbb{R}^2 : 0 \leq \theta \leq 2\pi\}$$

and

$$\partial\Omega_1 = \{(2r(\theta), \theta) \in \mathbb{R}^2 : 0 \leq \theta \leq 2\pi\}.$$

For $v = 1$, neglecting the gravity effects, the corresponding Navier-Stokes homogeneous system, in function of the variables (t, θ) where $t = r/r(\theta)$, is given by

$$L(u) - ud_1(u) - vd_2(u) - d_1(P) = 0, \tag{28.13}$$

$$L(v) - ud_1(v) - vd_2(v) - d_2(P) = 0, \tag{28.14}$$

$$d_1(u) + d_2(v) = 0, \tag{28.15}$$

where generically

$$L(u) = \nabla^2 u,$$

$$d_1(u) = \partial_x u,$$

and

$$d_2(u) = \partial_y u$$

will be specified in the next lines, in function of (t, θ).

Firstly, L is such that

$$L(u)\left(\frac{r(\theta)^2}{f_0(\theta)}\right) = \frac{\partial^2 u}{\partial t^2} + \frac{1}{t}f_2(\theta)\frac{\partial u}{\partial t}$$

$$+ \frac{1}{t}f_3(\theta)\frac{\partial^2 u}{\partial\theta\partial t} + \frac{f_4(\theta)}{t^2}\frac{\partial^2 u}{\partial\theta^2}, \tag{28.16}$$

in Ω. Here $f_0(\theta), f_2(\theta), f_3(\theta)$ and $f_4(\theta)$ are known functions.

More specifically, denoting

$$f_1(\theta) = \frac{-r'(\theta)}{r(\theta)},$$

we have

$$f_0(\theta) = 1 + f_1(\theta)^2,$$

$$f_2(\theta) = 1 + \frac{f_1'(\theta)}{1 + f_1(\theta)^2},$$

$$f_3(\theta) = \frac{2f_1(\theta)}{1 + f_1(\theta)^2},$$

and

$$f_4(\theta) = \frac{1}{1 + f_1(\theta)^2}.$$

Also d_1 and d_2 are expressed by

$$d_1 u = \hat{f}_5(\theta)\frac{\partial u}{\partial t} + (\hat{f}_6(\theta)/t)\frac{\partial u}{\partial \theta},$$

$$d_2 u = \hat{f}_7(\theta)\frac{\partial u}{\partial t} + (\hat{f}_8(\theta)/t)\frac{\partial u}{\partial \theta}$$

Where

$$\hat{f}_5(\theta) = \cos(\theta)/r(\theta) + \sin(\theta)r'(\theta)/r^2(\theta),$$
$$\hat{f}_6(\theta) = -\sin(\theta)/r(\theta),$$
$$\hat{f}_7(\theta) = \sin(\theta)/r(\theta) - \cos(\theta)r'(\theta)/r^2(\theta),$$
$$\hat{f}_8(\theta) = \cos(\theta)/r(\theta).$$

We also define

$$h_3(\theta) = \frac{f_0(\theta)}{r(\theta)^2},$$

$$f_5(\theta) = \left(\frac{r(\theta)^2}{f_0(\theta)}\right)\hat{f}_5(\theta),$$

$$f_6(\theta) = \left(\frac{r(\theta)^2}{f_0(\theta)}\right)\hat{f}_6(\theta),$$

$$f_7(\theta) = \left(\frac{r(\theta)^2}{f_0(\theta)}\right)\hat{f}_7(\theta),$$

and

$$f_8(\theta) = \left(\frac{r(\theta)^2}{f_0(\theta)}\right)\hat{f}_8(\theta),$$

Observe that $t \in [1,2]$ in Ω.

From equations (28.13) and (28.14) we may write

$$d_1(L(u) - ud_1(u) - vd_2(u) - d_1(P))$$
$$+d_2(L(v) - ud_1(v) - vd_2(v) - d_2(P)) = 0, \tag{28.1?}$$

From (28.15) we have

$$d_1[L(u)] + d_2[L(v)] = L(d_1(u) + d_2(v)) = 0,$$

and considering that

$$d_1(d_1(P)) + d_2(d_2(P)) = L(P),$$

from (28.17) we have

$$L(P) + d_1(u)^2 + d_2(v)^2 + 2d_2(u)d_1(v) = 0, \text{ in } \Omega.$$

Hence, in fact we solve the approximate system (this indeed is not exactly the Navier-Stokes one):

$$L(u) - ud_1(u) - vd_2(u) - d_1(P) = 0, \tag{28.1}$$

$$L(v) - ud_1(v) - vd_2(v) - d_2(P) = 0, \tag{28.1?}$$
$$L(P) + d_1(u)^2 + d_2(v)^2 + 2d_2(u)d_1(v) = 0, \text{ in } \Omega. \tag{28.2?}$$

Remark 28.1 Who taught me how to obtain this last approximate system was Professor Alvaro de Bortoli of Federal University of Rio Grande do Sul, UFRGS, Porto Alegre, RS-Brazil. ■

At this point, discretizing only in t (in N lines), defining $d = 1/N$ and $t_n = 1 + nd$, $\forall n \in \{1, \ldots, N-1\}$, we represent such a concerning system in partial finite differences.

Remark 28.2 In this text, we may generically consider the operators

$$\frac{\partial u_n}{\partial \theta} \quad \text{and} \quad \frac{\partial^2 u_n}{\partial \theta^2}$$

also in a finite differences context, so that in such a case we may also consider them as bounded operators. ■

Denoting

$$\hat{d}_1(u_n, u_{n-1}) = f_5(x) \frac{(u_n - u_{n-1})}{d} + \frac{f_6(x)}{t_n} \frac{\partial u_n}{\partial x},$$

and

$$\hat{d}_2(u_n, u_{n-1}) = f_7(x) \frac{(u_n - u_{n-1})}{d} + \frac{f_8(x)}{t_n} \frac{\partial u_n}{\partial x},$$

where x stands for θ, in partial finite differences, equation (28.18) stands for

$$\frac{u_{n+1} - 2u_n + u_{n-1}}{d^2} + \frac{f_2(x)}{t_n} \frac{u_n - u_{n-1}}{d}$$
$$+ \frac{f_3(x)}{t_n} \frac{\partial}{\partial x} \left(\frac{u_n - u_{n-1}}{d} \right) + \frac{f_4(x)}{t_n^2} \frac{\partial^2 u_n}{\partial x^2}$$
$$- u_n \hat{d}_1(u_n, u_{n-1}) - v_n \hat{d}_2(u_n, u_{n-1}) - \hat{d}_1(P_n, P_{n-1}) = 0. \tag{28.21}$$

Hence, denoting $\mathbf{u} = (u, v, P)$, we have

$$u_n = (T_1)_n(\mathbf{u}_{n+1}, \mathbf{u}_n, \mathbf{u}_{n-1}),$$

where

$$(T_1)_n(\mathbf{u}_{n+1}, \mathbf{u}_n, \mathbf{u}_{n-1})$$
$$= \left(u_{n+1} + u_n + u_{n-1} + \frac{f_2(x)}{t_n} (u_n - u_{n-1})d \right.$$
$$+ \frac{f_3(x)}{t_n} \frac{\partial}{\partial x}(u_n - u_{n-1})d + \frac{f_4(x)}{t_n^2} \frac{\partial^2 u_n}{\partial^2 x} d^2$$
$$\left. - u_n \hat{d}_1(u_n, u_{n-1})d^2 - v_n \hat{d}_2(u_n, u_{n-1})d^2 - \hat{d}_1(P_n, P_{n-1})d^2 \right) / 3.0.$$

Similarly, equation (28.19) stands for

$$\frac{v_{n+1} - 2v_n + v_{n-1}}{d^2} + \frac{f_2(x)}{t_n} \frac{v_n - v_{n-1}}{d}$$
$$+ \frac{f_3(x)}{t_n} \frac{\partial}{\partial x} \left(\frac{v_n - v_{n-1}}{d} \right) + \frac{f_4(x)}{t_n^2} \frac{\partial^2 v_n}{\partial x^2}$$
$$- u_n \hat{d}_1(v_n, v_{n-1}) - v_n \hat{d}_2(v_n, v_{n-1}) - \hat{d}_2(P_n, P_{n-1}) = 0. \tag{28.22}$$

Hence,
$$v_n = (T_2)_n(\mathbf{u}_{n+1}, \mathbf{u}_n, \mathbf{u}_{n-1}),$$

where

$$
\begin{aligned}
&(T_2)_n(\mathbf{u}_{n+1}, \mathbf{u}_n, \mathbf{u}_{n-1}) \\
&= \left(v_{n+1} + v_n + v_{n-1} + \frac{f_2(x)}{t_n}(v_n - v_{n-1})d \right. \\
&\quad + \frac{f_3(x)}{t_n}\frac{\partial}{\partial x}(v_n - v_{n-1})d + \frac{f_4(x)}{t_n^2}\frac{\partial^2 v_n}{\partial x^2}d^2 \\
&\quad \left. - u_n \hat{d}_1(v_n, v_{n-1})d^2 - v_n \hat{d}_2(v_n, v_{n-1})d^2 - \hat{d}_2(P_n, P_{n-1})d^2 \right)/3.0.
\end{aligned}
$$

Finally, (28.20) stands for

$$
\begin{aligned}
&\frac{P_{n+1} - 2P_n + P_{n-1}}{d^2} + \frac{f_2(x)}{t_n}\frac{P_n - P_{n-1}}{d} \\
&+ \frac{f_3(x)}{t_n}\frac{\partial}{\partial x}\left(\frac{P_n - P_{n-1}}{d}\right) + \frac{f_4(x)}{t_n^2}\frac{\partial^2 P_n}{\partial x^2} \\
&+ h_3(x)\left(\hat{d}_1(u_n, u_{n-1})^2 + \hat{d}_2(v_n, v_{n-1})^2 + 2\hat{d}_2(u_n, u_{n-1})\hat{d}_1(v_n, v_{n-1})\right) = 0.
\end{aligned}
\tag{28.23}
$$

Hence,
$$P_n = (T_3)_n(\mathbf{u}_{n+1}, \mathbf{u}_n, \mathbf{u}_{n-1}),$$

where

$$
\begin{aligned}
&(T_3)_n(\mathbf{u}_{n+1}, \mathbf{u}_n, \mathbf{u}_{n-1}) \\
&= \left(P_{n+1} + P_n + P_{n-1} + \frac{f_2(x)}{t_n}(P_n - P_{n-1})d \right. \\
&\quad + \frac{f_3(x)}{t_n}\frac{\partial}{\partial x}(P_n - P_{n-1})d + \frac{f_4(x)}{t_n^2}\frac{\partial^2 P_n}{\partial x^2}d^2 \\
&\quad \left. + h_3(x)\left(\hat{d}_1(u_n, u_{n-1})^2 d^2 + \hat{d}_2(v_n, v_{n-1})^2 d + 2\hat{d}_2(u_n, u_{n-1})\hat{d}_1(v_n, v_{n-1})d^2\right) \right)/3.0.
\end{aligned}
$$

Summarizing, we may write
$$\mathbf{u}_n = \hat{T}_n(\mathbf{u}_{n+1}, \mathbf{u}_n, \mathbf{u}_{n-1}),$$

where

$$
\begin{aligned}
&\hat{T}_n(\mathbf{u}_{n+1}, \mathbf{u}_n, \mathbf{u}_{n-1}) \\
&= \left((T_1)_n(\mathbf{u}_{n+1}, \mathbf{u}_n, \mathbf{u}_{n-1}), (T_2)_n(\mathbf{u}_{n+1}, \mathbf{u}_n, \mathbf{u}_{n-1}), (T_3)_n(\mathbf{u}_{n+1}, \mathbf{u}_n, \mathbf{u}_{n-1})\right),
\end{aligned}
\tag{28.24}
$$

$\forall n \in \{1, \dots, N-1\}$.

Therefore, for $n = 1$ we obtain

$$\mathbf{u}_1 = \hat{T}_1(\mathbf{u}_2, \mathbf{u}_1, \mathbf{u}_0).$$

We solve such an equation through the Banach fixed point theorem.

1. First set

$$(\mathbf{u}_1)^1 = \mathbf{u}_2.$$

2. In a second step, define $\{\mathbf{u}_1^k\}$ such that

$$\mathbf{u}_1^{k+1} = \hat{T}_1(\mathbf{u}_2, \mathbf{u}_1^k, \mathbf{u}_0), \ \forall k \in \mathbb{N}.$$

3. Finally obtain

$$\mathbf{u}_1 = \lim_{k \to \infty} \mathbf{u}_1^k \equiv F_1(\mathbf{u}_2, \mathbf{u}_0).$$

Now, reasoning inductively, having

$$\mathbf{u}_{n-1} = F_{n-1}(\mathbf{u}_n, \mathbf{u}_0),$$

we obtain $\mathbf{u}_n$ as indicated in the next lines.

1. First set

$$(\mathbf{u}_n)^1 = \mathbf{u}_{n+1}.$$

2. In a second step, define $\{\mathbf{u}_n^k\}$ such that

$$\mathbf{u}_n^{k+1} = \hat{T}_n(\mathbf{u}_{n+1}, \mathbf{u}_n^k, \mathbf{u}_0), \; \forall k \in \mathbb{N}.$$

3. Finally obtain

$$\mathbf{u}_n = \lim_{k \to \infty} \mathbf{u}_n^k \equiv F_n(\mathbf{u}_{n+1}, \mathbf{u}_0).$$

Thus, reasoning inductively we have obtained

$$\mathbf{u}_n = F_n(\mathbf{u}_{n+1}, \mathbf{u}_0), \; \forall n \in \{1, \dots, N-1\}.$$

In particular, for $n = N - 1$, we have $\mathbf{u}_N = \mathbf{u}_f = (u_f, v_f, P_f)$.
Therefore

$$\mathbf{u}_{N-1} = F_{N-1}(\mathbf{u}_N, \mathbf{u}_0) \equiv H_{N-1}(\mathbf{u}_f, \mathbf{u}_0).$$

From this we obtain

$$\mathbf{u}_{N-2} = F_{N-2}(\mathbf{u}_{N-1}, \mathbf{u}_0) \equiv H_{N-2}(\mathbf{u}_f, \mathbf{u}_0),$$

and so on, up to finding

$$\mathbf{u}_1 = F_1(\mathbf{u}_2, \mathbf{u}_0) \equiv H_1(\mathbf{u}_f, \mathbf{u}_0).$$

The problem is then solved.

With such results in mind, with a software similar to those presented in the previous last chapter, truncating the concerning series solutions for terms of order up to d^2 (in d), for the field of velocity u we have obtained the following expressions for the lines (here x stands for θ):

Line 1

$$
\begin{aligned}
u_1(x) \;=\; & -0.045 f_5(x) P_f(x) + 0.045 f_5(x) P_0(x) \\
& + 0.899 u_0(x) - 0.034 f_2(x) u_0(x) + 0.029 f_5(x) u_0(x)^2 \\
& + 0.029 f_7(x) u_0(x) v_0(x) - 0.011 f_6(x) P_f' \\
& - 0.022 f_6(x) P_0'(x) - 0.034 f_3(x) u_0'(x) \\
& - 0.016 f_6(x) u_0(x) u_0'(x) - 0.016 f_8(x) v_0(x) u_0'(x) \\
& + 0.018 f_4(x) u_0''(x)
\end{aligned}
$$

Line 2

$$
\begin{aligned}
u_2(x) \;=\; & -0.081 f_5(x) P_f(x) + 0.081 f_5(x) P_0(x) \\
& + 0.799 u_0(x) - 0.034 f_2(x) u_0(x) + 0.059 f_5(x) u_0(x)^2 \\
& + 0.048 f_7(x) u_0(x) v_0(x) - 0.022 f_6(x) P_f' \\
& - 0.036 f_6(x) P_0'(x) - 0.059 f_3(x) u_0'(x) \\
& - 0.025 f_6(x) u_0(x) u_0'(x) - 0.025 f_8(x) v_0(x) u_0'(x) \\
& + 0.028 f_4(x) u_0''(x)
\end{aligned}
$$

Line 3

$$
\begin{aligned}
u_3(x) \;=\; & -0.106 f_5(x) P_f(x) + 0.106 f_5(x) P_0(x) \\
& +0.698 u_0(x) - 0.075 f_2(x) u_0(x) + 0.060 f_5(x) u_0(x)^2 \\
& +0.060 f_7(x) u_0(x) v_0(x) - 0.031 f_6(x) P'_f \\
& -0.044 f_6(x) P'_0(x) - 0.075 f_3(x) u'_0(x) \\
& -0.029 f_6(x) u_0(x) u'_0(x) - 0.029 f_8(x) v_0(x) u'_0(x) \\
& +0.033 f_4(x) u''_0(x)
\end{aligned}
$$

Line 4

$$
\begin{aligned}
u_4(x) \;=\; & -0.121 f_5(x) P_f(x) + 0.121 f_5(x) P_0(x) \\
& +0.597 u_0(x) - 0.084 f_2(x) u_0(x) + 0.064 f_5(x) u_0(x)^2 \\
& +0.064 f_7(x) u_0(x) v_0(x) - 0.037 f_6(x) P'_f \\
& -0.046 f_6(x) P'_0(x) - 0.084 f_3(x) u'_0(x) \\
& -0.029 f_6(x) u_0(x) u'_0(x) - 0.029 f_8(x) v_0(x) u'_0(x) \\
& +0.034 f_4(x) u''_0(x)
\end{aligned}
$$

Line 5

$$
\begin{aligned}
u_5(x) \;=\; & -0.126 f_5(x) P_f(x) + 0.126 f_5(x) P_0(x) \\
& +0.497 u_0(x) - 0.086 f_2(x) u_0(x) + 0.062 f_5(x) u_0(x)^2 \\
& +0.062 f_7(x) u_0(x) v_0(x) - 0.041 f_6(x) P'_f \\
& -0.044 f_6(x) P'_0(x) - 0.086 f_3(x) u'_0(x) \\
& -0.026 f_6(x) u_0(x) u'_0(x) - 0.026 f_8(x) v_0(x) u'_0(x) \\
& +0.032 f_4(x) u''_0(x)
\end{aligned}
$$

Line 6

$$
\begin{aligned}
u_6(x) \;=\; & -0.121 f_5(x) P_f(x) + 0.121 f_5(x) P_0(x) \\
& +0.397 u_0(x) - 0.080 f_2(x) u_0(x) + 0.056 f_5(x) u_0(x)^2 \\
& +0.056 f_7(x) u_0(x) v_0(x) - 0.041 f_6(x) P'_f \\
& -0.031 f_6(x) P'_0(x) - 0.069 f_3(x) u'_0(x) \\
& -0.017 f_6(x) u_0(x) u'_0(x) - 0.017 f_8(x) v_0(x) u'_0(x) \\
& +0.022 f_4(x) u''_0(x)
\end{aligned}
$$

Line 7

$$
\begin{aligned}
u_7(x) \;=\; & -0.105 f_5(x) P_f(x) + 0.105 f_5(x) P_0(x) \\
& +0.297 u_0(x) - 0.069 f_2(x) u_0(x) + 0.045 f_5(x) u_0(x)^2 \\
& +0.045 f_7(x) u_0(x) v_0(x) - 0.037 f_6(x) P'_f \\
& -0.031 f_6(x) P'_0(x) - 0.069 f_3(x) u'_0(x) \\
& -0.017 f_6(x) u_0(x) u'_0(x) - 0.017 f_8(x) v_0(x) u'_0(x) \\
& +0.022 f_4(x) u''_0(x)
\end{aligned}
$$

Line 8

$$u_8(x) = -0.080f_5(x)P_f(x)+0.080f_5(x)P_0(x)$$
$$+0.198u_0(x)-0.051f_2(x)u_0(x)+0.032f_5(x)u_0(x)^2$$
$$+0.032f_7(x)u_0(x)v_0(x)-0.029f_6(x)P_f'$$
$$-0.021f_6(x)P_0'(x)-0.051f_3(x)u_0'(x)$$
$$-0.011f_6(x)u_0(x)u_0'(x)-0.011f_8(x)v_0(x)u_0'(x)$$
$$+0.015f_4(x)u_0''(x)$$

Line 9

$$u_9(x) = -0.045f_5(x)P_f(x)+0.045f_5(x)P_0(x)$$
$$+0.099u_0(x)-0.028f_2(x)u_0(x)+0.016f_5(x)u_0(x)^2$$
$$+0.016f_7(x)u_0(x)v_0(x)-0.017f_6(x)P_f'$$
$$-0.022f_6(x)P_0'(x)-0.012f_3(x)u_0'(x)$$
$$-0.006f_6(x)u_0(x)u_0'(x)-0.006f_8(x)v_0(x)u_0'(x)$$
$$+0.008f_4(x)u_0''(x)$$

For the field of velocity v, we have obtained the following expressions for the lines:

Line 1

$$v_1(x) = -0.045f_7(x)P_f(x)+0.045f_7(x)P_0(x)$$
$$+0.899v_0(x)-0.034f_2(x)v_0(x)+0.029f_5(x)u_0(x)v_0(x)$$
$$+0.029f_7(x)v_0(x)^2-0.011f_8(x)P_f'(x)$$
$$-0.022f_8(x)P_0'(x)-0.034f_3(x)v_0'(x)$$
$$-0.016f_6(x)u_0(x)v_0'(x)-0.016f_8(x)v_0(x)v_0'(x)$$
$$+0.018f_4(x)v_0''(x)$$

Line 2

$$v_2(x) = -0.081f_7(x)P_f(x)+0.081f_7(x)P_0(x)$$
$$+0.799v_0(x)-0.059f_2(x)v_0(x)+0.048f_5(x)u_0(x)v_0(x)$$
$$+0.048f_7(x)v_0(x)^2-0.022f_8(x)P_f'(x)$$
$$-0.036f_8(x)P_0'(x)-0.059f_3(x)v_0'(x)$$
$$-0.025f_6(x)u_0(x)v_0'(x)-0.025f_8(x)v_0(x)v_0'(x)$$
$$+0.028f_4(x)v_0''(x)$$

Line 3

$$v_3(x) = -0.106f_7(x)P_f(x)+0.106f_7(x)P_0(x)$$
$$+0.698v_0(x)-0.075f_2(x)v_0(x)+0.060f_5(x)u_0(x)v_0(x)$$
$$+0.060f_7(x)v_0(x)^2-0.031f_8(x)P_f'(x)$$
$$-0.044f_8(x)P_0'(x)-0.075f_3(x)v_0'(x)$$
$$-0.029f_6(x)u_0(x)v_0'(x)-0.029f_8(x)v_0(x)v_0'(x)$$
$$+0.033f_4(x)v_0''(x)$$

Line 4

$$v_4(x) = -0.121f_7(x)P_f(x) + 0.121f_7(x)P_0(x)$$
$$+0.597v_0(x) - 0.084f_2(x)v_0(x) + 0.064f_5(x)u_0(x)v_0(x)$$
$$+0.064f_7(x)v_0(x)^2 - 0.037f_8(x)P_f'(x)$$
$$-0.046f_8(x)P_0'(x) - 0.084f_3(x)v_0'(x)$$
$$-0.029f_6(x)u_0(x)v_0'(x) - 0.029f_8(x)v_0(x)v_0'(x)$$
$$+0.034f_4(x)v_0''(x)$$

Line 5

$$v_5(x) = -0.126f_7(x)P_f(x) + 0.126f_7(x)P_0(x)$$
$$+0.497v_0(x) - 0.086f_2(x)v_0(x) + 0.062f_5(x)u_0(x)v_0(x)$$
$$+0.062f_7(x)v_0(x)^2 - 0.041f_8(x)P_f'(x)$$
$$-0.044f_8(x)P_0'(x) - 0.086f_3(x)v_0'(x)$$
$$-0.026f_6(x)u_0(x)v_0'(x) - 0.026f_8(x)v_0(x)v_0'(x)$$
$$+0.032f_4(x)v_0''(x)$$

Line 6

$$v_6(x) = -0.121f_7(x)P_f(x) + 0.121f_7(x)P_0(x)$$
$$+0.397v_0(x) - 0.080f_2(x)v_0(x) + 0.056f_5(x)u_0(x)v_0(x)$$
$$+0.056f_7(x)v_0(x)^2 - 0.022f_8(x)P_f'(x)$$
$$-0.041f_8(x)P_0'(x) - 0.080f_3(x)v_0'(x)$$
$$-0.022f_6(x)u_0(x)v_0'(x) - 0.022f_8(x)v_0(x)v_0'(x)$$
$$+0.028f_4(x)v_0''(x)$$

Line 7

$$v_7(x) = -0.105f_7(x)P_f(x) + 0.105f_7(x)P_0(x)$$
$$+0.297v_0(x) - 0.069f_2(x)v_0(x) + 0.045f_5(x)u_0(x)v_0(x)$$
$$+0.045f_7(x)v_0(x)^2 - 0.037f_8(x)P_f'(x)$$
$$-0.031f_8(x)P_0'(x) - 0.069f_3(x)v_0'(x)$$
$$-0.017f_6(x)u_0(x)v_0'(x) - 0.017f_8(x)v_0(x)v_0'(x)$$
$$+0.022f_4(x)v_0''(x)$$

Line 8

$$v_8(x) = -0.080f_7(x)P_f(x) + 0.080f_7(x)P_0(x)$$
$$+0.198v_0(x) - 0.051f_2(x)v_0(x) + 0.032f_5(x)u_0(x)v_0(x)$$
$$+0.032f_7(x)v_0(x)^2 - 0.029f_8(x)P_f'(x)$$
$$-0.021f_8(x)P_0'(x) - 0.051f_3(x)v_0'(x)$$
$$-0.011f_6(x)u_0(x)v_0'(x) - 0.011f_8(x)v_0(x)v_0'(x)$$
$$+0.015f_4(x)v_0''(x)$$

Line 9

$$v_9(x) = -0.045f_7(x)P_f(x) + 0.045f_7(x)P_0(x)$$
$$+0.099v_0(x) - 0.028f_2(x)v_0(x) + 0.016f_5(x)u_0(x)v_0(x)$$
$$+0.016f_7(x)v_0(x)^2 - 0.017f_8(x)P'_f(x)$$
$$-0.011f_8(x)P'_0(x) - 0.029f_3(x)v'_0(x)$$
$$-0.057f_6(x)u_0(x)v'_0(x) - 0.057f_8(x)v_0(x)v'_0(x)$$
$$+0.008f_4(x)v''_0(x)$$

Finally, for the field of pressure P, we have obtained the following lines:

Line 1

$$P_1(x) = 0.101P_f(x) + 0.034f_2(x)P_f(x)$$
$$+0.899P_0(x) - 0.034f_2(x)P_0(x) + 0.046h_3(x)f_5(x)^2u_0(x)^2$$
$$+0.092h_3(x)f_5(x)f_7(x)u_0(x)v_0(x) + 0.046h_3(x)f_7(x)^2v_0(x)^2$$
$$+0.034f_3(x)P'_f(x) - 0.034f_3(x)P'_0(x)$$
$$-0.045h_3(x)f_5(x)f_6(x)u_0(x)u'_0(x) - 0.045h_3(x)f_5(x)f_8(x)v_0(x)u'_0(x)$$
$$+0.014h_3(x)f_6(x)^2u'_0(x)^2 - 0.045h_3(x)f_6(x)f_7(x)u_0(x)v'_0(x)$$
$$-0.045h_3(x)f_7(x)f_8(x)v_0(x)v'_0(x) + 0.027h_3(x)f_6(x)f_8(x)u'_0(x)v'_0(x)$$
$$+0.014h_3(x)f_8(x)^2v'_0(x)^2 + 0.008f_4(x)P''_f(x)$$
$$+0.018f_4(x)P''_0(x)$$

Line 2

$$P_2(x) = 0.201P_f(x) + 0.059f_2(x)P_f(x)$$
$$+0.799P_0(x) - 0.059f_2(x)P_0(x) + 0.082h_3(x)f_5(x)^2u_0(x)^2$$
$$+0.163h_3(x)f_5(x)f_7(x)u_0(x)v_0(x) + 0.082h_3(x)f_7(x)^2v_0(x)^2$$
$$+0.059f_3(x)P'_f(x) - 0.059f_3(x)P'_0(x)$$
$$-0.074h_3(x)f_5(x)f_6(x)u_0(x)u'_0(x) - 0.074h_3(x)f_5(x)f_8(x)v_0(x)u'_0(x)$$
$$+0.020h_3(x)f_6(x)^2u'_0(x)^2 - 0.074h_3(x)f_6(x)f_7(x)u_0(x)v'_0(x)$$
$$-0.074h_3(x)f_7(x)f_8(x)v_0(x)v'_0(x) + 0.041h_3(x)f_6(x)f_8(x)u'_0(x)v'_0(x)$$
$$+0.020h_3(x)f_8(x)^2v'_0(x)^2 + 0.015f_4(x)P''_f(x)$$
$$+0.028f_4(x)P''_0(x)$$

Line 3

$$P_3(x) = 0.302P_f(x) + 0.075f_2(x)P_f(x)$$
$$+0.698P_0(x) - 0.075f_2(x)P_0(x) + 0.107h_3(x)f_5(x)^2u_0(x)^2$$
$$+0.214h_3(x)f_5(x)f_7(x)u_0(x)v_0(x) + 0.107h_3(x)f_7(x)^2v_0(x)^2$$
$$+0.075f_3(x)P'_f(x) - 0.075f_3(x)P'_0(x)$$
$$-0.089h_3(x)f_5(x)f_6(x)u_0(x)u'_0(x) - 0.089h_3(x)f_5(x)f_8(x)v_0(x)u'_0(x)$$
$$+0.023h_3(x)f_6(x)^2u'_0(x)^2 - 0.089h_3(x)f_6(x)f_7(x)u_0(x)v'_0(x)$$
$$-0.089h_3(x)f_7(x)f_8(x)v_0(x)v'_0(x) + 0.045h_3(x)f_6(x)f_8(x)u'_0(x)v'_0(x)$$
$$+0.023h_3(x)f_8(x)^2v'_0(x)^2 + 0.021f_4(x)P''_f(x)$$
$$+0.033f_4(x)P''_0(x)$$

Line 4

$$
\begin{aligned}
P_4(x) &= 0.403P_f(x) + 0.084f_2(x)P_f(x) \\
&+ 0.597P_0(x) - 0.084f_2(x)P_0(x) + 0.122h_3(x)f_5(x)^2u_0(x)^2 \\
&+ 0.245h_3(x)f_5(x)f_7(x)u_0(x)v_0(x) + 0.122h_3(x)f_7(x)^2v_0(x)^2 \\
&+ 0.084f_3(x)P_f'(x) - 0.084f_3(x)P_0'(x) \\
&- 0.094h_3(x)f_5(x)f_6(x)u_0(x)u_0'(x) - 0.094h_3(x)f_5(x)f_8(x)v_0(x)u_0'(x) \\
&+ 0.022h_3(x)f_6(x)^2u_0'(x)^2 - 0.094h_3(x)f_6(x)f_7(x)u_0(x)v_0'(x) \\
&- 0.094h_3(x)f_7(x)f_8(x)v_0(x)v_0'(x) + 0.045h_3(x)f_6(x)f_8(x)u_0'(x)v_0'(x) \\
&+ 0.022h_3(x)f_8(x)^2v_0'(x)^2 + 0.025f_4(x)P_f''(x) \\
&+ 0.034f_4(x)P_0''(x)
\end{aligned}
$$

Line 5

$$
\begin{aligned}
P_5(x) &= 0.503P_f(x) + 0.086f_2(x)P_f(x) \\
&+ 0.497P_0(x) - 0.086f_2(x)P_0(x) + 0.127h_3(x)f_5(x)^2u_0(x)^2 \\
&+ 0.255h_3(x)f_5(x)f_7(x)u_0(x)v_0(x) + 0.127h_3(x)f_7(x)^2v_0(x)^2 \\
&+ 0.086f_3(x)P_f'(x) - 0.086f_3(x)P_0'(x) \\
&- 0.089h_3(x)f_5(x)f_6(x)u_0(x)u_0'(x) - 0.089h_3(x)f_5(x)f_8(x)v_0(x)u_0'(x) \\
&+ 0.020h_3(x)f_6(x)^2u_0'(x)^2 - 0.089h_3(x)f_6(x)f_7(x)u_0(x)v_0'(x) \\
&- 0.089h_3(x)f_7(x)f_8(x)v_0(x)v_0'(x) + 0.040h_3(x)f_6(x)f_8(x)u_0'(x)v_0'(x) \\
&+ 0.020h_3(x)f_8(x)^2v_0'(x)^2 + 0.026f_4(x)P_f''(x) \\
&+ 0.032f_4(x)P_0''(x)
\end{aligned}
$$

Line 6

$$
\begin{aligned}
P_6(x) &= 0.603P_f(x) + 0.080f_2(x)P_f(x) \\
&+ 0.397P_0(x) - 0.080f_2(x)P_0(x) + 0.122h_3(x)f_5(x)^2u_0(x)^2 \\
&+ 0.244h_3(x)f_5(x)f_7(x)u_0(x)v_0(x) + 0.122h_3(x)f_7(x)^2v_0(x)^2 \\
&+ 0.080f_3(x)P_f'(x) - 0.080f_3(x)P_0'(x) \\
&- 0.079h_3(x)f_5(x)f_6(x)u_0(x)u_0'(x) - 0.079h_3(x)f_5(x)f_8(x)v_0(x)u_0'(x) \\
&+ 0.017h_3(x)f_6(x)^2u_0'(x)^2 - 0.079h_3(x)f_6(x)f_7(x)u_0(x)v_0'(x) \\
&- 0.079h_3(x)f_7(x)f_8(x)v_0(x)v_0'(x) + 0.033h_3(x)f_6(x)f_8(x)u_0'(x)v_0'(x) \\
&+ 0.017h_3(x)f_8(x)^2v_0'(x)^2 + 0.026f_4(x)P_f''(x) \\
&+ 0.028f_4(x)P_0''(x)
\end{aligned}
$$

Line 7

$$
\begin{aligned}
P_7(x) \;=\; & 0.703 P_f(x) + 0.069 f_2(x) P_f(x) \\
& + 0.297 P_0(x) - 0.069 f_2(x) P_0(x) + 0.107 h_3(x) f_5(x)^2 u_0(x)^2 \\
& + 0.213 h_3(x) f_5(x) f_7(x) u_0(x) v_0(x) + 0.107 h_3(x) f_7(x)^2 v_0(x)^2 \\
& + 0.069 f_3(x) P_f'(x) - 0.069 f_3(x) P_0'(x) \\
& - 0.063 f_5(x) f_6(x) u_0(x) u_0'(x) - 0.063 f_5(x) f_8(x) v_0(x) u_0'(x) \\
& + 0.013 h_3(x) f_6(x)^2 u_0'(x)^2 - 0.063 h_3(x) f_6(x) f_7(x) u_0(x) v_0'(x) \\
& - 0.063 h_3(x) f_7(x) f_8(x) v_0(x) v_0'(x) + 0.025 h_3(x) f_6(x) f_8(x) u_0'(x) v_0'(x) \\
& + 0.013 h_3(x) f_8(x)^2 v_0'(x)^2 + 0.023 f_4(x) P_f''(x) \\
& + 0.022 f_4(x) P_0''(x)
\end{aligned}
$$

Line 8

$$
\begin{aligned}
P_8(x) \;=\; & 0.802 P_f(x) + 0.051 f_2(x) P_f(x) \\
& + 0.198 P_0(x) - 0.051 f_2(x) P_0(x) + 0.081 h_3(x) f_5(x)^2 u_0(x)^2 \\
& + 0.162 h_3(x) f_5(x) f_7(x) u_0(x) v_0(x) + 0.081 h_3(x) f_7(x)^2 v_0(x)^2 \\
& + 0.051 f_3(x) P_f'(x) - 0.051 f_3(x) P_0'(x) \\
& - 0.043 h_3(x) f_5(x) f_6(x) u_0(x) u_0'(x) - 0.043 h_3(x) f_5(x) f_8(x) v_0(x) u_0'(x) \\
& + 0.009 h_3(x) f_6(x)^2 u_0'(x)^2 - 0.043 h_3(x) f_6(x) f_7(x) u_0(x) v_0'(x) \\
& - 0.043 h_3(x) f_7(x) f_8(x) v_0(x) v_0'(x) + 0.017 h_3(x) f_6(x) f_8(x) u_0'(x) v_0'(x) \\
& + 0.009 h_3(x) f_8(x)^2 v_0'(x)^2 + 0.018 f_4(x) P_f''(x) \\
& + 0.015 f_4(x) P_0''(x)
\end{aligned}
$$

Line 9

$$
\begin{aligned}
P_9(x) \;=\; & 0.901 P_f(x) + 0.028 f_2(x) P_f(x) \\
& + 0.099 P_0(x) - 0.028 f_2(x) P_0(x) + 0.045 h_3(x) f_5(x)^2 u_0(x)^2 \\
& + 0.191 h_3(x) f_5(x) f_7(x) u_0(x) v_0(x) + 0.045 h_3(x) f_7(x)^2 v_0(x)^2 \\
& + 0.028 f_3(x) P_f'(x) - 0.028 f_3(x) P_0'(x) \\
& - 0.022 h_3(x) f_5(x) f_6(x) u_0(x) u_0'(x) - 0.022 h_3(x) f_5(x) f_8(x) v_0(x) u_0'(x) \\
& + 0.004 h_3(x) f_6(x)^2 u_0'(x)^2 - 0.022 h_3(x) f_6(x) f_7(x) u_0(x) v_0'(x) \\
& - 0.022 h_3(x) f_7(x) f_8(x) v_0(x) v_0'(x) + 0.009 h_3(x) f_6(x) f_8(x) u_0'(x) v_0'(x) \\
& + 0.004 h_3(x) f_8(x)^2 v_0'(x)^2 + 0.010 f_4(x) P_f''(x) \\
& + 0.008 f_4(x) P_0''(x)
\end{aligned}
$$

28.3.1 Numerical examples through the generalized method of lines

We consider some examples in which

$$
\Omega = \{(r, \theta) \mid 1 \le r \le 2,\ 0 \le \theta \le 2\pi\},
$$

$$
\partial \Omega_0 = \{(1, \theta) \mid 0 \le \theta \le 2\pi\},
$$

and

$$
\partial \Omega_1 = \{(2, \theta) \mid 0 \le \theta \le 2\pi\}.
$$

For such cases, the boundary conditions are

$$u = u_0(\theta), \quad v = v_0(\theta), \quad \text{on } \partial\Omega_0,$$
$$u = v = 0, \quad \text{on } \partial\Omega_1,$$

so that in these examples we do not have boundary conditions for the pressure.

Through the generalized method of lines, neglecting gravity effects and truncating the series up to the terms in d^2, where $d = 1/N$ is the mesh thickness concerning the discretization in r, we present numerical results for the following approximation of the Navier-Stokes system,

$$\begin{cases} \nu\nabla^2 u - u\partial_x u - v\partial_y u - \partial_x P = 0, & \text{in } \Omega, \\[2mm] \nu\nabla^2 v - u\partial_x v - v\partial_y v - \partial_y P = 0, & \text{in } \Omega, \\[2mm] \varepsilon\nabla^2 P + \partial_x u + \partial_y v = 0, & \text{in } \Omega, \end{cases} \tag{28.25}$$

where $\varepsilon > 0$ is a very small parameter. We highlight, since $\varepsilon > 0$ must be very small, the results obtained through the generalized of lines, as indicated in the last section, may have a relevant error. Anyway, we may use such a procedure to obtain the general line expressions, which we expect to be analytically suitable to obtain a numerical result by calculating numerically the concerning coefficients for such lines, through a minimization of the L^2 norm of equation errors, in a finite differences context.

Thus, from such a method, the general expression for the velocity and pressure fields on the line n, are given by (here x stands for θ and P_0 must be calculated numerically in the optimization process):

$$\begin{aligned} u_n(x) = \; & a_1[n]\cos(x) + a_2[n]u_0(x) + a_3[n]\cos[x]u_0(x)^2 \\ & a_4[n]\sin(x)u_0(x)v_0(x) + a_5[n]\sin(x)u_0(x)u_0'(x) + a_6[n]\cos(x)v_0(x)u_0'(x) \\ & a_7[n]u_0''(x) + a_8[n]\sin(x) + a_9[n]\sin(x)P_0(x) \\ & a_{10}[n]P_0(x)\cos(x) + a_{11}[n], \end{aligned} \tag{28.26}$$

$$\begin{aligned} v_n(x) = \; & b_1[n]\sin(x) + b_2[n]v_0(x) + b_3[n]\sin[x]v_0(x)^2 \\ & b_4[n]\cos(x)u_0(x)v_0(x) + b_5[n]\sin(x)v_0'(x)u_0(x) + b_6[n]\cos(x)v_0(x)u_0'(x) \\ & b_7[n]v_0''(x) + b_8[n]\cos(x) + b_9[n]\sin(x)P_0(x) \\ & b_{10}[n]P_0(x)\cos(x) + b_{11}[n], \end{aligned} \tag{28.27}$$

$$\begin{aligned} P_n(x) = \; & c_1[n] + c_2[n]\cos(x)u_0(x) + c_3[n]\sin(x)v_0(x) + c_4[n]\sin(x)u_0'(x) \\ & + c_5[n]\cos(x)v_0'(x) + c_6[n]u_0''(x) + c_7[n]v_0''(x) \\ & + c_9[n]P_0(x) + c_{10}[n]P_0''(x). \end{aligned} \tag{28.28}$$

■ **First example**: For the first example,

$$u_0(x) = -1.5\sin(x)$$

and

$$v_0(x) = 1.5\cos(x).$$

Denoting

$$\begin{aligned} J(u,v,P) = \; & \int_\Omega \left(\nu\nabla^2 u - u\partial_x u - v\partial_y u - \partial_x P\right)^2 \, d\Omega \\ & + \int_\Omega \left(\nu\nabla^2 v - u\partial_x v - v\partial_y v - \partial_y P\right)^2 \, d\Omega \\ & + \int_\Omega (\partial_x u + \partial_y v)^2 \, d\Omega, \end{aligned} \tag{28.29}$$

as, above mentioned, the coefficients $\{a_i[n]\}, \{b_i[n]\}, \{c_i[n]\}$ have been obtained through the numerical minimization of $J(\{u_n\}, \{v_n\}, \{P_n\})$, so that for the mesh in question, we have obtained

For this first example: $J(\{u_n\}, \{v_n\}, \{P_n\}) \approx 9.23 \ 10^{-12}$ for $v = 0.1$,

We have plotted the fields u, v and P, for the lines $n = 1$, $n = 5$, $n = 10$, $n = 15$ and $n = 19$, for a mesh 20×150 corresponding to 20 lines. Please, see the Figures from 28.1 to 28.6 for the case $v = 0.1$. For all graphs, please consider units in x to be multiplied by $2\pi/150$.

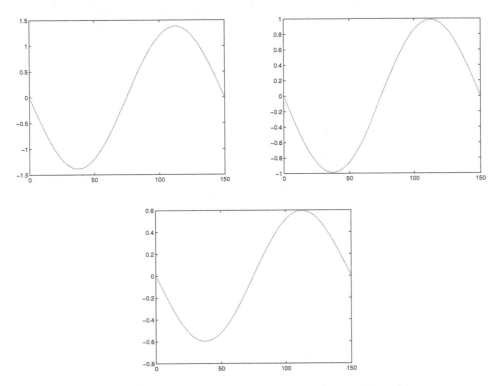

Figure 28.1: First example, from the left to the right, fields of velocity $u_1(x)$, $u_5(x)$, $u_{10}(x)$ for the lines $n = 1$, $n = 5$ and $n = 10$.

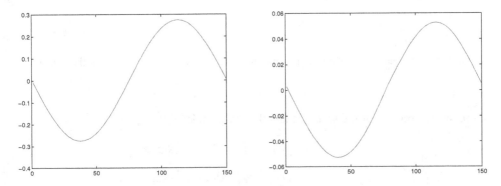

Figure 28.2: First example, from the left to the right, fields of velocity $u_{15}(x)$, $u_{19}(x)$ for the lines $n = 15$, and $n = 19$.

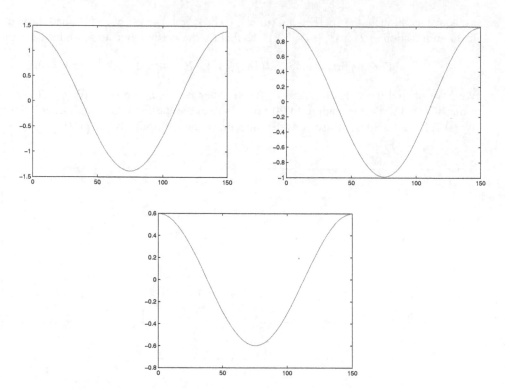

Figure 28.3: First example, from the left to the right, fields of velocity $v_1(x)$, $v_5(x)$, $v_{10}(x)$ for the lines $n = 1$, $n = 5$ and $n = 10$.

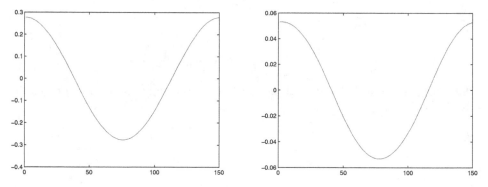

Figure 28.4: First example, from the left to the right, fields of velocity $v_{15}(x)$, $v_{19}(x)$ for the lines $n = 15$, and $n = 19$.

■ **Second example**: For the second example, we consider

$$u_0(x) = -3.0 \, \cos(x)\sin(x),$$

$$v_0(x) = -(-2.0\cos(x)^2 + \sin(x)^2)$$

and $v = 1.0$.

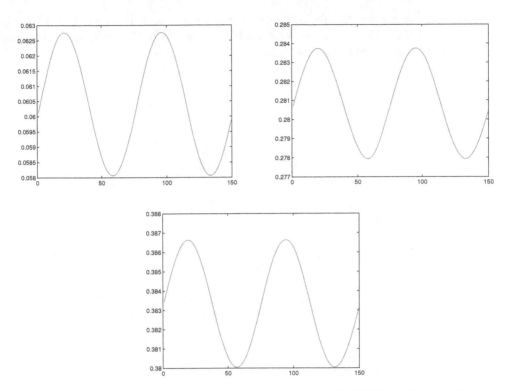

Figure 28.5: First example, from the left to the right, fields of pressure $P_1(x)$, $P_5(x)$, $P_{10}(x)$ for the lines $n = 1$, $n = 5$ and $n = 10$.

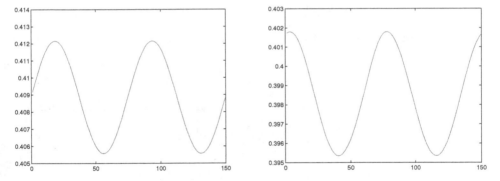

Figure 28.6: First example, from the left to the right, fields of pressure $P_{15}(x)$, $P_{19}(x)$ for the lines $n = 15$, and $n = 19$.

Again the coefficients $\{a_i[n]\}, \{b_i[n]\}, \{c_i[n]\}$ have been obtained through the numerical minimization of $J(\{u_n\}, \{v_n\}, \{P_n\})$, so that for the mesh in question, we have obtained

For this second example: $J(\{u_n\}, \{v_n\}, \{P_n\}) \approx 6.0 \ 10^{-7}$ for $v = 1.0$.

In any case, considering the values obtained for J, it seems we have got good first approximations for the concerning solutions.

For the second example, for the field of velocities u and v, and pressure field P, for the lines $n = 1$, $n = 5$, $n = 10$, $n = 15$ and $n = 19$, please see Figures 28.7 to 28.12. Once more, for all graphs, please consider units in x to be multiplied by $2\pi/150$.

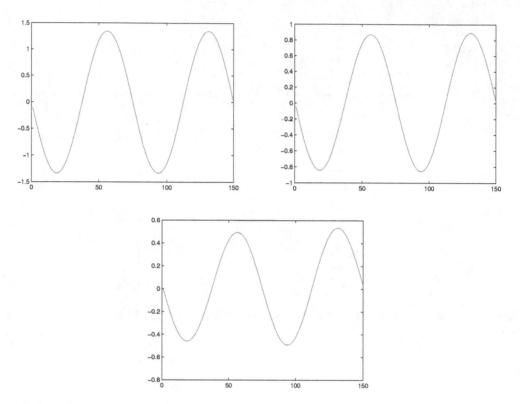

Figure 28.7: Second example, from the left to the right, fields of velocity $u_1(x)$, $u_5(x)$, $u_{10}(x)$ for the lines $n = 1$, $n = 5$ and $n = 10$.

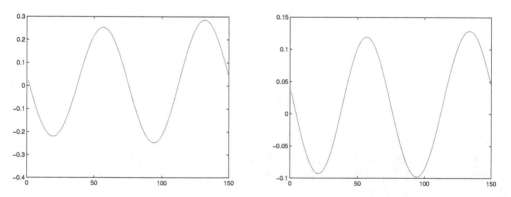

Figure 28.8: Second example, from the left to the right, fields of velocity $u_{15}(x)$, $u_{19}(x)$ for the lines $n = 15$, and $n = 19$.

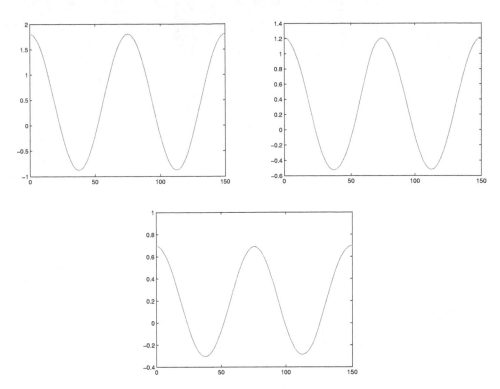

Figure 28.9: Second example, from the left to the right, fields of velocity $v_1(x)$, $v_5(x)$, $v_{10}(x)$ for the lines $n = 1$, $n = 5$ and $n = 10$.

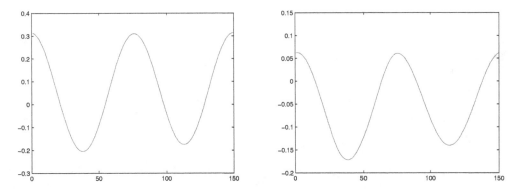

Figure 28.10: Second example, from the left to the right, fields of velocity $v_{15}(x)$, $v_{19}(x)$ for the lines $n = 15$, and $n = 19$.

28.4 Conclusion

In the first part of this chapter, we obtain a linear system whose solution solves the time-independent incompressible Navier-Stokes system. In the second part, we develop solutions for two-dimensional examples also for the time-independent incompressible Navier-Stokes system, through the generalized method of lines.

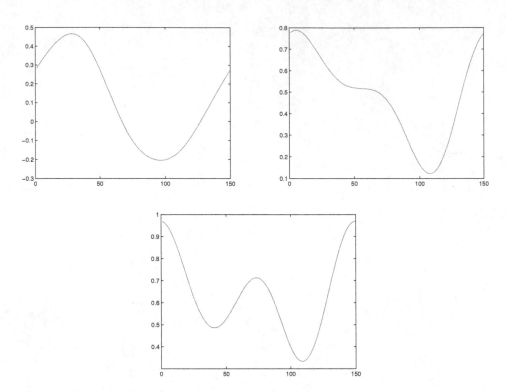

Figure 28.11: Second example, from the left to the right, fields of pressure $P_1(x)$, $P_5(x)$, $P_{10}(x)$ for the lines $n = 1$, $n =$ and $n = 10$.

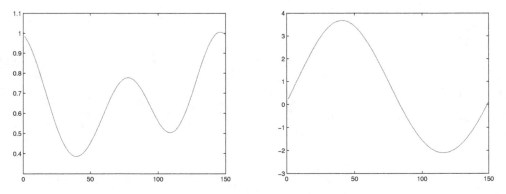

Figure 28.12: Second example, from the left to the right, fields of pressure $P_{15}(x)$, $P_{19}(x)$ for the lines $n = 15$, and $n = 1$

Considering the values for J obtained, we have got very good approximate solutions for the model in que tion, in a finite differences context. The extension of such results to $\mathbb{R}^3$, compressible and time depende cases is planned for a future work.

Chapter 29

A Numerical Method for an Inverse Optimization Problem through the Generalized Method of Lines

29.1 Introduction

In this chapter we develop a numerical method to compute the solution of an inverse problem through the generalized method of lines.

More specifically, we consider a Laplace equation on a domain $\Omega \subset \mathbb{R}^2$ with an internal boundary denoted by $\partial \Omega_0$ and an external one denoted by $\partial \Omega_1$. We prescribe boundary conditions for both $\partial \Omega_0$ and $\partial \Omega_1$ and a third boundary condition for $\partial \Omega_1$ (the external one) and consider the problem of finding the optimal shape for the internal boundary $\partial \Omega_0$ for which such a third boundary condition for the external boundary is satisfied.

The idea is to discretize the domain in lines (in fact curves) and write the solution of Laplace equation on these lines as functions of the unknown internal boundary shape, through the generalized method of lines.

The second step is to minimize a functional which corresponds to the L^2 norms of Laplace equation and concerning third boundary condition.

Remark 29.1   About the references, this and many other similar problems are addressed in [49]. The generalized method of lines has been originally introduced in [19], with additional results in [18, 14]. Moreover, about finite differences schemes we would cite [73]. Finally, details on the function spaces here addressed may be found in [1].   ∎

29.2 The mathematical description of the main problem

At this point we start to describe mathematically our main problem.

Let $\Omega \subset \mathbb{R}^2$ be a bounded, closed and connected set defined by

$$\Omega = \{(r, \theta) \in \mathbb{R}^2 \; : \; r(\theta) \leq r \leq R \; : \; 0 \leq \theta \leq 2\pi\}.$$

Consider a Laplace equation and concerning boundary conditions expressed by

$$\begin{cases} \nabla^2 u = 0, & \text{in } \Omega, \\ u = u_o, & \text{on } \partial\Omega_0, \\ u = u_f, & \text{on } \partial\Omega_1, \\ \nabla u \cdot \mathbf{n} = w, & \text{on } \partial\Omega_1, \end{cases} \tag{29.1}$$

where u_o, u_f and $w \in C^2([0, 2\pi])$ are known periodic functions with period 2π, $\mathbf{n}$ denotes the outward normal field to $\partial\Omega$,

$$\partial\Omega_0 = \{(r(\theta), \theta) : 0 \le \theta \le 2\pi\},$$

$$\partial\Omega_1 = \{(R, \theta) : 0 \le 0 \le 2\pi\},$$

and $R > 0$.

The main idea is to discretize the domain in lines and through the generalized method of lines to obtain the solution $u_n(r(\theta))$ on each line n as a function of $r(\theta)$. The final step is to compute the optimal $r(\theta)$ in order to minimize the cost functional $J(r(\theta))$ defined by

$$J(r(\theta)) = \|\nabla^2 u\|_{2,\Omega}^2 + K\|\nabla u \cdot \mathbf{n} - w\|_{2,\partial\Omega_1}^2,$$

where $K > 0$ is an appropriate constant to be specified.

29.3   About the generalized method of lines and the main result

At this point we start to describe the main result.

Consider firstly a Laplace equation in polar coordinates, that is,

$$\frac{\partial^2 u}{\partial r^2} + \frac{1}{r}\frac{\partial u}{\partial r} + \frac{1}{r^2}\frac{\partial^2 u}{\partial \theta^2} = 0.$$

In order to apply the generalized method of lines, we a define a variable t, through the equation

$$t = \frac{r - r(\theta)}{R - r(\theta)}$$

so that $t \in [0, 1]$ in the set Ω previously specified.

Hence, denoting $u(r, \theta) = \bar{u}(t, \theta)$, we obtain,

$$\begin{aligned} \frac{\partial u}{\partial \theta} &= \frac{\partial \bar{u}}{\partial t}\frac{\partial t}{\partial \theta} + \frac{\partial \bar{u}}{\partial \theta} \\ &= \frac{\partial \bar{u}}{\partial t}\left(\frac{t-1}{R - r(\theta)}\right)r'(\theta) + \frac{\partial \bar{u}}{\partial \theta} \\ &= \frac{\partial \bar{u}}{\partial t}(f_1(\theta) + t f_2(\theta)) + \frac{\partial \bar{u}}{\partial \theta}, \end{aligned} \tag{29.2}$$

where

$$f_1(\theta) = \frac{-r'(\theta)}{R - r(\theta)},$$

and

$$f_2(\theta) = \frac{r'(\theta)}{R - r(\theta)}.$$

Also,

$$
\begin{aligned}
\frac{\partial u}{\partial r} &= \frac{\partial \bar{u}}{\partial t}\frac{\partial t}{\partial r} \\
&= \frac{\partial \bar{u}}{\partial t}\frac{1}{R - r(\theta)},
\end{aligned}
\tag{29.3}
$$

so that

$$
\begin{aligned}
\frac{\partial^2 u}{\partial r^2} &= \frac{\partial^2 \bar{u}}{\partial t^2}\left(\frac{\partial t}{\partial r}\right)^2 \\
&= \frac{\partial^2 \bar{u}}{\partial t^2}\frac{1}{(R - r(\theta))^2} \\
&= f_3(\theta)\frac{\partial^2 \bar{u}}{\partial t^2},
\end{aligned}
\tag{29.4}
$$

where

$$
f_3(\theta) = \frac{1}{(R - r(\theta))^2}.
$$

Moreover, we may also obtain

$$
\begin{aligned}
\frac{\partial^2 u}{\partial \theta^2} &= f_4(t,\theta)\frac{\partial^2 \bar{u}}{\partial t^2} + f_5(t,\theta)\frac{\partial \bar{u}}{\partial t} \\
&\quad + f_6(t,\theta)\frac{\partial^2 \bar{u}}{\partial t \partial \theta} + \frac{\partial^2 \bar{u}}{\partial \theta^2},
\end{aligned}
\tag{29.5}
$$

where,

$$
f_4(t,\theta) = (f_1(\theta) + t f_2(\theta))^2,
$$
$$
f_5(t,\theta) = f_1'(\theta) + t f_2'(\theta) + f_2(\theta)(f_1(\theta) + t f_2(\theta)),
$$
$$
f_6(t,\theta) = 2(f_1(\theta) + t f_2(\theta)).
$$

Thus, for the new variables (t,θ), dropping the bar in $\bar{u}$, the Laplace equation is equivalent to

$$
\frac{\partial^2 u}{\partial t^2} + f_7(t,\theta)\frac{\partial u}{\partial t} + f_8(t,\theta)\frac{\partial^2 u}{\partial t \partial \theta} + f_9(t,\theta)\frac{\partial^2 u}{\partial \theta^2} = 0,
$$

in $\hat{\Omega}$ where

$$
\hat{\Omega} = \{(t,\theta) \in \mathbb{R}^2 \ : \ 0 \le t \le 1, \ 0 \le \theta \le 2\pi\},
$$
$$
r = t(R - r(\theta)) + r(\theta),
$$
$$
f_0(t,\theta) = \frac{1}{(R - r(\theta))^2} + \frac{f_4(t,\theta)}{r^2},
$$
$$
f_7(t,\theta) = \frac{\tilde{f}_7(t,\theta)}{f_0(t,\theta)},
$$
$$
f_8(t,\theta) = \frac{\tilde{f}_8(t,\theta)}{f_0(t,\theta)},
$$
$$
f_9(t,\theta) = \frac{\tilde{f}_9(t,\theta)}{f_0(t,\theta)}
$$

and where

$$\tilde{f}_7(t,\theta) = \left(\frac{1}{r(R-r(\theta))} + \frac{f_5(t,\theta)}{r^2} \right)$$

$$\tilde{f}_8(t,\theta) = \frac{f_6(t,\theta)}{r^2}$$

and

$$\tilde{f}_9(t,\theta) = \frac{1}{r^2}.$$

So, discretizing the interval $[0,1]$ into $N \in \mathbb{N}$ pieces, that is defining,

$$t_n = \frac{n}{N}, \ \forall n \in \{0,\ldots,N\}$$

and $d = 1/N$, in partial finite differences, the concerning equation stands for

$$\frac{u_{n+1} - 2u_n + u_{n-1}}{d^2} + f_7(t_n,\theta) \left(\frac{u_n - u_{n-1}}{d} \right)$$

$$+ f_8(t_n,\theta) \frac{\partial}{\partial \theta} \left(\frac{u_n - u_{n-1}}{d} \right) + f_9(t_n,\theta) \frac{\partial^2 u_n}{\partial \theta^2} = 0, \tag{29.6}$$

$\forall n \in \{1,\ldots,N-1\}$, where

$$u_0 = u_o(\theta)$$

and

$$u_N = u_f(\theta).$$

At this point we describe how to obtain the general expression for u_n corresponding to the line n, through the generalized method of lines.

For $n = 1$ in (29.6), we have

$$u_2 - 2u_1 + u_0 + f_7(t_1,\theta)(u_1 - u_0)d$$

$$+ f_8(t_1,\theta) \frac{\partial(u_1 - u_0)}{\partial \theta} d + f_9(t_1,\theta) \frac{\partial^2 u_1}{\partial \theta^2} d^2 = 0, \tag{29.7}$$

so that

$$u_1 = T_1(u_2,u_1,u_0),$$

where

$$T_1(u_2,u_1,u_0) = (u_2 + u_1 + u_0 + f_7(t_1,\theta)(u_1 - u_0)d$$

$$+ f_8(t_1,\theta) \frac{\partial(u_1 - u_0)}{\partial \theta} d + f_9(t_1,\theta) \frac{\partial^2 u_1}{\partial \theta^2} d^2 \right) /3. \tag{29.8}$$

To solve this equation we apply the Banach fixed point theorem as it follows:

1. Set $u_1^1 = u_2$.

2. Define

$$u_1^{k+1} = T_1(u_2,u_1^k,u_0), \ \forall k \in \mathbb{N}.$$

3. Obtain

$$u_1 = \lim_{k \to \infty} u_1^k \equiv F_1(u_2,u_0).$$

Reasoning inductively, having

$$u_{n-1} = F_{n-1}(u_n, u_0)$$

also from (29.6), for the line n, we have

$$u_n = T_n(u_{n+1}, u_n, u_0),$$

where

$$
\begin{aligned}
T_n(u_{n+1}, u_n, u_0) \quad = \quad & (u_{n+1} + u_n + F_{n-1}(u_n, u_0) + f_7(t_n, \theta)(u_n - F_{n-1}(u_n, u_0))d \\
& + f_8(t_n, \theta) \frac{\partial (u_n - F_{n-1}(u_n, u_0))}{\partial \theta} d + f_9(t_n, \theta) \frac{\partial^2 u_n}{\partial \theta^2} d^2 \Big) / 3.
\end{aligned}
\tag{29.9}
$$

Again, to solve this equation, we apply the Banach fixed point theorem, as it follows:

1. Set $u_n^1 = u_{n+1}$.

2. Define

$$u_n^{k+1} = T_n(u_{n+1}, u_n^k, u_0), \ \forall k \in \mathbb{N}.$$

3. Obtain

$$u_n = \lim_{k \to \infty} u_n^k \equiv F_n(u_{n+1}, u_0), \ \forall n\{2, \ldots, N-1\}.$$

In particular, for $n = N - 1$, we get

$$u_{N-1} = F_{N-1}(u_N, u_0) = F_{N-1}(u_f, u_0).$$

Similarly, for $n = N - 2$, we obtain

$$u_{N-2} = F_{N-2}(u_{N-1}, u_0),$$

and so on, up to finding

$$u_1 = F_1(u_2, u_0).$$

This means that we have obtained a general expression

$$u_n = F_n(u_{n+1}, u_0) \equiv H_n(u_f, u_0, r(\theta)), \ \forall n \in \{1, \ldots, N-1\},$$

Anyway, we remark to properly run the software we have to make the approximation $t_n = t, \forall n \in \{1, \ldots, N-1\}$. Thus, we have used the method above described just to find a general expression for u_n, which is approximately given by (here x stands for θ)

$$
\begin{aligned}
u_n(x) \quad \approx \quad & a_n[1]u_f(x) + a_n[2]u_0(x) + a_n[3] \ f_7[t_n, x]u_f(x) + a_n[4] \ f_7[t_n, x]u_0(x) \\
& + a_n[5] \ f_8[t_n, x]u_f'(x) + a_n[6] \ f_8[t_n, x]u_0'(x) \\
& + a_n[9] \ f_9[t_n, x]u_f''(x) + a_n[10] \ f_9[t_n, x]u_0''(x).
\end{aligned}
\tag{29.10}
$$

Indeed this a first approximation for the series representing the solution on each line obtained by considering terms up to order d^2 (in d).

Moreover the coefficients $a_n[k]$ and $r(\theta)$ expressed in finite differences are calculated through the minimization of $J(r(\theta), \{a_n[k]\})$ given by

$$J(r(\theta), \{a_n[k]\}) = \|\nabla^2 u\|_{2,\Omega}^2 + K\|\nabla u \cdot \mathbf{n} - w\|_{2,\partial\Omega_1}^2,$$

29.3.1 The numerical results

In a first step we solve, through the generalized method of lines, the following Laplace equation

$$\begin{cases} \nabla^2 u = 0, & \text{in } \Omega, \\ u = u_o, & \text{on } \partial\Omega_0, \\ u = u_f, & \text{on } \partial\Omega_1, \\ \nabla u \cdot \mathbf{n} = w, & \text{on } \partial\Omega_1, \end{cases} \qquad (29.11)$$

where

$$\partial\Omega_0 = \{(\theta, r(\theta)) \in \mathbb{R}^2 \, : \, r(\theta) = 5.5(1.4 + \cos(\theta))/1.5, \text{ and } 0 \le \theta \le 2\pi\},$$

and as above indicated

$$\partial\Omega_1 = \{(\theta, R) \in \mathbb{R}^2 \, : \, 0 \le \theta \le 2\pi\}.$$

Having obtained $u = \{u_n\}_{n=1}^N$ for $N = 26$ lines we define $W = (u_f - u_{n-1})/d$, where $d = 1/N$. We have considered $N_1 = 42$ nodes for the discretization in θ.

In a second step, we calculate the optimal solution $r(\theta)$, for the boundary conditions

$$u_o(x) = (0.5\sin(x) + 0.8)/3.0$$

$$u_f(x) = (0.5\cos(x) + 1.0)/3.0$$

which minimizes, as above indicated,

$$J(r(\theta), \{a_n[k]\}) = \|\nabla^2 u\|_{2,\Omega}^2 + K\|\nabla u \cdot \mathbf{n} - w\|_{2,\partial\Omega_1}^2.$$

We present numerical results for $R = 20$, $K = 100$. Please see Figure 29.1 for the exact solution and Figure 29.2 for the optimal shape which minimizes J, calculated through the generalized method of lines (we have used a very different initial solution for the iterative minimization procedure).

Finally, we have obtained,

$$\|(\nabla u) \cdot \mathbf{n} - w\|_\infty \approx 3.1218 \; 10^{-5}.$$

and

$$\|\nabla^2 u\|_\infty \approx 1.0626 \; 10^{-6}$$

These last results indicate the method proposed has been successful to compute such a problem.

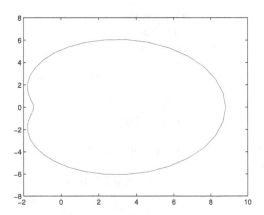

Figure 29.1: Exact solution for the internal boundary $\partial\Omega_0$ of Ω.

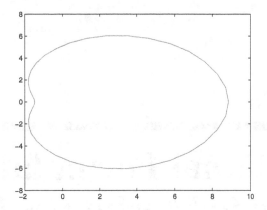

Figure 29.2: Optimal shape calculated through the generalized method of lines for the internal boundary $\partial\Omega_0$ of Ω.

29.4 Conclusion

In this chapter we have used the generalized method of lines to obtain an approximate solution for an inverse optimization problem.

We emphasize the numerical results obtained indicate the numerical procedure proposed is an interesting possibility to compute such a type of problem.

We also highlight the method here developed may be applied to a large class of similar models.

Chapter 30

A Variational Formulation for Relativistic Mechanics based on Riemannian Geometry and its Application to the Quantum Mechanics Context

This chapter has been published in a similar form by the Journal Ciência e Natura, from the Federal University of Santa Maria, Santa Maria, RS - Brazil (reference [23]).

30.1 Introduction

In this chapter we develop a variational formulation suitable for the relativistic quantum mechanics approach in a free particle context. The results are based on fundamental concepts of Riemannian geometry and suitable extensions for the relativistic case. Definitions such as vector fields, connection, Lie Bracket and Riemann tensor are addressed in the subsequent sections for the main energy construction.

Indeed, the action developed in this article, in some sense, generalizes and extends the one presented in the Weinberg book [81], in Chapter 12 at page 358. In such a book, the concerned action is denoted by $I = I_M + I_G$, where I_M, the matter action, for N particles with mass m_n and charge e_n, $\forall n \in \{1, \ldots, N\}$, is given by

$$
\begin{aligned}
I_M = & -\sum_{n=1}^{N} m_n \int_{-\infty}^{\infty} \sqrt{-g_{\mu\nu}(x(p))\frac{dx_n^{\mu}(p)}{dp}\frac{dx_n^{\nu}(p)}{dp}} \, dp \\
& -\frac{1}{4}\int_{\Omega} \sqrt{g}F_{\mu\nu}F^{\mu\nu} \, d^4x \\
& +\sum_{n=1}^{N} e_n \int_{-\infty}^{\infty} \frac{dx_n^{\mu}(p)}{dp}A_{\mu}(x(p)) \, dp,
\end{aligned}
\tag{30.1}
$$

where $\{x_n(p)\}$ is the position field with concerning metrics $\{g_{\mu\nu}(x(p))\}$ and

$$F_{\mu\nu} = \frac{\partial A_\nu}{\partial x_\mu} - \frac{\partial A_\mu}{\partial x_\nu}$$

represents the electromagnetic tensor field through a vectorial potential $\{A_\mu\}$.

Moreover, the gravitational action I_G is defined by

$$I_G = -\frac{1}{16\pi G} \int_\Omega R(x) \sqrt{g} \, d^4x,$$

where

$$R = g^{\mu\nu} R_{\mu\nu}.$$

Here,

$$R_{\mu\nu} = R^\sigma_{\mu\sigma\nu},$$

where

$$R^\eta_{\mu\sigma\nu}$$

are the components of the well known Riemann curvature tensor.

According to [81], the Euler-Lagrange equations for I correspond to the Einstein field equations,

$$R^{\mu\nu} - \frac{1}{2} g^{\mu\nu} R + 8\pi G T^{\mu\nu} = 0,$$

where the energy-momentum tensor $T^{\mu\nu}$ is expressed by

$$\begin{aligned} T^{\lambda\kappa} &= \sqrt{g} \sum_{n=1}^N \int_{-\infty}^\infty \frac{dx_n^\lambda(p)}{d\tau_n} \frac{dx_n^\kappa(p)}{d\tau_n} \delta^4(x - x_n) \, d\tau_n \\ &\quad + F_\mu^\lambda(x) F^{\mu\kappa}(x) - \frac{1}{4} g^{\lambda\kappa} F_{\mu\nu} F^{\mu\nu}. \end{aligned} \tag{30.2}$$

One of the main differences of our model from this previous one, is that we consider a possible variation in the density along the mechanical system.

Also, in our model, the motion of the system in question is specified by a four-dimensional manifold given by the function

$$\mathbf{r}(\hat{\mathbf{u}}(\mathbf{x},t)) = (ct, X_1(\mathbf{u}(\mathbf{x},t)), X_2(\mathbf{u}(\mathbf{x},t)), X_3(\mathbf{u}(\mathbf{x},t))),$$

with corresponding mass density

$$(\rho \circ \hat{\mathbf{u}}) : \Omega \times [0,T] \to \mathbb{R}^+,$$

where $\Omega \subset \mathbb{R}^3$ and $[0,T]$ is a time interval. At this point, we define $\phi(\hat{\mathbf{u}}(\mathbf{x},t))$ as a complex function such that

$$|\phi(\hat{\mathbf{u}}(\mathbf{x},t))|^2 = \frac{\rho(\hat{\mathbf{u}}(\mathbf{x},t))}{m},$$

where m denotes the total system mass at rest. We emphasize it seems to be clear that in the previous book the parametrization of the position field, through the parameter p, is one-dimensional.

In this work we do not consider the presence of electromagnetic fields.

Anyway, the final expression of the related new action here developed is given by

$$
\begin{aligned}
&J(\phi, \mathbf{r}, \hat{\mathbf{u}}, E) \\
&= \int_0^T \int_\Omega mc \sqrt{-g_{ij} \frac{\partial u_i}{\partial t} \frac{\partial u_j}{\partial t}} \, |\phi(\hat{\mathbf{u}}(\mathbf{x}, t))|^2 \sqrt{-g} \, |\det(\hat{\mathbf{u}}'(\mathbf{x}, t))| \, d\mathbf{x} \, dt \\
&\quad + \frac{\gamma}{2} \int_0^T \int_\Omega g^{jk} \frac{\partial \phi}{\partial u_j} \frac{\partial \phi^*}{\partial u_k} \sqrt{-g} \, |\det(\hat{\mathbf{u}}'(\mathbf{x}, t))| \, d\mathbf{x} dt \\
&\quad + \frac{\gamma}{4} \int_0^T \int_\Omega g^{jk} \left(\phi^* \frac{\partial \phi}{\partial u_l} + \phi \frac{\partial \phi^*}{\partial u_l} \right) \Gamma_{jk}^l \sqrt{-g} \, |\det(\hat{\mathbf{u}}'(\mathbf{x}, t))| \, d\mathbf{x} dt \\
&\quad + \frac{\gamma}{2} \int_0^T \int_\Omega |\phi|^2 g^{jk} \left(\frac{\partial \Gamma_{lk}^l}{\partial u_j} - \frac{\partial \Gamma_{jk}^l}{\partial u_l} + \Gamma_{lk}^p \Gamma_{jp}^l - \Gamma_{jk}^p \Gamma_{lp}^l \right) \sqrt{-g} \, |\det(\hat{\mathbf{u}}'(\mathbf{x}, t))| \, d\mathbf{x} dt \\
&\quad - \int_0^T E(t) \left(\int_\Omega |\phi(\hat{\mathbf{u}}(\mathbf{x}, t))|^2 \sqrt{-g} \, |\det(\hat{\mathbf{u}}'(\mathbf{x}, t))| \, d\mathbf{x} - 1 \right) dt. \quad (30.3)
\end{aligned}
$$

Observe that the action part

$$
\begin{aligned}
&\frac{\gamma}{2} \int_0^T \int_\Omega |\phi|^2 g^{jk} \left(\frac{\partial \Gamma_{lk}^l}{\partial u_j} - \frac{\partial \Gamma_{jk}^l}{\partial u_l} + \Gamma_{lk}^p \Gamma_{jp}^l - \Gamma_{jk}^p \Gamma_{lp}^l \right) \sqrt{-g} \, |\det(\hat{\mathbf{u}}'(\mathbf{x}, t))| \, d\mathbf{x} dt \\
&= \frac{\gamma}{2} \int_0^T \int_\Omega |\phi|^2 \hat{R} \sqrt{-g} \, |\det(\hat{\mathbf{u}}'(\mathbf{x}, t))| \, d\mathbf{x} dt, \quad (30.4)
\end{aligned}
$$

where

$$
\hat{R} = g^{jk} \hat{R}_{jk},
$$

$$
\hat{R}_{jk} = \hat{R}_{jlk}^l
$$

and

$$
\hat{R}_{ijk}^l = \frac{\partial \Gamma_{jk}^l}{\partial u_i} - \frac{\partial \Gamma_{ik}^l}{\partial u_j} + \Gamma_{jk}^p \Gamma_{ip}^l - \Gamma_{ik}^p \Gamma_{jp}^l
$$

represent the Riemann curvature tensor, corresponds tho the Hilbert-Einstein one, as specified in the subsequent sections.

In the last section, we show how such a formulation may result, as an approximation, the well known relativistic Klein-Gordon one and the respective Euler-Lagrange equations. We believe the main results obtained may be extended to more complex mechanical systems, including in some extent, the quantum mechanics approach.

Finally, about the references, details on the Sobolev Spaces involved may be found in [1, 14]. For standard references in quantum mechanics, we refer to [13, 48, 55] and the non-standard [12].

30.2 Some introductory topics on vector analysis and Riemannian geometry

In this section we present some introductory remarks on Riemannian geometry.

We start with the definition of surface in $\mathbb{R}^n$.

Definition 30.2.1 (Surface in $\mathbb{R}^n$, the respective tangent space and the dual one) *Let $D \subset \mathbb{R}^m$ be an open, bounded, connected set with a regular (Lipschitzian) boundary denoted by ∂D. We define a m dimensional C^1 class surface $M \subset \mathbb{R}^n$, where $1 \leq m < n$, as the range of a function $\mathbf{r} : D \subset \mathbb{R}^m \to \mathbb{R}^n$ where*

$$
M = \{\mathbf{r}(\mathbf{u}) : \mathbf{u} = (u_1, \ldots, u_m) \in D\}
$$

and

$$\mathbf{r}(\mathbf{u}) = \hat{X}_1(\mathbf{u})\mathbf{e}_1 + \hat{X}_2(\mathbf{u})\mathbf{e}_2 + \cdots + \hat{X}_n(\mathbf{u})\mathbf{e}_n,$$

where $\hat{X}_k : D \to \mathbb{R}$ is a C^1 class function, $\forall k \in \{1, \ldots, n\}$, and $\{\mathbf{e}_1, \ldots, \mathbf{e}_n\}$ is the canonical basis of $\mathbb{R}^n$.
Let $\mathbf{u} \in D$ and $p = \mathbf{r}(\mathbf{u}) \in M$. We also define the tangent space of M at p, denoted by $T_p(M)$, as

$$T_p(M) = \left\{ \alpha_1 \frac{\partial \mathbf{r}(\mathbf{u})}{\partial u_1} + \cdots + \alpha_m \frac{\partial \mathbf{r}(\mathbf{u})}{\partial u_m} : \alpha_1, \ldots, \alpha_m \in \mathbb{R} \right\}.$$

We assume

$$\left\{ \frac{\partial \mathbf{r}(\mathbf{u})}{\partial u_1}, \cdots, \frac{\partial \mathbf{r}(\mathbf{u})}{\partial u_m} \right\}$$

to be a linearly independent set $\forall \mathbf{u} \in D$.
Finally, we define the dual space to $T_p(M)$, denoted by $T_p(M)^*$, as the set of all continuous and linear functionals (in fact real functions) defined on $T_p(M)$, that is,

$$T_p(M)^* = \left\{ f : T_p(M) \to \mathbb{R} : f(\mathbf{v}) = \alpha \cdot \mathbf{v}, \text{ for some } \alpha \in \mathbb{R}^n, \forall \mathbf{v} = v_i \frac{\partial \mathbf{r}(\mathbf{u})}{\partial u_i} \in T_p(M) \right\}.$$

Theorem 30.2.2 *Let $M \subset \mathbb{R}^n$ be a m-dimensional C^1 class surface, where*

$$M = \{\mathbf{r}(\mathbf{u}) \in \mathbb{R}^n : \mathbf{u} \in D \subset \mathbb{R}^m\}.$$

Let $\mathbf{u} \in D$, $p = \mathbf{r}(\mathbf{u}) \in D$ and $f \in C^1(M)$.
Define $df : T_p(M) \to \mathbb{R}$ by

$$df(\mathbf{v}) = \lim_{\varepsilon \to 0} \frac{(f \circ \mathbf{r})(\{u_i\} + \varepsilon\{v_i\}) - (f \circ \mathbf{r})(\{u_i\})}{\varepsilon},$$

$\forall \mathbf{v} = v_i \frac{\partial \mathbf{r}(\mathbf{u})}{\partial u_i} \in T_p(M)$.
Under such hypotheses,

$$df \in T_p(M)^*.$$

Reciprocally, let $F \in T_p(M)^*$.
Under such assumption, there exists $f \in C^1(M)$ such that

$$F(\mathbf{v}) = df(\mathbf{v}), \forall \mathbf{v} \in T_p(M).$$

Proof 30.1   Let $\mathbf{v} = v_i \frac{\partial \mathbf{r}(\mathbf{u})}{\partial u_i} \in T_p(M)$.
Thus,

$$
\begin{aligned}
df(\mathbf{v}) &= \lim_{\varepsilon \to 0} \frac{(f \circ \mathbf{r})(\{u_i\} + \varepsilon\{v_i\}) - (f \circ \mathbf{r})(\{u_i\})}{\varepsilon} \\
&= \frac{\partial (f \circ \mathbf{r})(\mathbf{u})}{\partial \hat{X}_j} \frac{\partial \hat{X}_j(\mathbf{u})}{\partial u_i} v_i \\
&= \nabla f(\mathbf{r}(\mathbf{u})) \cdot \mathbf{v} \\
&= \alpha \cdot \mathbf{v},
\end{aligned}
\tag{30.5}
$$

where

$$\alpha = \nabla f(\mathbf{r}(\mathbf{u})),$$

so that $df \in T_p(M)^*$.
Reciprocally, assume $F \in T_p(M)^*$, that is, suppose there exists $\alpha \in \mathbb{R}^n$ such that

$$F(\mathbf{v}) = \alpha \cdot \mathbf{v},$$

$\forall \mathbf{v} = v_i \frac{\partial \mathbf{r}(\mathbf{u})}{\partial u_i} \in T_p(M)$.

Define $f : M \to \mathbb{R}$ by

$$f(\mathbf{w}) = \alpha \cdot \mathbf{w}, \ \forall \mathbf{w} \in M.$$

In particular,

$$f(\mathbf{r}(\mathbf{u})) = \alpha \cdot \mathbf{r}(\mathbf{u}) = \alpha_j \hat{X}_j(\mathbf{u}), \ \forall \mathbf{u} \in D.$$

For $p = \mathbf{r}(\mathbf{u}) \in M$ and $\mathbf{v} = v_i \frac{\partial \mathbf{r}(\mathbf{u})}{\partial u_i} \in T_p(M)$, we have

$$
\begin{aligned}
df(\mathbf{v}) &= \lim_{\varepsilon \to 0} \frac{(f \circ \mathbf{r})(\{u_i\} + \varepsilon\{v_i\}) - (f \circ \mathbf{r})(\{u_i\})}{\varepsilon} \\
&= \frac{\partial (f \circ \mathbf{r})(\mathbf{u})}{\partial \hat{X}_j} \frac{\partial \hat{X}_j(\mathbf{u})}{\partial u_i} v_i \\
&= \alpha_j \frac{\partial \hat{X}_j(\mathbf{u})}{\partial u_i} v_i \\
&= \alpha \cdot \mathbf{v},
\end{aligned}
\tag{30.6}
$$

Therefore,

$$F(\mathbf{v}) = df(\mathbf{v}), \ \forall \mathbf{v} \in T_p(M).$$

The proof is complete.

At this point, we present the tangential vector field definition, to be addressed in the subsequent results and sections.

Definition 30.2.3 (Vector field) *Let $M \subset \mathbb{R}^n$ be a m-dimensional C^1 class surface, where $1 \le m < n$. We define the set of C^1 class tangential vector fields in M, denoted by $\mathscr{X}(M)$, as*

$$\mathscr{X}(M) = \left\{ X = X_i(\mathbf{u}) \frac{\partial \mathbf{r}(\mathbf{u})}{\partial u_i} \in T(M) = \{T_p(M) \ : \ p = \mathbf{r}(\mathbf{u}) \in M\} \right\},$$

where $X_i : D \to \mathbb{R}$ is a C^1 class function, $\forall i \in \{1, \dots, m\}$.

Let $f \in C^1(M)$ and $X \in \mathscr{X}(M)$. We define the derivative of f on the direction X at $\mathbf{u}$, denoted by $(X \cdot f)(p)$, where $p = \mathbf{r}(\mathbf{u})$, as

$$
\begin{aligned}
(X \cdot f)(p) &= df(X(\mathbf{u})) \\
&= \lim_{\varepsilon \to 0} \frac{(f \circ \mathbf{r})(\{u_i\} + \varepsilon\{X_i(\mathbf{u})\}) - (f \circ \mathbf{r})(\{u_i\})}{\varepsilon} \\
&= \frac{\partial (f \circ \mathbf{r})(\mathbf{u})}{\partial u_i} X_i(\mathbf{u}).
\end{aligned}
\tag{30.7}
$$

The next definition is also very important for this work, namely, the connection one.

Definition 30.2.4 (Connection) *Let $M \subset \mathbb{R}^n$ be a m-dimensional C^1 class surface, where*

$$M = \{\mathbf{r}(\mathbf{u}) \in \mathbb{R}^n \ : \ \mathbf{u} \in D \subset \mathbb{R}^m\}$$

and

$$\mathbf{r}(\mathbf{u}) = \hat{X}_1(\mathbf{u})\mathbf{e}_1 + \cdots + \hat{X}_n(\mathbf{u})\mathbf{e}_n.$$

We define an affine connection on M, as a map $\nabla : \mathscr{X}(M) \times \mathscr{X}(M) \to \mathscr{X}(M)$ such that

1.

$$\nabla_{fX + gY} Z = f \nabla_X Z + g \nabla_Y Z,$$

2.
$$\nabla_X(Y+Z) = \nabla_X Y + \nabla_X Z$$

and

3.
$$\nabla_X(fY) = (X \cdot f)Y + f\nabla_X Y,$$

$\forall X, Y, Z \in \mathscr{X}(M)$, $f, g \in C^\infty(M)$.

About the connection representation, we have the following result.

Theorem 30.2.5 *Let $M \subset \mathbb{R}^n$ be a m-dimensional C^1 class surface, where*

$$M = \{\mathbf{r}(\mathbf{u}) \in \mathbb{R}^n : \mathbf{u} \in D \subset \mathbb{R}^m\}$$

and

$$\mathbf{r}(\mathbf{u}) = \hat{X}_1(\mathbf{u})\mathbf{e}_1 + \cdots + \hat{X}_n(\mathbf{u})\mathbf{e}_n.$$

Let $\nabla : \mathscr{X}(M) \times \mathscr{X}(M) \to \mathscr{X}(M)$ be an affine connection on M. Let $\mathbf{u} \in D$, $p = \mathbf{r}(\mathbf{u}) \in M$ and $X, Y \in \mathscr{X}(M)$ be such that

$$X = X_i(\mathbf{u})\frac{\partial \mathbf{r}(\mathbf{u})}{\partial u_i},$$

and

$$Y = Y_i(\mathbf{u})\frac{\partial \mathbf{r}(\mathbf{u})}{\partial u_i}.$$

Under such hypotheses, we have

$$\nabla_X Y = \sum_{i=1}^m \left(X \cdot Y_i + \sum_{j,k=1}^m \Gamma^i_{jk} X_j Y_k \right) \frac{\partial \mathbf{r}(\mathbf{u})}{\partial u_i} \in T_p(M), \tag{30.8}$$

where Γ^i_{jk} are defined through the relations,

$$\nabla_{\frac{\partial \mathbf{r}(\mathbf{u})}{\partial u_j}} \frac{\partial \mathbf{r}(\mathbf{u})}{\partial u_k} = \Gamma^i_{jk}(\mathbf{u})\frac{\partial \mathbf{r}(\mathbf{u})}{\partial u_i}.$$

Proof 30.2   Observe that

$$
\begin{aligned}
\nabla_X Y &= \nabla_{X_i \frac{\partial \mathbf{r}(\mathbf{u})}{\partial u_i}} \left(Y_j \frac{\partial \mathbf{r}(\mathbf{u})}{\partial u_j} \right) \\
&= X_i \nabla_{\frac{\partial \mathbf{r}(\mathbf{u})}{\partial u_i}} \left(Y_j \frac{\partial \mathbf{r}(\mathbf{u})}{\partial u_j} \right) \\
&= X_i \left(\frac{\partial \mathbf{r}(\mathbf{u})}{\partial u_i} \cdot Y_j \right) \frac{\partial \mathbf{r}(\mathbf{u})}{\partial u_j} + X_i Y_j \nabla_{\frac{\partial \mathbf{r}(\mathbf{u})}{\partial u_i}} \frac{\partial \mathbf{r}(\mathbf{u})}{\partial u_j} \\
&= \left(X_i \frac{\partial \mathbf{r}(\mathbf{u})}{\partial u_i} \cdot Y_j \right) \frac{\partial \mathbf{r}(\mathbf{u})}{\partial u_j} + X_i Y_j \Gamma^k_{ij} \frac{\partial \mathbf{r}(\mathbf{u})}{\partial u_k} \\
&= \sum_{i=1}^m \left(X \cdot Y_i + \sum_{j,k=1}^m \Gamma^i_{jk} X_j Y_k \right) \frac{\partial \mathbf{r}(\mathbf{u})}{\partial u_i}. \tag{30.9}
\end{aligned}
$$

The proof is complete.

Remark 30.1 If the connection in question is such that

$$\Gamma^i_{jk} = \frac{1}{2} g^{il} \left(\frac{\partial g_{kl}}{\partial u_j} + \frac{\partial g_{jl}}{\partial u_k} - \frac{\partial g_{jk}}{\partial u_l} \right)$$

such a connection is said to be the Levi-Civita one. In the next lines we assume the concerning connection is indeed the Levi-Civita one. ■

We finish this section with the Lie Bracket definition.

Definition 30.2.6 (Lie bracket) *Let $M \subset \mathbb{R}^n$ be a C^1 class m-dimensional surface where $1 \le m < n$.*
Let $X, Y \in \tilde{\mathscr{X}}(M)$, where $\tilde{\mathscr{X}}(M)$ denotes the set of the $C^\infty(M) = \cap_{k \in \mathbb{N}} C^k(M)$ class vector fields. We define the Lie bracket of X and Y, denoted by $[X, Y] \in \tilde{\mathscr{X}}(M)$, by

$$[X, Y] = (X \cdot Y_i - Y \cdot X_i) \frac{\partial \mathbf{r}(\mathbf{u})}{\partial u_i},$$

where

$$X = X_i \frac{\partial \mathbf{r}(\mathbf{u})}{\partial u_i}$$

and

$$Y = Y_i \frac{\partial \mathbf{r}(\mathbf{u})}{\partial u_i}.$$

30.3 A relativistic quantum mechanics action

In this section we present a proposal for a relativistic quantum mechanics action.

Let $\Omega \subset \mathbb{R}^3$ be an open, bounded, connected set with a C^1 class boundary denoted by $\partial \Omega$. Denoting by c the speed of light, in a free volume context, for a C^1 class function $\mathbf{r}$ and $\hat{\mathbf{u}} \in W^{1,2}(\Omega \times [0, T]; \mathbb{R}^4)$, let $(\mathbf{r} \circ \hat{\mathbf{u}}) : \Omega \times [0, T] \to \mathbb{R}^4$ be a particle position field where

$$\mathbf{r}(\hat{\mathbf{u}}(\mathbf{x}, t)) = (ct, X_1(\mathbf{u}(\mathbf{x}, t)), X_2(\mathbf{u}(\mathbf{x}, t)), X_3(\mathbf{u}(\mathbf{x}, t))),$$

with corresponding mass density

$$(\rho \circ \hat{\mathbf{u}}) : \Omega \times [0, T] \to \mathbb{R}^+,$$

where $[0, T]$ is a time interval.

We denote $\hat{\mathbf{u}} : \Omega \times [0, T] \to \mathbb{R}^4$ point-wise as

$$\hat{\mathbf{u}}(\mathbf{x}, t) = (u_0(\mathbf{x}, t), \mathbf{u}(\mathbf{x}, t)),$$

where

$$u_0(\mathbf{x}, t) = ct,$$

and

$$\mathbf{u}(\mathbf{x}, t) = (u_1(\mathbf{x}, t), u_2(\mathbf{x}, t), u_3(\mathbf{x}, t)),$$

$\forall (\mathbf{x}, t) = ((x_1, x_2, x_3), t) \in \Omega \times [0, T]$.

At this point, we recall to have defined $\phi(\hat{\mathbf{u}}(\mathbf{x}, t))$ as a complex function such that

$$|\phi(\hat{\mathbf{u}}(\mathbf{x}, t))|^2 = \frac{\rho(\hat{\mathbf{u}}(\mathbf{x}, t))}{m},$$

where m denotes the total system mass at rest. Also, we assume ϕ to be a C^2 class function and define

$$d\tau^2 = c^2 dt^2 - dX_1(\mathbf{u}(\mathbf{x}, t))^2 - dX_2(\mathbf{u}(\mathbf{x}, t))^2 - dX_3(\mathbf{u}(\mathbf{x}, t))^2,$$

so that the mass differential will be denoted by

$$
\begin{aligned}
dm &= \frac{\rho(\hat{\mathbf{u}}(\mathbf{x},t))}{\sqrt{1-\frac{v^2}{c^2}}}\sqrt{-g}\,|\det(\hat{\mathbf{u}}'(\mathbf{x},t))|\,d\mathbf{x} \\
&= \frac{m|\phi(\hat{\mathbf{u}}(\mathbf{x},t))|^2}{\sqrt{1-\frac{v^2}{c^2}}}\sqrt{-g}\,|\det(\hat{\mathbf{u}}'(\mathbf{x},t))|\,d\mathbf{x},
\end{aligned}
\tag{30.10}
$$

where $d\mathbf{x} = dx_1 dx_2 dx_3$ and $\hat{\mathbf{u}}'(\mathbf{x},t)$ denotes the Jacobian matrix of the vectorial function $\hat{\mathbf{u}}(\mathbf{x},t)$.

Also,

$$
\mathbf{g}_i = \frac{\partial \mathbf{r}(\hat{\mathbf{u}})}{\partial u_i}, \ \forall i \in \{0,1,2,3\},
$$

$$
g_{ij} = \mathbf{g}_i \cdot \mathbf{g}_j, \ \forall i,j \in \{0,1,2,3\},
$$

and

$$
g = \det\{g_{ij}\}.
$$

Moreover,

$$
\{g^{ij}\} = \{g_{ij}\}^{-1}.
$$

30.3.1 The kinetics energy

Observe that

$$
\begin{aligned}
c^2 - v^2 &= -\frac{d\mathbf{r}(\hat{\mathbf{u}})}{dt} \cdot \frac{d\mathbf{r}(\hat{\mathbf{u}})}{dt} \\
&= -\left(\frac{\partial \mathbf{r}(\hat{\mathbf{u}})}{\partial u_i}\frac{\partial u_i}{\partial t}\right) \cdot \left(\frac{\partial \mathbf{r}(\hat{\mathbf{u}})}{\partial u_j}\frac{\partial u_j}{\partial t}\right) \\
&= -\frac{\partial \mathbf{r}(\hat{\mathbf{u}})}{\partial u_i} \cdot \frac{\partial \mathbf{r}(\hat{\mathbf{u}})}{\partial u_j}\frac{\partial u_i}{\partial t}\frac{\partial u_j}{\partial t} \\
&= -g_{ij}\frac{\partial u_i}{\partial t}\frac{\partial u_j}{\partial t},
\end{aligned}
\tag{30.11}
$$

where the product in question is generically given by

$$
\mathbf{y}\cdot\mathbf{z} = -y_0 z_0 + \sum_{i=1}^{3} y_i z_i, \ \forall \mathbf{y} = (y_0,y_1,y_2,y_3), \ \mathbf{z} = (z_0,z_1,z_2,z_3) \in \mathbb{R}^4
$$

and

$$
v = \sqrt{\left(\frac{dX_1(\mathbf{u}(\mathbf{x},t))}{dt}\right)^2 + \left(\frac{dX_2(\mathbf{u}(\mathbf{x},t))}{dt}\right)^2 + \left(\frac{dX_3(\mathbf{u}(\mathbf{x},t))}{dt}\right)^2}.
$$

The semi-classical kinetics energy differential is given by

$$
\begin{aligned}
dE_c &= \frac{d\mathbf{r}(\hat{\mathbf{u}})}{dt} \cdot \frac{d\mathbf{r}(\hat{\mathbf{u}})}{dt}\,dm \\
&= -\left(\frac{d\tau}{dt}\right)^2 dm \\
&= -(c^2 - v^2)\,dm,
\end{aligned}
\tag{30.12}
$$

so that

$$
\begin{aligned}
dE_c &= -m \frac{(c^2 - v^2)}{\sqrt{1 - \frac{v^2}{c^2}}} |\phi(\hat{\mathbf{u}})|^2 \sqrt{-g} |\det(\hat{\mathbf{u}}'(\mathbf{x}, t))| \, d\mathbf{x} \\
&= -mc^2 \sqrt{1 - \frac{v^2}{c^2}} |\phi(\hat{\mathbf{u}})|^2 \sqrt{-g} |\det(\hat{\mathbf{u}}'(\mathbf{x}, t))| \, d\mathbf{x} \\
&= -mc \sqrt{c^2 - v^2} |\phi(\hat{\mathbf{u}})|^2 \sqrt{-g} |\det(\hat{\mathbf{u}}'(\mathbf{x}, t))| \, d\mathbf{x} \\
&= -mc \sqrt{-g_{ij} \frac{\partial u_i}{\partial t} \frac{\partial u_j}{\partial t}} |\phi(\hat{\mathbf{u}})|^2 \sqrt{-g} |\det(\hat{\mathbf{u}}'(\mathbf{x}, t))| \, d\mathbf{x},
\end{aligned} \tag{30.13}
$$

and thus, the semi-classical kinetics energy E_c is given by

$$
E_c = \int_0^T \int_\Omega dE_c \, dt,
$$

that is,

$$
E_c = -\int_0^T \int_\Omega mc \sqrt{-g_{ij} \frac{\partial u_i}{\partial t} \frac{\partial u_j}{\partial t}} |\phi(\hat{\mathbf{u}})|^2 \sqrt{-g} |\det(\hat{\mathbf{u}}'(\mathbf{x}, t))| \, d\mathbf{x} \, dt.
$$

30.3.2   The energy part relating the curvature and wave function

At this point we define an energy part, related to the Riemann curvature tensor, denoted by E_q, where

$$
E_q = \frac{\gamma}{2} \int_0^T \int_\Omega g^{jk} R_{jk} \sqrt{-g} |\det(\hat{\mathbf{u}}'(\mathbf{x}, t))| \, d\mathbf{x} \, dt.
$$

and

$$
R_{jk} = Re[R^i_{jik}(\phi)].
$$

Also, generically $Re[z]$ denotes the real part of $z \in \mathbb{C}$ and $R^l_{ijk}(\phi)$ is such that

$$
\begin{aligned}
&\nabla_{\left(\phi \frac{\partial \mathbf{r}(\hat{\mathbf{u}})}{\partial u_i}\right)} \nabla_{\frac{\partial \mathbf{r}(\hat{\mathbf{u}})}{\partial u_j}} \left(\phi^* \frac{\partial \mathbf{r}(\hat{\mathbf{u}})}{\partial u_k}\right) - \nabla_{\left(\phi \frac{\partial \mathbf{r}(\hat{\mathbf{u}})}{\partial u_j}\right)} \nabla_{\frac{\partial \mathbf{r}(\hat{\mathbf{u}})}{\partial u_i}} \left(\phi^* \frac{\partial \mathbf{r}(\hat{\mathbf{u}})}{\partial u_k}\right) \\
&- \nabla_{\left[\phi \frac{\partial \mathbf{r}(\hat{\mathbf{u}})}{\partial u_i}, \frac{\partial \mathbf{r}(\hat{\mathbf{u}})}{\partial u_j}\right]} \left(\phi^* \frac{\partial \mathbf{r}(\hat{\mathbf{u}})}{\partial u_k}\right) = R^l_{ijk}(\phi) \frac{\partial \mathbf{r}(\hat{\mathbf{u}})}{\partial u_l}.
\end{aligned} \tag{30.14}
$$

More specifically, we have

$$
\begin{aligned}
&\nabla_{\left(\phi\frac{\partial \mathbf{r}(\hat{\mathbf{u}})}{\partial u_i}\right)} \nabla_{\frac{\partial \mathbf{r}(\hat{\mathbf{u}})}{\partial u_j}} \left(\phi^* \frac{\partial \mathbf{r}(\hat{\mathbf{u}})}{\partial u_k}\right) \\
&= \nabla_{\left(\phi\frac{\partial \mathbf{r}(\hat{\mathbf{u}})}{\partial u_i}\right)} \left(\frac{\partial \phi^*}{\partial u_j} \frac{\partial \mathbf{r}(\hat{\mathbf{u}})}{\partial u_k} + \phi^* \Gamma_{jk}^l \frac{\partial \mathbf{r}(\hat{\mathbf{u}})}{\partial u_l}\right) \\
&= \phi \frac{\partial^2 \phi^*}{\partial u_i \partial u_j} \frac{\partial \mathbf{r}(\hat{\mathbf{u}})}{\partial u_k} + \phi \frac{\partial \phi^*}{\partial u_j} \Gamma_{ik}^p \frac{\partial \mathbf{r}(\mathbf{u})}{\partial u_p} \\
&\quad + \phi \frac{\partial \left(\phi^* \Gamma_{jk}^l\right)}{\partial u_i} \frac{\partial \mathbf{r}(\hat{\mathbf{u}})}{\partial u_l} \\
&\quad + |\phi|^2 \Gamma_{ij}^l \Gamma_{il}^p \frac{\partial \mathbf{r}(\hat{\mathbf{u}})}{\partial u_p} \\
&= \phi \frac{\partial^2 \phi^*}{\partial u_i \partial u_j} \delta_{kl} \frac{\partial \mathbf{r}(\hat{\mathbf{u}})}{\partial u_l} + \phi \frac{\partial \phi^*}{\partial u_j} \Gamma_{ik}^l \frac{\partial \mathbf{r}(\hat{\mathbf{u}})}{\partial u_l} \\
&\quad + \phi \frac{\partial (\phi^* \Gamma_{jk}^l)}{\partial u_i} \frac{\partial \mathbf{r}(\hat{\mathbf{u}})}{\partial u_l} \\
&\quad + |\phi|^2 \Gamma_{jk}^p \Gamma_{ip}^l \frac{\partial \mathbf{r}(\hat{\mathbf{u}})}{\partial u_l}
\end{aligned}
\tag{30.15}
$$

and similarly,

$$
\begin{aligned}
&\nabla_{\left(\phi\frac{\partial \mathbf{r}(\hat{\mathbf{u}})}{\partial u_j}\right)} \nabla_{\frac{\partial \mathbf{r}(\hat{\mathbf{u}})}{\partial u_i}} \left(\phi^* \frac{\partial \mathbf{r}(\hat{\mathbf{u}})}{\partial u_k}\right) \\
&= \phi \frac{\partial^2 \phi^*}{\partial u_i \partial u_j} \delta_{kl} \frac{\partial \mathbf{r}(\hat{\mathbf{u}})}{\partial u_l} + \phi \frac{\partial \phi^*}{\partial u_i} \Gamma_{jk}^l \frac{\partial \mathbf{r}(\hat{\mathbf{u}})}{\partial u_l} \\
&\quad + \phi \frac{\partial (\phi^* \Gamma_{ik}^l)}{\partial u_j} \frac{\partial \mathbf{r}(\hat{\mathbf{u}})}{\partial u_l} \\
&\quad + |\phi|^2 \Gamma_{ik}^p \Gamma_{jp}^l \frac{\partial \mathbf{r}(\hat{\mathbf{u}})}{\partial u_l}.
\end{aligned}
\tag{30.16}
$$

Moreover,

$$
\begin{aligned}
&\nabla_{\left[\phi\frac{\partial \mathbf{r}(\hat{\mathbf{u}})}{\partial u_i}, \frac{\partial \mathbf{r}(\hat{\mathbf{u}})}{\partial u_j}\right]} \left(\phi^* \frac{\partial \mathbf{r}(\hat{\mathbf{u}})}{\partial u_k}\right) \\
&= \nabla_{\left(-\frac{\partial \mathbf{r}(\hat{\mathbf{u}})}{\partial u_j} \cdot \phi\right) \frac{\partial \mathbf{r}(\hat{\mathbf{u}})}{\partial u_i}} \left(\phi^* \frac{\partial \mathbf{r}(\hat{\mathbf{u}})}{\partial u_k}\right) \\
&= -\nabla_{\left(\frac{\partial \phi}{\partial u_j} \frac{\partial \mathbf{r}(\hat{\mathbf{u}})}{\partial u_i}\right)} \left(\phi^* \frac{\partial \mathbf{r}(\hat{\mathbf{u}})}{\partial u_k}\right) \\
&= -\frac{\partial \phi}{\partial u_j} \nabla_{\frac{\partial \mathbf{r}(\hat{\mathbf{u}})}{\partial u_i}} \left(\phi^* \frac{\partial \mathbf{r}(\hat{\mathbf{u}})}{\partial u_k}\right) \\
&= -\frac{\partial \phi}{\partial u_j} \frac{\partial \phi^*}{\partial u_i} \frac{\partial \mathbf{r}(\hat{\mathbf{u}})}{\partial u_k} - \frac{\partial \phi}{\partial u_j} \phi^* \nabla_{\frac{\partial \mathbf{r}(\hat{\mathbf{u}})}{\partial u_i}} \frac{\partial \mathbf{r}(\hat{\mathbf{u}})}{\partial u_k} \\
&= -\frac{\partial \phi}{\partial u_j} \frac{\partial \phi^*}{\partial u_i} \frac{\partial \mathbf{r}(\hat{\mathbf{u}})}{\partial u_k} - \frac{\partial \phi}{\partial u_j} \phi^* \Gamma_{ik}^l \frac{\partial \mathbf{r}(\hat{\mathbf{u}})}{\partial u_l} \\
&= -\frac{\partial \phi}{\partial u_j} \frac{\partial \phi^*}{\partial u_i} \delta_{kl} \frac{\partial \mathbf{r}(\hat{\mathbf{u}})}{\partial u_l} - \frac{\partial \phi}{\partial u_j} \phi^* \Gamma_{ik}^l \frac{\partial \mathbf{r}(\hat{\mathbf{u}})}{\partial u_l}.
\end{aligned}
\tag{30.17}
$$

Thus,

$$
\begin{aligned}
R^l_{ijk}(\phi) &= \phi\,\frac{\partial^2 \phi^*}{\partial u_i \partial u_j}\,\delta_{kl} + \phi\,\frac{\partial \phi^*}{\partial u_j}\Gamma^l_{ik} + \phi\,\frac{\partial(\phi^*\Gamma^l_{jk})}{\partial u_i} + |\phi|^2\,\Gamma^p_{jk}\,\Gamma^l_{ip} \\
&\quad -\phi\,\frac{\partial^2 \phi^*}{\partial u_i \partial u_j}\,\delta_{kl} - \phi\,\frac{\partial \phi^*}{\partial u_i}\Gamma^l_{jk} - \phi\,\frac{\partial(\phi^*\Gamma^l_{ik})}{\partial u_j} - |\phi|^2\,\Gamma^p_{ik}\,\Gamma^l_{jp} \\
&\quad +\frac{\partial \phi}{\partial u_j}\,\frac{\partial \phi^*}{\partial u_i}\,\delta_{kl} + \frac{\partial \phi}{\partial u_j}\,\phi^*\Gamma^l_{ik}.
\end{aligned}
\tag{30.18}
$$

Simplifying this last result, we obtain

$$
\begin{aligned}
R^l_{ijk}(\phi) &= \phi\,\frac{\partial \phi^*}{\partial u_j}\Gamma^l_{ik} + \phi\,\frac{\partial(\phi^*\,\Gamma^l_{jk})}{\partial u_i} - \phi\,\frac{\partial \phi^*}{\partial u_i}\Gamma^l_{jk} - \phi\,\frac{\partial(\phi^*\Gamma^l_{ik})}{\partial u_j} \\
&\quad + |\phi|^2\left(\Gamma^p_{jk}\,\Gamma^l_{ip} - \Gamma^p_{ik}\,\Gamma^l_{jp}\right) + \frac{\partial \phi}{\partial u_j}\,\frac{\partial \phi^*}{\partial u_i}\,\delta_{kl} + \frac{\partial \phi}{\partial u_j}\,\phi^*\Gamma^l_{ik} \\
&= |\phi|^2\left(\frac{\partial \Gamma^l_{jk}}{\partial u_i} - \frac{\partial \Gamma^l_{ik}}{\partial u_j} + \Gamma^p_{jk}\,\Gamma^l_{ip} - \Gamma^p_{ik}\,\Gamma^l_{jp}\right) \\
&\quad + \frac{\partial \phi}{\partial u_j}\,\frac{\partial \phi^*}{\partial u_i}\,\delta_{kl} + \frac{\partial \phi}{\partial u_j}\,\phi^*\Gamma^l_{ik} \\
&= |\phi|^2 \hat{R}^l_{ijk} \\
&\quad + \frac{\partial \phi}{\partial u_j}\,\frac{\partial \phi^*}{\partial u_i}\,\delta_{kl} + \frac{\partial \phi}{\partial u_j}\,\phi^*\Gamma^l_{ik},
\end{aligned}
\tag{30.19}
$$

where

$$
\hat{R}^l_{ijk} = \frac{\partial \Gamma^l_{jk}}{\partial u_i} - \frac{\partial \Gamma^l_{ik}}{\partial u_j} + \Gamma^p_{jk}\,\Gamma^l_{ip} - \Gamma^p_{ik}\,\Gamma^l_{jp}
$$

represents the Riemann curvature tensor.

At this point, we recall to have defined this energy part by

$$
E_q = \frac{\gamma}{2}\int_0^T \int_\Omega R\,\sqrt{-g}\,|\det(\hat{\mathbf{u}}'(\mathbf{x},t))|\,d\mathbf{x}\,dt,
$$

where

$$
R = g^{jk}R_{jk},
$$

and as above indicated, $R_{jk} = Re[R^i_{jik}(\phi)]$.

Hence the final expression for the energy (action) is given by

$$
\begin{aligned}
J(\phi,\mathbf{r},\hat{\mathbf{u}},E) &= -E_c + E_q \\
&\quad - \int_0^T E(t)\left(\int_\Omega |\phi(\hat{\mathbf{u}}(\mathbf{x},t))|^2\,\sqrt{-g}\,|\det(\mathbf{u}'(\mathbf{x},t))|\,d\mathbf{x} - 1\right)dt,
\end{aligned}
\tag{30.20}
$$

where $E(t)$ is a Lagrange multiplier related to the total mass constraint.

More explicitly, the final action (the generalized Einstein-Hilbert one), would be given by

$$J(\phi, \mathbf{r}, \hat{\mathbf{u}}, E)$$

$$= \int_0^T \int_\Omega mc \sqrt{-g_{ij} \frac{\partial u_i}{\partial t} \frac{\partial u_j}{\partial t}} \, |\phi(\hat{\mathbf{u}}(\mathbf{x}, t))|^2 \sqrt{-g} \, |\det(\hat{\mathbf{u}}'(\mathbf{x}, t))| \, d\mathbf{x} \, dt$$

$$+ \frac{\gamma}{2} \int_0^T \int_\Omega g^{jk} \frac{\partial \phi}{\partial u_j} \frac{\partial \phi^*}{\partial u_k} \sqrt{-g} \, |\det(\hat{\mathbf{u}}'(\mathbf{x}, t))| \, d\mathbf{x} dt$$

$$+ \frac{\gamma}{4} \int_0^T \int_\Omega g^{jk} \left(\phi^* \frac{\partial \phi}{\partial u_l} + \phi \frac{\partial \phi^*}{\partial u_l} \right) \Gamma_{jk}^l \sqrt{-g} \, |\det(\hat{\mathbf{u}}'(\mathbf{x}, t))| \, d\mathbf{x} dt$$

$$+ \frac{\gamma}{2} \int_0^T \int_\Omega |\phi|^2 \, g^{jk} \left(\frac{\partial \Gamma_{lk}^l}{\partial u_j} - \frac{\partial \Gamma_{jk}^l}{\partial u_l} + \Gamma_{lk}^p \Gamma_{jp}^l - \Gamma_{jk}^p \Gamma_{lp}^l \right) \sqrt{-g} \, |\det(\hat{\mathbf{u}}'(\mathbf{x}, t))| \, d\mathbf{x} dt$$

$$- \int_0^T E(t) \left(\int_\Omega |\phi(\hat{\mathbf{u}}(\mathbf{x}, t))|^2 \sqrt{-g} \, |\det(\hat{\mathbf{u}}'(\mathbf{x}, t))| \, d\mathbf{x} - 1 \right) dt. \tag{30.21}$$

Where γ is an appropriate positive constant to be specified.

30.4 Obtaining the relativistic Klein-Gordon equation as an approximation of the previous action

In particular for the special case in which

$$\mathbf{r}(\hat{\mathbf{u}}(\mathbf{x}, t)) = \hat{\mathbf{u}}(\mathbf{x}, t) \approx (ct, \mathbf{x}),$$

so that

$$\frac{d\mathbf{r}(\hat{\mathbf{u}}(\mathbf{x}, t))}{dt} \approx (c, 0, 0, 0),$$

we would obtain

$$\mathbf{g}_0 \approx (1, 0, 0, 0), \ \mathbf{g}_1 \approx (0, 1, 0, 0), \ \mathbf{g}_2 \approx (0, 0, 1, 0) \text{ and } \mathbf{g}_3 \approx (0, 0, 0, 1) \in \mathbb{R}^4,$$

and $\Gamma_{ij}^k \approx 0$, $\forall i, j, k \in \{0, 1, 2, 3\}$.

Therefore, denoting $\phi(\hat{\mathbf{u}}(\mathbf{x}, t)) \approx \phi(ct, \mathbf{x})$ simply by a not relabeled $\phi(\mathbf{x}, t)$, we may obtain

$$E_q/c \approx \frac{\gamma}{2} \int_0^T \int_\Omega \left(-\frac{1}{c^2} \frac{\partial \phi(\mathbf{x}, t)}{\partial t} \frac{\partial \phi^*(\mathbf{x}, t)}{\partial t} \right.$$

$$\left. + \sum_{k=1}^3 \frac{\partial \phi(\mathbf{x}, t)}{\partial x_k} \frac{\partial \phi^*(\mathbf{x}, t)}{\partial x_k} \right) d\mathbf{x} dt, \tag{30.22}$$

and

$$E_c/c = m c^2 \int_0^T \int_\Omega |\phi|^2 \sqrt{1 - v^2/c^2} \sqrt{-g} |\det(\hat{\mathbf{u}}'(\mathbf{x}, t))| \, d\mathbf{x} \, dt/c \approx mc^2 \int_0^T \int_\Omega |\phi(\mathbf{x}, t)|^2 \, d\mathbf{x} dt.$$

Hence, we would also obtain

$$
J(\phi,\mathbf{r},\hat{\mathbf{u}},E)/c \;\approx\; \frac{\gamma}{2}\left(\int_0^T\int_\Omega -\frac{1}{c^2}\frac{\partial\phi(\mathbf{x},t)}{\partial t}\frac{\partial\phi^*(\mathbf{x},t)}{\partial t}\,d\mathbf{x}dt\right.
$$
$$
\left.+\sum_{k=1}^3\int_\Omega\int_0^T\frac{\partial\phi(\mathbf{x},t)}{\partial x_k}\frac{\partial\phi^*(\mathbf{x},t)}{\partial x_k}\,d\mathbf{x}dt\right)
$$
$$
+mc^2\int_0^T\int_\Omega|\phi(\mathbf{x},t)|^2\,d\mathbf{x}dt
$$
$$
-\int_0^T E(t)\left(\int_\Omega|\phi(\mathbf{x},t)|^2d\mathbf{x}-1\right)\,dt. \tag{30.23}
$$

The Euler Lagrange equations for such an energy are given by

$$
\frac{\gamma}{2}\left(\frac{1}{c^2}\frac{\partial^2\phi(\mathbf{x},t)}{\partial t^2}-\sum_{k=1}^3\frac{\partial^2\phi(\mathbf{x},t)}{\partial x_k^2}\right)
$$
$$
+mc^2\phi(\mathbf{x},t)-E(t)\phi(\mathbf{x},t)=0,\ \text{in}\ \Omega, \tag{30.24}
$$

where we assume the space of admissible functions is given by $C^1(\Omega\times[0,T];\mathbb{C})$ with the following time and spatial boundary conditions,

$$
\phi(\mathbf{x},0)=\phi_0(\mathbf{x}),\ \text{in}\ \Omega,
$$
$$
\phi(\mathbf{x},T)=\phi_1(\mathbf{x}),\ \text{in}\ \Omega,
$$
$$
\phi(\mathbf{x},t)=0,\ \text{on}\ \partial\Omega\times[0,T].
$$

Equation (30.24) is the relativistic Klein-Gordon one.

For $E(t)=E\in\mathbb{R}$ (not time dependent), at this point we suggest a solution (and implicitly related time boundary conditions) $\phi(\mathbf{x},t)=e^{-\frac{iEt}{\hbar}}\phi_2(\mathbf{x})$, where

$$
\phi_2(\mathbf{x})=0,\ \text{on}\ \partial\Omega.
$$

Therefore, replacing this solution into equation (30.24), we would obtain

$$
\left(\frac{\gamma}{2}\left(-\frac{E^2}{c^2\hbar^2}\phi_2(\mathbf{x})-\sum_{k=1}^3\frac{\partial^2\phi_2(\mathbf{x})}{\partial x_k^2}\right)+mc^2\phi_2(\mathbf{x})-E\phi_2(\mathbf{x})\right)e^{-\frac{iEt}{\hbar}}=0,
$$

in Ω.

Denoting

$$
E_1=-\frac{\gamma E^2}{2c^2\hbar^2}+mc^2-E,
$$

the final eigenvalue problem would stand for

$$
-\frac{\gamma}{2}\sum_{k=1}^3\frac{\partial^2\phi_2(\mathbf{x})}{\partial x_k^2}+E_1\phi_2(\mathbf{x})=0,\ \text{in}\ \Omega
$$

where E_1 is such that

$$
\int_\Omega|\phi_2(\mathbf{x})|^2\,d\mathbf{x}=1.
$$

Moreover, from (30.24), such a solution $\phi(\mathbf{x},t) = e^{-\frac{iEt}{\hbar}}\phi_2(\mathbf{x})$ is also such that

$$\frac{\gamma}{2}\left(\frac{1}{c^2}\frac{\partial^2\phi(\mathbf{x},t)}{\partial t^2} - \sum_{k=1}^{3}\frac{\partial^2\phi(\mathbf{x},t)}{\partial x_k^2}\right)$$
$$+mc^2\phi(\mathbf{x},t) = i\hbar\frac{\partial\phi(\mathbf{x},t)}{\partial t}, \text{ in } \Omega. \tag{30.25}$$

At this point, we recall that in quantum mechanics,

$$\gamma = \hbar^2/m.$$

Finally, we remark this last equation (30.25) is a kind of relativistic Schrödinger-Klein-Gordon equation.

30.5 A note on the Einstein field equations in the vacuum

In this section we obtain the Einstein field equations for a field of position in the vacuum.

Let $\Omega \subset \mathbb{R}^3$ be an open, bounded and connected set with a regular boundary denoted by $\partial\Omega$. Let $[0,T]$ be a time interval and consider the Hilbert-Einstein action given by $J : U \to \mathbb{R}$, where for an appropriate constant $\gamma > 0$,

$$J(\mathbf{r}) = \frac{\gamma}{2}\int_0^T\int_\Omega \hat{R}\sqrt{-g}\,d\mathbf{x}dt,$$

where again

$$\mathbf{u} = (u_0, u_1, u_2, u_3) = (t, x_1, x_2, x_3) = (x_0, x_1, x_2, x_3).$$

Also,

$$g_{jk} = \frac{\partial\mathbf{r}(\mathbf{u})}{\partial u_j}\cdot\frac{\partial\mathbf{r}(\mathbf{u})}{\partial u_k},$$
$$g = det\{g_{jk}\},$$
$$\hat{R} = g^{jk}R_{jk},$$
$$R_{ik} = R^j_{ijk},$$

and

$$R^l_{ijk} = \frac{\partial\Gamma^l_{jk}}{\partial u_i} - \frac{\partial\Gamma^l_{ik}}{\partial u_j} + \Gamma^p_{jk}\,\Gamma^l_{ip} - \Gamma^p_{ik}\,\Gamma^l_{jp}$$

represents the Riemann curvature tensor.

Finally, as above indicated,

$$\mathbf{r} : \Omega \times [0,T] \to \mathbb{R}^4$$

stands for

$$\mathbf{r}(\mathbf{u}) = (ct, X_1(\mathbf{u}), X_2(\mathbf{u}), X_3(\mathbf{u}))$$

and

$$U = \{\mathbf{r} \in W^{2,2}(\Omega;\mathbb{R}^4) : \mathbf{r}(0,u_1,u_2,u_3) = \mathbf{r}_1(u_1,u_2,u_3),$$
$$\mathbf{r}(cT,u_1,u_2,u_3) = \mathbf{r}_2(u_1,u_2,u_3) \text{ in } \Omega, \mathbf{r}|_{\partial\Omega} = \mathbf{r}_0 \text{ on } [0,T]\}. \tag{30.26}$$

Hence, already including the Lagrange multipliers, considering $\mathbf{r}$ and $\{g_{jk}\}$ as independent variables, such a functional again denoted by $J(\mathbf{r}, \{g_{jk}\}, \lambda)$ is expressed as

$$J(\mathbf{r}, \{g_{jk}\}, \lambda) = \frac{\gamma}{2}\int_0^T\int_\Omega \hat{R}\sqrt{-g}\,d\mathbf{x}dt + \int_0^T\int_\Omega \lambda_{jk}\left(\frac{\partial\mathbf{r}}{\partial u_j}\cdot\frac{\partial\mathbf{r}}{\partial u_k} - g_{jk}\right)d\mathbf{x}dt.$$

The variation of such a functional in g give us

$$\gamma \left(R_{jk} - \frac{1}{2} g_{jk} \hat{R} \right) \sqrt{-g} - \lambda_{jk} = 0, \text{ in } \Omega.$$

The variation in $\mathbf{r}$, provide us

$$\frac{\partial^2 X_l(\mathbf{u})}{\partial u_j \partial u_k} \lambda_{jk} + \frac{\partial X_l(\mathbf{u})}{\partial u_j} \frac{\partial \lambda_{jk}}{\partial u_k} = 0, \text{ in } \Omega, \tag{30.27}$$

so that

$$\frac{\partial^2 X_l(\mathbf{u})}{\partial u_j \partial u_k} \left([R_{jk} - \frac{1}{2} g_{jk} \hat{R}] \sqrt{-g} \right) + \frac{\partial X_l(\mathbf{u})}{\partial u_j} \frac{\partial [(R_{jk} - \frac{1}{2} g_{jk} \hat{R}) \sqrt{-g}]}{\partial u_k} = 0, \text{ in } \Omega, \tag{30.28}$$

$\forall l \in \{1, 2, 3\}$.

Observe that the condition $R_{jk} = 0$ in $\Omega \times [0, T]$, $\forall j, k \in \{0, 1, 2, 3\}$, it is sufficient to solve the system indicated in (30.28) but it is not necessary.

The system indicated in (30.28) is the Einstein field one. It is my understanding the actual variable for this system is $\mathbf{r}$ not $\{g_{jk}\}$.

However in some situations, it is possible to solve (30.28) through a specific metric $\{(g_0)_{jk}\}$, but one question remains, how to obtain a corresponding $\mathbf{r}$.

With such an issue in mind, given a specific metric $\{(g_0)_{jk}\}$, we suggest the following control problem,

$$\text{Find } \mathbf{r} \in U \text{ which minimizes } J_1(\mathbf{r}) = \sum_{j,k=0}^{3} \left\| \frac{\partial \mathbf{r}(\mathbf{u})}{\partial u_j} \cdot \frac{\partial \mathbf{r}(\mathbf{u})}{\partial u_k} - (g_0)_{jk} \right\|_2^2,$$

subject to

$$\frac{\partial^2 X_l(\mathbf{u})}{\partial u_j \partial u_k} \left([R_{jk} - \frac{1}{2} g_{jk} \hat{R}] \sqrt{-g} \right) + \frac{\partial X_l(\mathbf{u})}{\partial u_j} \frac{\partial [(R_{jk} - \frac{1}{2} g_{jk} \hat{R}) \sqrt{-g}]}{\partial u_k} = 0, \text{ in } \Omega, \tag{30.29}$$

$\forall l \in \{1, 2, 3\}$.

30.6   Conclusion

This work proposes an action (energy) suitable for the relativistic quantum mechanics context. The Riemann tensor represents an important part of the action in question, but now including the density distribution of mass in its expression. In one of the last sections, we obtain the relativistic Klein-Gordon equation as an approximation of the main action, under specific properly described conditions.

We believe the results obtained may be applied to more general models, such as those involving atoms and molecules subject to the presence of electromagnetic fields.

Anyway, we postpone the development of such studies for a future research.

References

[1] R.A. Adams, *Sobolev Spaces,* Academic Press, New York, 1975.

[2] R.A. Adams and J.F. Fournier, *Sobolev Spaces, second edition*, Elsevier, 2003.

[3] G. Allaire, *Shape Optimization by the Homogenization Method*, Springer-Verlag, New York, 2002.

[4] J.F. Annet, *Superconductivity, Superfluids and Condensates*, Oxford Master Series in Condensed Matter Physics, Oxford University Press, Reprint, 2010.

[5] H. Attouch, G. Buttazzo and G. Michaille, *Variational Analysis in Sobolev and BV Spaces*, MPS-SIAM Series in Optimization, Philadelphia, 2006.

[6] G. Bachman and L. Narici, *Functional Analysis*, Dover Publications, Reprint, 2000.

[7] J.M. Ball and R.D. James, *Fine mixtures as minimizers of energy,* Archive for Rational Mechanics and Analysis, 100: 15–52, 1987.

[8] M.P. Bendsoe and O. Sigmund, *Topology Optimization, Theory Methods and Applications*, Springer, Berlin, 2003.

[9] F. Bethuel, H. Brezis and F. Helein, *Ginzburg-Landau vortices*, Birkhäuser, Basel, 1994.

[10] W.R. Bielski, A. Galka and J.J. Telega, The Complementary Energy Principle and Duality for Geometrically Nonlinear Elastic Shells. I. Simple case of moderate rotations around a tangent to the middle surface. Bulletin of the Polish Academy of Sciences, Technical Sciences, 38: 7–9, 1988.

[11] W.R. Bielski and J.J. Telega, A contribution to contact problems for a class of solids and structures, Arch. Mech., 37(4-5): 303-320, Warszawa, 1985.

[12] D. Bohm, A suggested interpretation of the quantum theory in terms of hidden variables I, Phys. Rev. 85: 2, 1952.

[13] D. Bohm, *Quantum Theory* (Dover Publications INC., New York, 1989).

[14] F. Botelho, *Functional Analysis and Applied Optimization in Banach Spaces*, Springer Switzerland, 2014.

[15] F. Botelho, *Variational Convex Analysis*, Ph.D. thesis, Virginia Tech, Blacksburg, VA -USA, 2009.

[16] F. Botelho, *Variational Convex Analysis, Applications to non-Convex Models*, Lambert Academic Publishing, Berlin, June, 2010.

[17] F. Botelho, *Dual variational formulations for a non-linear model of plates*, Journal of Convex Analysis, 17(1): 131-158, 2010.

[18] F. Botelho, *Existence of solution for the Ginzburg-Landau system, a related optimal control problem and its computation by the generalized method of lines*, Applied Mathematics and Computation, 218: 11976–11989, 2012.

[19] F. Botelho, *Topics on Functional Analysis, Calculus of Variations and Duality*, Academic Publications, Sofia, 2011.

[20] F. Botelho, *On duality principles for scalar and vectorial multi-well variational problems*, Nonlinear Analysis, 75: 1904–1918, 2012.

[21] F. Botelho *On the Lagrange multiplier theorem in Banach spaces*, Computational and Applied Mathematics, 32: 135–144, 2013.

[22] F. Botelho, *A Classical Description of Variational Quantum Mechanics and Related Models*, Nova Science Publishers, New York, 2017.

[23] F.Botelho, A variational formulation for relativisticmechanics based on Riemannian geom-etry and its application to the quantum mechanics context, arXiv:1812.04097v2[math.AP], 2018.

[24] H. Brezis, *Analyse Fonctionnelle*, Masson, 1987.

[25] L.D. Carr, C.W. Clark and W.P. Reinhardt, *Stationary solutions for the one-dimensional nonlinear Schrödinger equation. I—Case of Repulsive Nonlinearity*, Physical Review A, Volume 62, 063610, 2000.

[26] I.V. Chenchiah and K. Bhattacharya, *The relaxation of two-well energies with possibly unequal moduli*, Arch. Rational Mech. Anal., 187: 409–479, 2008.

[27] M. Chipot, *Approximation and oscillations*, Microstructure and Phase Transition, The IMA Volumes in Mathematics and Applications, 54: 27–38, 1993.

[28] R. Choksi, M.A. Peletier and J.F. Williams, *On the Phase Diagram for Microphase Separation of Diblock Copolymers: an Approach via a Nonlocal Cahn-Hilliard Functional*, to appear in SIAM J. Appl. Math., 2009.

[29] P. Ciarlet, *Mathematical Elasticity*, Vol. I—Three Dimensional Elasticity, North Holland Elsivier, 1988.

[30] P. Ciarlet, *Mathematical Elasticity*, Vol. II—Theory of Plates, North Holland Elsevier, 1997.

[31] P. Ciarlet, *Mathematical Elasticity*, Vol. III—Theory of Shells, North Holland Elsevier, 2000.

[32] B. Dacorogna, *Direct methods in the Calculus of Variations*, Springer-Verlag, 1989.

[33] I. Ekeland and R. Temam, *Convex Analysis and Variational Problems,* North Holland, 1976.

[34] L.C. Evans, *Partial Differential Equations*, Graduate Studies in Mathematics, 19, AMS, 1998.

[35] U. Fidalgo and P. Pedregal, *A general lower bound for the relaxation of an optimal design problem in conductivity with a quadratic cost functional and a general linear state equation*, Journal of Convex Analysis 19(1): 281–294, 2012.

[36] N.B. Firoozye and R.V. Khon, *Geometric parameters and the relaxation for multiwell energies*, Microstructure and Phase Transition, the IMA volumes in mathematics and applications, 54: 85–110, 1993)

[37] I. Fonseca and G. Leoni, *Modern Methods in the Calculus of Variations, L^p Spaces*, Springer, New York 2007.

[38] D.Y. Gao and G. Strang, *Geometric nonlinearity: Potential energy, complementary energy and the gap function*, Quartely Journal of Applied Mathematics, 47: 487–504, 1989a.

[39] D.Y. Gao, *On the extreme variational principles for non-linear elastic plates.* Quarterly of Applied Mathematics, XLVIII(2): 361–370, June 1990.

[40] D.Y. Gao, *Finite deformation beam models and triality theory in dynamical post-Buckling analysis,* International Journal of Non-linear Mechanics 35: 103–131, 2000.

[41] D.Y. Gao, *Pure complementary energy principle and triality theory in finite elasticity,* Mech. Res. Comm., 26(1): 31–37, 1999.

[42] D.Y. Gao, *General analytic solutions and complementary variational principles for large deformation non-smooth mechanics,* Meccanica 34: 169–198, 1999.

[43] D.Y. Gao, *Duality Principles in Nonconvex Systems, Theory, Methods and Applications,* Kluwer, Dordrecht, 2000.

[44] M. Giaquinta and S. Hildebrandt, Calculus of variations I, a series of comprehensive studies in mathematics, vol. 310, Springer, 1996.

[45] M. Giaquinta and S. Hildebrandt, Calculus of variations II, a series of comprehensive studies in mathematics, vol. 311, Springer, 1996.

[46] T. Giorgi and R.T. Smits, *Remarks on the existence of global minimizers for the Ginzburg-Landau energy functional* Nonlinear Analysis, Theory Methods and Applications, 53, 147: 155, 2003.

[47] E. Giusti, *Direct Methods in the Calculus of Variations,* World Scientific, Singapore, Reprint, 2005.

[48] B. Hall, Quantum Theory for Mathematicians (Springer, New York, 2013).

[49] V. Isakov, Inverse Problems for Partial Differential Equations, Series Applied Mathematical Sciences 127, second edition, Springer, New York, 2006.

[50] A. Izmailov and M. Solodov *Otimização* Volume 1, second edition, IMPA, Rio de Janeiro, 2009.

[51] A. Izmailov and M. Solodov *Otimização* Volume 2, IMPA, Rio de Janeiro, 2007.

[52] R.D. James and D. Kinderlehrer *Frustration in ferromagnetic materials,* Continuum Mech. Thermodyn., 2: 215–239, 1990.

[53] H. Kronmuller and M. Fahnle, *Micromagnetism and the Microstructure of Ferromagnetic Solids,* Cambridge University Press, 2003.

[54] K. Ito and K. Kunisch, Lagrange Multiplier Approach to Variational Problems and Applications, Advances in Design and Control, SIAM, Philadelphia, 2008.

[55] L.D. Landau and E.M. Lifschits, *Course of Theoretical Physics, Vol. 5—Statistical Physics, part 1,* Butterworth-Heinemann, Elsevier, Reprint, 2008.

[56] E.M. Lifschits and L.P. Pitaevskii, *Course of Theoretical Physics, Vol. 9—Statistical Physics, part 2,* Butterworth-Heinemann, Elsevier, Reprint, 2002.

[57] D.G. Luenberger, *Optimization by Vector Space Methods,* John Wiley and Sons, Inc., 1969.

[58] G.W. Milton, *Theory of composites,* Cambridge Monographs on Applied and Computational Mathematics. Cambridge University Press, Cambridge, 2002.

[59] A. Molter, O.A.A. Silveira, J. Fonseca and V. Bottega, *Simultaneous piezoelectric actuator and sensor placement optimization and control design of manipulators with flexible links using SDRE method,* Mathematical Problems in Engineering, 2010: 1-23.

[60] P. Pedregal, *Parametrized measures and variational principles*, Progress in Nonlinear Differential Equations and their Applications, **30**, Birkhauser, 1997.

[61] P. Pedregal and B. Yan, *On two dimensional ferromagnetism*, Proceedings of the Royal Society of Edinburgh 139A: 575–594, 2009.

[62] M. Reed and B. Simon, *Methods of Modern Mathematical Physics, Volume I, Functional Analysis*, Reprint Elsevier (Singapore, 2003).

[63] S M. Robinson, *Strongly regular generalized equations*, Math. of Oper. Res., 5: 43–62, 1980.

[64] R.T. Rockafellar, *Convex Analysis*, Princeton Univ. Press, 1970.

[65] R.C. Rogers, *A nonlocal model of the exchange energy in ferromagnet materials*, Journal of Integral Equations and Applications, 3(1), Winter 1991.

[66] R.C. Rogers, *Nonlocal variational problems in nonlinear electromagneto-elastostatics*, SIAM J. Math, Anal., 19(6), November 1988.

[67] W. Rudin, *Functional Analysis*, second edition, McGraw-Hill, 1991.

[68] W. Rudin, *Real and Complex Analysis*, third edition, McGraw-Hill, 1987.

[69] H. Royden, *Real Analysis*, third edition, Prentice Hall, India, 2006.

[70] O. Sigmund, *A 99 line topology optimization code written in Matlab*, Struc. Muldisc. Optim., 21: 120–127 Springer-Verlag, 2001.

[71] J.J.A. Silva, A. Alvin, M. Vilhena, C.Z. Petersen and B. Bodmann, *On a closed form solution of the point cinetics equations with reactivity feedback of temprerature* International Nuclear Atlantic Conference-INAC-ABEN, Belo Horizonte, MG, Brazil, October 24–28, 2011, ISBN: 978-85-99141-04-05.

[72] E.M. Stein and R. Shakarchi, *Real Analysis*, Princeton Lectures in Analysis III, Princeton University Press, 2005.

[73] J.C. Strikwerda, *Finite Difference Schemes and Partial Differential Equations*, SIAM, second edition (2004).

[74] D.R.S. Talbot and J.R. Willis, *Bounds for the effective contitutive relation of a nonlinear composite*, Proc. R. Soc. Lond., 460: 2705–2723, 2004.

[75] R. Temam, *Navier-Stokes Equations,* AMS Chelsea, Reprint, 2001.

[76] J.J. Telega, *On the complementary energy principle in non-linear elasticity. Part I: Von Karman plates and three dimensional solids*, C.R. Acad. Sci. Paris, Serie II, 308: 1193–1198; Part II: Linear elastic solid and non-convex boundary condition. Minimax Approach, ibid, pp. 1313–1317, 1989.

[77] A. Galka and J.J. Telega *Duality and the complementary energy principle for a class of geometrically non-linear structures. Part I. Five parameter shell model; Part II. Anomalous dual variational priciples for compressed elastic beams*, Arch. Mech., 47(677-698): 699–724, 1995.

[78] J.F. Toland, *A duality principle for non-convex optimisation and the calculus of variations*, Arch. Rath. Mech Anal., 71(1): 41–61, 1979.

[79] J.L. Troutman, *Variational Calculus and Optimal Control*, second edition, Springer, 1996.

[80] N.X. Vinh, *Flight Mechanics of High Performance Aircraft*, Cambridge University Press, New York, 1993.

[81] S. Weinberg, *Gravitation and Cosmology*, Principles and Applications of the General Theory of Relativity Wiley and Sons, Cambridge, Massachusetts, 1972.

[82] E. Kreyszig, *Introductory Functional Analysis with Applications*, John Wiley and Sons, New York, 1989.

[83] D.Y. Gao and H.F. Yu, *Multi-scale modeling and canonical dual finite element method in phase transition in solids*, Int. J. Solids Struct. 45: 3660–3673, 2008.

Index

Fundamentals
of Electric Machines

Fundamentals
of Electric Machines

A Primer with MATLAB®

Warsame Hassan Ali
Samir Ibrahim Abood
Matthew N. O. Sadiku

CRC Press
Taylor & Francis Group
Boca Raton London New York

CRC Press is an imprint of the
Taylor & Francis Group, an **informa** business

CRC Press
Taylor & Francis Group
6000 Broken Sound Parkway NW, Suite 300
Boca Raton, FL 33487-2742

© 2019 by Taylor & Francis Group, LLC
CRC Press is an imprint of Taylor & Francis Group, an Informa business

No claim to original U.S. Government works

Printed on acid-free paper

International Standard Book Number-13: 978-0-367-25098-0 (Hardback)

Library of Congress Cataloging-in-Publication Data

Names: Ali, Warsame Hassan, author. | Sadiku, Matthew N. O., author. | Abood, Samir, author.
Title: Fundamentals of electric machines : a primer with MATLAB / Warsame Hassan Ali, Matthew N. O. Sadiku and Samir Abood.
Description: Boca Raton : Taylor & Francis, a CRC title, part of the Taylor & Francis imprint, a member of the Taylor & Francis Group, the academic division of T&F Informa, plc, 2019.
Identifiers: LCCN 2019007098 | ISBN 9780367250980 (hardback : acid-free paper) | ISBN 9780367250980 (ebook)
Subjects: LCSH: Electric machinery. | Electric machines. | MATLAB.
Classification: LCC TK2000 .A566 2019 | DDC 621.31/042--dc23
LC record available at https://lccn.loc.gov/2019007098

Visit the Taylor & Francis Web site at
http://www.taylorandfrancis.com

and the CRC Press Web site at
http://www.crcpress.com

My parents, Fadma Ibrahim Omer and Hassan Ali Hussein (Shigshigow), for their unconditional love and support, to my beloved wife, Engineer Shukri Mahdi Ali, and my children, Mohamed, Faduma, Dahaba, Hassan, and Khalid, for their patience, and to all my caring siblings.

Warsame Hassan Ali

My great parents, who never stop giving of themselves in countless ways, and my beloved brothers and sisters.

My dearest wife, who offered me unconditional love with the light of hope and support.

My beloved kids: Daniah, and Mustafa, whom I can't force myself to stop loving.

To all my family, the symbol of love and giving,

Samir Ibrahim Abood

My parents, Ayisat and Solomon Sadiku, and my wife, Kikelomo.

Matthew N. O. Sadiku

Contents

Preface

An electric machine is a device that converts mechanical energy into electrical energy or vice versa. It can take the form of an electric generator, electric motor, or transformer. Electric generators produce virtually all electric power we use all over the world.

Electric machine blends the three major areas of electrical engineering: power, control, and power electronics. This book presents the relation of power quantities for the machine as the current, voltage power flow, power losses, and efficiency. The control condition presents the methods of speed control and electrical drive. Power electronics is important to machine control and drive.

The purpose of this book is to provide a good understanding of the machine behavior and its drive. The book begins with the study of salient features of electrical DC and AC machines. Then it presents their applications in the different types of configurations in lucid detail. This book is intended for college students, both in community colleges and universities.

This book organized into 11 chapters. With a short review of the basic concept of magnetism in Chapter 1, it starts with a discussion of magnetism and electricity, history of magnetism, types of magnets, and magnetic materials. It also discusses the lines of magnetic forces and force generated in the field. Chapter 2 presents magnetic circuits and quantities. The description of electrical circuit elements, its connection in series and parallel, and its phasor diagram are discussed in Chapter 3 under the title of Alternating Current Power.

Chapter 4 deals with an electric transformer, installation of the transformer, core shape, principle of operation, and the operation of the transformer under no load and full load conditions. Chapter 5 deals with the transformer design techniques. It includes the conventional design of core and shell type for single- and three-phase transformers. It illustrates all the calculation that is required to design a transformer. The description details of all types of DC machine and related voltage, current, power, and efficiency are covered in Chapter 6. All types of AC machines are also covered in Chapter 7. The chapter also elaborates the related voltage, current, power, and efficiency.

In Chapter 8, the principles of conversion from AC to DC involving single-phase as well as three-phase are presented. DC choppers and the study of several applications of power electronics are also mentioned. Chapter 9 discusses electric drives in general and concept of DC drive in particular. Chapter 10 describes the basic principles of speed control techniques employed in three-phase induction motors using power electronics converters.

Chapter 11 introduces some machines that have special applications. The examples explained in this chapter include stepper motors, brushless DC motor, switched reluctance motor, servomotors, synchro motors, and resolvers.

Several problems are provided at the end of each chapter. The answers to odd-number problems appear in Appendix D.

It is not necessary that the reader has previous knowledge of MATLAB®. The material of this text can be learned without MATLAB. However, the authors highly recommend that the reader studies this material in conjunction with the MATLAB Student Version. Appendix C of this text provides a practical introduction to MATLAB.

MATLAB® and Simulink® are registered trademarks of The MathWorks, Inc. For product information, please contact:

The MathWorks, Inc.
3 Apple Hill Drive
Natick, MA 01760-2098 USA
Tel: 508 647 7000
Fax: 508-647-7001
E-mail: info@mathworks.com
Web: www.mathworks.com

Acknowledgments

We are indebted to Dr. John Fuller and Dr. Penrose Cofie who reviewed the manuscript. We are very grateful to Dr. Kelvin K. Kirby, the interim head of the Department of Electrical and Computer Engineering for his support. Dr. Pamela Obiomon, the Dean of the Roy G. Perry – College of Engineering, for providing a sound academic environment at Prairie View A&M University, We are very thankful to Dr. Ruth Simmons, the President of Prairie View A&M University, who is consistently encouraging faculties to write books. Special thanks to our graduate students, Nafisa Islam, Faduma Sheikh Yusuf, Yogita Akhare, Kenechukwu Victor Akoh, Chandan Reddy Chittimalle, and, Akeem Green, for going over the manuscript and pointing out some errors.

Finally, we express our profound gratitude to our wives and children, without whose cooperation this project would have been difficult if not impossible. We appreciate feedback from students, professors, and other users of this book. We can be reached at whali@pvamu.edu, sabood@student.pvamu.edu, and sadiku@ieee.org.

Authors

Warsame Hassan Ali received his BSc from King Saud University Electrical Engineering Department, Riyadh, Saudi Arabia, and his MS from Prairie View A & M University, Prairie View, Texas. He received his PhD in Electrical Engineering from the University of Houston, Houston, Texas. Dr. Ali was promoted to associate professor and tenured in 2010 and professor in 2017. Dr. Ali joined NASA, Glenn Research Center, in the summer of 2005, and Texas Instruments (TI) in 2006 as a faculty fellow.

Dr. Ali has given several invited talks and is also the author of more than 100 research articles in major scientific journals and conference. Dr. Ali has received several major National Science Foundation (NSF), Naval Sea Systems Command (NAVSEA), Air Force Research Laboratory (AFRL) and Department of Energy (DOE) awards. At present, he is teaching undergraduate and graduate courses in the Electrical and Computer Engineering Department at Prairie View A & M University. His main research interests are the application of digital PID Controllers, digital methods to electrical measurements, and mixed-signal testing techniques, power systems, High Voltage Direct Current (HVDC) power transmission, sustainable power and energy systems, power electronics and motor drives, electric and hybrid vehicles, and control system.

Samir Ibrahim Abood received his BSc and MSc from the University of Technology, Baghdad, Iraq in 1996 and 2001, respectively. From 1997 to 2001, he worked as an engineer at the same university. From 2001 to 2003, he was an assistant professor at the University of Baghdad and at AL-Nahrain University, and from 2003 to 2016, Mr. Abood was an assistant professor at Middle Technical University/Baghdad – Iraq. Presently, he is doing his PhD in an electrical power system in the Electrical and Computer Engineering Department at Prairie View A & M University. He is the author of 25 papers and 3 books. His main research interests are in the area of sustainable power and energy system, microgrid, power electronics and motor drives, and control system.

Matthew N. O. Sadiku received his BSc in 1978 from Ahmadu Bello University, Zaria, Nigeria, and his MSc and PhD from Tennessee Technological University. He was an assistant professor at Florida Atlantic University, where he did graduate work in computer science. From 1988 to 2000, he was at Temple University, Philadelphia, Pennsylvania, where he became a full professor. From 2000 to 2002, he worked with Lucent/Avaya, Holmdel, New Jersey, as a system engineer and with Boeing Satellite Systems as a senior scientist. At present, he is a professor at Prairie View A & M University.

He is the author of over 500 professional papers and over 80 books, including *Elements of Electromagnetics* (Oxford University Press, 7th ed., 2018), *Fundamentals of Electric Circuits* (McGraw-Hill, 6th ed., 2017, coauthored with C. Alexander), *Numerical Techniques in Electromagnetics with MATLAB®* (CRC Press, 3rd ed., 2009), and *Metropolitan Area Networks* (CRC Press, 1995). Some of his books have been translated into French, Korean, Chinese (and Chinese long form in Taiwan), Italian, Portuguese, and Spanish. He was the recipient of the McGraw-Hill/Jacob Millman Award in 2000 for an outstanding contribution in the field of electrical engineering. He was also the recipient of the Regents Professor award for 2012–2013 given by the Texas A&M University System.

His current research interests are in the area of computational electromagnetics and computer communication network. He is a registered professional engineer and a fellow of the Institute of Electrical and Electronics Engineers (IEEE) "for contributions to computational electromagnetics and engineering education." He was the IEEE Region 2 Student Activities Committee chairman. He was an associate editor for IEEE Transactions on Education. He is also a member of the Association for Computing Machinery (ACM).

1

Basic Concepts of Magnetism

I am a slow walker, but I never walk backwards.

Abraham Lincoln

Magnetism is a force generated in the matter by the motion of electrons within its atoms. Magnetism and electricity represent different aspects of the force of electromagnetism, which is one part of nature's fundamental magnetic force. The region in space that is penetrated by the imaginary lines of magnetic force describes a magnetic field. The strength of the magnetic field is determined by the number of lines of force per unit area of space. Magnetic fields are created on a large scale either by the passage of an electric current through magnetic metals or by magnetized materials called *magnets*. The elemental metals—iron, cobalt, nickel, and their solid solutions or alloys with related metallic elements—are typical materials that respond strongly to magnetic fields. Unlike the all-pervasive fundamental force field of gravity, the magnetic force field within a magnetized body, such as a bar magnet, is polarized—that is, the field is strongest and of opposite signs at the two poles of the magnet.

1.1 History of Magnetism

The history of magnetism was dated to earlier than 600 B.C., but it is only in the twentieth century that scientists have begun to understand it and develop technologies based on this understanding. Magnetism was most probably first observed in a form of the mineral magnetite called lodestone, which consists of an iron oxide—a chemical compound of iron and oxygen. The ancient Greeks were the first known to have used this mineral, which they called a magnet because of its ability to attract other pieces of the same material and iron.

The British physicist William Gilbert (1600 B.C.) explained that the earth itself is a giant magnet with magnetic poles that are somewhat distracted from its geographical poles. The German scientist Gauss then studied the nature of earth's magnetism, followed by the French scientist Koldem (1821 A.C.) known that the magnet is a ferrous material only.

Quantitative studies of magnetic phenomena initiated in the eighteenth century by Frenchman Charles Coulomb (1736–1806), who established the inverse square law of force, which states that the attractive force between two magnetized objects is directly proportional to the product of their individual fields and inversely proportional to the square of the distance between them. Danish physicist Hans Christian Oersted (1777–1851) first suggested a link between electricity and magnetism. Experiments involving the effects of magnetic and electric fields on one another were then conducted by Frenchman Andre Marie Ampere (1775–1836) and Englishman Michael Faraday (1791–1869), but it was the Scotsman, James Clerk Maxwell (1831–1879), who provided the theoretical foundation to the physics of electromagnetism in the nineteenth century by showing that electricity and magnetism represent different aspects of the same fundamental force field. Then, in the late 1960s, American Steven Weinberg (1933–) and Pakistani Abdus Salam (1926–), performed yet another act of theoretical synthesis of the fundamental forces by showing that electromagnetism is one part of the electroweak force.

The modern understanding of magnetic phenomena in condensed matter originates from the work of two Frenchmen: Pierre Curie (1859–1906), the husband and scientific collaborator of Madame Marie Curie (1867–1934) and Pierre Weiss (1865–1940). Curie examined the effect of temperature on magnetic materials and observed that magnetism disappeared suddenly above a certain critical temperature in materials like iron. Weiss proposed a theory of magnetism based on an internal molecular field proportional to the average magnetization that spontaneously aligns the electronic micro magnets in the magnetic matter. The present-day understanding of magnetism based on the theory of the motion and interactions of electrons in atoms (called quantum electrodynamics) stems from the work and theoretical models of two Germans, Ernest Ising (1900–) and Werner Heisenberg (1901–1976). Werner Heisenberg was also one of the founding fathers of modern quantum mechanics.

1.2 The Cause of Magnetism

The prime reason of magnetism is the movement of electrons in a particular object. The movement of electrons in specific orbits around the nucleus is as same as the motion of the planets around the earth and the rotation of the earth around its axis. In this case, there are two movements: orbital and rotational. Due to these movements, a magnetic moment is produced on each electron that behaves like an intricate magnet. In a particular matter, the numbers of all electrons of an atom are equal. In each orbit, there are many pairs of electrons that are usually remaining as many pairs in each orbit and each electron is rotating in two opposite direction. As a result, the total torque that was created by electrons in each orbit is zero.

The metallic bond is usually seen in pure metal and some metalloids. Most of the metal is symbolized by high electrical and thermal conductivity as well as by malleability, materials contain transition atoms. On the other hand, there are some metal that contains an odd number of an electron at their outer shell such as—iron, cobalt, and nickel. These metals have their own specific value for the sum of the torque. Some of which contain an odd number of electrons such as iron, cobalt, and nickel. The sums of torque in these orbits contain a specific value. The positively charged particles of the magnetic materials gain the magnetic effect generated by the electrons which are revolving around them. Positively charged ions or group of atoms (about 10^5 atoms) that are directed toward the same direction, is called a domain. Each field has its magnetic field with its northern and southern poles. Non-magnetized field of iron pieces is oriented in random, non-uniform directions. When placing the iron piece within a relatively weak external magnetic field, some iron fields will align with some and with the outer magnetic field. As the external magnetic field increases, lined and outward-facing fields will increase remarkably. As the strength of the outer field increases, we retain a situation in what all the fields of the iron segment are heading toward this area. Consequently, the object is perceived as being saturated magnetically, whereas all the fields have been lined up in the desired direction. In either case, lifting the effect of the magnetic field outside from the iron piece, the fields will return to the previous random states.

1.3 Types of Magnets

Magnets are divided into two types according to the composition method, namely, natural and artificial magnets. There are confined to the natural magnets in those present in the form of natural rocks such as the ancient Greeks and China, and the characteristics of these magnets to the rocks produced by nature and no human being. Natural rock or magnets are currently present in the United States of America, Sweden, and Norway, but it is not practical to use in known electrical appliances because of the ease and economy of making very powerful and efficient magnets. Synthetic magnets produce magnetic materials such as iron, nickel, and cobalt in various shapes and sizes.

To obtain very strong magnets, special alloys are used for iron or electrically magnetized steel. This is done by placing the magnetized iron piece inside a coil of insulated wire through which an electric current or pack of electrons pass through as shown in Figure 1.1, and the force that turns cutting iron into a magnet is a line of magnetic forces quite like the lines of the electric field. Synthetic magnets are classified into both temporary and permanent depending on their ability to retain magnetic properties after the magnetization is removed.

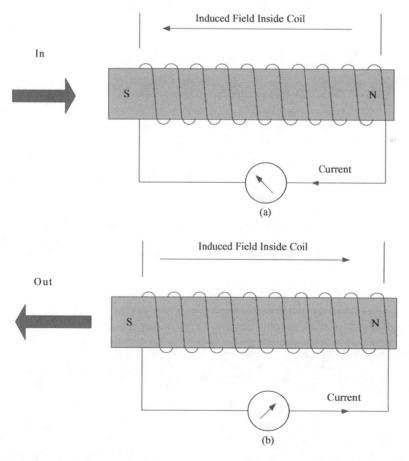

FIGURE 1.1
Magnetized piece of iron (a) When Magnetized piece of iron In direction case. (b) When Magnetized piece of iron Out direction case.

Magnetic materials such as iron, iron-silicon alloys, and nickel-iron alloys, used in the manufacture of transformers, motors, and generators have highly reluctance, which are easy to magnetize and give temporary magnets. However, these materials keep a small amount of magnetism after removing the external magnetization force. This magnetism is called *residual*, which is essential in the work of electric machines. Another criterion for measuring the susceptibility of a material to the formation of a permanent or temporary magnet is the ease with which these materials allow the lines of magnetic forces to be distributed within it. This is called *the permeability* of the material, so the permanent magnets have high impedance and low permeability while the temporary magnets low impedance and high permeability. The advantages of permanent magnets are: There is no operating cost, no maintenance, and high efficiency also doesn't need electrical power to operate, which excludes the possibility of failure of the feeding system, and they can be used in a hazardous environment, do not require electrical connections, and are not affected by shocks and vibrations. The disadvantage of permanent magnets is difficulty controlling their power. While the advantages of temporary magnets are ease of control over their ability, loadable, and withstand higher temperatures than permanent magnets, and can cut the feed and the field distance and edit the load easily. The disadvantages of temporary magnets, it needs a power source and exposure to the problems of potential disruption of this source, and that it can operate in a specific environment because it does not resist water, heat, and shocks.

1.4 Applications of Magnet

Magnets are widely used in power generation, transmission, and transmission systems. They are the basis for the work of the electrical stations that convert mechanical energy (in the form of heat to generate steam or waterfalls) to electrical power. In addition, it is the basis for the work of electrical transformers that convert the voltage from the amount supplied to other quantities needed by electrical and electronic devices. Magnets are found in electric motors which convert electrical energy into kinetic energy to move household, office, or industrial equipment.

Permanent magnets are used in sensitive and durable devices in sensitive and accurate devices such as loudspeakers, earphones, and measuring devices, in addition to vital and important uses in the field of magnetic recording and computer storage units and in audio-visual systems, see Figure 1.2.

1.5 Magnetic Materials

Magnetic materials are classified according to the properties they possess.

Ferromagnetism: These material exchange electronic forces in ferromagnetism are very large, thermal energy eventually overcomes the exchange and produces a randomizing effect. This occurs at a particular temperature called *Curie temperature*. In this process, magnetic fields are automatically directed in one direction at the Curie temperature to generate a magnetic field. At the Curie temperature, these magnetic fields are directed back to their random direction of

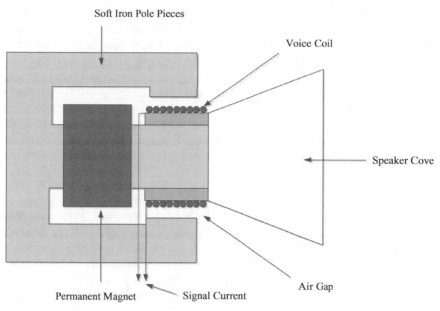

FIGURE 1.2
Magnet applications in the loudspeakers.

high temperature. The intensity of one field is equal in size and is the direction of one field parallel to some in the same direction. Also, a total torque of each field does not have to be in the direction of the other field's intensity itself. Thus, the piece of the electromagnetic material may not have a total magnetic torque, but when placed within an external magnetic field, the inhibition of the individual fields will go together in the same direction forming a total determination. So, the notable magnetic property of the ferromagnetic material is to turn into another kind of magnetic material at Curie temperature. The elements Fe, Ni, and Co are typical ferromagnetic materials which are characterized by the ability to attract and repulse with other magnetic objects.

Paramagnets: This substance that shows a positive response to the magnetic field, but it's a weak response. The amount of response is determined by a standard called *magnetic susceptibility*, which represents the ratio of the magnetic force of the material to the magnetic field strength, which is a quantity without units. The paramagnetic phenomenon is observed in substances whose atoms contain single electrons (while these electrons are usually conjugal and opposite), so its magnetic torque can't be zero. Paramagnetic materials include transient and rare substances in nature or generally substances whose atoms contain non-conjugate electrons. The magnetic properties of these substances depend on the temperature. Examples of these substances are sodium, potassium, and liquefied oxygen.

Diamagnetism: Is the material that is generated when placed within an external magnetic field, an opposition torque that contradicts the direction of the outer field and explains that as a result of the currents that are excited in atoms of the material with non-bilateral electrons are generated according to the law of Ampere of an opposite that contrasts with the field caused by this note that these materials contradict with magnets close to them. The relative permeability of these materials is slightly less than one, i.e., their magnetic intensity is negative. Examples of diamagnetism materials are copper, gold, carbon, diamonds, nitrogen gas, and carbon dioxide.

1.6 Lines of Magnetic Forces

A simple experiment is executed to acquire the properties or the behavior of magnetic materials. For the experiment, a permanent magnetic rod is used to put it on paper and scattered iron filings on the rod. It is noted that the distribution of these filings would be in such a way that remarks to the direction of the magnetic force lines generated by the magnetic bar as shown in Figure 1.3. The shape represents the nature of the magnetic field represented by the distribution of lines of forces that exit from the one terminals of the magnet and enter the other end of the so-called terminals. In this case of poles, the lines of forces involve the three-dimensional space surrounding the magnetic rod, while the experiment can show only two dimensions of them and poles include the entire length of the bar. But the intensity of the impact is concentrated at the terminals strongly and the party which the lines of force outcome is called *the north pole* and the party that enters the lines of forces called *the south pole* and symbolized with the letters N, and S, respectively. The magnetic field is called a dipole. This means that the magnetism of the bar (or any other magnet) is concentrated in two poles that are strongly equal to the opposite field in the direction of effect when cutting the bar magnetic to two parts each has a bipolar magnet, so that any number of bipolar magnets can be generated until one atom is reached, confirming that the atom is the primary source of the magnet. The magnetic field strength is measured by the number of lines of forces that leave a pole to interfere with the other, depending on the magnetic material and the magnetism and magnetic form. The magnets are usually in different shapes and sizes and are used in the form of discs, rods, or the horseshoe.

The latter is a rod-shaped like a letter (U) as shown in Figure 1.4. In this case, the magnet contains two poles, the lines of the forces of the pole, and interfere in the other exactly

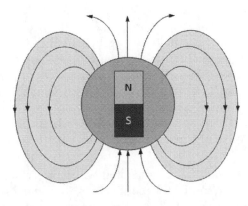

FIGURE 1.3
The distribution of lines of magnetic forces.

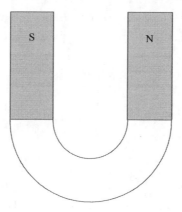

FIGURE 1.4
Magnets in the shape of a horse's suit.

how much in the case of the bar except that the magnetic field is more focused between the poles because they converge and direct lines between them in the air. When the magnets are placed on a magnetic rod or when the magnet is in the form of a loop, the circle of the path of the lines is closed and does not need to pass through the air, but will pass the majority in the metal body in the form of closed rings.

Some specialized devices are used to detect the lines of magnetic forces that can be sensed or influenced. These lines are surrounded through closed rings that emerge from the north pole to enter to mediate as shown in Figure 1.5. The speed of metal materials is useful to estimate the magnet forces such as iron when rounding from it. These lines belong to many properties:

1. Never intersect with each other.
2. All are identical and equal in intensity.
3. Take the shortest path with the least resistance between the poles.

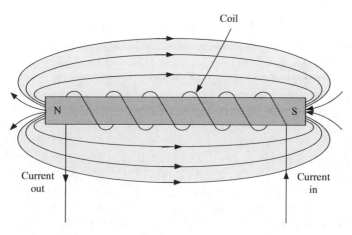

FIGURE 1.5
The lines of magnetic forces.

4. The intensity of the lines decreases as the distance between poles increases.

5. Heading from the south pole to the north pole inside the magnet and vice versa outside.

6. The static lines do not move and impose a trend as if they are moving.

1.7 Magnetic Force

There is a similarity between the magnetic and electric fields, and there are certain differences between them, and there is generally a very strong reciprocal relationship between them, and we know in our study of the field of electricity that the two different charges and the two similar charges are released because of the force affecting them according to the law of Coulomb. In the magnetic field, magnetism behaves in the same context as the conduction of charges in the electric field.

The principle of magnetic effect is to attract another object and convert them into another magnet when it is placed nearby. When two magnetic objects are close by, a certain force will affect them, and this force acts as proportional to the magnitude of the field and inversely proportional with the square of the distance between them. Either this force may be an attraction of the opposite poles or a repulsion force when the polarity of the poles is similar. If only one of the two bodies is magnetic, then the magnetic body strongly attract the other non-magnetic body to be a ferromagnetic material.

If it is a magnetic material, dissipation occurs weakly in terms of its principle. The magnetic force affects within a limited distance that is described as the field of force. Each magnetic pole generates an area around where each magnetic pole generates, turn sheds the force of the objects in this field. This field is conceived as the set of force lines such as in the case of the magnetic field. These lines converge near the poles and spread outward as they become more distant from the poles as they diverge from some, and their number in the unit area corresponds to the intensity of the field in the designated area.

1.8 The Direction of Magnetic Field Lines

A magnetic field is induced when an electric current pass through a conductor, and the generated magnetic field line will be and directed to the direction of current passing through the conductor. This phenomenon can be observed by taking an isolated straight conductor passing through the center of a carton plate as in Figure 1.6. When a current is passed in this conductor, the iron filings are scattered on the carton plate. Due to passing current, the generated magnetic field makes the iron bar take concentric rings surrounding the conductor. Compasses are placed to know the direction of these rings. All of which affect the same direction around the loop. When the direction of the current passing through the conductor changes, the direction of the compasses will change by 180°.

In general, the direction of the force lines generated by the current passing through the conductor can be determined by using the right-hand rule, which is to capture the conductor with the right hand so that the thumb is parallel to the axis of the conductor and pointing to the current. The four fingers holding on the conductor will indicate the direction of the force lines as shown in Figure 1.7.

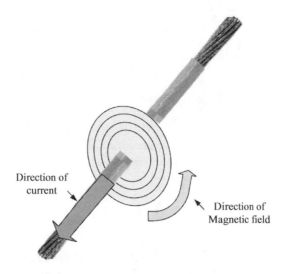

FIGURE 1.6
The direction of the generated field lines in the conductor.

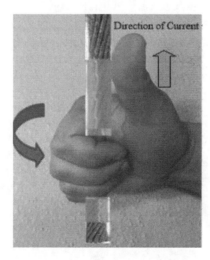

FIGURE 1.7
Directions of field lines.

The head of the arrow represents the direction of the current toward us, and the bottom of the arrow represents the direction of the current toward the page. Screwdriver method is used to detect the direction of the lines of the field. On Figure 1.8, screwdriver method is shown as well as the direction of currents and field. The spiral is in the direction of clockwise indicates the force lines, and the direction of the spiral is the direction of the current. When the coil is not plugged, the direction of the force lines is in the opposite direction of the anticlockwise. The characteristics of these lines are summarized as follows:

1. The magnetic field consists of regular concentric circles whose center corresponds to the axis of the conductor.
2. The direction of the lines depends on the direction of the current in the connector and the density of these lines on the current intensity.

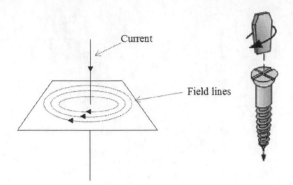

FIGURE 1.8
Apply the screwdriver method.

3. Field strength equal to zero in the axis of the connector, the maximum at the surface and less as move away from the surface of the connector.
4. All the rings are symmetrical relative to the axis of the connector, and if not, otherwise it is evidence of the presence of an influential external force that leads to it.
5. The lines of magnetic forces in two conductors carry a current remove each other, which constitutes the strength of attraction if the direction of the currents is similar, as in Figure 1.9, and support each other of the strength of repulsion if the direction of the two currents in contrast.

The magnetic field describes the most major functions is the lines of forces, that gives their number and total called *magnetic flux*, it symbolizes the letter (ϕ) and the unit is (Weber), and the flux is the flow of magnetic force represented by lines of force that move between the poles of the north and south of the part magnet. The properties of the magnetic field are given the density of the lines at a certain point in the field, and this is called *flux density*, it symbolizes the letter (B), and its unit is the Weber per the unit area (Wb/m^2) or (Tesla).

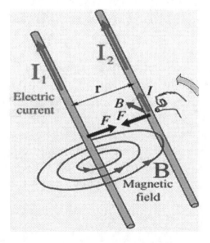

FIGURE 1.9
Two conductors carrying current. (http://WWW.phys4arab.net).

The one Weber is equal to one hundred million field lines (1 Wb = 10^8 field line) and given by the formula:

$$B = \phi/A, \tag{1.1}$$

where A is the area of the cross-section (m^2) where the density of the lines is to be found.

Example 1.1

What is the density of the flux in the area of the number of lines passing through a rectangle of 40 × 250 cm, which are 12 million lines?

Solution

We convert the number of lines to Weber (the unit of the flux), where the Weber equals one hundred million lines:

$$\Phi = 12 \times 10^6 / 10^8 = 0.12 \; Wb.$$

The area of the passage through which the lines of forces are:

$$A = 250 \times 40 \times 10^{-4} = 1 \; m^2$$

So, the density of the flux is equal:

$$B = 0.12/1 = 0.12 \; Wb/m^2.$$

1.9 Magnetic Field and Its Polarity

In 1820, Oersted discovered the relationship between magnetism and electrical when he showed that the current carrying wire generates its own magnetic field. When the wire forms a solenoid and a current is passed, the sphere of single coil rings accumulates to generate a strong magnetic field within the axis of the coil, as shown in Figure 1.10. In the case of putting a rod of iron or plastic material of ferromagnetic within this coil as shown in Figure 1.11, the magnetic field will increase several times. This arrangement is very

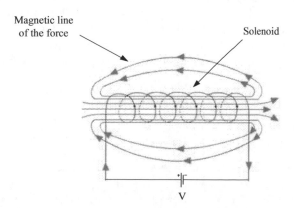

FIGURE 1.10
Field lines of the magnetic field through and around a current carrying solenoid.

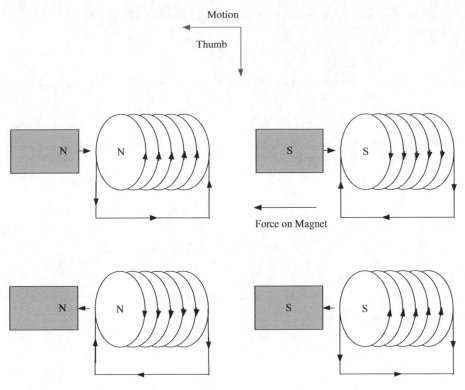

FIGURE 1.11
Change the direction of winding.

important and has many uses and is called *electromagnetic* that the force lines come out of the coil carrying the current on the north (N) pole and the enter in the other side of the south (S) pole.

The polarity of the solenoid is determined by using the right-hand rule, which states: Hold the coil with the right hand so that the four fingers represent the direction of the current passing through the coil. The parallel thumb of the coil axis, in this case, will affect the north pole of the magnetic field formed. When both the direction of the current and the direction of the winding of the coil as in (A), (B), or (C) and (D) in Figure 1.11, the polarity of the coil remains unchanged when the direction of the current is changed only, the polarity will change, and when the direction of the coil is changed only the polarity will change as shown in Figure 1.11, where there are four possibilities for these polar accordingly.

1.10 Magnetism of Magnetic Materials

Conversion of an ordinary piece of metal into a magnet is basically placed inside a spiral coil. This conversion takes place by regaining the magnetic properties and generating a magnetic moment in it. When the current passes through the solenoid, the magnetic field will pass its lines inside the iron, the magnetization of the iron piece by placing it within the magnetic field called indirect. The iron can be magnetized directly

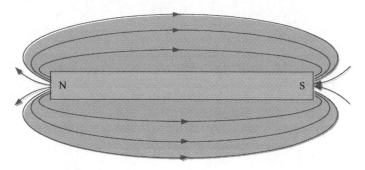

FIGURE 1.12
Field lines around a magnet bar.

by making it part of an electric circuit. The electric current passing through it is called *the magnetic current*, whose value determines the field strength of the object. The magnet can be directly cut off when it is rounded by permanent magnets. When the external magnetic field of the solenoid or permanent magnets is removed from the iron piece, its magnetic susceptibility will decay except for a small fraction called residual. When the length of the iron piece inside the solenoid is several times greater than the diameter, the lines of forces that pass through the iron piece are straight and parallel to the axis and denser within them much more than in the air surrounding this piece. This is because the air resistance to the passage of force lines is much more than that of the iron piece. When the lines of forces from the north pole of the magnetized piece are released, they are spread in the air in a large area because it can't carry this number of lines in the limited volume unit, as shown in Figure 1.12. The number of lines of magnetic forces or the flux (ϕ) generated by the solenoid is directly proportional to the current (I) and the number of coils (N) it consists of. The result of multiplying these two amounts is called *the magnetic motive force* (MMF) and is symbolized by letter (F) and its unit is ampere-turn (AT) is expressed mathematically as:

$$F = I * N. \tag{1.2}$$

1.11 Force Generated in the Field

When the coil is placed within an external magnetic field, its field will interact with the external field, and the nature of the reaction depends on the current of the current passing through the coil and on the polarity of the outer field or the direction of the force lines in it. To simplify the situation, we take one conductor carrying a current toward the page and put it within the field of external magnetic lines from the north pole to the south pole as shown in Figure 1.13. Note that the external field will emit the force of the conductor within this area because the lines of the outer areas, and to the connector to the right of the axis of the poles are in the same direction increasing density, while to the left of the axis of the poles and the opposite of each other, and become a few in this area.

The direction of force generated by the interaction between two fields can be determined using the rule of the left hand or using the filming rule as in Figure 1.14. The rule of the left-hand states that the left hand is extended within the magnetic field to the palm

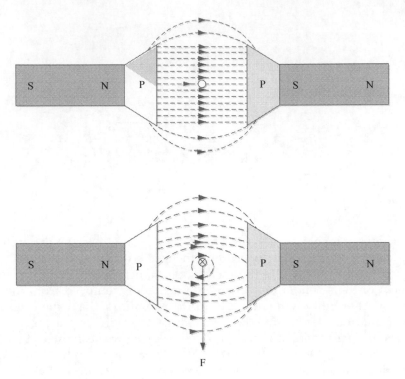

FIGURE 1.13
The departure of lines from the Arctic to the South Pole.

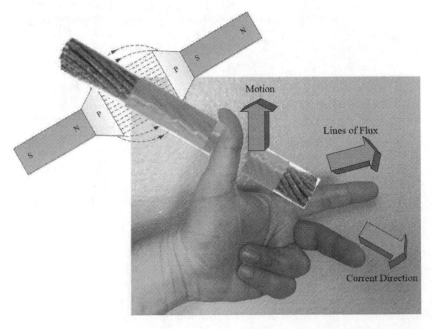

FIGURE 1.14
The rule of the left hand.

of the hand. The four fingers point to the direction of the current passing through the conductor. The orthogonal thumb with the four fingers indicates the direction of the expelling force (f). There are four possibilities for the direction of this force and depends on the direction of the lines of the outside field or its polarity and the direction of current passing through the conductor. The amount of force generated depends on the current of the current (I), the intensity of the force lines of field (B), and the effective length of the conductor within the field (L), which is usually equal to the length of the electrodes. This force is expressed as follows:

$$f = B * I * L. \tag{1.3}$$

The value of the force increases when the value of any amount on the right side of the equation increases. The length is also increased when making the conductor in the shape of a coil inside the magnetic field and often design the coil in the form of a frame that has two sides and when placed within the external magnetic field and make it moves freely around the axis, this carrier of the current will start circling around the axis. Because the direction of the upper and lower forces is one, the direction of the frame is identical to the clockwise movement. When the current is changed in the coil or the polarity of the external magnetic field changes, the direction of the frame rotation will be reversed. This phenomenon is one of the most important principles of the utilization of the transformation of electric energy into the kinetic energy of electric motors.

When the direction of the conductor movement within the magnetic field at a certain angle (θ), with the direction of the lines of the magnetic forces of this field, the expression of force becomes:

$$f = B * i * L * sin\ θ, \tag{1.4}$$

where: L distance of moving current i. The magnetic field affects the moving charge (Q) and the charge generates a moving current for a distance (L) in time (T) within the field, and the direction of the movement of the charge is at an angle (θ), the expression of force becomes:

$$f = B * Q * V * sin\ θ. \tag{1.5}$$

Example 1.2

When a 20 cm conductor is passed in which a current of 20 A is within a magnetic field, the force exerted on the conductor is 24 N. What is the density of the lines of this field?

Solution

The density of the field lines or the density of the magnetic flux is:

$$B = f/IL = 2.4/(20 \times 20 \times 10^{-2})$$

$$= 0.6\ Wb/m^2.$$

1.12 Hysteresis Loop

The (B-H) curve basically emphasizes the flux intensity and field strength of the material with additionally expresses the relationship between magnetic and non-magnetic materials and describes the former condition of it. The curve mentioned the relationship of a magnetic equation of the magnetizing and demagnetizing experiences which is drawn by a closed loop. In fact, it is also known as the most essential characteristics that represent the relationship of the field-density (B-H) of the material. It experiences a condition of material before the method of heat treatment and mechanical tension.

The magnetization curve can be drawn when a magnetic force is cast on a sample like steel or iron. When the magnetic force is applied, the curve starts to increase from zero to maximum, and it again goes to zero after removing the magnetization as shown in Figure 1.15. From the figure below, it can be seen that the field strength value goes higher when the magnetization field is lower which indicates the value of the flux intensity (B) is less than the field strength (H). After continuing to the value increase of magnetic force toward the negative direction, it goes again zero to maximum and again to zero as previous.

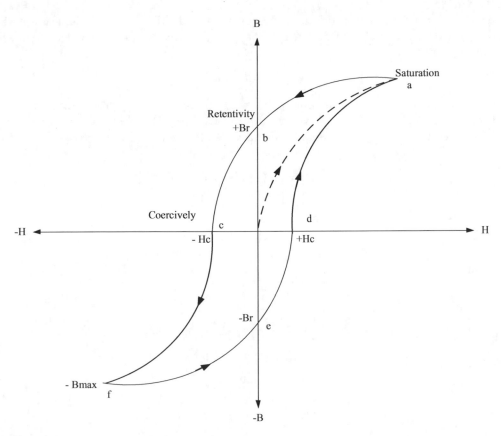

FIGURE 1.15
B-H curve.

From the figure, two very important values can be defined as below:

Residual Magnetism (B$_r$): is the amount of the intensity of the flux when the field strength is zero. It is symbolized as a residual.

Coercive Force (H$_c$): is the amount of field strength that makes the magnetism of (B) zero.

It is symbolized as a coercive force.

The hysteresis loop represents the properties of the magnetic material and the number of its basic quantities. The hinged ring may be wide. That is, the amount of (H$_c$) is roughly equal to (B$_r$) so that the magnetic material is low permeability and high impedance, and both the residual magnetism and coercive force are of high value. The loop may be tight, in other words, the amount of (B$_r$) is much higher than the amount of (H$_c$). The material is highly permeable and low impedance, with high residual magnetism and low coercive force.

Problems

1.1 What is the practical rule for passing the direction of the electric motive generated in a conductor moving within a magnetic field?

1.2 When the electric current passes in a coil, a magnetic field is formed. What are the polar potentials of this area and what do they depend on?

1.3 What are the factors in which the wide loop of hysteresis differs from those in the narrow hysteresis loop?

1.4 What are the common characteristics of all the ferromagnetic materials?

1.5 A conductor 0.32 m long with 0.025 Ω resistance is located within and normal to a uniform magnetic field of 1.3 T. Determine (a) the voltage drop across the conductor to cause a force of 120 N to be exerted on the conductor; (b) repeat part (s) assuming an angle of 65 degrees between the conductor and the magnetic field.

1.6 When a 25 cm conductor is passed in which a current of 30 A is within a magnetic field, the force exerted on the conductor is 20 N. What is the density of the lines of this field?

1.7 Find the linear velocity of a 0.54 m conductor that will generate 30.6 V when cutting flux in 0.86 T magnetic field.

1.8 What is the density of the flux in the area of the number of lines passing through a rectangle of 40 * 250 cm, which is 12 * 10^6 lines?

1.9 A coil of wire with 80 turns has a cross-sectional area of 0.04 m^2. A magnetic field of 0.6 T passes through the coil. Calculate the total magnetic flux passing through the coil.

2

Magnetic Circuit

Education is what remains after one has forgotten everything he learned in school.

Albert Einstein

Magnetic circuits are those parts of devices that employ magnetic flux due to inducing a voltage. Such devices include generators, transformers, motors, and other actuators as solenoid actuators and loudspeakers. In such devices, it is necessary to produce magnetic flit. This is usually done with pieces of ferromagnetic. In this sense, the magnetic circuits are like the electric circuits in which conductive material such as aluminum or copper has high electric conductivity and are used to guide electric current. The analogies between electric and magnetic circuits are two: the electric circuit quantity of current is analogous to magnetic circuit quantity flth. The electric circuit quantity of voltage or electromotive force (EMF) is analogous to the magnetic circuit quantity of magnetomotive force (MMF). EMF is the integral of electric field E, and MMF is the integral of magnetic field H.

2.1 Magnetic Quantities

The previous chapter discussed the various topics of many quantities describing the magnetic field. This section will define these quantities and give the mathematical expression and its units used to measure them as follows.

2.1.1 Flux Density

It represents the magnetic field near the poles of the magnet, where they emerge from the north pole and enter the south pole, and the total of these lines called *the magnetic flux* (ϕ), which is known as the total number of lines of forces in the magnetic field and the number of lines passing in the vertical area unit on the direction of the flux (B) and the flux unit is (Wb/m^2) or Tesla.

$$B = \phi/A \tag{2.1}$$

2.1.2 Permeability

It's characterized by the properties of magnetic materials, and the permeability is divided into absolute magnetic permeability, which is characterized by magnetic and non-magnetic materials and symbolized by the letter (μ), and relative permeability and symbolized by the letter (μ_r), which is the ratio of the absolute permeability of the material to the absolute permeability of air ($\mu_o = 4\,\pi * 10^{-7}$), and it's unit given in the Henry/meter of length, so the relative permeability, which is without unit, is equal to:

$$\mu_r = \mu/\mu_o \tag{2.2}$$

The relative permeability of air and non-magnetic materials ($\mu_r = 1$) and alloys of iron, nickel, and cobalt to tens of thousands.

Also, the absolute permeability of the material is known as changing the intensity of the flux (Λ B) relative to the corresponding change. The magnetic field intensity (Δ H):

$$\mu = \Delta B/\Delta H \tag{2.3}$$

Field intensity is the magnetic force of magnetic material or the amount of magnetic motive force in the unit of length necessary for the flow of magnetic flux:

$$H = F/L = I.N/L \tag{2.4}$$

Its unit is Amp. turn/meter or (A.T/m), and the density of the flux is approximately proportional to the field strength, and the relatively constant here is the permeability of the medium in which these lines of force pass:

$$B = \mu.H \tag{2.5}$$

The permeability of magnetic materials depends on the technique of manufacturer and arrangement of this material and gives the casting plants tables showing the permeability of their materials or the relationship between the density of the flux and the intensity of the field or may give the curves of the relationship (B-H), it's called *magnetization curves*.

2.1.3 Magnetic Reluctance

Is the resistance that faces the lines of forces passing in the middle and called *magnetic resistance* or *reluctance* and symbolized by the letter (S). The reluctance is opposing to the generation of magnetic flux and depends on the amount of the distance of the medium through which this flux and permeability of this medium can be derived expression through the flux on what follows:

$$\phi = B.A = F/S \tag{2.6}$$

This expression is like the Ohm's law (I = V/R) for this reluctance is equal:

$$S = \frac{L}{\mu A}, \left(\frac{\mu}{H/\mu \times \mu^2} = \frac{1}{H} \right) \tag{2.7}$$

Also:

$$S = \frac{F}{\phi} \tag{2.8}$$

So, the reluctance unit is a reciprocal Henry or amp-turn/Weber.

The magnetic motive force (F = I.N) is the source that generates the motive and drives it in the magnetic circuit. The passivity of this circuit is called *the permeance*, which is like the conductivity in the circuit. Which is equal to the reciprocal resistance where the equalization of the reciprocal reluctance in Henry:

$$\frac{1}{S}=\frac{\mu}{L}.A=\frac{\phi}{F} \tag{2.9}$$

The magnetic circuit is not very different from the electrical circuit, consisting of three main components, as in Figure 2.1:

- A voltage source (E)
- Connection wires as a conductor for the current
- Load (Lamp) consumes electricity and converts it to another form
- Resistance (R).

The magnetic circuit as shown in Figure 2.2 consisting of:

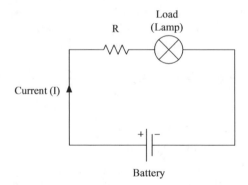

FIGURE 2.1
Simple electrical circuit.

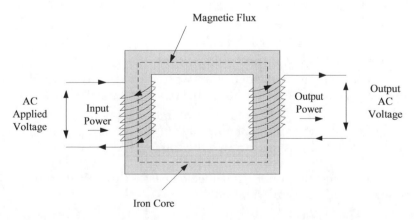

FIGURE 2.2
The magnetic circuit.

- A source of magnetic motive force (F) is simply a coil consisting of a number of turns (N_1) and passes by the current (I_1) because of the voltage applied (V_1).
- An arrangement of a magnetic material that is a path to the passage of the magnetic flux (ϕ).
- Load receives magnetic energy and transfers it into another form. It is here in the form of a second coil consisting of the winding (N_2), and where the ratio of the two voltages (V_2/V_1) as the ratio of the number of turns in the coils (N_2/N_1).

Magnetic path is one of the most important parts of the magnetic circuit and is usually made of a highly suitable ferromagnetic material for magnetization in the form of a closed loop, often different, for example, a circular ring or square or rectangular frame. The path may be simply consisting of a single series or complex consisting of several rings formed of more than two parallel branches. The magnetic path may be made without cutting. There may be parts in the closed, if empty, called *the air gap*, when the conduct of a single material with a specific reluctance and without an air gap, it is uniform, and the calculation of its coefficients is simple. When the path consists of two or more different objects with different reluctances or parts of which have a different sectioned area or when there is an air gap, the behavior is not uniform, and some complexity is complicated by the method of calculation. When calculating the magnetic circuit, you must know the dimensions and nature of the material of each part.

There are two ways to calculate the parameters of the magnetic circuit, depending on the information known and what is needed to find in them:

First: From the knowledge of finding the path and the current that passes through the coil required to find the magnetic flux and to achieve this follow the following steps:

1. Calculate the length of the wire (L) and the segment area (A).
2. Assume several values of the flux density (B) in the circuit and extract the corresponding flux ($\phi = B.A$).
3. We use the B–H of the path material to find the appropriate field strength (H) for each density of flux (B), and then according to the magnetic motive force of each amount ($F = H.l$).
4. Plot the relationship curve between ϕ and F and for the coil defined by its data, we find the amount ($F = I.N$).

Second: From the knowledge of the dimensions of the path, the coil, and the flux that passes through the path to be found and the amount of current that achieves it.

1. From the knowledge of magnetic flux, we find the density of the flux ($B = \phi/A$), and using the magnetization curve (B–H), we find the field strength (H).
2. From the knowledge of (L) and (H), we find the amount of magnetic motive force ($F = HL$), and from the knowledge of the number of turns, we find the amount of current passing through the coil ($I = F/N$).

As in the case of a continuous circuit, the same flux passes through all its parts. There is no branch or node that leads to its distribution.

Example 2.1

The coil has 200 turns and the flux of magnetization passing through it 1 A is placed on magnetic conduction equal to the same permeability 0.2. Calculate the value of the flux passing through the path with an average diameter of 10 cm and the diameter of the ring segment 4 cm.

Solution

First, the length of the ring:

$$L = \pi.D_a = \pi \times 10 \times 10^{-2} = 0.314 \text{ m.}$$

The cross section Area:

$$A = \pi.\left(\frac{D}{2}\right)^2 = \pi \times 2^2 \times 10^{-4} = 0.001256 \text{ m}^2.$$

The magnetic reluctance:

$$S = L/\mu.A = 0314/0.2 \times 0.001256 = 1250 \text{ AT/Wb.}$$

Magnetic motive force is:

$$F = I.N = 1 \times 200 = 200 \text{ AT.}$$

The magnetic flux passing through the path is:

$$\phi = F/S = 200/1250 = 0.16 \text{ Wb} = 160 \text{ mWb.}$$

Example 2.2

A medium diameter iron ring 20 cm is made of a metal rod with a diameter of 2 cm, with a coil number of turns 100 turns. Which is the necessary magnetic current to pass in the coil to generate a magnetic flux of 0.5 mWb? If the relative permeability of material 400 and for the air $4 \pi \times 10^{-7}$.

Solution

The length of the path of force lines:

$$L = \pi.D_A = \pi \times 20 \times 10^{-2} = 0.628 \text{ m.}$$

Wire cross section area:

$$A = \pi \, (2/2)^2 \times 10^{-4} = 3.14 \times 10^{-4} \text{ m}^2.$$

The magnetic flux density is equal to:

$$B = \phi/A = (0.5 \times 10^{-3})/(3.14 \times 10^{-4}) = 1.56 \text{ Wb/m}^2.$$

Field strength is equal to:

$$H = B/\mu = B/\mu_o = 1.56/4 \times 10^{-7} \times 400 = 310 \text{ A/m.}$$

The magnetization current is then equal (Tables 2.1 and 2.2):

$$I = H/N = 310/100 = 3.1 \text{ A.}$$

TABLE 2.1

Comparison of Electrical and Magnetic Quantities

Electrical Components		Magnetic Components	
The Amount	**Symbol**	**The Amount**	**Symbol**
Voltage or electric motive force	V or E	Magnetic motive force	$F = NI = HL$
Current	I	Flux	ϕ
Resistance	$R = \rho L/A = L/GA$	Reluctance	$S = L/\mu A$
Conductivity	$G = 1/\rho$	Permeability	μ

TABLE 2.2

Comparison between Electric and Magnetic Circuit

Electrical Circuit	Magnetic Circuit
The path traced by the current is known as the electric current.	The path traced by the magnetic flux is called as a magnetic circuit.
The current is actually flows, i.e., there is a movement of electrons.	Due to MMF, flux gets established and does not flow in the sense in which current flows.
There are many materials which can be used as insulators (air, PVC, synthetic resins, etc.) which current cannot pass.	There is no magnetic insulator as flux can pass through all the materials, even through the air as well.
Energy must be supplied to the electric circuit to maintain the flow of current.	Energy is required to create the magnetic flux, but is not required to maintain it.
The resistance and conductivity are independent of current density under constant temperature. But may change due to the temperature.	The reluctance, permanence, and permeability are dependent on the flux density.
Electric lines of flux are not closed. They start from a positive charge and end on a negative charge.	Magnetic lines of flux are closed lines. They flow from N pole to S pole externally while S pole to N pole internally.
There is continuous consumption of electrical energy.	Energy is required to create the magnetic flux and not to maintain it.
Kirchhoff current law and voltage law is applicable to the electric circuit.	Kirchhoff MMF law and flux law is applicable to the magnetic flux.
The current density.	The flux density.
EMF is the driving force in the electric circuit. The unit is volts.	MMF is the driving force in the magnetic circuit. The unit is ampere-turns.
There is a current I in the electric circuit which is measured in amperes.	There is flux Φ in the magnetic circuit which is measured in the Weber.
The flow of electrons decides the current in the conductor.	The number of magnetic lines of force decides the flux.

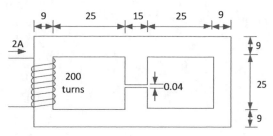

All dimensions in cm.

FIGURE 2.3
The magnetic circuit of Example 2.3.

Example 2.3

A magnetic core with three legs is shown in Figure 2.3. Its depth is 0.04 m, and there are 200 turns on the left most leg. The relative permeability of the core is 1500 with core lengths as shown. For the transformer circuit derive the following:

1. Equivalent circuit diagram with calculated values of magneto motive forces and reluctances. Neglect fringing at air gap
2. Flux Φg
3. Flux density in the left leg
4. Write MATLAB code to verify the answer in a, b, and c above.

Solution

$$\text{Length right} = \text{length left} = (L_r = L_l) = 2(0.075 + 0.25 + 0.045) + 0.09 + 0.25 = 1.08 \text{ m}$$

$$\text{Length of center path} = L_C = 0.34 \text{ m}$$

$$\text{Air gap length } (L_{ag}) = 0.0004 \text{ m}$$

a) $$R = \frac{L}{\mu_o \mu_r A}$$

$$R_{right} = R_{left} = \frac{1.08}{4\pi \times 10^{-7} \times 1500 \times 0.09 \times 0.04} = 159154.94 \text{ } AT/wb$$

$$R_{center} = \frac{0.34}{4\pi \times 10^{-7} \times 1500 \times 0.15 \times 0.04} = 30062.6 \text{ } AT/wb$$

$$R_{air} = \frac{0.0004}{4\pi \times 10^{-7} \times 0.15 \times 0.04} = 53051.65 \text{ } AT/wb$$

$$R_{total} = \frac{(53051.65 + 30062.6) \times 159154.94}{53051.65 + 30062.6 + 159154.94} + 159154.94 = 213755.54 \text{ } AT/wb$$

b) $$\varphi_{total} = \varphi_{left} = \frac{200 \times 2}{213755.54} = 1.8713 \text{ } mWb$$

$$\varphi_{right} = 1.8713 \times 10^{-3} \times \frac{53051.65 + 30062.6}{53051.65 + 30062.6 + 159154.94} = 0.6419 \ mWb$$

c)   *Flux Density$_{left}$* $= \dfrac{1.8713 \times 10^{-3}}{0.09 \times 0.04} = 0.5198 \ Wb/m^2$.

d)   MATLAB Code:

```
clc;
clear all;
N=200; I= 2;
ur= 1500; u0= (4*pi*10^-7);
Lr= 1.08; Ll= 1.08; Lc= 0.34; Lair= 0.0004;
Ar=(0.09 * 0.04)
Ac=(0.15 * 0.04)
Aair=(0.15 * 0.04)
Rright= ((Lr)/(ur*u0*Ar))
Rleft= ((Ll)/(ur*u0*Ar))
Rcenter= ((Lc)/(ur*u0*Ac))
Rair= ((Lair)/(u0*Ac))
Rtotal= ((Rcenter+Rair)*(Rleft)/(Rcenter+Rair+Rleft))+Rright
Fluxtotal= (N*I)/Rtotal
FluxLeft= Fluxtotal
Fluxright= Fluxtotal*((Rcenter+Rair)/(Rcenter+Rair+Rright))
Fluxdensityleft= Fluxtotal/(Ar)
```

MATLAB Output:

Ar = 0.0036

Ac = 0.0060

Aair = 0.0060

Rright = 1.5915e+05

Rleft = 1.5915e+05

Rcenter = 3.0063e+04

Rair = 5.3052e+04

Rtotal = 2.1376e+05

Fluxtotal = 0.0019

FluxLeft = 0.0019

Fluxright = 6.4198e−04

Fluxdensityleft = 0.5198

2.2 Electromagnetic Induction

The English scientist Faraday first developed the rules of the relationship between electricity and magnetism when he discovered the magnetic effect of the electric current, followed by Lenz several years later, which is one of the most important of the progress of human civilization is the basis of the design and work of transformers, motors, generators, and a lot of electrical equipment and appliances. The Faraday simple experiment is shown in Figure 2.4.

When moving a piece of a magnet near the coil, the electric motive force will shrink in this coil to verify the movement of the index galvanometer bound with him and that the placement of a conductor within the magnetic field fixed as in Figure 2.5 and move the conductor to intersect the lines of the field of electrodes, an electric motive force will be induced in this conductor. From these experiments, the following facts can be inferred:

1. The direction of the magnet movement in the first experiment or the conductor in the second experiment has a direct effect on the direction of the displacement of the galvanometer index or on the direction of the electric motive force being produced

2. The magnetism of the coil close to the coil has a direct effect on the direction of displacement of the galvanometer index or the direction of the electric motive force being generated

3. Do not induce electrical motive force and the galvanometer remains stationary when both the coil and magnets are stationary, that is, there must be a relative movement between them to induce the force

4. When the speed of the magnet movement is constant and constant, the amount of motion in the coil will be alternately alternating depending on the speed of movement.

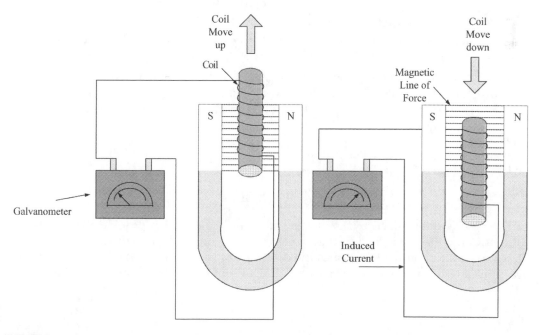

FIGURE 2.4
The Faraday experiment.

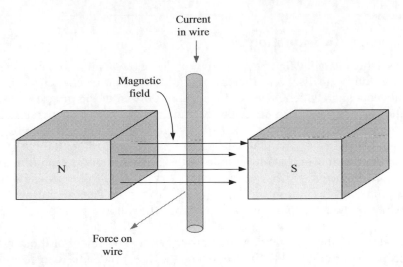

FIGURE 2.5
Place a conductor within a fixed magnetic field.

The amount of electrical motive force is dependent on the following:

1. The number of force lines that intersect the coil turns or the amount of the intensity of the flux (B) of the magnetic field
2. The number of coil turns because the total amount of electrical motive force generated in the coil is equal to the sum of the amount of electrical motive force produced in a single turn in the number of turns
3. The length of the coil (or conductor) that falls within the magnetic field and intersects its lines and is usually equal to the length of the electrodes (L)
4. The speed of movement of the magnet (or conductor) within the sphere (v), (the rate of the intersection of the force lines with the coil)
5. The direction of the motion of the magnet for the coil, or the conductor for the direction of the force lines, which is expressed in angle (θ) where equal zero when the direction of motion is parallel to the direction of force lines, and equal 90° when it is vertical.

In general, the amount of electrical motive force (e) induced by the following expression can be determined from:

$$e = B.L.N. \sin \theta \qquad (2.10)$$

Its direction is determined by using the right-hand rule, which provides the right-hand rest within the magnetic field so that the force lines enter it. If the orthogonal thumb with the rest of the four fingers indicates the direction of the movement of the conductor, the direction of the four fingers indicates the direction of the electric motive force, generally.

The direction in which a conductor is in a magnetic field is contrary to the direction of movement that caused it. This text is defined by the Lenz's law.

When the electric motive is driven in a loop-enclosed conductor, an electric current will pass through it, so that the magnetic effect of this current is counter to the change in the amount of flux.

This is consistent with the Lenz's law, which asserts that the direction of the current must be in such a way as to make the density of the field lines high in the path of the conductor to obstruct this movement. To illustrate this, we take Figure 2.6 where the current of the current in the conductor is within the magnetic field toward the page, and then the force F is generated to block the conductor movement within the field, the effect of the force generated in this case (F) helps in the movement of the conductor, which is not true because it is contrary to the Lenz's law. It is this conclusion that according to the right-hand rule referred to above that the direction of the current must match the direction which caused it to apply the Lenz's law to find the passing current in a spiraling coil due to the approximation of a magnet piece of it as in Figure 2.7a and b, the effect of the current must be obstructing the movement of the magnetic segment. This is done with the magnetic pole of the coil, like the magnet pole nearby. To determine the poles of the coil, hold it with the right hand so that the thumb is parallel to the axis of the coil and an indicator toward the north pole, the direction of the current toward the four fingers.

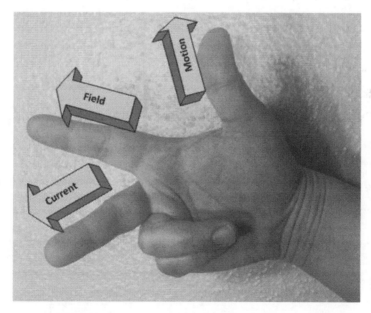

FIGURE 2.6
Direction of the current in the conductor.

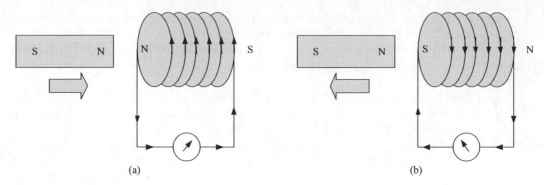

FIGURE 2.7
Direction of current passing in a spiraling coil. (a) Direction from S to N. (b) Direction from N to S.

2.3 Induced Electric Motive Force (EMF)

In summary of what has been mentioned, the conductor in the conductor moves within a magnetic field equal to the law of Faraday interchange rate of the flux with this conductor, or the coil of the N turn, this electric motive force will cause a current in the conductor to reverse the force of the conductor under the Lenz's law, if this force is in support of the conductor movement, we will have an acceleration and an increase in potential energy. This is not possible under the energy conservation law, that is, we can not get energy from anything. This is the rate of change of time over time. This is expressed mathematically as follows:

$$E = -\frac{\Delta \phi}{\Delta T} \tag{2.11}$$

The presence of a negative sign on the right side of the equation is an explanation of Lenz's law, which indicates that the induction of the object is to counteract the change in the magnetic flux. When a coil of N is placed within the magnetic field, Equation 2.11 becomes:

$$E = -N\frac{\Delta \phi}{\Delta T} \tag{2.12}$$

This can be explained by a conductor moving within a magnetic field in another way. A force will affect the positive and negative conductor of the nucleus and the electrons of the conductor atoms so that the direction of the effect of the two charges is reversed. Since the conductor is metallic, the free mobility of the electrons will lead to negative charges increase in the negative sign charges of the conductor and decrease in the other end, this leads to the formation of the voltage difference between the two ends of the conductor, and the voltage difference is the electric motive force.

Example 2.4

A conductor curved single-turn that moves vertically on the lines of an area of a density of a flux 0.04 T that decreases at a constant rate to zero during 20 seconds. Find the value of electrical motive force in the conductor if the cross-section area is 8 cm².

Solution

The field flux:

$$\phi = BA = 0.04 \times 8 \times 10^{-4} = 3.2 \times 10^{-5} \text{ Wb.}$$

The electrical motive force induced in a coil whose number of turns is N=1 is:

$$E = -N\frac{\Delta\phi}{\Delta T} = -\left(3.2 \times 10^{-5}/20\right) = -0.16 \times 10^{-5} \text{ V.}$$

The placing of the coil within a variable magnetic field, or moving the coil within a static magnetic field leads to the generation of electric motive force, and this is the principle of all generators. Also, when placing a metal object within the range of magnetic variable, the electric motive force will urge in the example of what happens in the coil or connector. This electric motive force will be stimulated in the metal body when moving within the static magnetic field. Considering the metal body as a closed circuit, a current will pass through it.

As shown in Figure 2.8, this current is called *the eddy current*, a normal harmful current because it converts part of the electric energy into useless heat energy. To reduce

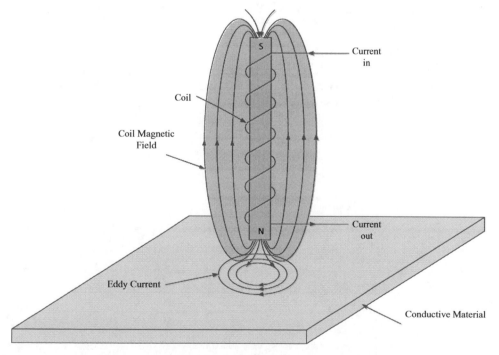

FIGURE 2.8
The eddy current.

the amount of this current increases the resistance of the body by adding silicon in metal casting, as well as to make the body a narrow segment isolated to reduce some of the total currents. It is worth mentioning that the currents are sometimes useful when used for industrial purposes such as modern furnaces.

2.4 Types of Inductance

There are two types of inductance, self and mutual.

2.4.1 Self-Inductance

When an electric current pass through a coil, it will generate a magnetic flux that intertwines with the coil turns and corresponds to the amount of this current. The proportional ratio between the flux and the current is called *inductance* and is symbolized by the letter L. The proportion is linear when the coil is wrapped alone in the air or on a non-magnetic object such as wood and paper and is not linear when the body is ferromagnetic. Inductance is expressed as:

$$L = \frac{N\phi}{I} = \frac{N}{I} \times \frac{NI}{S} = \frac{N^2}{S} \tag{2.13}$$

The inductance unit is Henry (H), and the only one is the inductance of the coil, which evokes an electric impulse of 1 volt when the current passing through the coil changes at a rate of 1 ampere per second.

Example 2.5

An iron ring with a coil on it of 200 turns with a current of 20 A. When the current decreases to 12 A, the flux changes from 1 mWb to 0.8 mWb. Find the amount of inductance of the coil.

Solution

Change the power supply:

$$\Delta I = 20 - 12 = 8\,A.$$

And magnetic flux change:

$$\Delta\phi = 1 \times 10^{-3} - 0.8 \times 10^{-3} = 0.2 \times 10^{-3}\,Wb.$$

The inductance is equal:

$$L = N.\phi/I = N.\frac{\Delta\phi}{\Delta I} = 200\,(0.2 \times 10^{-3}/8)$$

$$= 5 \times 10^{-3}\,H = 5mH$$

In Equation 1.18, the inductance is directly proportional to the number of coil turns and vice versa with reluctance (S = L/NA), the self-inductance is directly proportional to the permeability of the magnetic medium and, conversely, with the length of the coil. The change of the flux with the coil leads to the induction of electric motive force, which is expressed as follows:

$$E = -N.\Delta\phi / \Delta t \qquad (2.14)$$

From the knowledge that the flux of linkage is directly proportional to the current, the self-sustaining (e_L) in the coil is expressed as:

$$e_L = -\frac{\Delta(Li)}{\Delta t} = -L\frac{\Delta i}{\Delta t} \qquad (2.15)$$

This expression also indicates that the inductance is the property of the magnetic circuit in its induction when the current in its coil changes. When the coil is wrapped around a ferromagnetic body, the permeability of the magnetic medium changes with the current change.

Example 2.6

A resistance of the coil is 50 Ω and the number of turns 200 turns, connecting in a row with a resistive galvanometer 300 Ω, and placed in the magnetic field amount of 2 mWb. Find the value of current in the coil when moving during 0.2 sec to 0.6 mWb.

Solution

The change the magnetic flux:

$$\Delta\phi = 2\times10^{-3} - 0.6\times10^{-3} = 1.4\times10^{-3} \text{ Wb.}$$

The time when the flux changes:

$$\Delta t = 0.2 - 0 = 0.2 \text{ sec.}$$

The self-sustaining amount is:

$$E = N\frac{\Delta\phi}{\nabla t} = 200\left(1.4\times10^{-3}\right)/0.2 = 1.4 \text{ V}$$

The amount of current in the coil is equal:

$$I = E/R = 1.4/(300 + 50) = 4 \text{ mA.}$$

2.4.2 Mutual Inductance

When placing two adjacent coils coupled on a metal rod as shown in Figure 2.9, you will be prompted to close (or open) the first key (S) in the first coil (e_1). This coil is self-stimulated and is proportional to the current (e_1) in the first circuit, and the factor of proportionality is self-inductor (L). This is expressed as follows:

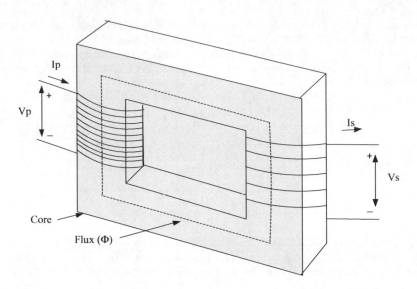

FIGURE 2.9
Mutual induction effect.

$$e_1 = -L1\frac{\Delta i1}{\Delta t} \tag{2.16}$$

The second one, which is generated in the second coil (e_2), is mutually induced by the change of current and therefore the magnetic flux in the first circuit. The expression of this unit is as:

$$E_{21} = -M_{21}\frac{\Delta i1}{\Delta t} \tag{2.17}$$

The self-sustaining factor is directly proportional to the rate of current change over time, and the proportionality factor is called mutual inductance and is denoted as letter M. In Equation 2.17, the mutual inductance (M_{21}) is the proportionality factor between the amounts of the object as reflected in the second coil due to the change in the amount of current in the first coil. The expression of the updated volume is also the rate of change of the first syllable of the coil with the second coil turns and the following:

$$E_{21} = -N_2\frac{\Delta \phi 1}{\Delta t} \tag{2.18}$$

The negative sign in this expression according to Lenz's law indicates that mutual induction seeks to create an opposite effect of the current change in the first coil when the second coil is closed. Thus, a current and therefore variable flux will consist of the intersection of the lines of the two coils. Two forces motives are energized: the first is self-contained in the second coil (e_2), the other is mutually induced in the first coil (e_{12}). In this case, M12 is the first coil inductance relative to the second coil. The mutual inductance unit is also Henry, and this is equal to 1 Henry when the current change in the rate of 1 ampere per second leads to a mutual exchange of 1 volt in the other coil. The amount of mutual inductance depends on the permeability of the magnetic medium and on the shape, dimensions, and position of the exchanged coils.

We conclude that the expression of the coupling factor between the two coils is:

$$K_c = \frac{M}{\sqrt{L_1 . L_2}} \qquad (2.19)$$

In the case where the coupling factor is ideal ($K_c = 1$), the mutual inductance is equal:

$$M = \sqrt{L_1 . L_2} \qquad (2.20)$$

Example 2.7

Two coils that are adjacent to the first current 0.5 A, and change at the rate of 0.01 sec. If the coefficient of the mutual inductance of both coils 0.1 H, calculate the electric motive force that arises in the second coil.

Solution

$$E_2 = M_{12} \frac{\Delta i}{\Delta t}$$

$$E_2 = 0.15 \times (0.5 / 0.01) = 7.5\,V$$

Example 2.8

A coil contains 200 turns with an electric current of 0.5 A, causing a magnetic flux of 1.5 mWb at 0.01 sec, place adjacent to a second coil with 50 turns and changing the current by 0.2 A causing a change in the flux by 2 mWb at a rate of 0.015 seconds. Calculate the mutual inductance and the electric motive force that arise in each of the two coils due to mutual induction if k = 0.9.

Solution

Induction coefficient in the first coil:

$$L_1 = N_1 \frac{\Delta\phi 1}{\Delta I1} = 200 \times \left(1.5 \times 10^{-3}\right) \big/ 0.5 = 0.6\,H$$

Induction coefficient in the second coil:

$$L_2 = N_2 \frac{\Delta\phi 2}{\Delta I2} = 50 \times \left(2 \times 10^{-3}\right) \big/ 0.2 = 0.5\,H$$

Mutual inductance between the two coils:

$$M = K \sqrt{L_1 . L_2} = 0.9 \times \sqrt{0.6 \times 0.5} = 0.9 \times \sqrt{0.3} = 0.493\,H$$

The induced EMF in the first coil:

$$E_1 = M_{12} \frac{\Delta i1}{\Delta t1} = 0.493 \times (0.5/0.01) = 24.65 \, \text{volt}$$

The induced EMF in second coil:

$$E_2 = M_{12} \frac{\Delta i2}{\Delta t2} = 0.493 \times (0.2/0.015) = 6.57 \, \text{volt}$$

2.5 Stored Energy

When placing a coil number (N) on a metal rod, the opening of the switch (S) or the circuit breaker, the magnetic field will be gradually reduced. This means the completion of work goes to convert the magnetic energy into electrical energy, all of which are dissipated to heat in the coil turns, and this work is accomplished if the magnetic field disappears. When the S switch is closed again, the current (I) will pass in the coil (L), and the electric energy will come back again, and the magnetic field is storage. This stored energy is expressed as follows:

$$W = \frac{1}{2} L I^2 \tag{2.21}$$

Example 2.9

A coil has 500 turns, connect to a constant current source. The amount of generated magnetic flux 0.1 mWb, when passing through the current 1 A and 0.001 seconds, calculate the energy stored in the coil.

Solution

$$W = \frac{1}{2} L I^2$$

$$L = \frac{N\phi}{I} \left(500 \times 0.1 \times 10^{-3}\right)/1 = 0.05 \, \text{H}$$

$$W = \frac{1}{2} \times 0.05 \times 1^2 = 0.025 \, \text{Joule}$$

Problems

2.1 What are the areas of comparison between the electrical and magnetic circuits?

2.2 How do you determine the total inductance of two components, (a) in series (b) in parallel, in the cases of mutually reinforcing and antagonistic?

2.3 Two conductors carrying two currents in the same direction 200 A and 300 A, respectively, and distance from some distance 180 cm. Find the amount of force affecting each meter of length for the two conductors.

2.4 A new ring with the relative permeability of 800, and its circumference rate 120 cm. It is made of iron bar diameter 4 cm and a coil with 400 turns. What is the magnetization current required generating flux with an excess of 0.5 Wb?

2.5 A metal ring with a relative permeability of 800 and a diameter of 50 cm. A 20 mm diameter rod is formed, which a piece of length 2 mm is cut, the coil is placed on it the number of turns is 200 turns. How much current is needed to pass the coil to generate a magnetic flux 0.2 mWb in the air gap?

2.6 What is the relative permeability of the magnetic circuit length 20 cm, and the area of its regular section 4 cm², and its coil consist of 200 turns, and when the passing of 2 A in this coil generates a flux of 6.3 mWb.

2.7 The coil has 300 turns and the flux of magnetization passing through it 2 A is placed on magnetic conduction equal to the same permeability 0.2. Calculate the value of the flux passing through the path with an average diameter of 10 cm and diameter of the ring segment 4 cm.

2.8 A medium diameter iron ring 25 cm is made of a metal rod with a diameter of 3 cm, with a coil number of turns 120 turns. Which is the necessary magnetic current to pass in the coil to generate a magnetic flux of 0.25 mWb? If the relative permeability of material 400 and for the air $4\,\pi \times 10^{-7}$.

2.9 A conductor curved single-turn that moves vertically on the lines of an area of a density of a flux 0.05 T that decreases at a constant rate to zero during 25 seconds. Find the amount of electrical motive force in the conductor if the cross-section area is 10 cm².

2.10 An iron ring with a coil on it of 250 turns with a current of 25 A. When the current decreases to 18 A, the flux changes from 1 mWb to 0.8 mWb. Find the amount of inductance of the coil.

2.11 A resistance of the coil is 100 Ω and the number of turns 250 turns, connecting in a series with a resistive galvanometer 300 Ω and placed in the magnetic field amount of 2 mWb. Find the value of current in the coil when moving during 0.5 sec to 0.8 mWb.

2.12 A coil contains 250 turns with an electric current of 1.5 A, causing a magnetic flux of 1.5 mWb at 0.01 sec, place adjacent to a second coil with 50 turns and changing the current by 0.2 A, causing a change in the flux by 2.5 mWb at a rate of 0.005 seconds, calculate the mutual inductance and the electric motive force that arise in each of the two coils due to mutual induction if k = 0.8.

3

Alternating Current Power

Setting goals is the first step in turning the invisible into the visible.

Anthony Robbins

Alternating Current (AC) is used in large areas and in all different life facilities, for ease of generating. AC generators are electrical machines that operate on the principle of electromagnetic induction to generate electric power. AC is defined as the current whose value and direction change continuously over a period on a wave called *the sinusoidal wave* and as shown in Figure 3.1.

The advantages of alternating current are as follows:

1. Changes in value and direction and the polarity are not fixed
2. Generates mechanical methods by cutting magnetic fields such as generators
3. Widely used
4. Low cost of production, especially when generated with large power
5. It can be changed to constant current using electrical components
6. It can be converted from low voltage to high voltage and vice versa using electric transformers
7. It can transfer to long distances using high voltage towers.

3.1 Sinusoidal Wave Cycle and Frequency

The complete cycle of the sinusoidal, or the sine wave, is called *the period*, and the number of oscillations per second is called *the frequency*, and it is denoted by the letter *F*. From Figure 3.2, the portion between the two points (A–B) is called *the full wave*, the part bound between (A–C) is called *half the positive wave*, and (C–B) is called *half the negative wave*. The relationship between the time of the wave *T* and the frequency *F* for one cycle is:

$$F = \frac{1}{T} \tag{3.1}$$

$$T = \frac{1}{F} \tag{3.2}$$

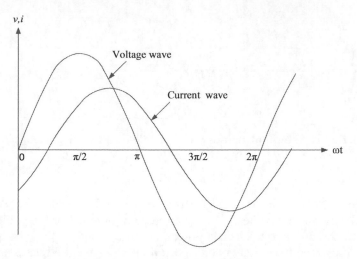

FIGURE 3.1
Voltage and current sine wave.

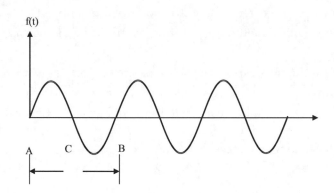

FIGURE 3.2
Sinusoidal waveform.

Example 3.1

If the one cycle is for a sine wave 10 msec, what is the frequency?

Solution

$$F = 1/T = 1/(10 \times 10^{-3}) = 1000/10 = 100 \text{ Hz}.$$

Example 3.2

Find the time of the one cycle if the frequency is 50 Hz.

Solution

$$T = 1/F = 1/50 = 0.02 \text{ sec}.$$

3.2 Electric Power Generation

Figure 3.3 shows the model of a generator that generates alternating current.

The above generator consists of a coil with copper wire in the form of a rectangular frame that connects its ends to isolated copper conductors installed on a rotational axis moving at a constant speed within two magnetic poles north and south.

Because of the conductor rotation and a side intersection with the lines of the magnetic forces in which an electric motive force is produced as in Figure 3.4, the value depends on the magnetic flux density (B), the length of the conductor within the magnetic field (L), and the velocity of the conductor (V). The angle formed by the direction of rotation with the direction of the lines of the magnetic field and the following expression for any moment is:

The instantaneous value of electric motive force.

$$e = B.L.V.sin\,\theta \tag{3.3}$$

As the coil revolves (n) of the cycles per minute, the frequency calculates by the following equation:

$$F = \frac{n}{60} \tag{3.4}$$

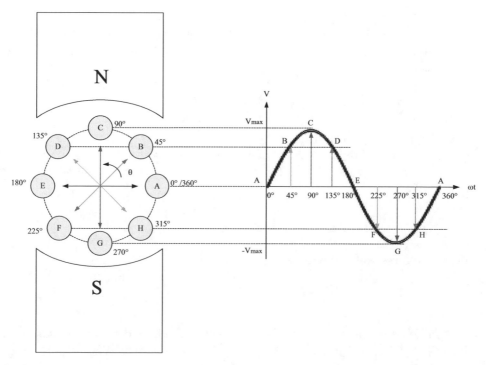

FIGURE 3.3
The model of a generator that produces alternating current.

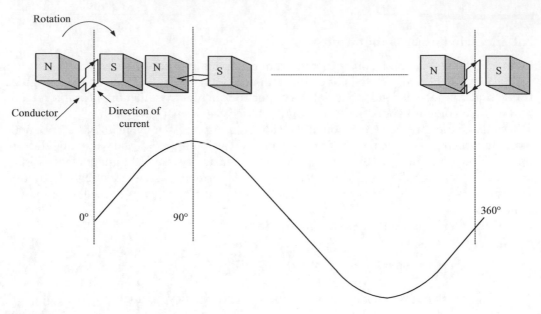

FIGURE 3.4
The process of generating (EMF) in the conductor.

3.3 Terms and Concepts

Each cycle generates one period of the Electric Motive Force (EMF) within two poles of the magnet for any moment, called *the instantaneous value* of the electric motive force as in the following equation:

$$E = 2\pi.\phi.f.N. \sin\theta \qquad (3.5)$$

where N is the number of coil turns.

When the direction of the movement of the coil is vertical on the lines of the magnetic field, i.e., the angle is equal to 90°, i.e., ($\theta = 90°$), so (sin 90° = 1). On this basis, the maximum value of the voltage is:

$$E_{max} = 2\pi\, \phi f N \qquad (3.6)$$

Therefore, the expression of the real value of the voltage is:

$$E = E_{max} \sin\theta \qquad (3.7)$$

The actual value of the electric voltage (V_{eff}) is the value indicated by the measuring instruments is equal to:

$$V_{eff} = (0.707)\, V_{max} \qquad (3.8)$$

Example 3.3

A coil containing 100 turns rotates at a speed of 1500 rpm within the two magnetic poles, calculate: (i) Frequency, (ii) Time, and (iii) Immediate electric motive force if you know that the angle between the movement of the conductor and the lines of the field is 30°, and the maximum value of voltage 3.14 V.

Solution

   i) $f = n/60$

        $= 1500/60 = 25$ Hz.

  ii) $T = 1/f$

        $= 1/25 = 0.04$ sec.

 iii) $e = E_{Max} \sin\theta$

        $= 3.14 \times \sin 30°$

        $= 3.14 \times 0.5 = 1.57\,V$

Example 3.4

A coil containing 50 turns rotates at a regular velocity inside a regular magnetic field. Calculate:

  (a) The speed at which the coil must be rotated, by the frequency of the electric motive force 25 Hz

  (b) A number of poles

  (c) The maximum value of electric motive force if the current value 15 V and the amount of angle 30°

  (d) The actual value of electric motive force.

Solution

  a) $F = n/60$

      $N = 60 \times F = 60 \times 35 = 1500$ rpm.

  b) $p = (120F)/n = (120 \times 25)/1500 = 2$ poles

  c) $e = E_{max} \cdot \sin\theta$

      $E_{max} = \dfrac{e}{\sin\theta} = \dfrac{e}{\sin 30°} = \dfrac{15}{0.5} = 30V.$

  d) $E_{eff} = 0.707\, E_{max}$

      $= 0.707 \times 30 = 12.21$ V.

3.4 AC Current Values

The alternating current changes in value and direction on a sine wave. Some important values of the AC current can be calculated as follows.

3.4.1 The Maximum Value of the Alternating Current

It is the highest value of the current direction, and arises in the case where the direction of the movement of the terminals of the coil is vertical on the lines of the magnetic field, and symbolizes it (I_{max}) and shall be calculated from the following formula:

$$I_{max} = \sqrt{2} \cdot I_{eff}$$

$$I_{max} = 1.414 \cdot I_{eff}$$

(3.9)

3.4.2 Average Value of Alternating Current (Mean Value)

The AC value of the sinusoidal wave can be calculated according to the following equation and symbolized by the symbol (I_{av}):

$$I_{av} = 0.637\, I_{max}$$

(3.10)

3.4.3 Actual AC Value

It is the reading that is indicated by the measuring instruments that are read on these devices, symbolized by the symbol (I_{eff}) from the following equation:

$$I_{eff} = 0.707\, I_{max}$$

(3.11)

3.4.4 The Instantaneous Value of Alternating Current

Which is the value we get at any moment of the current, and symbolized by the letter (i) and calculated from the following equation:

$$I = I_{max} \cdot sin\,\theta$$

(3.12)

Example 3.5

What is the maximum value voltage of AC source has a voltage 220 V?

Solution

$$V_{eff} = 0.707\ V_{max}.$$

$$V_{max} = V_{eff}/0.707.$$

$$V_{max} = 220/0.707 = 311\ V.$$

Example 3.6

An ammeter is connected to a circuit through which an alternating current passes. If the reading recorded by this device is 2 A, find the maximum value of the current.

Solution

$$I_{eff} = 0.707 \, I_{max}.$$

$$I_{max} = I_{eff}/0.707.$$

$$I_{max} = 2/0.707.$$

$$I_{max} = 2.82 \text{ A}.$$

Example 3.7

A wire rotates at a speed of 300 rpm in a uniform magnetic field. If the current of the circuit is at 50 Hz. Calculate the number of poles.

Solution

$$N = \frac{60 \ f}{P}.$$

$$P = \frac{60 \ f}{n} = \frac{60 \times 50}{3000}.$$

$$P = 3000/3000 = 1 \text{ pole pair or 2 poles.}$$

3.5 AC Circuits

3.5.1 AC Circuit Containing Pure Resistance

Figure 3.5a shows a pure resistance connected to an AC source, where the current wave is in phase with the voltage wave as shown in Figure 3.5b, both waves start from zero and reach the greatest value at 90° angle, then fall to zero at 180°, and so the waves are completed simultaneously, as in Figure 3.5b represents of the two waves with directional lines.

The resistance in the circuit shall be calculated as follows:

$$R = \frac{V}{I} \tag{3.13}$$

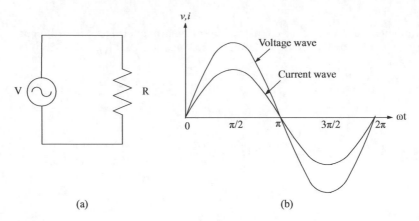

(a)   (b)

FIGURE 3.5
AC circuit containing only pure resistance. (a) Circuit diagram. (b) Voltage and current waveforms.

or of the following formula:

$$R = \frac{\rho L}{A} \tag{3.14}$$

where:
ρ = resistivity and its unit $\Omega.\text{mm}^2/\text{m}$
L = Conductor length in meters
A = cross section area in mm^2.

3.5.2 AC Circuit with Inductive Reactance

Figure 3.6 represents the connection of inductive in the circuit of the alternating current, as the current wave is delayed from the voltage wave at an angle of 90° because of the phenomenon of self-induction. As the back EMF at the moment of zero, the current wave passes until the voltage wave reaches the upper value at an angle of 90°. As the current wave starts to appear because the back EMF at this moment is equal to zero. Figure 3.7 represent the voltage and current waves in the circuit contain inductive load connected to the AC source.

The inductive reactance is represented by the symbol "X_L", and the unit is measured by the Ω (Ω). It is calculated from the following formula:

$$X_L = 2\pi \cdot f \cdot L \tag{3.15}$$

where:
f = frequency
L = Self inductive factor

$$\omega = 2\pi \cdot f \tag{3.16}$$

ω = Angler speed (rad/sec)

$$X_L = \omega \cdot L \tag{3.17}$$

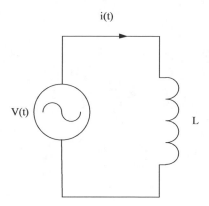

FIGURE 3.6
AC circuit containing inductive reactance.

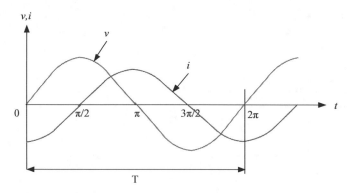

FIGURE 3.7
The voltage and current currents of the inductor.

From the above, scientifically, each coil has a natural resistivity ($R = \rho.L/A$), and magnetic resistance ($X_L = 2\pi.f.L$), and these two resistors are always connected to the series.

3.5.3 AC Circuit with Capacitive Reactance

Figure 3.8 represents a capacitor in the circuit of the alternating current, as the current wave in this case, ahead of the voltage wave at an angle 90°, for the passage of current in the capacitor to charge, and then show the voltage on the ends gradually, as shown in Figure 3.9a. Figure 3.9b represents the voltage and current waves with directional lines.

The capacitive reactance (Xc) and its unit of measurement (Ω), and its value depends on frequency (f) and capacitor (C) and is calculated from the following formula:

$$Xc = \frac{1}{2\pi \cdot F \cdot C}$$

(3.18)

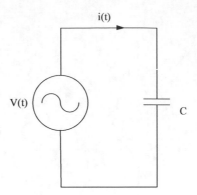

FIGURE 3.8
Capacitor in the circuit of the alternating current.

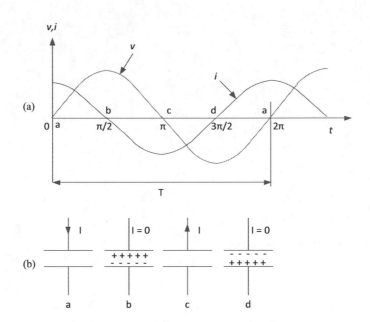

FIGURE 3.9
Voltage and current waveforms and the direction of the charges and current follow. (a) Voltage and current waveforms (b) Direction of the current.

where C = Capacitor capacity and calculated by F (Farad):

$$\omega = 2\pi \cdot f \tag{3.19}$$

$$X_C = \frac{1}{\omega C} \tag{3.20}$$

ω = angular speed in (rad/sec).

From the above, scientifically, where both resistance (R) and capacitive reactance $\left(X_C = \dfrac{1}{2\pi f c} \right)$ are always connect to each other.

3.6 Series Impedance Connection to the AC Circuit

3.6.1 R-L Series Circuit

Figure 3.10a represents a series connection of resistance and inductor with an AC voltage source. From Figure 3.10b, note that the voltage (V_R) on the resistance (R) is in phase with the angle of the current in the circuit, and the voltage across inductor (V_L) ahead of the current at an angle 90°, so the total voltage of the circuit (V) is the sum of both voltages.

From the right-angled triangle in Figure 3.10b, according to Pythagoras' theory, the total voltage is equal to:

$$V = \sqrt{V_R{}^2 + V_L{}^2}$$

Thus, the power factor can be calculated from this triangle:

$$\cos\theta = \frac{VR}{VL} \tag{3.21}$$

where $\cos\theta$ = power factor of the circuit.

The power factor can also be calculated from the impedance triangle Figure 3.11.

$$\cos\theta = \frac{R}{Z} \tag{3.22}$$

From Figure 3.10a, the current is equal in all parameters of the circuit, so the total impedance of the circuit is equal to:

$$Z = \frac{V}{I} \tag{3.23}$$

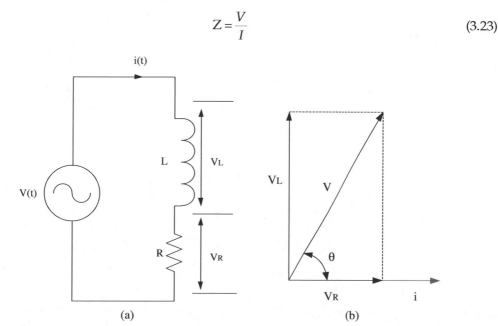

(a) (b)

FIGURE 3.10
Circuit and phasor diagram of R-L series circuit.

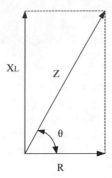

FIGURE 3.11
The impedance triangle.

where:
  Z = total impedance of the circuit and calculated by Ω
  V = the total voltage of the circuit and calculated by volts
  I = total circuit current and calculated in amperes,

and

$$R = V_R/I \tag{3.24}$$

$$X = V_L/I \tag{3.25}$$

Example 3.6

A self-inductance of 10 mH and its resistance 5 Ω connect to AC source of voltage 200 V, and frequency of 60 Hz, calculate the current in the circuit and voltage on both ends of resistance and reactance.

Solution

$$X_L = 2\pi \cdot f \cdot L$$

$$X_L = 2 \times 3.14 \times 60 \times 0.01$$

$$X_L = 3.77 \ \Omega$$

$$Z = \sqrt{R^2 + X_L^2} = \sqrt{5^2 + 3.77^2} = 6.262 \ \Omega$$

$$I = V/Z = 200/6.262 = 31.93 \ A$$

$$V_R = I \times R = 31.93 \times 5 = 159.7 \ V$$

$$V_L = I \times X_L = 33.9 \times 3.77 = 120.41 \ V.$$

3.6.2 R-C Series Circuit

Figure 3.12a represents a series connection of resistance and capacitor with an AC voltage source. From Figure 3.12b, the voltage on the resistance is aligned with the current, with

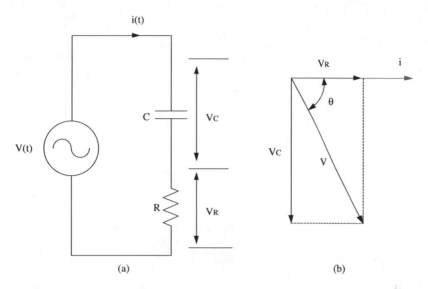

(a)   (b)

FIGURE 3.12
Resistors and capacitor in series. (a) Circuit diagram (b) Voltage and current directions.

the current of the circuit, and the voltage on the capacitive reactance is 90° below the current. Therefore, the total voltage of the circuit is the sum of both voltages.

Figure 3.12b represents the voltage triangle and can be found in the power factor formula:

$$\cos \theta = \frac{VR}{V} \tag{3.26}$$

As for the impedance triangle as in Figure 3.13, the power factor formula can be calculated:

$$\cos \theta = \frac{R}{Z} \tag{3.27}$$

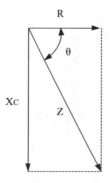

FIGURE 3.13
RC impedance triangle.

Example 3.7

A resistance of 30 Ω and a capacitor of 80 μF are connected in series to AC source of voltage 200 V, 50 Hz. Calculate the total current of the circuit and the voltage on the capacitor.

Solution

$$Xc = \frac{1}{2\pi fC} = \frac{1}{2 \times 3.14 \times 50 \times 80}$$

$$= 40\ \Omega$$

$$Z = \sqrt{R^2 + X_C^2}$$

$$= \sqrt{30^2 + 40^2}$$

$$= 50\ \Omega$$

$$I = \frac{V}{Z} = \frac{200}{50}$$

$$= 4\,A$$

$$V_C = I \times X_C = 4 \times 40$$

$$= 160\,V.$$

3.6.3 R-L-C Series Circuit

Figure 3.14a represents a series connection of resistance, inductor, and capacitor with an AC voltage source. From the phasor diagram shown in Figure 3.14b, we note that the voltage on the resistance is in phase with the current. The phase angle between the voltage on the inductance and the current is 90°, while the phase angle between the voltage on the capacitance and current is 90° as shown in Figure 3.14b. The voltage on the inductive reactance, the voltage on the capacitance reactance on one straightness, and at an angle of 180° is reversed in the direction.

From Figure 3.14b, we note:

$$V_{LC} = V_L - V_C$$

$$V = \sqrt{V_R^2 + V_{LC}^2}$$

$$V = \sqrt{V_R^2 + (V_L - V_C)^2}$$

$$I.Z = \sqrt{I^2.R^2 + (I.X_L - I.X_C)^2} \tag{3.28}$$

$$I.Z = I\sqrt{R^2 + (X_L - X_C)^2}$$

$$Z = \sqrt{R^2 + (X_L - X_C)^2} \quad \text{if } X_L > X_C$$

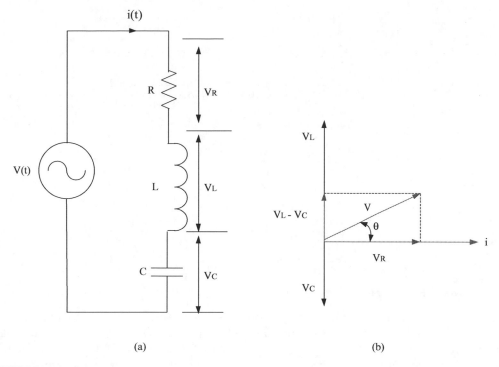

FIGURE 3.14
Circuit diagram and voltage vectors for the (R-C-L) circuit. (a) Circuit diagram (b) phasor diagram.

$$Z = \sqrt{R^2 + (X_C - X_L)^2} \quad \text{if } X_L < X_C \qquad (3.29)$$

$$\cos\theta = \frac{R}{Z} \qquad (3.30)$$

Example 3.8

An R-L-C circuit containing a resistance of 20 Ω, inductance of 0.16 H, and capacitor of 91 μF, has connected an alternating current source of 200 V, and its frequency 50 Hz. Calculate the total impedance and the total current in the circuit.

Solution

$$X_L = 2\pi.f.L$$

$$X_L = 2 \times 3.14 \times 50 \times 0.16 = 50 \, \Omega$$

$$X_C = 1/2\pi f C$$

$$X_C = 1/\left(2 \times 3.14 \times 50 \times 91 \times 10^{-6}\right)$$

$$X_C = 10^6/(91 \times 314) = 35\,\Omega$$

$$Z = \sqrt{R^2 + (X_L - X_C)^2}$$

$$Z = \sqrt{20^2 + (50-35)^2} = \sqrt{20^2 + 15^2} = \sqrt{400^2 + 225^2}$$

$$Z = 25\,\Omega$$

$$I = 200/25 = 8 \text{ A.}$$

3.7 Parallel Connection

3.7.1 Parallel R-L Circuit

Figure 3.15a represents a parallel connection of resistance and inductor with an AC voltage source. In Figure 3.15b, we note that the voltage (V) of the circuit is equally on terminals of both the resistance and inductance. The drawing of the directional lines shows the horizontal line representing the voltage of the circuit. With the voltage, the current passing through the resistance in phase with voltage while the current following in the reactance is late with the voltage by 90°.

From Figure 3.15b and according to the Pythagoras theory:

$$I^2_{T} = I^2_{R} + I^2_{L}$$

$$I_T = \sqrt{I_R{}^2 + I_L{}^2} \tag{3.31}$$

$$\sin\theta = I_L/I_T \tag{3.32}$$

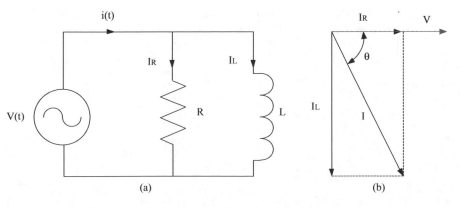

(a)   (b)

FIGURE 3.15
The R-L parallel circuit. (a) Circuit diagram. (b) Phasor diagram.

$$I_L = I_T . \sin\theta \tag{3.33}$$

$$\cos\theta = I_R / I_T \tag{3.34}$$

$$I_R = I_T . \cos\theta, \tag{3.35}$$

where:
  I_R = resistive current
  I_L = inactive current
  I_T = total current.

Example 3.9

A circuit with 50 Ω of resistance, parallel to the self-induction coefficient 89 mH, connected to the source of the alternating voltage 200 V and its frequency 50 Hz. Calculate the total current of the circuit.

Solution

$$I_R = V/R$$

$$I_R = 200/50 = 4 \text{ A}$$

$$X_L = 2\pi f L$$

$$X_L = 2 \times \pi \times 50 \times 0.089$$

$$X_L = 28\ \Omega$$

$$I_L = 200/28 = 7.12 \text{ A}$$

$$I_T = \sqrt{I_R^2 + I_L^2}$$

$$I = \sqrt{4^2 + 7.12^2} = 8.18\,\text{A}.$$

3.7.2 Parallel R-C Circuit

The electric circuit in Figure 3.16 represents R-C circuit connected in parallel to the voltage supply. The voltage of the circuit (V) is the same at terminals of both resistance and capacitance. The current in the resistance is in phase with the voltage and current in the capacitance ahead of the voltage by an angle 90°, and the sum of both currents is the total current (I) as in Figure 3.16b.

  From Figure 3.16b.

$$I^2_T = I^2_R + I^2_C$$

$$I_T = \sqrt{I_R^2 + I_C^2}, \tag{3.36}$$

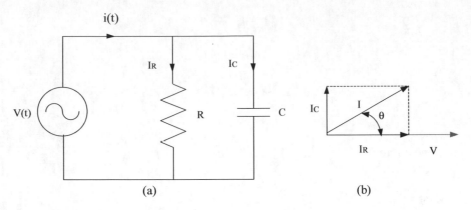

FIGURE 3.16
The voltage and current vectors of the parallel circuit R-C. (a) Circuit diagram (b) phasor diagram.

where:
  I_T = the total current
  I_R = the current in the resistance
  I_C = the current in the capacitance.

Example 3.10

A parallel R-C circuit containing 25 Ω of resistance, and a capacitor of 95.5 μF, connected to a voltage source of 200 V and a frequency of 50. Calculate the total current of the circuit.

Solution

$$X_C = \frac{1}{2\pi.F.C}$$

$$= 1/\left(2 \times 3.14 \times 50 \times 95.5 \times 10^{-6}\right)$$

$$= 10^6/29987$$

$$= 33.34\ \Omega$$

$$I_C = V/X_C$$

$$= 200/33.34 = 6\ \text{A}$$

$$I_R = V/R$$

$$= 200/25 = 8\ \text{A}$$

$$I_T = \sqrt{I_R{}^2 + I_C{}^2}$$

$$= \sqrt{8^2 + 6^2}$$

$$= \sqrt{64 + 36}$$

$$= 10\ \text{A}$$

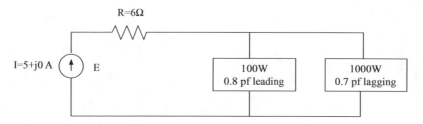

FIGURE 3.17
Circuit of Example 3.11.

Example 3.11

The following Figure 3.17, power system has two loads attached to a 6 Ω line with a supply current of values of $5\angle0°A$, if the loads data are given as:

Load 1: P = 100 W at 0.8 pf leading
Load 2: P = 1000 W at 0.7 pf lagging. Determine:
  a)   The value of the total watts, VAR, and VA for the circuit
  b)   Value of the supply voltage E
  c)   The power factor for the circuit
  d)   Type of element and their impedance in each box
  e)   Write a MATLAB program to verify the answers.

Solution

Given Load 1: $P_1 = 100$ W, $\cos\varphi_1 = 0.8$ pf leading.
  Reactive power is:

$$Q_1 = P_1 \tan \varphi_1 = 100 \times \tan (\cos^{-1}0.8) = 75 \text{ VAR (Volt Amper Reactive)}.$$

The total complex power for Load 1 is:

$$S_1 = P_1 - jQ_1 = 100 - j75 \text{ VA (Volt Amper)}.$$

For Load 2: $P_2 = 1000$ W, $\cos\varphi_2 = 0.7$ pf lagging.
  Reactive power is:

$$Q_2 = P_2 \tan \varphi_2 = 100 \times \tan (\cos^{-1}0.7) = 1020.2 \text{ VAR}.$$

The total complex power for Load 2 is:

$$S_2 = P_2 - jQ_2 = 1000 + j1020.2 \text{ VA}.$$

The P and Q of 6 ohm line is:

$$P_r = I^2 \times R = 5^2 \times 6 = 150 \text{ W}.$$

The reactive power of resistor is zero i.e., $Q_r = 0$ VAR.

  a) The total watts are:

$$P = P_1 + P_2 + P_3 = 150 + 100 + 1000 = 1250 \text{ W}.$$

The total VAR is:

$$Q = Q_2 - Q_1 = 1020.2 - 75 = 945.2 \text{ VAR}.$$

The total VA is:

$$S = P + jQ = 1250 + j945.2 \text{ VA} = 1567.13 \angle 37.1° \, VA$$

b) The supply voltage E:

$$E = \frac{S}{I^*} = \frac{1567.13 \angle 37.1°}{5 \angle 0°} = 313.43 \angle 37.1 \text{ V}.$$

c) The power factor for the circuit:

$$\text{pf} = \cos\left(\tan^{-1}\frac{945.2}{1250}\right) = 0.798 \text{ lagging}.$$

d) The voltage across loads is:

$$V = E - I \times R = 313.43 \angle 37.1° - 5 \angle 0° \times 6 = 290.06 \angle 40.67° V.$$

The current through Load 1 is:

$$I_1 = \left(\frac{S_1}{V}\right)^* = \left(\frac{100 - j75}{290.06 \angle 40.67°}\right)^* = 0.431 \angle 77.54 \text{ A}.$$

The impedance of Load 1 is:

$$Z_1 = \frac{V}{I_1} = \frac{290.06 \angle 40.67°}{0.431 \angle 77.54°} = 538.4 - j403.8 \; \Omega$$

Resistor and capacitor as load as pf is leading:
Current through Load 2 is:

$$I_2 = I - I_1 = 5 \angle 0° - 0.431 \angle 77.54° = 4.925 \angle - 4.9° \, A.$$

The impedance of Load 2 is:

$$Z_2 = \frac{V}{I_2} = \frac{290.06 \angle 40.67°}{4.925 \angle - 4.9°} = 41.23 - j42.06 \, \Omega$$

Resistor and inductor as load as pf is lagging.

e) MATLAB CODE
```
P1 = 100;
pf1 = 0.8;
Q1=P1*tan(acos(pf1));
P2 = 1000;
pf2 = 0.7;
Q2=P2*tan(acos(pf2));
I=5;
```

```
Pl=I^2 * 6;
P=Pl+P1+P2
Q=Q2-Q1
S=P+Q*i;
VA=abs(S)
E=(S/conj(I));
magE=abs(E)
angleE=angle(E)*180/pi
powerFactor=cos(atan(Q/P))
V=E-(I*6);
I1=conj((P1-Q1*i)/V);
Z_1=V/I1
I2=I-I1;
Z_2=V/I2
P1 = 100;
pf1 = 0.8;
Q1=P1*tan(acos(pf1));
P2 = 1000;
pf2 = 0.7;
Q2=P2*tan(acos(pf2));
I=5;
Pl=I^2 * 6;
P=Pl+P1+P2
Q=Q2-Q1
S=P+Q*i;
VA=abs(S)
E=(S/conj(I));
magE=abs(E)
angleE=angle(E)*180/pi
powerFactor=cos(atan(Q/P))
V=E-(I*6);
I1=conj((P1-Q1*i)/V);
Z_1=V/I1
I2=I-I1;
Z_2=V/I2
```

Ans:

$P = 1250$

$Q = 945.2041$

$VA = 1.5671e + 03$

$magE = 313.4269$

$angleE = 37.0952$

$powerFactor = 0.7976$

$Z\_1 = 5.3847e + 02 - 4.0385e + 02i$

$Z\_2 = 41.2269 + 42.0598i.$

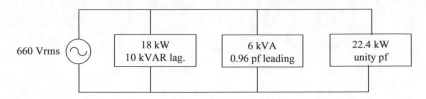

FIGURE 3.18
Circuit of Example 3.12.

Example 3.12

Three loads are connected in parallel across 660 V(rms), 60 Hz line as shown in Figure 3.18.

    Load 1: absorbs 18 kW and 10 kVAR with lagging power factor
    Load 2: absorbs 6 kVA at 0.96 leading power factor
    Load 3: absorbs 22.4 kW at unity power factor. Determine:
        a)   The value of the total watts, VAR, and VA for the circuit
        b)   The power factor for the circuit.

Solution

Given data:

        Load 1: absorbs 18 kW and 10 kVAR with lagging power factor
        Load 2: absorbs 6 kVA at 0.96 leading power factor
        Load 3: absorbs 22.4 kW at unity power factor

For Load 2, the reactive power is:

$$Q_2 = S_2 \sin \varphi_2 = 6 \times \sin (\cos^{-1} 0.96) = 5.76 \text{ kVAR}.$$

The total complex power for Load 2 is:

$$S_2 = P_2 + jQ_2 = 6 + j5.76 \text{ kVA}.$$

For Load 3: $P_3 = 22.4$ kW, $\cos\varphi_3 = 1$.
Reactive power is $Q_3 = 0$ VAR (unity power factor).
   The total complex power for Load 3 is:

$$S_3 = P_3 - jQ_3 = 22.4 + j0 \text{ kVA}.$$

  a) The total watts are:

$$P = P_1 + P_2 + P_3 = 18 + 5.76 + 22.4 = 46.16 \text{ kW}.$$

    The total VAR is:

$$Q = Q_1 + Q_2 + Q_3 = 10 + (-1.68) = 8.32 \text{ kVAR}.$$

    The total VA is:

$$S = P + jQ = 46.16 + j8.32 \text{ kVA} = 46.9 \angle 10.21° \text{ kVA}.$$

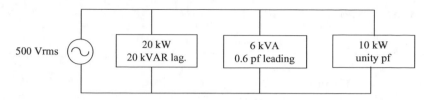

FIGURE 3.19
Circuit of Example 3.13.

b) The power factor for the circuit:

$$\text{p.f} = \cos\left(\frac{46.16}{46.9}\right) = 0.798 \, \text{lagging.}$$

Example 3.13

Three loads are connected in parallel across 500 V(rms), 60 Hz line as shown in Figure 3.19.

Load 1: absorbs 20 kW and 10 kVAR with lagging power factor
Load 2: absorbs 6 kVA at 0.6 leading power factor
Load 3: absorbs 10 kW at unity power factor. Determine:
  a)   The value of the total watts, VAR, and VA for the circuit
  b)   The power factor for the circuit
  c)   The value of capacitance C connected in parallel across the loads that will raise the power factor to unity.

Solution

Given data:

Load 1: absorbs 20 kW and 20 kVAR with lagging power factor
Load 2: absorbs 6 kVA at 0.6 leading power factor
Load 3: absorbs 10 kW at unity power factor.
For Load 2, the reactive power is:

$$Q_2 = S_2 \sin \varphi_2 = 6 \times \sin (\cos^{-1}0.6) = 4.8 \, \text{kVAR.}$$

The total complex power for Load 2 is:

$$S_2 = P_2 + jQ_2 = 6 + j4.8 \, \text{kVA.}$$

For Load 3: $P_3 = 10$ kW, $\cos\varphi_3 = 1$.
Reactive power is $Q_3 = 0$ VAR (unity power factor).
The total complex power for Load 3 is:

$$S_3 = P_3 - jQ_3 = 10 + j0 \, \text{kVA.}$$

a) The total watts are:

$$P = P_1 + P_2 + P_3 = 20 + 6 + 10 = 36 \, \text{kW.}$$

The total VAR is:

$$Q = Q_1 + Q_2 + Q_3 = 20 + (-4.8) = 15.2 \, \text{kVAR.}$$

The total VA is:

$$S = P + jQ = 36 + j15.2\,kVA = 39\angle 22.89°\,kVA.$$

b) The power factor for the circuit:
$$p.f = \cos(22.89) = 0.921 \text{ lagging.}$$

c) With required unity power factor $S = P = 36$ kVA, and $Q = 0$, so $Q_c = 15.2$ kVAR.

$$X_c = \frac{V}{Q_c} = \frac{500}{15.2 \times 10^3} = 0.033\,\Omega$$

$$C = \frac{1}{2\pi f X_c}$$

$$C = \frac{1}{2\pi \times 60 \times 0.03} = 88.42\ mF.$$

Problems

3.1   What are the advantages of alternating current?

3.2   What are the phase shift angles between voltage and current for series R-L, R-C, and R-L-C circuits?

3.3   Explain your answer to the diagram about the current and voltage relationship in the AC circuit containing:
a) pure resistance, b) inductance, and c) capacitance.

3.4   Define the following: alternating current, instantaneous current, wave time, and frequency.

3.5   Give the reason: the current wave is delayed by the voltage wave in the AC circuit containing only inductance.

3.6   Give the reason: the voltage wave is delayed by the current wave in the AC circuit containing only capacitance.

3.7   An electrical generator, four poles, and the speed of 1800 rpm. Calculate the value of frequency.

3.8   A copper wire length of 400 m, and a cross-section area 2 mm². Calculate the resistance if its resistivity of $\rho = 1.68 \times 10^{-8}\ \Omega.m$.

3.9   Calculate the speed of the coil revolving within the magnetic field arising from the 12 poles, and the frequency is 60 Hz.

3.10   Calculate the value of the reactance of the self-inductor of 0.1 H if it is to be connected to a circuit of frequency 60 Hz.

3.11   Calculate the capacitive reactance of a capacitor 31.8 µF if it is known to be connected to a 50 Hz voltage source.

3.12 Calculate the value of the reactance of a coil, connected with a resistance of 8 Ω, the circuit fed by a voltage source of 110 V, and the current passing through the circuit of 11 A.

3.13 A series circuit of 20 Ω resistance, and a capacitor of 100 µF. This circuit is connected to an AC source of 200 V at 50 Hz. Calculate:

    a. The total current of the circuit

    b. The voltage on both terminals of the resistance

    c. The voltage on both terminals of the capacitor

    d. The total impedance of the circuit

    e. Power factor.

3.14 Find the maximum value of a current 10 A.

3.15 Find power factor for an electric circuit with a resistance of 6 Ω, and connected in series, with a reactance of 8 Ω.

3.16 A reactance of 10 Ω, and this self-inductance of 150 mH calculated this frequency.

3.17 A reactance of 62.25 Ω, calculate its self-inductive factor if this coil connected with an AC voltage source of 240 V and the frequency of 25 Hz. The total resistance of the circuit is equal to 80 Ω.

3.18 A series circuit contains resistance of 10 Ω, and a self-inductance coefficient of 0.2 H, connected with an alternating voltage source of 64 V and frequency of 50 Hz., calculate:

    a. The total current of the circuit

    b. The current that passes through both resistance and the coil

    c. The voltage on the terminals of the coil

    d. The power factor of the circuit.

4

Transformers

Successful people are always looking for opportunities to help others. Unsuccessful people are always asking, "What's in it for me?"

Brian Tracey

An electric transformer is a device which converts electrical energy of a specific voltage into another voltage through a pair of electrical windings. It can also be defined as a static electromagnetic device which converts AC power from certain components to other components.

Transformers are efficient electrical machines that can convert 99.75% of the input power of the transformer to the output of the transformer. The manufacturing of electric transformers is done in different sizes ranging from thumb-sized transformers (where the microphone can consist of a pair of these transformers) to very large sizes transformers weighing hundreds of tons and used in power grids. The operation of all types of transformers is based on the same basic principles despite difference sizes. Different types and sizes of electrical transformers are given in Figure 4.1. In this chapter, the focus is on single-phase electrical transformers.

4.1 Installation of the Transformer

The transformer consists of two windings wires wound around an iron-core. The terminal connected to a source called *the primary*, while the associated end of the load is called *the secondary winding*, as shown in Figure 4.2a, i.e., the transformer can be considered to be composed of two circuits: a magnetic circuit consisting of the core, and the other is an electric circuit consisting of primary and secondary windings. The components of the transformer are placed inside a container to maintain the components of dust, moisture, and mechanical shocks.

The transformer is used to change the value of the voltage in the AC power transmission system where the transformer cannot operate in DC systems. If the secondary voltage is less than the primary voltage, the transformer is step-down voltage, and if the secondary voltage is higher than the primary voltage, the voltage step-up is shown in Figure 4.2.

The transformer windings must contain the following:

- High mechanical strength, enough to protect it from distortions which may result from short or excess currents

- Sufficient thermal strength so that the high temperature does not lead to thermal breakdown of the insulation material

FIGURE 4.1
Different sizes and shapes of electrical transformers. (https://www.udvavisk.com/cfd-analysis-transformer-room/; http://www.procontransformers.com/; http://www.mpja.com/24V-10A-Center-Tapped-12-0-12-Transformer/productinfo./)

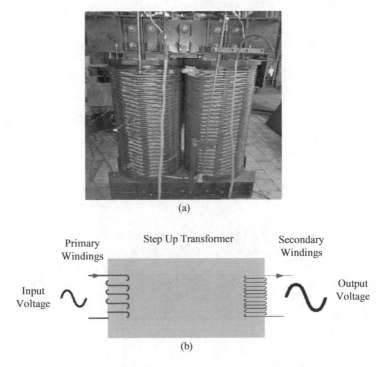

FIGURE 4.2
Two windings step-up transformer. (a) Two windings transformer. (b) Step-up voltage transformer.

- Enough electrical strength so that the insulation materials and insulation distances are sufficient to prevent electric breakdown or electric arc.

The transformer windings vary according to their currents and nominal voltages and are made of copper or aluminum wires with circular or rectangular sections.

4.2 Core Shape

The core's installation of the transformer depends on the following factors: voltages, current, and frequency. The size and cost of installation of the core are also considered, and the air-core is usually used for soft-core or steel-core. The selection of a particular type of core depends on the area of the transformer's use. Generally, air-cored transformers are used when the source voltage is high frequency, and the iron-core switches are usually used when the source voltage is low frequency, transmitting a better power than the air-core transformer. Soft iron-core transformers are very useful when the size of the transformer is small. The steel-cored transformer loses heat easily and thus is used to efficiently convert power. There are two types of slides that make up the steel-core of the transformers: the first is called *the core-shaped* transformer, as shown in Figure 4.3. The core is made of several steel strips, as shown (Figure 4.4):

The second shape of the core is the shell-core frame, which is the most common and efficient, as shown in Figure 4.5. Each layer or slice of the core consists of a part of letter E and the other a letter I, when combined produces a single slide shape, isolates these segments by isolation, and presses to be the core.

In this type of transformer, the windings are placed around the center column of the iron-core, as shown in Figure 4.6.

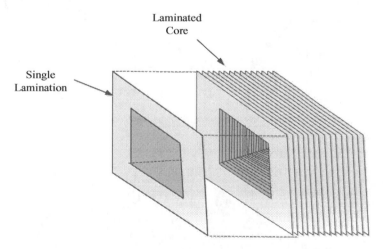

FIGURE 4.3
The core-shaped transformer.

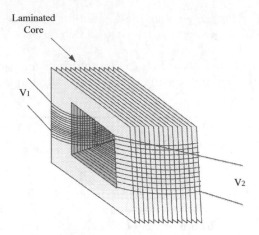

FIGURE 4.4
The location of converted coils around the core.

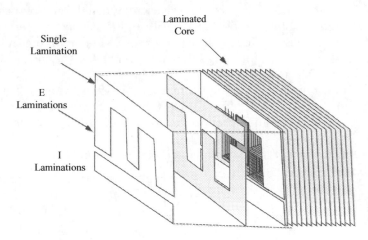

FIGURE 4.5
The shape of the frame-frame slides.

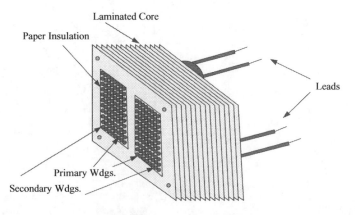

FIGURE 4.6
The location of converted coils with the frame.

4.3 Principle of Operation

The principle work of the electric transformer is based on Faraday's law of electromagnetic induction, which states that the value of the electric force (electric voltage) is directly proportional to the rate of change of the magnetic flux. For this reason, the transformer does not work in DC systems because the DC creates a static magnetic field. The amount of its changed is zero, so it is not possible to create an electrical voltage in a way of induction, the main reasons for why the preference of the current AC at the time when there is not even a practical and economical method to adjust the value of the voltage.

When an electric current I_1 is applied in the primary windings, a magnetic flux ϕ_{12} can be determined by the right-hand rule. When the right-hand fingers point to the direction of the coils, the thumb indicates the direction of the magnetic flux and, as there isn't contact between the primary and secondary terminals, the magnetic flux is applied in a magnetic circuit between the two ends. When the flux of the secondary party coils reaches the flux of the current in these coils I_2, the direction can be determined in the above manner. But this time by making the direction of the thumb first correspond to the direction of magnetic flux ϕ_{21}, pointing to the direction of current flux in the coils, as shown in Figure 4.7.

4.4 Ideal Transformer

The ideal transformer is a theoretical assumption only and is used to understand the real transformer. The ideal transformer assumes that there is no loss of energy where the energy moves from the primary coil circuit to the secondary coil circuit. The ideal transformer also assumes that the coils have no resistance to the flow of the current and there is no leakage in the magnetic flux. These hypotheses help to infer the different relationships. The ideal transformer is composed of two coils that have only an inductive impedance and are wound around an iron-core, as shown in Figure 4.8. If the primary coil is connected by a variable voltage source, it produces an alternating magnetic flux and frequency as well as the number of primary coil turns. This alternating flux is intertwined with the secondary coil, generating an alternating voltage dependent on the number of secondary coil turns.

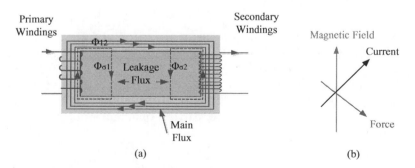

(a) (b)

FIGURE 4.7
The principle of the work of the electric transformer. (a) Direction of current (b) Direction of current, force and magnetic field.

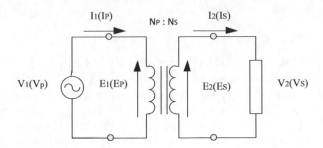

FIGURE 4.8
The composition of the ideal transformer and its components.

If we assume that the initial voltage is V_1, and the resulting magnetic flux is an inverse electric force (E_1) is generated in the primary coil given by:

$$E_1 = V_1 = N_1 \times \frac{d\Phi}{dt} \tag{4.1}$$

where:
  E_1 = the induce electric motive force in the primary windings (volt)
  N_1 = the number of primary windings turns
  Φ = the amount of magnetic flux (Wb)
  t = time (second).

The following of a variable electric current in the secondary windings causes the generation of an induced voltage according to Faraday's law equal to:

$$E_2 = V_2 = N_2 \times \frac{d\Phi}{dt} \tag{4.2}$$

where:
  E_2 = the induce electric motive force in the secondary windings (volt)
  N_2 = the number of secondary windings turns.

By dividing Equation 4.2 into Equation 4.1, it results in:

$$E_2/E_1 = V_2/V_1 = N_2/N_1. \tag{4.3}$$

The ideal transformer shown in Figure 4.8, in the case of connecting its secondary coil to the load, will follow a current through which the electric power will be transferred from the primary circuit to the secondary circuit. In the ideal case (by neglecting the losses), all incoming energy will be transferred from the primary circuit to the magnetic field and to the secondary circuit, so that the input power can be equal to the output power as given in the following equation:

$$P_{in} = I_1 . V_1 = P_{out} = I_2 . V_2. \tag{4.4}$$

That is, the ideal transformer equation is equal to:

$$V_2/V_1 = N_2/N_1 = I_1/I_2. \tag{4.5}$$

It is possible to calculate voltages on both ends of the primary windings in terms of voltages on both ends of the secondary windings by the following equation:

$$V_1 = (N_1/N_2) * V_2. \tag{4.6}$$

It is also possible to calculate the current following through the primary windings in terms of current following through the secondary windings by the following equation:

$$I_1 = (N_2/N_1) * I_2, \tag{4.7}$$

where:

V_P = the primary windings voltages, V_S: secondary windings voltages
N_P = the number of primary windings turns, N_S: the number of secondary windings turns
I_P = the primary windings current, and
I_S = the secondary windings current.

So, the ideal transformer equation is as follows:

$$V_S/V_P = N_S/N_P = I_P/I_S.$$

If the transformer is ($V_S > V_P$), then the current will be lower ($I_S < I_P$) by the same ratio, this ratio is called *transformation ratio* and symbolized by the symbol (α). The induced electrical motive force of the primary windings is calculated by the following equation:

$$E_1 = 4.44 \, \phi f N_1. \tag{4.8}$$

And the induced electrical motive force generated in the secondary windings is calculated by:

$$E_2 = 4.44 \, \phi f N_2. \tag{4.9}$$

The following assumptions are made in the analysis of an ideal transformer:

1. The transformer windings are perfect conductors, meaning that there is zero winding resistance.
2. The core permeability is infinite, meaning that the reluctance of the core is zero.
3. All magnetic flux is confined to the transformer core, meaning that no leakage flux has occurred.
4. The core losses are hypothetically assumed to be zero.

4.5 Transformer Rating

Transformers carry ratings related to the primary and secondary windings. The ratings refer to the power in kVA and primary/secondary voltages. A rating of 10 kVA, 1100/110 V means that the primary is rated for 1100 V, while the secondary is rated for 110 V ($\alpha = 10$).

The kVA rating gives the power information, with a kVA rating of 10 kVA and a voltage rating of 1100 V, the rated current for the primary is $10{,}000/1100 = 9.09$ A, while the secondary rated current is $10{,}000/110 = 90.9$ A.

Example 4.1

A transformer gives 500 A at 24 V when fed with a source voltage of 120 V. How many turns are required in the secondary side? When the number of turns of the primary side is 3000 turns, and how much is the primary current?

Solution

$$N_2/N_1 = V_2/V_1$$

$$N_2 = V_2/V_1 \times N_1$$

$$N_2 = (24/120) * 3000$$

$$N_2 = 600 \text{ Turns}$$

$$N_2/N_1 = I_1/I_2$$

$$I_1 = N_2/N_1 * I_2$$

$$I_1 = (600/3000) * 500 = 100 \text{ A.}$$

Example 4.2

In a step-down voltage transformer, if the number of high-voltage side is equal to 500 turns, while the number of turns of the low voltages 100 turns, the load current is 12 A, calculate:

  i. The transformation ratio (α)
  ii. The primary windings current.

Solution

  i. The transformation ratio $\alpha = N_1/N_2$

$$\alpha = 500/100$$

$$\alpha = 5$$

  ii. $N_2/N_1 = I_1/I_2$
  The primary windings current $I_1 = 100/500 * 12$

$$I_1 = 2.4 \text{ A.}$$

Example 4.3

A single-phase transformer operates at a frequency of 50 Hz. If the iron-core is a square shape, the length of its side 20 cm and the maximum magnetic flux density allowed follow in the iron-core 10000 line/cm², calculate the number of coils to be placed for both the primary and secondary windings to be a voltage transformation ratio 220/3000 V.

Solution

$$\phi = A \times B = 20 \times 20 \times 10^{-8} \times 10000 = 0.04 \text{ Wb}$$

$$E_1 = 4.44 \, \phi \, f \, N_1$$

$$N_1 = E_1/4.44 \, \phi \, f \, N_1 = 3000/4.44 \times 0.04 \times 50 = 338 \text{ turns}$$

$$E_2 = 4.44 \, \phi. \, f. \, N_2$$

$$N_2 = E_2/4.44 \, \phi \, f \, N_2 = 220/4.44 \times 0.04 \times 50 = 25 \text{ turns.}$$

4.6 Transformer Operation

A transformer is said to be on "no-load" when its secondary side winding is open circuited, in other words, nothing is attached and the transformer loading is zero. When an electrical load is connected to the secondary winding of a transformer and the transformer loading is, therefore, greater than zero, a current flows in the secondary winding and out to the load.

4.6.1 The Transformer Operation at No-Load

It is known that the theory of transformer operation depends on electromagnetic induction. When the transformer is connected to an alternating current source, a current follows through the primary winding called *no-load current* I_o. The flux of this current results in a variable magnetic flux followed by the current. This flux cut the primary and the secondary windings and generate in each of them an opposite electric motive force proportional to the number of turns and the rate of change in the flux to time as mentioned above.

The flow of the no-load current is the result of two currents:

1. The vertical component is called *the effective current*, which is the current responsible for the iron losses in the iron-core of the transformer (the heat produced in the iron-core when the transformer is working) and is symbolized by the symbol (I_e).
2. The horizontal component is called *the magnetism current*, which is the current responsible for the magnetic field that occurs in the iron-core and is symbolized by the symbol (I_m), as shown in Figure 4.9.

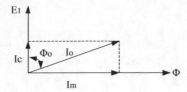

FIGURE 4.9
The flowing current and its two components.

The power dissipated to be carried out is the iron ΔP_{Fe}, which is calculated from the relationship:

$$\Delta P_{Fe} = V\, I_o\, \cos \phi_o \tag{4.10}$$

$$\Delta P_{Fe} = V\, i_e \tag{4.11}$$

$$I_e = \Delta P_{Fe}/V\,. \tag{4.12}$$

In applying Pythagoras' theorem to the right-angled triangle, we find that:

$$I_o{}^2 = I_e{}^2 + I_m{}^2$$

$$I_m = \sqrt{I_o{}^2 - I_e{}^2} \tag{4.13}$$

Example 4.4

A 250 V transformer takes a current of 0.5 mA with a power factor at no-load of 0.3 lagging. Calculate the magnetic current.

Solution

$$I_e = I_o \cos \theta_o$$

$$I_e = 0.5 \times 0.3$$

$$I_e = 0.15 \text{ A}$$

$$I_m = \sqrt{0.5^2 - 0.15^2}$$

$$I_m = 0.477 \text{ A}.$$

4.6.2 The Operation Transformer at Load

If the load is connected at the terminals of the secondary windings, a current followed called *the secondary windings current* I_2, which causes a magnetic flux and an induced electric motive force $-E_2$ to counteract the electric motive generated in the primary windings

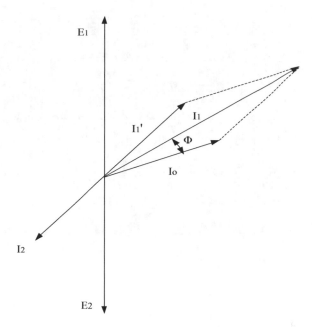

FIGURE 4.10
The current of the load current.

E_1. Thus, the current in the primary windings I_1' is equal to the secondary current I_2, which is equivalent to the amount and reverses it in the direction, as well as the current of the no-load I_o, so the current following through the primary windings with the load is large because it is the sum of two currents, I_0 and the load current of I_2 or I_1', as shown in Figure 4.10.

Transformers can be divided according to use into:

1. Power Transformers: Convert the voltage from one level to another and are used for power transmission and in various manufacturing areas, as shown in Figure 4.11 and in the household uses

2. Regulation Transformers: Used to obtain different voltage values in laboratories, research centers, and automated control

3. Transformers to change the number of phase currents (m): The alternating frequency (f) and pulse shape are mainly used in electronic devices, wired communications, and automatic control that does not exceed the capacity such as these transformers have several VA

4. Measurement Transformers: Such as serial current transformer and serial voltage transformer are used in electrical measurements and distribution boards and processes like feeding.

The transformers are divided in terms of the number of phases to:

1. Single-phase Transformers
2. Three-phase Transformers
3. Polyphase Transformers.

FIGURE 4.11
Power transformers.

Also divided in terms of conversion rates, which are:

1. Step-down Transformers: The high primary windings voltage V_1 is converted to a low voltage V_2 ($V_1 > V_2$)
2. Step-up Transformers: The primary windings voltage V_1 is converted to high voltage secondary V_2 ($V_1 < V_2$).

Also divided in terms of its cooling method into ways like following:

1. Dry Transformers: Cooled by natural or forced air and are usual transformers with small and medium capacities.
2. Transformers immersed in oil: Are cooled with oil as power transformers with medium and large capacity used in different power stations, and these transformers are characterized by the dangers of explosion and therefore provide advanced control circuits.
3. Transformers refrigeration with SF_6: Recently successfully tested and are now being used in indoor environments.

Figure 4.12 shows the types of electrical transformers.

4.7 Non-ideal Transformer Equivalent Circuits

The non-ideal transformer equivalent circuit in Figure 4.13 expresses the various ways in which all the loss terms that are neglected in the ideal transformer model. The individual loss terms in the equivalent circuit are:

R_1, R_2—primary and secondary winding resistances (losses in the windings due to the resistance of the wires)

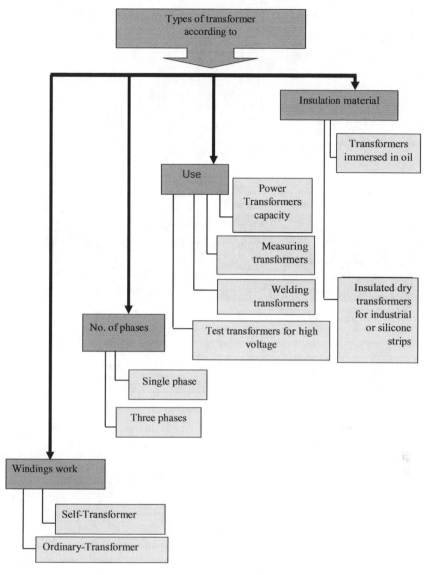

FIGURE 4.12
Types of electrical transformers.

X_1, X_2_primary and secondary leakage reactance's (losses due to flux leakage out of the transformer core)

R_c_core resistance (core losses due to hysteresis loss and eddy current loss)

X_m_magnetizing reactance (magnetizing current necessary to establish a magnetic flux in the transformer core).

Using the impedance reflection technique, all the quantities on the secondary side of the transformer can be reflected on the primary side of the circuit. The resulting equivalent

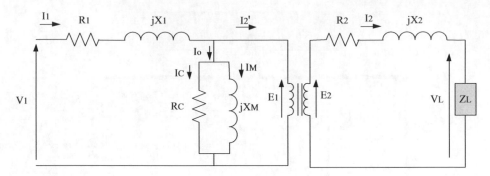

FIGURE 4.13
The non-ideal transformer equivalent circuit diagram.

circuit is shown below. The primed quantities represent those values that equal the original secondary quantity multiplied by voltages, divided by a current or multiplied by a^2 impedance components. Figure 4.14 shows the non-ideal transformer equivalent circuit diagram referred to the primary side and Figure 4.15 equivalent circuit for the secondary side of the transformer.

$$V_2' = a \cdot V_2 \tag{4.14}$$

$$I_2' = \frac{I_2}{a} \tag{4.15}$$

$$R_2' = a^2 \cdot R_2 \tag{4.16}$$

$$X_2' = a^2 \cdot X_2 \tag{4.17}$$

$$Z_2' = a^2 \cdot Z_2 \tag{4.18}$$

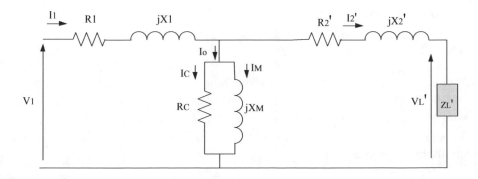

FIGURE 4.14
The non-ideal transformer equivalent circuit diagram referred to the primary side.

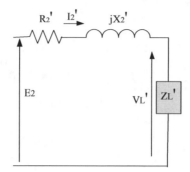

FIGURE 4.15
Equivalent circuit for the secondary side of the transformer.

4.8 Determination of Equivalent Circuit Parameters

To utilize the complete transformer equivalent circuit, the values of R_1, R_2, X_1, X_2, R_c, X_m, and a must be known. These values can be computed given the complete design data for the transformer including dimensions and material properties. The equivalent circuit parameters can also be determined by performing two simple test measurements. These measurements are the no-load (or open-circuit) test and the short-circuit test.

4.8.1 No-Load Test (Determine R_c and X_m)

The rated voltage at the rated frequency is applied to the high-voltage (HV) or low-voltage (LV) winding with the opposite winding open-circuits. Measurements of current, voltage, and real power are made on the input winding (most often the LV winding, for convenience). Figure 4.16 shows the equivalent circuit for no-load test.

$$P_L = \frac{V_L^2}{R_{CL}}$$ (4.19)

$$I_{CL} = \frac{V_L}{R_{CL}}$$ (4.20)

$$I_L = \sqrt{I_{CL}^2 + I_{ML}^2}$$ (4.21)

$$I_{ML} = \sqrt{I_L^2 - I_{CL}^2}$$ (4.22)

$$V_L = I_{ML} \cdot (jX_{ML})$$ (4.23)

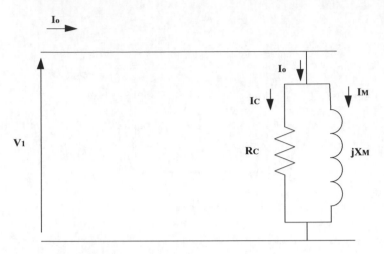

FIGURE 4.16
Equivalent circuit for no-load test.

4.8.2 Short-Circuit Test (Determine $R_{eq.H}$ and $X_{eq.H}$)

Either the LV or HV winding is short-circuited, and a voltage at the rated frequency is applied to the opposite winding, such that the rated current results. Measurements of current, voltage, and real power are made on the input winding (most often the HV winding, for convenience, since a relatively low-voltage is necessary to obtain rated current under short-circuit conditions). Figure 4.17 shows the equivalent circuit for the short-circuit test.

The values measured on the HV winding (primary) in the short-circuit test need to be referred to the LV side. Note that turns ratio is given by:

$$V_H' = \frac{V_H}{a} \tag{4.24}$$

$$I_H' = a \cdot I_H \tag{4.25}$$

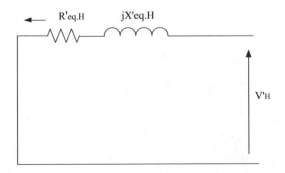

FIGURE 4.17
Equivalent circuit for short-circuit test.

$$R_{eq.L} = \frac{P_H}{I_H'^2} \tag{4.26}$$

$$V_H' = \left| R_{eq.H} + jX_{eq.H} \right| \cdot I_H' = Z_{eq.H} \cdot I_H' \tag{4.27}$$

and

$$X_{eq.H} = \sqrt{Z_{eq.H}^2 - R_{eq.H}^2} \tag{4.28}$$

4.9 Transformer Voltage Regulation

For a given input (primary) voltage, the output (secondary) voltage of an ideal transformer is independent of the load attached to the secondary. As seen in the transformer equivalent circuit, the output voltage of a realistic transformer depends on the load current. If the current through the excitation branch of the transformer equivalent circuit is small in comparison to the current that flows through the winding loss and leakage reactance components, the transformer approximate equivalent circuit referred to the primary is shown below. Note that the load on the secondary (Z_2) and the resulting load current (I_2) have been reflected the primary (Z_{N2}, I_{N2}).

The percentage voltage regulation (V_R) is defined as the percentage change in the magnitude of the secondary voltage as the load current changes from the no-load to the loaded condition.

$$V_R = \frac{|V_2|_{NL} - |V_2|_L}{|V_2|_L} \tag{4.29}$$

Example 4.5

A 15 kVA, 2400:240/120 V, 60 Hz, the two-winding transformer is to be reconnected as a 2400:2520 V step-up transformer. From test work on the two-winding transformer, it is known that its rated voltage core losses and coil losses are 280 and 300 W, respectively. For this autotransformer, (*a*) determine the apparent power rating and (*b*) the full-load efficiency if supplying 2520 V to a 0.8 PF lagging load.

Solution

a)

$$I_H = I_2 = \frac{15,000}{120} = 125 \text{ A}$$

$$V_H = V_1 + V_2 = 2400 + 120 = 2520$$

$$S_X = S_H = V_H I_H = (2520)(125) = 315 \text{ kVA}$$

b) The core and copper losses are unchanged from the two-winding transformer.

$$P_o = S_H PF = (315,000)(0.8) = 252 \text{ kW}$$

$$\eta = \frac{P_o(100)}{P_o + \text{losses}} = \frac{(252,000)(100)}{252,000 + 280 + 300} = 99.77\%.$$

Example 4.6

Determine the value of the coefficient of coupling (k) for the transformer of Example 4.5.

Solution

The turns ratio is:

$$a = \frac{V_1}{V_2} = \frac{240}{120} = 2$$

$$M = \frac{X_m}{\omega a} = \frac{400}{2\pi(60)(2)} = 0.5305 \text{ H}$$

$$L_1 = \frac{X_1}{\omega} + aM = \frac{0.18}{2\pi(60)} + 2(0.5305) = 1.0615 \text{ H}$$

$$L_2 = \frac{X_2}{\omega} + \frac{M}{a} = \frac{0.045}{2\pi(60)} + \frac{0.5305}{2} = 0.2654 \text{ H}$$

$$k = \frac{M}{\sqrt{L_1 L_2}} = \frac{0.5305}{\sqrt{(1.0615)(0.2654)}} = 0.999$$

Example 4.7

For the ideal transformer circuit of Figure 4.18, $R_p = 18 \ \Omega$, $R_L = 6 \ \Omega$, and $X_L = 0.5 \ \Omega$. If $\vec{V}_2 = 120\angle 0° \ V$ and $P_s = 5600 \text{ W}$, (a) determine the turns ratio a, (b) the source voltage $\vec{V}_S$, and (c) the input power factor.

(a) Equations

$$P_{R_L} = \frac{V_2^2}{R_L} = \frac{(120)^2}{6} = 2400 \text{ W}$$

$$P_{R_p} = P_S - P_{R_L} = 5600 - 2400 = 3200 \text{ W}$$

$$V_1 = \sqrt{P_{R_p} R_p} = \sqrt{(3200)(18)} = 240 \text{ V}$$

$$a = \frac{V_1}{V_2} = \frac{240}{120} = 2$$

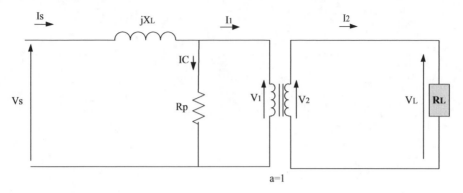

FIGURE 4.18
Circuit of Example 4.7.

(b) Equations

$$\bar{I}_2 = \frac{\bar{V}_2}{R_L} = \frac{120\angle0°}{6} = 20\angle0° \text{ A}$$

$$\bar{I}_1 = \frac{1}{a}\bar{I}_2 = \frac{1}{2}(20\angle0°) = 10\angle0° \text{ A}$$

$$\bar{I}_S = \bar{I}_1 + \frac{\bar{V}_1}{R_p} = 10\angle0° + \frac{240\angle0°}{18} = 23.33\angle0° \text{ A}$$

$$\bar{V}_S = Z_l\bar{I}_S + \bar{V}_1 = (0.5\angle90°)(23.33\angle0°) + 240\angle0° = 240.28\angle2.78° \text{ V}$$

(c) Equations

$$PF_S = \frac{P_S}{V_SI_S} = \frac{5600}{(240.28)(23.33)} = 0.999 \text{ lagging}$$

Example 4.8

Consider the circuit shown in Figure 4.19. The input to the circuit is the voltage of the voltage source $V_s(t)$. The output is the voltage across the 9 H inductor, $V_o(t)$. Determine the output voltage, $V_o(t)$. Write MATLAB program to verify the answers.

Given $V_s(t) = 75.5 \cos(4t + 26°)$ V, so $V_s = 75.5\angle26°$ V,
where frequency $\omega = 4$ rad/sec.
The impedance offered by inductor 9H is:

$$Z_L = J\omega.$$

$$L = J(4 \times 9) = J36 \ \Omega.$$

The total impedance referred to the primary side using turns ratio is given as:

$$Z = 8 + 30 \times \left(\frac{3}{2}\right)^2 + (J36) \times \left(\frac{3}{2}\right)^2$$

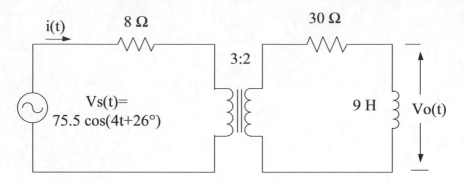

FIGURE 4.19
Circuit of Example 4.8.

$$Z = 8 + 67.5 + J81$$

$$Z = 75.5 + J81.$$

The current from the source is:

$$I_s = \frac{V_s}{Z} = \frac{75.5\angle 26°}{75.5 + J81} = 0.682\angle -21.01° A$$

Now current in the secondary is:

$$I_o = I_s \times \frac{3}{2} = 0.682\angle -21.01° = 1.023\angle -21.01° A$$

Now output voltage is:

$$V_o = I_o Z_L = 1.023\angle -21.01° \times J36 = 36.82\angle 68.99 V$$

$$V_o(t) = 36.82 \cos(4t + 68.99°) \text{ V}.$$

MATLAB CODE
```
W=4;
L=9;
Z_L=w*9*i;
Z=8+30*(3/2)^2+Z_L*(3/2)^2;
Vs=75.5*exp(i*26*pi/180);
Is=Vs/Z;
Io=Is*(3/2);
Vo=Io*Z_L;
magVo=abs(Vo)
phaseVo=angle(Vo)*180/pi
%The above matlab code executed in octave online compatible with matlab
below octave:5>=w;
L=9;
Z_L=w*9*i;
Z=8+30*(3/2)^2+Z_L*(3/2)^2;
Vs=75.5*exp(i*26*pi/180);
Is=Vs/Z;
Io=Is*(3/2);
Vo=Io*Z_L;
magVo=36.8191
phaseVo=68.9872
```

4.10 Three-Phase Transformers

A three-phase transformer may consist of three single-phase windings on the same core inside a single tank or three single-phase transformers wired externally in wye or delta. It is also common practice to construct valid three-phase transformation using only two single-phase transformers. This configuration is called "open-delta". In three-phase systems, the default voltage designation is the line-to-line value, $V_{L\text{-}L}$. Commercial and light industrial systems utilizing three-phase are usually served at 208 V (with 120 V line-to-neutrals). Larger industrial systems utilize 480 V (277 V line-to-neutral). Most industrial plants may use 4160 V (2400 V line-to-neutral) or even higher ranges. A high-voltage system will transmit a given amount of power at a lower current than a lower-voltage system. This is important for many reasons, not the least of which is I^2R losses (efficiency).

There are four ways to configure a standard three-phase transformer bank: delta-delta (Λ-Λ), wye-wye (Y-Y), delta-wye (Δ-Y), and wye-delta (Y-Δ). The most commonly used configuration is delta-wye (Δ-Y). When creating a delta winding using single-phase transformers, great care must be taken to ensure proper phasing and proper polarity. If a delta is completed with the wrong polarity, it is possible to create a dead short between phases, resulting in great damage to the transformer. Never close a delta transformer connection without measuring the voltage across the open corner. The life you save may be your own.

When a three-phase load is served by two single-phase transformers, the connection is called open-delta. Open-delta transformer banks will carry less than rated capacity since both transformers carry not only their own phase current, but also a portion of the current for the third phase. For two transformers of equal size (kVA rating), wired in open-delta, and serving a balanced three-phase load, the de-rating factor is 57% of a corresponding three-phase bank. For example, two 33.3 kVA transformers that are wired in open-delta would have a three-phase capacity of 58 kVA, whereas three transformers in the delta would have a capacity of 100 kVA.

4.10.1 Three-Phase Transformer Configuration

A three-phase transformer or 3-ϕ transformer can be constructed either by connecting together three single-phase transformers, thereby forming a so-called three-phase transformer bank or by using one pre-assembled and balanced three-phase transformer which consists of three pairs of single-phase windings mounted onto one single laminated core.

The advantages of building a single three-phase transformer are that for the same kVA rating it will be smaller, cheaper, and lighter than three individual single-phase transformers connected together because the copper and iron-core are used more effectively. The methods of connecting the primary and secondary windings are the same, whether using just one three-phase transformer or three separate single-phase transformers.

4.10.2 Three-Phase Transformer Connections

The primary and secondary windings of a transformer can be connected in a different configuration as shown to meet practically any requirement. In the case of three-phase transformer windings, three forms of connection are possible: "star" (wye), "delta" (mesh), and "interconnected-star" (zig-zag).

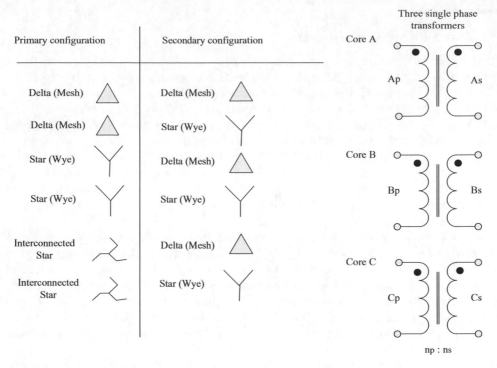

FIGURE 4.20
Three-phase transformer connections.

The combinations of the three windings may be with the primary delta-connected and the secondary star-connected, or star-delta, star-star, or delta-delta, depending on the transformers used. When transformers are used to provide three or more phases, they are generally referred to as a poly-phase transformer (Figure 4.20).

4.10.2.1 Three-Phase Transformer Star and Delta Configurations

A three-phase transformer has three sets of primary and secondary windings. How these sets of windings are interconnected determines whether the connection is a star (also known as wye) or delta (also known as mesh) configuration.

The three available voltages, which themselves are each displaced from the other by 120 electrical degrees, not only decide on the type of the electrical connections used on both the primary and secondary sides, but determine the flow of the transformers currents.

With three single-phase transformers connected together, the magnetic flux in the three transformers differs in phase by 120°. With a single three-phase transformer, there are three magnetic flux in the core differing in time-phase by 120°.

The standard method for marking three-phase transformer windings is to label the three primary windings with capital (upper case) letters A, B, and C used to represent the three individual phases of red, yellow, and blue. The secondary windings are labeled with small (lower case) letters a, b, and c. Each winding has two ends normally labeled 1 and 2 so that, for example, the second winding of the primary has ends which will be labeled B1 and B2, while the third winding of the secondary will be labeled c1 and c2, as shown (Figure 4.21).

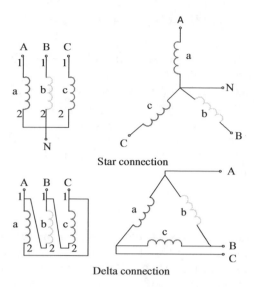

FIGURE 4.21
Transformer star and delta configurations.

4.10.2.2 Transformer Star and Delta Configurations

Symbols are generally used on a three-phase transformer to indicate the type or types of connections used with uppercase Y for star connected, Δ for delta connected, and Z for interconnected star primary windings, with lower case y, d, and z for their respective secondaries. Then, star-star would be labeled Yy, delta-delta would be labeled Dd, and interconnected star to interconnected star would be Zz for the same types of connected transformers.

Connection	Primary Winding	Secondary Winding
Delta	D	d
Star	Y	y
Interconnected	Z	z

4.10.2.3 Transformer Winding Identification

We now know that there are four different ways in which three single-phase transformers may be connected between their primary and secondary three-phase circuits. These four standard configurations are given as delta-delta (Δ-Δ), star-star (Y-Y), star-delta (Y-Δ), and delta-star (Δ-Y).

Transformers for high-voltage operation with the star connections have the advantage of reducing the voltage on an individual transformer, reducing the number of turns required, and an increase in the size of the conductors, making the coil windings easier and cheaper to insulate than delta transformers.

The delta-delta connection, nevertheless, has one big advantage over the star-delta configuration, in that, if one transformer of a group of three should become faulty or disabled, the two remaining ones will continue to deliver three-phase power with a capacity equal to approximately two-thirds of the original output from the transformer unit.

4.10.2.4 Transformer Delta and Delta Connections

In a delta connected (Δ-Δ) group of transformers, the line-voltage, V_L, is equal to the supply voltage, $V_L = V_S$. But the current in each phase winding is given as $1/\sqrt{3} \times I_L$ of the line current, where I_L is the line current.

One disadvantage of delta connected three-phase transformers is that each transformer must be wound for the full-line-voltage, for example, 120 V, as shown in Figure 4.22 and for 57.7%, line current. The greater number of turns in the winding, together with the insulation between turns, necessitates a larger and more expensive coil than the star connection. Another disadvantage with delta connected three-phase transformers is that there is no "neutral" or common connection.

4.10.2.5 Transformer Star and Star Connections

In the star-star arrangement (Y-Y), each transformer has one terminal connected to a common junction or neutral point with the three remaining ends of the primary windings connected to the three-phase mains supply. The number of turns in a transformer winding for star connection is 57.7% of that required for delta connection.

The star connection requires the use of three transformers, and if any one transformer becomes fault or disabled, the whole group might become disabled. Nevertheless, the star connected three-phase transformer is especially convenient and economical in electrical power distributing systems, in that, a fourth wire may be connected as a neutral point, (n), of the three-star connected secondary, as shown in Figure 4.23.

The voltage between any lines of the three-phase transformer is called *the line-voltage*, V_L, while the voltage between any line and the neutral point of a star connected transformer is called *the phase voltage*, V_P. This phase voltage between the neutral point and any one of the line connections is $1/\sqrt{3} \times V_L$ of the line-voltage. The primary side phase voltage, V_P, is given as:

$$V_P = \frac{1}{\sqrt{3}} \cdot V_L = \frac{1}{\sqrt{3}} \times 208 = 120 \, Volts.$$

The secondary current in each phase of a star-connected group of transformers is the same as that for the line current of the supply, then $I_L = I_S$. Then the relationship between line and phase voltages and currents in a three-phase system can be summarized as follows.

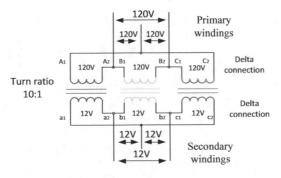

FIGURE 4.22
Transformer delta and delta connections.

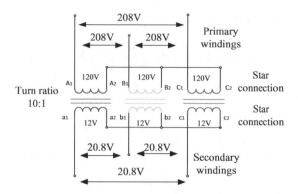

FIGURE 4.23
Transformer star and star connections.

4.10.3 Three-Phase Voltage and Current

Where again, V_L is the line-to-line-voltage and V_P is the phase-to-neutral voltage on either the primary or the secondary side. Other possible connections for three-phase transformers are star-delta Yd, where the primary winding is star-connected, and the secondary is delta-connected or delta-star (Δ-Y) with a delta-connected primary and a star-connected secondary.

Delta-star connected transformers are widely used in low power distribution with the primary windings providing a three-wire balanced load to the utility company while the secondary windings provide the required 4th-wire neutral or earth connection.

When the primary and secondary have different types of winding connections, star or delta, the overall turns ratio of the transformer becomes more complicated. If a three-phase transformer is connected as delta-delta (Δ-Δ) or star-star (Y-Y), then the transformer could potentially have a 1:1 turns ratio. That is the input and output voltages for the windings are the same.

However, if the three-phase transformer is connected in star-delta, (Y-Δ), each star-connected primary winding will receive the phase voltage, V_P, of the supply, which is equal to $1/\sqrt{3} \times V_L$. Then each corresponding secondary winding will then have this same voltage induced in it, and since these windings are delta-connected, the voltage $1/\sqrt{3} \times V_L$ will become the secondary line-voltage. Then with a 1:1 turns ratio, a star-delta connected transformer will provide a $\sqrt{3}$:1 step-down line-voltage ratio.

Connection	Phase Voltage	Line-Voltage	Phase Current	Line Current
Star	$V_P = V_L/\sqrt{3}$	$V_L = \sqrt{3} \times V_P$	$I_P = I_L$	$I_L = I_P$
Delta	$V_P = V_L$	$V_L = V_P$	$I_P = I_L/\sqrt{3}$	$I_L = \sqrt{3} \times I_P$

4.10.3.1 Star-Delta Turns Ratio

For a star-delta (Y-Δ) connected transformer, the turns ratio becomes:

$$\text{Transformation ratio} = \frac{N_P}{N_S} = \frac{V_P}{\sqrt{3}V_S} \tag{4.30}$$

4.10.3.2 Delta-Star Turns Ratio

Likewise, for a delta-star (Δ-Y) connected transformer, with a 1:1 turns ratio, the transformer will provide a 1:$\sqrt{3}$ step-up line-voltage ratio. Then for a delta-star connected transformer, the turns ratio becomes:

$$\text{Transformation ratio} = \frac{N_P}{N_S} = \frac{\sqrt{3}V_P}{V_S} \qquad (4.31)$$

Then for the four basic configurations of a three-phase transformer, we can list the transformers secondary voltages and currents with respect to the primary line-voltage, V_L, and its primary line current I_L, as shown in the following Table 4.1.

where n equals the transformers "turns ratio" of the number of secondary windings N_S, divided by the number of primary windings N_P, and V_L is the line-to-line-voltage with V_P being the phase-to-neutral voltage.

Example 4.8

The primary winding of a delta-star connected 50 kVA transformer is supplied with a 100 volt, 60 Hz, three-phase supply. If the transformer has 500 turns on the primary and 100 turns on the secondary winding, calculate the secondary side voltages and currents.

Solution

$$n = \frac{N_s}{N_p} = \frac{100}{500} = 0.2$$

$$V_{L(sec)} = \sqrt{3} \cdot n \cdot V_{L(pri)}$$

$$V_{L(sec)} = \sqrt{3} \times 0.2 \times 100$$

$$= 34.64\,\text{V}$$

$$V_{p(sec)} = \frac{V_{L(sec)}}{\sqrt{3}} = \frac{34.64}{\sqrt{3}} = 20\,V$$

$$I_{L(pri)} = \frac{V_A}{\sqrt{3} \cdot V_{L(pri)}} = \frac{50 \times 1000}{\sqrt{3} \times 100} = 288.67\,A$$

TABLE 4.1

Three-Phase Transformer Line-Voltage and Current

Primary-Secondary Configuration	Line-Voltage Primary or Secondary	Line-Current Primary or Secondary
Delta-Delta	$V_L \to nV_L$	$I_L \to \dfrac{I_L}{n}$
Delta-Star	$V_L \to \sqrt{3}nV_L$	$I_L \to \dfrac{I_L}{\sqrt{3}.n}$
Star-Delta	$V_L \to \dfrac{nV_L}{\sqrt{3}}$	$I_L \to \sqrt{3}.\dfrac{I_L}{n}$
Star-Star	$V_L \to nV_L$	$I_L \to \dfrac{I_L}{n}$

$$I_{sec} = \frac{V_{L(pri)}}{\sqrt{3} \cdot n} = \frac{288.67}{\sqrt{3} \times 0.2} = 833.33\,A$$

Then the secondary side of the transformer supplies a line-voltage, V_L, of about 35 V giving a phase voltage, V_P, of 20 V at 0.834 A.

4.10.4 Three-Phase Transformer Construction

We have said previously that the three-phase transformer is effectively three interconnected single-phase transformers on a single laminated core and considerable savings in cost, size, and weight can be achieved by combining the three windings onto a single magnetic circuit, as shown in Figure 4.24.

A three-phase transformer generally has the three magnetic circuits that are interlaced to give a uniform distribution of the dielectric flux between the high- and low-voltage windings. The exception to this rule is a three-phase shell-type transformer. In the shell-type of construction, even though the three cores are together, they are non-interlaced.

The three-limb core-type three-phase transformer is the most common method of three-phase transformer construction allowing the phases to be magnetically linked. The flux

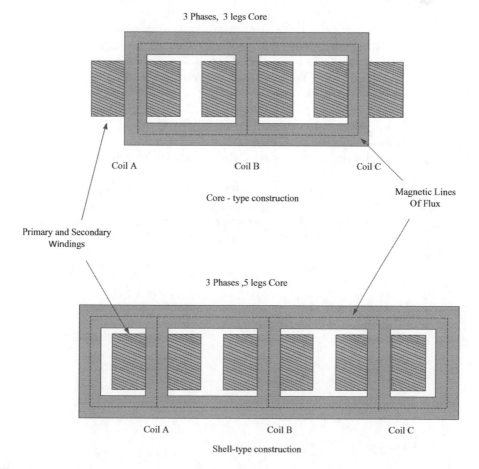

FIGURE 4.24
Three phase transformer construction.

of each limb uses the other two limbs for its return path with the three magnetic flux in the core generated by the line-voltages differing in time-phase by 120°. Thus, the flux in the core remains nearly sinusoidal, producing a sinusoidal secondary supply voltage.

The shell-type five-limb type three-phase transformer construction is heavier and more expensive to build than the core-type. Five-limb cores are generally used for very large power transformers as they can be made with reduced height. A shell-type transformers core material, electrical windings, steel enclosure, and cooling are much the same as for the larger single-phase types.

Problems

4.1   Explain the construction of electrical transformers.

4.2   What are the types of transformers in terms of iron-core shape?

4.3   Derive transformation ratio in electrical transformers.

4.4   A step-up transformer, if the number of high-voltage side is 500 turns, while the number of turns of the low-voltages sides 100 turns, calculate the transformation ratio (α).

4.5   A single-phase transformer of 25 kVA, number of primary turn 500 turns and secondary 40 turns, a voltage source connected to primary windings side of 3000 V, calculate primary windings current and secondary windings current at full load, secondary EMF voltage, and maximum magnetic flux in the magnetic circuit.

4.6   How can electric transformers be divided?

4.7   Explain the work of the transformer at no-load.

4.8   What is the sum of the currents in the transformer at full load?

4.9   Draw the phasor diagram of the transformer at full load.

4.10 What is the significance of stray losses, and should it be within some limits of total losses?

4.11 What is the best core material one should use to achieve minimum losses?

4.12 Consider the circuit shown in Figure 4.25. The input to the circuit is the voltage of the voltage source $V_s(t)$. The output is the voltage across the 3 H inductor, $V_o(t)$. Determine the output voltage, $V_o(t)$. Write MATLAB program to verify the answers.

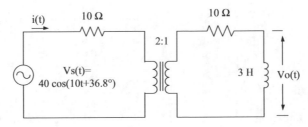

FIGURE 4.25
Circuit diagram for Problem 4.12.

5

Transformer Design

You must learn from your past mistakes, but not lean on your past successes.

Dennis Waitley

This chapter deals with the conventional design of core and shell type for single- and three-phase transformers. It provides all the calculations that are needed to design a transformer. It also gives a design using MATLAB program.

5.1 The Output Equations

It gives the relationship between the electrical rating and physical dimensions of the machines. Let:

V_1 = primary voltage say low voltage side Low Voltage (LV)

V_2 = secondary voltage say high voltage side High Voltage (HV)

I_1 = primary current

I_2 = secondary current

N_1 = primary no. of turns

N_2 = secondary no. of turns

a_1 = sectional area of LV conductors (m²)

$$= \frac{I_1}{\delta}$$

a_1 = sectional area of HV conductors (m²)

$$= \frac{I_2}{\delta}$$

δ = permissible current density (A/m²)

Q = rating in Kilo Volt Amper (KVA).

The transformer winding, the first half of low voltage side on one limb and rest half of LV on another limb to reduce leakage flux. So, the arrangement is low voltage side insulation, then half low voltage side turns, then high voltage side insulation, and then half high voltage side turn.

5.1.1 Single-Phase Core Type Transformer

In single-phase core type transformer, the core is made up of the magnetic core and built with laminations to form a rectangular frame and the windings are arranged concentrically with each other around the legs or limbs. The low voltage windings are wound near the core and high voltage windings are wound over low voltage winding away from the core in order to reduce the number of insulating materials required. Coupling primary and secondary windings are close together to reduce the leakage reactance. In this type of transformer, the rating is given by: (Figure 5.1)

$$Q = V_1 . I_1 \times 10^{-3} \quad \text{KVA}$$

$$= \left(4.44 f . \varphi_m . N_1\right). I_1 \times 10^{-3} \quad \text{KVA} \qquad \left(\because V_1 = 4.44 f . \varphi_m . N_1\right)$$

$$\because \quad \varphi_m = A_i . B_m$$

$$\therefore Q = \left(4.44 f . A_i . B_m . N_1\right). I_1 \times 10^{-3} \quad \text{KVA} \qquad (5.1)$$

where:

f = frequency
φ_m = maximum flux in the core
A_i = sectional area of core
B_m = maximum flux density in the core.

Window space factor:

$$K_w = \frac{\text{Actual CU section area of winding in window}}{\text{Window area } A_w}$$

$$= \frac{a_1 . N_1 + a_2 . N_2}{A_w}$$

$$= \frac{(I_1 / \delta) . N_1 + (I_2 / \delta) . N_2}{A_w} \qquad \left(\because a_1 = I /_1 \delta \,\&\, a_2 = I /_2 \delta\right)$$

$$= \frac{I_1 . N_1 + I_2 . N_2}{\delta . A_w}$$

$$= \frac{2 I_1 N_1}{\delta . A_W}.$$

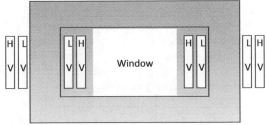

FIGURE 5.1
Phase core type transformer with concentric windings.

For ideal transformer, $I_1 N_1 = I_2 N_2$. So:

$$\left[N_1.I_1 = \frac{\delta.K_w.A_w}{2} \right] \tag{5.2}$$

Put the equation value of $N_1 I_1$ from Equation (5.2) to Equation (5.1):

$$Q = 4.44\,f.A_i.B_m \frac{\delta.K_w.A_w}{2} \times 10^{-3} \qquad KVA \tag{5.3}$$

$$Q = 2.22\,f.A_i.B_m.\delta.K_w.A_w \times 10^{-3} \qquad KVA$$

5.1.2 Single-Phase Shell Type Transformer

In single-phase shell type transformers, the windings are wrap around the central limb and the flux path is completed through two side limbs. The central limb carries total mutual flux while the side limbs forming a part of a parallel magnetic circuit carry half the total flux. The cross-sectional area of the central limb is twice that of each side limbs. In this type of transformer, the window space factor is (Figure 5.2):

$$
\begin{aligned}
K_W &= \frac{a_1.N_1 + a_2.N_2}{A_w} \\
&= \frac{(I_1/\delta).N_1 + (I_2/\delta).N_2}{A_w} \qquad (\because a_1 = I/_1\delta\ \&\ a_2 = I/_2\delta) \\
&= \frac{I_1.N_1 + I_2.N_2}{\delta.A_w} \\
&= \frac{2I_1 N_1}{\delta A_W}.
\end{aligned}
$$

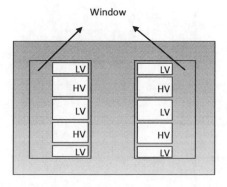

FIGURE 5.2
Phase shell type transformer with sandwich windings.

For ideal transformer $I_1 N_1 = I_2 N_2$. So:

$$N_1 . I_1 = \frac{\delta . K_w . A_w}{2} \tag{5.4}$$

Put the equation value of $N_1 I_1$ from Equation (5.4) to Equation (5.1):

$$Q = 4.44 f . A_i . B_m \frac{\delta . K_w . A_w}{2} \times 10^{-3} \quad KVA$$

$$\tag{5.5}$$

$$Q = 2.22 f . A_i . B_m . \delta . K_w . A_w \times 10^{-3} \quad KVA$$

Note, it is same as for 1-phase core type transformer, i.e., Equation (5.3).

5.1.3 Three-Phase Shell Type Transformer

In three-phase shell type transformer shown in Figure 5.3, the window space factor:

$$\begin{aligned} K_W &= \frac{a_1 . N_1 + a_2 . N_2}{A_w} \\ &= \frac{(I_1 / \delta) . N_1 + (I_2 / \delta) . N_2}{A_w} \quad (\because a_1 = I /_1 \delta \,\&\, a_2 = I /_2 \delta) \\ &= \frac{I_1 . N_1 + I_2 . N_2}{\delta . A_w} \\ &= \frac{2 I_1 N_1}{\delta A_W} . \end{aligned}$$

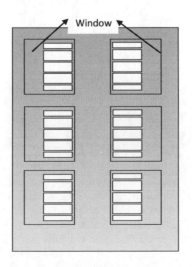

FIGURE 5.3
Three-phase shell type transformer with sandwich windings.

For ideal transformer $I_1N_1 = I_2N_2$. So:

$$N_1.I_1 = \frac{\delta.K_w.A_w}{2}$$ (5.6)

Put the equation value of N_1I_1 from Equation (5.6) to Equation (5.3):

$$Q = 3 \times 4.44 f.A_i.B_m \frac{\delta.K_w.A_w}{2} \times 10^{-3} \qquad KVA$$

$$Q = 6.66 f.A_i.B_m.\delta.K_w.A_w \times 10^{-3} \qquad KVA$$ (5.7)

5.2 Choice of Magnetic Loading (B_m)

The selection of magnetic loading depends on the service condition (i.e., distribution or transmission) and the material used for laminations of the core. The flux density decides the magnetic loading, area of the cross-section of core, and core loss.

1. Normal Si-steel $B_m = 0.9$ T–1.1 T
   35 m thickness, 1.5%–3.5% Si)
2. HRGO $B_m = 1.2$ T–1.4 T
   (Hot rolled grain oriented Si steel)
3. CRGO $B_m = 1.4$ T–1.7 T
   (Cold rolled grain oriented Si steel)
   (0.14 mm–0.28 mm thickness)

5.3 Choice of Electric Loading (Δ)

This factor depends upon the cooling method employed:

1. Natural cooling: 1.5 A/mm²–2.3 A/mm²
   AN   Air natural cooling
   ON   Oil natural cooling
   OFN   Oil forced circulated with natural air cooling
2. Forced cooling: 2.2 A/mm²–4.0 A/mm²
   AB   Air blast cooling
   OB   Oil blast cooling
   OFB   Oil forced circulated with air blast cooling

3. Water cooling: 5.0 A/mm^2–6.0 A/mm^2

                        OW   Oil immersed with circulated water cooling

                        OFW   Oil forced with circulated water cooling

5.4 Core Construction

The cross-sectional area of core type transformer may be rectangular, square, or stepped. When circular coils are required for distribution and power transformers, the square and stepped cores are used.

For shell type transformer, the cross-section may be rectangular. When rectangular cores are used the coils are also rectangular in shape. The rectangular core is suitable for small and low voltage transformers. Figure 5.4 shows a core construction.

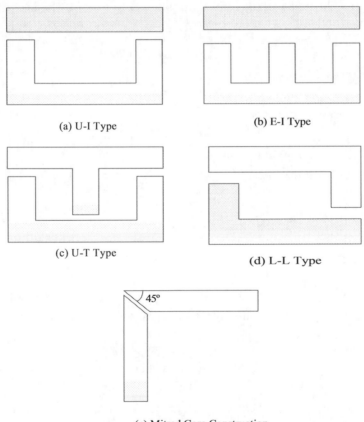

(a) U-I Type   (b) E-I Type

(c) U-T Type   (d) L-L Type

(e) Mitred Core Construction

FIGURE 5.4
Core constructions (a) U-I type, (b) E-I type, (c) U-T type, (d) L-L type, and (e) Mitered core construction.

5.5 Electric Motive Force (EMF) per Turn

$$V_1 = 4.44 f . \varphi_m . N_1 \tag{5.8}$$

$$\text{So EMF / Turn} \quad E_t = \frac{V_1}{N_1} = 4.44 f . \varphi_m \tag{5.9}$$

and

$$Q = V_1 . I_1 \times 10^{-3} \quad KVA \qquad \text{(Note: Take Q as per phase rating in KVA)}$$

$$= \left(4.44 f . \varphi_m . N_1 \right) . I_1 \times 10^{-3} \quad KVA$$

$$= E_t . N_1 . I_1 \times 10^{-3} \quad KVA \tag{5.10}$$

In the design, the ratio of total magnetic loading and electric loading may be kept constant:

$$\text{Magnetic loading} = \varphi_m$$

$$\text{Electric loading} = N_1 . I_1 .$$

$$\text{So} \quad \frac{\varphi_m}{N_1 I_1} = const.(say "r") \Rightarrow N_1 I_1 = \frac{\varphi_m}{r} \quad \text{put in Equation (5.10):}$$

$$Q = E_t \frac{\varphi_m}{r} \times 10^{-3} \quad KVA$$

$$\text{Or } Q = E_t . \frac{E_t}{4.44 f . r} \times 10^{-3} \quad KVA$$

using Equation (5.9):

$$E_t^2 = (4.44 f . r \times 10^{-3}) \times Q$$

$$\text{Or} \qquad E_t = K_t \sqrt{Q} \qquad \qquad Volts/Turn,$$

where:

$K_t = \sqrt{4.44 f . r \times 10^{-3}}$ is a constant and values are:
$K_t = 0.6–0.7$ for 3-phase core type power transformer
$K_t = 0.45$ for 3-phase core type distribution transformer
$K_t = 1.3$ for 3-phase shell type transformer
$K_t = 0.75–0.85$ for 1-phase core type transformer
$K_t = 1.0–1.2$ for 1-phase shell type transformer.

5.6 Estimation of Core X-Sectional Area A_i

We know:

$$E_t = K_t \sqrt{Q} \qquad (5.11)$$

$$E_t = 4.44 f \cdot \varphi_m$$

$$\text{or } E_t = 4.44 f \cdot A_i \cdot B_m \qquad (5.12)$$

$$\text{so} \qquad A_i = \frac{E_t}{4.44 f \cdot B_m} \qquad (5.13)$$

The core may be following types as shown in Figure 5.5,
where d = diameter of the circumscribed circle.
For square core:

$$\text{Gross Area} = \frac{d}{\sqrt{2}} \times \frac{d}{\sqrt{2}} = 0.5 d^2$$

Let stacking factor:

$$K_i = 0.9$$

Actual iron area:

$$A_i = 0.9 \times 0.5 \times d^2$$

$$= 0.45 \times d^2$$

(0.45 for square core and take "K" as a general case):

$$= K \cdot d^2$$

So $$A_i = K \cdot d^2.$$

Or $$d = \sqrt{\frac{A_i}{K}}$$

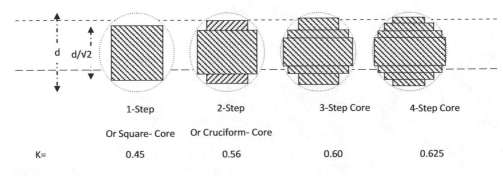

FIGURE 5.5
Types of the core.

5.7 Graphical Method to Calculate Dimensions of the Core

Consider 2 step core (Figure 5.6):

$$\theta = \frac{90^\circ}{n+1}, \qquad n = No \ of \ Steps$$

$$\text{i.e } n = 2 \qquad So \ \ a = d\cos\theta$$

$$\theta = \frac{90^\circ}{2+1} = 30^\circ \quad b = d\sin\theta$$

Percentage fill

$$= \frac{\text{Gross Area of Stepped core}}{\text{Area of circumcircle}} = \frac{K.d^2/K_i}{\pi.d^2/4}$$

$$= \frac{0.625 d^2 / 0.9}{\frac{\pi}{4}(d^2)} \qquad \text{for 4 Step core}$$

$$= 0.885 \text{ or } 88.5\%.$$

No of Steps	1	2	3	4	5	6	7	9	11
% Fill	63.7%	79.2%	84.9%	88.5%	90.8%	92.3%	93.4%	94.8%	95.8%

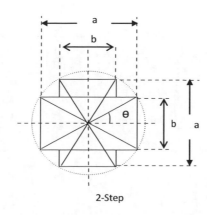

2-Step

FIGURE 5.6
2-step cruciform-core.

5.8 Estimation of Main Dimensions

Consider a 3-phase core type transformer shown in Figure 5.7.
 From the output equation:

$$Q = 3.33\, f \,.A_i \,.B_m \,.\delta \,.K_w \,.A_w \times 10^{-3} \quad KVA,$$

so, window area:

$$A_w = \frac{Q}{3.33\, f \,.A_i \,.B_m \,.\delta \,.K_w \times 10^{-3}} \quad m^2 \tag{5.14}$$

where K_w = window space factor.

$$K_w = \frac{8}{30 + HigherKV} \quad for\ upto\ 10\ KVA$$

$$K_w = \frac{10}{30 + HigherKV} \quad for\ upto\ 200\ KVA$$

$$K_w = \frac{12}{30 + HigherKV} \quad for\ upto\ 1000\ KVA.$$

For higher rating K_w = 0.15 to 0.20.
 Assume some suitable range for:

$$D = (1.7\ to\ 2)\ d.$$

Width of the window $W_w = D - d$.
 The height of the window:

$$L = \frac{A_w}{width\ of\ window(W_w)}$$

$$(\because L \times W_w = A_w).$$

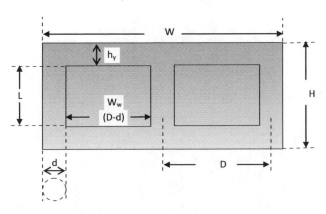

FIGURE 5.7
Three-phase core type transformer.

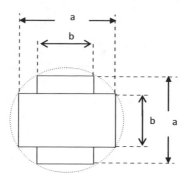

FIGURE 5.8
2-Step or cruciform-core.

Generally, $\frac{L}{W_w} = 2$ *to* 4.

Yoke area A_y is generally taken 10% to 15% higher than core section area (A_i), it is to reduce the iron loss in the yoke section. But if we increase the core section area (A_i), more copper will be needed in the windings and so more cost though we are reducing the iron loss in the core. A further length of the winding will increase resulting in higher resistance to more copper loss (Figure 5.8).

$$A_y = (1.10 \text{ to } 1.15) \; A_i$$

Depth of yoke $\qquad\qquad\qquad\qquad D_y = a$

Height of the yoke $\qquad\qquad\qquad h_y = A_y/D_y$

The width of the core:

$$W = 2 \times D + d$$

The height of the core:

$$H = L + 2 \times h_y$$

Flux density in the yoke

$$B_y = \frac{A_i}{A_y} B_m.$$

5.9 Estimation of Core Loss and Core Loss Component of No-Load Current I_c

$$\text{The volume of iron in core} = 3 \times L \times A_i \; m^3$$

$$\text{The weight of iron in core} = \text{density} \times \text{volume}$$

$$= \rho_i \times 3 \times L \times A_i \; Kg \qquad\qquad (5.15)$$

$$\rho_i = \text{density of iron (kg/m}^3)$$

$$= 7600 \text{ Kg/m}^3 \text{ for normal iron/steel}$$

$$= 6500 \text{ Kg/m}^3 \text{ for M-4 steel.}$$

From the graph we can find out specific iron loss, p_i (Watt/Kg) corresponding to flux density B_m in the core.

So:

the iron loss in core $= p_i \times \rho_i \times 3 \times L \times A_i$ Watt   (5.16)

Similarly:

the iron loss in yoke $= p_y \times \rho_i \times 2 \times W \times A_y$ Watt   (5.17)

where p_y = specific iron loss corresponding to flux density B_y in yoke.

The total iron loss P_i = iron loss in core + iron loss in the yoke.

The core loss component of no-load current

I_c = core loss per phase/primary voltage:

$$I_c = \frac{P_i}{3V_1}.$$

5.10 Estimation of Magnetizing Current of No-Load Current I_m

Find out magnetizing force H, $at_{core/m}$ corresponding to flux density B_m in the core and at_{yoke} corresponding to flux density in the yoke from B-H curve (Figure 5.9):

$$\left(B_m \Rightarrow at_{core} / m, \quad B_c \Rightarrow at_{yoke} / m \right).$$

So

Magnetic Motive Force (MMF) required for the core $= 3 \times L \times at_{core}$

Magnetic Motive Force (MMF) required for the yoke $= 2 \times W \times at_{yoke}.$

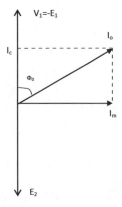

FIGURE 5.9
No-load phasor diagram.

We account for 5% of joints, so

$$Total\ MMF\ required\ = 1.05[MMF\ for\ core + MMF\ for\ yoke].$$

The peak value of the magnetizing current:

$$I_{m,peak} = \frac{Total\ MMF\ required}{3N_1}.$$

Root Mean Square (RMS) value of the magnetizing current:

$$I_{m,RMS} = \frac{I_{m,peak}}{\sqrt{2}}$$

$$I_{m,RMS} = \frac{Total\ MMF\ required}{3\sqrt{2}N_1}.$$

5.11 Estimation of No-Load Current and Phasor Diagram

The no-load current I_o:

$$I_o = \sqrt{I_c^{\,2} + I_m^{\,2}},$$

and the no-load power factor:

$$Cos\varphi_o = \frac{I_c}{I_o}.$$

The no-load current should not exceed 5% of the full the load current.

5.12 Estimation of Number of Turns on LV and HV Windings

$$\text{Primary no. of turns} \quad N_1 = \frac{V_1}{E_t}.$$

$$\text{Secondary no. of turns } N_2 = \frac{V_2}{E_t}.$$

5.13 Estimation of Sectional Area of Primary and Secondary Windings

$$\text{Primary current } I_1 = \frac{Q \times 10^{-3}}{3V_1}.$$

$$\text{Secondary current } I_2 = \frac{Q \times 10^{-3}}{3V_2} \quad OR \quad \frac{N_1}{N_2}I_1.$$

The sectional area of primary winding $a_1 = \dfrac{I_1}{\delta}$.

The sectional area of secondary winging $a_2 = \dfrac{I_2}{\delta}$,

where δ is the current density. Now we can use round conductors or strip conductors.

5.14 Determination of R_1, R_2, and Copper Losses

Let L_{mt} = length of mean turn.

The resistance of the primary winding:

$$R_{1,\,dc,\,75°} = 0.021 \times 10^{-6} \frac{L_{mt}.N_1(m)}{a_1(m^2)} \tag{5.18}$$

$$R_{1,\,ac,\,75°} = (1.15 \ to \ 1.20)\,R_{1,\,dc,\,75°}. \tag{5.19}$$

The resistance of the secondary winding:

$$R_{2,\,dc,\,75°} = 0.021 \times 10^{-6} \frac{L_{mt}.N_2(m)}{a_2(m^2)} \tag{5.20}$$

$$R_{2,\,ac,75°} = (1.15 \ to \ 1.20)\,R_{2,\,dc,75°} \tag{5.21}$$

$$\text{The copper loss in the primary winding} = 3I_1^2.R_1. \tag{5.22}$$

$$\text{The copper loss in the secondary winding} = 3I_2^2.R_2 \tag{5.23}$$

$$\begin{aligned}
\text{The total copper loss} &= 3I_1^2.R_1 + 3I_2^2.R_2 \\
&= 3I_1^2.(R_1 + R_2') \\
&= 3I_1^2.R_p
\end{aligned} \tag{5.24}$$

where R_{o1} is the total resistance:

$$R_{01} = R_p = R_1 + R_2' \tag{5.25}$$

Under no-load condition, there is a magnetic field connecting leads, which causes additional stray losses in the transformer tanks and other metallic parts. These losses may be taken as 7% to 10% of total copper losses.

5.15 Determination of Efficiency

The efficiency of the transformer can be calculated from:

$$\eta = \frac{Output\ Power}{Input\ Power}$$

$$\eta = \frac{Output\ Power}{Output\ Power + Losses} \tag{5.26}$$

$$\eta = \frac{Output\ Power}{Output\ Power + Iron\ Loss\ + Cu\ loss} \times 100\%$$

5.16 Estimation of Leakage Reactance

Assumptions:

1. Consider permeability of iron as infinity that is MMF is needed only for leakage flux path in the window
2. The leakage flux lines are parallel to the axis of the core

Consider an elementary cylinder of leakage flux lines of thickness dx at a distance x as shown in Figure 5.10.

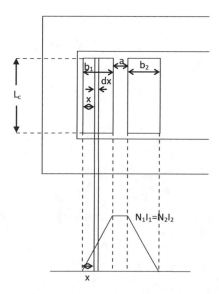

FIGURE 5.10
MMF distribution.

MMF at distance x:

$$M_x = \frac{N_1.I_1}{b_1}.x$$

The Permeance of this elementary cylinder:

$$= \mu_o \frac{A}{L}$$

$$= \mu_o \frac{L_{mt}.dx}{L_c} \ (L_c = \text{length of winding})$$

$$\left(\because S = \frac{1}{\mu_o} \frac{L}{A} \quad \& \quad Permeance = \frac{1}{S} \right).$$

The leakage flux lines associated with the elementary cylinder:

$$d\varphi_x = M_x \times Permeance$$

$$= \frac{N_1.I_1}{b_1} x \times \mu_o \frac{L_{mt}.dx}{L_c}.$$

The flux linkage due to this leakage flux:
$d\psi$ = number of turns which is associated. $d\varphi_x$

$$= \frac{N_1.I_1}{b_1} \times \frac{N_1.I_1}{b_1} x \times \mu_o \frac{L_{mt}.dx}{L_c}$$

$$\tag{5.27}$$

$$= \mu_o.N_1^2.\frac{L_{mt}}{L_c}.I_1.\left(\frac{x}{b_1} \right)^2.dx.$$

The flux linkages (or associated) with the primary winding:

$$\psi_1' = \mu_o.N_1^2.\frac{L_{mt}}{L_c}.I_1.\int_0^{b_1}\left(\frac{x}{b_1} \right)^2 dx = \mu_o.N_1^2.\frac{L_{mt}}{L_c}.I_1.\left(\frac{b_1}{3} \right) \tag{5.28}$$

The flux linkages (or associated) with space "a" between primary and secondary windings:

$$\psi_o = \mu_o.N_1^2.\frac{L_{mt}}{L_c}.I_1.a \tag{5.29}$$

We consider half of this flux linkage with primary and rest half with the secondary winding. So total flux linkages with the primary winding:

$$\psi_1 = \psi_1' + \frac{\psi_o}{2}$$

$$\tag{5.30}$$

$$\psi_1 = \mu_o.N_1^2.\frac{L_{mt}}{L_c}.I_1.\left(\frac{b_1}{3} + \frac{a}{2} \right)$$

Similarly, total flux linkages with the secondary winding:

$$\psi_2 = \psi_2' + \frac{\psi_o}{2} \tag{5.31}$$

$$\psi_2 = \mu_o.N_2^2.\frac{L_{mt}}{L_c}.I_2.\left(\frac{b_2}{3} + \frac{a}{2}\right) \tag{5.32}$$

The primary and secondary leakage inductance:

$$L_1 = \frac{\psi_1}{I_1} = \mu_o.N_1^2.\frac{L_{mt}}{L_c}.\left(\frac{b_1}{3} + \frac{a}{2}\right) \tag{5.33}$$

$$L_2 = \frac{\psi_2}{I_2} = \mu_o.N_2^2.\frac{L_{mt}}{L_c}.\left(\frac{b_2}{3} + \frac{a}{2}\right) \tag{5.34}$$

The primary and secondary leakage reactance:

$$X_1 = 2\pi.f.L_1 = 2\pi.f.\mu_o.N_1^2.\frac{L_{mt}}{L_c}.\left(\frac{b_1}{3} + \frac{a}{2}\right) \tag{5.35}$$

$$X_2 = 2\pi.f.L_2 = 2\pi.f\,\mu_o.N_2^2.\frac{L_{mt}}{L_c}.\left(\frac{b_2}{3} + \frac{a}{2}\right) \tag{5.36}$$

The total leakage reactance referred to the primary side:

$$X_{01} = X_P = X_1 + X_2' = 2\pi.f.\mu_o.N_1^2.\frac{L_{mt}}{L_c}.\left(\frac{b_1+b_2}{3} + a\right) \tag{5.37}$$

The total leakage reactance referred to the secondary side:

$$X_{02} = X_S = X_1' + X_2 = 2\pi.f.\mu_o.N_2^2.\frac{L_{mt}}{L_c}.\left(\frac{b_1+b_2}{3} + a\right) \tag{5.38}$$

It must be 5% to 8% or maximum 10%.

To control X_P: If increasing the window height (L), L_c will increase and the following will decrease b_1, b_2 & L_{mt} and so we can reduce the value of X_P.

5.17 Calculation of Voltage Regulation of Transformer

$$V.R. = \frac{I_2.R_{o2}.\text{Cos}\varphi_2 \;\pm\; I_2.X_{o2}.\text{Sin}\varphi_2}{E_2} \times 100 \tag{5.39}$$

$$= \frac{R_{o2}.\text{Cos}\varphi_2}{E_2 / I_2} \times 100 \pm \frac{X_{o2}.\text{Sin}\varphi_2}{E_2 / I_2} \times 100 \tag{5.40}$$

$$= \left(R_{o2}\cos\varphi_2 \pm X_{o2}\sin\varphi_2\right)\%.$$

5.18 Transformer Tank Design

The width of the transformer (tank) in Figure 5.11:

$$W_t = 2D + D_e + 2b \tag{5.41}$$

where:

D_e = external diameter of HV winding

b = clearance widthwise between HV and tank.

The depth of the transformer (tank):

$$l_t = D_e + 2a \tag{5.42}$$

where a = clearance depth wise between HV and tank.

The height of the transformer (tank):

$$H_t = H + h \tag{5.43}$$

where $h = h_1 + h_2$ = clearance height wise of top and bottom.

Figure 5.12 shows the transformer tank design outside.

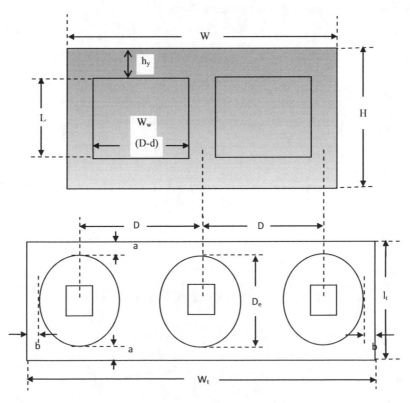

FIGURE 5.11

Transformer tank design.

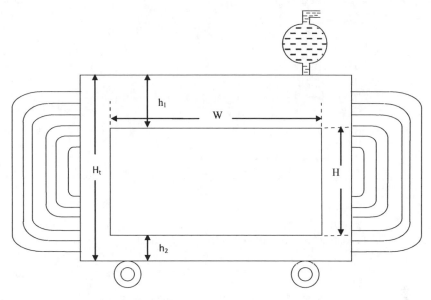

FIGURE 5.12
Transformer tank design outside (http://eed.dit.googlepages.com).

5.19 Calculation of Temperature Rise

The surface area of four vertical sides of the tank (Heat is considered to be dissipated from four vertical sides of the tank):

$$S_t = 2(W_t + l_t)\, H_t \text{ m}^2 \tag{5.44}$$

(Excluding area of top and bottom of the tank)
   Let:

$$\theta = \text{temperature rise of oil } (35°C–50°C)$$

$$12.5 S_t \theta = \text{total full load losses (iron loss + Cu loss)}$$

$$\text{So, the temperature rise in }°C\ \theta = \frac{\text{Total full load losses}}{12.5\ S_t}.$$

If the temperature rise so calculated exceeds the limiting value, the suitable no. of cooling tubes or radiators must be provided. Specific heat dissipation 6 Watt/m^2.°C by radiation.

5.20 Calculation Cooling Tubes Numbers

Let xS_t = surface area of all cooling tubes, then the losses to be dissipated by the transformer walls and cooling tube:
   = total losses

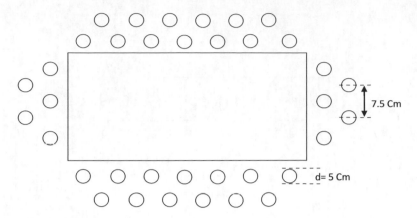

FIGURE 5.13
Tank and arrangement of cooling tubes (http://eed.dit.googlepages.com).

$$6 \text{ W-radiation} + 6.5 \text{ W} = 12.5 \text{ convection}$$

$$6.5 \times 1.35 \text{ W} \approx 8.5 \ (\approx 35\% \text{ more}) \text{ convection only.}$$

So from the above equation, we can find out the total surface of cooling tubes (xS_t) (Figure 5.13).

Normally we use 5 cm diameter tubes and keep them 7.5 cm apart.

$$A_t = \text{surface area of one cooling tube}$$

$$= \pi \times d_{tube} \times l_{tube,\,mean}.$$

Hence:

$$\text{No. of cooling tubes} = \frac{xS_t}{A_t}.$$

5.21 The Weight of Transformer

Let W_i = weight of iron in core and yoke (core volume × density + yoke volume × density).
And W_c = weight of copper in the winding (volume × density)

$$(\text{density of cu} = 8900 \text{ Kg/m}^3)$$

weight of oil = volume of oil × 880.
Add 20% of ($W_i + W_c$) for fittings, tank etc.
Total weight is equal to the weight of above all parts.

5.22 MATLAB Programs

5.22.1 Single-Phase Transformer Design Using MATLAB Program

Following a MATLAB program to design of single-phase transformer 125 kVA,11000/120 V:
clc

$$k = 0.75, S = 125, Q = 2 \times S, f = 60, Bm = 1.1, D = 0.34, V1 = 11000, V2 = 120,$$
$$del = 2.2, Kw = 0.29,$$

$$c1 = 0.42, c2 = .1, d1 = 0.7, d2 = 0.14, e1 = 0.9, e2 = .42$$

%main dimension

$$Et = k \times (S)^{\wedge}0.5 \text{ \%emf per turn}$$

$$Ai = Et/(4.44 \times f \times Bm) \text{ \%core area}$$

$$d = (Ai/k)^{\wedge}0.5$$

$$Aia = k \times ((d)^{\wedge}2) \text{ \%amended net core section}$$

$$fa = Aia \times Bm \text{ \%flux}$$

$$Et = 4.44 \times f \times fa \text{ \%emf per turn}$$

$$T2 = V2/Et \text{ \%no. secondary turn per phase}$$

$$T1 = T2 \times (V1/V2) \text{ \%no. primary turn per phase}$$

$$Aw = S/(2.22 \times f \times Ai \times Bm \times del \times Kw) \text{ \%window area}$$

$$L = Aw/(D-d) \text{ \%window length}$$

$$W = 2 \times D + 0.9 \times d \text{ \%window width}$$

%magnetic circuit

$$c11 = 2 \times (c1 \times c2) \times d^{\wedge}2$$

$$d11 = 2 \times (d1 \times d2) \times d^{\wedge}2$$

$$e11 = 1 \times (e1 \times e2) \times d^{\wedge}2$$

$$GC = c11 + d11 + e11$$

$$ai = 0.9 \times GC$$

$$netsec = 0.9 \times 2 \times c1 \times d \times e1 \times d$$

%core loss

$$cv = 3 \times ai \times L \text{ %core volume}$$

$$we = cv \times 7.55 \text{ %core weight}$$

$$pi0 = 2.7 \text{ %from curve 2.10}$$

$$loss = we \times pi0 \times 10^{\wedge}(-3) \text{ %loss of core}$$

$$Y = 2 \times netsec \times W \times 1000000 \text{ %yoke size}$$

$$weight = Y \times 7.55 \times 10^{\wedge}(-3)$$

$$den = 1.35 \times Aia/netsec$$

$$pi1 = 1.9$$

$$loss1 = weight \times pi1 \times 10^{\wedge}(-3)$$

$$losst = (loss + loss1 \times 1.075) \times 1000 \text{ %total core loss}$$

%magnetizing current

$$ATC = 3 \times L$$

$$ATY = 2 \times W \times 4 \times 100$$

$$ATT = ATC + ATY$$

$$vac = 31; vay = 10;$$

$$VA = we \times vac + weight \times vay$$

$$Im = VA/(V1)$$

$$I2 = S \times 1000/V2$$

$$del2 = 2.8$$

$$a2 = I2/del2$$

$$l2 = pi \times d \times T2$$

$$ro = 0.021 \times 10^{\wedge}(-6)$$

$$R2 = ro \times l2/(a2 \times 10^{\wedge}(-6))$$

$$Ploss2 = (I2^{\wedge}2) \times R2$$

%primary wdg

$$I1 = S \times 1000/V1$$

$$a1 = I1/del2$$

$$d1 = ((4 \times a2/pi)^{\wedge}0.5) \times 0.001$$

$$DIA = 2 \times d1 + d$$

$$l1 = DIA \times T1$$

$$R1 = ro \times l1/(a1 \times 10^{\wedge}-6)$$

$$Ploss1 = (I1^{\wedge}2) \times R1$$

$$Pc = (Ploss1 + Ploss2) \times 1.07 \ \%total\ copper\ loss$$

$$Plosstot = Pc + losst$$

$$eff = 1-(Plosstot/(Plosstot + S \times 1000))$$

$$x = (losst/Pc)^{\wedge}0.5$$

%Weight of transformer

$$weiron = we + weight$$

$$wecopper = (l1 \times a1 + l2 \times a2) \times 8900 \times (10^{\wedge}(-6))$$

$$wic = weiron + wecopper$$

$$copiro = weiron/wecopper$$

Results:

Design Values	d1 = 0.7000	c11 = 0.0032	losst = 1.1315e + 03	d1 = 0.0218
	d2 = 0.1400	d11 = 0.0075	ATC = 969.0345	DIA = 0.2389
	e1 = 0.9000	e11 = 0.0144	ATY = 684.6360	l1 = 313.3376
k = 0.7500	e2 = 0.4200	GC = 0.0251	ATT = 1.6537e + 03	R1 = 1.6213
S = 125	Et = 8.3853	ai = 0.0226	VA = 8.4790e + 03	Ploss1 = 209.3665
Q = 250	Ai = 0.0286	netsec = 0.0260	Im = 0.7708	Pc = 799.5532
f = 60	d = 0.1953	cv = 21.8945	I2 = 1.0417e + 03	Plosstot = 1.931e + 3
Bm = 1.1000	Aia = 0.0286	we = 165.3037	del2 = 2.8000	eff = 0.9848
D = 0.3400	fa = 0.0315	pi0 = 2.7000	a2 = 372.0238	x = 1.1896
V1 = 11000	Et = 8.3853	loss = 0.4463	l2 = 8.7817	weiron = 500.7625
V2 = 120	T2 = 14.3108	Y = 4.4432e + 4	ro = 2.1000e − 08	wecopper = 40.3941
del = 2.2000	T1 = 1.3118e + 3	weight = 335.4587	R2 = 4.9571e − 04	wic = 541.1566
Kw = 0.2900	Aw = 46.7308	den = 1.4881	Ploss2 = 537.8795	copiro = 12.3969
c1 = 0.4200	L = 323.0115	pi1 = 1.9000	I1 = 11.3636	
c2 = 0.1000	W = 0.8558	loss1 = 0.6374	a1 = 4.0584	

5.22.2 Three-Phase Transformer Design Using MATLAB Program

Following a MATLAB program to design three-phase transformer 300 kVA,6600/440 V:
  clc

$$k = 0.6, S = 300, Q = 2 \times S/3, f = 60, Bm = 1.35, D = 0.34, V1 = 6600,$$

$$V2 = 440,$$

$$del = 2.5, Kw = 0.29,$$

$$c1 = 0.42, c2 = .1, d1 = 0.7, d2 = 0.14, e1 = 0.9, e2 = .42$$

%main dimension

$$Et = k \times (Q)^{\wedge}0.5 \text{ \%emf per turn}$$

$$Ai = Et/(4.44 \times f \times Bm) \text{ \%core area}$$

$$d = (Ai/k)^{\wedge}0.5$$

$$Aia = k \times ((d)^{\wedge}2) \text{ \%amended net core section}$$

$$fa = Aia \times Bm \text{ \%flux}$$

$$Et = 4.44 \times f \times fa \text{ \%emf per turn}$$

$$T2 = (V2/(3)^{\wedge}0.5)/Et \text{ \%no. secondary turn per phase}$$

$$T1 = T2 \times (V1/V2) \text{ \%no. primary turn per phase}$$

$$Aw = S/(3.33 \times f \times Ai \times Bm \times del \times Kw) \text{ \%window area}$$

$$L = Aw/(D–d) \text{ \%window length}$$

$$W = 2 \times D + 0.9 \times d$$

%magnetic circuit

$$c11 = 2 \times (c1 \times c2) \times d^{\wedge}2$$

$$d11 = 2 \times (d1 \times d2) \times d^{\wedge}2$$

$$e11 = 1 \times (e1 \times e2) \times d^{\wedge}2$$

$$GC = c11 + d11 + e11$$

$$ai = 0.9 \times GC$$

$$netsec = 0.9 \times 2 \times c1 \times d \times e1 \times d$$

%core loss

$$cv = 3 \times ai \times L \text{ %core volume}$$

$$we = cv \times 7.55 \text{ %core weight}$$

$$pi0 = 2.7 \text{ %from curve 2.10}$$

$$loss = we \times pi0 \times 10^{\wedge}(-3) \text{ %loss of core}$$

$$Y = 2 \times netsec \times W \times 1000000 \text{ %yoke size}$$

$$weight = Y \times 7.55 \times 10^{\wedge}(-3)$$

$$den = 1.35 \times Aia/netsec$$

$$pi1 = 1.9$$

$$loss1 = weight \times pi1 \times 10^{\wedge}(-3)$$

$$losst = (loss + loss1 \times 1.075) \times 1000 \text{ %total core loss}$$

%magnetizing current

$$ATC = 3 \times L$$

$$ATY = 2 \times W \times 4 \times 100$$

$$ATT = ATC + ATY$$

$$vac = 31; vay = 10;$$

$$VA = we \times vac + weight \times vay$$

$$Im = VA/(3 \times V1)$$

$$s2 = S/3$$

$$I2 = s2 \times 1000/((V2/(3)^{\wedge}(.5)))$$

$$del2 = 2.8$$

$$a2 = I2/del2$$

$$l2 = pi \times d \times T2$$

$$ro = 0.021 \times 10^{\wedge}(-6)$$

$$R2 = ro \times l2/(a2 \times 10^{\wedge}(-6))$$

$$Ploss2 = 3 \times (I2^{\wedge}2) \times R2$$

%primary wdg

$$I1 = S \times 1000/(3 \times V1)$$

$$a1 = I1/del2$$

$$d1 = ((4 \times a2/pi)^{\wedge}0.5) \times 0.001$$

$$DIA = 2 \times d1 + d$$

$$l1 = DIA \times T1$$

$$R1 = ro \times l1/(a1 \times 10^{\wedge}-6)$$

$$Ploss1 = 3 \times (I1^{\wedge}2) \times R1$$

$$Pc = (Ploss1 + Ploss2) \times 1.07 \text{ \%total copper loss}$$

$$Plosstot = Pc + losst$$

$$eff = 1-(Plosstot/(Plosstot + S \times 1000))$$

$$x = (losst/Pc)^{\wedge}0.5$$

%Weight of transformer

$$weiron = we + weight$$

$$wecopper = 3 \times (l1 \times a1 + l2 \times a2) \times 8900 \times (10^{\wedge}(-6))$$

$$wic = weiron + wecopper$$

$$copiro = weiron/wecopper$$

$$besr = wecopper/S$$

$$\%Lmt = 0.5(Lmt1 + Lmt2)$$

$$AT = I2 \times T2$$

Results:

Design	d1 = 0.7000	c11 = 0.0033	Lost = 1.362e + 03
values	d2 = 0.1400	d11 = 0.0077	DIA = 0.2251
	e1 = 0.9000	e11 = 0.0149	l1 = 101.0678
k = 0.6000	e2 = 0.4200	GC = 0.0259	R1 = 0.3922
S = 300	Et = 8.4853	ai = 0.0233	Ploss1 = 270.1268
Q = 200	Ai = 0.0236	netsec = 0.0268	Pc = 1.6748e + 03
f = 60	d = 0.1983	cv = 32.0571	Plosstot = 3.0367e + 03
Bm = 1.3500	Aia = 0.0236	we = 242.0308	eff = 0.9900
D = 0.3400	fa = 0.0319	pi0 = 2.7000	x = 0.9018
V1 = 6600	Et = 8.4853	loss = 0.6535	weiron = 588.8583
V2 = 440	T2 = 29.9382	Y = 4.5937e + 04	wecopper = 84.6124
del = 2.5000	T1 = 449.0731	Weight = 46.8275	wic = 673.4707
Kw = 0.2900	Aw = 65.0213	den = 1.1905	copiro = 6.9595
c1 = 0.4200	L = 458.8672	pi1 = 1.9000	besr = 0.2820
c2 = 0.1000	W = 0.8585	loss1 = 0.6590	AT = 1.1785e + 04

5.22.3 Three-Phase Transformer Design Using MATLAB Program

Following a MATLAB program to design three-phase transformer 125 kVA, 2000/440 V:
clc

k = 1, S = 125, Q = 2 × S/3, f = 50, Bm = 1.1, D = 0.34, V1 = 2000, V2 = 440, del = 2.2, Kw = 0.29,

c1 = 0.42, c2 = .1, d1 = 0.7, d2 = 0.14, e1 = 0.9, e2 = .42

%main dimension

$$Et = k \times (S)^{0.5} \text{ \%emf per turn}$$

$$Ai = Et/(4.44 \times f \times Bm) \text{ \%core area}$$

$$d = (Ai/k)^{0.5}$$

$$Aia = k \times ((d)^{2}) \text{ \%amended net core section}$$

$$fa = Aia \times Bm \text{ \%flux}$$

$$Et = 4.44 \times f \times fa \text{ \%emf per turn}$$

$$T2 = V2/Et \text{ \%no. secondary turn per phase}$$

$$T1 = T2 \times (V1/V2) \text{ \%no. primary turn per phase}$$

$$Aw = S/(2.22 \times f \times Ai \times Bm \times del \times Kw) \text{ \%window area}$$

$$L = Aw/(D-d) \text{ \%window length}$$

$$W = 2 \times D + 0.9 \times d \text{ \%window width}$$

%magnetic circuit

$$c11 = 2 \times (c1 \times c2) \times d\text{\textasciicircum}2$$

$$d11 = 2 \times (d1 \times d2) \times d\text{\textasciicircum}2$$

$$e11 = 1 \times (e1 \times e2) \times d\text{\textasciicircum}2$$

$$GC = c11 + d11 + e11$$

$$ai = 0.9 \times GC$$

$$netsec = 0.9 \times 2 \times c1 \times d \times e1 \times d$$

%core loss

$$cv = 3 \times ai \times L \text{ \%core volume}$$

$$we = cv \times 7.55 \text{ \%core weight}$$

$$pi0 = 2.7 \text{ \%from curve 2.10}$$

$$loss = we \times pi0 \times 10\text{\textasciicircum}(-3) \text{ \%loss of core}$$

$$Y = 2 \times netsec \times W \times 1000000 \text{ \%yoke size}$$

$$weight = Y \times 7.55 \times 10\text{\textasciicircum}(-3)$$

$$den = 1.35 \times Aia/netsec$$

$$pi1 = 1.9$$

$$loss1 = weight \times pi1 \times 10\text{\textasciicircum}(-3)$$

$$losst = (loss + loss1 \times 1.075) \times 1000 \text{ \%total core loss}$$

%magnetizing current

$$ATC = 3 \times L$$

$$ATY = 2 \times W \times 4 \times 100$$

$$ATT = ATC + ATY$$

$$vac = 31; vay = 10;$$

$$VA = we \times vac + weight \times vay$$

$$Im = VA/(V1)$$

$$I2 = S \times 1000/V2$$

$$del2 = 2.8$$

$$a2 = I2/del2$$

$$l2 = pi \times d \times T2$$

$$ro = 0.021 \times 10^{\wedge}(-6)$$

$$R2 = ro \times l2/(a2 \times 10^{\wedge}(-6))$$

$$Ploss2 = (I2^{\wedge}2) \times R2$$

%primary wdg

$$I1 = S \times 1000/V1$$

$$a1 = I1/del2$$

$$d1 = ((4 \times a2/pi)^{\wedge}0.5) \times 0.001$$

$$DIA = 2 \times d1 + d$$

$$l1 = DIA \times T1$$

$$R1 = ro \times l1/(a1 \times 10^{\wedge}-6)$$

$$Ploss1 = (I1^{\wedge}2) \times R1$$

$$Pc = (Ploss1 + Ploss2) \times 1.07 \text{ %total copper loss}$$

$$Plosstot = Pc + losst$$

$$eff = 1-(Plosstot/(Plosstot + S \times 1000))$$

$$x = (losst/Pc)^{\wedge}0.5$$

%Weight of transformer

$$weiron = we + weight$$

$$wecopper = (l1 \times a1 + l2 \times a2) \times 8900 \times (10^{\wedge}(-6))$$

$$wic = weiron + wecopper$$

$$copiro = weiron/wecopper$$

Results:

Design	d1 = 0.7000	c11 = 0.0038	losst = 1.2994e+03	d1 = 0.0114
values	d2 = 0.1400	d11 = 0.0090	ATC = 834.2855	DIA = 0.2367
	e1 = 0.9000	e11 = 0.0173	ATY = 698.0590	l1 = 42.3427
k = 1	e2 = 0.4200	GC = 0.0301	ATT = 1.5323e+03	R1 = 0.0398
S =125	Et = 11.1803	ai = 0.0271	VA = 9.3986e+03	Ploss1 = 155.6093
Q = 83.3333	Ai = 0.0458	netsec = 0.0312	Im = 4.6993	Pc = 639.3490
f = 50	d = 0.2140	cv = 22.6200	I2 = 284.0909	Plosstot = 1.939e+03
Bm = 1.1000	Aia = 0.0458	we = 170.7809	del2 = 2.8000	eff = 0.9847
D = 0.3400	fa = 0.0504	pi0 = 2.7000	a2 = 101.4610	x = 1.4256
V1 − 2000	Et − 11.1803	loss − 0.4611	l2 = 26.4547	weiron = 581.2238
V2 = 440	T2 = 39.3548	Y = 5.4363e+04	ro = 2.1000e−08	wecopper = 32.3005
del = 2.2000	T1 = 178.8854	weight = 410.4429	R2 = 0.0055	wic = 613.5243
Kw = 0.2900	Aw = 35.0481	den = 1.9841	Ploss2 = 441.9131	copiro = 17.9943
c1 = 0.4200	L = 278.0952	pi1 = 1.9000	I1 = 62.5000	
c2 = 0.1000	W = 0.8726	loss1 = 0.7798	a1 = 22.3214	

Problems

5.1 A 2000 kVA, 3300/208 V, 60 Hz, single-phase core type transformer, maximum flux density of 1.75 Wb/m^2, current density of 2.5 A/mm^2, EMF voltage per turn = 25 V, and window factor = 0.3, calculate:

1. Core area A$_i$

2. Window area Aw.

5.2 A three-phase, Δ/Y transformer, core type, rated at 400 kVA, 3300/208 V, a suitable core with 3-step having a circumscribing circle of 0.25 m diameter, and a leg spacing of 0.45 m is available. EMF = 8.6 V/turn, δ = 2.55 A/mm^2, Kw = 0.3, and stacking factor S$_f$ = 0.8, determine the main dimensions.

5.3 A 250 kVA, 1100/120 V, 60 Hz, single-phase, core type transformer, maximum flux density of 1.25 Wb/m^2, current density of 2.5 A/mm^2, EMF voltage per turn = 8.2 V, window factor = 0.28, ratio of effective cross-section area of core to square of diameter of circumscribing circle is 0.75, and ratio of height to width of window is 2.5, calculate:

1. The dimensions of the core

2. The number of turns

3. The cross-section area of the conductor.

5.4 A 150 kVA, 2200/120 V, 60 Hz, single-phase, shell type transformer, the ratio of magnetic and electric loadings equal to 480 × 10^{-8}, Bm = 1.25 Wb/m^2, δ = 2.5 A/mm^2, Kw = 0.3, and stacking factor = 0.9, determine:

1. The dimensions of the core

2. The number of primary and secondary windings

3. The cross section area of conductor in the primary and secondary windings.

5.5 Design a single-phase transformer using MATLAB program with the following data:

123 kVA, V1 = 11000 V, V2 = 110 V, 60 Hz, k = 0.75, Kw = 0.29, D = 0.34, δ = 2.2, c_1 = 0.42, c_2 = .1, d_1 = 0.7, d_2 = 0.14, e_1 = 0.9, and e_2 = .42.

5.6 A 150 kVA, 2200/208 V, Y/Δ, 60 Hz, three-phase, shell type transformer, the ratio of magnetic and electric loadings equal to 480×10^{-8}, B_m = 1.1 Wb/m^2, δ = 2.5 A/mm^2, Kw = 0.3, and stacking factor = 0.9, determine:

1. The dimensions of the core

2. The number of primary and secondary windings

3. The cross-section area of the conductor in the primary and secondary windings.

5.7 Design a three-phase transformer using a MATLAB program with the following data:

600 kVA, 1100/208 V, 60 Hz, three-phase, core type transformer, maximum flux density of 1.25 Wb/m^2, current density of 2.5 A/mm^2, EMF voltage per turn = 8.2 V, window factor = 0.28, ratio of effective cross-section area of core to square of diameter of circumscribing circle is 0.75, and ratio of height to width of window is 2.5.

5.8 Design a three-phase transformer using a MATLAB program with the following data:

1200 kVA, 1000/200 V, 60 Hz, three-phase, core type transformer, maximum flux density of 1.2 Wb/m^2, current density of 2.5 A/mm^2, EMF voltage per turn = 5.2 V, window factor = 0.28, ratio of effective cross-section area of core to square of diameter of circumscribing circle is 0.75, and ratio of height to width of window is 2.5.

6

Direct Current Machines

You have to think anyway, so why not think big?

Donald Trump

In this chapter, the well-established theory of DC machines is set forth, and the dynamic characteristics of the separate excited and self-excited machines are illustrated. DC machines are generators that convert mechanical energy to DC electric energy and motors that convert DC electric energy into mechanical energy. This chapter will first explain the principles of DC machine operation by using simple examples, and then consider some of the complications that occur in real DC machines.

6.1 DC Machines

Despite the widespread use of AC machines, DC motors are still widespread in the industry for their simplicity of operation and ease of speed regulation. A DC machine, which converts mechanical energy into electricity, is called *the DC generator*, and a DC machine that converts electrical energy into mechanical is called *the DC motor*, and these machines are typically designed according to the nature of its operation.

Although DC generators can be operated as motors and vice versa, the design specification determines whether the machine is a generator or a motor. Figure 6.1 represents a DC machine.

6.2 DC Machine Parts

DC machine consists of the following basic parts (Figure 6.2).

6.2.1 Stator

The stator consists of the outer frame, which is made up of wrought iron or cast iron to generate the magnetic field. The stator is made from the steel plate set and acts as a path to complete the magnetic circuit, and the poles are installed on it as shown in Figure 6.3.

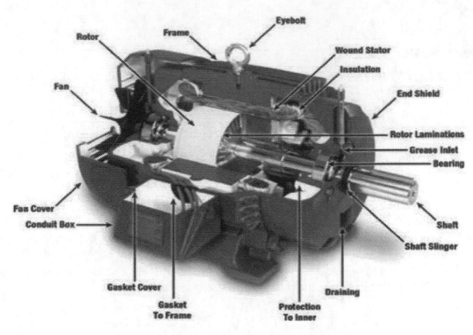

FIGURE 6.1
Direct current machine. (https://newmachineparts.blogspot.com).

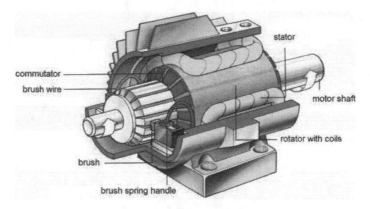

FIGURE 6.2
Parts of DC machine. (https://dumielauxepices.net/wallpaper-1106121).

The main poles are installed in the outer frame and made of steel sheets. The pole interface is called the pole chose. It is useful for attaching the magnetic pole coils and reducing the air gap between the rotor and the iron core of the pole magnetic.

Auxiliary poles are placed between the main poles as shown in Figure 6.4. It is useful to reduce the sparks between the carbon and commutator, and reduce the reaction effect in the armature and directly connect with the armature, and in the outer part of the stator part fix the carbon brushes holder, and the advantage of carbon brushes to transfer the current to and from the armature.

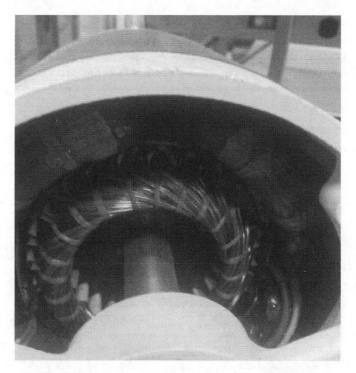

FIGURE 6.3
The stator of the DC machine.

FIGURE 6.4
Magnetic pole of the stator DC machine. (penang-electronic.blogspot.com/2008/06/motors-and-machines-animations-and-java.html).

6.2.2 Rotor

The rotor is called *armature*, it consists of the core which contains the slots where the armature wires are placed, and also has a commutator that alternates AC current into DC current. All the ends of the coils reach the commutator—a set of isolated copper pieces isolated between them and between the axis of rotation and the armature body completely isolated, separating the stator and the rotor the air gap. Figure 6.5 shows parts for DC machine.

6.2.3 Commutator

Commutator shall be cylindrical on the axis of armature rotation by a suitable holder, consisting of a set of copper pieces insulated from each other by mica or steel fiber, and isolated "well" from the axis of rotation as in Figure 6.6.

FIGURE 6.5
Rotor for DC machine. (https://slideplayer.com/slide/8758479/).

FIGURE 6.6
The commutator with the carbon brushes installed on it. (https://www.youtube.com/watch?v=cCB0WgpV8Iw).

6.2.4 Armature Coils

The armature is made up of copper or sometimes aluminum wire with different diameters according to the type of application. The wiring is placed in the slots of the armature, and each turn has the beginning and end, reaching the commutator terminals with the pieces of copper, and can be connected in two ways: lap winding and wave winding.

6.2.4.1 Lap Winding

Lap winding coil connects two adjacent pieces of the commutators. The number of parallel circuits will be equal to the number of poles. This type of winding is typically used in low voltage and high current machines. Refer to Figure 6.7 for a single turn lap winding.

2a—represents the number of parallel circuits

2p—represents the number of poles

2a = 2p.

6.2.4.2 Wave Winding

The beginning and end of the coil are connected to two pieces separated by commutator segment. Figure 6.8 shows a wave winding specified for a single turn. The number of parallel circuits is equal to two irrespective of the number of poles. This type of winding is typically used in high voltage and low current machines.

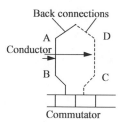

FIGURE 6.7
Lap winding.

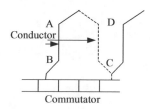

FIGURE 6.8
Wave winding.

6.3 DC Generator

DC generator converts mechanical energy into electrical energy. The operation principle of generators works on Faraday's theory that the Electric Motive Force (EMF) generated in the coil is of the sine waveform. According to Faraday's theory, the rotation of a coil within a magnetic field (magnetic poles) creates an electric motive force in the coil as a result of cutting the magnetic field lines.

6.3.1 Calculate the Motive Force Generated by the Generator (E.M.F)

The electric motive force generated is calculated as follows:

$$E = \frac{N}{60} \times \frac{z}{2a} \times \Phi \times 2p \qquad (6.1)$$

where:
  E = represents the electric motive force voltage (volt)
  N = the speed of the generator (rpm)
  Z = the number of conductors in the armature ducts
  Φ = the magnetic flux in Weber
  2p = the number of poles
  2a = number of parallel circuits.

In the case of a lap winding, the formula shall be as follows:

$$2a = 2p$$

$$E = \frac{NZ\Phi}{60}. \qquad (6.2)$$

And in the case of a wave winding, the formula shall be as follows:

$$2a = 2$$

$$E = \frac{NZ\Phi \times 2P}{60 \times 2} \qquad (6.3)$$

Example 6.1

A DC generator has eight poles, wound as lap then rewound as a wave, calculate the electrical motive force generated in each case if you know that the number of conductors in the armature slots is 240. The magnetic flux of each pole 0.04 Wb. It runs at 1200 rpm.

Solution

In the case of lap winding:

$$2P = 8, Z = 240, \phi = 0.04 \text{ Wb}, N = 1200 \text{ rpm}$$

$$E = \frac{NZ\Phi}{60} = \frac{1200 \times 240 \times 0.04}{60}$$

$$E = 192 \text{ V}.$$

In the case of wave winding:

$$2a = 2$$

$$E = \frac{NZ\Phi 2p}{60 \times 2} = \frac{1200 \times 240 \times 0.04 \times 8}{120}$$

$$E = 768 \text{ V}.$$

6.3.2 Method of Excitation of DC Machines

An external simulated field coil is required to generate magnetic motive force to obtain electric motive force when the field coils continue to generate flux either through an external source or internal source from the machine. The DC generators are divided in terms of methods excitation into two types: separately and self-excited.

6.3.2.1 Separately Excited Generator

The magnetic coils are fed by an external DC source (battery or any other source) as in Figure 6.9. This type of characteristic machine generator and the electric motive force generated depends on the amount of current feed, which needs to organize voltage such as (Leonard machines).

6.3.2.2 Self-Excited Generator

A self-excited generator requires the armature current feeds of the magnetic pole coils to generate a magnetic motive force. The coils are connected via carbon brushes to the armature circuit. There are three ways to connect these coils with the armature: series, shunt, and compound.

6.3.2.2.1 Series Generator

The field windings resistance R_f is connected in series with the armature windings, as shown in Figure 6.10, and have a large diameter and a few turns to carry the current and the current of the load is same armature current. Under no-load condition, the voltage at the terminals of the generator is zero due to an open circuit on the magnetic field, the voltage

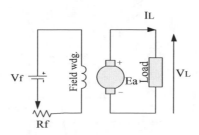

FIGURE 6.9
Electrical circuit of the separately excited generator.

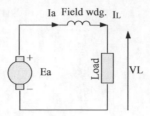

FIGURE 6.10
Series DC generator.

increases depending on the load increase and reaches the maximum value at full load. This type of generator is used as a voltage compensator for the DC power transmission networks (Figures 6.11 through 6.13). The induced electric motive force is calculated as follows:

$$E_a = V_L + I_a \left(R_a + R_f \right) \tag{6.4}$$

$$I_a = I_L = I_f \tag{6.5}$$

where:
  I_a = armature current (current produced by the alternator) (A)
  I_L = current load (A)
  I_f = current field (A)
  V_L = load voltage (V)
  E_a = the electric motive force (V)
  R_a = armature windings resistance (Ω)
  R_f = field windings resistance (Ω)
  R_L = load resistance (Ω).

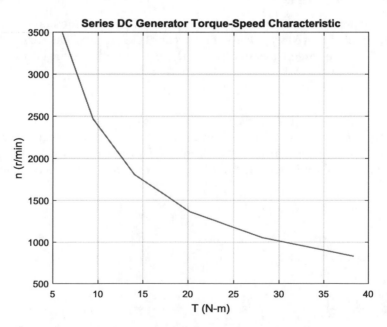

FIGURE 6.11
Input torque-speed characteristics of series DC generator.

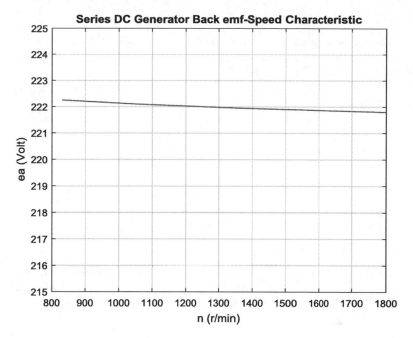

FIGURE 6.12
EMF-speed characteristics of series DC generator.

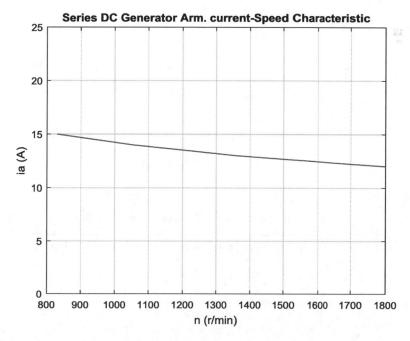

FIGURE 6.13
Armature current-speed characteristics of series DC generator.

Example 6.2

A DC generator feeding a load of current 20 A and voltage 220 V. Calculate the electrical motive force generated in the armature, if it is known that the resistance of the armature 0.02 Ω and the resistance of the field 0.01 Ω.

Solution

$$I_L = I_a = I_f = 20 \text{ A}$$

$$E_a = V_L + I_a (R_a + R_f)$$

$$E_a = 220 + 20 (0.02 + 0.01)$$

$$= 220.6 \text{ V.}$$

Example 6.3

A DC generator feeding a purely resistive load, if it is known that the current produced by the generator is 20 A, the resistance of the field 0.2 Ω, the resistance of the armature 0.1 Ω, and the electric motive force generated 230 V. Calculate load resistance.

Solution

$$I_a = I_L = 20 \text{ A}$$

$$E_a = V_L + I_a \cdot (R_a + R_f)$$

$$230 = V_L + 20 (0.2 + 0.1)$$

$$230 = V_L + 20 \times 0.3$$

$$230 = V_L + 6$$

$$V_L = 230 - 6 = 224 \text{ V}$$

$$V_L = I_L R_L$$

$$224 = 20 \times R_L$$

$$R_L = \frac{224}{20} = 11.2 \text{ Ω.}$$

Example 6.4

A DC generator feeds a load consisting of 22 lamps, each lamp capacity 100 W, 220 V, calculate the amount of electrical motive force generated if the resistance of the armature is known as 0.2 Ω while ignoring the resistance of the field winding.

Solution

$$\text{Total load power} = 22 \times 100 = 2200 \text{ W}$$

$$I_L = P_L/V_L = 2200/220 = 10 \text{ A}$$

$$I_a = I_L = I_f = 10 \text{ A}$$

$$E_a = V_L + I_a (R_a + R_f)$$

$$E_a = 220 + 10 \times 0.2$$

$$E_a = 220 + 2 = 222 \text{ V.}$$

Series DC Generator MATLAB Program

```
ea=[0 10 15 18 31 45 60 75 90 104 120 132 145 160 173 184 200 205 215 223
231];
n0 = 1000;
vt = 220;
ra = 0.15;
ia = 10:1:15;
ea = vt + ia * ra;
ea0 = interp1(if_values, ea_values, ia,'spline');
n1 = 1050;
Eao1 = interp1(if_values, ea_values,58,'spline');
Ea1 = vt + 15 * ra;
Eao2 = interp1(if_values, ea_values, ia,'spline');
n = ((ea./Ea1).* (Eao1./ Eao2)) * n1/100;
T = ea.* ia./ (n * pi / 30);
figure(1);
plot(T, n,'b-','LineWidth',2.0);
hold on;
xlabel('T (N-m)');
ylabel('n (r/min)');
title ('Series DC Generator Torque-Speed Characteristic');
grid on;
hold off;
figure(2);
plot(n, ea,'b-','LineWidth',2.0);
xlabel('n (r/min)');
ylabel('ea (Volt)');
title ('Series DC Generator Back emf-Speed Characteristic');
axis([800 1800 215 225]);
grid on;
%hold off;
figure(3);
plot(n, ia,'b-','LineWidth',2.0);
xlabel('n (r/min)');
ylabel('ia (A)');
title ('Series DC Generator Arm. current-Speed Characteristic');
axis([800 1800 0 25]);
grid on;
%hold off;
```

6.3.2.2.2 The Shunt Generator

The field windings resistance (R_f) is connected in parallel with the armature windings, as in Figure 6.14, and have a large number of turns and a small section area because of the current passing through it is relatively small, and the voltage on both ends of the generator in the case of no-load maximum. Because the circuit of the magnetic field is

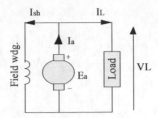

FIGURE 6.14
Shunt generator.

closed, so the voltage does not change on both ends of the generator in the case of a load or no-load. This type of generator is used in cases where the voltage is required for the vehicle and feeding AC generators (Figures 6.15 through 6.17). The electric motive force generated:

$$E_a = V_L + I_a R_a \tag{6.6}$$

$$V_f = V_L = I_f R_f \tag{6.7}$$

$$I_a = I_L + I_f. \tag{6.8}$$

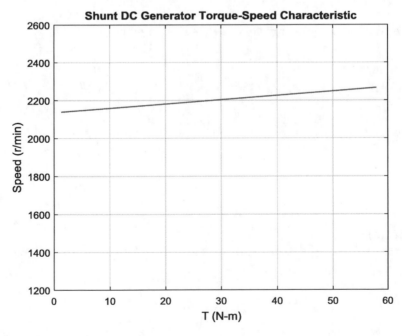

FIGURE 6.15
Input torque-speed characteristics of shunt DC generator.

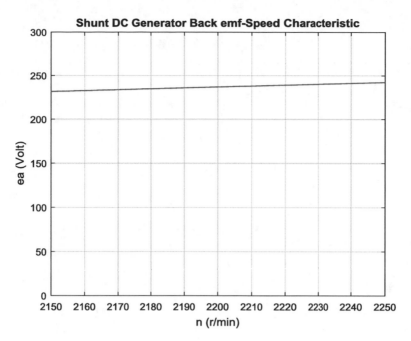

FIGURE 6.16
EMF-speed characteristics of shunt DC generator.

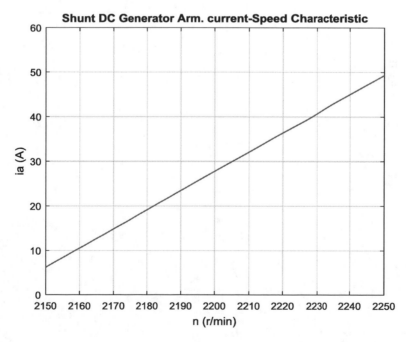

FIGURE 6.17
Armature current-speed characteristics of shunt DC generator.

Example 6.5

A shunt generator feeds a load with a current of 300 A at a voltage of 240 V. If the resistance of the armature windings is 0.02 Ω and the resistance of the field windings is 60 Ω, calculate the generated electrical motive force.

Solution

$$I_f = V/R_f = 300/60 = 4 \text{ A}$$

$$I_a = i_l + I_f = 300 + 4 = 304 \text{ A}$$

$$E_a = V_L + I_a R_a$$

$$E_a = 240 + 304 \times 0.02$$

$$E_a = 240 + 6.08$$

$$E_a = 246.08 \text{ V}.$$

Example 6.6

A shunt DC generator feeds a load of 4068 W, voltage 226 V, armature resistance 0.2 Ω, and the amount of electric motive force generated 230 V, calculate the resistance of field windings.

Solution

$$I_L = P_L/V_L = 4068/226 = 18 \text{ A}$$

$$E_a = V_L + I_a R_a$$

$$230 = 226 + I_a \times 0.2$$

$$4 = I_a \times 0.2$$

$$I_a = I_L + I_f$$

$$20 = 18 + I_f$$

$$I_f = 20 - 18 = 2 \text{ A}$$

$$V_L = V_f = I_f R_f$$

$$R_f = V/I_f = 226/2 = 113 \text{ Ω}.$$

DC Shunt Generator MATLAB Program

```
if_values=[0 0.1 0.2 0.3 0.4 0.5 0.6 0.7 0.8 0.9 1.0 1.1 1.2 1.3 1.4 1.5
1.6 1.7 1.8 1.9 2.0];
ea_values=[0 10 15 18 31.30 45.46 60.26 75.06 89.74 104.4 118.86 132.86
146.46 159.78 172.18 183.98 195.04 205.18...
```

```
214.52 223.06 231.2];
n0 = 1500;
vt = 230;
rf = 100;
radd = 75;
ra = 0.250;
il = 0:1:55;
nf = 2700;
far0 = 1500;
ia = il + vt / (rf + radd);
ea = vt + ia * ra;
ish= vt / (rf + radd);
ea0 = interp1(if_values, ea_values, ish);
n = (ea./ ea0) * n0;
T = ea.* ia./ (n * 2 * pi / 60);
figure(1);
plot(T, n,'b-','LineWidth',2.0);
xlabel('T (N-m)');
ylabel('Speed (r/min)');
title ('Shunt DC Generator Torque-Speed Characteristic');
axis([0 60 1200 2600]);
grid on;
%hold off;
figure(2);
plot(n, ea,'b-','LineWidth',2.0);
xlabel('n (r/min)');
ylabel('ea (Volt)');
title ('Shunt DC Generator Back emf-Speed Characteristic');
axis([2150 2250 0 300]);
grid on;
%hold off;
figure(3);
plot(n, ia,'b-','LineWidth',2.0);
xlabel('n (r/min)');
ylabel('ia (A)');
title ('Shunt DC Generator Arm. current-Speed Characteristic');
axis([2150 2250 0 60]);
grid on;
%hold off;
```

6.3.2.2.3 Compound Generator

The compound generator contains both series and shunt coils connected to armature windings. If the shunt coils connected directly with the ends of the armature, this method of connection is called *the short compound* as shown in Figure 6.18, or shunt coils are connected across the terminals of the outer circuit (the armature with the series coils), it is called *the long compound* as shown in Figure 6.19.

The self-excited generators polarity depends on the field coils connection with the armature, where the field current helps the remaining magnetism in the machine.

Any inverse connection to the field coils will cause the remaining magnetism to be eliminated, and the construction of potential is not generated on the terminals of the machine. The compound generator shall be composed of two types of short and long compound (Figure 6.20 through 6.22).

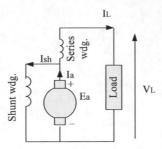

FIGURE 6.18
Short compound DC generator.

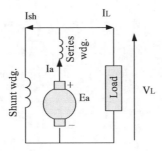

FIGURE 6.19
Long compound DC generator.

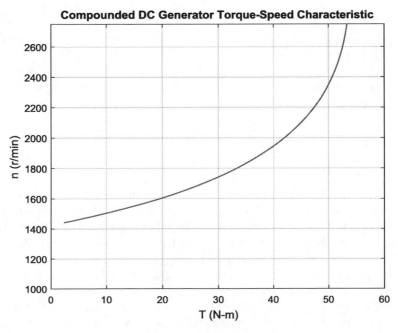

FIGURE 6.20
Input torque-speed characteristics of compound DC generator.

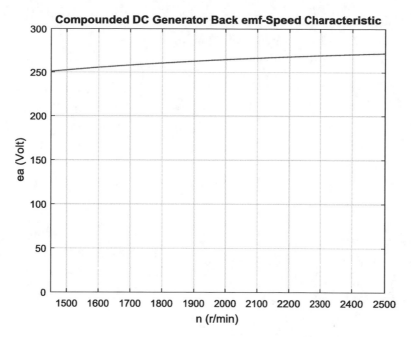

FIGURE 6.21
EMF-speed characteristics of compound DC generator.

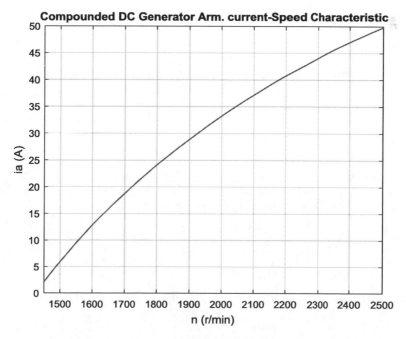

FIGURE 6.22
Armature current-speed characteristics of compound DC generator.

i. The short compound DC generator

From Figure 6.18, the general equations are:

$$E_a = V_L + I_a\,R_a + I_L\,R_{se} \tag{6.9}$$

$$I_a = I_L = I_{sh} \tag{6.10}$$

$$I_{se} = I_L \tag{6.11}$$

$$I_{sh} = \frac{Vl + Il\,Rse}{Rsh} \tag{6.12}$$

where:

Rse = series resistant (Ω)

R_{sh} = resistant of parallel coils (Ω)

I_{sh} = shunt current (A)

I_{se} = series windings current (A).

ii. The long compound DC generator

From Figure 6.19, the general equations are:

$$E_a = V_L + I_a\,(R_a + R_{se}) \tag{6.13}$$

$$I_a = i_L + I_{sh} \tag{6.14}$$

$$I_{se} = i_a \tag{6.15}$$

$$I_{sh} = V_L/R_{sh} \tag{6.16}$$

The voltage difference on both ends of the generator can be extracted as follows:

Voltage difference on both ends of the generator = $V_L + I_L\,R_{se}$.

Example 6.7

A long compound DC generator feeding a load of current 100 A, at a voltage of 230 V. Calculate the generated electrical motive force if the resistance of the armature is 0.04 Ω, the resistance of 0.01 Ω, and the resistance of the shunt coils 115 Ω.

Solution

$$E_a = V_L + I_a\,(R_a + R_{se})$$

$$E_a = 230 + I_a\,(0.04 + 0.01)$$

$$I_a = I_L + I_{sh}$$

$$I_{sh} = \frac{VL}{Rsh}$$

$$I_{sh} = \frac{230}{115} = 2A$$

$$I_a = 100 + 2 = 102$$

$$E_a = 230 + 102 \times 0.05$$

$$E_a = 230 + 501 = 235.1 \; volt.$$

Example 6.8

A short compound DC generator feeding power of 22 kW, 220 V, armature resistance of 0.2 Ω, the series resistance of 0.1 Ω, and shunt resistance of 120 Ω. Calculate the generated electric motive force.

$$E_a = V_L + I_a R_a + I_L R_{se}$$

$$P_L = 22 \times 1000 = 22000 \; W$$

$$I_L = P_L/V_L = 22000/220 = 100 \; A$$

$$I_{sh} = (V_L + I_L R_{se})/R_{sh} = (200 + 100 \times 0.2)/120 = 2 \; A$$

$$I_a = IL + Ish = 100 + 2 = 102 \; A$$

$$E_a = V_L + I_a R_a + I_L R_{se}$$

$$E_a = 220 + 102 \times 0.2 + 100 \times 0.1$$

$$E_a = 220 + 20.4 + 10 = 250.4 \; volt.$$

Compound DC Generator MATLAB Program

```
If=[0 0.1 0.2 0.3 0.4 0.5 0.6 0.7 0.8 0.9 1.0 1.1 1.2 1.3 1.4 1.5 1.6 1.7
1.8 1.9 2.0];
ea=[0 10 15 18 31.30 45.46 60.26 75.06 89.74 104.4 118.86 132.86 146.46
159.78 172.18 183.98 195.04 205.18...
214.52 223.06 231.2];
n0 = 1000;
vt = 250;
rf = 100;
radd = 75;
ra = 0.44;
il = 0:55;
nf = 2700;
nse = 27;
ia = il + vt / (rf + radd);
ea = vt + ia * ra;
ifi = vt / (rf + radd) - (nse / nf) * ia;
ea0 = interp1(if_values, ea_values, ifi);
n = (ea./ ea0) * n0;
T = ea.* ia./ (n * pi / 30);
figure(1);
plot(T, n,'b-','LineWidth',2.0);
xlabel('T (N-m)');
```

```
ylabel('n (r/min)');
title ('Compounded DC Generator Torque-Speed Characteristic');
axis([0 60 1000 2750]);
grid on;
figure(2);
plot(n, ea,'b-','LineWidth',2.0);
xlabel('n (r/min)');
ylabel('ea (Volt)');
title ('Compounded DC Generator Back emf-Speed Characteristic');
axis([1450 2500 0 300]);
grid on;
%hold off;
figure(3);
plot(n, ia,'b-','LineWidth',2.0);
xlabel('n (r/min)');
ylabel('ia (A)');
title ('Compounded DC Generator Arm. current-Speed Characteristic');
axis([1450 2500 0 50]);
grid on;
%hold off;
```

6.3.3 Losses in DC Generator

When the generator converts mechanical energy into electrical energy, the part of this energy is lost and is usually dissipated as heat in resistance. The generated heat in the machine may cause damage to insulation materials and a short circuit between the windings. Heat calculations must be done to reduce the loss in the machine so that the coefficient of quality (efficiency) high means reducing the cost of operation of the machine.

During the conversion of mechanical energy into electrical energy by the generator, a portion of the energy in the magnetic circuit and part of the electrical circuit is lost, as well as a part of the mechanical process (friction) when the rotor is rotated.

A. The Loss in the Magnetic Circuit (Iron Losses)

The magnetic **losses** including: eddy current loss and hysterics loss. It is considered to be a fixed loss.

Hysteresis Losses in DC Machine

Hysteresis losses occur in the armature winding due to the reversal of magnetization of the core. When the core of the armature exposed to a magnetic field, it undergoes one complete rotation of magnetic reversal. The portion of the armature which is under S-pole, after completing half electrical revolution, the same piece will be under the N-pole, and the magnetic lines are reversed in order to overturn the magnetism within the core. The constant process of magnetic reversal in the armature consume some amount of energy which is called hysteresis loss. The percentage of loss depends upon the quality and volume of the iron.

The Frequency of Magnetic Reversal

$$f = \frac{P.n}{120} \qquad (6.17)$$

where:

P = number of poles

n = speed in rpm.

The Steinmetz formula is used for the hysteresis loss calculation.

$$P_h = \eta.B_{max}^{1.6}.f.V \tag{6.18}$$

where:

η = Steinmetz hysteresis co-efficient

B_{max} = maximum flux density in armature winding

f = frequency

V = volume of the armature in m^3.

Eddy Current Loss in DC Machine

According to Faraday's law of electromagnetic induction, when an iron core rotates in the magnetic field, an EMF is also induced in the core. Similarly, when the armature rotates in a magnetic field, a small amount of EMF induced in the core which allows the flow of charge in the body due to the conductivity of the core. This current is useless for the machine. This loss of current is called eddy current. This loss is almost constant for the DC machines. It could be minimized by selecting the laminated core. The eddy current can calculate by using the formula:

$$P_e = K_h.B_{max}^2.f^2.t^2 \tag{6.19}$$

where:

t = lamination thickness.

B. **Losses in the Electrical Circuit (Copper Losses):**

It is caused by the passage of the current in the parts of the circuit in the windings of the rotor and the windings of the field, which is a variable loss and according to:

$I_a^2 R_a$ = armature copper losses

$I_{se}^2 R_{se}$ = series windings copper losses

$I_f^2 R_f$ = field windings copper losses.

C. **Mechanical Loss:**

The losses associated with the mechanical friction of the machine are called mechanical losses. These losses occur due to friction in the moving parts of the machine-like bearing, brushes, etc., and windage losses occur due to the air inside the rotating coil of the machine. These losses are usually very small about 15% of full load loss. Figure 6.23 shows the power stage diagram of the generator.

$$P_g = E_a.I_a$$

$$P_g = P_{in} - (P_{mech} + P_i)$$

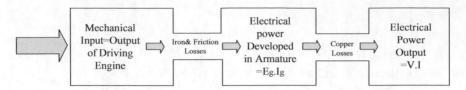

FIGURE 6.23
The power stage diagram of the generator.

$$Pg = \text{armature capacity generated in the air gap}$$

$$Pin = \text{input capacity (mechanical capacity) (horsepower)}$$

$$\text{Each 1 Hp} = 746 \text{ W}$$

$$O/P = \text{output power } (V_L.I_L).$$

6.3.4 Efficiency Calculation

With reference to the power path within the DC generator, the overall efficiency can be calculated as follows:

$$\eta = \frac{o/p}{i/p} = (V_L \, I_L) / HP \times 746 \tag{6.20}$$

$$\eta = P_{out} / (p_{out} + \text{Losses}) \tag{6.21}$$

$$\eta = (P_{in} - \text{losses}) / P_{in} = 1 - (\text{Losses} / P_{in}) \tag{6.22}$$

Example 6.9

A long compound DC generator, rotating at a speed of 1000 rpm, fed a load capacity of 22 kW at 220 V, if the resistance of the armature coils 0.02 Ω, the resistance of the series coils 0.01 Ω, and the resistance of the shunt coils 110 Ω, calculate generator efficiency if the iron and mechanical losses are 2500 W.

Solution

$$P_i + P_{mec} = 2500 \ W$$

$$I_L = P_{out} / VL = \frac{22 \times 1000}{220} = 100A$$

$$I_{sh} = V_{sh}/R_{sh} = V_L/R_{sh} = 220/110 = 2 \ A$$

$$I_a = I_l + i_{sh} = 100 + 2 = 102 \ A$$

$$P_{cu} = I_a^2 R_a + I_a^2 R_{se} + I_{sh}^2 R_{sh}$$

$$P_{cu} = (102)^2 \times 0.02 + (102)^2 \times 0.01 + (2)^2 \times 110$$

$$P_{cu} = 752.12 \text{ W}$$

$$\text{the power losses} = P_{cu} + (P_i + P_{mech.})$$

$$\text{so, power losses} = 752.1 + 2500 = 3252.1 \text{ W}$$

$$\eta = \frac{P_{out}}{P_{out} + losses} = \frac{22000}{22000 + 3252.1} = \frac{22000}{25252.1} = \eta = 0.87 = 87\%.$$

Example 6.10

A shunt DC generator feeds a load with a current of 20 A, 200 V. Calculate generator efficiency if the armature resistance is 0.2 Ω, the shunt resistance is 100 Ω, and the mechanical and iron losses are 201 W.

Solution

$$I_f = \frac{vf}{Rf} = \frac{VI}{Rf} = 200 / 100 = 2A$$

$$I_a = I_L + I_f = 20 + 2 = 22 \text{ A}$$

$$P_{cu} = I_a^2 R_a + I_f^2 R_f = (22)^2 \times 0.02 + (2)^2 \times 100 = 409.68$$

$$P_{cu} = 9.68 + 400 = 409.68 \text{ W}$$

$$\text{the power losses} = P_{cu} + (P_i + P_{mech}) = 409.68 + 201$$

$$= 610.68 \text{ W}$$

$$P_{out} = I_L \cdot V_L = 20 \times 200 = 4000 \text{ W}$$

$$\eta = \frac{P_{out}}{P_{out} + losses} = \frac{4000}{4000 + 610.68} = 86.7 = 86.7\%.$$

Example 6.11

A shunt DC generator has a shunt field resistance of 60 Ω. When the generator delivers 60 KW to a resistive load at a terminal voltage of V = 120 V, while the generated Generator voltage (Eg) is 135 V. Determine (a) the armature circuit resistance Ra and (b) determine the generated voltage Eg when the output is changed to 20 KW and the terminal voltage is V = 135 V. Armature consists of generated voltage Eg in series with armature resistance Ra. (c) Write MATLAB code program to repeat a and b above.

Solution

a) $R_{sh} = 60$ ohm
    the terminal voltage V = 120 V
    the field current in the winding $I_f = V/R_{sh} = 120/60 = 2A$
    when the connected load is 60 kW,
    $V \times I_L = 60000$
    $I_L = 60000/120 = 500A,$

for a shunt generator the
armature current = load current + field current

$$Ia = 500 + 2 = 502 \text{ A},$$

the generated EMF Eg = V + Ia. Ra

$$\Rightarrow 135 = 120 + 502 \times Ra$$

$$Ra = 0.0298 \; \Omega.$$

b) When the output is 20 kW and terminal voltage V = 135 V,

$$V \times I_l = 20000$$

$$I_l = 20000/135 = 148 \text{ A}$$

$$I_f = 135/60 = 2.25 \text{ A}$$

$$Ia = 148 + 2.25 = 150.25 \text{ A}$$

$$Eg = V + Ia. Ra$$

$$= 135 + 150.25 \times 0.0298 = 139.47 \text{ V}.$$

c) MATLAB code

```
R_sh=60;
Eg1=135;
V_l1=120;
i_f=V_l1/R_sh;
load1=60000;
i_l1=load1/V_l1;
i_a=i_f+i_l1;
R_a=(Eg1-V_l1)/i_a;
disp(R_a)
Load2=20000;
Vl2=135;
I_f2=Vl2/R_sh;
Il2=Load2/Vl2;
Ia2=Il2+I_f2;
Eg2=Vl2+Ia2*R_a;
disp(Eg2)
Ra=0.0299
Eg2=139.4940
```

6.4 DC Motors

The DC motor consists of the same parts as the DC generator.

The theory of operation is as follows:

When a wire is placed in a magnetic field, a mechanical force affects the wire, depending on the intensity of the magnetic flux, the length of the wire and the current strength, as in Figure 6.24.

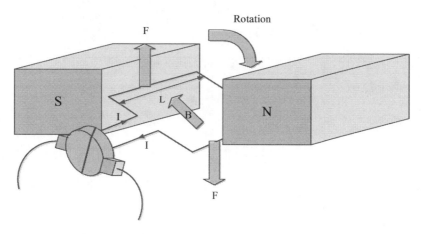

FIGURE 6.24
The diagram shows the effect of magnetic field forces.

$$F = B. L. I \qquad (6.23)$$

where:
F = mechanical force in Newton
B = flux density in Weber/m²
I = current in A.

When a conductor wire is placed in the form of a coil that carries an electric current, to rotate around a specific axis within a magnetic field with two poles, the direction of the magnetic lines around the wire is opposite to the direction of the lines resulting from the magnetic poles from one end of the coil, a mechanical force moves the coil in a direction that we can assign according to the left-hand rule as in Figure 6.25. The difference between the motor and the generator is illustrated in Figure 6.26.

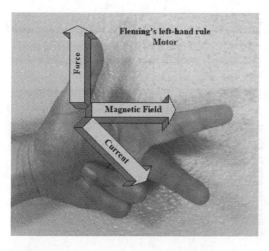
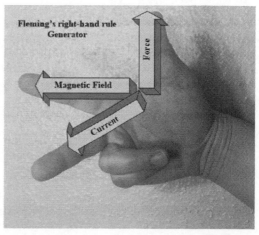

FIGURE 6.25
Fleming's hand rule for generator and motor.

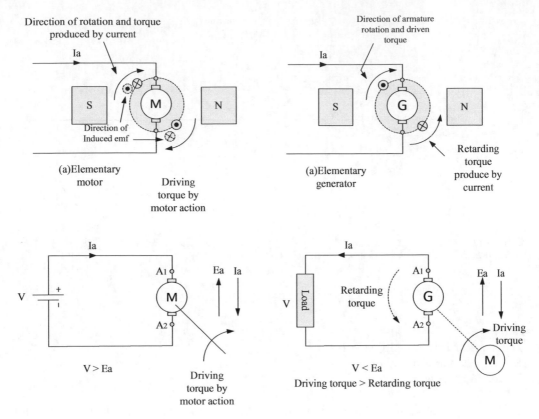

FIGURE 6.26
Comparisons between the generator and the motor, noting the direction of the current in each.

6.4.1 Types of DC Motors

6.4.1.1 Series Motor

The magnetic poles coils are connected to the armature coils, with a few turns and a thick section. The motor is used in cases where high torque is required, such as moving electric trains and cranes, starting the motor drive. The direction of rotation can be reversed as opposed to the ends of the magnetic poles windings (Figure 6.27).

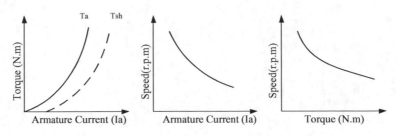

FIGURE 6.27
The relationship between armature current, speed, and torque.

It is noticeable from the characteristic curve that speed increases very significantly at no-load (T = 0). Therefore, it is preferable to not use the series motor at no-load because this causes mechanical problems related to speed up (Figures 6.28 through 6.30).

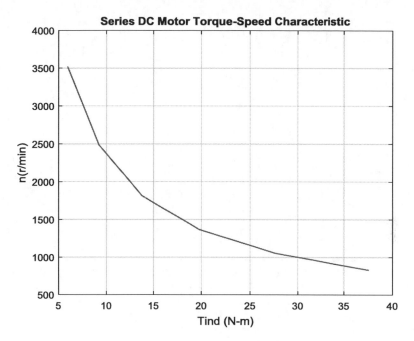

FIGURE 6.28
Torque-speed characteristics series DC motor.

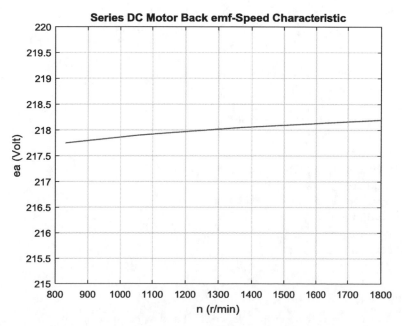

FIGURE 6.29
EMF-speed characteristics of series DC motor.

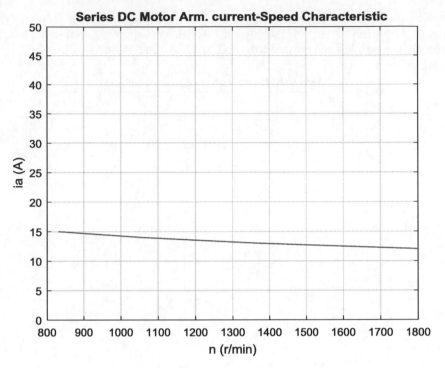

FIGURE 6.30
Armature current-speed characteristics of series DC motor.

The back electrical voltage is given as follows:

$$E_b = V_{in} - I_a \left(R_a + R_{se}\right) \tag{6.24}$$

$$I_a = I_{in} = I_{se} \tag{6.25}$$

where:
I_{in} = the current drawn by the motor (A)
I_{se} = the field current (A)
I_a = the armature current (A)
R_{se} = the series resistance (Ω)
R_a = the armature resistance (Ω)
E_b = the back electric motive force (V)
V_{in} = source voltage (V)
T   = torque (N. m)
n   = speed (rpm).

Example 6.12

A DC series motor, with series resistance 0.2 Ω and armature resistance 0.1 Ω, calculate the back-electric motive force if the value of the drawn current is 50 A and the voltage of the source is 220 V.

Solution

$$E_b = V_{in} - I_a (R_a + R_{se})$$

$$I_a = I_{in} = I_{se} = 50 \text{ A}$$

$$E_b = 220 - 50 (0.1 + 0.2)$$

$$E_b = 220 - 50 \times 0.3$$

$$E_b = 220 - 15$$

$$E_b = 205 \text{ V.}$$

Series DC Motor MATLAB Program

```
ea=[0 10 15 18 31 45 60 75 90 104 120 132 145 160 173 184 200 205 215 223
231];
n0 = 1000;
vt = 220;
ra = 0.15;
ia = 10:1:15;
ea = vt - ia * ra
ea0 = interp1(if_values, ea_values, ia,'spline');
n1 = 1050;
Eao1 = interp1(if_values, ea_values,58,'spline');
Ea1 = vt - 15 * ra;
Eao2 = interp1(if_values, ea_values, ia,'spline');
n = ((ea./Ea1).* (Eao1./ Eao2)) * n1/100;
tind = ea.* ia./ (n * pi / 30);
figure(1);
plot(tind, n,'b-','LineWidth',2.0);
hold on;
xlabel('Tind (N-m)');
ylabel('n(r/min)');
title ('Series DC Motor Torque-Speed Characteristic');
grid on;
hold off;
figure(2);
plot(n, ea,'b-','LineWidth',2.0);
xlabel('n (r/min)');
ylabel('ea (Volt)');
title ('Series DC Motor Back emf-Speed Characteristic');
axis([800 1800 215 220]);
grid on;
%hold off;
figure(3);
plot(n, ia,'b-','LineWidth',2.0);
xlabel('n (r/min)');
ylabel('ia (A)');
title ('Series DC Motor Arm. current-Speed Characteristic');
axis([800 1800 0 50]);
grid on;
%hold off;
```

6.4.1.2 Shunt Motor

The magnetic pole windings are connected in parallel to the armature by carbon brushes. The selection of magnetic pole windings have many turns and a small section area to obtain relatively high resistance. Therefore, the speed of the motor does not change with the change of load, and on this basis, it is used in situations requiring constant speed and when the load changes such as electric locomotives, elevators, printing machines, and papermaking machines. Figure 6.31 represents the equivalent electric circuit of the shunt motor, and Figure 6.32 represents the relationship between speed, torque, and armature current (Figures 6.33 through 6.35).

$$E_b = V_{in} - I_a R_a \tag{6.26}$$

$$I_a = I_{in} - I_{sh} \tag{6.27}$$

$$I_{sh} = V_{in}/R_{sh} \tag{6.28}$$

Motor speed,

$$n = \frac{V_{in}}{K_b \Phi} \tag{6.29}$$

$$K_b = \frac{z.2p}{60 \times 2a} \tag{6.30}$$

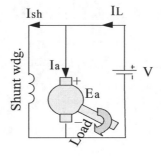

FIGURE 6.31
Equivalent electric circuit of the shunt motor.

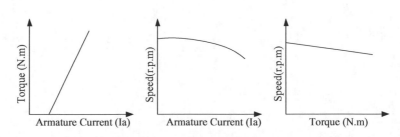

FIGURE 6.32
Relationship between speed, torque, and armature current.

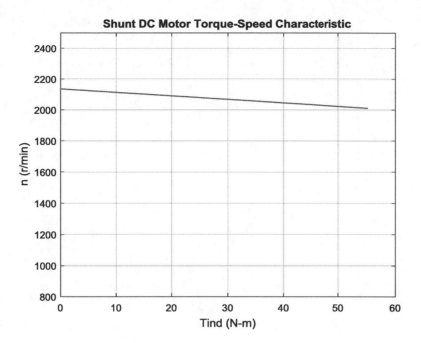

FIGURE 6.33
Torque-speed characteristics of shunt DC motor.

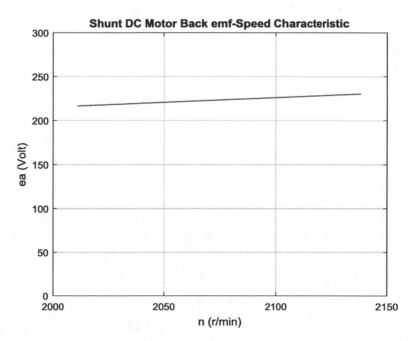

FIGURE 6.34
EMF-speed characteristics of shunt DC motor.

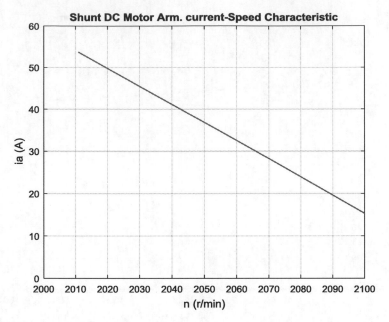

FIGURE 6.35
Armature current-speed characteristics of shunt DC motor.

Example 6.13

A shunt motor operates at 220 V and draws a current of 22 A. If the resistance of shunt 110 Ω and of armature 0.2 Ω, calculate the back-electric motive force.

Solution

$$I_{sh} = V_{in}/R_{sh} = 220/110 \ 2 \ A$$

$$I_a = I_{in} - I_{sh} = 22 - 2 = 20 \ A$$

$$E_b = V_{in} - I_a \, R_a$$

$$E_b = 220 - 20 \times 0.2 = 220 - 4 = 216 \ V.$$

Example 6.14

A four-pole DC shunt motor, 220 V, the number of conductors in the armature 1000, the typical winding coil, the motor draws 52 A, the magnetic flux per pole 0.02 Wb, the armature resistance 0.2 Ω, and shunt resistance 110 Ω. Calculate motor speed.

Solution

$$I_{sh} = V_{in}/R_{sh} = 220/110 = 2A$$

$$I_a = I_{in} - I_{sh} = 52 - 2 = 50 \ A$$

$$E_b = V_{in} - I_a \, R_a$$

$$E_b = 220 - 50 \times 0.2 = 210 \ V$$

$$E_b = \frac{nz\Phi 2p}{120}$$

$$210 = \frac{n \times 1000 \times 0.02 \times 4}{120}$$

$$n = \frac{210 \times 120}{1000 \times 0.02 \times 4} = 315 \text{ rpm}$$

Shunt DC Motor MATLAB Program

```
if_values=[0 0.1 0.2 0.3 0.4 0.5 0.6 0.7 0.8 0.9 1.0 1.1 1.2 1.3 1.4 1.5
1.6 1.7 1.8 1.9 2.0];
ea_values=[0 10 15 18 31.30 45.46 60.26 75.06 89.74 104.4 118.86 132.86
146.46 159.78 172.18 183.98 195.04 205.18...
214.52 223.06 231.2];
n0 = 1500;
vt = 230;
rf = 100;
radd = 75;
ra = 0.250;
il = 0:1:55;
nf = 2700;
far0 = 1500;
ia = il - vt / (rf + radd);
ea = vt - ia * ra;
ish= vt / (rf + radd);
ea0 = interp1(if_values, ea_values, ish);
n = (ea./ ea0) * n0;
Tind = ea.* ia./ (n * 2 * pi / 60);
figure(1);
plot(Tind, n,'b-','LineWidth',2.0);
xlabel('Tind (N-m)');
ylabel('n (r/min)');
title ('Shunt DC Motor Torque-Speed Characteristic');
axis([0 60 800 2500]);
grid on;
%hold off;
figure(2);
plot(n, ea,'b-','LineWidth',2.0);
xlabel('n (r/min)');
ylabel('ea (Volt)');
title ('Shunt DC Motor Back emf-Speed Characteristic');
axis([2000 2150 0 300]);
grid on;
%hold off;
figure(3);
plot(n, ia,'b-','LineWidth',2.0);
xlabel('n (r/min)');
ylabel('ia (A)');
title ('Shunt DC Motor Arm. current-Speed Characteristic');
axis([2000 2100 0 60]);
grid on;
%hold off;
```

6.4.1.3 Compound DC Motor

The compound motor is the basis of a shunt motor with series windings, in which the source current is passed in the short compound motor in Figure 6.36 or the current of the armature in the long compound motor as shown in Figure 6.37. In a certain direction so that the effect of the magnetic field given by these coils is in the magnetic field of shunt windings. Thus, the motor acquires certain properties for speed and torque. The compound motor can be used to obtain high torque and constant speed, which are not significantly affected by the change in load as in moving locomotives, electric buses, and printing machines (Figures 6.38 through 6.40).

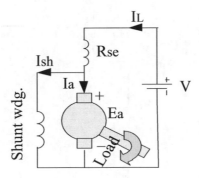

FIGURE 6.36
Short compound motor.

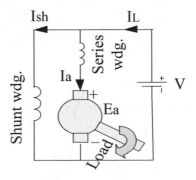

FIGURE 6.37
Long compound motor.

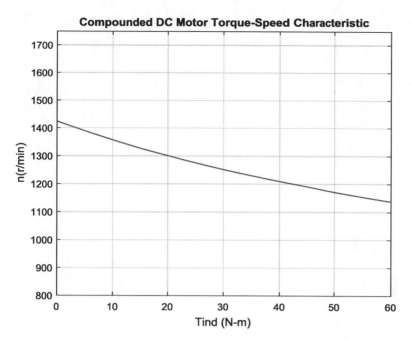

FIGURE 6.38
Torque-speed characteristics of compound DC motor.

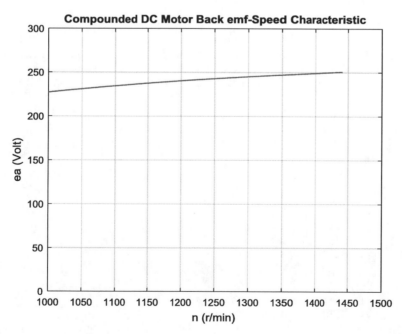

FIGURE 6.39
EMF-speed characteristics of compound DC motor.

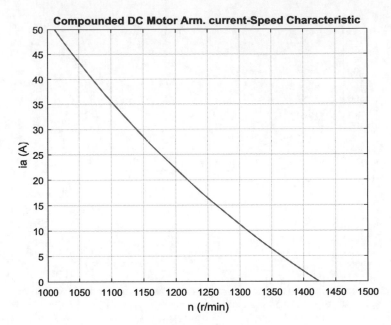

FIGURE 6.40
Armature current-speed characteristics of compound DC motor.

Compound DC Motor MATLAB Program

```
If=[0 0.1 0.2 0.3 0.4 0.5 0.6 0.7 0.8 0.9 1.0 1.1 1.2 1.3 1.4 1.5 1.6 1.7
1.8 1.9 2.0];
ea=[0 10 15 18 31.30 45.46 60.26 75.06 89.74 104.4 118.86 132.86 146.46
159.78 172.18 183.98 195.04 205.18 214.52 223.06 231.2];
n0 = 1000;
vt = 250;
rf = 100;
radd = 75;
ra = 0.44;
il = 0:55;
nf = 2700;
nse = 27;
ia = il - vt / (rf + radd);
ea = vt - ia * ra;
ifi = vt / (rf + radd) + (nse / nf) * ia;
ea0 = interp1(if_values, ea_values, ifi);
n = (ea./ ea0) * n0;
tind = ea.* ia./ (n * pi / 30);
figure(1);
plot(tind, n,'b-','LineWidth',2.0);
xlabel('Tind (N-m)');
ylabel('n(r/min)');
title ('Compounded DC Motor Torque-Speed Characteristic');
axis([0 60 800 1750]);
grid on;
figure(2);
plot(n, ea,'b-','LineWidth',2.0);
```

```
xlabel('n (r/min)');
ylabel('ea (Volt)');
title ('Compounded DC Motor Back emf-Speed Characteristic');
axis([1000 1500 0 300]);
grid on;
%hold off;
figure(3);
plot(n, ia,'b-','LineWidth',2.0);
xlabel('n (r/min)');
ylabel('ia (A)');
title ('Compounded DC Motor Arm. current-Speed Characteristic');
axis([1000 1500 0 50]);
grid on;
%hold off;
```

6.4.2 DC Motor Speed Control

When using DC motors for industrial purposes, it is necessary to control the start of its movement and regulate its speed to suit the purposes used.

The speed changes either by a resistance connected to the armature or by voltage exerted on the ends of the motor or by changing the magnetic flux through the circuit of the field. Both the shunt motor and the compound are like the speed control methods.

6.4.2.1 Speed Control of the Shunt Motor

The speed control methods of the shunt motor are:

1. By using variable resistance:

    Variable resistance is connected with the armature circuit, the speed changes by changing the value of the resistance by a switch that controls variable resistance values. One of the disadvantages of this method is to reduce the efficiency of the motor. As illustrated in Figure 6.41

2. Speed control by using voltage control:

    The speed of the shunt motor can be controlled by controlling the voltage applied to it, as in the case of Ward Leonard. However, this method is very expensive (Figure 6.42)

3. Speed control by field:

    This method is simple and low cost, as low-power field resistance is used to control the current of the field and then the magnetic flux (Figure 6.43).

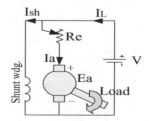

FIGURE 6.41
The speed regulation of a shunt motor using resistance with the armature.

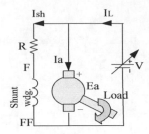

FIGURE 6.42
Speed control by the voltage control method.

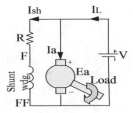

FIGURE 6.43
Speed control by field.

6.4.2.2 Speed Control of the Series Motor

The speed control methods of the series motor are:

1. By connecting series resistance with the motor circuit:

    The motor speed can be changed by adding resistance to the armature circuit (Figure 6.44).

2. Connect resistance in parallel with the field windings:

    Controlling the field current value can only be achieved by connecting resistance in parallel with the field windings. Thus, we can control the current of the field and thus the speed of the motor (Figure 6.45).

3. Speed control by using armature diverter:

    A diverter across the armature can be used for giving speeds lower than the normal speed as shown in Figure 6.46.

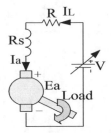

FIGURE 6.44
Connect series resistance with the motor circuits.

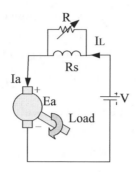

FIGURE 6.45
Resistance parallel with the field windings control the motor speed.

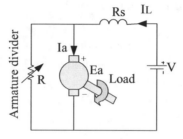

FIGURE 6.46
Armature diverter.

6.4.3 Starting Methods

The purpose of using different ways to start the DC motors is to reduce the current at the start, where this current is very high, and this is illustrated by the current equations in the below:

$$I_a = (V_{in} - E_b)/R_a \tag{6.31}$$

$$I_a = (V_{in} - E_b)/(R_a + R_{se}) \tag{6.32}$$

This means that the current will be very high because of the low resistance to its windings, so the resistance to start the movement until the arrival of the motor to 75% of its actual speed after up growth of the back-electric motive force, which reduces the resistance gradually until the value of zero.

A manual or automatic initiator is used, and the initiator of the motion is a set of conductors that are conductively connected and of which any number can be added or separated from the motor by a variable key.

Problems

6.1 What does the stator part of DC machines contain?

6.2 What are the ways to turn the armature in the DC machine?

6.3 What are the methods of feeding magnetic pole windings in DC generators?

6.4 What is the use of the commutator in the DC machine?

6.5 What is the benefit of the auxiliary poles in the DC machine?

6.6 Draw the DC generator circuit and the motor of the shunt type indicating where the direction of the current.

6.7 Why is it not allowed to operate a DC series motor without a load?

6.8 What types of power losses in DC generators?

6.9 Draw a diagram of the power path in the DC generators.

6.10 What are the ways to speed control of the DC motor?

6.11 How can reverse the rotation of DC motors?

6.12 What is the beneficial use of the resistance to start a movement in the DC motors?

6.13 What is the theory of running DC motors?

6.14 What is the beneficial use of carbon brushes in DC generators?

6.15 What is the effect of the feeding current in the DC motor rotation speed?

6.16 A DC generator have the number of poles four, wave winding, and the number of conductors in the armature 1000 conductor, if the electric motive force generated 200 V, the value of the magnetic flux 0.02 Wb calculated its speed.

6.17 A DC generator feeding a load of 10 Ω, 200 V, the series resistance of 0.2 Ω. Calculate the resistance of the armature.

6.18 A DC generator that feeds a load of 22 kW. The current in the load is 88 A. Calculate the current passing through the shunt windings if you know that the resistance is 125 Ω.

6.19 A short compound generator feeding a load with a current of 98 A, a shunt resistance of 100 Ω, a series resistance of 0.2 Ω, and the current passing through the shunt resistance 2 A. Calculate the armature resistance.

6.20 A short compound DC generator with eight poles, and a lap windings, the number of conductors is 1200, the speed is 600 rpm, the armature current 50 A, magnetic flux 0.02 Wb, resistance of armature windings 0.4 Ω, series resistance 0.1 Ω, shunt resistance 110 Ω, find efficiency if the mechanical and iron losses 1330 W.

6.21 A DC power motor with four poles, the armature has a typical coil, the number of conductors is 1000 conductor, the resistance of the series windings 0.2 Ω, resistance of the armature windings 0.4 Ω, calculate its speed if it is known that the magnetic flux generated in each pole 0.02 Wb, the current drawn 50 amp, and working on the source 230 V.

6.22 A DC motor is designed to reduce its speed, adding resistance in series with the shunt windings, calculate the value of this added resistance, if the armature resistance is known to be 0.5 Ω, and the shunt resistance is 10 Ω, the current is pulled by the motor 60 A, the current that passes the armature resistance is 10 A, and operates at 220 V.

6.23 A long compound DC generator that feeds a load of 2200 W, 220 V, armature resistance of 0.2 Ω, 44 Ω of parallel resistance, and 227.5 V generated electric motive force. Calculate the resistance of the series windings.

7

AC Motors

Your world is a living expression of how you are using and have used your mind.

Earl Nightingale

The machine that converts electrical energy into mechanical energy is the electric motor. Electric motors are characterized by the ability to classify them in sizes and capacities ranging from very small that can be placed in the clocks to the very large used in large cranes and in various industries. It is also characterized by different speeds and the possibility of controlling the speed of rotation, and the possibility of rotation in two directions, and the need for maintenance and no exhaust from them and use electric motors in various devices such as refrigerators, washing machines, vacuum cleaners, mixers, air conditioners, fans, and elevators and many devices. The alternative electric car is a vehicle that contains internal combustion motors that now exist that contribute to environmental pollution by a large amount.

7.1 Single-Phase Motor

The electric motors have been used in the operation of various types of modern devices and equipment that emerged with the emergence of these motors such as audio and video recorders, computer drives, printers, games, and industrial robots of uses.

The principle of the work of the electric motor is the opposite of the principle of the work of the generator when following an electric current in a wire located within a fixed magnetic field generates mechanical force affect the wire and push it to the movement, and thus could convert electricity to mechanical energy (Figure 7.1).

The single-phase electric motors are of many types depending on the nature of their work and design.

1. Induction-motors that decrease in speed with increased load
2. Synchronous motors that do not affect the speed when the load is changed on the axis of rotation
3. General motors (universal-motors).

7.1.1 Induction Motors

These machines are the most widely used and used in daily use for household, office, medical, and other purposes, because they are simple in design, low cost, good efficiency, low power, and rotate at different speeds (Figure 7.2).

FIGURE 7.1
Different types of single-phase motor.

(a)Squirrel cage rotor motor.   (b)Slip ring motor.

FIGURE 7.2
Single-phase induction motor.

7.1.1.1 Motor Construction

These motors consist of major parts that are present in all types, and additional parts are only present in some of them. The main parts are:

1. **The stator**: It consists of three basic parts:
   a. **Yoke (outer frame)**

      It is made up of steel (cast iron) and contains fins on the outer surface, which work to cool the coils during the air pump from the cooling fan and uses the frame to carry the sheets (the chips) of the core of the iron, as well as to install the side sinks. It lists motor specifications such as current, voltage, frequency, speed, capacity, and so on (Figure 7.3).

   b. **Iron core**

      It is made up of iron-silicon wafers or sheets, isolated from each other by varnishes and compressed to form the iron core.

      Figure 7.4a shows the iron core flakes of the stator and the rotor, and Figure 7.4b shows the iron core after the collection of chips.

FIGURE 7.3
The outer shape of the single-phase induction motor. (https://www.indiamart.com/proddetail/induction-motor-11915087030.html).

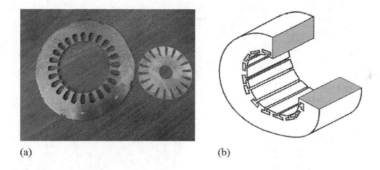

(a)   (b)

FIGURE 7.4
Iron core of single-phase induction motor. (a) The iron core chips of the stator and the rotor. (b) The iron core after collecting the chips (https://www.indiamart.com/proddetail/induction-motor-11915087030.html).

   c. **Running winding coils**

      Made of copper wires isolated by varnish, thick wires, called *operating coils*, and the number of turns proportional to the capacity of the motor and connect the AC source (Figure 7.5).

2. The rotor

The rotor is made up of a spindle, which is based on the side rails in parallel to the occupant. It presses on the shaft of iron sheets that are isolated from each other by varnishes. The outer perimeter has long, straight, or sloping ducts where copper or aluminum rods are placed. Aluminum is usually poured directly into these slots.

These coils are not fed by an electrical current. This kind of motor is called *squirrel cage motors*. Figure 7.6a shows the rotor in the induction motors in which the slots are slanted, and Figure 7.6b shows where the slots are fair inside the stator.

Figure 7.7 shows different sizes of the iron cores of the rotor and the stator.

From the observation of the construction of these motors that there is no electrical connection between the stator and rotor, and the only link is the electromagnetic bond, that is, this type of motor works in magnetic induction.

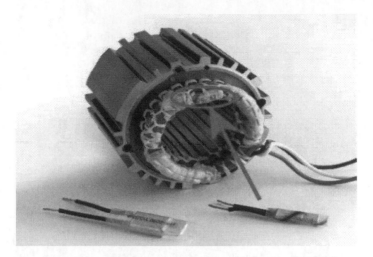

FIGURE 7.5
The stator coils of a single-phase induction motor.

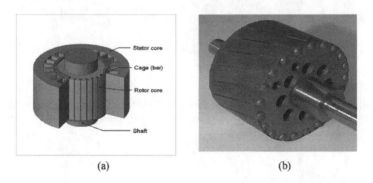

FIGURE 7.6
(a) The rotor in induction motors with skewed slots. (b) The slots are fair inside the stator.

FIGURE 7.7
Different sizes of the iron cores of the rotor and the stator.

FIGURE 7.8
The side cover of the motor.

3. The two sides cover

They are made of solid steel, i.e., of the metal frame itself, are fixed by bolts, and the central part of the centrifuge switch is placed in the front cover.

The two covers are mounted on the spindle, balancing the rotor, facilitating rotation, and making it in a position to move freely. Figure 7.8 shows the motor side cover.

4. Ventilation fan

It is an important part of aluminum or plastic and operates during the rotation of the motor, the air flux between the fins of the frame. So, the temperature resulting from following the current in the fixed iron core coils will reduce.

7.1.1.2 The Speed

In induction motors, the speed of the magnetic field of the stator is higher than the actual speed of the rotor. The reason is the air resistance of the rotor, and the friction between the axis of rotation and the supports, and this difference is called *slip*.

The speed of the magnetic field in the stator is proportional to the frequency of the source, and inversely with the number of poles of the motor (2P), i.e.:

$$N_S = 60 \times f/p \tag{7.1}$$

where:
N_S = the speed of the magnetic field and is measured in (rpm)
F = source frequency, measured in (Hz)
P = the number of pairs of poles
2P = the number of poles.

Example 7.1

A motor that has four poles connected to a frequency source 50 Hz, calculate its harmonic speed?

Solution

$$N_S = 60 \times f/P$$

$$N_S = 60 \times 50/2 = 1500 \text{ rpm.}$$

Example 7.2

An induction motor has the speed of 3000 rpm, connect to a source of frequency 50 Hz. Find the number of its poles.

Solution

$$N_S = 60 \, f/P$$

$$N_S \, P = 60 \, f$$

$$P = 60 \times f/N_s$$

$$P = 60 \times 50/3000 = 1 \text{ pole pair}$$

$$2p = 2 \text{ poles.}$$

7.1.1.3 The Theory of Work

When a single-phase alternating voltage is applying on the stator windings, the current following through it will generate a magnetic field that moves to the rotor windings through the air gap, then cuts it and electromotive force (EMF) voltage will generate, then lead to the passage of an electric current that constitutes another magnetic field of mutual interaction between the two magnetic fields of the stator and the rotor. Electromagnetic forces are composed of the sum of these forces, and as a result, the rotor's torque that is given on the rotor is zero.

To operate the motor, rotor rotation is required by creating a primary torque. This is done by displacing the magnetic field in the rotor from its position to create an angle between it and the magnetic field in the constant.

7.1.1.4 Starting Methods of the Motor

There are several ways to start single-phase induction motors, including:

1. Starting Windings Method

These windings are placed in parallel with the stator's windings (in the stator's slots) have a small section area and high resistance, occupy one-third of the total number of slots, reach parallel to the operating coils, and are usually designed for a limited period not exceeding several seconds. The centrifuge switch is connected sequentially to the auxiliary coils to separate it from the main circuit. When the motor speed reaches (75%) of the synchronous speed. The torque in this type is relatively low, efficiency and power factor are low.

The usefulness of these windings is when the motor is connected to the source. A current in the start coils passes at an angle from the current of the main operating coils and then two different fields occur, thus creating a rotary magnetic field that produces a primary torque for the motor. Figure 7.9 shows the connection of the circuit containing the auxiliary coils and the centrifuge switch. This method is not enough to give a high starting torque, so another factor must be added.

2. Capacitor Start Method

In this method, the capacitor is connected continuously with the start coils and the centrifuge switch. This circuit is connected in parallel with the main operation coils, and Figure 7.10 shows the circuit connection.

When the motor is connected to the source, a current follow through both the start and run windings, but the presence of the capacitor provides the flowing current in the start windings from the current in the operating windings at a given angle, thus creating a starting torque due to the two-phase differentials. The starting torque of the motor is high and very efficient and also reduces the consumption of current. This type of motors is used in washing machines and large refrigerators.

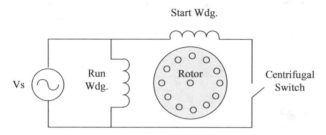

FIGURE 7.9
The connection of the circuit containing the auxiliary coils and the centrifuge switch.

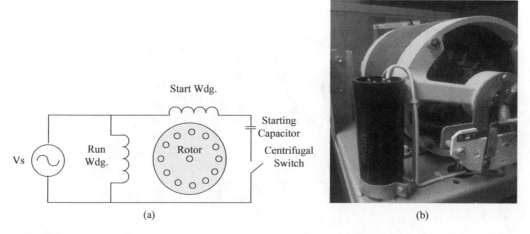

(a)   (b)

FIGURE 7.10
Capacitor start motor. (a) Circuit diagram. (b) Start capacitor.

The capacitor may remain connected to the starting coils to the source and continuously when the motor is operating. On this basis, the auxiliary coils are designed so that they are not damaged and, in this case, you do not need a centrifuge switch. The result is a low noise during operation and a power factor improvement. This type of motor is used in applications that require low noise during operation as in ceiling and desktop fans and called *capacitor run motor*. Figure 7.11 shows the connection of this type of motor.

There are motors that contain capacitors. The first is run capacitor, a small paper type saturated with oil and connected in parallel with the auxiliary coil circuit, and the second is the start capacitor which is connected in series with the auxiliary coils and the centrifuge switch.

Figure 7.12a shows the method of connecting the capacitors in the motor circuit, and Figure 7.12b shows the form of motor containing capacitors.

3. Shaded Pole Motor

In these motors the starting torque is low, and the stator is different in which the motors containing the starting motion coils. The stator is made of thin sheets of iron assembled together with prominent poles placed around the operating coils, and to obtain a magnetic field rotary to start the movement, a slit occurs on one side of the pole and placed around it a ring of copper wire is a bit thick and represent a restricted coil called *shaded pole*. This circuit is compensated for starting coils in single-phase motors. The rotor of these motors is of the type of squirrel cage.

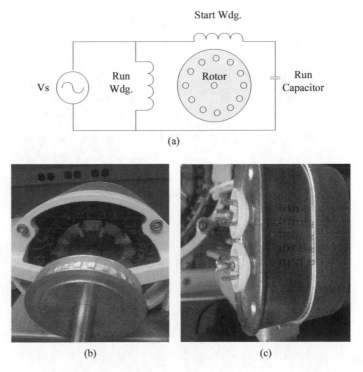

(a)

(b)   (c)

FIGURE 7.11
Capacitor run motor. (a) Circuit diagram. (b) Stator and rotor. (c) Run capacitor.

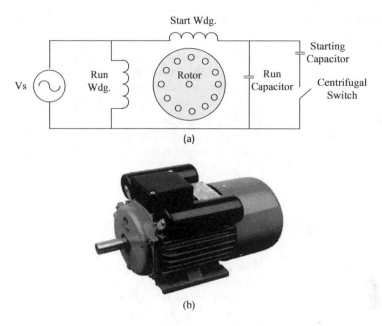

(a)

(b)

FIGURE 7.12
Capacitor start with capacitor run motor. (a) Capacitor start with capacitor run motor circuit. (b) Motor containing two capacitors.

The electric losses in the shaded ring are relatively large at the rated speed, making so the efficiency low, the motors are often designed at a low power of not more than 40 W. However, due to some improvements in motor design, common uses for this motor are up to 300 W. It also has a low power factor.

Despite the disadvantages of this type of motor, it also finds many uses because of its low cost and simplicity of design, and the noise level is very low due to the lack of slots, as well as the appropriate choice of structural parts and exact calculation of motor dimensions even it is possible to improve the power index and increase the starting torque and may be designed sometimes using two rings are short for each pole, and there are other designs intended to obtain the optimal rotational magnetic field. Figure 7.13 shows the forms of stator and rotor of the shaded pole motor.

FIGURE 7.13
Shaded pole motor. (https://www.indiamart.com/proddetail/shaded-pole-motor).

The working theory of shaded poles motor:

The main coil in the stator acts as an exciting coil in the DC motors, which is responsible for the formation of the magnetic field in the motor. When an alternating current in the coils creates a magnetic field that moves to the copper ring shorted, and as it differs from the main coils in terms of cross-section area for this generates high current as well as a magnetic field opposite to the original field and the latter at an angle, thus obtaining a difference angle between the two fields leads to the emergence of torque by the lag angle. The angle of lag between the two fields in this type of motor is very low, and this is the torque is low, and enough to run light loads.

7.1.2 Synchronous Motors

Synchronous motors are those which synchronize the speed of the magnetic field (N_s) in the stator with rotor speed (N_r), where there is no difference between them, that is, the two speeds are equal, and the slip is zero.

The power of synchronous motors ranges from 1 W to 10 MW or more and operates at any voltage (single or multi-phase) at speeds ranging from 125 to 3600 rpm. Synchronous motors are practical to operate at a speed of less than 500 rpm. When directly connected to the load, as in the case of pumps, mills, and foams, for example, with a power greater than 75 kW, the synchronous motor rated at this low speed is less expensive when compared with the induction motor of the same size. The speed of synchronous motors will not affect by any load. The synchronous motors can not start by itself, it needs an auxiliary external means to rotate them. As soon as their rotation speed reaches the synchronous speed, they remain at a constant speed. Figure 7.14 shows different types of synchronous motors.

The methods used to start the motor are:

1. Put a squirrel-cage coil in the rotor poles called *the starting coil*, so the motor starts as an induction motor until it reaches its synchronous speed. When the DC voltage is placed on the excitation coil, the motor is attracted to the synchronization state

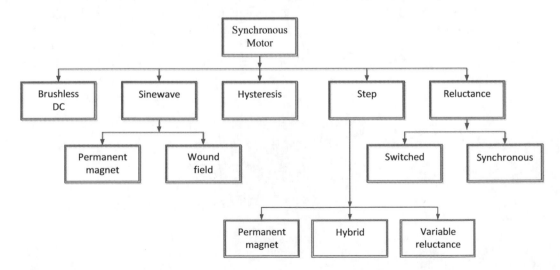

FIGURE 7.14
Types of synchronous motors.

2. Reduce the speed of rotation of the stator's field significantly by reducing the frequency of the voltage applied on the stator windings. Then start the rotor turn slowly to increase with increasing frequency.

7.1.2.1 Synchronous Motor Construction

1. Stator

It is like a stator in induction motors. The iron core is composed of iron chips and contains internal slots where the main windings connect to an AC source.

2. Rotor

The magnetic poles that are permanent magnetism, which consists of an alloy of different metals, the hysteresis loop or the power of de-magnetization are high and used in small machines, or coils fed by a direct current follow the slip-rings and carbon brushes.

The shape of the rotor varies according to the type of poles used, the number and the size of the machine, and its speed. They may be salient poles. These are suitable for slow and medium speed. In high-speed machines, the rotor is a hidden pole, which appears in a cylindrical shape containing slots, the magnetism that feeds from a constant current source through slip-rings and carbon brushes.

7.1.2.2 Synchronous Motor Operation Theory

When the alternating current flows from the source to the stator coils, when positive half-wave follow the polarity of the coils will be positive. Thus, there will be repulsion between the coils and the magnetic poles for the same polarity (positive and positive). This repulsion causes the poles to deviate from their position at a certain angle. In a moment when the half-wave changes to the negative wave, the polarity of the coils will be negative, and thus the polarization will take place between them and the poles. The magnetic poles then return to their original position. Therefore, the starting torque of these motors is zero. For starting torque tracking procedures already mentioned.

The efficiency of these motors is high, employing a high-power factor. Some of these motors are used in wall clocks, timers, and sensitive measuring devices, used in printing machines and textile laboratories. Figure 7.15 shows the stator and the rotor and how the current follows.

7.1.2.3 Synchronous Motor Features

1. Higher power factor.
2. High efficiency and overload capacity.
3. The speed of rotation is fixed within the field of regular load.

7.1.3 Universal Motor

A universal motor (or series AC motor) is a motor that can be operated in direct current, and alternating current about the same speed. It is frequently used in horsepower motors that are used in household uses such as mixers, drill bits, and sewing machines.

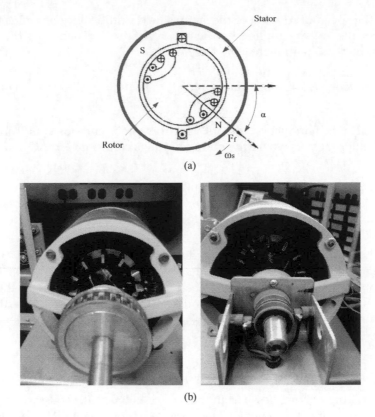

(a)

(b)

FIGURE 7.15
(a) The stator and the rotor and how the current follow. (b) Synchronous motor.

It has a high starting torque, which is variable speed, and its speed is high when it is not overloaded, and its speed decreases as the load increases. Figure 7.16 shows the shape of the motor and the internal structure of the motor.

7.1.3.1 Universal Motor Construction

1. Stator: It contains the salient magnetic poles placed around the coils of copper and reach the coils of magnetic poles, respectively, with the rotor by carbon brushes.
2. Rotor: Made of a shaft placed around it the chips of iron after the collection to form the iron core, which contains long slots from the outside, where the coils are placed of copper insulated, and reach the ends of the coils to the parts of the commutator according to the type of winding mentioned in Chapter six. Figure 7.17 shows the iron core of the stator and rotor.

7.1.4 Centrifuge Switches

One of the most common switches used in most motors is the one-phase centrifuge switch. Which consists of two main parts:

1. The first part contains two contact points and is fixed in one of the side covers or in the front of the stator.

(a) (https://en.wikipedia.org/wiki/Universal_motor).

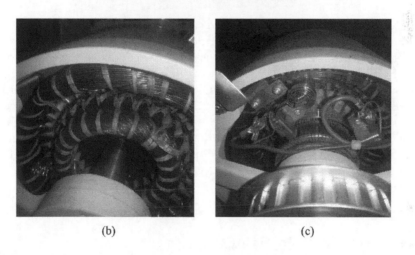

(b)   (c)

FIGURE 7.16
The shape of the motor and the internal structure of the universal motor. (a) The stator and rotor (b) The windings (c) The centrifuge switch

FIGURE 7.17
The iron core of the stator and rotor.

FIGURE 7.18
The construction of the centrifuge switch. (beyondelectric.com/Html/productshow.asp).

2. The second part is a movement on the spindle and is influenced by the centrifugal forces resulting from rotor rotation. It contains a set of springs that make the two points of contact touching when the motor is stationary.

The switch works to open and close the two contact points in the help coil circuit.

When the rotation starts, the two points are closed, completing the current circuit in the auxiliary coils, and after the spinner reaches 75% of its actual speed, the two contact points are opened by the centrifugal force influencing the moving part of the switch. When the motor is stopped, the moving part returns to its position, closes the contact points, and completes the auxiliary coil circuit. Figure 7.18 shows the construction of the centrifuge switch.

As well as the main types of motors, which mentioned that there are special types of motors designed to meet the need of some applications, the most important of which.

7.2 Three-Phase Induction Motor

Induction motors are the most widely used motors for appliances, industrial control, and automation. Hence, they are often called the *workhorse* of the motion industry. They are robust, reliable, and durable. When power is supplied to an induction motor at the recommended specifications, it runs at its rated speed. However, many applications need variable speed operations. For example, a washing machine may use different speeds for each wash cycle. Historically, mechanical gear systems were used to obtain variable speed. Recently, electronic power and control systems have matured to allow these components to be used for motor control in place of mechanical gears. These electronics not only control the motor's speed, but can improve the motor's dynamic and steady-state characteristics. In addition, electronics can reduce the system's average power consumption and noise generation of the motor. Induction motor control is complex due to its nonlinear characteristics.

7.2.1 Construction of Induction Machines

The induction machine is a very important AC machine. It is mostly used as a motor. The stator and the rotor are made of laminated steel sheets with stamped in slots. The stator slots contain one symmetrical three-phase winding, which can be connected to the three-phase network in a star or delta connection. The stator of a simple induction machine has six slots per pole pair, in each case one for the forward and one for the backward conductor for each phase winding. Generally, the winding is carried out with many pole pairs (p > 1) and distributed in different slots (q > 1).

We can distinguish induction machines according to the type of the rotor between squirrel cage motors and slip ring motors. The stator construction is the same in both motors.

7.2.1.1 Squirrel Cage Motor

Almost 90% of induction motors are squirrel cage motors. This is because the squirrel cage motor has a simple and rugged construction. The rotor consists of a cylindrical laminated core with axially placed parallel slots for carrying the conductors. Each slot carries a copper, aluminum, or alloy bar. If the slots are semi-closed, then these bars are inserted from the ends.

These rotor bars are permanently short-circuited at both ends by means of the end rings, as shown in Figure 7.1. This total assembly resembles the look of a squirrel cage, which gives the motor its name. The rotor slots are not exactly parallel to the shaft. Instead, they are given a skew for two main reasons:

1. To make the motor run quietly by reducing the magnetic hum
2. To help reduce the locking tendency of the rotor

Rotor teeth tend to remain locked under the stator teeth due to the direct magnetic attraction between the two. This happens if the number of stator teeth is equal to the number of rotor teeth.

7.2.1.2 Slip Ring Motors

The windings on the rotor are terminated to three insulated slip rings mounted on the shaft with brushes resting on them. This allows an introduction of an external resistor to the rotor winding. The external resistor can be used to boost the starting torque of the motor and change the speed-torque characteristic. When running under normal conditions, the slip rings are short-circuited, using an external metal collar, which is pushed along the shaft to connect the rings. So, in normal conditions, the slip ring motor functions like a squirrel cage motor (Figure 7.19 through 7.22).

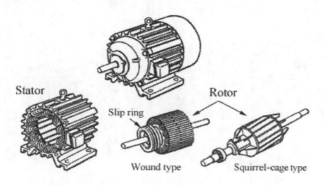

FIGURE 7.19
Rotor structure of induction machines. (https://www.ato.com/three-phase-induction-motor-construction).

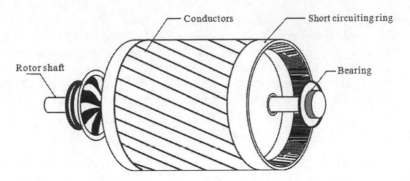

FIGURE 7.20
Typical squirrel cage rotor. (https://www.theengineeringprojects.com).

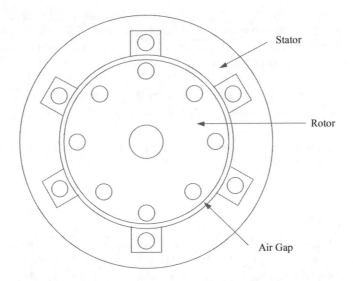

FIGURE 7.21
Schematic construction principle of an induction machine.

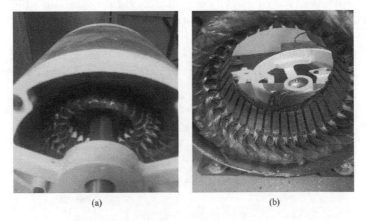

FIGURE 7.22
Three-phase induction motor. (a) Wound rotor. (b) Squirrel cage.

7.2.2 Operation

When the rated AC supply is applied to the stator windings, it generates a magnetic flux rotating at synchronous speed with a constant magnitude. The flux passes through the air gap, sweeps past the rotor surface, and through the stationary rotor conductors. An EMF is induced in the rotor conductors due to the relative speed differences between the rotating flux and stationary conductors. The voltage induced in the bars will be slightly out of phase with the voltage in the next one. Since the rotor bars are shorted at the ends, the EMF induced produces a current in the rotor conductors. The flux linkages will change in it after a short delay. If the rotor is moving at synchronous speed, together with the field, no voltage will be induced in the bars or the windings.

The frequency of the induced EMF is the same as the supply frequency. Its magnitude is proportional to the relative speed between the flux and the conductors. The direction of the rotor current opposes the relative speed between rotating flux produced by the stator and stationary rotor conductors according to Lenz's law.

To reduce the relative speed, the rotor starts rotating in the same direction as that of flux and tries to catch up with the rotating flux. But in practice, the rotor never succeeds in "catching up" to the stator field. So, the rotor runs slower than the speed of the stator field. This difference in speed is called *slip speed*. This slip speed depends upon the mechanical load on the motor shaft (Figure 7.23).

The frequency and speed of the motor, with respect to the input supply, is called *the synchronous frequency* and *synchronous speed*. Synchronous speed is directly proportional to the ratio of supply frequency and a number of poles in the motor. The rotor speed n_r of an induction motor is given by:

$$n_r = \frac{120f}{P}(1-S) \qquad (7.2)$$

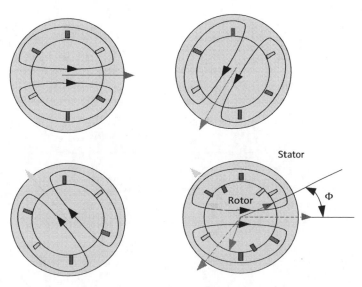

FIGURE 7.23
Concept of operation of three-phase induction motor.

7.2.3 Speed-Torque Characteristics of Induction Motor

Figure 7.24 shows the typical speed-torque characteristics of an induction motor. The *X*-axis shows speed and slip. The *Y*-axis shows the torque and current. The characteristics are drawn with rated voltage and frequency supplied to the stator.

During start-up, the motor typically draws up to seven times the rated current. This high current is a result of stator and rotor flux, the losses in the stator and rotor windings, and losses in the bearings due to friction. This high starting current overcomes these components and produces the momentum to rotate the rotor.

At start-up, the motor delivers 1.5 times the rated torque of the motor. This starting torque is also called *locked rotor torque* (LRT). As the speed increases, the current drawn by the motor reduces slightly (see Figure 7.24). The current drops significantly when the motor speed approaches to 80% of the rated speed. At base speed, the motor draws the rated current and delivers the rated torque.

At base speed, if the load on the motor shaft is increased beyond its rated torque, the speed starts dropping and slip increases. When the motor is running at approximately 80% of the synchronous speed, the load can increase up to 2.5 times the rated torque. This torque is called *breakdown torque*. If the load on the motor is increased further, it will not be able to take any further load and the motor will stall.

In addition, when the load is increased beyond the rated load, the load current increases following the current characteristic path. Due to this higher current flow in the windings, inherent losses in the windings increase as well. This leads to a higher temperature in the motor windings. Motor windings can withstand different temperatures, based on the class of insulation used in the windings and cooling system used in the motor. Some motor manufacturers provide the data on overload capacity and load over a duty cycle. If the motor is overloaded for longer than recommended, then the motor may burn out.

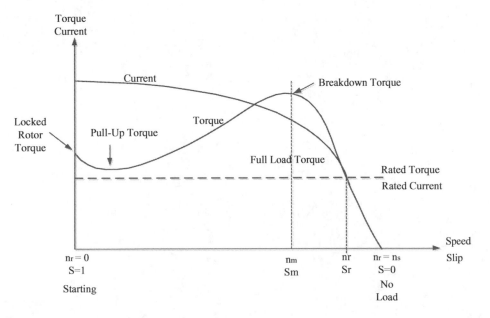

FIGURE 7.24
Speed-torque characteristics of induction motors.

As seen in the speed-torque characteristics, torque is highly nonlinear as the speed varies. In many applications, the speed needs to be varied, which makes the torque vary. We will discuss a simple open loop method of speed control called, variable voltage variable frequency (VVVF or V/*f*) in this application note.

7.2.4 Speed-Torque Characteristics of Induction Motors Using MATLAB Program

%Three phase induction motor

```
r1 = 0.200;
x1 = 0.250;
r2 = 0.10;
x2 = 0.250;
xm = 25.0;
v_phase = 208 / sqrt(3);
nsync = 1800;
wsync = 2*pi*nsync/60;
vth = v_phase * (xm / sqrt(r1^2 + (x1 + xm)^2));
zth = ((j*xm) * (r1 + j*x1)) / (r1 + j*(x1 + xm));
rth = real(zth);
xth = imag(zth);
s = (0:1:50) / 50;
s(1) = 0.001;
nm = (1 - s) * nsync;
for i = 1:51
td(i) = (3 * vth^2 * r2 / s(i)) / (wsync * ((rth + r2/s(i))^2 +
(xth + x2)^2));
end
figure(1);
plot(nm, td,'k-','LineWidth',2.0);
xlabel('n (r/m)');
ylabel('Td (N.m)');
title ('Induction Motor Torque-Speed Characteristics');
grid on;
```

7.2.5 Basic Equations and Equivalent Circuit Diagram

The stator and rotor of the induction machine both are equipped with a symmetrical three-phase winding. Because of the symmetry, it is sufficient to take only one phase into consideration. Every phase of the stator and the rotor winding has an active resistance of R_1 and R_2, as well as a self-inductance of L_1 and L_2. The windings of the stator and the rotor are magnetically coupled through a mutual inductance M. Since the current flowing in the stator winding has the frequency f_1 and the current flowing in the rotor winding has the frequency f_2, then at the rotor speed n,

- Currents induced from the stator into the rotor have $f = f_2$
- Currents induced from the rotor into the stator have $f = f_1$.

According to this, voltage equations for the primary and secondary sides can be derived. The equivalent circuit diagram after the conversion of the rotor parameters on the stator side is presented in Figure 7.25.

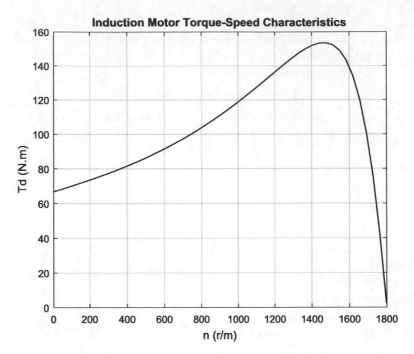

FIGURE 7.25
Speed-torque characteristics of induction motors using the MATLAB program.

To simplify the determination of torque and power equations from the induction machine equivalent circuit, the total real power per phase that crosses the air gap (the air gap power = $P_{\text{air gap}}$) and is delivered to the rotor is:

$$P_{airgap} = I_2'^2 \left[R_{w2}' + \frac{R_{w2}'}{S}(1-S) \right] = I_2'^2 \frac{R_{w2}'}{S} \tag{7.3}$$

The portion of the air gap power that is dissipated in the form of ohmic loss (copper loss $P_{r.cu}$) in the rotor conductors is:

$$P_{r.cu} = I_2'^2 R_{w2}' \tag{7.4}$$

The total mechanical power (P_{mech}) developed internally to the motor is equal to the air gap power minus the ohmic losses in the rotor which gives:

$$P_{mech} = P_{airgap} - P_{r.cu} = I_2'^2 \frac{R_{w2}'}{S} - I_2'^2 R_{w2}' = P_{airgap}(1-S)$$

$$P_{mech} = P_{airgap}(1-S) \tag{7.5}$$

$$P_{r.cu} = P_{airgap}S \tag{7.6}$$

According to the previous equations, of the total power crossing the air gap, the portion S goes to ohmic losses while the portion (1–S) goes to mechanical power. Thus, the induction machine is an efficient machine when operating at a low value of slip. Conversely, the induction machine is a very inefficient machine when operating at a high value of slip. The overall mechanical power is equal to the power delivered to the shaft of the machine plus losses (windage, friction).

The mechanical power (P_{mech}) is equal to torque (N-m) time's angular velocity (rad/s). Thus, we may write:

$$P_{mech} = T.\omega = I_2'^2 \frac{R_{w2}'}{S}(1-S) \tag{7.7}$$

where T is the torque and ω is the angular velocity of the motor in radians per second given by:

$$\omega = \frac{2\pi.n}{60} = \frac{2\pi.n_s}{60}.(1-S) = \omega_s.(1-S) \tag{7.8}$$

where ω_s is the angular velocity at synchronous speed. Using the previous equation, we may write:

$$(1-S) = \frac{\omega}{\omega_s} \tag{7.9}$$

Inserting this result into the equation relating torque and power gives:

$$P_{mech} = T.\omega = I_2'^2 \frac{R_{w2}'.\omega}{S.\omega_s} \tag{7.10}$$

Solving this equation for the torque yields:

$$T = \frac{I_2'^2 R_{w2}'}{S.\omega_s} = \frac{P_{air\ gap}}{\omega_s} \tag{7.11}$$

Returning to the Thevenin transformed equivalent circuit, we find:

$$I_2'^2 R_{w2}' = \left[\frac{V_{th}}{\left(R_{th} + \dfrac{R_{w2}'}{S}\right) + j\left(X_{th} + X_2'\right)} \right]^2 R_{w2}'.$$

Note that the previous equation is a phasor while the term in the torque expression contains the magnitude of this phasor. The complex numbers in the numerator and denominator may be written in terms of magnitude and phase to extract the overall magnitude term desired.

$$I_2'^2 R_{w2}' = \left[\frac{V_{th} \angle \theta 1}{\left(R_{th} + \dfrac{R_{w2}'}{S} \right)^2 + j\left(X_{th} + X_2' \right)^2 \angle \theta 2} \right]^2 R_{w2}'$$

The magnitude of the previous expression is:

$$I_2'^2 R_{w2}' = \left[\frac{V_{th} R_{w2}'}{\left(R_{th} + \dfrac{R_{w2}'}{S} \right)^2 + j\left(X_{th} + X_2' \right)^2} \right]$$

Inserting this result into the torque per phase equation gives:

$$T = \frac{1}{\omega_s} \cdot \left[\frac{V_{th}^2 \left(R_{w2}' / S \right)}{\left(R_{th} + \dfrac{R_{w2}'}{S} \right)^2 + \left(X_{th} + X_2' \right)^2} \right] \tag{7.12}$$

This equation can be plotted as a function of slip for a particular induction machine yielding the general shape curve shown in Figure 7.24. At low values of slip, the denominator term of $R_{w2}N/s$ is dominated and the torque can be accurately approximated by:

$$T = \frac{1}{\omega_s} \cdot \left[\frac{V_{th}^2 S}{R_{w2}'} \right] \tag{7.13}$$

where the torque curve is approximately linear in the vicinity of $S = 0$. At large values of slip ($S = 1$ or larger), the overall reactance term in the denominator of the torque equation is much larger than the overall resistance term such that the torque can be approximated. The torque is therefore inversely proportional to the slip for large values of slip. Between $S = 0$ and $S = 1$, a maximum value of torque is obtained. The maximum value of torque with respect to slip can be obtained by differentiating the torque equation with respect to S and setting the derivative equal to zero. The resulting maximum torque (called *the breakdown torque*) is:

$$T_{max} = \frac{1}{2\omega_s} \left[\frac{V_{th}^2}{R_{th} + \sqrt{R_{th}^2 + \left(X_{th} + X_2' \right)^2}} \right] \tag{7.14}$$

and the slip at this maximum torque is:

$$S_{max.T} = \left[\frac{R_{w2}'}{\sqrt{R_{th}^2 + \left(X_{th} + X_2' \right)^2}} \right] \tag{7.15}$$

If the stator winding resistance R_{w1} is small, then the Thevenin resistance is also small, so that the maximum torque and slip at maximum torque equations are approximated by:

$$T_{\max} = \frac{1}{2\omega_s}\left[\frac{V_{th}^2}{X_{th} + X_{12}'}\right]$$

$$S_{\max.T} = \frac{R_{w2}^2}{X_{th} + X_{12}'} \tag{7.16}$$

The efficiency of an induction machine is defined in the same way like that for a transformer. The efficiency (η) is the ratio of the output power (P_{out}) to the input power (P_{in}).

$$\eta = \frac{P_{out}}{P_{in}} \times 100\% \tag{7.17}$$

The input power is found using the input voltage and current at the stator. The output power is the mechanical power delivered to the rotor minus the total rotational losses.

$$P_{in} = 3V_\phi I_\phi \cos(\theta_v - \theta_i)$$

$$P_{out} = P_{mech} - P_{rot} = (1-S)P_{airgap} - P_{rot} \tag{7.18}$$

The *internal efficiency* (η_{int}) of the induction machine is defined as the ratio of the output power to the air gap power which gives:

$$\eta_{int} = \frac{P_{out}}{P_{airgap}} = (1-S) \tag{7.19}$$

The internal efficiency gives a measure of how much of the power delivered to the air gap is available for mechanical power.

7.2.6 No-Load Test and Blocked Rotor Test

The equivalent circuit parameters for an induction motor can be determined using specific tests on the motor, just as was done for the transformer.

7.2.6.1 No-Load Test

Balanced voltages are applied to the stator terminals at the rated frequency with the rotor uncoupled from any mechanical load. Current, voltage, and power are measured at the motor input. The losses in the no-load test are those due to core losses, winding losses, windage, and friction. In no-load test, the slip of the induction motor at no-load is very low. Thus, the value of the equivalent resistance:

$$\frac{R_{w2}'}{S}(1-S).$$

In the rotor branch of the equivalent circuit is very high. The no-load rotor current is then negligible and the rotor branch of the equivalent circuit can be neglected. The approximate equivalent circuit for the no-load test becomes as in Figure 7.26.

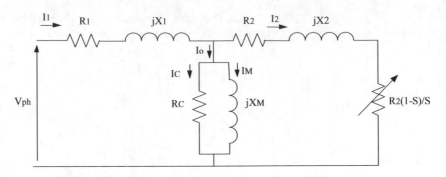

FIGURE 7.26
Equivalent circuit diagram of the induction machine.

Induction machine equivalent circuit for the no-load test note that the series resistance in the no-load test equivalent circuit is not simply the stator winding resistance. The no-load rotational losses (windage, friction, and core losses) will also be seen in the no-load measurement. This is why the additional measurement of the DC resistance of the stator windings is required. Given that the rotor current is negligible under no-load conditions, the rotor copper losses are also negligible. Thus, the input power measured in the no-load test is equal to the stator copper losses plus the rotational losses.

$$P_{NL} = P_{cu1} + P_{rot.} \tag{7.20}$$

where the stator copper losses are given by:

$$P_{cu1} = 3I_{NL}^2 R_{w1} \tag{7.21}$$

From the no-load measurement data (V_{NL}, I_{NL}, P_{NL}) and the no-load equivalent circuit, the value of R_{NL} is determined from the no-load dissipated power.

$$P_{NL} = 3I_{NL}^2 R_{NL} \rightarrow R_{NL} = \frac{P_{NL}}{3I_{NL}^2} \tag{7.22}$$

The ratio of the no-load voltage to current represents the no-load impedance which, from the no-load equivalent circuit, is:

$$\frac{V_{NL}}{I_{NL}} = Z_{NL} = \sqrt{R_{NL}^2 + \left(X_{l1} + X_{ml}\right)^2} \tag{7.23}$$

And the blocked rotor reactance sum $X_{l1} + X_{ml}$ is:

$$X_{l1} + X_{m1} = \sqrt{Z_{NL}^2 - R_{NL}^2} \tag{7.24}$$

Note that the values of X_{l1} and X_{m1} are not uniquely determined by the no-load test data alone (unlike the transformer no-load test). The value of the stator leakage reactance can be determined from the blocked rotor test. The value of the magnetizing reactance can then be determined.

7.2.6.2 Blocked Rotor Test

The rotor is blocked to prevent rotation and balanced voltages are applied to the stator terminals at a frequency of 25% of the rated frequency at a voltage where the rated current is achieved. Current, voltage, and power are measured at the motor input.

In addition to these tests, the DC resistance of the stator winding should be measured to determine the complete equivalent circuit.

The slip for the blocked rotor test is unity since the rotor is stationary. The resulting speed-dependent equivalent resistance:

$$\frac{R'_{w2}}{S}(1-S).$$

Goes to zero and the resistance of the rotor branch of the equivalent circuit becomes very small. Thus, the rotor current is much larger than the current in the excitation branch of the circuit such that the excitation branch can be neglected.

The resulting equivalent circuit for the blocked rotor test is shown in Figure 7.27.

The reflected rotor winding resistance is determined from the dissipated power in the blocked rotor test (Figure 7.28).

$$P_{BR} = 3I_{BR}^2(R_{w1} + R'_{w2}) \rightarrow R'_{w2} = \frac{P_{BR}}{3I_{BR}^2} - R_{w1} \tag{7.25}$$

The ratio of the blocked rotor voltage and current equals the blocked rotor impedance.

$$\frac{V_{BE}}{I_{BR}} = Z_{BR} = \sqrt{(R_{w1} + R'_{w2})^2 + (X_{l1} + X'_{l2})^2} \tag{7.26}$$

The reactance sum is:

$$X_{BR} = X_{l1} + X'_{l2} = \sqrt{Z_{BR}^2 - (R_{w1} + R'_{w2})^2} \tag{7.27}$$

Note that this reactance is that for which the blocked rotor test is performed. All reactances in the induction machine equivalent circuit are those at the stator frequency. Thus, all reactances computed based on the blocked rotor test frequency must be scaled according

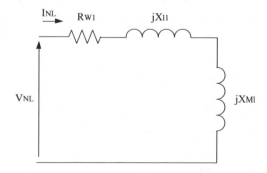

FIGURE 7.27
Equivalent circuit of induction motor at the no-load test.

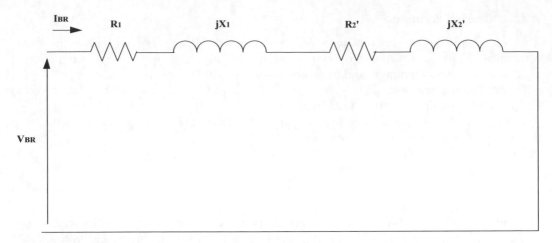

FIGURE 7.28
Equivalent circuit of induction motor at short-circuit test.

to relative frequencies (usually, a factor of 4 since T_{BR} is usually 25% of T_{NL}). The actual distribution of the total leakage reactance between the stator and the rotor is typically unknown, but empirical equations for different classes of motors (squirrel-cage motors) can be used to determine the values of X_{l1} and X_{l2}.

The following is a description of the four different classes of squirrel-cage motors.

> *Class A Squirrel-Cage Induction Motor*: characterized by normal starting torque, high starting current, low operating slip, low rotor impedance, good operating characteristics at the expense of high starting current, common applications include fans, blowers, and pumps.
>
> *Class B Squirrel-Cage Induction Motor*: characterized by normal starting torque, low starting current, low operating slip, higher rotor impedance than Class A, good general purpose motor with common applications being the same as Class A.
>
> *Class C Squirrel-Cage Induction Motor*: characterized by high starting torque, low starting current, higher operating slip than Classes A and B, common applications include compressors and conveyors.
>
> *Class D Squirrel-Cage Induction Motor*: characterized by high starting torque, high starting current, high operating slip, inefficient operation efficiency for continuous loads, common applications are characterized by an intermittent load such as a punch press.

Motor Reactance Distribution

Squirrel-cage Class A $X_{l1} = 0.5X_{BR}, X_{l2} = 0.5X_{BR}$
Squirrel-cage Class B $X_{l1} = 0.4X_{BR}, X_{l2} = 0.6X_{BR}$
Squirrel-cage Class C $X_{l1} = 0.3X_{BR}, X_{l2} = 0.7X_{BR}$
Squirrel-cage Class D $X_{l1} = 0.5X_{BR}, X_{l2} = 0.5X_{BR}$
And for Wound rotor $X_{l1} = 0.5 X_{BR}, X_{l2} = 0.5X_{BR}$.

Using these empirical formulas, the values of X_{l1} and X_{l2} can be determined from the calculation of X_{BR} from the blocked rotor test data. Given the value of X_{l1}, the magnetization reactance can be determined according to:

$$X_{m1} = \sqrt{Z_{NL}^2 - R_{NL}^2} - X_{l1} \qquad (7.28)$$

Problems

7.1 What is the motor? What types of motors? What is the principle of motor work?

7.2 What are the induction motors? And why it is called by that name?

7.3 An induction motor with 1800 rpm, connect to an AC source frequency 60 Hz, calculate the number of connections.

7.4 An Induction motor with a speed of 600 rpm and the number of poles 10, what is the frequency of the source?

7.5 What are the induction motor parts? What is the nameplate?

7.6 What types of rotor windings, and how to be placed in the iron core?

7.7 Enumerate the starting methods of the single-phase induction motor.

7.8 Describe the properties of the start windings, and how to connect with the running windings.

7.9 What is the advantage use of start capacitor in single-phase motors?

7.10 Is there another capacitor? What specifications?

7.11 What are the advantages and disadvantages of the shaded pole motor? Where do we use them?

7.12 Explain the theory of operation of the shaded pole motor.

7.13 Why are synchronous motors called by this name? Where do we use them?

7.14 What are the means to assist synchronous motor rotation?

7.15 What are synchronous motor parts?

7.16 Explain the theory of synchronous motor operation. With its properties mentioned.

7.17 What is the universal motor? Where do we use them?

7.18 Explain the universal motor construction.

7.19 Where is the centrifuge switch used? Are there ceiling fans?

7.20 Where do you use stepper motors? What are its components?

7.21 What are the uses of control motors, and what are their components?

8

Power Electronics

Formal education will make you a living; self-education will make you a fortune.

Jim Rohn

In this chapter, we cover the most important applications in the power electronics circuits, the most important of which are the rectifiers, uncontrolled and controlled circuits. Also, the DC chopper circuit will explain.

8.1 Rectifiers (AC-DC Converters)

One of the first and most widely used applications of power electronic devices have been in rectification. Rectification refers to the process of converting an AC voltage or current source to DC voltage and current. Rectifiers specially refer to power electronic converters where the electrical power flows from the AC side to the DC side.

8.1.1 Rectifier Types

1. Single-phase rectifier
2. Three-phase rectifier.

Classification of single-phase rectifiers (Figure 8.1)

1. Half-wave rectifiers
    a. Uncontrolled rectifier
    b. Controlled rectifier
2. Full-wave rectifiers
    a. Uncontrolled rectifier
    b. Controlled rectifier.

In the rectifier analysis, following simplifying assumptions will be made.

- The internal impedance of the AC source is zero
- Power electronic devices used in the rectifier are ideal switches.

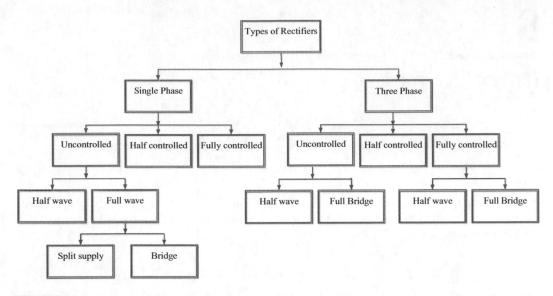

FIGURE 8.1
Classification of rectifiers.

Uncontrolled Rectifiers: Provide a fixed DC the output voltage for a given AC supply where diodes are used only.

Controlled Rectifiers: Provide an adjustable DC output voltage by controlling the phase at which the devices are turned on, where thyristors and diodes are used.

The full-wave rectifiers can divide into:

1. *Half-controlled*
   Allows electrical power flow from AC to DC (i.e., rectification only)
2. *Fully controlled*
   Allow power flow in both directions (i.e., rectification and inversion).

8.1.2 Performance Parameters

The performance of a rectifier can be evaluated in terms of the following parameters:

1. Output DC power

   Output DC power = average output voltage × average output current:

   $$P_{dc} = V_{dc}\, I_{dc} \tag{8.1}$$

2. Total power

   Total power = rms output voltage × rms output current:

   $$P_{total} = V_{rms}\, I_{rms} \tag{8.2}$$

3. The efficiency of a rectifier, *Rectification efficiency* $(\mathcal{J}) = \dfrac{P_{dc}}{P_{ac}} = \dfrac{I_{av} \cdot V_{av}}{I_{rms} \cdot V_{rms}}$.

Let f be the instantaneous value of any voltage or current associated with a rectifier circuit, then the following terms, characterizing the properties of f, can be defined.

Average (DC) value of f (F_{av}): Assuming f to be periodic over the time period T:

$$F_{av} \ or \ F_{mean} = \frac{1}{T} \int_0^T f(t) dt \tag{8.3}$$

4. Output AC component

The output voltage can be considered to have two components: including (1) DC value and (2) the AC components or ripple. The RMS value of the AC component of the output voltage is:

$$V_{ac} = \sqrt{V_{rms}^2 - V_{dc}^2} \tag{8.4}$$

5. Form factor, FF

RMS (effective) value of f (F_{rms}): For periodic current over the time period T:

$$F_{rms} = \sqrt{\frac{1}{T} \int_0^T f^2(t) dt} \tag{8.5}$$

The form factor of f (FF): It is a measure of the shape of the output voltage. It is defined as:

$$FF = \frac{F_{rms}}{F_{av}} \tag{8.6}$$

6. Ripple factor (RF)

It is a measure of the ripple content:

$$RF = \frac{V_{ac}}{V_{dc}} \tag{8.7}$$

or

$$RF = \sqrt{\left(\frac{V_{rms}}{V_{dc}}\right)^2 - 1} = \sqrt{FF^2 - 1} \tag{8.8}$$

7. Transformer utilization factor (TUF)

A transformer is most often used both to introduce galvanic isolation between the rectifier input and the AC mains and to adjust the rectifier AC input voltage to a level suitable for the required application. One of the parameters used to define the characteristics of the transformer is the TUF:

$$TUF = \frac{P_{dc}}{V_S I_S} \tag{8.9}$$

where:

V_S = RMS voltage of the transformer secondary, and

I_S = RMS current of the transformer secondary.

8. Displacement factor (*DF*):

$$DF = \text{Cos } \Phi \qquad (8.10)$$

where Φ is the phase angle between the fundamental of the input current and voltage.

9. Harmonic factor (HF)

$$HF = \sqrt{\frac{I_S^2 - I_1^2}{I_1^2}} \qquad (8.11)$$

where I_1 is the fundamental RMS component of the input current.

10. Power factor (PF)

$$PF = \frac{V_S I_1 \text{ Cos } \Phi}{V_S I_S} = \frac{I_1}{I_S} \text{ Cos } \Phi \qquad (8.12)$$

= Distortion Factor. Displacement Factor

8.1.3 Uncontrolled Rectifiers

The rectifiers converted AC to DC using power diodes and called uncontrolled rectifiers because they give constant out-voltage and fixed value if the value of the input voltage (AC voltage) is constant. The diode is an appropriate component of uncontrolled rectifiers circuits because of one-way conducted properties and classifications of circuits are based on:

1. A number of phases: One phase and three phases
2. Waveform shape: Half-wave or full-wave.

In this chapter, we will look at single-phase circuits, if the diode has ideal properties, i.e., its resistance to the current is zero if the diode is forward biased, and its resistance is very high if it is reverse biased.

8.1.3.1 *Half-Wave Uncontrolled Rectifiers*

8.1.3.1.1 Single-Phase Half-Wave Rectifier with Resistive Load

This circuit is one of the simplest circuits of the rectifier. During the positive half of the input voltage wave, the rectifier is in the forward bias state and allows the current to pass through to the load resistance. In the case of an ideal diode used, the value of the voltage drops on both ends of the rectifier is zero. Consequently, the voltage on both ends of the load resistance is the same as the positive half the input voltage as shown in Figure 8.2.

There are two advantages of connecting using the transformer. First, it allows raising and reducing the source voltage as needed. Second, it achieves electrical insulation between the source of the alternating current and rectifier, to prevent sudden electric shocks in the secondary coil circuit.

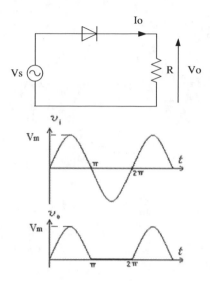

FIGURE 8.2
Half-wave uncontrolled rectifier.

The continuous output voltage can be calculated from the following formula:

$$V_{out} = V_{dc} = \frac{Vm}{\pi} = \frac{\sqrt{2}}{\pi} \times V_{rms} \qquad (8.13)$$

The output current is calculated from the following formula:

$$I_{out} = I_{dc} = \frac{V_{dc}}{R} = \frac{V_m}{\pi R} \qquad (8.14)$$

where:
V_m = Represents the maximum value of the source voltage, and
V_{rms} = Effective source voltage.

To calculate the average value of the output voltage of the one-half-wave is the value measured by the continuous voltmeter and can be calculated from:

$$V_{av} = V_m/\pi \qquad (8.15)$$

where V_{av} is the average value of the output voltage.

Example 8.1

A single-phase single wave has a purely resistive load of 8 Ω and the source voltage of 100 V at 60 Hz. Calculate the output voltage and current.

Solution

$$V_o = V_{dc} = V_m/\pi$$

$$= (\sqrt{2} \times 100)/\pi = 45 \text{ V}$$

$$I_o = I_{dc} = V_{dc}/R = 45/8$$

$$= 5.63 \text{ A.}$$

Example 8.2

Find the V_{av} of the combined voltage half-wave if the maximum value of the source voltage is 50 V.

Solution

$$V_{av} = V_m/\pi$$

$$= 50/3.14$$

$$= 15.9 \text{ V.}$$

Formula derivative:

$$V_{av} \quad \text{or} \quad V_{dc} \quad \text{or} \quad V_{lmean} = \frac{1}{T}\int_0^T f(t)dt$$

$$= \frac{1}{2\pi}\int_0^\pi V_{s\,max} sin\theta d\theta$$

$$= \frac{V_{s\,max}}{2\pi}\int_0^\pi sin\theta d\theta$$

$$= -\frac{V_{s\,max}}{2\pi}[cos\pi - cos0]$$

$$v_{s\,rms} = \frac{V_{s\,max}}{\sqrt{2}}$$

$$\therefore V_{lmean} = \frac{V_{s\,max}}{\pi}$$

$$\therefore I_{lmean} = \frac{V_{lmean}}{R} \quad \text{or} \quad I_{lmean} = \frac{1}{2\pi}\int_0^\pi I_{s\,max}\sin\theta d\theta$$

$$v_{o\,rms} = \sqrt{\frac{1}{T}\int_0^\pi V^2(t)dt}$$

$$v_{0\ rms} = \sqrt{\frac{1}{2\pi} \int_0^\pi V_{smax}^2 sin^2\theta d\theta}$$

$$= V_{smax} \sqrt{\frac{1}{2\pi} \int_0^\pi \frac{1}{2}(1 - cos2\theta)d\theta}$$

$$= V_{smax} \sqrt{\frac{1}{4\pi} \int_0^\pi (1 - cos2\theta)d\theta}$$

$$= V_{smax} \sqrt{\frac{1}{4\pi} \int_0^\pi 1 d\theta - \frac{1}{2} \int_0^\pi cos2\theta d\theta}$$

$$= V_{smax} \sqrt{\frac{1}{4\pi} \left[\theta - \frac{sin2\theta}{2} \right]_0^\pi}$$

$$= V_{smax} \sqrt{\frac{1}{4\pi} \left[\pi - \frac{sin2\pi}{2} - 0 + \frac{sin2\times 0}{2} \right]}$$

$$= V_{smax} \sqrt{\frac{1}{4\pi} [\pi]}$$

$$\therefore i_{0\ rms} = \frac{v_{o\ rms}}{R} \text{ or } i_{0\ rms} = \sqrt{\frac{1}{2\pi} \int_0^\pi I_{smax}^2 sin^2\theta d\theta}$$

$$P_{dc} = \frac{\left(\dfrac{Vm}{\pi}\right)^2}{R} = \frac{V_m^2}{\pi^2 R}$$

$$P_{total} = \frac{V_{rms}^2}{R} = \frac{(V_m / 2)^2}{R} = \frac{1}{4}\frac{V_m^2}{R}.$$

1. $\dfrac{P_{dc}}{P_{total}} = \dfrac{1}{\pi^2 R}\ 4R = 0.405\ (40.5\%).$

2. $FF = \dfrac{V_{rms}}{V_{dc}} = \dfrac{\dfrac{Vm}{2}}{Vm\Big/\pi} = \dfrac{\pi}{2} = 1.57\ (157\%).$

3. $RF = \sqrt{FF^2 - 1}$

    $= \sqrt{1.57^2 - 1}$

    $= 1.21\ (121\%).$

$$4.\ TUF = \frac{P_{dc}}{V_S.I_S} = \frac{\frac{V_m^2}{\pi^2 R}}{\left(\frac{V_m}{\sqrt{2}}\right) * \left(\frac{V_m}{2R}\right)} = \frac{2\sqrt{2}}{\pi^2} = 0.287\ \#.$$

Example 8.3

For the single-phase half-wave uncontrolled rectifier circuit shown in Figure 8.3, $R = 5\Omega$: $V_s - 150\ sinwt$. Calculate V_{Lmean}, I_{Lmean}, $v_{s_{r.m.s}}$, $v_{0_{r.m.s}}$, $i_{0_{r.m.s}}$ and RF.

Solution

$$V_{lmean} = \frac{V_{s\ max}}{\pi} = \frac{150}{3.14} = 47.7\ V$$

$$I_{lmean} = \frac{V_{lmean}}{R} = \frac{47.7}{5} = 9.5 A$$

$$v_{s\ rms} = \frac{V_{s\ max}}{\sqrt{2}} = \frac{150}{\sqrt{2}} = 106.06 V$$

$$v_{o\ rms} = \frac{V_{s\ max}}{2} = \frac{150}{2} = 75 V$$

$$i_{o\ rms} = \frac{v_{o\ rms}}{R} = \frac{75}{5} = 15\ A$$

$$RF = \sqrt{\frac{V_{o\ rms}^2 - V_{lmean}^2}{V_{lmean}^2}} = \sqrt{\frac{(75)^2 - (47.7)^2}{(47.7)^2}} = 1.213.$$

8.1.3.1.2 *Single-Phase Half-Wave Rectifier with Inductive Load*

Due to the inductive load, the conduction period of the diode will extend beyond 180° until the current becomes zero. Consider the case of the resistance-inductance (R-L) load as shown in Figure 8.4. The voltage source, is a sine wave, the positive half period when $0 < \omega t < \pi$, and the negative half period when $\pi < \omega t < 2\pi$. When the voltage source starts becoming positive, the diode starts conducting and the source keeps the diode in conduction till ωt reaches π radians. At that instant defined by $\omega t = \pi$ radians, the current through the circuit is not zero and there is some energy stored in the inductor.

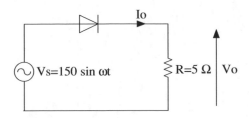

FIGURE 8.3
Single-phase half-wave uncontrolled rectifier circuit diagram.

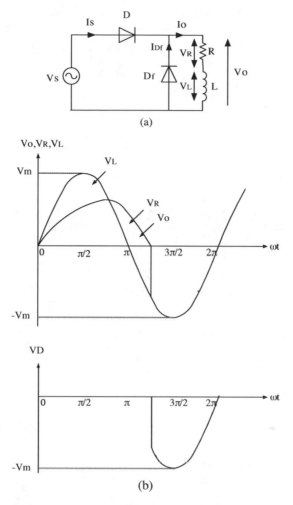

FIGURE 8.4
Single-phase half-wave rectifier with an inductive load. (a) Circuit diagram. (b) Waveforms.

The voltage across an inductor is positive when the current through it is increasing, and it becomes negative when the current through it tends to fall. When the voltage across the inductor is negative, it is in such a direction as to forward-bias the diode. The polarity of the voltage across the inductor is as shown in the waveforms shown in Figure 8.3. When source voltage changes from a positive to a negative value, the voltage across the diode changes its direction and there is current through the load at the instant $\omega t = \pi$ radians and the diode continues to conduct till the energy stored in the inductor becomes zero. After that, the current tends to flow in the reverse direction and the diode blocks conduction. The entire applied voltage now appears across the diode as a reverse bias voltage. An expression for the current through the diode can be obtained by solving the differential equation representing the circuit.

During diode conduction:

$$L\frac{di}{dt} + iR = V_m \sin \omega t.$$

A solution of this differential equation is:

$$i = \frac{V_m}{Z}\left[Sin(\omega t - \Phi) + Sin\,\Phi\, e^{\frac{-\omega t}{\tan\Phi}} \right]; \quad 0 \le \omega t \le \beta$$

(8.16)

$$Z = \sqrt{R^2 + \omega^2 L^2}\,; \tan\Phi = \frac{\omega L}{R}$$

At $\omega t = \beta, i - 0$

$$\therefore 0 = Sin(\beta - \Phi) + \sin\Phi\, e^{\frac{-\beta}{\tan\Phi}}.$$

This is a transcendental equation and can be solved by iterative techniques. The extinction angle can be determined for a given load impedance angle Φ.

The average output voltage is:

$$V_{dc} = \frac{V_m}{2\pi}\int_0^\beta \sin\omega t\, d\omega t$$

$$= \frac{V_m}{2\pi}(1 - \cos\beta)$$

(8.17)

The average output current is:

$$I_{dc} = \frac{V_m}{2\pi R}(1 - \cos\beta)$$

(8.18)

Example 8.4

For the single-phase half-wave uncontrolled rectifier circuit shown in Figure 8.5, $R = 10\,\Omega, L = 9\,mH, V_s = 150\sin wt$. Calculate and trace V_{Lmean}, I_{Lmean}, $v_{s_{r.m.s}}$ and FF.

Solution

$$\varphi = \tan^{-1}\frac{\omega L}{R} = \tan^{-1}\frac{2 \times 3.14 \times 50 \times 9 \times 10^{-3}}{10} = 15.7^0$$

$$V_{lmean} = \frac{1}{2\pi}\int_0^{\pi + \varphi} V_{s\,max}\sin\theta d\theta$$

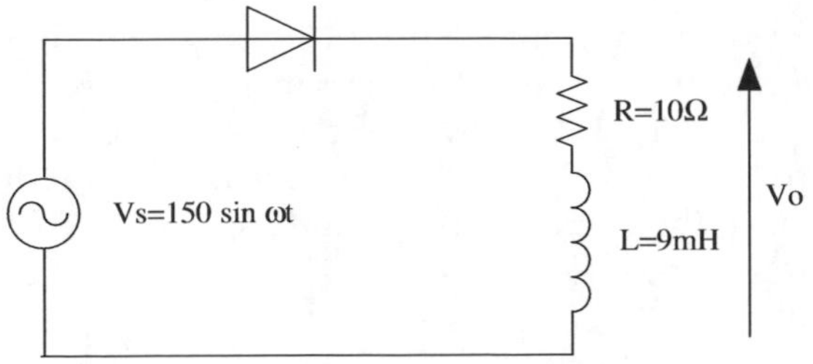

FIGURE 8.5
Circuit of Example 8.4.

$$V_{lmean} = \frac{V_{smax}}{2\pi}[1 + \cos\varphi]$$

$$= \frac{150}{2\pi}[1 + \cos 15.7] = 46.8 \text{ V}$$

$$\text{so } I_{lmean} = \frac{V_{lmean}}{R} = \frac{46.8}{10} = 4.68 \text{ A}$$

$$V_{s\,rms} = \frac{V_{s\,max}}{\sqrt{2}} = \frac{150}{\sqrt{2}} = 106.06 \text{ V}$$

$$v_{0\,rms} = V_{smax}\sqrt{\frac{1}{4\pi}\left[\pi + \varphi - \frac{sin2(\pi + \varphi)}{2}\right]}$$

$$= 150\sqrt{\frac{1}{4 \times 3.14}\left[3.14 + 15.7^0 \times \frac{\pi}{180} - \frac{sin2(180 + 15.7)}{2}\right]}$$

$$= 150\sqrt{0.08\left[3.14 + 0.27 - \frac{0.52}{2}\right]} \approx 75.2\,V$$

$$Form\,factor\,(FF) = \frac{v_{0\,rms}}{V_{lmean}} = \frac{75.2}{46.8} = 1.6$$

8.1.3.2 Full-Wave Rectifiers

Although the rectifier half-wave has some applications, the use of a rectifier full-wave is more prevalent in the sources of power for DC, and the difference between full-wave and the half-wave rectifier is that the rectifier full-wave allows one-way current through the load in each half cycle. While a single half-wave allows the current to pass through the positive half of the wave only, and as a result, we get the positive pulse number from the output of the full-wave rectifier equal to twice the number of positive pulses we get from a single half-wave output over the same time period. Figure 8.6 shows a full-wave rectifier.

FIGURE 8.6
Full-wave rectifier (https://www.grainger.com/product).

8.1.3.2.1 Full-Wave Center Tapped Rectifier with Resistive Load

Figure 8.7 shows a complete wave rectifier circuit using an adapter with an intermediate junction point. In this circuit, when the sinusoidal wave is in the positive half, the rectifier D_1 is in the front bias position, so the rectifier (D_2) is in reverse bias state.

In the case of a sinusoidal wave in the negative half, the rectifier (D_1) is in reverse bias, and the D_2 is in the front bias state. As a result, it conducts the current and thus the half of the positive and negative wave appears on the output. Figures 8.8a and b show the stages in which the full-wave are obtained.

The average value of the voltage, in this case, shall be calculated from the following formula:

$$V_{av} = 2V_m/\pi \qquad\qquad (8.19)$$

Note that the average value of the voltage (V_{av}) in this case (full-wave) is twice the value obtained in a half-wave state.

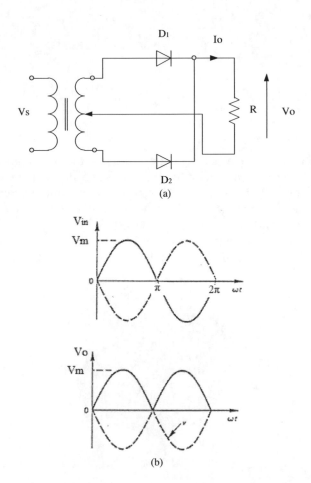

FIGURE 8.7
Full-wave center tapped rectifier. (a) Circuit diagram. (b) Waveforms.

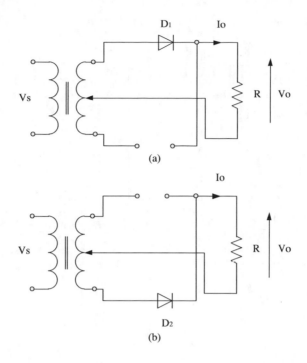

FIGURE 8.8
The stages in which the full-wave is obtained. (a) Diode 1 ON. (b) Diode 2 ON.

Example 8.5

Find the V_{av} of the load voltage a full-wave if the maximum value of the output voltage is 15 V, and create the load current if the load resistance is 3 Ω.

Solution

$$V_{av} = 2\,Vm/\pi$$

$$= 2 \times 15/3.14$$

$$= 9.55\ v$$

$$I_{dc} = V_{av}/R$$

$$= 9.55/3$$

$$= 3.18\ A.$$

Example 8.6

The rectifier shown in Figure 8.9 has a purely resistive load of R. Calculate:
    i. The rectifier efficiency.
   ii. Form factor.
  iii. Ripple factor.
  iv. Peak inverse voltage (PIV) for diode D_1.

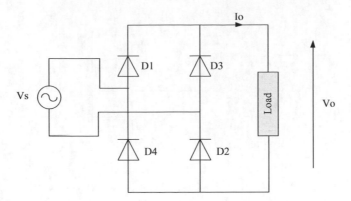

FIGURE 8.9
Single-phase full-wave (bridge) rectifier.

Solution

i. $\eta = \dfrac{V_{dc} \cdot I_{dc}}{V_{rms} \cdot I_{rms}}$

$= \dfrac{\dfrac{2V_m}{\pi} \cdot \dfrac{2V_m}{\pi \cdot R}}{\dfrac{V_m}{\sqrt{2}} \cdot \dfrac{V_m}{\sqrt{2} \cdot R}} = 81.05\%.$

ii. $F.F = \dfrac{V_{rms}}{V_{dc}}$

$= \dfrac{\dfrac{V_m}{\sqrt{2}}}{\dfrac{2V_m}{\pi}} = 111\%.$

iii. $R.F = \dfrac{V_{ac}}{V_{dc}} = \sqrt{F.F^2 - 1}$

$= \sqrt{1.11^2 - 1} = 48\%$

iv. $PIV = 2.Vm$

8.1.3.2.2 Single-Phase Full-Wave (Bridge) Rectifier

8.1.3.2.2.1 Single-Phase Full-Wave (Bridge) Rectifier with a Resistive Load   This rectifier needs four diodes to form the circuit. as shown in Figure 8.9.

The diodes (D_1, D_2) are in a state of ON during the positive part of the input voltage and the diodes (D_3, D_4) are reverse biased. The current follow through the diode (D_1, D_2), and during the negative half of the input voltage wave through D_3 and D_4 were ON and Diodes (D_1, D_2) are in reverse bias. The current follow through the diode in (D_3, D_4) (Figure 8.10).

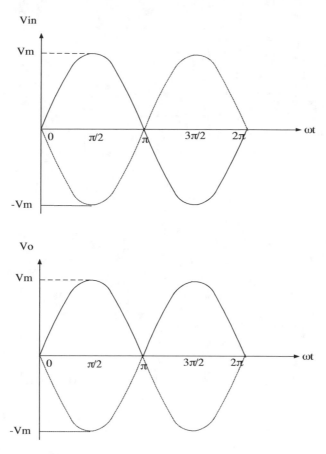

FIGURE 8.10
Input and output voltage of single-phase bridge rectifier.

The voltage and current calculation equations are same used in the full-wave rectifier with the center tap transformer, but should be noted that the diode in the rectifier of the bridge rectifier is subjected to a reverse voltage equal to half of the voltage exerted by the diode in the rectifier with the center tap transformer, which reduces rating of the diode. Figures 8.11a and b show the stages of the rectifier of a full-wave using four diodes.

Formula Derivative:

$$V_{Imean} = \frac{1}{2\pi} \int_0^\pi V_{s\,max} \sin\theta d\theta \times 2$$

$$= \frac{V_{s\,max}}{\pi} \int_0^\pi \sin\theta d\theta$$

$$= \frac{V_{s\,max}}{\pi} \left[-\cos\theta\right]_0^\pi$$

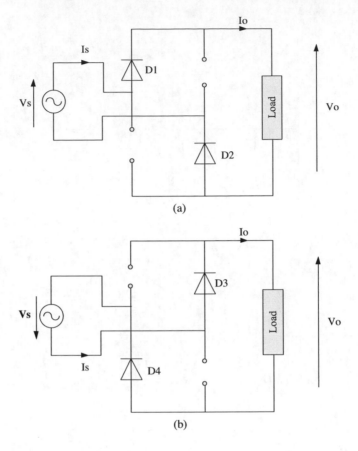

FIGURE 8.11
The stages of operation using a single-phase bridge rectifier. (a) D_1 and D_2 are ON. (b) D_3 and D_4 are ON.

$$= -\frac{V_{smax}}{\pi}[cos\pi - cos0]$$

$$\therefore V_{lmean} = \frac{2V_{smax}}{\pi}$$

$$\therefore I_{lmean} = \frac{V_{lmean}}{R}$$

$$\text{or } I_{lmean} = \frac{1}{2\pi}\int_{0}^{\pi} I_{smax} sin\theta d\theta \times 2$$

$$v_{srms} = \frac{V_{smax}}{\sqrt{2}}$$

$$v_{0rms} = \sqrt{\frac{1}{2\pi}\int_{0}^{\pi} V_{smax}^{2} sin^{2}\theta d\theta \times 2}$$

$$= V_{smax} \sqrt{\frac{1}{\pi} \int_0^\pi \frac{1}{2}(1-cos2\theta)d\theta}$$

$$= V_{smax} \sqrt{\frac{1}{2\pi} \int_0^\pi (1-cos2\theta)d\theta} = V_{smax} \sqrt{\frac{1}{2\pi} \int_0^\pi 1d\theta - \frac{1}{2}\int_0^\pi cos2\theta d\theta}$$

$$= V_{smax} \sqrt{\frac{1}{2\pi}\left[\pi - \frac{sin2\pi}{2} - 0 + \frac{sin2 \times 0}{2}\right]}$$

$$= V_{smax} \sqrt{\frac{1}{2\pi}[\pi]}$$

$$v_{o\,rms} = \frac{V_{smax}}{\sqrt{2}}$$

$$\therefore i_{o\,rms} = \frac{v_{o\,rms}}{R} \text{ or } i_{o\,rms} = \sqrt{\frac{1}{2\pi}\int_0^\pi I_{smax}^2 \sin^2\theta d\theta \times 2}.$$

Example 8.7

For the full-wave uncontrolled rectifier (bridge) circuit shown in Figure 8.12 below, $R = 10\Omega$, Calculate $V_{Lmean}, I_{Lmean}, v_{sr.m.s}. v_{0r.m.s}$, and $i_{0r.m.s}$.

Solution

$$V_{L_{mean}} = \frac{1}{2\pi} \int_0^\pi V_{max} \sin\theta d\theta \times 2 = \frac{2V_{max}}{\pi} = 127.3\,V$$

$$I_{L_{mean}} = \frac{V_{L_{mean}}}{R} = \frac{127.3}{10} = 12.73\,A$$

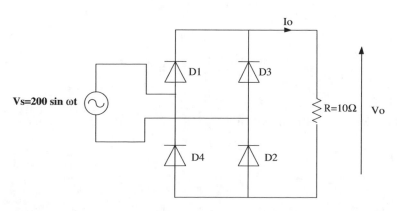

FIGURE 8.12
Circuit of Example 8.7.

$$v_{s_{rms}} = \frac{V_{smax}}{\sqrt{2}} = \frac{200}{\sqrt{2}} = 141.42\,V$$

$$v_{o\,rms} = \sqrt{\frac{1}{2\pi}\int_0^\pi V_{smax}^2 sin^2\theta d\theta \times 2} = \frac{V_{max}}{\sqrt{2}} = \frac{200}{\sqrt{2}} = 141.42\,V$$

$$i_{0\,rms} = i_{s\,rms} = \frac{v_{o\,rms}}{R} = \frac{141.42}{10} = 14.14\,A.$$

8.1.3.2.2.2 Single-Phase Full-Wave (Bridge) Rectifier with Inductive Load For an RL series-connected load in Figure 8.13, the method of analysis is like that for the half-wave rectifier with the freewheeling diode discussed previously. After a transient that occurs during start-up, the load current $i_{o(t)}$ reaches a periodic steady-state condition. For the bridge circuit, current is transferred from one pair of diodes to the other pair when the source changes polarity. The voltage across the R-L load is a full-wave rectified sinusoid, as it was for the resistive load. The full-wave rectified sinusoidal voltage across the load can be expressed as a Fourier series consisting of a DC term and the even harmonics.

$$V_{dc} = \frac{2V_m}{\pi}; \quad V_{rms} = \frac{V_m}{\sqrt{2}}$$

$$RF = \sqrt{\left(\frac{V_{rms}}{V_{dc}}\right)^2 - 1} = \sqrt{\left(\frac{\pi}{2\sqrt{2}}\right) - 1} = 0.483.$$

This is significantly less than the half-wave rectifier. With a highly inductive load, which is the usual practical application, virtually constant load current flows. The bridge diode currents are then square blocks of current with magnitude I_{dc}, in this application:

$$\text{Average current in each diode} = I_{D,av} = \frac{I_{dc}}{2}.$$

$$\text{R.M.S. current in each diode} = I_{D,rms} = \frac{I_{dc}}{\sqrt{2}}.$$

$$\text{Diode current, } FF = \frac{I_{D,rms}}{I_{D,av}} = \sqrt{2}.$$

Example 8.8

Repeat Example 8.7 with a heavily inductive load, also find RF and efficiency.

Solution

$$V_{L_{mean}} = \frac{2V_{S_{max}}}{\pi} = \frac{2\times 200}{\pi} = 127.3\,V$$

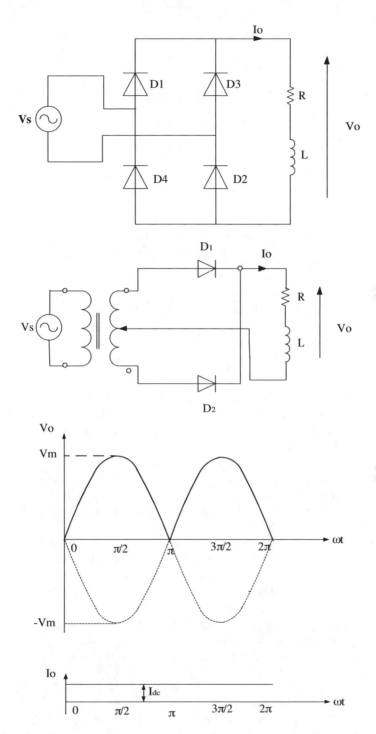

FIGURE 8.13
Single-phase full-wave (bridge and center tap) rectifier with an inductive load and its related waveforms.

$$I_{L_{mean}} = \frac{V_{L_{mean}}}{R} = \frac{127.3}{10} = 12.73\,A$$

$$v_{S_{rms}} = v_{0_{rms}} = \frac{V_{smax}}{\sqrt{2}} = \frac{200}{\sqrt{2}} = 141.42\,V$$

$$i_{s\,rms} = i_{o\,rms} = \sqrt{\frac{1}{2\pi}\int_0^\pi I_{lmean}{}^2\theta d\theta \times 2} = I_{lmean} = 12.73\,A$$

$$Form\ factor\ (FF) = \frac{v_{o\,rms}}{V_{lmean}} = \frac{141.42}{127.3} = 1.11$$

$$Ripple\ Factor\,(RF) = \sqrt{FF^2 - 1} = \sqrt{(1.11)^2 - 1} = 0.48$$

$$efficiency\,(\eta) = \frac{P_{av}}{P_{ac}} = \frac{V_{lmean} \times I_{lmean}}{v_{o\,rms} \times i_{o\,rms}} = \frac{127.3}{141.42} = 90\%.$$

8.1.4 Rectifiers with Filter Circuits

The previous rectifier circuits that the output is a rectifier voltage, the direction of the variable value in the form of pulses, and to reduce the value of the ripples in the voltage, we should be using some types of filters that apply to the output of the rectifier circuits. Either capacitor (C) or inductance and capacitance (LC).

For a given load, a larger capacitor will reduce ripple, but will cost more and will create higher peak currents in the transformer secondary and in the supply feeding it. In extreme cases where many rectifiers are loaded onto a power distribution circuit, it may prove difficult for the power distribution authority to maintain a correctly shaped sinusoidal voltage curve.

For a given tolerable ripple, the required capacitor size is proportional to the load current and inversely proportional to the supply frequency and the number of output peaks of the rectifier per input cycle. The load current and the supply frequency are generally outside the control of the designer of the rectifier system, but the number of peaks per input cycle can be affected by the choice of rectifier design.

A half-wave rectifier will only give one peak per cycle and for this and other reasons are only used in very small power supplies. A full-wave rectifier achieves two peaks per cycle and this is the best that can be done with single-phase input. For three-phase inputs, a three-phase bridge will give six peaks per cycle and even higher numbers of peaks can be achieved by using transformer networks placed before the rectifier to convert to a higher phase order.

To further reduce this ripple, a capacitor-input filter can be used. This complements the reservoir capacitor with an inductor and a second filter capacitor so that a steadier DC output can be obtained across the terminals of the filter capacitor. The inductor presents a high impedance to the ripple current. http://en.wikipedia.org/wiki/Rectifier - cite_note-dcp-0#cite_note-dcp-0 Figure 8.14 shows the filtration process.

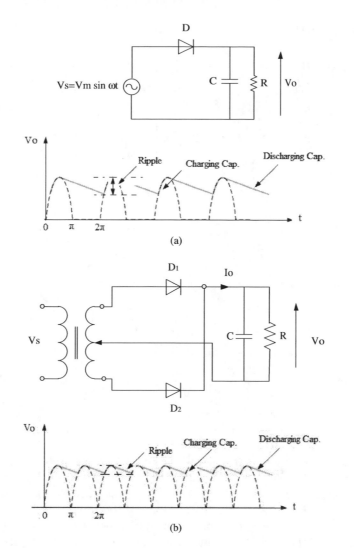

FIGURE 8.14
Filtration process. (a) Half-wave rectifier. (b) Full-wave rectifier.

8.1.5 Controlled Rectifiers

These rectifiers use the thyristors and the delay angle to control the output voltage and current.

8.1.5.1 Thyristor Firing Circuits

The thyristor becomes a conductor if a positive voltage is found between the anode and the cathode, as well as a pulse trigger on the gate. The trigger is a signal on the gate in the form of a pulse that takes a certain amount of time to operate the thyristor.

Shockley diodes are curious devices, but rather limited in application. Their usefulness may be expanded, however, by equipping them with another means of latching. In doing

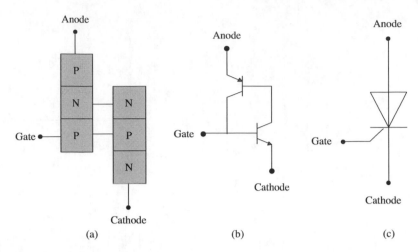

FIGURE 8.15
The silicon controlled rectifier. (a) Physical diagram. (b) Equivalent schematic. (c) Schematic symbol.

so, they become true amplifying devices (if only in an on/off mode), and we refer to them as silicon-controlled rectifiers or SCRs.

The progression from Shockley diode to SCR is achieved with one small addition, nothing more than a third wire connecting to the existing PNPN structure (Figure 8.15).

If SCR's gate is left floating (disconnected), it behaves exactly as a Shockley diode. It may be latched by brake overvoltage or by exceeding the critical rate of voltage rise between anode and cathode, just as with the Shockley diode. Dropout is accomplished by reducing current until one or both internal transistors fall into the cutoff mode, also like the Shockley diode. However, because the gate terminal connects directly to the base of the lower transistor, it may be used as an alternative means to latch the SCR. By applying a small voltage between gate and cathode, the lower transistor will be forced on by the resulting base current, which will cause the upper transistor to conduct, which then supplies the lower transistor's base with the current so that it no longer needs to be activated by a gate voltage. The necessary gate current to initiate latch-up, of course, will be much lower than the current through the SCR from the cathode to anode, so the SCR does achieve a measure of amplification.

This method of securing SCR conduction is called triggering, and it is by far the most common way that SCRs are latched in actual practice. In fact, SCRs are usually chosen so that their brake overvoltage is far beyond the greatest voltage expected to be experienced from the power source so that it can be turned on only by an intentional voltage pulse applied to the gate.

8.1.5.2 The Use of a Thyristor in the Controlled Rectifier Circuits

The thyristors are used in the circuits of controlled rectifier to convert the alternating current to a constant current and voltage that can be controlled by the angle of the trigger of the thyristor (α) and called (trigger angle). The control rectifiers are characterized by simplicity and high efficiency, as well as the low cost of manufacturing, which requires a voltage change to control the speed of the motors.

8.1.5.3 Single-Phase Half-Wave Controlled Rectifier with Resistive Load

The circuit in Figure 8.16 is one of the simplest applications on the controlled rectifier circuits, consisting of alternating current source and thyristor and resistance load.

During the positive half of the wave, the anode voltage is higher than the cathode voltage, and the thyristor is in the forward bias state. When the thyristor is triggered by the current of the gate at the firing angle, the input voltage will appear on the load. When the current follow in the load to be zero the thyristor becomes off, and when the negative half of the input wave will be the anode voltage is lower than the cathode voltage, and the thyristor is in reverse bias and repeated with each cycle (2 π). Figure 8.17 shows the input

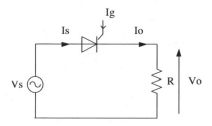

FIGURE 8.16
Half-wave controlled rectifier.

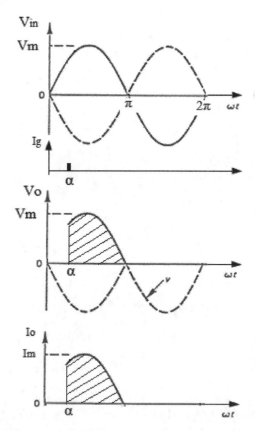

FIGURE 8.17
Input, output voltage, and output current waveforms.

wave and the gate current, load current and voltage. The average value of the mean output voltage can be found in the following equation:

$$V_o = \frac{V_m}{2\pi}(1 + \cos \alpha) \tag{8.20}$$

8.1.5.4 Single-Phase Half-Wave Control Rectifier with R-L Load

The importance of this circuit lies in the fact that most industrial loads are an inductive load containing resistance and inductive.

Figure 8.18 shows a circuit with a resistive and inductive load, and Figure 8.19 showing the input and output currents of the voltages and current of this circuit.

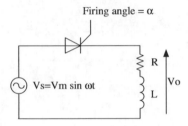

FIGURE 8.18
Single-phase half-wave control rectifier with R-L load circuit.

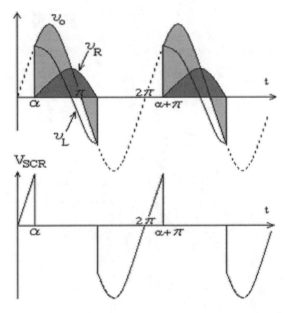

FIGURE 8.19
Output voltage and current and SCR voltage of single-phase half-wave control rectifier with the R-L load.

The current will start to flow in the thyristor when it is triggered, and this current is lagging the negative voltage, due to the transient state of the current caused by the inductor and sometimes freewheeling diode to get rid of the negative part of the load voltages.

The average value of the output voltage without the freewheeling diode can be found in the following formula:

$$V_o = \frac{V_m}{2\pi}(cos \propto - cos \beta) \tag{8.21}$$

where $\beta =$ is the angle at which the thyristor turns OFF.

Example 8.9

A single-phase half-wave controlled rectifier is used to feed a pure resistance load of $R = 10 \, \Omega$, and AC voltage of 220 V at frequency $f = 60$ Hz. Calculate:

    a. Output voltages at delay angle $\alpha = 90°$.
    b. The output current and the thyristor current at the angle of the $\alpha = 90°$.

Solution

    (a)

$$V_o = \frac{Vm}{2\pi}(1 + cos \propto)$$

$$= \frac{220}{2\pi}\sqrt{2}(1 + cos \ 90°)$$

$$V_o = 49.51 \text{ V}.$$

    (b)

$$I_o = V_o/R = 49.51/10 = 4.951 \text{ A}$$

$$I_{av} = I_o$$

$$= 4.951 \text{ A}.$$

Example 8.10

A single-phase half-wave controlled rectifier with a resistive load and high inductance with a supply voltage of 120 V and frequency of 60 Hz.

    a. Calculate the output voltage at $\alpha = 60°$ and $\beta = 200°$.
    b. Calculate the output current at the two angles above.

Solution

(a)

$$V_o = (Vm/2\pi) \cdot (\cos \propto - \cos \beta)$$
$$= ((\sqrt{2} * 120)/2\pi) \cdot (\cos 60° - \cos 200°)$$
$$V_o = 38.88 \text{ V}.$$

(b)

$$I_o = V_o/R = 38.88/5$$
$$= 7.77 \text{ A}.$$

Example 8.11

A single-phase half-wave controlled rectifier with a resistance R = 2 Ω in series with a very high inductance. The supply voltage of 200 V and f = 60 Hz. Firing angle for each thyristor $\alpha = 60°$. Calculate:

a. The output voltage V_o.
b. The output current I_o.

Solution

(a)

$$V_o = (2.Vm/\pi) \cos \propto$$
$$= ((2\sqrt{2} \times 200)/\pi) \cdot \cos 60°$$
$$= 90 \text{ V}.$$

(b)

$$I_o = V_o/R = 90/2 = 45 \text{ A}.$$

Example 8.12

For the single-phase half-wave-controlled rectifier circuit shown in Figure 8.20 has a purely resistive load of *R* and firing angle α = 50% of total power. Determine:

i. Derive an expression for load voltage in term of α.
ii. Efficiency.
iii. FF.
iv. RF.

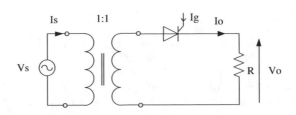

FIGURE 8.20
Circuit of Example 8.12.

Solution

i.

$$V_{lmean} = \frac{1}{2\pi} \int_{\alpha}^{\pi} V_{smax} \sin\theta \, d\theta$$

$$= \frac{V_{smax}}{2\pi} \int_{\alpha}^{\pi} \sin\theta \, d\theta$$

$$= \frac{V_{smax}}{2\pi} \left[-\cos\theta \right]_{\alpha}^{\pi}$$

$$= -\frac{V_{smax}}{2\pi} \left[\cos\pi - \cos\alpha \right]$$

$$\therefore V_{lmean} = \frac{V_{smax}}{2\pi} \left[1 + \cos\alpha \right].$$

ii.

$$V_{lmean} = \frac{V_{smax}}{2\pi} \left[1 + \cos 90 \right] = \frac{V_{smax}}{2\pi} = 0.1592 V_{smax}$$

$$\text{Average power } (P_{av}) = \frac{V_{lmean}^2}{R} = \frac{(0.1592 V_{smax})^2}{R}$$

$$v_{o\,rms} = \sqrt{\frac{1}{2\pi} \int_{\alpha}^{\pi} V_{smax}^2 \sin^2\theta \, d\theta}$$

$$= V_{smax} \sqrt{\frac{1}{2\pi} \int_{\alpha}^{\pi} \frac{1}{2} (1 - \cos 2\theta) \, d\theta}$$

$$= V_{smax} \sqrt{\frac{1}{4\pi} \int_{\alpha}^{\pi} (1 - \cos 2\theta) \, d\theta}$$

$$= V_{smax} \sqrt{\frac{1}{4\pi} \int_{\alpha}^{\pi} 1 \, d\theta - \frac{1}{2} \int_{\alpha}^{\pi} \cos 2\theta \, d\theta}$$

$$= V_{smax} \sqrt{\frac{1}{4\pi} \left[\theta + \frac{\sin 2\theta}{2} \right]_{\alpha}^{\pi}}$$

$$= V_{smax} \sqrt{\frac{1}{4\pi} \left[\pi - \frac{\sin 2\pi}{2} - \alpha + \frac{\sin 2\alpha}{2} \right]}$$

$$= V_{smax} \sqrt{\frac{1}{4\pi} \left[\pi - \frac{\sin 2\pi}{2} - \frac{\pi}{2} + \frac{\sin 2 \times \frac{\pi}{2}}{2} \right]}$$

$$= V_{smax}\sqrt{\frac{1}{4\pi}\left[\frac{\pi}{2}\right]} = V_{smax}\sqrt{\frac{1}{8}} = 0.3536\,V_{smax}$$

$$\text{AC power}\,(P_{ac}) = \frac{v_{0\,rms}{}^2}{R} = \frac{(0.3536V_{s\,max})^2}{R}$$

$$efficiency\,(\eta) = \frac{P_{av}}{P_{ac}} = \frac{\dfrac{\left(0.1592V_{s\,max}\right)^2}{R}}{\dfrac{(0.3536V_{s\,max})^2}{R}} = 20.27\%$$

iii.

$$Form\,factor\,(FF) = \frac{v_{o\,rms}}{V_{lmean}} = \frac{0.3536\,V_{s\,max}}{0.1592\,V_{s\,max}} = 2.221.$$

iv.

$$Ripple\,Factor\,(RF) = \sqrt{FF^2 - 1} = \sqrt{(2.221)^2 - 1} = 1.9831.$$

8.1.5.5 Single-Phase Full-Wave Control Rectifier with R-L Load

The fully controlled rectifier contains four thyristors. We will assume that the load here contains resistance and high inductance to filter the output current.

The thyristors (T_1, T_2) are connected to the positive part of input voltage after being triggered at angle α, and the work continues to the next firing angle for (T_3, T_4) at $\alpha + \pi$. The occurrence of the thyristors (T_3, T_4) in the negative part of the input wave due to the transient condition of the presence of inductance in the load, as shown in Figure 8.21. Figure 8.22 shows input and output voltages and source current. The average output voltage can be found by the following equation (Figure 8.23):

$$V_o = \frac{2V_m}{\pi}\cos\alpha \tag{8.22}$$

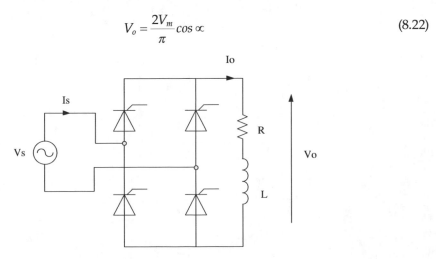

FIGURE 8.21
Single-phase full-wave control rectifier with R-L load circuit.

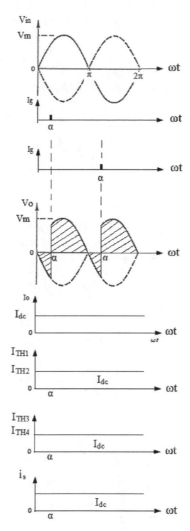

FIGURE 8.22
Input, output voltages, load, SCR, and source currents.

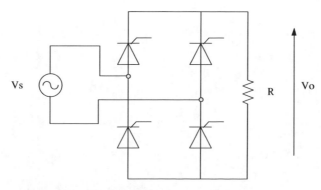

FIGURE 8.23
Single-phase full-wave control rectifier with R load circuit.

Formula Derivative:

a. resistive load:

$$V_{lmean} = \frac{1}{2\pi} \int_{\alpha}^{\pi} V_{s\,max} sin\theta\, d\theta \times 2$$

$$= \frac{V_{s\,max}}{\pi} \int_{\pi}^{\pi} sin\theta\, d\theta$$

$$= \frac{V_{s\,max}}{\pi} \left[-cos\theta \right]_{\alpha}^{\pi}$$

$$= -\frac{V_{s\,max}}{\pi} \left[cos\pi - cos\alpha \right]$$

$$\therefore V_{lmean} = \frac{V_{s\,max}}{\pi} \left[1 + cos\alpha \right]$$

$$\therefore I_{lmean} = \frac{V_{lmean}}{R}$$

$$\text{or} \quad I_{lmean} = \frac{1}{2\pi} \int_{\alpha}^{\pi} I_{s\,max} \sin\theta\, d\theta \times 2$$

$$v_{s\,rms} = \frac{V_{s\,max}}{\sqrt{2}}$$

$$v_{0\,rms} = \sqrt{ \frac{1}{2\pi} \int_{\alpha}^{\pi} V_{smax}{}^2 sin^2\theta\, d\theta \times 2 }$$

$$= V_{smax} \sqrt{ \frac{1}{\pi} \int_{\alpha}^{\pi} \frac{1}{2}(1 - cos2\theta) d\theta }$$

$$= V_{smax} \sqrt{ \frac{1}{2\pi} \int_{\alpha}^{\pi}(1 - cos2\theta) d\theta } = V_{smax} \sqrt{ \frac{1}{2\pi} \int_{\alpha}^{\pi} 1 d\theta - \frac{1}{2} \int_{\alpha}^{\pi} cos2\theta\, d\theta }$$

$$v_{0rms} = V_{smax} \sqrt{ \frac{1}{2\pi} \left[\pi - \frac{sin2\pi}{2} - \alpha + \frac{sin2\alpha}{2} \right] }$$

$$\therefore i_{0rms} = \frac{v_{orms}}{R}$$

Example 8.13

For the fully controlled full-wave rectifier circuit shown in Figure 8.24, firing angle $\alpha = 90°$, calculate the efficiency.

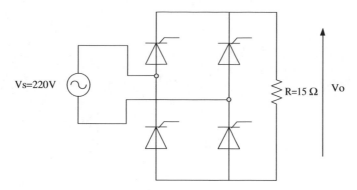

FIGURE 8.24
Circuit of Example 8.13.

Solution

$$V_{lmean} = \frac{V_{s\,max}}{\pi}\left[1 + \cos\alpha\right] = \frac{220 \times \sqrt{2}}{\pi}\left[1 + \cos 90°\right] = 99.08\,V$$

$$I_{lmean} = \frac{V_{lmean}}{R} = \frac{99.08}{15} = 6.6\,A$$

$$v_{0\,rms} = \sqrt{\frac{1}{2\pi}\int_{\alpha}^{\pi}V_{smax}^{2}\sin^{2}\theta d\theta \times 2} = V_{smax}\sqrt{\frac{1}{2\pi}\left[\pi - \frac{sin2\pi}{2} - \alpha + \frac{sin2\alpha}{2}\right]} = 155.5\,V$$

$$i_{orms} = \frac{v_{orms}}{R} = \frac{155.5}{15} = 10.36\,A$$

$$efficiency\,(\eta) = \frac{P_{av}}{P_{ac}} = \frac{V_{lmean} \times I_{lmean}}{v_{o\,rms} \times i_{o\,rms}} = \frac{99.08 \times 6.6}{155.5 \times 10.36} = 40.59\%.$$

8.1.5.6 Single-Phase Full-Wave Half Control Rectifier with R-L Load

When a converter contains both diode and thyristors, it is called *half-controlled*. These three circuits produce identical load waveforms, neglecting any differences in the number type of semiconductor voltage drops. The power to the load is varied by controlling the angle α, at which the load voltage from going negative, extend the conduction period, and reduce the AC ripple (Figure 8.25).

$$Average\ output\ voltage = V_{dc} = \frac{1}{\pi}\int_{\alpha}^{\pi}V_{m}\sin\omega t\,d\omega t$$

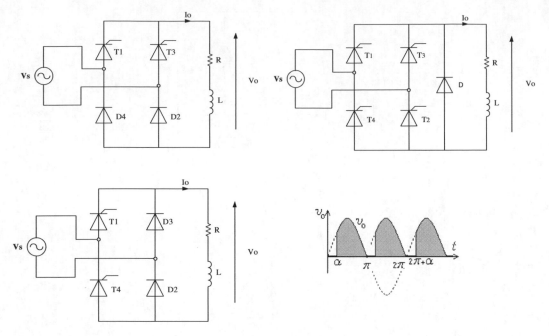

FIGURE 8.25
Half-controlled rectifiers and output voltage.

$$= \frac{V_m}{\pi}(1 + cos\alpha)$$

$$RMS\ output\ voltage = V_{rms} = \sqrt{\frac{1}{\pi}\int_{\alpha}^{\pi}(V_m sin\omega t)^2\,d\omega t}$$

$$= V_m\sqrt{\frac{\pi - \alpha + \dfrac{1}{2}sin2\alpha}{2\pi}}.$$

Inversion is not possible.

Example 8.14

For the fully controlled full-wave rectifier circuit shown in Figure 8.26, firing angle $\alpha = 45$. Calculate efficiency.

Solution

$$\varphi = \tan^{-1}\frac{\omega L}{R} = \tan^{-1}1\frac{2\pi \times 50 \times 5 \times 10^{-3}}{10} = 8.9°$$

$$V_{lmean} = \frac{V_{s\,max}}{\pi}\left[cos\ \varphi + cos\ \alpha\right] = \frac{150}{3.14}\left[cos\ 8.9° + cos\ 45°\right] = 81V$$

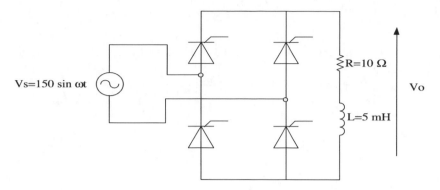

FIGURE 8.26
Fully controlled full-wave rectifier circuit of Example 8.14.

$$I_{lmean} = \frac{V_{lmean}}{R} = \frac{81}{10} = 8.1\,A$$

$$v_{0\,rms} = \sqrt{\frac{1}{2\pi} \int_{\alpha}^{\pi+\varphi} V_{smax}{}^2 sin^2\theta d\theta \times 2} \approx 101.1\,V$$

$$i_{orms} = \frac{v_{orms}}{R} = \frac{101.1}{10} = 10.11\,A$$

$$efficiency\,(\eta) = \frac{P_{av}}{P_{ac}} = \frac{V_{lmean} \times I_{lmean}}{v_{0\,rms} \times i_{0\,rms}} = \frac{81 \times 8.1}{101.1 \times 10.11} = 64.25\%$$

Example 8.15
For the fully controlled full-wave rectifier circuit shown in Figure 8.27, firing angle $(\alpha) = 45°$. Calculate the efficiency.

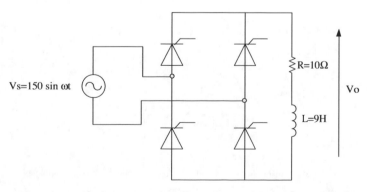

FIGURE 8.27
Fully controlled full-wave rectifier circuit of Example 8.15.

Solution

$$\varphi = \tan^{-1}\frac{\omega L}{R} = \tan^{-1}\frac{2\pi \times 50 \times 9}{10} = 89.79°$$

$$V_{lmean} = \frac{2V_{s\,max}}{\pi}\left[\cos\alpha\right] = \frac{2}{3.14}\frac{150}{}\left[\cos45°\right] = 67.5\,V$$

$$I_{lmean} = \frac{V_{lmean}}{R} = \frac{67.5}{10} = 6.75\,A$$

$$v_{0\,rms} = \sqrt{\frac{1}{2\pi}\int_{\alpha}^{\pi+\alpha}V_{smax}^2\sin^2\theta d\theta \times 2} \approx 105.8\,V$$

$$i_{0\,rms} = \sqrt{\frac{1}{2\pi}\int_{\alpha}^{\pi+\alpha}I_{lmean}^2\,\theta d\theta \times 2} = I_{lmean} = 67.5\,A$$

$$efficiency\,(\eta) = \frac{P_{av}}{P_{ac}} = \frac{V_{lmean} \times I_{lmean}}{v_{0\,rms} \times i_{0\,rms}} = \frac{67.5 \times 6.75}{105.8 \times 6.75} = 63.8\%.$$

8.2 Power Electronics Circuits with MATLAB Program

8.2.1 MATLAB Simulation of Single-Phase Half-Wave Uncontrolled Rectifier

1. Open MATLAB Program, and open the *Simulink Library Browser*. From which, open new module (Figure 8.28)
2. From the library, Simulink opens the branch **Sinks**, from which get the block **Scope**. Set the following: Number of Axis = 2. Also get the block **Display**
3. From the library, Simulink opens the branch **Signal Routing** and choose **Mux** block

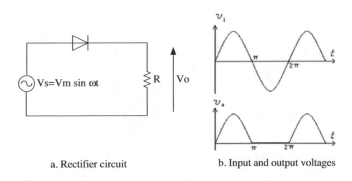

a. Rectifier circuit   b. Input and output voltages

FIGURE 8.28
Single-phase half-wave uncontrolled rectifier. (a) Rectifier circuit (b) Input and output voltages.

4. Open the library **Sim Power Systems**, and then open the branch *Electrical Sources*. From which get the block **AC Voltage Source**. Set the following: Peak Amplitude Voltage = 100 V, Frequency = 50, Sample Time = 0.0001

5. From the same library open the *Elements* and get the block **Ground (Input)**

6. Open the branch *Elements* and get the Block **Series RLC branch**. Set the following: Resistance = 10 Ω, Inductance = 0, Capacitance reactance = inf, or choose brunch type on only resistance

7. From the branch, *measurements* get the block **Voltage Measurement and current measurement**

8. From the *Power, Electronic* branch gets the block **Diode**. Set the following: Snubber resistance = $1e^9$

9. Connect the circuit in Figure 8.29.

NOTE: a. Add power graphical user interface (GUI) block in Figure 8.29.

b. From the menu *simulation*, choose *configuration parameters*, and from which set the type of the solver to the following: ode23tb (stiff/TR-BDF2)

Now, run the module and draw V_o and V_{Imean}. As shown in Figures 8.30 and 8.31. Controlled *Half-Wave* Rectifier with a Resistive Load.

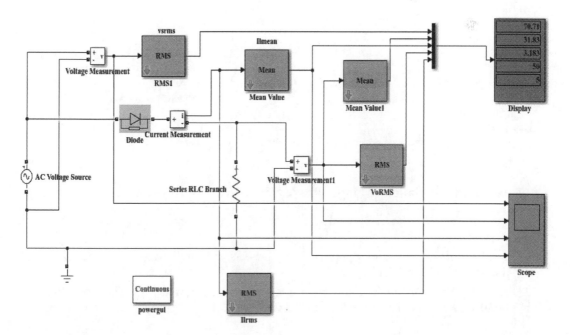

FIGURE 8.29
Single-phase half-wave uncontrolled rectifier Simulink.

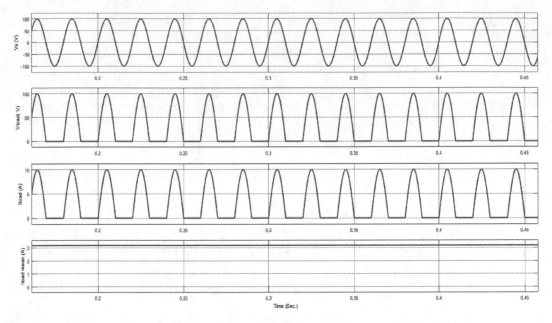

FIGURE 8.30
Waveforms.

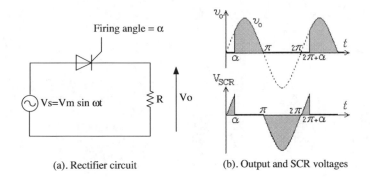

(a). Rectifier circuit   (b). Output and SCR voltages

FIGURE 8.31
Single-phase half-wave controlled rectifier. (a) Rectifier circuit (b) Output and SCR voltages.

8.2.2 MATLAB Simulation of Single-Phase Half-Wave Controlled Rectifier

1. Open MATLAB Program, and open the *Simulink Library Browser*. From which, open new module

2. From the *library, Simulink* open the branch *Sinks*, from which get the block **Scope**. Also get the block **Display**

3. From the *library, Simulink* open the branch *Sources*, from which get the block **Pulse Generator**. Set the Period = 0.05, Pulse width % = 1

4. Open the library *Sim Power Systems,* and then open the branch *Electrical Sources.* From which get the block *AC Voltage Source.* Set the following: Peak Amplitude Voltage = 200 V, Frequency = 50 Hz, Sample Time = 0.0001

5. From the same library open the branch *Connectors* and get the block *Ground (Input)*

6. Open the branch *Elements* and get the Block *Series RLC branch.* Set the following: Resistance = 10 Ω, Inductance = 0, Capacitance reactance = inf

7. From the branch *measurements* get the block *Voltage Measurement*

8. From the *Power, Electronic* branch gets the block *Detailed Thyristor.* Set the following: Snubber resistance = 1e9

9. Connect the circuit in Figures 8.32 and 8.33

10. Now, from the menu *simulation,* choose *configuration parameters,* and from which set the type of the solver to the following: ode23tb (stiff/TR-BDF2)

11. Open the block *Pulse Generator* and change the **Phase delay** to the values **0.001666, 0.0025, 0.00333,** and **0.005.** These values are equivalent to 30°, 45°, 60°, and 90°, respectively

$$\text{Firing angle} = \frac{\alpha}{360} \times \frac{1}{\text{freq}}.$$

12. Draw the waveform of *Vo* and V_{Imean}.

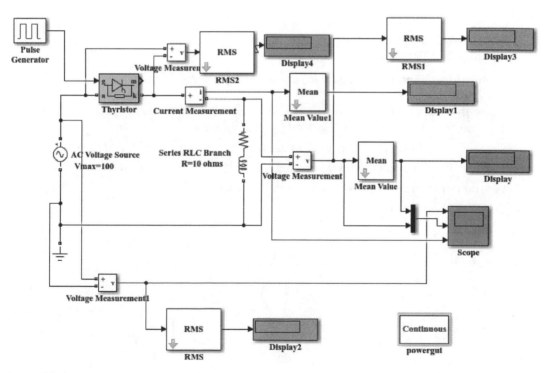

FIGURE 8.32
Single-phase half-wave uncontrolled rectifier Simulink.

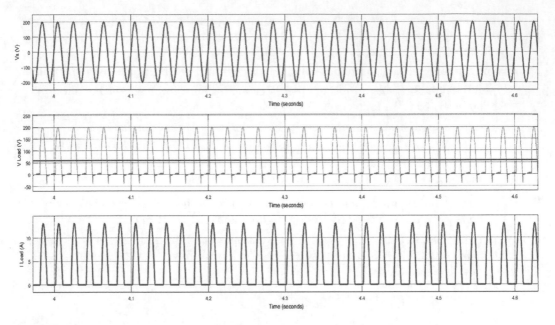

FIGURE 8.33
Waveforms.

8.2.3 MATLAB Simulation of Single-Phase Half-Wave Controlled Rectifier with an Inductive Load

To implement the diagram in Figure 8.34 using MATLAB/Simulink

1. Copy the branch *Elements* and get the Block ***Series RLC branch***. Set the following:
   Resistance = 10 Ω, Inductance = 9mH, Capacitance reactance = inf

2. Connect the circuit as in Figure 8.35 and run the module

3. Use the Simulink library browser as Figure 8.36

4. Set the parameters as shown in Figures 8.36 through 8.43

5. Draw the waveforms of V_o, V_R, and V_L.

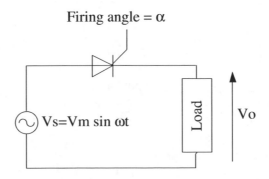

FIGURE 8.34
Single phase half wave-controlled rectifier.

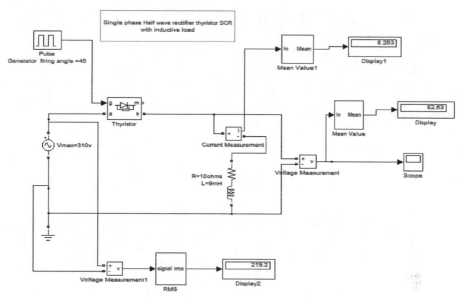

FIGURE 8.35
Single-phase half-wave controlled rectifier with an inductive load Simulink.

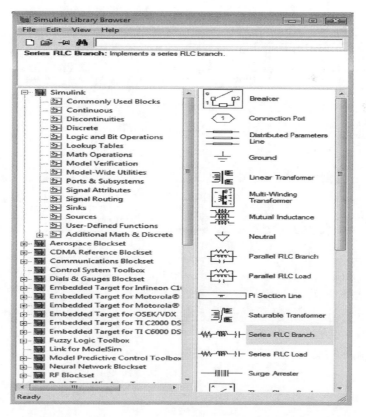

FIGURE 8.36
Simulink library browser.

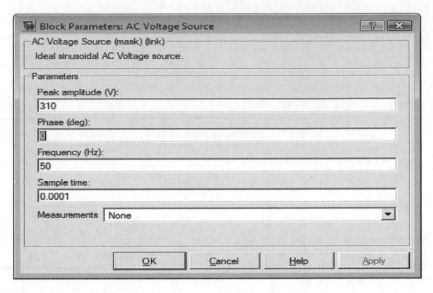

FIGURE 8.37
Block parameters of AC voltage source.

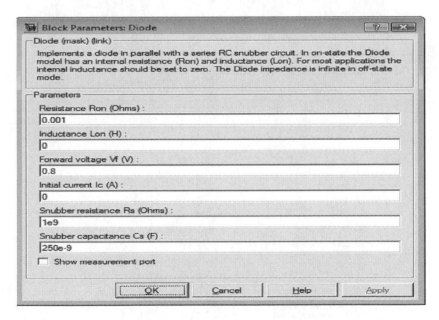

FIGURE 8.38
Block parameters of the diode.

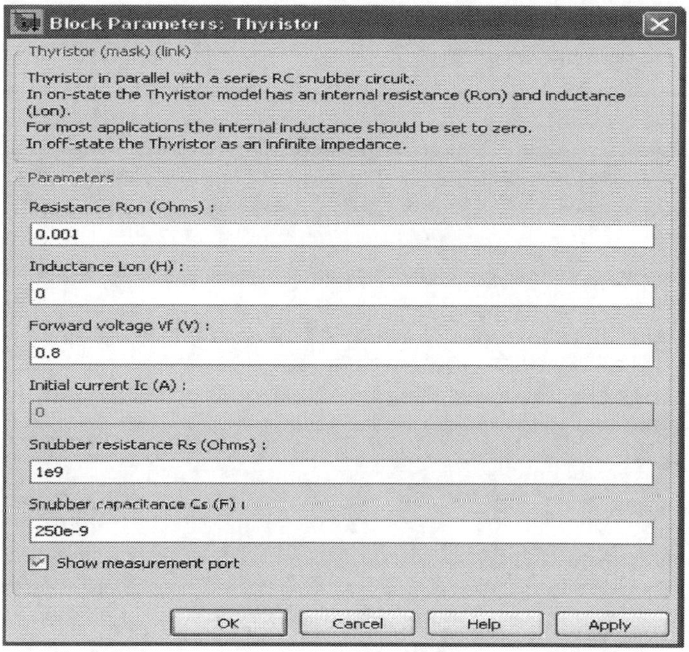

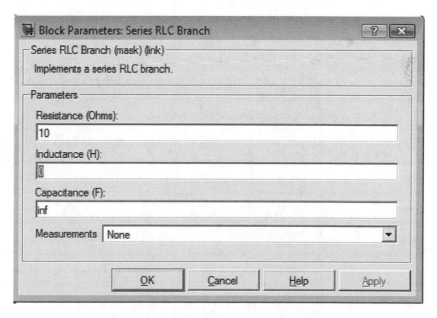

FIGURE 8.39
Block parameters of SCR.

FIGURE 8.40
Block parameters of series RLC branch.

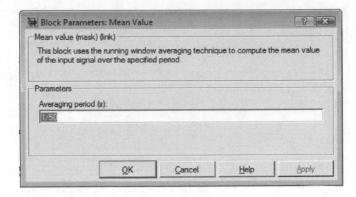

FIGURE 8.41
Mean value block parameters.

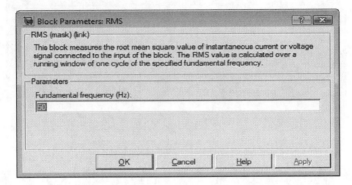

FIGURE 8.42
RMS block parameters.

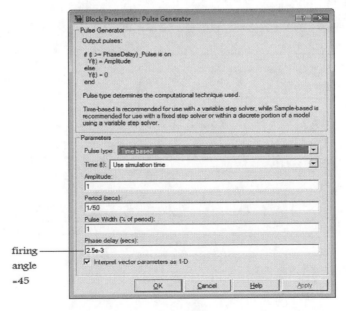

FIGURE 8.43
Pulse generator block parameters.

Example 8.16

For the full-wave uncontrolled rectifier circuit shown in Figure 8.26, $R = 10\,\Omega$, $V_{max} = 200\,V$ and firing angle $(\alpha) = 45°$. Trace and calculate V_{Lmean}, I_{Lmean}, and $v_{S_{r.m.s}}$ (Figures 8.44 and 8.45).

Solution

$$V_{L_{mean}} = \frac{1}{2\pi} \int_0^\pi V_{max} sin\theta\, d\theta \times 2 = \frac{2V_{max}}{\pi} = 127.3\,volts$$

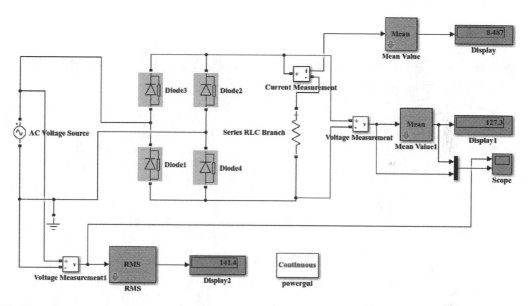

FIGURE 8.44
Full-wave uncontrolled rectifier circuit Simulink.

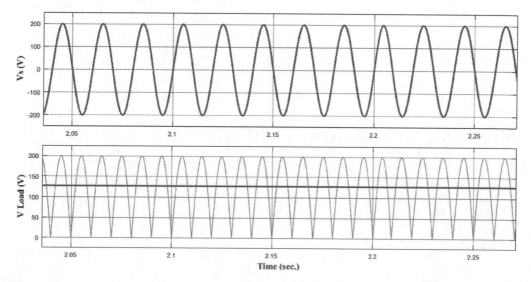

FIGURE 8.45
Waveforms.

$$I_{L_{mean}} = \frac{V_{L_{mean}}}{R} = \frac{127.3}{15} = 8.49\,Amper$$

$$V_{S_{rms}} = \frac{V_{max}}{\sqrt{2}} = \frac{200}{\sqrt{2}} = 141.42\,volts$$

Example 8.17

For the full-wave controlled rectifier circuit shown in Figure 8.46, $R = 10\Omega$, $L = 5mH$, $V_{max} = 150$ volt, and firing angle $(\alpha) - 30°$. Trace and calculate V_{Lmean}, I_{Lmean}, and $v_{S_{r,m,s}}$ (Figure 8.47).

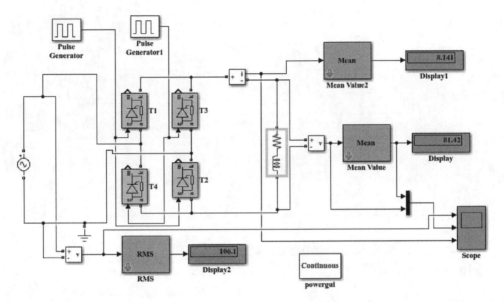

FIGURE 8.46
Full-wave controlled rectifier circuit Simulink.

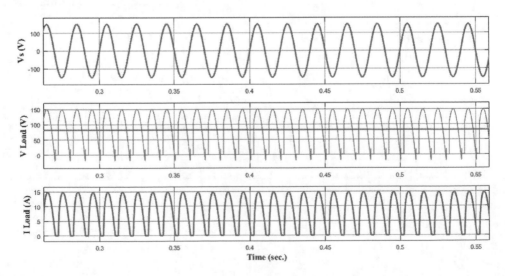

FIGURE 8.47
Waveforms.

Solution

$$\phi = tan^{-1}\frac{\omega L}{R} = 8.9°$$

$$V_{L_{mean}} = \frac{1}{2\pi}\int_{\alpha}^{\pi+f} V_{max}sin\theta d\theta \times 2 = \frac{V_{max}}{\pi}\left[cos\phi + cos\alpha\right] = 88.5\, volts$$

$$I_{L_{mean}} = \frac{V_{L_{mean}}}{R} = \frac{88.5}{10} = 8.85\, Amper$$

$$V_{S_{RMS}} = \frac{V_{max}}{\sqrt{2}} = \frac{150}{\sqrt{2}} = 106.01\, volts.$$

8.3 DC-DC Converter Basics

A DC-DC converter is a device that accepts a DC input voltage and produces a DC output voltage. Typically, the output produced is at a different voltage level than the input. In addition, DC-DC converters are used to provide noise isolation, power bus regulation, etc. This section considers a summary of some of the popular DC-DC converter topologies.

8.3.1 Step-Down (Buck) Converter

The circuit in Figure 8.48, the transistor turning ON will put voltage V_{in} on one end of the inductor. This voltage will tend to cause the inductor current to rise. When the transistor is OFF, the current will continue flowing through the inductor, but now flowing through the diode. We initially assume that the current through the inductor does not reach zero, thus the voltage at V_x will now be only the voltage across the conducting diode during the full OFF time. The average voltage at V_x will depend on the average ON time of the transistor provided the inductor current is continuous as in Figure 8.49.

To analyze the voltages of this circuit let us consider the changes in the inductor current over one cycle. From the relation:

$$V_x - V_0 = L\frac{di}{dt} \tag{8.23}$$

The change of current satisfies:

$$di = \int_{ON}\left(V_x - V_0\right)dt + \int_{OFF}\left(V_x - V_0\right)dt \tag{8.24}$$

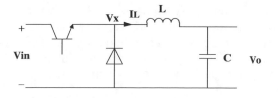

FIGURE 8.48
Buck converter.

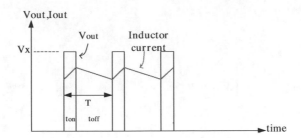

FIGURE 8.49
Voltage and current changes.

For steady-state operation, the current at the start and end of a period T will not change. To get a simple relation between voltages, we assume no voltage drop across transistor or diode while ON and a perfect switch change. Thus, during the ON time $V_x = V_{in}$ and in the OFF $V_x = 0$.

So

$$0 = di = \int_0^{t_{on}} \left(V_{in} - V_0\right)dt + \int_{t_{on}}^{T} \left(-V_0\right)dt \tag{8.25}$$

which simplifies to:

$$\left(V_{in} - V_0\right).t_{on} + \left(-V_0\right).t_{off} = 0$$

or

$$\frac{V_o}{V_{in}} = \frac{t_{on}}{T} \tag{8.26}$$

and defining "duty ratio" as:

$$D = \frac{t_{on}}{T} \tag{8.27}$$

The voltage relationship becomes $V_o = D\, V_{in}$ since the circuit is lossless and the input and output powers must match on the average $V_o \times I_o = V_{in} \times I_{in}$. Thus, the average input and output current must satisfy $I_{in} = D\, I_o$. These relations are based on the assumption that the inductor current does not reach zero.

8.3.1.1 Transition between Continuous and Discontinuous

When the current in the inductor L remains always positive then either the transistor T_1 or the diode D_1 must be conducting. For continuous conduction, the voltage V_x is either V_{in} or 0. If the inductor current ever goes to zero then the output voltage will not be forced to either of these conditions. At this transition point, the current just reaches zero as seen in Figure 8.50. During the ON time $V_{in}-V_{out}$ is across the inductor thus:

$$I_L\left(peak\right) = \left(V_{in} - V_{out}\right).\frac{t_{on}}{L} \tag{8.28}$$

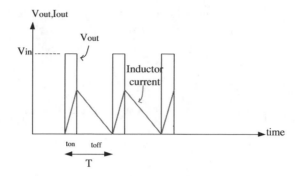

FIGURE 8.50
Buck converter at the boundary.

The average current which must match the output current satisfies:

$$I_L(av) = \frac{I_L(peak)}{2} = (V_{in} - V_{out}).\frac{dT}{2L} \tag{8.29}$$

If the input voltage is constant, the output current at the transition point satisfies:

$$I_L(av) = I_{out} = V_{in}.\frac{(1-d)}{2L}.T \tag{8.30}$$

8.3.1.2 Voltage Ratio of Buck Converter (Discontinuous Mode)

As for the continuous conduction analysis, we use the fact that the integral of the voltage across the inductor is zero over a cycle of switching T. The transistor OFF time is now divided into segments of diode conduction $\delta_d T$ and zero conduction $\delta_o T$ as shown in Figure 8.51. The inductor average voltage thus gives:

$$(V_{in} - V_o)\,DT + (-V_o)\,\delta.dT = 0 \tag{8.31}$$

$$\frac{V_{out}}{V_{in}} = \frac{d}{d+d_\delta} \tag{8.32}$$

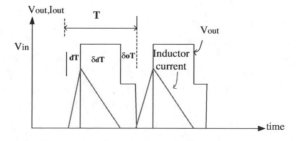

FIGURE 8.51
Buck converter—discontinuous conduction.

for the case $d + \delta_d < 1$. To resolve the value of δ_d, consider the output current which is half the peak when averaged over the conduction times $d + \delta_d$:

$$I_{out} = \frac{I_{L(peak)}}{2}.d + \delta_d \tag{8.33}$$

Considering the change of current during the diode conduction time:

$$I_{L(peak)} = \frac{V_o(\delta_d T)}{L} \tag{8.34}$$

Thus, from equations 8.33 and 8.34, we can get:

$$I_{out} = \frac{V_o \delta_d T(d + d_\delta)}{2L} \tag{8.35}$$

using the relationship in equation 8.32:

$$I_{out} = \frac{V_{in}dTd_\delta}{2L} \tag{8.36}$$

And solving for the diode conduction:

$$\delta_d = \frac{2LI_{out}}{V_{in}dT} \tag{3.37}$$

The output voltage is thus given as:

$$\frac{V_{out}}{V_{in}} = \frac{d^2}{d^2 + \left(\dfrac{2LI_{out}}{V_{in}T}\right)} \tag{3.38}$$

Defining $k^* = 2L/(V_{in} T)$, we can see the effect of discontinuous current on the voltage ratio of the converter. As seen in Figure 8.52, once the output current is high enough, the voltage ratio depends only on the duty ratio "d." At low currents, the discontinuous operation tends to increase the output voltage of the converter toward V_{in}.

8.3.2 Step-Up (Boost) Converter

The schematic in Figure 8.53 shows the basic boost converter. This circuit is used when a higher output voltage than input is required.

   While the transistor is ON $V_x = V_{in}$ and the OFF state the inductor current flows through the diode giving $V_x = V_o$. For this analysis, it is assumed that the inductor current always remains flowing (continuous conduction). The voltage across the inductor is shown in Figure 8.54, and the average must be zero for the average current to remain in a steady-state:

$$V_{in}.t_{on} + (V_{in} - V_o)\, t_{off} = 0$$

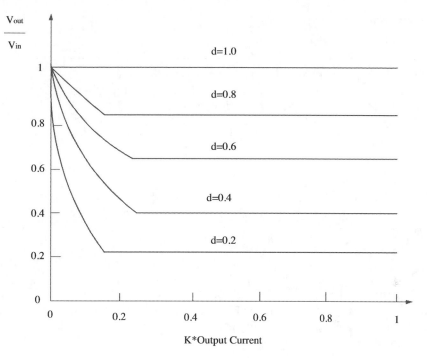

FIGURE 8.52
Output voltage versus current.

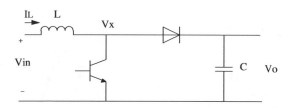

FIGURE 8.53
Boost converter circuit.

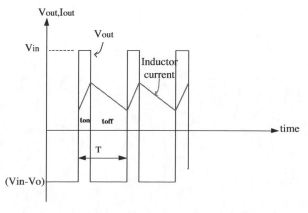

FIGURE 8.54
Voltage and current waveforms (boost converter).

This can be rearranged as:

$$\frac{V_{out}}{V_{in}} = \frac{T}{t_{off}} = \frac{1}{(1-D)} \qquad (8.39)$$

And for a lossless circuit, the power balance ensures:

$$\frac{I_o}{I_{in}} = (1-D) \qquad (8.40)$$

Since the duty ratio "D" is between 0 and 1, the output voltage must always be higher than the input voltage in magnitude. The negative sign indicates a reversal of sense of the output voltage.

8.3.3 Buck-Boost Converter

With continuous conduction for the buck-boost converter in Figure 8.55, $V_x = V_{in}$ when the transistor is ON and $V_x = V_o$ when the transistor is OFF. For zero net current change over a period, the average voltage across the inductor is zero.

From Figure 8.56:

$$V_{in} \cdot t_{on} + V_o \cdot t_{off} = 0,$$

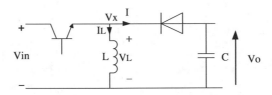

FIGURE 8.55
Schematic for the buck-boost converter.

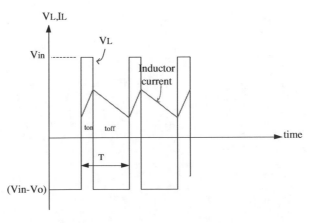

FIGURE 8.56
Waveforms for the buck-boost converter.

which gives the voltage ratio:

$$\frac{V_{out}}{V_{in}} = \frac{T}{t_{off}} = \frac{D}{(1-D)} \tag{8.41}$$

and the corresponding current:

$$\frac{I_{out}}{I_{in}} = \frac{(1-D)}{D} \tag{8.42}$$

Since the duty ratio D is between 0 and 1, the output voltage can vary between lower or higher than the input voltage in magnitude. The negative sign indicates a reversal of sense of the output voltage.

8.3.4 Converter Comparison

The voltage ratios achievable by the DC-DC converters are summarized in Figure 8.57. Notice that only the buck converter shows a linear relationship between the control (duty ratio) and output voltage. The buck-boost can reduce or increase the voltage ratio with rectifier gain for a duty ratio of 50%.

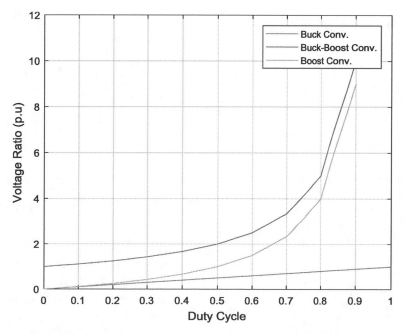

FIGURE 8.57
Comparison of voltage ratio.

MATLAB program

```
clc
clear all
D=0:0.1:1;
X1=D; %X1=Vo/Vin Buck converter
X2 = 1./(1-D); %X2-Vo/Vin Boost converter
X3=D./(1-D); %X3=Vo/Vin Buck-Boost converter
figure
plot(D, X1,D, X2,D, X3)
title('Comparison of voltage ratio')
xlabel('Duty Cycle')
ylabel('Voltage Ratio (p.u)')
grid
```

Example 8.18

A step-down (buck) converter is used to supply an inductive load of $R = 3\Omega$ and $L = 9$ mH, with a DC voltage of 60V as shown in Figures 8.58 and 8.59. If $D = 0.4$, $F_s = 2$ KH$_z$, calculate:

a. I_{Lmean} and V_{Lmean}.
b. Trace the output voltage and current.

a. $Ts = 1/2000 = 5 \times 10^{-4}$

$$D = \frac{a}{Ts}$$

$$a = D \times Ts = 0.4 \times 5 \times 10^{-4} = 2 \times 10^{-4}$$

$$Ts = a + b$$

$$b = Ts - a = 5 \times 10^{-4} - 2 \times 10^{-4} = 3 \times 10^{-4}$$

$$I_{max} = \frac{Ed}{R} \left[\frac{e^{\frac{RTs}{L}} - e^{\frac{Rb}{L}}}{e^{\frac{RTs}{L}} - 1} \right] = 8.4A$$

$$I_{min} = I_{max} e^{\frac{-Rb}{L}} = 7.6A$$

$$I_{Lmean} = \frac{I_{max} + I_{min}}{2} = 7.8A$$

b. $V_{max} = I_{max} \times R = 25.2V$

$V_{min} = I_{min} \times R = 22.8V$

$V_{Lmean} = \dfrac{V_{max} + V_{min}}{2} = 24V$

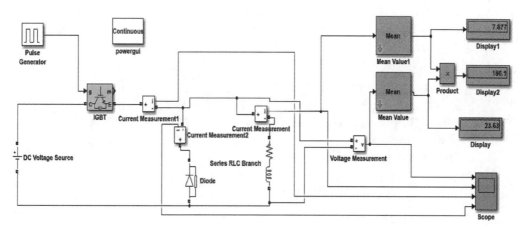

FIGURE 8.58
Step-down (buck) converter of Example 8.18.

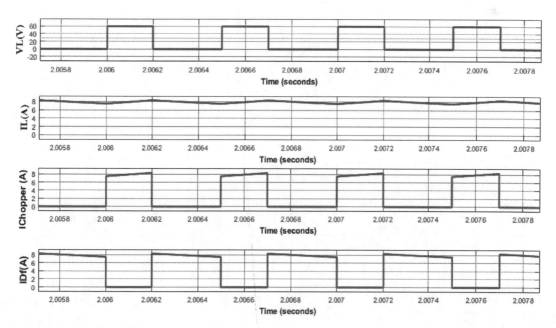

FIGURE 8.59
Waveforms.

Problems

8.1   A single-phase half-wave controlled rectifier the thyristor turn ON at an angle
$\alpha = 90°$. If the load resistance of R = 10 Ω and supply voltage of 120 V.

Required:

1.   Draw input voltage, output voltage, and thyristor current

2.   Draw the input voltage, output voltage, and thyristor current if $\alpha = 0$.

8.2   A single-phase full-wave control rectifier, all the thyristor works at delay angle
$\alpha = 90°$ supplied bus AC voltage of 120 V. Draw the input voltage, output voltage
output current if load:

1.   Pure resistance

2.   Resistance with high-value inductance.

8.3   A single-phase full-wave bridge rectifier operate at an angle of $\alpha = 90°$, load A pure
resistance of R = 20 Ω is supplied by an AC source of V = 20 sin ωt at frequency
f = 50 Hz. Calculate:

1.   Output voltage

2.   Output current.

8.4   The circuit is shown in Figure 8.60.

A.   Draw the waveform of the load voltages of the thyristor at $\alpha = 60°$

B.   Draw the voltages of the load pure resistance at $\alpha = 90°$

C.   Draw the output current I_o for A and B above.

8.5   The circuit is shown in Figure 8.61. If the rated load is 10 Ω and the source voltage
is 120 volt. Required:

a.   Derive an expression for load and thyristor current

b.   Use an equation for load voltage. Calculate the load current at $\alpha = 90°$.

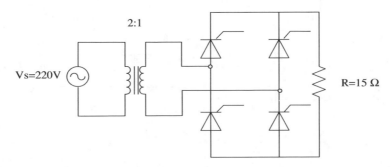

FIGURE 8.60
Circuit diagram for Problem 8.4.

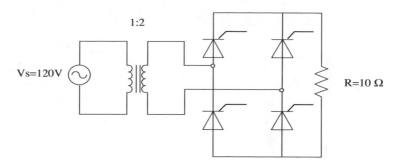

FIGURE 8.61
Circuit diagram for Problem 8.5.

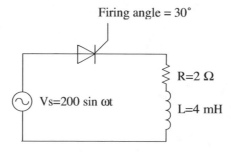

FIGURE 8.62
Circuit diagram for Problem 8.7.

8.6 A step-down (buck) converter is used to supply an inductive load of $R = 10\Omega$ and $L = 5mH$, with a D.C voltage of 80V, if $D = 0.6$, $F_s = 20\ KH_z$, and input DC voltage, calculate:

1. I_{Lmean} and V_{Lmean}
2. Trace the output voltage and current.

8.7 For the single-phase half-wave controlled rectifier circuit shown in Figure 8.62, $R = 2\Omega$, L = 4mH, $V_s = 200\sin\omega t$ volt, and firing angle $(\alpha) = 30°$.

   i. Calculate V_{Lmean}, I_{Lmean}, and $v_{Sr.m.s}$

   ii. Repeat the circuit at placed free-wheeling diode (FWD) and calculate V_{Lmean}, I_{Lmean}, $v_{Sr.m.s}$ and RF.

8.8 For the full-wave half-controlled rectifier circuit shown in Figure 8.63, $R = 17\Omega, L = 8H$, $V_{max} = 150$ volt and firing angle $(\alpha) = 30°$. Calculate and trace V_{Lmean}, I_{Lmean}, and $v_{Sr.m.s}$.

8.9 The DC chopper in Figure 8.64 has an inductive load of R = 0.5 Ω and L = 0.5 mH. The input voltage value is Vs = 120 V, f = 5 KHz, E = 0 V, and duty cycle = 0.4. Calculate the I_{min} and I_{max}, the average load current I_{av}, the average value of the

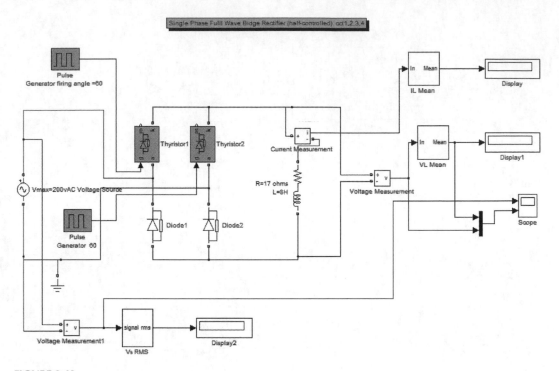

FIGURE 8.63
Full-wave half-controlled rectifier circuit Simulink.

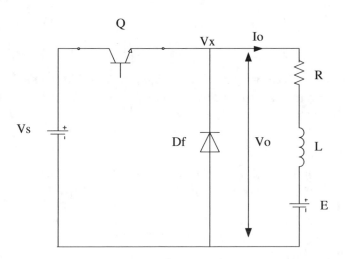

FIGURE 8.64
Circuit diagram for Problem 8.9.

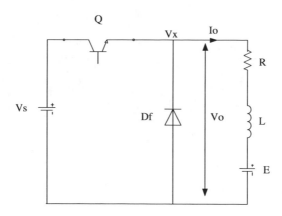

FIGURE 8.65
Circuit diagram for Problem 8.10.

diode current I_D, and the effective input resistance R_i. Draw the voltage across the chopper, the output voltage, and the current in the freewheeling diode.

8.10 The DC chopper in Figure 8.65 has an inductive load of $R = 1.2\ \Omega$ and $L = 0.5$ mH. The input voltage value is $Vs = 140$ V, $f = 5$ KHz, $E = 10$ V, and duty cycle = 0.75. Calculate the I_{min} and I_{max}, the average load current I_{av}, the average value of the diode current I_D, and the effective input resistance R_i. Draw the voltage across the chopper, the output voltage, and the current in the freewheeling diode.

9

Concept of DC Drive

In imagination, there's no limitation.

Mark Victor Hansen

A DC motor is used to drive a mechanical load. Variable DC drives have been used to control DC motors longer than variable frequency drives have been used to control AC motors. The first motor-speed control used DC motors because of the simplicity of controlling the voltage to the armature and field of a DC motor.

9.1 DC Motors Drive

The brushed DC motor is one of the earliest motor designs. Today, it is the motor of choice in most of the variable speed and torque control applications.

9.1.1 Advantages

- Easy to understand the design
- Easy to control the speed
- Easy to control torque
- Simple, cheap drive design.

9.1.1.1 Easy to Understand the Design

The design of the brushed DC motor is quite simple. A permanent magnetic field is created in the stator by either of two means:

- Permanent magnets
- Electromagnetic windings.

If the field is created by permanent magnets, the motor is said to be a "permanent-magnet DC motor." (PMDC). If created by electromagnetic windings, the motor is often said to be a "shunt wound DC motor." (SWDC). Today, because of cost-effectiveness and reliability, the PMDC is the motor of choice for applications involving fractional horsepower DC motors, as well as most applications up to about 3 horsepower.

At 5 horsepower and greater, various forms of the shunt wound DC motor are most commonly used. This is because the electromagnetic windings are more cost effective than permanent magnets in this power range.

The section of the rotor where the electricity enters the rotor windings is called *the commutator*. The electricity is carried between the rotor and the stator by conductive graphite-copper brushes (mounted on the rotor), which contact rings on the stator. In most DC motors, several sets of windings or permanent magnets are present to smooth out the motion.

9.1.1.2 Easy to Control the Speed

Controlling the speed of a brushed DC motor is simple. The higher the armature voltage, the faster the rotation. This relationship is linear to the motor's maximum speed. The maximum armature voltage which corresponds to a motor's rated speed (these motors are usually given a rated speed and a maximum speed, such as 1750/2000 rpm) is available in certain standard voltages, which roughly increase in conjunction with horsepower. Thus, the smallest industrial motors are rated 90 and 180 V. Larger units are rated at 250 V and sometimes higher. Specialty motors for use in mobile applications are rated 12, 24, or 48 V. Other tiny motors may be rated 5 V.

Most industrial DC motors will operate reliably over a speed range of about 20:1 down to about 5%–7% of base speed. This is a much better performance than the comparable AC motor. This is partly due to the simplicity of control, but is also partly due to the fact that most industrial DC motors are designed with variable speed operation in mind and have added heat dissipation features which allow lower operating speeds.

9.1.1.3 Easy to Control Torque

In a brushed DC motor, torque control is also simple since output torque is proportional to current. If you limit the current, you have just limited the torque which the motor can achieve. This makes this motor ideal for delicate applications such as textile manufacturing.

9.1.1.4 Simple, Cheap Drive Design

The result of this design is that variable speed or variable torque electronics are easy to design and manufacture. Varying the speed of a brushed DC motor requires little more than a large enough potentiometer. In practice, these have been replaced for all but sub-fractional horsepower applications by the Silicon Controlled Rectifier (SCR) and pulse-width modulation (PWM) drives, which offer relatively precisely control voltage and current.

Large DC drives are available up to hundreds of horsepower. However, over about 10 horsepower careful consideration should be given to the price/performance trade-offs with AC inverter systems since the AC systems show a price advantage in the larger systems. (But they may not be capable of the application's performance requirements.)

9.1.2 Disadvantages

DC motors have the following disadvantages:

1. Expensive to produce
2. Not can reliably control at lowest speeds
3. Physically larger
4. High maintenance.

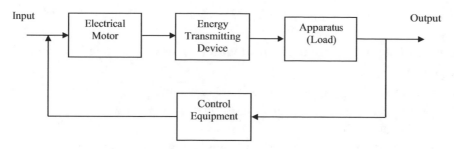

FIGURE 9.1
A block diagram of an electrical drive.

A drive consists of three main parts: prime mover; energy transmitting device; and actual apparatus (load), which perform the desired job. The function of the first two parts is to impart motion and operate the third one. In electrical drives, the prime mover is an electric motor or an electromagnet. A block diagram of an electrical drive shown in Figure 9.1.

The electromagnetic forces or torque developed by the driving motor tend to propagate motion of the drive system. This motion may be uniform translational or rotational motion or non-uniform as in case of starting, braking, or changing the load.

In the uniform motion, the motor torque (Figure 9.2) must be in a direction opposite to that of the load and equal to it, taken on the same shaft (motor or load shaft) the steady-state condition has a uniform motion at ω_m = Constant:

$$T_m = T_L$$

$$T_m - T_L = 0 \tag{9.1}$$

For a uniform motion $T_m \neq T_L$, therefore the system will either in acceleration if $T_m > T_L$ or deceleration when $T_m < T_L$ and the dynamic equation relating them is given by:

$$T_m - T_L = J \frac{d\omega}{dt} \tag{9.2}$$

where J = moment of inertia of all rotating part transformed to the motor shaft.

Types of the load: There are two main loads type:

Active load: which provides active torque (gravitational), deformation in elastic bodies, springs forces, compressed air, …), these load may cause motion of the system.

Passive loads: are that loads which have a torque all the times opposing the motion such as frictional load and shearing loads.

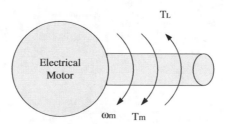

FIGURE 9.2
Motion of the motor drive and torque direction.

Also, loads may be sub-divided into:

1. *Constant loads*: which are unchanged with time or with a variation of speed
2. *Linear varying loads*:
   a. First order, $T_L = a + b\omega$ where a, b are constant
   b. Second order (second varying load), $T_L = a + b\,\omega + c\,\omega^2$, where a, b, c are constant and ω is the speed.
3. *The frictional load*: may be considered constant, which pump load consider linear, and compressor loads may consider the second order.

The electrical drive may operate in one of three main modes of operation:

1. *Continuous mode*: by which the motor is started and operate for enough time to reach steady-state temperature then when stopped (Figure 9.3), it must be left for enough to reach the initial temperature (room temperature).

   In such an operation and to avoid thermal stresses:

   $$T_L \leq T_{rated} \text{ (torque of the motor)}.\theta$$

   $$T_s > T_L \text{ for the normal motor operation,}$$

   where $T_s = $ starting torque.

2. *Interruptive operation:* In such operation, the motor operates and stops for approximately equal intervals by which at any interval, the motor will not reach steady-state temperature when started and not return to initial temperature when stopped as shown in Figure 9.4. In such operation and according to the ON and OFF intervals $T_L < T_{rated}$ of the motor (depending on the number of ON and OFF intervals).

   For each interval $T_s > T_L$ to ensure starting. The number of intervals is limited by the temperature reached which must not exceed the designed temperature limit θ_{max}. Such operation of drives may be found in the control system

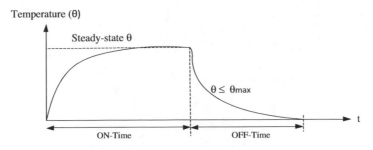

FIGURE 9.3
Continuous mode.

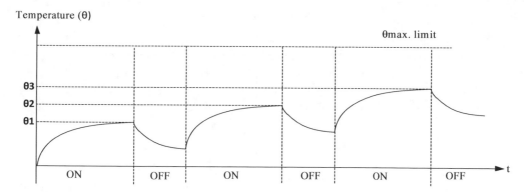

FIGURE 9.4
Interruptive operation.

3. *Short time operation*: in such operation, the motor is started and operate for a small interval of time, then switch-OFF for a very long period as shown in Figure 9.5.

   Θ may reach θ_{Max}, but the long OFF time will be sufficient to return the temperature to its initial value $t_{ON} \ll t_{OFF}$.

   The motor may be overloaded during the ON time so the temperature θ may reach θ_{Max} at small times, but the long OFF period will return it back to its initial value.

In such drive, $T_L > T_{rated}$ motor so multi-successive ON (starting) may damage the motor. Such derive may be found in cars starting system.

For quadrants operation of drives: By the effect of active and passive loads, and according to the relationship and directions of w, T_m, and T_L the electrical drive may operate in any quadrant of 4-quadrants diagram whose axis is $\pm T$ and $\pm \omega$ as shown in Figure 9.6.

The first quadrant (I) represents motor operation by which the motor torque and speed are in the same direction (+ve) and opposing the load torque T_L.

The speed of the motor and its torque is positive (counter clockwise direction). The speed may be changed from 0 to ω_o, where ω_o = no-load angular speed and the motor torque

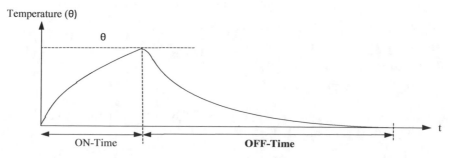

FIGURE 9.5
Short time operation.

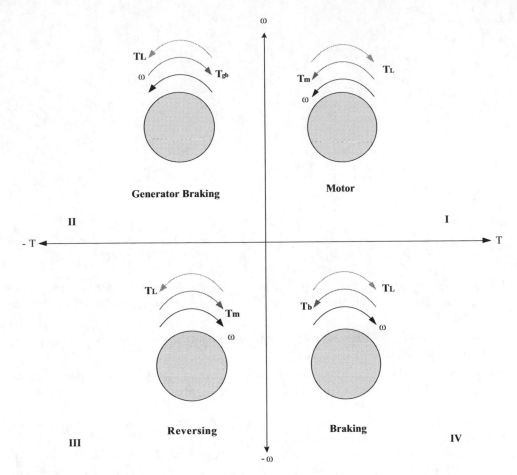

FIGURE 9.6
Quadrants operation of drives.

change from 0 at no-load to T_s at $\omega = 0$ (starting torque). For normal motor operation, the following relation must be satisfied:

$T_r \geq T_L \geq 0$ for the normal thermal operating condition

$0 < \omega_L \leq \omega_o$ according to the selective method of speed variation

$Ts > T_L$ for all motor operation to ensure starting

T_r = rated torque of the motor (N·m)

T_L = load torque of the motor shaft (N·m)

T_s = starting torque of the motor (N·m).

The normal steady-state operation is when $T_m = T_L$ which make the dynamic torque $(T_m - T_L) = 0$. This means that:

$$T_m - T_L = J\frac{d\omega}{dt} = 0$$

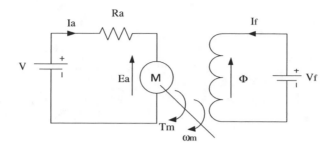

FIGURE 9.7
Equivalent circuit for separate excited DC motor.

or ω = constant = operating speed.

If for any reason the motor speed exceeds ω_o, then the motor back EMF I will be genera-tor then the supply voltage (V). From equivalent circuit for separately excited DC motor in Figure 9.7, the armature current:

$$I_a = \frac{V-E}{R_a} \tag{9.3}$$

will be negative (return to the supply) and produce a torque opposing the speed (brake).

Therefore, moving the second quadrant will change the operation from motor to genera-tor braking. At generator braking, ω is positive, but greater than ω_o and the motor torque opposes the motion (the load torque in the same direction as the motion).

At no-load speed ω_o, E = V, $I_a = 0$, $T_m = 0$ (Y-axis) the 3rd quadrant represent also motor operation, but in the reverse direction with respect to that in the first quadrant (i.e., rota-tion in the clockwise direction and T_m also in the clockwise direction and opposing T_L).

The 4th quadrant represents brake operation by which the motor takes current from the supply and produce positive torque, but opposing the motion.

The motion is started by the effect of the load which is greater than T_s (active) and hence the motor move opposite to its electrical direction causing E to change its direction. So,

$$I_a = \frac{V-E}{R_a} \rightarrow I_a = \frac{V+E}{R_a}$$

Since E is negative.

And hence T_m increase as ω increase until at certain ω:
$T_m = T_L$, and the load will brake at speed = ω $\omega_o > \omega > 0$.

9.2 Torque-Speed Characteristics

In many applications, the rotational speed of the motor may be much higher than the required speed or vice versa. In other applications, multi-speeds are required. To change the load speed, there are mechanical and electrical methods as well as their combination. In mechanical method, the use of pullies, the gear may make the load speed equal to the required speed from given motor rotational speed.

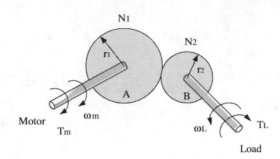

FIGURE 9.8
A simple two gears.

A simple two gears may have radius r_1 and r_2 and number of tooth N_1 and N_2 as shown in Figure 9.8, correspondingly, may give the following relationship. Since teeth pitch is the same for both gears:

So when gear A moves one revolution, $(2\pi r_1)$, gear B must move the same length $(2\pi r_2)$ according to the radius.

i.e.
$$1 \times 2\pi r_1 = i(2\pi r_2) \tag{9.4}$$

where:

$$i = \frac{Load\ speed}{Motor\ speed} = \frac{\omega_L}{\omega_m} \tag{9.5}$$

$$i = \frac{I_1'}{I_2'} = \frac{N_1}{N_2} \tag{9.6}$$

$$\omega_L = I \cdot \omega_m \tag{9.7}$$

$I > 1$ if $r_1 > r_2$ or $N_1 > N_2$.

If the gears have the efficiency of 100%:

So,
$$P_{in} = P_{out}.$$

Or
$$T_m \cdot \omega_m = T_L \cdot \omega_L,$$

$$\therefore T_L = T_m \cdot \frac{\omega_M}{\omega_L} = \frac{T_m}{i} \tag{9.8}$$

And if the efficiency $\eta \neq 100\%$ (losses) $1 > \eta > 0$:

So,
$$P_{in} = P_{out}/\eta \tag{9.9}$$

Or

$$T_m \cdot \omega_m = T_L \cdot \omega_L/\eta \tag{9.10}$$

$$\therefore T_L = \eta \times T_m \cdot \frac{\omega_M}{\omega_L} = \eta \times \frac{T_m}{i} \qquad (9.11)$$

If there are multi-stages of gears with gear ratio $= i_1, i_2, i_3, \ldots$
So,

$$\omega_L = (i_1 \times i_2 \times i_3 \times \ldots)\omega_m \qquad (9.12)$$

For ideal case:

$$T_L = \frac{T_m}{i_1 \times i_2 \times i_3 \times \ldots\ldots} \qquad (9.13)$$

For actual case:

$$T_L = \frac{(\eta_1 \times \eta_2 \times \eta_3 \times \ldots)T_m}{i_1 \times i_2 \times i_3 \times \ldots\ldots} \qquad (9.14)$$

where η_n represents the efficiency of stage n.

When the load is transferred to the motor shaft, and hence its speed will be ω_m, so the load torque at the motor shaft will be:

$$P_L = \omega_L \times T_L = \omega_m \times T'_L. \qquad (9.15)$$

For ideal case:

$$T'_L = \frac{\omega_L}{\omega_m} \cdot T_L = i \cdot T_L \qquad (9.16)$$

For actual case:

$$T'_L = \frac{\omega_L}{\omega_m \cdot \eta} \cdot T_L = \frac{i \cdot T_L}{\eta} \qquad (9.17)$$

The equivalent system will be as shown (Figure 9.9):
So steady-state operation when:

$$T'_L = \frac{i \cdot T_L}{\eta} \qquad (9.18)$$

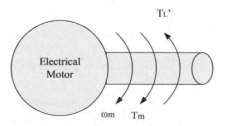

FIGURE 9.9
Equivalent system for a motor drive with the load.

Electrical methods: there are many electrical methods which are used to make the load operation at the required speed by changing ω_m. To select any method, the following points are taken into consideration:

1. The direction of speed variation: UP; DOWN; and UP and DOWN with respect to operating speed at natural characteristics
2. The dynamic range of speed variation, i.e., the possible ratio of maximum to minimum speed achieved:

$$D = \text{Dynamic range of speed} = \frac{\omega_m}{\omega_{min}}.$$

ω_{min} is taken to be at least 10% of rated speed
3. The stiffness coefficient of mechanical characteristics (β) which is equal to:

$$\beta = -\frac{dt}{d\omega} \tag{9.19}$$

This factor represents the sensitivity of the motor to the load variation. The ideal motor must keep constant speed when its load changes, i.e., $\beta = \infty$, and as this factor decrease, it means the motor will be much sense to load variation
4. The value of the load torque
5. The efficiency, simplicity, cost, ….

Different motors have a different electrical method of speed variation; therefore DC and AC motors will take each alone with their possible speed variation methods.

9.3 DC Motors Parametric Methods

9.3.1 Separate Excited and Shunt Motor

The main equations for the separately excited DC motor are:

$$I_a = \frac{V - E}{R_a} \tag{9.20}$$

$$E = k\phi\omega \tag{9.21}$$

$$T = k\phi Ia \tag{9.22}$$

$$K = constant = \frac{Z.P}{2\pi \cdot a} \tag{9.23}$$

where:

Z = conductors/armature
P = no. of poles
A = no. of parallel paths
A = 2 for wave connected, and a = P for lap connected.

The mechanical characteristics ($\omega = f(T)$) can be found from Equations 9.20 through 9.22 as:

$$\omega = \frac{V}{K\Phi} - \frac{R_a}{(K\Phi)^2} \cdot T \tag{9.24}$$

For separate or shunt motor, inter poles and compensating windings ϕ can be considered = constant independent of load.

So for natural characteristics ϕ = constant = ϕ_{nom}.

The mechanical characteristics will be a straight line since V, k, ϕ, R_a are constants.

If $V = V_{nom}$, $\phi = \phi_{nom}$, and no additional elements are inserted, the mechanical characteristics are natural characteristics. From Figure 9.10:

At no-load ($T = I_a = 0$)

$$\omega_o = \frac{V}{K\Phi} \quad \left(\text{no load speed}; E = V\right) \tag{9.25}$$

At $\omega = 0$, $T = T_{so}$ = starting condition is given by:

$$T_{so} = K\Phi \cdot \frac{V}{R_a} \quad (\text{when } E = 0) \tag{9.26}$$

If the motor is operating directly with the load at its natural characteristics:

So, $\qquad\qquad\qquad T_r \geq T_L \geq 0$, and $\omega_0 \geq \omega_L \geq \omega_r$.

So for any operating load torque, the operating point will lay on the mechanical characteristics between points A and B, for example at T_L, ω_L (point C).

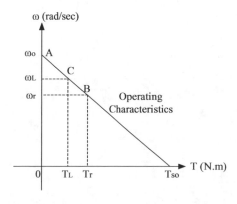

FIGURE 9.10
Torque-speed characteristics.

The stiffness coefficient:

$$\beta = -\frac{dT}{d\omega} = \frac{(K\Phi)^2}{R_a} \tag{9.27}$$

9.3.1.1 Adding Resistance to the Armature

From the equivalent circuit for adding resistance to the armature shown in Figure 9.11, the mechanical characteristics will be represented by:

$$\omega_i = \frac{V_{nom}}{K\Phi_{nom}} - \frac{R_a + R_d}{(K\Phi_{nom})^2} \cdot T \tag{9.28}$$

which gives different speed ω_i.

At different added R_d for given load torque T (Figure 9.12).

At no-load ($T = I_a = 0$):

$$\omega_i = \frac{V_{nom}}{K\Phi_{nom}} = \omega_o = \text{constant independent of } R_d.$$

At $\omega = 0$, $T = T_{si}$ = starting condition, and $E = 0$.

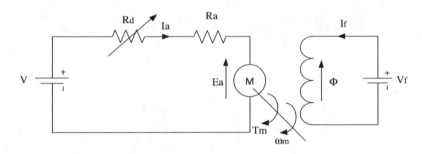

FIGURE 9.11
Adding resistance to the armature.

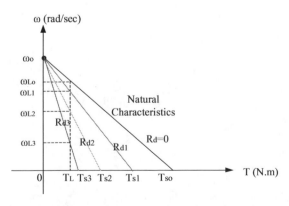

FIGURE 9.12
Torque-speed characteristics when adding resistance to the armature.

$$T_{si} = K\Phi_{nom} \cdot \frac{V_{nom}}{R_a + R_d}$$

$$T_{si} \downarrow as R_d \uparrow$$

$$\beta = -\frac{dT}{d\omega} = \frac{(K\Phi_{nom})^2}{R_a + R_d} \downarrow as R_d \uparrow.$$

If $\qquad\qquad 0 < R_{d1} < R_{d2} < R_{d3}.$

So $\qquad\qquad T_{so} > T_{s1} > T_{s2} > T_{s3}$

$$\omega_{Lo} > \omega_{L1} > \omega_{L2} > \omega_{L3}$$

$$\beta_o > \beta_1 > \beta_2 < \beta_3.$$

The method can decrease the speed only to ω_{Lo} and the stiffness coefficient will be as R_d increase. The method is simple with low efficiency (high losses) due to $I_a^2 (R_d + R)$, therefore it can be used only for small motors. Any operating condition can be achieved by a certain R_d which make the developed characteristics pass through that point. For example, it is required to operate a load of T_L at speed ω_L. Given natural characteristics, such motor can drive such load as shown in Figure 9.13.

If

$$T_L \le T_m \quad at \quad rated$$

$$T_r = \frac{P_r}{\dfrac{2\pi N_r}{60}} \qquad\qquad (9.29)$$

If the speed is given in rpm, the natural characteristics are not suitable if:

$$\omega = \frac{V_{nom}}{K\Phi_{nom}} - \frac{R_a}{(K\Phi_{nom})^2} \cdot T_L \qquad\qquad (9.30)$$

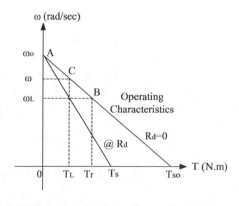

FIGURE 9.13
Torque-speed characteristics when adding resistance to the armature for two values of resistance.

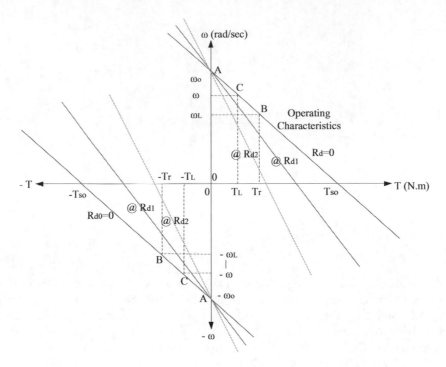

FIGURE 9.14
Torque-speed characteristics when adding resistance to the armature in two operation mode.

It is greater than ω_L required.

To find R_d such that the load T_L operates at ω_L, the characteristics must have R_d so that it passes through the point (T_L, ω_L) (Figure 9.14). R_d can be found if the point in the 1st quadrant (motor operation) either by using the equation:

$$\omega_L = \frac{V_{nom}}{K\Phi_{nom}} - \frac{R_a + R_d}{\left(K\Phi_{nom}\right)^2} \cdot T_L \tag{9.31}$$

So R_d can be found if the other parameters are known, or by similarity:

$$\frac{\omega_o}{T_s} = \frac{\omega_0 - \omega_L}{T_L} \tag{9.32}$$

So,

$$T_s = \frac{\omega_o}{\omega_0 - \omega_L} T_L \tag{9.33}$$

$$T_s = K\Phi_{nom} \cdot \frac{V_{nom}}{R_a + R_d} \tag{9.34}$$

For the motor to operate in the fourth quadrant, each developed characteristics can be extended to the other quadrant.

So, for a brake operation, ω_L will be negative will T_L is positive. For generator braking, ω_L is positive $> \omega_o$, but T_L is negative.

For reverse operation:

$$-\omega_L = \frac{V}{K\Phi} - \frac{R_a - R_d}{(K\Phi)^2} \cdot (-T_L) \tag{9.35}$$

9.3.1.2 Changing the Armature Supply Voltage

If the armature supply voltage can vary,
so $V_{nom} \geq V > 0$ will give a set of mechanical characteristics:

$$\omega = \frac{V}{K\Phi_{nom}} - \frac{R_a}{(K\Phi_{nom})^2} \cdot T_L \tag{9.36}$$

At any selected voltage, the characteristics are a straight line parallel to each other given by:

$$\omega_{0i} = \frac{V_{noim}}{K\Phi_{nom}} \tag{9.37}$$

$$T_{si} = K\Phi_{nom} \cdot \frac{V_i}{R_a} \tag{9.38}$$

As $V_i \downarrow$, $\omega_{oi} \downarrow$ and $T_{si} \downarrow$

$$\beta = -\frac{dT}{d\omega} = \frac{(K\Phi)^2}{R_a} = \text{constant.}$$

So the mechanical characteristics are parallel to each other as shown in Figure 9.15.

$$V_{nom} > V_1 > V_2$$

$$\omega_o > \omega_{o1} > \omega_{o2}$$

$$T_{so} > T_{s1} > T_{s2}.$$

At constant load T_L:

$$\omega_L > \omega_{L1} > \omega_{L2}.$$

So the speed variation only down. By the use of power electronics, the armature supply voltage can be changed. The efficiency of this method is better than that with R_d and the stiffness coefficient remains constant.

9.3.1.3 Changing the Field Flux

By adding resistance to the field circuit or by inserting series winding, the flux of the motor can be changed over and under ϕ_{nom} of the single winding as shown in Figure 9.16.

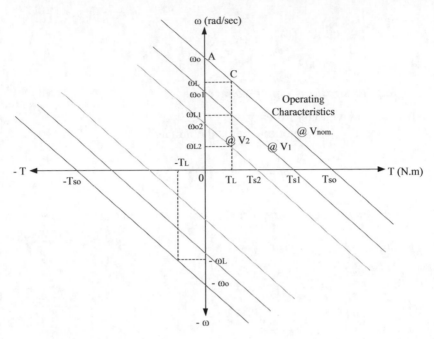

FIGURE 9.15
Torque-speed characteristics changing the armature supply voltage in two operation mode.

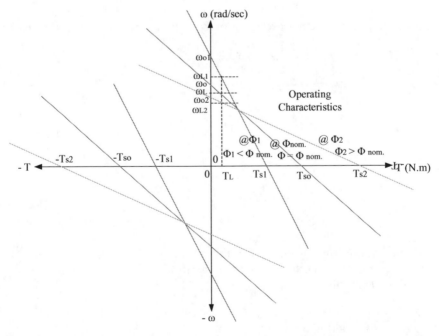

FIGURE 9.16
Torque-speed characteristics changing the field flux in two operation mode.

$$\omega_i = \frac{V_{nom}}{K\Phi_i} - \frac{R_a}{\left(K\Phi_i\right)^2} \cdot T_L.$$

$$\text{As } \Phi_i \uparrow, \omega_{oi} = \frac{V_{nom}}{K\Phi_i} \downarrow \quad and \quad T_{si} = \frac{V_{nom}}{R_a} \uparrow.$$

So for $\phi_1 < \phi_{nom}$

$$\omega_1 > \omega_o$$

$$T_{s1} > T_{so}$$

And for $\phi_2 > \phi_{nom}$

$$\omega_{o2} < \omega_o$$

$$T_{s2} > T_{so}.$$

According to the value of T_L, increasing or decreasing of the flux may give different results of speed (uncertainty), for example, decreasing the flux from ϕ_1 to ϕ_2 will give increasing of speed if the load torque is T_1 from ω_1 to ω_2. If the load torque is T_2 the change from the flux from ϕ_1 to ϕ_2 will not change the speed.

At load torque T_3 the changing from the flux from ϕ_1 to ϕ_2 will decrease the speed from ω_3 to ω_4 as shown in Figure 9.17.

The method affects the stiffness coefficient:

$$\beta = -\frac{dT}{d\omega} = \frac{\left(K\Phi\right)^2}{R_a} \downarrow$$

When $\phi_i \downarrow$, but the method is more efficient than that with R_d in the armature.

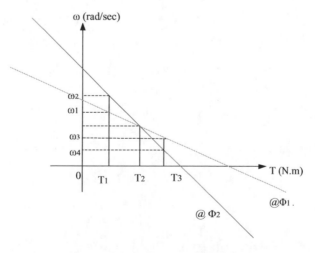

FIGURE 9.17
Torque-speed characteristics changing the field flux for two values of flux.

Example 9.1

A separate excited, 3 kW, 150 Volt, 1610 rpm, DC motor has an armature resistance = 1 Ω and is to be used in the drive through a reducer of Nm/NL = 10. If the load torque is 150 N·m, calculate the required R_d added to the armature to operate the load at 1 rad/sec? What voltage reduction is used for such operation if no R_d is added?

Solution

$$\omega_r = \frac{2\pi \cdot n_r}{60} = \frac{2\pi \times 1610}{60} = 168.6 \, \text{rad/sec}$$

$$T'_L = \frac{150}{10} = 15 \, \text{N} \cdot \text{m}$$

$$T_{mr} = \frac{P_r}{\omega_r} = \frac{3 \times 10^3}{168.6} = 17.79 \, \text{N} \cdot \text{m}$$

(such motor can drive the load 15 N·m).

For natural characteristics,

$$\omega_r = \frac{V}{K\Phi} - \frac{R_a}{(K\Phi)^2} \cdot T_r$$

$$168.6 = \frac{150}{K\Phi} - \frac{1}{\left(K\Phi\right)^2} \times 17.79$$

Solving:

$$168.6(K\Phi)^2 - 150 \, (K\Phi) + 17.79 = 0$$

$$K\Phi = 0.7487. \; Or \; K\Phi = 0.1409$$

$$\omega_o = \frac{V}{K\Phi}$$

$$\omega_o = \frac{150}{0.7487} = 200.34 \, rad/sec$$

Or:

$$\omega_o = \frac{150}{0.1409} = 1064.58 \; rad/sec$$

By neglecting the high value of ω_o.

$$\text{So } k\Phi = 0.7487.$$

To find R_d required:

$$10 = 200.34 - \frac{1 + R_d}{\left(0.7487\right)^2} \cdot 15$$

$$R_d = 6.11 \ \Omega$$

$$\text{For } R_d = 0 \ \Omega$$

$$10 = \frac{V}{0.7487} - \frac{1}{\left(0.7487\right)^2} \cdot 15$$

$$V = 27.52 \ \text{V}.$$

So the reduction of supply voltage = 150 − 27.52 = 122.478 Volt.

9.3.2 Series Motor

For the series motor, the value of R_f (field resistance) is added to armature total resistance and the change of load (torque) will change also ϕ.

So ϕ is no longer constant, but the function of the armature current, for example, (Figure 9.18):

$$K\Phi = \frac{A \cdot I_a^2 + B \cdot I_a + C}{D \cdot I_a} \tag{9.39}$$

where A, B, C, and D are constants whose values give a linear relationship at small I_a and constant $k\phi$ at very high I_a (saturation). Figure 9.19 shows torque-speed characteristics for the series motor.

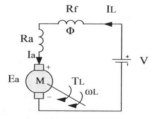

FIGURE 9.18
Equivalent circuit for series motor.

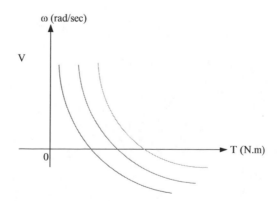

FIGURE 9.19
Torque-speed characteristics for the series motor.

At no-load ideally $\phi = 0$ and $\omega_o = \infty$, therefore no-load or high load operation of the series motor is not allowed since the resultant speed is dangerously high. For the same reason, a movement from 1st quadrant (generator braking) is not allowed also. The mechanical characteristics are given by the same equation:

$$\omega = \frac{V}{K\Phi} - \frac{R_a}{(K\Phi)^2} \cdot T \text{ with } \Phi \text{ changing with T.}$$

Adding the resistance to the armature or changing the supply voltage will change the ϕ also. The characteristics are a no-longer straight line.

Example 9.2

A series motor 120 Volt, 2.2 KW, 850 rpm, has $R_a = 0.5\ \Omega$ and $R_F = 0.3\ \Omega$. The field is linearly proportional to armature current. Find the required additional armature resistance to drive a load of 20 N·m at speed of 250 rpm.

Solution

Let $k\phi = A \cdot I_a$,
    where A = constant.

$$\omega_r = \frac{2\pi \cdot n_r}{60} = \frac{2\pi \times 850}{60} = 89.011 \text{rad / sec}$$

$$T_{mr} = \frac{P_r}{\omega_r} = \frac{2.2 \times 10^3}{89.011} = 24.716 \text{N.M}$$

$$T = K\Phi \cdot I_a = A \cdot I_a^2$$

$$\omega_r = \frac{V}{K\Phi} - \frac{R_a + R_F}{(K\Phi)^2} \cdot T_r$$

$$89.011 = \frac{120}{K\Phi} - \frac{0.5 + 0.3}{(K\Phi)^2} \cdot 24.716.$$

Solving:

$$89.011(K\Phi)^2 - 120(K\Phi) + 19.7728 = 0.$$

$$\text{Either } K\Phi = 1.156$$

$$\text{or} = 0.192.$$

$$\omega_o = \frac{V}{K\Phi}$$

$$\omega_o = \frac{120}{1.156} = 103.8 \text{ rad/sec}$$

or

$$\omega_o = \frac{120}{0.192} = 625 \ \text{rad/sec.}$$

By neglecting the high value of ω_o,

so

$$\therefore k\phi = 1.156 = A \cdot I_a$$

$$T_r = A \cdot I_a^2 = 24.716$$

$$\therefore \frac{T_r}{K\Phi} = I_a = \frac{24.716}{1.156} = 21.38 \, A$$

$$1.156 = A \cdot 21.38,$$

$$\text{so } A = 0.054$$

To find R_d required:

$$10 = 200.34 - \frac{1 + R_d}{(0.7487)^2} \cdot 15,$$

$$R_d = \ 6.11 \ \Omega.$$

$$\text{For } R_d = 0 \ \Omega$$

$$T_L = A \cdot I_a^2$$

$$20 = 0.054 \, I_a^2$$

$$I_a = 19.24 \ \text{A.}$$

$$\text{New } k\phi = A \cdot I_a = 0.054 \times 19.24 = 1.039$$

$$\omega_L = \frac{2\pi n_L}{60}$$

$$\omega_L = \frac{2\pi \times 250}{60} = 26.18 \, rad/sec$$

$$\omega_L = \frac{V}{K\Phi} - \frac{R_a + R_F + R_d}{(K\Phi)^2} \cdot T_L$$

$$26.18 = \frac{120}{1.039} - \frac{0.5 + 0.3 + R_d}{(1.039)^2} \cdot 20,$$

$$R_d = 4.021 \ \Omega.$$

9.4 DC Drive Circuits

The DC drives are classified under:

1. Single-phase DC drives
2. Three-phase DC drives
3. Chopper drives.

First, the basic operating characteristics of DC motor are presented and then power electronics circuits strategies as mentioned in previous chapters are described. In this section, the single-phase rectifier drive and DC chopper drive will discuss.

9.4.1 DC Drive Rectifier Circuits

9.4.1.1 *Single-Phase Half-Wave Converter Drives a Separately Excited DC Motor*

Assume that this circuit fed through single-phase half-wave converter is shown in Figure 9.20. Motor field circuit is fed through a single-phase semi converter in order to reduce the ripple content in the field circuit. Single-phase half-wave converter feeding a DC motor offers one-quadrant drive, Figure 9.21a. The waveforms for source voltage V_s, armature terminal voltage V_t, armature current i_a, source current i_s, and freewheeling diode current i_{fd}. Note that thyristor current $i_T = i_s$: The armature current is assumed ripple free.

For single-phase half-wave converter, the average output voltage of the converter, V_o = armature terminal voltage, V_t is given by:

$$V_0 = V_t = \frac{V_m}{2\pi}(1+cos\alpha) \qquad For\, 0 < \alpha < \pi, \tag{9.40}$$

where V_m = maximum value of source voltage. For single-phase semi converter in the field circuit, the average output voltage is given by:

$$V_f = \frac{V_m}{\pi}(1+cos\alpha_1) \qquad For\, 0 < \alpha_1 < \pi. \tag{9.41}$$

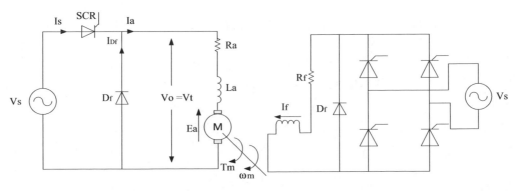

FIGURE 9.20
Circuit diagram of single-phase half-wave converter drive.

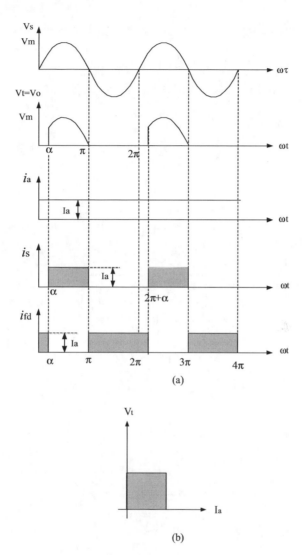

FIGURE 9.21
Waveforms single-phase half-wave converter drive (a) Voltages and currents waveforms (b) V-I characteristics.

It is seen from the waveforms of Figure 9.21a that RMS value of armature current, $I_{ar} = I_{aRMS}$ value of source or Thyristor current, and the relation between armature current and terminal voltage given in Figure 9.21b.

$$I_{sr} = \sqrt{I_a^2 \frac{\pi - \alpha}{2\pi}} = I_a \left(\frac{\pi - \alpha}{2\pi} \right)^{1/2}. \tag{9.42}$$

The RMS value of freewheeling-diode current,

$$I_{sr} = \sqrt{I_a^2 \frac{\pi + \alpha}{2\pi}} = I_a \left(\frac{\pi + \alpha}{2\pi} \right)^{1/2}. \tag{9.43}$$

$$Apparent\ input\ power = (rms\ source\ voltage) \cdot (rms\ source\ current)$$

$$= V_s \cdot I_{sr}. \tag{9.44}$$

$$\text{Power delivered to motor} = E_a I_a + I_a^2 \cdot r_a = (E_a + I_a r_a)\, I_a = V_t \cdot I_a \tag{9.45}$$

Input supply power factor:

$$pf = \frac{E_a I_a + I_a^2 r_a}{V_s \cdot I_{sr}} \tag{9.46}$$

Example 9.3

A separately excited DC motor has $k\varphi = 1.44$, $R_a = 0.86$ Ω. When it is operating at 150 rad/sec, the armature current is $I_a = 40A$. The terminal voltage V_t is held constant under all conditions.

1. Compute V_t
2. Compute the no-load speed in rad/sec.

Solution

1.

$$E_a = k\varphi w_m = (1.44) \cdot (150) = 216V$$

$$V_t = E_a + I_a r_a = 216 + (40) \cdot (0.86).$$
$$= 250.4V$$

2. "No-load" speed occurs when T = 0. Therefore,

$$\omega = \frac{V_t}{k\varphi} - \frac{R_a T}{(k\varphi)^2} = \frac{V_t}{k\varphi} \ . \ \text{When } T = 0$$

$$\omega_0 = \frac{250.4}{1.44} = 173.9\,\text{rad/sec.}$$

Example 9.4

A separately excited DC motor is supplied from 220 V, 60 Hz source through a single-phase half-wave controlled converter. Its field is fed through i-phase semi converter with zero degrees firing-angle delay. Motor resistance $R_a = 0.1$ Ω, and motor constant = 0.6 V-sec/rad. For rated load torque of 15 N·m at 1200 rpm and for continuous ripple free currents, determine:

1. The firing angle delay of the armature converter
2. The RMS value of thyristor and freewheeling diode currents
3. The input power factor of the armature converter.

Solution

1. The average current = 15/0.5 = 30 A

$$\omega_m = \frac{2\pi \cdot n_m}{60}$$

$$\omega_m = \frac{2\pi \cdot 1200}{60} = 125.66\, rad\,/\,sec.$$

Motor EMF $E_a = K_m \cdot \omega_m = 0.6 \times 125.66 = 75.4$ V.
  For 1-phase half-wave converter feeding a DC motor,

$$V_0 = V_t = \frac{V_m}{2\pi}(1 + \cos\alpha) = E_a + I_a r_a$$

$$V_0 = V_t = \frac{220 \cdot \sqrt{2}}{2\pi}(1 + \cos\alpha) = 75.4 + 30 \times 0.1.$$

Solving for α to get $\alpha = 54.32° = 0.948$ rad.

  2. RMS value of thyristor current is:

$$I_{sr} = \sqrt{I_a^2 \frac{\pi - \alpha}{2\pi}} = I_a\left(\frac{\pi - \alpha}{2\pi}\right)^{1/2}$$

$$I_{sr} = 30\left(\frac{\pi - 0.948}{2\pi}\right)^{1/2} = 17.725\,A,$$

the RMS value of freewheeling diode current is:

$$I_{sr} = \sqrt{I_a^2 \frac{\pi + \alpha}{2\pi}} = I_a\left(\frac{\pi + \alpha}{2\pi}\right)^{1/2}$$

$$I_{sr} = 30\left(\frac{\pi + 0.948}{2\pi}\right)^{1/2} = 24.2\,A.$$

  3. The input power factor of armature converter:

$$pf = \frac{V_t \cdot I_a}{V_s \cdot I_{sr}}$$

$$pf = \frac{78.4 \times 30}{220 \times 24.2} = 0.441\,lagging.$$

Example 9.5

Draw the speed response for separate excited DC motor drive: A 5 hp, 240 V separately excited DC motor with parameters given as follows in Figure 9.22.

Solution

The Simulink of the circuit as shown in Figure 9.23, and the related waveforms given in Figure 9.24.

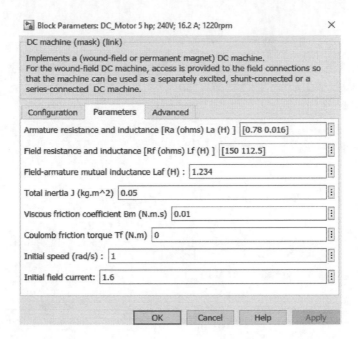

FIGURE 9.22
DC machine mask.

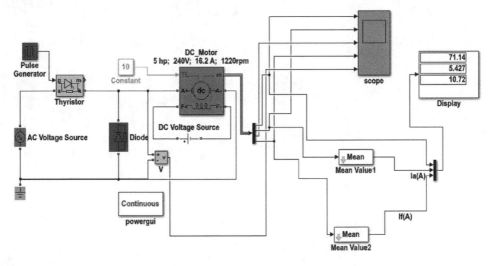

FIGURE 9.23
Separate excited DC motor drive Simulink.

9.4.1.2 *Single-Phase Full-Wave Converter Drives a Separately Excited DC Motor*

Example 9.6

A separately excited DC motor drives a rated load torque of 40 Nm at 1000 rpm as shown in Figure 9.25. The field circuit resistance is 200 Ω and armature circuit resistance is 0.05 Ω. The field winding, connected to single-phase, 120 V source, is fed through 1-phase full converter with zero degrees firing angle. The armature circuit is also fed through another full converter from the same converter single-phase, 120 V source.

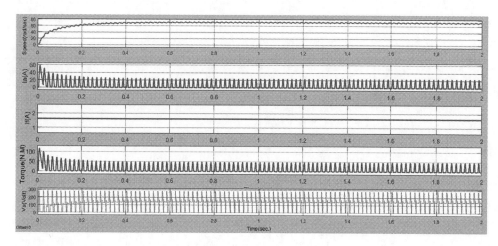

FIGURE 9.24
Waveforms.

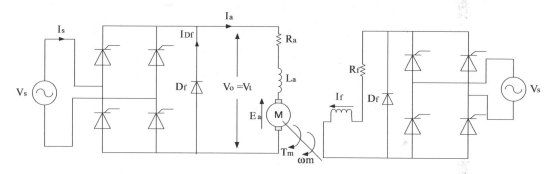

FIGURE 9.25
Separately excited DC motor drives.

With magnetic saturation neglected, the motor constant is 0.8 V-sec/A-rad. For ripple free armature and field currents. Determine:

1. Rated armature current
2. Firing angle delay of armature converter at rated load
3. Speed regulation at full load
4. Input power factor (PF) of the armature converter and the drive at rated load.

Solution

1. For field converter, angle delay $\alpha = 0°$.

The field voltage,

$$V_f = \frac{2V_m}{\pi} = \frac{2\sqrt{2} \times 120}{\pi} = 108.03\,V$$

$$I_f = \frac{V_f}{r_f} = \frac{108.03}{200} = 0.54\,A.$$

With magnetic saturation neglected, $\phi_1 = K_1 I_f$

$$E_a = K_a \phi \omega_m = K_a \cdot K_1 I_f \omega_m = K \cdot I_f \omega_m$$

and the torque $T = K_a \phi I_a = K_a \cdot K_1 I_f I_a = K \cdot I_f I_a$.
The rated armature current, I_a:

$$I_a = \frac{T}{K \cdot I_f}$$

$$I_a = \frac{40}{0.8 \times 0.54} = 92.59 \, A$$

2.

$$V_0 = V_t = \frac{2V_m}{\pi}(\cos\alpha) = E_a + I_a r_a$$

$$V_0 = V_t = \frac{2\sqrt{2} \times 120}{\pi}(\cos\alpha) = 0.8 \times 0.54 \times \frac{2\pi \times 1000}{60} + 92.59 \times 0.05$$

$$\alpha = 62.51°.$$

3. At the same firing angle of $\alpha = 62.51°$ motor EMF at no-load,

$$E_a = K_a \phi \omega_m = K_a \cdot K_1 I_f \omega_m = K \cdot I_f \omega_m = V_t = 49.87 \text{ V.}$$

The speed at no-load:

$$\omega_{mo} = \frac{E_a}{K \cdot I_f} = \frac{49.87}{0.8 \times 0.54} = 115.43 \, rad/sec.$$

Or 1102.37 rpm.
Speed regulation at full load:

$$N_R\% = \frac{No\,load\,speed - Full\,load\,speed}{Full\,load\,speed}$$

$$N_R\% = \frac{1102.37 - 1000}{1000} = 10.23\%.$$

4. The input power factor of the armature converter

$$pf = \frac{V_t \cdot I_a}{V_s \cdot I_{ar}}$$

$$pf = \frac{49.87 \times 92.59}{120 \times 92.59} = 0.415 \quad lagging$$

RMS value of current in armature converter,

$$I_{ar} = 92.59 \text{ A.}$$

RMS value of current in field circuit,

$$1_{fT} = I_f = 0.54 \text{ A}.$$

Total current taken from the source,

$$I_{sr} = \sqrt{I_{ar}^2 + I_{fr}^2}$$

$$I_{sr} = \sqrt{92.59^2 + 0.54^2} = 92.591 A.$$

Input S = $V_s \cdot I_{sr}$ = 120 × 92.591 = 11.11 kVA.
   With no loss in the converters, total power input to motor and field:

$$= V_t I_a + V_f I_f = 49.87 \times 59.03 + 108.03 \times 0.54 = 3 \text{ KW}.$$

$$Input\, power\, factor = \frac{P}{S} = \frac{3}{11.11} = 0.27 \,lagging.$$

Example 9.7

Draw the speed response for DC shunt motor drive: A 5 hp, 240 V, 1750 rpm, separate excited DC motor with parameters given as follows:

$$R_a = 2.581 \text{ Ω}, L_a = 0.028 \text{ H}, R_f = 281.3 \text{ Ω}, L_f = 156 \text{ H}, J = 0.02215 \text{ kg·m}^2.$$

Solution

The Simulink of the circuit as shown in Figure 9.26, and the related waveforms given in Figure 9.27.

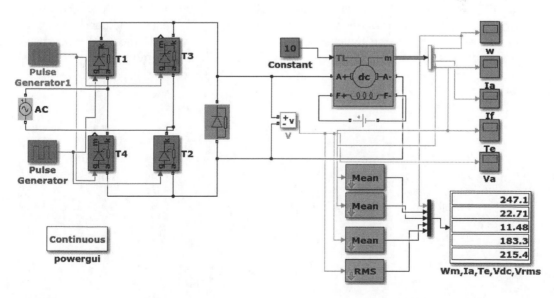

FIGURE 9.26
Circuit diagram using MATLAB/Simulink.

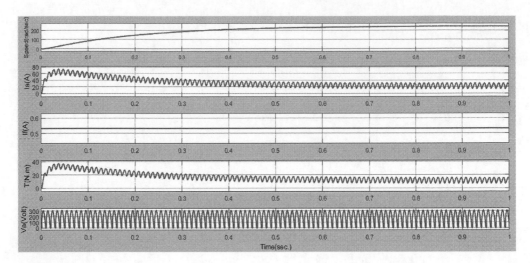

FIGURE 9.27
Waveforms.

9.5 DC Chopper Drive

The speed of a DC Motor is directly proportional to the line voltage applied to it. Given a fixed DC source, VS, and a power Insulated Gate Bi-polar Transistor (IGBT) to act as a switch, it is possible to control the average voltage applied to the motor using a technique called PWM.

In the circuit shown in Figure 9.28 below, the source voltage, V_s, is "chopped" to produce an average voltage somewhere between 0% and 100% of V_s. Thus, the average value of the voltage applied to the motor, V_m, is controlled by closing and opening the "switch," Q1. To close the switch, a firing signal is delivered to the gate of the IGBT, causing it to conduct between source and drain. To open the switch, the firing signal is removed and the IGBT is self-biased to stop conducting. In PWM, the switch is closed and opened every modulation period.

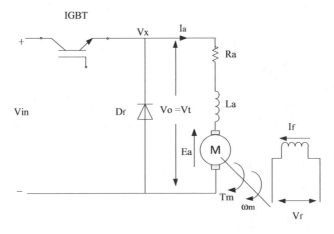

FIGURE 9.28
DC chopper drive circuit.

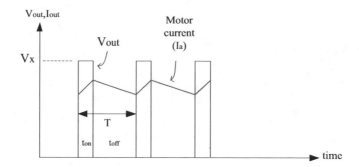

FIGURE 9.29
Waveforms of DC chopper drive circuit.

In discussing the period of modulation, let time be divided into uniform periods of 1 millisecond each and let a period be called T, the modulation period. During T, there is a time, t_0 to t_1, during which the MOSFET Q1 is on, and a time, t_1 to t_2, during which it is off, as indicated in Figure 9.29. This is true for each period and therefore Q1 turns on and off 1000 times every second when T = 1 ms.

When Q1 is on, V_s volts are applied to the motor for t_1 milliseconds. When Q1 is off, zero volts are applied to the motor. However, the armature current, I_a, is still allowed to circulate through the diode. The magnitude of the armature current will diminish between t_1 and t_2 as losses in the motor dissipate energy.

The voltage V_m seen by the motor can be expressed in terms of the source voltage V_s and the "ON" time t_1 and the period of modulation T. The equation is:

$$V_m = \alpha\, V_s \tag{9.47}$$

where:

$$\alpha = t_1/T \tag{9.48}$$

The symbol α is called *the duty cycle*. As the duty cycle is increased from 0% to 100%, the average voltage applied to the motor increases from 0 to V_s volts and the motor speeds up.

It is sometimes more useful to think of the frequency of modulation as opposed to its period. The modulation frequency is just the inverse of the period of modulation, $f_m = 1/T$. State of the art PWM converters have modulation frequencies as high as 200 k, although some people argue that frequencies above 20 k lead to unnecessary expense.

Example 9.8

Draw speed, armature current, field current, and torque for the DC chopper drive circuit shown in Figure 9.30 using MATLAB/Simulink.

Solution

The Simulink of the circuit as shown in Figure 9.30 and the related waveforms given in Figure 9.31.

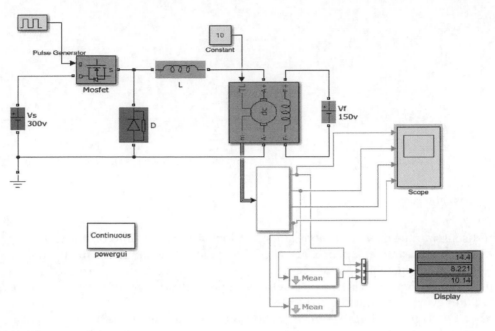

FIGURE 9.30
DC chopper drive circuit using MATLAB/Simulink.

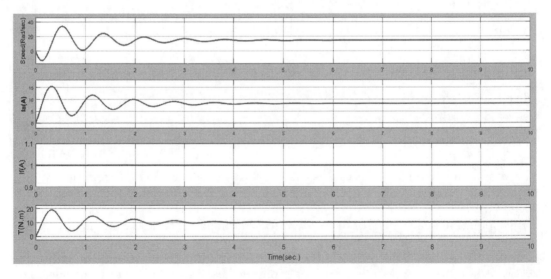

FIGURE 9.31
Waveforms.

Example 9.9

A step-down (buck) converter in Figure 9.32 is used to supply 100 V DC motor and an inductive load of $L = 4mH, R_a = 0$, if $F_s = 5\ KHz$, $T = 50 N \cdot m$, and input DC voltage an is 300 V, calculate:

1. V_{Lmean} and speed of duty cycle range 0.4–0.8
2. Trace the output voltage and current of case D = 0.7.

Solution

The Simulink of the circuit as shown in Figure 9.33 and the related waveforms given in Figure 9.34.

1.

Duty Cycle (D)	I_{Lmean} (A)	V_{Lmean} (V)	Speed (Rad/sec)
0.4	67.7	120	104.8
0.5	67.4	150	140.5
0.6	67.7	180.2	176.5
0.7	68.04	210.5	212.6
0.8	68.35	240.4	248.1

2.

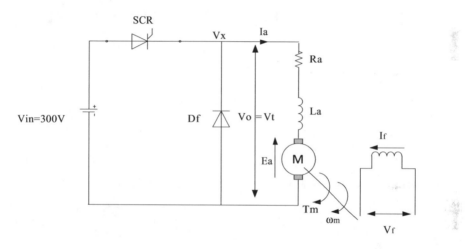

FIGURE 9.32
Step-down (buck) converter.

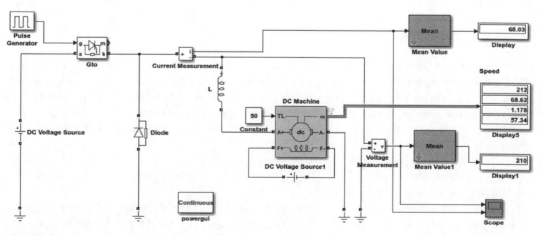

FIGURE 9.33
Step-down (buck) converter MATLAB/Simulink.

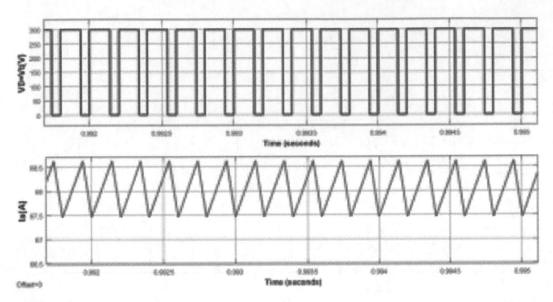

FIGURE 9.34
Waveforms.

9.6 Electrical Braking of Separate Excited DC Motor

The separately excited DC motor can be braking by one of the following methods:

1. Generator braking
2. Supply reverse braking
3. Dynamic braking.

9.6.1 Generator Braking

When the speed of a separately excited motor exceeds its no-load speed, the motor will exert a torque oppose the motion and the direction of the armature current is reversed.

The current will move from the machine to the supply (the machine will operate as a generator), and the produced torque is opposing the motion, hence, it is called generator braking or regenerative braking. Figure 9.35 shows torque-speed characteristics in the case of generator braking.

Suppose a motor drive a load T_L at speed ω_L according to the mechanical characteristics as shown in Figure 9.35. If the voltage is reduced or ϕ is increased such that the new mechanical characteristics have $\omega_o < \omega_L$ as shown.

Then at $t = 0$ when V is changed, the motor will produce a torque $= T_{gb}$ opposing the motion. As ω reduced from ω_L to ($\omega_{o\ new}$), this torque will change from T_{gb} to zero. For the operating characteristics, $\omega_o > \omega_L$ and the machine operate as a motor with $E = (k\phi \cdot \omega_L)$ is less than V, the armature current is:

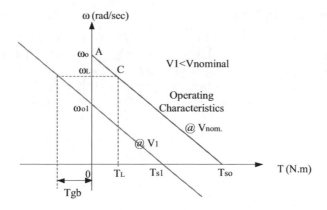

FIGURE 9.35
Torque-speed characteristics in case generator braking.

$$I_a = \frac{V-E}{R_a} = \frac{V-K\Phi\omega}{R_a}.$$

From the supply to the motor producing a torque:

$$T = K\Phi \cdot I_a = K\Phi\frac{V-E}{R_a} = K\Phi\frac{V-K\Phi\omega}{R_a}.$$

In the same direction as that of the speed (motor operation). When the voltage V is changed to V_1, where $V_1 < V$ such that:

$$\omega_{o\,new} = \frac{V_1}{K\Phi} < \omega_L$$

$$I_{a\,new} = \frac{V_1-E}{R_a} = \frac{V_1-K\Phi\omega}{R_a}$$

$$\omega_{o\,new} = \frac{V_1}{K\Phi} < \omega_L$$

$$\therefore I_{a\,new} = \frac{V_1-E}{R_a} = \frac{V_1-K\Phi\omega_L}{R_a} = -ve. \tag{9.49}$$

So new $(k\phi \cdot \omega_o) = V_1$.

For example, the current will return to the supply and produced torque will be $T_{gb} = k\phi$. $I_{a\,new} = k\phi(-I_{a\,new}) = $ negative torque, i.e., oppose the motion, so the motor will brake.

Similar results when ϕ increase to give $\omega_{o\,new} < \omega_L$. Notice that the flux remains constant in value and direction. If the motors are shunt motor, similar operation can be activated. Series motor can never be operated in such mode of operation since its speed will be dangerously high.

9.6.2 Supply Reversing Braking

A separately excited DC motor operates at certain mechanical characteristics at point (T_L, ω_L) as shown in Figure 9.36. By changing the polarity of the armature supply voltage, the motor will produce a very large brake torque as shown in Figure 9.37.

Suppose point A is the operating point on the selected characteristics $A(T_L, \omega_L)$.

The armature current will be from the supply to the machine (motor) and equal to:

$$I_a = \frac{V - E}{R_a}$$

$$E = \text{brake EMF} = k\phi \cdot \omega_L, V > E.$$

When the supply polarity is reversed (ϕ kept constant in amplitude and direction), so $V \rightarrow -V$ (in the same direction as E which remain the same at $t = 0$).

$$I_a = \frac{-V - E}{R_a}$$

$$I_a = -\frac{V + E}{R_a} \tag{9.50}$$

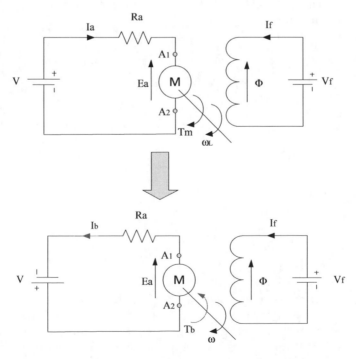

FIGURE 9.36
Supply reversing braking.

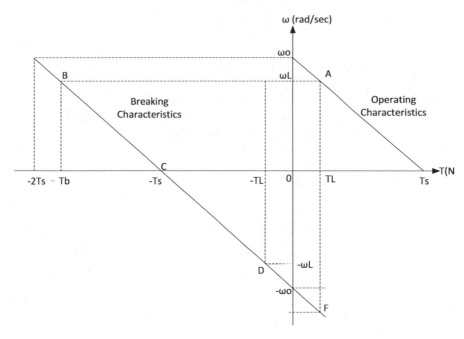

FIGURE 9.37
Torque-speed characteristics in case supply reversing braking.

Passing through the armature from the supply, but in opposite direction to that for the motor, (For motor from A1 to A2 while for supply reversing from A2 to A1). So, the resulting torque:

$$T = K\Phi I_a = -K\Phi \frac{V+E}{R_a} = -ve \tag{9.51}$$

which mean opposing the motion (brake).
So, the resultant brake torque:

$$T_b = K\Phi I_a = K\Phi \frac{V+E}{R_a} \tag{9.52}$$

At t = 0, and as this brake will reduce the speed quickly to zero, E = kφ·ω will reduce and the braking torque through the braking process will reduce too.

The maximum possible brake torque in such method is the case of braking by supply reversing at no-load. For example, if the motor is at no-load, then V = E, $I_a = 0$. When V is reversed then:

$$I_a = \frac{-V-V}{R_a}$$

$$I_a = -2\frac{V}{R_a}$$

$$I_a = -2I_s \tag{9.53}$$

$$Tb = k\phi \cdot Ia = -2\ k\phi \cdot Is = -2\ Ts \text{ (negative means brake).} \tag{9.54}$$

So, if the operating point at $(0, \omega_o)$, the brake torque at $t = 0 = 2T_s$ and the armature current $= 2\ I_s$. So, $2T_s \geq T_b \geq T_s$ depending on the position of the operating point A.

At $t = 0$, the corresponding point on the braking characteristic is B and the corresponding brake torque is T_b. As ω_L reduced $E \downarrow$, $T_b \downarrow$ and the point will move on the brake characteristics from B to C.

At point C the load speed is zero, therefore if the armature supply V is switched OFF the system will stop and the time taken for stopping is very short (with respect to the previous method and which stopped due to frictional force only). If the supply is not switched OFF, then the motor will have a torque:

$$T = -T_s \left(K\Phi \frac{-V}{R_a} \right).$$

Therefore, the motor will be starting to operate in reverse direction. The final operating point such case will be point $D(-T_L, -\omega_L)$ if the load torque T_L is passive torque (frictional), otherwise, the operating point will be the point (F), $F(T_L, -\omega_L)$, and o $\omega_L > \omega$ (generator braking).

An example of such case is braking a lift move upward by supply reversing. Supply voltage reversing is applied to shunt or series motor as shown in Figure 9.38.

The motor will not have braked because the supply polarity reversing cause the armature current and the field current both to change their direction then ϕ will change to $-\phi$ and I_a will change also to $-I_a$ which give:

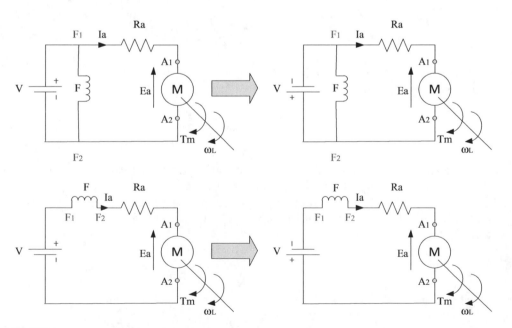

FIGURE 9.38
Supply voltage reversing is applied to shunt or series motor.

$$T = k\phi I_a \rightarrow k(-\phi)(-I_a) = k\phi I_a$$

For example, the shunt and series motor will not have affected by supply reversing. Therefore, for shunt and series motor, supply reversing braking can be done only if either the direction of armature current or field current is kept the same (unchanged). This can be done using two similar fields winding one wound opposite the other and used alternatively when the supply polarity reversing.

In such case, the current of the motor operation passes through the armature from A1 to A2 (Figure 9.39) and through the upper field winding from F1 to F2 causing a motor torque T_L in the same direction as ω_L. When the supply polarity is reversed, the upper field winding is disconnected and the lower field winding is connected, so the armature current direction will be reversed (from A2 to A1) which means $-I_a$ with respect to the first case, but the field current in the lower field winding is still from F1 to F2 which means ϕ is not changed. So,

$$T = k\phi I_a \Rightarrow T_b = k\phi(-I_a)$$

9.6.3 Dynamic Braking

For the separately excited motor, operate with a load T_L at speed ω_L at point A of the selective mechanical characteristics can be braking by switching OFF the supply armature voltage and closing the armature terminal through a short circuit or through R_d as shown with field kept constant (unchanged).

For motor operation and from Figure 9.40:

$$I_a = \frac{V - E}{R_a}$$

$$E = K\Phi \cdot \omega_L$$

and $T_L = k\phi \cdot I_a$.

When $V = 0$ and the terminal of the armature is short-circuited:

$$I_a = \frac{0 - E}{R_a} = -\frac{E}{R_a} \tag{9.55}$$

In the reverse direction, since ϕ is kept the same (constant)

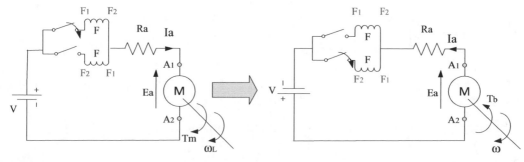

FIGURE 9.39
Supply voltage reversing is applied to the series motor.

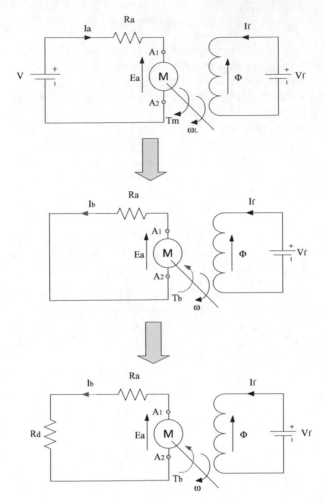

FIGURE 9.40
Motor with dynamic braking.

so,

$$T = k\phi\, I = k\phi\, E/Ra \tag{9.56}$$

As $\omega_L \downarrow$, $E \downarrow$ and $T_b \downarrow$.

For terminating the armature through R_d:

$$I_b = \frac{0 - E}{R_a + R_d} = -\frac{E}{R_a + R_d} \tag{9.57}$$

In the opposite direction, but less than that with short circuit:

$$T_b = K\Phi\, I_b = K\Phi\left(-\frac{E}{R_a + R_d}\right) \tag{9.58}$$

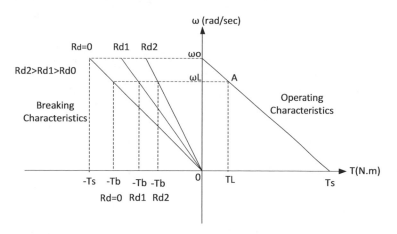

FIGURE 9.41
Torque-speed characteristics in case dynamic braking.

As $\omega\downarrow$, $I_a\downarrow$, $T_b\downarrow$.

Figure 9.41 shows torque-speed characteristics in case dynamic braking.

At $\omega = 0$, $I_a = 0$, and $T_b = 0$. The maximum possible:

$$I_b = \frac{V}{R_a} \qquad (9.59)$$

At no-load, since $E = V$, and the brake torque is (T_s), as $R_d\uparrow$, $I_a\downarrow$, and $T_b\downarrow$.

If the load is passive load, the final position is the origin, but if the load is active load the final operating point will move to the 4th quadrant at $(T_L,-\omega_L)$.

From Figure 9.42, operating in position (1) represented motor operation in a certain direction, while operating in position (2) represented also motor operation in the reverse direction.

Changing from position (1) to position (2) or vice versa through operation will represent supply reversing braking.

Changing from position (1) to position (4) or from (2) to (3) will represent dynamic braking through R_d which can be selected as required.

Example 9.10

A separate excited 150 Volt, 3 KW, 960 rpm DC motor has a total armature resistance of 1 Ω is used to lift a load of 0.8 Tr at speed of 10% of its no-load speed. Find:

1. The required additional resistance for such operation
2. If the supply is reverse braking is used, what will be the initial brake torque and what will be the final operation?
3. If the supply is switched OFF and the armature is terminated through resistance = 3 Ω, what will be the initial brake torque and what will be the final position?

Solution

$$\omega_r = \frac{2\pi\, n_r}{60} = \frac{2\pi \times 960}{60} = 100.53\, rad/sec$$

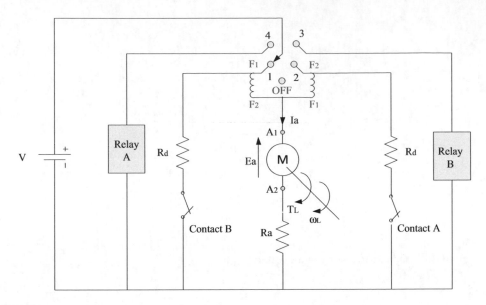

FIGURE 9.42
Dynamic braking modes.

$$T_r = \frac{P_r}{\omega_r} = \frac{3 \times 10^3}{100.53} = 29.84 \, N \cdot m$$

$$\omega_L = 10\% \, \omega_o$$

$$\omega_L = 10\% \times 100.53 = 10.053 \, rad/sec$$

$$T_L = 0.8 T_r$$

$$T_L = 0.8 \times 29.84 = 23.87 \text{ N·m}$$

$$\omega_r = \frac{V}{K\Phi} - \frac{R_a}{\left(K\Phi\right)^2} \cdot T_r$$

$$100.53 = \frac{150}{K\Phi} - \frac{1.2}{\left(K\Phi\right)^2} \cdot 29.84$$

$$100.53 \, (K\Phi)^2 - 150 \, K\Phi + 29.84 = 0$$

$$K\Phi = 1.255 \text{ or } K\Phi = 0.236 \text{ (neglecting very high } \omega_o \text{ and } T_s \approx T_r)$$

$$\omega_o = \frac{V}{K\Phi} = \frac{150}{1.255} = 119.52 \, rad/sec$$

$$\omega_L = 12 \, rad/sec.$$

To find R_d:

$$\omega_L = \frac{V}{K\Phi} - \frac{R_a + R_d}{(K\Phi)^2} \cdot T_L$$

$$12 = \frac{150}{1.255} - \frac{1 + R_d}{(1.255)^2} \cdot 23.87$$

$$R_d = 6\ \Omega.$$

For supply reversing the initial braking is represented by point B (T_b):

$$I_b = \frac{-V - E}{R_{at}}$$

$$I_b = \frac{-150 - 1.255 \times 12}{7} = 23.58\ A$$

(In the opposite direction to that for motor operation.).

So, $T_b = k\phi \cdot I_b = 1.25 \times 23.58 = 29.6$ N·m,

Since the load is the potential load, therefore, the final operating position will be represented by point C (general braking for reverse operation).

$$-\omega_L = \frac{V}{K\Phi} + \frac{R_{at}}{(K\Phi)^2} \cdot T_L$$

$$-\omega_L = 120 + \frac{7}{(1.255)^2} \cdot 23.87 = 226.93\ rad/sec.$$

For dynamic braking through 3 Ω:

$$I_b = \frac{0 - 1.255 \times 12}{7 + 3} = -1.506\ A.$$

In the opposite direction to motor operation.

So, $T_b = l\phi I_b = 1.255 \times 1.5 = 1.89$ N·m. (Point D) (Figure 9.43 torque-speed characteristics).

The final operating position will also be generator braking (pot. load).

$$-\omega_L = \frac{V}{K\Phi} + \frac{R_{at} + 3}{(K\Phi)^2} \cdot T_L$$

$-\omega_L = 120 + 7 + 3/(1.255)^2 \cdot 23.87 = 271.55\ rad/sec$ at point (E) (see Figure 9.44 torque-speed characteristics).

Example 9.11

The separately excited DC motor is shown in Figure 9.45 was operated at no-load, and the following data were recorded $\omega_m = 1000\pi/30$ rad/sec, $I_a = 0.95$ A, V = 240 V, and $V_\beta = 150$ V. The field voltage V_s is unchanged, but the motor is loaded so that it supplies an output power P = 10 HP at 1000 rpm to a coupled mechanical load. At this load

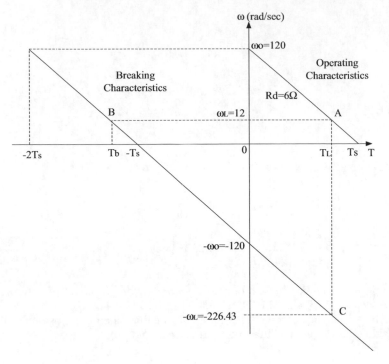

FIGURE 9.43
Torque-speed characteristics Example 9.10 i.

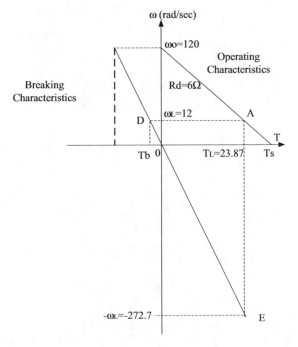

FIGURE 9.44
Torque-speed characteristics Example 9.10 ii.

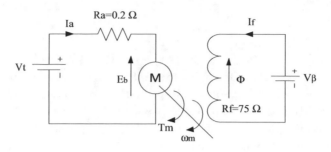

FIGURE 9.45
Motor circuit for Example 9.11.

point, determine the rotational losses, armature current, the no-load terminal voltage, efficiency. Also, calculate braking current when adding 3 Ω resistance in series with armature circuit. Neglect armature reaction.

Solution

Given data at no-load:
Speed = 1000 × Pi/30 rad/sec,

$$I_a = 0.95 \text{ A}, V_t = 240 \text{ V}, V_\beta = 150 \text{ V}.$$

Generated back EMF at no-load $E_b = V_t - I_a . R_a = 240 - 0.95 \times 0.2 = 239.81$ V.
No-load copper loss = $I_a{}^2 R_a = 0.952 \times 0.2 = 0.1805$ W.
Total input power at no-load = $V \times I_a = 240 \times 0.95 = 228$ W.
No-load losses (friction and windage losses) = $228 - 0.1805 = 227.8195$ W.
At given load:
Speed and field are constants. It means back EMF is constant.

$$\therefore E_b = 239.81 \text{ V}.$$

Power output = $E_b \times I_a$ + friction and windage losses = 10×746 W

$$E_b \times I_a = 7460 - 227.8195 = 7687.81 \text{ W}.$$

$$\text{Therefore, } I_a = 7687.81/(239.81) = 32.05 \text{ A}$$

$$\text{copper losses} = 32.052 \times 0.2 = 205.4405 \text{ W}$$

$$\text{input power} = V_s \times I_a = 240 \times 32.05 = 7693.91 \text{ W}$$

$$\text{efficiency } \eta = P_{out}/P_{input}$$

$$\eta = 7460/7693.91$$

$$\eta = 96.95\%$$

$$I_b = \frac{-V_t - E_b}{R_{at}} = \frac{-240 - 239.81}{3.2} = -149.94 \, A.$$

Problems

9.1   A separate excited, 3 kW, 175 Volt, 1600 rpm, DC motor has an armature resistance $= 1 \,\Omega$ is to be used in the drive through a reducer of $Nm/N_L=10$. If the load torque is 140 N·m, find the required R_d added to the armature to operate the load at 1 rad/sec? What voltage reduction is used for such operation if no R_d is added?

9.2   A series 120 Volt, 2.2 KW, 850 rpm motor has $R_a = 0.5 \,\Omega$ and $RF = 0.3 \,\Omega$. The field is linearly proportional to armature current. Find the required additional armature resistance to drive a load of 20 N·m at speed of 250 rpm.

9.3   A separately excited DC motor has $K\Phi = 1$, $R_a = 1.2 \,\Omega$. When it is operating at 150 rad/sec, the armature current is $I_a = 45$ A. The terminal voltage V_t is held constant under all conditions.

   1.   Compute V_t

   2.   Compute the no-load speed in rad/sec.

9.4   A separately excited DC motor is supplied from 220 V, 50 sources through a single-phase half-wave controlled converter. Its field is fed through i-phase semi converter with zero degrees firing-angle delay. Motor resistance $R_a = 0.1 \,\Omega$, and motor constant $= 0.6$ V-sec/rad. For rated load torque of 15 N·m at 1200 rpm and for continuous ripple free currents, determine:

   1.   Firing-angle delay of the armature converter

   2.   The RMS value of thyristor and freewheeling diode currents

   3.   Input power factor of the armature converter. Solution: (a) motor constant $= 0.5$ V-sec/rad $= 0.5$ Nm/A $= K_m$, but motor torque, $T_e = K_m \cdot I_a$.

9.5   A separately excited DC motor drives in Figure 9.46, a rated load torque of 40 Nm at 1200 rpm. The field circuit resistance is 200 Ω and armature circuit resistance is 0.05 Ω. The field winding, connected to single-phase, 120 V source, is fed through 1-phase full converter with zero degrees firing angle. The armature circuit is also fed through another full converter from the same converter single-phase, 120 V source. With magnetic saturation neglected, the motor constant is 1.2 V-sec /A-rad. For ripple free armature and field currents, determine:

   1.   Rated armature current

   2.   Firing angle delay of armature converter at rated load

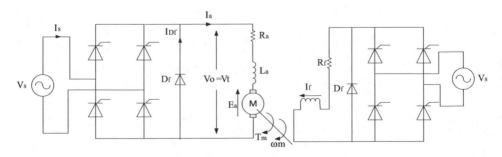

FIGURE 9.46
Separately excited DC motor drives of Problem 9.5.

3. Speed regulation at full load

4. Input PF of the armature converter and the drive at rated load.

9.6 A separate excited 150 Volt, 3 kW, 960 rpm DC motor has a total armature resistance of 1 Ω is used to lift a load of 0.8 Tr at speed of 10% of its no-load speed. Find:

1. The required additional resistance for such operation

2. If the supply is reverse braking is used, what will be the initial brake torque and what will be the final operation?

3. If the supply is switched OFF and the armature is terminated through resistance = 3 Ω, what will be the initial brake torque, and what will be the final position?

9.7 A step-down (buck) converter shown in Figure 9.47 is used to supply 80 V DC motor and an inductive load of $L = 5\text{mH}$, if $f_s = 6$ kHz, torque $= 40\text{N} \cdot \text{m}$, and input DC voltage is 200 V, calculate:

1. V_{Lmean} and speed of duty cycle range $0.5 - 1$.

2. Trace the output voltage and current of case $D = 0.7$.

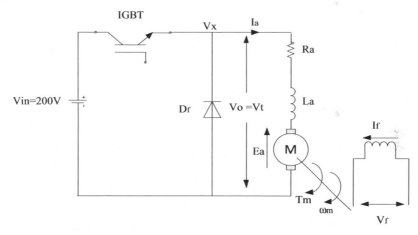

FIGURE 9.47
Step-down (buck) converter of Problem 9.7.

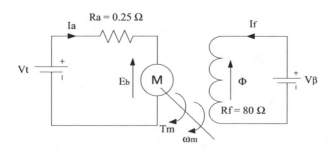

FIGURE 9.48
Motor circuit for Problem 9.8.

9.8 The separately excited DC motor is shown in Figure 9.48 was operated at no-load and the following data were recorded $\omega_m = 1000\pi/30$ rad/sec, $I_a = 1.1$ A, $V = 250$ V, and $V_\beta = 160$ V. The field voltage V_s is unchanged, but the motor is loaded so that it supplies an output power $P = 10$ HP at 1000 rpm to a coupled mechanical load. At this load point, determine the rotational losses, armature current, the no-load terminal voltage, efficiency. Also, calculate braking current when adding 2 Ω resistance in series with armature circuit. Neglect armature reaction.

10

AC Drives

Everyone enjoys doing the kind of work for which he is best suited.

Napoleon Hill

Three-phase induction motors are admirably suited to fulfill the demand of loads requiring substantially constant speed. Several industrial applications, however, need adjustable speeds for their efficient operation. The objective of the present chapter is to describe the basic principles of speed control techniques employed in three-phase induction motors using power electronics converters.

10.1 Advantages of AC Drives

1. AC motors are lighter in weight as compared to DC motors in the same rating.
2. AC motors are low maintenance as well as less expensive in comparison to equivalent DC motors.
3. AC motors are more effective in a hazardous environment like chemical, petrochemical conditions, but DC motors are unsuitable for such environments because of commutation sparking.

10.2 Disadvantages of AC Drives

1. AC drives detect some complication to control speed, voltage, as well as the whole controls, compared to DC motors.
2. Power converters for AC drives are more expensive then DC motors for more horsepower ratings.
3. Power converters for AC drives generate harmonics in the supply system of a load circuit.

10.3 Speed Control of Three-Phase Induction Motor

The induction motor (IM) is required to drive various kinds of loads, each of them having a different torque versus speed characteristics among a number of methods that exist for this purpose Figure 10.1, some commonly used ones are:

1. Stator voltage control
2. Stator frequency control
3. Stator current control
4. Controlling the stator voltage and the frequency Vs/f kept constant
5. Controlling the induced voltage and the frequency Es/f kept constant
6. Rotor resistance control
7. Slip energy recovery scheme.

In the modern industrial application and with the help of the development of semiconductor devices, changing the supply frequency (f) continuously to change the speed of the induction motor is now convincible. It is also manageable to change the supplied voltage of stator windings with the continuously changed supply frequency. Nowadays, the induction motor is progressively interchanging the variable speed DC drive—in the industry.

The magnitude of the alternator voltage can be controlled by controlling the alternator field excitation, but at constant excitation, this voltage is proportional to the frequency.

The induced back Electric Motive Force (EMF) equation of an induction motor is:

$$E = V_{ph} = 2.22 \ f\Phi Z_{ph} \tag{10.1}$$

where:
 f = supply frequency
 ϕ = flux per pole
 Z_{ph} = a number of series conductors per phase.

Hence, if the only f is varied keeping the applied voltage constant, the air-gap flux varies. If f is decreased from rated value, φ increase and the machine get saturated and the magnetizing current increase.

Hence, it is usual to keep the air-gap flux at its normal value. For this purpose, we must keep $E1/f$, i.e., $V1/f$ ratio constant, so that flux approximately remain constant. Let $K = (f/f_1) \leq 1$, where (f) is the operating frequency and (f_1) the normal rated frequency for which the machine is designed (i.e., 50). The synchronous speed is now $K\ n_s$, the applied voltage KV_1, and all the reactance KX. Substituting there in torque equation and simplifying, we get:

$$T = \frac{1}{2\pi \dfrac{n_s}{60}} \cdot \frac{3V_1^2}{\left(\dfrac{R_1}{K} + \dfrac{R_2'}{KS}\right)^2 + \left(X_1 + X_2'\right)^2} \cdot \frac{R_2'}{KS} \tag{10.2}$$

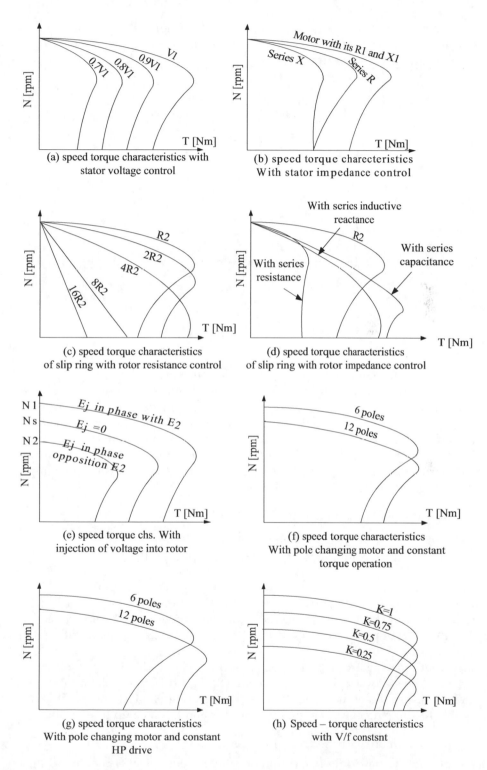

FIGURE 10.1

Torque-speed characteristics of an induction motor for a different method of speed control.

It may be seen that this expression is of the same as the original torque equation, but resistances have become larger by a factor K. Similar results can be obtained for starting torque, maximum torque, and maximum slip:

$$T_{st} = \frac{3}{2\pi \frac{n_s}{60}} \cdot \frac{V_1^2 \frac{R_2'}{K}}{\left(\frac{R_1}{K} + \frac{R_2'}{K}\right)^2 + \left(X_1 + X_2'\right)^2} \tag{10.3}$$

$$T_{max} = \frac{3}{2\pi \frac{n_s}{60}} \cdot \frac{V_1^2}{2\left[\frac{R_1}{K} + \sqrt{\left(\frac{R_1}{K}\right)^2 + \left(X_1 + X_2'\right)^2}\right]} \tag{10.4}$$

The slip at which maximum torque occurs becomes:

$$S_{max} = \frac{\frac{R_2'}{K}}{\sqrt{\left(\frac{R_1}{K}\right)^2 + \left(X_1 + X_2'\right)^2}} \tag{10.5}$$

The slip at which maximum torque occurs when the frequency becomes larger. When the operating frequency decreases and the maximum torque decreases vaguely. In the beginning, torque starts to increase for a slight decrease in frequency, after reaching the maximum point, the torque starts to decrease even with the further decreasing frequency. This can be happened due to an increase of resistance of the machine and decrease of air-gap flux which is ignored in the torque equation above. And, the reduction in air gap resulted from the voltage drop in the stator impedance.

The speed-torque characteristics can be modified by:

1. By varying the supply voltage [Figure 10.1a]
2. Varying the stator parameters (R_1, X_1) [Figure 10.1b]
3. Varying the rotor parameters (R_2, X_2) [Figure 10.1c and d]

The last method of speed control in the case of slip ring induction motors.

Another method of speed control in the case of slip ring induction motor is by injection of voltage into rotor circuit by an external frequency converter or by special construction as in the case of Schrage motor [Figure 10.1e]. These methods require additional machine or special construction.

Another method of controlling the speed of squirrel cage induction motor is by changing the poles, i.e., the stator winding may be connected for different pole number and hence the synchronous speed of the motor can be changed (Figure 10.1f and g). This method is not possible in slip ring induction motor. This method has its own limitation:

1. The speed can be controlled only in steps.
2. It is possible to have only two different synchronous speed for one winding.
3. In cascade control, we can have speeds corresponding to P_1, P_2, and $P_1 \pm P_2$ pole numbers, but this method requires two machines mechanically coupled.

And the last method of controlling the speed of the induction motor is by keeping the V/f ratio constant as in Figure 10.1h.

10.4 Methods of Control Techniques

Three-phase induction motors are more commonly employed in adjustable-speed drives than three-phase synchronous motors. There are two types of three-phase inductor motors: squirrel-cage induction motors (SCIMs) and slip-ring (or wound-rotor) induction motors (SRIMs). The rotor of SCIM is built of copper or aluminum bars which are short-circuited by two end rings. The rotor of SRIM carries three-phase winding connected to three slip rings on the rotor shaft. The speed control of three-phase induction motors different according to the type of motor as follows:

10.4.1 Speed Control of Three-Phase Induction Motors

There are several methods of speed control through semiconductor devices are given below:

1. Stator voltage control
2. Stator frequency control
3. Stator voltage and frequency control
4. Stator current control
5. Static rotor-resistance control
6. Slip-energy recovery control.

SCIMs and SRIMs methods are valid to (1) to (4) both SCIMs and SRIMs, but methods (5) and (6) can only be valid for SRIMs. These methods are now described in what follows.

The angular speed in rad/s of an induction AC machine mechanical speed is given by:

$$\omega_m = (1-S).\omega_s \tag{10.6}$$

From the equation, it can be seen that there are two possibilities for speed regulation of an induction motor. One can be achieved by modifying the slips (typical <5%) or the other can be achieved by changing synchronous speed ω_s. When the motor is connected to the main circuit with the constant frequency, the speed can be found at the intersection of the motor and load torque (Figure 10.2).

To get the valid estimation for motor drives, we may ignore the stator resistance and the magnetizing impedance in the equivalent circuit model of an induction motor. After an effective calculation of motors, simplification of an equation for the shaft torque in N_m.

$$T_e = {}^2T_{emax} \frac{S_m S}{S_m + S} \tag{10.7}$$

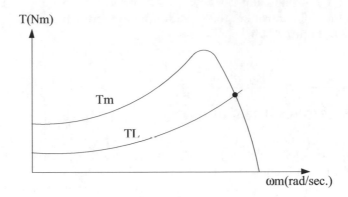

FIGURE 10.2
Typical motor torque-load characteristics.

where the maximal torque and corresponding slip are, respectively, given by:

$$T_{emax} = \frac{3|V_1|^2}{2\omega_s X_\sigma} \tag{10.8}$$

$$S_m = \frac{R_2'}{X_\sigma} \tag{10.9}$$

The variation of the slip S is usually small. Since the slip from no load to full load is very small which is about 0.05, the speed of an induction motor is almost constant specifically when frequency and number of poles are constant.

There is another method of speed control which is known as in the case of Schrage motor, in the case of slip ring induction motor. These schemes need some added machine configurations. In these methods, the rotor circuit is required to inject voltages by the external frequency converter.

- *Voltage control*

  As the voltage decreases, the torque decreases (the torque is developed in an induction motor is proportional to the square of the terminal voltage). Practically, this is confined to 80%–100% control. Unfortunately, this is not an effective control.

- *Frequency control*

  Frequency control or speed control is the most effective control among all the control. However, machine saturation is not acceptable. Since the flux is proportional to V/f, this control has to assure that the magnitude of the voltage is proportional to the speed. Power electronic circuits are best suited for this kind of control.

- *Vector control*

  In vector control, the magnetizing current (I_d) and the current that produced by torque (I_q) are controlled in two different control loops. As I_d lags voltage by 90° and I_q are always in phase with voltage, the two vectors I_d and I_q which are always 90° apart. Afterward, then vector sum is sent to the modulator which modifies the vector information into a rotating Pulse Width Modulation (PWM) modulated 3-phase system with the correct frequency and voltage. Consequently, this phenomenon helps to get the fast-dynamic response for the induction motor by reducing torque pulsation and robust control.

10.4.1.1 Stator Voltage Control

Stator voltage control of an induction motor is used generally for three purposes: (a) to control the speed of the motor, (b) to control the starting and braking behavior of the motor, and (c) to maintain optimum efficiency in the motor when the motor load varies over a large range. Here, we are discussing speed control of the motor. At a given load, if the voltage applied to the motor decreases, keeping the frequency constant, air-gap flux decreases. This result in a reduction in torque or power developed. As the rotor speed decreases, increasing the value of slip at which the torque developed will balance the load torque. Thus, the variation of voltage results in a variation of slip frequency and speed control of the motor. For speed control in a reasonably wide range, the rotor should have a large resistance. The torque equation is given by:

$$T_e = \frac{3}{\omega_s} \cdot \frac{(K.V)^2}{\left(r_1 + \dfrac{r_2}{S}\right)^2 + (x_1 + x_2)^2} \cdot \frac{r_2}{S} \tag{10.10}$$

The equation of torque defines the value of torque is dependent on the square of the supplied voltage. When supply voltage decreases, the motor torque and the speed will be reduced accordingly. To control the speed, changed voltages are applied to a 3-phase IM and 3-phase voltage controller is mostly installed to implement the purpose. Figure 10.3 shows a 3-phase voltage controller feeding a 3-phase IM. From the figure, it can be seen that the thyristor is anti-parallelly connected with every phase. Therefore, the root mean square (RMS) value of the voltage can be regulated with the adjustment of the firing angle of the thyristor. Also, the motor torque and the speed of the machine is controlled in. The torque-speed characteristics of the three-phase induction motors for varying supply voltage and also for the fan load are shown in Figure 10.4.

   This process holds few shortcomings in the field of speed control of the device. It controls speed only when a speed control below the normal rated speed as the operation of the voltages. Because the higher rated voltage is not allowable to execute the whole process. This method is suitable where the alternating operation of the drive is required and also for the fan and pump drives. In fan and pump drives, the load torque changes as the square of the speed. These types of drives required low torque at lower speeds. This condition can be attained by applying a lower voltage without exceeding the motor current. Three pairs of thyristor are needed in this induction where each pair consisting of two

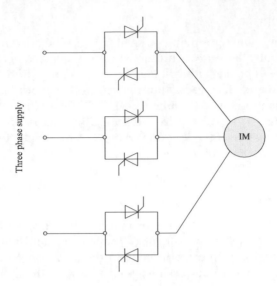

FIGURE 10.3
Thyristor voltage controller drive.

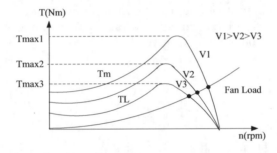

FIGURE 10.4
Torque-speed characteristics for different values of the supply voltage.

thyristors. The diagram below illustrates that the stator voltage is controlled by the thyristor voltage controller, speed control is acquired by varying the firing angle of the triac. These controllers are identified as *solid-state fan regulators*. As the solid-state regulators are more compact and efficient as compared to the conventional variable regulator. Thus, they are preferred over the normal regulator.

The variable voltage for speed control of small size motors mainly for single phase can be obtained by the following methods given below.

- By connecting an external resistance in the stator circuit of the motor
- By using an autotransformer
- By using a thyristor voltage controller
- By using a triac controller.

Nowadays, the thyristor voltage control method is selected for varying the firing angle of the thyristor. Triac is used to control speed variation, this method is also known as the phase angle control method. Additionally, energy consumption can also be controlled by controlling AC power.

10.4.1.2 Stator Frequency Control

By changing the supply frequency, motor synchronous speed can be altered and thus torque and speed of a 3-phase induction motor can be controlled. For a three-phase induction motor, the per-phase supply voltage is:

$$E_1 = \sqrt{2}\pi f_1 . N_1 \Phi k_{w1}$$ (10.11)

In this equation, if the frequency is reduced with constant V_1, then the air-gap flux increases. As a result, the induction motor magnetic circuit gets saturated. Furthermore, the motor parameters will change leading to inaccurate speed-torque characteristics. Thus, at low frequency, the reactance will flow high leading currents. Consequently, it increases loss and makes the system ineffective. In particular, speed control with constant supply voltage and reduced supply frequency are rarely suitable. But with the constant supply voltage and the increased supply frequency, the synchronous speed and therefore motor speed rises. But, with an increase in frequency and decreased flux and torque, IM performance at constant voltage and increase: frequency can be obtained by neglecting magnatizem reactance (Xm) and r_1 from the equivalent circuit induction motor. This assumption is not going to introduce any noticeable error as magnetizing current at high frequency is quite small. Thus:

$$I_2 = \frac{V_1}{\left[(r_2/S)^2 + (x_1 + x_2)^2 \right]^{1/2}}$$ (10.12)

Synchronous speed,

$$\omega_s = \frac{4\pi . f_1}{P} = \frac{2\omega_1}{P}$$ (10.13)

Motor torque:

$$T_e = \frac{3}{\omega_s} . I_2^2 . \frac{r_2}{S}$$ (10.14)

$$T_e = \frac{3P}{\omega_s} . \frac{(V_1)^2}{(r_2/S)^2 + (x_1 + x_2)^2} . \frac{r_2}{S}$$ (10.15)

$$S = \frac{f_2}{f_1} = \frac{\omega_2}{\omega_1}$$ (10.16)

or

$$\omega_2 = S.\omega_1$$

Here, f_2 and ω_2 are the rotor frequencies in and rad/sec, respectively. Substituting the value ω of slip $s = \omega_2/\omega_1$ in Equation 10.15, to get:

$$T_e = \frac{3P}{2\omega_1^2} . \frac{(V_1)^2.\omega_2}{r_2^2 + \omega_2^2.(l_1 + l_2)^2} . r_2$$ (10.17)

Slip at which maximum torque occurs is given as:

$$S_m = \frac{r_2}{x_1 + x_2} \tag{10.18}$$

Rotor frequency in rad/sec at which maximum torque occurs is given by:

$$\omega_{2m} = S_m.\omega_1 = \frac{\omega_1.r_2}{\omega_1.(l_1 + l_2)} = \frac{r_2}{(l_1 + l_2)} \tag{10.19}$$

Note that ω_{2m} does not depend on the supply frequency ω_1. Substituting $r_2 = \omega_{2m}^2.(l_1 + l_2)$ in Equation 10.19:

$$T_{e.m} = \frac{3P}{2\omega_1^2} \cdot \frac{(V_1)^2.\omega_{2m}^2.(l_1 + l_2)}{\omega_{2m}^2.(l_1 + l_2)^2 + \omega_{2m}^2.(l_1 + l_2)^2} \tag{10.20}$$

$$T_{e.m} = \frac{3P}{4\omega_1^2} \cdot \frac{V_1^2}{l_1 + l_2} \tag{10.21}$$

Equation 10.21 indicates that $T_{e.m}$ is inversely proportional to supply-frequency squared. Also,

$$T_{e.m}.\omega_1^2 = \frac{3P}{4} \cdot \frac{V_1^2}{l_1 + l_2} \tag{10.22}$$

At given source voltage V1, $\frac{3P}{4} \cdot \frac{V_1^2}{l_1 + l_2}$ is constant, therefore, (Tem. ω^2) is also constant. As the operating frequency ω_1 is increased, $T_{e.m}.\omega_1^2$ remains constant, but maximum torque at increased frequency ω_1 gets reduced as shown in Figure 10.5. Such type of IM behavior is

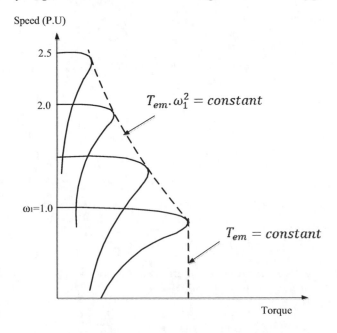

FIGURE 10.5
Speed torque characteristics of a three-phase IM with stator frequency control with constant supply voltage.

similar to the working of DC series motors. With constant voltage and increased-frequency operation, air-gap flux gets reduced, therefore, during this control, IM is said to be working in field-weakening mode. Constant voltage and variable frequency control of Figure 10.5 can be obtained by feeding 3-phase IM through three-phase.

Example 10.1

A 3-phase, 208 V, 20 kW, 1450 rpm, 60 Hz, star-connected induction motor has rotor leakage impedance of 0.4 + J1.6 Ω. Stator leakage impedance and rotational losses are assumed negligible.

If this motor is energized from 120 Hz, 400 V, 3-phase source, then calculate:

1. The motor speed at rated load stator frequency control with constant supply voltage
2. The slip at which maximum torque occurs, and
3. The maximum torque.

Solution

1.

$$T_e = \frac{P}{\omega_m} = \frac{60 \times 20000}{2\pi \times 1450} = 131.71 N.m$$

$$\omega_s = \frac{4\pi.f_1}{P} = \frac{2\omega_1}{P} = \frac{4\pi \times 120}{4} = 377 \, rad/sec$$

$$Z_{120\,Hz} = 0.4 + j1.5 \times \frac{120}{60} = 0.4 + j3$$

$$131.71 = \frac{3}{377} \cdot \frac{(208/\sqrt{3})^2}{(3)^2} \cdot \frac{0.4}{S}$$

$$S = 0.0387$$

2.

$$S_m = \frac{r_2}{x_2} = \frac{0.4}{3} = 0.1333$$

3.

$$T_{e.m} = \frac{3}{2\omega_s} \cdot \frac{V_1^2}{x_2} = \frac{3}{2\times 377} \cdot \frac{\left(\frac{208}{\sqrt{3}}\right)^2}{3} = 19.12 N.m$$

Example 10.2 For the MATLAB/Simulink shown in Figure 10.6, draw the speed response.

Solution

Figure 10.7 shows a speed response of three-phase IM with stator frequency control with constant supply voltage using MATLAB/Simulink circuit shown in Figure 10.6.

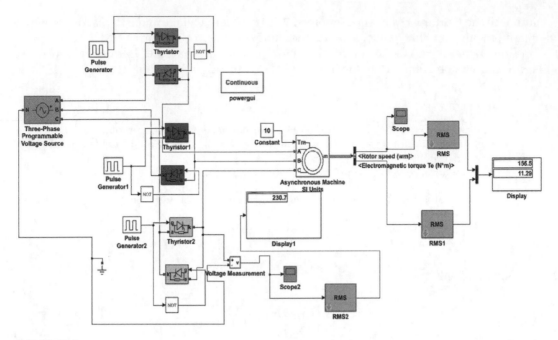

FIGURE 10.6
Thyristor voltage controller Simulink drive.

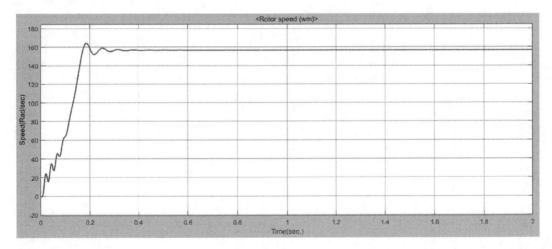

FIGURE 10.7
Speed response of Example 10.2.

10.4.1.3 Stator Voltage and Frequency Control

Synchronous speed can be controlled by varying the supply frequency. The voltage induced in the stator is $E_1 \propto \Phi.f$, where Φ is the air-gap flux and f is the supply frequency. As we can neglect the stator voltage drop, we obtain terminal voltage $V_1 \propto \Phi.f$. Thus, reducing the frequency without changing the supply voltage will lead to an increase in the air-gap flux which is undesirable. Hence, whenever frequency is varied in order to control speed, the terminal voltage is also varied so as to maintain the V/f ratio constant. Thus, by maintaining a constant V/f ratio, the maximum torque of the motor becomes constant for changing speed.

V/f control is the most popular and has found widespread use in industrial and domestic applications because of its ease-of-implementation. However, it has an inferior dynamic performance compared to vector control. Thus, in areas where precision is required, V/f control is not used. The various advantages of V/f control are as follows:

1. It provides a good range of speed.
2. It gives good running and transient performance.
3. It has a low starting current requirement.
4. It has a wider stable operating region.
5. Voltage and frequencies reach rated values at base speed.
6. The acceleration can be controlled by controlling the rate of change of supply frequency.
7. It is cheap and easy to implement.

Induction motors are the most widely used motors for appliances, industrial control, and automation. Hence, they are often called *the workhorse* of the motion industry.

They are robust, reliable, and durable. When power is supplied to an induction motor at the recommended specifications, it runs at its rated speed. However, many applications need variable speed operations. For example, a washing machine may use different speeds for each wash cycle. Historically, mechanical gear systems were used to obtain variable speed. Recently, electronic power and control systems have matured to allow these components to be used for motor control in place of mechanical gears. These electronics not only control the motor's speed, but can improve the motor's dynamic and steady-state characteristics. In addition, electronics can reduce the system's average power consumption and noise generation of the motor.

Induction motor control is complex due to its nonlinear characteristics. While there are different methods for control, variable voltage variable frequency (VVVF) or *V/f* is the most common method of speed control in open loop. This method is most suitable for applications without position control requirements or the need for high accuracy of speed control. Examples of these applications include heating, air conditioning, fans, and blowers. *V/f* control can be implemented by using low-cost peripheral interface controllers (PIC), rather than using costly digital signal processors (DSPs).

Many PIC micro microcontrollers have two hardware PWM, one less than the three required to control a 3-phase induction motor. In this application, we generate a third PWM in software, using a general purpose timer and an Input/Output (I/O) pin resource that are readily available on the PIC micro microcontroller.

10.4.1.4 V/f Control Theory

In the speed-torque characteristics, the induction motor draws the rated current and delivers the rated torque at the base speed. When the load is increased, the speed drops, and the slip increases. As we have discussed in the earlier section, the motor can take up to 2.5 times the rated torque with around 20% drop in the speed. Any further increase of load on the shaft can stall the motor.

The torque established by the motor is directly proportional to the magnetic field produced by the stator. So, the voltage applied to the stator is directly proportional to the product of stator flux and angular velocity. This makes the flux produced by the stator proportional to the ratio of applied voltage and frequency of supply.

By varying the frequency, the speed of the motor can be varied. Therefore, by varying the voltage and frequency by the same ratio, flux, and hence, the torque can be kept constant throughout the speed range.

$$\text{Stator voltage} = E_1 \approx V_1 \propto \Phi.2\pi.f \tag{10.23}$$

So:

$$\Phi \propto \frac{V}{f} \tag{10.24}$$

Above relation presents V/f as the most common speed control. Figure 10.8 indicates the torque curve between voltage and frequency. Figure 10.8 validates the voltage and frequency are reached up to the base speed. At base speed, the voltage and frequency reach the rated values as listed in the nameplate. It can be possible to run the motor beyond base speed by increasing the frequency further. However, there is a limit to reach the voltage, it cannot across the rated voltage. As a result, the only parameter can be increased that is frequency. When the motor runs above base speed, the torque becomes complex as the friction and windage losses increase drastically changes due to high speeds. Therefore, the torque curve gives a nonlinear pattern with respect to speed or frequency.

10.4.1.5 Static Rotor-Resistance Control

In a SRIM, a three-phase variable resistor R_2 can be inserted in the rotor circuit as shown in Figure 10.9a. By varying the rotor circuit resistance R_2. the motor torque can be controlled as shown in Figure 10.9b. The starting torque and starting current can also be varied by controlling the rotor circuit resistance, Figure 10.9b and c. This method of speed control is used when speed drop is required for short time, as for example, in overhead cranes, in load equalization.

The disadvantages of this method of speed control are:

1. Reduced efficiency at low speeds
2. Speed changes vary widely with load variation
3. Unbalances in voltages and currents if rotor circuit resistances are not equal

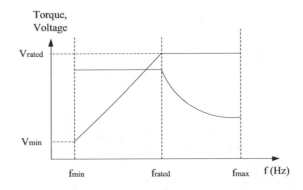

FIGURE 10.8
Relation between the voltage and torque versus frequency.

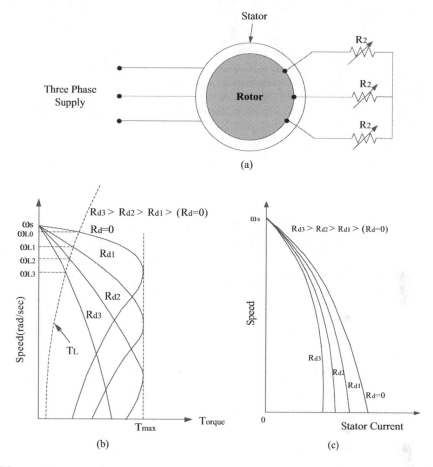

FIGURE 10.9
Three-phase IM speed control by rotor resistance. (a) Circuit arrangement. (b) Effect on developed torque. (c) Effect on stator current.

The three-phase resistor of Figure 10.9a may be replaced by a three-phase diode rectifier, chopper, and one resistor as shown in Figure 10.10. In this figure, the function of inductor L_d is to smoothen the current I_d. A Turn ON OFF (GTO) chopper allows the effective rotor circuit resistances to be varied for the speed control of SRIM. Diode rectifier converts slip frequency input power to DC at its output terminals. When chopper is on, $V_{dc} = V_d = 0$ and resistance R gets short-circuited. When chopper is off, $V_{dc} = V_d$ and resistance in the rotor circuit are R. This is shown in Figure 10.11.

c) From this figure, effective external resistance Re is:

$$Re = R. \, T_{off}/T$$

$$= R. \, (T\text{-}Ton)/T$$

$$= R. \, (1\text{--}k),$$

where k = Ton/T = duty cycle of chopper.

The equivalent circuit for 3-phase IM, diode rectifier, and chopper circuit is shown in Figure 10.12. If stator and rotor leakage impedances are neglected as compared to inductor

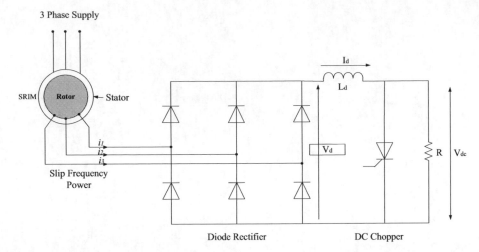

FIGURE 10.10
Static rotor-resistance control circuit drive.

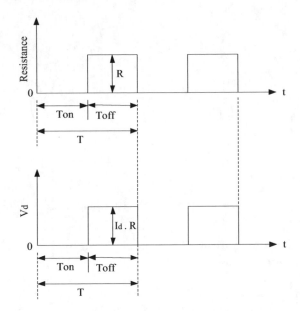

FIGURE 10.11
Resistance and rectifier voltage.

L_d, the equivalent circuit of Figure 10.12b is obtained. Stator voltage V_1, when referred to rotor circuit, gives slip-frequency voltage as:

$$s. \, V_1 \, N_1/N_2 = s. \, \alpha. \, V_1 = s. \, E_2,$$

where E_2 = rotor induced EMF per phase at the standstill.

V_1 = stator voltage per phase.

α = (rotor effective turns, N_2)/(Stator effective turns, N_1) = per phase turns ratio from rotor to stator.

Voltage $S.E_2 = S.\alpha.V_1$, after rectification by a three-phase diode, appears as V_d (rectifier output voltage).

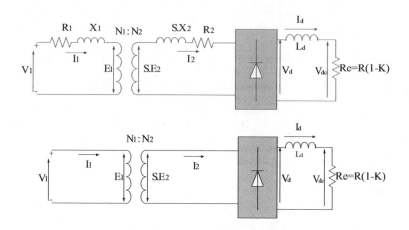

FIGURE 10.12
Equivalent circuit for 3-phase IM, diode rectifier, and chopper circuit.

$$V_d = \frac{3V_m}{\pi} = \frac{3.\sqrt{3}V_{mph}}{\pi} = 2.339\,S\,a\,V_1 \tag{10.25}$$

For lossless rectifier:

$$\text{Total slip power} = 3S\ .P_g = V_d.I_d \tag{10.26}$$

$$P_m = (1-S).P_d \tag{10.27}$$

$$P_m = (1-S).\frac{V_d.I_d}{3S} \tag{10.28}$$

And:

$$P_m = T_e.\omega_r = T_e.\omega_s.(1-S) \tag{10.29}$$

So:

$$P_m = (1-S).\frac{V_d.I_d}{3S} = T_e.\omega_s.(1-S)$$

$$I_d = T_e.\omega_s.3S/V_d$$

Or

$$I_d = \frac{T_L.\omega_s}{2.339\,a\,V_1} \tag{10.30}$$

And:

$$V_d = I_d.R.(1-k)$$

So:

$$V_d = 2.339 \; S \; a \; V_1 = I_d.R.(1-k)$$

$$\therefore S = \frac{I_d.R.(1-S)}{2.339a \; .V_1} \tag{10.31}$$

And $\omega_m = \omega_s(1-s)\omega_m = \omega_s(1-s)$.
   So the motor speed:

$$\omega_m = \omega_s \left[1 - \left(\frac{T_L.\omega_s.R.(1-k)}{(2.339 a V_1)^2} \right) \right] \tag{10.32}$$

The relationship of the torque-speed characteristic with a variable rotor resistance is given in Figure 10.13, the critical slip happens when motor produces maximum torque is proportional to the rotor resistance. As the rotor resistance increases, the critical slip value increase. As we can see from the diagram, for example: assume a constant load torque (T_L), the motor operating points shift from A to B to C with the change of resistance, respectively. According to the changes of resistances, the changed speeds are N_1, N_2, and N_3 with $N_1 > N_2 > N3$. In this way, the speed of the induction motor decreases with increases in the rotor resistance.

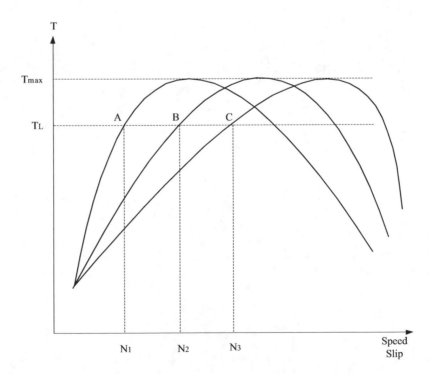

FIGURE 10.13
Torque-speed characteristics with a variable rotor resistance.

10.4.1.6 Slip-Energy Recovery Control

Due to output, the slip power losses are induced at the terminal of slip ring induction motor. Also, slip power losses are more often due to using variable resistances. To control these losses or to recovery the slip power, the power losses at the terminal are needed to be reduced.

In the recent days, there are two methods have introduced to control the speed of the slip ring. One of them is static rotor-resistance control and another one is slip-energy recovery control. In between two of them, the slip-energy recovery control method is developed to reduce the slip energy losses and to improve overall poor drive efficiency of the system. External resistance control by using three rheostats is a primitive method which is easy to control. In this method, the speed of the wound rotor of the induction motor is varied with the external resistance at the terminal. When the external resistance value is zero, the slip rings start to act as a short circuit. At the rated load, the internal torque slip curve of the machine starts to provide speed at this point. As the external resistance is getting started to increase, the curve would start to look flat. Eventually, at the high resistance, there will be no speed detected. When output slip terminals are connected to each other with any coil and the motor work as a full load, then we can use a variable rheostat at the terminal to control the speed. The speed of the motor can be controlled by varying the rheostat. As the rotor circuit needs energy and the power losses occur, the slip control is not an effective method to control slip (Figure 10.14).

Changing the resistance by using a diode like bridge rectifier and the chopper is more efficient rather than changing it mechanically. Normally, a bridge rectifier, the chopper is directly connected to power supply, whereas the slip voltage is rectified to DC by diode rectifier in the rotor circuit. With the connection of series inductor, the DC voltage is converted to the current source then it is used to feed to IGBT shunt chopper with shunt resistance shown. When the IGBT is off, the DC link current flows through it. On the contrary, when the device is ON, the resistance is short-circuited and the current is bypassed through it. Consequently, the established torque and the speed can be controlled by the variation of the duty cycle of the chopper. This can be assumed that electronic control of rotor resistance is definitely advantageous compared to rheostat controlled. Nevertheless, the problem of poor drive efficiency cannot be changed. This specific structure is effective in limited range like in intermittent speed-controlled application.

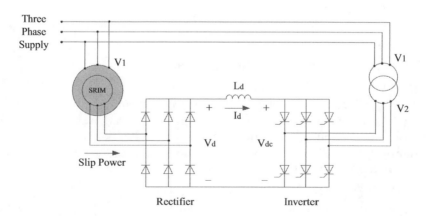

FIGURE 10.14
Slip power recovery drive.

To summarize, the speed control mechanism for high-power SRIM and low-power SRIM is analyzed as an extension and nonslip recovery mechanism to the available, respectively. The mechanism can be employed either at the rotor side or stator side. The mechanism that has engaged with a single thyristor with ON/OFF state at the rotor side of the slip ring induction motor. The basic projected outline here is to control the high-power machine by directly varying the load current or the rotor side current by the single thyristor after converting at the available load side current into the DC current with the help of three-phase rectifier bridge. As the three-phase rectifier is made of six diodes, respectively, all the AC power is converted into simple DC current and is controlled by the thyristor ON/OFF state time. The switch ON/OFF state time contribute the needed speed control as the rotor current is directly proportional to speed and torque in an increased power scenario as in this case. Although, the losses due to nonslip power recovery can be ignored because of the machines high-power status. In this scheme, we are able to simplify control of a high-power SRIM by just changing the amount of ON/OFF time of the thyristor at the rotor end that operates efficiently. Moreover, it is the best suited for very high inertia loads where a pull-out torque requires to be zero speed and accelerates to full speed with minimum current drawn in a very short time period.

Problems

10.1 List out some examples of prime movers.

10.2 List out some advantages of electric AC drives.

10.3 Give some examples of electric drives.

10.4 What are the advantages of static Kramer system, over static Scherbius system?

10.5 What is the function of a conventional Kramer system?

10.6 What is meant by slip power?

10.7 A 3-phase, 208 V, 25 kW, 1650 rpm, 60 Hz, star-connected induction motor has rotor leakage impedance of $0.4 + J1.6\ \Omega$. Stator leakage impedance and rotational losses are assumed negligible. If this motor is energized from 120 Hz, 400 V, 3-phase source, then calculate:

 1. The motor speed at rated load stator frequency control with constant supply voltage

 2. The slip at which maximum torque occurs

 3. The maximum torque.

10.8 A 3-phase, 208 V, 60 Hz, 6 poles, Y connected round-rotor synchronous motor has $Zs = 0 + J2\ \Omega$. Load torque, proportional to speed squared, is 300 Nm at rated synchronous speed. The speed of the motor is lowered by keeping V/f constant and maintaining unity Power Factor (pf) by field control of the motor. For the motor operation at 600 rpm, calculate:

 1. The supply voltage

 2. The armature current

 3. The excitation voltage

4. The load angle

5. The pullout torque, neglect rotational losses.

10.9 A static Kramer drive is used for the speed control of a 4-pole SRIM fed from 3-phase, 300 V, 60 Hz supply. The inverter is connected directly to the supply. If the motor is required to operate at 1400 rpm, find the firing advance angle of the inverter. The voltage across the open-circuited slip rings at stand-still is 750 V. Allow a voltage drop of 0.7 V and 1.5 V across each of the diodes and thyristors, respectively. Inductor drop is neglected.

11

Special Machines

Live your life and forget your age.

Norman Vincent Peale

There are some special motors with their electrically special applications in the modern era. This chapter delivery a brief fundamental to electrical special machines which have special applications. Stepper motors, brushless *direct current* DC motor, and switched reluctance motor are the examples of some special machines that are mostly used. Additionally, this chapter has a short description of servomotors, synchro motors, and resolvers.

Among all the motors, electric motor plays a vital role in our daily life. The main mechanism of this kind of motor is basically to convert electrical energy into mechanical energy. There are several devices in our daily life whose movement is produced by the electric motor. Hair dryer, Video Cassette Recorder (VCR), and a disk drive in a computer are few examples that are using the mechanism of the electric motor to run the specific applications.

There are various types of an electric motor according to the nature of the supplied current and sizes, namely, *alternating current (AC)* electric motors and *DC* electric motors. According to the name, DC electric motor will not run when AC current supplied. On the other hand, AC electric motor will not operate with the DC supplied current. Furthermore, AC electric motors are also subdivided into *single phase* and *three-phase* motors. Single phase AC electrical supply is what is typically supplied in a home and three-phase electrical power is commonly only available in a factory setting.

11.1 Stepper Motors

In the early 1960s, stepper motors were developed and introduced as a low-cost alternative to position servo systems in the emerging computer peripheral industry. To achieve the accurate position control without knowing the position feedback, the stepper motor is effective to use. However, it reduces the cost of a position control system by running "open-loop." Also, stepper motors apply a doubly salient topology, which means they have "teeth" on both the rotor and stator. Magnetizing stator teeth are used to generate electrical torque, and the permanent-magnet rotor teeth are utilized to line up with the stator teeth. There are many different configurations of stepper motors and even more different ways to drive them. There is a most common stator configuration which consists of two coils which are arranged around the circumference of the stator in such a way that if they are driven with square waves which have a quadrature phase relationship between them, the motor will rotate. To rotate the rotor motor in the opposite direction, simply reverse the phase relationship between the two coils signals. This alteration of either square wave causes the rotor to move by a small amount or a "step." Therefore, the motor is itself known as a *stepper motor*. The size of this step depends on the teeth arrangement of the motor, but a common value is 1.8° or 200 steps per revolution. Speed control is achieved by simply varying the frequency of the square-waves. As stepper motors can be driven with square waves, they are easily controlled by inexpensive digital circuitry and do not even require PWM (pulse width modulation). For this reason, stepper motors have often been inappropriately referred to as "digital motors." However, the quadrature changes into sine and cosine waveforms by utilizing power modulation techniques which may provide more resolution. This is termed as "micro-stepping," where each discrete change in the sine and cosine levels constitutes one micro-step.

A stepper motor, also known as a stepping motor, a pulse motor, or digital motor, is an electromechanical device which rotates a discrete step angle when energized electrically. Stepper motors are synchronous motors in which rotor's positions depend directly on driving signal.

The main difference between the stepping motor and a general motor is that the stepping motor only powered by a fixed driving voltage that does not rotate. Also, stepping motor displays excellent functions such as accurate driving, rapid stopping, and rapid starting. The stepper motor provides controllable speed or position in response to input step pulses commonly applied from an appropriate control circuit. Stepping motors are driven by a pulse signal. When a digital pulse signal is used as an input into the stepping motor, the rotor of the stepping motor is rotated by a fixed angle, that is, a well-known stepping angle. Since the stepper motor increments in a precise amount with each step pulse, it converts digital information, as represented by the input step pulses, to the corresponding incremental rotation. By increasing the rate of the step pulses, it is possible to increase the speed of the motor.

11.1.1 Step Angle

The angle through which the motor shaft rotates for each command pulse is called *the step angle β*. The parameter step angle can determine the number of steps per revolution and the accuracy of the position. It is observed, the smaller the step angle, the greater the number of steps per revolution and higher the resolution or accuracy of positioning obtained. The step angles can be as small as 0.72° or as large as 90°. But the most common step sizes are 1.8°, 2.5°, 7.5°, and 15°.

The value of step angle can be expressed either in terms of the rotor and stator poles (teeth) Nr and Ns, respectively, or in terms of the number of stator phases (m) and the number of rotor teeth.

$$\beta = \frac{(N_s - N_r)}{N_s \cdot N_r} \times 360° \tag{11.1}$$

$$\beta = \frac{360°}{m \cdot N_r} = \frac{360°}{\text{No. of stator phases} \times \text{No. of rotor teeth}} \tag{11.2}$$

For example, if Ns = 10 and Nr = 6, β = (10−6) × 360°/8 × 6=30°.

The resolution is given by the number of steps needed to complete one revolution of the rotor shaft. The higher the resolution, the greater the accuracy of positioning of objects by the motor.

$$\text{Resolution} = \text{No. of steps/revolution} = 360°/\beta \tag{11.3}$$

A stepping motor also possess extraordinary ability to operate at very high stepping rates up to 20,000 steps per second in some motors and yet to remain fully in synchronism with the command pulses. When the pulse rate is high the shift rotation seems continuous and operation at a high rate is called *slewing*. When in the slewing range, the motor generally emits an audible whine having a fundamental frequency equal to the stepping rate. If f is the stepping frequency (or pulse rate) in pulses per second (pps) and β is the step angle, when motor shaft speed is given by:

$$n = \frac{\beta.f}{360°} = \text{pulse frequncy revolution} \qquad (11.4)$$

If the stepping rate is increased too quickly, the motor loses synchronism and stops. The same thing happens if when the motor is slewing, command pulses are suddenly stopped instead of being progressively slowed.

Stepping motors are designed to operate for long periods with the rotor held in a fixed position and with rated current flowing in the stator windings. It means that stalling is no problem for such motors, whereas for most of the other motors, stalling results in the collapse of back Electric Motive Force (EMF) (E_b) and a very high current which can lead to a quick burn-out.

Example 11.1

A hybrid variable reluctance (VR) stepping motor has eight main poles which have been castellated to have five teeth each. If the rotor has 60 teeth, calculate the stepping angle.

Solution

$$Ns = 8 \times 5 = 40; \text{ and } Nr = 60$$

$$\beta = \frac{(N_s - N_r)}{N_s.N_r} \times 360°$$

$$\beta = \frac{(N_r - N_s)}{N_s.N_r} \times 360° = \frac{(60 - 40)}{40.60} \times 360° = 3°$$

Example 11.2

A stepper motor has a step angle of 2.5°. Determine (a) resolution, (b) number of steps required for the shaft to make 25 revolutions, and (c) shaft speed, if the stepping frequency is 3600 pps.

Solution

  a. resolution = 360°/β = 360°/2.5° = 144 steps/revolution
  b. steps/revolution = 144. Hence, steps required for making 25 revolutions = 144 × 25 = 3600
  c. n = β × f/360° = 2.5 × (3600/360°) = 25 rps.

11.1.2 How Stepper Motors Work

Stepper motors consist of a permanent-magnet rotating shaft, called *the rotor* and electro-magnets on the stationary portion that surrounds the motor called *the stator*. Figure 11.1 illustrates one complete rotation of a stepper motor. At position 1, we can see that the rotor

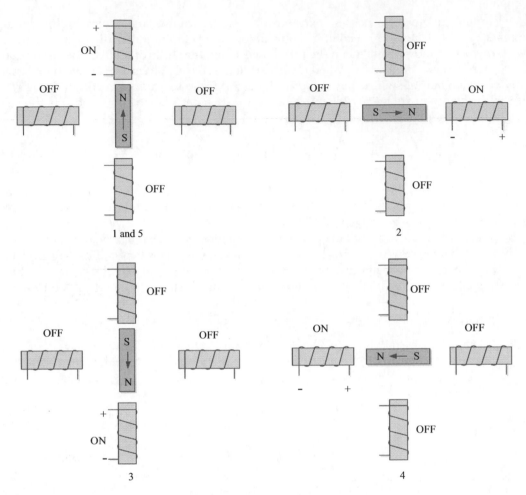

FIGURE 11.1
One complete rotation of a stepper motor. Rotor position Case 1&5, Rotor position Case 2, Rotor position Case 3, Rotor position Case 4.

is beginning at the upper electromagnet, which is currently active (has voltage applied to it). To move the rotor clockwise (CW), the upper electromagnet is deactivated and the right electromagnet is activated, causing the rotor to move 90°CW, aligning itself with the active magnet. This process is repeated in the same manner at the south and west electromagnets until we once get the starting position again.

In the above example, we used a motor with a resolution of 90° for demonstration purposes. In reality, this would not be a very practical motor for most applications. The average stepper motor's resolution—a number of degrees rotated per pulse—is much higher than this. For example, a motor with a resolution of 5° would move its rotor 5° per step, thereby requiring 72° pulses (steps) to complete a full 360° rotation.

You may perhaps double the resolution of some motors by a method known as "half-stepping." Instead of switching the next electromagnet in the rotation on one at a time, you can turn on both electromagnets by the help of half stepping phenomena that results in an equal attraction between thereby doubling the resolution. As you can see in Figure 11.2, in the first position only the upper electromagnet is active, and the rotor

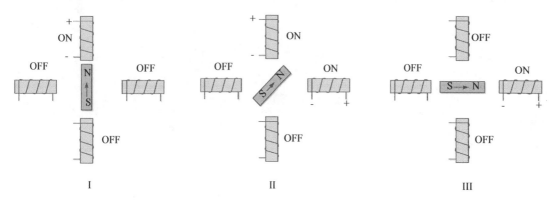

FIGURE 11.2
Rotation of stepper motor.

is drawn completely to it. In position 2, both the top and right electromagnets are active, causing the rotor to position itself between the two active poles. Finally, in position 3, the top magnet is deactivated and the rotor is drawn all the way right. This process can then be repeated for the entire rotation.

Among all types of stepper motors—4-wire stepper motors basically contain only two electromagnets. In fact, the operation is more complicated than the motors which contain three or four magnets because the driving circuit must be able to reverse the current after each step. For our purposes, we will be using a 6-wire motor. For example, motors which rotated 90° per step, real-world motors employ a series of mini-poles on the stator and rotor to increase resolution. Although this may seem to add more complexity to the process of driving the motors, the operation is identical to the simple 90° motor we used in our example. An example of a multipole motor can be seen in Figure 11.3. In position 1, the north pole of the rotor's permanent magnet is aligned with the south pole of the stator's

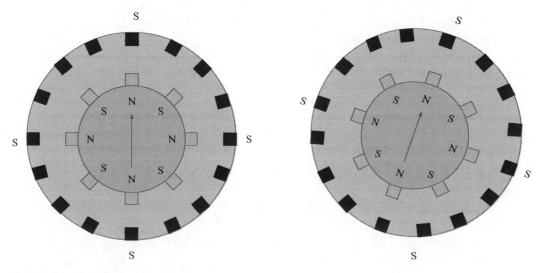

FIGURE 11.3
An example of a multipole motor.

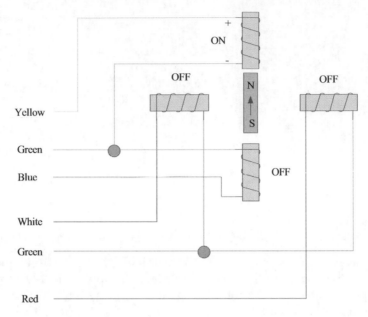

FIGURE 11.4
The electrical equivalent of the stepper motor.

electromagnet. Also, it is noted that multiple positions are aligned at once. In position 2, the upper electromagnet is deactivated and the next one to its immediate left is activated, causing the rotor to rotate a precise amount of degrees. In this example, after eight steps the sequence repeats.

The specific stepper motor we are using for our experiments (5° per step) has six wires coming out of the casing. If we follow Figure 11.4, the electrical equivalent of the stepper motor, we can see that three wires go to each half of the coils and that the coil windings are connected in pairs. This is true for all four-phase stepper motors.

11.1.3 DC Motors versus Stepper Motors

- Stepper motors are operated open-loop, where most of the DC motors are operated closed loop.
- Stepper motors are easy to control with microprocessors, however, logic and drive electronics are more complex.
- Stepper motors are brushless and brushes contribute several problems, e.g., wear, sparks, electrical transients.
- DC motors have a continuous displacement and can be accurately positioned, whereas stepper motor motion is incremental, and its resolution is limited to the step size.
- Stepper motors can slip if became overloaded and the error can remain undetected (A few stepper motors use closed-loop control.).
- Feedback control with DC motors gives a much faster response time compared to stepper motors.

11.1.4 Advantages of Stepper Motors

- The resulted position error is noncumulative, high accuracy of motion is possible at even under open-loop control
- Huge savings in the sensor (measurement system) and a controller which cost are affordable when the open-loop mode is used
- Because of the incremental nature of command and motion, stepper motors are easily adaptable to digital control applications
- No serious stability problems exist, even under open-loop control
- Torque capacity and power requirements can be optimized and the response can be controlled by electronic switching
- Brushless construction has obvious advantages.

11.1.5 Disadvantages of Stepper Motors

- They have low torque capacity (typically less than 2,000 oz-in) compared to DC motors
- They have limited speed (limited by torque capacity and by pulse-missing problems due to faulty switching systems and drive circuits)
- Large errors and oscillations can result when a pulse is missed under open-loop control
- They have high vibration levels due to the stepwise motion.

11.1.6 Specification of Stepping Motor Characteristics

In this section, some technical terms are used for specifying the characteristics of a stepping motor are discussed.

11.1.6.1 Static Characteristics

The characteristics relating to stationary motors are called *static characteristics.*

11.1.6.1.1 T/Θ Characteristics

Firstly, the stepping motor is kept stationary at a rest (equilibrium) position by supplying a current in a specified mode of excitation with the assumption with single-phase or two-phase excitation. If an external torque is applied to the shaft, an angular displacement will occur. The relation between the external torque and the displacement may be plotted as in Figure 11.5. This curve is conventionally called *the* T/θ *characteristic curve,* and the maximum of static torque is termed the "holding torque," which occurs at $\theta = \theta_M$ Figure 11.6. At displacements larger than θ_M, the static torque does not act in a direction toward the original equilibrium position, but in the opposing direction toward the next equilibrium position. The holding torque is rigorously defined as "the maximum static torque that can be applied to the shaft of an excited motor without causing continuous motion." The angle at which the holding torque is produced is not always separated from the equilibrium point by one step angle.

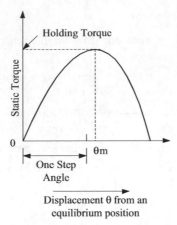

FIGURE 11.5
T/θ characteristics.

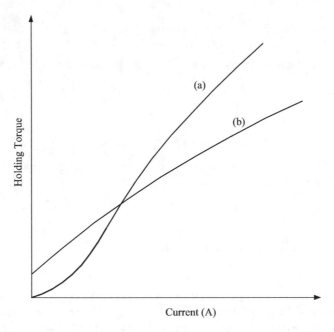

FIGURE 11.6
Examples of T/I characteristics. (a) a 1.8° four phase variable reluctance motor; and (b) a 1.8° four phase hybrid motor.

11.1.6.1.2 *T/I Characteristics*

The holding torque increases with current, and this relation is conventionally referred to as Torque/Amper (T/I) characteristics. Figure 11.6 compares the T/I characteristics of a typical hybrid motor with those of a variable-reluctance motor, the step angle of both being 18°. The maximum static torque appearing in the hybrid motor with no current is the detent torque, which is defined as the maximum static torque that can be applied to the shaft of an unexcited motor without causing continuous rotation.

11.1.6.2 Dynamic Characteristics

The characteristics relating to motors which are in motion or about to start are called *dynamic characteristics*.

11.1.6.2.1 Pull-In Torque Characteristics

These are alternatively called *the starting characteristics* and refer to the range of frictional load torque at which the motor can start and stop without losing steps for various frequencies in a pulse train. The number of pulses in the pulse train used for the test is 100 or so. The reason why the word "range" is used here, instead of "maximum," is that the motor is not capable of starting or maintaining a normal rotation at small frictional loads in certain frequency ranges as indicated in Figure 11.8. When the pull-in torque is measured or discussed, it is also necessary to identify the driving circuit clearly, the method of measuring, the method of coupling, and the inertia to be coupled to the shaft. Generally, the self-starting range decreases with the increasing inertia.

11.1.6.2.2 Pull-Out Torque Characteristics

This characteristic is also alternatively called *the slewing characteristic*. After the test motor is started by a specified driver in the specified excitation mode in self-starting range, the pulse frequency is gradually increased, the motor will eventually run out of synchronism. The relation between the frictional load torque and the maximum pulse frequency with which the motor can synchronize is called *the pull-out characteristic* (see Figure 11.7). The pull-out curve is greatly affected by the driver circuit, coupling, measuring instruments, and other conditions.

11.1.6.2.3 The Maximum Starting Frequency

This is defined as the maximum control frequency at which the unloaded motor can start and stop without losing steps.

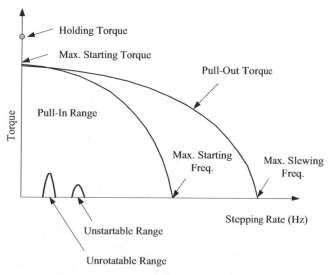

FIGURE 11.7
Dynamic characteristics.

11.1.6.2.4 Maximum Pull-Out Rate

This is defined as the maximum frequency (stepping rate) at which the unloaded motor can run without losing steps and is alternatively called *the maximum slewing frequency.*

11.1.6.2.5 Maximum Starting Torque

This is alternatively called "maximum pull-in torque" and is defined as the maximum frictional load torque with which the motor can start and synchronize with the pulse train of a frequency as low as 10.

11.1.7 Steady State Phasor Analysis

Figure 11.8 illustrates the equivalent circuit in the time domain of a phase winding. The circuit equation for one phase excitation can be written as:

$$V = R.i + L\frac{di}{dt} + \frac{dL(\theta)}{d\theta} \cdot \omega_r i + \frac{d\lambda_m}{dt} \tag{11.5}$$

where L is the stator winding inductance, λ_m the stator winding flux linkage due to the permanent magnet, and:

$$e = \frac{dL(\theta)}{d\theta} \cdot \omega_r i + \frac{d\lambda_m}{dt} \tag{11.6}$$

11.1.7.1 Phasor Expression of Variable Reluctance Stepping Motor

For a variable reluctance stepping motor, we have $\lambda_m = 0$,
   and:

$$L(\theta) = L_o + L_1 sin(N_r\theta) \tag{11.7}$$

Since the unipolar drive is employed, we may express the fundamental components of the voltage and current in the stator phase winding as:

$$v(t) = V_o + V_1 cos(\omega t) \tag{11.8}$$

and

$$i(t) = I_o + I_1 cos(\omega t - \delta - \alpha) \tag{11.9}$$

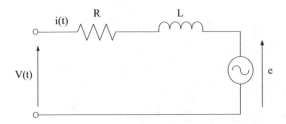

FIGURE 11.8
Per phase equivalent circuit in the time domain.

Substituting and neglecting the high-frequency terms, we obtain the voltage and current relations as:

$$V_o = R.I_o \tag{11.10}$$

$$V_1 \cos(\omega t) = RI_1 \cos(\omega t - \delta - \alpha) - \omega L_o I_1 \sin(\omega t - \delta - \alpha) + \omega L_1 I_o \cos(\omega t - \delta) \tag{11.11}$$

In phasor expression, the above voltage-current relationship becomes:

$$V = RI + j\omega L_o I + E \tag{11.12}$$

$$E = \omega \ L_1 \ I_o \angle -\delta \tag{11.13}$$

11.1.7.2 Phasor Expression of PM and Hybrid Stepping Motors

For permanent magnet (PM) and hybrid motors, L can be considered as independent of the rotor position.

The fundamental component of the voltage and current can be expressed as:

$$v(t) = V \cos(\omega t) \tag{11.14}$$

and

$$i(t) = I \cos(\omega t - \delta - \alpha) \tag{11.15}$$

Assuming the flux linkage of the stator winding due to the permanent magnet is:

$$\lambda_m = \lambda'_m \sin(\omega t - \delta) \tag{11.16}$$

$$V_1 \cos(\omega t) = RI \cos(\omega t - \delta - \alpha) - \omega LI \sin(\omega t - \delta - \alpha) + \omega \lambda_m \cos(\omega t - \delta) \tag{11.17}$$

In phasor expression, the above voltage-current relationship becomes:

$$V = RI + j\omega LI + E \tag{11.18}$$

where

$$E = \omega \lambda'_m \angle -\delta \tag{11.19}$$

11.1.7.3 Equivalent Circuit in Frequency Domain

A common phasor expression for all stepping motors is:

$$V = R \ I + j \ \omega \ L \ I + E \tag{11.20}$$

where

$$E = \omega \ L_1 \ I_o \ \angle -\delta \tag{11.21}$$

For a VR stepping motor, and:

$$E = \omega \lambda'_m \angle -\delta \tag{11.22}$$

and

$$\beta = tan^{-1} \frac{X_L}{R}$$

For a PM or hybrid stepping motor, Figure 11.9 shows the corresponding phasor diagram.

11.1.7.4 Pull-Out Torque Expression

From the phasor diagram, it can be derived that the electromagnetic torque of a stepping motor can be expressed as:

$$T = \frac{pmEI \cos(\beta - \delta)}{\omega\sqrt{R^2 + \omega^2 L^2}} - \frac{pmE^2 R}{\omega(R^2 + \omega^2 L^2)} \tag{11.23}$$

m is the number of phases, and $p = N_r/2$ the pole pairs of the motor.

The pull-out torque is the maximum torque for a certain speed and can be determined by letting $\delta = \beta$. Therefore:

$$T_{max} = \frac{pmEI}{\omega\sqrt{R^2 + \omega^2 L^2}} - \frac{pmE^2 R}{\omega(R^2 + \omega^2 L^2)} \tag{11.24}$$

Figure 11.10 plots the predicted pull-out torque against the rotor speed by equation (11.24), where:

$$K = \frac{L_1 V_o}{L_0 V_1} \tag{11.25}$$

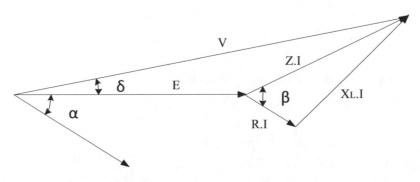

FIGURE 11.9
Phasor diagram of stepping motors.

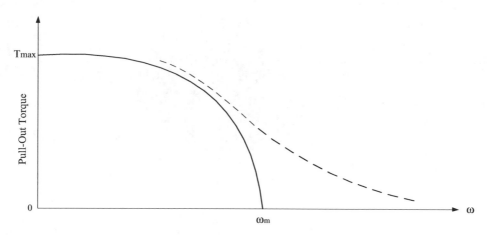

FIGURE 11.10
Predicted pull-out torque against rotor speed.

For a VR stepping motor, and:

$$K = \frac{\lambda'_m}{VL / R} \tag{11.26}$$

For a PM or hybrid stepping motor.

11.1.8 Applications

Stepper motors can be found almost anywhere. Most of us use every day without even realizing it. As for instance, steppers power "analog" wristwatches, disc drives, printers, robots, cash points, machine tools, CD players, profile cutters, plotters, and much more. Unlike other electric motors, they do not simply rotate smoothly when switched on. Every revolution is divided into a number of steps (typically 200), and the motor must be sent a separate signal for each step. It can only take one step at a time, and the size of each step is the same. Therefore, step motors may be considered a digital device.

11.2 Permanent-Magnet DC Motor

A PMDC motor is similar to an ordinary DC shunt motor except that its field is provided by permanent magnets instead of a salient-pole wound-field structure. Figure 11.11a shows 2-pole PMDC motor ,whereas Figure 11.11b shows a 4-pole wound-field DC motor for comparison purposes.

11.2.1 Construction

As shown in Figure 11.11a, the permanent magnets of the PMDC motor are supported by a cylindrical steel stator which also serves as a return path for the magnetic flux. The rotor (i.e., armature) has winding slots, commutator segments, and brushes as in conventional machines.

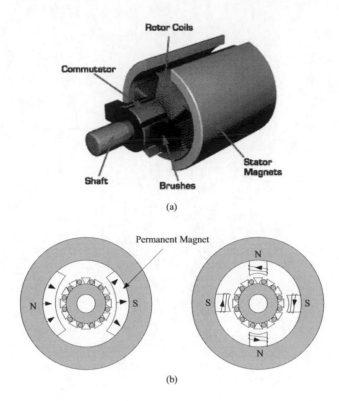

FIGURE 11.11
Permanent-magnet DC motor (a) Motor construction (b) Direction of field and current. (https://www.electrical4u.com/permanent-magnet-dc-motor-or-pmdc-motor/)

There are three types of permanent magnets used for such motors. The materials used have residual flux density and high coercively.

1. Alnico magnets: These magnets which are in having the ratings in the range of 1 KW to 150 KW used in motors
2. Ceramic (ferrite) magnets: This material is much more economical in fractional kilowatt motors
3. Rare-earth magnets: These magnets are made of samarium cobalt and neodymium iron cobalt which have the highest energy product. Such magnetic materials are costly, but are the best economic choice for small as well as large motors

Another form of the stator construction is the one in which permanent-magnet material is cast in the form of a continuous ring instead of in two pieces as shown in Figure 11.11b.

11.2.2 Working

Most of these motors usually run on 6 V, 12 V, or 24 V DC supply obtained either from batteries or rectified alternating current. In such motors, torque is produced by the interaction between the axial current-carrying rotor conductors and the magnetic flux produced by the permanent magnets.

11.2.3 Performance

The speed-torque curve is a straight line which makes this motor ideal for a servomotor. Moreover, input current increases linearly with load torque. The efficiency of such motors is higher as compared to wound-field DC motors because, in their case, there is no field copper loss.

11.2.4 Speed Control

Since flux remains constant, the speed of a PMDC motor cannot be controlled by using flux control method. The only way to control its speed is to vary the armature voltage with the help of an armature rheostat or electronically by using choppers. Consequently, such motors are found in systems where speed control below base speed only is required.

11.2.5 Advantages

1. In very small ratings, use of permanent-magnet excitation results in lower manu-facturing cost
2. In many cases, a PMDC motor is smaller in size than a wound-field DC motor of equal power rating
3. Since field excitation current is not required, the efficiency of these motors is gen-erally higher than that of the wound-field motors
4. Low-voltage PMDC motors produce less air noise
5. When designed for low-voltage (12 V or less), these motors produce very little radio and TV interference.

11.2.6 Disadvantages

1. Since their magnetic field is active at all times even when the motor is not being used, these motors are made totally enclosed to prevent their magnets from col-lecting magnetic junk from the neighborhood. Hence, as compared to wound-field motors, their temperature tends to be higher. However, it may not be much of a disadvantage in situations where the motor is used for short intervals
2. A more serious disadvantage is that the permanent magnets can be demagnetized by armature reaction Magnetic Motive Force (MMF) causing the motor to become inoperative. Demagnetization can result from (a) improper design, (b) excessive armature current caused by a fault or transient or improper connection in the armature circuit, (c) improper brush shift and, (d) temperature effects.

11.2.7 Applications

1. Small, 12-V PMDC motors are used for driving automobile heater and air conditioner blowers, windshield wipers, windows, fans, radio antennas, etc. They are also used for electric fuel pumps, marine motor starters, wheelchairs, and cordless power tools
2. Toy industry uses millions of such motors which are also used in other appli-ances such as the toothbrush, food mixer, ice crusher, portable vacuum cleaner, and shoe polisher and also in portable electric tools such as drills, saber saws, hedge trimmers, etc.

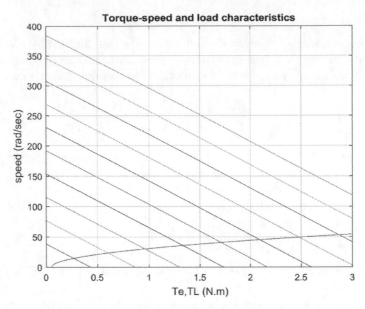

FIGURE 11.12
Permanent-magnet DC motor torque-speed and load characteristics.

MATLAB program for permanent-magnet DC motor.

```
Ra=1.5; ka=0.13;
Te=0:0.01:3;
for Va=5:5:50;
wr=Va/ka-(Ra/ka^2)*Te;
wrl=0:1:100;
Tl=0.05 + 0.001*wrl.^2;
Plot(Te, wr,'-',Tl, wrl,'-');
hold on;
axis([0, 3, 0, 400]);
end;
grid
```

Results see Figure 11.12.

11.3 Low-Inertia DC Motors

These motors are so designed as to make their armature mass very low. This permits them to start, stop and change direction, and speed very quickly making them suitable for instrumentation applications. The two common types of low-inertia motors are:

1. shell-type motor and
2. printed circuit (PC) motor.

11.3.1 Shell-Type Low-Inertia DC Motor

The armature of this kind of DC motor is made up of flat aluminum or copper coils bonded together to form a hollow cylinder. This hollow cylinder is not basically attached physically to its iron core which is stationary and is located inside the shell-type rotor. Since iron does not form part of the rotor, the rotor inertia is very small.

11.3.2 Printed-Circuit (Disc) DC Motor

It is a low-voltage DC motor which has its armature (rotor) winding and commutator printed on a thin disk of nonmagnetic insulating material. This disk-shaped armature contains no iron and etched copper conductors are printed on its both sides. It uses permanent magnets to produce the necessary magnetic field. The magnetic circuit is completed through the flux-return plate which also supports the brushes. Brushes mounted in an axial direction bear directly on the inner parts of the armature conductors which thus serve as a commutator. Since the number of armature conductors is very large, the torque produced is uniform even at low speeds. Typical sizes of these motors are in the fractional and sub-fractional horsepower ranges.

The speed can be controlled by varying either the applied armature voltage or current. Because of their high efficiency, fan cooling is not required in many applications. But the motor brushes need periodic inspection and replacement. The rotor disk which carries the conductors and commutator are very thin has a limited life. Hence, it requires replacing after some time.

The main features of this motor are:

1. Very low-inertia
2. High overload current capability
3. Linear speed-torque characteristic
4. Smooth torque down to near-zero speed
5. Very suitable for direct drive control applications
6. High torque/inertia ratio.

Advantages
1. High efficiency
2. Simplified armature construction
3. Being of low-voltage design produces a minimum radio and TV interference.

Disadvantages
1. Restricted to low-voltages only
2. Short armature life
3. Suited for intermittent duty cycle only because motor overheats in a very short time since there is no iron to absorb excess heat
4. Liable to burn out if stalled or operated with the wrong supply voltage.

Applications

These low-inertia motors have been developed specifically to provide high performance characteristics when used in direct-drive control applications. Examples are:

1. High-speed paper tape readers
2. Oscillographs
3. *X-Y* recorders
4. Layer winders
5. Point-to-point tool positioners, i.e., as positioning servomotors
6. With in-built optical position encoder, it competes with stepping motor
7. In high rating is being manufactured for heavy-duty drives such as lawn mowers, battery-driven vehicles, etc.

11.4 Servo Motors

The servo motors are used to turn the electrical signal to mechanical displacement of the rotor. The electrical signal is basically the voltage applied to the primary coil. These motors are extensively used in control systems.

These motors are specially established in control systems to measure devices, and in all types of controls, and their response is quicker in terms of change in speed for any change in voltage where the motor is mechanically connected to the load axis directly or through the gears which it is moving. This motor helps some electric devices that are unable to generate sufficient torque help move the load and do the job instead. Figure 11.13 illustrates different forms of service motors.

FIGURE 11.13
Different forms of servo motors. (https://www.efxkits.us/different-types-servo-motor-applications/).

It is a two-phase inductive drive and it contains two windings in the fixed between the 90° angle, and each coil is independent of the each other where the source of the voltage is subjected to fixed value and frequency on one of the coils called (excitation windings), and variable voltage can be controlled on the other coil, called (control windings).

The rotor has the windings of the squirrel cage, usually a small diameter, and a drop much less than its length to reduce the torque of the inertia and to be responsive.

There are several types of servo motors, but in this section just deal with a simple DC type here. If you take a normal DC motor that can be bought at Radio Shack, it has one coil (two wires). If you attach a battery to those wires the motor will spin. This is very different from a stepper already. Reversing the polarity will reverse the direction. Attach that motor to the wheel of a robot and watch the robot move to note the speed. When adding a heavier payload to the robot, the robot will slow down due to the increased load. The computer inside of the robot would not know this happened unless there was an encoder on the motor keeping track of its position.

So, in a DC motor, the speed and current draw are affected by the load. For applications that the exact position of the motor must be known, a feedback device like an encoder must be used. The control circuitry to perform good servo control of a DC motor is much more complex than the circuitry that controls a stepper motor.

One of the main differences between servo motors and stepper motors is that servo motors, by definition, run using a control loop and require feedback of some kind. A control loop uses feedback from the motor to help the motor get to the desired state (position, velocity, and so on). There are many different types of control loops. Generally, the PID (Proportional, integral, and derivative) control loop is used for servo motors.

When using a control loop such as PID, you may need to tune the servo motor. Tuning is the process of making a motor response in a desirable way. Tuning a motor can be a very difficult and tedious process, but is also an advantage in that it lets the user have more control over the behavior of the motor.

Since servo motors have a control loop to check what state they are in, they are generally more reliable than stepper motors. When a stepper motor misses a step for any reason, there is no control loop to compensate in the move. The control loop in a servo motor is constantly checking to see if the motor is on the right path and, if it is not, it makes the necessary adjustments.

In general, servo motors run more smoothly than stepper motors except when microstepping is used. Also, as speed increases, the torque of the servo remains constant, making it better than the stepper at high speeds (usually above 1000 rpm).

Some of the advantages of servo motors over stepper motors are as follows:

1. High intermittent torque
2. High torque to inertia ratio
3. High speeds
4. Work well for speed control
5. Available in all sizes
6. Quiet.

Some of the disadvantages of servo motors compared with stepper motors are as follows:

1. More expensive than stepper motors
2. Cannot work open-loop—feedback is required
3. Require tuning of control loop parameters
4. More maintenance due to brushes on brushed DC motors.

11.4.1 Mathematical Model of Servo Motor

The differential equations and transfer functions for DC servo motors are more compli-cated than those of AC servo motors. Since DC servo motors have time lags because of both the armature inductance and the winding, while AC motors have only a single time constant. DC servo motors are described by three differential equations: The developed torque [T(t)] is described by:

$$T(t) = K_2 . i_{f(t)} \qquad (11.27)$$

where $i_f(t)$ is the current through the field and K_2 is constant. The field voltage [$V_f(t)$] is described by:

$$V_f(t) = R_f . i_f(t) + L_F \frac{d}{dt} i_{f(t)} \qquad (11.28)$$

where R_f is the field resistance and L_f is the field inductance. Lastly, the mechanical torque [T(t)] is described by:

$$T(t) = J \frac{d^2}{dt^2} \theta(t) + B . \frac{d}{dt} \theta(t) \qquad (11.29)$$

where J is the motor's moment of inertia, B is the motor's viscous damping, and $\theta(t)$ is the motor's angular position. By assuming zero initial conditions and then Laplace transform-ing each of these equations, s-domain equations are reached. The developed torque is now:

$$T(s) = K_2 . I_f(s) \qquad (11.30)$$

the field voltage is now:

$$V_f(s) = (L_f . s + R_f) . I_f(s) \qquad (11.31)$$

and the mechanical torque is now:

$$T(s) = (J . s^2 + R_f . s) . \Theta(s) \qquad (11.32)$$

where all of the constants have the same meaning as in the time-domain differential equa-tions, I_f is used in place of i_f and Θ is used in place of θ. By substituting and solving, the transfer function of the motor is found to be:

$$\frac{\Theta(s)}{V_f(s)} = \frac{K_m}{s . (\tau_m . s + 1) . (\tau_e . s + 1)} \qquad (11.33)$$

where:

$$\tau_m = \frac{J}{B} \tag{11.34}$$

is the mechanical time constant of the motor:

$$\tau_e = \frac{L_f}{R_f} \tag{11.35}$$

is the electrical time constant of the motor, and:

$$K_m = \frac{K_2}{B.R_f} \tag{11.36}$$

is another constant. This is the transfer function that will be used for the control analysis of the DC servo motor in the next section.

The differential equations and transfer functions for AC servo motors are considerably less complicated than those for DC servo motors because AC servo motors only have a single time constant while DC servo motors have two. AC motors are described by two differential equations: The torque [T(t)] is described by:

$$T(t) = K.v(t) - m.\frac{d}{dt}\theta(t) \tag{11.37}$$

where K is a constant, v(t) is the voltage provided to the motor, θ(t) is the angular position of the motor, and m is described by:

$$m = \frac{\text{stall torque (at rated voltage)}}{\text{no-load speed (at rated voltage)}}$$

where stall torque (at rated voltage) and no-load speed (at rated voltage) are characteristics of any specific AC motor. The torque is also described by:

$$T(t) = J\frac{d^2}{dt^2}\theta(t) + B.\frac{d}{dt}\theta(t) \tag{11.38}$$

which is identical to the third differential equation that describes DC motors and has the same meaning. By equating the two AC motor equations, assuming zero initial conditions, and then taking the Laplace transform of the resultant equation, the transfer function of an AC motor is found to be:

$$\frac{\Theta(s)}{V(s)} = \frac{K_m}{s.(\tau.s+1)} \tag{11.39}$$

where:

$$K_m = \frac{K}{m+B} \tag{11.40}$$

is a constant, and:

$$\tau = \frac{J}{m+B} \tag{11.41}$$

is the time constant of the motor. This is the transfer function that will be used for the control analysis of the AC motor in the next section.

11.4.2 The Difference between Stepper Motors and Servos Motor

Stepper motors are less expensive and typically easy to use rather than a servo motor of a similar size. Because of moving into discrete steps, this kind of motor is called *stepper motor*. Therefore, a stepper motor needs a stepper drive and control to control the motor. Controlling a stepper motor requires a stepper drive and a controller.

The drive then interprets these signals and drives the motor. Stepper motors can be run in an open-loop configuration (no feedback) and are good for low-cost applications. In general, a stepper motor will have high torque at low speeds, but low torque at high speeds.

Movement at low speeds is also choppy unless the drive has the micro-stepping capability. At higher speeds, the stepper motor is not as choppy, but it does not have as much torque. When idle, a stepper motor has a higher holding torque than a servo motor of similar size, since current is continuously flowing in the stepper motor windings.

A stepper motor's shaft has permanent magnets attached to it. Around the body of the motor is a series of coils that create a magnetic field that interacts with the permanent magnets. When these coils are turned on and off the magnetic field cause the rotor to move. As the coils are turned on and off in sequence the motor will rotate forward or reverse. This sequence is called *the phase pattern* and there are several types of patterns that will cause the motor to turn. Common types are a full-double phase, full-single phase, and half step.

To make a stepper motor rotate, you must constantly turn on and off the coils. If you simply energize one coil the motor will just jump to that position and stay there resisting change. This energized coil pulls full current even though the motor is not turning. The stepper motor will generate a lot of heat at standstill. The ability to stay put at one position rigidly is often an advantage of stepper motors. The torque at standstill is called *the holding torque*.

Because steppers can be controlled by turning coils on and off, they are easy to control using digital circuitry and micro-controller chips. The controller simply energizes the coils in a certain pattern and the motor will move accordingly. At any given time the computer will know the position of the motor since the number of steps given can be tracked. This is true only if some outside force of greater strength than the motor has not interfered with the motion.

An optical encoder could be attached to the motor to verify its position, but steppers are usually used open-loop (without feedback). Most stepper motor control systems will have a home switch associated with each motor that will allow the software to determine the starting or reference "home" position.

Some of the advantages of stepper motors over servo motors are as follows:

1. Low cost
2. Low maintenance (brushless)
3. Excellent holding torque (eliminated brakes/clutches)
4. Can work in an open-loop (no feedback required)
5. Excellent torque at low speeds

6. Very rugged—any environment

7. Excellent for precise positioning control

8. No tuning required.

Some of the disadvantages of stepper motors in comparison with servo motors are as follows:

1. Rough performance at low speeds unless you use micro-stepping. Consume current regardless of load

2. Limited sizes available

3. High noisy

4. Torque decreases with speed (you need an oversized motor for higher torque at higher speeds)

5. Stepper motors can stall or lose position running without a control loop.

11.5 Brushed DC Motors

Brushed DC motors are widely used in applications ranging from toys to push-button adjustable car seats. Brushed DC (BDC) motors are inexpensive, easy to drive, and are readily available in all sizes and shapes.

This section will discuss how a BDC motor works, how to drive a BDC motor, and how a drive circuit can be interfaced to a peripheral interface controller (PIC) microcontroller.

The last decade or two, servomotors have evolved from largely brush types to brushless. This has been driven by lower maintenance and higher reliability of brushless motors. As brushless motors have become more prevalent during this period, the circuit and system techniques used to drive them have evolved as well. The variety of control schemes has led to a similar variety of buzzwords that describe them.

Most high-performance servo systems employ an inner control loop that regulates torque. This inner torque loop will then be enclosed in outer velocity and position loops to attain the desired type of control. While the designs of the outer loops are largely independent of motor type, the design of the torque loop is inherently specific to the motor being controlled.

The torque produced by a brush motor is fairly easy to control because the motor commutates itself. Torque is proportional to the DC current into the two terminals of the motor, irrespective of speed. Torque control can, therefore, be implemented by a Proportional-Integration (P-I) feedback loop which adjusts the voltage applied to the motor in order to minimize the error between requested and measured motor currents.

11.5.1 Stator

The stator generates a stationary magnetic field that surrounds the rotor. This field is generated by either permanent magnets or electromagnetic windings. The different types of BDC motors are distinguished by the construction of the stator or the way the electromagnetic windings are connected to the power source.

11.5.2 Rotor

The rotor, also called *the armature*, is made up of one or more windings. When these windings are energized they produce a magnetic field. The magnetic poles of this rotor field will be attracted to the opposite poles generated by the stator, causing the rotor to turn. As the motor turns, the windings are constantly being energized in a different sequence so that the magnetic poles generated by the rotor do not overrun the poles generated in the stator. This switching of the field in the rotor windings is called *commutation*.

11.5.3 Brushless Motor Basics

A brushless DC motor consists of a permanent magnet, which rotates (the rotor) and surrounded by three equally spaced windings which are fixed (the stator). The flowing current in each winding produces a magnetic field vector, which sums up the fields from the other windings. By controlling currents in the three windings, a magnetic field of an arbitrary direction and magnitude can be produced by the stator. Torque is then produced by the attraction or repulsion between this net stator field and the magnetic field of the rotor.

For any position of the rotor, there is an optimal direction of the net stator field, which maximizes torque, there is also a direction, which will produce no torque. If the permanent magnet rotor in the same direction as the field produces the net stator field, no torque is produced. The fields interact to produce a force, but because the force is in line with the axis of rotation of the rotor, it only serves to compress the motor bearings, not to cause rotation. On the other hand, if the stator field is orthogonal to the field produced by the rotor, the magnetic forces work to turn the rotor and torque is maximized as shown in Figure 11.14.

11.5.4 Advantage and Disadvantage of the Brushless DC Motor

The most noticeable advantage of the brushless configuration is being without brushes, as they reducing the brush maintenance and emits any other problems regarding the brushes. As for example, brushes tend to yield radiofrequency interference and the sparking associated with them which are a source of ignition in inflammable environments.

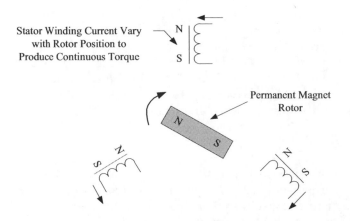

FIGURE 11.14
Brushless DC motor.

Additionally, brushless motors are potentially cleaner, faster, make less noise, and more reliable than induction motors. The rotor losses are very low, and the stator easily cooled because of the fine slot structure and the proximity of the outside air.

Their main disadvantages are: (i) the need for shaft position sensing and (ii) increased complexity in the electronic controller. There are few more disadvantages of brush DC motors which include inadequate heat dissipation, high rotor inertia, low-speed range due to limitations imposed by the brushes, and electromagnetic interference (EMI) generated by brush arcing. Brushless DC motors (BLDC) motors have a number of advantages over their brush brothers.

Problems

11.1 A hybrid VR stepping motor has ten main poles which have been castellated to have five teeth each. If the rotor has 80 teeth, calculate the stepping angle.

11.2 A hybrid VR stepping motor has 12 main poles which have been castellated to have six teeth each. If the rotor has 80 teeth, calculate the stepping angle.

11.3 A stepper motor has a step angle of 3°. Determine (a) resolution, (b) number of steps required for the shaft to make 25 revolutions, and (c) shaft speed, if the stepping frequency is 3600 pps.

11.4 A stepper motor has a step angle of 2.8°. Determine (a) resolution, (b) number of steps required for the shaft to make 30 revolutions, and (c) shaft speed, if the stepping frequency is 3600 pps.

11.5 What are the parts of brushed DC motors.

11.6 What are the advantage and disadvantage of the brushless DC motor.

11.7 Compare between stepper motors, servos motor, and brushless DC motor.

11.8 Give the mathematical model of the servo motor.

Appendix A: Mathematical Formula

This appendix, by no means exhaustive, serves as a handy reference. It does contain all the formulas needed to solve problems in this book.

A.1 Quadratic Formulas

The roots of the quadratic equation $ax^2 + bx + c = 0$:

$$x_1, x_2 = \frac{-b \pm \sqrt{b^2 - 4ac}}{2a}$$

A.2 Trigonometric Identities

$$\sin(-x) = -\sin x$$

$$\cos(-x) = \cos x$$

$$\sec x = \frac{1}{\cos x}, \qquad \csc x = \frac{1}{\sin x}$$

$$\tan x = \frac{\sin x}{\cos x}, \qquad \cot x = \frac{1}{\tan x}$$

$$\sin(x \pm 90°) = \pm \cos x$$

$$\cos(x \pm 90°) = \mp \sin x$$

$$\sin(x \pm 180°) = -\sin x$$

$$\cos(x \pm 180°) = -\cos x$$

$$\cos^2 x + \sin^2 x = 1$$

$$\frac{a}{\sin A} = \frac{b}{\sin B} = \frac{c}{\sin C} \qquad \text{(law of sines)}$$

347

$$a^2 = b^2 + c^2 - 2bc\cos A \qquad \text{(law of cosines)}$$

$$\frac{\tan\frac{1}{2}(A-B)}{\tan\frac{1}{2}(A+B)} = \frac{a-b}{a+b} \qquad \text{(law of tangents)}$$

$$\sin(x \pm y) = \sin x \cos y \pm \cos x \sin y$$

$$\cos(x \pm y) = \cos x \cos y \mp \sin x \sin y$$

$$\tan(x \pm y) = \frac{\tan x \pm \tan y}{1 \mp \tan x \tan y}$$

$$2\sin x \sin y = \cos(x-y) - \cos(x+y)$$

$$2\sin x \cos y = \sin(x+y) - \sin(x-y)$$

$$2\cos x \cos y = \cos(x+y) - \cos(x-y)$$

$$\sin 2x = 2\sin x \cos x$$

$$\cos 2x = \cos^2 x - \sin^2 x = 2\cos^2 x - 1 = 1 - 2\sin^2 x$$

$$\tan 2x = \frac{2\tan x}{1 - \tan^2 x}$$

$$\sin^2 x = \frac{1}{2}(1 - \cos 2x)$$

$$\cos^2 x = \frac{1}{2}(1 + \cos 2x)$$

$$a\cos x + b\sin x = K\cos(x+\theta), \text{ where } K = \sqrt{a^2 + b^2} \text{ and } \theta = \tan^{-1}\left(\frac{-b}{a}\right)$$

$$e^{\pm jx} = \cos x \pm j\sin x \qquad \text{(Euler's formula)}$$

$$\cos x = \frac{e^{jx} + e^{-jx}}{2}$$

$$\sin x = \frac{e^{jx} - e^{-jx}}{2j}$$

$$1\text{rad} = 57.296°$$

A.3 Hyperbolic Functions

$$\sinh x = \frac{1}{2}\left(e^x - e^{-x}\right)$$

$$\cosh x = \frac{1}{2}\left(e^x + e^{-x}\right)$$

$$\tanh x = \frac{\sinh x}{\cosh x}$$

$$\coth x = \frac{1}{\tanh x}$$

$$\csc hx = \frac{1}{\sinh x}$$

$$\sec hx = \frac{1}{\cosh x}$$

$$\sinh\left(x \pm y\right) = \sinh x \ \cosh y \pm \cosh x \ \sinh y$$

$$\cosh\left(x \pm y\right) = \cosh x \ \cosh y \pm \sinh x \ \sinh y$$

$$\tan\left(x \pm y\right) = \frac{\tan x \pm \tan y}{1 \mp \tan x \ \tan y}$$

A.4 Derivatives

If $U = U(x)$, $V = V(x)$, and a = constant:

$$\frac{d}{dx}\left(aU\right) = a\frac{dU}{dx}$$

$$\frac{d}{dx}\left(UV\right) = U\frac{dV}{dx} + V\frac{dU}{dx}$$

$$\frac{d}{dx}\left(\frac{U}{V}\right) = \frac{V\frac{dU}{dx} - U\frac{dV}{dx}}{V^2}$$

$$\frac{d}{dx}\left(aU^n\right) = naU^{n-1}$$

$$\frac{d}{dx}\left(a^U\right) = a^U \, 1na \frac{dU}{dx}$$

$$\frac{d}{dx}\left(e^U\right) = e^U \frac{dU}{dx}$$

$$\frac{d}{dx}(\sin U) = \cos U \frac{dU}{dx}$$

$$\frac{d}{dx}(\cos U) = -\sin U \frac{dU}{dx}$$

$$\frac{d}{dx}\tan U = \frac{1}{\cos^2 U}\frac{dU}{dx}$$

A.5 Indefinite Integrals

If $U = U(x)$, $V = V(x)$, and $a = $ constant:

$$\int a\,dx = ax + C$$

$$\int U\,dV = UV - \int V\,dU \qquad \text{(integration by parts)}$$

$$\int U^n dU = \frac{U^{n+1}}{n+1} + C, \qquad n \neq 1$$

$$\int \frac{dU}{U} = 1nU + C$$

$$\int a^U dU = \frac{a^U}{1na} + C, \qquad a > 0, a \neq 1$$

$$\int e^{ax} dx = \frac{1}{a}e^{ax} + C$$

$$\int xe^{ax} dx = \frac{e^{ax}}{a^2}\left(ax - 1\right) + C$$

$$\int x^2 e^{ax} dx = \frac{e^{ax}}{a^3}\left(a^2 x^2 - 2ax + 2\right) + C$$

$$\int \ln x\, dx = x\ln x - x + C$$

$$\int \sin ax\, dx = -\frac{1}{a}\cos ax + C$$

$$\int \cos ax\, dx = \frac{1}{a}\sin ax + C$$

$$\int \sin^2 ax\, dx = \frac{x}{2} - \frac{\sin 2ax}{4a} + C$$

$$\int \cos^2 ax\, dx = \frac{x}{2} + \frac{\sin 2ax}{4a} + C$$

$$\int x\sin ax\, dx = \frac{1}{a^2}\left(\sin ax - ax\cos ax\right) + C$$

$$\int x\cos ax\, dx = \frac{1}{a^2}\left(\cos ax + ax\sin ax\right) + C$$

$$\int x^2 \sin ax\, dx = \frac{1}{a^3}\left(2ax\sin ax + 2\cos ax - a^2 x^2 \cos ax\right) + C$$

$$\int x^2 \cos ax\, dx = \frac{1}{a^3}\left(2ax\cos ax - 2\sin ax + a^2 x^2 \sin ax\right) + C$$

$$\int e^{ax}\sin bx\, dx = \frac{e^{ax}}{a^2 + b^2}\left(a\sin bx - b\cos bx\right) + C$$

$$\int e^{ax}\cos bx\, dx = \frac{e^{ax}}{a^2 + b^2}\left(a\cos bx + b\sin bx\right) + C$$

$$\int \sin ax\sin bx\, dx = \frac{\sin(a-b)x}{2(a-b)} - \frac{\sin(a+b)x}{2(a+b)} + C, \qquad a^2 \neq b^2$$

$$\int \sin ax\cos bx\, dx = -\frac{\cos(a-b)x}{2(a-b)} - \frac{\cos(a+b)x}{2(a+b)} + C, \qquad a^2 \neq b^2$$

$$\int \cos ax\cos bx\, dx = \frac{\sin(a-b)x}{2(a-b)} + \frac{\sin(a+b)x}{2(a+b)} + C, \qquad a^2 \neq b^2$$

$$\int \frac{dx}{a^2 + x^2} = \frac{1}{a}\tan^{-1}\frac{x}{a} + C$$

$$\int \frac{x^2 dx}{a^2 + x^2} = x - a\tan^{-1}\frac{x}{a} + C$$

$$\int \frac{dx}{\left(a^2 + x^2\right)^2} = \frac{1}{2a^2}\left(\frac{x}{x^2 + a^2} + \frac{1}{a}\tan^{-1}\frac{x}{a}\right) + C$$

A.6 Definite Integrals

If *m* and *n* are integers:

$$\int_0^{2\pi} \sin ax\, dx = 0$$

$$\int_0^{2\pi} \cos ax\, dx = 0$$

$$\int_0^{\pi} \sin^2 ax\, dx = \int_0^{\pi} \cos^2 ax\, dx = \frac{\pi}{2}$$

$$\int_0^{\pi} \sin mx\ \sin nx\, dx = \int_0^{\pi} \cos mx\ \cos nx\, dx = 0, \qquad m \neq n$$

$$\int_0^{\pi} \sin mx\ \cos nx\, dx = \begin{cases} 0, & m+n = even \\ \dfrac{2m}{m^2 - n^2}, & m+n = odd \end{cases}$$

$$\int_0^{2\pi} \sin mx \sin nx\, dx = \int_{-\pi}^{\pi} \sin mx\ \sin nx\, dx = \begin{cases} 0, & m \neq n \\ \pi, & m \neq n \end{cases}$$

$$\int_0^{\infty} \frac{\sin ax}{x}\, dx = \begin{cases} \dfrac{\pi}{2}, & a > 0 \\ 0, & a = 0 \\ -\dfrac{\pi}{2}, & a < 0 \end{cases}$$

$$\int_0^{\infty} \frac{\sin^2 x}{x}\, dx = \frac{\pi}{2}$$

$$\int_0^\infty \frac{\cos bx}{x^2 + a^2}\, dx = \frac{\pi}{2a} e^{-ab}, \qquad a > 0, b > 0$$

$$\int_0^\infty \frac{x \sin bx}{x^2 + a^2}\, dx = \frac{\pi}{2} e^{-ab}, \qquad a > 0, b > 0$$

$$\int_0^\infty \sin cx\, dx = \int_0^\infty \sin c^2 x\, dx = \frac{1}{2}$$

$$\int_0^\pi \sin^2 nx\, dx = \int_0^\pi \sin^2 x\, dx = \int_0^\pi \cos^2 nx\, dx = \int_0^\pi \cos^2 x\, dx = \frac{\pi}{2}, \qquad n = \text{an integer}$$

$$\int_0^\pi \sin mx\, \sin nx\, dx = \int_0^\pi \cos mx\, \cos nx\, dx = 0, \qquad m \neq n, m, n \text{ integers}$$

$$\int_0^\pi \sin mx\, \cos nx\, dx = \begin{cases} \dfrac{2m}{m^2 - n^2}, & m + n = \text{odd} \\ 0, & m + n = \text{even} \end{cases}$$

$$\int_{-\infty}^\infty e^{\pm j2\pi tx}\, dx = \delta(t)$$

$$\int_0^\infty x^n e^{-ax}\, dx = \frac{n!}{a^{n+1}}$$

$$\int_0^\infty e^{-a^2 x^2}\, dx = \frac{\sqrt{\pi}}{2a}, \qquad a > 0$$

$$\int_0^\infty x^{2n} e^{-ax^2}\, dx = \frac{1 \bullet 3 \bullet 5\ \bullet\bullet\bullet (2n-1)}{2^{n+1} a^n} \sqrt{\frac{\pi}{a}}$$

$$\int_0^\infty x^{2n+1} e^{-ax^2}\, dx = \frac{n!}{2a^{n+1}}, \qquad a > 0$$

A.7 L'Hopital's Rule

If $f(0) = 0 = h(0)$, then:

$$\lim_{x \to 0} \frac{f(x)}{h(x)} = \lim_{x \to 0} \frac{f'(x)}{h'(x)},$$

where the prime indicates differentiation.

A.8 Taylor and Maclaurin Series

$$f(x) = f(a) + \frac{(x-a)}{1!} f'(a) + \frac{(x-a)^2}{2!} f''(a) + \dots$$

$$f(x) = f(0) + \frac{x}{1!} f'(0) + \frac{x^2}{2!} f''(0) + \dots$$

where the prime indicates differentiation.

A.9 Power Series

$$e^x = 1 + x + \frac{x^2}{2!} + \frac{x^3}{3!} + \dots + \frac{x^n}{n!} + \dots$$

$$\sin x = x - \frac{x^3}{3!} + \frac{x^5}{5!} - \frac{x^7}{7!} + \dots$$

$$\cos x = 1 - \frac{x^2}{2!} + \frac{x^4}{4!} - \frac{x^6}{6!} + \frac{x^8}{8!} - \dots$$

$$\tan x = x + \frac{x^3}{3} + \frac{2x^5}{15} + \frac{17x^7}{315} + \dots$$

$$(1+x)^n = 1 + nx + \frac{n(n+1)}{2!} x^2 + \frac{n(n-1)(n-2)}{3!} x^3 + \dots + \binom{n}{k} x^k + \dots + x^n$$

$$\approx 1 + nx, \qquad |x| = 1$$

$$\frac{1}{1-x} = 1 + x + x^2 + x^3 + \dots, \qquad |x| < 1$$

$$Q(x) = \frac{e^{-x^2/2}}{x\sqrt{2\pi}} \left(1 - \frac{1}{x^2} + \frac{1 \cdot 3}{x^4} - \frac{1 \cdot 3 \cdot 5}{x^6} + \dots \right)$$

$$J_n(x) = \frac{1}{n!} \left(\frac{x}{2} \right)^n - \frac{1}{(n+1)!} \left(\frac{x}{2} \right)^{n+2} + \frac{1}{2!(n+2)!} \left(\frac{x}{2} \right)^{n+4} - \dots$$

$$J_n(x) \approx \sqrt{\frac{2}{\pi x}} \cos \left(x - \frac{\pi}{4} - \frac{n\pi}{2} \right), \qquad x < 1$$

A.10 Sums

$$\sum_{k=1}^{N} k = \frac{1}{2} N(N+1)$$

$$\sum_{k=1}^{N} k^2 = \frac{1}{6} N(N+1)(2N+1)$$

$$\sum_{k=1}^{N} k^3 = \frac{1}{4} N^2 (N+1)^2$$

$$\sum_{k=0}^{N} a^k = \frac{a^{N+1} - 1}{a - 1} \qquad a \neq 1$$

$$\sum_{k=M}^{N} a^k = \frac{a^{N+1} - a^M}{a - 1} \qquad a \neq 1$$

$$\sum_{k=0}^{N} \binom{N}{k} a^{N-k} b^k = (a+b)^N, \quad \text{where} \quad \binom{N}{k} = \frac{N!}{(N-k)!k!}$$

A.11 Logarithmic Identities

$$\log xy = \log x + \log y$$

$$\log \frac{x}{y} = \log x - \log y$$

$$\log x^n = n \log x$$

$$\log_{10} x = \log x \quad \text{(common logarithm)}$$

$$\log_e x = \ln x \quad \text{(natural logarithm)}$$

A.12 Exponential Identities

$$e^x = 1 + x + \frac{x^2}{2!} + \frac{x^3}{3!} + \frac{x^4}{4!} + \cdots$$

where e = 2.7182

$$e^x e^y = e^{x+y}$$

$$(e^x)^n = e^{nx}$$

$$\ln e^x = x$$

A.13 Approximations

$$\sin x = x \qquad or \qquad \lim_{x \to 0} \frac{\sin x}{x} = 1$$

Appendix B: Complex Numbers

The ability to handle complex numbers is important in signals and systems. Although calculators and computer software packages such as MATLAB are now available to manipulate complex numbers, it is advisable that students be familiar with how to handle them by hand.

B.1 Representation of Complex Numbers

A complex number z may be written in *rectangular form* as:

$$z = x + jy \tag{B.1}$$

where $j = \sqrt{-1}$; x is the real part of z, while y is the imaginary part, that is:

$$x = \text{Re}(z), \qquad y = \text{Im}(z) \tag{B.2}$$

The complex number z is shown plotted in the complex plane in Figure B.1. Since $j = \sqrt{-1}$:

$$
\begin{aligned}
\frac{1}{j} &= -j \\[4pt]
j^2 &= -1 \\[4pt]
j^3 &= j \cdot j^2 = -j \\[4pt]
j^4 &= j^2 \cdot j^2 = 1 \\[4pt]
j^5 &= j \cdot j^4 = j \\
&\;\;\vdots
\end{aligned}
\tag{B.3}
$$

A second way of representing the complex number z is by specifying it magnitude r and angle θ it makes with the real axis, as shown in Figure B.1. This is known as the *polar form*. It is given by:

$$z = |z| \angle \theta = r \angle \theta \tag{B.4}$$

where:

$$r = \sqrt{x^2 + y^2}, \qquad \theta = \tan^{-1}\frac{y}{x} \tag{B.5a}$$

or

$$x = r \cos \theta, \qquad y = r \sin \theta \tag{B.5b}$$

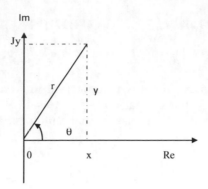

FIGURE B.1
Graphical representation of a complex number.

that is,

$$z = x + jy = r \angle \theta = r \cos \theta + jr \sin \theta \tag{B.6}$$

In converting from rectangular to polar form using Eq. (B.5), we must exercise care in determining the correct value of θ. These are the four possibilities:

$$z = x + jy, \qquad \theta = \tan^{-1} \frac{y}{x} \qquad \text{(1st quadrant)}$$

$$z = -x + jy, \qquad \theta = 180^{\circ} - \tan^{-1} \frac{y}{x} \qquad \text{(2nd quadrant)}$$

$$z = -x - jy, \qquad \theta = 180^{\circ} + \tan^{-1} \frac{y}{x} \qquad \text{(3rd quadrant)} \tag{B.7}$$

$$z = x - jy, \qquad \theta = 360^{\circ} - \tan^{-1} \frac{y}{x} \qquad \text{(4th quadrant)}$$

assuming that x and y are positive.

The third way of representing the complex number x is the *exponential form*:

$$z = re^{j\theta} \tag{B.8}$$

This is almost the same as the polar form because we use the same magnitude r and the angle θ.

The three forms of representing a complex number are summarized as follows.

$$z = x + jy, \qquad (x = r \cos \theta, y = r \sin \theta) \qquad \text{Rectangular form}$$

$$z = r \angle \theta, \qquad \left(r = \sqrt{x^2 + y^2}, \theta = \tan^{-1} \frac{y}{x} \right) \qquad \text{Polar form}$$

$$z = re^{j\theta}, \qquad \left(r = \sqrt{x^2 + y^2}, \theta = \tan^{-1} \frac{y}{x} \right) \qquad \text{Exponential form} \tag{B.9}$$

B.2 Mathematical Operations

Two complex numbers $z_1 = x_1 + jy_1$ and $z_2 = x_2 + jy_2$ are equal if and only their real parts are equal, and their imaginary parts are equal, that is:

$$x_1 = x_2, \qquad y_1 = y_2 \tag{B.10}$$

The complex conjugate of the complex number $z = x + jy$ is:

$$z^* = x - jy = r \angle -\theta = re^{-j\theta} \tag{B.11}$$

Thus, the complex conjugate of a complex number is found by replacing every j by –j.

Given two complex numbers $z_1 = x_1 + jy_1 = r_1 \angle \theta_1$ and $z_2 = x_2 + jy_2 = r_2 \angle \theta_2$, their sum is:

$$z_1 + z_2 = (x_1 + x_2) + j(y_1 + y_2) \tag{B.12}$$

and their difference is

$$z_1 - z_2 = (x_1 - x_2) + j(y_1 - y_2) \tag{B.13}$$

While it is more convenient to perform addition and subtraction of complex numbers in rectangular form, the product and quotient of two complex numbers are best done in polar or exponential form. For their product:

$$z_1 z_2 = r_1 r_2 \angle \theta_1 + \theta_2 \tag{B.14}$$

Alternatively, using the rectangular form:

$$z_1 z_2 = (x_1 + jy_1)(x_2 + jy_2)$$
$$= (x_1 x_2 - y_1 y_2) + j(x_1 y_2 + x_2 y_1) \tag{B.15}$$

For their quotient,

$$\frac{z_1}{z_2} = \frac{r_1}{r_2} \angle \theta_1 - \theta_2 \tag{B.16}$$

Alternatively, using the rectangular form:

$$\frac{z_1}{z_2} = \frac{x_1 + jy_1}{x_2 + jy_2} \tag{B.17}$$

We rationalize the denominator by multiplying both the numerator and denominator by z_2^*:

$$\frac{z_1}{z_2} = \frac{(x_1 + jy_1)(x_2 - jy_2)}{(x_2 + jy_2)(x_2 - jy_2)} = \frac{x_1 x_2 + y_1 y_2}{x_2^2 + y_2^2} + j\frac{x_2 y_1 - x_1 y_2}{x_2^2 + y_2^2} \tag{B.18}$$

B.3 Euler's Formula

Euler's formula is an important result in complex variables. We derive it from the series expansion of e^x, $\cos\theta$, and $\sin\theta$. We know that:

$$e^x = 1 + x + \frac{x^2}{2!} + \frac{x^3}{3!} + \frac{x^4}{4!} + \dots \tag{B.19}$$

Replacing x by $j\theta$ gives:

$$e^{j\theta} = 1 + j\theta - \frac{\theta^2}{2!} - j\frac{\theta^3}{3!} + \frac{\theta^4}{4!} + \dots \tag{B.20}$$

Also:

$$\cos\theta = 1 - \frac{\theta^2}{2!} + \frac{\theta^4}{4!} - \frac{\theta^6}{6!} + \dots \tag{B.21a}$$

$$\sin\theta = \theta - \frac{\theta^3}{3!} + \frac{\theta^5}{5!} - \frac{\theta^7}{7!} + \dots \tag{B.21b}$$

so that:

$$\cos\theta + j\sin\theta = 1 + j\theta - \frac{\theta^2}{2!} - j\frac{\theta^3}{3!} + \frac{\theta^4}{4!} + j\frac{\theta^5}{5!} - \dots \tag{B.22}$$

Comparing eqs. (B.20) and (B.22), we conclude that:

$$e^{j\theta} = \cos\theta + j\sin\theta \tag{B.23}$$

This is known as Euler's formula. The exponential form of representing a complex number as in Eq. (B.8) is based on Euler's formula. From Eq. (B.23), notice that:

$$\cos\theta = \mathrm{Re}\ (e^{j\theta}), \qquad \sin\theta = \mathrm{Im}\ (e^{j\theta}), \tag{B.24}$$

and that

$$|e^{j\theta}| = \sqrt{\cos^2\theta + \sin^2\theta} = 1 \tag{B.25}$$

Replacing θ by $-\theta$ in eq. (B.23) gives:

$$e^{-j\theta} = \cos\theta - j\sin\theta \tag{B.26}$$

Adding eqs. (B.23) and (B.26) yields:

$$\cos\theta = \frac{1}{2}\left(e^{j\theta} + e^{-j\theta}\right) \tag{B.27}$$

Subtracting eq. (B.26) from eq. (B.23) yields:

$$\sin\theta = \frac{1}{2j}\left(e^{j\theta} - e^{-j\theta}\right) \tag{B.28}$$

The following identities are useful in dealing with complex numbers. If $z = x + jy = r\angle\theta$, then:

$$zz^* = |z|^2 = x^2 + y^2 = r^2 \tag{B.29}$$

$$\sqrt{z} = \sqrt{x + jy} = \sqrt{r}e^{j\theta/2} = \sqrt{r} \angle \theta/2 \tag{B.30}$$

$$z^n = (x + jy)^n = r^n \angle n\theta = r^n(\cos n\theta + j\sin n\theta) \tag{B.31}$$

$$z^{1/n} = (x + jy)^{1/n} = r^{1/n} \angle \theta/n + 2\pi k/n, \quad k = 0,1,2,\ldots,n-1 \tag{B.32}$$

$$\ln(re^{j\theta}) = \ln r + \ln e^{j\theta} = \ln r + j\theta + j2\pi k \quad (k = \text{integer}) \tag{B.33}$$

$$\begin{aligned} e^{\pm j\pi} &= -1 \\ e^{\pm j2\pi} &= 1 \\ e^{j\pi/2} &= j \\ e^{-j\pi/2} &= -j \end{aligned} \tag{B.34}$$

$$\mathrm{Re}\left(e^{(\alpha + j\omega)t}\right) = \mathrm{Re}\left(e^{\alpha t}e^{j\omega t}\right) = e^{\alpha t}\cos\omega t$$

$$\mathrm{Im}\left(e^{(\alpha + j\omega)t}\right) = \mathrm{Im}\left(e^{\alpha t}e^{j\omega t}\right) = e^{\alpha t}\sin\omega t \tag{B.35}$$

MATLAB handles complex numbers quite easily as real numbers.

Example B.1: Evaluate the complex numbers:

a. $z_1 = \dfrac{j(3 - j4)^*}{(-1 + j6)(2 + j)^2}$

b. $z_2 = \left[4 \angle 30° + 2 - j + 6e^{j\pi/4}\right]^{1/2}$.

Solution

a. This can be solved in two ways: working with z in rectangular form or polar form.

Method 1 (working in rectangular form):

$$\text{Let } z_1 = \frac{z_3 z_4}{z_5 z_6}$$

where

$$z_3 = j$$

$$z_4 = (3 - j4)^* \quad = \text{ the complex conjugate of } (3 - j4)$$

$$= (3 + j4)$$

$$z_5 = -1 + j6$$

$$z_6 = (2 + j)^2 = 4 + j4 - 1 = 3 + j4$$

Hence,

$$z_1 = \frac{j(3 + j4)}{(-1 + j6)(3 + j4)} = \frac{j3 - 4}{-3 - j4 + j18 - 24}$$

$$= \frac{-4 + j3}{-27 + j14}$$

Multiplying and dividing by $-27 - j14$ (rationalization), we have:

$$z_1 = \frac{(-4 + j3)(-27 - j14)}{(-27 + j14)(-27 - j14)} = \frac{150 - j25}{27^2 + 14^2}$$

$$= 0.1622 - j0.027 = 0.1644 \angle -9.46°$$

Method 2 (working in polar form)

$$z_3 = j = 1\angle 90°$$

$$z_4 = (3 - j4)^* = (5\angle -53.13°)^* = 5\angle 53.13°$$

$$z_5 = (-1 + j6) = \sqrt{37} \angle 99.46°$$

$$z_6 = (2 + j)^2 = (\sqrt{5} \angle 26.56°)^2 = 5\angle 53.13°$$

Hence,

$$z_1 = \frac{z_3 z_4}{z_5 z_6} = \frac{(1\angle 90°)(5\angle 53.13°)}{(\sqrt{37} \angle 99.46°)(5\angle 53.13°)} = \frac{1}{\sqrt{37}} \angle(90° - 99.46°)$$

$$= 0.1644 \angle -9.46° = 0.1622 - j0.027$$

b. Let

$$z_7 = 4\angle 30° = 4\cos 30° + j4\sin 30° = 3.464 + j2$$

$$z_8 = 2 - j$$

$$z_9 = 6e^{j\pi/4} = 6\cos 45° + j6\sin 45° = 4.243 + j4.243$$

Then,

$$z_2 = \left[z_7 + z_8 + z_9 \right]^{1/2}$$
$$= \left[3.464 + j2 + 2 - j + 4.243 + j4.243 \right]^{1/2}$$
$$= (9.707 + j5.243)^{1/2} = (11.03 \angle 28.374°)^{1/2}$$
$$= 3.32 \angle 14.19°$$

Practice Problem B.1 Evaluate the following complex numbers:

(a) $j^3 \left[\dfrac{1+j}{2-j} \right]^2$

(b) $6\angle 30° + j5 - 3 + e^{j45°}$

Answer: (a) $0.24 + j0.32$, (b) $2.03 + j8.707$

Appendix C: Introduction to MATLAB®

MATLAB has become a powerful tool for technical professionals worldwide. The term MATLAB is an abbreviation for MATrix LABoratory implying that MATLAB is a computational tool that employs matrices and vectors/arrays to carry out numerical analysis, signal processing, and scientific visualization tasks. Because MATLAB uses matrices as its fundamental building blocks, one can write mathematical expressions involving matrices just as easily as one would on paper. MATLAB is available for Macintosh, Unix, and Windows operating systems. A student version of MATLAB is available for Personal Computers (PCs). A copy of MATLAB can be obtained from:

The Mathworks, Inc.

3 Apple Hill Drive

Natick, MA 01760-2098

Phone:(508) 647-7000

Website: http://www.mathworks.com

A brief introduction to MATLAB is presented in this appendix. What is presented is sufficient for solving problems in this book. Other information on MATLAB required in this book is provided on the chapter-to-chapter basis as needed. Additional information about MATLAB can be found in MATLAB books and from online help. The best way to learn MATLAB is to work with it after one has learned the basics.

C.1 MATLAB Fundamentals

The Command window is the primary area where you interact with MATLAB. A little later, we will learn how to use the text editor to create M-files, which allow executing sequences of commands. For now, we focus on how to work in the Command window. We will first learn how to use MATLAB as a calculator. We do so by using the algebraic operators in Table C.1.

To begin to use MATLAB, we use these operators. Type commands to MATLAB prompt ">>" in the Command window (correct any mistakes by backspacing) and press the <Enter> key. For example,

```
» a=2;  b=4;  c=-6;
» dat = b^2 - 4*a*c
dat =
   64
» e=sqrt(dat)/10
e =
   0.8000
```

The first command assigns the values 2, 4, and −6 to the variables *a*, *b*, and *c*, respectively. MATLAB does not respond because this line ends with a colon. The second command sets *dat* to $b^2 - 4ac$ and MATLAB return the answer as 64. Finally, the third line sets *e* equal to

TABLE C.1

Basic Operations

Operation	MATLAB Formula	
Addition	a + b	
Division (right)	a/b	(means $a \div b$)
Division (left)	a\b	(means $b \div a$)
Multiplication	a × b	
Power	a^b	
Subtraction	a − b	

the square root of *dat* and divides by 10. MATLAB prints the answer as 0.8. As function *sqrt* is used here, other mathematical functions listed in Table C.2 can be used. Table C.2 provides just a small sample of MATLAB functions. Others can be obtained from the online help. To get help, type:

```
>> help
```

[a long list of topics come up]
and for a specific topic, type the command name. For example, to get help on a *log to base 2*, type:

TABLE C.2

Typical Elementary Math Functions

Function	Remark
abs(x)	Absolute value or complex magnitude of x
acos, acosh(x)	Inverse cosine and inverse hyperbolic cosine of x in radians
acot, acoth(x)	Inverse cotangent and inverse hyperbolic cotangent of x in radians
angle(x)	Phase angle (in radian) of a complex number x
asin, asinh(x)	Inverse sine and inverse hyperbolic sine of x in radians
atan, atanh(x)	Inverse tangent and inverse hyperbolic tangent of x in radians
conj(x)	Complex conjugate of x
cos, cosh(x)	Cosine and hyperbolic cosine of x in radian
cot, coth(x)	Cotangent and hyperbolic cotangent of x in radian
exp(x)	Exponential of x
fix	Round toward zero
imag(x)	Imaginary part of a complex number x
log(x)	Natural logarithm of x
log2(x)	Logarithm of x to base 2
log10(x)	Common logarithms (base 10) of x
real(x)	Real part of a complex number x
sin, sinh(x)	Sine and hyperbolic sine of x in radian
sqrt(x)	Square root of x
tan, tanh	Tangent and hyperbolic tangent of x in radian

```
>> help log2
```

[a help message on the log function follows]

Note that MATLAB is case sensitive so that sin(a) is not the same as sin(A).

Try the following examples:

```
>> 3^(log10(25.6))
>> y=2* sin(pi/3)
>>exp(y+4-1)
```

In addition to operating on mathematical functions, MATLAB easily allows one to work with vectors and matrices. A vector (or array) is a special matrix with one row or one column. For example:

```
>> a = [1 -3 6 10 -8 11 14];
```

is a row vector. Defining a matrix is similar to defining a vector. For example, a 3×3 matrix can be entered as:

```
>> A = [1 2 3; 4 5 6; 7 8 9]
```

or as:

```
>> A = [1 2 3
4 5 6
7 8 9]
```

In addition to the arithmetic operations that can be performed on a matrix, the operations in Table C.3 can be implemented.

Using the operations in Table C.3, we can manipulate matrices as follows.

```
» B = A'
B =
   1 4 7
   2 5 8
   3 6 9
» C = A + B
```

TABLE C.3

Matrix Operations

Operation	Remark
A'	Finds the transpose of matrix A
det(A)	Evaluates the determinant of matrix A
inv(A)	Calculates the inverse of matrix A
eig(A)	Determines the eigenvalues of matrix A
diag(A)	Finds the diagonal elements of matrix A
expm(A)	Exponential of matrix A

```
C =
  2 6 10
  6 10 14
  10 14 18
» D = A^3 - B*C

D =
  372 432 492
  948 1131 1314
  1524 1830 2136
» e= [1 2; 3 4]

e =
  1 2
  3 4
» f=det(e)

f =
  -2
» g = inv(e)

g =
  -2.0000 1.0000
  1.5000 -0.5000
» H = eig(g)

H =
  -2.6861
  0.18611
```

Note that not all matrices can be inverted. A matrix can be inverted if and only if its determinant is nonzero. Special matrices, variables, and constants are listed in Table C.4. For example, type:

```
>> eye(3)
```

TABLE C.4

Special Matrices, Variables, and Constants

Matrix/Variable/Constant	Remark
eye	Identity matrix
ones	An array of ones
zeros	An array of zeros
i or j	Imaginary unit or sqrt(–1)
pi	3.142
NaN	Not a number
inf	Infinity
eps	A very small number, 2.2e–16
rand	Random element

```
ans=
   1  0  0
   0  1  0
   0  0  1
```

to get a 3 × 3 identity matrix.

C.2 Using MATLAB to Plot

To plot using MATLAB is easy. For two-dimensional plot, use the plot command with two arguments as:

```
>> plot(xdata, ydata),
```

where *xdata* and *ydata* are vectors of the same length containing the data to be plotted.

For example, suppose we want to plot $y = 10 \times \sin(2 \times pi \times x)$ from 0 to $5 \times pi$, we will proceed with the following commands:

```
>> x = 0:pi/100:5*pi;    % x is a vector, 0 <= x <= 5*pi, increments of pi/100
>> y = 10*sin(2*pi*x);   % create a vector y
>> plot(x, y);           % create the plot.
```

With this, MATLAB responds with the plot in Figure C.1.

MATLAB will let you graph multiple plots together and distinguish with different colors.

This is obtained with the command plot (xdata, ydata, "color"), where the color is indicated by using a character string from the options listed in Table C.5.

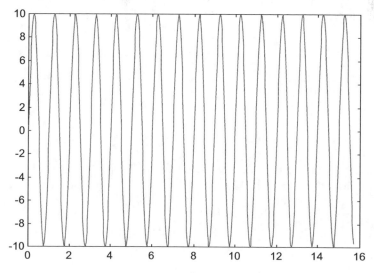

FIGURE C.1
MATLAB plot of $y = 10 \times \sin(2 \times pi \times x)$.

TABLE C.5

Various Color and Line Types

y	yellow	.	point
m	magenta	o	circle
c	cyan	x	x-mark
r	red	+	plus
g	green	-	solid
b	blue	*	star
w	white	:	dotted
k	black	-.	dashdot
		--	dashed

For example,

```
>> plot(x1, y1, 'r', x2,y2, 'b', x3,y3, '--');
```

will graph data (x1,y1) in red, data (x2,y2) in blue, and data (x3,y3) in dashed line all on the same plot.

MATLAB also allows for logarithm scaling. Rather that the **plot** command, we use:

```
loglog       log(y) versus log(x)
semilogx     y versus log(x)
semilogy     log(y) versus x.
```

Three-dimensional plots are drawn using the functions *mesh* and *meshdom* (mesh domain). For example, draw the graph of $z = x \times \exp(-x^2 - y^2)$ over the domain $-1 < x, y < 1$, we type the following commands:

```
>> xx = -1:.1:1;
» yy = xx;
» [x, y] = meshgrid(xx, yy);
» z=x.*exp(-x.^2 -y.^2);
» mesh(z);.
```

(The dot symbol used in x. and y. allows element-by-element multiplication.) The result is shown in Figure C.2.

Other plotting commands in MATLAB are listed in Table C.6. The **help** command can be used to find out how each of these is used.

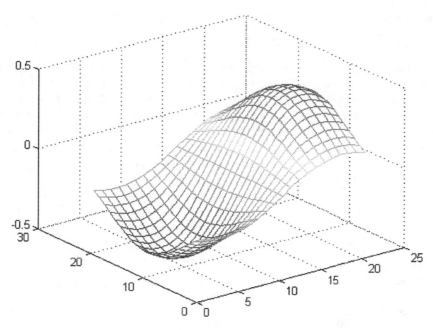

FIGURE C.2
A three-dimensional plot.

TABLE C.6

Other Plotting Commands

Command	Comments
Bar(x, y)	A bar graph
Contour(z)	A contour plot
Errorbar(x, y, l, u)	A plot with error bars
Hist(x)	A histogram of the data
Plot3(x, y, z)	A three-dimensional version of plot()
Polar(r, angle)	A polar coordinate plot
Stairs(x, y)	A stairstep plot
Stem(n, x)	Plots the data sequence as stems
Subplot(m, n, p)	Multiple (m-by-n) plots per window
Surf(x, y, x, c)	A plot of 3-D colored surface

C.3 Programming with MATLAB

So far MATLAB has been used as a calculator, you can also use MATLAB to create your own program. The command line editing in MATLAB can be inconvenient if one has several lines to execute. To avoid this problem, one creates a program which is a sequence of statements to be executed. If you are in Command window, click **File/New/M-files** to open a new file in the MATLAB Editor/Debugger or simple text editor. Type the program and save the program in a file with an extension.m, say filename.m; it is, for this reason, called an M-file. Once the program is saved as an M-file, exit the Debugger window. You are now back in Command window. Type the file without the extension.m to get results. For example, the plot that was made above can be improved by adding title and labels and typed as an M-files called example1.m.

```
x = 0:pi/100:5*pi;        % x is a vector, 0 <= x <= 5*pi, increments of pi/100
y = 10*sin(2*pi*x);       % create a vector y
plot(x, y);               % create the plot
xlabel('x (in radians)'); % label the x-axis
ylabel('10*sin(2*pi*x)'); % label the y-axis
title('A sine functions'); % title the plot
grid                      % add grid.
```

Once it is saved as *example1.m* and we exit text editor, type:

```
>> example1,
```

in the Command window, and hit <Enter> to obtain the result shown in Figure C.3.

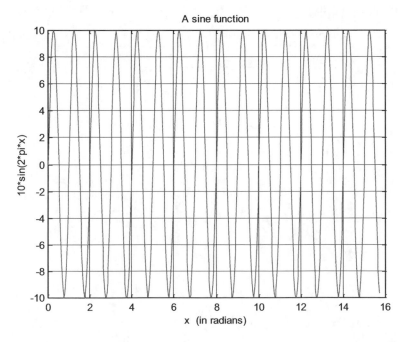

FIGURE C.3
MATLAB plot of $y = 10 \times \sin(2 \times pi \times x)$ with title and labels.

TABLE C.7

Relational and Logical Operators

Operator	Remark
<	Less than
< =	Less than or equal
>	Greater than
> =	Greater than or equal
= =	Equal
~ =	Not equal
&	And
|	Or
~	Not

To allow flow control in a program, certain relational and logical operators are necessary. They are shown in Table C.7. Perhaps the most commonly used flow control statements are *for* and *if*. The *for* statement is used to create a loop or a repetitive procedure and has the general form:

```
for x = array
```

    [commands]

```
End.
```

The *if* statement is used when certain conditions need be met before an expression is executed. It has the general form:

    if expression
    [commands if the expression is True]

```
else,
```

    [commands if the expression is False]

```
end.
```

For example, suppose we have an array y(x) and we want to determine the minimum value of y and its corresponding index x. This can be done by creating an M-file as shown below.

```
% example2.m
% This program finds the minimum y value and its corresponding x index
x = [1 2 3 4 5 6 7 8 9 10]; %the nth term in y
y = [3 9 15 8 1 0 -2 4 12 5];
min1 = y(1);
for k=1:10
  min2=y(k);
  if(min2 < min1)
    min1 = min2;
    xo = x(k);
```

```
  else
    min1 = min1;
  end
end
diary
min1, xo
diary off.
```

Note the use of *for* and *if* statements. When this program is saved as example2.m, we execute it in the Command window and obtain the minimum value of y as –2 and the corresponding value of x as 7, as expected.

```
» example2
min1 =
      -2
xo =
    7
```

If we are not interested in the corresponding index, we could do the same thing using the command:

```
>> min(y).
```

The following tips are helpful in working effectively with MATLAB:

- Comment your M-file by adding lines beginning with a % character
- To suppress output, end each command with a semi-colon (;), you may remove the semi-colon when debugging the file
- Press up and down arrow keys to retrieve previously executed commands
- If your expression does not fit on one line, use an ellipse (…) at the end of the line and continue on the next line. For example, MATLAB considers:

$$y = \sin (x + \log10(2x + 3)) + \cos(x + \ldots$$

$$\log10(2x + 3));$$

as one line of expression
- Keep in mind that variable and function names are case sensitive.

C.4 Solving Equations

Consider the general system of n simultaneous equations as:

$$a_{11}x_1 + a_{12}x_2 + \bullet\bullet\bullet + a_{1n}x_n = b_1$$

$$a_{21}x_1 + a_{22}x_2 + \bullet\bullet\bullet + a_{2n}x_n = b_2$$

$$\ldots \qquad \ldots \qquad \ldots$$

$$a_{n1}x_1 + a_{n2}x_2 + \bullet\bullet\bullet + a_{nn}x_n = b_n$$

or in matrix form:

$$AX = B$$

where:

$$A = \begin{bmatrix} a_{11} & a_{12} & \bullet\bullet\bullet & a_{1n} \\ a_{21} & a_{22} & \bullet\bullet\bullet & a_{2n} \\ \bullet\bullet\bullet & \bullet\bullet\bullet & \bullet\bullet\bullet & \bullet\bullet\bullet \\ a_{n1} & a_{n2} & a_{n3} & a_{nn} \end{bmatrix}, \quad X = \begin{bmatrix} x_1 \\ x_2 \\ \bullet\bullet\bullet \\ x_n \end{bmatrix}, \quad B = \begin{bmatrix} b_1 \\ b_2 \\ \bullet\bullet\bullet \\ b_n \end{bmatrix}$$

A is a square matrix and is known as the coefficient matrix, while X and B are vectors. X is the solution vector we are seeking to get. There are two ways to solve for X in MATLAB. First, we can use the backslash operator ($\backslash$) so that:

$$X = A \backslash B.$$

Second, we can solve for X as:

$$X = A^{-1}B,$$

which in MATLAB is the same as:

$$X = inv(A) \times B.$$

We can also solve equations using the command **solve**. For example, given the quadratic equation $x^2 + 2x - 3 = 0$, we obtain the solution using the following MATLAB command:

```
>> [x]=solve('x^2 + 2*x - 3 = 0')
x =
[-3]
[1].
```

Indicating that the solutions are x = –3 and x = 1. Of course, we can use the command **solve** for a case involving two or more variables. We will see that in the following example.

Example C.1 Use MATLAB to solve the following simultaneous equations

$$25x_1 - 5x_2 - 20x_3 = 50$$

$$-5x_1 + 10x_2 - 4x_3 = 0$$

$$-5x_1 - 4x_2 + 9x_3 = 0.$$

Solution

We can use MATLAB to solve this in two ways:

Method 1:

The given set of simultaneous equations could be written as:

$$\begin{bmatrix} 25 & -5 & -20 \\ -5 & 10 & -4 \\ -5 & -4 & 9 \end{bmatrix} \begin{bmatrix} x_1 \\ x_2 \\ x_3 \end{bmatrix} = \begin{bmatrix} 50 \\ 0 \\ 0 \end{bmatrix} \quad or \quad AX = B$$

We obtain matrix A and vector B and enter them in MATLAB as follows.

```
» A = [25 -5 -20; -5 10 -4; -5 -4 9]
A =
  25 -5 -20
  -5 10 -4
  -5 -4 9
» B = [50 0 0]'
B =
  50
  0
  0
» X = inv(A)*B
X =
  29.6000
  26.0000
  28.0000
» X = A\B
X =
  29.6000
  26.0000
  28.0000.
```

Thus, $x_1 = 29.6$, $x_2 = 26$, and $x_3 = 28$.

Method 2:

Since the equations are not many in this case, we can use the command **solve** to obtain the solution of the simultaneous equations as follows:

$[x_1, x_2, x_3]$ = solve('25 × x_1–5 × x_2–20 × x_3 = 50',
    '–5 × x_1 + 10 × x_2–4 × x_3 = 0', '–5 × x_1–4 × x_2 + 9 × x_3 = 0')

x_1 =

148/5

$$x_2 =$$

26

$$x_3 =$$

28,

which is the same as before.

Practice Problem C.1 Solve the problem the following simultaneous equations using MATLAB:

$$3x_1 - x_2 - 2x_3 = 1$$

$$-x_1 + 6x_2 - 3x_3 = 0$$

$$-2x_1 - 3x_2 + 6x_3 = 6.$$

Answer: $x_1 = 3 = x_3$, $x_2 = 2$.

C.5 Programming Hints

A good program should be well documented, of reasonable size, and capable of performing some computation with reasonable accuracy within a reasonable amount of time. The following are some helpful hints that may make writing and running MATLAB programs easier.

- Use the minimum commands possible and avoid execution of extra commands. This is particularly true of loops
- Use matrix operations directly as much as possible and avoid *for, do,* and/or *while* loops if possible
- Make effective use of functions for executing a series of commands over several times in a program
- When unsure about a command, take advantage of the help capabilities of the software
- It takes much less time running a program using files on the hard disk than on a floppy disk
- Start each file with comments to help you remember what it is all about later
- When writing a long program, save frequently. If possible, avoid a long program, break it down into smaller subroutines.

C.6 Other Useful MATLAB Commands

Some common useful MATLAB commands which may be used in this book are provided in Table C.8.

TABLE C.8

Other Useful MATLAB Commands

Command	Explanation
Diary	Save screen display output in text format
Mean	Mean value of a vector
Min(max)	Minimum (maximum) of a vector
Grid	Add a grid mark to the graphic window
Poly	Converts a collection of roots into a polynomial
Roots	Finds the roots of a polynomial
Sort	Sort the elements of a vector
Sound	Play vector as sound
Std	Standard deviation of a data collection
Sum	Sum of elements of a vector

Appendix D: Answer to Odd-Numbered Problems

Chapter 1

1.5   (a) 7.21 V, (b) 7.987 V

1.7   65.89 m/sec

1.9   1.92 Wb

Chapter 2

2.3 $6.67 \times 10 - 3$ N

2.5   6.655 A

2.7 $\phi = 0.48$ Wb

2.9 $E = -0.2 \times 10^{-5}$ V

2.11 $I = 1.5$ mA

Chapter 3

3.7 $f = 60$ Hz

3.9   600 rpm

3.11   100 Ω

3.13   (i) 5.32 A, (ii) 10.6.4 V, (iii) 167.5 V, (iv) 37.6 Ω, and (v) 0.53 leading

3.15   0.6 lagging

3.17   0.318 Henry

Chapter 4

4.5   8.33 A, 104.13 A, 0.027 wb

Chapter 5

5.1 (i) Core area $A_i = 0.0536 * 10^6 \, mm^2$

(ii) Window area $A_w = 0.3735 * 10^6 \, mm^2$

5.3 (i) Ai = 0.0246 m², d = 0.1812 m, Aw = 0.0872 m², Ww = 0.1867 m

(ii) Tp = 134 turns, Ts = 15 turns

(iii) Ap = 90 mm², As = 833.33 mm²

Chapter 6

6.17 Ra = 0.8 Ω

6.19 Ra = 1.8 Ω

6.21 N = 300 rpm

6.23 Rse = 0.3 Ω

Chapter 7

7.3 2p = 2

Chapter 8

8.3 6.366 V, 0.31 A

8.5 10.8 A

8.7 i.$V_{lmean} = 54.9 \, V$, $I_{lmean} = 27.45 \, A$, $v_{s\,rms} = 141.42 \, V$

8.8 ii.$V_{lmean} - 59.59 \, V$, $I_{lmean} = 29.8 \, A$, $v_{o\,rms} = 98.61 \, V, RF = 1.318$

8.9 $I_{min} = 92.28$ A, $I_{max} = 101.79$ A, the average load current $I_{av} = 96$ A, the average value of the diode current $I_D = 57.9$, the effective input resistance $R_i = 1.25$

Chapter 9

9.1 $R_d = 10 \, \Omega$, V = 24.35 Volts

9.3 Vt = 279 V, $\omega_o = 186$ rad/sec

9.5 $I_a = 61.72$ A, $\alpha = 38.53°$

$N_R \% = 3.78\%$, pf = 0.704 lagging

Input power factor $= \dfrac{P}{S} = 0.712$ lagging

9.7

i.

Duty Cycle (D)	I_{Lmean} (A)	V_{Lmean} (V)	Speed (Rad/sec)
0.5	54.96	83.25	87.82
0.6	55.17	99.94	111.7

ii.

Duty Cycle (D)	I_{Lmean} (A)	V_{Lmean} (V)	Speed (Rad/sec)
100%	56	200	206.9

iii.

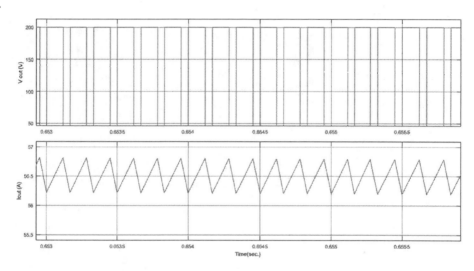

Chapter 10

10.7 (i) Te = 144.68 Nm, ω_s = 377 rad / sec

   (ii) S_m = 0.08

   (iii) $T_{e.m}$ = 11.47 N.m

10.9 Firing advance angle of inverter = 17.95°

Chapter 11

11.1 β = 2.7°

11.3 (a) Resolution = 120 steps/revolution

   (b) The steps required for making 25 revolutions = 3000

   (c) n = 20 rps

Selected Bibliography

1. *"Stepper Motor Controller,"* Texas Instruments Incorporated.
2. Z. Qi, *"Design of a Driver of Two-Phase Hybrid Stepper Motor Based on THB6064H,"* 2017 2nd Asia Conference on Power and Electrical Engineering IOP.
3. X. H. WAN, *Electrical Machinery.* Beijing, China: China Machine PRESS, 2009.
4. N. Q. Le, and J. W. Jeon, "An open-loop stepper motor driver-based on FPGA," *International Conference on Control, Automation and Systems,* 2007 October 17–20, 2007 In COEX, Seoul, Korea.
5. AC Motors, (Http://Www. Allaboutcircuits.Com/Vol_2/Chpt_13/5.Html).
6. Hard Disk Drives, (Http://Www.Storagereview.Com/Guide2000/Ref/Hdd/Op/Actactuator. Html).
7. OSR Journal of Electrical And Electronics Engineering (IOSR-JEEE) Www.Iosrjournals.Org, International Conference On Advances In Engineering & Technology, 2014.
8. P. C. Krause, O. Wasynczuk, and S. D. Sudhoff, *"Analysis of Electric Machinery and Drive Systems,"* 2nd ed., Hoboken, NJ: Wiley Interscience, A John Wiley & Sons, INC. Publication.
9. A. Hughes, *"Electric Motors and Drives, Fundamentals, Types, and Applications,"* 3rd ed., New York: Elsevier, 2002.
10. P. S. Bimibhra, *"Power Electronics,"* 4th ed., 2007, New Delhi, India: Khanna Publishers.
11. "Introduction to power electronics," A Tutorial Burak Ozpineci Power Electronics and Electrical Power Systems Research Center, The U.S. Department of Energy.
12. R. W. Erickson, *"Fundamentals of Power Electronics "* University of Colorado, Boulder, 1997, New York: Chapman & Hall.
13. C. W. Lander, *"Power Electronics,"* 3rd ed., New York: McGraw-Hill, 1993.
14. J. Zhao, and Y. Yu. "Brushless DC Motor Fundamentals Application Note," 2011.
15. A. Nafees, Http://Eed.Dit.Googlepages.Com.
16. V. Larin, and D. Matveev "Analysis of transformer frequency response deviations using white-box modeling," *Conference: CIGRE Study Committee A2 COLLOQUIUM,* Cracow, Poland, October 2017.
17. J. C. Stephen, *"Electric Machinery Fundamentals,"* 5th ed., New York: McGraw-Hill, 2012.
18. A. E. Fitzgerald, C. Kingsley, and D. Stephen, *"Electric Machinery,"* 6th ed., London: McGraw-Hill, 2012.
19. P.C. Krause, O. Wasynczuk, and S. D. Sudhoff, *"Analysis of Electrical Machinery and Drive Systems,"* 2nd ed., Hoboken, NJ: John Wiley & Sons, 2002.
20. A. Hughes, *"Electric Motors and Drives Fundamentals, Types and Applications,"* 3rd ed., Elsevier, 2006.
21. M. H. Rashid, *"Power Electronics, Circuits, Devices, and Applications",* Upper Saddle River, NJ: Prentice, 2003.
22. P. Krause, O. Wasynczuk, S. Sudhoff, and S. Pekarek, *"Analysis of Electric Machinery and Drive Systems",* 3rd ed., Hoboken, NJ: John Wiley & Sons, 2013.

Index

Note: Page numbers in italic and bold refer to figures and tables, respectively.